Earth

Portrait of a Planet

Third Edition

STEPHEN MARSHAK

University of Illinois

W. W. NORTON & COMPANY

NEW YORK   LONDON

W. W. Norton & Company has been independent since its founding in 1923, when William Warder Norton and Mary D. Herter Norton first published lectures delivered at the People's Institute, the adult education division of New York City's Cooper Union. The Nortons soon expanded their program beyond the Institute, publishing books by celebrated academics from America and abroad. By mid-century, the two major pillars of Norton's publishing program—trade books and college texts—were firmly established. In the 1950s, the Norton family transferred control of the company to its employees, and today—with a staff of four hundred and a comparable number of trade, college, and professional titles published each year—W. W. Norton & Company stands as the largest and oldest publishing house owned wholly by its employees.

Composition by TSI Graphics
Manufacturing by Courier Companies, Inc.
Illustrations for the Second and Third Editions by Precision Graphics

Editor: Jack Repcheck
Senior project editor: Thomas Foley
Production manager: Christopher Granville
Copy editor: Barbara Curialle
Managing editor, college: Marian Johnson
Science media editor: April Lange
Associate editor, science media: Sarah England
Photography editors: Kelly Mitchell and Michelle Riley
Editorial assistant: Mik Awake
Geotour spreads designed by Precision Graphics
Developmental editor for the First Edition: Susan Gaustad

978–0-393-93036-8

W. W. Norton & Company, Inc., 500 Fifth Avenue, New York, N.Y. 10110
wwnorton.com

W. W. Norton & Company Ltd., Castle House, 75/76 Wells Street, London W1T 3QT

3 4 5 6 7 8 9 0

BRIEF CONTENTS

CONTENTS

Imagine a desert canyon at dawn. Stark cliffs of red rock descend like a staircase down to the gravelly bed of a dry stream on the canyon floor. Mice patter among dry shrubs and cactus. Suddenly, the sound of a hammer cracking rock rises from below. Some hours later, a sweating geologist—a scientist who studies the Earth—scales back up the cliffs, carrying a backpack filled with heavy rock samples that he will eventually take to a lab. Why? By closely examining natural exposures of rocks and sediments in the field (such as those in the canyon just described), as well as by studying samples in a laboratory, analyzing satellite imagery, and developing complex computer models, geologists can answer a number of profound and fascinating questions about the character and history of our planet: How do rocks form? What do fossils tell us about the evolution of life? Why do earthquakes shake the ground and why do volcanoes erupt? What causes mountains to rise? Has the map of the Earth always looked the same? Does climate change through time? How do landforms develop? Where do we dig or drill to find valuable resources? What kinds of chemical interactions occur among land, air, water, and life? How did the Earth originate? Does our planet resemble others? The modern science of geology (or geoscience), the study of the Earth, addresses these questions and more. Indeed, a look at almost any natural feature leads to a new question, and new questions drive new research. Thus, geology remains as exciting a field of study today as it was when the discipline originated in the eighteenth century.

Before the mid-twentieth century, geologists considered each of the above questions as a separate issue, unrelated to the others. But since 1960, two paradigm-shifting advances have unified thinking about the Earth and its features. The first, *the theory of plate tectonics*, shows that the Earth's outer shell, rather than being static, consists of discrete plates that constantly move very slowly relative to each other, so that the map of our planet constantly changes. We now understand that plate interactions cause earthquakes and volcanoes, build mountains, provide gases for the atmosphere, and affect the distribution of life on Earth. The second advance establishes the concept that our planet is a complex system—*the Earth System*—in which water, land, air, and living inhabitants are dynamically interconnected in ways that allow materials to cycle constantly among various living and nonliving reservoirs. With the Earth System concept in mind, geologists now realize that the history of life links intimately to the physical history of our planet.

Earth: Portrait of a Planet is an introductory geoscience textbook that weaves the theory of plate tectonics and the concept of Earth System science into its narrative from the first page to the last. The book strives to create a modern, coherent image—a portrait—of the very special sphere on which we all live. As such, the book helps students understand the origin of the Earth and its internal structure, the nature of plate movement, the diversity of Earth's landscapes, the character of materials that make up the Earth, the distribution of resources, the structure of the air and water that surround our planet, the evolution of continents during the Earth's long history, and the nature of global change through time. The story of our planet, needless to say, is interesting in its own right. But knowledge of this story has practical applications as well. Students reading *Earth: Portrait of a Planet* and studying this book's multitude of drawings and photos will gain insight that can help address practical and political issues too. Is it safe to build a house on a floodplain or beach? How seriously should we take an earthquake prediction? Is global warming for real, and if so, should we worry about its impacts? Which candidate has a more realistic energy and environmental policy? Should your town sell permits to a corporation that wants to extract huge amounts of water from the ground beneath the town? The list of such issues seems endless.

NARRATIVE THEMES

To understand a subject, students must develop an appreciation of fundamental concepts and by doing so create a mental "peg board" on which to hang and organize observations and ideas. In the case of *Earth: Portrait of a Planet*, these concepts define "narrative themes" that are carried throughout the book, as discussed more fully in the Prelude.

1. The Earth is a complex system in which rock, oceans, air, and life are interconnected. This system is unique in the Solar System.

2. Internal energy (due to the make-up and processes occurring in our planet's interior) drives the motion

of plates, and the interactions among plates, in turn, drive a variety of geologic phenomena, such as the uplift of mountain ranges, the eruption of volcanoes, the vibration of earthquakes, and the drift of continents. But what plate tectonics builds, other Earth phenomena tear down. Specifically, gravity causes materials at the tops of cliffs to slip down to lower elevations. And external energy (provided by the Sun), along with gravity, drives the flow of water, ice, and wind on the Earth's surface—this flow acts like a rasp, capable of eventually grinding away even the highest mountain.

3. The Earth is a planet, formed like other planets from a cloud of dust and gas. Because of its location and history, the Earth differs greatly from its neighbors.

4. Our planet is very old—about 4.57 billion years old. During this time, the map of the planet has changed, surface landscapes have developed and disappeared, and life has evolved.

5. Natural features and processes on Earth can be a hazard—earthquakes, volcanic eruptions, floods, hurricanes, and landslides can devastate societies. But understanding these features can help prevent damage and save lives.

6. Energy and material resources come largely from the Earth. Geologic knowledge can help find them and can help people understand the consequences of using them.

7. Geology ties together ideas from many sciences, and thus the study of geology can increase science literacy in chemistry, physics, and biology.

ORGANIZATION

Topics covered in *Earth: Portrait of a Planet* have been arranged so that students can build their knowledge of geology on a foundation of basic concepts. The book's parts group chapters so that interrelationships among subjects are clear. Part I introduces the Earth from a planetary perspective. It includes a discussion of cosmology and the formation of the Earth and introduces the architecture and composition of our planet, from surface to center. With this background, students are ready to delve into plate tectonics theory. Plate tectonics theory appears early in this book, a departure from traditional textbooks, so that students will be able to relate the contents of all subsequent chapters to this theory. Understanding plate tectonics, for example, helps students to understand the chapters of Part II, which introduce Earth materials (minerals and rocks). A familiarity with plate tectonics and Earth materials together, in turn, provides a basis for the study of volcanoes, earthquakes, and mountains (Part III). And with this background, students have sufficient preparation to understand the fundamentals of Earth history and the character of natural resources (Parts IV and V).

The final part of this book, Part VI, addresses processes and problems occurring at or near the Earth's surface, from the unstable slopes of hills, down the course of rivers, to the icy walls of glaciers, to the shores of the sea and beyond. This part also includes a summary of atmospheric science and concludes with a topic of growing concern—global change. As we think about the future of the planet, concerns about the warming of the climate and the contamination of the environment loom large.

TEACHING PHILOSOPHY

Students learn best by actively engaging in the learning process, by basing learning on the formulation of questions, and by linking clear explanations to visual images. With these concepts as a foundation, the Third Edition of *Earth: Portrait of a Planet* provides a variety of new active-learning and inquiry-based teaching tools, as well as an even broader array of outstanding illustrations, all couched in a highly readable narrative. Most notably, each chapter provides a Geotour, which uses the magical *Google Earth™* to take students to field localities worldwide in order to *see for themselves* what geology looks like, first-hand. Related questions on the book's website allow students to apply their new knowledge to the real world—instantly. This book also helps students to switch into inquiry mode right from the start, by introducing each chapter with a question—a geopuzzle—that prompts students to pursue answers while they read. To encourage students to pause and register the essence of the subject before moving on, each section ends with a take-home message. And in addition to the paper and ink of the book itself, the book's website provides students with access to a variety of resources including 2-D and 3-D animations (to help with visualization of how geologic features evolve over time), practice questions, links to web resources, videos, and even geo-crossword puzzles.

With the advent of smart classrooms, laptop computers, the web, PowerPoint™, Quicktime™, and a myriad of pedagogical tools, students enjoy ever-widening opportunities to absorb and understand the subjects they study. But while this boggling array of diverse learning options helps makes education fun, it does harbor a risk.

Specifically, students may miss seeing the unifying threads that hold a subject together, and may find themselves longing for a clear, straightforward explanation of a topic. To meet this need, *Earth: Portrait of a Planet* holds to the concept that a well illustrated, well written, and well organized textbook helps provide a framework for understanding a subject. This book, therefore, offers thousands of carefully designed drawings and carefully selected photographs, with annotations to highlight key features. It includes interpretative sketches that help students to understand *what a geologist sees* when looking at a photograph of a landscape or outcrop. And to ensure that the illustrations mean something, this book provides carefully crafted explanations and discussion in an accessible, sequential, narrative form. The book treats each topic with sufficient thoroughness that students should be able to understand a topic by reading the book alone. The approach has stood the test of time—the repeated refrain by previous users of this book is that it is "easy to read" and "makes concepts very understandable."

SPECIAL FEATURES

Broad Coverage

Earth: Portrait of a Planet provides complete coverage of the topics in traditional Physical Geology or Introductory General Geology courses. But increasingly, first-semester courses in geology incorporate aspects of historical geology and of Earth System science. Therefore, this book also provides chapters that address Earth history, the atmosphere, the oceans, and global change. Finally, to reflect the international flavor of geoscience, the book contains examples and illustrations from around the world.

Flexible Organization

Though the sequence of chapters in *Earth: Portrait of a Planet* was chosen for a reason, it has been structured to be flexible, so that instructors can rearrange chapters to fit their own strategies for teaching. Geology is a nonlinear subject—individual topics are so interrelated that there is not a single best way to order them. Thus, each chapter is largely self-contained, repeating background material where necessary for the sake of completeness. Readers will note that the book includes a Prelude and several Interludes. These treat shorter subjects in a coherent way that would not be possible if they were simply

subsections within a larger chapter. Finally, the book includes two Appendices. The first reviews basic physics and chemistry, and as such can be used as an introduction to minerals, if students lack the necessary science background. The second provides full-page versions of important charts and maps.

A Connection to Societal Issues

Geology's practical applications are addressed in chapters on volcanic eruptions, earthquakes, energy resources, mineral resources, global change, and mass wasting. Here, students learn that natural features can be hazardous, but that with a little thought, danger may be lessened. In addition, where relevant, *Earth: Portrait of a Planet* introduces students to some of the ways in which geologic understanding can be applied to environmental issues. Case studies show how geologists have used their knowledge to solve practical problems.

Boxed Inserts

Throughout the text, boxes expand on specific topics. "The Human Angle" boxes introduce links between geologic phenomena and the human experience. "Science Toolboxes" provide background scientific data. And "The Rest of the Story" boxes give additional interesting, but optional, detail. "Case Studies" boxes provide specific examples of geologic phenomena that impact society.

Superb Artwork

It's hard to understand features of the Earth System without being able to see them. To help students visualize topics, *Earth: Portrait of a Planet* is lavishly illustrated—the book contains over 200 more illustrations than competing texts! The author has worked closely with the artists to develop an illustration style that conveys a realistic context for geologic features without overwhelming students with extraneous detail. The talented artists who worked on the figures have pushed the envelope of modern computer graphics, and the result is the most realistic pedagogical art ever produced for a geoscience text.

In addition to line art, *Earth* features photographs from all continents. Many of the pictures were taken by the author and provide interesting alternatives to the stock images that have appeared for many years in introductory books. Where appropriate, photographs are accompanied by annotated sketches labeled "What a geologist sees," which help students see the key geologic features in the photos.

In the past, students would need to go to a museum to see bold, colorful paintings of geologic features. Now, they need only flip through the pages of *Earth: Portrait of a Planet*. Famed British painter Gary Hincks has provided spectacular two-page synoptic paintings that illustrate key concepts introduced in the text and visually emphasize the relationships among components of the Earth System.

CHANGES IN THE THIRD EDITION

The Third Edition of *Earth: Portrait of a Planet* contains a number of major changes and updates. Key changes include:

1. *Geotours*: Each chapter of this book contains a Geotour, which provides the coordinates, a thumbnail photo, and a description of several examples of geologic features that illustrate concepts from the book. By entering the coordinates in *Google Earth™*, the student instantly flies to the site and can view it from any altitude and perspective, in 3-D. Geotours are, in effect, guides that take students around the world to see for themselves what geologic features look like. To ensure ease of use, and to provide an inquiry-based approach to using geotours, the book's website provides buttons that take students to the sites at the click of a mouse, and questions that prompt active learning.

2. *New Pedagogical Features*: This edition contains three new teaching aids, in addition to Geotours. First, each chapter starts with a *geopuzzle*, a question that prompts students to pursue the concepts contained in the chapter. The answer to the puzzle, provided at the end of the chapter, serves to synthesize the key points of the chapter. Second, every section ends with a *Take-Home Message*, which allows students to pause and take stock of what they've learned. Third, a set of questions, under the heading *On Further Thought*, has been added to the end of each chapter, to encourage critical thinking.

3. *Incorporation of New Data and New Events*: Students relate to geology in the news, so every effort has been made to incorporate examples of geologic events that have made global headlines. For example, the Third Edition contains complete coverage of Hurricane Katrina and the Indian Ocean tsunami. Chapters covering rapidly evolving subjects have been revised to incorporate the latest data. For example, the chapter on global change reflects the conclusions of the 2007 Intergovernmental Panel on Climate Change. This book provides the most complete and up-to-date coverage of geoscience available at the introductory level.

4. *New Figures and Photos*: Close to 300 figures and photos have been revised or updated for the Third Edition. At just about every point where a students might be thinking, "What does this look like?" the book has an illustration.

SUPPLEMENTS

Geotours

Earth, Third Edition, is the first textbook to utilize *Google Earth™* to emphasize active student learning. Using *Google Earth*'s spectacular 3-D maps of the planet's surface, Stephen Marshak has created 23 Geotours that take students on "virtual" geology field trips where they can apply textbook lessons to real-world geologic features. Appearing in the book as two-page spreads in each chapter, Geotours can be incorporated into engaging lectures that work seamlessly with the book. From the free StudySpace student website, students can access the *Google Earth™* Geotours file and Worksheet Activities, developed with Scott Wilkerson of DePauw University. These tools make it easy to use the *Google Earth™* software to explore the locations illustrated in the text, and to assign these explorations as homework or as a quiz.

Earth: Videos of a Planet Instructor's DVD, Version 3.0

This outstanding lecture resource features 24 short film clips carefully selected from the U.S. Geological Survey archives for use in physical geology lectures. The DVD also includes 10 spectacular 3-D art animations that focus on the geologic processes that are the hardest to understand.

Animations

Students and instructors will have access to over 60 original animations, including 20 animations new to the Third Edition. All have been developed by Stephen Marshak to illustrate dynamic Earth processes. Conveniently accessible from the free StudySpace student website, offline versions of the animations are also available on the Instructor's CD-ROM, linked from PowerPoint slides and easily enlargeable for classroom display with VCR-like controls that allow instructors to control the pace of the animation during lecture.

Zoomable Art

Explore Gary Hincks's spectacular synoptic paintings in vivid detail using the zoomable art feature. Ideal for lecture use, these figures are included among the resources on the Instructor's CD-ROM.

ebook

Earth, Third Edition, is also available as a value-priced electronic edition—same great book, half the price! Go to norton**ebooks**.com for more information.

FOR INSTRUCTORS:

1. Norton Media Library Instructor's CD-ROM

This instructor CD-ROM offers a wealth of easy-to-use multimedia resources, all structured around the text and designed for use in lecture presentations, including:

- editable PowerPoint lecture outlines for each chapter by Ron Parker of Earlham College
- links, from PowerPoint, to the *Google Earth™* Geotours file. These shortcuts make it easy to "fly to" relevant geologic localities identified in the textbook
- all of the art and most of the photographs from the text
- zoomable art versions of Gary Hincks's spectacular synoptic paintings from the text
- additional photographs from Stephen Marshak's own archives
- over sixty animations unique to *Earth: Portrait of a Planet*, Third Edition
- multiple-choice GeoQuiz Clicker *Questions for each chapter*

2. Norton Resource Library Instructor's Website

www.wwnorton.com/nrl

Maintained as a service to our adopters, this password-protected instructor website offers book-specific materials for use in class or within WebCT, Blackboard, or course websites. Resources include:

- editable PowerPoint lecture outlines by Ron Parker of Earlham College
- over sixty animations unique to *Earth: Portrait of a Planet*, Third Edition
- multiple-choice GeoQuiz Clicker *Questions for each chapter*
- Test Bank questions in *ExamView® Assessment Suite*, WebCT, and BlackBoard-ready formats.

- JPEG versions of all drawn art in the textbook
- instructor's manual and test bank in PDF format

3. Instructor's Manual and Test Bank

Prepared by John Werner of Seminole Community College, Terry Engelder of Pennsylvania State University, and Stephen Marshak, this manual offers useful material to help instructors as they prepare their lectures and includes over 1,200 multiple-choice and true-false test questions. The test bank is available in printed form and electronically in *ExamView® Assessment Suite*, WebCT, and Blackboard-ready formats.

4. Transparencies

Approximately 200 figures from the text are available as color acetates.

5. Supplemental Slide Set

This collection of 35mm slides supplements the photographs in the text with additional images from Stephen Marshak's own photo archives. These images are also available as PowerPoint slides on the Norton Media Library CD-ROM.

FOR STUDENTS:

1. StudySpace Student Website

www.wwnorton.com/studyspace

Developed specifically for *Earth: Portrait of a Planet*, Third Edition, and free and open for students, StudySpace provides a wealth of materials to help students organize, learn, and connect the concepts they are learning.

- **Study Plans** highlight the learning tools built into each chapter of the textbook and its associated media resources.
- **Guides to Reading** prepared by Rita Leafgren of the University of Northern Colorado provide an overview of the major ideas introduced in each chapter.
- **GeoTours** with direct links to *Google Earth™* locations and worksheet Activities that can be printed out for self study or assigned as homework or quizzes.
- **Diagnostic Quizzes** for each chapter, prepared by Rita Leafgren of the University of Northern Colorado, help students identify which parts of the text they need to review further and give them instant feedback on right and wrong answers.
- **Animations** developed for *Earth*, Third Edition.
- **FlashCards** that help students review key terms in each chapter.

- **Feature Articles** by Stephen Marshak that comment on particular topics of interest.
- **Geology in the News** featuring newsfeeds to the most geology news, updated regularly.
- **Ebook** direct links that allow students using the ebook to go directly to the relevant sections mentioned in the Study Plan.
- **Norton Gradebook** that allows students to submit [email] their Diagnostic Quiz results and Geotours Worksheets directly to instructors' gradebook.

ACKNOWLEDGMENTS

I am very grateful for the assistance of many people in bringing this book from the concept stage to the shelf in the first place, and for helping to provide the momentum needed to make the Third Edition take shape. First and foremost, I wish to thank my family. My wife, Kathy, has helped throughout in the overwhelming task of keeping track of text and figures and of handling mailings. In addition, she helped edit text, read proofs, shuttle artwork, and provide invaluable advice. My daughter, Emma, helped locate and scan photos, and my son, David, helped me keep the project in perspective and highlighted places where the writing could be improved. During the initial development of the First Edition, I greatly benefited from discussions with Philip Sandberg, and during later stages in the development of the First Edition, Donald Prothero contributed text, editorial comments, and end-of-chapter material.

The publisher, W. W. Norton & Company, has been incredibly supportive of my work and has been very generous in their investment in this project. Steve Mosberg signed the First Edition, and Rick Mixter put the book on track. Jack Repcheck bulldozed aside numerous obstacles and brought the First Edition to completion. He has continued to be a fountain of sage advice and an understanding friend throughout the development of the Second and Third Editions. Jack has provided numerous innovative ideas that have strengthened the book and brought it to the attention of the geologic community. Under Jack's guidance, this book has been able to reach a worldwide audience.

April Lange has expertly coordinated development of the ancillary materials. She has not only managed their development, but also introduced innovative approaches and wrote part of the material. Her contributions have set a standard of excellence. JoAnn Simony did a superb job of managing production of the First Edition and of doing the page makeup. Thom Foley and Chris Granville have expertly and efficiently handled the task of managing production for this Third Edition. They have calmly handled all the back and forth involved in developing a book and in keeping it on schedule. Susan Gaustad did an outstanding job of copy editing the First Edition. This tradition continued through the efforts of Alice Vigliani on the Second Edition and Barbara Curialle on the Third Edition. Kelly Mitchell has taken over and greatly modernized photo research for the Third Edition and has done a great job. Michelle Riley has been invaluable in obtaining permissions and maintaining the credit list for the photos, and Mik Awake, editorial assistant, was a great help in tying up any and all loose ends.

Production of the illustrations has involved many people. I am particularly grateful to Joanne Bales and Stan Maddock, who helped create the overall style of the figures, produced most of them for all editions, and have worked closely with me on improvements. I would also like to thank Becky Oles, who helped create the new art for this edition. Terri Hamer and Kristina Seymour have done an excellent job as production managers at Precision Graphics, coordinating the art team. Jennifer Wasson skillfully designed all of the Geotours and generously helped in the challenging task of fitting the material into the space available. Jon Prince creatively programmed all of the 2-D animations, while Simon Shak, Dan Whitaker, Jeff Griffin, and Andrew Troutt have done a wonderful job of developing all of the new 3-D animations. It has been a delight to interact with the artists, production staff, and management of Precision Graphics over the past several years. It has also been a great pleasure to work with Scott Wilkerson on the Geotours.

It has been great fun to interact with Gary Hincks, who painted the incredible two-page spreads, in part using his own designs and geologic insights. Some of Gary's paintings appeared in *Earth Story* (BBC Worldwide, 1998) and were based on illustrations jointly conceived by Simon Lamb and Felicity Maxwell, working with Gary. Others were developed specifically for *Earth: Portrait of a Planet*. Some of the chapter quotes were found in *Language of the Earth*, compiled by F. T. Rhodes and R. O. Stone (Pergamon, 1981).

As this book has evolved, I have benefited greatly from input by expert reviewers for specific chapters, by general reviewers of the entire book, and by comments from faculty and students who have used the earlier editions and were kind enough to contact me by e-mail. The list of people whose comments were incorporated includes:

Jack C. Allen, Bucknell University
David W. Anderson, San Jose State University
Philip Astwood, University of South Carolina
Eric Baer, Highline University
Victor Baker, University of Arizona
Keith Bell, Carleton University

Mary Lou Bevier, University of British Columbia
Daniel Blake, University of Illinois
Michael Bradley, Eastern Michigan University
Sam Browning, Massachusetts Institute of Technology
Rachel Burks, Towson University
Peter Burns, University of Notre Dame
Sam Butler, University of Saskatchewan
Katherine Cashman, University of Oregon
George S. Clark, University of Manitoba
Kevin Cole, Grand Valley State University
Patrick M. Colgan, Northeastern University
John W. Creasy, Bates College
Norbert Cygan, Chevron Oil, retired
Peter DeCelles, University of Arizona
Carlos Dengo, ExxonMobil Exploration Company
John Dewey, University of California, Davis
Charles Dimmick, Central Connecticut
    State University
Robert T. Dodd, State University of New York at
    Stony Brook
Missy Eppes, University of North Carolina, Charlotte
Eric Essene, University of Michigan
James E. Evans, Bowling Green State University
Leon Follmer, Illinois Geological Survey
Nels Forman, University of North Dakota
Bruce Fouke, University of Illinois
David Furbish, Vanderbilt University
Grant Garvin, John Hopkins University
Christopher Geiss, Trinity College, Connecticut
William D. Gosnold, University of North Dakota
Lisa Greer, William & Mary College
Henry Halls, University of Toronto at Mississuaga
Bryce M. Hand, Syracuse University
Judy Harden, University of South Florida
Anders Hellström, Stockholm University
Tom Henyey, University of South Carolina
Jim Hinthorne, Central Washington University
Paul Hoffman, Harvard University
Neal Iverson, Iowa State University
Donna M. Jurdy, Northwestern University
Thomas Juster, University of Southern Florida
Dennis Kent, Lamont Doherty/Rutgers

Jeffrey Knott, California State University, Fullerton
Ulrich Kruse, University of Illinois
Lee Kump, Pennsylvania State University
David R. Lageson, Montana State University
Robert Lawrence, Oregon State University
Craig Lundstrom, University of Illinois
John A. Madsen, University of Delaware
Jerry Magloughlin, Colorado State University
Paul Meijer, Utrecht University (Netherlands)
Alan Mix, Oregon State University
Robert Nowack, Purdue University
Charlie Onasch, Bowling Green State University
David Osleger, University of California, Davis
Lisa M. Pratt, Indiana University
Mark Ragan, University of Iowa
Bob Reynolds, Central Oregon Community College
Joshua J. Roering, University of Oregon
Eric Sandvol, University of Missouri
William E. Sanford, Colorado State University
Matthew Scarborough, University of Cape Town
    (South Africa)
Doug Shakel, Pima Community College
Angela Speck, University of Missouri
Tim Stark, University of Illinois (CEE)
Kevin G. Stewart, University of North Carolina at
    Chapel Hill
Don Stierman, University of Toledo
Barbara Tewksbury, Hamilton College
Thomas M. Tharp, Purdue University
Kathryn Thornbjarnarson, San Diego State University
Basil Tickoff, University of Wisconsin
Spencer Titley, University of Arizona
Robert T. Todd, State University of New York at
    Stony Brook
Torbjörn Törnqvist, Tulane University
Jon Tso, Radford University
Alan Whittington, University of Missouri
Lorraine Wolf, Auburn University
Christopher J. Woltemade, Shippensburg University

I apologize if anyone was inadvertently not included on
the list.

ABOUT THE AUTHOR

Professor Stephen Marshak is the head of the Department of Geology at the University of Illinois, Urbana-Champaign. He holds an A.B. from Cornell University, an M.S. from the University of Arizona, and a Ph.D. from Columbia University. Steve's research interests lie in the fields of structural geology and tectonics. Over the years, he has explored geology in the field on several continents. Since 1983 Steve has been on the faculty of the University of Illinois, where he teaches courses in introductory geology, structural geology, tectonics, and field geology and has won the university's highest teaching awards. In addition to research papers and *Earth: Portrait of a Planet*, Steve has authored or co-authored *Essentials of Geology, Earth Structure: An Introduction to Structural Geology and Tectonics*, and *Basic Methods of Structural Geology*.

THANKS!

I am very grateful to the faculty who selected the earlier editions of this book for use in their classes and to the students who engaged so energetically with it. I continue to welcome your comments and can be reached at: smarshak@uiuc.edu.

Stephen Marshak

To see the world in a grain of sand
and heaven in a wild flower.
To hold infinity in the palm of the hand
and eternity in an hour.

—William Blake (British poet, 1757–1827)

And Just What Is Geology?

Geopuzzle

Tourists might look at this photo and see a beautiful view. What do geologists see in this landscape?

Geology students exploring the Earth System in the Wasatch Mountains, Utah. Here, air, water, rock, and life all interact to produce a complex and fascinating landscape.

Civilization exists by geological consent, subject to change without notice.

> —**Will Durant (American historian, 1885–1981)**

P.1 IN SEARCH OF IDEAS

Our Hercules transport plane rose from a smooth ice runway on the frozen sea surface at McMurdo Station, Antarctica, and headed south. We were off to spend a month studying unusual rocks exposed on a cliff about 250 km (kilometers) away. As we climbed past the smoking summit of Mt. Erebus, Earth's southernmost volcano, we had one nagging thought: no aircraft had ever landed at our destination, so the ground conditions there were unknown—if deep snow covered the landing site, the massive plane might get stuck and would not be able to return to McMurdo. Because of this concern, the flight crew had added a crate of rocket canisters to the pile of snowmobiles, sleds, tents, and food in the plane's cargo hold. "If the props can't lift us, we can clip a few canisters to the tail, light them, and rocket out of the snow," they claimed.

For the next hour, we flew along the Transantarctic Mountains, a ridge of rock that divides the continent into two parts, East Antarctica and West Antarctica (▶Fig. P.1a). A vast ice sheet, in places over 3 km thick, covers East Antarctica—the surface of this ice sheet forms a high plain known as the Polar Plateau. From the plane's window, we admired glaciers, rivers of ice cutting valleys through the Transantarctic Mountains as they slowly flow from the Polar Plateau to the Ross Ice Shelf, until suddenly, we heard the engines slow.

As the plane descended, it lowered its ski-equipped landing gear. The loadmaster shouted an abbreviated reminder of the emergency alarm code: "If you hear three short blasts of the siren, hold on tight!" Roaring toward the ground, the plane touched the surface of our first choice for a landing spot, the ice at the base of the rock cliff we wanted to study. *Wham, wham, wham, wham*!!!! Sastrugi (frozen snow drifts) rippled the ice surface, and as the skis slammed into them at about 180 km an hour, it seemed as though a fairy-tale giant was shaking the plane. Seconds later, the landing aborted, we were airborne again, looking for a softer runway above the cliff. Finally, we landed in a field of deep snow, unloaded, and bade farewell to the plane (▶Fig. P.1b). The Hercules trundled for kilometers through the snow before gaining enough speed to take off, but fortunately did not need to use the rocket canisters. When the

(a)

(b)

(c)

FIGURE P.1 (a) Map of Antarctica. **(b)** Geologists unloading a cargo of tents, sleds, and snowmobiles from the tail of a C-130 Hercules transport plane that has just landed in a snowfield. Note the large skis over the wheels. **(c)** Geologists sledding to a field area in Antarctica. The sleds carry a month's worth of food, sample bags, rock hammers, and notebooks as well as tents and clothes (and a case of frozen beer).

FIGURE P.2 A geologist studying exposed rocks on a mountain slope on the desert island of Zabargad, in the Red Sea, off the coast of Egypt.

plane passed beyond the horizon, the silence of Antarctica hit us—no trees rustled, no dogs barked, and no traffic rumbled in this stark land of black rock and white ice. It would take us a day and a half to haul our sleds of food and equipment down to our study site (►Fig. P.1c). All this to look at a few dumb rocks?

Geologists, scientists who study the Earth, explore remote regions such as Antarctica almost routinely. Such efforts often strike people in other professions as a strange way to make a living. The Scottish poet Walter Scott (1771–1832), when describing geologists at work, said: "Some rin uphill and down dale, knappin' the chucky stones to pieces like sa' many roadmakers run daft. They say it is to see how the warld was made!" Indeed—to see how the world was made, to see how it continues to evolve, to find its valuable resources, to prevent contamination of its waters and soils, and to predict its dangerous movements. That is why geologists spend months at sea drilling holes in the ocean floor, why they scale mountains (►Fig. P.2), camp in rain-drenched jungles, and trudge through scorching desert winds. That is why geologists use electron microscopes to examine the atomic structure of minerals, use mass spectrometers to define the composition of rock and water, and use supercomputers to model the paths of earthquake waves. For over two centuries, geologists have pored over the Earth—in search of ideas to explain the processes that form and change our planet.

P.2 WHY STUDY GEOLOGY?

Geology, or geoscience, is the study of the Earth. Not only do geologists address academic questions, such as the formation and composition of the Earth, the causes of earthquakes (►Fig. P.3) and ice ages, the history of mountain building, and the evolution of life, they also address practical problems, such as how to prevent groundwater contamination, how to find oil and minerals, and how to stabilize slopes. And in recent years, geologists have contributed to the study of global climate change. When news reports begin with "Scientists say . . ." and then continue with "an earthquake occurred today off Japan," or "landslides will threaten the city," or "contaminants from the proposed toxic waste dump will destroy the town's water supply," or "there's only a limited supply of oil left," the scientists referred to are geologists. Because geologists address so many different kinds of problems, it's convenient to divide geology into many different specialities, just as it's convenient to divide medicine into many specialities (cardiology, psychiatry, hematology, and so on). ►Table P.1 lists some of the many subdiscipines of geology.

The fascination of geology attracts many people to careers in this science. Thousands of geologists work for oil, mining, water, engineering, and environmental companies, while a smaller number work in universities, government geological surveys, and research laboratories. Nevertheless, since the majority of students reading this book will not become professional geologists, it's fair to ask the question, Why, in general, should people study geology?

First, geology may be one of the most practical subjects you can learn, for geologic phenomena and issues affect our daily lives, sometimes in unexpected ways. Think about the following questions:

- Do you live in a region threatened by landslides, volcanoes, earthquakes, or floods (Fig. P.3)? These are geologic natural hazards that destroy property and take lives.

- Are you worried about the price of energy or about whether there will be a war in an oil-supplying country? Oil, coal, and uranium are energy resources whose distribution is controlled by geologic processes.

- Do you ever wonder about where the copper in your home's wires come from? Metals come from geologic materials—ore deposits—found by geologists.

- Have you seen fields of green crops surrounded by desert and wondered where the water to irrigate the crops comes from? Most likely, the water comes from underground, where it fills cracks and pores in geologic materials.

- Would you like to buy a dream house on a coastal sandbar (a ridge of sand just offshore)? The surroundings look beautiful, but geologists suggest that, on a time scale of centuries, sandbars are temporary landforms, and your investment might disappear in the next storm.

FIGURE P.3 Human-made cities cannot withstand the vibrations of a large earthquake. These apartment buildings collapsed during an earthquake in Turkey.

Clearly, all citizens of the twenty-first century, not just professional geologists, will need to make decisions concerning Earth-related issues. And they will be able to make more reasoned decisions if they have a basic understanding of geologic phenomena. History is full of appalling stories of people who ignored geologic insight and paid a horrible price for their ignorance. Your knowledge of geology may help you to avoid building your home on a hazardous floodplain or fault zone, on an unstable slope, or along a rapidly eroding coast. With a basic understanding of groundwater, you may be able to save money when drilling an irrigation well, and with knowledge of the geologic controls on resource distribution, you may be able to invest more wisely in the resource industry.

Second, the study of geology gives you a perspective on the planet that no other field can. As you will see, the Earth is a complicated system—its living organisms, climate, and solid rock interact with one another in a great variety of ways. Geologic study reveals Earth's antiquity (it's about 4.57 billion years old) and demonstrates how the planet has changed profoundly during its existence. What was the center of the Universe to our ancestors becomes, with the development of geologic perspective, our "island in space" today, and what was an unchanging orb originating at the same time as humanity becomes a dynamic planet that existed long before people did.

Third, the study of geology puts human achievements and natural disasters in context. On the one hand, our cities seem to be no match for the power of an earthquake, and a rise in sea level may swamp all major population centers. But on the other hand, we are now changing the face of the land worldwide at rates that far exceed those resulting from natural geologic processes. By studying geology, you can develop a frame of reference for judging the extent and impact of changes.

Finally, when you finish reading this book, your view of the world will be forever colored by geologic curiosity. When you walk in the mountains, you will think of the many forces that shape and reshape the Earth's surface. When you hear about a natural disaster, you will have insight into the processes that brought it about. And when you next go on a road trip, the rock exposures next to the highway will no longer be gray, faceless cliffs, but will

TABLE P.1   Principal Subdisciplines of Geology (Geoscience)

Name	Subject of Study
Engineering geology	The stability of geologic materials at the Earth's surface, for such purposes as controlling landslides and building tunnels.
Environmental geology	Interactions between the environment and geologic materials, and the contamination of geologic materials.
Geochemistry	Chemical compositions of materials in the Earth and chemical reactions in the natural environment.
Geochronology	The age (in years) of geologic materials, the Earth, and extraterrestrial objects.
Geomorphology	Landscape formation and evolution.
Geophysics	Physical characteristics of the whole Earth (such as Earth's magnetic field and gravity field) and of forces in the Earth.
Hydrogeology	Groundwater, its movement, and its reaction with rock and soil.
Mineralogy	The chemistry and physical properties of minerals.
Paleontology	Fossils and the evolution of life as preserved in the rock record.
Petrology	Rocks and their formation.
Sedimentology	Sediments and their deposition.
Seismology	Earthquakes and the Earth's interior as revealed by earthquake waves.
Stratigraphy	The succession of sedimentary rock layers.
Structural geology	Rock deformation in response to the application of force.
Tectonics	Regional geologic features (such as mountain belts) and plate movements and their consequences.
Volcanology	Volcanic eruptions and their products

present complex puzzles of texture and color telling a story of Earth's history. Have you ever been bored while driving?

Instead of feeling miserable and confined, feel the bones of the earth as you ride past the exposed evidence of the planet's history. That's roadside geology, road food for the mind and eye.
—*James Gorman* (**New York Times**, *Nov. 16, 2001*)

P.3 WHAT ARE THE THEMES OF THIS BOOK?

A number of narrative themes appear (and reappear) throughout this text. Consider these themes, listed below, to be this book's "take-home message."

- *The Earth is a unique, evolving system*. Geologists increasingly recognize that the Earth is a complicated system; its interior, solid surface, oceans, atmosphere, and life forms interact in many ways to yield the landscapes and environment in which we live. Within this **Earth System**, chemical elements cycle between different types of rock, between rock and sea, between sea and air, and between all of these entities and life. Aside from material addded to the Earth by the impact of a meteorite, all the material involved in these cycles originates in the Earth itself—our planet is truly an island in space.

- *Plate tectonics explains many Earth processes*. Like other planets, Earth is not a homogeneous ball, but rather consists of concentric layers: from center to surface, Earth has a core, mantle, and crust. We live on the surface of the crust, where it meets the atmosphere and the oceans. In the 1960s, geologists recognized that the crust, together with the uppermost part of the underlying mantle, forms a 100- to 150-km-thick semi-rigid shell. Large cracks separate this shell into discrete pieces, called **plates**, which move very slowly relative to each other (▶Fig. P.4). The theory that describes this movement and its consequences is now known as the **theory of plate tectonics**, and it is the foundation for understanding most geologic phenomena. Although plates move very slowly, generally less than 10 cm (centimeters) a year, their movements yield earthquakes, volcanoes, and mountain ranges, and cause the distribution of continents to change over time.

- *The Earth is a planet*. The subtitle of this book, *Portrait of a Planet*, highlights the view that despite the uniqueness of Earth's system and inhabitants, Earth fundamentally can be viewed as a planet, formed like the other planets of the Solar System from dust and gas that encircled the newborn Sun. Though Earth resembles the other inner planets (Mercury, Venus, and

FIGURE P.4 Simplified map of the Earth's principal plates. The arrow on each plate indicates the direction the plate moves, and the length of the arrow indicates the plate's velocity (the longer the arrow, the faster the motion). We discuss the types of plate boundaries in Chapter 4.

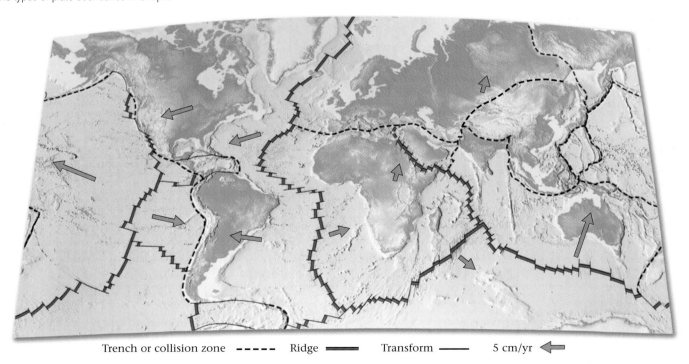

Trench or collision zone ----- Ridge ▬▬ Transform ▬▬ 5 cm/yr ⇐

Mars), it differs from them in having plate tectonics, an oxygen-rich atmosphere and liquid-water ocean, and abundant life. Further, because of the dynamic interactions among various aspects of the Earth System, our planet is constantly changing; the other inner planets are static.

- **The Earth is very old.** Geologic data indicate that the Earth formed 4.57 billion years ago—plenty of time for geologic processes to generate and destroy features of the Earth's surface, for life forms to evolve, and for the map of the planet to change. Plate-movement rates of only a few centimeters per year, if continuing for hundreds of millions of years, can move a continent thousands of kilometers. In geology, we have time enough to build mountains and time enough to grind them down, many times over! To define intervals of this time, geologists developed the **geologic time scale** (▶Fig. P.5). Geologists call the last 542 million years the Phanerozoic Eon, and all time before that the Precambrian. They further divide the Precambrian into three main intervals named, from oldest to youngest, the Hadean, the Archean, and the Proterozoic Eons. The Phanerozoic Eon is also divided into three main intervals named, from oldest to youngest, the Paleozoic, the Mesozoic, and the Cenozoic Eras. (Chapter 13 provides further details about geologic time.)

- **Internal and external processes interact at the Earth's surface.** Internal processes are those phenomena driven by heat from inside the Earth. Plate movement is an example, and since plate movements cause mountain building, earthquakes, and volcanoes, we call all of these phenomena internal processes as well. External processes are those phenomena driven by heat supplied by radiation coming to the Earth from the Sun. This heat drives the movement of air and water, which grinds and sculpts the Earth's surface and transports the debris to new locations, where it accumulates. The interaction between internal and external processes forms the landscapes of our planet. As we'll see, gravity plays an important role in both internal and external processes.

- **Geologic phenomena affect our environment.** Volcanoes, earthquakes, landslides, floods, and even more subtle processes such as groundwater flow and contamination or depletion of oil and gas reserves are of vital interest to every inhabitant of this planet. Therefore, throughout this book we emphasize linkages between geology and the environment.

- **Physical aspects of the Earth System are linked to life processes.** All life on this planet depends on physical features such as the minerals in soil; the temperature, humidity, and composition of the atmosphere; and the

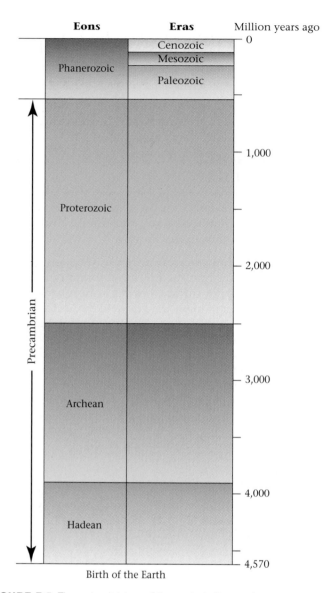

FIGURE P.5 The major divisions of the geologic time scale.

flow of surface and subsurface water. And life in turn affects and alters these same physical features. For example, the atmosphere's oxygen comes primarily from plant photosynthesis, a life activity. This oxygen in turn permits complex animals to survive, and affects chemical reactions among air, water, and rock. Without the physical Earth, life could not exist, but without life, this planet's surface might have become a frozen wasteland like that of Mars, or enshrouded in acidic clouds like that of Venus.

- **Science comes from observation, and people make scientific discoveries.** Science is not a subjective guess or an arbitrary dogma, but rather a consistent set of objective statements resulting from the application of the

scientific method (▶Box P.1). Every scientific idea must be constantly subjected to testing and possible refutation, and can be accepted only when supported by documented observations. Further, scientific ideas do not appear out of nowhere, but are the result of human efforts. Wherever possible, this book shows where geologic ideas came from and tries to answer the question, How do we know that?

- *The study of geology can increase general science literacy*. Studying geology provides an ideal opportunity to learn basic concepts of chemistry and physics, because these concepts can be applied directly to understanding tangible phenomena. Thus, in this book, where appropriate, basic concepts of physical science are introduced in boxed features called "Science Toolboxes." Also, Appendix A provides a systematic introduction to matter and energy, for those readers who have not learned this information previously or who need a review.

As you read this book, please keep these themes in mind. Don't view geology as a list of words to memorize, but rather as an interconnected set of concepts to digest. Most of all, enjoy yourself as you learn about what may be the most fascinating planet in the Universe.

Key Terms

Archean (p. 6)
Cenozoic (p. 6)
Earth System (p. 5)
geologic time scale (p. 6)
geologists (p. 3)
geology (p. 3)
Hadean (p. 6)
hypothesis (p. 8)
Mesozoic (p. 6)
Paleozoic (p. 6)
Phanerozoic (p. 6)

plates (p. 5)
Precambrian (p. 6)
Proterozoic (p. 6)
science (p. 7)
scientific law (p. 8)
scientific method (p. 7)
scientists (p. 7)
shatter cones (p. 8)
theory (p. 8)
theory of plate tectonics (p. 5)

Geopuzzle Revisited

A glance at the dirt (soil) beneath their feet makes geologists think of the processes that break rocks into a mass of loose grains in which plants can root. By studying the shape of the mountains, geologists imagine an earlier time when glaciers (rivers of ice) flowed slowly down the mountains, rasping and ripping at the underlying rock. By studying the rock itself, geologists picture an even earlier time when the present land surface lay kilometers underground, beneath a chain of erupting volcanoes.

BOX P.1 SCIENCE TOOLBOX

The Scientific Method

Sometime during the past 200 million years, a large block of rock or metal, which had been orbiting the Sun, crossed the path of Earth's orbit. In seconds, it pierced the atmosphere and slammed into our planet (thereby becoming a meteorite) at a site in what is now the central United States, today a landscape of flat cornfields. The impact released more energy than a nuclear bomb—a cloud of shattered rock and dust blasted skyward, and once-horizontal layers of rock from deep below the ground sprang upward and tilted on end in the gaping hole left by the impact. When the dust had settled, a huge crater, surrounded by debris spread over fractured land, marked the surface of the Earth at the impact site. Later in Earth history, running water and blowing wind wore down this jagged scar. Some 15,000 years ago, sand, gravel, and mud carried by a vast gla-

cier buried what remained, hiding it entirely from view (▶Fig. P.6a, b). Wow! So much history beneath a cornfield. How do we know this? It takes scientific investigation.

The movies often portray science as a dangerous tool, capable of creating Frankenstein's monster, and scientists as warped or nerdy characters with thick glasses and poor taste in clothes. In reality, **science** is simply the use of observation, experiment, and calculation to explain how nature operates, and **scientists** are people who study and try to understand natural phenomena. Scientists carry out their work using the **scientific method**, a sequence of steps for systematically analyzing scientific problems in a way that leads to verifiable results. Let's see how geologists employed the steps of the scientific method to come up with the meteorite-impact story.

1. *Recognizing the problem*: Any scientific project, like any detective story, begins by identifying a mystery. The cornfield mystery came to light when water drillers discovered limestone, a rock typically made of shell fragments, just below the 15,000-year-old glacial sediment. In surrounding regions, the rock at this depth consists of sandstone, made of cemented-together sand grains. Since limestone can be used to build roads, make cement, and produce the agricultural lime used in treating soil, workers stripped off the glacial sediment and built a quarry to excavate the limestone. They were amazed to find that rock layers exposed in the quarry tilted steeply and had been shattered by large cracks. In the surrounding regions, all rock layers are horizontal, like

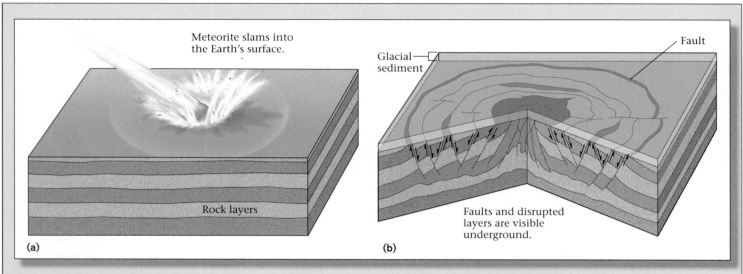

Meteorite slams into
the Earth's surface.

Rock layers

(a)

Glacial
sediment

Fault

Faults and disrupted
layers are visible
underground.

(b)

FIGURE P.6 (a) The site of an ancient meteorite impact in the American Midwest, during impact. Note the horizontal layers of rock below the ground surface. The thin lines represent boundaries between the successive layers. During impact, a large crater, surrounded by debris, formed. **(b)** The site of the impact today. The crater and the surface debris were eroded away. Relatively recently, the area was buried by gravel and sand brought by glaciers. Underground, the impact disrupted layers of rock by tilting them and by generating faults (fractures on which sliding occurs).

the layers in a birthday cake, the limestone layer lies underneath the sandstone, and the rocks contain relatively few cracks. Curious geologists came to investigate and soon realized that the geologic features of the land just beneath the cornfield presented a problem to be explained: What phenomena had brought limestone up close to the Earth's surface, tilted the layering in the rocks, and shattered the rocks?

2. *Collecting data*: The scientific method proceeds with the collection of observations or clues that point to an answer. Geologists studied the quarry and determined the age of its rocks, measured the orientation of rock layers, and documented (made a written or photographic record of) the fractures that broke up the rocks.

3. *Proposing hypotheses*: A scientific **hypothesis** is merely a possible explanation, involving only naturally occurring processes, that can explain a set of observations. Scientists propose hypotheses during or after their initial data

collection. The geologists working in the quarry came up with two alternative hypotheses. First, the features in this region could result from a volcanic explosion; and second, they could result from a meteorite impact.

4. *Testing hypotheses*: Since a hypothesis is no more than an idea that can be either right or wrong, scientists must put hypotheses through a series of tests to see if they work. The geologists at the quarry compared their field observations with published observations made at other sites of volcanic explosions and meteorite impacts, and studied the results of experiments designed to simulate such events. They learned that if the geologic features visible in the quarry were the result of volcanism, the quarry should contain rocks formed by the freezing of molten rock erupted by a volcano. But no such rocks were found. If, however, the features were the consequence of an impact, the rocks should contain **shatter cones**, small, cone-shaped cracks (▶Fig. P.7). Shatter cones can easily be overlooked, so the geologists returned to the quarry specifically to search for them, and found them in abundance. The impact hypothesis passed the test!

Theories are scientific ideas supported by an abundance of evidence; they have passed many tests and have failed none. Scientists have much more confidence in a theory than they do in a hypothesis. Continued study in the quarry eventually yielded so much evidence for impact that the

impact hypothesis came to be viewed as a theory. Scientists continue to test theories over a long time. Successful theories withstand these tests and are supported by so many observations that they come to be widely accepted. (As you will discover in Chapters 3 and 4, geologists consider the idea that continents drift around the surface of the Earth to be a theory, because so much evidence supports it.) However, some theories may eventually be disproven, to be replaced by better ones.

Some scientific ideas must be considered absolutely correct, for if they were violated, the natural Universe as we know it would not exist. Such ideas are called **scientific laws**; examples include the law of gravity.

FIGURE P.7 Shatter cones in limestone. These cone-shaped fractures, formed only by severe impact, open up in the direction away from the impact. At this locality, the cones open up downward, indicating that the impact came from above.

See for yourself...

How to Use *Google Earth*™ to See Geologic Features

In earlier editions of this book, we could provide photos of landscapes, but could not convey a 3-D image of a region from a variety of perspectives. Fortunately, web-based computer tools such as *Google Earth*™, *NASA World Wind*, or *Microsoft Virtual Earth*, now permit you to tour our planet at speeds faster than a rocket. You can visit millions of locations and *see for yourself* what geologic features look like in context. These tools provide a compilation of imagery in a format that permits you to go to locations quickly and view them from any elevation, perspective, or direction. You can examine boulders on a hillside, zoom to airplane height to see the landform that the boulders rest on, and then zoom to astronaut height to see where the landform lies in a continent. You can look straight down or obliquely, and you can stay in one position, circle around a location, or fly like a bird over a landscape.

To help you use computer-visualization programs to understand geology, each chapter of this book includes a *Geotour*, with locations and descriptions of geologic features relevant to the chapter. Because only *Google Earth*™ can be used on both PC and Mac computers (as of this writing), we key the Geotours to the *Google Earth*™ format. Below, we provide a brief introduction to the use of *Google Earth*™ in the context of this book. There's not room for a complete tutorial, but you'll find that the program is so easy to use that you will be proficient with it simply by working with it for a few minutes. The thumbnail images provided on this page are only to help identify tour sites. Go to wwnorton.com/studyspace to experience this flyover tour.

Opening *Google Earth*™

To use a Geotour, start by opening *Google Earth*™. As the program initializes, an image of the globe set in a background of stars appears in a window. The toolbar across the top provides a window icon. Click on this icon and a left sidebar appears (Image GP.1)—click on the icon again and the sidebar disappears so that the image becomes larger. The toolbar also provides a pushpin tool for marking locations, and a ruler tool for measuring distances. The sidebar contains information about locations and provides options for adding information to the screen image. For example, when you click on "borders," "roads," and "Populated Places" in the Layers panel, political boundaries, highways, and city names appear to provide a visual reference frame (Image GP.2). Any location you have marked with a pushpin appears in the sidebar within the Places panel. Clicking on the location in the sidebar will make the pushpin appear and will take you to the location. If you type a location in the Search panel at the top, a click on the magnifying glass icon will fly you to the location.

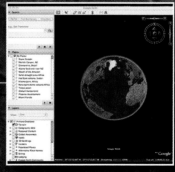

GP.1

GP.2

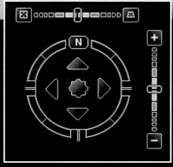

GP.3

GP.4

GP.5

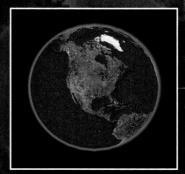

GP.6

GP.7

GP.8

The Basic Tools of *Google Earth*™

In the upper right corner of the screen (for *Google Earth*™ 4.X) are three tools (Image GP.3): (1) The horizontal bar tilts the image. You can either move the sliding bar, or click and hold on one of the Xs. Double clicking the upright X provides a view looking straight down, whereas double clicking on the tilted X provides a view looking horizontally. Note that your eye elevation Indicated in the lower right corner of the window) changes as you change your tilt. You can see this contrast by comparing Image GP.4 (vertical view) to Image GP.5 (oblique view, looking NW)—these images show a site in the Andes Mountains of Bolivia (at Lat 18°21'36.08"S, Long 66°0'13.95") from an altitude of 10 km (6 miles). Because of the nature of the imagery used by *Google Earth*™, steep cliffs may appear distorted. (2) The vertical bar zooms you up or down. You can either move the sliding bar or click on the end of the bar—the + end takes you to a lower elevation and enlarges the image, whereas the − end takes you to a higher elevation and reduces the image. The sliding bar offers better control. Using the "View" menu bar, you can display a bar scale in the lower left portion of the screen. (3) The compass tool allows you to rotate the image (if you are looking straight down) or fly around an image (if you are looking at it obliquely). Double click on the N button to reorient the image with north at the top.

Within the window showing the Earth, you can see a hand-shaped cursor. When you drag the cursor across the screen, the image moves. If you quickly drag the hand cursor, while holding down on your mouse, and then let go, the movement will continue. Thus, you can set the globe spinning (from a distant view) or get the feeling that you are flying across the landscape (at lower elevations).

Working with Location and Elevation
(Lat 43°8'18.84"N, Long 77°34'18.36"W)

On the bottom rule of the window, you will see three information items. The location of the point just beneath the hand-shaped cursor on the screen is specified on the left in terms of latitude (degrees, minutes, and seconds north or south of the equator) and longitude (degrees, minutes, and seconds east or west of the of the prime meridian). Just to the right of the Lat/Long information, a number indicates the elevation of the ground surface just below the hand-shaped cursor. The next number to the right tells you how much of the image has streamed onto your computer—an image starts out blurry, and then as streaming approaches 100%, it becomes clearer. The number on the far right ("Eye alt") indicates your viewing elevation.

To practice using these tools, enter the latitude and longitude provided above and zoom to a viewpoint 25,000 km (15,500 miles) out in space. You will see all of North America (Image GP.6). If you zoom down to 1800 m (5900 ft), you can see the details of Cobb's Hill Park in Rochester, New York (Image GP.7). Tilt your image so you just see the horizon, and rotate the image so you are looking SW (Image GP.8). With this perspective, you realize that the bean-shaped reservoir sits on top of an elongate ridge.

The Issue of Resolution
(Lat 32°10'58.91"S, Long 137°57'4.96"E)

The sharpness or clarity of images on *Google Earth*™ depends on the resolution of the image that is available in the database. High-resolution images are clear at high magnification, whereas low-resolution images appear grainy or pixilated at high magnification. Because the globe that you see is a compilation of many separate images merged by computer, some regions look like a patchwork. To see this effect, fly to the coordinates provided in South Australia, and zoom to an altitude of 2000 km (1242 miles) (Image GP.9). On this image, dark greenish land areas are from a low-resolution image set, whereas light brown areas are from a high-resolution set. Zoom down to an elevation of 5 km (3 miles) at this location. The view straddles a resolution boundary, with low resolution on the left and high resolution on the right (Image GP.10).

GP.9

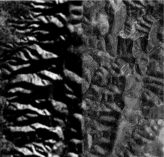

GP.10

Finding Locations Using Latitude and Longitude
(Lat 48°52'25.26"N, Long 2°17'42.18"E)

You can find towns, parks, and landmarks on the Earth using *Google Earth*™ by entering the name in the search panel in the upper left corner. For example, enter Arc de Triomphe, hit the return button, and you fly to this well known landmark in the center of Paris, France (Image GP.11). Many of the places you will visit in Geotours of this book are not, however, near a well-known landmark. Thus, to get you to a locality, we provide a latitude and longitude in the form:

Lat 48°52'25.44"N, Long 2°17'42.31"E

The first number is the latitude in degrees, minutes (60' = 1°), and seconds (60.00" = 1'). Simply enter the latitude and longitude on your screen. When typing numbers into the space provided on the program, you can abbreviate as:

48 52 25.44N, 2 17 42.31E

Note that instead of typing in latitude and longitude, you can simply copy the ".kmz" files provided in this book's website onto your computer. When you double-click on a file, it will open *Google Earth*™ and add all the locations referred to in the book to the "Temporary Places" folder. Then you simply need to click on the particular image referred to in a Geotour and the program will fly you straight there.

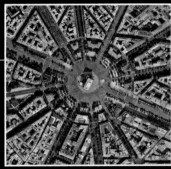

GP.11

Enjoy!

Please take advantage of the Geotours provided in this book. They provide you with an understanding of geology that simply can't appear on a printed page, and they are, arguably, as fun as a video game. Take control of the *Google Earth*™ tools, change your elevation and perspective, and fly around the landscape. Geology will come alive for you, and exploration of the Earth may become your hobby.

Our Island in Space

When you look out toward the horizon from a mountain top, the Earth seems endless, and before the modern era, many people thought it was. But to astronauts flying to the Moon, the Earth is merely a small, shining globe—they can see half the planet in a single glance. From the astronauts' perspective, it appears that we are riding on a small island in space.

Earth may not be endless, but it is a very special planet: its temperature and composition, unlike those of the other planets in the Solar System, make it habitable. In Part I of this book, we first learn scientific ideas about how the Earth, and the Universe around it, came to be. Then we take a quick tour of the planet to get a sense of its composition and its various layers. With this background, we're ready to encounter the twentieth-century revolution in geology that yielded the set of ideas we now call the theory of plate tectonics. We'll see that this theory, which proposes that the outer layer of the Earth is divided into plates that move with respect to each other, provides a rational explanation for a great variety of geologic features—from the formation of continents to the distribution of fossils. In fact, geologists now recognize that plate interactions even led to the formation of gases from which the atmosphere and oceans formed, and without which life could not exist.

A photograph of the Earth as seen by the Apollo 17 astronauts. This image emphasizes that our planet is a sphere with finite limits. But it's a very special sphere because the air, land, water, and life on our planet all interact in ways found no where else in the Solar System.

Cosmology and the Birth of Earth

Geopuzzle

How did the Solar System (including the Earth) form, and what is the source of the material from which it formed?

When the Hubble Space Telescope looks into what, to the naked eye, appears to be the black void of the night sky, it reveals a spectacle of disks and spirals of hazy light. Each of these is a distant galaxy, a cluster of as many as 300 billion stars. This is the fabric of space.

This truth within thy mind rehearse,

That in a boundless Universe

Is boundless better, boundless worse.

—Alfred, Lord Tennyson (British poet, 1809–1892)

1.1 INTRODUCTION

Sometime in the distant past, more than 100,000 generations ago, humans developed the capacity for complex, conscious thought. This amazing ability, which distinguishes our species from all others, brought with it the gift of curiosity, an innate desire to understand and explain the workings of ourselves and all that surrounds us—our **Universe**. Questions that we ask about the Universe differ little from questions a child asks of a playmate: Where do you come from? How old are you? Such musings first spawned legends in which heroes, gods, and goddesses used supernatural powers to sculpt the landscape. Increasingly, science, the systematic analysis of natural phenomena, has provided insight into these questions. However, the development of **cosmology**, the study of the overall structure and evolution of the Universe, has proven to be a rough one, booby-trapped with tempting but flawed approaches and cluttered with misleading prejudices.

In this chapter, we begin with a brief historical sketch of cosmological thought. For brevity, we restrict the discussion to the Western tradition, though an equally rich history of ideas developed in other cultures. We then look at currently accepted ideas of modern scientific cosmology and the key discoveries that led to scientific ideas about how our planet fits into the fabric of a changing Universe. The chapter concludes with a description of Earth's formation as it may have occurred about 4.57 billion years ago.

1.2 AN EVOLVING IMAGE OF THE EARTH'S POSITION AND SHAPE

Three thousand years ago (1000 B.C.E.; "before the Common Era"), the Earth's human population totaled only several million, the pyramids of Egypt had already been weathering in the desert for 1,600 years, and Homer, the great Greek poet, was compiling the *Iliad* and the *Odyssey*. In Homer's day, astronomers of the Mediterranean region knew the difference between stars and planets. They had observed that the positions of stars remained fixed relative to each other but that the whole star field slowly revolved around a fixed point (►Box 1.1), while the planets moved relative to the

stars and to each other, etching seemingly complex paths across the night sky. In fact, the word "planet" comes from the Greek word *planēs*, which means "wanderer." Despite their knowledge of the heavens, people of Homer's day did not realize fully that Earth itself is a planet. Some envisioned the Earth to be a flat disk, with land toward the center and water around the margins, that lay at the center of a celestial sphere, a dome to which the stars were attached. This disk supposedly lay above an underworld governed by the fearsome god Hades. Placing the Mediterranean region at the center of the Universe must have made people of that region feel quite important indeed! Philosophers also toyed with numerous explanations for the Sun: to some, it was a burning bowl of oil, and to others, a ball of red-hot iron. Most favored the notion that movements of celestial bodies represented the activities of gods and goddesses.

Beginning around 600 B.C.E., philosophers in the Mediterranean region began to argue about the structure of the Universe. Some advocated a **geocentric model** (►Fig. 1.3a), in which the Earth sits motionless at the center of the Universe while the Sun and all planets follow perfectly circular orbits around it. Others advocated a **heliocentric model**, in which all planets, including the Earth, orbit the Sun (►Fig. 1.3b). The geocentric model came to be favored by most people, perhaps because it appealed to human vanity—it placed the Earth at the most important point in the Universe and implied that humans were the Universe's most important creatures. The model gained credibility when Ptolemy (100–170 C.E.), an Egyptian mathematician, used it to develop equations that appeared to predict the future positions of planets. Ptolemy's calculations were so influential that the geocentric model became religious dogma in Europe for the next 1,400 years. During this period, known as the Middle Ages, anyone who disagreed with Ptolemy risked charges of heresy.

Then came the Renaissance. The very word means "rebirth" or "revitalization," and in fifteenth-century Europe, bold thinkers spawned a new age of exploration and discovery. As the Renaissance dawned, Nicolaus Copernicus (1473–1543) reintroduced and justified the heliocentric concept in a book called *De revolutionibus* (Concerning the Revolutions) but, perhaps fearing the wrath of officials, published the book just days before he died (Fig. 1.3b). *De revolutionibus* did indeed spark a bitter battle that pitted astronomers such as Johannes Kepler (1571–1630) and Galileo (1564–1642) against the establishment. Kepler showed that the planets follow elliptical, not circular, orbits, and thus demonstrated that Ptolemy's calculations, based on the assumption that orbits are circular, were wrong. Galileo, using the newly invented telescope, observed that Venus has phases like our own Moon (a characteristic that could only be possible if Venus orbited the Sun) and that Jupiter has its own moons. His discoveries demonstrated that all heavenly bodies do not revolve

BOX 1.1 THE REST OF THE STORY

How Do We Know That Earth Rotates?

As you sit reading this book, you probably aren't conscious that you are moving. But in fact, because of the Earth's spin, you actually are moving rapidly around Earth's axis (the imaginary line that connects the North and South Poles). In fact, a person sitting on the equator is hurtling along at about 1,674 km/h (1,040 mph)—faster than the speed of sound! How do we know that the Earth spins around its axis? The answer comes from observing the apparent motion of the stars (▶Fig. 1.1).

If you gaze at the night sky for a long time, you'll see that the stars move in a circular path around the North Star. Curiously,

it was not until the middle of the nineteenth century that Jean-Bernard-Léon Foucault (1819–1868), a French physicist, proved that the Earth spins on its axis. He made this discovery by setting a heavy pendulum, attached to a long string, in motion. As the pendulum continued to swing, Foucault noted that the plane in which it oscillated (a plane perpendicular to the Earth's surface) appeared to rotate around a vertical axis (a line perpendicular to the Earth's surface). If Newton's first law of motion—objects in motion remain in motion, objects at rest remain at rest—was correct, then this phenomenon required that the Earth rotate

under the pendulum while the pendulum continued to swing in the same plane (▶Fig. 1.2a, b). Foucault displayed his discovery beneath the great dome of the Pantheon in Paris, to much acclaim.

We now know that, in fact, the Earth's spin axis is not fixed in orientation; rather, it wobbles. This wobble, known as **precession**, is like the wobble of a toy top as it spins. We'll see later in this book that the precession of the Earth's axis, which takes 23,000 years, may affect the planet's climate.

FIGURE 1.1 Time exposure of the night sky over an observatory. Note that the stars appear to be fixed relative to each other, but that they rotate around a central point, the North Star. This motion is actually due to the rotation of the Earth on its axis.

FIGURE 1.2 Foucault's experiment. **(a)** An oscillating pendulum at a given time. **(b)** The same pendulum at a later time. The pendulum stays in the same plane, but the Earth, and hence the frame, rotates.

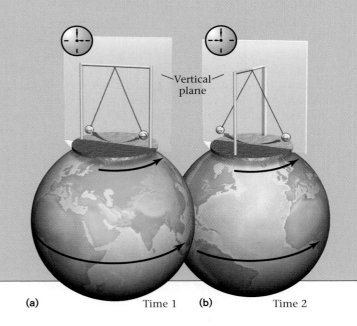

Vertical plane

(a)   Time 1   (b)   Time 2

around the Earth. So Galileo's observations also contradicted the geocentric hypothesis. But Europe was not quite ready for Galileo. Officials charged him with heresy, and he spent the last ten years of his life under house arrest.

In the year of Galileo's death, Isaac Newton (1642–1727), perhaps the greatest scientist of all time, was born in England. Newton derived mathematical laws governing gravity and basic mechanics (the movement of

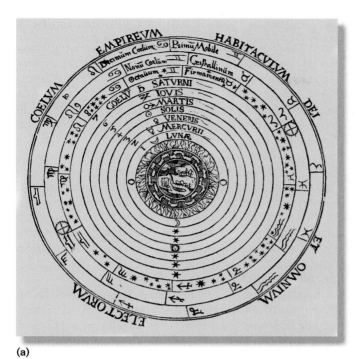

(a)

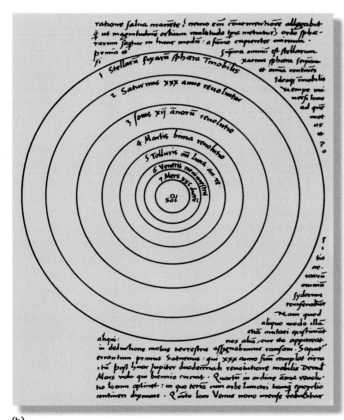

(b)

FIGURE 1.3 **(a)** The geocentric model of the Universe. Earth, at the center, is surrounded by air and fire and the Moon, Mercury, Venus, the Sun, Mars, Jupiter, and Saturn. Everything lies within the globe of the stars. **(b)** The heliocentric model of the Universe, as illustrated in this woodcut from Copernicus's *De revolutionibus*. "Sol" is the sun.

objects in space) and thus provided the tools for explaining how physical processes operate in nature. Newton was able to use his laws of motion and gravity to show why planetary orbits are elliptical, as Copernicus proposed. After Newton, support for a geocentric model of the Universe vanished. Thus, as the seventeenth century came to a close, people for the first time possessed a clear image of the movements of planets. Unfortunately for human self-esteem, Earth had been demoted from its place of prominence at the center of the Universe and it became merely one of many planets circling the Sun.

> **Take-Home Message**
>
> The Sun lies at the center of the Solar System and is but one star at the edge of one galaxy in a Universe of hundreds of billions of galaxies.

As we noted earlier, the ancient Greeks, like people of many other cultures, originally considered the Earth to be a flat disk. But by the time of Aristotle (c. 257–180 B.C.E.), many philosophers realized that the Earth had to be a sphere, because they could observe sailing ships disappear progressively from base to top as the ships moved beyond the horizon, and they could see that the Earth cast a curved shadow on the Moon during an eclipse. In fact, Ptolemy had developed the concepts of latitude and longitude to define locations on a spherical Earth. Thus, though a few clerics espoused the flat Earth view through the Middle Ages, it is likely that nearly everyone had rejected the idea as the Renaissance began. When Ferdinand Magellan successfully circumnavigated the Earth in 1520, the image of the Earth as a globe became firmly established.

1.3 A SENSE OF SCALE

We use enormous numbers to describe the size of the Earth, the distance from the Earth to the Sun, the distance between stars, and the distance between galaxies. Where do these numbers come from?

How Can We Calculate the Circumference of the Earth?

The Greek astronomer Eratosthenes (c. 276–194 B.C.E.) served as chief of the library in Alexandria, Egypt, one of the great ancient centers of learning in the Mediterranean region. One day, while filing papyrus scrolls, he came across a report noting that in the southern Egyptian city of Syene, the Sun lit the base of a deep vertical well precisely at noon on the first day of summer. Eratosthenes deduced that the Sun's rays at noon on this day must be exactly perpendicular to the Earth's surface at Syene, and that if the Earth was spherical, then the Sun's rays could *not* simultaneously be perpendicular to the Earth's surface at Alexandria, 800 km to the north (►Fig. 1.4). So on the first day of summer, Eratosthenes measured the shadow cast by a tower in Alexandria at noon. The angle between the tower and the Sun's rays, as indicated by the shadow's length, proved to be 7.2°. He then commanded a servant to pace out a straight line from Alexandria to Syene. The sore-footed servant found the distance to be 5,000 stadia (1 stadium = 0.1572 km). Knowing that a circle contains 360°, Eratosthenes then calculated the Earth's circumference as follows:

$$\frac{360°}{x} = \frac{7.2°}{5,000 \text{ stadia}}$$

$$x = \frac{360° \times 5,000 \text{ stadia}°}{7.2°} = 250,000 \text{ stadia}$$

$$250,000 \text{ stadia} \times 0.1572 \text{ km/stadium}$$
$$= 39,300 \text{ km} = 24,421 \text{ miles}$$

Thus, twenty-two centuries ago, Eratosthenes determined the circumference of the Earth to within 2% of today's accepted value (40,008 km, or 24,865 miles) without the aid of any sophisticated surveying equipment—a truly amazing feat.

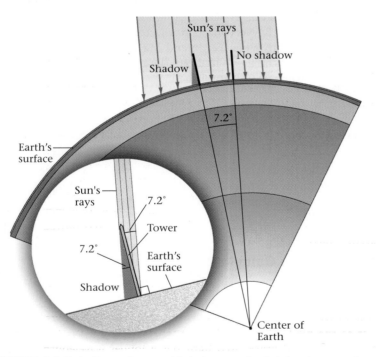

FIGURE 1.4 Eratosthenes discovered that at noon on the first day of summer, the Sun's rays were perpendicular to the Earth's surface (that is, parallel to the radius of the Earth) at Syene, but made an angle of 7.2° with respect to a vertical tower at Alexandria. Thus, the distance between Alexandria and Syene represented 7.2°/360° of the Earth's circumference. Knowing the distance between the two cities, therefore, allowed him to calculate the circumference of the Earth.

The Distance from Earth to Celestial Objects

Around 200 B.C.E., Greek mathematicians, using ingenious geometric calculations, determined that the distance to the Moon was about thirty times the Earth's diameter, or 382,260 km. This number comes close to the true distance, which on average is 381,555 km (about 237,100 miles). But it wasn't until the seventeenth century that astronomers figured out that the mean distance between the Earth and the Sun is 149,600,000 km (about 93,000,000 miles). As for the stars, the ancient Greeks realized that they must be much farther away than the Sun in order for them to appear as a fixed backdrop behind the Moon and planets, but the Greeks had no way of calculating the actual distance. Our modern documentation of the vastness of the Universe began in 1838, when astronomers found that the nearest star to Earth, Alpha Centauri, lies 40.85 trillion km away.

Since it's hard to fathom the distances to planets and stars without visualizing a more reasonably sized example, imagine that the Sun is the size of an orange. At this scale, the Earth would be a grain of sand at a distance of 10 meters (m) (30 feet) from the orange. Alpha Centauri would lie 2,000 km (about 1,243 miles) from the orange.

When astronomers realized that light travels at a constant (i.e., unchanging) speed of about 300,000 km (about 186,000 miles) per second, they realized that they had a way to describe the huge distances between objects in space conveniently. They defined a large distance by stating how long it takes for light to traverse that distance. For example, it takes light about 1.3 seconds to travel from the Earth to the Moon, so we can say that the Moon is about 1.3 light seconds away. Similarly, we can say that the Sun is 8.3 light minutes away. A **light year**, the distance that light travels in one Earth year, equals about 9.5 trillion km (about 6 trillion miles). When you look up at Alpha Centauri, 4.3 light years distant, you see light that started on its journey to Earth about 4.3 years ago.

Astronomers didn't develop techniques for measuring the distance to very distant stars and galaxies until the twentieth century. With these techniques (see an astronomy book for details), they determined that the farthest celestial objects that can be seen with the naked eye are over 2.2 million light years away. Powerful telescopes allow us to see much farther. The edge of the *visible* Universe lies over 13 billion light years away, which means that light traveling to Earth from this location began its journey about 9 billion years before the Earth even existed. When such numbers became available by the middle of the twentieth century, people came to the realization that the dimensions of the Universe are truly staggering.

> **Take-Home Message**
>
> The Universe is immense. Galaxies at the outer edge of the Universe lie over 13 billion light years away (i.e., over 120 trillion km, or 78 trillion miles, away).

1.4 THE MODERN IMAGE OF THE UNIVERSE

We've seen that the burst of discovery during the Renaissance forced astronomers to change their view of Earth's central place in the Universe. Eventually, they realized that the Earth is but one of many objects in the Solar System (the Sun and all objects that travel around it). They also learned that stars are not randomly scattered through the Universe; gravity pulls them together to form immense systems, or groups, called **galaxies**. The Sun is but one of over 300 billion stars that together form the Milky Way, and the Milky Way is but one of more than 100 billion galaxies constituting the visible Universe (see chapter opening photo). Galaxies are so far away that, to the naked eye, they look like stars in the night sky. The nearest galaxy to ours, Andromeda, lies over 2.2 million light years away.

If we could view the Milky Way from a great distance, it would look like a flattened spiral, 100,000 light years across, with great curving arms gradually swirling around a glowing, disk-like center (▶Fig. 1.5). Presently, our Solar System lies near the outer edge of one of these arms and rotates around the center of the galaxy about once every 250 million years. We hurtle through space, relative to an observer standing outside the galaxy, at about 200 km per second.

FIGURE 1.5 An image of what the Milky Way might look like if viewed from outside. Note that the galaxy consists of spiral arms around a central cluster. Our Sun lies at the edge of one of these arms.

1.5 HOW DID THE UNIVERSE FORM?

Do galaxies move with respect to other galaxies? Does the Universe become larger or smaller with time? Has the Universe always existed? Answers to these fundamental questions came from an understanding of a phenomenon called the Doppler effect. Though the term may be unfamiliar, the phenomenon it describes is an everyday experience. After introducing the Doppler effect, we show how an understanding of it leads to a theory of Universe formation.

How Can We Determine if a Star is Moving? The Doppler Effect

When a train whistle screams, the sound you hear has moved through the air from the whistle to your ear in the form of sound waves. (Waves are disturbances that transmit energy from one point to another by causing periodic motions.) As each wave passes, air alternately compresses, then expands. The pitch of the sound, meaning its note in the musical scale, depends on the frequency of the sound waves, meaning the number of waves that pass a point in a given time interval. Now imagine that as you are standing on the station platform, the train moves toward you. The sound of the

whistle gets louder as the train approaches, but its pitch remains the same. Then, the instant the train passes, the pitch abruptly changes; it sounds like a lower note in the musical scale. An Austrian physicist, C. J. Doppler (1803–1853), first interpreted this phenomenon, and thus it is now known as the **Doppler effect**. When the train moves toward you, the sound has a higher frequency (the waves are closer together), because the sound source, the whistle, has moved slightly closer to you between the instant that it emits one wave and the instant that it emits the next (▶Fig. 1.6a, b). When the train moves away from you, the sound has a lower frequency (the waves are farther apart), because the whistle has moved slightly farther from you between the instant it emits one wave and the instant it emits the next.

Light energy also moves in the form of waves. In shape, light waves somewhat resemble water waves. Visible light comes in many colors—the colors of the rainbow. The color of light you see depends on the frequency of the light waves, just as the pitch of a sound you hear depends on the frequency of sound waves. Red light has a longer wavelength (lower frequency) than blue light (▶Fig. 1.7a, b). The Doppler effect also applies to light but is noticeable only if the light source moves very fast (e.g., at least a few percent of the speed of light). If a light source moves away from you, the light you see becomes redder (as the light shifts to lower frequency). If the source moves toward you, the light you see becomes bluer (as the light shifts to higher frequency). We call these changes the red shift and the blue shift, respectively (▶Fig. 1.7c).

Does the Size of the Universe Change?

In the 1920s, astronomers such as Edwin Hubble (after whom the Hubble Space Telescope was named) braved many a frosty night beneath the open dome of a mountaintop observatory in order to aim telescopes into deep space. These researchers had begun a search for distant galaxies. At first, they documented only the location and shape of newly discovered galaxies. But then, one astronomer began an additional project to study the wavelength of light produced by the distant galaxies. The results yielded a surprise that would forever change humanity's perception of the Universe.

Astronomers found, to their amazement, that the light of distant galaxies displayed red shifts relative to the light of nearby stars. Hubble pondered this mystery and, around 1929, realized that the red shifts must be a consequence of the Doppler effect—and thus that the distant galaxies must be moving away from Earth at an immense velocity. At the

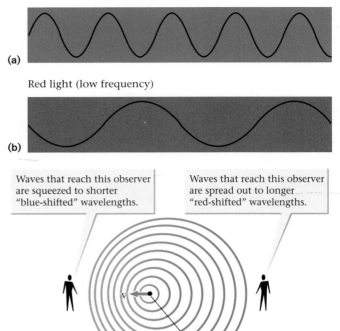

FIGURE 1.7 Light waves resemble ocean waves in shape, but physically they are quite different. **(a)** Blue light has a relatively short wavelength (higher frequency). **(b)** Red light has a relatively long wavelength (lower frequency). **(c)** The shift in light frequency that an observer sees depends on whether the source is moving toward or away from the observer.

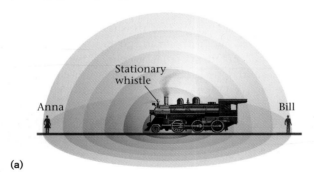

FIGURE 1.6 The spacing of waves is wavelength. Wavelength, and, therefore, frequency (the number of waves passing a point in an interval of time) are affected by the speed of the source. Frequency determines pitch. **(a)** Sound emanating from a stationary source has the same wavelength in all directions (the circular shells represent the waves), and Anna and Bill hear the same pitch. **(b)** If the source is moving toward Anna, she hears a shorter-wavelength sound than does Bill. Therefore, Anna hears a higher-pitched (higher-frequency) sound than does Bill.

time, astronomers thought the Universe had a fixed size, so Hubble initially assumed that if some galaxies were moving away from Earth, others must be moving toward Earth. But this was not the case. On further examination, Hubble realized that the light from *all* distant galaxies, regardless of their direction from Earth, exhibits a red shift. In other words, *all* distant galaxies are moving rapidly away from us.

How can all galaxies be moving away from us, regardless of which direction we look? Hubble puzzled over this question and finally recognized the solution. The whole Universe must be expanding! To picture the expanding Universe, imagine a ball of bread dough with raisins scattered throughout. As the dough bakes and expands into a loaf, each raisin moves away from its neighbors, in every direction (▶Fig. 1.8a). This idea came to be known as the **expanding Universe theory**.

Hubble's expanding Universe theory marked a revolution in thinking. No longer could we view the Universe as being fixed in dimension, with galaxies locked in position. Now we see the Universe as an expanding bubble, in which galaxies race away from each other at incredible speeds.

This image immediately triggers the key question of cosmology: Did the expansion begin at some specific time in the past? If it did, then this instant would mark the beginning of the Universe, the beginning of space and time.

The Big Bang

Most astronomers have concluded that expansion did indeed begin at a specific time, with a cataclysmic explosion called the **big bang**. According to the big bang theory, all matter and energy—everything that now constitutes the Universe—was initially packed into an infinitesimally small point. For reasons that no one fully understands, the point exploded, according to current estimates, 13.7 (± 1%) billion years ago. Since the big bang, the Universe has been continually expanding (▶Fig. 1.8b).

> **Take-Home Message**
>
> According to the big bang theory, the Universe started in a cataclysmic explosion and has been expanding ever since. Distant galaxies move away from Earth at immense speeds.

1.6 MAKING ORDER FROM CHAOS

Aftermath of the Big Bang

Of course, no one was present at the instant of the big bang, so no one actually saw it happen. But by combining clever calculations with careful observations, researchers have developed a consistent model of how the Universe evolved, beginning an instant after the explosion (▶Fig. 1.9).

According to the contemporary model of the big bang, profound change happened at a fast and furious rate at the outset. During the first instant of existence, the Universe was so small, so dense, and so hot that it consisted entirely of energy—atoms, or the smallest subatomic particles that make up atoms, could not even exist. But within a few seconds, it had cooled sufficiently for the smallest atoms, hydrogen atoms, to form. By the time the Universe reached an age of 3 minutes, its temperature had fallen below 1 billion degrees, and its diameter had grown to about 100 billion km (60 billion miles). Under these conditions, nuclei of new atoms began to form through

(a)

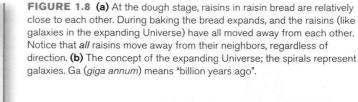

FIGURE 1.8 (a) At the dough stage, raisins in raisin bread are relatively close to each other. During baking the bread expands, and the raisins (like galaxies in the expanding Universe) have all moved away from each other. Notice that *all* raisins move away from their neighbors, regardless of direction. **(b)** The concept of the expanding Universe; the spirals represent galaxies. Ga (*giga annum*) means "billion years ago".

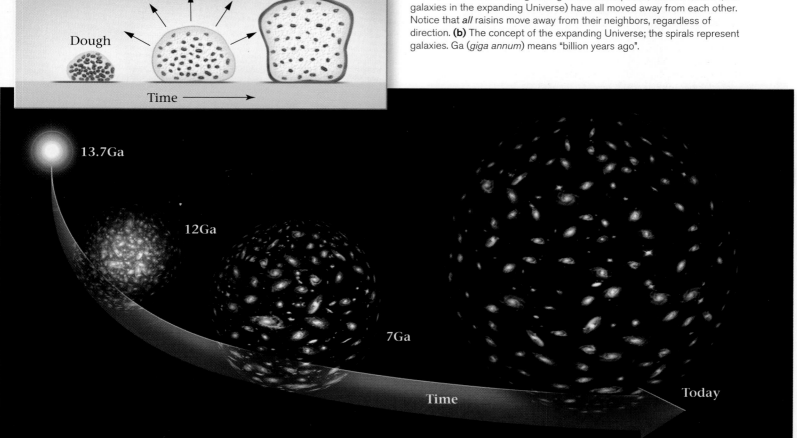

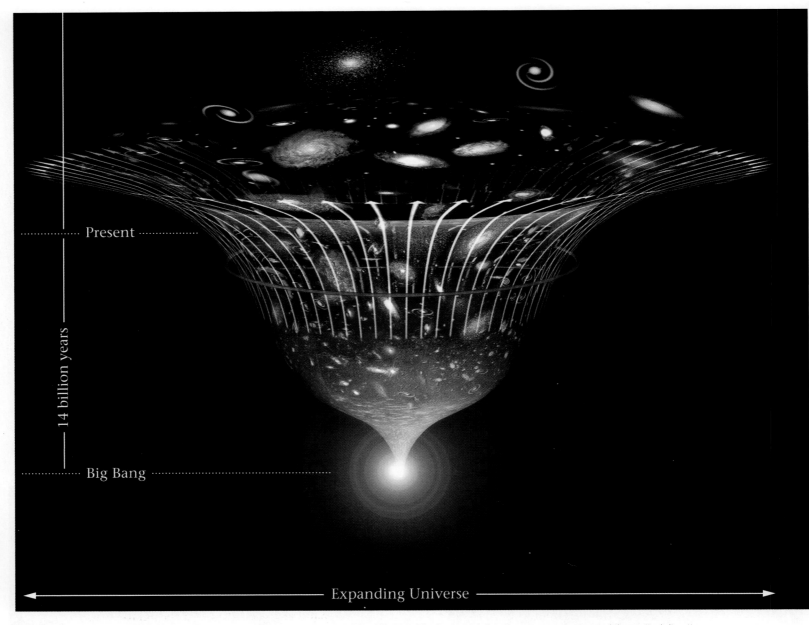

Present

14 billion years

Big Bang

Expanding Universe

FIGURE 1.9 An artist's rendition of the big bang, followed by expansion of the Universe. The horizontal direction represents size, and the vertical direction represents time. Recent work suggests that the rate of expansion has changed as time has passed.

the collision and fusion (sticking together) of hydrogen atoms (see Appendix A). Formation of new nuclei by fusion reactions at this time is called "big bang nucleosynthesis" because it happened *before* any stars existed. Big bang nucleosynthesis could produce only small atoms (such as helium), meaning ones containing a small number of protons, and it happened very rapidly. In fact, virtually all of the new atomic nuclei that would form by big bang nucleosynthesis had formed by the end of 5 minutes.

Why did nucleosynthesis stop? When it reached an age of 5 minutes, the Universe had expanded so much that its average density had decreased to 0.1 kg/m^3, a value one-tenth that of water, meaning, the atoms were so far apart that they rarely collided.

For the next interval of time, the Universe consisted of nuclei dispersed in a turbulent sea of wandering electrons. Physicists refer to such a material as a plasma. After a few hundred thousand years, temperature dropped below a few thousand degrees. Under these conditions, neutral atoms (in which negatively charged electrons orbit a positively charged nucleus) could form. Eventually, the Universe became cool enough for chemical bonds to bind atoms of certain elements together in molecules—most notably, two hydrogen atoms could join to form molecules of H_2. As the Universe continued to expand and cool further, atoms and molecules slowed down and accumulated into **nebulae**, patchy clouds of gas. Gas making up the earliest nebulae of the Universe consisted entirely of the smallest atoms, namely, hydrogen (74%), helium (25%), and traces of lithium, beryllium, and boron.

Forming the First Stars

When the Universe reached its 200 millionth birthday, it contained immense, slowly swirling, dark nebulae separated by vast voids of empty space (▶Fig. 1.10). The Universe could not remain this way forever, however, because of the invisible but persistent pull of gravity. Eventually, gravity began to remold the Universe pervasively and permanently.

All matter exerts gravitational pull—a type of force—on its surroundings, and as Isaac Newton first pointed out, the amount of pull depends on the amount of mass. Somewhere in the young Universe, the gravitational pull of an initially denser region of a nebula began to pull in surrounding gases and, in a grand example of "the rich getting richer," grew in mass and, therefore, density. As this denser region sucked in progressively more gas, more matter compacted into a smaller region, and the initial swirling movement of gas transformed into a rotation around an axis that became progressively faster and faster. (The same phenomenon occurs when a spinning ice skater pulls her arms inward and speeds up.) Because of rotation, the condensing portion of the nebula evolved into a spinning disk-shaped mass of gas called an **accretion disk** (see art spread, pp. 26–27). Eventually, the gravitational pull of the accretion disk became great enough to trigger wholesale inward collapse of the surrounding nebula. With all the additional mass available, gravity aggressively pulled the inner portion of the accretion disk into a dense ball. The energy of motion (kinetic energy), of gas falling into this ball, transformed into heat (thermal energy) when it landed on the ball. (The same phenomenon happens when you drop a plate and it shatters—if you measured the temperature of the pieces immediately after breaking, they would be slightly warmer.) Moreover, the squeezing together of jostling gas atoms and molecules in the ball increased the gas's temperature still further. (The same phenomenon happens in the air that you compress in a bicycle pump.) Eventually, the central ball of the accretion disk became hot enough to glow, and at this point it became a **protostar**.

FIGURE 1.10 Gases clump to form distinct nebulae, which look like clouds in the sky. In this Hubble Space Telescope picture, new stars are forming at the top of the nebula on the left. Stars that have already formed backlight the nebulae.

A protostar continues to grow, by pulling in successively more mass, until its core becomes very dense and its temperature reaches about 10 million degrees. Under such conditions, fusion reactions begin to take place; hydrogen nuclei in a protostar join, in a series of steps, to form helium nuclei. Such fusion reactions produce huge amounts of energy and make a star into a fearsome furnace. When the first nuclear fusion reactions began in the first protostar, the body "ignited" and the first true star formed. When this happened, the first starlight pierced the newborn Universe. This process would soon happen again and again, and many first-generation stars began to burn.

First-generation stars tended to be very massive (e.g., 100 times the mass of the Sun) because compared with the nebulae of today, nebulae of the very young Universe contained much more matter. Astronomers have shown that the larger the star, the hotter it burns and the faster it runs out of fuel and "dies." A huge star may survive from only a few million years to a few tens of millions of years before it dies by violently exploding to form a **supernova**.[1] Thus, not long after the first generation of stars formed, the Universe began to be peppered with the first generation of supernova explosions.

1. The name "supernova" comes from the Latin word *nova*, which means "new"; when the light of a supernova explosion reaches Earth, it looks like a very bright new star in the sky.

1.7 WE ARE ALL MADE OF STARDUST

Where Do Elements Come From?

Nebulae from which the first-generation stars formed consisted entirely of small atoms (elements with atomic numbers smaller than 5), because only these small atoms were generated by big bang nucleosynthesis. In contrast, the Universe of today contains ninety-two naturally occurring elements (see Appendix A). Where do the other eighty-seven elements come from? In other words, how did elements such as carbon, sulfur, silicon, iron, gold, and uranium form? These elements, which are common on Earth, have large atomic numbers. (For example, carbon has an atomic number of 6, and uranium has an atomic number of 92.) Physicists now realize that these elements didn't form during or immediately after the big bang. Rather, they formed later, during the life cycle of stars, by the process of *stellar nucleosynthesis*. Because of stellar nucleosynthesis, we can consider stars to be element factories, constantly fashioning larger atoms out of smaller atoms.

The specific reactions that take place during stellar nucleosynthesis depend on the mass of the star, because the temperature and density of a more massive star are greater than those of a less massive star; as temperature increases, the velocity of particles increases so larger nuclei can be driven together. Low-mass stars, like our Sun, burn slowly and may survive for 10 billion years. Nuclear reactions in these stars produce elements up to an atomic number of 6 (carbon). As we have seen, high-mass stars (10 to 100 times the mass of the Sun) burn quickly, and may survive for only 20 million years. They produce elements up to an atomic number of 26 (iron). Very large atoms (atoms with atomic numbers greater than that of iron) require even more violent circumstances to form than can occur within even a high-mass star. These atoms form most efficiently during a supernova explosion, though some form in massive stars.

Now you can understand why we call stars and supernova explosions "element factories." They fashion larger atoms—new elements—that had not formed during or immediately after the big bang. What happens to these atoms? Some escape from a star into space during the star's lifetime, simply by moving fast enough to overcome the star's gravitational pull. The stream of atoms emitted from a star during its lifetime is a **stellar wind** (▶Fig. 1.11). Escape also happens at the death of a star. Specifically, a low-mass star (like our Sun) releases a large shell of gas when it dies (as we will discuss in Chapter 23), whereas a high-mass star blasts matter into space during a supernova explosion (▶Fig. 1.12). Once in space, atoms form new nebulae or mix back into existing nebulae.

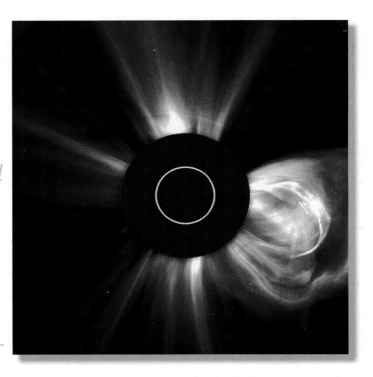

FIGURE 1.11 In this image, a black disk hides the Sun so we can see the stellar wind that our Sun produces. The white circle indicates the diameter of the Sun. Note that some of the particles shoot into space in long jets. Occasionally, archlike fountains suddenly erupt.

FIGURE 1.12 Very heavy elements form during supernova explosions. Here we see the rapidly expanding shell of gas ejected into space that appeared in 1054 C.E. This shell is called the Crab Nebula.

In effect, when the first generation of stars died, they left a legacy of new elements that mixed with residual gas from the big bang. A second generation of stars and associated planets formed out of the new, compositionally more diverse nebulae. Second-generation stars lived and died, and contributed elements to third-generation stars. Succeeding generations of stars and planets contain a greater proportion of heavier elements. Because different stars live for varied periods of time, at any given moment the Universe contains many different generations of stars, including small stars that have been living for a long time and large stars that have only recently arrived on the scene. The mix of elements we find on Earth includes relics of primordial gas from the big bang as well as the disgorged guts of dead stars. Think of it—the elements that make up your body once resided inside a star!

> **Take-Home Message**
>
> Small atoms (H and He) formed during the big bang. Larger atoms form in stars, and the largest during explosions of stars. Thus, we all contain atoms that were once inside stars.

The Nature of Our Solar System

We've just discussed how stars form from nebulae. Can we apply this model to our own Solar System—that is, the Sun plus all the objects that orbit it? The answer is yes, but before we do so, let's define the components that make up our Solar System.

Our Sun does not sail around the Milky Way in isolation. In its journey it holds, by means of gravitational "glue," many other objects. Of these, the largest are the eight planets: Mercury, Venus, Earth, Mars, Jupiter, Saturn, Uranus, and Neptune (named in order from the closest to the Sun to the farthest). A **planet** is a large, spherical solid object orbiting a star, and it may itself travel with a moon or even many moons (▶Box 1.2). By definition, a **moon** is an object locked in orbit around a planet; all but two of the planets have them. For example, Earth has one moon (the Moon), whereas Jupiter has at least sixteen, of which four are as big as or bigger than the Moon. In the past several years, research has documented planets in association with dozens of other stars. At least 150 of these so-called exoplanets have been found to date. In addition to planets and moons, our Solar System also includes a belt of asteroids (relatively small chunks of rock and/or metal) between the orbit of Mars and the orbit of Jupiter, and perhaps a trillion comets (relatively small blocks of "ice" orbiting the Sun; ice, in this context, means the solid version of materials that could be gaseous under Earth's surface conditions) in clouds that extend very far beyond the orbit of Pluto (see Box 1.2). Even though there are many objects in the Solar System, 99.8% of the Solar System's mass resides in the Sun. The next largest object in the Solar System—the planet Jupiter—accounts for about 71% of all nonsolar mass in the Solar System.

Of the eight planets in our Solar System, the four closest to the Sun (Mercury, Venus, Earth, and Mars) are called the inner planets, or the **terrestrial planets** (Earth-like planets), because they consist of a shell of

BOX 1.2 THE REST OF THE STORY

Discovering and Defining Planets

Before the invention of the telescope, astronomers recognized five other planets besides Earth, for these planets (Mercury, Venus, Mars, Jupiter, and Saturn) can be seen with the naked eye. Telescopes allowed astronomers to find the next farthest planet, Uranus, in 1781. It is interesting to note that Uranus did not follow its expected orbit exactly. The discrepancy implied that the gravity of yet another planet must be tugging on Uranus, and this prediction led to the discovery of Neptune in 1846. Discrepancies in Neptune's orbit, in turn, prompted a race to find yet another planet, leading to the discovery of Pluto in 1930. Pluto, however, proved to be a very different sort of planet—it's much smaller than the others, consists mostly of ice, and fol-

lows an orbit that does not lie in the same plane as the orbits of other planets.

In 1992, astronomers found that millions of icy objects, similar in composition to Pluto, occupy the region between the orbit of Neptune and a distance perhaps ten times the radius of Neptune's orbit. These objects together comprise the Kuiper Belt, named for the astronomer who predicted the objects' existence. As the twenty-first century dawned, astronomers learned that, though most Kuiper Belt objects are tiny, some are comparable in size to Pluto. In fact, Eris, an object found in 2003, is 20% larger than Pluto. Clearly, scientists needed to reconsider the traditional concept of a planet, or we could eventually have

dozens or hundreds of planets. Thus, in August 2006, the International Astronomical Union proposed a new definition of the word *planet*. This definition states that a planet is a celestial body that orbits the Sun, has a nearly spherical shape, and has cleared its neighborhood of other objects. The last phrase means that the object has either collided with and absorbed other objects in its orbit, has captured them to make them moons, or has gravitationally disturbed their orbits sufficiently to move them elsewhere. According to this definition, only the eight classical planets discovered by 1846 have the honor of being considered full-fledged planets. Pluto and Eris are now called "dwarf planets."

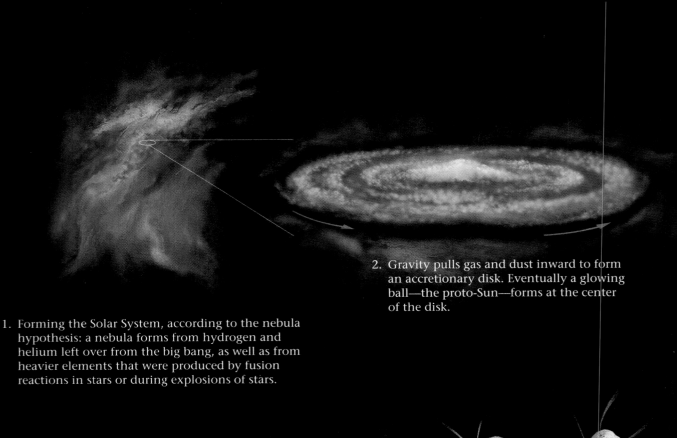

2. Gravity pulls gas and dust inward to form an accretionary disk. Eventually a glowing ball—the proto-Sun—forms at the center of the disk.

1. Forming the Solar System, according to the nebula hypothesis: a nebula forms from hydrogen and helium left over from the big bang, as well as from heavier elements that were produced by fusion reactions in stars or during explosions of stars.

6. Gravity reshapes the proto-Earth into a sphere. The interior of the Earth separates into a core and mantle.

5. Forming the planets from planetesimals: Planetesimals grow by continuous collisions. Gradually, an irregularly shaped proto-Earth develops. The interior heats up and becomes soft.

7. Soon after Earth forms, a small planet collides with it, blasting debris that forms a ring around the Earth.

8. The Moon forms from the ring of debris.

3. "Dust" (particles of refractory materials) concentrates in the inner rings, while "ice" (particles of volatile materials) concentrates in the outer rings. Eventually, the dense ball of gas at the center of the disk becomes hot enough for fusion reactions to begin. When it **ignites**, it becomes the Sun.

4. Dust and ice particles collide and stick together, forming planetesimals.

Forming the Planets and the Earth-Moon System

9. Eventually, the atmosphere develops from volcanic gases. When the Earth becomes cool enough, moisture condenses and rains to create the oceans. Some gases may be added by passing comets.

rock surrounding a core of iron alloy, as does the Earth. The next four have traditionally been called the outer planets, the Jovian planets (Jupiter-like planets), or the **gas-giant planets**. The adjective "giant" certainly seems appropriate for these planets (▶Fig. 1.13). Jupiter, for example, has a mass that is about 318 times that of Earth and a diameter 11.2 times larger. The inner two Jovian planets (Jupiter and Saturn) differ significantly from the outer two (Uranus and Neptune). Jupiter and Saturn have an elemental composition similar to the Sun's and thus consist predominantly of hydrogen and helium. Uranus and Neptune, in contrast, appear to consist predominantly of "ice" (solid methane, hydrogen sulfide, ammonia, and water). Note that even though Jupiter and Saturn have the same composition as the Sun, they did not ignite like the Sun because their masses are so small that their interiors never

became hot enough for fusion reactions to commence. Jupiter would have to be 80 times bigger for it to start to burn.

Forming the Solar System

Earlier in this chapter we presented ideas about how the first stars formed, during the early history of the Universe. To keep things simple, we didn't mention the formation of planets, moons, asteroids, or comets in that discussion. Now, let's develop a model to explain where they came from (see art, pp. 26–27).

Our Solar System formed at about 4.57 Ga ("Ga" means "giga annum" or "billion years ago"), over 9 billion years after the big bang, and thus, our Sun is probably a third- or fourth- or fifth-generation star (no one can say for sure) created from a nebula that contained all ninety-two elements. The materials in this nebula could be divided into two classes. Volatile materials—such as hydrogen, helium, methane, ammonia, water, and carbon monoxide—are ones that could exist as gas at the Earth's surface. In the pressure and temperature conditions of space, some volatile materi-

> **Take-Home Message**
>
> Our Solar System consists of the Sun (99% of the mass), four small Earth-like planets, and four gas-giant planets. It also contains rocky or metallic asteroids, and icy Kuiper Belt and Oort belt objects. Pluto is now considered to be a Kuiper Belt object.

FIGURE 1.13 (a) The relative sizes of the planets of our Solar System. Pluto no longer qualifies as a planet, as of 2006, so it does not appear here. **(b)** A diagram of the Solar System indicates that all of the classical planets have orbits that lie in the same plane. A belt of asteroids, rocky and metallic planetesimals that never coalesced into a planet, lies between Mars and Jupiter. The Kuiper Belt of icy objects (not shown) lies outside the orbit of Neptune. Pluto, a planet until its reclassification in 2006, has an orbit that lies oblique to the plane of the Solar System. Pluto is probably a Kuiper Belt object whose orbit has been changed in response to the gravitational pull of the planets.

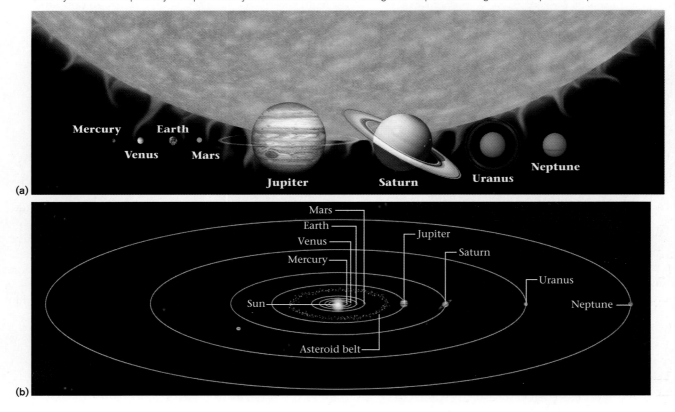

(a)

(b)

FIGURE 1.14 **(a)** Photograph, taken with a scanning electron microscope, showing a speck of interplanetary "dust." This speck is 0.01 mm across. By comparison, a pinhead is 1.0 mm across. **(b)** The grainy interior of this meteorite (a fragment of solid material that fell from space and landed on Earth) may resemble the texture of a small planetesimal. The sample is about 15 cm long.

als remain in gaseous form, but others condense or freeze to form different kinds of "ice." Refractory materials are ones that melt only at high temperatures; they condense to form solid soot-sized particles of "dust" in the coldness of space (►Fig. 1.14a). Thus, when astronomers refer to dust, they mean specks of solid rocky or metallic material, or clumps of the molecules that make up rock or metal.

We saw earlier that the first step in the formation of a star is the development of an accretion disk. When our Solar System formed, this accretion disk contained not only hydrogen and helium gas, but also other gases, as well as ice and dust. Such an accretion disk can also be called a **protoplanetary disk**, because it contains the raw materials from which planets form. With time, the central ball of our protoplanetary disk developed into the proto-Sun, and the remainder evolved into a series of concentric rings. A protoplanetary disk is hotter toward its center than toward its rim. Thus, the warmer inner rings of the disk ended up with higher concentrations of dust, whereas the cooler outer rings ended up with higher concentrations of ice.

Even before the proto-Sun ignited, the material of the surrounding rings began to coalesce, or accrete. Recall that ac-

cretion is the process by which smaller pieces of matter clump and bind together to form larger pieces. First, soot-sized particles merged to form sand-sized grains. Then, these grains clumped together to form grainy basketball-sized blocks (►Fig. 1.14b), which in turn collided. The fate of these blocks depended on the speed of the collision—if the collision was slow, blocks stuck together or simply bounced apart. If the collision was fast, one or both of the blocks shattered, producing smaller fragments that had to recombine all over again.

Eventually, enough blocks accreted to form **planetesimals**, bodies whose diameter exceeded about 1 km. Because of their mass, planetesimals exert enough gravity to attract and pull in other objects that are nearby. Figuratively, planetesimals acted like vacuum cleaners, sucking debris—small pieces of dust and ice, as well as smaller planetesimals—into their orbit, and in the process, they grew progressively larger. Eventually, victors in the competition to attract mass grew into **protoplanets**, bodies almost the size of today's planets. Once the protoplanets had succeeded in incorporating virtually all the debris near their orbits, so that their growth nearly ceased and they were the only inhabitants of their orbits, they became the planets that exist today. Some planetesimals still exist (►Box 1.3).

Early stages in the accretion process probably occurred very quickly—some computer models suggest that it may have taken only a few hundred thousand years to go from the dust stage to the large planetesimal stage. Estimates for the growth of planets from planetesimals range from 10 million years (m.y.) to 200 m.y. Recent studies favor faster growth and argue that planet formation in our Solar System was essentially completed by 4.558 Ga.

The nature of the planet resulting from accretion depended on its distance from the proto-Sun. In the inner orbits, where the protoplanetary disk consisted mostly of dust, small terrestrial planets composed of rock and metal formed. In the outer part of the Solar System, where in addition to gas and dust significant amounts of ice existed, larger protoplanets—as big as 15 times the size of Earth—formed. These, in turn, pulled in so much gas and ice that they grew into the gas-giant planets. Rings of dust and ice orbiting these planets became their moons.

When the Sun ignited, toward the end of the time when planets were forming, it generated a strong stellar wind (in this case, the solar wind) that blew any remaining gases out of the inner portion of the new-born Solar System. But the wind was too weak to blow away the gases of the gas-giant planets, for the gravitational pull of these planets was too strong.

> **Take-Home Message**
>
> The nebula theory states that stars and planets form when gravity pulls gas, dust, and ice together to form a swirling disk. The center of the disk becomes a star. Rings around the star condense into planetesimals which combine to form planets.

BOX 1.3 THE REST OF THE STORY

Comets and Asteroids—The Other Stuff of the Solar System

The process of planet formation, all in all, was very efficient. Almost all of the dust, ice, and gas that formed the accretionary disk around the proto-Sun eventually became incorporated into the planets. But some material escaped this fate. This material now constitutes two distinct classes of solid bodies—asteroids and comets. Meteorites that strike Earth provide samples of these bodies.

Asteroids are small bodies of solid rock or metal that orbit the Sun. Most reside in a belt, called the asteroid belt, between the orbits of Mars and Jupiter. Some asteroids are small rocky planetesimals that were never incorporated into planets, whereas others are fragments of once-larger planetesimals that collided with each other and disintegrated very early in the history of the Solar System. The debris in the asteroid belt never merged to form a planet because it is constantly churned by Jupiter's gravitational pull. Asteroids are too small for their own gravity to reshape them into spheres, so they are irregular, pockmarked masses (▶Fig. 1.15). Astronomers have found 1,000 asteroids with diameters greater than 30 km and estimate that there may be 10 million more with diameters greater than 1 km. Though asteroids are numerous, taken together their combined mass only equals that of Earth's Moon.

A **comet** is an icy planetesimal whose highly elliptical orbit brings it sufficiently close to the Sun that, during part of its journey, the comet evaporates and releases gas and dust to form a glowing tail. Comets that take less than 200 years to orbit the Sun originate from a disk-shaped region of icy fragments called the Kuiper Belt, extending from the orbit of Neptune out to a distance of about 50 times the radius of Earth's orbit. Those with longer orbits originate from a diffuse spherical region of icy fragments called the Oort Cloud, which extends out to a distance of about 100 times the radius of Earth's orbit. All told, there could be a trillion objects in the Oort Cloud and the Kuiper Belt, with a combined mass that may exceed the mass of Jupiter. Objects from the Oort Cloud or the Kuiper Belt become comets when gravity tugs on them and sends them on a trajectory into the inner Solar System.

In recent decades, researchers have sent spacecraft to observe comets. During an approach to Halley's comet in 1986, *Giotto* photographed jets of gas and dust spurting from the comet's surface (▶Fig. 1.16a, b). *Stardust* visited a comet in 2004 and returned to Earth with samples, and in 2005, *Deep Impact* dropped a copper ball on a comet and analyzed the debris ejected by the impact. Such studies confirm that comets consist of frozen water (H_2O), carbon dioxide (CO_2), methane (CH_4), ammonia (NH_3), and other volatile compounds, along with a variety of organic chemicals and dust (tiny rocky or metallic particles). Considering these components, astronomers often refer to comets as "dirty snowballs."

Some comet paths cross the orbits of planets, so collisions between comets and planets can and do occur. In 1994, astronomers observed four huge impacts between a fragmented comet and Jupiter. One of the impacts resulted in a 6 million megaton explosion. This would be equivalent to blowing up 600 times the entire nuclear arsenal on Earth all at once! Comets have collided with the Earth during historic time. For example, one exploded in the atmosphere above Tunguska, Siberia, in 1908 and flattened trees over an area of 2,150 square kilometers. Even larger comet impacts in the geologic past may have been responsible for disrupting life on Earth, as we discuss later in this book. Researchers speculate that comets may have added significant water to the Earth over its history and perhaps even seeded the Earth with life-related chemicals.

FIGURE 1.15 Photograph of the asteroid Ida, a body that is about 56 km long.

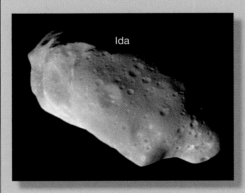

Ida

FIGURE 1.16 (a) Photograph of comet Hale-Bopp, which approached the Earth in 1997. The head of this comet is about 40 km across. **(b)** A close-up photo of Halley's comet, in 1986. On the side facing the Sun, jets of gas and dust spew into space. The solid nucleus is 14 km long.

(a)

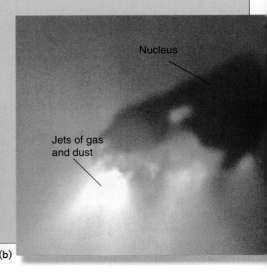

Nucleus

Jets of gas and dust

(b)

The overall model that we've just described is called the **nebular theory** of Solar System formation. Astronomers like the nebular theory because it explains why the ecliptic (the elliptical, or oval, plane traced out by a planet's orbit) of each planet is nearly the same, and why all planets orbit the Sun in the same direction. These observations make sense if all the planets formed out of a flattened disk of gas revolving in the same direction around a central mass.

Differentiation of the Earth and Formation of the Moon

When they first developed, larger planetesimals and protoplanets had a fairly uniform distribution of material throughout, because the smaller pieces from which they formed all had much the same composition and collected together in no particular order. But large planetesimals did not stay homogeneous for long. As they formed, they began to heat up. The heat came primarily from two sources: the transformation of kinetic energy into thermal energy during collisions, and the decay of radioactive elements.[2] In bodies whose temperature rose sufficiently to cause melting, denser iron alloy separated out and sank to the center of the body, whereas lighter rocky materials remained in a shell surrounding the center. By this process, called **differentiation**, protoplanets and large planetesimals developed internal layering early in their history. As we will see in Chapter 2, the central ball of iron alloy constitutes the body's core and the outer shell constitutes its mantle. Eventually, even partially molten planetesimals cooled and largely solidified.

In the early days of the Solar System, planets continued to be bombarded by **meteorites** (solid objects falling from space that land on a planet) even after the Sun had ignited and differentiation had occurred. Heavy bombardment in the early days of the Solar System (perhaps peaking at about 3.9 Ga) pulverized the surfaces of the planets and eventually left huge numbers of craters. It also contributed to heating the planets (▶Geotour 1).

In the case of the Earth, a particularly large collision early in Solar System history profoundly changed the planet and generated the Moon (see art spread, pp. 26–27). Constrained by the age of Moon rocks, geologists have concluded that at about 4.53 Ga, a Mars-sized protoplanet slammed into Earth. In the process, the colliding body disintegrated, along with a large part of the Earth's mantle. As much as 65% of the objects melted; some of their mass may even have vaporized. A ring of debris formed around the Earth, which at the time would have been covered with a sea of molten rock, and quickly accreted to form the Moon. Because of the manner in which the Moon formed, its overall composition resembles that of Earth's mantle.

Why Is the Earth Round?

Planetesimals were jagged or irregular in shape, and asteroids today have irregular shapes. Planets, on the other hand, are essentially spheres. Why are planets spherical? Simply put, when a protoplanet gets big enough, gravity can change its shape. To picture how, take a block of cheese outside on a hot summer afternoon and place it on a table. As the Sun warms the cheese, it gets softer and softer, and eventually gravity causes the cheese to spread out in a pancake-like blob on the table. This model shows that gravitational force alone can cause material to change shape if the material is soft enough. Now let's apply this model to planetary growth.

The rock composing a small planetesimal is cool and strong enough so that the force of gravity is not sufficient to cause the rock to flow. But once a planetesimal grows beyond a certain critical size, the insides of the planet become warm and the rock becomes soft enough to flow in response to gravity; also, the gravitational force becomes stronger. As a consequence, protrusions are pulled inward toward the center, and the planetesimal re-forms into a shape that permits the force of gravity to be the same at all points on its surface (see art, pp. 26–27). This special shape is a sphere because in a sphere the distribution of mass around the center has evened out.

Chapter Summary

- A geocentric model of the Universe placed the Earth at the center of the Universe, with the planets and Sun orbiting around the Earth within a celestial sphere speckled with stars. The heliocentric model placed the Sun at the center.

- The heliocentric model did not gain wide popularity until the Renaissance.

- Eratosthenes was able to measure the size of the Earth in ancient times, but it was not until fairly recently that astronomers accurately determined the distances to the Sun, planets, and stars. Distances in the Universe are so large that they must be measured in light years.

- The Earth is one of eight planets orbiting the Sun, and this Solar System lies on the outer edge of a slowly revolving galaxy, the Milky Way, which is composed of

> ### Take-Home Message
> The Moon probably formed from the debris of a collision between Earth and a large planetesimal. Earth was fairly homogeneous at first, but when iron sank to the center, it differentiated into a metallic core surrounded by a rocky mantle.

---

2. Radioactive elements are ones that spontaneously transform, or "decay," to form other elements by the releasing of one or more subatomic particles from the nucleus, or by the splitting of the nucleus into fragments. The process releases energy.

See for yourself . . .
Meteorite Impact Sites on Earth

Look at the surface of our Moon or that of Mars. You will find that the surfaces of the bodies are covered with craters, circular depressions formed during impact. On Earth, relatively few craters pock-mark the surface, but there are still some to be seen, as you can now *see for yourself.* The thumbnail images provided on this page are only to help identify tour sites. Go to wwnorton.com/studyspace to experience each flyover tour.

Extraterrestrial Craters

Earth's Moon contains craters in a great range of sizes. Some of the Moon's craters have very sharp rims (Image G1.1). Huge aprons of debris surround some of the craters of Mars (Image G1.2). Mercury's surface was also scarred by impacts. Here, we see the 45 km (250 mile)-wide Degas crater, from which light-colored rays of debris emanate (Image G1.3).

G1.1

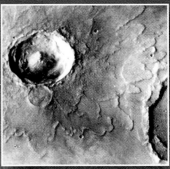
G1.2

G1.3

Meteor Crater, Arizona
(Lat 35° 1'38.03"N, Long 111° 1'21.64"W)

Click on "Fly To" and enter the coordinates of Meteor Crater. Once you have reached the location, zoom to an elevation of about 400 km (250 miles). You'll just barely see Meteor Crater, at the center of your image, but you can see the Grand Canyon in the upper left corner of your image. Zoom lower and the crater becomes clear. If you zoom down to an elevation of about 45 km (15,000 feet), you can see the crater clearly, but it's a bit blurry because of low resolution (Image G1.4). Descend a bit further and tilt the image so that you just barely see the Earth's horizon at the top of your screen, and use the compass tool to fly around the crater (Image G1.5). Notice the uplifted rim and the steep slopes down to the crater floor. This crater, formally known as Barringer Crater but commonly called Meteor Crater, formed when an iron-nickel meteorite slammed into Earth at a speed of about 12 km/s (28,000 mph), fifty thousand years ago. Calculations suggest that the meteorite was about 50 m (150 feet) across and weighed 300,000 tons. Hardly anything remains of the meteor because it exploded on impact, but because of the dry climate of northern

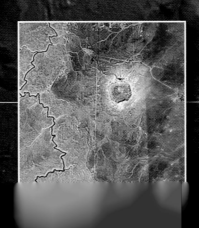

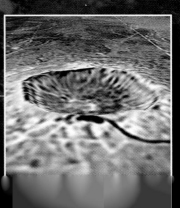

Manicouagan Crater, Quebec, Canada
(Lat 51°23'58.50"N, Long 68°41'44.11"W)

Fly to the coordinates of the crater and zoom to an elevation of 800 km (500 miles). Note that the crater is really obvious even from this elevation. Now zoom to a lower elevation. At what elevation does the crater fill the field of view? Tilt the field of view to see the horizon and, using the compass tool, fly around the crater to see its context (Image G1.6). If you descend to 4,570 m (15,000 feet), do you even realize that you are within a crater? Notice that the interior of the crater has risen, relative to the rim. This is due to the crust rebounding after the crater was excavated by impact. Manicouagan Crater formed between 206 and 214 million years ago. The preserved portion of the crater is about 70 km (43.5 miles) in diameter, but before erosion, it was probably 100 km (62 miles) in diameter. Due to the construction of a dam, the depressed outer edge of the crater filled with a lake.

G1.6

Chesapeake Bay Crater, Maryland
(Lat 37°15'01.92"N, Long 76°00'33.05"W)

Fly to the location and zoom to an elevation of 400 km (250 miles) and you can see all of Chesapeake Bay (Image G1.7). Now, zoom down to an elevation of 31 km (50 miles) and you'll find yourself on the east shore of Chesapeake Bay, near its mouth. At about 35 Ma, a meteorite struck the Earth at this spot. The impact fractured the continent at least to a depth of 8 km (4.97 miles) below the surface and produced an 85 km (53 miles)-wide crater—the sixth largest on Earth. After impact, the crater filled with a 1.2 km (0.75 miles)-thick layer of rubble (crater breccia), which geologists sampled by drilling in 2006. You may be wondering—where's the crater? You can't see it because about 450 m (1,476 feet) of sediment (sand and silt) buried it subsequent to its formation (Image G1.8a,b). The weight of these overlying sedimentary layers has compacted the underlying crater breccia—the same phenomenon happens when you push down on a sponge. As a consequence, the land surface at the mouth of Chesapeake Bay has been sinking faster than at any other place along the East Coast of North America. Because of this sinking, water of the Atlantic Ocean has submerged the outlets of the Susquehanna and Potomac rivers, thereby forming Chesapeake Bay.

G1.7

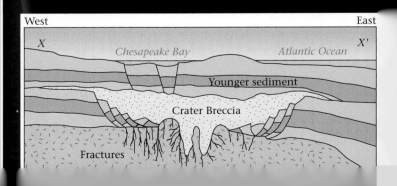

about 300 billion stars. The Universe contains at least hundreds of billions of galaxies.

- The red shift of light from distant galaxies, a manifestation of the Doppler effect, indicates that all distant galaxies are moving away from the Earth. This observation leads to the expanding Universe theory. Most astronomers agree that this expansion began after the big bang, a cataclysmic explosion about 13.7 billion years ago.

- The first atoms (hydrogen and helium) of the Universe developed soon after the big bang. These atoms formed vast gas clouds, called nebulae.

- According to the nebular theory of planet formation, gravity caused clumps of gas in the nebulae to coalesce into revolving balls. As these balls of gas collapsed inward, they evolved into flattened disks with bulbous centers. The protostars at the center of these disks eventually became dense and hot enough that fusion reactions began in them. When this happened, they became true stars, emitting heat and light.

- Heavier elements form during fusion reactions in stars; the heaviest are mostly made during supernova explosions. The Earth and the life forms on it contain elements that could only have been produced during the life cycle of stars. Thus, we are all made of stardust.

- Planets developed from the rings of gas and dust, the planetary nebulae, that surrounded protostars. The gas condensed into planetesimals that then clumped together to form protoplanets, and finally true planets. The rocky and metallic balls in the inner part of the Solar System did not acquire huge gas coatings; they became the terrestrial planets. Outer rings grew into gas-giant planets.

- The Moon formed from debris ejected when a Mars-sized planet collided with the Earth early during the history of the Solar System.

- A planet assumes a near-spherical shape when it becomes so soft that gravity can smooth out irregularities.

Geopuzzle Revisited

Geologists conclude that the Solar System formed from atoms generated by the big bang, and from atoms produced in stars or during the explosion of stars. Gravity pulled all this material together into a bulbous disk whose central ball became the Sun. The remainder of the disk condensed into planetesimals, which in turn coalesced to form planets.

Key Terms

accretion disk (p. 23)	moon (p. 25)
asteroids (p. 30)	nebula (p. 23)
big bang (p. 21)	nebular theory (p. 29)
comet (p. 30)	planet (p. 25)
cosmology (p. 15)	planetesimals (p. 29)
differentiation (p. 31)	precession (p. 16)
Doppler effect (p. 19)	protoplanetary disk
expanding Universe theory	(p. 29)
(p. 20)	protoplanets (p. 29)
galaxies (p. 19)	protostar (p. 23)
gas-giant planets (p. 28)	stellar wind (p. 24)
geocentric model (p. 15)	supernova (p. 23)
heliocentric model (p. 15)	terrestrial planets (p. 25)
light year (p. 18)	Universe (p. 15)
meteorite (p. 31)	

Review Questions

1. Why do the planets appear to move with respect to the stars?

2. Contrast the geocentric and heliocentric Universe concepts.

3. How did Galileo's observations support the heliocentric Universe concept?

4. Describe how Foucault's pendulum demonstrates that the Earth is rotating on its axis.

5. How did Eratosthenes calculate the circumference of the Earth?

6. Imagine you hear the main character in a low-budget science-fiction movie say he will "return ten light years from now." What's wrong with his usage of the term "light year"? What are light years actually a measure of?

7. Describe how the Doppler effect works.

8. What does the red shift of the galaxies tell us about their motion with respect to the Earth?

9. Briefly describe the steps in the formation of the Universe and the Solar System.

10. How is a supernova different from a normal star?

11. Why do the inner planets consist mostly of rock and metal, but the outer planets mostly of gas?

12. Why are all the planets in the Solar System orbiting the Sun in the same direction and in the same plane?

13. Describe how the Moon was formed.

14. Why is the Earth round?

On Further Thought

1. Look again at Figure 1.1. The North Star, a particularly bright star, lies just about at the center of the circles of light tracked out by other stars. (a) What does this mean about the position of the North Star relative to Earth's spin axis? Why is it called the "North Star"? (b) Consider the wobble of Earth's axis. Will the North Star be in the same position in a photograph taken from the same location as Figure 1.1 in about 10,000 years? Why?

2. When Copernicus reintroduced the heliocentric Universe model, he thought that the Sun sat at the *center of the whole Universe* and that planets had perfectly circular orbits. Consider the modern image of the Universe. Which aspects of Copernicus's model are still thought to be correct, and which are not?

3. The horizon is the line separating sky from the Earth's surface. Consider the shape of the Earth. How does the distance from your eyes to the horizon change as your elevation above the ground increases? To answer this question, draw a semicircle to represent part of the Earth's surface, then draw a vertical tower up from the surface. With your ruler, draw a line from various elevations on the tower to where the line is tangent to the surface of the Earth. (A *tangent* is a line that touches a circle at one point and is perpendicular to a radius.)

4. Astronomers discovered that more distant galaxies move away from the Earth more rapidly than do nearer ones. Why? To answer this question, make a model of the problem by drawing three equally spaced dots along a line; the dot at one end represents the Earth, and the other two represent galaxies. "Stretch" the line by drawing the line and dots again, but this time make the line twice as long. This stretching represents Universe expansion. Notice that the dots are now farther apart. Recall that: velocity = distance × time. If you pretend that it took 1 second to stretch the line (so "time" = 1 second), measurement of the distance that each galaxy moved relative to the Earth allows you to calculate velocity.

5. Consider that the deaths of stars eject quantities of heavier elements into space, and that these elements then become incorporated in nebulae from which the next generation of stars forms. Do you think that the ratio of heavier to lighter elements in, say, a sixth-generation star is larger or smaller than the ratio in a second-generation star? Why?

6. List the ways in which the three craters described in Geotour 1 differ from each other. Why do you think that these differences exist? With your answer in mind, why do you think that the Moon's craters are so well preserved and so numerous?

Suggested Reading

Allegre, C. 1992. *From Stone to Star*. Cambridge, Mass.: Harvard University Press.

Canup, R. M., and K. Righter, eds. 2000. *Origin of the Earth and Moon*. Tucson: University of Arizona Press.

Freedman, R. A., and W. J. Kaufmann, III. 2001. *Universe*, 6th ed. New York: Freeman.

Hawking, S., and L. Miodinow 2005. *A Briefer History of Time*. New York: Bantam.

Hester, J., et al. 2007. *21st Century Astronomy*, 2nd edition. New York: W. W. Norton.

Hoyle, F., G. Burbidge, and J. W. Narlikar. 2000. *A Different Approach to Cosmology: From a Static Universe through the Big Bang towards Reality*. Cambridge: Cambridge University Press.

Keel, W. C. 2002. *The Road to Galaxy Formation*. New York: Springer.

Kirshner, R. P. 2002. *The Extravagant Universe: Exploding Stars, Dark Energy, and the Accelerating Cosmos*. Princeton: Princeton University Press.

Liddle, A. 2003. *An Introduction to Modern Cosmology*, 2nd ed. New York: John Wiley & Sons.

Mackenzie, D. 2003. *The Big Splat, or How Our Moon Came to be*. Hoboken, N.J.: John Wiley & Sons.

Silk, J. 2006. *The Infinite Cosmos: Questions from the Frontiers of Cosmology*. New York: Oxford University Press USA.

Weinberg, S. 1993. *The First Three Minutes*. New York: Basic Books.

THE VIEW FROM SPACE A group of new born stars occurs in a cluster 12 billion light years from Earth, as viewed by the Hubble Space Telescope. The light from the stars makes the gas and dust surrounding them glow. Our Solar System may have formed from such a cloud.

Journey to the Center of the Earth

The Black Canyon of the Gunnison, in Colorado, is an 829 m (2,722-foot-) deep gash into the ancient rock comprising North America. The floor of the canyon almost always lies in shadow. But despite the awesome height of its sheer walls, the canyon is a mere scratch on Earth's surface—its depth is only 0.04% of the way to our planet's center.

The Earth is not a mere fragment of dead history, stratum upon stratum like the leaves of a book . . . but living poetry like the leaves of a tree.

—Henry David Thoreau (1817–1862)

2.1 INTRODUCTION

For most of human history, people perceived the other planets of our Solar System to be nothing more than bright points of light that moved in relation to each other and in relation to the stars. In the seventeenth century, when Galileo first aimed his telescope skyward, the planets became hazy spheres, and through the nineteenth and early twentieth centuries, with the advent of powerful telescopes, our image of the planets continued to improve. Now that we have actually sent space probes out to investigate them, we have exquisitely detailed pictures displaying the landscapes of planetary surfaces, as well as basic data about planetary composition (▶Fig. 2.1).

What if we turned the tables and became explorers from outside our Solar System undertaking a visit to Earth for the first time? What would we see? Even without touching the planet, we could detect its magnetic field and atmosphere, and could characterize its surface. We could certainly distinguish regions of land, sea, and ice. We could also get an idea of the nature of Earth's interior, though we could not see the details. In the first part of this chapter, we imagine rocketing to Earth to study its external characteristics. In the second part, we build an image of Earth's interior, based on a variety of data. (Of course, no one can see the interior firsthand, because high pressures and temperatures would crush and melt any visitor.) This high-speed tour of Earth will provide a frame of reference for the remainder of the book.

2.2 WELCOME TO THE NEIGHBORHOOD

Let's begin our journey to Earth from interstellar space, somewhere beyond the edge of the heliosphere. Astronomers define the heliosphere as the region of space whose diameter is more than 200 times that of the Earth's orbit and where the few atoms present came from the solar wind. Compared with the air we breath at sea level, interstellar space is a profound vacuum, for it contains an average of less than one atom per liter. In comparison, air at sea level contains 27,000,000,000,000,000,000,000 (or, in scientific notation, 2.7×10^{22}) atoms per liter. Notably, the two

FIGURE 2.1 Satellite studies emphasize that the surfaces of Mercury, Venus, Earth, and Mars differ markedly from each other. All four planets have mountains, valleys, and plains. But only Earth has distinct continents and ocean basins. Surface features provide clues to the nature of geologic processes happening inside a planet.

Voyager spacecraft, which left Earth in 1976 to explore the outer planets, will cross the edge of the heliosphere in about 2020.

Once we are within the heliosphere, our journey toward Earth remains uneventful until we begin to detect the ice fragments of the Oort Cloud and the Kuiper Belt, some of which have been slung into orbits that carry them into the inner Solar System, where they become comets (see Box 1.3). As we continue to fly toward Earth, we eventually pass within the orbit of Neptune and enter interplanetary space. This region is still a profound vacuum, but less so than interstellar space—between planets we may detect up to 5,000 atoms per liter. Our journey takes us past the orbits of Uranus, Saturn, Jupiter, the asteroid belt, and Mars, until finally we see the Earth.

As our rocket nears the Earth, its instruments detect the planet's magnetic field, like a signpost shouting, Approaching Earth! A **magnetic field**, in a general sense, is the region affected by the force emanating from a magnet. This force, which grows progressively stronger as you approach the magnet, can attract or repel another magnet and can cause charged particles (ions or subatomic particles with an electrical charge; see Appendix A) to move. Earth's magnetic field, like the familiar magnetic field around a bar magnet, is largely a **dipole**, meaning it has a North Pole and a South Pole. We can portray the magnetic field by drawing magnetic field lines, the paths along which magnets would align, or charged particles would flow, if placed in the field (▶Fig. 2.2).

The solar wind interacts with Earth's magnetic field, distorting it into a huge teardrop pointing away from the Sun. Fortunately, the magnetic field deflects most of this

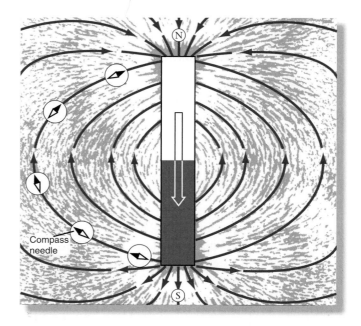

FIGURE 2.2 The magnetic field around a bar magnet can be displayed by sprinkling iron filings on a sheet of paper lying over the magnet. The filings define curving trajectories—these are the field lines. Charged particles placed in the field would flow along these lines, and compass needles would align with them. Note that the magnet has a north pole and a south pole.

wind, so that most of the particles in the wind do not reach the Earth's surface. In this way, the magnetic field acts like a shield against the solar wind; the region inside this magnetic shield is called the magnetosphere (▶Fig. 2.3). Without this shield, the solar wind would bathe the Earth in lethal radiation. Even with the shield, particularly strong

FIGURE 2.3 The magnetic field of the Earth interacts with the solar wind—the wind distorts the field so that it tapers away from the Sun, and the field isolates the Earth from most of the wind. Note the Van Allen belts near the Earth.

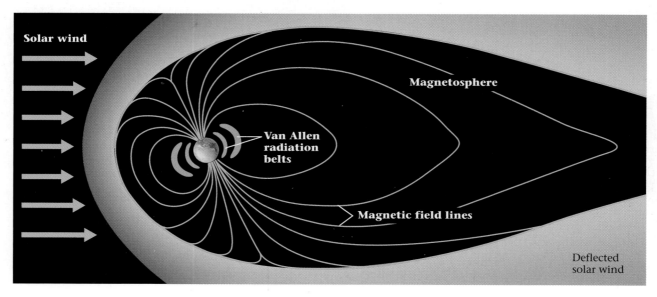

bursts of solar wind can disrupt the electronics on spacecraft and can even affect electric grids on Earth.

Though it protects the Earth from the solar wind, the magnetic field does not stop our rocket ship, and we continue to speed toward the planet. At distances of about 10,500 km and 3,000 km out from the Earth, we encounter the Van Allen radiation belts, named for the physicist who first recognized them in 1959. These consist of solar wind particles, as well as cosmic rays (nuclei of atoms emitted from supernova explosions), that were moving so fast they were able to penetrate the weaker outer part of the magnetic field and were then trapped by the stronger magnetic field closer to the Earth. By trapping cosmic rays, the Van Allen belts protect life on Earth from dangerous radiation. Some charged particles make it past the Van Allen belts and are channeled along magnetic field lines to the polar regions of Earth. When these particles interact with gas atoms in the upper atmosphere they cause the gases to glow, like the gases in neon signs, creating spectacular aurorae (►Fig. 2.4a, b).

> **Take-Home Message**
>
> The Solar System includes the Sun, the planets, and many tiny icy and rocky fragments. Earth has a magnetic field that shields our planet's surface from solar wind and cosmic rays.

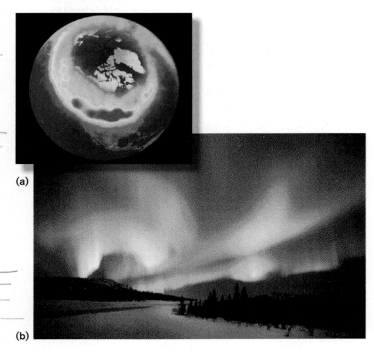

(a)

(b)

FIGURE 2.4 **(a)** Satellite view showing the glowing ring of the aurora borealis as it appears superimposed on a map of North America. **(b)** The aurora as seen from the ground in Alaska.

FIGURE 2.5 **(a)** Sunset view from a space shuttle. The gases and dust of the atmosphere reflect light and absorb certain wavelengths of light, creating a glowing palette of color. The vacuum of space is always black. **(b)** Nitrogen and oxygen constitute most of the gas in the atmosphere.

2.3 THE ATMOSPHERE

As we descend farther, we enter Earth's **atmosphere**, an envelope of gas consisting overall of 78% nitrogen (N_2) and 21% oxygen (O_2), with minor amounts (1% total) of argon, carbon dioxide (CO_2), neon, methane, ozone, carbon monoxide, and sulfur dioxide (►Fig. 2.5a, b). Other terrestrial planets have atmospheres, but none of them are like Earth's. For example, Venus's atmosphere, dense enough to hide the planet's surface, consists almost entirely of carbon dioxide. Mercury has only a trace of an atmosphere, because the planet's high temperatures allowed the atmosphere to escape into space long ago. Mars has a thin atmosphere that, like Venus's, consists almost entirely of carbon dioxide. (We'll discuss these atmospheres further in Chapter 20.) The density of gas constituting the Earth's atmosphere gradually decreases with altitude until it's the same as that of interplanetary space at about 10,000 km from Earth. But 99% of the gas in the atmosphere lies below 50 km, and most of the remaining 1% lies between 50 and 500 km.

The weight of overlying air (the mixture of gases in the atmosphere) squeezes on the air below, and thus pushes gas molecules in the air below closer together (►Fig. 2.6a, b). Thus, both the density (mass per unit volume) of air and the air pressure (the amount of push that

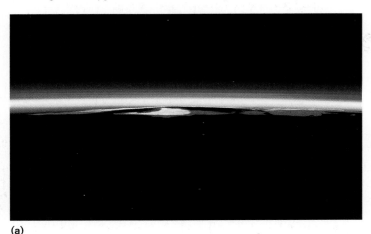

(a)

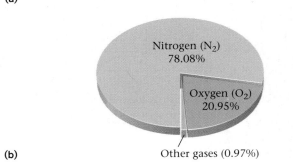

Nitrogen (N_2)
78.08%

Oxygen (O_2)
20.95%

(b)

Other gases (0.97%)

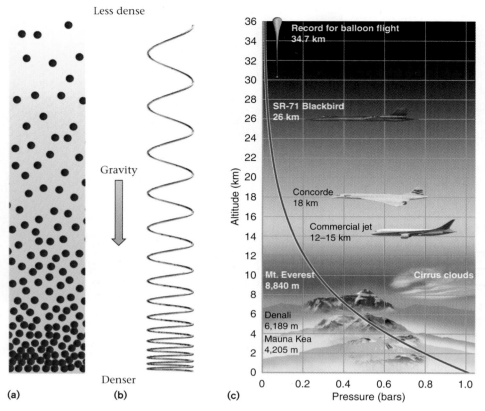

FIGURE 2.6 **(a)** Atmospheric density increases toward the base of the atmosphere because the weight of the upper atmosphere squeezes together gas molecules in the lower atmosphere. **(b)** By analogy, if you place a spring on a table in a gravity field, the weight of the upper part of the spring pushes down on the lower part and causes it to squeeze together. **(c)** A graph displaying the variation of air pressure with elevation shows that by an elevation of 30 km, atmospheric pressure is less than 1% of the atmospheric pressure at sea level.

2.4 LAND AND OCEANS

Now, imagine that we've gone into orbit around the Earth, and we've set about mapping the planet—what obvious features should we put on the map? Earth has two distinctly different types of surface: about 30% consists of dry land (continents and islands) and about 70% consists of surface water (oceans, lakes, and streams). Geologists refer to surface water on Earth, along with groundwater (the water that fills cracks and holes in rock or sediment underground), as the **hydrosphere**. Ice covers significant areas of land and sea. Taken together, the ice-covered regions comprise the **cryosphere**.

While orbiting the Earth, we can clearly see that its land surface is not flat. In other words, **topography**—physical features of the land surface represented by changes in elevation—varies dramatically from place to place (▶Fig. 2.7). For example, the lowest dry land today lies along the Dead Sea, 400 m (1,300 feet) below sea level, and the highest, as we have seen, lies at the summit of Mt. Everest, 8.85 km (29,035 feet) above sea level. Elevation differences, coupled with regional contrasts in vegetation due to variations in rainfall and atmospheric temperature, lead to an immense variety of landscapes on Earth (▶Geotour 2). Notably, Earth's surface displays relatively few meteorite impact craters, in comparison to the pockmarked surfaces of the Moon and Mars, for most of Earth's craters have been destroyed by various processes discussed later in this book. But a few impact structures can still be found (see Geotour 1). Our instruments also tell us that the sea floor is not flat. We recognize submarine plains, oceanic ridges, and deep trenches, or troughs.

A graph, called a hypsometric curve, plotting surface elevation on the vertical axis and the percentage of the Earth's surface on the horizontal axis shows that a relatively small proportion of the Earth's surface occurs at very high elevations (mountains) or at great depths

the air exerts on material beneath it) decrease with increasing elevation (▶Fig. 2.6c). Technically, we specify pressure in units of force, or push, per unit area. Such units include atmospheres (abbreviated atm) and bars. 1 atm = 1.04 kilograms per square centimeter, or 14.7 pounds per square inch. Atmospheres and bars are almost the same, for 1 atm = 1.01 bars. At sea level, average air pressure on Earth is 1 atm, whereas on the peak of Mt. Everest, 8.85 km above sea level, air pressure is only 0.3 atm. Where the space shuttle orbits the Earth, an altitude of about 400 km (about 250 miles), air pressure is only 0.0000001 atm. Humans cannot live for long at elevations greater than about 4.5–5.5 km.

> ### Take-Home Message
>
> Earth's atmosphere consists mostly of nitrogen and oxygen. Ninety-nine percent of the atmosphere's gas lies below an elevation of 50 km. So, relative to our planet's diameter, the atmosphere is very thin indeed.

> ### Take-Home Message
>
> Most land lies less than 1 km above sea level, but the tallest mountain reaches a height of 8.85 km. Most sea floor lies 4 to 5 km below sea level, but the deepest point reaches a depth of 11 km.

FIGURE 2.7 This map of the Earth shows variations in elevation of both the land surface and the sea floor. Darker blues are deeper water in the ocean. Greens are lower elevation and brown and reds are higher elevation on land.

(deep trenches). In fact, most of the land surface lies just within a kilometer of sea level, and most of the sea floor is between 4 and 5 km deep (►Fig. 2.8). A slight change in sea level would dramatically change the amount of dry land.

2.5 WHAT IS THE EARTH MADE OF?

Elemental Composition

At this point, we leave our fantasy space voyage and turn our attention to the materials that make up the solid Earth, which we need to know about before proceeding to the Earth's interior. In Chapter 1, we learned that the atoms that make up the Earth consist of a mixture of elements left over from the big bang, as well as elements produced by fusion reactions in stars and during supernova explosions. During the birth of the Solar System, solar wind blew volatile materials away from the region in which Earth was forming, like wind separating chaff from wheat. Earth and other terrestrial planets formed from the materials left behind. As a consequence, iron (35%), oxygen (30%), silicon (15%), and magnesium (10%) make up most of Earth's mass (►Fig. 2.9). The remaining 10% consists of the other eighty-eight naturally occurring elements.

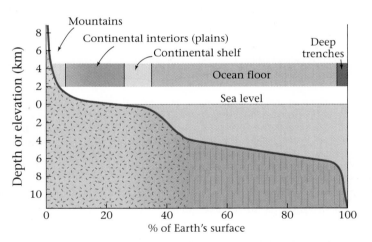

FIGURE 2.8 This graph shows a hypsometric curve, indicating the proportions of the Earth's solid surface at different elevations. Two principal zones—the continents and adjacent continental shelf areas (the submerged margins of continents) and the ocean floor—account for most of Earth's area. Mountains and deep trenches cover relatively little area.

Categories of Earth Materials

The elements making up the Earth combine to form a great variety of materials. We can organize these into several categories.

See for yourself…
The Variety of Earth's Surface

Google Earth™ and *NASA World Wind* allow you to sense what a space traveler orbiting the Earth would see if there were no clouds and almost no sea ice, and if features of the sea floor were visible. Let's orbit the Earth, and then look more closely at examples of its surface. The thumbnail images provided on this page are only to help identify tour sites. Go to wwnorton.com/studyspace to experience each flyover tour.

Distant View of the Earth

Zoom to view the planet from an elevation of about 12,000 km (about 8,000 miles). Turn on the "Lat/Lon Grid" to see lines of latitude and lines of longitude (Image G2.1). Orient the image so that the equator is horizontal and examine the numbering scheme of these lines relative to the equator and the prime meridian. Click and drag from right to left and then let go to simulate the Earth's rotation and see the distribution of land and sea. Note that different shades of blue distinguish between deeper and shallower water, as you can see in the western Caribbean and Bahamas region, from an elevation of 3,000 km (1,864 miles) (Image G2.2).

G2.1

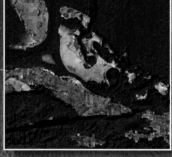

G2.2

G2.3

Eastern Greenland
(Lat 74°55'21.75"N, Long 22°5'32.62"W)

Ice covers the land surface of Greenland and Antarctica. Fly to this locality, then zoom to 3,200 km (2,000 miles) above sea level (Image G2.3). You can see that a vast sheet of ice covers almost all of Greenland.

Zoom to 160 km (100 miles) above sea level. Now you can see that the ice sheet drains to the sea via slowly moving "rivers of ice" (Image G2.4). These valley glaciers were once longer and they carved deep valleys. Rising sea level filled the valleys to form fingers of the ocean called fjords. Note that the water in these fjords has frozen to form sea ice. Next, zoom down to 16 km (10 miles) and tilt the image so you just see the Earth's horizon (Image G2.5). Fly along the coast to see spectacular valleys and cliffs.

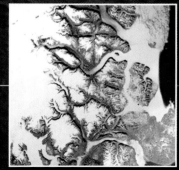

G2.4

G2.5

G2.6

G2.7

G2.8

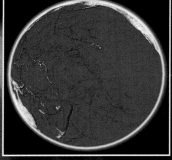

G2.9

Southern Alps, New Zealand
(Lat 43°30'45.61"S, Long 170°50'41.97"E)

Zoom to an elevation of 20 km (about 12 miles) and you can see an example of rugged, mountainous topography and sediment-choked streams (Image G2.6). Next, zoom to 16 km (10 miles), tilt so that you can see the horizon, and fly northwest up the river valley (Image G2.7). Turn 180° and fly down the river. Eventually, the river leaves the mountains and crosses the plains. Here, farmers have divided the land into fields (Image G2.8). The river eventually reaches the coast and drains into the Pacific Ocean.

Pacific Ocean
(Lat 3°43'48.84"N, Long 163°51'15.32"W)

Zoom out to an elevation of 6,700 km (4,200 miles) above sea level. The Pacific Ocean almost fills your field of view (Image G2.9). You can see a few islands and can note that the floor of the sea is not perfectly flat—it has seamounts, trenches, and oceanic ridges.

Midwestern United States
(Lat 39°51'02.69"N, Long 87°24'30.44"W)

Zoom to 9,000 m (30,000 feet) to see the view from a jet plane (Image G2.10). The image displays the Wabash River, farm fields, wooded areas along small streams, and a town (Newport, Indiana). What percentage of this landscape has been changed by human hands?

G2.10

A Sand Sea in Saudi Arabia
(Lat 28°55'2.02"N, Long 39°36'59.07°E)

Zoom to 35 km (22 miles). You see golden sand, blown by the wind into ridges called dunes. Tilt the image so that the horizon just appears, and then fly slowly north (Image G2.11). You'll cross the occasional barren rocky hill without a tree or shrub in sight.

G2.11

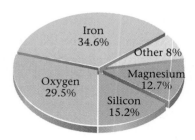

FIGURE 2.9 The proportions of major elements making up the mass of the whole Earth. Note that iron and oxygen account for most of the mass.

- *Organic chemicals*: Carbon-containing compounds that either occur in living organisms, or have characteristics that resemble those of molecules in living organisms, are called **organic chemicals**. Examples include oil, protein, plastic, fat, and rubber. Certain simple carbon-containing materials, such as pure carbon (C), carbon dioxide (CO_2), carbon monoxide (CO), and calcium carbonate ($CaCO_3$), are *not* considered organic.

- *Minerals*: A solid substance in which atoms are arranged in an orderly pattern is called a **mineral**. (We provide a more detailed definition in Chapter 5.) Almost all minerals are inorganic (not organic). Minerals grow either by freezing of a liquid or by precipitation out of a water solution. Precipitation occurs when atoms that had been dissolved in water come together and form a solid. For example, solid salt forms by precipitation out of seawater when the water evaporates. A single coherent sample of a mineral that grew to its present shape and has smooth, flat faces is a crystal. An irregularly shaped sample, or a fragment derived from a once-larger crystal or group of crystals, is a grain.

- *Glasses*: A solid in which atoms are not arranged in an orderly pattern is called **glass**. Glass forms when a liquid freezes so fast that atoms do not have time to organize into an orderly pattern.

- *Rocks*: Aggregates of mineral crystals or grains, and masses of natural glass, are called **rocks**. Geologists recognize three main groups of rocks. (1) Igneous rocks develop when hot molten (melted) rock cools and freezes solid. (2) Sedimentary rocks form from grains that break off preexisting rock and become cemented together, or from minerals that precipitate out of a water solution; an accumulation of loose mineral grains (grains that have not stuck together) is called **sediment**. (3) Metamorphic rocks are created when preexisting rocks undergo changes, such as the growth of new minerals in response to heat and pressure.

- *Metals*: Solids composed of metal atoms (such as iron, aluminum, copper, and tin) are called **metals**. In a metal, outer electrons are able to flow freely (see Chapter 15). An **alloy** is a mixture containing more than one type of metal atom (e.g., bronze is a mixture of copper and tin).

- *Melts*: **Melts** form when solid materials become hot and transform into liquid. Molten rock is a type of melt. Geologists distinguish between magma, which is molten rock beneath the Earth's surface, and lava, molten rock that has flowed out onto the Earth's surface.

- *Volatiles*: Materials that easily transform into gas at the relatively low temperatures found at the Earth's surface are called **volatiles**.

Note that the most common minerals in the Earth contain silica (SiO_2) mixed in varying proportions with other elements (typically iron, magnesium, aluminum, calcium, potassium, and sodium). These minerals are called silicate minerals, and, no surprise, rocks composed of silicate minerals are silicate rocks. Geologists distinguish four classes of igneous silicate rocks based, in essence, on the proportion of silicon to iron and magnesium. In order, from greatest to least proportion of silicon to iron and magnesium, these classes are: felsic (or silicic), intermediate, mafic, and ultramafic. As the proportion of silicon in a rock increases, the density (mass per unit volume) decreases. Thus, felsic rocks are less dense than mafic rocks.

Within each class are many different rock types, each with a name, that differ from the others in terms of composition (chemical makeup) and crystal size. These will be discussed in detail in Part II. But for now, we need to recognize the following four rock names: **granite** (a felsic rock with large grains), **basalt** (a mafic rock with small grains), **gabbro** (a mafic rock with large grains), and **peridotite** (an ultramafic rock with large grains).

> **Take-Home Message**
>
> The Earth consists mostly of silicate rock (e.g., granite, basalt, gabbro, peridotite) and iron alloy. Different types of silicate rock can be distinguished from each other by their composition (proportion of silicon to iron and magnesium) and on grain size. Composition affects rock density.

2.6 HOW DO WE KNOW THAT THE EARTH HAS LAYERS?

The world's deepest mine shaft penetrates gold-bearing rock that lies about 3.5 km (2 miles) beneath South Africa. Though miners seeking this gold must begin their workday by plummeting straight down a vertical shaft for almost ten minutes aboard the world's fastest elevator, the shaft is little more than a pinprick on Earth's surface when compared with the planet's radius (the distance from the center to the surface is 6,371 km). Even the deepest well ever drilled, a 12-km-deep hole in northern Russia, penetrates only the upper 0.03% of the Earth. We literally live on the thin skin of our planet, its interior forever inaccessible to our wanderings.

People have wondered about the Earth's interior since ancient times. What is the source of incandescent lavas spewed from volcanoes, of precious gems and metals, of sparkling spring water, and of the mysterious shaking that topples buildings? Without the ability to observe the Earth's interior firsthand, pre-twenty-first-century authors dreamed up fanciful images of it. For example, the English poet John Milton (1608–1674) described the underworld as a "dungeon horrible, on all sides round, as one great furnace flamed" (▶Fig. 2.10). Perhaps his image was inspired by volcanoes in the Mediterranean. In the eighteenth and nineteenth centuries, some European writers thought that the Earth's interior resembled a sponge, containing open caverns variously filled with molten rock, water, or air. In this way, the interior could provide both the water that bubbled up at springs *and* the lava that erupted at volcanoes. In fact, in the French author Jules Verne's popular 1864 novel *Journey to the Center of the Earth*, three explorers find a route through interconnected caverns to the Earth's center. Today, we picture the Earth's interior as having distinct layers. This image is the end product of many clues found over the past two hundred years.

FIGURE 2.10 A literary image of the Earth's insides: *The Fallen Angels Entering Pandemonium, from* Milton's "*Paradise Lost,*" *Book 1,* by English painter John Martin (1789–1854).

Clues from Measuring Earth's Density

The first key to understanding the Earth's interior came from studies that provided an estimate of the planet's density (mass per unit volume). To determine Earth's density, one must first determine the amount of matter making up the Earth. In 1776, the British Royal Astronomer, Nevil Maskelyne provided the first realistic estimate of Earth's mass. Maskelyne postulated that he could weigh the Earth by examining the deflection of a plumb bob attached to a surveying instrument. The angle of deflection (ß) of the plumb bob caused by the gravitational attraction of a mountain indicates the magnitude of gravitational attraction exerted by the mountain's mass relative to the gravitational attraction of the Earth's mass. Maskelyne tested his hypothesis at Schiehallion Mountain in Scotland (▶Fig. 2.11). His results led to an estimate that the Earth's average density is 4.5 times the density of water (i.e., 4.5g/cm^3 in the modern metric system). In 1778, another physicist using a different method arrived at a density estimate of 5.45 g/cm^3, fairly close to modern estimates. Significantly, typical rocks (such as granite and basalt) at the surface of the Earth have a density of only $2.2\text{–}2.5 \text{ g/cm}^3$, so the *average* density of Earth exceeds that of its surface rocks. Certainly, the open voids that Jules Verne described could not exist!

Clues from Measuring Earth's Shape

Once they had determined that the density of the Earth's interior was greater than that of its surface rocks, nineteenth-century scientists asked, "Does this density increase *gradually* with depth, or does the Earth consist of a less dense shell surrounding a much denser core?" If the Earth's density increased only gradually with depth, most of its mass would lie

FIGURE 2.11 A surveyor noticed that the plumb line he was using to level his surveying instrument did not hang exactly vertically near a mountain; it was deflected by an angle ß, owing to the gravitational attraction of the mountain. The angle of deflection represents the ratio between the mass of the mountain and the mass of the whole Earth.

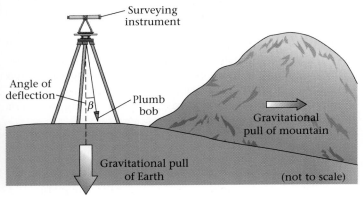

far from the center, and the planet's spin would cause the Earth to flatten into a disk. Since this doesn't happen, scientists concluded that much of Earth's mass must be concentrated close to the center. The dense center came to be known as the *core*. Eventually, geologists determined that the density of this core approaches 13 g/cm³.

To understand further the nature of Earth's interior, geologists measured tides, the rise and fall of the Earth's surface in response to the gravitational attraction of the Moon and Sun. If the Earth were composed of a liquid surrounded by only a thin solid crust, then the surface of the land would rise and fall daily, like the surface of the sea. We don't observe such behavior, so the Earth's interior must be largely solid.

By the end of the nineteenth century, geologists had recognized that the Earth resembled a hard-boiled egg, in that it had three principal layers: a not-so-dense **crust** (like an eggshell, but composed of rocks such as granite, basalt, and gabbro), a denser, solid **mantle** in between (the white, but composed of a then-unknown material), and a very dense **core** (the yolk, but also composed of a then-unknown material). Clearly, many questions remained. How thick are the layers? Are the boundaries between layers sharp or gradational? And what exactly are the layers composed of?

Clues from the Study of Earthquakes: Refining the Image

One day in 1889, a physicist in Germany noticed that the pendulum in his lab began to move without having been touched. He reasoned that the pendulum was actually standing still, because of its inertia (the tendency of an object at rest to remain at rest, and of an object in motion to remain in motion), and that the Earth was moving under it. A few days later, he read in a newspaper that a large **earthquake** (ground shaking due to the sudden breaking of rocks in the Earth) had taken place in Japan minutes before the movement of his pendulum began. The physicist deduced that vibrations due to the earthquake had traveled through the Earth from Japan and had jiggled his laboratory in Germany. The energy in such vibrations moves in the form of waves, called either seismic waves or earthquake waves, that resemble the shock waves you feel with your hands when you snap a stick (▶Fig. 2.12). The breaking of rock during an earthquake either produces a new fracture on which sliding occurs or causes sliding on a pre-existing fracture. A fracture on which sliding occurs is called a **fault**.

Geologists immediately realized that the study of seismic waves traveling through the Earth might provide a tool for exploring the Earth's insides (much as doctors today use ultrasound to study a patient's insides). Specifically, laboratory measurements demonstrated that

earthquake waves travel at different velocities (speeds) through different materials. Thus, by detecting the depths at which velocities suddenly change, geoscientists pinpointed the boundaries between layers and even recognized subtler boundaries within layers. (We'll explain how in Interlude D, after we've had a chance to describe earthquakes in more detail.)

Pressure and Temperature Inside the Earth

In order to keep underground tunnels from collapsing under the pressure created by the weight of overlying rock, mining engineers must design sturdy support structures. It is no surprise that deeper tunnels require stronger supports: the downward push from the weight of overlying rock increases with depth, simply because the mass of the overlying rock layer increases with depth. At the Earth's center, pressure probably reaches about 3,600,000 atm.

Temperature also increases with depth in the Earth. Even on a cool winter's day, miners who chisel away at gold veins exposed in tunnels 3.5 km below the surface swelter in temperatures of about 53°C (127°F). We refer to the rate of change in temperature with depth as the

FIGURE 2.12 When the rock inside the Earth suddenly breaks and slips, forming a fracture called a fault, it generates shock waves that pass through the Earth and shake the surface (creating an earthquake), much as the sound waves from a stick snapping travel to you and make your eardrum vibrate.

Earthquake wave

Fault plane

(not to scale)

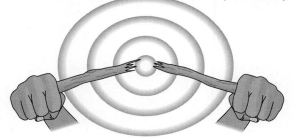

geothermal gradient. In the upper part of the crust, the geothermal gradient averages between 15° and 50°C per km. At greater depths, the rate decreases to 10°C per km or less. Thus, 35 km below the surface of a continent, the temperature reaches 400° to 700°C. No one has ever directly measured the temperature at the Earth's center, but recent calculations suggest it may reach 4,700°C, only about 800° less than the temperature at the surface of the Sun.

2.7 WHAT ARE THE LAYERS MADE OF?

We saw earlier that the material composing the Earth's insides must be much denser than familiar surface rocks such as granite and basalt. To discover what this material consists of, geologists

- conducted laboratory experiments to determine what kinds of materials inside the Earth could be a source of magma;
- studied unusual chunks of rock that may have been carried up from the mantle in magma;
- conducted laboratory experiments to measure densities in samples of known rock types, so that they could compare these with observed densities in the Earth; and
- estimated which elements would be present in the Earth if the Earth had formed out of planetesimals similar in composition to meteorites (chunks of rock and/or metal alloy that fell from space and landed on Earth; ▶Box 2.1).

As a result of this work, we now have a pretty clear sense of what the layers inside the Earth are made of, though this picture is constantly being adjusted as new findings become available. Let's now look at the properties of individual layers, starting with the Earth's surface.

The Crust

When you stand on the surface of the Earth, you are standing on the top of its outermost layer, the crust. The crust is our home and the source of all our resources. How thick is this all-important layer? Or, in other words, what is the depth of the crust-mantle boundary? An answer came from the studies of Andrija Mohorovičić, a researcher working in Zagreb, Croatia. In 1909, Mohorovičić discovered that the velocity of earthquake waves suddenly increased at a depth of about 50 km beneath the Earth's surface, and he suggested that this increase was caused by an abrupt change in the properties of rock (see Interlude D for further detail). Later studies showed that this change can be found most everywhere around our planet, though it actually occurs at different depths in different locations—it's deeper beneath continents than beneath oceans. Geologists now consider the change to be the crust-mantle boundary, and they refer to it as the **Moho** in Mohorovičić's honor. The relatively shallow depth of the Moho (7–70 km, depending on location), compared with the radius of the Earth (6,371 km), emphasizes that the crust is very thin indeed. The crust is only about 0.1% to 1.0% of the Earth's radius, so if the Earth were the size of a balloon, the crust would be about the thickness of the balloon's skin.

Geologists distinguish between two fundamentally different types of crust—oceanic crust, which underlies the sea floor, and continental crust, which underlies continents (▶Fig. 2.13a). The crust is not simply cooled mantle, like the skin on chocolate pudding, but rather consists of a variety of rocks that differ in composition (chemical makeup) from mantle rock.

Oceanic crust is only 7 to 10 km thick. At highway speeds (100 km per hour), you could drive a distance equal to the thickness of the oceanic crust in about five minutes. (It would take sixty-three hours, driving nonstop, to reach the Earth's center.) We have a good idea of what oceanic crust looks like in cross section, because geologists have succeeded in drilling down through its top few kilometers and have found places where slices of oceanic crust have been incorporated in mountains and therefore have been exposed on dry land. Studies of such examples show that oceanic crust consists of fairly uniform layers. At the top, we find a blanket of sediment, generally less than 1 km thick, composed of clay and tiny shells that have settled like snow. Beneath this blanket, the oceanic crust consists of a layer of basalt and, below that, a layer of gabbro.

Most continental crust is about 35 to 40 km thick—about four to five times the thickness of oceanic crust—but its thickness varies much more than does oceanic crust. In regions called rifts, continents have stretched and thinned to become only 25 km thick, whereas in some mountain belts, continents have squashed and thickened to become up to 70 km thick. In contrast to oceanic crust, continental crust contains a great variety of rock types, ranging from mafic to felsic in composition, but on average, continental crust is less mafic than oceanic crust—it has a felsic to intermediate composition—so a block of average continental crust weighs less than a same-size block of oceanic crust (▶Fig. 2.13b).

BOX 2.1 THE REST OF THE STORY

Meteors and Meteorites

During the early days of the Solar System, the Earth collided with and incorporated almost all of the planetesimals and smaller fragments of solid material lying in its path. Intense bombardment ceased about 3.9 Ga, but even today, collisions with space objects continue, and over 1,000 tons of material (rock, metal, dust, and ice) on average, fall to Earth every year. The vast majority of this material consists of fragments derived from the Oort Cloud, Kuiper Belt, or asteroid belt (see Box 1.2). Some of the material, however, consists of chips of the Moon or Mars, ejected into space when large objects collided with those bodies.

Astronomers refer to any object from space that enters the Earth's atmosphere as a meteoroid. Meteoroids move at speeds of up to 75 km/s, so fast that when they reach an altitude of about 150 km, friction with the atmosphere causes them to begin to evaporate, leaving a streak of bright, glowing gas. The glowing streak, an atmospheric phenomenon, is a **meteor** (also known colloquially, though incorrectly, as a "falling star"). Most visible meteoroids completely evaporate by an altitude of about 30 km. But dust-sized ones may slow down sufficiently to float to Earth, and larger ones (fist-sized or bigger) can survive the heat of entry to reach the surface of the planet. Objects that strike the Earth are called **meteorites**. Most are asteroidal or planetary fragments, for the icy material of small cometary bodies is too fragile to survive the fall. In some cases, the meteoroids explode in brilliant fireballs; such particularly luminous objects are also called bolides.

Scientists did not realize that meteors were the result of solid objects falling from space until 1803, when a spectacular meteor shower (the occurrence of a large number of meteors during a short time) lit the sky over Normandy, France, and over 3,000 meteorites were subsequently recovered on the ground. In the succeeding two centuries, many meteorites have been collected and studied in detail. On the basis of this work, researchers recognize three basic classes of meteorites: iron (made of iron-nickel alloy), stony (made of silicate rock), and stony iron (rock embedded in a matrix of metal). Of all known meteorites, about 93% are stony and 6% are iron. Studying their composition, researchers have concluded that some meteors (a special subcategory of stony meteorites called carbonaceous chondrites, because they contain carbon and small pea-sized balls called chondrules) are asteroids derived from planetesimals that never underwent differentiation into a core and mantle. Other stony meteorites and all iron meteorites are asteroids derived from planetesimals that differentiated into a metallic core and a rocky mantle early in Solar System history but later shattered into fragments during collisions with other planetesimals. Most meteorites appear to be about 4.54 Ga, but carbonaceous chondrites are as old as 4.56 Ga, the oldest known material ever measured.

Meteors over Hong Kong in 2001.

Meteorites are important to geologists because they consist of matter unchanged since the earliest days of the Solar System. Rocks that formed on Earth, in contrast, have been modified since they first formed.

Although almost all meteors are small and have not caused notable damage on Earth in historic time, a few have smashed through houses, dented cars, and bruised people. During the longer term of Earth history, however, some catastrophic collisions have left huge craters (see Geotour 1). As we will see later in this book, the largest collisions probably caused mass extinctions of life forms on our planet. It is likely that such collisions may happen again in the future—a large asteroid, for example, passed within 3 million miles of Earth in 2004. Researchers are just beginning to think about ways to deflect objects that are on a collision course with our planet.

Geologists have been able to calculate the overall chemical composition of the crust (▶Fig. 2.14). A glance at Figure 2.14 shows that regardless of whether you consider percentage by weight, percentage by volume, or percentage of atoms, *oxygen is by far the most abundant element in the crust!* This observation may surprise you, because most people picture oxygen as the colorless gas that we inhale when we breathe the atmosphere, not as a rock-forming chemical. But oxygen, when bonded to other elements, forms a great variety of minerals, and these minerals in turn make up the bulk of the rock in the Earth's crust. Because oxygen atoms are relatively large in comparison with their mass, oxygen actually occupies about 93% of the crust's volume. If you compare the composition of the crust to that of the whole Earth (see Fig. 2.9), you'll notice that the composition of the crust differs markedly from that of the whole Earth. That's because the composition of the entire Earth takes into account the core and mantle, which (as we discuss next) do not have the same composition as the crust.

Finally, it is important to note that most rock in the crust contains pores (tiny open spaces). In much of the upper several kilometers of the crust, the pores are filled with liquid water. This subsurface water, or groundwater, is what farmers pump out of wells for irrigation and that cities pump out for their water supplies.

The Mantle

The mantle of the Earth forms a 2,885-km-thick layer surrounding the core. In terms of volume, it is the largest part of the Earth. In contrast to the crust, the mantle consists entirely of an ultramafic rock called peridotite. This means that peridotite, though rare at the Earth's surface, is actually the most abundant rock in our planet! Overall, density in the mantle increases from about 3.5 g/cm³ at the top to about 5.5 g/cm³ at the base. On the basis of the occurrence of changes in the velocity of earthquake waves, geoscientists divide the mantle into two sublayers: the upper mantle, down to a depth of 660 km, and the lower mantle, from 660 km down to 2,900 km. The bottom part of the upper mantle, the interval lying between 400 km and 660 km deep, is also called the transition zone because here the character of the mantle undergoes a series of abrupt changes (see Interlude D).

Almost all of the mantle is solid rock. But even though it's solid, mantle rock below a depth of 100 to 150 km is so hot that it's soft enough to flow extremely slowly—at a rate of less than 15 cm a year. "Soft" here does not mean liquid; it simply means that over long periods of time mantle rock can change shape, like soft wax, without

FIGURE 2.13 (a) This simplified cross section illustrates the differences between continental crust and oceanic crust. Note that the thickness of continental crust can vary greatly. (b) Oceanic crust is denser than continental crust.

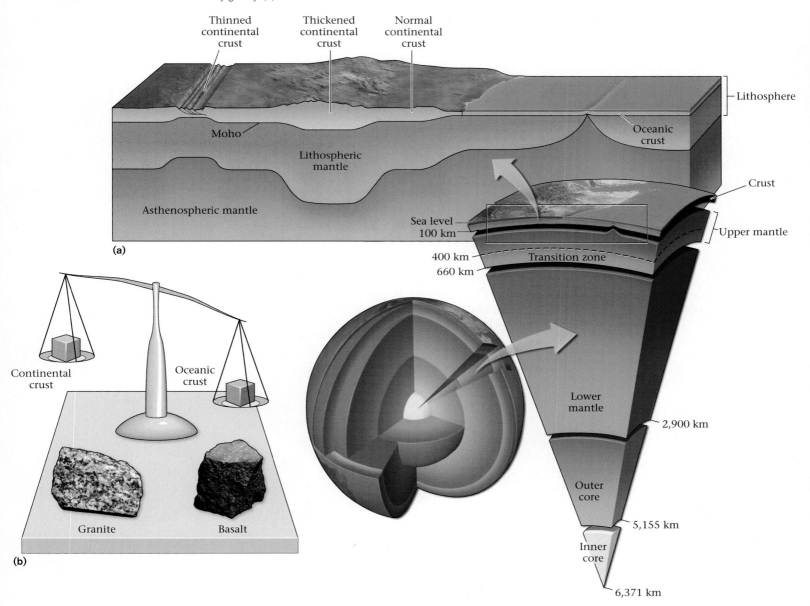

The Earth, from Surface to Center

If we could remove all the air that hides much of the solid surface from view, we would see that both the land areas and the sea floor have plains and mountains.

If we could break open the Earth, we would see that its interior consists of a series of concentric layers, called (in order from the surface to the center) the crust, the mantle, and the core. The crust is a relatively thin skin (7–10 km beneath oceans, 25–70 km beneath the land surface). Oceanic crust consists of basalt (mafic rock), while the average continental crust is intermediate to silicic. The mantle, which overall has the composition of ultramafic rock, can be divided into three layers: upper mantle, transition zone, and lower mantle. The core can be divided into an outer core of liquid iron alloy and an inner core of solid iron alloy. Temperature increases progressively with depth, so at the Earth's center the temperature may approach that of the Sun's surface.

Within the mantle and outer core, there is swirling, convective flow. Flow within the outer core generates the Earth's magnetic field.

When discussing plate tectonics, it is convenient to call the outer part of the Earth, a relatively rigid shell composed of the crust and uppermost mantle, the lithosphere and the underlying warmer, more plastic portion of the mantle the asthenosphere. These are not shown in this painting.

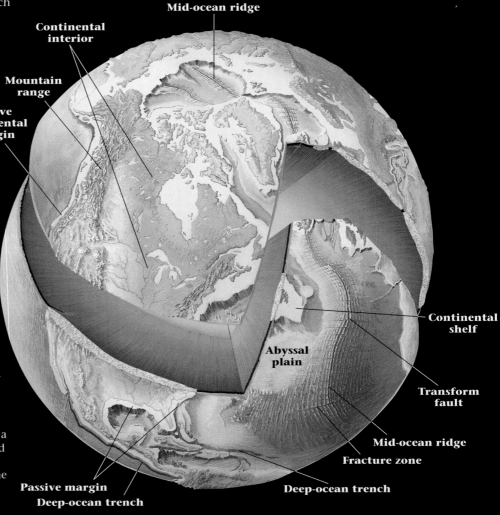

Mid-ocean ridge

Continental interior

Mountain range

Active continental margin

Continental shelf

Abyssal plain

Transform fault

Mid-ocean ridge

Fracture zone

Deep-ocean trench

Passive margin
Deep-ocean trench

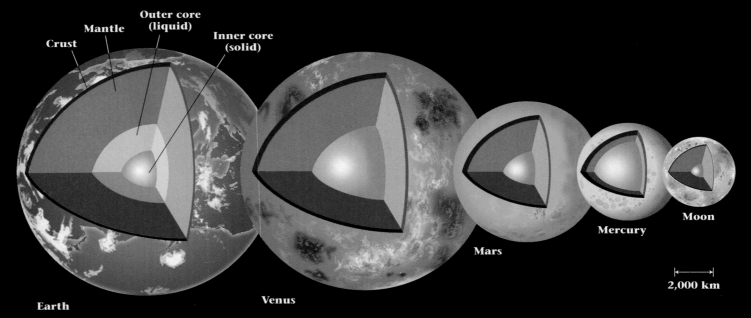

Crust
Mantle
Outer core (liquid)
Inner core (solid)

Earth

Venus

Mars

Mercury

Moon

2,000 km

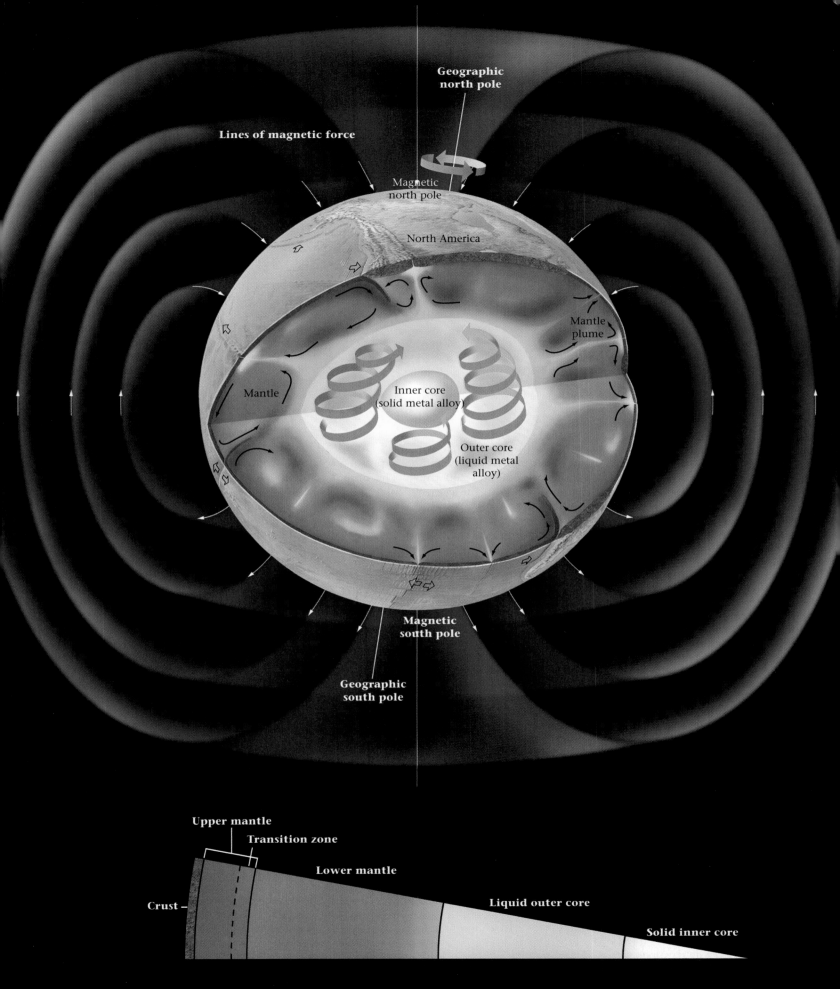

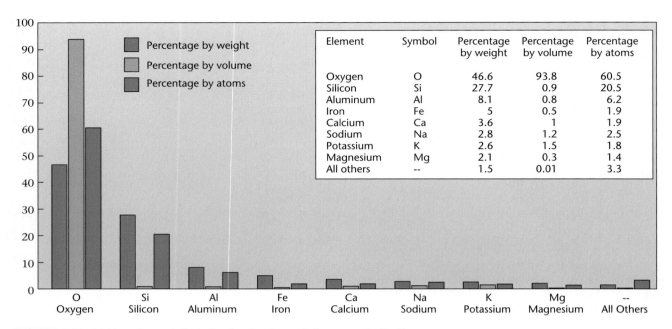

Element	Symbol	Percentage by weight	Percentage by volume	Percentage by atoms
Oxygen	O	46.6	93.8	60.5
Silicon	Si	27.7	0.9	20.5
Aluminum	Al	8.1	0.8	6.2
Iron	Fe	5	0.5	1.9
Calcium	Ca	3.6	1	1.9
Sodium	Na	2.8	1.2	2.5
Potassium	K	2.6	1.5	1.8
Magnesium	Mg	2.1	0.3	1.4
All others	--	1.5	0.01	3.3

FIGURE 2.14 A table and a graph illustrating the abundance of elements in the Earth's crust.

breaking. Note that we said *almost* all of the mantle is solid—in fact, up to a few percent of the mantle has melted in a layer that lies at depths of between 100 and 200 km beneath the ocean floor. This melt causes seismic waves to slow down, so geologists refer to this partly molten layer as the low-velocity zone.

Though overall the temperature of the mantle increases with depth, it varies significantly with location even at the same depth. The warmer regions are less dense, while the cooler regions are denser. The blotchy pattern of warmer and cooler mantle indicates that the mantle convects like water in a simmering pot. Warm mantle gradually flows upward, while cooler, denser mantle sinks.

The Core

Early calculations suggested that the core had the same density as gold, so for many years people held the fanciful hope that vast riches lay at the heart of our planet. Alas, geologists eventually concluded that the core consists of a far less glamorous material, iron alloy (iron mixed with smaller amounts of other elements). They arrived at this conclusion, in part, by comparing the properties of the core with the properties of metallic (iron) meteorites (see Box 2.1).

Studies of how earthquake waves bend as they pass through the Earth, along with the discovery that certain types of seismic waves cannot pass through the outer part of the core (see Interlude D), led geoscientists to divide the core into two parts, the **outer core** (between 2,900 and 5,155 km deep) and the **inner core** (from a depth of 5,155 km down to the Earth's center at 6,371 km). The outer core is a *liquid* iron alloy (composed of iron, nickel, and some other elements including silicon [Si], oxygen [O], and sulphur [S]) with a density of 10 to 12 g/cm³. It can exist as a liquid because the temperature in the outer core is so high that even the great pressures squeezing the region cannot lock atoms into a solid framework. Because it is a liquid, the iron alloy of the outer core can flow (see art, pp. 50–51); this flow generates Earth's magnetic field.

The inner core, with a radius of about 1,220 km and a density of 13 g/cm³, is a *solid* iron-nickel alloy, which may reach a temperature of over 4,700 °C. Even though it is hotter than the outer core, the inner core is a solid because it is deeper and subjected to even greater pressure. The pressure keeps atoms from wandering freely, so they pack together tightly in very dense materials. The inner core probably grows through time at the expense of the outer core, as the Earth slowly cools and the deeper part of the outer core solidifies. Recent data suggest the inner core rotates slightly faster than the rest of the Earth because of the force applied to it by the Earth's magnetic field.

> **Take-Home Message**
> Continental crust is 25 to 70 km thick and, on average, resembles granite in composition, whereas oceanic crust is about 7 km thick, and consists of basalt. The base of the crust is called the Moho. Earth's mantle is much thicker than the crust and consists of very dense rock. The core consists of iron alloy. Its outer part is liquid, and its inner part is solid.

2.8 THE LITHOSPHERE AND ASTHENOSPHERE

So far, we have divided the insides of the Earth into layers (crust, mantle, and core) based on the velocity at which earthquake waves travel through the layers. The three major layers (crust, mantle, and core) differ in composition from each other. An alternate way of thinking about Earth layers comes from studying the degree to which the material making up a layer can flow. In this context, we distinguish between "rigid" materials, which can bend but cannot flow, and "plastic" materials, which are relatively soft and can flow.

Let's apply this concept to the outer portion of the Earth's shell. Geologists have determined that the outer 100 to 150 km of the Earth is relatively rigid; in other words, the Earth has an outer shell composed of rock that cannot flow easily. This outer layer is called the **lithosphere**, and it consists of the crust plus the uppermost part of the mantle. We refer to the portion of the mantle within the lithosphere as the lithospheric mantle. Note that the terms *lithosphere* and *crust* are not synonymous—the crust is just part of the lithosphere. The lithosphere lies on top of the **asthenosphere**, which is the portion of the mantle in which rock can flow. Notice that the asthenosphere is entirely in the mantle and lies below a depth of 100 to 150 km. We can't assign a specific depth to the base of the asthenosphere because all of the mantle below 150 km can flow, but for convenience, some geologists place the base of the asthenosphere at the upper mantle/transition zone boundary. One final point: even though the asthenosphere can flow, do not think of it as a liquid. It is not. Rather, the asthenosphere is largely solid—a small amount of melt occurs in the low-velocity zone. At its fastest, the asthenosphere flows at rates of 10 to 15 cm/year.

Oceanic lithosphere and continental lithosphere are somewhat different (▶Fig. 2.15). Oceanic lithosphere, topped by oceanic crust, generally has a thickness of about 100 km. In contrast, continental lithosphere, topped by continental crust, generally has a thickness of about 150 km.

The boundary between the lithosphere and asthenosphere occurs where the temperature is about 1,280°C, for at this temperature mantle rock becomes soft enough to flow. To see how temperature affects the ability of a material to flow, take a cube of candle wax and place it in the freezer. The wax becomes very rigid and can maintain its shape for long periods of time; in fact, if you were to drop the cold wax, it would shatter.

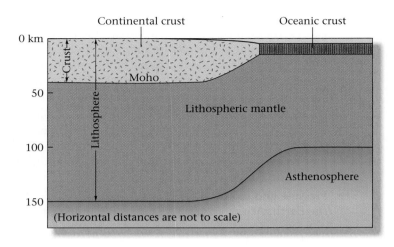

FIGURE 2.15 A cross section of the lithosphere, emphasizing the difference between continental and oceanic lithosphere.

But if you take another block of wax and place it in a warm (not hot) oven, it becomes soft, so that you can easily mold it into another shape. In fact, the force of gravity alone may cause the warm wax to slowly assume the shape of a pancake. Rock behaves somewhat similarly to the wax blocks. When rock is cool, it is quite rigid; but at high temperatures, rock becomes soft and can flow, though much more slowly than wax. This ability to flow slowly can occur at a temperature much lower than is necessary to cause rocks to melt. Rock of the lithosphere is cool enough to behave rigidly, whereas rock of the asthenosphere is warm enough to flow easily.

Now, with an understanding of Earth's overall architecture at hand, we can discuss geology's grand unifying theory—plate tectonics. The next two chapters introduce this key topic.

Take-Home Message

The crust and outermost part of the mantle make up a 100- to 150-km-thick layer called the lithosphere that behaves rigidly and cannot flow. It overlies the asthenosphere, the region of the mantle that is soft enough to flow (though very slowly).

Chapter Summary

- The Earth has a magnetic field that shields it from solar wind. Closer to Earth, the field creates the Van Allen belts, which trap cosmic rays.

- A layer of gas surrounds the Earth. This atmosphere consists of 78% nitrogen, 21% oxygen, and 1% other gases. Air pressure decreases with elevation, so 50% of the gas in the atmosphere resides below 5.5 km.

- The surface of the Earth can be divided into land (30%) and ocean (70%). Most of the land surface lies within 1 km of sea level. Earth's land surface has a great variety of landscapes because of variations in elevation and climate.

- The Earth consists of organic chemicals, minerals, glasses, rocks, metals, melts, and volatiles. Most rocks on Earth contain silica (SiO_2) and thus are called silicate rocks. We distinguish between felsic, intermediate, mafic, and ultramafic rocks based on the proportion of silica.

- The Earth's interior can be divided into three compositionally distinct layers, named in sequence from the surface down: the crust, the mantle, and the core. The first recognition of this division came from studying the density and shape of the Earth.

- Pressure and temperature both increase with depth in the Earth. At the center, pressure is 3.6 million times greater than at the surface, and the temperature reaches over 4,700 °C. The rate at which temperature increases as depth increases is the geothermal gradient.

- Studies of seismic waves have revealed the existence of sublayers in the core (liquid outer core and solid inner core) and mantle (upper mantle, transition zone, and lower mantle).

- The crust is a thin skin that varies in thickness from 7–10 km (beneath the oceans) to 25–70 km (beneath the continents). Oceanic crust is mafic in composition, whereas average continental crust is felsic to intermediate. The mantle is composed of ultramafic rock. The core is made of iron alloy and consists of two parts—the outer core is liquid, and the inner core is solid. Flow in the outer core generates the magnetic field.

- The crust plus the upper part of the mantle constitute the lithosphere, a relatively rigid shell up to 150 km thick. The lithosphere lies over the asthenosphere, mantle that is capable of flowing.

Geopuzzle Revisited

We can't see the interior of the Earth, so we have to deduce its nature from a variety of different kinds of measurements. From outside in, a slice through the Earth would reveal: a thin atmosphere; a thin crust of relatively low-density rock, with rocks of the continental crust being less dense, overall, than rocks of the oceanic crust; a thick mantle of very dense rock; and a core of iron alloy. Significantly, the crust plus the outermost part of the mantle together act as a rigid shell. Beneath this shell, the mantle is soft, but still mostly solid.

Key Terms

alloy (p. 44)
asthenosphere (p. 53)
atmosphere (p. 39)
basalt (p. 44)
core (p. 46)
crust (p. 46)
cryosphere (p. 40)
dipole (p. 38)
earthquake (p. 46)
fault (p. 46)
gabbro (p. 44)
geothermal gradient (p. 47)
glass (p. 44)
granite (p. 44)
hydrosphere (p. 40)
inner core (p. 52)
lithosphere (p. 53)

lower mantle (p. 49)
magnetic field (p. 38)
mantle (p. 46)
melts (p. 44)
metals (p. 44)
meteor (p. 48)
meteorites (p. 48)
mineral (p. 44)
Moho (p. 47)
organic chemicals (p. 44)
outer core (p. 52)
periodotite (p. 44)
rocks (p. 44)
sediment (p. 44)
topography (p. 40)
transition zone (p. 49)
upper mantle (p. 49)
volatiles (p. 44)

Review Questions

1. Why do astronomers consider the space between planets to be a vacuum, in comparison with the atmosphere near sea level?

2. What is the Earth's magnetic field? Draw a representation of the field on a piece of paper; your sketch should illustrate the direction in which charged particles would flow if placed in the field.

3. How does the magnetic field interact with solar wind? Be sure to consider the magnetosphere, the Van Allen radiation belts, and the aurorae.

4. What is Earth's atmosphere composed of, and how does it differ from the atmospheres of Venus and Mars? Why would you die of suffocation if you were to eject from a fighter plane at an elevation of 12 km without taking an oxygen tank with you?

5. What is the proportion of land area to sea area on Earth? From studies of the hypsometric curve, approximately what proportion of the Earth's surface lies at elevations above 2 km?

6. What are the two most abundant elements in the Earth? Describe the major categories of materials constituting the Earth.

7. What are silicate rocks? Give four examples of such rocks, and explain how they differ from one another in terms of their component minerals.

8. How did researchers first obtain a realistic estimate of Earth's average density? From this result, did they

conclude that the inside of the Earth is denser or less dense than rocks exposed at the surface? What observations led to the realization that the Earth is largely solid and that the Earth's mass is largely concentrated toward the center?

9. What are earthquake waves? Does the velocity at which an earthquake wave travels change or stay constant as the wave passes through the Earth? What are the principal layers of the Earth? What happens to earthquake waves when they reach the boundary between layers?

10. How do temperature and pressure change with increasing depth in the Earth? Be sure to explain the geothermal gradient.

11. What is the Moho? How was it first recognized? Describe the differences between continental crust and oceanic crust. Approximately what percentage of the Earth's diameter is within the crust?

12. What is the mantle composed of? What are the three sublayers within the mantle? Is there any melt within the mantle?

13. What is the core composed of? How do the inner core and outer core differ from each other? We can't sample the core directly, but geologists have studied samples of materials that are probably very similar in composition to the core. Where do these samples come from?

14. What is the difference among a meteoroid, a meteor, and a meteorite? Are all meteorites composed of the same material? Explain your answer.

15. What is the difference between lithosphere and asthenosphere? Be sure to consider material differences and temperature differences. Which layer is softer and flows easily? At what depth does the lithosphere/asthenosphere boundary occur? Is this above or below the Moho?

On Further Thought

1. (a) Recent observations suggest that the Moon has a very small, solid core that is less than 3% of its mass. In comparison, Earth's core is about 33% of its mass. Explain why this difference might exist. (*Hint*: Recall the model for Moon formation that we presented in Chapter 1.) (b) The Moon has virtually no magnetosphere. Why? (*Hint*: Remember what causes Earth's magnetic field.)

2. There is hardly any hydrogen or helium in the Earth's atmosphere, yet most of the nebula from which the Solar System formed consisted of hydrogen and helium. Where did all this gas go?

3. Popular media sometimes imply that the crust floats on a "sea of magma." Is this a correct image of the mantle just below the Moho? Explain your answer.

4. Why are meteorites significantly older than the oldest intact rock on Earth?

5. As you will see later in this book, emplacement of a huge weight (e.g., a continental ice sheet) causes the surface of lithosphere to sink, just as your weight causes the surface of a trampoline to sink. Emplacement of such a weight does not, however, cause a change in the thickness of the lithosphere. How is this possible? (*Hint*: Think about the nature of the asthenosphere.)

Suggested Reading

Bolt, B. A. 1982. *Inside the Earth*. San Francisco: Freeman.

Brown, G. C., and A. E. Mussett. 1993. *The Inaccessible Earth*. London: Chapman and Hall.

Fothergill, A., and Attenborough, D., 2007. *Planet Earth: As You've Never Seen It Before*. Berkeley, Calif.: University of California Press.

Freedman, R. A., and W. J. Kaufmann III. 2001. *Universe*, 6th ed. New York: Freeman.

Helffrich, G. R., and B. J. Wood. 2001. The Earth's mantle. *Nature* 412: 501–507.

Karato, S. I. 2003. *The Dynamic Structure of the Deep Earth*. Princeton: Princeton University Press.

Merrill, R. T., et al., 1998. *The Magnetic Field of the Earth*. Burlington, Mass.: Academic Press.

Sobel, D. 2006. *The Planets*. New York: Penguin.

Stein, S., and M. Wysession. 2003. *An Introduction to Seismology, Earthquakes, and Earth Structure*. London: Blackwell.

Vita-Finzi, C. 2006. *Planetary Geology*. UK: Terra Publishing.

Drifting Continents and Spreading Seas

Geopuzzle

At first glance, it looks like the continents on either side of the Atlantic Ocean could once have fit together quite nicely, like the pieces of a jigsaw puzzle. Do continents really move? Or, to put it another way, does the map of Earth's surface change over time?

Fossil leaves of *Glossopteris* from an exposure in Australia. The presence of this fossil on many continents was one of the observations that led to the proposal of continental drift.

It is only by combing the information furnished by all the Earth sciences that we can hope to determine "truth" here.

—Alfred Wegener (1880–1930)

3.1 INTRODUCTION

In September 1930, fifteen explorers led by a German meteorologist, Alfred Wegener, set out across the endless snowfields of Greenland to resupply two weather observers stranded at a remote camp (▶Fig. 3.1). The observers were planning to spend the long polar night recording wind speeds and temperatures on Greenland's polar plateau. At the time, Wegener was well known, not only to researchers studying climate but also to geologists. Some fifteen years earlier, he had published a small book, *The Origin of the Continents and Oceans*, in which he had dared to challenge geologists' long-held assumption that the continents had remained fixed in position through geologic time (the time since the formation of the Earth). Wegener proposed, instead, that the present distribution of continents and ocean basins had evolved. According to Wegener, the continents had once fit together like pieces of a giant jigsaw puzzle, to make one vast supercontinent. He suggested that this supercontinent, which he named **Pangaea** (pronounced Pan-jee-ah; Greek for "all land"), later fragmented into separate continents that then drifted apart, moving slowly to their present positions (▶Fig. 3.2). This idea came to be known as the **continental drift** hypothesis.

Wegener presented many observations in favor of the hypothesis, but he was met with strong resistance. Drifting continents? Absurd! Or so proclaimed the leading geologists of the day. At a widely publicized 1926 geology conference in New York City, a phalanx of celebrated American professors scoffed: "What force could possibly be great enough to move the immense mass of a continent?" Wegener's writings didn't provide a good answer,

FIGURE 3.1 Alfred Wegener, the German meteorologist who proposed a comprehensive model of continental drift and presented geologic evidence in support of the idea.

so despite all the supporting observations he had provided, most of the meeting's participants rejected continental drift.

Now, four years later, Wegener faced his greatest challenge. As he headed into the interior of Greenland, the weather worsened and most of his party turned back. But Wegener felt he could not abandon the isolated observers, and with two companions he trudged forward. On October 30, 1930, Wegener reached the observers and dropped off enough supplies to last the winter. Wegener and one companion set out on the return trip the next day, but they never made it home.

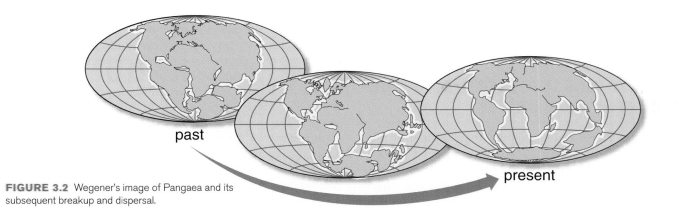

past

present

FIGURE 3.2 Wegener's image of Pangaea and its subsequent breakup and dispersal.

Had Wegener survived to old age, he would have seen his hypothesis become the foundation of a scientific revolution. Today, geologists accept Wegener's ideas and take it for granted that the map of the Earth constantly changes. Continents indeed waltz around our planet's surface, variously combining and breaking apart through geologic time. The revolution began in 1960, when Harry Hess, a Princeton University professor, proposed that continents drift apart because new ocean floor forms between them by a process that his contemporary Robert Dietz also described and labeled **sea-floor spreading**. Hess and others realized that in order for the circumference of the Earth to remain constant through time, ocean floor must eventually sink back into the mantle. Geologists now refer to this sinking process as **subduction** and recognize that when subduction consumes the ocean floor between two continents, the continents move toward one another.

By 1968, geologists had developed a fairly complete model of how continental drift, sea-floor spreading, and subduction all take place. In this model, Earth's lithosphere (its outer, relatively rigid shell) consists of about twenty distinct pieces, or plates, that move relative to each other as sea-floor spreading and subduction slowly take place. This model is now known as the theory of **plate tectonics** or, more simply, as "plate tectonics." The English word "tectonics" comes from the Greek word *tekton*, which means builder; plate movements "build" regional geologic features. Geologists view plate tectonics as the grand unifying theory of geology, because it successfully explains a great many geologic phenomena and features.

> **Take-Home Message**
>
> Alfred Wegener proposed that continents drifted apart following the breakup of a supercontinent that he called Pangaea. But only a few geologists agreed with his proposal at the time, because Wegener couldn't explain *how* drift occurred.

In this chapter, we examine the observations that led Wegener to propose his continental drift hypothesis. Then we learn how various key observations made by geologists during the mid-twentieth century led Harry Hess to propose the concept of sea-floor spreading. In Chapter 4, we will build on these concepts and describe the details of modern plate tectonics theory.

3.2 WHAT WAS WEGENER'S EVIDENCE FOR CONTINENTAL DRIFT?

Before Wegener, geologists viewed the continents and oceans as immobile—fixed in position throughout geologic time. According to Wegener, however, the positions of continents change through time. He suggested that a vast supercontinent, Pangaea, existed until the Mesozoic Era (the interval of geologic time, commonly known as the "age of dinosaurs," that lasted from 251 to 65 million years ago). During the Mesozoic, Pangaea broke apart to form the continents we see today; these continents then drifted away from each other. (Geologists now realize that supercontinents have formed and dispersed at least a few times during Earth's history. The name Pangaea applies only to the most recent supercontinent.) Let's look at some of Wegener's arguments and see why he came to this conclusion.

The Fit of the Continents

Almost as soon as maps of the Atlantic coastlines became available in the 1500s, scholars noticed the fit of the continents. The northwestern coast of Africa could tuck in against the eastern coast of North America, and the bulge of eastern South America could nestle cozily into the indentation of southwestern Africa. Australia, Antarctica, and India could all connect to the southeast of Africa; Greenland, Europe, and Asia could pack against the northeastern margin of North America (see Fig. 3.2). In fact, all the continents could be joined like the pieces of a jigsaw puzzle, with remarkably few overlaps or gaps, to create Pangaea. (Modern plate tectonics theory can now even explain the misfits.) Wegener concluded that the fit was too good to be coincidence (▶Geotour 3).

Locations of Past Glaciations

Wegener was an Arctic meteorologist by training, so it is no surprise that he had a strong interest in glaciers, rivers or sheets of ice that slowly flow across the land surface. He realized that glaciers form mostly at high latitudes, and thus that by studying the past locations of glaciers, he might be able to determine the past locations of continents. We'll look at glaciers in detail in Chapter 22, but we need to know something about them now to understand Wegener's arguments.

When a glacier moves, it scrapes sediment (pebbles, boulders, sand, silt, and mud) off the ground and carries it along. The sediment freezes into the base of the glacier, so the glacier becomes like a rasp and grinds exposed rock beneath it. In fact, rocks protruding from the base of the ice carve striations (scratches) into the underlying rock, and these striations indicate the direction in which the ice flowed. When the glacier eventually melts, the sediment collects on the ground and creates a distinctive layer of sediment called glacial till, a mixture of mud, sand, pebbles, and larger rocks. Later on, the till may be buried and preserved. Today, glaciers are found only in polar regions and in high mountains. But by studying the distribution and age of ancient till, geoscientists have determined that at several times during Earth's history, glaciers covered large areas of continents. We refer to these times as ice ages. One of the major ice ages occurred about 260 to 280 million years ago, near the end of the Paleozoic Era (the interval of geologic time between 542 and 251 million years ago).

Why was the study of ancient glacial deposits important to Wegener? When he plotted the locations of Late Paleozoic till, he found that glaciers of this time interval occurred in southern South America, southern Africa, southern India, Antarctica, and southern Australia. These places are all now widely separated from one another and, with the exception of Antarctica, do not currently lie in cold polar regions (▶Fig. 3.3a). To Wegener's amazement, all the late Paleozoic glaciated areas lie adjacent to each other on a map of Pangaea (▶Fig. 3.3b). Furthermore, when he plotted the orientation of glacial striations, they all pointed roughly outward from a location in southeastern Africa, just as we would expect if an ice sheet comparable to the present-day Antarctic polar ice cap had developed in southeastern Africa and had spread outward from its origin. In other words, Wegener determined that the distribution of glaciers at the end of the Paleozoic Era could easily be explained if the continents had been united in Pangaea, with the southern part of Pangaea located over the South Pole, but could not be explained if the continents had always been in their present positions.

The Distribution of Equatorial Climatic Belts

If the southern part of Pangaea had straddled the South Pole at the end of the Paleozoic Era, then during this same time interval southern North America, southern Europe, and northwestern Africa would have straddled the equator and would have had tropical or subtropical climates. Wegener searched for evidence for this configuration by studying sedimentary rocks that were formed at this time, for the material making up these rocks can reveal clues to the climate. Specifically, in the swamps and jungles of tropical regions, thick deposits of plant material accumulate, and when deeply buried, this material transforms into coal. And, in the clear shallow seas of tropical regions, large reefs built from the shells of marine organisms develop offshore. Finally, subtropical regions on either side of the tropical belt contain deserts, an environment in which sand dunes form and salt from evaporating seawater or salt lakes accumulates. Wegener thought that the distribution of late Paleozoic coal, sand-dune deposits, and salt deposits could define climate belts on Pangaea.

Sure enough, in the belt of Pangaea that Wegener expected to be equatorial, late Paleozoic sedimentary rock layers include abundant coal and the relics of reefs; and in the portions of Pangaea that Wegener predicted would be subtropical, late Paleozoic sedimentary rock layers include relics of desert dunes and of salt (▶Fig. 3.4). On a present-day map of our planet, these deposits are scattered around the globe at a variety of latitudes—including high latitudes, where they could not have formed. However, in Wegener's Pangaea, the deposits align in continuous bands that occupy appropriate latitudes.

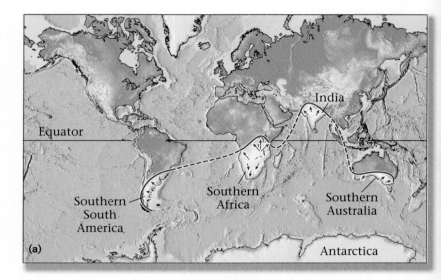

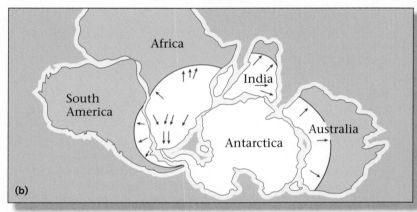

FIGURE 3.3 (a) The distribution of late Paleozoic glacial deposits on a map of the present-day Earth. The arrows indicate the orientation of striations. **(b)** The distribution of these glacial deposits on a map of the southern portion of Pangaea. Note that the glaciated areas fit together to define a polar ice cap.

The Distribution of Fossils

Today, different continents provide homes for different species. Kangaroos, for example, live only in Australia. Similarly, many kinds of plants grow only on one continent and not on others. Why? Because land-dwelling species of animals and plants cannot swim across vast oceans, and thus evolve independently on different continents. During a period of Earth history when all continents were in contact, however, land animals and plants conceivably could have migrated easily, so the same species could have lived on many continents.

With this concept in mind, Wegener plotted locations of fossils of land-dwelling species that lived during the late Paleozoic and early Mesozoic Eras (between about 300 and 210 million years ago) and found that they indeed existed

See for yourself...
Wegener's Evidence

Alfred Wegener could not measure plate motions directly, but he did gain insight simply from looking at the map of Earth's surface. Specifically, he recognized the "fit of the continents" and found landforms whose shape seemed to be the result of continental movement. The thumbnail images provided on this page are only to help identify tour sites. Go to wwnorton.com/studyspace to experience this flyover tour.

Matching Coastlines

View the planet from 12,000 km (roughly 8,000 miles) and orient the image so that you see South America on the left and Africa on the right (Image G3.1). Do you notice the similarity of the coastlines on opposite sides of the ocean? To see other examples, rotate the globe and compare the coast of eastern United States with the coast of northwest Africa (Image G3.2). Finally, compare the south coast of Australia with the nearest coast of Antarctica (Image G3.3).

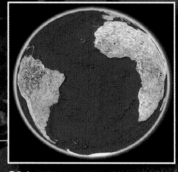
G3.1

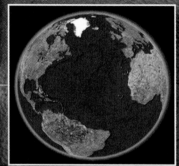
G3.2

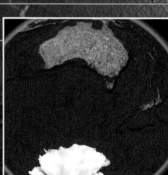
G3.3

The Scotia Arc
(Lat 57°44'06.79"S, Long 46°25'43.11"W)

Go to the Scotia Sea and zoom to an elevation of about 5,000 km (3,100 miles). Wegener was impressed that the southern tip of South America and the northern tip of the Antarctic Peninsula both curve to the east where they border the Scotia Sea, and he wondered whether the curves meant that the land masses bent as they drifted westward (Image G3.4). Modern studies suggest that such bending did indeed happen. Move to Lat 53° 56'8.91"S Long 70° 34'12.99"W, zoom to 1,950 km (1.2 miles), and tilt the image so you are looking north. You will see the curve of southern South America in greater detail (Image G3.5).

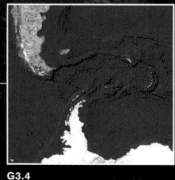
G3.4

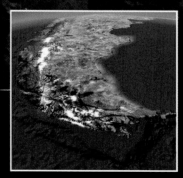
G3.5

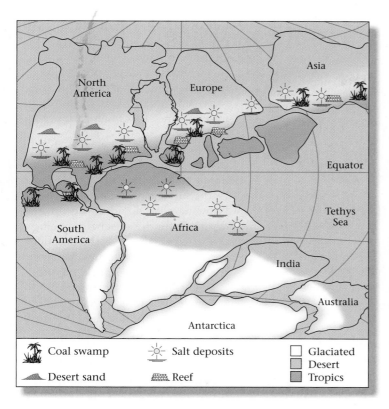

FIGURE 3.4 Map of Pangaea, showing the distribution of coal deposits and reefs (indicating tropical environments), and sand-dune deposits and salt deposits (indicating subtropical environments). Note how deposits now on different continents align in distinct belts.

Captain Robert Scott and his party of British explorers, who reached the South Pole in 1912, confirmed this proposal. On their return trip, the party died of starvation and cold, only 11 km from a food cache. When their bodies were found, their sled loads included *Glossopteris* fossils that they had hauled for hundreds of kilometers, in the process burning valuable calories that could possibly have kept them alive long enough to reach the cache. In 1969, paleontologists found fossils of *Lystrosaurus* in Antarctica, providing further confirmation that Wegener was right and that the continents had once been connected.

Matching Geologic Units

In the same way that an art historian can identify a Picasso painting and an architect a Victorian design, a geologist can identify a distinctive group of rocks. Wegener found that the same distinctive Precambrian (the interval of geologic time between Earth's formation and 542 million years ago) rock assemblages occurred on the eastern coast of South America and the western coast of Africa, regions now separated by an ocean (▶Fig. 3.6a). If the continents had been joined to create Pangaea in the past, then these matching rock groups would have been adjacent to one another, and thus could have composed continuous blocks. Wegener also noted that belts of rocks in the Appalachian Mountains of the United States and Canada closely resembled belts of rocks in mountains of southern Greenland, Great Britain,

on several continents (▶Fig. 3.5). For example, an early Mesozoic land-dwelling reptile called *Cynognathus* lived in both southern South America and southern Africa. *Glossopteris*, a species of seed fern, flourished in regions that now constitute South America, Africa, India, Antarctica, and Australia (see the chapter opening photo). *Mesosaurus*, a freshwater reptile, inhabited portions of what is now South America and Africa. *Lystrosaurus*, another land-dwelling reptile, wandered through present-day Africa, India, and Antarctica. None of these species could have traversed a large ocean. Thus, Wegener argued, the distribution of these species required the continents to have been adjacent to one another in the late Paleozoic and early Mesozoic Eras.

Considering that paleontologists found fossils of species such as *Glossopteris* in Africa, South America, and India, Wegener suspected that they might also be found in Antarctica. The tragic efforts of

FIGURE 3.5 This map shows the distribution of terrestrial (land-based) fossil species. Note that creatures such as *Lystrosaurus* could not have swum across the Atlantic to reach Africa. Sample locations are approximate.

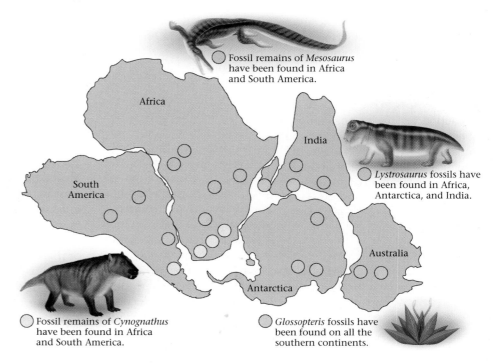

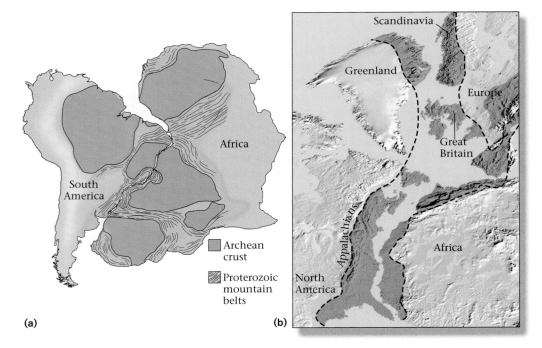

FIGURE 3.6 (a) Distinctive areas of rock assemblages on South America link with those on Africa, as if they were once connected and later broke apart. "Archean" is the older part of the Precambrian, and "Proterozoic" is the younger part. **(b)** If the continents are returned to their positions in Pangaea by closing the Atlantic, mountain belts (shown in brown) of the Appalachians lie adjacent to similar-age mountain belts in Greenland, Great Britain, Scandinavia, and Africa.

Scandinavia, and northwestern Africa (▶Fig. 3.6b), regions that would have lain adjacent to each other in Pangaea. Wegener thus demonstrated that not only did the coastlines of continents match—their component rocks did too.

Criticism of Wegener's Ideas

Wegener's model of a supercontinent that later broke apart explained the distribution of glaciers, coal, sand dunes, distinctive rock assemblages, and fossils we find today. Clearly, he had compiled a strong case for continental drift. But Wegener, as noted earlier, could not adequately explain how or why continents drifted. In his writings, Wegener suggested that the force created by the rotation of the Earth could cause a supercontinent centered at a pole to break up into pieces that would move toward equatorial latitudes. He proposed that the continental crust (he didn't refer to the lithosphere, which includes the crust and the uppermost part of the mantle) moved by "plowing" through oceanic crust as a ship plows through water. But other geologists of the time found his explanation wholly unsatisfac-

> **Take-Home Message**
>
> In support of his proposal of continental drift, Wegener noted that coastlines on opposite sides of oceans matched, and that the observed distribution of ancient glaciations, climate belts, fossils, and rock units make better sense if Pangaea existed.

tory. Experiments showed that the relatively weak rock making up continents cannot plow through the relatively strong rock making up the ocean floor, and that the force generated by Earth's spin is a million times too small to move a continent.

Wegener left on his final expedition to Greenland having failed to convince his peers, and he died in the icy wasteland never knowing that his ideas would lie dormant for decades before being reborn as the basis of the broader theory of plate tectonics. During these decades, a handful of iconoclasts continued to champion Wegener's notions. Among these was Arthur Holmes, a highly respected British geologist who argued that huge convection cells existed inside the Earth, slowly transporting hot rock from the deep interior up to the surface. Holmes suggested that continents might be split and the pieces dragged apart in response to convective flow in the mantle. But in general, geologists retreated to their subspecialties and remained indifferent to the possibility that a single bold idea could unify their work.

3.3 SETTING THE STAGE FOR THE DISCOVERY OF SEA-FLOOR SPREADING

Geologic work did not, of course, come to a halt following the death of Wegener. Researchers refined new techniques and instruments and completed countless studies of geologic features during the middle decades of the twentieth century. In particular, new fossil discoveries strengthened Wegener's argument that, prior to Mesozoic time, land animals had dispersed across all continents. The new technique of radiometric dating defined the age of rocks in years (see Chapter 12). New maps displayed the distribution of rock units on continents. New detection methods provided a clearer picture of Earth's interior. But arguably, the discoveries that were most influential in proving continental drift and in setting the stage for the proposal of sea-floor spreading came from research on a phenomenon called paleomagnetism and from exploration of the sea floor. Let's look at these discoveries.

Paleomagnetism and Apparent Polar Wander: The Basics

Rocks that contain tiny grains of magnetic minerals behave, overall, like very weak magnets. Researchers learned that some rocks develop their magnetization—their ability to produce a magnetic field (see Chapter 2)—at the time that the rocks themselves formed. Such rocks, in effect, preserve a record of the Earth's magnetic field at known times in the past: this record is called **paleomagnetism**. We can think of the orientation of the paleomagnetic field preserved in a rock specimen as an imaginary compass needle that points to where the Earth's magnetic pole was at the time the rock formed. The subject of paleomagnetism is fascinating, though somewhat complex. We introduce the basics of the topic here, but for students who want to learn more, we provide additional detail in Interlude A.

When researchers developed instruments sensitive enough to measure paleomagnetism in rocks, they asked the following question: Can the study of paleomagnetism detect changes in the position of the Earth's magnetic poles, relative to the continents, over geologic time? In the first attempt to answer this question, researchers measured the paleomagnetism in a collection of rock samples from Britain. Each sample in the collection had formed at a different time during the last 600 million years. Simply put, researchers found that the imaginary compass needle representing the paleomagnetism of each sample pointed to a different place on a map of the Earth. These results, announced in 1954, *implied* that the position of Earth's magnetic poles had indeed changed, relative to Britain, through geologic time. The researchers referred to this change as "polar wander" and represented what they thought was the progressive change in pole position over time by drawing a "polar-wander path" on a map.

If you're sitting in a car and see a person pass by, it's fair to wonder whether the person is standing still and the car is moving, or if the car is standing still and the person is moving. Researchers faced the same dilemma when studying paleomagnetism. Did observations of polar wander for Britain mean that Earth's magnetic poles have been moving relative to Britain, or do they instead mean that Britain has been moving (or drifting) relative to Earth's magnetic poles? To answer this question, researchers measured paleomagnetism in rock specimens from other continents and plotted polar-wander paths for other continents. They found that each continent has a different polar-wander path and realized that this result can mean only one thing: *Earth's magnetic poles do not move with respect to fixed continents. Rather, continents move relative to each other while the Earth's magnetic poles stay roughly fixed* (see Interlude A for illustrations). Because of this conclusion, the change over time in the apparent position of the Earth's magnetic poles relative to a continent came to be known as the **apparent polar-wander path** for the continent. The fact that each continent has a different apparent polar wander path proves that continents do move, and thus that Wegener's drift hypothesis was right after all. In the wake of this realization, geologists quickly turned their attention to the question that Wegener could not answer: *How* does drift occur?

New Images of Sea-Floor Bathymetry

Before World War II, we knew less about the shape of the ocean floor than we did about the shape of the Moon's surface. After all, we could at least see the surface of the Moon and could use a telescope to map its craters. But our knowledge of sea-floor **bathymetry** (the shape of the sea-floor surface) came only from scattered soundings of the sea floor. To sound the ocean depths, a surveyor let out a length of cable with a heavy weight attached. When the weight hit the sea floor, the length of the cable indicated the depth of the floor. Needless to say, it took many hours to make a single measurement, and not many could be made. Nevertheless, soundings carried out between 1872 and 1876 by the world's first oceanographic research vessel, the H.M.S. *Challenger*, did hint at the existence of submarine mountain ranges and deep troughs.

Military needs during World War II gave a boost to sea-floor exploration, for as submarine fleets grew, navies required detailed maps showing variations in the depth of the sea floor. The invention of echo sounding (sonar) permitted such maps to be made. Echo sounding works on the same principle that a bat uses to navigate and find insects. A sound pulse emitted from a ship travels down through the water, bounces off the sea floor, and returns up as an echo through the water to a receiver on the ship. Since sound waves travel at a known velocity, the time between the sound emission and the detection of the echo indicates the distance between the ship and the sea floor (velocity = distance/time, so distance = velocity × time). As the ship moves, echo sounding permits observers to obtain a continuous record of the depth of the sea floor; the resulting cross section showing depth plotted against location is called a bathymetric profile (▶Fig. 3.7a, b). By cruising back and forth across the ocean many times, investigators obtained a series of bathymetric profiles from which they constructed maps of the sea floor. Bathymetric maps revealed several important features of the ocean floor.

- *Mid-ocean ridges:* The floor beneath all major oceans includes two provinces: **abyssal plains**, the broad, relatively flat regions of the ocean that lie at a depth of about 4–5 km below sea level; and **mid-ocean ridges**, elongate submarine mountain ranges whose

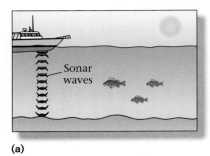

FIGURE 3.7 **(a)** To make a bathymetric profile, researchers use sonar. **(b)** An east-west bathymetric profile of the Atlantic Ocean. The inset shows a map of the ocean's floor.

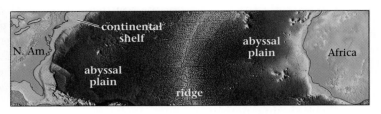

(a)

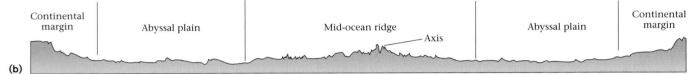

(b)

peaks lie only about 2–2.5 km below sea level (▶Figs. 3.8, 3.9a). Geologists call the crest of the mid-ocean ridge the ridge axis. All mid-ocean ridges are roughly symmetrical—bathymetry on one side of the axis is nearly a mirror image of bathymetry on the other side. Some, like the Mid-Atlantic Ridge, include steep escarpments (cliffs) as well as a distinct axial trough, a narrow valley that runs along the ridge axis.

- *Deep-ocean trenches*: Along much of the perimeter of the Pacific Ocean, and in a few other localities as well, the ocean floor reaches astounding depths of 8–12 km—deep enough to swallow Mount Everest. These deep areas define elongate troughs that are now referred to as **trenches**. Trenches border **volcanic arcs**, curving chains of active volcanoes.

- *Seamount chains*: Numerous volcanic islands poke up from the ocean floor: for example, the Hawaiian Islands lie in the middle of the Pacific. In addition to islands that rise above sea level, echo sounding has detected many **seamounts** (isolated submarine mountains),

FIGURE 3.8 The mid-ocean ridges, fracture zones, and principal deep-ocean trenches of today's oceans.

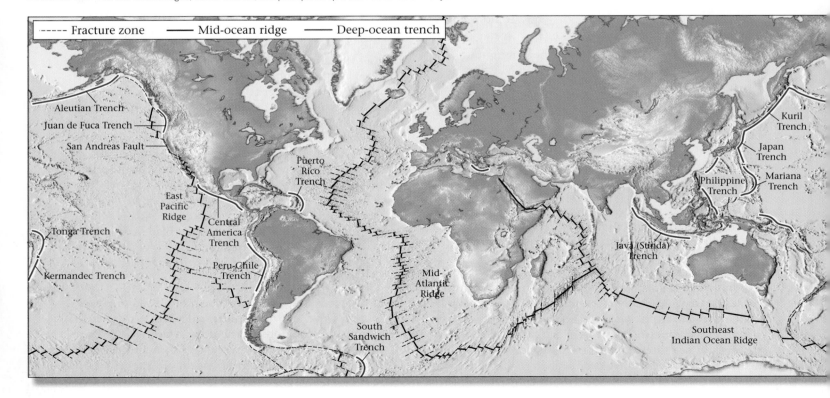

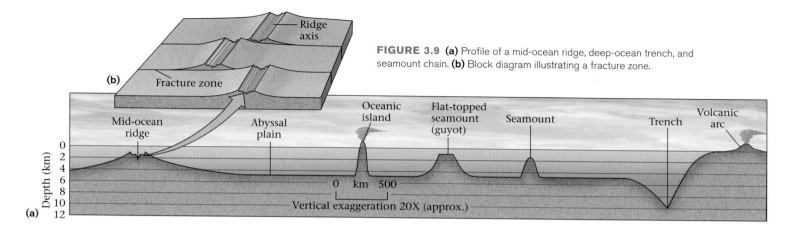

FIGURE 3.9 (a) Profile of a mid-ocean ridge, deep-ocean trench, and seamount chain. **(b)** Block diagram illustrating a fracture zone.

which were once volcanoes but no longer erupt. Oceanic islands and seamounts typically occur in chains, but in contrast to the volcanic arcs that border deep-ocean trenches, only one island at the end of a seamount chain is actively erupting today.

- *Fracture zones*: Surveys reveal that the ocean floor is diced up by narrow bands of vertical fractures. These **fracture zones** lie roughly at right angles to mid-ocean ridges, effectively segmenting the ridges into small pieces (▶Fig. 3.9b).

New Observations on the Nature of Oceanic Crust

By the mid-twentieth century, geologists had discovered many important characteristics of the sea-floor crust. These discoveries led them to realize that oceanic crust is quite different from continental crust, and further, that bathymetric features of the ocean floor provide clues to the origin of the crust. Specifically:

- A layer of sediment composed of clay and the tiny shells of dead plankton covers much of the ocean floor. This layer becomes progressively thicker away from the mid-ocean ridge axis. But even at its thickest, the sediment layer is too thin to have been accumulating for the entirety of Earth history.

Take-Home Message

The study of paleomagnetism showed that continents moved relative to the Earth's magnetic poles and thus proved that drift occurred. Studies of the sea floor and of the distribution of earthquakes set the stage for the discovery of sea-floor spreading.

- By dredging up samples, geologists learned that oceanic crust contains no granite and no metamorphic rock, common rock types on continents. Rather, oceanic crust contains only basalt and gabbro. Thus, it is fundamentally different in composition from continental crust.

- Heat flow, the rate at which heat rises from the Earth's interior up through the floor of the ocean, is not the same everywhere in the oceans. Rather, more heat seems to rise beneath mid-ocean ridges than elsewhere (▶Fig. 3.10). This observation led geologists to speculate that magma might be rising into the crust just below the mid-ocean ridge axis, because this hot molten rock could bring heat into the crust.

- When maps showing the distribution of earthquakes in oceanic regions became available in the years after World War II, geologists realized that earthquakes in these regions do not occur randomly, but rather define distinct belts (▶Fig. 3.11). Some belts follow trenches, some follow mid-ocean ridge axes, and others lie along portions of fracture zones. Since earthquakes define locations where rocks break and move, geologists realized that these bathymetric features are places where movements of the crust take place.

Now let's see how Harry Hess used these observations to come up with the hypothesis of sea-floor spreading.

FIGURE 3.10 In a mid-ocean ridge, heat from the mantle flows up through the crust; heat flow decreases away from the ridge axis.

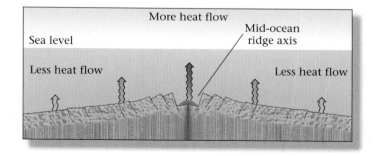

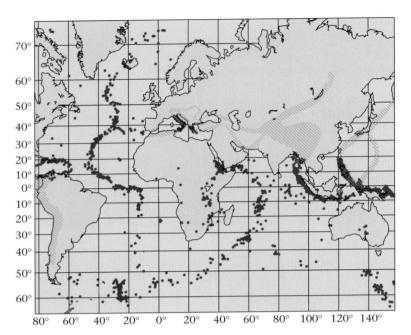

FIGURE 3.11 A 1953 map showing the distribution of earthquake locations in the ocean basins. Note that earthquakes occur in belts.

3.4 HARRY HESS AND HIS "ESSAY IN GEOPOETRY"

In the late 1950s, after studying the observations described above, Harry Hess realized that the overall thinness of the sediment layer on the ocean floor meant that the ocean floor might be much younger than the continents, and that the progressive increase in thickness of the sediment away from mid-ocean ridges could mean that the ridges themselves were younger than the deeper parts of the ocean floor. If this was so, then somehow new ocean floor must be forming at the ridges, and thus an ocean could be getting wider with time. But how? The association of earthquakes with mid-ocean ridges suggested to him that the sea floor was cracking and splitting apart at the ridge. The discovery of high heat flow along mid-ocean ridge axes provided the final piece of the puzzle, for it suggested the presence of molten rock beneath the ridges. In 1960, Hess suggested that molten rock rose upward beneath mid-ocean ridges and that this material solidified to create oceanic crust. The new sea floor then moved away from the

ridge, a process we now call sea-floor spreading. Hess realized that old ocean floor must be consumed somewhere, or the Earth's circumference would have to grow. He suggested that deep-ocean trenches might be places where the sea floor sank back into the mantle, and that earthquakes at trenches were evidence of this movement, but he didn't understand how the movement took place (▶Fig. 3.12).

Hess and his contemporaries realized that the sea-floor-spreading hypothesis instantly provided the long-sought explanation of how continental drift occurs. Rather than plowing through oceanic crust as Wegener suggested, continents passively move apart as the sea floor between them spreads at mid-ocean ridges, and they passively move together as the sea floor between them sinks back into the mantle at trenches. Further, it is lithosphere that moves, not just the crust. Thus, sea-floor spreading proved to be an important step on the route to plate tectonics—the idea seemed so good that Hess referred to his description of it as "an essay in geopoetry." But other key discoveries would have to take place before the whole theory of plate tectonics came together.

Take-Home Message

Ocean basins get wider with time due to the process of sea-floor spreading, and old ocean floor can sink back into the mantle by the process of subduction. As ocean basins get wider, continents drift apart. Subduction allows continents to move toward each other.

3.5 MARINE MAGNETIC ANOMALIES: EVIDENCE FOR SEA-FLOOR SPREADING

For a hypothesis to earn the status of theory, there must be proof. The proof of sea-floor spreading emerged from two discoveries. First, geologists found that the measured strength of Earth's magnetic field is not the same everywhere

FIGURE 3.12 Harry Hess's basic concept of sea-floor spreading. New sea floor forms at the mid-ocean ridge axis. As a result, the ocean grows wider. Old sea floor sinks into the mantle at a trench. Earthquakes occur at ridges and trenches. Hess implied, incorrectly, that only the crust moved. We will see in Chapter 4 that this sketch is an oversimplification.

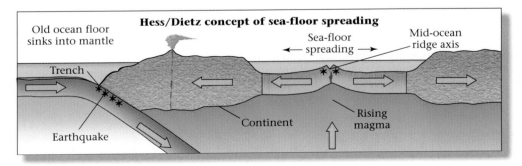

in the ocean basins; the variations are now called marine magnetic anomalies. Second, they found that Earth's dipole (the imaginary arrow inside the Earth that points from one pole to the other; see Chapter 2) reverses direction every now and then; such sudden reversals of the Earth's polarity are now called magnetic reversals. To understand why geologists find the concept of sea-floor spreading so appealing, we first need to learn about anomalies and reversals.

Marine Magnetic Anomalies

Geologists can measure the strength of Earth's magnetic field with an instrument called a magnetometer. At any given location on the surface of the Earth, the magnetic field that you measure includes two parts: one that is created by the main dipole of the Earth (which is caused in turn by the flow of liquid iron in the outer core; see Interlude A) and another that is created by the magnetism of near-surface rock. A **magnetic anomaly** is the difference between the expected strength of the Earth's main field at a certain location and the actual measured strength of the magnetic field at that location. Places where the field strength is stronger than expected are positive anomalies, and places where the field strength is weaker than expected are negative anomalies.

On continents, the pattern of magnetic anomalies is very irregular because continental crust contains many dif-ferent rock types. But magnetic anomalies on the sea floor yield a surprisingly different pattern. Geologists towed magnetometers back and forth across the ocean to map variations in magnetic field strength. As a ship cruised along its course, the magnetometer's gauge would first detect strong signals (a positive anomaly) and then weak signals (a negative anomaly). A graph of signal strength versus distance along the traverse, therefore, has a sawtooth shape (▶Fig. 3.13a). When data from many cruises was compiled on a map, these **marine magnetic anomalies** defined distinctive, alternating bands. And if we color positive anomalies dark and negative anomalies light, the map pattern made by the anomalies resembles the stripes on a candy cane (▶Fig. 3.13b). The mystery of the marine magnetic anomaly pattern, however, remained unsolved until geologists recognized the existence of magnetic reversals.

Magnetic Reversals

Soon after geologists began to study the phenomenon of paleomagnetism, they decided to see if the magnetism of rocks changed as time passed. To do this, they measured the paleomagnetism of many successive rock layers that represented a long period of time. To their surprise, they found that the polarity (which end of a magnet points north and which end points south; see Interlude A) of the

FIGURE 3.13 **(a)** A ship sailing through the ocean dragging a magnetometer detects first a positive anomaly and then a negative one, then a positive one, then a negative one. **(b)** Magnetic anomalies on the sea floor off the northwestern coast of the United States. The dark bands are positive anomalies, the light bands negative anomalies. Note the distinctive stripes of alternating anomalies. A positive anomaly overlies the crest of the Juan de Fuca Ridge (a small mid-ocean ridge).

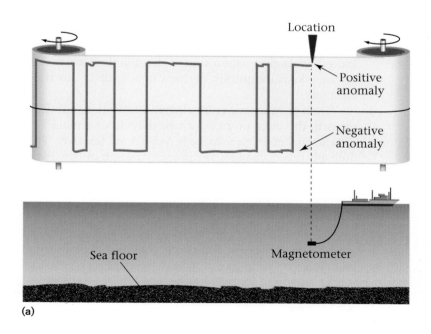

(a)

(b)

paleomagnetic field of some layers was the same as that of Earth's present magnetic field, whereas in other layers it was the opposite. Recall that Earth's magnetic field can be represented by an arrow, representing the dipole, that points from the magnetic north pole to the magnetic south pole; in some of the rock layers, the paleomagnetic dipole pointed south (these layers have normal polarity), but in others the dipole pointed north (these layers have reversed polarity) (▶Fig. 3.14).

At first, observations of reversed polarity were largely ignored, thought to be the result of lightning strikes or of chemical reactions between rock and water. But when repeated measurements from around the world revealed a systematic pattern of alternating normal and reversed polarity in rock layers, geologists realized that reversals were a global, not a local, phenomenon. At various times during Earth history, the polarity of Earth's magnetic field has suddenly reversed! In other words, sometimes the Earth has normal polarity, as it does today, and sometimes it has reversed polarity (▶Fig. 3.15a, b). Times when the Earth's field flips from normal to reversed polarity, or vice versa, are called **magnetic reversals**. When the Earth has reversed polarity, the south magnetic pole lies near the north geographic pole, and the north magnetic pole lies near the south geographic pole. If you were to use a compass during periods when the Earth's magnetic field was reversed, the north-seeking end of the needle would point to the south geographic pole.

Note that magnetic reversals are not related to apparent polar wander. Also, magnetic reversals are not related to the slight migration of the dipole with respect to Earth's

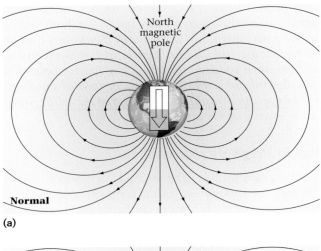

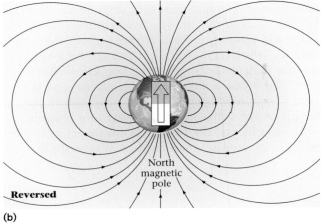

FIGURE 3.15 The magnetic field of the Earth has had reversed polarity at various times during Earth history. **(a)** If the dipole points from north to south, Earth has normal polarity. **(b)** If the dipole points from south to north, Earth has reversed polarity.

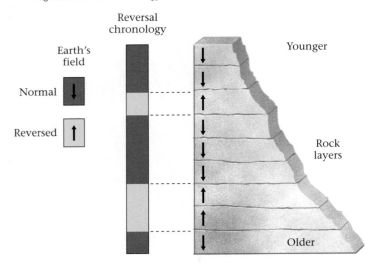

FIGURE 3.14 In a succession of rock layers on land, different flows exhibit different polarity (indicated here by whether the arrow points up or down). When these reversals are plotted on a time column, we have a magnetic-reversal chronology.

rotational axis, which occurs constantly (see Interlude A). Indeed, geologists have found evidence indicating that reversals take place quickly, perhaps in as little as one thousand years.

Though magnetic reversals have now been well documented, the mechanism by which they occur remains uncertain and continues to be a subject of research. They probably reflect changes in the configuration of flow in the outer core. Recent computer models of the Earth's magnetic field show that the field will flip spontaneously, without any external input, simply in response to the pattern of convection in the outer core. During the relatively short periods of time during which a reversal takes place, the magnetic field becomes disorganized and cannot be represented by a dipole (▶Fig. 3.16a–c).

In the 1950s, about the same time geologists discovered polarity reversals, they developed a technique that permitted them to define the age of a rock in years, by measuring the

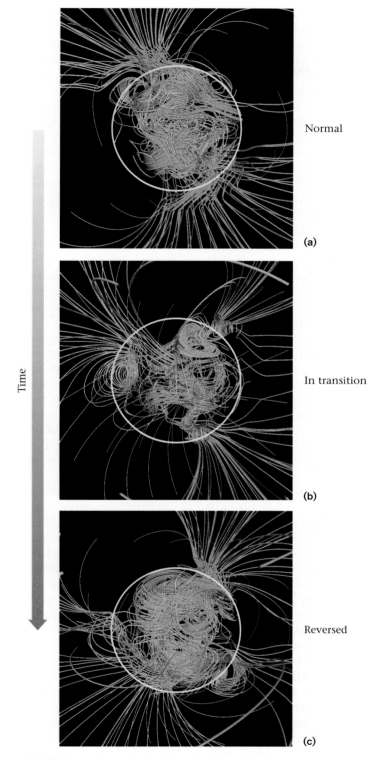

FIGURE 3.16 Images illustrating Earth's magnetic changes during a reversal, as calculated by a computer model. The colored lines represent magnetic field lines. The dipole points from yellow to blue. The white circle represents the outline of the core-mantle boundary inside the Earth. **(a)** Normal polarity. **(b)** Polarity during transition. **(c)** Reversed polarity.

rate of decay of radioactive elements in the rock. The technique is called radiometric dating. (It will be discussed in detail in Chapter 12.) Geologists applied the technique to determine the ages of rock layers from which they obtained their paleomagnetic measurements, and thus determined *when* the magnetic field of the Earth reversed. With this information, they constructed the history of magnetic reversals, now called the magnetic-reversal chronology. A diagram representing the Earth's magnetic-reversal chronology (▶Fig. 3.17) shows that reversals do not occur regularly, so the lengths of different polarity chrons, the time intervals between reversals, are different. For example, we have had a normal-polarity chron for about the last 700,000 years. Before that, there was a reversed-polarity chron. Geologists named the youngest four polarity chrons (Brunhes, Matuyama, Gauss, and Gilbert) after scientists who had made important contributions to the study of rock magnetism. As more measurements became available, investigators realized that there were some short-duration reversals (less than 200,000 years long) within the chrons; they called these shorter reversals polarity subchrons. Radiometric dating methods are not accurate enough to date reversals that happened about 4.5 million years ago.

FIGURE 3.17 Radiometric dating of lava flows allows us to determine the age of magnetic reversals during the past 4 million years. Major intervals of a given polarity are referred to as polarity chrons, and are named after scientists who contributed to the understanding of Earth's magnetic field. Shorter-duration reversals are called subchrons.

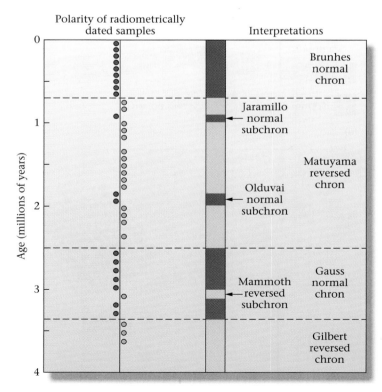

The Interpretation of Marine Anomalies

Why do marine magnetic anomalies exist? A graduate student in England, Fred Vine, working with his adviser, Drummond Matthews, and a Canadian geologist, Lawrence Morley (working independently), discovered a solution to this riddle. The three suggested that a positive anomaly occurs over areas of sea floor where the paleomagnetism preserved in ocean-floor basalt has normal polarity. In these areas, the magnetic force produced by the basalt *adds* to the force produced by Earth's dipole and creates a stronger magnetic signal than expected, as measured by the magnetometer. A negative anomaly occurs over regions of sea floor where basalt has reversed polarity. Here, the magnetic force of the basalt *subtracts* from the force produced by the dipole and results in a weaker magnetic signal (▶Fig. 3.18a).

FIGURE 3.18 **(a)** The explanation of marine anomalies. The sea floor beneath positive anomalies has the same polarity as Earth's magnetic field and therefore adds to it. The sea floor beneath negative anomalies has reversed polarity and thus subtracts from Earth's magnetic field. **(b)** The symmetry of the magnetic anomalies measured across the Mid-Atlantic Ridge south of Iceland. Note that individual anomalies are somewhat irregular, because the process of forming the sea floor, in detail, happens in discontinuous pulses along the length of the ridge.

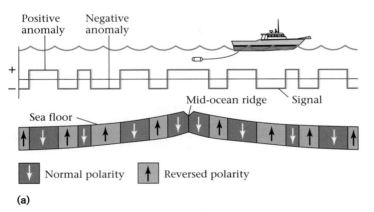

(a)

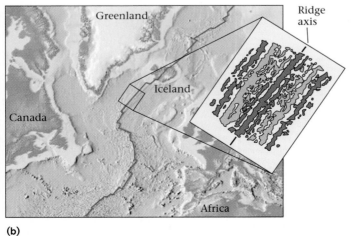

(b)

Vine, Matthews, and Morley pointed out that marine magnetic anomalies would be an expected consequence of sea-floor spreading—sea floor yielding positive anomalies developed at times when the Earth had normal polarity, whereas sea floor yielding negative anomalies formed when the Earth had reversed polarity. If this was so, then the seafloor-spreading hypothesis implies that the pattern of anomalies should be symmetric with respect to the axis of a mid-ocean ridge. With this idea in mind, geologists set sail to measure magnetic anomalies near mid-ocean ridges (▶Fig. 3.18b). By 1966, the story was complete. In the examples studied, the magnetic anomaly pattern on one side of a ridge was indeed a mirror image of the anomaly pattern on the other.

Let's look more closely at how marine magnetic anomalies are formed. Please refer to ▶Figure 3.19a. At Time 1 (sometime in the past), a time of normal polarity, the dark stripe of sea floor forms. The tiny dipoles of magnetite grains in basalt making up this stripe align with the Earth's field. As it forms, the rock in this stripe migrates away from the ridge axis, half to the right and half to the left. Later, at Time 2, the field has reversed, and the light-gray stripe forms with reversed polarity. As it forms, it too moves away from the axis, and still younger crust begins to develop along the axis. As the process continues over millions of years, many stripes form. A positive anomaly exists along the ridge axis today, because at the ridge axis is sea floor that has developed during the most recent interval of time, a chron of normal polarity. The magnetism of the rock along the ridge adds to the magnetism of the Earth's field.

Geologists realized that if the anomalies on the sea floor formed by sea-floor spreading as the Earth's magnetic field flipped between normal and reversed polarity, then the anomalies should correspond to the magnetic reversals that had been discovered and radiometrically dated in basalt layers on land. By relating the stripes on the sea floor to magnetic reversals found in dated basalt (▶Fig. 3.19b), geologists dated the sea floor back to an age of 4.5 million years, and they found that the relative widths of anomaly stripes on the sea floor exactly corresponded to the relative durations of polarity chrons in the magnetic-reversal chronology.

The relationship between anomaly-stripe width and polarity-chron duration provides the key for determining the rate (velocity) of sea-floor spreading, for it indicates that the rate of spreading has been fairly constant for the last 4.5 million years. Remember that velocity = distance/time. In the North Atlantic Ocean, 4.5-million-year-old sea floor lies 45 km away from the ridge axis. Therefore, the velocity (v) at which the sea floor moves away from the ridge axis can be calculated as follows:

$$v = \frac{45 \text{ km}}{4,500,000 \text{ years}} = \frac{4,500,000 \text{ cm}}{4,500,000 \text{ years}} = 1 \text{ cm/y}$$

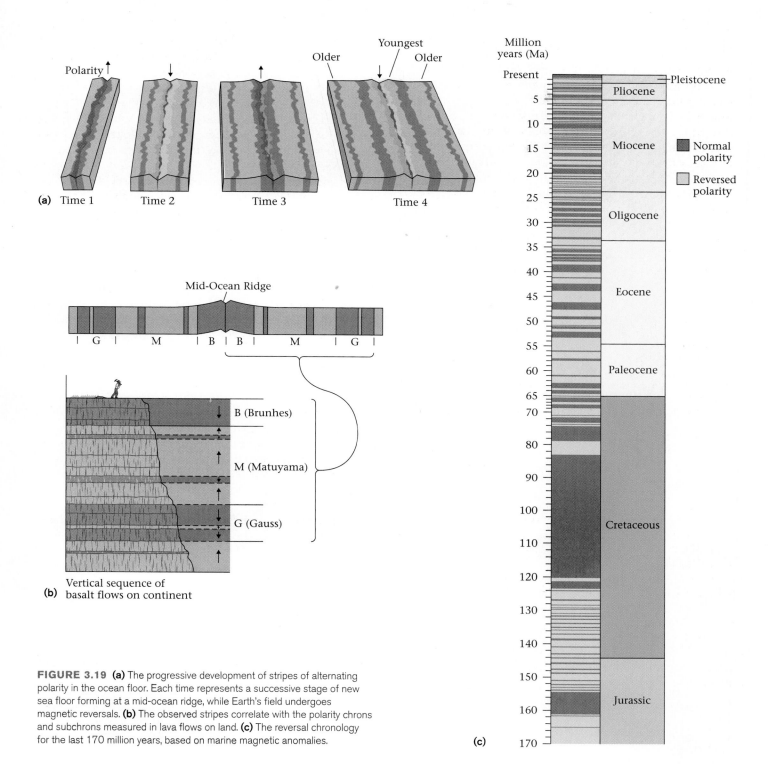

FIGURE 3.19 (a) The progressive development of stripes of alternating polarity in the ocean floor. Each time represents a successive stage of new sea floor forming at a mid-ocean ridge, while Earth's field undergoes magnetic reversals. **(b)** The observed stripes correlate with the polarity chrons and subchrons measured in lava flows on land. **(c)** The reversal chronology for the last 170 million years, based on marine magnetic anomalies.

This means that the crust moves away from the Mid-Atlantic Ridge axis at a rate of 1 cm per year, or that a point on one side of the ridge moves away from a point on the other side by 2 cm per year. We call this number the **spreading rate**. In the Pacific Ocean, sea-floor spreading occurs at the East Pacific Rise. (Geographers named this a "rise" because it is not as rough and jagged as the Mid-Atlantic Ridge.) The anomaly stripes bordering the East Pacific Rise are much wider, and 4.5-million-year-old sea floor lies about 225 km from the rise axis. This requires the sea floor to move away from the rise at a rate of about 5 cm per year, so the spreading rate for the East Pacific Rise is about 10 cm per year.

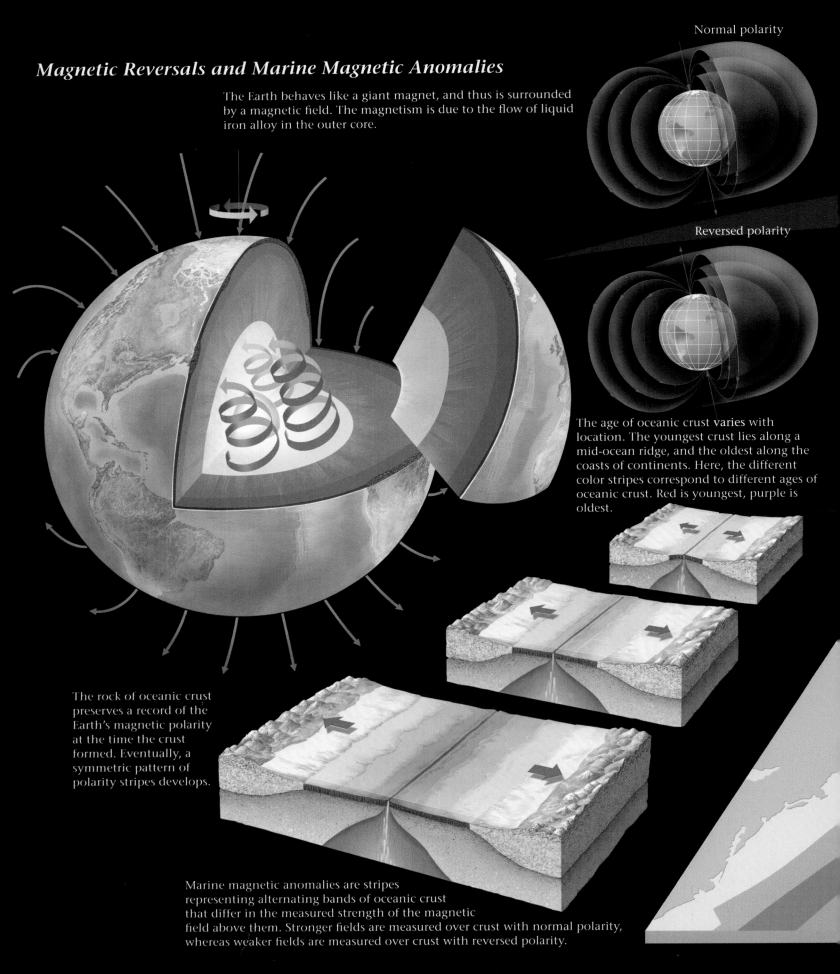

Magnetic Reversals and Marine Magnetic Anomalies

The Earth behaves like a giant magnet, and thus is surrounded by a magnetic field. The magnetism is due to the flow of liquid iron alloy in the outer core.

Normal polarity

Reversed polarity

The age of oceanic crust varies with location. The youngest crust lies along a mid-ocean ridge, and the oldest along the coasts of continents. Here, the different color stripes correspond to different ages of oceanic crust. Red is youngest, purple is oldest.

The rock of oceanic crust preserves a record of the Earth's magnetic polarity at the time the crust formed. Eventually, a symmetric pattern of polarity stripes develops.

Marine magnetic anomalies are stripes representing alternating bands of oceanic crust that differ in the measured strength of the magnetic field above them. Stronger fields are measured over crust with normal polarity, whereas weaker fields are measured over crust with reversed polarity.

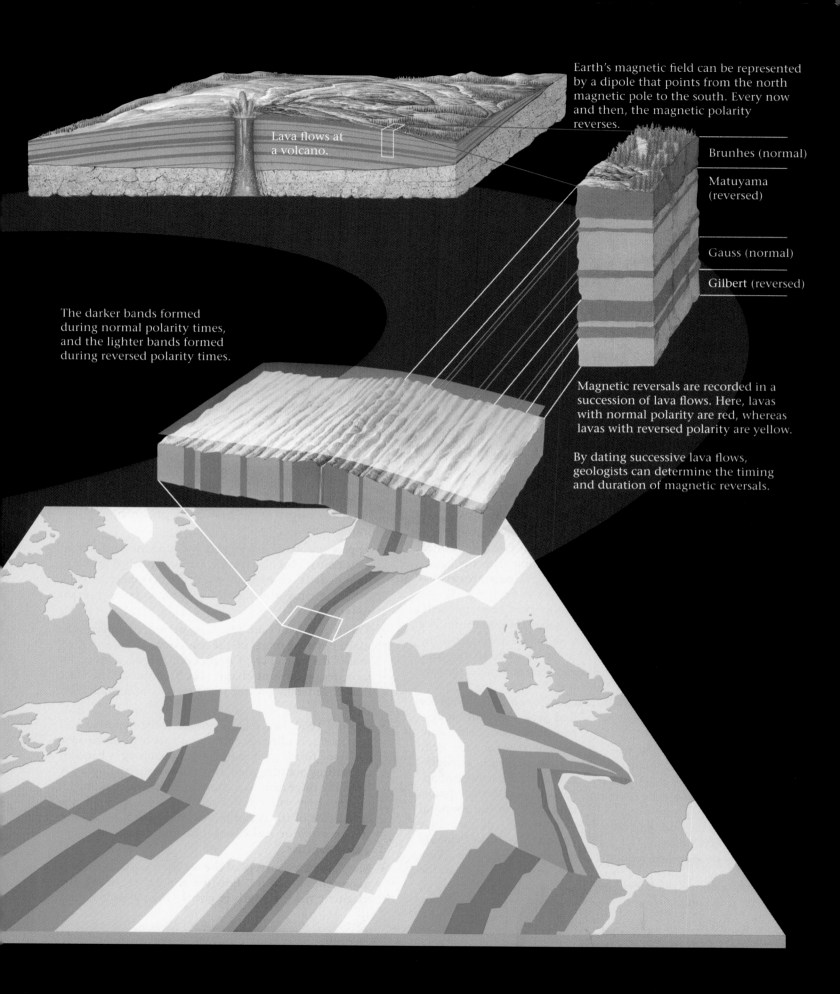

Earth's magnetic field can be represented by a dipole that points from the north magnetic pole to the south. Every now and then, the magnetic polarity reverses.

Lava flows at a volcano.

Brunhes (normal)

Matuyama (reversed)

Gauss (normal)

Gilbert (reversed)

The darker bands formed during normal polarity times, and the lighter bands formed during reversed polarity times.

Magnetic reversals are recorded in a succession of lava flows. Here, lavas with normal polarity are red, whereas lavas with reversed polarity are yellow.

By dating successive lava flows, geologists can determine the timing and duration of magnetic reversals.

Geologists also realized that if they could assume that the rate of sea-floor spreading has remained fairly constant for a long time, then they could date the ages of magnetic-field reversals further back in Earth history, simply by measuring the distance of successive magnetic anomalies from the ridge axis (time = distance/velocity). Such analysis eventually defined the magnetic-reversal chronology back to about 170 million years ago (▶Fig. 3.19c).

The spectacular correspondence between the record of marine magnetic anomalies and the magnetic-reversal chronology can be explained only by the sea-floor-spreading hypothesis. Thus, the discovery and explanation of marine magnetic anomalies was proof of sea-floor spreading and allowed geologists to measure rates of spreading. At a rate of 5 cm per year, sea-floor spreading produces a 5,000-km-wide ocean in 100 million years.

3.6 DEEP-SEA DRILLING: FURTHER EVIDENCE

Soon after geologists around the world began to accept the idea of sea-floor spreading, an opportunity arose to really put the concept to the test. In the late 1960s, a drilling ship called the *Glomar Challenger* set out to sail around the ocean drilling holes into the sea floor. This amazing ship could lower enough drill pipe to drill in 5-km-deep water and could continue to drill until the hole reached a depth of about 1.7 km (1.1 miles) below the sea floor. Drillers brought up cores of rock or sediment that geoscientists then studied on board.

On one of its early cruises, the *Glomar Challenger* drilled a series of holes through sea-floor sediment to the basalt layer. These holes were spaced at progressively greater distances from the axis of the Mid-Atlantic Ridge. If the model of sea-floor spreading was correct, then the sediment layer should be progressively thicker away from the axis, and the age of the oldest sediment just above the basalt should be progressively older away from the axis. When the drilling and the analyses were complete, the predictions were confirmed.

So by the early 1960s, it had become clear that Wegener had been right all along— continents do drift. But, though the case for drift had been greatly strengthened by the discovery of apparent polar-wander paths, it really took the proposal and proof of sea-floor spreading to make believers of most geologists. Very quickly, as we will see in the next chapter, these ideas became the basis of the theory of plate tectonics.

Take-Home Message

Marine magnetic anomalies form because reversals of the Earth's magnetic polarity take place while sea-floor spreading occurs. The discovery and interpretation of these anomalies proved the sea-floor-spreading hypothesis.

Chapter Summary

- Alfred Wegener proposed that continents had once been joined together to form a single huge supercontinent (Pangaea) and had subsequently drifted apart. This idea is the continental drift hypothesis.

- Wegener drew from several different sources of data to support his hypothesis: (1) coastlines on opposite sides of the ocean match up; (2) the distribution of late Paleozoic glaciers can be explained if the glaciers made up a polar ice cap over the southern end of Pangaea; (3) the distribution of late Paleozoic equatorial climatic belts is compatible with the concept of Pangaea; (4) the distribution of fossil species suggests the existence of a supercontinent; (5) distinctive rock assemblages that are now on opposite sides of the ocean were adjacent on Pangaea.

- Despite all the observations that supported continental drift, most geologists did not initially accept the idea, because no one could explain *how* continents could move.

- Rocks retain a record of the Earth's magnetic field that existed at the time the rocks formed. This record is called paleomagnetism. By measuring paleomagnetism in successively older rocks, geologists found that the apparent position of the Earth's magnetic pole relative to the rocks changes through time. Successive positions of the pole define an apparent polar-wander path.

- Apparent polar-wander paths are different for different continents. This observation can be explained by continental drift: continents move with respect to each other, wheras the Earth's magnetic poles remain roughly fixed.

- The invention of echo sounding permitted explorers to make detailed maps of the sea floor. These maps revealed the existence of mid-ocean ridges, deep-ocean trenches, seamount chains, and fracture zones. Heat flow is generally greater near the axis of a mid-ocean ridge.

- Around 1960, Harry Hess proposed the hypothesis of sea-floor spreading. According to this hypothesis, new sea floor forms at mid-ocean ridges, above a band of upwelling mantle, then spreads symmetrically away from the ridge axis. As a consequence, an ocean can get progressively wider with time, and the continents on either side of the ocean basins drift apart. Eventually, the ocean floor sinks back into the mantle at deep-ocean trenches.

- Magnetometer surveys of the sea floor revealed marine magnetic anomalies. Positive anomalies, where the magnetic field strength is greater than expected, and negative anomalies, where the magnetic field strength is less than expected, are arranged in alternating stripes.

- During the 1950s, geologists documented that the Earth's magnetic field reverses polarity every now and then. The record of reversals, dated by radiometric techniques, is called the magnetic-reversal chronology.

- The proof of sea-floor spreading came from the interpretation of marine magnetic anomalies. Sea floor that forms when the Earth has normal polarity results in positive anomalies, and sea floor that forms when the Earth has reversed polarity results in negative anomalies. Anomalies are symmetric with respect to a mid-ocean ridge axis, and their widths are proportional to the duration of polarity chrons, observations that can be explained only by sea-floor spreading. Study of anomalies allows us to calculate the rate of spreading.

- Drilling of the sea floor confirmed its age and was another proof of sea-floor spreading.

Geopuzzle Revisited

Many lines of evidence indicate that the position of continents indeed changes over time, and thus that the map of the Earth's surface is not fixed. Not only do coastlines match, but the observed distribution of rock units, fossils, and climate belts, and evidence of past glaciations all point to the occurrence of continental drift. Drift occurs because ocean basins grow wider by the process of sea-floor spreading, or get narrower by the process of subduction. The documentation and interpretation of marine magnetic anomalies proved that sea-floor spreading does happen.

Key Terms

abyssal plains (p. 63)
apparent polar-wander path (p. 63)
bathymetry (p. 63)
continental drift (p. 57)
fracture zones (p. 65)
magnetic anomaly (p. 67)
magnetic reversals (p. 68)
marine magnetic anomalies (p. 67)

mid-ocean ridges (p. 63)
Pangaea (p. 57)
paleomagnetism (p. 63)
plate tectonics (p. 58)
sea-floor spreading (p. 58)
seamounts (p. 64)
spreading rate (p. 71)
subduction (p. 58)
trenches (p. 64)
volcanic arcs (p. 64)

Review Questions

1. What was Wegener's continental drift hypothesis?

2. How does the fit of the coastlines around the Atlantic support continental drift?

3. Explain the distribution of glaciers as they occurred during the Paleozoic.

4. How does the evidence of equatorial climatic belts support continental drift?

5. Was it possible for a dinosaur to walk from New York to Paris when Pangaea existed? Explain your answer.

6. Why were geologists initially skeptical of Wegener's continental drift hypothesis?

7. Describe how the angle of inclination of the Earth's magnetic field varies with latitude. How could paleomagnetic inclination be used to determine the ancient latitude of a continent?

8. Describe the basic characteristics of mid-ocean ridges, deep-ocean trenches, and seamount chains.

9. Describe the hypothesis of sea-floor spreading.

10. How did the observations of heat flow and seismicity support the hypothesis of sea-floor spreading?

11. How were the reversals of the Earth's magnetic field discovered? How did they corroborate the sea-floor-spreading hypothesis?

12. What is a marine magnetic anomaly? How is it detected?

13. Describe the pattern of marine magnetic anomalies across a mid-ocean ridge. How do geologists explain the pattern?

14. How did geologists calculate rates of sea-floor spreading?

15. Did drilling into the sea floor contribute further proof of sea-floor spreading? If so, how?

On Further Thought

The following questions will be answered, in large part, by Chapter 4. But by thinking about them now, you can get a feel for the excitement of discovery that geologists enjoyed in the wake of the proposal of sea-floor spreading.

1. Alfred Wegener's writings implied that all continents had been linked to form Pangaea from the formation of the Earth until Pangaea's breakup in the Mesozoic. Modern geologists do not agree. Geologic evidence suggests that Pangaea itself was formed by the late Paleozoic collision and suturing together of continents that had been separate during most of the Paleozoic, and that other supercontinents had formed and broken up prior to the Paleozoic. What geologic evidence led geologists to this conclusion? (*Hint*: Keep in mind that modern geologists,

unlike Wegener, understand that mountain belts such as the Appalachians form when two continents collide, and that modern geologists, unlike Wegener, are able to determine the age of rocks using radiometric dating.)

2. Dating methods indicate that the oldest rocks on continents are almost 4 billion years old, whereas the oldest ocean floor is only 200 million years old. Why? (*Hint*: Ocean crust, when subducted, is denser than the asthenosphere, but continental crust is not.)

3. The geologic record suggests that when supercontinents break up, a pulse of rapid evolution, with many new species appearing and many existing species becoming extinct, takes place. Why might this be? (*Hint*: Consider how the environment, both global and local, might change as a result of breakup, and keep in mind the widely held idea that competition for resources drives evolution.)

4. Why are the marine magnetic anomalies bordering the East Pacific Rise in the southeastern Pacific Ocean wider than those bordering the Mid-Atlantic Ridge in the South Atlantic Ocean?

Suggested Reading

Butler, R. F. 1992. *Paleomagnetism: Magnetic Domains to Geologic Terranes*. Boston: Blackwell.

Campbell, W. H. 2001. *Earth Magnetism: A Guided Tour through Magnetic Fields*. New York: Harcourt/Academic Press.

Condie, K. C. 2001. *Mantle Plumes & Their Record in Earth History*. Cambridge, UK: Cambridge University Press.

Condie, K. C. 2005. *Earth as an Evolving Planetary System*. Burlington, Mass.: Academic Press.

Cox, A., and R. B. Hart. 1986. *Plate Tectonics: How It Works*. Palo Alto, Calif.: Blackwell.

Erikson, J. 1992. *Plate Tectonics: Unraveling the Mysteries of the Earth*. New York: Facts on File.

Glen, W. 1982. *The Road to Jaramillo: Critical Years of the Revolution in Earth Sciences*. Palo Alto, Calif.: Stanford University Press.

Kearey, P., and F. J. Vine, 1996. *Global Tectonics*, 2nd ed. Cambridge, Mass.: Blackwell.

McFadden, P. L., and M. W. McElhinny. 2000. *Paleomagnetism: Continents and Oceans*, 2nd ed. San Diego: Academic Press.

McPhee, J. A. 1998. *Annals of the Former World*. New York: Farrar, Straus and Giroux.

Oreskes, N., ed. 2003. *Plate Tectonics: An Insider's History of the Modern Theory of the Earth*. Boulder: Westview Press.

Sullivan, W. 1991. *Continents in Motion: The New Earth Debate*. 2nd ed. New York: American Institute of Physics.

THE VIEW FROM SPACE This image was produced by Christoph Hormann, using computer rendering techniques. It shows the Caucasus Mountains, between the Black Sea and the Caspian Sea. This range is forming due to the collision between two continental masses.

Paleomagnetism and Apparent Polar-Wander Paths

A.1 INTRODUCTION

In 1853, an Italian physicist noticed that volcanic rock behaved like a very weak magnet, and proposed that it became magnetic when it solidified from melt. In the 1950s, instruments became available that could routinely measure such weak magnetization, so researchers in England began to study magnetization in *ancient* rocks. Their work showed that rocks preserve a record of Earth's past magnetic field. The record of ancient magnetism preserved in rock is called **paleomagnetism**. In Chapter 3, we briefly introduced paleomagnetism and noted that its discovery led to the discovery of apparent polar-wander paths whose existence proved continental drift. In this Interlude, we provide the background needed to understand paleomagnetism more thoroughly, and provide additional detail about apparent polar-wander paths and their interpretation.

A.2 BACKGROUND ON MAGNETS AND ON EARTH'S FIELD

Some Fundamentals of Magnetism

If you hold a magnet over a pile of steel paper clips, it will lift the paper clips against the force of gravity. The magnet exerts an attractive force that pulls on the clips. A magnet can also create a repulsive force that pushes an object away. For example, when oriented appropriately, one magnet can levitate another. The push or pull exerted by a magnet is a **magnetic force**; this force creates an invisible **magnetic field** around the magnet. Magnetic forces can be created by a permanent magnet, a special material that behaves magnetically for a long time all by itself. Magnetic forces can also be produced by an electric current passing through a wire. An electrical device that produces a magnetic field is an electromagnet. The stronger the magnet, the greater its **magnetization**. When other magnets, special materials (such as iron), or electric charges enter a magnetic field, they feel a magnetic force. The strength of the pull that an object feels when placed in a magnet's field depends on the magnet's magnetization and on the distance of the object from the magnet.

Compass needles are simply magnetic needles that can pivot freely and that align with Earth's magnetic field. Recall from Chapter 2 that you can symbolically represent a magnetic field by a pattern of curving lines, known as magnetic field lines. You can see the form of these lines by sprinkling iron filings on a sheet of paper placed over a bar magnet; each filing acts like a tiny magnetic compass needle and aligns itself with the magnetic field lines (see Fig. 2.2).

All magnets have two **magnetic poles**, a north pole at one end and a south pole at the other. Opposite poles attract, but like poles repel. The imaginary line through the magnet that connects one pole to another represents the magnet's dipole. Physicists specify the dipole by an arrow that points from the north to the south pole. The **polarity** of a magnet refers to the direction the arrow points; the dipoles of magnets with opposite polarity are represented by arrows with arrowheads at opposite ends. Because of the dipolar nature of magnetic fields, we can draw arrowheads on magnetic field lines oriented to form a continuous loop through the magnet.

An electron, which is a spinning, negatively charged particle that orbits the nucleus of an atom, behaves like a tiny electromagnet because its movement produces an electric current. Most of the magnetism is due to the electron's spin, but a little may come from its orbital motion (▶Fig. A.1a). Each atom, therefore, can be pictured as a little dipole (▶Fig. A.1b). But even though all materials consist of atoms, not all materials behave like strong, permanent magnets. In fact, most materials (wood, plastic, glass, gold, tin, etc.) are

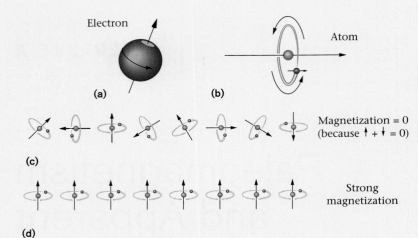

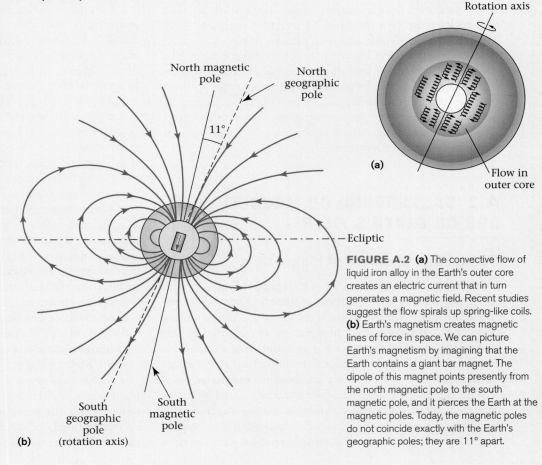

FIGURE A.1 **(a)** A spinning electron creates an electric current. **(b)** The magnetic dipole of an atom can be represented by an arrow that points from north to south. **(c)** In a nonmagnetic material, atoms tilt all different ways, so the dipoles cancel each other out, yielding a net magnetization of 0. **(d)** In a permanent magnet, the dipoles lock into alignment, so that they add to each other and produce a strong magnetization.

essentially nonmagnetic. That's because the atomic dipoles in the materials are randomly oriented, so overall the dipoles of the atoms cancel each other out (▶Fig. A.1c). In a permanent magnet, however, all atomic dipoles lock into alignment with one another. When this happens, the magnetization of each atom adds to that of its neighbor, so the material as a whole becomes magnetic (▶Fig. A.1d).

Earth's Magnetic Field, Revisited

In Chapter 2, we learned that Earth has a magnetic field that deflects the solar wind and traps cosmic rays. Why does this field exist? Geologists do not yet have a complete answer, but they have hypothesized that the field results from the circulation of liquid iron alloy, an electrical conductor, in the Earth's outer core—in other words, the outer core behaves like an electromagnet (▶Fig. A.2a; ▶Box A.1). For convenience, however, we can picture the planet as a giant bar magnet producing a magnetic field (▶Fig. A.2b). In the terminology of physics, magnetic field lines point toward the south end of this magnet, and away from the north end. By geologic convention, the north magnetic pole is near the north end of the Earth, and the south magnetic pole is near the south end. This convention arises because the north-seeking end of a compass points toward the north magnetic pole, while the south-seeking end points toward the south magnetic pole. We define the **dipole** of the Earth as an imaginary arrow that points from the north magnetic pole to the south magnetic pole and passes through the planet's center.

Presently, Earth's dipole tilts at about 11° to the planet's rotational axis (the imaginary line through the center of the Earth around which Earth spins). Therefore, the geographic poles of the planet, the places where the rotational axis intersects the Earth's surface, do not coincide exactly with the magnetic poles. For example, the north magnetic pole currently lies in arctic Canada. As a consequence, the north-seeking end of a compass needle in New York points about 14° west of north. The angle between the direction that a compass needle points at a given location and the direction to "true"

FIGURE A.2 **(a)** The convective flow of liquid iron alloy in the Earth's outer core creates an electric current that in turn generates a magnetic field. Recent studies suggest the flow spirals up spring-like coils. **(b)** Earth's magnetism creates magnetic lines of force in space. We can picture Earth's magnetism by imagining that the Earth contains a giant bar magnet. The dipole of this magnet points presently from the north magnetic pole to the south magnetic pole, and it pierces the Earth at the magnetic poles. Today, the magnetic poles do not coincide exactly with the Earth's geographic poles; they are 11° apart.

Generating Earth's Magnetic Field

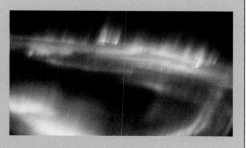

The magnetic field causes an aurora, here viewed from space.

Chinese scientists first studied Earth's magnetic field in 1040 C.E., yet almost a millennium later, Albert Einstein noted that the question of why Earth has a magnetic field remained one of the great physics questions of all times. Space exploration shows that planets, in fact, do not have to have magnetic fields. Neither Venus nor Mars presently has a significant field—so what is so special about the Earth that causes it to have a strong field? The path toward an answer became clear in 1926, when researchers proved that the Earth's outer core consists of liquid iron alloy. The flow of this liquid metal, presumably, can generate an electric current, which in turn can generate a magnetic field. In other words, the flow of iron alloy makes the Earth's outer core an electromagnet.

To better understand the generation of Earth's magnetic field, let's first consider how an electric power plant works. In a power plant, water or wind power spins a wire coil (an electrical conductor) around an iron bar (a permanent magnet). This apparatus is a **dynamo.** The motion of the wire in the bar's magnetic field generates an electric current in the wire, which in turn generates more magnetism. Applying this concept to the Earth, we can picture the flow of the outer core as playing the role of a spinning wire coil. But what plays the role of the permanent magnet in the Earth? There can't be a permanent magnet in the core, because at the very high temperatures found in the core, thermal agitation causes atoms to vibrate and tumble so much that their atomic

dipoles cannot lock into parallelism with each other—and without locked-in parallelism of atomic dipoles, permanent magnets can't exist (see Fig. A.1d). Thus, researchers suggest that the Earth is a self-exiting dynamo. Somehow, in Earth's earlier history, flow in the outer core took place in the presence of a magnetic field. This flow generated an electric current. Once the current existed, it generated a magnetic field. Continued flow in the presence of this generated magnetic field produced more electric current, which in turn produced more magnetic field. Once started, the system perpetuated itself.

Flow of iron alloy in the outer core must take place for a self-existing dynamo to exist in the Earth. What causes this flow, and how does flow result in the geometry of the field that we measure today? This topic remains an area of active research, but recent work provides some possible answers.

Calculations suggest that as the Earth cools, the inner core is growing in diameter, at a rate of 0.1 to 1 mm per year. Growth occurs as new crystals of solid iron form along the surface of the inner core. (An interesting observation is that from the present rate of growth, it appears that the inner core started forming only 1 to 2 billion years ago.) Solid iron crystals do not have room for lighter elements, such as silicon, sulfur, hydrogen, carbon, or oxygen, which had been contained in the liquid iron alloy of the outer core. Thus, these elements migrate into the base of the outer core as the inner core grows. The relatively high concentration of

these lighter elements makes the base of the outer core less dense than the top. As a result, the base of the outer core is buoyant; like a block of Styrofoam floating to the top of a pool, the outer core begins to rise, and this rise causes flow. We can consider the flow to be convection. But unlike the familiar thermal convection that takes place in a pot of water on your stove, in which differences in density are caused by differences in temperature (warmer water is less dense and, thus, is buoyant), convection in the outer core is largely chemical convection caused by contrasts in composition.

Calculations suggest that convective motion in the outer core results in the flow of iron alloy in columnar spirals (resembling the coils of a spring), whose axes roughly parallel the spin axis of the Earth. That's because the spin of the Earth influences the geometry of convective flow in the mantle. Possibly for this reason, the magnetic dipole of the Earth roughly parallels the spin axis of the Earth. Because it is so hot, iron alloy in the outer core may flow at rates of up to 20 km per year.

(geographic) north is called the **magnetic declination** (▶Fig. A.3). Measurements over the past couple of centuries show that magnetic poles migrate very slowly through time, probably never straying more than about 15° of latitude from the geographic pole. In fact, the magnetic declination of a compass changes by 0.2° to 0.5° per year. Notably, when averaged over about 10,000 years, the magnetic poles are thought to coincide with the geographic poles.

▶Figure A.4 illustrates the magnetic field lines in space around the Earth, as seen in cross section (without the warping caused by solar wind). Note that close to the Earth, the lines parallel Earth's surface at the equator; the lines tilt at an angle to the surface at mid-latitudes, and the lines are perpendicular to the surface at the magnetic poles. Thus, if we traveled to the equator and set up a magnetic needle such that it could pivot up and down freely, the needle

would be horizontal. If we took the needle to mid-latitudes, it would tilt at an angle to Earth's surface; and at a magnetic pole, the needle would point straight down. The needle's angle of tilt (which, as Fig. A.4 shows, depends on latitude) is called the **magnetic inclination**. Note that a regular compass needle does not indicate inclination, because it cannot tilt—a compass needle aligns parallel to the projection of the magnetic field lines on the Earth's surface. (You can think of the projection as the shadow of a magnetic field line on Earth's surface.)

How Do Rocks Develop Paleomagnetism?

More than 1,500 years ago, Chinese sailors discovered that an elongated piece of lodestone suspended from a thread "magically" pivots until it points north, and thus that this

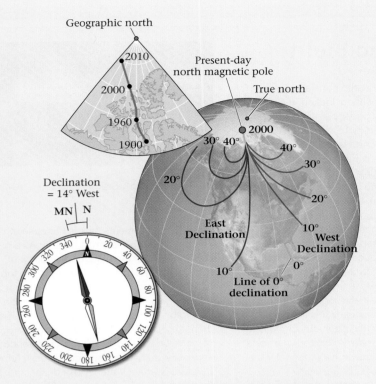

FIGURE A.3 The projection of lines of constant declination in North America at present. Recall that lines of longitude run north-south, so in most places a compass needle will not parallel longitude. For example, a compass needle at New York would make an angle of about 14° to the west of true north. Note that along the circumference that passes through both magnetic north and geographic north, the magnetic declination = 0°. See Appendix B for magnetic declination maps for the United States and for the world.

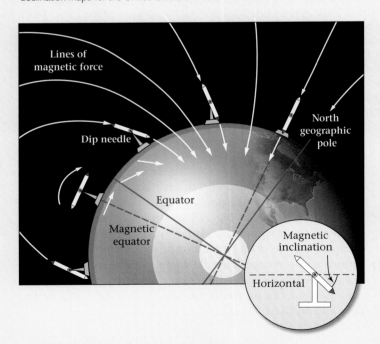

FIGURE A.4 An illustration of magnetic inclination. A magnetic needle that is free to rotate around a horizontal axis aligns with magnetic field lines (here depicted in cross section). Because magnetic field lines curve in space, this needle is horizontal at the equator, tilts at an angle at mid-latitudes, and is vertical at the magnetic pole. Therefore, the angle of tilt depends on the latitude.

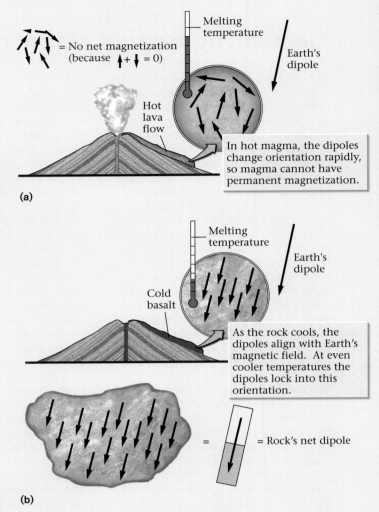

FIGURE A.5 The formation of paleomagnetism. **(a)** At high temperatures (greater than 350°–550°C), thermal vibration causes atoms to have random orientations; the dipoles thus cancel each other out, and the sample has no overall magnetic dipole. **(b)** As the sample cools to below 350°–550°C, the atoms slow down and their dipoles lock into alignment with the Earth's field.

rock could help guide their voyages. We now know that lodestone exhibits this behavior because it consists of magnetite, an iron-rich mineral that acts like a permanent magnet. Small crystals of magnetite or other magnetic minerals occur in many rock types. Each crystal produces a tiny magnetic force. The sum of the magnetic forces produced by all the crystals makes the rock, as a whole, weakly magnetic.

To see how magnetic rocks preserve a record of Earth's past magnetic field, let's examine the development of magnetization in one type of rock, basalt. Basalt is dark-colored, magnetite-containing igneous rock that forms when lava, flowing out of a volcano, cools and solidifies. When lava first comes out of a volcano, it is very hot (up to about 1,200°C), and thermal energy makes its atoms wobble and tumble chaotically. Each atom acts like

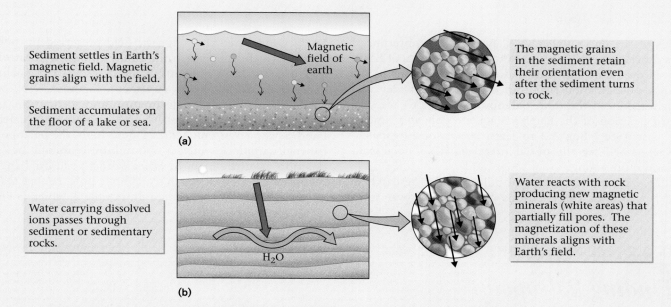

FIGURE A.6 **(a)** Paleomagnetism can form during the settling of sediments. **(b)** Paleomagnetism can also form when iron-bearing minerals precipitate out of groundwater passing through sediment.

a mini-dipole, but the mini-dipoles of the wildly dancing atoms point in all different directions. When this happens, the magnetic force exerted by one atom cancels out the force of another with an oppositely oriented dipole, so the lava as a whole is not magnetic (▶Fig. A.5a). However, as the temperature of the lava decreases to below the melting temperature (about 1,000 °C), basalt rock starts to solidify. As the magnetite crystals in the basalt form and cool (i.e., as thermal energy decreases), their iron atoms slow down. The dipoles of all the atoms gradually become parallel with each other and with the Earth's magnetic field lines at the location where the basalt cools. Finally, at temperatures below 350°–550°C, well below the melting temperature, the dipoles lock into position, pointing in the direction of the magnetic pole, and the basalt becomes a permanent magnet (▶Fig. A.5b). Since this alignment is permanent, the basalt provides a record of the orientation of the Earth's magnetic field lines, relative to the rock, at the time the rock cooled. This record is paleomagnetism.

Basalt is not the only rock to preserve a good record of paleomagnetism. Certain kinds of sedimentary rocks also can preserve a record of ancient magnetism. In some cases, the record forms when magnetic sedimentary grains align with the Earth's magnetic field as they settle to form a layer; this orientation can be preserved when the layer turns to rock (▶Fig. A.6a). Paleomagnetism may also develop when magnetic minerals (magnetite or another iron-bearing mineral, hematite) grow in the spaces between grains after the sediment has accumulated. These minerals form from ions that had been dissolved in groundwater passing through the sediment (▶Fig. A.6b).

Interpreting *Apparent* Polar-Wander Paths: Evidence for Continental Drift

When geologists measured paleomagnetism in samples of basalt that had formed millions of years ago, they were surprised to find that the dipole representing this paleomagnetism did not point to the present-day magnetic

FIGURE A.7 A rock sample can maintain paleomagnetization for millions of years. In this example, the dipole representing the paleomagnetism in this rock sample, from a village on the equator, does not parallel the Earth's present field. Note that I (inclination) is not 0°, as it would be today for rock forming near the equator. (See Fig. A.8a, b.)

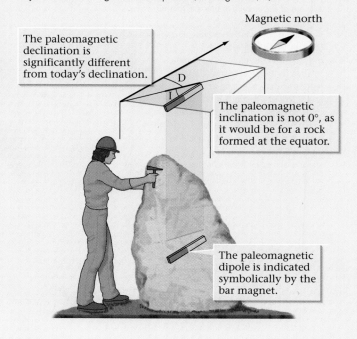

poles of the Earth (▶Fig. A.7). At first, they interpreted this observation to mean that the magnetic poles of Earth moved over time and thus had different locations in the past, a phenomenon that they called polar wander, and they introduced the term **paleopole** to refer to the supposed position of the Earth's magnetic pole at a time in the past. (See ▶Box A.2 to learn about locating paleopoles.) As we'll see, this initial interpretation of polar wander was quite wrong.

To explore the concept of polar wander, researchers decided to track the positions of paleopoles over time.

So they measured the magnetic field preserved in rocks of many different ages from about the same location. ▶Figure A.9 shows how this would be done for a locality called X on an imaginary continent. Figure A.9a shows that the paleomagnetic declination and inclination are different in different layers. The paleopoles can be calculated, as shown in Box A.2. In Figure A.9b, the paleopole for rock sample 1, which is 600 million years old, lies at position 1 on the map. In other words, position 1 indicates the location of Earth's magnetic pole, *relative to locality X*, 600 million years ago. In sample 2, which is 500

BOX A.2 THE REST OF THE STORY

Finding Paleopoles

How do you find a paleopole position from the orientation of a paleomagnetic dipole in a rock sample? Keep in mind that the paleomagnetic dipole points to the relative location of the magnetic pole at the time the rock sample cooled. The horizontal projection of the dipole arrow on the Earth's surface (picture the projection as the shadow cast on the Earth's surface by the arrow if the Sun were directly overhead) is like a

compass needle that points in the direction of the paleopole; that is, the projection defines an imaginary circle around the Earth that passes through the paleopole and the sample (▶Fig. A.8a, b). Note that when drawing the circle, we assume that the declination at the time the sample was magnetized equals 0 (because we assume that averaged over time, the magnetic pole coincides with the geographic pole). To find the

specific position of the paleopole on this circle, we must look at the inclination of the paleomagnetic dipole in the rock. Recall that inclination depends on latitude (Fig. A.4). Thus, the inclination of the *paleomagnetic* dipole defines the *paleolatitude* of the sample with respect to the paleopole, and paleolatitude simply represents the distance (measured in degrees) from the pole along the circle to where the sample formed.

FIGURE A.8 (a) Relative to present-day magnetic north, the sample of Figure A.7 has a declination angle D in map view, and an inclination angle I in a vertical plane. (b) The paleomagnetic dipole preserved in the rock indicates that, relative to the sample site, the north magnetic pole sat at point P when the rock formed. The declination defines the orientation of a circumference (dashed line) that passes between the sample site and the paleopole. The distance between the two along this line is defined by the inclination. It represents the paleolongitude.

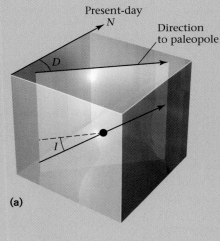

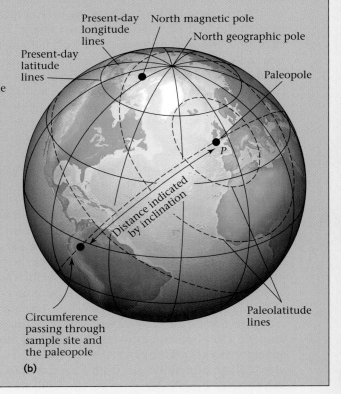

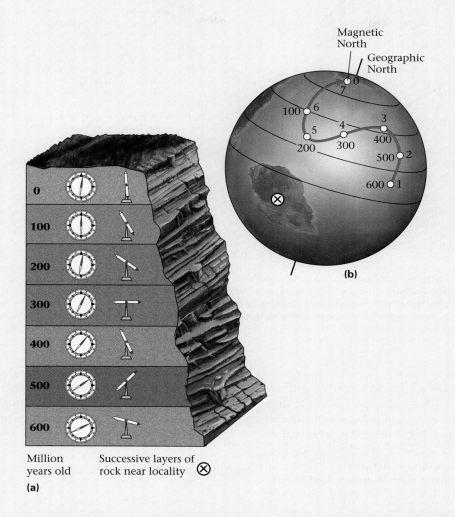

FIGURE A.9 (a) A cliff at location ✗ exposes a succession of dated lava flows. A geologist measures the orientation of the dipole in the rock. (The arrowheads aren't shown, as they don't matter here.) The paleopoles are the places where the dipole intersects the surface of the Earth. (b) The succession of paleopole positions through time for location ✗ defines the polar-wander path for the location. The path ends at the position of the present-day magnetic pole, near the North Pole.

million years old, the paleopole lies at position 2 on the map, and so on. When all the points are plotted, the resulting curving line shows the progressive change in the position of the Earth's magnetic pole, relative to locality X, *assuming* that the position of X on Earth has been fixed through time. This curve was called a polar-wander path. Note that the polar-wander path ends near the present North Pole, because recent rocks became magnetized when Earth's magnetic field was close to its position today.

In the early 1950s, geologists determined what they thought was the polar-wander path for Europe. When this path was first plotted, they did not accept the notion of continental drift, and assumed that the position of the continents was fixed. Thus, they *interpreted* the path to represent how the position of Earth's north magnetic pole migrated over time. Were they in for a surprise! When geologists then determined the polar-wander path for North America, they found that North America's path differed from Europe's (▶Fig. A.10a). In fact, when paths were

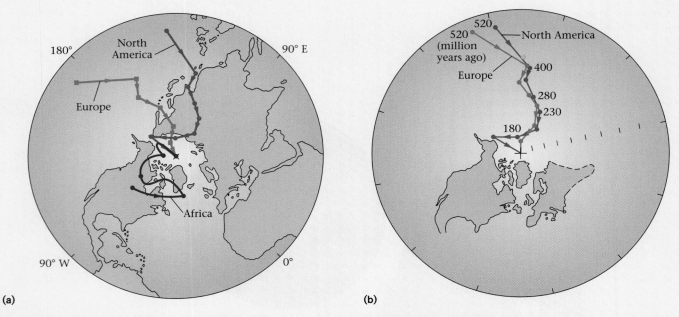

(a)   (b)

FIGURE A.10 **(a)** Apparent polar-wander paths of North America, Europe, and Africa for the past several hundred million years, plotted on a present-day map of the Earth. **(b)** The apparent polar-wander paths for North America and Europe would have coincided with each other from about 280 to 180 million years ago, because Europe and North America moved together as a unit when both were part of Pangaea. When Pangaea broke up, the two began to develop separate apparent polar-wander paths.

plotted for all continents, they all turned out to be different from each other (▶Fig. A.10b). The hypothesis that the continents are fixed and the magnetic poles move simply *cannot* explain this observation. If the magnetic poles really moved while the continents stayed fixed, then all continents should have the same polar-wander paths.

Geologists suddenly realized that they were looking at polar-wander paths in the wrong way. *It's not the pole that moves with respect to fixed continents, but rather the continents that move with respect to a fixed pole* (▶Fig. A.11a, b)! Further, if the pole is fixed, then in order for each continent to have its own unique polar-wander path, the continents must move, or "drift," in relation to each other. Poles do *not* wander, but the continents do drift while the pole stays fixed. To emphasize this point, we now call a curve like those in Figure A.9 an **apparent polar-wander path**.

Having finally understood the meaning of apparent polar-wander paths, geologists looked once again at the paths for Europe and North America and realized that they had the same shape between about 280 and 180 million years ago, the period during which the continents supposedly linked together to form part of Pangaea (Fig. A.10b). This makes sense—during times when continents move with respect to one another, they develop different apparent polar-wander paths; but when continents are stuck together, they develop the same apparent polar-wander path.

The discovery that apparent polar-wander paths for different continents differ from one another led to a renewal of support for Wegener's model of continental drift.

FIGURE A.11 The two alternative explanations for an apparent polar-wander path. **(a)** In a "true polar-wander" model, the continent is fixed. If this model is to explain polar-wander paths, the magnetic pole must move substantially. (In reality, the magnetic pole does move a little, but it never strays very far from the geographic pole.) **(b)** In a continental drift model, the magnetic pole is fixed near the geographic pole, and the continent drifts relative to the pole.

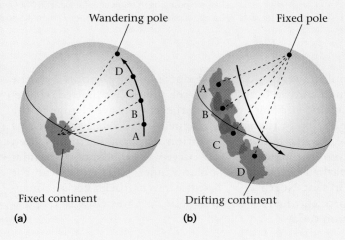

(a)   (b)

The Way the Earth Works: Plate Tectonics

The Alps formed during the Cenozoic as a consequence of the collision between a "microplate" (including Italy) and the Eurasian plate. Such continental collisions build complex mountain belts. Because of the high elevation of its peaks, much of the Alps remains snow covered all year. Some Alpine valleys still contain glaciers.

Geopuzzle

Why do earthquakes, volcanoes, and mountain belts occur where they do? And why does Earth's surface differ so much from those of other planets?

4.1 INTRODUCTION

Thomas Kuhn, an influential historian of science working in the 1960s, argued that scientific thought evolves in fits and starts. According to Kuhn, scientists base their interpretation of the natural world on an established line of reasoning, a scientific paradigm. But once in a while, a revolutionary thinker or group of thinkers proposes a radically new point of view that invalidates an old paradigm. Almost immediately, the scientific community scraps old hypotheses and formulates others consistent with the new paradigm. Kuhn called such abrupt changes in thought a scientific revolution. Some new paradigms work so well that they become scientific law and will never be replaced.

In physics, Isaac Newton's mathematical description of moving objects established the paradigm that natural phenomena obey physical laws; older paradigms suggesting that natural phenomena followed the dictates of Greek philosophers had to be scrapped. In biology, Charles Darwin's proposal that species evolve by natural selection required biologists to rethink hypotheses based on the older paradigm that species never change. And in geology, a scientific revolution in the 1960s yielded the new paradigm that the outer layer of the Earth, the lithosphere, consists of separate pieces, or plates, that move with respect to each other. This idea, which we now call the theory of plate tectonics, or simply **plate tectonics**, required geologists to cast aside hypotheses rooted in the paradigm of fixed continents, and thus led to a complete restructuring of how geologists think about Earth history. Compare this book with a geology textbook from the 1950s, and you will instantly see the difference.

Alfred Wegener planted the seed of plate tectonics theory with his proposal of continental drift in 1915, but until 1960 this seed lay dormant while geoscientists focused on collecting new data about the Earth. Discoveries about the ocean floor and about apparent polar wander led to the germination of the seed in 1960, with Harry Hess and Robert Dietz's proposal of sea-floor spreading. The roots took hold three years later when marine magnetic anomalies supplied proof of sea-floor spreading. During the next five years, the study of geoscience turned into a feeding frenzy, as many investigators dropped what they'd been doing and turned their attention to examining the broader implications of sea-floor spreading. Thanks primarily to the work of at least two dozen different investigators, by 1968 the sea-floor-spreading hypothesis had bloomed into plate tectonics theory. Geologists clarified the concept of a plate, described the types of plate boundaries, calculated plate motions, related plate tectonics to earthquakes and volcanoes, showed how plate motions generate mountain belts and seamount chains, and defined the history of past plate motions. After excited investigators had pre-sented their new ideas to standing-room-only audiences at many conferences between 1968 and 1970, the geoscience community, with few exceptions, embraced plate tectonics theory and has built on it ever since.

To begin our explanation of the key elements of plate tectonics theory, we first learn about lithosphere plates, the three types of plate boundaries, and the nature of geologic activity that occurs at each boundary. We then look at hot spots and other special locations on plates. Finally, we see how continents break apart and how they collide, and we learn about what makes plates move. Because plate tectonics theory is geology's grand unifying theory, it is now an essential foundation for the discussion of all geology.

4.2 WHAT DO WE MEAN BY PLATE TECTONICS?

The Concept of a Lithosphere Plate

As we learned in Chapter 2, geoscientists divide the interior of the Earth into layers. If we want to distinguish layers according to chemical composition, we speak of the crust, mantle, and core. We can define the boundaries between these layers by abrupt changes in the speed of seismic waves. But if, instead, we want to distinguish layers according to whether they are rigid or can flow relatively easily, we use the names *lithosphere* and *asthenosphere*. Let's now clarify the definitions of these important terms.

The **lithosphere** consists of the crust *plus* the uppermost (coolest) part of the upper mantle. It behaves rigidly and somewhat elastically, meaning that when a force pushes or pulls on it, it does not flow overall, but rather bends and flexes, or breaks (▶Fig. 4.1a). The lithosphere floats on a relatively soft, or "plastic," layer called the **asthenosphere**, composed of warmer (>1,280°C) mantle that can flow (though very slowly) when acted on by force. Therefore, the asthenosphere can undergo convection, like water in a pot, but the lithosphere cannot.

Continental lithosphere and oceanic lithosphere differ in thickness. On average, continental lithosphere has a thickness of 150 km, whereas old oceanic lithosphere has a thickness of about 100 km; for reasons discussed later in this chapter, *new* oceanic lithosphere at a mid-ocean ridge is only 7 to 10 km thick. Recall that the crustal part of continental lithosphere ranges from 25 to 70 km thick and consists of relatively low-density rock. In contrast, the crustal part of oceanic lithosphere is only 7 to 10 km thick and consists of relatively high-density rock.

The surface of continental lithosphere lies at a higher elevation than the surface of oceanic lithosphere. To

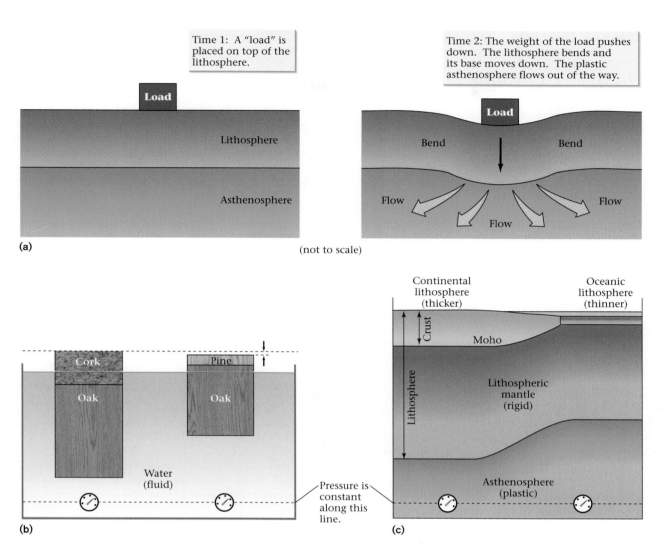

FIGURE 4.1 **(a)** Lithosphere bends when a load is placed on it, whereas asthenosphere flows. **(b)** We can picture continental lithosphere as a thick oak block (lithospheric mantle) overlaid by a layer of cork (continental crust), and oceanic lithosphere as a thinner block of oak overlaid by a layer of pine (oceanic crust). The pine layer is thinner than the cork layer. If both blocks float in a tub of water, the surface of the thick cork/oak block lies at a higher elevation than that of the pine/oak block. **(c)** Similarly, the ocean floor lies 4–5 km below the surface of the continents, on average, because lithosphere, like the wood blocks, floats on the asthenosphere.

picture why, imagine that we have two blocks of oak (a high-density wood), one 15 cm thick and one 10 cm thick. On top of the thicker block, we glue a 4-cm-thick layer of cork (a low-density wood), and on top of the thinner block, we glue a 1-cm-thick layer of pine (a medium-density wood). Now we place the two blocks in water (▶Fig. 4.1b). The total mass of the cork-covered block exceeds the total mass of the pine-covered block, so the base of the cork-covered block sinks deeper into the water. But because the cork-covered block is thicker and has a lower overall density, it floats higher. In our analogy, the cork-covered block represents continental lithosphere, with its

thick crust of low-density rock, whereas the pine-covered block represents oceanic lithosphere, with its thin crust of dense rock. The oak represents the very high-density rock constituting the mantle part of the lithosphere, thicker for the continent than for the ocean (▶Fig. 4.1c). Our analogy emphasizes that ocean floors are low, and thus fill with water to form oceans, because continental lithosphere is more buoyant and floats higher than oceanic lithosphere (▶Box 4.1, ▶Fig. 4.2).

The lithosphere forms the Earth's relatively rigid shell. But unlike the shell of a hen's egg, the lithosphere shell contains a number of major "breaks," which separate the

BOX 4.1 SCIENCE TOOLBOX

Archimedes' Principle of Buoyancy

Archimedes (c. 287–212 B.C.E.), a Greek mathematician and inventor, left an amazing legacy of discoveries. He described the geometry of spheres, cylinders, and spirals; introduced the concept of a center of gravity; and was the first to understand buoyancy. **Buoyancy** is the upward force acting on an object immersed or floating in a fluid. According to legend, Archimedes recognized this concept suddenly, while bathing in a public bath, and was so inspired that he jumped out of the bath and ran home naked, shouting "Eureka!"

Archimedes realized that when you place a solid object in water, the object displaces a volume of water equal in mass to the object (►Fig. 4.2). An object denser than water, such as a stone, sinks through the water, because even when completely submerged, the stone's mass exceeds the mass of the water it displaces. When submerged, however, the stone weighs less than it does in air. (For this reason, a scuba diver can lift a heavy object underwater.) An object less dense than water, such as an iceberg, sinks only until the mass of the water displaced equals the total mass of the iceberg. This condition happens while part of the iceberg still protrudes up into the air. Put another way, an object placed in a fluid feels a "buoyancy force" that tends to push it up. If the object's weight is less than the buoyancy force, the object floats, but if its weight is greater than the buoyancy force, the object sinks.

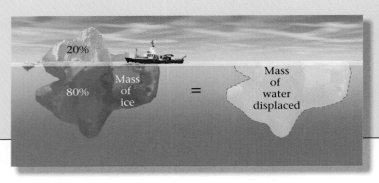

FIGURE 4.2 According to Archimedes' principle of buoyancy, an iceberg sinks until the total mass of the water displaced equals the total mass of the whole iceberg. Since water is denser, the volume of the water displaced is less than the volume of the iceberg, so only 20% of the iceberg protrudes above the water.

lithosphere into distinct pieces.[1] We call the pieces lithosphere plates, or simply **plates**, and we call the breaks **plate boundaries**. Geoscientists distinguish twelve major plates and several microplates. Some plates have familiar names (the North American Plate, the African Plate), whereas some do not (the Cocos Plate, the Juan de Fuca Plate).

Some plate boundaries follow continental margins, the boundary between a continent and an ocean, but others do not. For this reason, we distinguish between **active margins**, which are plate boundaries, and **passive margins**, which are not plate boundaries. Along passive margins, continental crust is thinner than normal (►Fig. 4.3). (As we'll discuss in Section 4.7, this thinning occurs during the initial formation of the ocean; during this process, the upper part breaks into wedge-shaped slices.) Thick (10–15 km) accumulations of sediment cover this thinned crust. The surface of this sediment layer is a broad, shallow (less than 500 m deep) region called the **continental shelf**, home to the major fisheries of the world. Note that some plates consist entirely of oceanic lithosphere or entirely of continental lithosphere, whereas some plates consist of both. For example, the Nazca Plate is made up entirely of ocean floor, whereas the North American Plate consists of North America plus the western half of the North Atlantic Ocean.

The Basic Premise of Plate Tectonics

We can now restate plate tectonics theory concisely as follows. The Earth's lithosphere is divided into 15 to 20 plates (►Fig. 4.4) that move relative to each other and relative to the underlying asthenosphere. Plate movement occurs at rates of about 1 to 15 cm per year. As a plate moves, its internal area remains largely rigid and intact, but rock along the plate's boundaries undergoes deformation (cracking, sliding, bending, stretching, and squashing) as the plate grinds or scrapes against its neighbors or pulls away from them. As plates move, so do the continents that form part of the plates, resulting in continental drift. Because of plate tectonics, the map of Earth's surface constantly changes.

---

[1]Note that the above definition equates the base of a plate with the base of the lithosphere. This definition works well for oceanic plates, because the asthenosphere directly beneath oceanic lithosphere is particularly weak. But some geologists now think that the upper 100 to 150 km of the asthenosphere beneath continents actually moves with the continental lithosphere. Thus, the base of continental plates—the base of the layer that moves during plate motion—may actually lie within the asthenosphere. To simplify our discussion, we ignore this complication.

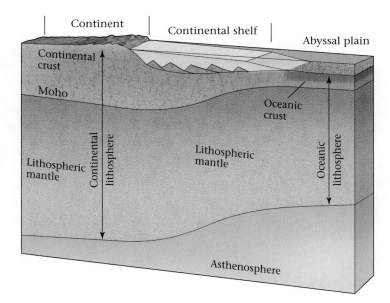

Continent | Continental shelf | Abyssal plain

Continental crust
Moho
Oceanic crust
Lithospheric mantle
Continental lithosphere
Lithospheric mantle
Oceanic lithosphere
Asthenosphere

FIGURE 4.3 In this block diagram of a passive margin, note that the continental crust thins along the boundary (see Section 4.7, "How Do Plate Boundaries Form and Die?"). The sediment pile that accumulates over this thinned crust underlies the continental shelf.

Identifying Plate Boundaries

How do we recognize the location of a plate boundary? The answer becomes clear from looking at a map showing the locations of earthquakes (►Fig. 4.5). Recall from Chapter 2 that earthquakes are vibrations caused by shock waves that are generated where rock breaks and suddenly shears (slides) along a fault (a fracture on which sliding occurs). The hypocenter (or focus) of the earthquake is the spot where the fault begins to slip, and the epicenter marks the point on the surface of the Earth directly above the focus. Earthquake epicenters do not speckle the Earth's surface randomly, like buckshot on a target. Rather, the majority occur in relatively narrow, distinct belts. These earthquake belts define the position of plate boundaries, because the fracturing and slipping that occur along plate boundaries as plates move generate earthquakes. (We will learn more about this process in Chapter 10.) Plate interiors, regions away from the plate boundaries, remain relatively earthquake free because they are stronger and do not accommodate much movement. Note that earthquakes occur frequently along active continental margins, such as the Pacific coast of the Americas, but not along passive continental margins, such as the eastern coast of the Americas.

Although earthquakes are the most definitive indicators of a plate boundary, other prominent geologic features also develop along plate boundaries. By the end of this chapter, we will see that each type of plate boundary is associated with a diagnostic group of geologic features such as volcanoes, deep-ocean trenches, or mountain belts.

Geologists define three types of plate boundaries, simply on the basis of the relative motions of the plates on either side of the boundary (►Fig. 4.6a–c). A boundary at which two plates move apart from one another is called a **divergent boundary**. A boundary at which two plates move toward one another so that one plate sinks beneath the other is called a **convergent boundary**. And a boundary at which one plate slips along the side of another plate is called a **transform boundary**. Each type looks different and behaves differently from the others, as we will now see.

> **Take-Home Message**
>
> The Earth's lithosphere, its rigid shell, consists of about twenty plates that move relative to each other. The distribution of earthquakes delineates the boundaries between plates. Geologists recognize three distinct types of plate boundaries.

4.3 DIVERGENT PLATE BOUNDARIES AND SEA-FLOOR SPREADING

At divergent boundaries, or spreading boundaries, two oceanic plates move apart by the process of sea-floor spreading. Note that an open space does not develop between diverging plates. Rather, as the plates move apart, new oceanic lithosphere forms along the divergent boundary (►Fig. 4.7). This process takes place at a submarine mountain range called a **mid-ocean ridge** (such as the Mid-Atlantic Ridge, the East Pacific Rise, and the Southeast Indian Ocean Ridge), which rises 2 km above the adjacent abyssal plains of the ocean. Thus, geologists also commonly call a divergent boundary a mid-ocean ridge, or simply a ridge.

Characteristics of a Mid-Ocean Ridge

To understand a divergent boundary better, let's look at one mid-ocean ridge, the Mid-Atlantic Ridge, in more detail (►Figs. 4.8, 4.9). The Mid-Atlantic Ridge extends from the waters between northern Greenland and northern Scandinavia southward across the equator to the latitude opposite the southern tip of South America. For most of its length, the elevated area of the ridge is about 1,500 km wide.

Geologists have mapped segments of the Mid-Atlantic Ridge in detail, using sonar from ships and from research submarines. They have found that the formation of new sea floor takes place only across a remarkably narrow band—less than a few kilometers wide—along the axis (centerline) of the ridge. The axis lies at water depths of about 2 to 2.5 km. Along ridges, like the Mid-Atlantic, where sea-floor spreading occurs slowly, the axis lies in a narrow trough about 500 m deep and less than 10 km

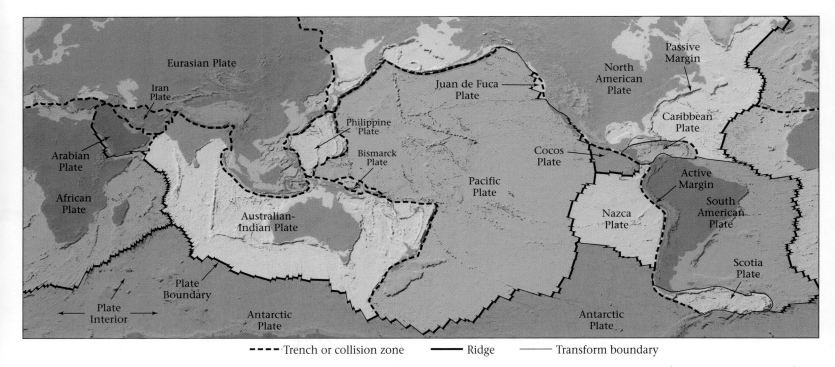

--- Trench or collision zone   ━━━ Ridge   ── Transform boundary

FIGURE 4.4 Simplified map showing the major plates making up the lithosphere. Note that some plates are all ocean floor, whereas some contain both continents and oceans. Thus, some plate boundaries lie along continental margins (coasts), but others do not. For example, the eastern border of South America is not a plate boundary, but the western edge is a plate boundary.

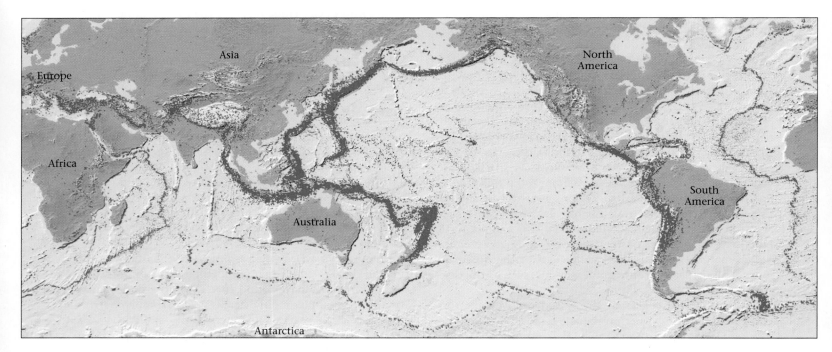

FIGURE 4.5 The locations of most earthquakes fall in distinct bands, or belts. These earthquake belts define the positions of the plate boundaries. Compare this map with the plate boundaries on Figure 4.3. For more detailed earthquake maps, see Appendix B.

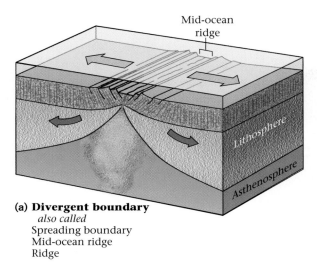

(a) Divergent boundary
also called
Spreading boundary
Mid-ocean ridge
Ridge

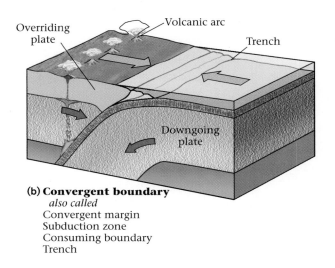

(b) Convergent boundary
also called
Convergent margin
Subduction zone
Consuming boundary
Trench

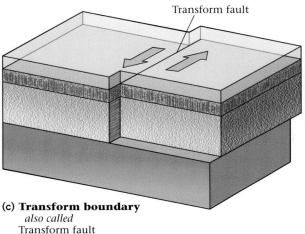

(c) Transform boundary
also called
Transform fault

FIGURE 4.6 Geologists recognize three types of plate boundary on the basis of the nature of relative movement at the boundary. **(a)** At a divergent boundary (its other names are listed below), two oceanic plates move away from one another. The lithosphere thickens with increasing distance from the ridge. **(b)** At a convergent boundary, one oceanic plate bends and sinks into the mantle beneath another plate. **(c)** At a transform boundary, two plates slide past each other along a vertical fault surface.

wide, bordered on either side by steep cliffs. In rough terms, the Mid-Atlantic Ridge is symmetrical—its eastern half looks like a mirror image of its western half. As illustrated by Figure 4.9, along its length, the ridge consists of short segments (tens to hundreds of km long) linked by breaks called transform faults, which we will discuss later.

Not all mid-ocean ridges look just like the Mid-Atlantic. For example, ridges at which spreading occurs rapidly, such as the East Pacific Rise, do not have the axial trough we see along the Mid-Atlantic Ridge. Also, the region of elevated sea floor of faster-spreading ridges is much wider.

The Formation of Oceanic Crust at a Mid-Ocean Ridge

As noted above, sea-floor spreading does not create an open space between diverging plates. Rather, as each increment of spreading occurs, new sea floor develops in the space. How does this happen?

As sea-floor spreading takes place, hot asthenosphere (the soft, flowable part of the mantle) rises beneath the ridge (Fig. 4.8). As this asthenosphere rises, it begins to melt, producing molten rock, or magma. Magma has a lower density than solid rock, so it behaves buoyantly and rises. It eventually accumulates in the crust below the ridge axis. The lower part of this region is a mush of crystals, above which magma pools in a fairly small magma chamber. Some of the magma solidifies along the side of the chamber to make a coarse-grained, mafic igneous rock called gabbro. Some of the magma rises still higher to fill vertical cracks, where it solidifies and forms wall-like sheets, or dikes, of basalt. Finally, some magma rises all the way to the surface of the sea floor at the ridge axis and spills out of small submarine volcanoes. The resulting lava cools to form a layer of basalt blobs, called pillow basalt, on the sea floor. We can't easily see the submarine volcanoes because they occur at depths of more than 2 km beneath sea level, but they have been observed

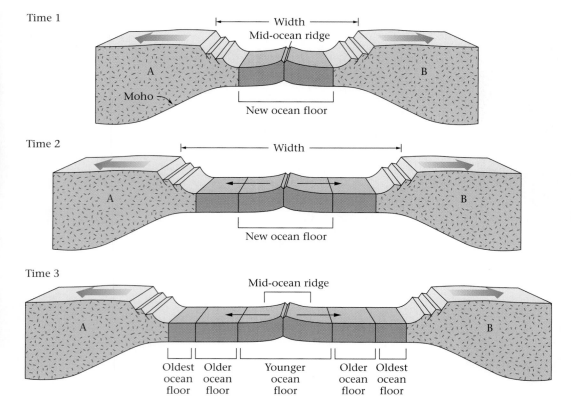

Time 1

Width
Mid-ocean ridge

A

Moho

New ocean floor

B

Time 2

Width

A

New ocean floor

B

Time 3

Mid-ocean ridge

A

B

Oldest ocean floor | Older ocean floor | Younger ocean floor | Older ocean floor | Oldest ocean floor

FIGURE 4.7 These sketches depict successive stages in sea-floor spreading along a divergent boundary (mid-ocean ridge); only the crust is shown. The top figure represents an early stage of the process, after the mid-ocean ridge formed but before the ocean grew very wide. With time, as seen in the next two figures, the ocean gets wider and continent A drifts way from continent B. Note that the youngest ocean crust lies closest to the ridge.

FIGURE 4.8 How new lithosphere forms at a mid-ocean ridge. Rising hot asthenosphere partly melts underneath the ridge axis. The molten rock, magma, rises to fill a magma chamber in the crust. Some of the magma solidifies along the sides of the chamber to make coarse-grained mafic rock called gabbro. Some magma rises still farther to fill cracks, solidifying into basalt that forms wall-like sheets of rock called dikes. Finally, some magma rises all the way to the surface of the sea floor at the ridge axis. This magma, now called lava, spills out and forms a layer of basalt. As sea-floor spreading continues, the oceanic crust breaks along faults. Also, as a plate moves away from a ridge axis and cools, the lithospheric mantle thickens.

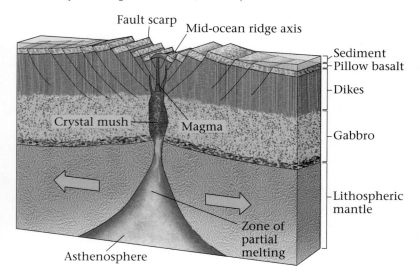

Fault scarp   Mid-ocean ridge axis

Sediment
Pillow basalt

Dikes

Crystal mush   Magma

Gabbro

Lithospheric mantle

Zone of partial melting

Asthenosphere

by the research submarine *Alvin*. *Alvin* has also detected chimneys spewing hot, mineralized water that rose through cracks in the sea floor after being heated by magma below the surface. These chimneys are called **black smokers** because the water they emit looks like dark smoke (▶Fig. 4.10).

As soon as it forms, new oceanic crust moves away from the ridge axis. As this happens, more magma rises from below, so still more crust forms. In other words, magma from the mantle rises to the Earth's surface at the ridge like a vast, continuously moving conveyor belt. Then it solidifies to form oceanic crust, and finally moves laterally away from the ridge. Because all sea floor forms at mid-ocean ridges, the youngest sea floor occurs on either side of the ridge, and sea floor becomes progressively older away from the ridge (Fig. 4.7). In the Atlantic Ocean, the oldest sea floor lies adjacent to the passive continental margins on either side of the ocean (▶Fig. 4.11). The oldest ocean floor on our planet is in the western Pacific Ocean; this crust formed 200 million years ago.

As spreading takes place, the tension (stretching force) applied to newly formed solid crust breaks this new crust, resulting in the formation of faults. Slip on the faults causes divergent-boundary earthquakes and creates numerous cliffs, or scarps, that parallel the ridge axis.

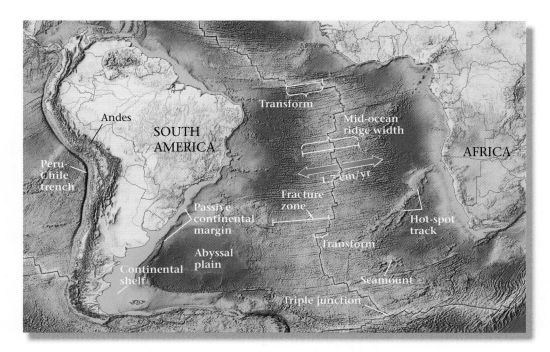

FIGURE 4.9 A map showing the bathymetry of the Mid-Atlantic Ridge in the south Atlantic Ocean. The lighter colors are shallower depths. The map also shows the trench along the west coast of South America.

The Formation of Lithospheric Mantle at a Mid-Ocean Ridge

So far, we've seen how oceanic crust forms at mid-ocean ridges. What about the formation of the mantle part of the oceanic lithosphere? Remember that this part consists

FIGURE 4.10 A column of superhot water gushing from a vent (known as a "black smoker") along the mid-ocean ridge. The water has been heated by magma (molten rock) just below the surface. The cloud of "smoke" actually consists of tiny mineral grains; the elements making up these minerals had been dissolved in the hot water, but when the hot water mixes with cold water of the sea, they precipitate. Many exotic species of life, such as giant worms, live around these vents.

of the cooler uppermost area of the mantle, in which temperatures are less than about 1,280°C. At the ridge axis, such temperatures occur almost at the base of the crust because of the presence of rising hot asthenosphere and hot magma, so the lithospheric mantle beneath the ridge axis effectively doesn't exist. But as the newly formed oceanic crust moves away from the ridge axis, the crust and the uppermost mantle directly beneath it gradually cool as they lose heat to the ocean above. As soon as mantle rock cools below 1,280°C, it becomes, by definition, part of the lithosphere.

As oceanic lithosphere continues to move away from the ridge axis, it continues to cool. Thus, the lithospheric mantle, and therefore the oceanic lithosphere as a whole, grow progressively thicker away from the ridge (▶Fig. 4.12a). This process doesn't change the thickness of the oceanic crust, for the crust formed entirely at the ridge axis. The rate at which cooling and thickening occur decreases with distance from the ridge axis. In fact, by the time the lithosphere is about 80 million years old, it has just about reached its maximum thickness (▶Fig. 4.12c).

The Reason Mid-Ocean Ridges Are High

Why does the surface of the sea floor rise to form a mid-ocean ridge along divergent plate boundaries? The answer comes from considering the buoyancy of oceanic lithosphere (see Box 4.1). As sea floor ages, the asthenosphere

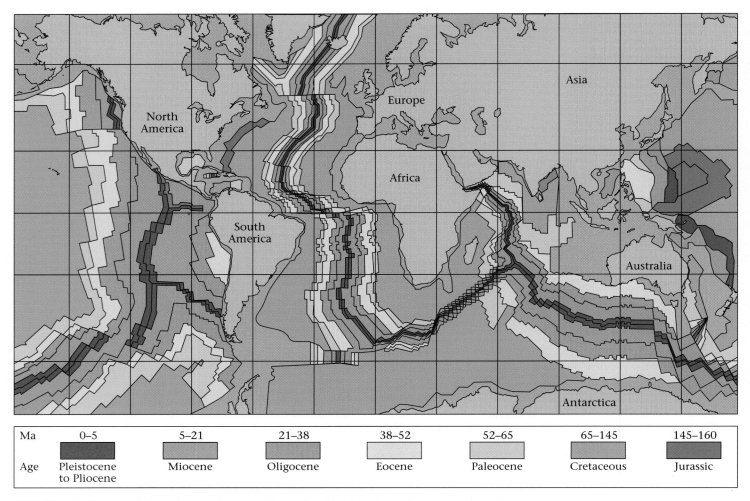

Ma	0–5	5–21	21–38	38–52	52–65	65–145	145–160
Age	Pleistocene to Pliocene	Miocene	Oligocene	Eocene	Paleocene	Cretaceous	Jurassic

FIGURE 4.11 This map of the world shows the age of the sea floor. Note how the sea floor grows older with increasing distance from the ridge axis (Ma = million years ago).

below cools enough to become part of the lithosphere, and the lithospheric mantle thickens. Cooler rock is denser than warmer rock, so the process of cooling and thickening the lithosphere, like adding ballast to a ship, causes the lithosphere to sink deeper into the asthenosphere (▶Fig. 4.12c). Hot young lithosphere is less dense and floats higher; this high-floating lithosphere constitutes the mid-ocean ridge. Because lithosphere cools and thickens as it grows older, the depth of the sea floor depends on its age (Fig. 4.12c).

> **Take-Home Message**
>
> Sea-floor spreading occurs at divergent plate boundaries. Through this process, new oceanic plates form and move apart. These plate boundaries are delineated by mid-ocean ridges, along which submarine volcanoes erupt.

4.4 CONVERGENT PLATE BOUNDARIES AND SUBDUCTION

At convergent plate boundaries, (or convergent margins), two plates, at least one of which is oceanic, move toward each other. But rather than butting each other like angry rams, one oceanic plate bends and sinks down into the asthenosphere beneath the other plate. Geologists refer to the sinking process as **subduction**, so convergent margins are also known as subduction zones. Because subduction at a convergent margin consumes old ocean lithosphere and thus closes (or "consumes") oceanic basins, geologists also refer to convergent margins as consuming boundaries. Because they are delineated by deep-ocean trenches, margins are also sometimes simply called **trenches** (Fig. 4.6b).

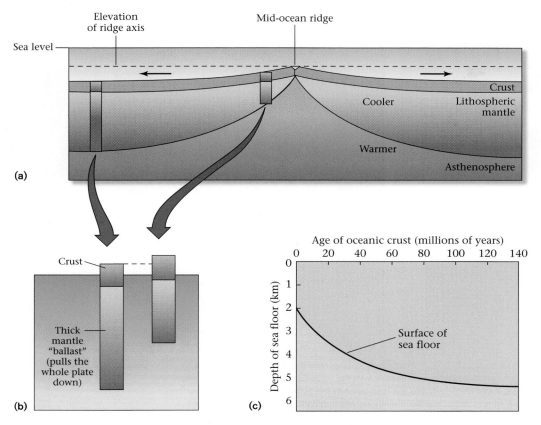

FIGURE 4.12 **(a)** As sea floor ages, the dense lithospheric mantle thickens. **(b)** Like the ballast of a ship, thicker lithosphere sinks deeper into the mantle. **(c)** The depth of the sea floor increases as a plate moves away from the ridge and grows older.

The amount of oceanic plate consumption worldwide equals the amount of sea-floor spreading worldwide, so the surface area of the Earth remains constant through time.

Subduction occurs for a simple reason: once oceanic lithosphere has aged at least 10 million years, it is denser than asthenosphere and thus can sink down through the asthenosphere. When it lies flat on the surface of the asthenosphere, oceanic lithosphere doesn't sink because the resistance of the asthenosphere to flow is too great; however, once the end of the convergent plate bends down and slips into the mantle, it begins to sink like an anchor falling to the bottom of a lake (▶Fig. 4.13a, b). As the lithosphere sinks, asthenosphere flows out of the way, just as water flows out of the way of an anchor. But even though it is relatively soft and plastic, the asthenosphere resists flow so oceanic lithosphere can sink only very slowly, at a rate of less than 10 to 15 cm per year.

Note that the downgoing plate (or slab), the plate that has been subducted, *must* be composed of oceanic lithosphere. The overriding plate (or slab), which does not sink, can consist of either oceanic or continental lithosphere. Continental crust cannot be subducted because it is too buoyant; the low-density rocks of continental crust act like a life preserver, keeping the continent afloat. If continental crust moves into a convergent margin, subduction stops. Because of subduction, all ocean floor on the planet is less than about 200 million years old. Because continental crust cannot subduct, some continental crust has persisted at the surface of the Earth for over 3.8 billion years.

Earthquakes and the Fate of Subducted Plates

At convergent plate boundaries, the downgoing plate grinds along the base of the overriding plate, a process that generates large earthquakes. These earthquakes occur fairly close to the Earth's surface, so some of them trigger massive destruction in coastal cities. But earthquakes also happen in downgoing plates at greater depths, deep below the

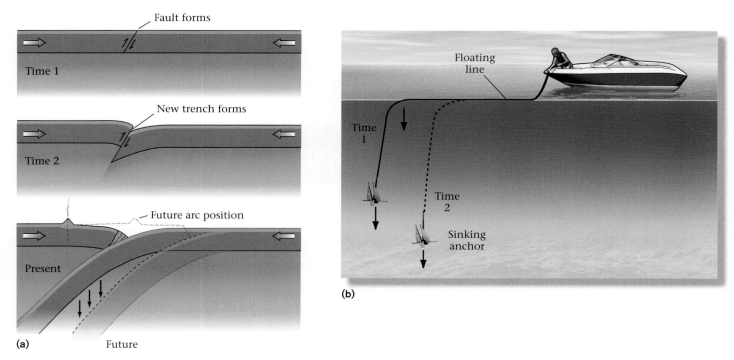

(a)

(b)

FIGURE 4.13 The concept of subduction. **(a)** A plate bends, and one piece pushes over the other. The red arrows indicate the relative motion of the two plates. Oceanic lithosphere is denser than the underlying asthenosphere, but when it lies flat on the surface of the asthenosphere, it can't sink because the resistance of the asthenosphere to flow is too great. However, once the end of the plate is pushed into the mantle, the lithosphere begins to sink. **(b)** The process of sinking is like an anchor pulling a floating anchor line down. As a consequence, the bend in the plate (or in the anchor line) progressively moves with time.

FIGURE 4.14 **(a)** The Wadati-Benioff zone is a band of earthquakes that occur in subducted oceanic lithosphere. The discovery of these earthquakes led to the proposal of subduction. **(b)** A hypothesis for the ultimate fate of subducted lithosphere. In this hypothesis, the lower mantle contains two regions (shallower and deeper), which differ in density and possibly composition. D″ is the name for a hot region just above the core-mantle boundary.

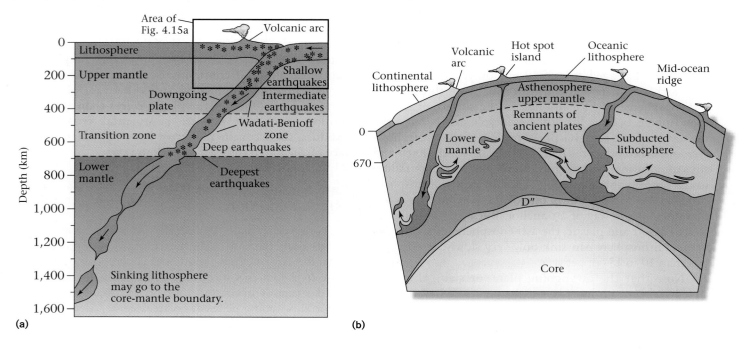

(a)

(b)

overriding plate. In fact, geologists have detected earthquakes within downgoing plates to a depth of 660 km; the belt of earthquakes in a downgoing plate is called a Wadati-Benioff zone, after its two discoverers (▶Fig. 4.14a).

At depths greater than 660 km, conditions leading to earthquakes evidently do not occur. Recent evidence, however, indicates that some downgoing plates do continue to sink below a depth of 660 km—they just do so without generating earthquakes. In fact, some studies suggest that the lower mantle may be a graveyard for old subducted plates (▶Fig. 4.14b).

Geologic Features of a Convergent Boundary

To become familiar with the various geologic features that occur along a convergent plate boundary, let's look at an example, the boundary between the western coast of the South American Plate and the eastern edge of the Nazca Plate (a portion of the Pacific Ocean floor). A deep-ocean trench, the Peru-Chile Trench, delineates this boundary (Fig. 4.9). Such trenches form where the plate bends as it starts to sink into the asthenosphere.

In the Peru-Chile Trench, as the downgoing plate slides under the overriding plate, sediment (clay and plankton) that had settled on the surface of the downgoing plate, as well as sand that fell into the trench from the shores of South America, gets scraped up and incorporated in a wedge-shaped mass known as an accretionary prism (▶Fig. 4.15a). An accretionary prism forms in basically the same way as a pile of snow in front of a plow, and like the snow, the sediment tends to be squashed and contorted during the formation of the prism (▶Fig. 4.15b).

A chain of volcanoes known as a **volcanic arc** develops behind the accretionary prism (▶Geotour 4). As we will see in Chapter 6, the magma that feeds these volcanoes forms at or just above the surface of the downgoing plate when the plate reaches a depth of about 150 km below the Earth's surface. If the volcanic arc forms where an oceanic plate subducts beneath continental lithosphere, the resulting chain of volcanoes grows on the continent and forms a continental volcanic arc. In some cases, the plates push together,

> **Take-Home Message**
>
> At a convergent plate boundary, an oceanic plate sinks into the mantle beneath the edge of another plate. This process allows the two plates to move toward each other and ocean basins to close. Trenches and volcanic arcs delineate convergent boundaries.

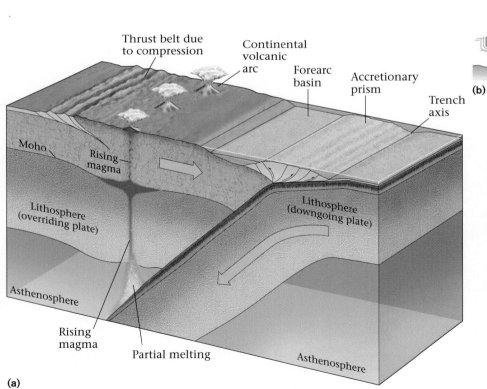

(a)

FIGURE 4.15 (a) This model shows the geometry of subduction along an active continental margin. The trench axis (lowest part of the trench) roughly defines the plate boundary. Numerous faults form in the accretionary prism, which is composed of material scraped off the sea floor. Behind the prism lies a basin (a forearc basin) of trapped sediment. A volcanic arc is created from magma that forms at or just above the surface of the downgoing plate. Here, the plate subducts beneath continental lithosphere, so the chain of volcanoes is called a continental arc. Faulting occurs on the backside of the arc. The Andes in South America and the Cascades in the United States are examples of such continental arcs. **(b)** The action of a bulldozer pushing snow or soil is similar to the development of an accretionary prism.

See for yourself...
Plate Boundaries

Where do you go if you want to see a plate boundary for yourself? First, it's important to realize that a plate boundary is not a simple line on the surface of the Earth, but is a zone perhaps 40 to 200 km wide. Still, there are places where you can go to study plate-boundary characteristics in person. Visit the following localities and you'll see for yourself. The thumbnail images provided on this page are only to help identify tour sites. Go to wwnorton.com/studyspace to experience each flyover tour.

Divergent Plate Boundaries

Most mid-ocean ridges that define the trace of a divergent plate boundary lie submerged below 2 km (1.2 miles) of water. Thus, to get close to see the faults and volcanic vents of a ridge, geologists descend in research submersibles. *Google Earth*™, however, can help you see the shape of a mid-ocean ridge without the submersible, by taking you to Iceland, one of the few places on this planet where a ridge rises above sea level.

The Mid-Atlantic Ridge (Overall View)

Zoom out to about 8,000 km (5,000 miles), and orient your view to show the north Atlantic Ocean (Image G4.1). See how the coastline of northwest Africa matches that of eastern North America? Now focus on the ocean floor between these coastlines and you'll see an image of the Mid-Atlantic Ridge—its trace follows the curve of the coastlines. Though this image has only low resolution, you can see the segmentation of the ridge axis and can recognize the oceanic fracture zones that link the end of one segment to the end of the next. Fracture zones appear to extend beyond the junctions with ridge segments. But only the portion of a fracture zone between two ridge segments is an active transform fault.

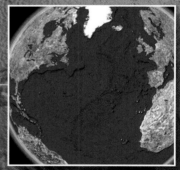

G4.1

The Mid-Atlantic Ridge in Iceland
(Lat 64°15'11.95"N, Long 20°56'12.66"W)

Trace the ridge northward to Iceland, an island that straddles the ridge. Zoom in to an elevation of 35 km (22 miles) at the above coordinates, tilt, and look northeast (Image G4.2). Along the south coast, east of Reykjavik, you can see a set of northeast trending ridges whose faces are fault scarps along the ridge axis, and if you look around, you'll see some volcanoes. A glacier covers part of the ridge—after all, it is Iceland!

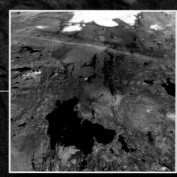

G4.2

Convergent Plate Boundaries

These boundaries, along which subduction occurs, are marked by a deep-sea trench and a bordering volcanic arc. Some volcanic arcs are island arcs, such as the Marianas arc in the Pacific, whereas some are continental arcs, such as the Cascade arc in Washington and Oregon.

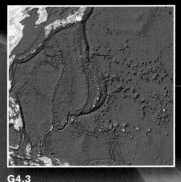

G4.3

G4.4

The Mariana Trench
(Lat 16°20'49.97"N, Long 145°41'48.16"E)

Fly to the coordinates given above, and you'll find yourself over the western Pacific Ocean. Zoom to an elevation of 6,000 km (3,700 miles), and you will be looking at a curving band of dark blue, south of Japan (Image G4.3). This is the Mariana Trench, whose floor—the deepest point in the ocean—marks the boundary where the Pacific Plate subducts beneath the Philippine Plate. The curving chain of islands (the Marianas) to the west of the trench is the volcanic island arc on the edge of the Philippine Plate.

Zoom down to an elevation of 12 km (7.5 miles). You will be looking down on the volcano of Anatahan (Image G4.4). The central portion of the island has collapsed to form a depression called a caldera.

G4.5

Cascade Volcanic Arc
(Lat 46°12'25.47"N, Long 121°29'25.80"W)

Fly to the coordinates given (Mt. Adams volcano), and zoom to an elevation of 45 km (28 miles). You'll be looking down on the peak of Mt. Adams, a volcano in the Cascade volcanic arc, a continental arc. Tilt your image so you just see the northern horizon, and use the compass to reorient your image to look along the volcanic chain (Image G4.5). You can see the spacing between the volcanoes.

Transform Plate Boundaries

Many major earthquakes happen on the San Andreas Fault. All along its length, slip on this continental transform plate boundary has affected the landscape—valleys, elongate ponds, and narrow ridges follow the fault. Also, the fault has offset stream channels.

G4.6

San Andreas Transform Fault
(Lat 34°30'43.93"N, Long 118°01'06.19"W)

Fly to these coordinates and zoom to an elevation of 12 km (7.5 miles). You are looking down on the San Andreas Fault, southeast of Palmdale, California. This is the transform boundary between the Pacific Plate and the North American Plate. The trace of the fault is the boundary between flat land to the northeast and the hilly area to the southwest (Image G4.6). At this locality, you can see a gravelly river channel that has been offset at the fault. Zoom closer, tilt the image, and rotate the view so you are looking along the fault. Follow its trace and see all the buildings, canals, and roads that cross it.

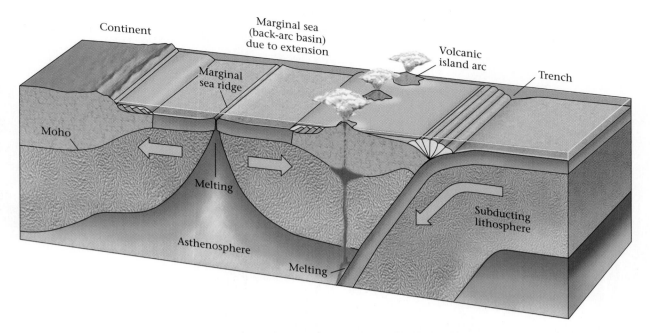

FIGURE 4.16 Subduction along an island arc. Here, the volcanoes build on the sea floor. Behind some island arcs, a marginal sea forms. This sea resembles a small ocean basin, with a spreading ridge that is created when the plate behind the arc moves away from the arc.

causing mountains to rise (Fig. 4.15a). If, however, the volcanic arc forms where one oceanic plate subducts beneath another oceanic plate, the resulting volcanoes form a chain of islands known as a volcanic island arc. A marginal sea (or back-arc basin), the small ocean basin between an island arc and the continent, forms either in cases where subduction happens to begin offshore, trapping ocean lithosphere behind the arc, or where stretching of the lithosphere behind the arc leads to the formation of a small spreading ridge between the arc and the continent (▶Fig. 4.16).

4.5 TRANSFORM PLATE BOUNDARIES

We saw earlier that the spreading axis of a mid-ocean ridge consists of short segments. The ends of these segments are linked to each other by narrow belts of broken and irregular sea floor, known as fracture zones (see Fig. 4.9). Fracture zones lie roughly at right angles to the ridge segments and extend beyond the ends of the segments (▶Fig. 4.17a). The geometric relationship of fracture zones to ridge segments, and evidence indicating that fracture zones are made of broken-up crust, originally led geoscientists to conclude that fracture zones were faults. They then incorrectly assumed that sliding on faults in fracture zones broke an originally continuous ridge into segments and displaced

the segments sideways (▶Fig. 4.17b, c). This interpretation implies that one segment moves with respect to its neighbor, as shown by the arrows in Figure 4.17c. But soon after Harry Hess proposed his model of sea-floor spreading in 1960, a Canadian named J. Tuzo Wilson realized that if sea-floor spreading really occurred, then the notion that fracture zones offset an originally continuous ridge could not be correct.

In Wilson's alternative interpretation, the fracture zone formed *at the same time* as the ridge axis itself, and thus the ridge consisted of separate segments to start with. These segments were *linked* (not offset) by fracture zones. With this idea in mind, he drew a sketch map showing two ridge-axis segments linked by a fracture zone, and he drew arrows to indicate the direction that ocean crust was moving, relative to the ridge axis, as a result of sea-floor spreading (▶Fig. 4.17d). Look at Wilson's arrows. Clearly, the movement direction on the fracture zone must be opposite to the movement direction that geologists originally thought occurred on the structure. Further, in Wilson's model, slip occurs *only* along the segment of the fracture zone between the two ridge segments. Plate A moves with respect to plate B as sea-floor spreading occurs on the mid-ocean ridge. This movement results in slip along the segment of the fracture zone between points X and Y. But to the west of point X, the fracture zone continues merely as a boundary between two different parts of plate A. The portion of plate A at point 1, just to the north of the boundary (▶Fig. 4.17e), must be

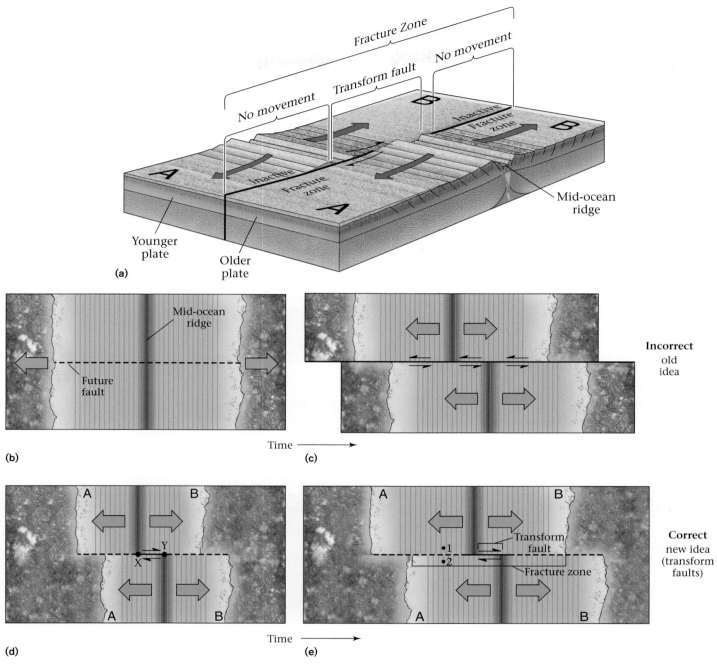

FIGURE 4.17 (a) The fracture zone beyond the ends of the transform fault does not slip, and thus is not a plate boundary. It does, however, mark the boundary between portions of the plate that are different in age. **(b)** In this *incorrect* interpretation of an oceanic fracture zone, the fault forms and cuts across an originally continuous ridge. **(c)** After slip on the fault, indicated by the arrows, the ridge consists of two segments. **(d)** In Wilson's *correct* interpretation, the ridge initiates at the same time as the transform fault, and thus was never continuous. Note that the way in which the fault slips (along the fracture zone between points X and Y) makes sense if sea-floor spreading takes place, but contrasts with the slip in (c). **(e)** Even though the ocean grows, the transform fault can stay the same length. Point 1 on plate A is younger than point 2 because it lies closer to the ridge axis.

younger than the portion at point 2 just to the south, because point 1 lies closer to the ridge axis; but since points 1 and 2 move at the same speed, this segment of the fracture zone does not slip and is not a plate boundary.

Wilson introduced the term **transform fault** for the actively slipping segment of a fracture zone between two ridge segments, and he pointed out that transform faults made a third type of plate boundary. Geologists now also call them transform boundaries, or simply "transforms." At a transform boundary, one plate slides sideways past another, but no new plate forms and no old plate is consumed. Transform boundaries are therefore defined by a vertical fault on which slip parallels the Earth's surface (Fig. 4.17a).

Not all transforms link ridge segments. Some, such as the Alpine Fault of New Zealand, link trenches, whereas others link a trench to a ridge segment. Further, not all transform faults occur in oceanic lithosphere; a few cut across continental lithosphere. The San Andreas Fault, for example, which cuts across California, defines part of the plate boundary between the North American Plate and the Pacific Plate—the portion of California that lies to the west of the fault (including Los Angeles) is part of the Pacific Plate, while the portion that lies to the east of the fault is part of the North American Plate (▶Fig. 4.18a, b). On average, the Pacific Plate moves about 6 cm north, relative to North America, every year. If this motion continues, Los Angeles will become a suburb of Anchorage, Alaska, in about 100 million years (see Geotour 4).

> **Take-Home Message**
>
> At transform plate boundaries, one plate slips sideways past another. Most transform boundaries link segments of mid-ocean ridges. But some, such as the San Andreas Fault, cut across continental crust.

The grinding of one plate past another along a transform fault generates frequent earthquakes, including some huge ones. Fortunately, most of these earthquakes occur out in the ocean basin, far from people. But large earthquakes along the transform faults that cut across continental crust can be very destructive. A huge earthquake on the San Andreas Fault, for example, in conjunction with the ensuing fire, destroyed much of San Francisco in 1906.

FIGURE 4.18 (a) The San Andreas Fault is a transform plate boundary between the Pacific Plate to the west and the North American Plate to the east. At its southeastern end, the San Andreas connects to spreading ridge segments in the Gulf of California. **(b)** The San Andreas Fault, where it cuts across a dry landscape. The fault trace is the narrow valley running the length of the photo. The land has been pushed up slightly, along the fault; streams have cut small side canyons into this uplifted land.

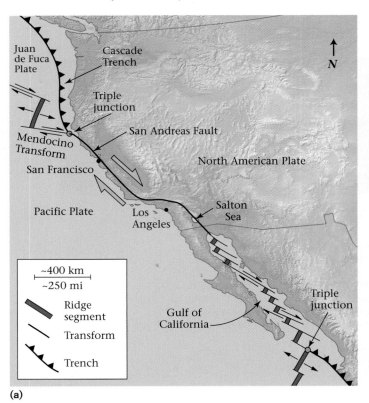

(a)

(b)

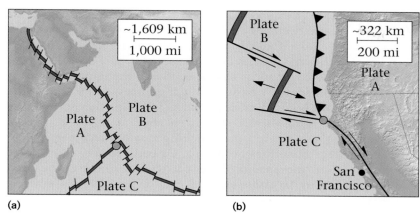

(a)

(b)

FIGURE 4.19 **(a)** A ridge-ridge-ridge triple junction (at the dot). **(b)** A trench-transform-transform triple junction (at the dot).

4.6 SPECIAL LOCATIONS IN THE PLATE MOSAIC

Triple Junctions

So far, we've focused attention on boundaries—divergent (mid-ocean ridge), convergent (trench), and transform—between *two* plates. But in several places, *three* plate boundaries intersect at a point. Geologists refer to these points as **triple junctions**. We name triple junctions after the types of boundaries that intersect. For example, the triple junction formed where the Southwest Indian Ocean Ridge intersects two arms of the Mid–Indian Ocean Ridge (this is the triple junction of the African, Antarctic, and Australian plates) is a ridge-ridge-ridge triple junction (▶Fig. 4.19a). The triple junction north of San Francisco is a trench-transform-transform triple junction (▶Fig. 4.19b).

Hot Spots

Most subaerial (above sea level) volcanoes are situated in the volcanic arcs that border trenches, and most submarine (underwater) volcanoes lie hidden along mid-ocean ridges. The volcanoes of volcanic arcs and mid-ocean ridges are plate-boundary volcanoes, in that they formed as a consequence of movement along the boundary. Not all volcanoes on Earth, however, are plate-boundary volcanoes. Geoscientists have identified about 50 to 100 volcanoes that exist as isolated points and appear to be independent of movement at a plate boundary; these are called hot-spot volcanoes, or simply **hot spots** (▶Fig. 4.20). Most hot spots are located in the interiors of plates, away from the boundaries, but some straddle mid-ocean ridges.

FIGURE 4.20 The dots represent the locations of selected hot-spot volcanoes. The tails represent hot-spot tracks. The most recent volcano (dot) is at one end of this track. Some of these are extinct, indicating that the plume no longer exists. Some hot spots are fairly recent and do not have tracks. Dashed tracks indicate places where a track was broken by sea-floor spreading.

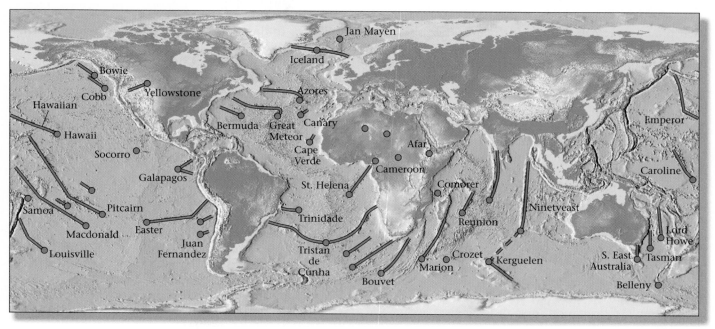

What causes hot-spot volcanoes? The question remains the subject of lively debate. One idea, called the deep-mantle plume model, has been widely (though not universally) accepted for over 30 years. Here, we first discuss the deep-mantle plume model and then we introduce alternatives.

In 1963, the Canadian geologist J. Tuzo Wilson noted that many active hot-spot volcanoes lie at the end of a long chain of extinct volcanic islands and seamounts. (An *active* volcano is one that is erupting, or has erupted relatively recently; a volcano that has died and will never erupt again is *extinct*; see Chapter 9.) With the brand-new concept of moving plates in mind, Wilson proposed that the heat source causing a hot-spot volcano lies in the asthenosphere beneath the plate, and that this *heat source remains relatively fixed in position* while the plate moves over it. Plate movement slowly carries the volcano off the heat source, and eventually, the volcano dies and a new volcano forms over the heat source. The process continues, producing a succession of extinct volcanoes along a line that parallels the plate movement. This line came to be known as a **hot-spot track**. If the heat source persists for a long time, the hot-spot track can be hundreds or even thousands of kilometers long. But at any given time, only the volcano currently over the heat source is active. In contrast, all volcanoes making up the volcanic arc of a convergent plate margin are active at more or less the same time. Wilson's model requires the extinct volcanoes along a hot-spot track to be progressively older the farther they are from the active volcano.

Several years after Wilson's proposal, an American geologist named Jason Morgan suggested that the heat source beneath a hot spot is a **mantle plume**, a column of very hot rock that flows upward until it reaches the base of the lithosphere (▶Fig. 4.21a). Morgan suggested that plumes originate deep in the mantle, just above the core-mantle boundary. In this model, such *deep-mantle* plumes form because heat rising from the Earth's core is warming rock at the base of the mantle. The heated rock expands and becomes less dense, eventually becoming buoyant enough to rise like a hot-air balloon through the overlying mantle. When rock in the plume reaches the base of the lithosphere, it partly melts (for reasons described in Chapter 6) and produces magma that seeps up through the lithosphere and erupts at the Earth's surface (▶Fig. 4.21b). In the context of the deep-mantle plume concept, a hot-spot track forms when the overlying plate moves over a fixed plume. The movement slowly carries the volcano off the top of the plume. The volcano then becomes extinct and a new volcano grows over the plume.

Although the deep-mantle plume model of hot-spot volcanism seems reasonable, and computer models can easily simulate formation, not all geologists accept the model because plumes are hard to see, even using the best techniques available for making images of the mantle. Some geologists suggest alternative models in which either the plumes beneath some (or all?) hot-spot volcanoes originate at *shallow* depths in the upper mantle, or plumes don't exist at all. If plumes don't exist, what process could cause hot spots? Perhaps they form where the lithosphere cracks open above a region of special asthenosphere whose melting can produce particularly large quantities of magma. Or perhaps they form where plate movements drive particularly intense, localized flow in the asthenosphere.

The Hawaiian chain is a classic example of a hot-spot track (Fig. 4.21a). Currently, volcanic eruptions happen only on the big island of Hawaii and in a submarine volcano just to the southeast. The other islands of Hawaii are extinct volcanoes; of these, the oldest (Kauai), lies farthest from Hawaii (Fig. 4.21a). Other, smaller extinct volcanic islands lie to the northwest of Hawaii as far as Midway Island. To the northwest of Midway, the extinct volcanoes are submerged and thus are properly called seamounts. (The submergence of the extinct volcanoes occurs partly because erosion and submarine landslides transfer material from higher to lower elevations, partly because the sea floor beneath the extinct volcanoes sinks as it ages, and partly because the weight of the volcano gradually pushes down the surface of the plate.) (▶Fig. 4.21c) The Hawaiian seamount chain extends another 1,100 km to the northwest of Midway to a point where it links to the Emperor seamount chain, which trends north-northwest for another 1,500 km. Rocks dredged from the seamount at the junction between the two seamount chains erupted at 43 Ma. If the trend of a hot-spot track indicates the direction of plate movement, relative to a point fixed in the deep mantle, then the existence of this "bend" in the Hawaiian-Emperor seamount chain implies that the movement direction of the Pacific Plate changed 43 million years ago (▶Fig. 4.22).

Some hot spots lie within continents. For example, several have been active in the interior of Africa, and one now underlies Yellowstone National Park. The famous geysers (natural steam and hot-water fountains) of Yellowstone exist because hot magma, formed above the Yellowstone hot spot, lies not far below the surface of the park. Yellowstone lies at the northeastern end of the Snake River Plain, a valley covered with the products of volcanic eruptions in the past. In the plume model, the Snake River

> **Take-Home Message**
>
> Three plate boundaries join at a triple junction. Hot spots are places where volcanoes exist as isolated points that are not necessarily a direct consequence of movement at plate boundaries. The magma at hot spots may form by melting at the top of mantle plumes.

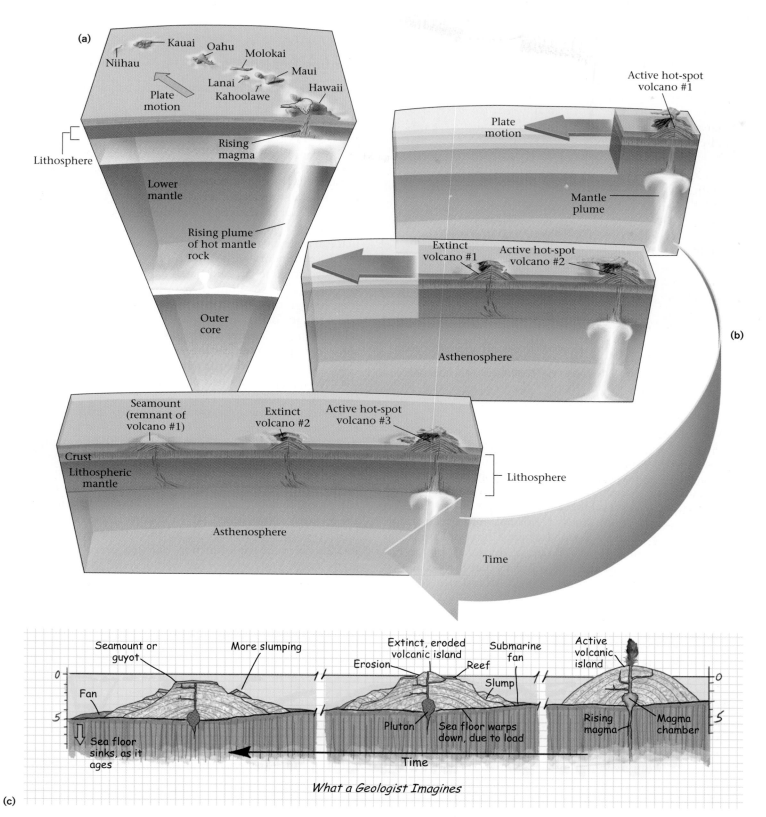

FIGURE 4.21 A plume-generated hot-spot model for Hawaii. **(a)** In this model, the plume that forms Hawaii rises from the base of the mantle. **(b)** A hot spot at the base of a plate leads to the growth of a volcano on the surface of the plate. As the plate moves, the volcano is carried off the hot spot; it then dies (becomes extinct), and a new volcano forms above the hot spot. As the process continues, a chain of extinct volcanoes develops, with the oldest one farthest from the hot spot. The extinct volcanoes gradually sink below sea level and become seamounts. Only the volcano above the hot spot erupts. The chain of islands is a hot-spot track. **(c)** A geologist's sketch emphasizes stages in the evolution of a hot-spot island volcano. When the volcano dies, erosion, slumping, sagging of the sea floor due to the weight of the volcano, and gradual aging (and sinking) of the underlying lithosphere all cause the island to sink below sea level.

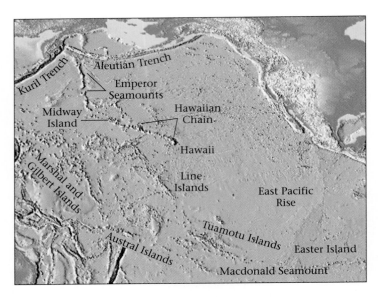

FIGURE 4.22 Bathymetric map showing hot-spot tracks in the Pacific Ocean. Note that the chains have a 40° bend in them, resulting from a change in the direction of motion of the Pacific Plate about 40 million years ago.

Plain volcanics represent the hot-spot track left as North America drifted westward (see Chapter 6).

As mentioned earlier, many hot spots lie on mid-ocean ridges. Where this happens, a volcanic island protrudes above sea level, because the hot spot produces far more magma than does a normal mid-ocean ridge. Iceland, for example, formed where a hot spot underlies the Mid-Atlantic Ridge. The extra volcanism of the hot spot built up the island of Iceland so that it rises almost 3 km above other places on the Mid-Atlantic Ridge.

4.7 HOW DO PLATE BOUNDARIES FORM AND DIE?

The configuration of plates and plate boundaries visible on our planet today has not existed for all of geologic history and will not exist indefinitely into the future. Because of plate motion, oceanic plates form and are later consumed, while continents merge and later split apart. How does a new divergent boundary come into existence, and how does a convergent boundary cease to exist? Most new divergent boundaries form when a continent splits and separates into two continents. We call this process continental **rifting**. A convergent boundary ceases to exist when a piece of buoyant lithosphere, such as a continent or an island arc, moves into the subduction zone. We call this process **collision**.

Continental Rifting

A **continental rift** is a linear belt in which continental lithosphere pulls apart (▶Fig. 4.23). The lithosphere stretches horizontally, so it thins vertically, much as when you pull a piece of taffy between your fingers. Near the surface of the continent, where the crust is cold and brittle, stretching causes rock to break and faults to develop. On these faults, blocks of crust slip down, leading to the formation of low areas that gradually become buried by sediment. Deeper in the crust, and down in the lithospheric mantle, rock is warmer and softer, so stretching takes place in a plastic manner without breaking the rock.

As continental lithosphere thins, hot asthenosphere rises beneath the rift and partly melts. This molten rock erupts at volcanoes along the rift. If rifting continues for a long enough time, the continent breaks in two, a new mid-ocean ridge forms, and sea-floor spreading begins. The relict of the rift evolves into a passive margin (Fig. 4.3). In some cases, however, rifting stops before the continent splits in two. Then, the rift remains as a permanent scar in the crust, defined by a belt of faults, volcanic rocks, and a thick layer of sediment.

Perhaps the most spectacular example of a rift today occurs in eastern Africa; geoscientists aptly refer to this structure as the East African Rift (▶Fig. 4.24a). To astronauts in orbit, the rift looks like a giant gash in the crust. On the ground, it consists of a deep trough bordered on both sides by high cliffs formed by faulting. Along the length of the rift, several major volcanoes smoke and fume; these include the formerly snow-crested Mt. Kilimanjaro, towering over 6 km above the savannah. At its north end, the rift joins the Red Sea ridge at a triple junction. The Red Sea ridge dies out at its north end in the Gulf of Suez rift (▶Fig. 4.24b). Another major rift, known as the Basin and Range Province, breaks up the landscape of the western United States between Salt Lake City, Utah, and Reno, Nevada (▶Fig. 4.25). Here, movement on numerous faults tilted blocks of crust to form narrow mountain ranges, while sediment that eroded from the blocks filled the adjacent basins (the low areas between the ranges).

Collision

India was once a small, separate continent that lay far to the south of Asia. But subduction consumed the ocean between India and Asia, and India moved northward, finally slamming into the southern margin of Asia about 40 to 50 million years ago. Continental crust, unlike oceanic crust, is too buoyant to subduct. So when India collided with Asia, the attached oceanic plate broke off and sank down into the deep mantle. India pushed hard into Asia, squashing the rocks and sediment that once lay between the two continents into the 8-km-high

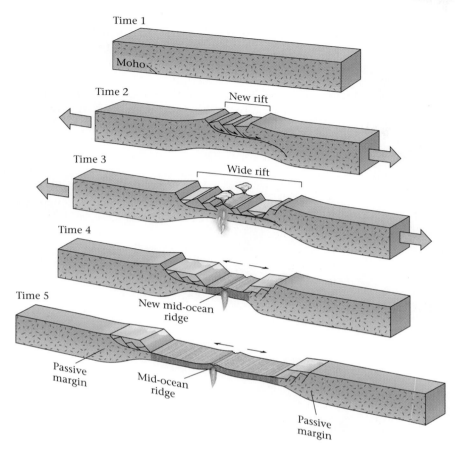

Time 1

Moho

Time 2

New rift

Time 3

Wide rift

Time 4

New mid-ocean
ridge

Time 5

Passive
margin

Mid-ocean
ridge

Passive
margin

FIGURE 4.23 When continental lithosphere stretches during continental rifting, the upper part of the crust breaks up into a series of faults. The lower part of the crust and the lithospheric mantle stretches more like soft plastic. The region that has stretched is the rift. With continued stretching, the crust becomes much thinner, and the asthenosphere that rises beneath the rift partly melts. As a consequence, volcanoes form in the rift. Eventually, the continent breaks in two, and a new mid-ocean ridge forms. With time, an ocean develops. The relicts of the stretched and broken crust of the rift underlie the thick sediment wedge of the passive margins. In this figure, we do not show the lithosphere mantle.

welt that we now know as the Himalayan Mountains. During this process, not only did the surface of the Earth rise, but the crust became thicker. The crust beneath a collisional mountain range can be up to 60 to 70 km thick, about twice the thickness of normal continental crust.

Geoscientists refer to the process during which two buoyant pieces of lithosphere converge and squash together as collision (▶Fig. 4.26a, b). Some collisions involve two continents; some involve continents and an island arc. When a collision is complete, the convergent plate boundary that once existed between the two colliding pieces ceases to exist. Collisions yield some of the most spectacular mountains on the planet, such as the Himalayas and the Alps. They also yielded major mountain ranges in the

past, which subsequently eroded away so that today we see only their relicts. For example, the Appalachian Mountains in the eastern United States were formed as a consequence of three collisions. After the last one, a collision between Africa and North America around 280 million years ago, North America became part of the Pangaea supercontinent.

4.8 WHAT DRIVES PLATE MOTION?

As we discussed in Chapter 2, the mantle consists of solid rock. But beneath the base of the lithosphere, this rock is so hot that it behaves somewhat plastically, and it can flow at very slow rates—1 to 15 cm per year. Because of its ability to flow, the mantle beneath the lithosphere convects, like a vast vat of simmering chocolate. During convection, hotter mantle from greater depths rises, while cooler mantle at shallow levels sinks.

When geoscientists first proposed plate tectonics, they thought the process occurred simply because convective flow

Take-Home Message

Rifting can split a continent in two and can lead to the formation of a new divergent plate boundary. When two continents come together at a convergent plate boundary, they collide, a mountain belt forms, and subduction ceases.

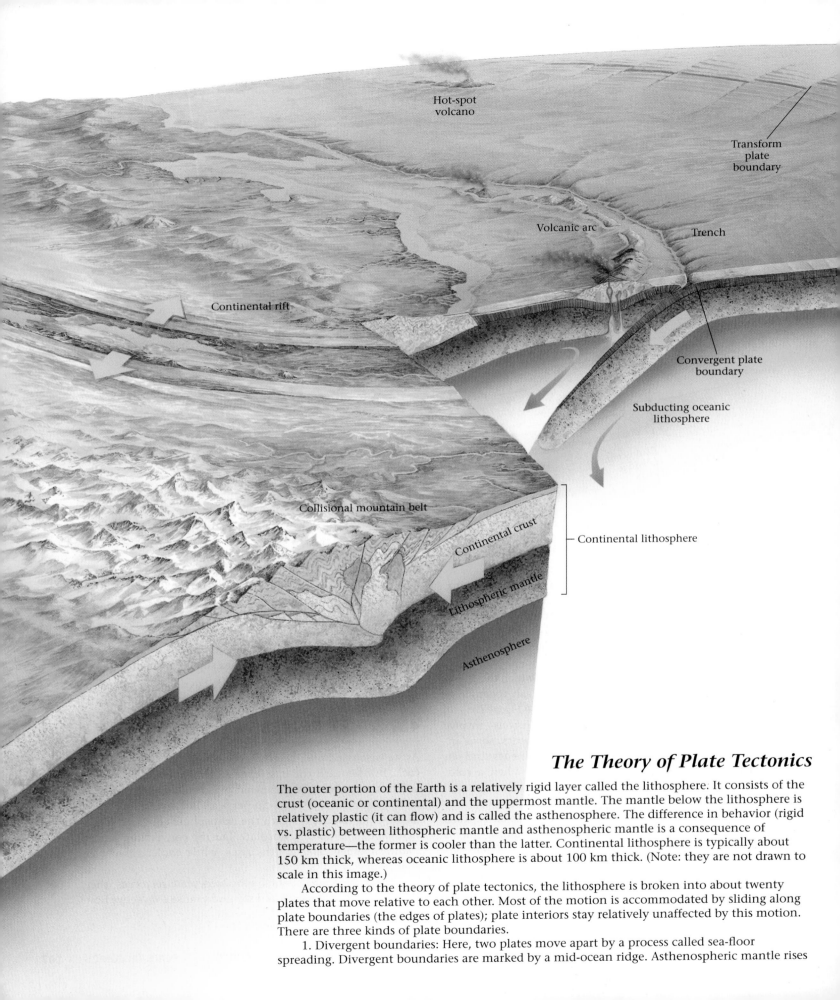

Hot-spot
volcano

Transform
plate
boundary

Volcanic arc

Trench

Continental rift

Convergent plate
boundary

Subducting oceanic
lithosphere

Collisional mountain belt

Continental crust

Continental lithosphere

Lithospheric mantle

Asthenosphere

The Theory of Plate Tectonics

The outer portion of the Earth is a relatively rigid layer called the lithosphere. It consists of the crust (oceanic or continental) and the uppermost mantle. The mantle below the lithosphere is relatively plastic (it can flow) and is called the asthenosphere. The difference in behavior (rigid vs. plastic) between lithospheric mantle and asthenospheric mantle is a consequence of temperature—the former is cooler than the latter. Continental lithosphere is typically about 150 km thick, whereas oceanic lithosphere is about 100 km thick. (Note: they are not drawn to scale in this image.)

According to the theory of plate tectonics, the lithosphere is broken into about twenty plates that move relative to each other. Most of the motion is accommodated by sliding along plate boundaries (the edges of plates); plate interiors stay relatively unaffected by this motion. There are three kinds of plate boundaries.

1. Divergent boundaries: Here, two plates move apart by a process called sea-floor spreading. Divergent boundaries are marked by a mid-ocean ridge. Asthenospheric mantle rises

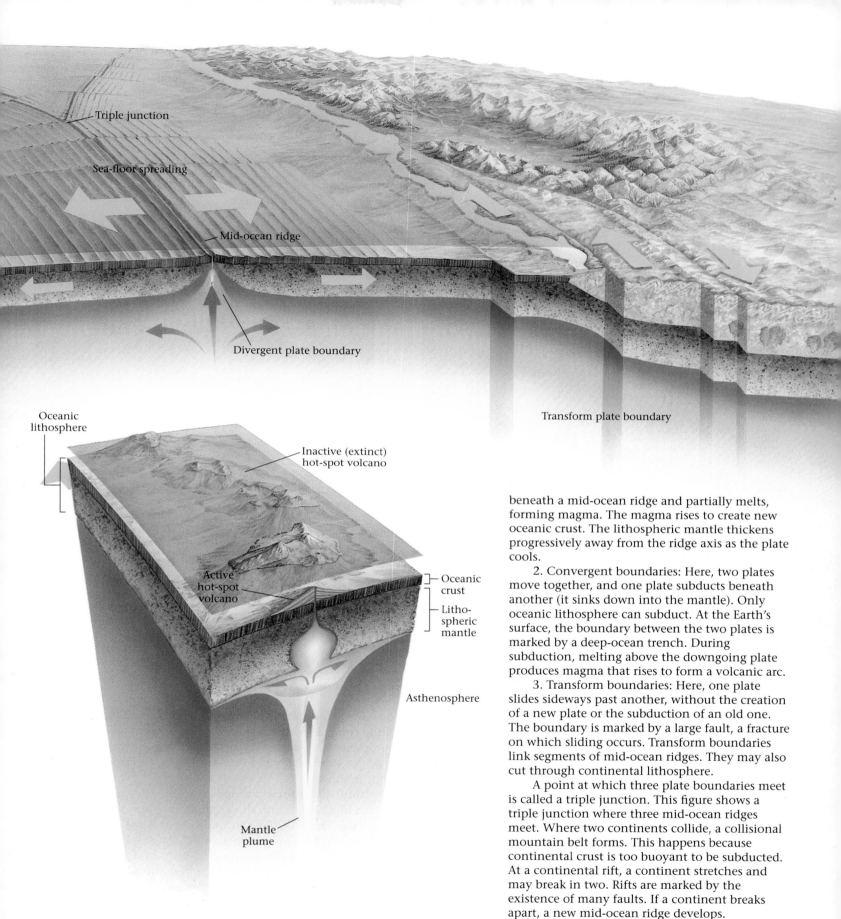

Triple junction

Sea-floor spreading

Mid-ocean ridge

Divergent plate boundary

Transform plate boundary

Oceanic lithosphere

Inactive (extinct) hot-spot volcano

Active hot-spot volcano

Oceanic crust

Lithospheric mantle

Asthenosphere

Mantle plume

beneath a mid-ocean ridge and partially melts, forming magma. The magma rises to create new oceanic crust. The lithospheric mantle thickens progressively away from the ridge axis as the plate cools.

2. Convergent boundaries: Here, two plates move together, and one plate subducts beneath another (it sinks down into the mantle). Only oceanic lithosphere can subduct. At the Earth's surface, the boundary between the two plates is marked by a deep-ocean trench. During subduction, melting above the downgoing plate produces magma that rises to form a volcanic arc.

3. Transform boundaries: Here, one plate slides sideways past another, without the creation of a new plate or the subduction of an old one. The boundary is marked by a large fault, a fracture on which sliding occurs. Transform boundaries link segments of mid-ocean ridges. They may also cut through continental lithosphere.

A point at which three plate boundaries meet is called a triple junction. This figure shows a triple junction where three mid-ocean ridges meet. Where two continents collide, a collisional mountain belt forms. This happens because continental crust is too buoyant to be subducted. At a continental rift, a continent stretches and may break in two. Rifts are marked by the existence of many faults. If a continent breaks apart, a new mid-ocean ridge develops.

Hot-spot volcanoes form above plumes of hot mantle rock that rise from near the core-mantle boundary. As a plate drifts over a hot spot, it leaves a chain of extinct volcanoes.

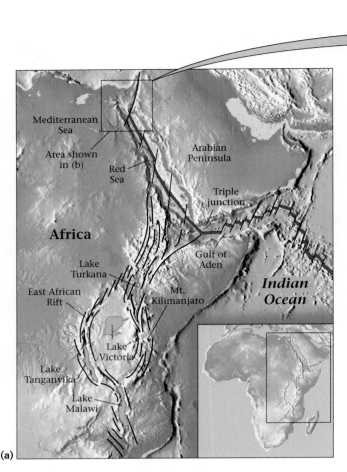

(b)

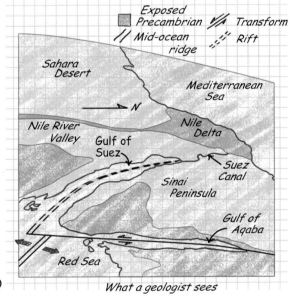

(c)

FIGURE 4.24 **(a)** If the East African Rift were to continue growing, part of Africa would break off, forming a continental fragment. Note that the East African Rift intersects the Red Sea and the Gulf of Aden at a triple junction. The Red Sea and the Gulf of Aden started as rifts but are now narrow oceans, bisected by new mid-ocean ridges. Many of the most important fossils of early humans have come from rocks within the East African Rift; the region's lakes, formed because the axis of the rift drops down, provided an environment in which these hominids could survive. **(b)** Satellite image of the northern Red Sea and Gulf of Suez. **(c)** A geologist's interpretation of the satellite image.

in the asthenosphere actively dragged plates along, as if the plates were rafts on a flowing river. Thus, early images depicting plate motion showed simple convection cells—elliptical flow paths of convecting asthenosphere—beneath mid-ocean ridges (▶Fig. 4.27). At first glance, this hypothesis looked pretty good, but on closer examination, it failed. Among other reasons, it is impossible to draw a global arrangement of convection cells that can explain the complex geometry of plate boundaries on Earth. Gradually, geoscientists came to the conclusion that convective flow within the asthenosphere does occur, but does not directly drive plate motion. In other words, hot asthenosphere does rise in some places and sink in others because of temperature con-

trasts, but the specific directions of this flow do not necessarily coincide with the directions of plate motion. Today, geoscientists favor the hypothesis that two forces—ridge-push force and slab-pull force—strongly influence individual plate motion.

Ridge-push force develops because mid-ocean ridges lie at a higher elevation than the adjacent abyssal plains of the ocean (▶Fig. 4.28a). To understand ridge-push force, imagine you have a glass containing a layer of water over a layer of honey. By tilting the glass momentarily and then returning it to its upright position, you can create a temporary slope in the boundary between these substances. While the boundary has this slope, gravity causes the ele-

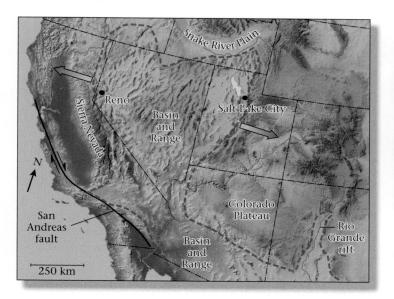

FIGURE 4.25 The Basin and Range Province or Rift is the broad region between Reno, Nevada and Salt Lake City, Utah. Note the narrow north-south trending mountain ranges, the tips of fault-bounded blocks.

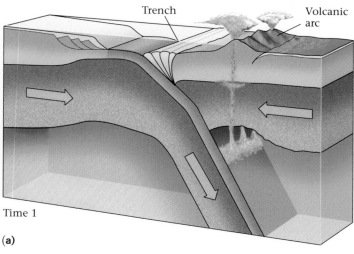

(a)

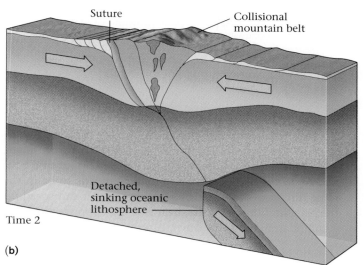

(b)

FIGURE 4.26 (a) Before a continental collision takes place, subduction consumes an oceanic plate until it collides with another plate. Here, a passive continental margin collides with a continental volcanic arc. **(b)** After the collision, the oceanic plate detaches and sinks into the mantle. Rock caught in the collision zone gets broken, bent, and squashed, and forms a mountain range. Slivers of oceanic crust may be trapped along the boundary, or suture, between what once was two continents. As the crust squashes horizontally, it thickens vertically.

vated honey to push against the glass adjacent to the side where the honey surface lies at lower elevation. If the glass were to suddenly disappear, this force would push the honey out over the table. The geometry of a mid-ocean ridge resembles this situation. The surface of the sea floor is higher along a mid-ocean ridge axis than in adjacent abyssal plains. Thus, the surface of the sea floor overall slopes away from the ridge axis. Gravity causes the elevated lithosphere at the ridge axis to push on the lithosphere that lies farther from the axis (much as the tilted honey layer pushes on the side of the glass), making it move away. As lithosphere moves away from the ridge axis, new hot asthenosphere rises to fill the gap; it then moves away, cools, and itself becomes lithosphere. Note that the upward movement of asthenosphere beneath a mid-ocean ridge is a *consequence* of sea-floor spreading, not the cause.

> **Take-Home Message**
>
> Though convective movement in the mantle may contribute to plate motion, it probably isn't the dominant force acting on a plate. The details of plate motions appear to be related to ridge-push and slab-pull forces.

Slab-pull force, the force that subducting plates (also called downgoing slabs) apply to oceanic lithosphere at a convergent margin, arises simply because lithosphere that was formed more than 10 million years ago is denser than asthenosphere, so it can sink into the asthenosphere (▶Fig. 4.28b). Thus, once an oceanic plate starts to sink

down into the mantle, it gradually pulls the rest of the plate along behind it, like an anchor pulling down the anchor line. This "pull" is the slab-pull force.

Now let's summarize our discussion of forces that drive plate motions. Plates move away from ridges—in other words, sea-floor spreading occurs—in response to the ridge-push force. Subducting lithosphere generates a slab-pull force that tows the rest of the plate along with it. But ridge push and slab pull are not the only forces acting on the plate.

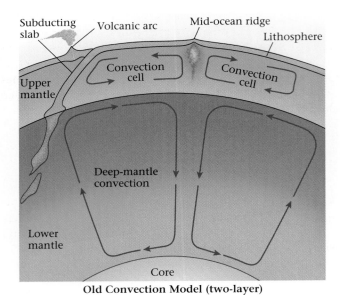

Old Convection Model (two-layer)

FIGURE 4.27 In an early hypothesis for the forces that drive plates, convection cells in the asthenosphere, indicated by the arrows that show the presumed flow direction in the asthenosphere, dragged the plates along. Although convection certainly occurs, such simple geometries are now thought to be unlikely.

The asthenosphere does convect, and the flow of the asthenosphere probably exerts a force on the base of the plate, just as flowing water exerts a force on the bottom of a boat tied to a dock. If this force, or shear, happens to be in the same direction the plate is already moving, it can speed up the plate motion, but if the shear is in the opposite direction, it might slow the plate down. Also, where one plate grinds

against another, as occurs along a transform fault or at the base of an overriding plate at a convergent margin, friction (the force that resists sliding on a surface) may slow the plate down.

4.9 THE VELOCITY OF PLATE MOTIONS

How fast do plates move? It depends on your frame of reference. To illustrate this concept, imagine two cars speeding in the same direction down the highway. From the viewpoint of a tree along the side of the road, car A zips by at 100 km an hour, while car B moves at 80 km an hour. But relative to car B, car A moves at only 20 km an hour. Likewise, geologists use two different frames of reference for describing plate velocity (velocity = distance/time). If we describe the movement of plate A with respect to plate B, then we are talking about **relative plate velocity**. But if we describe the movement of both plates relative to a fixed point in the mantle, then we are speaking of **absolute plate velocity**.

We've already seen one method of determining relative plate motions. Geoscientists measure the distance of a known magnetic anomaly from the axis of a mid-ocean ridge, and they calculate the velocity of a plate relative to the ridge axis by applying the equation: plate velocity equals the distance from the anomaly to the ridge axis divided by the age of anomaly. The velocity of the plate on one side of the ridge relative to the plate on the other is twice this value. We've also seen a way to estimate absolute plate motions. If we *assume* that the position of a mantle plume does not change much for a long time, then the track of hot-spot volcanoes on the plate moving over the plume provides a

FIGURE 4.28 **(a)** A simplified profile (not to scale) of a mid-ocean ridge. Note that along the flanks of the ridge, the sea floor slopes. The elevation of the ridge causes an outward ridge-push force that drives the lithosphere plate away from the ridge. A similar situation exists in a glass containing honey and water. If the boundary between the honey and water tilts, the honey exerts an outward force at its base. **(b)** In this cross section illustrating slab-pull force, the oceanic plate is denser than the asthenosphere, so it sinks into the asthenosphere like a stone into water, only much more slowly.

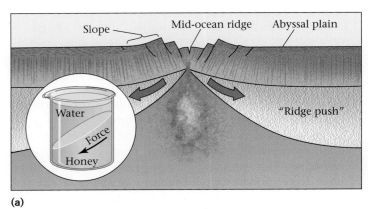

(a)

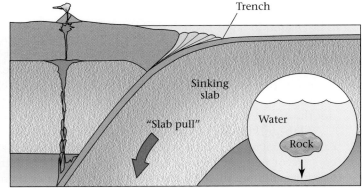

(b)

record of the plate's absolute velocity and indicates the direction of movement (▶Fig. 4.29). (In reality, plumes are not immovably fixed; geologists must use other, more complex methods to calculate absolute plate motions.) The Hawaiian-Emperor seamount chain, for example, may define the absolute velocity of the Pacific Plate. Note that the Hawaiian chain runs northwest, whereas the Emperor chain curves north-northwest (Fig. 4.22). Radiometric dates of volcanic rocks from the bend indicate that they formed about 43 million years ago. Thus, the direction in which the Pacific Plate moved changed significantly at this time.

Working from the calculations described above, geologists have determined that relative plate motions on Earth today occur at rates of about 1 to 15 cm per year. But these rates, though small, can yield large displacements given the immensity of geologic time. At a rate of 10 cm/y, a plate can move 100 km in a million years. Can we detect such slow rates? Until the last decade, the an-

Take-Home Message

Plates move at 1 to 15 cm/y, about the rate that your fingernails grow. We can describe the "relative motion" of one plate with respect to another, or the "absolute motion" of a plate with respect to the underlying asthenosphere. GPS can now detect plate motion.

swer was no. Now the answer is yes. Satellites orbiting the Earth are providing us with the **global positioning system (GPS)**. Automobile drivers can use a GPS receiver to find their destinations, and geologists can use an array of GPS receivers to monitor plate displacements of millimeters per year (▶Fig. 4.30). In other words, we can now see the plates move!

4.10 THE DYNAMIC PLANET

Now, having completed our two-chapter introduction to plate tectonics, we can see more easily why plate tectonics holds the key to understanding most of geology. To start with, plate tectonics explains the origin and distribution of earthquakes, major sea-floor features (mid-ocean ridges, deep-ocean trenches, seamount chains, and fracture zones), and volcanoes (▶Fig. 4.31). It also tells us why mountain belts form. Finally, plate tectonics explains the drift of continents and why the distribution of land changes with time, a change that significantly affects the evolution of life on Earth (▶Fig. 4.32). In coming chapters, we will explore these consequences and others in more detail.

FIGURE 4.29 Relative plate velocities: the blue arrows show the rate and direction at which the plate on one side of the boundary is moving with respect to the plate on the other side. Outward-pointing arrows indicate spreading (divergent boundaries), inward-pointing arrows indicate subduction (convergent boundaries), and parallel arrows show transform motion. The length of an arrow represents the velocity. Absolute plate velocities: the red arrows show the velocity of the plates with respect to a fixed point in the mantle.

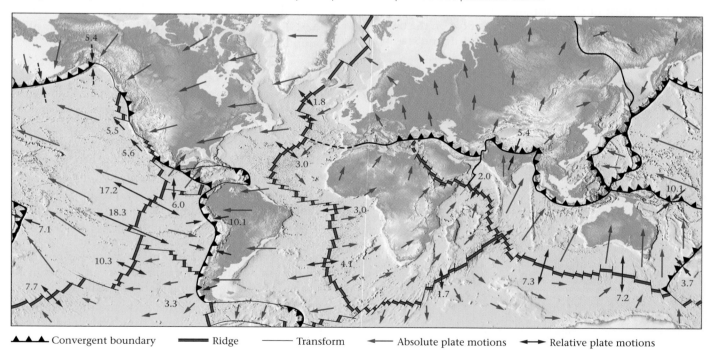

◄▲▲▲ Convergent boundary ▬▬ Ridge ▬▬ Transform ◄— Absolute plate motions ◄—► Relative plate motions

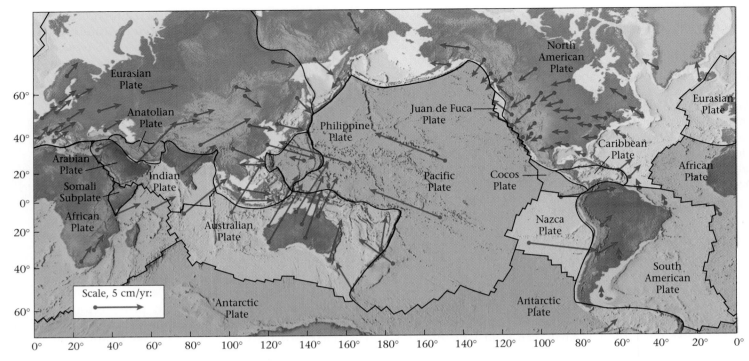

FIGURE 4.30 The Global Positioning System (GPS) is used to measure plate motions at many locations on Earth. The velocities shown here are determined for stations that continuously record GPS data.

Chapter Summary

- The lithosphere, the rigid outer layer of the Earth, is broken into discrete plates that move relative to each other. Plates consist of the crust and the uppermost (cooler) mantle. Lithosphere plates effectively float on the underlying soft asthenosphere. Continental drift and sea-floor spreading are manifestations of plate movement.

FIGURE 4.31 Plate tectonics involves the transfer of material from the mantle to the surface and back down again. The insides and surface of our dynamic planet are in constant motion.

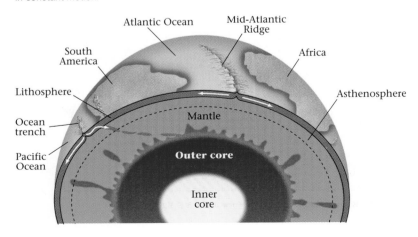

- Some continental margins are plate boundaries, but many are not. A single plate can consist of continental lithosphere, oceanic lithosphere, or both.

- Most plate interactions occur along plate boundaries; the interior of plates remains relatively rigid and intact. Earthquakes delineate the position of plate boundaries.

- There are three types of plate boundaries—divergent, convergent, and transform—distinguished from each other by the movement the plate on one side of the boundary makes relative to the plate on the other side.

- Divergent boundaries are marked by mid-ocean ridges. At divergent boundaries, sea-floor spreading takes place, a process that forms new oceanic lithosphere.

- Convergent boundaries, also called convergent margins or subduction zones, are marked by deep-ocean trenches and volcanic arcs. At convergent boundaries, oceanic lithosphere of the downgoing plate is subducted beneath an overriding plate. The overriding plate can consist of either continental or oceanic lithosphere. An accretionary prism forms out of sediment scraped off the downgoing plate as it subducts.

- Subducted lithosphere sinks back into the mantle. Its position can be tracked down to a depth of about 670 km by a belt of earthquakes known as the Wadati-Benioff zone.

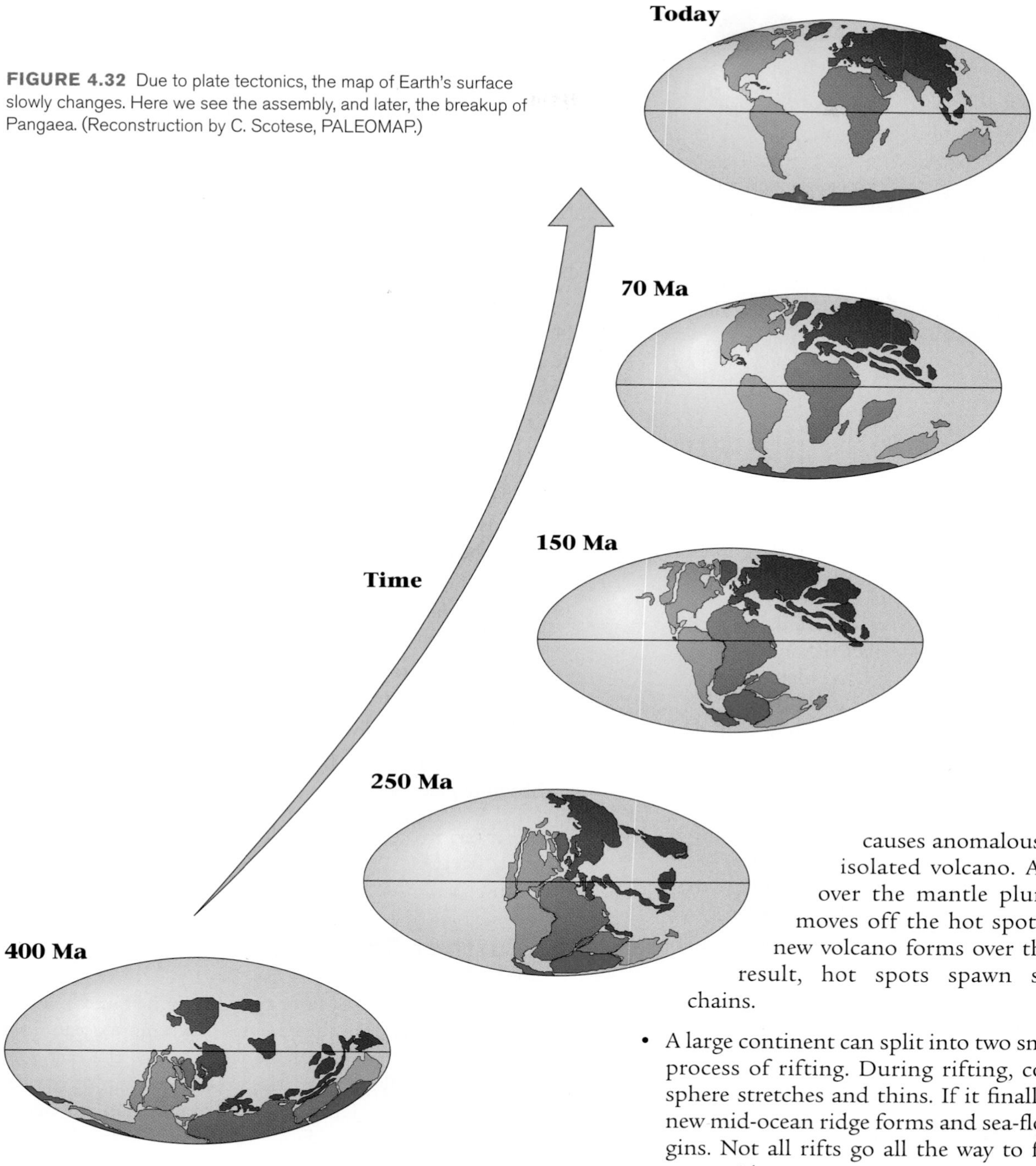

FIGURE 4.32 Due to plate tectonics, the map of Earth's surface slowly changes. Here we see the assembly, and later, the breakup of Pangaea. (Reconstruction by C. Scotese, PALEOMAP.)

Today

70 Ma

150 Ma

Time

250 Ma

400 Ma

causes anomalous volcanism at an isolated volcano. As a plate moves over the mantle plume, the volcano moves off the hot spot and dies, and a new volcano forms over the hot spot. As a result, hot spots spawn seamount/island chains.

- A large continent can split into two smaller ones by the process of rifting. During rifting, continental lithosphere stretches and thins. If it finally breaks apart, a new mid-ocean ridge forms and sea-floor spreading begins. Not all rifts go all the way to form a new mid-ocean ridge.

- Convergent plate boundaries cease to exist when a buoyant piece of crust (a continent or an island arc) moves into the subduction zone. When that happens, collision occurs. The collision between two continents yields large mountain ranges.

- Ridge-push force and slab-pull force drive plate motions. Plates move at rates of about 1 to 15 cm per year. Modern satellite measurements can detect these motions.

- Transform boundaries, also called transform faults, are marked by large faults at which one plate slides past another. No new plate forms and no old plate is consumed at a transform boundary.

- Triple junctions are points where three plate boundaries intersect.

- Hot spots are places where a plume of hot mantle rock rises from just above the core-mantle boundary and

Key Terms

absolute plate velocity (p. 112)
active margins (p. 88)
asthenosphere (p. 86)
black smokers (p. 92)
buoyancy (p. 88)
collision (p. 106)
continental rift (p. 106)
continental shelf (p. 88)
convergent boundary (p. 89)
divergent boundary (p. 89)
global positioning system (p. 113)
hot spot (p. 103)
hot-spot track (p. 104)
lithosphere (p. 86)
mantle plume (p. 104)

mid-ocean ridge (p. 89)
passive margins (p. 88)
plate (p. 88)
plate boundaries (p. 88)
plate tectonics (p. 86)
relative plate velocity (p. 112)
ridge-push force (p. 110)
rifting (p. 106)
slab-pull force (p. 111)
subduction (p. 94)
transform boundary (p. 89)
transform fault (p. 102)
trenches (p. 94)
triple junctions (p. 103)
volcanic arc (p. 97)

Review Questions

1. What is a scientific revolution? How is plate tectonics an example of a scientific revolution?

2. What are the characteristics of a lithosphere plate?

3. How does oceanic crust differ from continental crust in thickness, composition, and density?

4. Describe how Archimedes' principle of buoyancy can be applied to continental and oceanic lithosphere.

5. Contrast active and passive margins.

6. What are the basic premises of plate tectonics?

7. How do we identify a plate boundary?

8. Describe the three types of plate boundaries.

9. How does crust form along a mid-ocean ridge?

10. What happens to the mantle beneath the mid-ocean ridge?

11. Why are mid-ocean ridges high?

12. Why is subduction necessary on a nonexpanding Earth with spreading ridges?

13. What is a Wadati-Benioff zone, and how does it help to define the location of subducting plates?

14. Describe the major features of a convergent boundary.

15. Why are transform plate boundaries required on an Earth with spreading and subducting plate boundaries?

16. What are two examples of famous transform faults?

17. What is a triple junction?

18. Explain the processes that form a hot spot.

19. How is a hot-spot track produced, and how can hot-spot tracks be used to plot the past motions of the overlying plate?

20. Describe the characteristics of a continental rift, and give examples of where this process is occurring today.

21. Describe the process of continental collision, and give examples of where this process has occurred.

22. Discuss the major forces that move lithosphere plates.

23. Explain the difference between relative plate velocity and absolute plate velocity.

24. Can we measure present-day plate motions directly?

On Further Thought

1. Look at the map of sea-floor ages shown in Figure 4.11. Explain the observation that there is much more ocean floor to the west of the East Pacific Rise than to the east, and that the ocean floor along the western margin of the Pacific (southeast of Japan) is much older than the ocean floor on the eastern side of the Pacific (west of central South America).

2. The Pacific Plate moves north relative to the North American Plate at a rate of 6 cm per year. How long will it take Los Angeles (a city on the Pacific Plate) to move northwards by 480 km, the present distance between Los Angeles and San Francisco?

3. Look at a map of the western Pacific Ocean, and examine the position of Japan with respect to mainland Asia. Japan's older crust contains rocks similar to those of east-

ern Asia. Presently, there are many active volcanoes along the length of Japan. With these facts in mind, explain how the Japan Sea (the region between Japan and the mainland) formed.

Suggested Reading

Butler, R. F. 1992. *Paleomagnetism: Magnetic Domains to Geologic Terranes.* Boston: Blackwell.

Condie, K. C. 2005. *Earth as an Evolving Planetary System.* Burlington, Mass.: Academic Press.

Condie, K. C. 2001. *Mantle Plumes and Their Record in Earth History.* Cambridge: Cambridge University Press.

Cox, A., and R. B. Hart. 1986. *Plate Tectonics: How It Works.* Palo Alto, Calif.: Blackwell.

Glen, W. 1982. *The Road to Jaramillo: Critical Years of the Revolution in Earth Sciences.* Palo Alto, Calif.: Stanford University Press.

Kearey, P., and F. J. Vine. 1996. *Global Tectonics,* 2nd ed. Cambridge, Mass.: Blackwell.

McFadden, P. L., and M. W. McElhinny. 2000. *Paleomagnetism: Continents and Oceans,* 2nd ed. San Diego: Academic Press.

McPhee, J. A. 1998. *Annals of the Former World.* New York: Farrar, Straus, and Giroux.

Moores, E. M., and R. J. Twiss. 1995. *Tectonics.* New York: Freeman.

Oreskes, N., ed. 2003. *Plate Tectonics: An Insider's History of the Modern Theory of the Earth.* Boulder: Westview Press.

Sullivan, W. 1991. *Continents in Motion: The New Earth Debate,* 2nd ed. New York: American Institute of Physics.

THE VIEW FROM SPACE This computer-generated image by Christoph Hormann shows the active continental margin that accommodates the relative motion between the North American plate and the Pacific Plate. In the United States, the margin is a transform fault in most of California, and a convergent boundary in Oregon and Washington. The complex topography of the Cordillera holds the record of earlier subduction events, and more recently, of rifting.

Earth Materials

What is the Earth made of? There are four basic components: the solid Earth (the crust, mantle, and core), the biosphere (living organisms), the atmosphere (the envelope of gas surrounding the planet), and the hydrosphere (the liquid and solid water at or near the ground surface). In this part of the book, we focus on the materials that make up the crust and mantle of the solid Earth. We will find that these consist primarily of rock. Most rock, in turn, contains minerals, so minerals are, in effect, the building blocks of our planet.

We therefore begin in Chapter 5 by learning about minerals and how they grow. Then we see, in Interlude B, how geologists distinguish three categories of rock—igneous, sedimentary, and metamorphic—based on how the rocks form. In each of the next three chapters (6, 7, and 8), we look at one of these rock categories. Finally, Interlude C shows us how materials in the Earth System pass through a rock cycle, as atoms constituting one rock type may end up being incorporated into a succession of other rock types.

The Pinnacles in Nambung National Park in Western Australia are columns of sandstone and limestone capped by a resistant layer of calcrete (a concrete-like material formed in soil). They formed when wind eroded away weaker rock in between. A great variety of different Earth materials give variety to our planet's surface.

Patterns in Nature:
Minerals

This photo is real, not a computer collage! We're seeing the world's largest known mineral
crystals jutting from the walls of a cave near Chihuahua, Mexico. The crystals are of the
mineral gypsum; they formed by precipitation from water solutions.

I died a mineral, and became a plant. I died as plant and rose to animal, I died as animal and I was Man. Why should I fear?

—*Jalal-Uddin Rumi (Persian mystic and poet, 1207–1273)*

5.1 INTRODUCTION

Zabargad Island rises barren and brown above the Red Sea, about 70 km off the coast of southern Egypt. Nothing grows on Zabargad except for scruffy grass and a few shrubs, so no one lives there now. But in ancient times many workers toiled on this 5-square-km patch of desert, gradually chipping their way into the side of its highest hill, seeking glassy green, pea-sized pieces of peridot, a prized gem. Carefully polished peridots were worn as jewelry by ancient Egyptians and were buried with them when they died. Eventually, some of the gems appeared in Europe, where jewelers set them into crowns and scepters (▶Fig. 5.1). These peridots now glitter behind glass cases in museums, millennia after first being pried free from the Earth, and perhaps 10 million years after first being formed by the bonding together of still more ancient atoms.

Peridot is one of about 4,000 minerals that have been identified on Earth so far, and it fascinates collectors and geologists alike. Fifty to one hundred new minerals are rec-

ognized every year. Each different mineral has a name. Some names come from Latin, Greek, German, or English words describing a certain characteristic (e.g., "albite" comes from the Latin word for white, orthoclase comes from the German words meaning splits at right angles, and olivine is olive-colored); some honor a person (sillimanite was named for Benjamin Silliman, a famous nineteenth-century mineralogist); some indicate the place where the mineral was first recognized (illite was first identified in rocks from Illinois); and some reflect a particular element in the mineral (chromite contains chromium). Several minerals have more than one name—for example, peridot is the gem-quality version of olivine, a common mineral. Although the vast majority of mineral types are rare, forming only under special conditions, many are quite common and occur in a variety of rock types at Earth's surface.

Though ancient Greek philosophers pondered minerals and medieval alchemists puttered with minerals, true scientific study of minerals did not begin until 1556, when Georgius Agricola, a German physician, published *De Re Metallica*, in which he discussed mining and gave basic descriptions of minerals.[1] In 1669, more than a century after Agricola's work, Nicholas Steno, a Danish monk, discovered important geometric characteristics of minerals. Steno's work became the basis for systematic descriptions of minerals, a task that occupied many researchers during the next two centuries. These researchers were the first **mineralogists**, people who specialize in the study of minerals.

The study of minerals with an optical microscope began in 1828, but though such studies helped in mineral identification, they could not reveal the arrangement of atoms inside minerals. That understanding had to wait until 1912, when Max von Laue of Germany proposed that X-rays, electromagnetic radiation whose wavelength is comparable to the distance between atoms in a mineral, could be used to study the internal structure of minerals. A father-and-son team, W. H. and W. L. Bragg of England, published the first X-ray study of a mineral, work for which they shared the 1915 Nobel Prize in physics. In subsequent decades, researchers developed progressively more complex instruments to aid their study of minerals. For example, in the 1960s, mineralogists began to use electron microscopes to obtain actual images of the internal structure of minerals, and electron microprobes to analyze the chemical composition of grains that are almost too small to see.

Why study minerals? Without exaggeration, we can say that minerals are the building blocks of our planet. To a geologist, almost any study of Earth materials depends on an understanding of minerals, for minerals make up the rocks

FIGURE 5.1 A royal crown containing a variety of valuable jewels. The large gemstone near the base of the crown is a green peridot.

---

[1]Agricola wrote his book in Latin. It's interesting to note that the book's first English translation was completed in 1912 by Herbert Hoover and his wife, Lou. At the time, Hoover was a successful geological engineer—he became president of the United States seventeen years later.

and sediments that make up the Earth and its landscapes. Minerals are also important from a practical standpoint. Industrial minerals serve as the raw materials for manufacturing chemicals, concrete, and wallboard. Ore minerals are the source of valuable metals such as copper and gold and provide energy resources such as uranium (▶Fig. 5.2a, b). Certain forms of minerals, gems, delight the eye as jewelry. Unfortunately, not all minerals are beneficial; some pose environmental hazards. No wonder **mineralogy**, the study of minerals, fascinates professionals and amateurs alike.

The word *mineral* has a broader meaning in everyday English than it does in geology. Nutritionists talk about the "vitamins and minerals" in various types of foods—to them, a mineral is a metallic compound. In geology, however, a mineral is a special kind of substance with certain distinctive characteristics. In this chapter, we begin by discussing the geologic definition of a mineral. Then we look at how minerals form and examine the main characteristics that enable us to identify them. Finally, we note the basic scheme that geologists use to classify minerals. This chapter assumes that you understand the fundamental concepts of matter and energy, especially the nature of atoms, molecules, and chemical bonds. If you are rusty on these topics, please review Appendix A. Basic terms from chemistry are summarized in ▶Box 5.1, for your convenience.

FIGURE 5.2 **(a)** Museum specimen of malachite, a bright-green mineral containing copper. (Its formula is $Cu_2[CO_3][OH]_2$.) Malachite is an ore mineral mined to produce copper, but because of its beauty, it is also used for jewelry. **(b)** Copper pennies made by the processing of malachite and other ore minerals of copper.

(a)

(b)

5.2 WHAT IS A MINERAL?

To a geologist, a **mineral** is a naturally occurring solid, formed by geologic processes, that has a crystalline structure and a definable chemical composition, and in general is inorganic. Let's pull apart this mouthful of a definition and examine its meaning in detail.

- *Naturally occurring*: Minerals are produced in nature, not in factories. We need to emphasize this point because in recent decades industrial chemists have learned how to synthesize materials that have characteristics virtually identical to those of real minerals. These materials are not minerals in a geologic sense, even though they are commonly referred to in the commercial world as "synthetic minerals."

- *Solid*: A solid is a state of matter that can maintain its shape indefinitely, and thus will not conform to the shape of its container. Liquids (such as oil or water) and gases (such as air) are not minerals.

- *Formed by geologic processes*: Minerals, as we see later in this chapter, can form by the freezing of molten rock, by precipitation out of a water solution, or by chemical reactions within or on the surface of preexisting rocks. All of these processes are considered to be "geologic" since they occur naturally on or in the Earth. This component of the definition has a caveat, however. Some substances are identical in character to minerals produced by geologic processes, but are a byproduct of living organisms—the calcite in a clam shell is an example. Such materials are minerals, but they are called **biogenic minerals** to emphasize their origin. Significantly, biogenic substances that cannot also be formed by geologic processes are *not* considered to be minerals.

- *Definable chemical composition*: This part of the definition simply means that you can write a chemical formula for a mineral (see Box 5.1). Some minerals contain only one element, but most are compounds of two or more elements. For example, diamond and graphite both have the formula C, because they consist

BOX 5.1 SCIENCE TOOLBOX

Some Basic Definitions from Chemistry

To describe minerals, we need to use several terms from chemistry (for a more in-depth discussion, see Appendix A). To avoid confusion, terms are listed in an order that permits each successive term to utilize previous terms.

- **Element**: a pure substance that cannot be separated into other elements.

- **Atom**: the smallest piece of an element that retains the characteristics of the element. An atom consists of a nucleus surrounded by a cloud of orbiting electrons; the nucleus is made up of protons and neutrons (except in hydrogen, whose nucleus contains only one proton and no neutrons). Electrons have a negative charge, protons have a positive charge, and neutrons have a neutral charge. An atom that has the same number of electrons as protons is said to be "neutral," in that it does not have an overall electrical charge.

- **Atomic number**: the number of protons in an atom of an element.

- **Atomic weight**: the approximate number of protons plus neutrons in an atom of an element.

- **Ion**: an atom that is not neutral. An ion that has an excess negative charge (because it has more electrons than protons) is an **anion**, whereas an ion that has an excess positive charge (because it has more protons than electrons) is a **cation**. We indicate the charge with a superscript. For example, Cl$^-$ (chlorine) has a single excess electron; Fe^{2+} is missing two electrons.

- **Chemical bond**: an attractive force that holds two or more atoms together. For example, *covalent bonds* form when atoms share electrons. *Ionic bonds* form when a cation and anion (ions with opposite charges) get close together and attract each other. In materials with **metallic bonds**, some of the electrons can move freely.

- **Molecule**: two or more atoms bonded together. The atoms may be of the same element or of different elements.

- **Compound**: a pure substance that can be subdivided into two or more elements. The smallest piece of a compound that retains the characteristics of the compound is a molecule.

- **Chemical**: a general name used for a pure substance (either an element or a compound).

- **Chemical formula**: a shorthand recipe that itemizes the various elements in a chemical and specifies their relative proportions. For example, the formula for water, H$_2$O, indicates that water consists of molecules in which two hydrogens bond to one oxygen.

- **Chemical reaction**: a process that involves the breaking or forming of chemical bonds. Chemical reactions can break molecules apart or create new molecules and/or isolated atoms.

- **Mixture**: a combination of two or more elements or compounds that can be separated without a chemical reaction. For example, a cereal composed of bran flakes and raisins is a mixture—you can separate the raisins from the flakes without destroying either.

- **Solution**: a type of material in which one chemical (the solute) dissolves (becomes completely incorporated) in another (the solvent). In solutions, a solute may separate into ions during the process. For example, when salt (NaCl) dissolves in water, it separates into sodium (Na$^+$) and chlorine (Cl$^-$) ions. In a solution, atoms or molecules of the solvent surround atoms, ions, or molecules of the solute.

- **Precipitate**: (noun) a compound that forms when ions in liquid solution join together to create a solid that settles out of the solution; (verb) the process of forming solid grains by separation and settling from a solution. For example, when saltwater evaporates, solid salt crystals precipitate and settle to the bottom of the remaining water.

---

entirely of carbon. Quartz has the formula SiO$_2$, meaning that it consists of the elements silicon and oxygen in a proportion of one silicon atom for every two oxygen atoms. Some minerals have complicated formulas. For example, a flaky black mineral called biotite has this formula: K(Mg,Fe)$_3$(AlSi$_3$O$_{10}$)(OH)$_2$. This formula indicates that biotite can contain magnesium, iron, or both in varying proportions.

- *Orderly arrangement of atoms*: The atoms that make up a mineral are not distributed randomly and cannot move around easily. Rather, they are fixed in a specific pattern that repeats itself over a very large region relative to the size of atoms. (To picture the contrast between a random arrangement and a fixed pattern, compare the distribution of people at a casual party with the distribution of people in a military regiment at attention. At the party, clusters of two or three people stand around chatting, and people or groups of people move around the room. But in the regiment at attention, everyone stands aligned in orderly rows and columns, and no one dares to move.) A material in which atoms are fixed in an orderly pattern is called a crystalline solid. Mineralogists refer to the pattern itself (the imaginary framework representing the arrangement of atoms) as a **crystal lattice** (▶Fig. 5.3a, b).

- *Inorganic, in general*: To explain what this statement means, we must first distinguish between organic and inorganic chemicals. Organic chemicals are molecules containing carbon-hydrogen bonds. Although some organic chemicals contain only carbon and hydrogen, others also contain oxygen, nitrogen, and other elements in varying quantities. Sugar ($C_{12}H_{22}O_{11}$), for example, is an organic chemical. (The name *organic* came into use during the eighteenth century, when chemists thought that such chemicals could form only by life processes. But that restriction no longer applies because countless examples of organic chemicals—plastics, for example—have been synthesized in the laboratory, outside of living organisms.) Almost all minerals are inorganic. Thus, sugar and protein are not minerals. But, we have to add the qualifier "in general" because mineralogists do consider about thirty organic substances formed by "the action of geologic processes on organic materials" to be minerals. Examples include the crystals that grow in ancient deposits of bat guano. We do not discuss such examples further.

With these definitions in mind, we can make an important distinction between a mineral and glass. Both minerals and glasses are solids, in that they can retain their shape indefinitely (see Appendix A). But a mineral is crystalline, and glass is not. Whereas atoms, ions, or molecules in a mineral are ordered into a crystal lattice, like soldiers standing in formation, those in a glass are arranged in a semichaotic way, like a crowd of people at a party, in small clusters or chains that are neither oriented in the same way nor spaced at regular intervals (▶Fig. 5.3c, d). Note that the chemical compound silica (SiO_2) forms the mineral quartz when arranged in a crystalline lattice, but forms common window glass when arranged in a semichaotic way.

If you ever need to figure out whether a substance is a mineral or not, just check it against the criteria listed above. Is motor oil a mineral? No—it's a liquid. Is table salt a mineral? Yes—it's a solid crystalline compound with the formula NaCl.

5.3 BEAUTY IN PATTERNS: CRYSTALS AND THEIR STRUCTURE

What Is a Crystal?

The word *crystal* brings to mind sparkling chandeliers, elegant wine goblets, and shiny jewels. But, as is the case with the word *mineral*, geologists have a more precise definition for crystal. A **crystal** is a single, continuous (uninterrupted) piece of a crystalline solid bounded by flat surfaces called **crystal faces** that grew naturally as the mineral formed. The word comes from the Greek *krystallos*, meaning ice. Many crystals have beautiful shapes that look as if they belong in the pages of a geometry book. These shapes fascinated Nicholas Steno, who discovered that for a given mineral, the angle between two adjacent crystal faces of one specimen is identical to the angle between the corresponding faces of another specimen. For example, a perfectly formed quartz crystal looks like an obelisk (▶Fig. 5.4a). The angle between the faces of the columnar part of a quartz crystal is exactly 120°. This rule holds regardless of whether the whole crystal is big or small and regardless of whether all of the faces are the same size (▶Fig. 5.4b).

Crystals come in a great variety of shapes including cubes, trapezoids, pyramids, octahedrons, hexagonal

FIGURE 5.3 (a) Internally, this quartz crystal contains an orderly arrangement of atoms. (b) This gridwork of scaffolding surrounding the Washington Monument in Washington, D.C., provides an analogy for the fixed arrangements of atoms in a mineral. (c) Disordered atoms, as occur in glass, do not define a regular pattern. (d) Ordered atoms like these are found in a mineral.

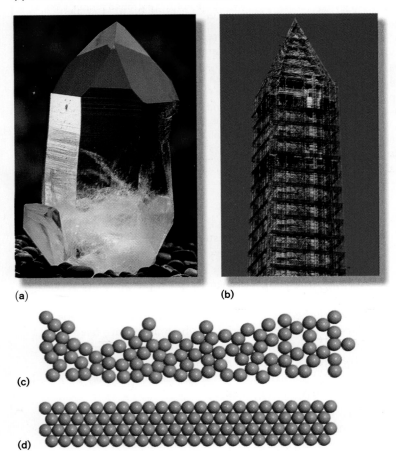

(a)

(b)

(c)

(d)

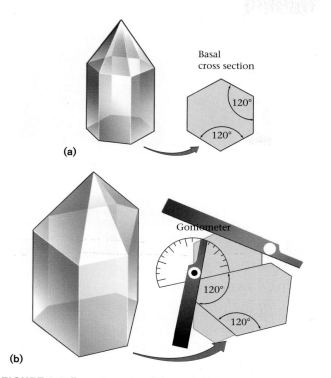

(a)

Basal cross section

120°

120°

Goniometer

120°

120°

(b)

FIGURE 5.4 For a given mineral, the angle between two adjacent crystal faces in one specimen is the same as the angle between corresponding faces in another. **(a)** A small crystal of quartz whose vertical crystal faces happen to be the same size. The intersection between crystal faces makes an angle of 120°, as shown by the cross-section slice through the crystal. **(b)** A large crystal of quartz whose vertical crystal faces are not all the same size. Even though the dimensions of the faces differ from each other, the angle between the faces is still 120°, as measured by a goniometer (an instrument that measures angles).

In recent years, it has become fashionable for people to wear crystals around their necks, suspend them over their heads at night, or put them in prominent places in their homes in the belief that crystals can somehow "channel" the "life force" of the Universe into a person's soul. As far as the vast majority of geologists are concerned, crystals have no demonstrable effect on health or mood. For millennia, crystals have inspired awe because of the way they sparkle, but such behavior is simply a consequence of how crystal structures interact with light.

What's Inside a Crystal?

What do the insides of a mineral actually look like? We can picture atoms in minerals as tiny balls packed together tightly and held in place by chemical bonds. The way in which atoms are packed defines the **crystal structure** of the mineral. As we will see in Section 5.4, the physical properties of a mineral (for example, the shape of its crystals, how hard it is, how it reacts chemically with other substances) depend both on the identity of the elements making up the mineral and on the way these elements are arranged and bonded in a crystal structure.

Because of its importance, we now look a little more closely at the nature of chemical bonding in minerals. Chemists recognize five different types of bonds (covalent, ionic, metallic, van der Waals', and hydrogen) that are based on the way in which atoms stick or link to each other. For example, in covalently bonded materials, atoms stick to each other by sharing electrons, whereas in ionically bonded materials, atoms either add electrons to become negative ions

columns, blades, needles, columns, and obelisks. All the faces of some crystals have the same shape (▶Fig. 5.5a, b, e), whereas on others, different faces have different shapes (▶Fig. 5.5c, d, f, g, h).

Because crystals have a regular geometric form, people have always considered crystals to be special, and many cultures have attributed magical powers to them. For example, shamans commonly relied on talismans or amulets made of crystals, which supposedly brought power to their wearer or warded off evil spirits. Even in modern fantasy and science-fiction stories, crystals play a special role—in the TV series *Star Trek*, the starship *Enterprise* required a "dilithium crystal" to achieve "warp speed." (In reality, no such crystal exists.)

FIGURE 5.5 Crystals come in all kinds of shapes. Some are double pyramids, some are cubes, and some have blade shapes. Some crystals terminate at a point, and some terminate in a chisel-like wedge. **(a)** Halite, **(b)** diamond, **(c)** staurolite, **(d)** quartz, **(e)** garnet, **(f)** stibnite, **(g)** calcite, and **(h)** kyanite.

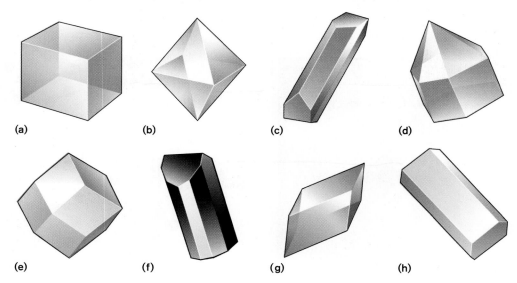

(a)   **(b)**   **(c)**   **(d)**

(e)   **(f)**   **(g)**   **(h)**

(anions) or lose electrons to become positive ions (cations)—the two kinds of ions stick to each other because opposite charges attract. (Appendix A illustrates these bonds and discusses the other types as well.) Not all minerals have the same kind of bonding, and in some minerals, more than one type of bonding occurs. The type of bonding, the ease with which bonds form or are broken, and the geometric arrangement of bonds play an important role in determining the characteristics of minerals. As your intuition might suggest, bonds are stronger in harder minerals and in minerals with higher melting temperatures. In some minerals, the nature and strength of bonding vary with direction in the mineral. If bonds form more easily in one direction than another, a crystal will grow faster in one direction than another. And if a mineral has weak bonds in one direction and strong bonds in another direction, it will break more easily in one direction than in the other.

To illustrate crystal structures, we look at a few examples. Halite (rock salt) is an ionically bonded mineral in that it consists of oppositely charged ions that stick together because opposite charges attract. In halite, the anions are chloride (Cl^-) and the cations are sodium (Na^+). In halite, six chloride ions surround each sodium ion, producing an overall arrangement of atoms that defines the shape of a cube (▶Fig. 5.6a, b). Diamond, by contrast, is a

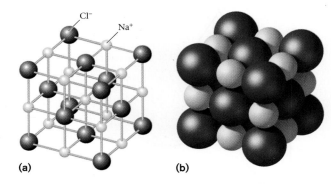

FIGURE 5.6 Minerals are composed of atoms that stick together by chemical bonds. **(a)** A ball-and-stick model of halite, composed of ionically bonded ions of sodium and chloride. The sticks represent bonds, and the balls represent atoms. **(b)** A ball model of halite. The different sizes of balls represent the relative sizes of ions. Note that the smaller sodium ions fit in between the larger chloride ions.

mineral made entirely of carbon. In diamond, each atom is covalently bonded to four neighbors. The atoms are arranged in the form of a tetrahedron, so some naturally formed diamond crystals display this shape (▶Fig. 5.7a, b). Covalent bonds are very strong, so diamond is very hard. Graphite, another mineral composed entirely of carbon,

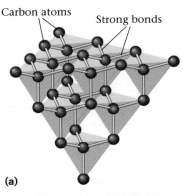

FIGURE 5.7 (a) A ball-and-stick model of diamond, composed of covalently bonded atoms of carbon arranged in a tetrahedron. **(b)** A photograph of a diamond crystal. **(c)** A ball-and-stick model of graphite. Note that the carbon atoms are arranged in sheets of hexagons; the sheets are held together by weak bonds. **(d)** A photograph of a graphite crystal.

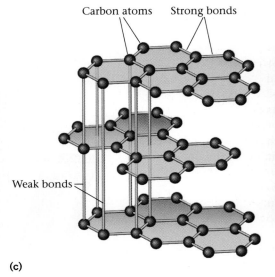

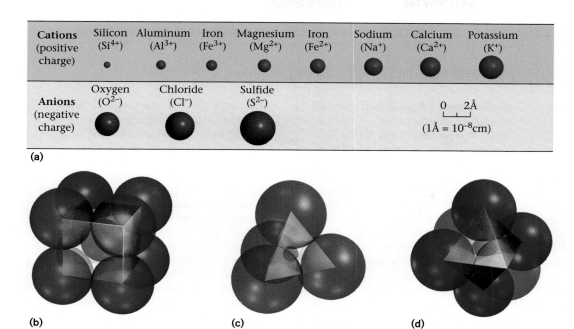

Cations (positive charge)	Silicon (Si^{4+})	Aluminum (Al^{3+})	Iron (Fe^{3+})	Magnesium (Mg^{2+})	Iron (Fe^{2+})	Sodium (Na^+)	Calcium (Ca^{2+})	Potassium (K^+)

Anions (negative charge)	Oxygen (O^{2-})	Chloride (Cl^-)	Sulfide (S^{2-})

0   2Å
($1Å = 10^{-8}$cm)

(a)

(b)   (c)   (d)

FIGURE 5.8 (a) Relative sizes of ions that are common in minerals. **(b)** The packing of atoms in a cube. **(c)** The packing of atoms in a tetrahedron. **(d)** The packing of atoms in an octahedron.

behaves very differently from diamond. In contrast to diamond, graphite is so soft that it can be used as the "lead" in a pencil; as you move a pencil across paper, tiny flakes of graphite peel off the pencil point and adhere to the paper. This behavior occurs because the carbon atoms in graphite are not arranged in tetrahedra, but rather occur in sheets (▶Fig. 5.7c, d). The sheets are bonded to each other by weak bonds (van der Waals' bonds) and thus can separate from each other easily. Note that two different minerals (such as diamond and graphite) that have the same composition but have different crystal structures are called **polymorphs**.

What determines how atoms pack together in a crystal? The size of an ion depends on the number of electrons orbiting the nucleus (▶Fig. 5.8a); so, since anions have extra electrons, they tend to be bigger than cations. Thus, cations nestle snugly in the spaces between anions in many crystal structures, and as many anions will try to fit around a cation as there is room for. Depending on the identity of an ion, different geometries of packing can occur (▶Fig. 5.8b–d).

Note that in halite, as described above, each ion is a single atom. In many ionically bonded minerals, the ions building the minerals consist of more than one atom. For example, the mineral calcite ($CaCO_3$) consists of calcium (Ca^{2+}) cations and carbonate (CO_3^{2-}) anions—each carbonate anion, or anionic group, consists of four atoms and thus is quite large.

The orderly arrangement of atoms inside a crystal—its crystal structure—provides one of nature's most spectacular

examples of a pattern. The pattern on a sheet of wallpaper may be defined by the regular spacing of, say, clumps of flowers. Similarly, the pattern in a crystal is defined by the regular spacing of atoms (▶Fig. 5.9a, b; ▶Box 5.2). If the crystal contains more than one type of atom, the atoms alternate in a regular pattern. The orderly arrangement controls the outward shape, or morphology, of crystals. For example, if the atoms in a mineral are packed into the shape of a cube, a crystal of the mineral will have faces that intersect at 90° angles. The pattern of atoms or ions in a mineral displays **symmetry**, meaning that the shape of one part of a mineral is a mirror image of the shape of another part. For example, if you were to cut a halite crystal in half and place one half against a mirror, the crystal would look whole again (▶Fig. 5.11a, b).

The Formation and Destruction of Minerals

New mineral crystals can form in five ways:

1. They can form by the solidification of a melt, meaning the freezing of a liquid; for example, ice crystals, a type of mineral, are made by freezing water.

FIGURE 5.9 (a) Repetition of a flower motif in wallpaper illustrates a regular pattern. **(b)** On the face of a crystal of galena (a type of lead ore), lead and sulfur atoms pack together in a regular array.

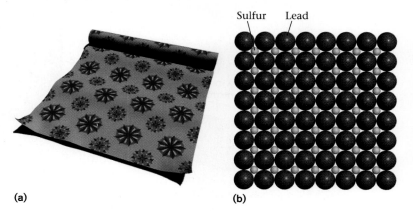

Sulfur   Lead

(a)   (b)

BOX 5.2 SCIENCE TOOLBOX

How Do We "See" the Arrangement of Atoms in a Crystal?

We can see crystals and hold them in our hands. But atoms are so small that the human eye cannot possibly distinguish them, even with the strongest optical microscope (a microscope using light that passes through lenses). So how do mineralogists come up with the models that depict atoms arranged in crystals? To explain, we must first provide some background about waves.

The energy of light, radio signals, and X-rays moves from one place to another in the form of waves. For simplicity, we picture such waves as having the same form as familiar water waves, with crests (high parts) and troughs (low parts). A set of waves moving in a given direction is called a wave train, and the distance between adjacent crests (or troughs) is the wavelength. When a train of straight waves strikes a wall that contains a small opening whose width is comparable to the wavelength, the opening acts as a new point source of waves (just as a pebble striking a pond surface is a point source of water), and a set of curving waves propagates outward from the opening. This phenomenon is called **diffraction**. If the obstacle

contains many equally spaced openings, the new waves emitting from each opening interfere with each other (▶Fig. 5.10a). This means that in some places, crests from one point opening overlap crests from another, thereby producing larger crests (manifested as a stronger signal). In other places, crests from one point opening overlap crests from another trough, thereby canceling each other out (manifested as a weaker signal).

In 1912, Max von Laue proposed that an X-ray passing through a crystal would undergo diffraction if the atoms were arranged in orderly columns and rows and if the spacing between columns was comparable to the wavelength of an X-ray. He surmised that the space between two adjacent columns would act as a point source of waves, that the curving waves emitting from adjacent spaces would interfere, and that because of this interference, the diffracted beams would produce a distinctive pattern of strong and weak signals, which would appear as spots on a photographic plate (▶Fig. 5.10b). (This phenomenon doesn't happen with light waves because the wavelengths of light are

too big.) When von Laue's students tested this proposal, they indeed produced diffraction patterns. The existence of such X-ray diffraction (XRD) patterns requires that the atoms in crystals have an orderly arrangement. From the patterns, researchers can now deduce the geometric arrangement of atoms in a crystal.

More recently, investigators have been able to use transmission electron microscopes (TEM) to "see" the arrangement of atoms in a mineral directly. A TEM shoots a beam of electrons at a thin slice of a crystal. Since electrons are much smaller than the spaces between atoms, the electrons can pass through spaces between atoms and can strike a detector, forming a light spot. Electrons that interact with atoms bounce off and scatter in all directions, and therefore don't reach the detector (▶Fig. 5.10c). As a result, a dark spot (somewhat like a shadow) remains under a column of atoms. The overall pattern of dark and light spots on the detector represents the distribution of atoms (▶Fig. 5.10d).

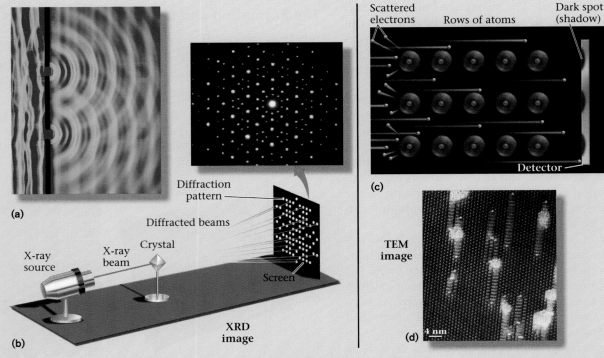

FIGURE 5.10 (a) A photo of water waves moving from left to right through two gaps in a barrier. Diffraction occurs to the right of the barriers. **(b)** A single X-ray beam undergoes diffraction when it passes through a crystal. Interaction of diffracted beams produces a pattern of dots on a photographic plate. This is an X-ray diffraction (XRD) image. X-rays interact this way with a crystal because of the crystal's orderly arrangement of atoms. **(c)** Some incoming electrons pass through a material, whereas some undergo scattering. **(d)** This phenomenon produces a transmission electron microscope (TEM) image.

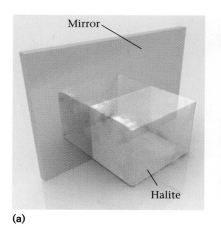

FIGURE 5.11 **(a)** Crystals have symmetry: one half of a halite crystal is a mirror image of the other half. **(b)** Snowflakes, crystals of ice, are symmetrical hexagons.

2. They can form by precipitation from a solution, meaning that atoms, molecules, or ions dissolved in water bond together and separate out of the water—salt crystals, for example, develop when you evaporate saltwater.

3. They can form by solid-state diffusion, the movement of atoms or ions through a solid to arrange into a new crystal structure; this process takes place very slowly. (In Chapter 8, we'll discuss the importance of diffusion during the formation of minerals in metamorphic rocks.)

4. They can form at interfaces between the physical and biological components of the Earth system by a process called biomineralization. Biomineralization occurs when living organisms cause minerals to precipitate either within or on their bodies, or immedi-

ately adjacent to their bodies. For example, clams and other shelled organisms extract ions from water to produce mineral shells (a clamshell consists of two minerals: calcite and its polymorph, aragonite), and the metabolism of certain species of cyanobacteria produces chemicals that change the acidity of the water they live in and cause calcite crystals to precipitate.

5. They can form directly from a vapor. This process, called *fumerolic mineralization*, occurs around volcanic vents or around geysers. At such locations, volcanic gases or steam enter the atmosphere and cool, so certain elements cannot remain in gaseous form. Some of the bright yellow sulfur deposits found in volcanic regions form in this way.

The first step in forming a crystal is the chance formation of a seed, or an extremely small crystal (▶Fig. 5.12a–c). Once the seed exists, other atoms in the surrounding material attach themselves to the face of the seed. As the crystal grows, crystal faces move outward from the center of the seed but maintain the same orientation. Thus, the youngest part of the crystal is always its outer edge (▶Fig. 5.13a, b).

In crystals formed by the solidification of a melt, atoms begin to attach to the seed when the melt becomes so cool that thermal vibrations can no longer break apart the attraction between the seed and the atoms in the melt. Crystals formed by precipitation from a solution develop when the solution becomes saturated, meaning that the number of dissolved ions per unit volume of solution becomes so great that they can get close enough to each other to bond together. If a solution is not saturated, dissolved ions are surrounded by solvent molecules, which shield the ions from the attractive forces of their neighbors. Sometimes crystals formed by precipitation from a solution grow from the walls of the

FIGURE 5.12 **(a)** New crystals nucleate (begin to form) in a water solution. They grow inward from the walls of the container. **(b)** At a later time, the crystals have grown larger. **(c)** On a crystal face, atoms in the solution are attracted to the surface and latch on.

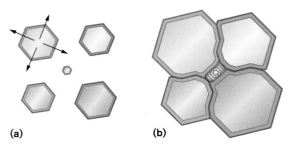
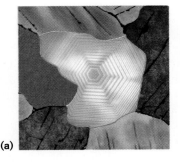
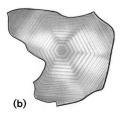

(a)   (b)

FIGURE 5.13 **(a)** Crystals grow outward from the central seed. **(b)** Crystals maintain their shape until they interfere with each other. When that happens, the crystal shapes can no longer be maintained.

(a)   (b)

FIGURE 5.15 A crystal growing in a confined space is anhedral, meaning its surface is not composed of crystal faces. **(a)** A crystal stops growing when it meets the surfaces of other grains and continues growing to fill in gaps. **(b)** The resulting mineral grain, if it were to be separated from other grains, would have an anhedral shape.

solution's "container" (e.g., a crack or pore in a rock). This process can form a spectacular **geode**, a mineral-lined cavity in rock (▶Fig. 5.14a).

As crystals grow, they develop their particular crystal shape, based on the geometry of their internal structure. The shape is defined by the relative dimensions of the crystal (needle-like, sheet-like, etc.) and the angles between crystal faces. If a mineral's growth is uninhibited so that it displays well-formed crystal faces, then it is a euhedral crystal (▶Fig. 5.14b). Typically, however, the growth of minerals is restricted in one or more directions because existing crystals act as obstacles. In such cases, minerals grow to fill the space that is available, and their shape is controlled by the shape of their surroundings. Minerals without well-formed crystal faces are anhedral grains (▶Fig. 5.15a, b). In rocks created by the solidification of melts, many crystals grow at about the same time, competing with each other for space. As a consequence, these minerals grow into each other, forming anhedral grains that interlock like pieces of a jigsaw puzzle.

A mineral can be destroyed by melting, dissolving, or some other chemical reaction. Melting involves heating a mineral to a temperature at which thermal vibration of the atoms or ions in the lattice can break the chemical bonds holding them to the lattice; the atoms or ions then separate, either individually or in small groups, to move around freely again. Dissolution occurs when you immerse a mineral in a solvent (such as water). Atoms or ions then separate from the crystal face and are surrounded by solvent molecules. Some minerals, such as salt, dissolve easily, but most do not dissolve much at all. Chemical reactions can destroy a mineral when it comes in contact with reactive materials. For example, iron-bearing minerals react with air and water to form rust. The action of microbes in the environment can also destroy minerals. In effect, microbes can "eat" certain minerals, using the energy stored in the chemical bonds that hold the atoms of the mineral together as a source of energy for metabolism.

> **Take-Home Message**
>
> In a crystalline material, atoms are arranged in a very regular pattern. Minerals are crystalline materials that form by several geologic processes, including solidification from a melt, precipitation from a solution, and diffusion in solids.

FIGURE 5.14 **(a)** A geode, in which euhedral crystals of purple quartz grow from the wall into the center. **(b)** An enlargement of a euhedral crystal, showing that the surfaces are crystal faces.

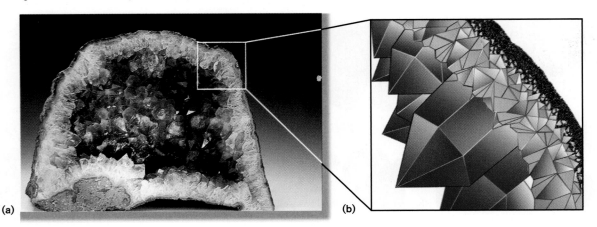

(a)   (b)

5.4 HOW CAN YOU TELL ONE MINERAL FROM ANOTHER?

Amateur and professional mineralogists alike get a kick out of recognizing minerals. They'll be the show-offs in a museum who hover around the display case, naming specimens without bothering to look at the labels. How do they do it? The trick lies in learning to recognize the basic **physical properties** (visual and material characteristics) that distinguish one mineral from another. Some physical properties, such as shape and color, can be seen from a distance—these are the properties that enable the show-offs to recognize specimens isolated behind the glass of a display case. Other properties, such as hardness and magnetization, can be determined only by handling the specimen or by performing an identification test on it. Identification tests include scratching the mineral against another object, placing it near a magnet, weighing it, tasting it, or placing a drop of acid on it. Let's examine some of the physical properties most commonly used in mineral identification. (Appendix B provides charts for identifying minerals on the basis of their physical properties.)

- *Color*: **Color** results from the way a mineral interacts with light. Sunlight contains the whole spectrum of colors; each color has a different wavelength. A mineral absorbs certain wavelengths, so the color you see when looking at a specimen represents the wavelengths the mineral does not absorb. Certain minerals always have the same color (galena is always gray, for example), but many show a range of colors. For example, quartz can be clear, white, purple, gray (smoky), red (rose), or just about anything in between (►Fig. 5.16). Purple quartz is known as amethyst (Fig. 5.14a). Color variations in a mineral reflect the presence of tiny amounts of impurities. For example, a trace of iron may produce a reddish or purple tint, and manganese may produce a pinkish tint.

FIGURE 5.16 The range of colors of quartz, displayed by different crystals: milky, clear, and rose quartz.

FIGURE 5.17 A streak plate, showing the red streak of hematite.

- *Streak*: The **streak** of a mineral refers to the color of a powder produced by pulverizing the mineral. You can obtain a streak by scraping the mineral against an unglazed ceramic plate (►Fig. 5.17). The color of a mineral powder tends to be less variable than the color of a whole crystal, and thus provides a fairly reliable clue to a mineral's identity. Calcite, for example, always yields a white streak even though pieces of calcite may be white, pink, or clear.

- *Luster*: **Luster** refers to the way a mineral surface scatters light. Geoscientists describe luster simply by comparing the appearance of the mineral with the appearance of a familiar substance. For example, minerals that look like metal have metallic luster, whereas those that do not have nonmetallic luster—the adjectives are self-explanatory (►Fig. 5.18a, b). Types of nonmetallic luster may be further described, for example, as silky, glassy, satiny, resinous, pearly, or earthy.

- *Hardness*: **Hardness** is a measure of the relative ability of a mineral to resist scratching, and therefore represents the resistance of bonds in the crystal structure to being broken. The atoms or ions in crystals of a hard mineral are more strongly bonded than those in a soft mineral. Hard minerals can scratch soft minerals, but soft minerals cannot scratch hard ones. Diamond is the hardest mineral known—it can scratch anything, which is why it is used to cut glass. In the early 1800s, a mineralogist named Friedrich Mohs listed some minerals in order of their relative hardness; a mineral with a hardness of 5 can scratch all minerals with a hardness of 5 or less. This list, now called the **Mohs hardness scale**, helps in mineral identification. When you use the scale (►Table 5.1), it might help to compare the hardness of a mineral with a common item such as your fingernail, a penny, or a glass plate. Note that not all of the minerals in Table 5.1 are common or familiar. Also, it's important to realize that the numbers on the

(a)

(b)

FIGURE 5.18 **(a)** This specimen of pyrite looks like a piece of metal because of its shiny gleam; we call this metallic luster. **(b)** These specimens of feldspar have a nonmetallic luster. The white one on the left is plagioclase, and the pink one on the right is orthoclase (potassium feldspar, or "K-spar").

TABLE 5.1 **Mohs Hardness Scale**

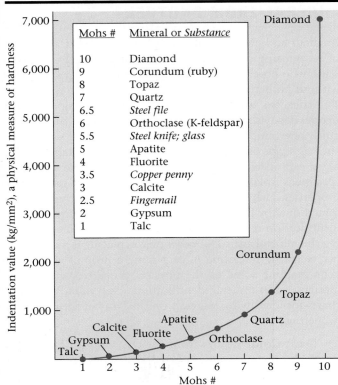

Mohs #	Mineral or *Substance*
10	Diamond
9	Corundum (ruby)
8	Topaz
7	Quartz
6.5	*Steel file*
6	Orthoclase (K-feldspar)
5.5	*Steel knife; glass*
5	Apatite
4	Fluorite
3.5	*Copper penny*
3	Calcite
2.5	*Fingernail*
2	Gypsum
1	Talc

Mohs hardness scale do not specify the true relative differences in hardness of minerals. For example, on the Mohs scale, talc has a hardness of 1 and quartz has a hardness of 7. But this does not mean that quartz is 7 times harder than talc. Careful tests show that quartz is actually about 100 times harder than talc, as indicated by how difficult it is to make an indentation in the mineral (see Table 5.1).

• *Specific gravity*: **Specific gravity** represents the density of a mineral, as specified by the ratio between the weight of a volume of the mineral and the weight of an equal volume of water at 4°C. For example, one cubic centimeter of quartz has a weight of 2.65 grams, whereas one cubic centimeter of water has a weight of

1.00 gram. Thus, the specific gravity of quartz is 2.65. Divers use lead weights to help them sink to great depths because lead is extremely heavy—it has a specific gravity of 11. In practice, you can develop a feel for specific gravity by hefting minerals in your hands. For example, a piece of galena (lead ore) "feels" heavier than a similar-sized piece of quartz.

• *Crystal habit*: The **crystal habit** of a mineral refers to the shape (morphology) of a single crystal with well-formed crystal faces, or to the character of an aggregate of many well-formed crystals that grew together as a group (▶Fig. 5.19). The habit depends on the internal arrangement of atoms in the crystal, for the arrangement, in turn, controls the geometry of crystal faces (e.g., triangular, square, rectangular, parallelogram) and the angular relationships among the faces. Mineralogists use a great many adjectives when describing habit. For example, a crystal may be compared with a geometric shape by using adjectives such as cubic or prismatic. A description of habit generally includes adjectives that define relative dimensions of the crystal. Crystals that are roughly the same length in all directions are called equant or blocky, those that are much longer in one dimension than in others are

(a)

(b)

(c)

FIGURE 5.19 Crystal habit refers to the shape or character of the mineral. **(a)** Kyanite, which generally occurs in clusters of blades. **(b)** Prismatic crystals. **(c)** A spray of needlelike, or fibrous, crystals.

columnar or needle-like, those shaped like sheets of paper are platy, and those shaped like knife blades are bladed. The relative dimensions depend on relative rates of crystal growth in different directions. A crystal that grows equally fast in three directions will be blocky, one that grows rapidly in two directions and slowly in the third direction will be platy, and one that grows rapidly in one direction but slowly in the other two directions will be needle-like. If the crystal grows as part of an aggregate, other adjectives can be used. For example, a group of needle-like crystals is a fibrous array, whereas a group of plates oriented in different directions is a rosette because it resembles a rose.

- *Fracture and cleavage*: Different minerals fracture (break) in different ways, depending on the internal arrangement of atoms. If a mineral breaks to form distinct planar surfaces that have a specific orientation in relation to the crystal structure, then we say that the mineral has **cleavage**, and we refer to each surface as a cleavage plane (▶Fig. 5.20a–e). Cleavage forms in directions where the bonds holding atoms together in the crystal are the weakest. Some minerals have one direction of cleavage. For example, mica has very weak bonds in one direction but strong bonds in the other two directions. Thus, it easily splits into parallel sheets; the surface of each sheet is a cleavage plane. Other minerals have two or three directions of cleavage that intersect at a specific angle. For example, halite has three sets of cleavage planes that intersect at right angles, so halite crystals break into little cubes. Specimens with good cleavage break easily along cleavage planes, whereas specimens with fair cleavage not only break on cleavage planes but

may break on other surfaces as well. Materials that have no cleavage at all (because bonding is equally strong in all directions) break either by forming irregular fractures or by forming conchoidal fractures. **Conchoidal fractures** are smoothly curving, clamshell-shaped surfaces; they typically form in quartz. Note that glass, which contains no crystal structure, can develop beautiful conchoidal fractures (▶Fig. 5.21).

One final note about cleavage: because both cleavage planes and crystal faces reflect light, it's easy to confuse them. Cleavage forms by breaking, so you can recognize a cleavage plane if it is one of several parallel planes arranged like steps. In a given orientation, there can be numerous parallel cleavage planes, but only one crystal face (▶Fig. 5.22a, b).

- *Special properties*: Some minerals have distinctive properties that readily distinguish them from other minerals. For example, calcite ($CaCO_3$) reacts with dilute hydrochloric acid (HCl) to produce carbon dioxide (CO_2) gas (▶Fig. 5.23); dolomite ($CaMg(CO_3)_2$) also reacts with acid, but not so strongly; graphite makes a gray mark on paper (it's the "lead" in pencils); magnetite attracts a magnet; halite tastes salty; and plagioclase has striations (thin parallel corrugations or stripes) on cleavage planes.

> **Take-Home Message**
>
> The physical characteristics of minerals (such as color, crystal shape, hardness, cleavage, and luster) are a manifestation of the crystal structure and chemical composition of minerals. You can identify a mineral by examining physical properties.

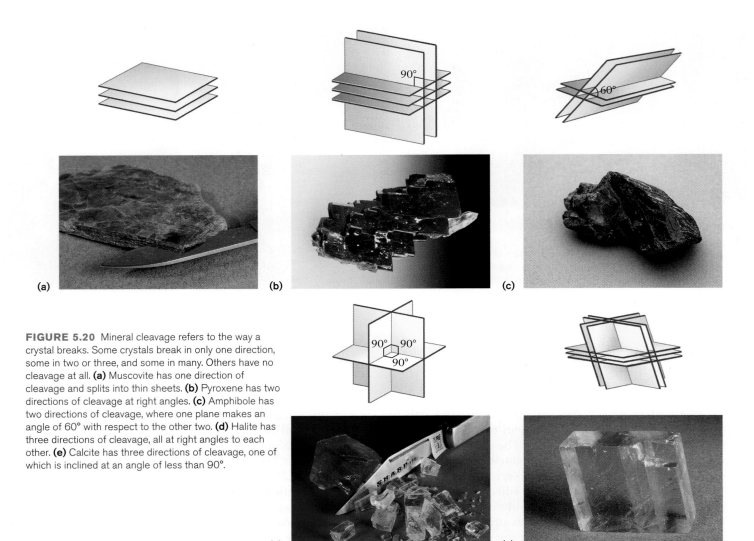

FIGURE 5.20 Mineral cleavage refers to the way a crystal breaks. Some crystals break in only one direction, some in two or three, and some in many. Others have no cleavage at all. **(a)** Muscovite has one direction of cleavage and splits into thin sheets. **(b)** Pyroxene has two directions of cleavage at right angles. **(c)** Amphibole has two directions of cleavage, where one plane makes an angle of 60° with respect to the other two. **(d)** Halite has three directions of cleavage, all at right angles to each other. **(e)** Calcite has three directions of cleavage, one of which is inclined at an angle of less than 90°.

FIGURE 5.21 Minerals without cleavage break on random fractures. Some materials, such as quartz and glass, break on conchoidal fractures, which have a curving, scoop shape. The photo shows concoidal fracture surfaces on volcanic glass.

FIGURE 5.22 You can distinguish between cleavage planes and crystal faces because **(a)** cleavage planes can be repeated, like a series of steps or terraces, whereas **(b)** a crystal face is a single surface. Note that there are no repetitions of the crystal face within a crystal.

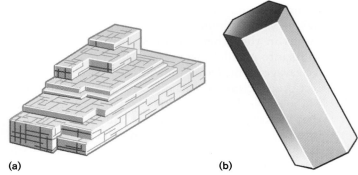

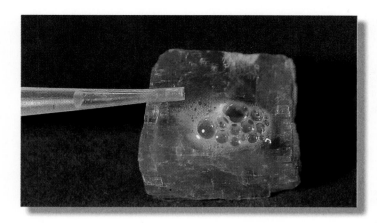

FIGURE 5.23 Calcite reacts with hydrochloric acid to produce carbon dioxide gas.

5.5 ORGANIZING OUR KNOWLEDGE: MINERAL CLASSIFICATION

Although there are close to 4,000 minerals already known, and 50 to 100 new minerals are identified every year, they can be separated into a relatively small number of groups, or mineral classes. You may think, "Why bother?" but classification schemes are useful because they help organize information and streamline discussion. Biologists, for example, classify animals into groups on the basis of how they feed their young and on the architecture of their skeletons; botanists classify plants according to the way they reproduce and by the shape of their leaves. In the case of minerals, a good means of classification eluded researchers until it became possible to determine accurately the chemical makeup of

minerals. A Swedish chemist, Baron Jöns Jacob Berzelius (1779–1848), analyzed the chemicals making up minerals and noted chemical similarities among many of them. Berzelius, along with his students, then established that most minerals can be classified by specifying the principal anion (negative ion) within the mineral. (Note again that some anions consist of single atoms, while others consist of a group of atoms that act as a unit; see Box 5.1.) We now take a look at these mineral groups, focusing especially on silicates, the class that constitutes most of the rock in the Earth.

The Mineral Classes

Mineralogists distinguish several principal classes of minerals on the basis of their chemical composition. Here are some of the major ones.

- *Silicates*: The fundamental component of most silicates in the Earth's crust is the SiO_4^{4-} anionic group, a silicon atom surrounded by four oxygen atoms that are arranged to define the corners of a tetrahedron, a pyramid-like shape with four triangular faces (▶Fig. 5.24a–d). Mineralogists commonly refer to this anionic group as the **silicon-oxygen tetrahedron**. We can identify a huge variety of silicate minerals that differ from each other in the way the tetrahedra link and in the cations present in the mineral. Olivine, a common example, has the formula $(Mg,Fe)_2SiO_4$. Another well-known example, quartz (Fig. 5.16), has the formula SiO_2. We will learn more about silicates in the next section.

- *Oxides*: Oxides consist of metal cations bonded to oxygen anions. Typical oxide minerals include magnetite (Fe_3O_4) and hematite (Fe_2O_3). Some oxides contain a

FIGURE 5.24 (a) The basic building block of a silicate mineral is the SiO_4^{4-} anionic group, the silicon-oxygen tetrahedron. In this compound, the oxygens occupy the corners of a tetrahedron, which can be depicted in different ways. **(b)** In this ball model, the small silicon atom is hidden, surrounded by larger oxygen atoms. **(c)** A tetrahedron has been superimposed on this ball-and-stick model. **(d)** Simple geometric sketches of quadrilatera.

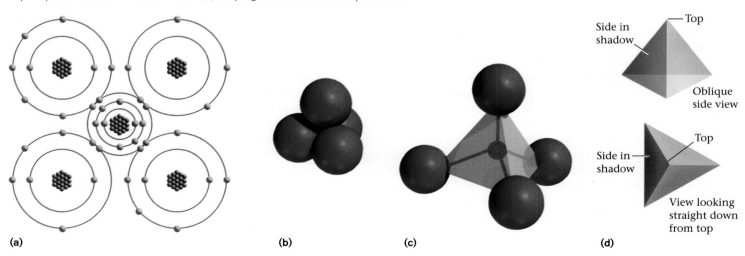

(a)   (b)   (c)   (d)

relatively high proportion of metal atoms, and thus are ore minerals.

- *Sulfides*: Sulfides consist of a metal cation bonded to a sulfide anion (such as S^{2-}). Examples include galena (PbS) and pyrite (FeS_2; Fig. 5.18a). As with oxides, the metal forms a high proportion of the mineral, so many sulfides are considered ore minerals. Many sulfide minerals have a metallic luster.

- *Sulfates*: Sulfates consist of a metal cation bonded to the SO_4^{2-} anionic group. Many sulfates form by precipitation out of water at or near the Earth's surface. An example is gypsum ($CaSO_4 \cdot 2H_2O$), in which water molecules bond to the calcium-sulfate molecules. Pulverized gypsum mixed with water can be spread out in thin sheets that harden when they dry. Contractors use these sheets as wallboard (Sheetrock) in houses.

- *Halides*: The anion in a halide is a halogen ion (such as chlorine [Cl^-] or fluorine [F^-]), an element from the second column from the right in the periodic table (see Appendix A). Halite, or rock salt (NaCl; Fig. 5.20d), and fluorite (CaF_2), a source of fluoride, are common examples.

- *Carbonates*: In carbonates, the molecule CO_3^{2-} serves as the anionic group. Elements like calcium or magnesium bond to this group. The two most common carbonates are calcite ($CaCO_3$; Fig. 5.20e) and dolomite ($CaMg[CO_3]_2$).

- *Native metals*: Native metals consist of pure masses of a single metal. The metal atoms have metallic bonds. Copper and gold, for example, may occur as native metals. A gold nugget is a mass of native gold that has been broken out of a rock.

Silicates: The Major Rock-Forming Minerals

Silicate minerals compose over 95% of the continental crust. Rocks making up oceanic crust and the Earth's mantle consist almost entirely of silicate minerals. Thus, silicate minerals are the most common minerals on Earth.

As noted earlier, most silicate minerals in the crust consist of combinations of a fundamental building block called the silicon-oxygen tetrahedron. (Silicate minerals in the mantle, where pressures are very high, have an octahedral structure.) The tetrahedra can link together, forming larger molecules, by sharing oxygen atoms. Silicate minerals are divided into seven groups, of which five are described below. Groups are distinguished from each other on the basis of how tetrahedra in the mineral link together. The number of links determines how many oxygen atoms are shared between tetrahedra and, therefore, the ratio of Si to O in the mineral. Each group includes several distinct minerals that differ from each other primarily in terms of the cations they contain.

- *Independent tetrahedra*: In this group, the tetrahedra are independent and do not share any oxygen atoms (▶Fig. 5.25a). The mineral is held together by the attraction between the tetrahedra and positive ions (metal cations, such as iron or magnesium). This group includes olivine, a glassy green mineral that usually occurs in small, sugarlike crystals. Garnet is also a member of this group.

- *Single chains*: In a single-chain silicate, the tetrahedra link to form a chain by sharing two oxygen atoms each (▶Fig. 5.25b). The most common of the many different types of single-chain silicates are pyroxenes, a group of black or dark-green minerals that occur in elongate crystals with two cleavage directions at 90° to one another (Fig. 5.20b).

- *Double chains*: In a double-chain silicate, the tetrahedra link to form a double chain by sharing two or three oxygen atoms (▶Fig. 5.25c). Amphiboles are the most common type; these typically are black or dark-brown elongate crystals, with two cleavage directions. You can distinguish amphiboles from pyroxenes because the cleavage planes of amphiboles lie at about 60° to one another (Fig. 5.20c).

- *Sheet silicates*: The tetrahedra in this group all share three oxygen atoms and therefore link to form two-dimensional sheets (▶Fig. 5.25d). Other ions and, in some cases, water molecules fit between the sheets in some sheet silicates. Because of their structure, sheet silicates have a single strong cleavage in one direction, and they occur in books of very thin sheets. In this group we find micas, a type of sheet silicate including muscovite (light-brown or clear mica; Fig. 5.20a) and biotite (black mica). Clay minerals are also sheet silicates; they have a crystal structure similar to that of mica, but clay occurs only in extremely tiny flakes.

- *Framework silicates*: In a framework silicate, each tetrahedron shares all four oxygen atoms with its neighbors, forming a 3-D structure (▶Fig. 5.25e). Examples include feldspar and quartz. The two most common types of feldspar are plagioclase, which tends to be white, gray, or blue; and orthoclase (also called potassium feldspar, or K-feldspar), which tends to be pink (Fig. 5.18b). Feldspars contain aluminum (which substitutes for silicon in the tetrahedra), as well as varying proportions of other elements, such as calcium, sodium, and potassium. Quartz, in contrast, contains only silicon and oxygen. As the ratio of silicon to oxygen in quartz is 1:2, the mineral has the familiar formula SiO_2.

> **Take-Home Message**
>
> Mineralogists classify minerals into a number of different classes (such as silicates, carbonates, oxides, and sulfides) on the basis of their chemical composition. Silicates are the most abundant. Inside, they consist of arrangements of silicon-oxygen tetrahedra.

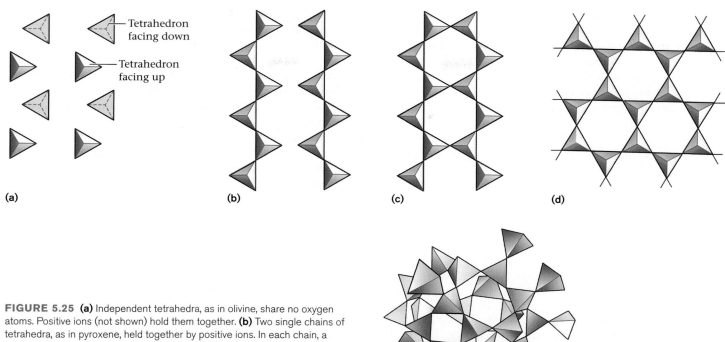

FIGURE 5.25 **(a)** Independent tetrahedra, as in olivine, share no oxygen atoms. Positive ions (not shown) hold them together. **(b)** Two single chains of tetrahedra, as in pyroxene, held together by positive ions. In each chain, a tetrahedron shares two oxygens. **(c)** A double chain of tetrahedra, as in amphibole. Here, two single chains link by sharing oxygens. Some tetrahedra share two oxygens, some share three. **(d)** A sheet of tetrahedra, as in mica. Each tetrahedron shares three oxygens. **(e)** A 3-D network (framework) of tetrahedra. Note that within the framework, each tetrahedron shares all four oxygens with its neighbors.

(e)

5.6 SOMETHING PRECIOUS—GEMS!

Mystery and romance follow famous gems. Consider the stone now known as the Hope Diamond, recognized by name the world over. No one knows who first dug it out of the ground. Was it mined in the 1600s, or was it stolen off an ancient religious monument? What we do know is that in the 1600s, a French trader named Jean Baptiste Tavernier obtained a large (112.5 carats, where 1 carat ≈ 200 milligrams), rare blue diamond in India, perhaps from a Hindu statue, and carried it back to France. King Louis XIV bought the diamond and had it fashioned into a jewel of 68 carats. This jewel vanished in 1762 during a burglary. Perhaps it was lost forever—perhaps not. In 1830, a 44.5-carat blue diamond mysteriously appeared on the jewel market for sale. Henry Hope, a British banker, purchased the stone, which then became known as the Hope Diamond (▶Fig. 5.26). It changed hands several times until 1958, when the famous New York jeweler Harry Winston donated it to the Smithsonian Institution in Washington, D.C., where it now sits behind bulletproof glass in a heavily guarded display. Legend has it that whoever owns the stone suffers great misfortune.

What makes stones like the Hope Diamond so special that people risk life and fortune to obtain them? What is the difference between a gemstone, a gem, and any other mineral? A gemstone is a mineral that has

FIGURE 5.26 The Hope Diamond, now on display in the Smithsonian Institution, Washington, D.C.

special value because it is rare and people consider it beautiful. A **gem** is a cut and finished stone ready to be used in jewelry. Jewelers distinguish between precious stones (e.g., diamond, ruby, sapphire, emerald), which are particularly rare and expensive, and semiprecious stones (e.g., topaz, tourmaline, aquamarine, garnet), which are less rare and less expensive. All the stones mentioned so far are transparent crystals, though most have some color (see ▶Table 5.2). The category of semiprecious stones also includes opaque or translucent minerals such as lapis, malachite (Fig. 5.2a), and opal.

In everyday language, pearls and amber may also be considered gemstones. Unlike diamonds and garnets, which form inorganically in rocks, pearls form in living oysters when the oyster extracts calcium and carbonate ions from water and precipitates them around an impurity, such as a sand grain, embedded in its body. Thus, they are a result of biomineralization. Most pearls used in jewelry today are "cultured" pearls, made by artificially introducing round sand grains into oysters in order to stimulate round pearl production. Amber is also formed by organic processes—it consists of fossilized tree sap. But because amber consists of organic compounds that are not

arranged in a crystal structure, it does not meet the definition of a mineral.

Rare means hard to find, and some gemstones are indeed hard to find. Many diamond localities, for example, occur in isolated regions of Congo, South Africa, Brazil, Canada, Russia, India, and Borneo (▶Box 5.3). In some cases, it is not the mineral itself but rather the "gem-quality" versions of the mineral that are rare. For example, garnets are found in many rocks in such abundance that people use them as industrial abrasives. But most garnets are quite small and contain inclusions (specks of other minerals and/or bubbles) or fractures, so they are not particularly beautiful. Gem-quality garnets—clean, clear, large, unfractured crystals—are unusual. In some cases, gemstones are merely pretty and rare versions of more common minerals. For example, ruby is a special version of the common mineral corundum, and emerald is a special version of the common mineral beryl (▶Fig. 5.28). As for the beauty of a gemstone, this quality lies basically in its color and, in the case of transparent gems, its "fire"—the way the stone bends and internally reflects the light passing through it and disperses the light into a spectrum. Fire makes a diamond sparkle more than a similarly cut piece of glass.

TABLE 5.2   Precious and Semi-Precious Materials

Gem Name	Material/Formula	Comments
Amber	Fossilized tree sap	Composed of organic chemicals; amber is not strictly a mineral.
Amethyst	Quartz/SiO_2	The best examples precipitate from water in openings in igneous rocks; a deep-purple version of quartz.
Aquamarine	Beryl/$Be_3Al_2Si_6O_{18}$	A bluish version of emerald.
Diamond	Diamond/C	Brought to the surface from the mantle in igneous bodies called diamond pipes; may later be mixed in deposits of sediment.
Emerald	Beryl/$Be_3Al_2Si_6O_{18}$	Occurs in coarse igneous rocks (pegmatites) (see Chapter 6).
Garnet	Garnet/(e.g., $Mg_3Al_2[SiO_4]_3$)	A variety of types differ in composition (Ca, Fe, Mg, and Mn versions); occurs in metamorphic rocks (see Chapter 8).
Jade	Jadeite/$NaAlSi_2O_6$ Nephrite/$Ca_2(Mg,Fe)_5Si_8O_{22}(OH)_2$	Jade can be one of two minerals, jadeite (a pyroxene) or nephrite (an amphibole); both occur in metamorphic rocks.
Opal	Composed of microscopic spheres of hydrated silica packed together	Most opal comes from a single mining district in central Australia; occurs in bedrock that has reacted with water near the surface.
Pearl	Aragonite/$CaCO_3$	Formed by oysters, which secrete coatings around sand grains that are accidentally embedded in the soft parts of the organism. Cultured pearls are formed the same way, but the impurity is a spherical bead that is intentionally introduced.
Ruby	Corundum/Al_2O_3	The red color is due to chromium impurities; found in coarse igneous rocks called pegmatites and as a result of contact metamorphism (see Chapters 6 and 8).
Sapphire	Corundum/Al_2O_3	A blue version of ruby.
Topaz	$Al_2SiO_4(F,OH)_2$	Found in igneous rocks, and as a result of the reaction of rock with hot water.
Tourmaline	$Na(Mg,Fe)_3Al_6(BO_3)_3(Si_6O_{18})(OH,F)_4$	Forms in igneous and metamorphic rocks.
Turquoise	$CuAl_6(PO_4)_4(OH)_8 \cdot 4H_2O$	Found in copper-bearing rocks; a popular jewelry gem in the American Southwest.

BOX 5.3 THE REST OF THE STORY

Where Do Diamonds Come From?

As we saw earlier, diamond consists of the element carbon. Accumulations of carbon develop in a variety of ways: soot (pure carbon) results from burning plants at the surface of the Earth; coal (which consists mostly of carbon) forms from the remains of plants buried to depths of up to 15 km; and graphite develops from coal or other organic matter buried to still greater depths (15–70 km) in the crust during mountain building. Experiments demonstrate that the temperatures and pressures needed to form diamond are so extreme that, in nature, they generally occur only at depths of around 150 km below the Earth's surface—that is, in the mantle. Under these conditions, the carbon atoms that were arranged in hexagonal sheets in graphite rearrange to form the much stronger and more compact structure of diamond. (Note that engineers can duplicate these conditions in the laboratory; corporations manufacture several tons of diamonds a year.)

How does carbon get down into the mantle, where it transforms into diamond? Geologists speculate that the process of subduction provides the means. Carbon-containing rocks and sediments in oceanic lithosphere plates at the Earth's surface can be carried into the mantle at a convergent plate boundary. This carbon transforms into diamond, some of which becomes trapped in the lithospheric mantle beneath continents. But if diamonds form in the mantle, then how do they return to the surface? One possibility is that the process of rifting cracks the continental crust and causes a small part of the underlying lithospheric mantle to melt. Magma generated during this process rises to the surface, bringing the diamonds with it. Near the surface, the magma cools and so-lidifies to form a special kind of igneous rock called kimberlite (named for Kimberley, South Africa, where it was first found). Diamonds brought up with the magma are frozen into the kimberlite. Kimberlite magma contains a lot of dissolved gas and thus froths to the surface very rapidly. Kimberlite rock commonly occurs in carrot-shaped bodies 50 to 200 m across and at least 1 km deep that are called kimberlite pipes (▶Fig. 5.27a).

Controversial measurements suggest that many of the diamonds that sparkle on engagement rings today were created when subduction carried carbon into the mantle 3.2 billion years ago. The diamonds sat at depths of 150 km in the Earth until two rifting events, one of which took place in the late Precambrian and the other during the late Mesozoic, released them to the surface, like genies out of bottles. The Mesozoic rifting event led to the breakup of Pangaea.

In places where diamonds occur in solid kimberlite, they can be obtained only by digging up the kimberlite and crushing it, to separate out the diamonds (▶Fig. 5.27b). But nature can also break diamonds free from the Earth. In places where kimberlite has been exposed at the ground surface for a long time, the rock chemically reacts with water and air (a process called weathering; see Chapter 7). These reactions cause most minerals in kimberlite to disintegrate, creating sediment that washes away in rivers. Diamonds are so strong that they remain as solid grains in river gravel. Thus, many diamonds have been obtained simply by separating them from recent or ancient river gravel.

Diamond-bearing kimberlite pipes are found in many places around the world, particularly where very old continental litho-sphere exists. Southern and central Africa, Siberia, northwestern Canada, India, Brazil, Borneo, Australia, and the U.S. Rocky Mountains all have pipes. Rivers and glaciers, however, have transported diamond-bearing sediments great distances from their original sources. In fact, diamonds have even been found in farm fields of the midwestern United States. Not all natural diamonds are valuable: value depends on color and clarity. Diamonds that contain imperfections (cracks, or specks of other material), or are dark gray in color, won't be used for jewelry. These stones, called *industrial diamonds*, are used instead as abrasives, for diamond powder is so hard (10 on the Mohs hardness scale) that it can be used to grind away any other substance.

Gem-quality diamonds come in a range of sizes. Jewelers measure diamond size in carats, where one "carat" equals 200 milligrams (0.2 grams)—one ounce equals 142 carats. (Note that a carat measures gemstone weight, whereas a "karat" specifies the purity of gold. Pure gold is 24 karat; 18-karat gold is an alloy containing 18 parts of gold and 6 parts of other metals.) The largest diamond ever found, a stone called the Cullinan Diamond, was discovered in South Africa in 1905. It weighed 3,106 carats (621 grams) before being cut into nine large gems (the largest weighing 516 carats) as well as many smaller ones. In comparison, the diamond on a typical engagement ring weighs less than 1 carat. Diamonds are rare, but not so rare as their price suggests. A worldwide consortium of diamond producers stockpiles the stones so as not to flood the market and drive the price down.

(a)

(b)

FIGURE 5.27 (a) A mine in a kimberlite pipe. **(b)** A raw diamond still imbedded in kimberlite.

See for yourself . . .
Diamond Mines

Diamonds fetch such a high price that prospectors seek them even in very remote areas. In the 1860s, attention focused on Kimberley, South Africa—at the time, a very remote area. In recent years, new deposits have been found in Arctic Canada. The thumbnail images provided on this page are only to help identify tour sites. Go to wwnorton.com/studyspace to experience this flyover tour.

Kimberley, South Africa
(Lat 28°44'17.06"S, Long 24°46'30.77"E)

Fly to the coordinates provided and zoom out to an elevation of about 15 km (9 miles). You will see the town of Kimberley in the dry interior of South Africa. Zoom down to an elevation of 3 km (2 miles). You will be hovering over a large circular pit (Image G5.1). This is one of the diamond mines for which Kimberley is famous.

G5.1

Mine near Yellowknife, Canada
(Lat 64°43'14.74"N, Long 110°37'32.76"W)

At these coordinates, you will see a remote region of Arctic Canada in which prospectors found diamond pipes in the early 1990s, after a 20-year search. If you zoom to an altitude of 200 km (125 miles), you can get a feel for the tundra landscape of scrubby vegetation and frigid lakes (Image G5.2). From an altitude of 10 km (6 miles), you can see one of the mines where a pipe is being excavated (Image G5.3). For the story of this discovery, please see Krajick, K., 2001, *Barren Lands*; the complete reference is at the end of the chapter.

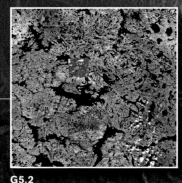

G5.2

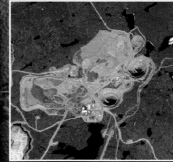

G5.3

Diamond Prospect, Diamantina, Brazil
(Lat 18°15'4.35"S, Long 43°34'57.21"W)

Fly to the coordinates, and hover at 15 km (9 miles). You are looking at the town of Diamantina, nestled in the Espinhaço Range of eastern Brazil (Image G5.4). The town's name means "diamond," and for a good reason. Prospectors (known as *Bandeirantes*) discovered diamonds at the locale in the 1720s, and the stones have been mined ever since. These diamonds are detrital—in the Precambrian, they weathered out of kimberlites and then were transported as sediment. Eventually, they accumulated with quartz grains and pebbles to form a thick unit of quartzite and conglomerate. These rocks form the ridges of the Espinhaço. Weathering of this rock frees the diamonds once again. Zoom to 3 km (2 miles), tilt, and look north, and you see a small prospect pit (Image G5.5)

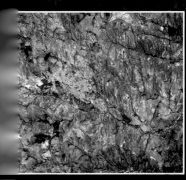

G5.4

G5.5

Gemstones form in many ways. Some solidify from a melt along with other minerals of igneous rock; some form by diffusion in a metamorphic rock; some precipitate out of a water solution in cracks; and some are a consequence of the chemical interaction of rock with water near the Earth's surface. Many gems come from pegmatites, particularly coarse-grained igneous rocks formed by the solidification of a steamy melt.

Most gems used in jewelry are "cut" stones. This does not mean that they have been sliced by a knife. Rather, the smooth **facets** on a gem are ground and polished surfaces made with a faceting machine (▶Fig. 5.29a). Facets are not the natural crystal faces of the mineral, nor are they cleavage planes, though gem cutters sometimes make the facets parallel to cleavage directions and will try to break a large gemstone into smaller pieces by splitting it on a cleavage plane. A faceting machine consists of a doping arm, a device that holds a stone in a specific orientation, and a lap, a rotating disk covered with a wet paste of grinding powder and water. The gem cutter fixes a gemstone to the end of the doping arm and positions the arm so that it holds the stone against the moving lap. The movement of the lap grinds a facet. When the facet is complete, the gem cutter rotates the arm by a specific angle, lowers the stone, and grinds another facet. The geometry of the facets defines the cut of the stone. Different cuts have different names, such as "brilliant," "French," "star," "pear," and "kite." Grinding facets is a lot of work—a typical engagement-ring diamond with a brilliant cut has fifty-seven facets (▶Fig. 5.29b)!

Some mineral specimens have special value simply because their geometry and color before cutting are beautiful. Prize specimens exhibit shapes and colors reminiscent of fine art and may sell for tens of thousands of dollars (▶Fig. 5.30). It's no wonder that mineral "hounds" risk their necks looking for a cluster of crystals protruding from the dripping roof of a collapsing mine or hidden in a crack near the smoking summit of a volcano.

> ### Take-Home Message
>
> Gems are minerals that have special value because of their beauty and rarity. Some gems are just particularly good specimens of more common minerals. The "cut" gems that you find at a jeweler's have been faceted by grinding.

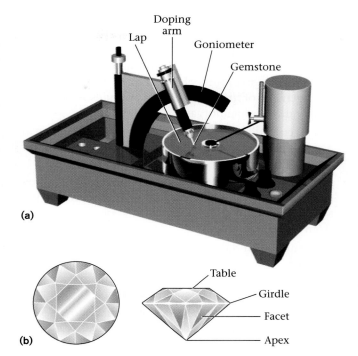

(a)

(b)

FIGURE 5.29 The shiny faces on gems in jewelry are made by a faceting machine. **(a)** In this faceting machine, the gem is held against the face of the spinning lap. **(b)** Top and side views show the many facets of a brilliant-cut diamond, and names for different parts of the stone.

FIGURE 5.30 A spectacular museum specimen of a mineral cluster. The arrangement of colors and shapes is like abstract art.

FIGURE 5.28 Beryl crystals in rock.

Chapter Summary

- Minerals are homogeneous, naturally occurring, solid substances with a definable chemical composition and an internal structure characterized by an orderly arrangement of atoms, ions, or molecules in a lattice. Most minerals are inorganic.

- In the crystalline lattices of minerals, atoms occur in a specific pattern—one of nature's finest examples of ordering.

- Minerals can form by the solidification of a melt, precipitation from a water solution, diffusion through a solid, the metabolism of organisms, or precipitation from a gas.

- There are about 4,000 different known types of minerals, each with a name and distinctive physical properties (color, streak, luster, hardness, specific gravity, crystal form, crystal habit, and cleavage).

- The unique physical properties of a mineral reflect its chemical composition and crystal structure. By observing these physical properties, you can identify minerals.

- The most convenient way of classifying minerals is to group them according to their chemical composition. Mineral classes include the following: silicates, oxides, sulfides, sulfates, halides, carbonates, and native metals.

- The silicate minerals are the most common on Earth. The silicon-oxygen tetrahedron, a silicon atom surrounded by four oxygen atoms, is the fundamental building block of silicate minerals.

- There are several groups of silicate minerals, distinguished from each other by the ways in which the silicon-oxygen tetrahedra that constitute them are linked.

- Gems are minerals known for their beauty and rarity. The facets on cut stones used in jewelry are made by grinding and polishing the stone with a faceting machine.

Geopuzzle Revisited

The geologic definition of a mineral is much narrower than the definition used in everyday conversation. Just because something is not or was not alive doesn't mean that it's a mineral. Minerals have an orderly internal crystalline structure and must have formed by geologic processes.

Key Terms

biogenic minerals (p. 122)
cleavage (p. 133)
color (p. 131)
conchoidal fractures (p. 133)
crystal (p. 124)
crystal faces (p. 124)
crystal habit (p. 132)
crystal lattice (p. 123)
crystal structure (p. 125)
diffraction (p. 128)
facets (p. 141)
gem (p. 138)
geode (p. 130)
hardness (p. 131)

luster (p. 131)
mineral (p. 122)
mineralogists (p. 121)
mineralogy (p. 122)
Mohs hardness scale (p. 131)
physical properties (p. 131)
polymorphs (p. 127)
silicate minerals (p. 136)
silicon-oxygen tetrahedron (p. 135)
specific gravity (p. 132)
streak (p. 131)
symmetry (p. 127)

Review Questions

1. What is a mineral, as geologists understand the term? How is this definition different from the everyday usage of the word?

2. Why is glass not a mineral?

3. Salt is a mineral, but the plastic making an inexpensive pen is not. Why not?

4. Diamond and graphite have an identical chemical composition (pure carbon), yet they differ radically in physical properties. Explain in terms of their crystal structure.

5. In what way does the arrangement of atoms in a mineral define a pattern? How can X-rays be used to study these patterns?

6. Describe the several ways that mineral crystals can form.

7. Why do some minerals contain beautiful euhedral crystals, whereas others contain anhedral grains?

8. List and define the principal physical properties used to identify a mineral.

9. Give the chemical formulas of the following important minerals: quartz, halite, calcite.

10. What holds atoms together in a mineral?

11. Discuss the shape of crystals, including angular relations between crystal faces. What factors control crystal shape?

12. How can you determine the hardness of a mineral? What is the Mohs hardness scale?

13. How do you distinguish cleavage surfaces from crystal faces on a mineral? How does each type form?

14. What is the prime characteristic that geologists use to separate minerals into classes?

15. What is the principal anionic group in most familiar silicate minerals? On what basis are silicate minerals further divided into distinct groups?

16. What is the relationship between the way in which silicon-oxygen tetrahedra bond in micas and the characteristic cleavage of micas?

17. How do sulfate minerals differ from sulfides?

18. Why are some minerals considered gems? How do you make the facets on a gem?

On Further Thought

1. The mineral olivine can exist in the crust and the upper mantle. In the very deep mantle, the elements that make up olivine rearrange, in effect, to form a different crystal structure whose atoms are packed more tightly together than in olivine. Why? How do you think this change affects the density of the mantle?

2. Compare the chemical formula of magnetite with that of biotite. Why is magnetite mined as iron ore, but biotite is not?

3. Imagine that you are given two milky white crystals, each about 2 cm across. You are told that one of the crystals is composed of plagioclase and the other of quartz. How can you determine which is which?

4. Could you use crushed calcite to grind and form facets on a diamond? Why or why not?

Suggested Reading

Campbell, G. 2002. *Blood Diamonds: Tracing the Deadly Path of the World's Most Precious Stones*. Boulder: Westview Press.

Ciprianni, C., A. Borelli, and K. Lyman, eds. 1986. *Simon and Schuster's Guide to Gems and Precious Stones*. New York: Simon & Schuster.

Deer, W. A., J. Zussman, and R. A. Howie. 1996. *An Introduction to the Rock-Forming Minerals*, 2nd ed. Boston: Addison-Wesley.

Hart, M. 2002. *Diamond: A Journey to the Heart of an Obsession*. New York: Dutton/Plume.

Hibbard, J. J. 2001. *Mineralogy: A Geologist's Point of View*. New York: McGraw-Hill.

Klein, C., C. S. Hurlbut, and J. D. Dana. 2001. *The Manual of Mineral Sources*, 22nd ed. New York: John Wiley & Sons.

Krajick, K. 2001. *Barren Lands: An Epic Search for Diamonds in the North American Arctic*. New York: Henry Holt and Co.

Kurlansky, M. 2003. *Salt: A World History*. New York: Penguin USA.

Matlins, A. L., and A. C. Bonanno. 2003. *Gem Identification Made Easy*, 2nd ed. Woodstock, Vt.: GemStone Press.

Nesse, W. D. 2000. *Introduction to Mineralogy*. New York: Oxford University Press.

——. 2003. *Introduction to Optical Mineralogy*, 3rd ed. Oxford: Oxford University Press.

Perkins, D. 2001. *Mineralogy*, 2nd ed. Upper Saddle River, N.J.: Pearson Education.

Perkins, D., and K. R. Henke. 1999. *Minerals in Thin Sections*. Upper Saddle River, N. J.: Pearson Education.

ANOTHER VIEW Sapphires come in many colors, and can be cut into many shapes.

Rock Groups

It took years of back-breaking labor for nineteenth-century workers to chisel and chip ledges and tunnels through the hard rock of the Sierra Nevada in their quest to run a rail line across this rugged range. In the process, the workers became very familiar with the nature of rock.

B.1 INTRODUCTION

During the 1849 gold rush in the Sierra Nevada Mountains of California, only a few lucky individuals actually became rich. The rest of the "forty-niners" either slunk home in debt or took up less glamorous jobs in new towns such as San Francisco. These towns grew rapidly, and soon the American west coast was demanding large quantities of manufactured goods from East Coast factories. Making the goods was no problem, but getting them to California meant either a stormy voyage around the southern tip of South America or a trek with stubborn mule teams through the deserts of Nevada or Utah. The time was ripe to build a railroad linking the east and west coasts of North America, and, with much fanfare, the Central Pacific line decided to punch one right through the peaks of the Sierras. In 1863, while the Civil War raged elsewhere in the United States, the company transported six thousand Chinese laborers across the Pacific in the squalor of unventilated cargo holds and set them to work chipping ledges and blasting tunnels. Foremen measured progress in terms of feet per day—if they were lucky. Along the way, untold numbers of laborers died of frostbite, exhaustion, mistimed blasts, landslides, or avalanches.

Through their efforts, the railroad laborers certainly gained an intimate knowledge of how rock feels and behaves—it's solid, heavy, and hard! They also found that some rocks seemed to break easily into layers whereas others did not, and that some rocks were dark-colored while others were light. They realized, like anyone who looks closely at rock exposures, that rocks are not just gray, featureless masses, but rather come in a great variety of colors and textures. Why are there so many distinct types of rocks? The answer is simple: rocks can form in many different ways, and rocks can be made out of many different materials. Because of the relationship between rock type and the process of formation, *rocks provide a historical record of geologic events and give insight into interactions among components of the Earth System.* The next few chapters are devoted to a discussion of rocks and a description of how rocks form; this interlude is a general introduction. Here, we learn what the term *rock* means to geologists, what rocks are made of, and how to distinguish the three principal groups of rocks. We also look at how geologists study rocks.

B.2 WHAT IS ROCK?

To geologists, **rock** is a coherent, naturally occurring solid, consisting of an aggregate of minerals or, less commonly, a mass of glass. Now let's take this definition apart.

- *Coherent*: A collection of unattached grains (for example, the sand on a beach) does not constitute a rock. A rock holds together, and it must be broken to separate it into pieces. As a result of its coherence, rock can form cliffs and can be carved into sculptures.

- *Naturally occurring*: Geologists consider only naturally occurring materials formed by geologic processes to be rocks. Thus, manufactured materials such as concrete and brick do not qualify. A minor point: the term *stone* usually refers to rock used as a construction material.

- *An aggregate of minerals or a mass of glass*: The vast majority of rocks consist of an aggregate (a collection) of many mineral grains, or crystals, stuck or grown together. (Note that a grain is any fragment or piece of mineral, rock, or glass. A crystal is a piece of a mineral that grew into its present shape. In casual discussion, geologists may use the word *grain* to include crystals.) Technically, a single mineral crystal is simply a "mineral specimen," not a rock, even if it is meters long. Some rocks contain only one kind of mineral, whereas others contain several different kinds. A few of the rock types that form at volcanoes (see Chapter 6) consist of glass, which may occur either as a homogeneous mass or as an accumulation of tiny flakes.

What holds rock together? Grains in nonglassy rock stick together to form a coherent mass either because they are bonded by natural **cement**, mineral material that precipitates from water and fills the space between grains (▶Fig. B.1a–c), or because they interlock with each other

(a)

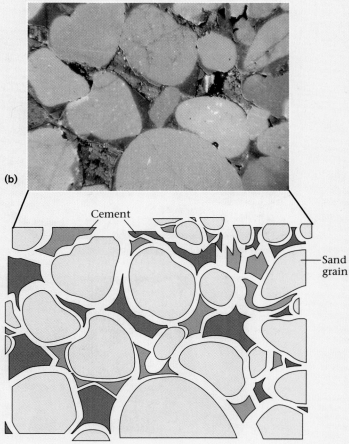

(b)

(c)

FIGURE B.1 (a) Hand specimens of sandstone. (b) A magnified image of the sandstone shows that it consists of round white sand grains, surrounded by cement. (c) This exploded image of the thin section emphasizes how the cement surrounds the sand grains. We depict the cement using two different colors because the cement contains two components formed at different times.

like pieces in a jigsaw puzzle (►Fig. B.2a–d). Rocks whose grains are stuck together by cement are called **clastic**; rocks whose crystals interlock with each other are called **crystalline**. Glassy rocks hold together either because they originate as a continuous mass (that is, they have no separate grains) or because glassy grains welded together while still hot.

All rocks, in the most basic sense, are just masses of chemicals bonded in molecules of varying size and complexity. But not all rocks contain the same chemicals. For example, granite—a rock commonly used for gravestones, building facades, and kitchen counters—contains oxygen, silicon, aluminum, calcium, iron, magnesium, and potassium. Marble—a rock favored by fine sculptors—contains oxygen, carbon, and calcium. Note, as we pointed out in Chapter 2 (Fig. 2.14), that the elements oxygen and silicon are the most common elements in the Earth's crust; indeed, oxygen constitutes 93.8% of the volume of the crust. It is no surprise, therefore, that most of the rock in the crust as a whole consists of silicate minerals (minerals containing the silicon-oxygen tetrahedron). Very close to the

Earth's surface, however, the activity of life plays a role in rock formation (see Chapter 7), so a significant proportion of bedrock exposed at the surface of the Earth consists of carbonate minerals (minerals containing the CO_3^- ion) extracted from water to form shells. Other minerals (such as oxides, sulfides, sulfates) are important as resources for metals and industrial materials, but they constitute only a small percentage of rocks in the crust.

B.3 ROCK OCCURRENCES

At the surface of the Earth, rock occurs either as broken chunks (pebbles, cobbles, or boulders; see Chapter 7) that have moved from their point of origin by falling down a slope or by being transported in ice, water, or wind; or as **bedrock**, which is still attached to the Earth's crust. Geologists refer to an exposure of bedrock as an **outcrop**. An outcrop may appear as a rounded knob out in a field, as a ledge along a cliff or ridge, on the face of a stream cut

FIGURE B.2 **(a)** A hand specimen of granite, a rock formed when melt cools underground. **(b)** A photomicrograph (a photo taken through a microscope) shows that the texture of granite is different from that of sandstone. In granite, the grains interlock with each other, like pieces of a jigsaw puzzle. **(c)** An artist's sketch emphasizes the irregular shapes of grains and how they interlock. **(d)** This exploded image highlights individual grain shapes.

(where a river has cut down into bedrock), or along human-made road cuts and excavations (▶Fig. B.3a–d).

To people who live in cities or forests or on farmland, outcrops of bedrock may be unfamiliar, because the outcrops may be covered by vegetation, sand, mud, gravel, soil, water, asphalt, concrete, or buildings. Outcrops are particularly rare in regions such as the midwestern United States, where, during the past million years, ice-age glaciers melted and left behind thick deposits of debris (see Chapter 22).

These deposits completely buried preexisting valleys and hills, so today the bedrock surface lies as deep as 100 m below the ground. The depth of bedrock plays a key role in urban planning, because architects prefer to set the foundations of large buildings on bedrock rather than on loose sand or mud. Because of this preference, the skyscrapers of New York City rise in two clusters on the island of Manhattan, one at the south end and the other in the center, locations where bedrock lies close to the surface.

FIGURE B.3 **(a)** Outcrops (natural rock exposures) in a field along the coast of Scotland. **(b)** A stream cut, which is an outcrop that forms when a stream's flow removes overlying soil and vegetation. Note that dense forest covers most of the adjacent hills along this stream cut in Brazil, obscuring outcrops that may be exposed there. **(c)** Road cuts, such as this one along a highway near Kingston, New York, are made by setting off dynamite placed at the bottom of drill holes. Note that the layers of rock exposed in this road cut are curved—such a bend is called a fold (see Chapter 11). **(d)** Mountain cliffs provide immense exposures of rock. These cliffs, in the Grand Teton Mountains of Wyoming, rise above a lowland in which bedrock has been covered by a layer of sediment, which hosts fields of sagebrush.

(a)

(b)

(c)

(d)

B.4 THE BASIS OF ROCK CLASSIFICATION

In the eighteenth century, geologists struggled to develop a sensible way to classify rocks, for they realized, as did miners from centuries past, that not all rocks are the same. One of the earliest classification schemes divided rocks into three groups—primary, secondary, and tertiary—on the basis of a perception (later proved incorrect) that the groups had formed in succession. In light of this concept, a German mineralogist named Abraham Werner proposed that a "universal ocean" containing dissolved and suspended minerals once had covered the Earth. According to Werner, the earliest rocks formed by precipitation from this solution; later, as sea level dropped, the action of rivers, waves, and wind wore down exposed rocks and produced debris that consolidated to form younger rocks. Werner was an influential teacher, and his followers came to be known as the Neptunists, after the Roman god of the sea.

At about the same time that Werner was developing his ideas, a Scottish gentleman farmer and doctor named James Hutton began exploring the outcrops of his native land. Hutton was a keen intellect who lived in Edinburgh, a hotbed of intellectual argument during the Age of Enlightenment where everything from political institutions to scientific paradigms became fodder for debate. He associated with prominent philosophers and scientists in Edinburgh and, like them, was open to new ideas. Hutton began to ponder the issue of how rocks formed; rather than force his perceptions to fit established dogma, he developed alternative ideas based on his own observations. For example, he watched sand settle on a beach and realized that some rocks could have formed from cementation of clasts. He examined exposures in which bodies of certain crystalline rocks appeared to have pushed into other rocks, heating the other rocks in the process, and concluded that some rocks could have formed by solidification from a melt. He also noticed that rocks adjacent to bodies of now solid melts had somehow been altered, and he proposed that these rocks formed by change of preexisting rocks as a result of what he referred to as "subterranean heat." Hutton, like Werner, attracted followers—Hutton's group came to be known as the Plutonists, after the Roman god of the underworld, because they favored the idea that the formation of certain rocks involved melts that had risen from deeper within the Earth.

In the last decades of the eighteenth century, as the armed rebellions that led to the formation of the United States and the Republic of France raged, a battle of ideas concerning the origin of rocks rattled the infant science of geology. This battle, pitting the Neptunists against the Plutonists, lasted for years. In the end the Plutonists won, for they demonstrated beyond a shadow of a doubt that certain crystalline rocks must have been in molten form when emplaced. As a consequence, it became clear that different rocks formed in different ways, and that as a starting point, rocks can best be *classified on the basis of how they formed*. This principle became the basis of the modern system of classification of rocks. For this contribution, among many others that we will describe later in the book, modern geologists revere Hutton as the "father of geology."

Hutton's scheme of rock classification is a "genetic classification," because it focuses on the genesis—the origin—of the rock. In modern terminology, geologists recognize three basic groups: (1) **igneous rocks**, which form by the freezing (solidification) of molten rock, or melt (▶Fig. B.4a); (2) **sedimentary rocks**, which form either by the cementing together of fragments (grains) broken off preexisting rocks or by the precipitation of mineral crystals out of water solutions at or near the Earth's surface (▶Fig. B.4b); and (3) **metamorphic rocks**, which form when preexisting rocks change into new rocks in response to a change in pressure and temperature conditions, and/or as a result of squashing, stretching, or shear (▶Fig. B.4c). Metamorphic change occurs in the "solid state," which means that it does not require melting. Each of the three groups contains many different individual rock types, distinguished from each other by physical characteristics such as

- *grain size*: The dimensions of individual grains (using the word here in a general sense to mean crystals, fragments of minerals, or fragments of preexisting rocks) in a rock may be measured in millimeters or centimeters. Some grains are so small that they can't be seen without a microscope; others are as big as a fist or larger. Some grains are **equant**, meaning that they have the same dimensions in all directions; some are **inequant**, meaning that their dimensions are not the same in all directions (▶Fig. B.5a–c). In some rocks, all the grains are the same size, but other rocks contain a variety of different-sized grains.

- *composition*: As we stated earlier, a rock is a mass of chemicals. These chemicals may be ordered into mineral grains or, less commonly, may be disordered and constitute glass. The term *rock composition* refers to the proportions of different chemicals making up the rock. The proportion of chemicals, in turn, affects the proportion of different minerals constituting the rock. As you will see, however, chemical composition does not completely control the minerals present in a rock. For example, two rocks with exactly the same chemical composition can have totally different assemblages of minerals, if each rock formed under different pressure and temperature conditions. That's because the process of mineral formation is affected by environmental factors such as pressure and temperature.

- *texture*: This term refers to the arrangement of grains in a rock; that is, the way grains connect to each other

(a)

(b)

(c)

FIGURE B.4 Examples of the major rock groups. **(a)** Lava (molten rock) is in the process of freezing to form basalt, an igneous rock. The molten tip of the flow still glows red. **(b)** Sand, deposited on a beach, eventually becomes buried to form layers of sandstone, a sedimentary rock, such as those exposed in the cliffs behind the beach. **(c)** When preexisting rocks become buried deeply during mountain building, the increase in temperature and pressure transforms them into metamorphic rocks. The lichen-covered outcrop in the foreground was once deeply buried beneath a mountain range, but was later exposed when glaciers and rivers stripped off the overlying rocks of the mountain.

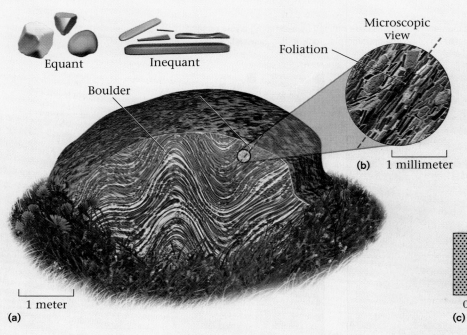

Equant

Inequant

Microscopic view

Foliation

Boulder

1 meter

(a)

(b)　1 millimeter

FIGURE B.5 **(a)** This boulder of metamorphic rock is an aggregate of mineral grains. **(b)** At high magnification, we can see that the rock consists of both equant and inequant grains. The inequant grains are aligned parallel to each other. **(c)** Using this comparison chart, we can measure the size of the grains, in millimeters.

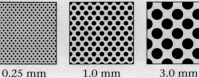

0.25 mm　　1.0 mm　　3.0 mm　　7.0 mm

(c)

and whether or not inequant grains are aligned parallel to one another. The concept of rock texture will become easier to grasp as we look at different examples of rocks in the following chapters.

- *layering*: Some rock bodies appear to contain distinct layering, defined either by bands of different compositions or textures, or by the alignment of inequant grains so that they trend parallel to one another. Different types of layering occur in different kinds of rocks. For example, the layering in sedimentary rocks is called **bedding**, whereas the layering in metamorphic rocks is called **metamorphic foliation** (▶Fig. B.6a, b).

Each individual rock type has a name. Names come from a variety of sources. Some come from the dominant component making up the rock, some from the region where the rock was first discovered or is particularly abundant, some from a root word of Latin or Greek origin, and some from a traditional name used by people in an area where the rock is found. All told, there are hundreds of different rock names, though in this book we introduce only about thirty.

B.5 STUDYING ROCK

Outcrop Observations

The study of rocks begins by examining a rock in an outcrop. If the outcrop is big enough, such an examination will reveal relationships between the rock you're interested in and the rocks around it, and will allow you to detect layering. Geologists carefully record observations about an outcrop, then break off a **hand specimen**, a fist-sized piece, which they can examine more closely with a hand lens (magnifying glass). Observation with a hand lens enables geologists to identify sand-sized or larger mineral grains, and may enable them to describe the texture of the rock.

Thin-Section Study

Geologists often must examine rock composition and texture in minute detail in order to identify a rock and develop a hypothesis for how it formed. To do this, they take a specimen back to the lab and make a very thin slice (about 3/100 mm thick, the thickness of a human hair). They study such a **thin section** (▶Fig. B.7a–c) with a petrographic microscope (*petro* comes from the Greek word for rock; geologists who specialize in the study of rocks are called petrologists). A petrographic microscope differs from an ordinary microscope in that it illuminates the thin section with transmitted polarized light. This means that the illuminating light beam first passes through a special filter that makes all the light waves in the beam vibrate in the same orientation, and then passes up through the thin section; an observer can then look through the thin section as if it were a window.

When illuminated with transmitted polarized light, each type of mineral grain displays a unique suite of colors. The specific color the observer sees depends on both the identity of the grain and its orientation with respect to the waves of polarized light, for a crystal interacts with polarized light and allows only certain colors to pass through.

The brilliant colors and strange shapes in a thin section rival the beauty of an abstract painting or stained glass. By

FIGURE B.6 Layering in rocks. **(a)** Bedding in a sedimentary rock, in this case defined by alternating layers of sand and gravel exposed in a cliff along the coast of Oregon. **(b)** Foliation in a metamorphic rock, in this case defined by alternating light and dark layers.

(a)

(b)

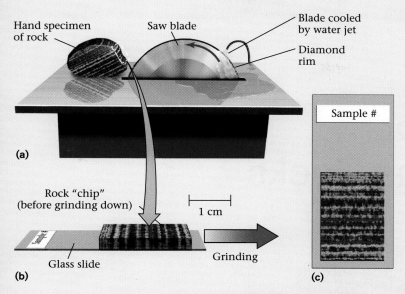

FIGURE B.7 **(a)** To prepare a thin section, a geologist cuts a brick-shaped chip out of rock, using a rock saw (a rotating circular blade with a diamond-studded rim). **(b)** The chip is glued to a glass slide and then ground down until it's paper thin. **(c)** The thin section is then labeled and ready to examine.

examining a thin section with a petrographic microscope, geologists can identify most of the minerals constituting the rock and can describe the way in which the grains connect to one another (▶Fig. B.8a). A photograph taken through a petrographic microscope is called a **photomicrograph**.

Note that to make a thin section, a geologist first uses a special rock saw with a spinning, diamond-studded

blade to cut a small rectangular block, or chip, of rock. (Rock saws work by grinding their way through rock—the hard diamonds embedded in the saw blade scratch and pulverize minerals as the saw blade rubs against the rock.) The geologist then cements the chip to a glass microscope slide with an epoxy adhesive (glue). By using a lap, a spinning plate coated with abrasive, the geologist can grind the chip down until only a thin slice, still cemented to the glass slide, remains, ready for examination with a petrographic microscope.

High-Tech Analytical Equipment

Beginning in the 1950s, high-tech electronic instruments became available that enabled petrologists to examine rocks at an even finer scale than can be done with a petrographic microscope. Modern research laboratories typically boast instruments such as electron microprobes, which can focus a beam of electrons on a small part of a grain to create a signal that defines the chemical composition of the mineral (▶Fig. B.8b); mass spectrometers, which analyze the proportions of atoms with different atomic weights contained in a rock; and X-ray diffractometers, which identify minerals by looking at the way X-ray beams diffract as they pass through crystals in a rock (see Box 5.2). Such instruments, in conjunction with optical examination, can provide petrologists with highly detailed characterizations of rocks, which in turn help them understand how the rocks formed and where the rocks came from. This information enables geologists to use the study of rocks as a basis for deciphering Earth history.

FIGURE B.8 **(a)** A rock photographed through a petrographic microscope. The colors are caused by the interaction of polarized light with the crystals. The long dimension of this photo is 2 mm. **(b)** An electron microprobe uses a beam of electrons to analyze the chemical composition of minerals.

(a)

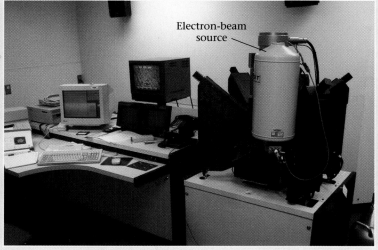

(b)

Up from the Inferno: Magma and Igneous Rocks

A river of molten rock (lava) weaves across a stark terrain of already solidified igneous rock. The volcano, the vent from which the lava has spilled out onto Earth's surface from the interior, can be seen in the distance.

Geopuzzle

Where does the red-hot molten rock that spills and/or blasts out of a volcano come from, and what does it turn into?

6.1 INTRODUCTION

Every now and then, an incandescent liquid—molten rock, or melt—begins to fountain from a crater (pit) or crack on the big island of Hawaii. Hawaii is a **volcano**, a vent at which melt from inside the Earth spews onto the planet's surface. Such an event is a volcanic eruption. Some of the melt, which is called **lava** once it has reached the Earth's surface during an eruption, pools around the vent, while the rest flows down the mountainside as a viscous (syrupy) red-yellow stream called a lava flow. Near its source, the flow moves swiftly, cascading over escarpments at speeds of up to 60 km per hour (▶Fig. 6.1a). At the base of the mountain, the lava flow slows but advances nonetheless, engulfing any roads, houses, or vegetation in its path. At the edge of the flow, beleaguered plants incinerate in a burst of flames. As the flow cools, it slows down, and its surface darkens and crusts over, occasionally breaking to reveal the hot, sticky

mass that remains within (▶Fig. 6.1b). Finally, it stops moving entirely, and within days to months the once red-hot melt has become a hard, black solid through and through (▶Fig. 6.1c). New **igneous rock**, made by the freezing of a melt, has formed. Considering the fiery heat of the melt from which igneous rocks develop, the name *igneous*—from the Latin *ignis*, meaning fire—makes sense. Igneous rocks are very common on Earth. They make up all of the oceanic crust and much of the continental crust.

It may seem strange to speak of freezing in the context of forming rock, for most people think of freezing as the transformation of liquid water to solid ice when the temperature drops below 0°C (32°F). Nevertheless, the freezing of liquid melt to form solid igneous rock represents the same phenomenon, except that igneous rocks freeze at high temperatures—between 650°C and 1,100°C. To put such temperatures in perspective, remember that home ovens attain a maximum temperature of only 260°C (500°F).

FIGURE 6.1 **(a)** Lava fountains in this crater of a volcano on Hawaii, and a river of lava streams out of a gap in its side. As the lava moves rapidly away from the crater, it cools, and a black crust forms on the surface. **(b)** Farther down the mountain, the surface of the lava has completely crusted over with newborn rock, while the insides of the flow remain molten, allowing it to creep across the highway (in spite of the stop sign). Smoke comes from burning vegetation. **(c)** Eventually the flow cools through and through, and a new layer of basalt rock has formed. This rock is only a few weeks old.

(b)

Lava fountain

Lava flow

(a)

(c)

Although some igneous rocks solidify at the surface during volcanic eruptions, a vastly greater volume results from solidification of melt underground, out of sight. Geologists refer to melt that exists below the Earth's surface as **magma**, and melt that has erupted from a volcano at the surface of the Earth as lava. Rock made by the freezing of magma underground, after it has pushed its way (intruded) into preexisting rock of the crust, is **intrusive igneous rock**, and rock that forms by the freezing of lava above ground, after it spills out (extrudes) onto the surface of the Earth and comes into contact with the atmosphere or ocean, is **extrusive igneous rock** (▶Fig. 6.2). Extrusive igneous rock includes both solid lava flows, formed when streams or mounds of lava solidify on the surface of the Earth, and deposits of **pyroclastic debris** (from the Greek word *pyro*, meaning fire). Some of the debris forms when clots of lava fly into the air in lava fountains and then freeze to form solid chunks before hitting the ground. Some forms when the explosion of a volcano shatters preexisting rock and ejects the fragments over the countryside. And some debris forms when an explosion blasts a fine spray of lava into the air—the fine spray of lava instantly freezes to form fine particles of glass called **ash**. Some of the ash billows up into the atmosphere, eventually drifting down from the sky as an ash fall. But some ash rushes down the side of the volcano in a scalding avalanche called an ash flow (see Chapter 9 for further detail).

Note that a significant amount of the material making up a volcano actually consists of pyroclastic debris accumu-lations that move downslope *after* being initially deposited. Volcanic debris may slip in a landslide; it may be carried away by streams and redeposited as stream sediment; or it may mix with water to form a slurry of mud and debris, called a volcanic debris flow, that oozes down the mountain like wet concrete. If the volcano is an island or seamount, the mud and debris move under water.

A great variety of igneous rocks exist on Earth. To understand why and how these rocks form, and why there are so many different kinds, we discuss why magma forms, why it rises, why it sometimes erupts as lava, and how it freezes in intrusive and extrusive environments. We then look at the scheme that geologists use to classify igneous rocks.

6.2 WHY DOES MAGMA FORM?

The Earth is hot inside, but the popular image that the planet's solid crust "floats on a sea of molten rock" is *not* correct. Magma only forms in special places—elsewhere, the crust and mantle of the Earth are solid. In this section, we first examine the source of Earth's heat and then consider the special conditions that trigger melting.

Why Is It Hot Inside the Earth?

Clearly, if the Earth were not hot inside, igneous processes would not take place. Where does our planet's internal heat come from? Much of the heat is left over from the Earth's early days. According to the nebula theory, this planet formed from the collision and merging of countless planetesimals. Every time a collision occurred, the kinetic energy (energy of motion) of the colliding planetesimals transformed into heat energy. (You can simulate this phenomenon by banging a hammer repeatedly on a nail—the head of the nail becomes quite warm.) As the Earth grew, gravity pulled matter inward until eventually the weight of overlying material squeezed the matter inside tightly together. Such compression made the Earth's insides even hotter, just as compressing a gas with a piston makes it hotter. Even after the Earth had grown to become a planet, intense bombardment continued to add heat energy. Eventually, the Earth became hot enough for iron to melt. The iron sank to the center to form the core. Friction between the sinking iron and its surroundings generated still more heat, just as rubbing your hands together generates heat. (Note that this process transformed gravitational potential energy into heat; see Appendix A.) Soon after Earth's formation, but probably after differentiation (see Chapter 1), a Mars-sized object collided with the Earth. This collision generated vast amounts of heat. Taken together, collisions and differentiation made the early Earth so hot that it was at least partially molten throughout, and its surface may have been an ocean of lava.

FIGURE 6.2 Extrusive igneous materials, namely ash and lava, form above the Earth's surface, whereas intrusive rocks develop below. Melt that erupts from a volcano is lava; underground melt is magma.

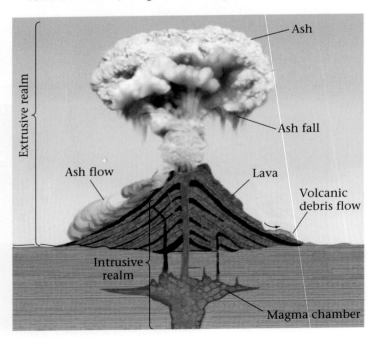

Extrusive realm

Ash

Ash fall

Ash flow

Lava

Volcanic debris flow

Intrusive realm

Magma chamber

Ever since the heat-producing catastrophes of its early days, the Earth has radiated heat into space and thus has slowly cooled. Eventually, the sea of lava solidified and formed igneous rock, the first rock on Earth. If no heat had been added to the Earth after the end of intense bombardment (at about 3.9 Ga), the Earth might be too cold by now for igneous activity to take place. This hasn't happened because of the presence of radioactive elements, primarily in the crust. Decay of a single radioactive atom produces only a tiny amount of heat, but the cumulative effect of radioactive decay throughout the Earth has been sufficient to slow the cooling of this planet. Thus, Earth remains very hot today, with temperatures at the base of the lithosphere reaching almost 1,300°C and temperatures at the planet's center exceeding 4,700°C.

Causes of Melting

Magma forms both in the upper part of the asthenosphere and in the lower crust. Here we discuss the physical conditions that lead to this melting. Later, we'll consider the geologic settings in which these conditions develop.

Melting as a result of a decrease in pressure (decompression). The variation in temperature with depth can be expressed on a graph by a curving line, the **geotherm**. Beneath typical oceanic crust, temperatures comparable with those of lava (650°–1,100°C) generally occur in the upper mantle (▶Fig. 6.3). But even though the upper mantle is very hot, its rock stays solid because it is also under high pressure from the weight of overlying rock. To put it simply, pressure at great depth prevents atoms from breaking free of solid mineral crystals.

Because pressure prevents melting, a decrease in pressure can permit melting. Specifically, if the pressure affecting hot mantle rock decreases while the temperature remains unchanged, a magma forms. This kind of melting, called decompression melting (▶Box 6.1), occurs where hot mantle rock rises to shallower depths in the Earth, because pressure decreases toward the surface and rock is such a good insulator that it doesn't lose much heat as it rises (▶Fig. 6.4a).

Melting as a result of the addition of volatiles. Magma also forms at locations where chemicals called volatiles mix with hot mantle rock. Volatiles are elements or molecules, such as water (H_2O) and carbon dioxide (CO_2), that evaporate easily and can exist in gaseous forms at the Earth's surface. When volatiles mix with hot rock, they help break chemical bonds, so that if you add volatiles to a solid, hot, dry rock, the rock begins to melt (▶Fig. 6.4b). In effect, adding volatiles decreases a rock's melting temperature. (Melting due to addition of volatiles is sometimes called flux melting.) Of the common volatiles, water plays the most important role in influencing melting.

We can understand the effect of volatiles by looking at the contrast between the melting curve (or solidus), which is the line defining the range of temperatures and pressures at which a rock starts to melt, for wet basalt and for dry basalt (▶Fig. 6.4c). Note that wet basalt (basalt containing volatiles) melts at much lower temperatures than dry (volatile-free) basalt. In fact, adding volatiles to rock to cause melting is somewhat like sprinkling salt on ice to make it melt.

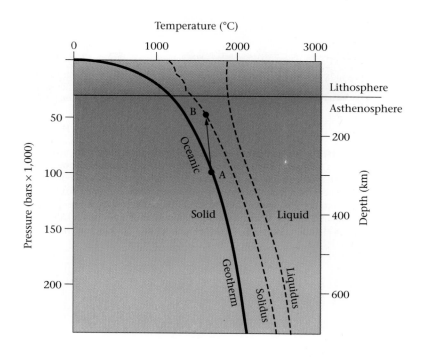

FIGURE 6.3 The graph plots the Earth's geotherm (solid line), which specifies the temperature at various depths below oceanic lithosphere, as well as the "liquidus" and "solidus" (dashed lines) for peridotite, the ultramafic rock that makes up the mantle. The solidus represents conditions of pressure and temperature at which a rock *begins to melt*, whereas the liquidus represents the conditions of pressure and temperature at which the *last solid disappears*. The region of the graph between the liquidus and solidus represents conditions under which there can be a mixture of solid and melt. Note that the geothermal gradient (the rate of change in temperature) decreases with greater depths; if it were constant, the geotherm would be a straight line. A rock that starts at pressure and temperature conditions indicated by point A, and then rises to point B, undergoes a significant decrease in pressure without much change in temperature. When it reaches the conditions indicated by point B, it begins to melt. This process is called decompression melting. Note that asthenosphere cools only slightly as it rises, because rock is such a good insulator.

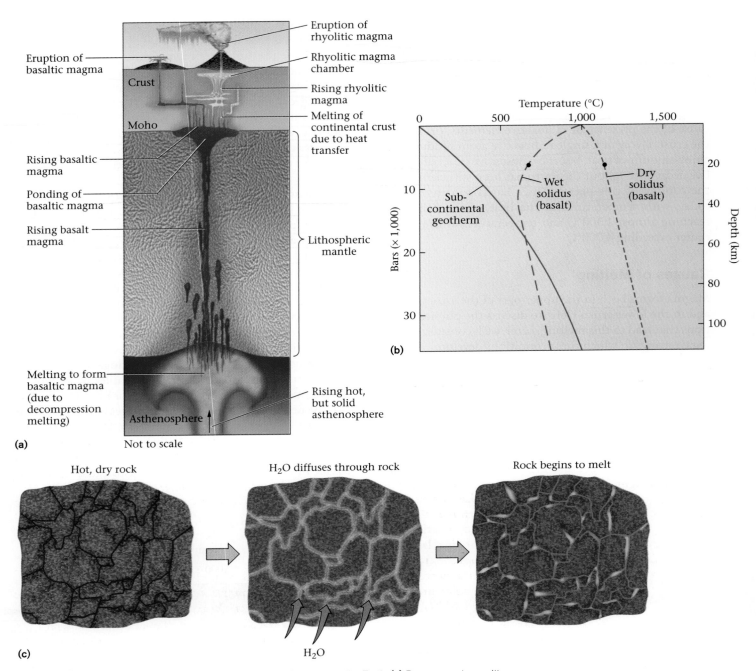

FIGURE 6.4 The three main causes of melting and magma formation in the Earth. **(a)** Decompression melting occurs when a hot rock rises to a shallow depth, where the pressure is lower. Melting as a result of heat transfer happens when hot magma rises into rock that has a lower melting temperature. For example, hot basaltic magma rising from the mantle can make the surrounding intermediate-composition crust melt. **(b)** The addition of volatiles decreases the melting temperature. For example, at a depth of 20 km, the melting temperature of wet basalt (basalt that contains volatiles) is about 500°C lower than the melting temperature of dry basalt. **(c)** Melting as a result of the addition of volatiles occurs when H_2O percolates into a solid hot rock. It's as if an "injection" of water triggers the melting.

BOX 6.1 THE REST OF THE STORY

Understanding Decompression Melting

Look again at Figure 6.3. The horizontal axis on the graph represents temperature, in degrees Centigrade, and the left vertical axis represents pressure. Since pressure in the Earth results from the weight of overlying rock, we can calibrate the vertical axis either in units of depth (km below the surface) or in units of pressure, such as bars. (Note: 1 bar approximately equals 1 atm, where 1 atm, or atmosphere, is the air pressure at sea level.)

The solid line on this graph is the geotherm, which defines the temperature as a function of depth in the Earth. Notice that the rate of increase in temperature, a quantity called the **geothermal gradient** (expressed in degrees per km), decreases with increasing depth. The dashed lines represent the solidus and liquidus for mantle rock (peridotite). The solidus defines the conditions of pressure and temperature at which mantle rock begins to melt, as determined by laboratory measurements. Values to the left of the solidus indicate pressures and temperatures for which the rock stays entirely solid. The liquidus represents conditions at which all solid disappears and only melt remains.

To see what happens during decompression, imagine a mantle rock at a depth of 300 km (point A on the graph). According to the graph, the pressure ≈ 95,000 bars and the temperature ≈ 1,700°C at this depth. Now imagine that the rock moves closer to the Earth's surface without cooling, as may occur in a rising mantle plume, and reaches point B, where the pressure is only about 47,000 bars. It has undergone decompression. Notice that point B lies on the solidus, so the rock begins to melt—the thermal vibration of atoms in the rock, no longer countered by pressure, can cause the atoms to break free of crystals. Also notice that decompression melting takes place without additional new heat. In fact, because the rock expands as it rises, it actually has cooled slightly.

Melting as a result of heat transfer from rising magma. When rock, or magma, from the mantle rises up into the crust, it brings heat with it. This heat flows into and raises the temperature of the surrounding crustal rock. In some cases, the rise in temperature may be sufficient to melt part of the crustal rock (see Fig. 6.4a). Imagine injecting hot fudge into ice cream; the fudge transfers heat to the ice cream, raises its temperature, and causes it to melt. We call such melting heat-transfer melting, because it results from the transfer of heat from a hotter material to a cooler one. Since mantle-derived magmas are very hot (over 1,100°C) and rocks of the crust melt at temperatures of about 650° to 850°C, when mantle-derived magma intrudes into the crust, it can raise the temperature of the surrounding crust enough to melt it.

> **Take-Home Message**
>
> Much of Earth's internal heat is a relict of the planet's formation. But the Earth would have become much cooler were it not for heat from radioactivity. Changes in pressure, volatile content, and temperature trigger melting in the upper mantle and lower crust.

grouped instead in clusters or short chains, relatively free to move with respect to each other.

"Dry" magmas contain no volatiles. "Wet" magmas, in contrast, include up to 15% dissolved volatiles such as water, carbon dioxide, nitrogen (N_2), hydrogen (H_2), and sulfur dioxide (SO_2). These volatiles come out of the Earth at volcanoes in the form of gas. Usually water constitutes about half of the gas erupting at a volcano. Thus, magma not only contains the elements that constitute solid minerals in rocks, but also can contain molecules that become water and air.

The Major Types of Magma

Imagine four pots of molten chocolate simmering on a stove. Each pot contains a different type of chocolate. One pot contains white chocolate, one milk chocolate, one semisweet chocolate, and one baker's chocolate. All the pots contain chocolate, but not all the chocolates are the same—they differ from each other in terms of proportions of sugar, cocoa butter, and milk. It's no surprise that different kinds of molten chocolate yield different kinds of solid chocolate; each type differs from the others in terms of taste and color.

Like molten chocolate, not all molten rock is the same. Specifically, magmas differ from one another in terms of the chemicals they contain. Geologists have found that the easiest way to describe the chemical makeup of magma is by specifying the proportions, in weight percent, of oxides. (By "weight percent" we mean the percentage of the magma's weight that consists of a given component, and by "oxide" we mean a combination of a cation with oxygen.) Common

6.3 WHAT IS MAGMA MADE OF?

All magmas contain silicon and oxygen, which bond to form the silicon-oxygen tetrahedron. But magmas also contain varying proportions of other elements such as aluminum (Al), calcium (Ca), sodium (Na), potassium (K), iron (Fe), and magnesium (Mg). Because magma is a liquid, its atoms do not lie in an orderly crystalline lattice but are

oxides include silica (SiO_2), iron oxide (FeO or Fe_2O_3), and magnesium oxide (MgO). To distinguish categories of magma from each other, geologists simply specify the relative proportions of silica (SiO_2) the magmas contain. The magma types are as follows. (1) **Felsic magma** contains about 66% to 76% silica. The name reflects the occurrence of felsic minerals (feldspar and quartz) in rocks formed from this magma. Because of its relatively high silica content, some geologists use the adjective "silicic" instead of "felsic" for this magma. (2) **Intermediate magma** contains about 52% to 66% silica. The name "intermediate" indicates that these magmas have a composition between that of felsic magma and mafic magma. (3) **Mafic magma** contains about 45% to 52% silica. Mafic magma is so named because it produces rock containing abundant mafic minerals, that is, minerals with a relatively high proportion of MgO and FeO or Fe_2O_3. Recall that the "ma" in the word mafic stands for magnesium, and the suffix "-fic" stands for iron (from the Latin *ferric*). (4) **Ultramafic magma** contains only 38% to 45% silica.

Magma properties depend on magma compositions. For example, the **viscosity** (resistance to flow) of magma reflects its silica content, for silica tends to polymerize, meaning it links up to form long, chainlike molecules whose presence slows down the flow. Thus, felsic magmas have higher viscosity (i.e., are stickier and flow less easily) than mafic magmas. The density of magma also reflects its composition, for SiO_2 is less dense than MgO or FeO. Thus, felsic magmas are less dense than mafic magmas. Finally, different magmas have different temperatures. Felsic magma can remain liquid at temperatures of only 650° to 800°C, whereas ultramafic magmas may reach temperatures of up to 1,300°C.

Why are there so many different kinds of magma? There are several factors.

Source rock composition. When you melt ice, you get water, and when you melt wax, you get liquid wax. There is no way to make water by melting wax. Clearly, the composition of a melt reflects the composition of the solid from which it was derived. Not all magmas form from the same source rock, so not all magmas have the same composition; magmas formed from crustal sources don't have the same composition as magmas formed from mantle sources, because the crust and mantle have different compositions to start with (see Chapter 2).

Partial melting. The melting of rock differs markedly from the melting of pure water ice, in that ice contains only one kind of mineral (crystals of H_2O), whereas most rocks contain a variety of different minerals. Because not all minerals melt by the same amount under given conditions, and because chemical reactions take place during melting, the magma that forms as a rock begins to melt does *not* have the

same composition as the original rock from which it formed. Typically, the magma contains more silica than did the original rock. Because magma can flow, it moves out of the original rock long before the rock completely melts. In the process, the melt carries away silica. The process during which only part of a rock melts to form a magma that then moves away is called **partial melting**.

Geologists find that, commonly, about 2% to 30% of a rock melts to produce a magma. The melt forms films on grain surfaces, and collects in little pockets between grains, until it eventually migrates away (▶Fig. 6.5a). "Crystal mush" forms if there is enough melt to keep the remaining crystals suspended (not in contact with each other) in the melt. Note that the solid rock left behind as partial melting occurs will be more mafic than the original rock. Thus, magma formed during later stages of melting contains less silica than magma formed during earlier stages (Fig. 6.5a).

Let's consider the implications of partial melting. Because melts formed by partial melting tend to be richer in silica than the rock from which they were derived, partial melting of ultramafic rock can produce mafic magma. Similarly, partial melting of intermediate rock produces a felsic magma.

Assimilation. As magma sits in a magma chamber before completely solidifying, it may incorporate chemicals derived from the wall rocks of the chamber. This process of **assimilation** (▶Fig. 6.5b) takes place when rocks fall into the magma and then partially melt, or when heat from the magma partially melts the walls of the chamber. In some cases, selected elements migrate out of the wall and into the magma without the wall melting. This process may be accelerated if hot water circulates through the wall rock, for the water may dissolve elements and carry them into the magma. Geologists do not agree about how much assimilation takes place during magma formation.

Magma mixing. Different magmas formed in different locations from different sources may come in contact within a magma chamber prior to freezing. When this happens, the originally distinct magmas may mix to create a new, different magma. For example, thoroughly mixing a felsic magma with a mafic magma in equal proportions produces an intermediate magma. Sometimes different magmas that come in contact do not mix together completely; when this occurs, blobs of rock formed from one magma remain suspended within a mass formed from the other magma after solidification occurs.

Fractional crystallization. Consider an imaginary magma that contains only two kinds of atoms (X and Y). The magma starts to cool, and crystals of a mineral that contains mostly X atoms begin to form. The solid crystals are denser than the magma, so they sink, and by doing so preferentially

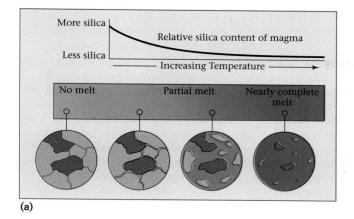

(a)

FIGURE 6.5 (a) The concept of partial melting. Rock does not all melt at once; at lower temperatures, only part of the rock melts. The first melt tends to be more silicic than the later-formed melt, as the graph shows. When the rock first starts to melt, the molten rock makes films around still-solid grains. Grains that melt at lower temperatures melt first, while grains that melt at higher temperatures remain. With further melting, a "crystal mush" develops, with relict solid grains surrounded by melt. **(b)** The concept of magma contamination. Blocks of rock fall into a magma, melt, and become mixed with the magma. Also, wall rock begins partially to melt and contributes new magma to the rising magma column. **(c)** The concept of fractional crystallization. The highest-melting-temperature (mafic) minerals begin to crystallize. These early-formed minerals sink to the bottom of the magma body. Elements incorporated in these minerals, therefore, are extracted from the melt. The remaining melt becomes more silicic.

Magma from partial melt of wall rock mixes with magma from below; this process is magma contamination.

Blocks of rock fall into magma and dissolve; this process is assimilation.

Heat transfer from deep magma melts wall rock and creates another magma source.

Deep magma rises

(b)

X atoms ●
Y atoms ●

Time 1

Time 2

Magma chamber

← Decreasing temperature →

In the original magma, at higher temperature, all atoms are dispersed through the magma. In this example there are equal quantities of X and Y atoms.

As the magma cools, crystals form and incorporate X atoms. When the crystals settle out due to gravity, they remove X atoms and leave the remaining magma enriched in Y atoms.

(c)

remove X atoms from the magma (see ▶Fig. 6.5c). As a result, the remaining magma contains a higher proportion of Y atoms. Such a process, in which magma changes composition as it cools because formation and sinking of crystals preferentially remove certain atoms from the magma, is called **fractional crystallization**.

Let's apply this concept to the real world. Imagine that a mafic magma intrudes into the crust and starts to cool. Not all minerals form at the same temperature. While the magma is still quite hot, crystals of olivine and then pyroxene form and then sink. These minerals contain relatively large amounts of iron and magnesium and remove these elements from the magma. Thus, over time the magma becomes enriched in silica. As the magma cools and changes composition, other minerals begin to crystallize and sink. If the process of fractional crystallization continues, the remaining magma evolves to become so silica rich that when it finally freezes, a felsic rock forms. In other words, fractional crystallization can result in the evolution of a mafic magma into a felsic magma. (In detail, the process is more complex than just described—see ▶Box 6.2 for additional discussion.)

> **Take-Home Message**
>
> Magma is a very hot (1,280°C to 650°C) liquid made up of varying proportions of Si, O, Fe, Mg, Ca, and other chemicals. The composition of a magma depends on its source, on how it interacts with its surroundings, and on whether crystals sink from it as they form.

6.4 MOVING MAGMA AND LAVA

Why Does Magma Rise?

If magma stayed put once it formed, new igneous rocks would not develop in or on the crust. But it doesn't stay put. Magma tends to move upward, away from where it formed. In some cases, it reaches the Earth's surface. Magma rises slowly through the crust, because it has to work its way up

BOX 6.2 THE REST OF THE STORY

Bowen's Reaction Series

During the 1920s, Norman L. Bowen, a geologist at the Carnegie Institution in Washington, D.C., and later at the University of Chicago, began a series of laboratory experiments designed to determine the sequence in which silicate minerals crystallize from a melt. Bowen first melted powdered mafic igneous rock in a sealed crucible by raising its temperature to about 1,280°C. Then he cooled the melt just enough to cause part of it to solidify, and he "quenched" the remaining melt by submerging it quickly in mercury. Quenching, during which melt undergoes a sudden cooling to form a solid, caused any remaining liquid to turn into glass, a material without a crystalline structure. The early-formed crystals were trapped in the glass. Bowen identified mineral crystals formed before quenching by examining thin sections with a microscope. Then he analyzed the composition of the remaining glass.

After repeated experiments at different temperatures, Bowen found that as new crystals form during crystallization, they extract certain chemicals preferentially from the melt. Thus, the chemical composition of the remaining melt progressively changes as the melt cools. Further, once formed, crystals continue to react with the remaining melt. Bowen described the specific sequence of mineral-producing reactions that take place in a cooling, initially mafic, magma. This sequence is now called **Bowen's reaction series** in his honor.

In a cooling melt, olivine and calcium-rich plagioclase form first. The Ca-plagioclase reacts with the melt to form more plagioclase, but the newer plagioclase contains more sodium (Na). Meanwhile, some olivine crystals react with the remaining melt to produce pyroxene, which may encase (i.e., surround) olivine crystals or even replace them. However, some of the olivine and Ca-plagioclase crystals settle out of the melt, taking iron, magnesium, and calcium atoms with them. By this process, the remaining melt becomes enriched in silica. As the melt continues to cool, plagioclase continues to form, with later-formed plagioclase having progressively more Na than earlier-formed plagioclase. Pyroxene crystals react with melt to form amphibole, and then amphibole reacts with the remaining melt to form biotite. All the while, crystals continue to settle out, so the remaining melt becomes more felsic. At temperatures of between 650°C and 850°C, only about 10% melt remains, and this melt has a high silica content. At this stage, the final melt freezes, yielding felsic minerals such as quartz, K-feldspar, and muscovite.

On the basis of the description just provided, Bowen realized that there are two tracks to the reaction series. The "discontinuous reaction series" refers to the sequence olivine, pyroxene, amphibole, biotite, K-feldspar/muscovite/quartz: each step yields a different class of silicate mineral. The "continuous reaction series" refers to the progressive change from calcium-rich to sodium-rich plagioclase: the steps yield different versions of the same mineral (▶ Fig. 6.6a, b). In the discontinuous reaction series, the first mineral to form is composed of isolated silicon-oxygen tetrahedra; the second contains single chains of tetrahedra; the third, double chains of tetrahedra; the fourth, 2-D sheets of tetrahedra; and the last, 3-D network silicates.

It's important to note that not all minerals listed in the series appear in all igneous rock. For example, a mafic magma may completely crystallize before felsic minerals such as quartz or feldspar have a chance to form.

Bowen's studies provided a remarkable demonstration of how laboratory experiments can help us understand processes that take place in locations (such as a deep magma chamber) that no human can visit firsthand.

FIGURE 6.6 **(a)** Bowen's reaction series. Minerals that crystallize at higher temperatures are at the top of the series. On the left is the discontinuous reaction series, consisting of a succession of different minerals. On the right is the continuous reaction series, consisting of progressively changing plagioclase compositions. Rocks formed from minerals at the top of the series are ultramafic and mafic, whereas rocks made from minerals at the bottom of the series are felsic. The short arrows indicate reactions. **(b)** With decreasing temperature, crystallization begins and magma composition changes.

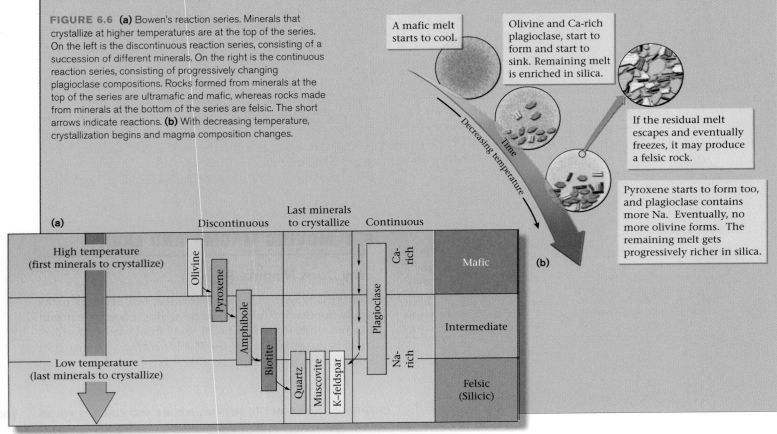

through cracks and narrow conduits. Friction between the magma and the walls of the conduits slows it down. In a broad sense, the movement of magma from deep within the Earth up to the crust or even onto the surface transfers both matter and heat from the interior of the Earth upward. This movement is a key component of the Earth System, because it provides the raw material for new rocks and for the atmosphere and ocean. But why does magma rise?

First, magma rises because it is less dense than surrounding rock, and thus is buoyant. Magma is less dense both because rock expands as it melts and because magma tends to contain a smaller proportion of heavy elements (such as iron) than does its source. Buoyancy drives magma upward just as it drives a wooden block up through water. Note that when volatile-rich magma rises to a shallower depth, the volatiles dissolved within it come out of solution and form bubbles, much as carbon dioxide makes bubbles in soda water when you pop the bottle cap off and release the pressure. The presence of gas bubbles within a magma further decreases its density and creates additional buoyancy force to drive the magma upward. The increase in volume due to the formation of gas bubbles can even propel melt out of the volcano at high velocity.

Second, magma rises because the weight of overlying rock creates a pressure at depth that literally squeezes magma upward. The same process happens when you step into a puddle barefoot and mud squeezes up between your toes.

What Controls the Speed of Flow?

Viscosity—the resistance to flow—determines how fast magmas or lavas move. Magmas with low viscosity flow more easily than those with high viscosity, just as water flows more easily than molasses. Magma viscosity depends on temperature, volatile content, and silica content. Hotter magma is less viscous than cooler magma, just as hot tar is less viscous than cool tar, because thermal energy breaks bonds and allows atoms to move more easily. Similarly, magmas or lavas containing more volatiles are less viscous than dry (volatile-free) magmas, because the volatile atoms also tend to break apart bonds. And mafic magmas are less viscous than felsic magmas, because silicon-oxygen tetrahedra tend to link together in the magma to create long chains that can't move past each other easily. Thus, hotter mafic lavas have relatively low viscosity and flow in thin sheets over wide regions, whereas cooler felsic lavas are highly viscous and clump at the volcanic vent (▶Fig. 6.7a, b).

> ### Take-Home Message
>
> Magma rises from its source because it is buoyant and because of pressure due to overlying rocks. As it moves, magma pushes into existing cracks or forces open new cracks. Its resistance to flow (viscosity) depends on its composition, temperature, and gas content.

(a)

(b)   Not to scale

FIGURE 6.7 The behavior of erupting lava reflects its viscosity. **(a)** Viscous lava (silica-rich lava) forms a blob or mound-like dome at the volcano's vent. **(b)** Nonviscous lava (hot mafic lava) spreads out in a thin flow and can travel far from the vent; it can also fountain.

6.5 HOW DO EXTRUSIVE AND INTRUSIVE ENVIRONMENTS DIFFER?

Recall that igneous rocks form in two environments. If lava erupts at the Earth's surface and freezes in contact with the atmosphere or the ocean, then the rock it forms is called extrusive igneous rock. The term implies that the melt was extruded from (it flowed or exploded out of) a vent in a volcano. In contrast, if magma freezes underground, the rock it forms is called intrusive igneous rock, implying that the magma pushed—intruded—into preexisting rock of the crust.

Extrusive Igneous Settings

Not all volcanic eruptions are the same, so not all extrusive rocks are the same (as will be discussed further in Chapter 9). Some volcanoes erupt streams of low-viscosity lava that run down the flanks of the volcanoes and then spread over the countryside. When this lava freezes, it forms a sheet of igneous rock also known as a lava flow. In contrast, some volcanoes erupt viscous masses of lava that pile into domes; and still others erupt explosively, sending clouds of volcanic ash and debris skyward, or avalanches of ash—ash flows—that tumble down the sides of the volcano. The ash and debris that billows into the sky (an ash cloud) cools and falls to Earth, creating an ash or debris fall that blankets the countryside (▶Fig. 6.8a). Ash flows remain hot, eventually settling and welding together (▶Fig. 6.8b).

Which type of eruption occurs depends largely on a magma's composition and volatile content. As noted above, mafic lavas tend to have low viscosity and spread in broad, thin flows. Volatile-rich felsic lavas tend to erupt explosively and form thick ash deposits. We discuss the products of extrusive settings in more detail in Chapter 9.

Intrusive Igneous Settings

Magma rises and intrudes into preexisting rock by slowly percolating upward between grains or by forcing open cracks. The magma that doesn't make it to the surface freezes solid underground in contact with preexisting rock and becomes intrusive igneous rock. Geologists commonly refer to the preexisting rock into which magma intrudes as country rock, or wall rock, and the boundary between wall rock and an intrusive igneous rock as an intrusive contact

(▶Fig. 6.9a, b). If the wall rock was cold to begin with, then heat from the intrusion "bakes" and alters it in a narrow band along an intrusive contact (see Chapter 8).

Geologists distinguish between different types of intrusions based on their shape. Tabular intrusions, or sheet intrusions, are planar and are of roughly uniform thickness; they range in thickness from millimeters to tens of meters, and can be traced for meters to tens or, in a few cases, hundreds of kilometers. In places where tabular intrusions cut across rock that does not have layering, a nearly vertical, wall-like tabular intrusion is called a **dike**, whereas a nearly horizontal, tabletop-shaped tabular intrusion is a **sill**. In places where tabular intrusions cut across rock that has layering (bedding or foliation), dikes are defined as intrusions that cut across layering, whereas sills are intrusions that are parallel to layering (▶Figs. 6.10a, b; 6.11a–g). Spectacular groups of dikes cut across the countrysides of interior Canada and western Britain. A large sill, the Palisades Sill, makes up the cliff along the western bank of the Hudson River opposite New York City. Another sill forms the ledge on which Hadrian's Wall, which bisects Britain, was built. Some intrusions start to inject between layers but then dome

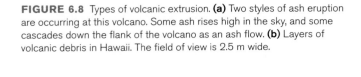

FIGURE 6.8 Types of volcanic extrusion. **(a)** Two styles of ash eruption are occurring at this volcano. Some ash rises high in the sky, and some cascades down the flank of the volcano as an ash flow. **(b)** Layers of volcanic debris in Hawaii. The field of view is 2.5 m wide.

(a)

(b)

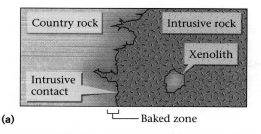

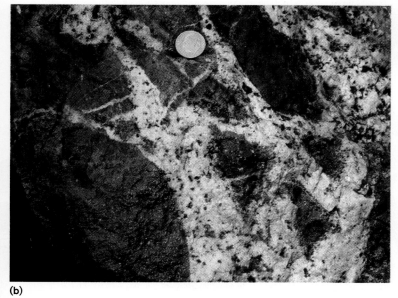

FIGURE 6.9 (a) An intrusive contact, showing the baked zone, blocks of country rock, fingers of the intrusion protruding into the country rock, and a xenolith. **(b)** A close-up photo of light-colored intrusive rock (granite) within dark-colored country rock. The coin indicates scale.

upwards, creating a blister-shaped intrusion known as a **laccolith** (▶Fig. 6.12a).

Plutons are irregular or blob-shaped intrusions that range in size from tens of meters across to tens of kilometers across (▶Figs. 6.12a–c; 6.13a, b). The intrusion of numerous plutons in a region produces a vast composite body that may be several hundred kilometers long and over 100 km wide; such immense masses of igneous rock are called **batholiths**. The rock making up the Sierra Nevada Mountains of California comprises a batholith created from plutons that intruded between 145 and 80 million years ago (▶Fig. 6.14a–c). Keep in mind that batholiths do not extend all the way down to the base of the crust. They are probably only a few kilometers to perhaps 10 km thick.

Where does the space for intrusions come from? Geologists continue to debate this issue, and over the past century they have suggested several models. Dikes form in regions where the crust is being stretched (for example, in a rift). Thus, as the magma that makes a dike forces its way up through a crack (sometimes causing the crack to form in the first place), the crust stretches sideways (▶Fig. 6.15a). Intrusion of sills occurs near the surface of the Earth, so the pressure of the magma effectively pushes up the rock over the sill, leading to uplift of the Earth's surface (▶Fig. 6.15b).

How do plutons form? This question remains a topic of active research and much controversy (▶Fig. 6.15c). Some geologists propose that a pluton is a frozen diapir, a lightbulb-shaped blob of magma that pierced overlying rock and pushed it aside as it rose. Others suggest that pluton formation involves **stoping**, a process during which magma assimilates wall rock, and blocks of wall rock break off and sink into the magma (▶Fig. 6.15d). (If a stoped block does not melt entirely, but rather becomes surrounded by new igneous rock, it can be called a **xenolith**,

FIGURE 6.10 (a) Dikes and sills are vertical or horizontal bands, respectively, on the face of an outcrop. **(b)** If we were to strip away the surrounding rock, dikes would look like walls, and sills would look like tabletops.

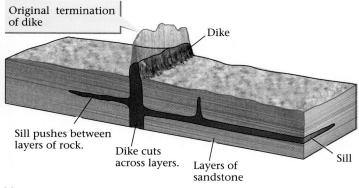

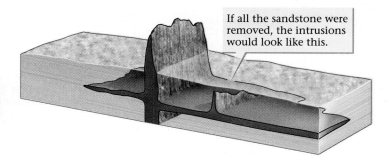

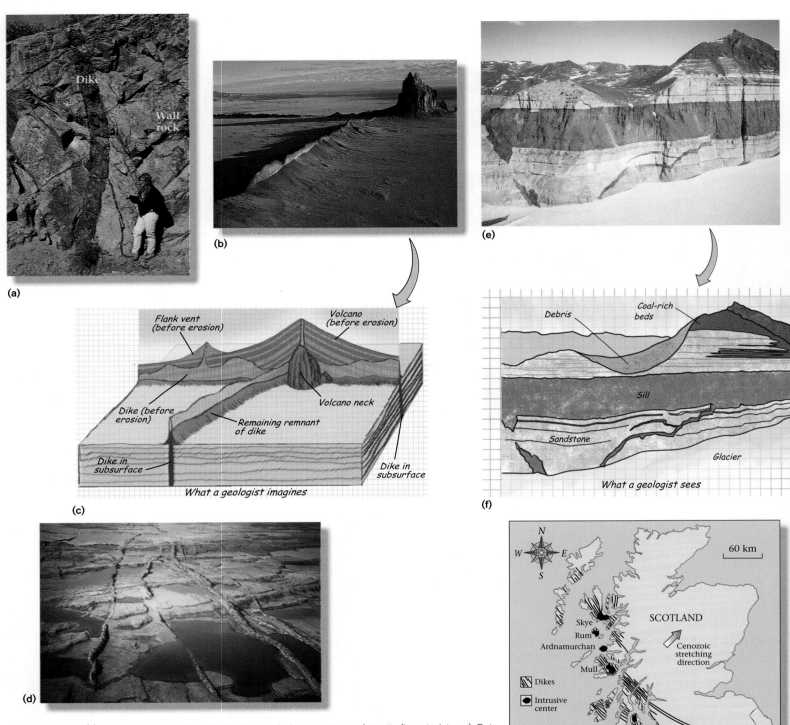

FIGURE 6.11 **(a)** A basalt dike looks like a black stripe painted on an outcrop of granite (here, in Arizona). But the dike actually intrudes, wall-like, into the outcrop. **(b)** At this ancient volcano called Shiprock, in New Mexico, ash and lava flows have eroded away, leaving a volcanic neck (the solid igneous rock that cooled in a magma chamber within the volcano). Large dikes radiate outward from the center, like spokes of a wheel. The softer rocks that once surrounded the dikes have eroded away, leaving wall-like remnants of the dikes exposed. **(c)** Shiprock was once in the interior of a volcano or below a volcano. **(d)** These Precambrian dikes exposed in the Canadian Shield formed when the region underwent stretching over a billion years ago; at that time, numerous cracks in the crust filled with magma. The dikes are about 100 m apart. **(e)** This dark sill, exposed on a cliff in Antarctica, is basalt; the white rock is sandstone. **(f)** This geologist's sketch shows the cliff face as viewed face on. **(g)** Map showing Cenozoic dikes in Great Britain and Ireland. Note that the dikes radiate from intrusive centers.

FIGURE 6.12 (a) While a volcano is active, a magma chamber exists underground; dikes, sills, and laccoliths intrude; and lava and ash erupt at the surface. Here we see that the active magma chamber, the one currently erupting, is only the most recent of many in this area. Earlier ones are now solidified. Each mass is a pluton. A composite of many plutons is a batholith. **(b)** Later, the bulbous magma chamber freezes into a pluton. The soft parts of the volcano erode, leaving wall-like dikes and column-like volcanic necks. Hard lava flows create resistant plateaus. **(c)** With still more erosion, volcanic rocks and shallow intrusions are removed, and we see plutonic intrusive rocks.

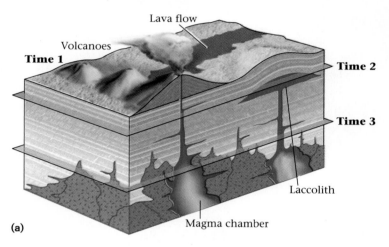
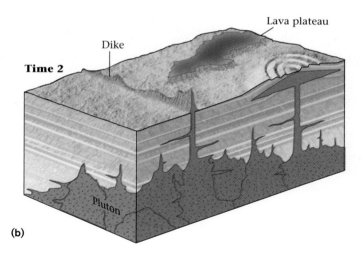

(a)

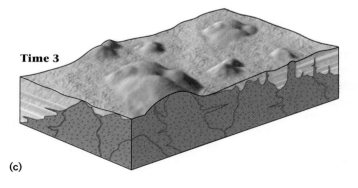

(b)

(c)

(a)

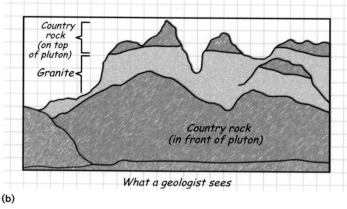

(b)

FIGURE 6.13 (a) Torres del Paines is a spectacular group of mountains in southern Chile. The light rock is a granite pluton, and the dark rock is the remains of the country rock (wall rock) into which the pluton intruded. A screen of country rock (in the lower half) hides the front of the pluton. **(b)** A geologist's sketch with the two major rock units labeled.

after the Greek *xeno*, meaning foreign; ▶Fig. 6.15e; see also Fig. 6.9.) More recently, geologists have proposed that plutons form by injection of several superimposed dikes or sills, which coalesce to become a single larger intrusion. Finally, a few geologists speculate that plutons grow when chemical exchanges between magma and the wall rock slowly transform the rock into granite.

If intrusive igneous rocks form beneath the Earth's surface, why can we see them exposed today? The answer comes from studying the dynamic activity of the Earth. Over long periods of geologic time, mountain building, driven by plate interactions, slowly uplifts huge masses of rock. Moving water, wind, and ice eventually strip away great thicknesses of overlying rock and expose the intrusive rock that had formed below.

Take-Home Message

Volcanoes erupt lava and ash. But not all magma erupts. Some solidifies underground either in tabular intrusions (dikes and sills) or blob-like intrusions (plutons). In places, immense volumes of intrusive igneous rock form, producing batholiths.

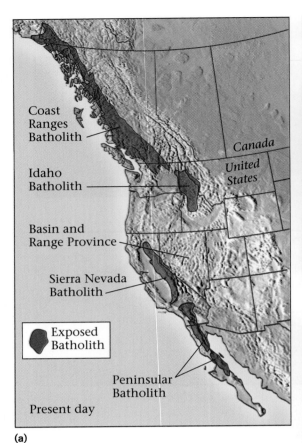

(b)

(c)

FIGURE 6.14 **(a)** The batholiths of western North America today. These formed between 145 and 80 Ma, ago when the west coast was a convergent plate boundary. **(b)** The Sierra Nevada Batholith as exposed today. The rounded, light-colored hills are all composed of granite-like intrusive igneous rock. These rocks formed several kilometers beneath a chain of volcanoes. **(c)** At this locality in the Mojave Desert of California, we see the top of a small pluton (light rock) where it is in contact with darker country rock.

6.6 TRANSFORMING MAGMA INTO ROCK

What makes magma freeze? In nature, two phenomena lead to the formation of solid igneous rock from a magma. Magma may freeze if the volatiles dissolved within it bubble out, for removal of volatiles (H_2O and CO_2) makes magma freeze at a higher temperature. Magma can also freeze simply when it cools below its freezing temperature and crystals start to grow. For cooling to occur, magma must move to a cooler environment. Because temperatures decrease toward the Earth's surface, magma automatically enters a cooler environment when it rises. The cooler environment may be cool country rock, if the magma intrudes underground, or it may be the atmosphere or ocean if the magma extrudes as lava at the Earth's surface.

The time it takes for a magma to cool depends on how fast it is able to transfer heat into its surroundings. To understand why, think about the process of cooling coffee. If you pour hot coffee into a thermos bottle and seal it, the coffee stays hot for hours; because of insulation, the coffee in the thermos loses

> **Take-Home Message**
>
> Magma freezes when it enters a cooler environment. The rate at which it freezes depends on the environment. For example, magma that solidifies deep in the crust cools more slowly than magma that erupts at the surface.

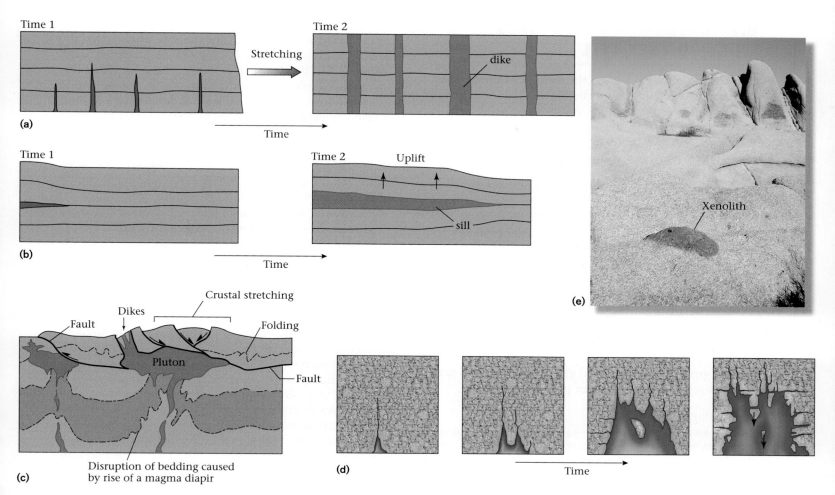

FIGURE 6.15 **(a)** Cross sections showing how the crust stretches sideways to accommodate dike intrusion. **(b)** Cross sections showing how intrusion of a sill may raise the surface of the Earth. **(c)** Ways in which crust accommodates emplacement of a pluton. **(d)** A magma stoping into country rock, gradually breaking off and digesting blocks as it moves. **(e)** Xenolith in a granite outcrop in the Mojave Desert. Note the coin for scale.

heat to the air outside only very slowly. Like the thermos bottle, rock acts as an insulator in that it transports heat away from a magma very slowly, so magma underground (in an intrusive environment) cools slowly. In contrast, if you spill coffee on a table, it cools quickly because it loses heat to the cold air. Similarly, lava that erupts at the ground surface cools more quickly because it is surrounded by air or water, which remove heat quickly. (Once a crust of rock forms on a lava flow, however, the interior of the flow cools more slowly, because the crust is an insulator.)

Three factors control the cooling time of magma that intrudes below the surface.

- *The depth of intrusion*: Intrusions deep in the crust cool more slowly than shallow intrusions, because warm country rock surrounds deep intrusions, whereas cold country rock surrounds shallow intru-

sions, and warmer country rock keeps the intrusion warmer for a longer time.

- *The shape and size of a magma body*: Heat escapes from an intrusion at the intrusion's surface, so the greater the surface area for a given volume of intrusion, the faster it cools. Thus, a pluton cools more slowly than a tabular intrusion with the same volume (because a tabular intrusion has a greater surface area across which heat can be lost), and a large pluton cools more slowly than a small pluton of the same shape. Similarly, droplets of lava cool faster than a lava flow, and a thin flow of lava cools more quickly than a thick flow (▶Fig. 6.16).

- *The presence of circulating groundwater*: Water passing through magma absorbs and carries away heat, much like the coolant that flows around an automobile engine.

6.7 IGNEOUS ROCK TEXTURES: WHAT DO THEY TELL US?

In Interlude B, we introduced the concept of a rock texture. When we talk about an igneous rock, a description of texture tells us whether the rock consists of glass, crystals, or fragments. A description of texture may also indicate the size of the crystals or fragments. Geologists distinguish among three types of texture:

- *Glassy texture*: A rock made of a solid mass of glass, or of tiny crystals surrounded by glass, has a glassy texture. Rocks with this texture are **glassy igneous rocks** (▶Fig. 6.17a). They reflect light as glass does, and they tend to break along conchoidal fractures.

- *Interlocking texture*: Rocks that consist of mineral crystals that intergrow when the melt solidifies, and thus fit together like pieces of a jigsaw puzzle, have an interlocking texture (▶Fig. 6.17b). Rocks with an interlocking texture are called **crystalline igneous rocks**. The interlocking of crystals in these rocks occurs because once some grains have developed, they interfere with the growth of later-formed grains. Later-formed grains fill irregular spaces between preexisting grains.

FIGURE 6.16 The cooling time of an intrusion increases with greater depth (because country rock is hotter at greater depth). The cooling time also depends on the shape of the intrusion (a thin sheet cools faster than a sphere of the same volume) and on the size of the intrusion (a small intrusion cools faster than a large one). Thus, ash (formed when droplets of lava come in contact with air) cools fastest, followed by a thin sheet of lava, a shallow sill, a deep sill, and a deep pluton.

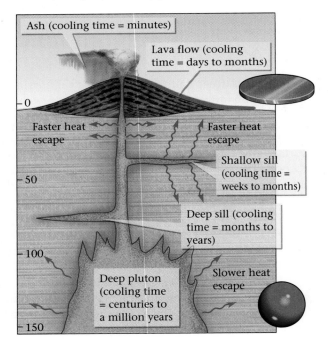

Geologists distinguish subcategories of crystalline igneous rocks according to the size of the crystals. Coarse-grained (phaneritic) rocks have crystals large enough to be identified with the naked eye. Typically, crystals in phaneritic rocks range in size from a couple of millimeters across to several centimeters across. Fine-grained (aphanitic) rocks have crystals too small to be identified with the naked eye. Porphyritic rocks have larger crystals surrounded by mass of fine crystals. In a porphyritic rock, the larger crystals are called phenocrysts, whereas the mass of finer crystals is called groundmass.

- *Fragmental texture*: Rocks with a fragmental texture consist of igneous fragments that are packed together, welded together, or cemented together after having solidified. Rocks with this texture are called **fragmental igneous rocks**. The fragments can consist of glass, individual crystals, bits of crystalline rock, or a mixture of all of these.

Geologists study igneous textures carefully, because the texture provides a clue to the way in which the rock formed. For example, cooling time plays an important role in determining texture. Specifically, the presence of glass indicates that cooling happened so quickly that the atoms within a lava didn't have time to arrange into crystal lattices. Crystalline rocks form when a melt cools more slowly, and in crystalline rocks, grain size depends on cooling time. A melt that cools rapidly, but not rapidly enough to make glass, forms fine-grained (aphanitic) rock, because crystals do not have time to grow large, whereas a melt that cools very slowly forms a coarse-grained (phaneritic) rock, because crystals do have time to grow large.

Because of the relationship between cooling time and texture, lava flows, dikes, and sills tend to be composed of fine-grained igneous rock. In contrast, plutons tend to be composed of coarse-grained rock. Plutons that intrude into hot country rock at great depth cool very slowly and thus are coarser grained than plutons that intrude into cool country rock at shallow depth, where they cool relatively rapidly. Porphyritic rocks form when a melt cools in two stages. First, the melt cools slowly at depth, so that large phenocrysts form. Then the melt erupts and the remainder cools quickly, so that fine-grained groundmass forms around the phenocrysts.

One important crystalline igneous rock type, **pegmatite**, doesn't quite fit the grain size–cooling time scheme just described. Pegmatite, a very coarse-grained igneous rock, contains crystals up to tens of centimeters

> **Take-Home Message**
>
> Igneous rocks come in a variety of textures—crystalline (made of interlocking crystals), glassy (made of a solid mass of glass), and fragmental (made of broken-up fragments of rock and/or ash). The grain size depends on the rate of cooling.

across and occurs in tabular intrusions called pegmatite dikes. A variety of precious gemstones are found in pegmatites. Because pegmatite occurs in dikes, which generally cool quickly, the coarseness of the rock may seem surprising. Researchers have shown that pegmatites are coarse because they form from *water-rich* melts in which atoms can move around so rapidly that large crystals can grow very quickly.

6.8 CLASSIFYING IGNEOUS ROCKS

Because melt can have a variety of compositions and can freeze to form igneous rocks in many different environments above and below the surface of the Earth, we observe a wide spectrum of igneous rock types. We classify these according to their texture and composition (▶Fig. 6.17c–j). Studying a rock's texture tells us about the rate at which it cooled and therefore the environment in which it formed, whereas studying its composition tells us about the original source of the magma and the way in which the magma evolved before finally solidifying.

Glassy Igneous Rocks

Glassy texture develops more commonly in felsic igneous rocks, because the presence of a high concentration of silica inhibits the easy growth of crystals. But basaltic and intermediate lavas can form glass if they cool rapidly enough. In some cases, a lava that cools to form obsidian contains a high concentration of gas bubbles—when the rock freezes, these bubbles remain as open holes known as **vesicles**. Geologists distinguish among several different kinds of glassy rocks.

- **Obsidian** is a mass of solid, felsic glass. It tends to be black or brown. Because it breaks conchoidally, sharp-edged pieces split off its surface when you hit a sample with a hammer. Preindustrial peoples worldwide used such pieces for arrowheads, scrapers, and knife blades.

- **Tachylite** is a bubble-free mass consisting of more than 80% mafic glass. This rock is relatively rare, in comparison with obsidian.

- **Pumice** is a glassy, felsic volcanic rock that contains abundant open pores (vesicles), giving it the appearance of a sponge. A preserved bubble is called a vesicle. Pumice forms by the quick cooling of frothy lava that resembles the foam head in a glass of beer. In some cases, pumice contains so many air-filled pores that it can actually float on water, like styrofoam. Ground-up pumice makes the grainy abrasive that blue-jean manufacturers use to "stonewash" jeans. Pumice tends to be light gray to tan in color.

- **Scoria** is a glassy, mafic volcanic rock that contains abundant vesicles (more than about 30%). Generally, the bubbles in scoria are bigger than those in pumice, and the rock overall is a medium gray to dark gray color.

Crystalline Igneous Rocks

The classification scheme for the principal types of crystalline igneous rocks is quite simple. The different compositional classes are distinguished on the basis of silica content—ultramafic, mafic, intermediate, or felsic—whereas the different textural classes are distinguished according to whether or not the grains are coarse or fine. The chart in ▶Figure 6.18 gives the texture and composition of the most commonly used rock names. As a rough guide, the color of an igneous rock reflects its composition: mafic rocks tend to be black or dark gray, intermediate rocks tend to be lighter gray or greenish gray, and felsic rocks tend to be light tan to pink or maroon. Unfortunately, color can be a misleading basis for rock identification, so geologists use a petrographic microscope to confirm their identifications. Different types of porphyritic rocks (these are not listed in Figure 6.18) are distinguished from each other according to their overall composition. For example, andesite porphyry is an andesite containing phenocrysts; generally, the phenocrysts consist of plagioclase.

Note that according to Figure 6.18, rhyolite and granite have the same chemical composition but differ in grain size. Which of these two rocks will develop from a melt of felsic composition depends on the cooling rate. A felsic lava that solidifies quickly at the Earth's surface, or in a thin dike or sill, turns into fine-grained rhyolite, but the same magma, if solidifying slowly at depth in a pluton, turns into coarse-grained granite. A similar situation holds for mafic lavas—a mafic lava that cools quickly in a lava flow forms basalt, but a mafic magma that cools slowly forms gabbro.

Fragmental Igneous Rocks

Geologists distinguish among different kinds of fragmental igneous rocks according to the size of the fragments and the way in which they stick together. Fragmental rocks form when a flow shatters into pieces during its movement and then the pieces weld together; when a fountain of lava sends droplets of lava into the air, forming bombs or cinders that accumulate around the vent; or when ash, crystals, and pre-existing volcanic rock blast

> **Take-Home Message**
>
> Geologists classify igneous rocks on the basis of texture and composition. For a given rock composition (proportion of silica), there is one name for a fine-grained version and another for a coarse-grained version. Other names are used for glassy and fragmental rocks.

FIGURE 6.17 **(a)** Thin section of a glassy, fine-grained volcanic rock. The light-colored grains are crystals, and the dark matrix (the region between the crystals) is glass. **(b)** This thin section of granite shows relatively large, interlocking crystals. **(c)** Rhyolitic welded tuff, **(d)** granite, **(e)** basalt, **(f)** gabbro, **(g)** porphyritic andesite, **(h)** pumice, **(i)** obsidian. **(j)** A pegmatite dike cutting granite.

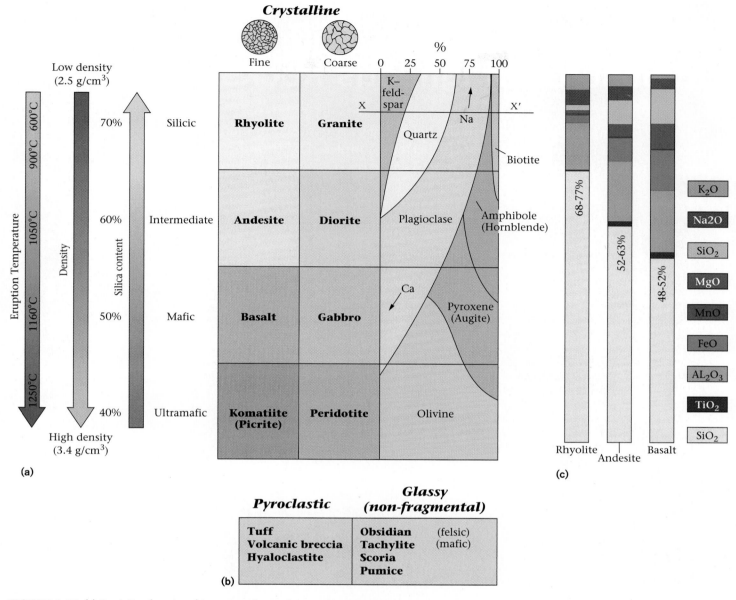

FIGURE 6.18 **(a)** Crystalline (nonglassy) igneous rocks are distinguished from each other by their grain size and composition. The right side of the chart shows the percentages of different minerals in the different rock types. To read this chart, draw a horizontal line next to a rock name; the minerals that the line crosses are the minerals found in that rock. (For example, granite [line X–X′] includes K-feldspar, quartz, plagioclase, amphibole, and biotite.) **(b)** The principal types of glassy and fragmental igneous rocks can be separated into several categories. **(c)** Graph illustrating the relative proportions of different oxides in igneous rocks.

from a volcano during an eruption and then settle down to the ground. Fragmental rocks composed of debris that has been blasted out of a volcano or thrown out in a fountain are commonly called **pyroclastic rocks** (from the Greek *pyro* for fire, and from the word *clastic*, which signifies composed of grains that were already solid when they stuck together). (Geologists use another term, volcaniclastic rock, in a general sense for rock made of volcanic debris that moved and was redeposited, commonly in

water, before becoming rock.) Because of the way in which they form, the fragments in pyroclastic rocks may themselves be glassy.

• **Tuff** is a fine-grained pyroclastic igneous rock composed of volcanic ash and/or fragmented lava and pumice. Tuff forms either from material that settles from the air (in an ash fall) and then is cemented together or sticks together, or from material that

See for yourself . . .
Exposures of Igneous Rocks

What do igneous rocks look like in the field? They are exposed in many places around the world. Here, we take you on tour to see a few of the better examples. The thumbnail images provided on this page are only to help identify tour sites. Go to wwnorton.com/studyspace to experience this flyover tour.

G6.1

Yosemite National Park, California
(Lat 37°47'25.70"N, Long 119°29'16.74"W)

Fly to the coordinates provided and zoom up to an elevation of 7 km (4 miles). You can see a vast expanse of whitish-gray outcrop in Yosemite National Park in the Sierra Nevada (Image G6.1). This outcrop consists of granite and similar rocks formed by slow cooling of felsic magma many kilometers beneath the surface. The magma that became these rocks formed when Pacific Ocean floor subducted beneath North America, along a convergent plate boundary, about 80 to 100 million years ago. Subsequent uplift and erosion stripped away volcanic rocks that once lay above the granite. During the last twenty thousand years, glacial erosion polished the outcrops.

Now tilt the image so you just see the horizon, and rotate so that you are looking southwest. This points you downstream along Merced Canyon (Image G6.2). If you fly slowly down the canyon, you will pass famous mountains—Half Dome (on your left) and eventually El Capitan on the right—whose hard surfaces appeal to mountain climbers.

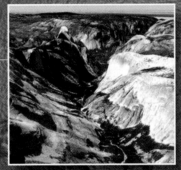

G6.2

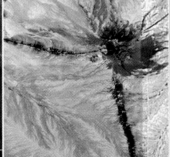
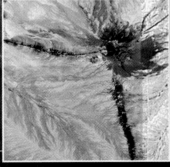

G6.3

Shiprock, New Mexico
(Lat 36°41'17.03"N, Long 108°50'09.31"W)

Fly to the coordinates provided and zoom out to an elevation of 10 km (6 miles). You are looking down on Shiprock. This is the eroded remnant of an explosive volcano that last erupted about 30 Ma (Image G6.3). Erosion stripped away layers of lava and ash that once formed the volcano, leaving behind dark intrusive rock that froze inside and just below the volcano. Shiprock consists of a rock type similar to basalt; this rock shattered into fragments as it approached the ground surface. Three dikes cut into the countryside, emanating from Shiprock like spokes from a wheel. These dikes are the remnants of wall-like intrusions.

Zoom down to 5 km (3 miles), tilt the image so you just see the horizon, and rotate so you're looking NW (Image G6.4). You can see the steep-sided mountain and two of the dikes. The mountain is 500 m (1,600 feet) in diameter and 600 m (2,000 feet) high.

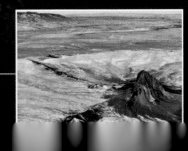

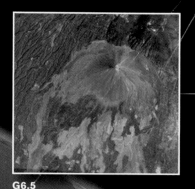

G6.5

G6.6

Izalco Volcano, El Salvador
(Lat 13°48'50.40"N, Long 89°37'57.74"W)

This volcano was active almost continuously from 1770 to 1958. Zoom to an elevation of 10 km (6 miles). Basalt lava flows flooded down Izalco's flanks and are still clearly visible (Image G6.5). Younger flows are darker colored than older flows, because the younger flows have had less time to react with the atmosphere and water to undergo weathering (see Chapter 7). The gray, evenly distributed material closer to the summit consists of tiny pellets of cooled lava. Zoom down to 7 km (4 miles), and tilt the image so that you are looking north and just see the horizon (Image G6.6). You can see nearby volcanoes. The summit of one, Santa Ana, has collapsed to form a circular depression called a caldera.

G6.7

Dikes, Western Australia
(Lat 22°50'13.69"S, Long 117°23'59.74"E)

Fly to this locality and zoom down to an elevation of 3.5 km (2 miles). You are hovering over the desert of northwestern Australia, an area where extensive areas of Precambrian rocks are exposed. Here, a number of dikes stand out in relief, because they are harder than the surrounding rock, which has eroded away. Tilt the image, look north, and you can see the dikes more clearly (Image G6.7).

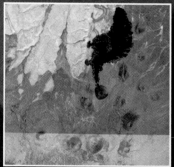

G6.8

G6.9

Cinder Cones, Arizona
(Lat 35°34'56.58"N, Long 111°37'55.10"W)

Fly to these coordinates and zoom to 19 km (12 miles) (Image G6.8). You can see many of the cinder cones that spot the landscape north of San Francisco Peak, a stratovolcano near Flagstaff, Arizona. The darkest spot is SP Crater, which erupted about 71 Ka. The black apron to the north of SP Crater is a 30 m-thick basalt flow. Descend to 2.5 km (1.5 miles), tilt and rotate so you are looking south for a better view (Image G6.9). At an elevation of 10 km (6 miles), fly 16.5 km (10 miles) to the SSE, to find Sunset Crater, which erupted an apron of bright orange tephra at 1 Ka. At this site (Lat 35°21'51.82"N, Long 111°30'10.87"W), zoom to 4 km, tilt the view, and look east (Image G6.10).

avalanches down the side of a volcano (an ash flow) and is still so hot when it settles that the glass shards weld together. Tuff that settles from air is called air-fall tuff, and tuff formed by the welding together of hot shards is called welded tuff.

- **Volcanic breccia** consists of larger fragments of volcanic debris. The fragments either fall through the air and accumulate, or form when a flow breaks into pieces that accumulate.

- **Hyaloclasite** is formed from lava that erupts under water or ice, and cools so quickly that it shatters into glassy fragments that then weld or cement together.

6.9 WHERE DOES IGNEOUS ACTIVITY OCCUR, AND WHY?

If you look at a map showing the distribution of igneous activity—the formation, movement, and in some cases eruption of molten rock—around the world (▶Fig. 6.19), you'll see that igneous activity occurs in four settings: (1) in volcanic arcs bordering deep-ocean trenches, (2) at isolated hot spots, (3) within continental rifts, and (4) along mid-ocean ridges. As evident from this list, most igneous activity takes place at established or newly forming plate boundaries. (Hot-spot volcanoes, however, which erupt in the interiors of plates, violate this rule.) Most volcanic activity along mid-ocean ridges happens underwater, at submarine volcanoes. Most volcanic activity in rifts, along convergent margins, and at hot spots takes place under the air, at subaerial volcanoes.

The Formation of Igneous Rocks at Volcanic Arcs—the Product of Subduction

Most subaerial volcanoes on Earth occur in long, curving chains, called volcanic arcs (or just arcs), adjacent to the deep-ocean trenches that mark convergent plate boundaries. Some of these arcs, such as the volcanoes of the Andes Mountains in South America, fringe the edge of a continent and are called continental arcs. Others, such as the volcanoes of the Aleutian Islands in Alaska, form oceanic islands and are called island arcs. They are called "arcs" because in many locations the volcanic chain defines a curve on a map. Recall that at convergent plate boundaries, where volcanic

FIGURE 6.19 The distribution of submarine and subaerial volcanoes worldwide. Note that volcanic activity occurs all along mid-ocean ridges, though most is submerged beneath the water and can't be seen. Most subaerial volcanoes lie in volcanic arcs bordering convergent plate boundaries. Others are found along continental rifts and at hot spots. Subaerial volcanoes are ones that rise above sea level.

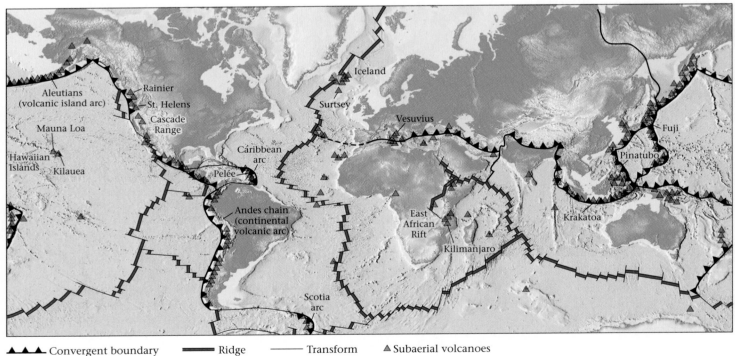

arcs form, oceanic lithosphere subducts and sinks into the mantle. How does this process trigger volcanic activity?

The oceanic lithosphere that subducts at convergent boundaries is made up of oceanic crust and the underlying lithospheric mantle. Some of the minerals constituting the oceanic crust contain volatiles. When the subducting crust sinks into the mantle to a depth of about 150 km, it is heated by the surrounding mantle to such a degree that its volatile compounds separate and enter the adjacent hot asthenosphere. The addition of volatiles, mainly water, causes the ultramafic rock of the asthenosphere to partially melt and produce a basaltic magma. This magma either rises directly, to erupt as basaltic lava, or undergoes fractional crystallization before erupting, becoming intermediate or felsic lava. Whether the crust on the downgoing slab partially melts and contributes to the volcanic system remains a subject of debate. If crustal melting makes a contribution, it is small.

In continental volcanic arcs, not all the mantle-derived basaltic magma rises directly to the surface; some gets trapped at the base of the continental crust, and some in magma chambers deep in the crust. When this happens, heat transfers into the continental crust and causes partial melting of this crust. Because much of continental crust is mafic to intermediate in composition to start with, the resulting magmas are intermediate to felsic in composition. This magma rises, leaving the basalt behind, and either cools higher in the crust to form plutons (▶Fig. 6.20), or rises to the surface and erupts. For this reason, granitic plutons and andesite lavas form at continental arcs.

Subduction now occurs along 60% of the margin of the present Pacific Plate, so volcanic arcs border 60% of the margin of the Pacific Ocean; geologists refer to the Pacific rim, therefore, as the "Ring of Fire." The volcanic arcs of the Ring of Fire include the Andes of South America, the Cascades of the northwestern United States, the Aleutian Islands of Alaska, the Kuril Islands off the eastern coast of Russia, Japan, and several arcs in the southwestern Pacific.

The Formation of Igneous Rocks at Hot Spots—a Surprise of Nature

Hawaii and other South Pacific island volcanoes are **hot-spot volcanoes**, isolated volcanoes that are independent of plate-boundary interactions. There are about 50 to 100 currently active hot-spot volcanoes scattered around the world (see Figs. 4.20–4.22). Oceanic hot-spot volcanoes erupt in the interior of oceanic plates away from convergent or divergent boundaries. Some, such as Iceland, sit astride a divergent boundary. (Geoscientists associate Iceland with a hot spot because its volcanoes generate far

FIGURE 6.20 Mt. Rushmore, a pluton of granite in the Black Hills of South Dakota, has been carved to display the colossal heads of four U.S. presidents. The sculptor Gutzon Borglum was able to carve these heads because of the uniformity and coherence of the pluton.

more lava than normal mid-ocean ridge volcanoes do.) And some hot-spot volcanoes grow on continents.

Continental hot-spot volcanoes erupt in the interior of continents. A continental hot-spot volcano produced the stunning landscape of Yellowstone National Park in northwestern Wyoming and adjacent states. The "yellow stone" exposed in the park consists of sulfur- and iron-stained layers of volcanic ash.

As we learned in Chapter 4, researchers think that many hot-spot volcanoes form above plumes of hot mantle rock that rise from the core-mantle boundary. Recent work suggests that some plumes may originate at shallower depths, and that there may be nonplume explanations for some hot spots.

According to the plume hypothesis, the rise of plumes occurs by the slow flow of solid rocks through the Earth's mantle—in other words, a plume does not consist of magma. But when the hot rock of a plume reaches the base of the lithosphere, decompression causes the rock (peridotite) of the plume to undergo partial melting, a process that generates mafic magma. The mafic magma then rises through the lithosphere, pools in a magma chamber in the crust, and eventually erupts at the surface, forming a volcano. In the case of oceanic hot spots, mostly mafic magma erupts. In the case of continental hot spots, some of the mafic magma erupts to form basalt, but some transfers heat to the continental crust, which then partially melts itself, producing felsic magmas that erupt to form rhyolite. The volume of magma erupted above a plume may change with time. When the top of a plume first arrives at the base of the lithosphere, it may have a bulbous head (like a lightbulb) in which a great deal of magma forms. As time passes, the head disappears, and only a narrow stalklike plume remains, in which less magma forms.

Mantle plume and
a hot-spot volcano

Subduction yields
a volcanic arc.

Melting occurs beneath
a mid-ocean ridge.

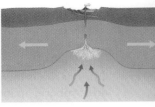

Melting occurs beneath
a continental rift.

Large volumes of magma erupt at a hot-spot volcano on
the oceanic crust, creating an oceanic island.

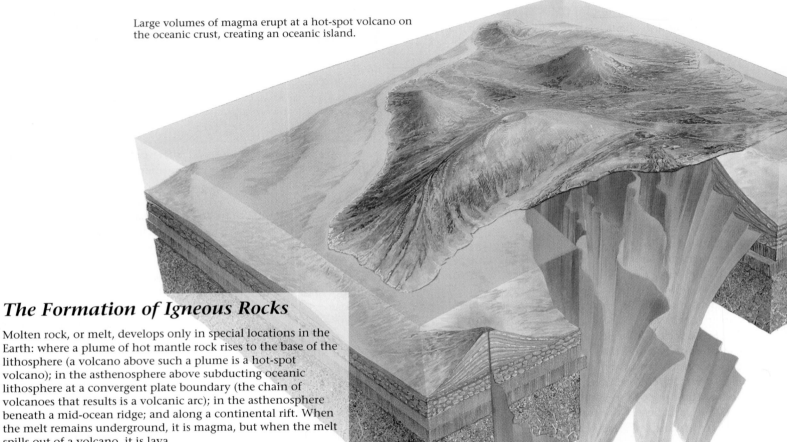

The Formation of Igneous Rocks

Molten rock, or melt, develops only in special locations in the
Earth: where a plume of hot mantle rock rises to the base of the
lithosphere (a volcano above such a plume is a hot-spot
volcano); in the asthenosphere above subducting oceanic
lithosphere at a convergent plate boundary (the chain of
volcanoes that results is a volcanic arc); in the asthenosphere
beneath a mid-ocean ridge; and along a continental rift. When
the melt remains underground, it is magma, but when the melt
spills out of a volcano, it is lava.

When magma or lava cools, different minerals form in
sequence until the melt solidifies (freezes) and igneous rock
forms. The composition of a melt depends on its origin and
cooling history. For example, partial melting of the mantle
results in basaltic magma. Basaltic magma is very hot, so when it
rises into the continental crust, it can transfer heat and cause
partial melting of the crust, yielding rhyolitic magma. Lava or
magma that cools quickly tends to be fine grained or glassy,
whereas lava or magma that cools slowly tends to be coarse
grained.

Igneous rock that forms by the solidification of magma
underground is intrusive rock. Blob-like intrusions are plutons;
sheet-like intrusions are dikes if they cut across preexisting
layers, and sills if they intrude parallel to preexisting layers. In
rock containing no preexisting layers, dikes are vertical and sills
are horizontal. Intrusions that are shaped like a blister are called
laccoliths.

Volcanoes erupt both lava flows and pyroclastic debris (ash
and other fragmental material ejected explosively). Igneous rock
that forms by the extrusion of lava or the accumulation of
pyroclastic debris is extrusive rock.

Less silica
(basalt)

More silica
(granite)

Granite forms from the cooling of a felsic melt, such as in a
continental pluton, whereas basalt results from the cooling of
a mafic melt, as at an oceanic hot-spot volcano.

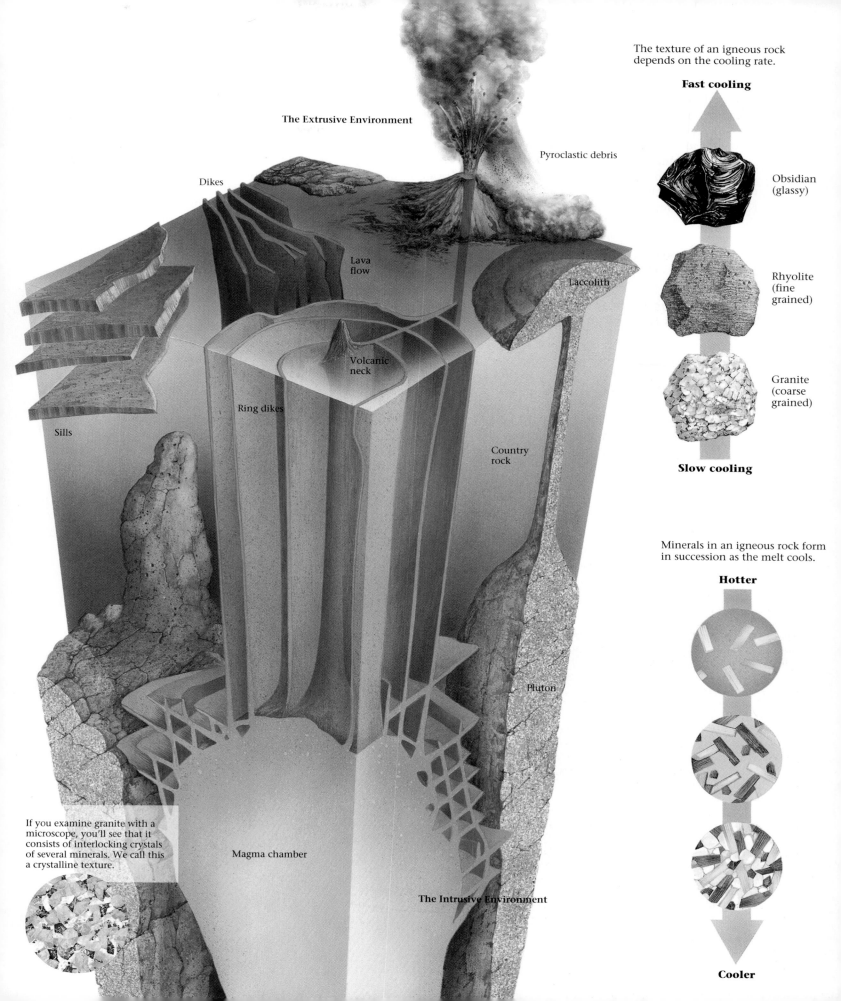

The Extrusive Environment

Dikes

Pyroclastic debris

Lava flow

Laccolith

Volcanic neck

Ring dikes

Country rock

Sills

Pluton

The texture of an igneous rock depends on the cooling rate.

Fast cooling

Obsidian (glassy)

Rhyolite (fine grained)

Granite (coarse grained)

Slow cooling

Minerals in an igneous rock form in succession as the melt cools.

Hotter

Cooler

If you examine granite with a microscope, you'll see that it consists of interlocking crystals of several minerals. We call this a crystalline texture.

Magma chamber

The Intrusive Environment

The position of a hot-spot plume remains relatively fixed with respect to the moving plates, so in the case of plate-interior hot spots, the drift of the plates causes a volcano that grew during one interval of time eventually to move off the hot spot. When this happens, the volcano dies and a new volcano forms. As a consequence, active hot-spot volcanoes commonly occur at the end of a chain of dead volcanoes, and this chain is called a **hot-spot track**. Hawaii, for example, is at the end of a track consisting of the Emperor seamount chain and the Hawaiian Islands (see Fig. 4.22). Similarly, Yellowstone Park lies at the end of a track manifested by a chain of now dead volcanoes along the Snake River Plain in Idaho.

Large Igneous Provinces (LIPs) and Their Effects on the Earth System

In many places on Earth, *particularly voluminous* quantities of mafic magma have erupted and/or intruded (▶Fig. 6.21). Some of these regions occur along the margins of continents, some in the interior of oceanic plates, and some in the interior of continents. The largest of these, the Ontong Java Oceanic Plateau of the western Pacific, covers an area of about 5,000,000 km² of the sea floor and has a volume of about 50,000,000 km³. Such provinces also occur on land. It's no surprise that these huge volumes of igneous rock are called **large igneous provinces** (LIPs).

The volume of rock that erupted during the formation of an LIP is much greater than the amount being erupted at even the most productive hot-spot volcano today. In fact, when active, the volume of material that erupted at a large LIP may have exceeded the amount that erupted along the Earth's entire mid-ocean ridge system during the same time. Thus, it seems that the eruption of an LIP is a special event in Earth history. Geologists suggest that LIPs may be a consequence of the formation of **superplumes** in the mantle—plumes that bring up vastly more hot asthenosphere than do normal plumes.

In the context of understanding the Earth System, it is important to keep in mind that eruption of an LIP may have a profound impact on the environment and may even affect the evolution of life. For example, the growth of a large undersea basalt plateau displaces seawater, causing a rise in sea level and causing seawater to flood the interiors of continents. The formation of the plateau could also change the geometry of ocean currents, which (as we will see in Chapter 18) play a major role in regulating climate. Eruption of volcanic gas and ash during production of an LIP could change global atmospheric temperature and clarity and could increase the supply of nutrients to the sea. A change in nutrient supply, in turn, could increase the amount of plankton that grows in seawater. Changes in ocean chemistry, global temperature, and atmospheric clarity triggered by an LIP eruption could even cause extinction of life forms.

FIGURE 6.21 A map showing the distribution of large igneous provinces (LIPs) on Earth. The red areas are underlain by immense volumes of basalt.

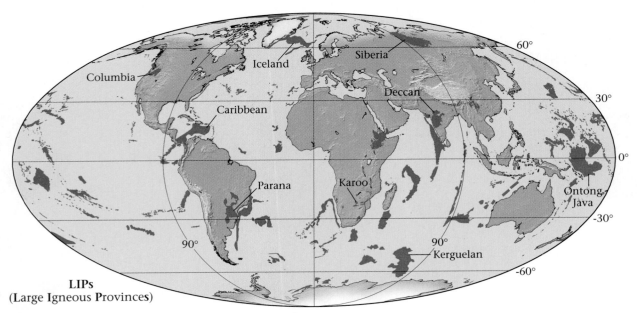

LIPs
(Large Igneous Provinces)

The Formation of Igneous Rocks at Rifts—a Record of Continental Breakup

According to the theory of plate tectonics (Chapter 4), rifts are places where continental lithosphere is being stretched and thinned. Successful rifting splits a continent in two and gives birth to a new mid-ocean ridge. As the continental lithosphere thins (before the continent splits), the weight of rock overlying the asthenosphere decreases, so pressure in the asthenosphere decreases and decompression melting takes place, producing basaltic magma, which rises into the crust. Some of this magma makes it to the surface, following the cracks that appear in the crust as a consequence of the stretching and breaking that accompany rifting, and erupts as basalt. However, some of the magma gets trapped in the crust and transfers heat to the crust. The resulting partial melting of the crust yields felsic (silicic) magmas that erupt as rhyolite or mix with basaltic magma to form andesitic magma. Thus, a sequence of volcanic rocks in a rift may include basaltic flows, sheets of rhyolitic ash, and even andesitic flows.

The most famous active rift, the East African Rift, presently forming a 4,000-km-long gash in the crust of Africa, has produced numerous volcanoes, including Mount Kilimanjaro. Recent rifting in North America has yielded the Basin and Range Province of Utah, Nevada, Arizona, and southeastern California. Though no currently erupting volcanoes exist in this region, the abundance of recent volcanic deposits suggests that igneous activity could occur again. Geoscientists are now monitoring the Mono Lakes volcanic area of California along the western edge of the Basin and Range, because of the possibility that volcanoes in this area may erupt in the very near future.

In some cases, the bulbous head of a mantle plume underlies a rift. More partial melting can occur in a plume head than in normal asthenosphere, because temperatures are higher in a plume head. Thus, an unusually large quantity of unusually hot magma forms where a rift overlies a plume head, so when volcanic eruptions begin in the rift, huge quantities of basaltic lava spew out of the ground, forming an LIP. The particularly hot basaltic lava that erupts at such localities has such low viscosity that it can flow tens to hundreds of kilometers across the landscape. Geoscientists refer to such flows as **flood basalts**. Flood basalts make up the bedrock of the Columbia River Plateau in Oregon and Washington (▶Fig. 6.22a), the Paraná Plateau in southeastern Brazil (▶Fig. 6.22b), the Karoo region of southern Africa, and the Deccan region of southwestern India.

The Formation of Igneous Rocks at Mid-Ocean Ridges—Hidden Plate Formation

Most igneous rocks at the Earth's surface form at mid-ocean ridges, that is, along divergent plate boundaries. Think about it—the entire oceanic crust, a 7- to 10-km-thick layer of basalt and gabbro that covers 70% of the Earth's surface, forms at mid-ocean ridges. And this entire volume gets subducted and replaced by new crust, over a period of about 200 million years.

Igneous magmas form at mid-ocean ridges for much the same reason they do at hot spots and rifts. As sea-floor spreading occurs and oceanic lithosphere plates drift away from the ridge, hot asthenosphere rises to fill the resulting space. As this asthenosphere rises, it undergoes decompression, which leads to partial melting and the generation of basaltic magma. As noted in Chapter 4, this magma rises into the crust and pools in a shallow magma chamber. Some cools slowly along the margins of the magma chamber to form massive gabbro, while some intrudes upward to fill vertical cracks that appear as newly formed crust splits apart (see Fig. 4.8). Magma that cools in the cracks forms basalt

FIGURE 6.22 (a) Flood basalts underlie the Columbia River Plateau in Washington and Oregon, the dark area on this map. (b) Iguazu Falls, on the Brazil-Argentina border. The falls flow over the huge flood basalt sheet (the black rock) of the Paraná Plateau. Flood basalt underlies all of the region in view.

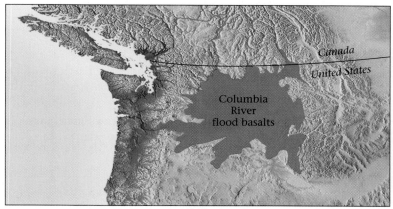

Columbia River flood basalts

Canada
United States

(a)

(b)

dikes, and magma that makes it to the sea floor and extrudes as lava forms basalt flows. The basalt flows of the sea floor don't look like those that erupt on land, because the seawater cools the lava so rapidly that it can't flow very far before solidifying into a pillow-shaped blob with a glassy rind. Eventually, the pressure of the lava inside a pillow breaks the glassy rind, and another pillow extrudes. Thus, sea-floor basalt is made up of a pile of pillows, known by geologists as **pillow basalt** (▶Fig. 6.23a–c).

In this chapter, we've focused on the diversity of igneous rocks, and why and where they form. We see that extrusive rocks develop at volcanoes. There's a lot more to say about volcanoes—eruptions have the potential to cause great harm. Proceed directly to Chapter 9 if you want to consider volcanic eruptions in detail at this point in your course.

> **Take-Home Message**
>
> Most igneous activity on Earth occurs at divergent and convergent plate boundaries, but igneous activity also occurs at hot spots and rifts. At large igneous provinces (LIPs), unusually large volumes of igneous rock have erupted.

cumstances—when the pressure decreases (decompression), when volatiles (such as water or carbon dioxide) are added to hot rock, and when heat is transferred by magma rising from the mantle into the crust.

- Magma occurs in a range of compositions: felsic (silicic), intermediate, mafic, and ultramafic. The composition of magma is determined in part by the original composition of the rock from which the magma formed and in part by the way the magma evolves, by such processes as assimilation and fractional crystallization.

- During partial melting, only part of the source rock melts to form magma. Magma tends to be more silicic than the rock from which it was extracted.

- Magma rises from the depth because of its buoyancy and because the pressure caused by the weight of overlying rock squeezes magma upward.

Chapter Summary

- Magma is liquid rock (melt) under the Earth's surface. Lava is melt that has erupted from a volcano at the Earth's surface.

- Magma forms when hot rock in the Earth partially melts. This process only occurs under certain cir-

(b)

(c)

FIGURE 6.23 (a) The formation of pillow basalt. **(b)** This pillow basalt forms part of an ophiolite, a slice of sea floor that was pushed up onto the surface of a continent during mountain building. **(c)** A cross section through a single pillow shows the glassy rind, with a more crystalline center.

Older pillows

New pillow forming

(a)

- Magma viscosity (its resistance to flow) depends on its composition. Felsic magma is more viscous than mafic magma.
- Geologists distinguish two types of igneous rocks. Extrusive igneous rocks form from lava that erupts out of a volcano and freezes in contact with air or the ocean. Intrusive igneous rocks develop from magma that freezes inside the Earth.
- Lava may solidify to form flows or domes, or it may explode into the air to form ash.
- Intrusive igneous rocks form when magma intrudes into preexisting rock (country rock) below Earth's surface. Blob-shaped intrusions are called plutons. Sheet-like intrusions that cut across layering in country rock are dikes, and sheet-like intrusions that form parallel to layering in country rock are sills. Huge intrusions, made up of many plutons, are known as batholiths.
- The rate at which intrusive magma cools depends on the depth at which it intrudes, the size and shape of the magma body, and whether circulating groundwater is present. The cooling time is reflected in the texture of an igneous rock.
- Crystalline (nonglassy) igneous rocks are classified according to texture and composition. Glassy igneous rocks are classified according to texture (a solid mass is obsidian; ash that has cemented or welded together is a tuff).
- The origin of igneous rocks can readily be understood in the context of plate tectonics. Magma forms at continental or island volcanic arcs along convergent margins, mostly because of the addition of volatiles to the asthenosphere above the subducting slab. Igneous rocks form at hot spots, owing to the decompression melting of a rising mantle plume. Igneous rocks form at rifts as a result of decompression melting of the asthenosphere below the thinning lithosphere. Igneous rocks form along mid-ocean ridges because of decompression melting of the rising asthenosphere.

Geopuzzle Revisited

Molten rock, or magma, forms in the upper mantle and lower crust, but only at special localities—along convergent and divergent plate boundaries, at hot spots, and in rifts. Most magma freezes underground, but some erupts as lava or ash at volcanoes. When melt cools and solidifies, it becomes igneous rock.

Key Terms

ash (p. 154)
assimilation (p. 158)
batholith (p. 163)
Bowen's reaction series (p. 160)
crystalline igneous rocks (p. 168)
dike (p. 162)
extrusive igneous rock (p. 154)
felsic magma (p. 158)
flood basalts (p. 179)
fractional crystallization (p. 159)
fragmental igneous rocks (p. 168)
geotherm (p. 155)
geothermal gradient (p. 157)
glassy igneous rocks (p. 168)
hot-spot track (p. 178)
hot-spot volcanoes (p. 175)
hyaloclasite (p. 174)
igneous rock (p. 153)
intermediate magma (p. 158)
intrusive igneous rock (p. 154)

laccolith (p. 163)
large igneous province (p. 178)
lava (p. 153)
mafic magma (p. 158)
magma (p. 154)
obsidian (p. 169)
partial melting (p. 158)
pegmatite (p. 168)
pillow basalt (p. 180)
plutons (p. 163)
pumice (p. 169)
pyroclastic debris (p. 154)
pyroclastic rocks (p. 171)
scoria (p. 169)
sill (p. 162)
stoping (p. 163)
superplumes (p. 178)
tachylite (p. 169)
tuff (p. 171)
ultramafic magma (p. 158)
vesicles (p. 169)
viscosity (p. 158)
volcanic breccia (p. 174)
volcano (p. 153)
xenolith (p. 163)

Review Questions

1. How is the process of freezing magma similar to that of freezing water? How is it different?
2. What is the source of heat in the Earth? How did the first igneous rocks on the planet form?
3. Describe the three processes that are responsible for the formation of magmas.
4. Why are there so many different types of magmas?
5. Why do magmas rise from depth to the surface of the Earth?
6. What factors control the viscosity of a melt?
7. What factors control the cooling time of a magma within the crust?
8. How does grain size reflect the cooling time of a magma?
9. What does the mixture of grain sizes in a porphyritic igneous rock indicate about its cooling history?
10. Describe the way magmas are produced in subduction zones.

11. What processes in the mantle may be responsible for causing hot-spot volcanoes to form?

12. Describe how magmas are produced at continental rifts.

13. What is a large igneous province (LIP)? How might the formation of LIPs have affected the Earth System?

14. Why does melting take place beneath the axis of a mid-ocean ridge?

On Further Thought

1. If you look at the Moon, even without a telescope, you see broad areas where its surface appears relatively darker and smoother. These areas are called *mare* (plural: *maria*), from the Latin word for "sea." The term is misleading, for they are not bodies of water, but rather plains of igneous rock formed after huge meteors struck the Moon and formed very deep craters. These impacts occurred early in the history of the Moon, when its insides were warmer. With this background information in mind, propose a cause for the igneous activity, and suggest the type of igneous rock that fills the mare. (*Hint*: Think about how the presence of a deep crater affects pressure in the region below the crater, and think about the viscosity of a magma that could spread over such a broad area.)

2. The Cascade volcanic chain of the northwestern United States is only about 800 km long (from the southernmost volcano in California to the northernmost one in Washington State). The volcanic chain of the Andes is several thousand kilometers long. Look at a map showing the Earth's plate boundaries, and explain why the Andes volcanic chain is so much longer than the Cascade volcanic chain.

3. A 250-m-high cliff, known as the Palisades, forms the western shore of the Hudson River at the latitude of New York City (see adjacent photo). This cliff exposes a sill of dark igneous rock that intruded into cool sedimentary rock between 186 and 192 Ma, when what is now the east coast of North America was an active rift. The rock in the sill is not homogeneous. At its top and bottom, the sill consists of several meters of basalt. The interior of the sill contains layers of different minerals. The bottom layer consists mostly of olivine, the next layer mostly of pyroxene, and the top layer mostly of plagioclase (see adjacent cross section). It is significant that the composition of the plagioclase lower in the sill contains more Ca, and the plagioclase at the top contains more Na. Explain the internal character of the Palisades sill.

Suggested Reading

Best, M. G., and E. H. Christiansen. 2001. *Igneous Petrology*. 2nd ed. Oxford: Blackwell Science.

Faure, G. 2000. *Origin of Igneous Rocks: The Isotopic Evidence*. New York: Springer-Verlag.

LeMaitre, R. E., ed. 2002. *Igneous Rocks: A Classification and Glossary of Terms*, 2nd ed. Cambridge: Cambridge University Press.

Leyrit, H., C. Montenat, and P. Bordet, eds. 2000. *Volcaniclastic Rocks, from Magmas to Sediments*. London: Taylor & Francis.

Mackenzie, W. S. 1982. *Atlas of Igneous Rocks and Their Textures*. New York: Wiley.

Middlemost, E. A. K. 1997. *Magmas, Rocks and Planetary Development: A Survey of Magma/Igneous Rock Systems*. Boston: Addison-Wesley.

Philpotts, A. R. 2003. *Petrography of Igneous and Metamorphic Rocks*. Long Grove, IL: Waveland Press.

Thorpe, R., and G. Brown. 1985. *The Field Description of Igneous Rocks*. New York: Wiley.

Winter, J. D. 2001. *Introduction to Igneous and Metamorphic Petrology*. Upper Saddle River, N.J.: Prentice-Hall.

Young, D. A. 2003. *Mind over Magma*. Princeton, N.J.: Princeton University Press.

(a)

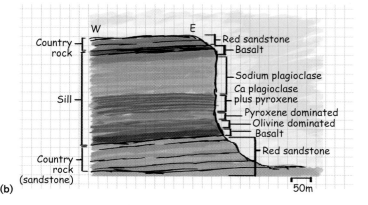

(b)

A Surface Veneer: Sediments, Soils, and Sedimentary Rocks

These cliffs, near Bryce Canyon (Utah), expose beds of sedimentary rock deposited in lakes, and by streams, over 40 million years ago. Present-day erosion has produced aprons of debris at the base of the cliffs.

Geopuzzle

Why do the walls of the Grand Canyon display spectacular layers of different-colored rock? Why do some layers form vertical cliffs while others do not?

In every out-thrust headland, in every curving beach, in every grain of sand there is the story of the Earth.
—*Rachel Carson (writer and ecologist, 1907–1964)*

7.1 INTRODUCTION

In the 1950s, the government of Egypt decided to build the Aswan High Dam to trap the water of the Nile River in a huge reservoir. To identify a good site for the dam's foundation, geologists drilled holes into the ground to find the depth to bedrock. They discovered that the present-day Nile River flows on the surface of a 1.5-km-thick layer of loose debris (gravel, sand, and mud) that fills a canyon that was once as large as the Grand Canyon (▶Fig. 7.1). The carving and subsequent filling of this canyon baffled geologists, because today the river flows along a plain almost at sea level. How could the river have carved a canyon 1.5 km deep, and why did the canyon later fill with sediment?

The origin of the Nile "canyon" remained a mystery until the summer of 1970, when geologists drilled holes into the floor of the Mediterranean Sea to find out what lay beneath. They expected the sea floor to be covered with shells of plankton (tiny floating organisms) that had settled out of the water, or with clay that rivers had carried to the sea. To their surprise, however, they found that, in addition to clay and plankton shells, a 2-km-thick layer of halite and gypsum lies beneath the floor of the Mediterranean Sea. These minerals form when seawater dries up, allowing the salt in the seawater to precipitate (see Chapter 5). The researchers realized that to yield a layer that is 2 km thick, the entire Mediterranean would have had to dry up completely several times, with the sea refilling after each drying event. This discovery solved the mystery of the pre-Nile canyon. When the Mediterranean Sea dried up, the Nile River was able to cut a canyon down to the level of the dry sea floor; and when the sea refilled with water, this canyon flooded and filled with sand and gravel.

Why did the Mediterranean Sea dry up? Only 10% of the water in the Mediterranean enters the sea from rivers, and since the sea lies in a hot, dry region, ten times that amount of water evaporates from its surface each year. Thus, most of the water in the Mediterranean enters through the Strait of Gibraltar from the Atlantic Ocean. If this flow stops, the Mediterranean Sea evaporates. About 6 million years ago, the northward-drifting African Plate collided with the European Plate, forming a natural dam separating the Mediterranean from the Atlantic. When global sea level dropped, the dam emerged above sea level and stopped the flow of water from the Atlantic, and the Mediterranean evaporated. All the salt that had been dissolved in its water accumulated as a solid deposit of halite and gypsum on the floor of the resulting basin. When sea level rose above the dam level, a gigantic flood rushed from the Atlantic into the Mediterranean, filling the basin again. This process was repeated many times. About 5.5 million years ago, the Mediterranean rose to its present level, and gravel, sand, and mud carried by the Nile River filled the Nile canyon.

Geologists refer to the kinds of deposits just described—sand, mud, gravel, halite and gypsum accumulates, shell fragments—as sediment. **Sediment**, in general, consists of loose fragments of rocks or minerals broken off bedrock, mineral crystals that precipitate directly out of water, and shells (formed when organisms extract ions from water). Much of what we know about the history of the Earth (including the amazing story of the Mediterranean Sea) comes from studying sediments—not only those that remain "unconsolidated" (loose and not connected), but also those that have been bound together into sedimentary rock. Formally defined, **sedimentary rock** is rock that forms at or near the surface of the Earth by the precipitation of minerals from water solutions, by the growth of skeletal material in organisms, or by the cementing together of shell fragments or of loose grains derived from preexisting rock. Layers of sediment and sedimentary rock are like the pages of a book, recording tales of ancient events and ancient environments on the ever-changing face of the Earth.

Sediments and sedimentary rocks only occur in the upper part of the crust—in effect, they form a surface veneer,

FIGURE 7.1 The present Nile River overlies a deep canyon, now filled with layers of sand and mud. The bottom of the canyon lies 1.5 km below sea level.

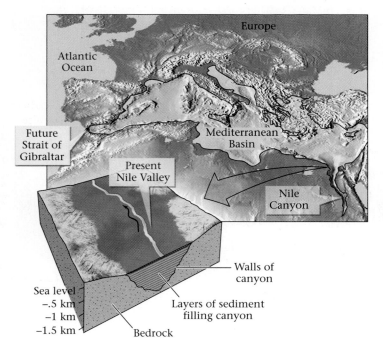

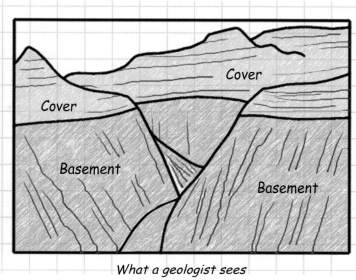

What a geologist sees

FIGURE 7.2 Near the bottom of the Grand Canyon, we can see the boundary between the sedimentary veneer, or cover (here, a succession of horizontal layers), and the older basement (here, the steep cliff of dark metamorphic rock that goes down to the river). The Colorado River flows along the floor of the canyon. A geologist's sketch emphasizes the contact, or boundary, between cover and basement.

or "cover," on older igneous and metamorphic rocks, which make up the "basement" of the crust (▶Fig. 7.2). This veneer ranges from nonexistent, in places where igneous and metamorphic rocks crop out at the Earth's surface, to several kilometers thick in regions called sedimentary basins. Though sediments and sedimentary rocks cover more than 80% of the Earth's surface, they actually constitute less than 1% of the Earth's mass. Nevertheless, they represent a uniquely important rock type, both because they contain a historic record and because they contain the bulk of the Earth's energy resources, as we'll see in Chapter 14. Further, some sediments transform into soil, essential for life. Let's now look at how sediments, soils, and sedimentary rocks form, and what these materials can tell us about the Earth System.

7.2 HOW DOES WEATHERING LEAD TO SEDIMENT FORMATION?

The Mountains Crumble

If you ever have the chance to hike or drive through granitic mountains, such as the Sierra Nevada of California or the Coast Mountains of Canada, you may notice that in some outcrops the granite surface looks hard and smooth and contains shining crystals of feldspar, biotite, and quartz, whereas in other outcrops the granite surface looks grainy and rough—the feldspar crystals appear dull, the biotite flakes have spots of rust, and the rock may peel apart like an onion. Why are these two types of outcrop dif-

ferent? The first type exposes "fresh" or unweathered rock whose mineral grains have kept their original composition and shape, while the second type exposes weathered rock that has reacted with air and/or water at or near the Earth's surface and has thus been weakened (▶Fig. 7.3).

FIGURE 7.3 This outcrop shows the contrast between fresh and weathered granite. The rock below the notebook is fresh—the outcrop face is a fairly smooth fracture. The rock above the notebook is weathered—the outcrop face is crumbly, breaking into grains that have fallen and collected on the ledge.

Weathered granite

Fresh granite

Weathering refers to the processes that break up and corrode solid rock, eventually transforming it into sediment. All mountains and other features on the Earth's surface sooner or later crumble away because of weathering. Geologists distinguish two types of weathering: physical weathering and chemical weathering. Just as a plumber can unclog a drain by using physical force (with a plumber's snake) or by causing a chemical reaction (with a dose of liquid drain opener), nature can attack rocks in two ways.

Physical Weathering

Physical weathering, sometimes referred to as mechanical weathering, breaks intact rock into unconnected grains or chunks, collectively called debris or detritus. Each size range of grains has a name (the measurements are grain diameters):

- *boulders*   more than 256 millimeters (mm)
- *cobbles*   between 64 mm and 256 mm
- *pebbles*   between 2 mm and 64 mm
- *sand*   between 1/16 mm and 2 mm
- *silt*   between 1/256 mm and 1/16 mm
- *mud*   less than 1/256 mm

For convenience, geologists refer to boulders, cobbles, and pebbles as coarse-grained sediment; sand as medium-grained sediment; and silt and mud as fine-grained sediment. Many different phenomena contribute to physical weathering.

Jointing. Rocks buried deep in the Earth's crust endure enormous pressure because the overburden (overlying rock) weighs a lot and presses down on the buried rock. Rocks at depth are also warmer than rocks nearer the surface, because of Earth's geothermal gradient (see Chapter 6). Over long periods, moving water, air, and ice at the Earth's surface grind away and remove overburden, so rock formerly at depth rises closer to the Earth's surface. As a result, the pressure squeezing this rock decreases, and the rock becomes cooler. A change in pressure causes rock to change shape slightly, for the same reason that a rubber ball changes shape when you squeeze it and then let go. Similarly, a change in temperature causes rock to change shape for the same reason that a baked apple changes shape when it is removed from the oven and cools. But unlike a rubber ball or a soft apple, hard rock may break into pieces when it changes shape (▶Fig. 7.4a, b). Natural cracks that form in rocks due to removal of overburden or due to cooling (and for other reasons as well; see Chapter 11) are known as **joints**.

Almost all rock outcrops contain joints. Some joints are fairly planar, some curving, and some irregular. The spacing between adjacent joints varies from less than a centimeter to tens of meters. Joints can break rock into large or small rectangular blocks, onion-like sheets, irregular chunks, or pillar-like columns. Typically, large granite plutons split into onion-like sheets along joints that lie parallel to the mountain face. This process is called exfoliation (▶Fig. 7.5a). Sedimentary rock layers tend to break into rectangular blocks (▶Fig. 7.5b). The for-

FIGURE 7.4 **(a)** The weight of overburden creates pressure on rocks at depth. Removal of the overburden by erosion allows once-deep rocks to be exposed at the Earth's surface. **(b)** The exposure of once-deep rocks causes them to crack. Different rock types crack in different ways. Here, the granite pluton develops exfoliation joints as well as vertical joints, while the sedimentary rock layers develop mostly vertical joints. Joint-bounded blocks break off the outcrop.

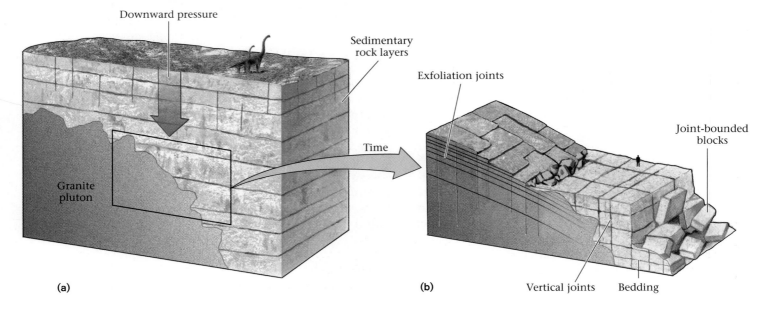

(a)

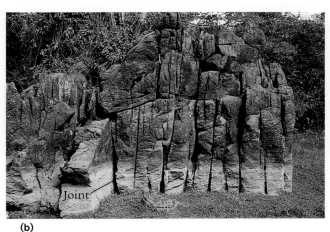

Joint

(b)

(c)

FIGURE 7.5 **(a)** Exfoliation joints in the Sierra Nevada, California. **(b)** Vertical joints in sedimentary rock (Brazil). **(c)** Talus has accumulated at the base of these cliffs near Mt. Snowdon in Wales.

mation of joints turns formerly intact bedrock into loose blocks. Eventually, these blocks fall from the outcrop at which they formed. After a while, they may collect in an apron of **talus**, rock rubble at the base of a slope (▶Fig. 7.5c).

Frost wedging. Freezing water bursts pipes and shatters bottles, because water expands when it freezes and pushes the walls of the container apart. The same phenomenon happens in rock. When the water trapped in a joint freezes, it forces the joint open and may cause the joint to grow. Such frost wedging helps break blocks free from intact bedrock (▶Fig. 7.6a). Of course, frost wedging is most common where water periodically freezes and thaws, as occurs in temperate climates or at high altitudes in mountains.

Root wedging. Have you ever noticed how the roots of an old tree can break up a sidewalk? Even though the wood of roots doesn't seem very strong, as roots expand they apply pressure to their surroundings. Tree roots that grow into joints can push those joints open in a process known as root wedging (▶Fig. 7.6b). Even the roots of small plants, fungi, and lichen get into the act by splitting open small cracks and pores.

Salt wedging. In arid climates, dissolved salt in groundwater precipitates and grows as crystals in open pore spaces in rocks. This process, called salt wedging, pushes apart the surrounding grains and so weakens the rock that when exposed to wind and rain, the rock disintegrates into separate grains. The same phenomenon happens in coastal areas, where salt spray percolates into surface rock and then dries (▶Fig. 7.6c).

Thermal expansion. When the heat of an intense forest fire bakes a rock, the outer layer of the rock expands. On cooling, the layer contracts. This change creates forces in the rock sufficient to make the outer part of the rock spall, or break off in sheet-like pieces.

Animal attack. Animal life also contributes to physical weathering: burrowing creatures, from earthworms to gophers, push open cracks and move rock fragments. And in the past century, humans have become perhaps the most energetic agents of physical weathering on the planet. When we excavate quarries, foundations, mines, or roadbeds by digging and blasting, we shatter and displace rock that might otherwise have remained intact for millions of years more.

FIGURE 7.6 Examples of processes contributing to physical weathering. **(a)** During the summer, cracks are closed. During the winter, water in the cracks freezes and forces rocks apart. Ice can even lift blocks up. **(b)** The roots of this old pine tree in Zion National Park, Utah, originally grew in exfoliation joints. Eventually, the roots pried the rock above the joints free. Thus, the roots are now exposed. **(c)** These gravestones, near the ruin of a medieval abbey on the seacoast near Whitby, England, absorbed salt from the sea spray. Salt wedging has resulted in honeycomb-like weathering.

Chemical Weathering

Up to now we've taken the "plumber's-snake approach" to breaking up rock; now let's look at the "liquid-drain-opener approach." **Chemical weathering** refers to the chemical reactions that alter or destroy minerals when rock comes in contact with water solutions or air. Because many of these reactions proceed more quickly in warm, wet conditions, chemical weathering takes place much faster in the tropics than it does in deserts, or near the poles. Chemical weathering in warm, wet climates can produce a layer of rotten rock, called **saprolite**, over 100 m thick. Common reactions involved in chemical weathering include the following.

Dissolution. Chemical weathering during which minerals dissolve into water is called dissolution. Dissolution primarily affects salts and carbonate minerals, but even quartz dissolves slightly (▶Fig. 7.7a, b). Some minerals, such as

halite, can dissolve rapidly in pure rainwater. But some, such as calcite, dissolve rapidly only when the water is acidic, meaning that it contains an excess of hydrogen ions (H^+). Acidic water reacts with calcite to form a solution and bubbles of CO_2 gas (see Chapter 5).

How does the water in rock near the surface of the Earth become acidic? As rainwater falls, it dissolves carbon dioxide gas in the atmosphere, and as the water sinks down through soil containing organic debris, it reacts with the debris. Both processes yield carbonic acid. Because of the solubility of calcite, limestone and marble (two types of rock composed of calcite) dissolve, widening joints and leading to the formation of caverns (▶Fig. 7.7c; see Chapter 19).

Hydrolysis. During hydrolysis, water chemically reacts with minerals and breaks them down. (*Lysis* means "loosen" in Greek.) Hydrolysis works faster in slightly acidic water. For example, potassium feldspar, a common mineral in gran-

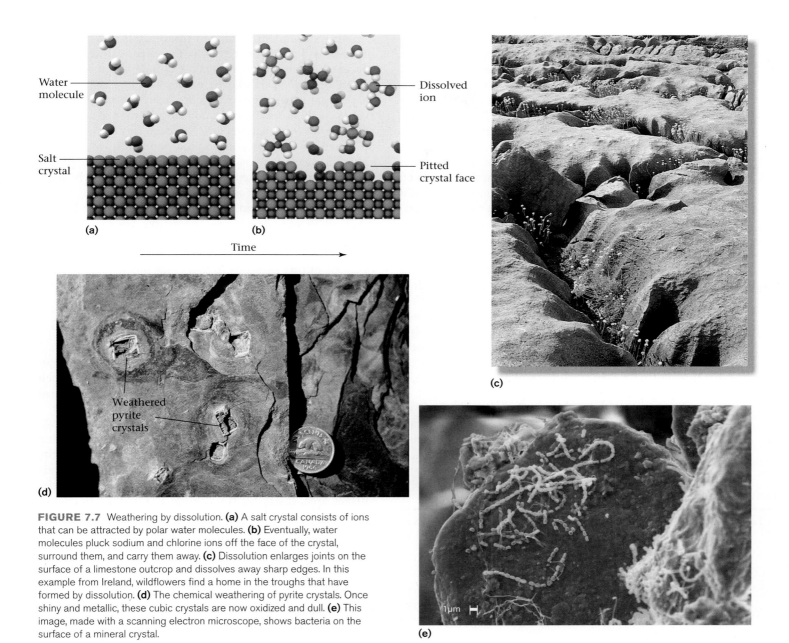

FIGURE 7.7 Weathering by dissolution. **(a)** A salt crystal consists of ions that can be attracted by polar water molecules. **(b)** Eventually, water molecules pluck sodium and chlorine ions off the face of the crystal, surround them, and carry them away. **(c)** Dissolution enlarges joints on the surface of a limestone outcrop and dissolves away sharp edges. In this example from Ireland, wildflowers find a home in the troughs that have formed by dissolution. **(d)** The chemical weathering of pyrite crystals. Once shiny and metallic, these cubic crystals are now oxidized and dull. **(e)** This image, made with a scanning electron microscope, shows bacteria on the surface of a mineral crystal.

ite, reacts with acidic water to produce kaolinite (a type of clay) and other dissolved ions.

Hydrolysis reactions break down not only feldspars, but many other silicate minerals as well—amphiboles, pyroxenes, micas, and olivines all react slowly and transform into various types of clay. Quartz also undergoes hydrolysis, but does so at such a slow rate that it survives weathering in most climates.

Oxidation. Chemists refer to a reaction during which an element loses electrons as an oxidation reaction, because such a loss commonly takes place when elements combine with oxygen. The oxidation, or rusting, of iron, is a familiar process example of oxidation.

Oxidation reactions in rocks transform iron-bearing minerals (such as biotite and pyrite) into a rusty-brown mixture of various iron-oxide and iron-hydroxide minerals, such as hematite and goethite (►Fig. 7.7d). Reactions such as these made the surface of Mars red.

Hydration. Hydration, the absorption of water into the crystal structure of minerals, causes some minerals, such as certain types of clay, to expand.

The relative stability of minerals during chemical weathering. Not all minerals undergo chemical weathering at the same rates. Some weather in a matter of months or years, whereas others remain unweathered for millions of years. In humid climates, for example, halite and calcite weather faster than most silicate minerals. Of the silicate minerals, those that crystallize at the highest temperatures are generally less stable under the cool temperatures of the Earth's surface than those that crystallize at lower temperatures (▶Table 7.1). The difference depends partly on crystal structure and partly on chemical composition. Specifically, minerals with fewer linkages between silicon-oxygen tetrahedra tend to have weaker structures and thus weather faster than do minerals with stronger structures. And minerals containing iron, magnesium, sodium, potassium, and aluminum tend to weather faster than minerals without these elements. Thus, quartz (pure SiO_2), which consists of a 3-D network with strong bonds in all directions, is very stable. When a granite (which contains quartz, mica, and feldspar) undergoes chemical weathering everything but quartz transforms to clay. That's why beaches typically consist of quartz sand; quartz is the only mineral left after the other minerals have turned to clay and washed away.

Chemical weathering produced by organisms. Until fairly recently, geoscientists tended to think of chemical weathering as strictly an inorganic chemical reaction occurring entirely independently of life forms. But it is now clear that organisms play a major role in the chemical-weathering process. For example, the roots of plants, fungi, and lichens secrete organic acids that help dissolve minerals in rocks; these organisms extract nutrients from the minerals. Microbes, such as bacteria, are amazing in that they literally eat minerals for lunch (▶Fig. 7.7e). Bacteria can metabolize an incredible range of compounds, depending on the environment they are living in. They pluck off molecules from minerals, and use the energy from the molecules' chemical bonds to supply their own life force. Mineral-eating bacteria live at depths of up to a few kilometers in the Earth's crust; at greater depths, temperatures are too high for them to survive. If microbes can live off the minerals below the surface of the Earth, can they do so beneath the surface of Mars? Future missions to Mars may provide the answer.

Physical and Chemical Weathering Working in Concert

So far we've looked at the processes of chemical and physical weathering separately, but in the real world they happen together, aiding each other in disintegrating rock to form sediment.

Physical weathering speeds up chemical weathering. To understand why, keep in mind that chemical-weathering reactions take place at the surface of a material, so the overall rate at which chemical weathering occurs depends on the ratio of surface area to volume—the greater the surface area, the faster the volume as a whole can chemically weather. When jointing (physical weathering) breaks a large block of rock into smaller pieces, the surface area increases, so chemical weathering happens faster (▶Fig. 7.8).

Similarly, chemical weathering speeds up physical weathering, because chemical weathering—by dissolving away grains or cements that hold a rock together, by transforming hard minerals (such as feldspar) into soft minerals (such as clay), or by causing minerals to absorb water and expand—makes the rock weaker, so it can disintegrate more easily (▶Fig. 7.9a–d). If you drop a block of fresh granite on the ground, it will most likely stay intact, but if you drop a block of chemically weathered granite on the ground, it will crumble into a pile of sand and clay.

Note that weathering happens faster at edges, and even faster at the corners of broken blocks. This is because weathering attacks a flat face from only one direction, an edge from two directions, and a corner from three directions. Thus, with time, edges of blocks become blunt and corners become rounded (▶Fig. 7.10a). In rocks such as granite, which do not contain layering that can affect

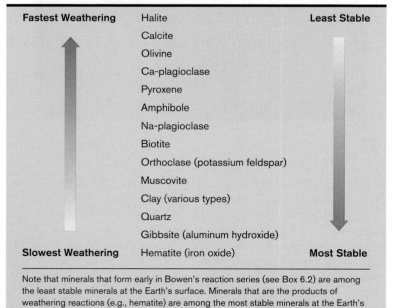

TABLE 7.1 **Relative Stability of Minerals at the Earth's Surface**

Fastest Weathering		Least Stable
	Halite	
	Calcite	
	Olivine	
	Ca-plagioclase	
	Pyroxene	
	Amphibole	
	Na-plagioclase	
	Biotite	
	Orthoclase (potassium feldspar)	
	Muscovite	
	Clay (various types)	
	Quartz	
	Gibbsite (aluminum hydroxide)	
Slowest Weathering	Hematite (iron oxide)	Most Stable

Note that minerals that form early in Bowen's reaction series (see Box 6.2) are among the least stable minerals at the Earth's surface. Minerals that are the products of weathering reactions (e.g., hematite) are among the most stable minerals at the Earth's surface. Mafic minerals weather by oxidation, felsic minerals by hydrolysis, carbonates and salts by dissolution, and oxide minerals don't weather at all.

FIGURE 7.8 The surface area per unit volume of a block increases every time you break the block into more pieces. For example, a coherent 1 m^3 block has a surface area of 6 m^2. Divide the block into eight pieces, and the surface area increases to 12 m^2, but the volume stays the same. The rate of chemical weathering increases as the surface area increases, because the weathering reactions occur at the surface. (To picture this, think about how fast granular sugar dissolves as compared with a solid cube of sugar.)

Fewer cracks, less surface area

More cracks, more surface area

Surface area = 6 m^2

Surface area = 12 m^2

Surface area = 60 m^2

weathering rates, rectangular blocks transform into spheroidal shapes (▶Fig. 7.10b).

Under a given set of environmental conditions, not all rock types weather at the same rate. When different rocks in an outcrop undergo weathering at different rates, we say that the outcrop has undergone differential weathering. Because of differential weathering, cliffs composed of a variety of rock layers take on a stair-step or sawtooth-like shape (▶Fig. 7.10c). Weak layers may weather away beneath a more resistant layer, creating an overhang. Similarly, the rate at which the land surface weathers depends on the rock type, so valleys tend to develop over weak rocks, while strong rocks hold up hills.

You can easily see the consequences of differential weathering if you walk through a graveyard. The inscriptions on some headstones are sharp and clear, whereas those on other stones have become blunted or have even disappeared (▶Fig. 7.11a, b). That's because the minerals in these different stones have different resistances to weathering. Granite, an igneous rock with a high quartz content, retains inscriptions the longest. But marble, a metamorphic rock composed of calcite, dissolves away relatively rapidly in acidic rain.

> **Take-Home Message**
>
> Rocks at or near the surface of the Earth undergo weathering. During chemical weathering, minerals dissolve and/or transform into new minerals (such as clay and iron oxide). During physical weathering, rock breaks down into smaller pieces.

FIGURE 7.9 Chemical weathering aids physical weathering by weakening the attachments between grains. **(a)** This rock is solid. **(b)** Susceptible minerals have started to weather. **(c)** The rock crumbles. **(d)** Weaker minerals break up or react to form clay and wash away.

Intact rock

Feldspar

Quartz

Biotite

Minerals weather; grains break apart

Rock has broken into loose grains; feldspar has turned into clay

Clay

Clay washes away; quartz grains become rounded

(a)   (b)   (c)   (d)

Time

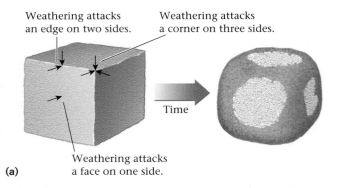

Weathering attacks an edge on two sides.

Weathering attacks a corner on three sides.

Weathering attacks a face on one side.

Time

(a)

(b)

(c)

FIGURE 7.10 **(a)** Weather attacks more vigorously at edges and most vigorously at corners, resulting in a rounded block. **(b)** Spheroidal weathering of granite blocks in Joshua Tree National Monument, California. **(c)** Sawtooth shape of an outcrop of weathered sedimentary rock, in New Mexico. Weak shale layers are softer than sandstone layers, so the sandstone layers stick out relative to the shale.

(a)

(b)

FIGURE 7.11 **(a)** Inscriptions in a granite headstone remain sharp for centuries. This example dates from 1856. **(b)** Inscriptions in a marble headstone weather away fairly rapidly. This example, from the same cemetery, dates from 1872.

7.3 SOIL: SEDIMENT INTERWOVEN WITH LIFE

I bequeath myself to the dirt to grow from the grass I love;
if you want me again, look for me under your boot soles.
　　　　　　　　　　　　—Walt Whitman (1819–1892)

Walt Whitman, an American poet of the nineteenth century, reveled in the natural world, and in his poetry he evoked life's cycle of growth, death, and rebirth. As the above quote from Whitman's masterpiece, *Leaves of Grass*, indicates, Whitman recognized that "dirt"—or what more technically can be called soil—plays an essential role in giving life, and that life, in turn, plays an essential role in generating soil. **Soil** consists of rock and sediment that has been modified by physical and chemical interaction with organic material and rainwater, over time, to produce a substrate that can support the growth of plants. Soil is one type of **regolith** (from the Greek *rhegos*, which means cover), a name that geologists use for any kind of unconsolidated debris that covers bedrock. By unconsolidated, in this context, we mean loose or unconnected—regolith includes both soil and accumulations of sediment that have not evolved into soil.

Geologists distinguish between residual soil, which forms directly from underlying bedrock, and transported soil, which forms from sediment that has been carried in from elsewhere. Transported soils include those formed from deposits left by rivers, glaciers, or wind. Much of the dense rainforest of Brazil grows on a residual soil formed on deeply weathered Precambrian bedrock, whereas some of the heavily farmed soil of the American Midwest is a transported soil in that it developed at the end of the ice age on thick accumulations of fine silt deposited by strong winds.

Soil is one of our planet's most valuable resources, for without it there could be no agriculture, forestry, ranching, or even home gardening. Because it takes special conditions and time to make soil, it is a resource that must be conserved—as we will see, misuse of soil can lead to its loss. We now look at how soil forms, and at the characteristics of soil.

Formation of Soil and Soil Horizons

Three processes taking place at or just below the surface of the Earth contribute to soil formation. First, chemical and physical weathering produces loose debris, new mineral grains (e.g., clay), and ions in solution.

Second, rainwater percolates through the debris and carries dissolved ions and clay flakes downward. The region in which this downward transport occurs is the **zone of leaching**, because leaching means "extracting and absorbing." Farther down, new mineral crystals precipitate directly out of the water or form when the water reacts with debris, and the water leaves behind its load of fine clay. The region in which new minerals and clay collect is the **zone of accumulation** (▶Fig. 7.12a, b).

FIGURE 7.12 During the formation of soil, the downward percolation of water creates a zone of leaching and a zone of accumulation. **(a)** In soil, the percolating water carries ions and clay downward. Soil formation also involves the metabolism of microbes and fungi and the addition of organic matter at the surface and underground. **(b)** The same process happens when you pour hot water through coffee grounds or tea leaves into a pot containing bread crumbs. Elements in the coffee or tea dissolve in the water and are carried down and collect in the bread crumbs; coffee eventually leaks from the pot.

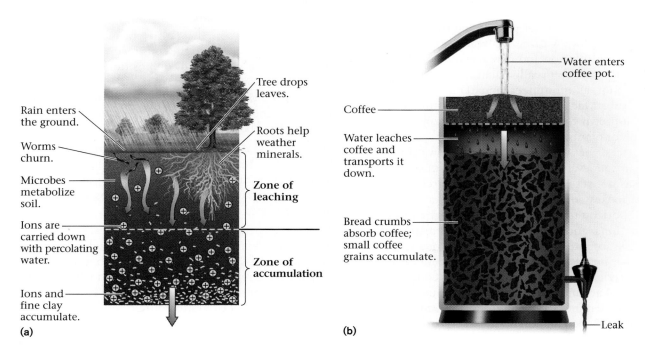

(a)

Tree drops leaves.

Rain enters the ground.

Worms churn.

Microbes metabolize soil.

Ions are carried down with percolating water.

Ions and fine clay accumulate.

Roots help weather minerals.

Zone of leaching

Zone of accumulation

(b)

Water enters coffee pot.

Coffee

Water leaches coffee and transports it down.

Bread crumbs absorb coffee; small coffee grains accumulate.

Leak

Third, microbes, fungi, plants, and animals interact with sediment by producing acids that weather grains, by absorbing nutrient atoms (e.g., K, Ca, Mg), and by leaving behind waste and remains. Plant roots and burrowing animals (insects, worms, and gophers) churn and break up the soil, and microbes metabolize mineral grains and release chemicals.

As a consequence of the above processes, regolith and rock evolve into soil—the soil's character (texture and composition) becomes very different from that of the starting material. Note that the biologic and physical components of the Earth System interact profoundly in the soil. Indeed, soils serve as home for a remarkable number of organisms—a single cubic centimeter of moist soil in a warm region hosts over 1 billion bacteria and 1 million protozoans. Over 1.5 million earthworms wriggle through each acre of such soil.

Because different soil-forming processes operate at different depths, soils typically develop distinct zones, known as **horizons**, arranged in a vertical sequence called a **soil profile** (▶Fig. 7.13a). Not all soils have the same horizons or the same degree of horizon development, because soil-forming processes vary with climate and with the length of time during which the soil has been forming. Nevertheless, to illustrate the concept of soil horizons, we can look at an idealized soil profile, from top to bottom, using a soil formed in a temperate forest as our example.

The highest horizon is the O-horizon (the prefix stands for "organic"), so called because it consists almost entirely of organic matter and contains barely any mineral matter. At the ground surface this material consists of "litter," undecomposed leaves and twigs. Deeper down the litter transforms into humus, organic material that has been decomposed by the action of insects, microbes, and fungi.

Below the O-horizon we find the A-horizon, in which humus has decayed further and has mixed with mineral grains (clay, silt, and sand). Water percolating through the A-horizon causes chemical-weathering reactions to occur and produces ions in solution and new clay materials. The downward-moving water eventually leaches and carries soluble chemicals (iron, aluminum, carbonate) and fine clay deeper into the subsurface. The A-horizon constitutes dark gray to blackish-brown topsoil, the fertile portion of soil that farmers till for planting crops.

In some places, the A-horizon grades downward into the E-horizon, a soil level that has undergone substantial leaching but has not yet mixed with organic material. Organic material makes soil dark; because it lacks organic material, the E-horizon tends to be noticeably lighter than the A-horizon. Ions and clay leached and transported down from above accumulate in the B-horizon, or subsoil. As a result, new minerals form, and clay fills open spaces. If the parent material from which the soil forms contained iron, the B-horizon attains a deep red color, because of the growth of iron-oxide minerals and the lack of organic matter. Note, from our description, that the O-, A-, and E-horizons make up the zone of leaching, whereas the B-horizon makes up the zone of accumulation.

Finally, at the base of a soil profile we find the C-horizon, which consists of material derived from the substrate that's been chemically weathered and broken apart, but has not yet undergone leaching or accumulation. In a residual soil the C-horizon grades downward into unweathered bedrock, whereas in a transported soil the soil grades into unweathered sediment.

The Variety of Soils: A Consequence of Many Factors

As farmers, foresters, and ranchers well know, the soil in one locality can differ greatly from the soil in another, in both composition and thickness (see the world soil map in Appendix B). And crops that grow well in one type of soil may wither and die in another. Such diversity exists because the makeup of a soil depends on several soil-forming factors (▶Fig. 7.13b–d).

- *Climate*: The total rainfall, the distribution of rainfall during the year, and the range and average of temperature during the year determine the rate and amount of chemical weathering and leaching that take place at a given location. Large amounts of rainfall and warm temperatures accelerate chemical weathering and cause most of the soluble elements to be leached. In regions with small amounts of rainfall and cooler temperatures, soils take a long time to develop and can retain unweathered minerals and soluble components. Climate seems to be the single most important factor in determining the nature of soils that develop.

- *Substrate composition*: Some soils form on basalt, some on granite, some on volcanic ash, and some on recently deposited quartz silt. These different substrates consist of different materials, so the soils formed on them end up with different chemical compositions. For example, a soil formed on basalt tends to be richer in iron than a soil formed on granite. Also, soils tend to develop faster on unconsolidated material (ash or sediment) than on hard bedrock.

- *Slope steepness*: A thick soil can accumulate under land that lies flat. But on a steep slope, regolith may wash away before it can evolve into a soil. Thus, all other factors being equal, soil thickness increases as the slope angle decreases.

- *Drainage*: Depending on the details of local topography and on the depth to the water table (the depth underground below which pores are filled with water), some localities in a region may be well drained whereas others may be saturated with water. Soils formed from saturated sediment tend to contain more organic material than do soils formed from dry sediment.

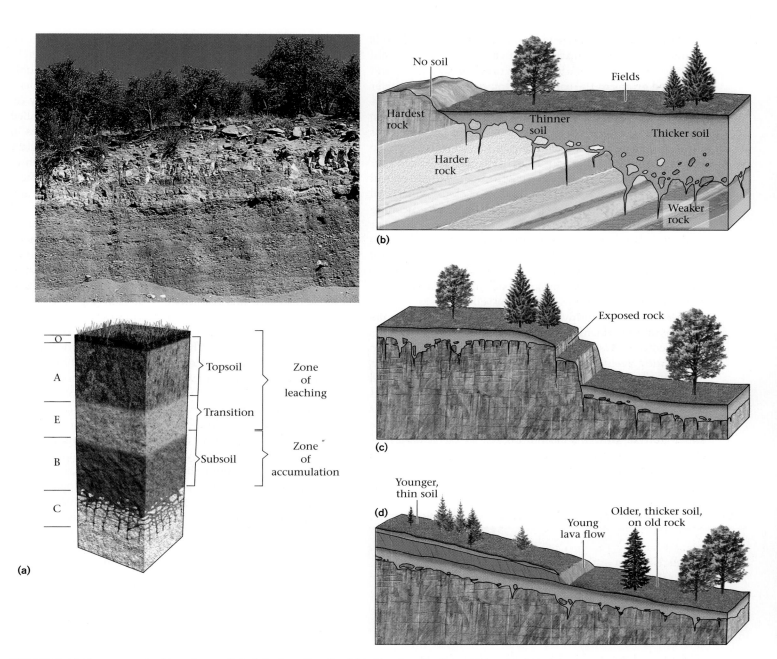

FIGURE 7.13 The process of soil formation results in distinctive soil profiles. **(a)** In this soil exposed on a cliff face, the dark layer (horizon) at the top is the organic-rich layer. Because of the redistribution of elements, the different horizons have different colors. The thickness of a soil at a given latitude depends on **(b)** the composition of the substrate (base), because less resistant underlying rocks weather more deeply; **(c)** the steepness of a slope, because soil washes or slides off steep slopes; and **(d)** the duration of soil formation; young soil is thinner than an old soil.

- *Time*: Because soil formation is an evolutionary process, a young soil tends to be thinner and less developed than an old soil. The rate of soil formation varies greatly with location. In a protected, moist, warm region soils may develop over the course of a few years to a few decades. But in an exposed, cold, dry region, soils may take thousands of years or more to develop. In temperate regions, soil forms at a rate of 0.02 to 0.20 mm per year, thereby producing 1 meter of soil in about 10,000 years.

- *Vegetation type*: Different kinds of plants extract or add different nutrients and quantities of organic matter to a soil. Also, some plants have deeper root systems than others and help prevent soil from washing away.

Depending on their evolution and composition, soils come in a variety of textures, structures, and colors. Soil texture reflects the relative proportions of sand, silt, and clay-sized grains in the soil. For many crops, farmers prefer to sow in **loam**, a type of soil consisting of about 10 to 30% clay and the rest silt and sand. In loam, pores (open spaces) can remain between grains so that water and air can pass through and roots can easily penetrate. In soils with too much clay, the clay packs together and prevents water movement. Soil structure refers to the degree to which soil grains clump together to form lumps or clods, which soil scientists refer to as "peds" (from the Latin *pedo*, meaning soil). The structure changes as a soil develops, because structure depends on clay content and organic content, both of which change with time. These materials give soil its stickiness. Soil color reflects its composition: organic-rich soil is gray or black, organic-poor and calcite-rich soil is whitish, and iron-rich soil is red or yellow.

Soil scientists worldwide have struggled mightily to develop a rational scheme for classifying soils. Not all schemes are based on the same criteria, and even today there is no worldwide agreement on which works best. An older scheme, which worked reasonably well in the United States, divided soils into categories primarily on the basis of the elements that accumulates in the B-horizon (▶Fig. 7.14a–c). In this scheme, pedalfer soil forms in temperate climates from a substrate that contains aluminum (*al*) and iron (*fer*); they have well-defined horizons, including an O-horizon and an organic-rich A-horizon (Fig. 7.14a). Pedocal soils form in arid climates and tend to be thin. Such soils do not have an O-horizon, and their A-horizon contains unweathered minerals, rock fragments, and a relatively high concentration of soluble minerals such as calcite, but very little organic matter. Calcite in a pedocal soil accumulates in the B-horizon and may cement the soil together, creating a solid mass sometimes called **caliche** or calcrete (Fig. 7.14b). Because evaporation rates in desert regions are high, water sometimes moves by capillary action upward through pedocal soil, contributing to accumulation of salt and calcite in horizons near the ground surface.

Laterite forms in tropical regions where abundant rainfall drenches the land during the rainy season, and the soil dries during the dry season. Because so much percolating water passes down through the soil, just about all mineral components get leached out of the soil until only insoluble iron and/or aluminum oxide remain. In fact, so much water passes through the soil that accumulation cannot take place, even at depth, so this soil has no B-horizon (Fig. 7.14c). During the dry season, capillary action brings water upward, contributing to the supply of oxide minerals. Iron oxide gives laterite a brick-red color. When dried, laterite can form a solid mass that can be used for construction—in fact, the word laterite comes from the Latin word *later*, meaning "brick." If laterite forms from iron-rich rock (e.g., mafic volcanics) or sediment, the concentration of iron in the soil is so great that the soil can be used as iron ore. Laterite soil derived from felsic rock (e.g., granite) may contain so much aluminum hydroxide that it can be quarried to provide aluminum ore (see Chapter 15). Aluminum-rich laterite is called bauxite.

In recent years, soil scientists have developed more complex and comprehensive classification schemes. One of these, the U.S. Comprehensive Soil Classification System, which distinguishes among twelve major "orders" of soil, is based on both physical characteristics and environment of formation (▶Table 7.2). There are literally thousands of suborders, known only to specialists. Different orders typically form in different environments. For example, an aridisol (≈ pedocal) is a soil that forms in arid climates, contains very little organic matter, and commonly contains caliche, whereas an alfisol (≈ pedalfer) develops in moist forests, has well-developed horizons, and contains abundant nutrients. Canadians use a different scheme focusing only on soils that develop north of the 40th parallel.

As we noted earlier, climate plays a major role in controlling the development of soil. Climate, in turn, depends in part on latitude and elevation. Thus, we can correlate the character and thickness of a soil with latitude and elevation (▶Fig. 7.14d).

Soil Use and Misuse

Though new techniques may someday provide an alternative to growing plants in soil, it almost goes without saying that soil remains essential for life on Earth. Soil is the substrate

TABLE 7.2   Soil Orders (U.S. Comprehensive Soil Classification System)

Alfisol	Gray/brown, has subsurface clay accumulation and abundant plant nutrients. Forms in humid forests.
Andisol	Forms in volcanic ash.
Aridisol	Low in organic matter, has carbonate horizons. Forms in arid environments.
Entisol	Has no horizons. Formed very recently.
Gelisol	Underlaid with permanently frozen ground.
Histosol	Very rich in organic debris. Forms in swamps and marshes.
Inceptisol	Moist, has poorly developed horizons. Formed recently.
Mollisol	Soft, black, and rich in nutrients. Forms in subhumid to subarid grasslands.
Oxisol	Very weathered, rich in aluminum and iron oxide, low in plant nutrients. Forms in tropical regions.
Spodosol	Acidic, low in plant nutrients, ashy, has accumulations of iron and aluminum. Forms in humid forests.
Ultisol	Very mature, strongly weathered soils, low in plant nutrients.
Vertisol	Clay-rich soils capable of swelling when wet, and shrinking and cracking when dry.

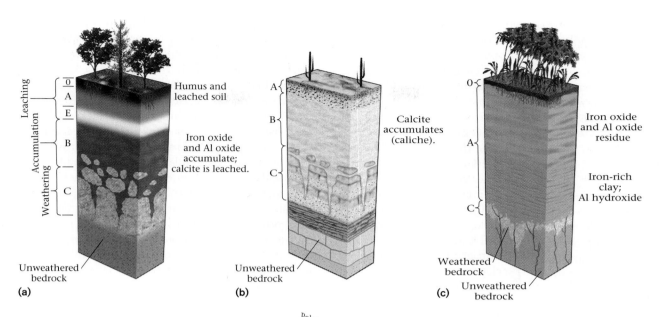

(a)

(b)

(c)

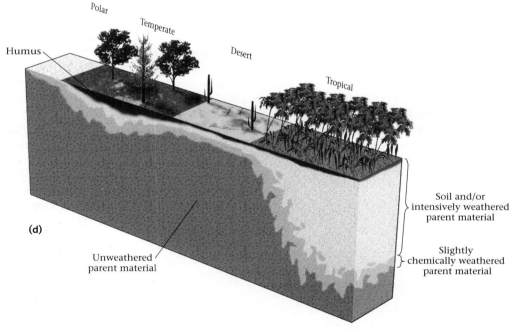

(d)

FIGURE 7.14 **(a)** In a pedalfer soil (alfisol) formed in a temperate climate, an O-horizon forms on top. Because of the moderate amount of rainfall, materials leached from the A-horizon can accumulate in the B-horizon. In this example, the C-horizon happens to consist of weathered granite. **(b)** Because of low rainfall, a thin pedocal soil (aridisol) in a desert has only a thin A-horizon. Soluble minerals, specifically calcite, that would be washed away in a temperate climate can accumulate in the B-horizon, creating calcrete. In this example, the C-horizon consists of weathered limestone. **(c)** In a thick tropical laterite (oxisol), so much water percolates down from the heavy rainfall that all reactive minerals dissolve or break down and get carried away. This leaves only a residue of iron oxide and/or aluminum oxide. (These are very stable.) There is no real zone of accumulation, but at depth, iron-rich clays collect. Here, the C-horizon is weathered metamorphic rock. **(d)** Soil thickness varies with latitude because of variations in temperature, rainfall, and vegetation.

for forests and fields, without which animals—including humans—could not survive. The life-giving character of soil comes from its ability to exchange key elements with plant roots. Roots are truly amazing. They contribute to weathering by prying apart grains physically and by providing acids that pluck ions off mineral grains. In addition, roots absorb water and nutrients (K, Ca, Mg, Na, Fe) out of the soil and supply them to plants. Soil also plays an important role in filtering water. The charged surfaces of clay flakes in soil can efficiently hold on to contaminants, such as mercury and uranium, and keep them from entering water supplies.

As we have seen, soils take time to form, so soils capable of supporting agriculture or forests should be considered a natural resource worthy of protection. However, practices such as agriculture, overgrazing, and clear-cutting, which remove the cover of vegetation that protects soil, have led to the destruction of soil. Humans have not been the best of custodians in maintaining soil supplies. First, crops rapidly remove nutrients from soil, so if they are not replaced (either by allowing fields to go fallow or by providing fertilizer), the soil will not contain sufficient nutrients to maintain plant life. Second, when the natural plant cover disappears, the surface of the soil becomes exposed to wind and water. Such actions, for example, the impact of falling raindrops or the rasping of a plow, break up the soil at the surface, with the result that it can wash away in water or blow away as dust. When this happens, **soil erosion**, the removal of soil by running water or by wind, takes place (▶Fig. 7.15). In some cases, almost 6 tons of soil may be lost from an acre of land per year, leading to removal of the A-horizon at a rate of about 0.04 cm per year. In extreme cases, the eroded soil chokes rivers and blackens the sky. Since clay, the finest-grained sediment, tends to be most easily moved, soil erosion makes soil sandier with time and less capable of retaining nutrients; the surfaces of clay flakes play an important role in holding on to nutrient atoms until plant roots can absorb them. Human activities increase rates of soil erosion by 10 to 100 times, so that it far exceeds the rate of soil formation. Droughts exacerbate the situation. For example, during the 1930s a succession of droughts killed off so much vegetation in the American plains that wind stripped the land of soil and caused devastating dust storms. Large numbers of people were forced to migrate away from the Dust Bowl of Oklahoma and adjacent areas.

> **Take-Home Message**
>
> If weathering products remain in place for a period of time, they transform into soil by interaction with rainwater and living organisms. Different types of soils form in different environments, and soil evolves over time at a location. Erosion removes soil.

The consequences of rainforest destruction on soil are particularly profound. Considering the lushness of rainforests, you might expect a thick humus (O-horizon) on a laterite soil. But, in fact, organic matter in a tropical climate decays so rapidly that the O-horizon remains fairly thin. Nevertheless, in an established rainforest, lush growth provides sufficient organic debris so that trees can grow. But if the forest is logged or cleared for agriculture, the humus rapidly disappears, leaving laterite that contains few nutrients. Crop plants use whatever nutrients there are so rapidly that the soil becomes infertile after only a year or two, useless for agriculture and unsuitable for regrowth of rainforest trees.

Soil erosion is but one of several problems that face society. The overuse of fertilizers, pesticides, and herbicides, as well as spills of a great variety of toxic chemicals, have contaminated soils. Too much irrigation in arid climates can make soils too saline for plant growth, for irrigation water contains trace amounts of salts. Fortunately, people have begun to realize the fragility of soil and have been working on ways to conserve soil. Most countries now have soil-conservation agencies.

7.4 INTRODUCING SEDIMENTARY ROCKS

So far, we've learned that weathering attacks bedrock and breaks it down to form dissolved ions and loose sediment grains. What happens next? The products of weathering may become components of soil, as we have seen. They can also become buried and transformed into sedimentary rock. Sedimentary rock develops from a variety of materials in a variety of environments, so there are many different kinds of sedimentary rock. Geologists divide sedimentary rocks into four major classes, based on their mode of origin. (1) **Clastic sedimentary rocks** consist of cemented-together solid fragments and grains derived from preexisting rocks (clastic comes from the Greek *klastos*, meaning "broken"). (2) **Biochemical sedimentary rocks** are made up of the shells of organisms. (3) **Organic sedimentary rocks** consist of carbon-rich relicts of plants. And (4) **chemical sedimentary rocks** are made up of minerals that precipitate directly from water solutions.

In some situations, it is also useful to distinguish among different kinds of sedimentary rocks on the basis of their predominant mineral composition. Siliceous rocks contain quartz, argillaceous rocks contain clay, and carbonate rocks contain calcite or dolomite. It's hard to estimate relative proportions of different types of sedimentary rocks, but by some estimates 70 to 85% of all the sedimentary rocks on Earth are siliceous or argillaceous clastic rocks, whereas 15 to 25% are carbonate biochemical or chemical rocks. Other kinds of sedimentary rocks occur only in minor quantities.

FIGURE 7.15 In this image, we can see that the lack of natural plant coverage has led to severe soil erosion by wind. Similar conditions created the Dust Bowl of the 1930s.

7.5 CLASTIC SEDIMENTARY ROCKS

Formation

Nine hundred years ago, a thriving community of Native Americans inhabited the high plateau of Mesa Verde, Colorado. In the hollows beneath huge overhanging ledges, they built multistory stone-block buildings that have survived to this day. Clearly, the blocks are solid and durable—they are, after all, rock. But if you were to rub your thumb along one, it would feel gritty, and small grains of quartz would break free and roll under your thumb, for the block consists of quartz sand grains cemented together. Geologists call such rock a **sandstone**.

Sandstone is an example of clastic sedimentary rock, rock created from solid grains (**clasts**, or detritus) stuck together to form a solid mass. The grains can consist of individual minerals (grains of quartz or flakes of clay) or fragments of rock (for example, pebbles of granite). The loose grains of sediment transform into clastic sedimentary rock by the following five steps (▶Fig. 7.16a, b).

- *Weathering*: Detritus forms by the disintegration of bedrock that happens in response to physical and chemical weathering.

- *Erosion*: **Erosion** is the combination of processes that separate rock or regolith from its substrate and carry it away. Erosion involves abrasion, plucking, scouring, and dissolution, and is caused by air, water, or ice.

- *Transportation*: Moving air, water, or ice transports sediment from one location to another. The ability of a medium to carry sediment depends on its viscosity and velocity. Solid ice can carry sediment of any size, regardless of how slowly the ice moves. Very fast moving, turbulent water can transport coarse fragments (cobbles and boulders), moderately fast moving water can carry only sand and gravel, and slowly moving water carries only silt and mud. Strong winds can move sand and dust, but gentle breezes carry only dust.

- *Deposition*: **Deposition** is the process by which sediment settles out of the transporting medium. When the ice of a glacier melts, its sediment load settles on the ground. Sediment settles out of wind or moving water when these fluids slow, because as the velocity decreases, the fluid no longer has the ability to transport sediment.

- *Lithification*: Geologists refer to the transformation of loose sediment into solid rock as **lithification**. The formation of clastic sedimentary rocks generally requires

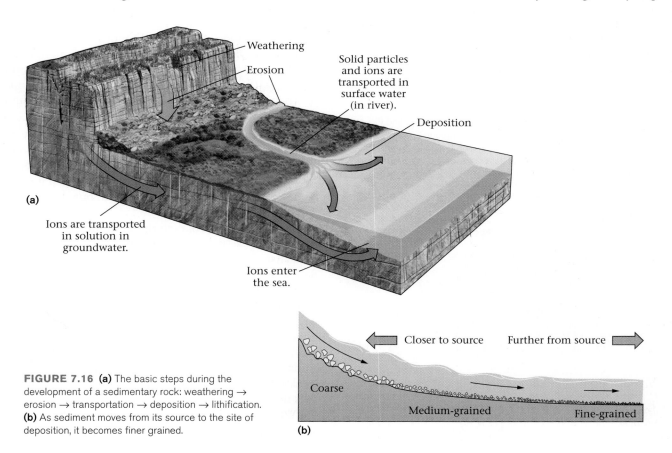

FIGURE 7.16 (a) The basic steps during the development of a sedimentary rock: weathering → erosion → transportation → deposition → lithification. **(b)** As sediment moves from its source to the site of deposition, it becomes finer grained.

the burial of the sediment by more sediment. When the sediment has been buried, pressure caused by the overburden squeezes out water and air that had been trapped between clasts, and the clasts press together tightly. Geologists refer to this process as **compaction** (▶Fig. 7.17). Mud, a mixture of clay and water, compacts (decreases in volume) by 50 to 80% when buried. Sand, on the other hand, compacts by only 10 to 20%. Compacted sediment may then be bound together to make coherent sedimentary rock by the process of **cementation**. Cement consists of minerals (commonly quartz or calcite) that precipitate from groundwater and partially or completely fill the spaces between clasts to attach each grain to its neighbors. Effectively, cement acts like glue and holds detritus together.

How Do We Describe and Classify Clastic Sedimentary Rocks?

Say you pick up a clastic sedimentary rock and want to describe it sufficiently so that, from your words alone, another person can picture the rock. What characteristics should you mention? Geologists find the following characteristics most useful:

- *Clast size*. This refers to the diameter of clasts making up a rock. Names used for clast size, listed in order from coarsest to finest, are: boulder, cobble, pebble, sand, silt, and clay (see Interlude B). Geologists infor-

mally use the term *gravel* for an accumulation of pebbles and cobbles, and *mud* for wet clay. In this context, clay refers to extremely small clasts—grains of this size typically consist of clay minerals (varieties of sheet silicate that occur in tiny flakes; see Chapter 5), but specks of quartz may be included.

- *Clast composition*. This refers to the makeup of clasts in sedimentary rock. Larger clasts (pebbles or larger) typically consist of rock fragments, meaning the clasts themselves are an aggregate of many mineral grains, whereas smaller clasts (sand or smaller) typically consist of individual mineral grains. In some cases, chips of fine-grained rock may be mixed in with sand grains. Such chips are called lithic clasts. Some sedimentary rocks contain only one clast of one composition, but others contain a variety of different kinds of clasts.

- *Angularity and sphericity*. The angularity of clasts indicates the degree to which grains have smooth or angular corners and edges. Sphericity, in contrast, refers to the degree to which a clast is equidimensional (i.e., resembles a sphere; ▶Fig. 7.18a).

- *Sorting*. **Sorting** of clasts indicates the degree to which the clasts in a rock are all the same size or include a variety of sizes (▶Fig. 7.18b). Well-sorted sediment consists entirely of sediment of the same size, whereas poorly sorted sediment contains a mixture of more than one grain size. If a sedimentary rock contains larger clasts surrounded by much smaller clasts (for example, cobbles surrounded by sand), then the mass of smaller grains constitutes the matrix of the rock.

- *Character of cement*. Not all clastic sedimentary rocks have the same kind of cement. In some, the cement consists predominantly of silica (quartz), whereas in others, it consists predominantly of calcite. Other kinds of mineral cements do occur, but they are rare.

With the above characteristics in mind, we can distinguish among several common types of clastic sedimentary rocks, listed in ▶Table 7.3 (▶Fig. 7.19a–l). Note that no single characteristic serves as a comprehensive basis for describing clastic rocks, but grain size proves to be the most important basis for classification. Geologists further distinguish among different kinds of sandstone (quartz sandstone, arkose, wacke) on the basis of clast composition and/or sorting, and they distinguish between shale and mudstone on the basis of the way in which the rock breaks. (Shale splits into thin sheets, whereas mudstone does not.)

What do the characteristics of a sedimentary rock tell us about the source of the sediment and about the environment of deposition? Following the fate of rock

FIGURE 7.17 The process of lithification. As sediment is buried, it becomes compacted (expelling the water between the grains), and the grains pack tightly together. Groundwater passing through the rock precipitates ions to form mineral cements that bind the grains together. If there is clay in the rock, weak chemical bonds may cause the clay grains to stick to each other.

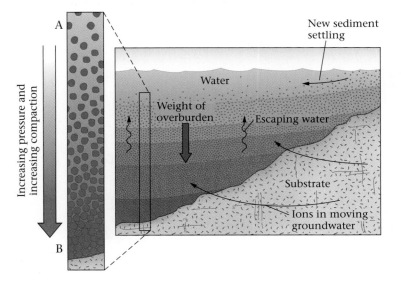

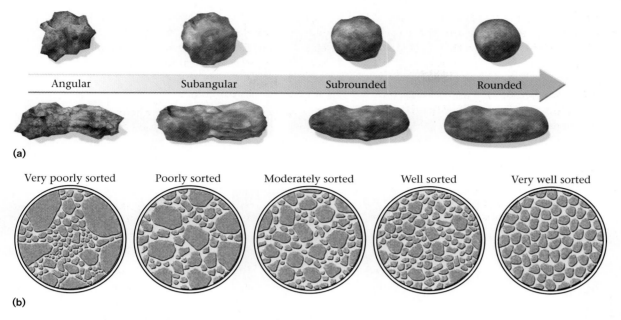

(a)

Angular   Subangular   Subrounded   Rounded

Very poorly sorted   Poorly sorted   Moderately sorted   Well sorted   Very well sorted

(b)

FIGURE 7.18 **(a)** A grain with high sphericity (top row) has roughly the same length in all directions, whereas one with low sphericity is elongate or flattened (bottom row). Sphericity is independent of angularity, which refers to whether the grain has sharp corners or edges or not. Grains on the left are more angular than grains on the right. **(b)** In a poorly sorted sediment, there is a great variety of different clast sizes, whereas in a well-sorted sediment, all the clasts are the same size.

fragments as they gradually move from a cliff face in the mountains via a river to the seashore provides some clues (see Fig. 7.16). Different kinds of sediment develop along the route and each of these types, if buried and lithified, yields a different kind of sedimentary rock.

To start, imagine that some large blocks of granite tumble off a cliff and slam into other blocks already at the bottom. The impact shatters the blocks, producing a pile of angular fragments with sharp edges. If these fragments were to be cemented together, the resulting rock would be **breccia** (Fig. 7.19a). Later, a storm causes the fragments (clasts) to slide downslope into a turbulent river. In the river, clasts bang into each other and into the bed of the stream, a process that shatters them into still smaller

TABLE 7.3 **Common Types of Sedimentary Rock**

Clast Size*	Clast Character	Rock Name (Alternate Name)
Coarse to very coarse	Rounded pebbles and cobbles	**Conglomerate**
	Angular clasts	**Breccia**
	Large clasts in muddy matrix	**Diamictite**
Medium to coarse	Sand-sized grains	**Sandstone**
	• quartz grains only	• **quartz sandstone (quartz arenite)**
	• quartz and feldspar sand	• **arkose**
	• sand-sized lithic clasts	• **lithic sandstone**
	• sand and lithic clasts in a clay-rich matrix	• **wacke** (informally called **graywacke**)
Fine	Silt-sized clasts	**Siltstone**
Very fine	Clay and/or very fine silt	**Shale** (if it breaks into platy sheets)
		Mudstone (if it doesn't break into platy sheets)

*For precise diameters, see Fig. B.5c on p. 149.

FIGURE 7.19 Examples of sediments and sedimentary rocks. **(a)** A sedimentary breccia. **(b)** Recently deposited stream gravel. **(c)** A conglomerate made of stream gravel that was later cemented together. **(d)** Hand samples of arkose. The whitish fragments consist of feldspar. **(e)** A photomicrograph (photograph of a thin section) showing quartz grains in a sandstone. The field of view is 3 mm. **(f)** A sandy desert in Australia. **(g)** A thick sandstone bed forms an overhang that protects ancient Native American dwellings on Mesa Verde, Colorado. **(h)** A hand specimen of sandstone. **(i)** Mud along a dirt road in central Australia makes a slippery obstacle for drivers. **(j)** Thin shale beds beneath a sandstone bed in Pennsylvania. Note the coin for scale. **(k)** A photograph taken with a scanning electron microscope shows flakes of clay that are less than 2 microns across. **(l)** A photomicrograph of graywacke. Note the large grains in a finer-grained matrix.

pieces and breaks off their sharp edges. Angular clasts gradually become rounded clasts. When the storm abates and the river water slows, pebbles and cobbles stop moving and form a mound or bar of gravel. Burial and lithification of these rounded clasts would produce **conglomerate** (Fig. 7.19b, c).

If the gravel stays put for a long time, it undergoes chemical weathering. As a consequence, cobbles and pebbles break apart into individual mineral grains, eventually producing a mixture of quartz, feldspar, and clay. If another storm causes the river to rise and flow faster, these sediments start to wash downstream. Clay is so fine that it may remain suspended in the water and be carried far downstream. Sand, however, may drop out along the stream bed or stream banks when the flow slows between storms. In the resulting sand bars, not far from the source, we find a mixture of quartz and some feldspar grains—this sediment, if buried and lithified, would become **arkose** (Fig. 7.19d). Over time, feldspar grains in sand continue to weather into clay so that gradually, during successive events that wash the sediment farther downstream, the sand loses feldspar and ends up being composed almost entirely of durable quartz grains. This sediment, when buried and lithified, becomes **quartz sandstone** (Fig. 7.19e–h). Some of the sand may make it to the sea, where waves carry it to beaches. Meanwhile, silt and clay may accumulate in the flat areas bordering the stream (regions called floodplains; see Chapter 17) that become inundated only during floods, or in a wedge of sediment, called a delta, that accumulates in the sea at the mouth of the river. Some of the silt and mud may be collected in lagoons or mud flats along the shore. The silt,

when lithified, becomes **siltstone**, and the mud, when lithified, becomes **shale** or **mudstone** (Fig. 7.19i–k).

Two of the rock names in Table 7.3 do not appear in the above narrative, because they don't form in the depositional settings just described. Diamictite forms either from debris flows (slurries consisting of mud mixed with larger clasts) both on land and under water, or in glacial settings where ice deposits clasts of all sizes. Wacke typically forms from the deposits of submarine avalanches (Fig. 7.19l). (Most wacke has a grayish color, and thus geologists informally refer to it as graywacke). Note that diamictites and wackes are poorly sorted.

You may have sensed from this narrative that as sediment moves downstream, grains overall become smaller and rounder; grains composed of minerals that are susceptible to chemical weathering progressively break down and disappear; and sediment becomes better sorted. Geologists use the term *sediment maturity* to refer to the degree to which a sediment has evolved from being just a crushed-up version of the original rock to a sediment that has lost its easily weathered components and has become well sorted and rounded. Thus, we can say that an immature sandstone is one that contains angular clasts, both durable and easily weathered minerals, and is poorly sorted. In contrast, a mature sandstone is one that contains only well-sorted grains of resistant minerals (▶Fig. 7.20).

> ### Take-Home Message
> Clastic sedimentary rocks consist of grains that were broken off preexisting rock and transported to a new location where they were deposited, buried, and lithified. We classify sedimentary rock on the basis of grain size and composition.

FIGURE 7.20 As sediments are transported progressively farther, weatherable sediments such as feldspar break down and convert to clay, which washes away, so the proportion of sediment consisting of resistant minerals such as quartz increases. Further, the physical bouncing and grinding that accompanies the transport of sediment progressively rounds the quartz grains and sorts them.

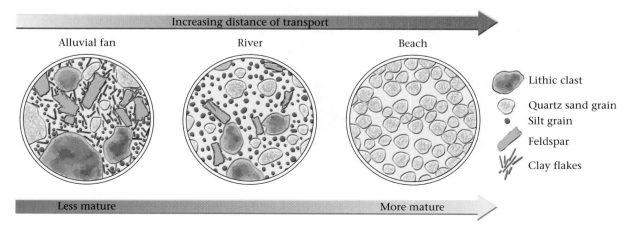

7.6 BIOCHEMICAL AND ORGANIC SEDIMENTARY ROCKS: BYPRODUCTS OF LIFE

The Earth System involves interactions between living organisms and the physical planet. In this chapter, we've already seen how living organisms play a role in chemically weathering rocks by secreting chemicals or by physically forcing open cracks. Here, we learn how some living organisms play a role in creating the materials constituting sedimentary rocks.

Numerous organisms have developed the ability to extract dissolved ions from seawater to make solid shells. Some organisms construct their shells out of calcium (Ca^{2+}) and carbonate (CO_3^{2-}) ions, which they merge to make the mineral calcite ($CaCO_3$) or its polymorph, aragonite, whereas other organisms make their shells out of dissolved silica (SiO_2). When the organisms die, the solid material in their shells turns into sediment that eventually becomes incorporated in the class of sedimentary rocks called biochemical sedimentary rocks. Since carbonate shells are much more common than silica shells, most biochemical rocks are carbonates. Plants, algae, bacteria, and plankton also yield materials that can be incorporated in sedimentary rocks. The rocks formed from this material contain organic chemicals and pure carbon derived from organic chemicals. Such rocks are called organic sedimentary rocks.

Biochemical Limestone

A snorkeler gliding above the Great Barrier Reef of Australia sees an incredibly diverse community of coral and algae, around which creatures such as clams, oysters, snails (gastropods), and lampshells (brachiopods) live, and above which plankton floats (▶Fig. 7.21a). Though they all look so different from one another, many of these organisms share an important characteristic: they make solid shells of calcite (or its polymorph, aragonite). When the organisms die, their skeletons may stay in place, as is the case with reef builders like coral; settle out of the water, like snowflakes; or be moved by currents or waves to another location, where they eventually settle out. During transport, shells may break up into small fragments. Rocks formed from the calcite or aragonite skeletons of organisms are the biochemical version of **limestone**, a type of carbonate rock (▶Fig. 7.21b, c). Note that geologists use the name *limestone* for any rock composed of calcite (and/or aragonite), regardless of origin; later, we'll discuss chemical limestone, a type that precipitates directly out of water.

There are many different kinds of limestones. They differ from each other according to the material from which they formed. Some retain the internal structure of coral colonies; some consist of large, angular shell fragments; some are made up of rounded grains of calcite that rolled about in the surf; some consist of lime mud (very tiny grains of calcite); and some consist of shells from plankton (particularly, microscopic animals called foraminifera). Three common types are fossiliferous limestone, consisting of identifiable shells and shell fragments; micrite, consisting of lime mud; and chalk, consisting of plankton shells. (Specialists use a variety of other names for types of limestone.)

Typically, ancient limestone is a massive light-gray to dark-bluish-gray rock that breaks into chunky blocks—it doesn't look much like a pile of shell fragments (▶Fig. 7.21d). That's because several processes change the texture of the rock over time. For example, organisms living in the depositional environment burrow into recently formed or deposited shells and break them up, and may even convert some to lime mud. Later, water passing through the rock precipitates new cement and also dissolves some carbonate grains and causes new ones to grow. Thus, original crystals may be replaced by new ones. Typically, all aragonite transforms into calcite, a more stable mineral, and smaller crystals of calcite are replaced by larger ones.

Biochemical Chert

If you walk beneath the northern end of the Golden Gate Bridge in Marin County, California, north of San Francisco, you will find outcrops of reddish, almost porcelain-like rock occurring in 3- to 15-cm-thick layers, one on top of another in a sequence with a total thickness of hundreds of meters (▶Fig. 7.22a). Hit it with a hammer, and the rock cracks almost like glass, creating smooth, spoon-shaped (conchoidal) fractures. Geologists call this rock biochemical chert; it's made from cryptocrystalline quartz (*crypto* is Greek for hidden), meaning quartz grains that are too small to be seen without the extreme magnification of an electron microscope. The chert beneath the Golden Gate Bridge formed from the shells of plankton (particularly, microscopic animals called radiolaria and diatoms). The shells accumulate on the sea floor as a silica-rich ooze. Gradually, after burial, the shells dissolve, forming a silica-rich solution. Chert then precipitates from this solution.

By the way, how did this chert end up beneath the Golden Gate Bridge? About 145 million years ago, the western United States was a convergent plate boundary. As the Pacific Ocean floor slipped beneath North America during subduction, the chert was scraped off the downgoing plate, like snow in front of a plow, and became incorporated in an accretionary prism that grew between the deep-ocean trench and the Sierra Nevada volcanic arc. The foundation of the Golden Gate Bridge stands on the remnants of this accretionary prism.

FIGURE 7.21 **(a)** In this modern coral reef in Australia, corals produce shells of calcite or aragonite. If buried and preserved, these become limestone. **(b)** A quarry face in Vermont shows the typical gray color of limestone; this rock is over 400 million years old. The thinly laminated layers are lime mud; the white mounds are relicts of small reefs. **(c)** This specimen of fossiliferous limestone consists entirely of small fossil shells and shell fragments. Not all fossiliferous limestones contain such a high proportion of fossils. **(d)** A roadcut exposing tilted beds of limestone in Pennsylvania.

Organic Rocks: Coal and Oil Shale

The Industrial Revolution of the nineteenth century, which transformed the world's economy from an agricultural to an industrial base, depended on power provided by steam engines. After decimating forests to provide fuel for these engines, industrialists turned to coal. **Coal** is a black, combustible rock consisting of over 50% carbon, and so differs markedly from the other sedimentary rocks discussed so far. The carbon of coal occurs as pure carbon or as an element in organic chemicals, not in minerals. Still, we consider coal a sedimentary rock because it is made up of detritus (of plants) deposited in layers (▶Fig. 7.22b). We'll look more at coal formation in Chapter 14. Here, we simply need to know that the carbon and the organic chemicals that make up coal come from the remains of plant material that died and accumulated on the floor of a forest or swamp. The remains were buried deeply, and the heat and pressure at depth compacted the plant material and drove off volatiles (hydrogen, water, carbon dioxide, ammonia), leaving a concentration of carbon.

Not all organic rocks come from plant material. Organic material, in the form of chemicals derived from fats and proteins that made up the flesh of plankton or algae, can mix with mud and be incorporated in shale. The organic material, which tends to color shale black, may gradually transform into oil. A fine-grained clastic rock containing the organic precursor of oil is called oil shale.

> ### Take-Home Message
>
> The origin of some sedimentary rocks involves the activity of living organisms. For example, some limestones consist of calcite shells or shell fragments, and coal consists of plant debris that was buried and altered.

(a)   (b)

FIGURE 7.22 **(a)** This bedded chert, which crops out near the northern foundation of the Golden Gate Bridge in Marin County, California, north of San Francisco, developed on the deep sea floor by the deposition of forms of plankton that secrete silica shells. The bends in the layers, called folds (see Chapter 11), formed when the layers were squeezed and wrinkled as they were scraped off the sea floor. **(b)** Coal is deposited in layers just like other kinds of sedimentary rocks. Here, we see a coal seam (a miner's term for a coal layer) between layers of sandstone and shale.

7.7 CHEMICAL SEDIMENTARY ROCKS

The colorful terraces, or mounds, around the vents of hot-water springs; the immense layers of salt that underlie the floor of the Mediterranean Sea; the smooth, sharp point of an ancient arrowhead—these materials all have something in common. They all consist of rock formed primarily by the precipitation of minerals out of water solutions. We call such rocks chemical sedimentary rocks. They typically have a crystalline texture, partly formed during the original precipitation and partly as the result of later recrystallization.

Evaporites: The Products of Saltwater Evaporation

In 1965, two daredevil drivers in jet-powered cars battled to be the first to break the land-speed record of 600 mph. On November 7, Art Arfons, in the *Green Monster*, peaked at 576.127 mph; but eight days later, Craig Breedlove, driving the *Spirit of America*, reached 600.601 mph. Traveling at such speeds, a driver must maintain an absolutely straight course; any turn will catapult the vehicle out of control, because its tires simply can't grip the ground. Thus, high-speed trials take place on extremely long and flat racecourses. Not many places can provide such conditions—the Bonneville Salt Flats, near the Great Salt Lake of central Utah, do.

How did this vast salt plain come into existence? Like all streams, streams bringing water from Utah's Wasatch Mountains into the Salt Lake basin carry trace amounts of dissolved ions, provided to the water by chemical weathering. Most lakes have an outlet, so the water in them constantly flushes out and the ion concentration stays low. But the Great Salt Lake has no such outlet, so water escapes from the lake only by evaporating. Evaporation removes just the water; dissolved ions stay behind, so over time, the lake water has become a concentrated solution of dissolved ions—in other words, very salty (▶Fig. 7.23a). In the past, when the region had a wetter climate, the Great Salt Lake was larger and covered the region of the Bonneville Salt Flats; this larger ancient lake was Lake Bonneville. Along its shores, water dried up and salt precipitated. When the lake shrank to its present dimension, the vast extent of the Bonneville Salt Flats was left high and dry, and covered with salt (▶Fig. 7.23b). Such salt precipitation occurs wherever there is saturated saltwater—along desert lakes with no outlet (e.g., the Dead Sea) and along margins of restricted seas (e.g., the Persian Gulf). For thick deposits of salt to form, large volumes of water must evaporate (▶Fig. 7.23c, d). This may happen when plate tectonic movements temporarily cut off arms of the sea (as we saw in the case of the Mediterranean Sea) or during continental rifting, when seawater first begins to spill into the rift valley.

Because salt deposits form as a consequence of evaporation, geologists refer to them as **evaporites**. The specific type of salt constituting an evaporite depends on the amount of evaporation. When 80% of the water evaporates, gypsum forms; when 90% of the water evaporates, halite precipitates. If seawater were to evaporate entirely, the resulting evaporite would consist of 80% halite, 13% gypsum, and the remainder of other salts and carbonates.

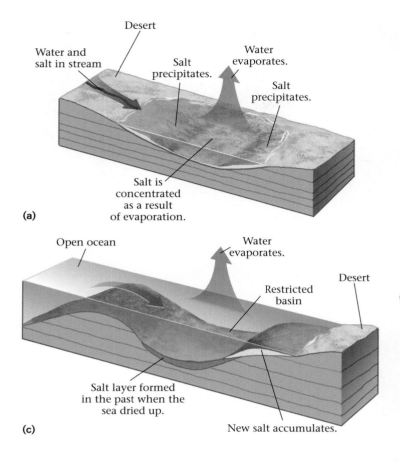

(a)

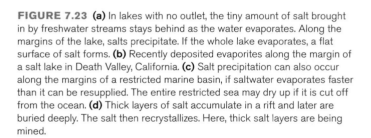

(c)

(b)

(d)

FIGURE 7.23 **(a)** In lakes with no outlet, the tiny amount of salt brought in by freshwater streams stays behind as the water evaporates. Along the margins of the lake, salts precipitate. If the whole lake evaporates, a flat surface of salt forms. **(b)** Recently deposited evaporites along the margin of a salt lake in Death Valley, California. **(c)** Salt precipitation can also occur along the margins of a restricted marine basin, if saltwater evaporates faster than it can be resupplied. The entire restricted sea may dry up if it is cut off from the ocean. **(d)** Thick layers of salt accumulate in a rift and later are buried deeply. The salt then recrystallizes. Here, thick salt layers are being mined.

Travertine (Chemical Limestone)

Travertine is a rock composed of crystalline calcium carbonate (calcite and/or aragonite) formed by chemical precipitation out of groundwater that has seeped out at the ground surface (in hot- or cold-water springs) or on the walls of caves. What causes this precipitation? It happens, in part, when the groundwater degasses, meaning that some of the carbon dioxide that had been dissolved in the groundwater bubbles out of solution. Dissolved carbon dioxide makes water more acidic and better able to dissolve carbonate, so removal of carbon dioxide decreases the ability of the water to hold dissolved carbonate. Precipitation also occurs when water evaporates and leaves behind dis-

solved ions, thereby increasing the concentration of carbonate. Various kinds of microbes live in the environments in which travertine accumulates, so biologic activity may also contribute to the precipitation process.

Travertine produced at springs forms terraces and mounds that are meters or even hundreds of meters thick. Spectacular terraces of travertine grew at Mammoth Hot Springs in Yellowstone National Park (▶Fig. 7.24a). Amazing column-like mounds of travertine grew up from the floor of Mono Lake, California, where hot springs seeped into the cold water of the lake (▶Fig. 7.24b); the columns are now exposed because the water level of the lake has been lowered. Travertine also grows on the walls of caves

(a)

(b)

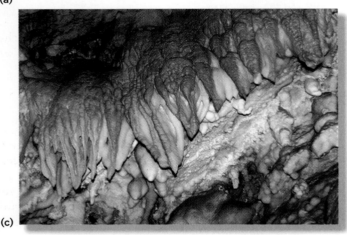

(c)

FIGURE 7.24 **(a)** A travertine buildup at Mammoth Hot Springs. Note the terraces. **(b)** Mounds of travertine forming at hot springs in Mono Lake, California. The material in these mounds is also called tufa. **(c)** A travertine buildup on the wall of a cave in Utah. The field of view is 2 m.

where groundwater seeps out. In cave settings, travertine builds up beautiful and complex growth forms called speleothems (▶Fig. 7.24c; see Chapter 19).

Travertine has been quarried for millennia to make building stones and decorative stones. The rock's beauty comes in part because in thin slices it is translucent, and in part because it typically displays growth bands. Bands develop in response to changes in the composition of groundwater, or in the environment into which the water drains. Some travertines (a type called tufa) contain abundant large pores.

Dolostone: Replacing Calcite with Dolomite

Dolostone differs from limestone in that it contains the mineral dolomite ($CaMg[CO_3]_2$). Most dolostone forms by a chemical reaction between solid calcite and magnesium-bearing groundwater. Much of the dolostone you may find in an outcrop actually originated as limestone but later re-

crystallized so that dolomite replaced the calcite. This recrystallization may take place beneath lagoons along a shore soon after the limestone formed, or a long time later, after the limestone has been buried deeply.

Replacement and Precipitated Chert

A tribe of Native Americans, the Onondaga, once inhabited the eastern part of New York State. In this region, outcrops of limestone contain layers of a black chert (▶Fig. 7.25a). Because of the way it breaks, artisans could fashion sharp-edged tools (arrowheads and scrapers) from this chert, so the Onondaga collected it for their own toolmaking industry and for use in trade with other tribes. Unlike the deep-sea (biochemical) chert described earlier, the chert collected by the Onondaga formed when cryptocrystalline quartz gradually replaced calcite crystals within a body of limestone long after the limestone was deposited; geologists thus call it replacement chert.

FIGURE 7.25 **(a)** Replacement chert occurring as nodules in a limestone. Chert forms the black band in these tilted layers. **(b)** A thin slice of agate, lit from the back. Note the growth rings.

Chert comes in many colors (black, white, red, brown, green, gray), depending on the impurities it contains. Black chert, or flint, made the tools of the Onondaga. Red chert, or jasper, which like all chert takes on a nice polish, makes beautiful jewelry. Petrified wood is chert that's made when silica-rich sediment, such as ash from a volcanic eruption, buries a forest. The silica dissolves in groundwater that then passes into the wood. Dissolved silica precipitates as cryptocrystalline quartz within wood, gradually replacing the wood's cellulose. The chert retains the shape of the wood and even its growth rings. Some chert, known as agate, precipitates in concentric rings inside hollows in a rock and ends up with a striped appearance, caused by variations in the content of impurities while precipitation took place (▶Fig. 7.25b).

> **Take-Home Message**
>
> Some sedimentary rocks form by precipitation of minerals directly out of solutions. For example, evaporite precipitates from salt water, and travertine from hot springs. Dolomite and replacement chert form by reaction of preexisting rock with groundwater.

7.8 SEDIMENTARY STRUCTURES

In the photo of a stark outcrop in Figure 7.10c, note the distinct lines across its face. In 3-D, we see that these lines are the traces of individual surfaces that separate the rock into sheets. In fact, sedimentary rocks in general contain distinctive layering. The layers themselves may have a characteristic internal arrangement of grains or distinctive markings on their surface. We use the term **sedimentary structure** for the layering of sedimentary rocks, surface features on layers formed during deposition, and the arrangement of grains within layers. Here, we examine some of the more important types.

Bedding and Stratification

Geologists have jargon for discussing sedimentary layers. A single layer of sediment or sedimentary rock with a recognizable top and bottom is called a **bed**; the boundary between two beds is a bedding plane; several beds together constitute **strata**; and the overall arrangement of sediment into a sequence of beds is bedding, or stratification. From the word *strata*, we derive other words, such as "stratigrapher" (a geologist who specializes in studying strata) and stratigraphy (the study of the record of Earth history preserved in strata).

When you examine strata in a region with good exposure, the bedding generally stands out clearly. Beds appear as bands across a cliff face (Fig. 7.10c). Typically, a contrast in rock type distinguishes one bed from adjacent beds. For example, a sequence of strata may contain a bed of sandstone, overlain by a bed of shale, overlain by a bed of limestone. Each bed has a definable thickness (from a couple of centimeters to tens of meters) and some contrast in composition, color, and/or grain size, which distinguishes it from its neighbors. But in many examples, adjacent beds all appear to have the same composition. In such cases, bedding may be defined by subtle changes in grain size, by surfaces that represent interruptions in deposition, or by cracks that have formed parallel to bed surfaces.

Why does bedding form? To find the answer, we need to think about how sediment is deposited. Typically, a succession of discrete beds forms, because deposition of a particular type of sediment at a location does not occur as a continuous, uninterrupted process, but rather during

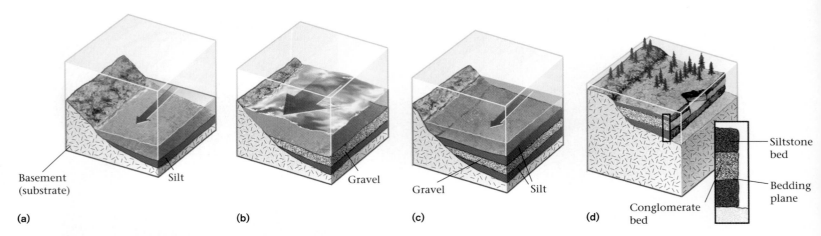

FIGURE 7.26 Bedding forms as a result of changes in the environment. **(a)** During a normal river flow, a layer of silt is deposited. **(b)** During a flood, turbulent water brings in a layer of gravel. **(c)** When the river returns to normal, another layer of silt is deposited. **(d)** Later, after lithification, uplift, and exposure, a geologist sees these layers as beds on an outcrop.

discrete intervals when conditions are appropriate for deposition. After the interval, some time may pass during which no sediment accumulates, and the surface of the just-deposited bed has time to weather a bit, or if conditions change, a different type of sediment starts to accumulate. Changes in the source of sediment, climate, or water depth control the type of sediment deposited at a location at a given time. For example, on a normal day a slow-moving river may carry only silt, which collects on the riverbed. During a flood, the river carries sand and pebbles, so a layer of sandy gravel forms over the silt layer. Then, when the flooding stops, more silt buries the gravel. If these sediments become lithified and exposed for you to see, they appear as alternating beds of siltstone and sandy conglomerate (►Fig. 7.26a–d).

Bedding is not always well preserved. In some environments, burrowing organisms disrupt the layering. Worms, clams, and other creatures churn sediment and may leave behind burrows. This process is called bioturbation.

During geologic time, long-term changes in a depositional environment can take place. Thus, a sequence of beds may differ markedly from sequences of beds above or below. If a sequence of strata is distinctive enough to be traced across a fairly large region, geologists call it a **stratigraphic formation**, or simply a formation (►Fig. 7.27). For example, a region may contain a succession of alternating sandstone and shale beds deposited by rivers, overlain by beds of marine limestone deposited later when the region was submerged by the sea. A stratigrapher might identify the sequence of sandstone and shale beds as one formation and the sequence of limestone beds as another. Formations are often named after the locality where they were first found and studied. For example, the Schoharie Formation was recognized and described from exposures near Schoharie, New York.

FIGURE 7.27 A particularly thick bed, a sequence of beds of the same composition, or a sequence of beds of alternating rock types can be called a stratigraphic formation, if the sequence is distinctive enough to be traced across the countryside. In this photo of the Grand Canyon, we can see five formations. Formations that consist primarily of one rock type may take the rock-type name (e.g., Kaibab Limestone), but a formation containing more than one rock type may just be called a "formation." The Supai Group is a group because it consists of several related formations, which are too thin to show here. Formations and groups are examples of stratigraphic units. Note that each formation consists of many beds, and that beds vary greatly in thickness. The boundaries between units are called contacts.

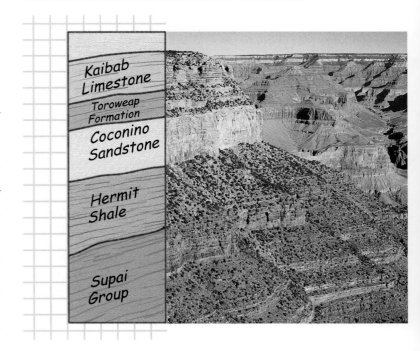

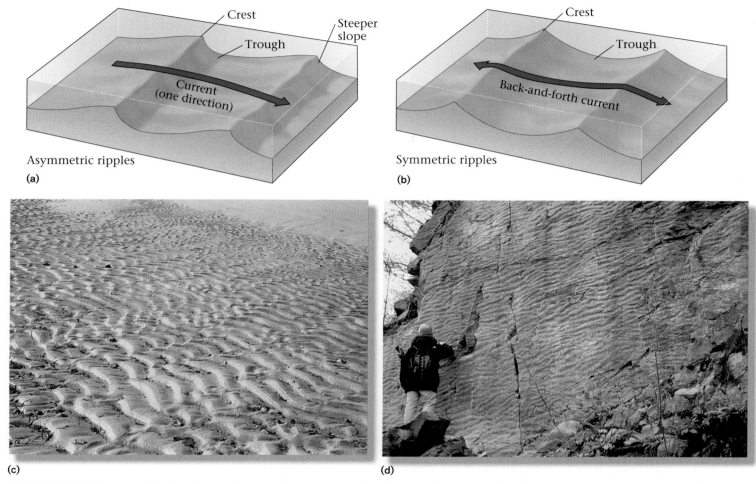

FIGURE 7.28 **(a)** A current that always flows in the same direction, as occurs in a stream, produces asymetric ripples. **(b)** A current that moves back and forth, as occurs on a wave-washed beach, produces symmetric ripples. **(c)** Modern ripples forming in the salt on a beach. **(d)** Ancient ripples preserved in a layer of quartzite that's over 1.5 billion years old. This outcrop occurs in Wisconsin.

Ripples, Dunes, and Cross Bedding: A Consequence of Deposition in a Current

Many clastic sedimentary rocks accumulate in moving fluids (wind, rivers, or waves). The movement of the fluids creates fascinating sedimentary structures at the interface between the sediment and the fluid—these structures are called bedforms. The bedforms that develop at a given location reflect factors such as the velocity of the flow and the size of the clasts. Though there are many types of bedforms, we'll focus on only two—ripples and dunes. The growth of both produces cross bedding, a special type of lamination within beds.

Ripples (or ripple marks) are relatively small (generally no more than a few centimeters high), elongated ridges that form on a bed surface at right angles to the direction of current flow (▶Fig. 7.28a–d). If the current always flows in the same direction, the ripple marks are asymmetric, with a steeper slope on the downstream (lee) side (Fig. 7.28a).

Along the shore, where water flows back and forth due to wave action, ripples tend to be symmetric. The crest (the high ridge) of a symmetric ripple is a sharp ridge, whereas the trough between adjacent ridges is a smooth, concave-up curve (Fig. 7.28b).

Dunes look like ripples, but they are much larger. For example, dunes on the bed of a stream may be tens of centimeters high, whereas wind-formed dunes occurring in deserts may be tens to over 100 meters high. Small ripples often form on the surfaces of dunes.

If you slice into a ripple or dune and examine it in cross section, you will find distinct internal laminations that are inclined at an angle to the boundary of the main sedimentary layer. Such laminations are called **cross beds**. Cross bedding forms as a direct consequence of the evolution of ripples or dunes. To see how, imagine a current of air or water moving uniformly in one direction (▶Fig. 7.29a). The current erodes and picks up clasts from the upstream part

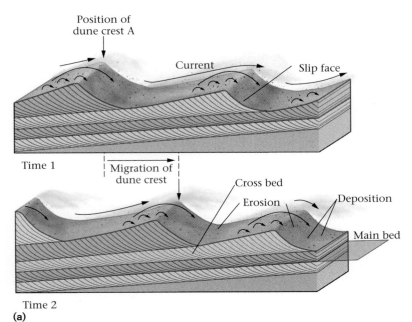

FIGURE 7.29 **(a)** Cross beds form as sand blows up the windward side of a dune and then accumulates on the slip face. At a later time, we see that dunes migrate, and eventually bury the layers below. **(b)** Successive layers, or master beds, of cross-bedded strata can be seen on this cliff face of sandstone in Zion National Park. We are looking at the remnants of ancient sand dunes. Cross beds indicate the wind direction during deposition.

of the bedform (because here the fluid moves quickly) and deposits them on the downstream or leeward part of the bedform (because here the fluid moves more slowly). The face of the downstream side of the bedform is called the slip face, for the accumulation of sediment allows this face to become so steep that gravity causes sediment to slip downward. The upper part of the slip face becomes steeper than its base, so the slip face becomes curved, with a concave-up

shape. The process repeats as more sediment builds up on the leeward side of the bedform and then slips down, so with time the leeward side of the bedform builds in the downstream direction. The curving surface of the slip face establishes the shape of the cross beds. During slippage events, heavier clasts (of denser minerals or larger grains) remain stranded along the slip face; these mark the visible boundaries between successive cross beds. Typically, erosion clips off the top of the cross bed, so only the bottom half or two-thirds will be preserved and buried.

With time, a new cross-bedded layer builds out over a preexisting one. The boundary between two successive layers is called the main bedding, and the internal curving surfaces within the layer constitute the cross bedding (▶Fig. 7.29b). Note that the shape of the cross bed indicates both the direction in which the current was flowing during deposition and the direction in a stratigraphic sequence in which beds are younger.

Turbidity Currents and Graded Beds

Sediment deposited on a submarine slope probably will not stay in place permanently. For example, an earthquake or storm might disturb this sediment and cause it to slip downslope. If the sediment is loose enough, it mixes with water to create a murky, turbulent cloud. This cloud is denser than clear water, and thus flows downslope like an underwater avalanche (▶Fig. 7.30a). We call this moving submarine suspension of sediment a **turbidity current**. Turbidity currents can be powerful enough to snap undersea phone cables and displace shipwrecks.

Eventually, in deeper water where the slope becomes gentler, or if the turbidity current spreads out, the turbidity current slows. When this happens, the sediment that it has carried starts to settle out. Larger grains sink faster through a fluid than do finer grains, so the coarsest sediment settles out first. Progressively finer grains accumulate on top, with the finest sediment (clay) settling out last. This process forms a **graded bed**—that is, a layer of sediment in which grain size varies from coarse at the bottom to fine at the top (▶Fig. 7.30b).

Typically, turbidity currents flow down submarine canyons—in fact, their flow contributes to scouring and deepening the canyon. The graded beds thus form an apron, called a submarine fan, at the mouth of the canyon (see Fig. 7.30a). Successive turbidity currents deposit successive graded beds, creating a sequence of strata called a **turbidite**.

Bed-Surface Markings

A number of features appear on the surface of a bed as a consequence of events that happen during deposition or soon after, while the sediment layer remains soft. These bed-surface markings include the following.

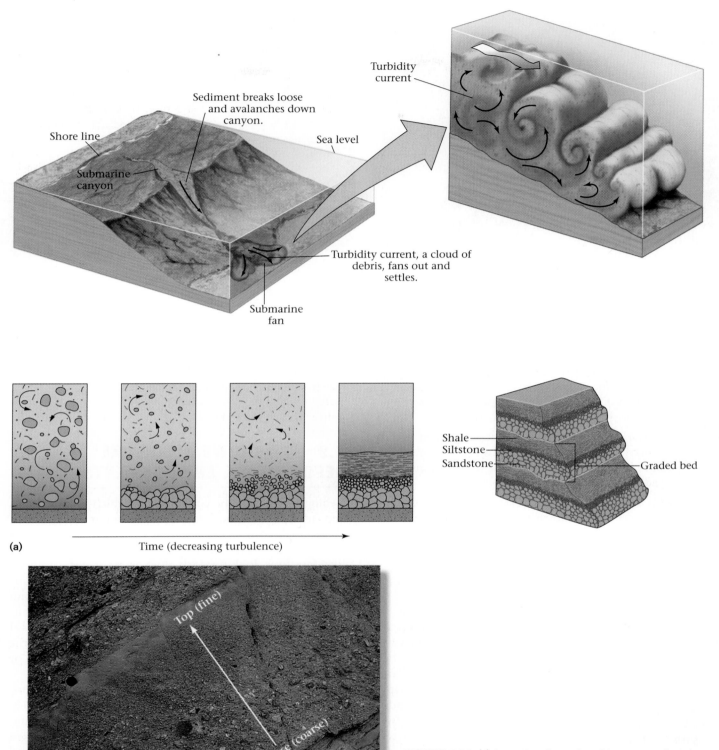

Shore line

Sediment breaks loose and avalanches down canyon.

Sea level

Submarine canyon

Turbidity current

Turbidity current, a cloud of debris, fans out and settles.

Submarine fan

(a)

Time (decreasing turbulence)

Shale
Siltstone
Sandstone
Graded bed

Top (fine)

Base (coarse)

(b)

FIGURE 7.30 **(a)** An earthquake or storm triggers an underwater avalanche (turbidity current), which mixes sediment of different sizes together. When the current slows, the larger grains settle faster, gradually creating a graded bed. **(b)** In this example of a graded bed, pebbles lie at the bottom of the bed, and silt at the top. The bed was tilted after deposition.

(a)

(b)

FIGURE 7.31 **(a)** Mud cracks in dried red mud, from Utah, as viewed from above. Note how the edges of the mud cracks curl up. **(b)** Mud cracks visible on the bottom of a 410-million-year-old bed, exposed in eastern New York. The photo was taken looking up at the base of an overhang.

- *Mud cracks*: If a mud layer dries up after deposition, it cracks into roughly hexagonal plates that typically curl up at their edges. We refer to the openings between the plates as mud cracks. Later, these fill with sediment and can be preserved (▶Fig. 7.31a, b).

- *Scour marks*: As currents flow over a sediment surface, they may scour out small troughs called scour marks parallel to the current flow. These indentations can be buried and preserved.

- *Fossils*: Fossils are relicts of past life. Some fossils are shell imprints or footprints on a bedding surface (see Interlude E).

The Value of Studying Sedimentary Structures

Sedimentary structures are not just a curiosity, but rather provide key clues that help geologists understand the environment in which clastic sedimentary beds were deposited. For example, the presence of ripple marks and cross bedding indicates that layers were deposited in a current. Graded beds indicate deposition by turbidity currents. The presence of mud cracks indicates that the sediment layer was exposed to the air on occasion. And fossil types can tell us whether sediment was deposited along a river or in the sea, for different species of organisms live in different environments. In the next section of this chapter, we examine these environments in greater detail.

7.9 HOW DO WE RECOGNIZE DEPOSITIONAL ENVIRONMENTS?

Geologists refer to the conditions in which sediment was deposited as the **depositional environment**. Examples include beach environments, glacial environments, and river environments. To identify these environments, geologists, like detectives, look for such clues as grain size, composition, sorting, and roundness of clasts, which can tell us how far the sediment has traveled from its source and whether it was deposited by the wind, by a fast-moving current, or from a stagnant body of water. Clues such as fossil content and sedimentary structures can tell us whether the sediments were deposited subaerially, just off the coast, or in the deep sea.

Now let's look at some examples of different depositional environments and the sediments deposited in them by imagining that we are taking a journey from the mountains to the sea, examining sediments as we go. We begin with terrestrial sedimentary environments, those formed on dry land, and end with marine sedimentary environments, those formed along coasts and under the waters of the ocean. (See art, pp. 216–17.) Of note, sediments deposited in terrestrial environments may oxidize (rust) when undergoing lithification in oxygen-bearing water, or if in contact with air. If this happens, the sedimentary beds de-

> **Take-Home Message**
>
> Sedimentary rocks occur in beds, because deposition takes place in discrete episodes. A sequence of beds makes up a stratigraphic formation. Sedimentary structures, such as ripple marks, cross beds, and graded beds, give clues to the environment of deposition.

velop a reddish color and can be called **redbeds**. The red comes from a film of iron oxide (hematite) that forms on grain surfaces.

Terrestrial (Nonmarine) Sedimentary Environments

Glacial environments. We begin high in the mountains, where it's so cold that more snow collects in the winter than melts away, so glaciers—rivers of ice—develop and slowly flow downslope. Because ice is a solid, it can move sediment of any size. So as a glacier moves down a valley in the mountains, it carries along *all* the sediment that falls on its surface from adjacent cliffs or gets plucked from the ground at its base. At the end of the glacier, where the ice finally melts away, it drops its load and makes a pile of glacial till (▶Fig. 7.32a). Till is unsorted and unstratified—it contains clasts ranging from clay size to boulder size all mixed together, with large clasts distributed through a matrix of silt and clay. Thus, in a sequence of strata, a layer of diamictite would be the record of an ancient episode of glaciation.

Mountain-stream environments. As we walk down beyond the end of the glacier, we enter a realm where turbulent streams rush downslope in mountain valleys. This fast-moving water has the power to carry large clasts; in fact, during floods, boulders and cobbles tumble down the stream bed. Between floods, when water flow slows, the largest clasts settle out to form gravel and boulder beds, while the stream carries finer sediments such as sand and mud away (▶Fig. 7.32b). Sedimentary deposits of a mountain stream would, therefore, include coarse conglomerate.

Alluvial-fan environments. Our journey now takes us to the mountain front, where the fast-moving stream empties onto a plain. In arid regions, where water is insufficient for the stream to flow continuously, the stream deposits its load of sediment right at the mountain front, creating a large, wedge-shaped apron called an alluvial fan (▶Fig. 7.32c). Deposition takes place here because when the stream pours from a canyon mouth and spreads out over a broader region, friction with the ground causes the water to slow down, and slow-moving water does not have the power to

(a)

(b)

(c)

FIGURE 7.32 **(a)** Glacial till at the end of a melting glacier in France. **(b)** Coarse boulders deposited by a flooding mountain stream in California. **(c)** An alluvial fan in California. Note the road for scale.

Glacial environment

Estuary

Beach

Bar

Continental shelf

Coastal
erosion

Turbidity
current

Submarine fan

Deep-sea current

Layers of sedimentary
rock accumulate.

Forming an unconformity

Mountain building folds
the rock layers.

The mountains are eroded; the
folded layers are submerged.

New sedimentary layers
accumulate.

The Formation of Sedimentary Rocks

Categories of sedimentary rocks include clastic sedimentary rocks, chemical sedimentary rocks (formed from the precipitation of minerals out of water), and biochemical sedimentary rocks (formed from the shells of organisms). Clastic sedimentary rocks develop when grains (clasts) break off preexisting rock by weathering and erosion and are transported to a new location by

wind, water, or ice; the grains are deposited to create sediment layers, which are then cemented together. We distinguish among types of clastic sedimentary rocks on the basis of grain size.

The character of a sedimentary rock depends on the composition of the sediment and on the environment in which it accumulated. For example, glaciers carry sediment of all sizes, so

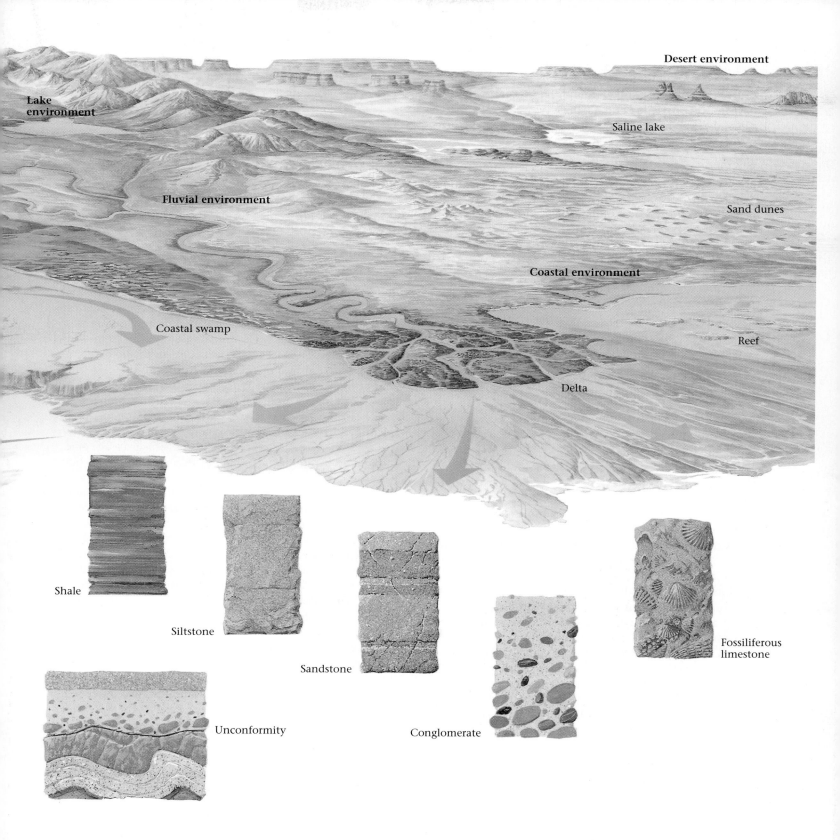

Desert environment

Lake environment

Saline lake

Fluvial environment

Sand dunes

Coastal environment

Coastal swamp

Reef

Delta

Shale

Siltstone

Sandstone

Conglomerate

Fossiliferous limestone

Unconformity

they leave deposits of poorly sorted (different-sized) till; streams deposit coarser grains in their channels and finer ones on floodplains; a river slows down at its mouth and deposits an immense pile of silt in a delta. Fossiliferous limestone develops on coral reefs. In desert environments, sand accumulates into dunes, and evaporates precipitate in saline lakes. Offshore, submarine canyons channel avalanches of sediment, or turbidity currents, out to the deep-sea floor.

Sedimentary rocks tell the history of the Earth. For example, the layering, or bedding, of sedimentary rocks is initially horizontal. So where we see layers bent or folded, we can conclude that the layers were deformed during mountain building. Where horizontal layers overlie folded layers, we have an unconformity: for a time, sediment was not deposited, and/or older rocks were eroded away.

move coarse sediment. Alluvial fans are so close to the source that the sand still contains feldspar grains, for these have not yet broken up and have not yet weathered into clay. Alluvial-fan sediments, when later buried and transformed into sedimentary rock, become arkose and conglomerate.

Sand-dune environments. In deserts, relatively few plants grow, so the ground lies exposed to the wind. The strongest winds can transport sand. As a result, large sand dunes, of well-sorted sand accumulate (Fig. 7.19f–h). Thus, thick layers of well-sorted sandstone, in which we see large (meters-high) cross beds, are relicts of desert sand-dune environments (Fig. 7.29b).

Lake environments. From the dry regions, we continue our journey into a temperate realm, where water remains at the surface throughout the year. Some of this water collects in lakes, in which relatively quiet water is unable to move coarse sediment; any coarse sediment brought into the lake by a stream settles out along the shore. Only fine clay makes it out into the center of the lake, where it eventually settles to form mud on the lake bed. Thus, lake sediments (also called lacustrine sediments) typically consist of finely laminated shale (▶Fig. 7.33a).

River environments. The lake drains into a stream that carries water onward toward the sea. As we follow the stream, it merges with tributaries to become a large river, winding back and forth across a plain. Rivers transport sand, silt, and mud. The coarser sediments tumble along the bed in the river's channel, while the finer sediments drift along, suspended in the water (▶Fig. 7.33b). This fine sediment settles out along the banks of the river, or on the floodplain, the flat region on either side of the river that is covered with water only during floods. Since river sediment is deposited in a current, the sediment surface develops ripple marks, and the sediment layers have small, internal cross beds. On the floodplain, mud layers dry out between floods, leading to the formation of mud cracks.

Because the river has transported sediment a great distance, the minerals making up the sediment have undergone chemical weathering. As a result, very little feldspar remains— most of it has changed into clay. Thus, river sediments (also called fluvial sediments, from the Latin word for river) lithify to form sandstone, siltstone, and shale. Typically, channels of coarser sediment (sandstone) are surrounded by layers of fine-grained floodplain deposits; in cross section, the channel has a lens-like shape (▶Fig. 7.33c, d). Note that if oxidized iron precipitates in pores during cementation, redbeds form (▶Fig. 7.33e).

Marine Sedimentary Environments

Marine delta deposits. After following the river downstream for a long distance, we reach its mouth, where it empties into the sea. Here, the river builds a delta of sediment out into the sea. Deltas were so named because the map shape of some deltas (e.g., the Nile Delta of Egypt) resembles the Greek letter *delta* (Δ), as we will discuss further in Chapter 17. River water stops flowing when it enters the sea, so sediment settles out.

In 1885, an American geologist named G. K. Gilbert studied small deltas that formed where mountain streams (carrying gravel, sand, and silt) emptied into lakes. He showed that the deltas contained three components (▶Fig. 7.34a): nearly horizontal topset beds composed of gravel, sloping foreset beds of gravel and sand (deposited on the sloping face of the delta), and nearly horizontal, silty bottomset beds, formed at depth on the floor of the water body. Gilbert's model makes intuitive sense but does not adequately describe the complexity of large river deltas. In large river deltas, there are many different sedimentary environments, ranging from fluvial and marsh environments to deeper-water marine environments (▶Fig. 7.34b). In addition, storms may cause masses of sediment to slip down the seaward-sloping face of the delta, creating mudflows (slurries of mud) or turbidity currents. Finally, sea-level changes may cause the positions of the different environments to move with time. Nevertheless, deposits of a delta can be identified in the stratigraphic record, as thick sequences in which deeper-water (offshore) sediments of a given age grade progressively into fluvial sediments in a shoreward direction.

Coastal beach sands. Now we leave the delta and wander along the coast. Oceanic currents transport sand along the coastline. The sand washes back and forth in the surf (▶Fig. 7.34c), so it becomes well sorted (waves winnow out mud and silt) and well rounded, and because of the back-and-forth movement of ocean water over the sand, the sand surface may become rippled. Thus, if you find well-sorted, medium-grained sandstone, perhaps with ripple marks, you may be looking at the remnants of a beach environment.

Shallow-marine clastic deposits. From the beach, we proceed offshore. Wave action transports coarser sediment shoreward, so in deeper water, where wave energy does not stir the sea floor, finer sediment accumulates. Also, finer sediment gets washed out to sea by the waves. As the water here may be only meters to a few tens of meters deep, geologists refer to this depositional setting as a shallow-marine environment. Clastic sediments that accumulate in this environment tend to be fine-grained, well-sorted, well-rounded silt, and they are inhabited by a great variety of organisms such as mollusks and worms. Thus, if you see smooth beds of siltstone and mudstone containing marine fossils, you may be looking at shallow-marine clastic deposits.

Shallow-water carbonate environments. In shallow-marine settings far from the mouth of a river, where relatively little clastic sediment (sand and mud) enters the water, the warm,

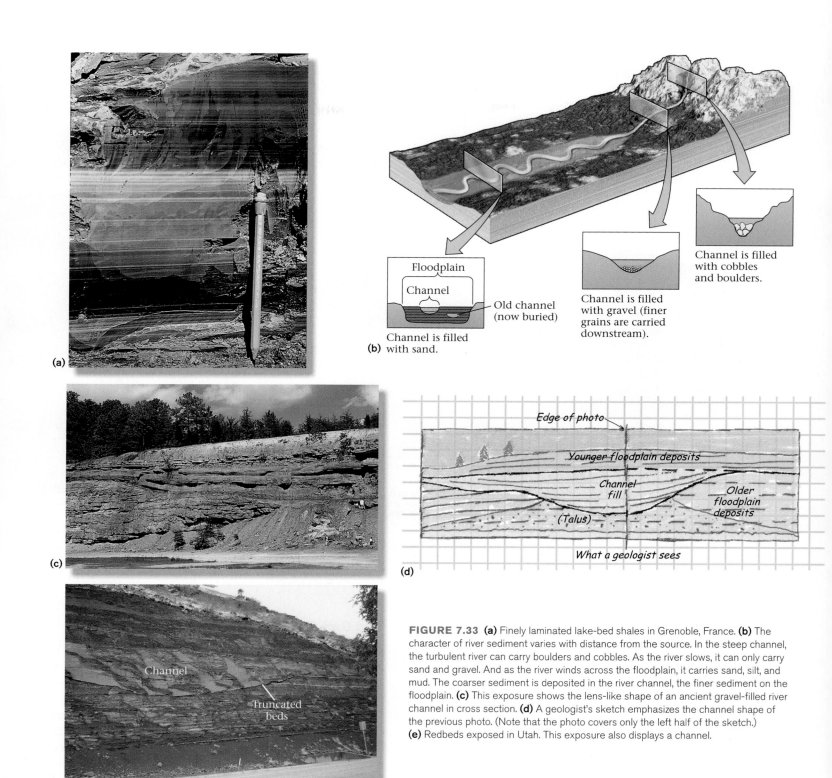

FIGURE 7.33 **(a)** Finely laminated lake-bed shales in Grenoble, France. **(b)** The character of river sediment varies with distance from the source. In the steep channel, the turbulent river can carry boulders and cobbles. As the river slows, it can only carry sand and gravel. And as the river winds across the floodplain, it carries sand, silt, and mud. The coarser sediment is deposited in the river channel, the finer sediment on the floodplain. **(c)** This exposure shows the lens-like shape of an ancient gravel-filled river channel in cross section. **(d)** A geologist's sketch emphasizes the channel shape of the previous photo. (Note that the photo covers only the left half of the sketch.) **(e)** Redbeds exposed in Utah. This exposure also displays a channel.

clear, and nutrient-rich water hosts an abundant number of organisms. Their shells, which consist of carbonate minerals, make up most of the sediment that accumulates, so we call such environments carbonate environments. The margins of tropical islands, away from the clastic debris of land, provide ideal carbonate environments (▶Fig. 7.35a). In carbonate environments, the nature of sediment depends on the water depth. Beaches collect sand composed of shell fragments,

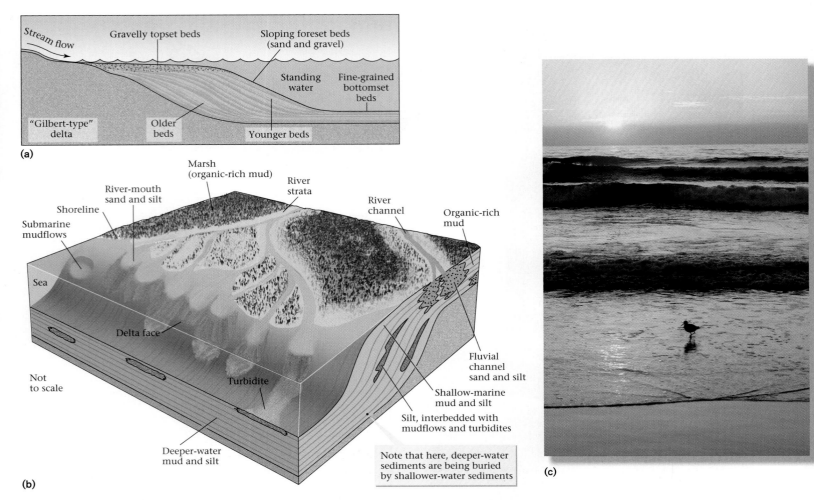

FIGURE 7.34 **(a)** A simple "Gilbert-type" delta formed where a small stream carrying gravel, sand, and silt enters a standing water body. The delta contains topset, foreset, and bottomset beds. **(b)** A larger river delta is very complex and doesn't fit the simple Gilbert-type model. The great variety of local depositional environments in a delta setting are labeled. Note that as time passes, the delta builds out seaward. **(c)** Waves wash the sand along the California coast.

lagoons (quiet water) are sites where lime mud accumulates, and reefs consist of coral and coral debris. Farther offshore of a reef, we can find a sloping apron of reef fragments (▶Fig. 7.35b). Shallow-water carbonate environments transform into sequences of fossiliferous limestone and micrite.

Deep-marine deposits. We conclude our journey by sailing offshore. Along the transition between coastal regions and the deep ocean, turbidity currents deposit turbidites (Fig. 7.30). Farther offshore, in the deep-ocean realm, only fine clay and plankton provide a source for sediment. The clay eventually settles out onto the deep sea floor, forming deposits of finely laminated mudstones, and plankton shells settle to form chalk (from calcite shells; ▶Fig. 7.36a, b) or chert (from siliceous shells). Thus, deposits of mudstone, chalk, or bedded chert indicate a deep-marine origin.

You Can Be a Sedimentary Detective!

Now you should be able to look at most sedimentary rocks in quarries, road cuts, and cliffs and take a pretty good guess as to what ancient environments they represent. From now on, when you see fossiliferous limestone in a quarry, it's not just a limestone, it's the record of a tropical reef. And a sandstone cliff—think ancient dune or beach. Coarse conglomerates should scream "alluvial fan" or "mountain stream," and a shale should bring to mind a floodplain, a lake bed, or the floor of the deep sea. Every sequence of strata has a story to tell.

> **Take-Home Message**
>
> Different types of sedimentary rocks accumulate in different terrestrial and marine depositional environments. By examining rock types and sedimentary structures, geologists can deduce the depositional environment in which the sediment accumulated.

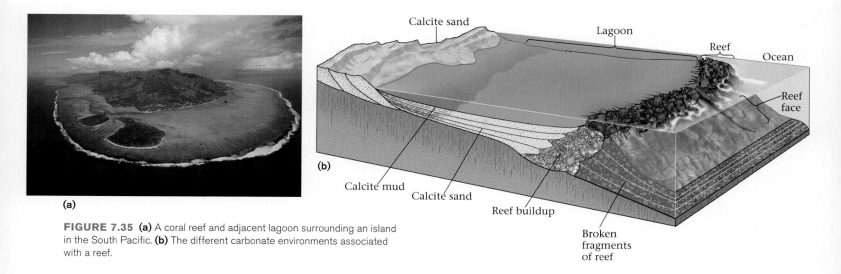

FIGURE 7.35 (a) A coral reef and adjacent lagoon surrounding an island in the South Pacific. **(b)** The different carbonate environments associated with a reef.

7.10 SEDIMENTARY BASINS

The sedimentary veneer on the Earth's surface varies greatly in thickness. If you stand in central Siberia or south-central Canada, you will find yourself on igneous and metamorphic basement rocks that are over a billion years old—there are no sedimentary rocks anywhere in sight. Yet if you stand along the southern coast of Texas, you would have to drill through over 15 km of sedimentary beds before reaching igneous and metamorphic basement. Thick accumulations of sediment form only in regions where the surface of the Earth's lithosphere sinks as sedi-

ment collects. Geologists use the term **subsidence** to refer to the sinking of lithosphere, and the term **sedimentary basin** for the sediment-filled depression. In what geologic settings do sedimentary basins form? An understanding of plate tectonics theory provides some answers.

Types of Sedimentary Basins in the Context of Plate Tectonics Theory

Geologists distinguish among different kinds of sedimentary basins on the basis of the region of a lithosphere plate in which they formed. Let's consider a few examples.

FIGURE 7.36 (a) These plankton shells, which make up deep-marine sediment, are so small (0.003 mm in diameter) that they could pass through the eye of a needle. This image was obtained with a scanning electron microscope. **(b)** The chalk cliffs of Dover, England. These were originally deposited on the sea floor and later uplifted.

See for yourself . . .

Sedimentary Rocks and Environments

You can see dramatic exposures of sedimentary rocks at many localities across the planet. Stratification in these exposures tends to be better where climate is drier and vegetation sparse. With a little searching, you can also find many examples of places where sediment is now accumulating. The thumbnail images provided on this page are only to help identify tour sites. Go to wwnorton.com/studyspace to experience this flyover tour.

Grand Canyon, Arizona
(Lat 36°08'00.34"N, Long 112°13'38.95"W)

Fly to these coordinates and zoom out to an elevation of 50 km (31 miles). You can see the entire width of the Grand Canyon, in northern Arizona (Image G7.1). Here, the Canyon is about 20 km (12 miles) wide. It cuts into the Colorado Plateau, a region of dominantly flat-lying Paleozoic and Mesozoic strata. Zoom down to about 6 km (4 miles), and tilt the image to see the horizon (Image G7.2). You now get a sense of the rugged topography of the canyon, and can see how the different layers of sedimentary rock stand out. More resistant rock units hold up cliffs. Fly down the Colorado River to develop a sense of the Canyon's majesty.

G7.1

G7.2

Lewis Range, Montana
(Lat 47°48'02.31"N, Long 112°45'30.97"W)

Fly to these coordinates and zoom out to an elevation of 25 km (15 miles). You will be looking at a series of north–south-trending cuestas underlain by tilted layers of late Paleozoic and Cretaceous strata (Image G7.3). A cuesta is an asymmetric ridge—one side is a cliff cutting across bedding, and the other side is a slope parallel to the tilted bedding. Faulting has caused units to be repeated. Thus, some of the adjacent cuestas contain the same strata. Tilt your field of view so the horizon just appears, and you get a clear sense of how bedding can control topography. Zoom down to 6 km (4 miles) and you'll be able to pick out individual beds.

G7.3

Death Valley, California
(Lat 36°20'49.15"N, Long 116°50'8.14"W)

Death Valley includes the lowest point in North America. It's a hot, dry desert area. When rain does fall, brief but intense floods carry sediment out of the bordering mountains and deposit it in alluvial fans at the foot of the mountains. Water that makes it out into the valley eventually evaporates and leaves behind salt. Fly to the coordinates given, zoom to an elevation of 10 km (6 miles), and tilt the image (Image G7.4). You can see contemporary sites of alluvial-fan and evaporite deposition.

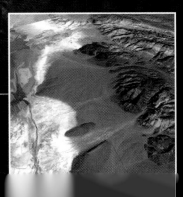

G7.5

Great Exhuma, Bahamas
(Lat 23°27'03.08"N, Long 75°38'15.68"W)

Here, in the warm waters of the Bahamas, you can see numerous examples of carbonate depositional environments. If you fly to the coordinates given, zoom to an elevation of 12 km (7 miles), and then tilt the image, you'll see reefs, beaches, lagoons, and deeper-water settings (Image G7.5). The white sand consists of broken shell fragments.

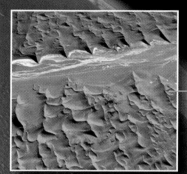

G7.6

G7.7

Sand Dunes, Namibia
(Lat 24°44'40.42"S, Long 15°30'5.34"E)

Fly to these coordinates and zoom to an elevation of about 18 km (11 miles). Tilt your image so you gain some 3-D perspective (Image G7.6). You are looking at a sea of sand piled in huge dunes by the wind. In the field of view, you can also see the deposits of a stream that occasionally washes through the area. Were the sand dunes to be buried and transformed into rock, they would become thick layers of cross-bedded sandstone. Zoom to 5.5 km (3 miles), tilt, and look NW (Image G7.7). You can determine the wind direction.

G7.8

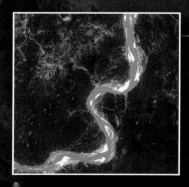

Niger Delta, Nigeria
(Lat 5°24'54.48"N, Long 6°31'26.25"E)

Head to the west coast of Africa, near the equator. The Niger River drains into the Atlantic and has built a 400 km (249 mile)-wide delta. Enter the coordinates given and zoom up to an elevation of 450 km (280 miles) to see the entire delta at once (Image G7.8). The present surface of the delta sits on top of a sedimentary accumulation that is many kilometers thick.

Now zoom down to 35 km (22 miles). You can see sandbars accumulating at bends in the river (Image G7.9). If the abundant vegetation of the dense surrounding jungle was preserved and buried, it would transform into coal.

- *Rift basins*: These form in continental rifts, regions where the lithosphere has been stretched. During the early stages of rifting, the surface of the Earth subsides simply because crust becomes thinner as it stretches. (To picture this process, imagine pulling on either end of a block of clay with your hands—as the clay stretches, the central region of the block thins and sinks lower than the ends.) As the rift grows, slip on faults drops blocks of crust down, creating low areas bordered by narrow mountain ridges. Alluvial-fan deposits form along the base of the mountains, and salt flats or lakes develop in the low areas between the mountains.

  Thinning is not the only reason that rifted lithosphere subsides. During rifting, warm asthenosphere rises beneath the rift and heats up the thin lithosphere. When rifting ceases, the rifted lithosphere then cools, thickens, and becomes denser. This heavier lithosphere sinks down, causing more subsidence, just as the deck of a tanker ship drops to a lower elevation when the ship is filled with ballast. Sinking due to cooling of the lithosphere is called thermal subsidence.

- *Passive-margin basins:* These form along the edges of continents that are *not* plate boundaries. They are underlain by stretched lithosphere, the remnants of a rift whose evolution ultimately led to the formation of a mid-ocean ridge (see Chapter 4). Passive-margin basins form because thermal subsidence of stretched lithosphere continues long after rifting ceases and sea-floor spreading begins. They fill with sediment carried to the sea by rivers and with carbonate rocks formed in coastal reefs. Sediment in a passive-margin basin can reach an astounding thickness of 15 to 20 km.

- *Intracontinental basins:* These develop in the interiors of continents, initially because of thermal subsidence over an unsuccessful rift. They may continue to subside for discrete episodes of time, even hundreds of millions of years after they first formed, for reasons that are not well understood. Illinois and Michigan are each underlain with an intracontinental basin (the Illinois basin and the Michigan basin, respectively) in which up to 7 km of sediment has accumulated. Most of this sediment is fluvial, deltaic, or shallow marine. At times, extensive swamps formed along the shoreline in these basins. The plant matter of these swamps was buried to form coal.

- *Foreland basins*: These form on the continent side of a mountain belt because as the mountain belt grows, large slices of rock are pushed up and out onto the surface of the continent. Such movement takes place by slip along large faults. The weight of these slices pushes down on the surface of the lithosphere, creating a wedge-shaped depression adjacent to the mountain range that fills with sediment eroded from the range. Fluvial and deltaic strata accumulate in foreland basins.

As indicated by the above descriptions, different assemblages of sedimentary rocks form in different sedimentary basins. So geologists may be able to determine the nature of the basin in which ancient sedimentary deposits accumulated by looking at the character of the deposits.

Transgression and Regression

Sea-level changes control the succession of sediments that we see in a sedimentary basin. At times during Earth history, sea level has risen by as much as a couple of hundred meters, creating shallow seas that covered the interiors of continents; there have also been times when sea level has fallen by a couple hundred meters, exposing even the continental shelves to air. Sea-level changes may be due to a number of factors, including climate changes, which control the amount of ice stored in polar ice caps and changes in the volume of ocean basins.

When sea level rises, the coast migrates inland—we call this process **transgression**. As the coast migrates, the sandy beach migrates with it, and the site of the former beach gets buried by deeper-water sediment. Thus, as transgression occurs, an extensive layer of beach sand eventually forms. This layer may look like a blanket of sand that was deposited all at once, but in fact the sand deposited at one location differs in age from the sand deposited at another location. When sea level falls, the coast migrates seaward—we call this process **regression** (▶Fig. 7.37). Typically, the record of a regression will not be well preserved, because as sea level drops, areas that had been sites of deposition become exposed to erosion. A succession of strata deposited during a cycle of transgression and regression is called a depositional sequence.

> **Take-Home Message**
>
> In order for a thick deposit of sediment to accumulate, the surface of the Earth must sink (subside) and form a depression called a sedimentary basin. Sediment fills a basin as it subsides. The deepest basins form where lithosphere stretches and then cools.

7.11 DIAGENESIS

Earlier in this chapter we discussed the process of lithification, by which sediment hardens into rock. Lithification is an aspect of a broader phenomenon called diagenesis. Geologists use the term **diagenesis** for all the physical, chemical, and biological processes that transform sediment into sedimentary rock and that alter characteristics of sedimentary rock once the rock has formed.

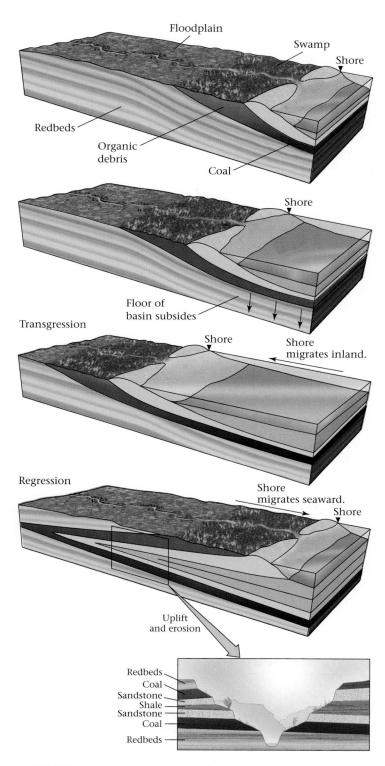

Floodplain

Swamp

Shore

Redbeds

Organic
debris

Coal

Shore

Transgression

Floor of
basin subsides

Shore

Shore
migrates inland.

Regression

Shore
migrates seaward.

Shore

Uplift
and erosion

Redbeds
Coal
Sandstone
Shale
Sandstone
Coal
Redbeds

FIGURE 7.37 The concept of transgression and regression. As sea level rises and the shore migrates inland, coastal sedimentary environments overlap terrestrial environments. Eventually, deeper-water environments overlap shallower ones. Thus, a regionally extensive layer does not all form at the same time. During regression, sea level falls and the shore moves seaward.

In the depositional environment, diagenesis includes bioturbation, growth of minerals in pore spaces and around grains, and replacement of existing crystals with new crystals. In buried sediment, diagenesis involves the compaction of sediment and the growth of cement that leads to complete lithification. Note that pressure due to the weight of overburden may cause a process called pressure solution, during which the faces of grains dissolve where they are squeezed against neighboring grains. In fully lithified sedimentary rocks, diagenesis continues both as a result of chemical reactions between the rock and groundwater passing through the rock, and as a result of increases in temperature and pressure. These reactions may dissolve existing cement and/or form new cement, and may grow new minerals. Though diagenesis may alter the texture and mineral composition of sedimentary rock, it usually does not destroy all sedimentary structures in the rock.

As temperature and pressure increase still deeper in the subsurface, the changes that take place in rocks become more profound. At sufficiently high temperature and pressure, a whole new assemblage of minerals forms, and/or mineral grains become aligned parallel to one another. Geologists consider such changes to be examples of metamorphism. The transition between diagenesis and metamorphism in sedimentary rocks is gradational. Most geologists consider changes that take place in rocks at temperatures of below about 150°C to be clearly diagenetic reactions, and those that occur in rocks at temperatures above about 300°C to be clearly metamorphic reactions. In the temperature range between 150°C and 300°C, whether diagenesis or metamorphism takes place depends on rock type. In the next chapter, we enter the true realm of metamorphism.

> **Take-Home Message**
>
> Once deposited and buried, sediment transforms into sedimentary rock. Even after lithification is complete, fluids passing through the rock precipitate additional cement, react with the rock to form new minerals, and may dissolve grains.

Chapter Summary

- Sediment consists of detritus (mineral grains and rock fragments derived from preexisting rock), mineral crystals that precipitate directly out of water, and shells (formed when organisms extract ions from water).

- Rocks at the surface of the Earth undergo physical and chemical weathering. During physical weathering, intact rock breaks into pieces. During chemical weathering, rocks react with water and air to produce new minerals such as clay, and ions in solution.

- The covering of loose rock fragments, sand, gravel, and soil at the Earth's surface is regolith. Soil differs from other types of regolith in that it has been changed by the activities of organisms, by downward-percolating rainwater, and by the mixing in of organic matter. Distinct horizons can be identified in soil. The type of soil that forms depends on factors such as climate and source material.

- Geologists recognize four major classes of sedimentary rocks. Clastic (detrital) rocks form from cemented-together detritus (mineral grains and rock fragments) that were first produced by weathering, then were transported, deposited, and lithified. Biochemical rocks develop from the shells of organisms. Organic rocks consist of plant debris or of altered plankton remains. Chemical rocks, such as evaporites, precipitate directly from water.

- Sedimentary structures include features such as bedding, cross bedding, graded bedding, ripple marks, and dunes. Their presence provides clues to depositional settings.

- Glaciers, mountain streams and fronts, sand dunes, lakes, rivers, deltas, beaches, shallow seas, and deep seas each accumulate a different assemblage of sedimentary strata. Thus, by studying sedimentary rocks, we can reconstruct the characteristics of past environments.

- Thick piles of sedimentary rocks accumulate in sedimentary basins, regions where the lithosphere sinks, creating a depression at the Earth's surface.

- Sea level changes with time. Transgressions occur when sea level rises and the coastline migrates inland. Regressions occur when sea level falls and the coastline migrates seaward.

- Diagenesis involves processes that lead to lithification and processes that alter sedimentary rock once it has formed.

Geopuzzle Revisited

The Grand Canyon cuts down through an over 1.5 km-thick succession of sedimentary strata, recording a long history of deposition in a variety of environments during successive transgressions and regressions of the sea. Contrasting layers consist of contrasting rock types—each rock type formed in a different depositional environment. Not only are different layers different colors (in part due to the amount of oxidized iron in the rock), but they also have different grain sizes and bedding thicknesses. Stronger rock units (sandstone and limestone) form steep cliffs, whereas weaker units (shale) form gentler slopes.

Key Terms

arkose (p. 203)
bed (p. 209)
biochemical sedimentary rocks (p. 198)
breccia (p. 201)
caliche (calcrete) (p. 196)
cementation (p. 200)
chemical sedimentary rocks (p. 198)
chemical weathering (p. 188)
clastic sedimentary rocks (p. 198)
clasts (p. 199)
coal (p. 205)
compaction (p. 200)
conglomerate (p. 203)
cross beds (p. 211)
deposition (p. 199)
depositional environment (p. 214)
diagenesis (p. 224)
dolostone (p. 208)
dunes (p. 211)
erosion (p. 199)
evaporites (p. 206)
graded bed (p. 212)
horizons (p. 194)
joints (p. 186)
laterite (p. 196)
limestone (p. 204)
lithification (p. 199)
loam (p. 196)
mudstone (p. 203)
organic sedimentary rocks (p. 198)

physical (mechanical) weathering (p. 186)
quartz sandstone (p. 203)
redbeds (p. 215)
regolith (p. 193)
regression (p. 224)
ripples (ripple marks) (p. 211)
sandstone (p. 199)
saprolite (p. 188)
sediment (p. 184)
sedimentary basin (p. 221)
sedimentary rock (p. 184)
sedimentary structure (p. 209)
shale (p. 203)
siltstone (p. 203)
soil (p. 193)
soil erosion (p. 198)
soil profile (p. 194)
sorting (p. 200)
strata (p. 209)
stratigraphic formation (p. 210)
subsidence (p. 221)
talus (p. 187)
transgression (p. 224)
travertine (p. 207)
turbidite (p. 212)
turbidity current (p. 212)
weathering (p. 186)
zone of accumulation (p. 193)
zone of leaching (p. 193)

Review Questions

1. Explain the circumstances that allowed the Mediterranean Sea to dry up.

2. How does physical weathering differ from chemical weathering?

3. Describe the processes that produce joints in rocks.

4. Feldspars are among the most common minerals in igneous rocks, but they are relatively rare in sediments. Why are they more susceptible to weathering, and what common sedimentary minerals are produced from weathered feldspar?

5. What types of minerals tend to weather more quickly?

6. Describe the different horizons in a typical soil profile.

7. What factors determine the nature of soils in different regions?

8. Describe how a clastic sedimentary rock is formed from its unweathered parent rock.

9. Clastic and chemical sedimentary rocks are both made of material that has been transported. How are they different?

10. Explain how biochemical sedimentary rocks form.

11. Describe how grain size and shape, sorting, sphericity, and angularity change as sediments move downstream.

12. Describe the two different kinds of chert. How are they similar? How are they different?

13. What kinds of conditions are required for the formation of evaporites?

14. What minerals precipitate out of seawater first? next? last? What does this suggest when geologists find huge volumes of pure gypsum in the Earth's crust?

15. How is dolostone different from limestone, and how does it form?

16. Describe how cross beds form. How can you read the current direction from cross beds?

17. Describe how a turbidity current forms and moves. How does it produce graded bedding?

18. Compare the deposits of an alluvial fan with those of a typical river environment and with those of a deep-marine deposit.

19. Why don't sediments accumulate everywhere? What types of tectonic conditions are required to create basins?

20. What happens during diagenesis, and how does diagenesis differ from metamorphism?

On Further Thought

1. Recent exploration of Mars by robotic vehicles suggests that layers of sedimentary rock cover portions of the planet's surface. On the basis of examining images of these layers, some researchers claim that the layers contain cross bedding and relicts of gypsum crystals. At face value, what do these features suggest about depositional environments on Mars in the past? (*Note*: Interpretation of the images remains controversial.)

2. The Gulf Coast of the United States is a passive-margin basin that contains a very thick accumulation of sediment. Drilling reveals that the base of the sedimentary succession in this basin consists of redbeds. These are overlain by a thick layer of evaporite. The evaporite, in turn, is overlain by deposits composed dominantly of sandstone and shale. In some intervals, sandstone occurs in channels and contains ripple marks, and the shale contains mud cracks. In other intervals, the sandstone and shale contains fossils of marine organisms. The sequence contains hardly any conglomerate or arkose. Be a sedimentary detective, and explain the succession of sediment in the basin.

3. Examine the Bahamas with Google Earth or NASA World Wind. (You can find a high-resolution image at Lat 23°58'40.98''N Long 77°30'20.37''W). Note that broad expanses of very shallow water surround the islands, that white sand beaches occur along the coast of the islands, and that small reefs occur offshore. What does the sand consist of, and what rock will it become if it eventually becomes buried and lithified? Compare the area of shallow water in the Bahamas area with the area of Florida. The bedrock of Florida consists mostly of shallow marine limestone. What does this observation suggest about the nature of the Florida peninsula in the past? Keep in mind that sea level on Earth changes over time. Presently, most of the land surface of Florida lies at less than 50 m (164 feet) above sea level.

Suggested Reading

Boggs, S., Jr. 2003. *Petrology of Sedimentary Rocks*. Caldwell, N.J.: Blackburn Press.

——. 2000. *Principles of Sedimentology and Stratigraphy*, 3rd ed. Upper Saddle River, N.J.: Pearson Education.

Brady, N. C., and R. R. Weil. 2001. *The Nature and Properties of Soils*, 13th ed. Upper Saddle River, N.J.: Prentice-Hall.

Einsele, G. 2000. *Sedimentary Basins: Evolution, Facies, and Sediment Budget*. New York: Springer-Verlag.

Harpstead, M. I., et al. 2001. *Soil Science Simplified*, 4th ed. Ames: Iowa State University Press.

Leeder, M. R. 1999. *Sedimentology and Sedimentary Basins: From Turbulence to Tectonics*. Oxford: Blackwell.

Prothero, D. R., and F. L. Schwab. 2003. *Sedimentary Geology*, 2nd ed. New York: Freeman.

Reading, H. G., ed. 1996. *Sedimentary Environments: Processes, Facies, and Stratigraphy*. Oxford: Blackwell.

Tucker, M. E. 2001. *Sedimentary Petrology*, 3rd ed. Oxford: Blackwell.

Metamorphism:
A Process of Change

Geopuzzle

Marble, the rock from which Michelangelo carved his sculptures, contains the same chemicals as limestone, a sedimentary rock. But grains in marble interlock, and the form of layering in marble suggests that the rock once flowed. The rock can't be igneous because its composition is unlike that of any known magma. So, how do rocks like marble form?

An outcrop of Precambrian metamorphic rock, exposed in the Wasatch Mountains in Utah. The layering, or foliation, formed during metamorphism and later was bent into a Z-shape.

Nothing in the world lasts, save eternal change.

—Honorat de Bueil (1589–1650)

8.1 INTRODUCTION

Cool winds sweep across Scotland for much of the year. In this blustery climate, vegetation has a hard time taking hold, so the landscape features countless outcrops of barren rock. During the latter half of the eighteenth century, James Hutton became fascinated with the Earth and examined these outcrops, hoping to learn how rock formed. Hutton found that many features in the outcrops resembled the products of present-day sediment deposition or of volcanic activity, and soon he came to an understanding of how sedimentary and igneous rocks form. But Hutton also found rock that contained minerals and textures quite different from those in sedimentary and igneous samples. He described this puzzling rock as "a mass of matter which had evidently formed originally in the ordinary manner . . . but which is now extremely distorted in its structure . . . and variously changed in its composition."

The rock that so puzzled Hutton is now known as metamorphic rock, from the Greek words *meta*, meaning beyond or change, and *morphe*, meaning form. In modern terms, a **metamorphic rock** is a rock that forms from a preexisting rock, or **protolith**, that undergoes mineralogical and textural changes in response to modification of its physical or chemical environment. This means that during **metamorphism**, the process of forming metamorphic rock, new minerals may grow at the expense of old ones, and/or the shape, size, and arrangement of grains in the rock may change. These changes occur when the protolith is subjected to heat, pressure, differential stress (a push, pull, or shear), and/or hydrothermal fluids (hot-water solutions). Keep in mind that when we speak of metamorphism, we do not mean all changes that happen in a rock after it forms—geologists do not consider weathering, diagenesis, or melting to be metamorphic changes. Because metamorphism does not involve melting, we say that *metamorphism is a solid-state process*.

Hutton did more than just note the existence of metamorphic rock—he also tried to understand why metamorphism takes place. Because he found metamorphic rocks adjacent to igneous intrusions, he concluded that metamorphism takes place when heat from an intrusion "cooks" the rock into which it intrudes. And because he realized that metamorphic rocks also occur over broad regions, in the absence of intrusions, he speculated that metamorphism also takes place when rock becomes deeply buried, for he knew that the Earth gets warmer with depth.

From Hutton's day to the present, geologists have undertaken field studies, laboratory experiments, and theoretical calculations to develop an understanding of metamorphism. In this chapter, we present the results of this work. We begin by explaining how metamorphism takes place. Then we describe the causes of metamorphism and the basis for classifying metamorphic rocks. We conclude by discussing the geologic settings in which these rocks form. As you will see, Hutton's speculations on the cause of metamorphism were basically correct, but they were only part of the story—the rest of the story could not take shape until the theory of plate tectonics came along.

8.2 WHAT HAPPENS DURING METAMORPHISM?

If someone were to put a rock on a table in front of you, how would you know that it is metamorphic? First, metamorphic rocks can have a **metamorphic texture**, meaning that the grains in the rock have grown in place and interlock. Second, metamorphic rocks can possess **metamorphic minerals**, new minerals that only grow under metamorphic temperatures and pressures—in fact, metamorphism can produce a group of minerals called a metamorphic mineral assemblage. And third, metamorphic rocks can have metamorphic foliation, defined by the parallel alignment of platy minerals (e.g., mica) and/or the presence of alternating light-colored and dark-colored layers. When these characteristics develop, a metamorphic rock becomes as different from its protolith as a butterfly is from a caterpillar. For example, metamorphism of red shale can yield a metamorphic rock consisting of aligned mica flakes and brilliant garnet crystals (▶Fig. 8.1a, b). Metamorphism of fossiliferous limestone can yield a metamorphic rock consisting of large interlocking crystals of calcite (▶Fig. 8.1c, d). Metamorphism of granite radically changes the rock's texture (▶Fig. 8.1e, f).

The formation of metamorphic textures and minerals happens very slowly—it may take thousands to millions of years—and it involves several processes, which sometimes occur alone and sometimes together. Let's consider the most common processes:

- *Recrystallization*: This process changes the shape and size of grains without changing the identity of the mineral constituting the grains. For example, during recrystallization of sandstone, quartz sand grains grow into larger quartz crystals that fit together tightly like pieces of a mosaic (▶Fig. 8.2a).

- *Phase change*: This process transforms a grain of one mineral into a grain of another mineral with the same composition but a different crystal structure. For

Protolith ———————▶ **Metamorphic rock**

(a)

(b)

(c)

(d)

(e)

(f)

FIGURE 8.1 (a) Hand specimens of red shale, consisting of clay flakes, quartz, and iron oxide (hematite). **(b)** Hand specimen of metamorphic rock (gneiss) containing biotite, quartz, feldspar, and bright purple garnets. A rock similar to the shale could have been the protolith of this gneiss. This hand specimen is about 10 cm wide. **(c)** This thin section of a Devonian limestone shows that the rock consists of small fossil shells and shell fragments that have been cemented together. The field of view is about 3 mm. **(d)** This thin section of marble shows how new crystals of metamorphic calcite grew to form an interlocking texture. This photo is taken with polarized light; the color and darkness of an individual grain depends on its orientation with respect to the light waves. **(e)** Granite has randomly oriented grains. **(f)** Metamorphosed granite has flattened and aligned grains.

example, the transformation of quartz into a rare mineral called coesite represents a phase change, for these minerals have the same formula (SiO_2) but different crystal structures. At an atomic scale, phase change involves rearrangement of atoms.

- *Metamorphic reaction*, or *neocrystallization* (from the Greek *neos*, new): This process results in the growth of new mineral crystals that differ from those of the protolith (▶Fig. 8.2b). During neocrystallization, one or more chemical reactions effectively digest minerals (reactants) of the protolith to produce new minerals (products) of the metamorphic rock. For this process to take place, atoms must migrate, or diffuse, through solid crystals, a very slow process, and/or dissolve and reprecipitate at grain boundaries, sometimes with the aid of hydrothermal fluids.

- *Pressure solution*: This process happens when a rock is squeezed more strongly in one direction than in others at relatively low pressures and temperatures, in the presence of water. Mineral grains dissolve where their surfaces are pressed against other grains, producing ions that migrate through the water to precipitate elsewhere. Precipitation may take place on faces where the grains are squeezed together less strongly, so pressure solution can cause grains to become shorter in one direction and longer in another (▶Fig. 8.2c). Pressure solution takes place under both non-metamorphic and metamorphic conditions.

> **Take-Home Message**
>
> Metamorphism changes an original rock (protolith) into a new metamorphic rock. The process involves one or more of the following: neocrystallization (growth of new minerals), recrystallization, plastic deformation, phase change, and pressure solution.

Protolith   **Metamorphic rock**

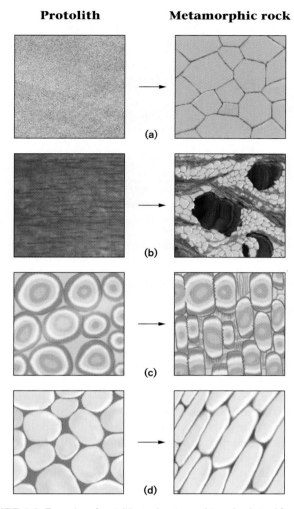

(a)

(b)

(c)

(d)

FIGURE 8.2 Examples of protoliths and metamorphic rocks derived from them, illustrating different mechanisms of metamorphism. Each box contains a sketch of a thin section, with the field of view about 1 mm wide. **(a)** A protolith of siltstone recrystallizes to form metamorphic rock made of larger quartz crystals of the same mineral. **(b)** Metamorphic reactions (neocrystallization) in a protolith of silty shale will form a rock formed of quartz, mica, large garnets, and other minerals. **(c)** A protolith of oolitic limestone (an oolite is a tiny snowball-like sphere of calcite with internal concentric rings) undergoes pressure solution so that grains have dissolved on two sides. **(d)** A protolith of quartz sandstone deforms plastically to produce a metamorphic rock in which the quartz grains have been flattened into wavy pancakes.

- *Plastic deformation*: This process happens at elevated temperatures and pressures, conditions that permit some minerals to behave like soft plastic in that if they are squeezed or stretched, they become flattened or elongate without breaking (▶Fig. 8.2d). Such deformation takes place without changing either the composition or the crystal structure of the mineral. The atomic-scale processes causing plastic deformation are complex, so we must defer an explanation of them to more advanced books.

8.3 WHAT CAUSES METAMORPHISM?

Caterpillars undergo metamorph*osis* because of hormonal changes in their bodies. Rocks undergo metamorph*ism* when they are subjected to heat, pressure, differential stress, and/or hydrothermal fluids. Let's now consider the details of how these agents of metamorphism operate.

Metamorphism Due to Heating

When you bake cake batter, the batter transforms into a new material—cake. Similarly, when you heat a rock, its ingredients transform into a new material—metamorphic rock. Why? Think about what happens to atoms in a mineral grain as the grain warms. Heat causes the atoms to vibrate rapidly, stretching and bending the chemical bonds that lock atoms to their neighbors. If bonds stretch too far and break, atoms detach from their original neighbors, move slightly, and form new bonds with other atoms. Repetition of this process leads to rearrangement of atoms within grains, or to migration of atoms into or out of grains. As a consequence, recrystallization and/or neocrystallization take place, enabling a metamorphic mineral assemblage to grow in solid rock.

Metamorphism takes place at temperatures between those at which diagenesis occurs and those that cause melting. Roughly speaking, this means that most metamorphic rocks you find in outcrops on continents formed at temperatures between 200°C and 850°C. However, melting temperature depends on composition and water content (see Chapter 6), so the upper limit of the metamorphic realm actually ranges between 650°C and 1,200°C, depending on rock composition and water content. The depth in the Earth at which metamorphic temperatures occur depends on the geothermal gradient, which, in turn, reflects the geologic setting. For example, near a hot, igneous intrusion, a temperature of 500°C can occur at very shallow depths. But in the upper part of *average* continental crust, away from intrusions, a temperature of 500°C occurs at a depth of about 20 to 25 km.

Metamorphism Due to Pressure

As you swim underwater in a swimming pool, water squeezes against you equally from all sides—in other words, your body feels pressure. Pressure can cause a material to collapse inward. For example, if you pull an air-filled balloon down to a depth of 10 m in a lake, the balloon becomes significantly smaller. Pressure can have the same effect on minerals. Near the Earth's surface, minerals with relatively open crystal structures (i.e., with relatively large spaces between atoms) are stable. However, if you subject these minerals to extreme pressure, denser minerals tend to form. Such transformations involve phase changes and/or neocrystallization.

Most metamorphic rocks that occur in outcrops on continents were metamorphosed at pressures of less than about 12 kbar (= 12,000 bars, which is about 12,000 times the pressure at Earth's surface for 1 bar ≈ 1 atm). Pressure increases at about 270 to 300 bars/km due to the weight of overlying rock, so a pressure of 12 kbar occurs at depths of about 40 km. However, in a few locations geologists have found ultra-high-pressure metamorphic rocks, which appear to have formed at pressures of up to 29 kbar, meaning depths of about 80 to 100 km. Rocks subjected to ultra-high-pressure contain grains of coesite, a phase of SiO_2 that is much denser than familiar quartz. In fact, some of these rocks contain tiny grains of diamond, a phase of carbon that only forms under very high pressure.

Changing Pressure and Temperature Together

So far, we've considered changes in pressure and temperature as separate phenomena. But in the Earth, pressure and temperature change together with increasing depth. For example, at a depth of 8 km, temperature in rock is about 200°C and pressure is about 2.3 kbar. If a rock slowly becomes buried to a depth of 20 km, as can happen during mountain building, temperature in the rock increases to 500°C, and pressure to 5.5 kbar. Experiments and calculations show that the stability (ability to last) of certain minerals and mineral assemblages depends on *both* pressure and temperature. As a result, a metamorphic rock formed at 8 km does not contain the same minerals as one formed at 20 km.

We can illustrate the relationship of mineral stability to pressure and temperature by studying the behavior of Al_2SiO_5 (aluminum silicate) as portrayed on a phase diagram, a graph with temperature indicated by one axis and pressure indicated by the other (▶Fig. 8.3). Al_2SiO_5 can exist as three different minerals: kyanite, andalusite, and sillimanite. Each of these minerals exists only under a specific range of temperatures and pressures, indicated by an area called a stability field, on the phase diagram. If a protolith containing the elements necessary to produce Al_2SiO_5 is taken to a depth in the Earth where the pressure is 2 kbar and the temperature is 450°C (point X), then andalusite grows. If the temperature stays at 450° but pressure on the rock increases to 5 kbar (point Y), then andalusite becomes unstable and kyanite grows. And if the pressure stays at 5 kbar but the temperature increases to 650°C (point Z), then sillimanite grows.

Differential Stress and Development of Preferred Mineral Orientation

Imagine that you have just built a house of cards and, being in a destructive mood, you step on it. The structure collapses because the downward push you apply with

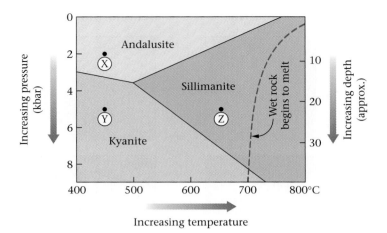

FIGURE 8.3 The stability fields for three metamorphic minerals (kyanite, andalusite, and sillimanite) that are polymorphs of Al_2SiO_5 (aluminum silicate) can be depicted on a phase diagram. At a pressure of 2 kbar and a temperature of 450°C (point X), andalusite is stable. At 5 kbar and 450°C (point Y), kyanite is stable. At 5 kbar and 650°C (point Z), sillimanite is stable.

your foot is greater than the push provided by air in other directions. If a material is squeezed (or stretched) unequally from different sides, we say that it is subjected to **differential stress**. In other words, under conditions of differential stress, the push or pull in one direction differs in magnitude from the push or pull in another direction (▶Fig. 8.4a). We distinguish two kinds of differential stress:

- *Normal stress*: Normal stress pushes or pulls perpendicular to a surface. We call a push *compression* and a pull *tension*. Compression flattens a material (▶Fig. 8.4b), whereas tension stretches a material.

- *Shear stress*: Shear stress, or shear, moves one part of a material sideways, relative to another. If, for example, you place a deck of cards on a table, then set your hand on top of the deck and move your hand parallel to the table, you shear the deck (▶Fig. 8.4c).

When rocks are subjected to differential stress at elevated temperatures and pressures (i.e., under metamorphic conditions), they can change shape without breaking. For example, a piece of rock that is squeezed or sheared during metamorphism may slowly become flatter. As it changes shape, the internal texture of the rock also changes, typically resulting in the development of **preferred mineral orientation**. By this, we mean that platy (pancake-shaped) grains lie parallel to each other, and elongate (cigar-shaped) grains align in the same direction. Both platy and elongate grains are inequant grains, meaning that the length of the grain is not the same in all directions; in contrast, equant grains have roughly the same

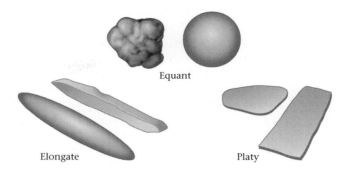

FIGURE 8.5 Some basic shapes in nature. Equant grains have roughly the same dimensions in all directions. Inequant grains can be either elongate (cigar shaped) or platy (pancake shaped).

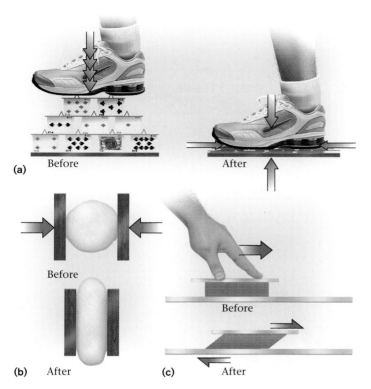

FIGURE 8.4 The concept of differential stress. **(a)** Before you step on it, a house of cards feels only air pressure, equal from all sides. As you step on the cards, they feel a differential stress because the vertical push by your foot (large arrows) is greater than the horizontal push by air (small arrows), so the house flattens. **(b)** A normal stress (in this case, compression) applied to a ball of dough flattens the ball into a pancake. **(c)** A shear stress smears out a pack of cards parallel to the table.

dimensions in all directions (▶Fig. 8.5). The preferred orientation of inequant minerals in a rock gives the rock a planar fabric, or layering. Such planar fabric is a type of metamorphic foliation.

How does preferred orientation form? In wet rocks at relatively low temperatures, pressure solution dissolves on the faces perpendicular to the direction of compression. Commonly, precipitation of the dissolved minerals takes place on faces where compression is less. So as a result of pressure solution, grains become shorter in one direction and longer in another (▶Fig. 8.6a). At relatively high temperatures, grains flatten in response to differential stress by means of plastic deformation (▶Fig. 8.6b). As a rock undergoes flattening, relatively rigid, inequant grains distributed throughout a soft matrix may rotate into parallelism as the overall rock changes shape, much as logs scattered in a flowing river align with the current (▶Fig. 8.6c). Shear can have the same effect (▶Fig. 8.6d). Finally, grain growth (by neocrystallization) may produce preferred orientation, because certain minerals grow faster in the direction in which a rock is stretching than in other directions (▶Fig. 8.6e).

The Role of Hydrothermal Fluids

Metamorphic reactions usually take place in the presence of hydrothermal fluids, because water occurs throughout the crust. We initially defined hydrothermal fluids simply as "hot water solutions." In fact, they actually can include hot water, steam, and so-called supercritical fluid. A supercritical fluid is a substance that forms under high temperatures and pressures and has characteristics of both liquid and gas. (Supercritical fluids permeate rock like a gas, seeping into every conceivable opening, and react with rock like a liquid.) Hydrothermal fluids are *chemically active*, in that they are able to react chemically with rock. For example, hydrothermal fluids can dissolve certain minerals. As a consequence, the fluids do not consist of pure water, but rather are solutions.

The water constituting hydrothermal fluids comes from several sources. Some originates as groundwater that entered the crust at the Earth's surface and then sank down, some was released from magma when the magma rose, and some originates as the product of metamorphic reactions themselves. To illustrate how metamorphic reactions produce hydrothermal fluids, consider the metamorphism of muscovite and quartz at high temperature:

$$KAl_3Si_3O_{10}(OH)_2 + SiO_2 \rightarrow KAlSi_3O_8 + Al_2SiO_5 + H_2O$$

muscovite   quartz   K-feldspar   sillimanite   water

This formula indicates that muscovite and quartz of the protolith decompose while new crystals of K-feldspar and sillimanite grow and new water molecules form.

Hydrothermal fluid plays many roles during metamorphism. First, fluids accelerate metamorphic reactions, for atoms involved in the reactions can migrate faster through a fluid than they can through a solid. Second, fluids provide water that can be absorbed during metamorphic reactions. Third, fluids passing through a rock may pick up some

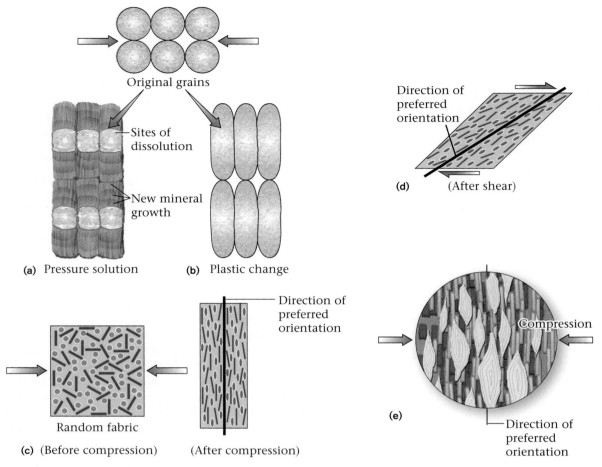

FIGURE 8.6 Squeezing or shearing a rock under metamorphic conditions can result in preferred mineral orientation in five ways. **(a)** Grains can undergo pressure solution, during which they dissolve on the side where the stress is greatest; the dissolved ions migrate through a water film and precipitate on the side where the stress is least. Commonly, the dissolved side becomes jagged in cross section. **(b)** At higher temperatures, grains can undergo plastic deformation. **(c)** Inequant grains distributed through a soft matrix may rotate into parallelism as the rock changes shape in response to compression. **(d)** Shear also will cause a preferred orientation to form. **(e)** Inequant grains can grow with a preferred orientation.

dissolved ions and drop off others (just as a bus on its route through a city picks up some passengers and drops off others) and thus can change the overall chemical composition of a rock during metamorphism. The process by which a rock's chemical composition changes because of a reaction with hydrothermal fluids is called **metasomatism**. Dissolved ions carried away by hydrothermal fluids either react with rocks elsewhere, reach the Earth's surface and wash away, or precipitate to form veins. A **vein** is a mineral-filled crack that cuts across preexisting rock (▶Fig. 8.7).

> **Take-Home Message**
>
> Metamorphism takes place in response to changes in temperature, pressure, application of differential stress (squashing, stretching, or shearing), and/or reaction with hydrothermal fluids. Application of differential stress during metamorphism aligns mineral grains.

FIGURE 8.7 These milky-white quartz veins cutting across a metamorphic rock formed by the precipitation of silica from hydrothermal fluids.

8.4 HOW DO WE CLASSIFY METAMORPHIC ROCKS?

Thousands of years ago, massive glaciers cut deep, steep-sided valleys into the western coast of Norway. When the glaciers melted away, sea level rose and the valleys became long, narrow arms of the sea known as fjords (see Chapter 18). Because fjords are so deep and their sides are so steep, cruise ships in the fjords of Norway can sail almost within spitting distance of spectacular cliffs. A tourist on deck will be treated to a spectrum of metamorphic rocks that differ in color, grain size, mineral content, and texture. Coming up with a way to classify and name the great variety of such rocks hasn't been easy. In the end, geologists have found it most convenient to divide metamorphic rocks into two fundamental classes: foliated rocks and nonfoliated rocks. Each class contains several rock types.

Foliated Metamorphic Rocks

To understand this class of rocks, we first need to discuss the nature of **foliation** in more detail. The word comes from the Latin word *folium*, for leaf. Geologists use *foliation* to refer to the repetition of planar surfaces or layers in a metamorphic rock. Some layers are indeed as thin as a leaf, or thinner, but some may be over a meter thick. Foliation can give metamorphic rocks a striped or streaked appearance in an outcrop, and/or give them the ability to split into thin sheets. A foliated metamorphic rock has foliation because it contains inequant mineral crystals that are aligned parallel to each other, defining preferred mineral orientation, and/or because the rock has alternating dark-colored and light-colored layers. As noted earlier, foliation develops in response to the application of differential stress during metamorphism.

Foliated metamorphic rocks can be distinguished from each other according to their composition, their grain size, and the nature of their foliation. The most common types include:

- **Slate**, the finest-grained foliated metamorphic rock, forms by the metamorphism of shale (a sedimentary rock consisting of clay) under relatively low pressures and temperatures. The foliation, or slaty cleavage, in slate results from the development of a strong preferred orientation of clay and chlorite grains, for these grains are shaped like extremely tiny sheets of paper (▶Fig. 8.8a). Slate tends to split on slaty cleavage planes into thin, impermeable sheets that serve as convenient roofing material (▶Fig. 8.8b, c).

  Slaty cleavage forms in response to differential stress. Typically, cleavage planes form perpendicular to the direction of compression. For example, end-on compression of a sequence of horizontal shale beds produces vertical slaty cleavage (▶Fig. 8.9a, b). The development of aligned clay is primarily a consequence of pressure solution and recrystallization—grains lying at an angle to the cleavage plane dissolve, whereas grains parallel to the cleavage plane grow. In addition, during this process, less soluble grains may passively rotate into the plane of cleavage, and new grains may grow.

- **Phyllite** is a fine-grained metamorphic rock with a foliation caused by the preferred orientation of very fine-grained white mica and, in some cases, chlorite. The

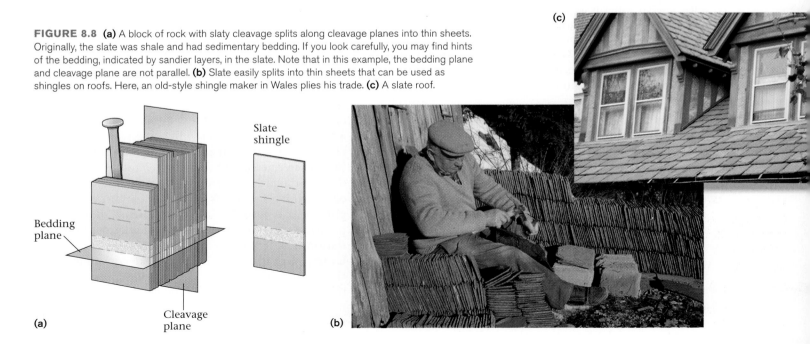

FIGURE 8.8 **(a)** A block of rock with slaty cleavage splits along cleavage planes into thin sheets. Originally, the slate was shale and had sedimentary bedding. If you look carefully, you may find hints of the bedding, indicated by sandier layers, in the slate. Note that in this example, the bedding plane and cleavage plane are not parallel. **(b)** Slate easily splits into thin sheets that can be used as shingles on roofs. Here, an old-style shingle maker in Wales plies his trade. **(c)** A slate roof.

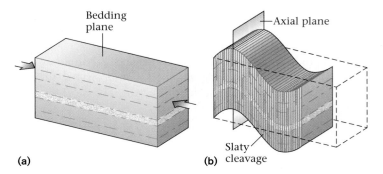

Bedding plane

Axial plane

(a)

(b) Slaty cleavage

FIGURE 8.9 **(a)** The end-on compression of a bed will create slaty cleavage at an angle perpendicular to the bedding. **(b)** Commonly, the rock folds (bends into curves) at the same time cleavage forms. Cleavage tends to be parallel to the axial plane of the fold, the imaginary plane that divides the fold in half (see Chapter 11). The dashed lines indicate the original shape of the rock body that was deformed.

word *phyllite* comes from the Greek word *phyllon*, for leaf, as does the word *phyllo*, the flaky dough in Greek pastry. The parallelism of translucent fine-grained mica gives phyllite a silky sheen, known as phyllitic luster (▶Fig. 8.10a). Phyllite forms by the metamorphism of slate at a temperature high enough to cause neocrystallization; metamorphic reactions produce a new assemblage of minerals (fine-grained mica and chlorite) out of clay. The formation of foliation in phyllite is due to differential stress during metamorphism.

• *Flattened-clast conglomerate*: Under the metamorphic conditions that transform shale to slate or to phyllite, a protolith of conglomerate also undergoes changes. Specifically, pebbles or cobbles flatten and become pancake shaped, and the alignment of these inequant clasts defines a foliation (▶Fig. 8.10b). The flattening of clasts occurs through a combination of plastic deformation and pressure solution. Geologists refer to conglomerates composed of flattened clasts as flattened-clast conglomerate, meta-conglomerate, or stretched-pebble conglomerate.

• **Schist** is a medium- to coarse-grained metamorphic rock that possesses a type of foliation, called **schistosity**, that is defined by the preferred orientation of large mica (e.g., muscovite and/or biotite) flakes (▶Fig. 8.10c). Again, the parallelism of the mica flakes develops in response to differential stress resulting from shearing and shortening during metamorphism—the flakes grow parallel to the foliation. Schist forms at a higher temperature than phyllite, and it differs from phyllite in that the mica grains are larger. Typically, schists also contain other minerals such as quartz, feldspar, kyanite, garnet, and amphibole—the specific minerals that grow depend on the chemical composition of the protolith. Schist can form from a shale but

also from a great variety of other protoliths, as long as the protolith contains the appropriate elements to make mica.

In many cases, certain mineral grains in schists grow to be much larger than surrounding minerals. For example, garnet crystals in schist may become many times larger than those of other minerals (see Fig. 8.2d). Especially large crystals that grow in a metamorphic rock are called porphyroblasts. Smaller grains surrounding porphyroblasts constitute a rock's matrix.

• **Gneiss** is a compositionally layered metamorphic rock, typically composed of alternating dark-colored and light-colored layers or lenses that range in thickness from millimeters to meters. Compositional layering, or gneissic banding, gives gneiss a striped appearance (▶Fig. 8.11). The contrasting colors respresent contrasting compositions. Light-colored layers contain predominantly felsic minerals such as quartz and feldspar, whereas the dark-colored layers contain predominantly mafic minerals such as amphibole, pyroxene, and biotite. If gneiss contains mica, the mica-rich layers have schistosity. Gneiss that formed at very high temperatures, however, does not contain mica, because at high temperatures, mica reacts to form other minerals.

How does the banding in gneiss form? There are many ways, but in this book, we can only introduce a few of the more common ones. Banding in some examples of gneiss evolved directly from the original bedding in a rock. For example, metamorphism of a protolith consisting of alternating beds of sandstone and shale produces a gneiss consisting of alternating beds of quartzite and mica schist. It is more common for gneissic banding to form when the protolith undergoes an extreme amount of shearing under conditions in which the rock can flow like soft plastic. Such flow stretches, folds, and smears out any kind of pre-existing compositional contrasts in the rock, and transforms them into aligned sheets (▶Fig. 8.12a–d). To picture this process, imagine slowly stirring vanilla batter in which there are blobs of chocolate batter— eventually you will see thin, alternating layers of dark and light batter. Similarly, imagine what happens if you take a rolling pin and flatten a ball consisting of two different colors of dough, then fold it in half, and flatten it again—you end up with thin parallel layers of contrasting colors. Finally, banding in some gneisses develops by an incompletely understood process called metamorphic differentiation. Differentiation may involve dissolution of minerals in some layers, migration of the chemical components of those minerals to other layers, where new minerals then grow. The process happens in such a way that mafic minerals and felsic minerals segregate in alternating layers (▶Fig. 8.12e).

(a)

(b)

(c)

FIGURE 8.10 (a) The sheen in this phyllite comes from the reflection of light off the tiny mica flakes that constitute the rock. The accordion-like crinkles (crenulations) on the surface are due to compression. (b) In a flattened-clast conglomerate (a type of metaconglomerate), the pebbles and cobbles are squashed into pancake-like shapes that align with each other to define a foliation. The larger clasts in this photo are about 20 cm (5 inches) long. (c) A hand specimen of schist, with coarse mica crystals.

FIGURE 8.11 An outcrop of Precambrian gneiss in Brazil. Note the camera lens cap for scale.

- *Migmatite*: At very high temperatures, or if hydrothermal fluids enter the rock and lower its melting temperature, gneiss begins to partially melt, producing magma that is enriched in silica. In some cases, this melt does not move very far before freezing to form a light-colored (felsic) igneous rock. Lenses of this new igneous rock are surrounded by bands of relict gneiss, which consists of minerals left behind when the felsic melt seeped out; the relict gneiss tends to be dark-colored (mafic). The resulting mixture of igneous and relict metamorphic rock is called a migmatite (▶Fig. 8.13).

Nonfoliated Metamorphic Rocks

A nonfoliated metamorphic rock contains minerals that recrystallized, or new minerals that grew during metamorphism, but it has no foliation. The lack of foliation may mean that metamorphism occurred in the absence of differential stress, or simply that all the new crystals are equant. We list below some of the common rock types that commonly occur without foliation. But we must add the following caveat: some of these rock types *can* develop foliation if the protolith was subjected to significant differential stress during metamorphism, or if the protolith contained bedding.

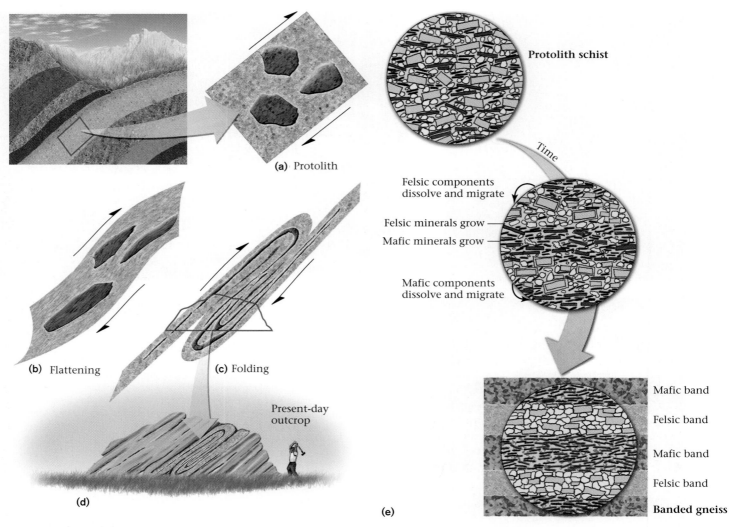

(a) Protolith

Protolith schist

Time

Felsic components dissolve and migrate

Felsic minerals grow

Mafic minerals grow

Mafic components dissolve and migrate

(b) Flattening

(c) Folding

Present-day outcrop

(d)

(e)

Mafic band

Felsic band

Mafic band

Felsic band

Banded gneiss

FIGURE 8.12 A schematic model illustrating one of the ways in which gneissic banding forms. **(a)** The protolith, in this case an intrusive igneous rock, contains patches that are more mafic than the surrounding felsic rock. **(b)** Shear stretches and flattens the rock. The mafic patches stretch and flatten too. While this is happening, recrystallization and neocrystallization are taking place throughout the rock, and a preferred mineral orientation develops. **(c)** The layer is folded back on itself in response to continued shear. **(d)** A present-day outcrop of this rock displays mafic bands separated by felsic bands. **(e)** During metamorphic differentiation, felsic minerals dissolve in mafic layers and grow in felsic layers, while mafic minerals dissolve in felsic layers and grow in mafic layers.

- *Hornfels*: Rock that undergoes metamorphism because of heating in the absence of differential stress becomes **hornfels**. Foliation does not appear in these rocks because the crystals grow in random orientations. The specific mineral assemblage in a hornfels depends on the composition of the protolith and on the temperature and pressure of metamorphism.

- *Amphibolite*: Metamorphism of mafic rocks (basalt or gabbro) can't produce quartz and muscovite when metamorphosed, for these rocks don't contain the right mix of chemicals to yield such minerals. Rather, they transform into amphibolite, a dark-colored meta-

morphic rock containing predominantly hornblende (a type of amphibole) and plagioclase (a type of feldspar) and, in some cases, biotite (▶Fig. 8.14). Where subjected to differential stress, amphibolites can develop a foliation, but the foliation tends to be poorly developed because the rock contains very little mica.

- *Quartzite*: Quartzite forms by the metamorphism of quartz sandstone. During metamorphism, preexisting quartz grains recrystallize, creating new, generally larger grains. In the process, the distinction between cement and grains disappears, open pore space disappears, and the grains become interlocking. In fact, recrystallization

(a)

FIGURE 8.13 An outcrop of migmatite in northern Michigan contains both light-colored (felsic) igneous rock and dark-colored (mafic) metamorphic rock. The mixture of the two rock types makes migmatite resemble marble cake.

makes the grains of a quartzite weld together into a tight mosaic, so that when quartzite cracks, the fracture cuts across grain boundaries. In contrast, fractures in sandstone curve around grains. Quartzite looks glassier than sandstone and does not have the grainy, sandpaper-like surface characteristic of sandstone (▶Fig. 8.15a, b). Depending on the impurities contained in the quartz, quartzite can vary in color from white to gray, purple, or green. Most quartzite is nonfoliated because it does not

FIGURE 8.14 A black amphibolite from the island of Zabargad, Egypt. The fold defined by the streaks of light-colored gneiss indicates that the rock, overall, was intensely sheared. But the pure amphibolite away from the light streaks (e.g., in the region above the jackknife) does not have an obvious foliation.

(b)

FIGURE 8.15 (a) A sandstone protolith, with a grainy surface. (b) A quartzite with a glassy surface. Both *a* and *b* consist predominantly of quartz, but they have different textures.

contain aligned mica or compositional layering. In some cases, however, quartz grains deform plastically and become pancake shaped. The alignment of "pancakes" yields a foliation. To avoid confusion, quartzite with this texture should be called foliated quartzite.

- *Marble*: The metamorphism of limestone yields marble. During the formation of marble, calcite composing the protolith recrystallizes; as a consequence, original sedimentary features such as fossil shells become hard to recognize (if they remain at all), pore space disappears, and the distinction between grains and cement disappears. Thus, marble typically consists of a fairly uniform mass of interlocking calcite crystals. It may also contain other, less familiar minerals formed from

(a)

(c)

(b)

FIGURE 8.16 **(a)** What appears to be "snow" on the mountain is actually white rock–Carrara marble. **(b)** The marble in this unfinished sculpture by Michelangelo is fairly soft and easy to carve, but it does not crumble. **(c)** Color bands in a marble floor tile of a staircase.

the reaction of calcite with quartz, clay, and iron oxide, if these minerals existed in the protolith.

Sculptors love to work with marble because the rock is relatively soft (only a 3 on the Mohs hardness scale; see Chapter 5), and has a uniform texture that gives it the cohesiveness and homogeneity needed to fashion large, smooth, highly detailed sculptures. Marble comes in a variety of colors—white, pink, green, and black—depending on the impurities it contains. Michelangelo, one of the great Italian Renaissance artists, sought large, unbroken blocks of creamy white marble from the quarries in the Italian Alps for his

masterpieces (▶Fig. 8.16a, b). This marble began as a pure limestone, formed from the shells of tiny marine organisms, and metamorphosed to marble during the mountain-building event that produced the Alps.

Like quartzite, marble commonly has no foliation, for it contains mostly equant grains. But impurities, such as iron oxide or graphite, may create beautiful color banding in marble, making it a prized decorative stone (▶Fig. 8.16c). The banding in such foliated marble typically started out as bedding. But because marble is a relatively weak rock, it flows like soft plastic under metamorphic conditions, and this flow can smear out different-colored portions of marble into beautiful contorted, curving bands.

Taking Chemical Composition into Account in Classifying Metamorphic Rocks

Up to this point, we've emphasized the importance of temperature and pressure on determining the mineral assemblage in a metamorphic rock. Let's not forget that composition plays a key role in determining which minerals form as well. For example, it's impossible to form a biotite-rich schist from a pure quartz sandstone, because biotite contains elements, such as iron, that do not occur in quartz.

To distinguish among different compositions of metamorphic rock, geologists use the following terms. (1) *Pelitic metamorphic rocks* form from sedimentary protoliths such as shale, which contain a rela-

> **Take-Home Message**
>
> Geologists divide metamorphic rock into two general classes based on whether or not the rock contains foliation. Types of foliated rocks (slate, phyllite, schist, and gneiss) differ from each other in terms of the nature of foliation, which in turn reflect metamorphic conditions.

tively high proportion of aluminum. Metamorphism of these rocks produces aluminum-rich metamorphic minerals such as muscovite. (2) *Basic (or mafic) metamorphic rocks* contain relatively little silica and an abundance of iron and magnesium. During metamorphism, minerals such as biotite and hornblende grow. (3) *Calcareous metamorphic rocks* form from calcium-rich sedimentary rocks (i.e., limestone) and contain calcite ($CaCO_3$). (4) *Quartzo-feldspathic metamorphic rocks* form from protoliths (e.g., granite) that contained mostly quartz and feldspar. These metamorphic rocks contain mostly the same minerals as the protolith, because quartz and feldspar remain stable under metamorphic conditions, but during metamorphism the quartz tends to flow and transform into thin ribbons that define a foliation (see Fig. 8.2d).

8.5 DESCRIBING THE INTENSITY OF METAMORPHISM

Metamorphic Grade

Not all metamorphism occurs under the same physical conditions. For example, rocks carried to a great depth beneath a mountain range undergo more intense metamorphism than do rocks closer to the surface. Geologists use the term **metamorphic grade**, in a somewhat informal way, to indicate the intensity of metamorphism, meaning the amount or degree of metamorphic change (▶Fig. 8.17a). (To provide a more complete indication of the intensity of metamorphism, geologists use the concept of metamorphic facies; ▶Box 8.1). Classification of metamorphic grade depends primarily on temperature, because temperature plays the dominant role in determining the extent of recrystallization and neocrystallization during metamorphism; the concept of grade doesn't apply to rocks formed due to high pressures at low temperatures. Metamorphic rocks that form under relatively low temperatures (between about 200°C and 320°C) are low-grade rocks, and rocks that metamorphose under relatively high temperatures (over 600°C) are high-grade rocks. Intermediate-grade rocks form under temperatures between these two extremes. Different grades of metamorphism yield different groups of metamorphic minerals (▶Fig. 8.17b).

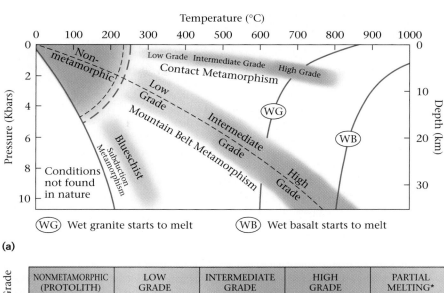

(a)

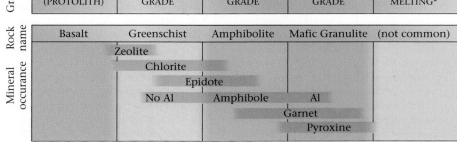

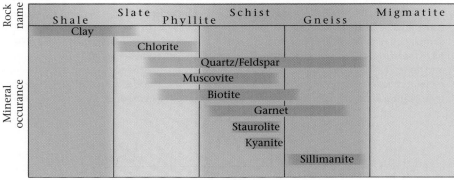

(b)

*Note: The temperature at which partial melting depends on rock composition and water content. Mafic rocks begin to melt at higher temperatures than do pelitic rocks. Wet rocks melt at lower temperatures than do dry rocks.

FIGURE 8.17 The concept of metamorphic grade. **(a)** A schematic graph showing the approximate conditions of various grades. At low temperatures, only diagenesis takes place. At progressively higher temperatures, a rock passes from low to intermediate to high grade. The presence of water may allow rocks under high-grade conditions to partially melt. The three colored bands represent typical ranges of temperature and pressure conditions in the crust. The top band is found where rocks are in contact with magma intrusions at shallow depths. The central band represents conditions beneath a mountain belt. The blue band represents conditions in an accretionary prism at a subduction zone. **(b)** The metamorphic minerals that form in a given rock depend on grade *and* composition. This chart contrasts important metamorphic minerals that form, at different grades, from a basalt protolith with those formed from a shale protolith.

BOX 8.1 THE REST OF THE STORY

Metamorphic Facies

In the early years of the twentieth century, geologists working in Scandinavia, where erosion by glaciers has left beautiful, nearly unweathered exposures, came to realize that metamorphic rocks, in general, do not consist of a hodgepodge of minerals formed at different times and in different places. Rather, these rocks consist of a distinct assemblage of minerals that grew in association with each other at a certain pressure and temperature. It seemed that the mineral assemblage present in a rock more or less represented a condition of chemical equilibrium, meaning that the chemicals making up the rock had organized into a group of mineral grains that were—to anthropomorphize a bit—comfortable with each other and their surroundings, and thus did not feel the need to change further.

These geologists also realized that the specific mineral assemblage in a rock depends on pressure and temperature conditions, and on the composition of the protolith. For example, a basalt metamorphosed at low pressures and temperatures doesn't contain the same minerals as a basalt metamorphosed under high temperatures and pressures, even though the two rocks have the same chemical composition. Similarly, a basalt and a shale metamorphosed under the same conditions do not contain the same assemblages of minerals, because they do not have the same chemical composition. But— and here's the key observation—a basalt metamorphosed at a given pressure and temperature at one location (location X) will

have the same mineral assemblage as a basalt metamorphosed under the same pressure and temperature at a different location (location Y). Likewise, a shale metamorphosed at location X will have the same assemblage as a shale metamorphosed at location Y. So if you determine the assemblage of minerals in a basalt at a given locality, you can predict the assemblage of minerals in a shale at the same locality.

The above discovery led the geologists to propose the concept of **metamorphic facies**. A metamorphic facies is a set of metamorphic mineral assemblages indicative of a certain range of pressure and temperature. Each specific assemblage in a facies reflects a specific protolith composition. According to this definition, a given metamorphic facies includes several different kinds of rocks that differ from each other in terms of composition (i.e., mineral content)—but all the rocks of a given facies formed under roughly the same temperature and pressure conditions.

Geologists recognize several major facies, of which the major ones are zeolite, hornfels, greenschist, amphibolite, blueschist, eclogite, and granulite. The names of the different facies are based on a distinctive feature or mineral found in some of the rocks of the facies. For example, some zeolite-facies rocks contain a class of minerals called zeolite, and greenschist-facies rocks contain a green, flaky mineral called chlorite. Some amphibolite-facies rocks contain a type of amphibole called hornblende, and some

blueschist-facies rocks contain a bluish amphibole called glaucophane.

We can represent the approximate conditions under which metamorphic facies formed by using a pressure-temperature graph (▶Fig. 8.18). Each area on the graph, labeled with a facies name, represents the approximate range of temperatures and pressures in which mineral assemblages characteristic of that particular facies form. For example, a rock subjected to the pressure and temperature at point A (4.5 kbar and 400°C) develops a mineral assemblage characteristic of the greenschist facies. As the graph implies, the boundaries between facies cannot be precisely defined, and there are likely broad transitions between facies.

We can also portray the geothermal gradients of different crustal regions on the graph. Beneath mountain ranges, for example, the geothermal gradient passes through the zeolite, greenschist, amphibolite, and granulite facies. In contrast, the geothermal gradient in an accretionary prism created at a subduction zone passes into the blueschist facies, because temperatures in the prism remain relatively cool, even at high pressure. Hornfels facies rocks form in the wall rocks of igneous intrusions where temperature is high but pressure is low.

"Grade" and "facies" are related terms in that they are both used to distinguish among rocks formed under different metamorphic conditions. But geologists use "grade" to give an approximate sense of metamorphic temperature, and "facies" to emphasize the mineral assemblage in the rock. Roughly speaking, zeolite and lower-greenschist facies rocks are low grade, upper-greenschist facies through lower-amphibolite facies are intermediate grade, and upper-amphibolite through granulite facies rocks are high grade.

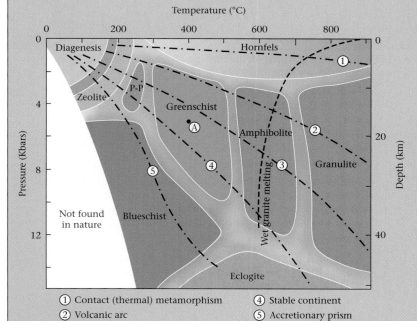

① Contact (thermal) metamorphism
② Volcanic arc
③ Collisional mountain belt
④ Stable continent
⑤ Accretionary prism

FIGURE 8.18 The common metamorphic facies. The boundaries between the facies are depicted as wide bands because they are gradational and approximate, for the various metamorphic reactions that transform minerals in one facies into minerals of another don't all occur at exactly the same pressure and temperature conditions. Note that some amphibolite-facies rocks and all granulite-facies rocks form at pressure-temperature conditions to the right of the melting curve for wet granite. Thus, such metamorphic rocks develop only if the protolith is dry. The dotted lines indicate approximations of various geothermal gradients found on Earth. One of the facies depicted on the graph is not mentioned in the text: specifically, P-P = prehnite-pumpellyite facies, named for two unusual metamorphic minerals. The hornfels facies includes several subfacies, each formed at a different temperature range; the dashed line is the wet granite melting line.

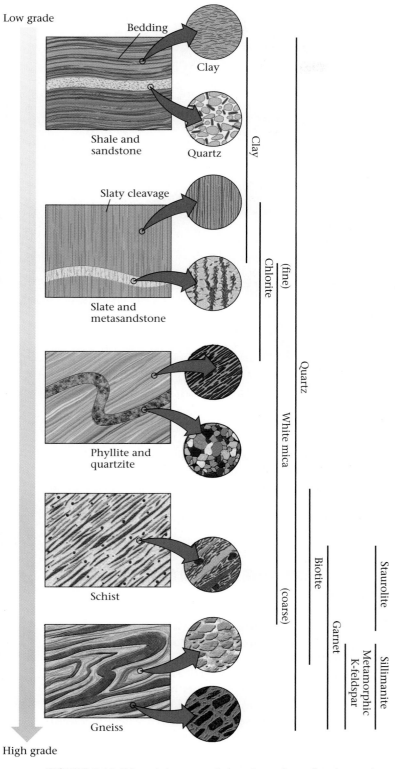

FIGURE 8.19 When shale progressively metamorphoses from low grade to high grade, it first becomes slate, then phyllite, then schist, then gneiss. In many cases, gneiss and schist can form under the same conditions. The side graph shows the stability range of various minerals.

Metamorphism that occurs while temperature and pressure progressively *increase* is called prograde metamorphism. During prograde metamorphism, recrystallization and neocrystallization produce coarser grains and new mineral assemblages that are stable at higher temperatures and pressures. As grade increases, metamorphic reactions release water, so high-grade rocks tend to be "drier" than low-grade rocks. This means that minerals in high-grade rocks do not contain minerals with -OH in their formula, whereas minerals in lower-grade rocks can.

To understand prograde metamorphism, consider the changes that a shale undergoes when it starts near the Earth's surface and ends up at great depth beneath a mountain range (▶Fig. 8.19). The clay flakes in shale lie more or less parallel to the bedding. Under low-grade metamorphic conditions and differential stress, shale transforms into slate. In slate, the clay flakes are a bit larger, are better formed, and align parallel to cleavage. As metamorphic grade increases a bit more, new crystals of fine-grained white mica, as well as new crystals of chlorite and quartz form, transforming the rock into phyllite.

Under intermediate-grade conditions, the minerals in phyllite react and decompose, yielding atoms that combine to produce large crystals of mica (such as muscovite and biotite), as well as other minerals such as garnet. The reactions also release water, which escapes. In our example, this metamorphism is taking place under differential stress, so the micas grow with a preferred orientation and the rock becomes a schist.

During metamorphic reactions under high-grade conditions, yet another assemblage of minerals forms. High-grade rocks do not contain much mica, if any, because mica contains -OH and tends to decompose and release water at high temperatures. In fact, high-grade rock typically includes water-free minerals, such as feldspar, quartz, pyroxene, and garnet. As mica disappears, the rock loses its schistosity but can develop gneissic layering.

Metamorphism that takes place when temperatures and pressures progressively decrease is known as retrograde metamorphism. For retrograde metamorphism to occur, water must be added back to the rock. Thus, retrograde metamorphism does not happen unless hydrothermal fluids enter the rock. Under cold and dry conditions, retrograde metamorphic reactions cannot proceed. It is for this reason that high-grade rocks formed early during Earth history have survived and can be exposed at the surface of the Earth today.

We can represent the concept of prograde and retrograde metamorphism as a path plotted on a pressure-temperature graph (▶Fig. 8.20). When a rock gets buried progressively deeper, temperature and pressure increase; the rock follows a prograde path on the graph. (In this particular example the rock was buried quickly, so pressure increased faster than temperature; rocks are good insulators and thus heat up very slowly.) Later, when the

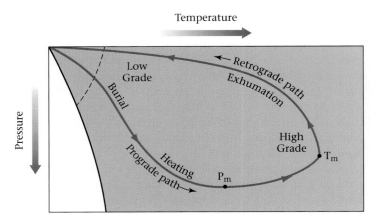

FIGURE 8.20 The metamorphic history of a rock can be portrayed on a graph showing variations in temperature and pressure. This graph shows one of many possible paths. As a rock experiences increased heating and pressure, it undergoes prograde metamorphism. As the temperature and pressure decrease, the rock undergoes retrograde metamorphism, if water can be added back. P_m is the peak pressure, and T_m is the peak temperature. In this example, the rock was buried so quickly that it reached its peak pressure before it reached its peak temperature.

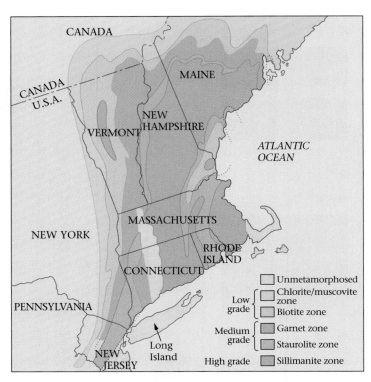

FIGURE 8.21 Metamorphic zones as portrayed on a map of New England (U.S.A.). Isograds, defined by the first appearance of an index mineral, separate the zones.

rock moves back toward the Earth's surface, because uplift raises the rock and erosion strips away overlying rock, it follows the retrograde path. Recently, geologists have been able to determine the times at which a rock reached particular locations along the pressure-temperature path. This information defines a "P-T-t path" (pressure-temperature-time path) for the rock. Knowledge of these factors help geologists interpret the geologic history of the rock.

Index Minerals and Metamorphic Zones

Geologists have discovered that the presence of certain minerals, known as index minerals, in a rock can indicate the approximate metamorphic grade of the rock. Geologists can determine the locations in a region where a particular index mineral first appears. The line on a map along which an index mineral first appears is called an isograd (from the Greek *iso*, meaning equal). All points along an isograd have approximately the same metamorphic grade. **Metamorphic zones** are the regions between two isograds; zones are named after an index mineral that was not present in the previous, lower-grade zone.

To compare rocks of different grades, you could take a hike from central New York State eastward into central Mass-

> **Take-Home Message**
>
> The type of metamorphic rock (low grade vs. high grade) that forms at a location depends on the conditions of metamorphism. For example, a given metamorphic mineral assemblage forms under a definable range of temperature and pressure.

achusetts in the northeastern United States. Your path starts in a region where rocks were not metamorphosed, and it takes you into the internal part of the Appalachian mountain belt, where rocks were intensely metamorphosed. As a consequence, you cross several metamorphic zones (▶Fig. 8.21).

8.6 WHERE DOES METAMORPHISM OCCUR?

By this point in the chapter, we've discussed the nature of changes that occur during metamorphism, the agents of metamorphism (heat, pressure, differential stress, and hydrothermal fluids), the rock types that form as a result of metamorphism, and the concepts of metamorphic grade and metamorphic facies. With this background, let's now examine the geologic settings on Earth where metamorphism takes place, as viewed from the perspective of plate tectonics theory (see art spread, pp. 250–251). Because of the wide range of possible metamorphic environments, metamorphism occurs at a wide range of conditions in the Earth. You will see that the range of conditions under which metamorphism occurs is not the same in all geologic settings. That's because the geothermal gradient

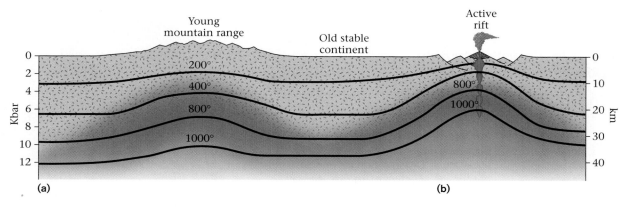

(a) **(b)**

FIGURE 8.22 The change in temperature with depth at different locations in a continent. The solid lines are isotherms—they connect locations where the temperature has the same value (along the 300°C isotherm, the temperature is 300°C). **(a)** Beneath a young collisional mountain range and **(b)** beneath an active rift. Note that isotherms rise to shallower depths beneath mountain ranges because plutons carry heat with them up to shallower depths. The crust is fairly cool beneath old, stable continents.

(meaning the rate of change of temperature with depth; ▶Fig. 8.22a, b), the extent to which rocks endure differential stress during metamorphism, and the extent to which rocks interact with hydrothermal fluids all depend on the geologic environment.

Thermal or Contact Metamorphism: Heating by an Igneous Intrusion

Imagine a hot magma that rises from great depth beneath the Earth's surface and intrudes into cooler country rock at a shallow depth. Heat flows from the magma into the country rock, for heat always flows from hotter to colder materials. As a consequence, the magma cools and solidifies while the country rock heats up. In addition, hydrothermal fluids circulate through both the intrusion and the country rock. As a consequence of the heat and hydrothermal fluids, the country rock undergoes metamorphism, with the highest-grade rocks forming immediately adjacent to the pluton, where the temperatures were highest, and progressively lower-grade rocks forming farther away. The distinct belt of metamorphic rock that forms around an igneous intrusion is called a **metamorphic aureole**, or contact aureole (▶Fig. 8.23). The width of an aureole depends on the size and shape of the intrusion, and on the amount of hydrothermal circulation: larger intrusions create wider aureoles.

The local metamorphism caused by igneous intrusion can be called either **thermal metamorphism** (▶Box 8.2), to emphasize that it develops in response to heat without a change in pressure and without differential stress, or **contact metamorphism**, to emphasize that it develops adjacent to an intrusion. Because this metamorphism takes place without application of differential stress, aureoles contain hornfels, a nonfoliated metamorphic rock.

You can see a classic example of contact metamorphism by traveling to the state of Maine in the northeastern United States. Here you will find a 14-km-long by 4-km-wide granitic pluton, named the Onawa Pluton, which formed about 400 million years ago when an 850°C

FIGURE 8.23 In a metamorphic aureole bordering an igneous intrusion, the highest-grade thermally metamorphosed rocks directly border the intrusion. The grade decreases away from the pluton. The gradation is analogous to the gradation from clay to pottery to porcelain, obtained by firing clay in an oven.

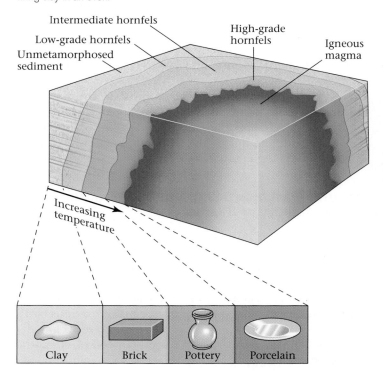

BOX 8.2 THE HUMAN ANGLE

Pottery Making—An Analog for Thermal Metamorphism

A crumbling brick in the wall of an adobe house, an earthenware pot, a stoneware bowl, and a translucent porcelain teacup may all be formed from the same lump of soft clay, scooped from the surface of the Earth and shaped by human hands. This pliable and slimy muck is a mixture of very fine clay minerals and quartz grains formed during the chemical weathering of rock and water. Fine potters' clay for making white china contains a particular clay mineral called kaolinite ($Al_2Si_2O_5[OH]_4 \cdot 2H_2O$), named after the locality in China (called *Kauling*, meaning high ridge) where it was originally discovered.

People in arid climates make adobe bricks by forming damp clay into blocks, which they then dry in the sun. Such bricks can be used for construction *only* in arid climates, because if it rains heavily, the bricks will rehydrate and turn back into sticky muck. Drying clay in the sun does not change the structure of the clay minerals.

To make a more durable material, brick makers place clay blocks in a kiln and bake ("fire") them at high temperatures. This process makes the bricks hard and impervious to water. Potters use the same process to make jugs. In fact, fired clay jugs that were used for storing wine and olive oil have been found intact in sunken Greek and Phoenician ships that have rested on the floor of the Mediterranean Sea for thousands of years! Clearly, the firing of a clay pot fundamentally and permanently changes clay in a way that makes it physically different (see Fig. 8.23). In other words, firing causes a thermal metamorphic change in the mineral assemblage that composes pottery. The extent of the transformation depends on the kiln temperature, just as the grade of metamorphic rock depends on temperature. Potters usually fire earthenware at about 1,100°C and stoneware (which is harder than a knife or fork) at about 1,250°C. To produce porcelain—fine china—the clay is partially melted at even higher temperatures. Just as it begins to melt, the potter cools ("quenches") it quickly. Quenching of the melt creates glass, which gives porcelain its translucent, vitreous (glassy) appearance.

magma intruded into country rock composed of 300°C slate, several kilometers below the surface of the Earth. Heat from the magma transformed the slate into hornfels in an aureole that reaches a maximum width of 2 km. Subsequently, erosion stripped off overlying rock, so that outcrops of the granite and hornfels can be seen today (▶Fig. 8.24a–d).

Contact metamorphism occurs anywhere that the intrusion of plutons occurs. According to plate tectonics theory, plutons intrude into the crust at convergent plate boundaries, in rifts, and during the mountain building that takes place when continents collide.

Burial Metamorphism: Due to Deep Burial in Sedimentary Basins

As sediment gets buried in a subsiding sedimentary basin, the pressure increases due to the weight of overburden, and the temperature increases due to the geothermal gradient. In the upper few kilometers, the temperatures and pressures are low enough that the changes taking place in the sediment can be considered to be manifestations of diagenesis. But at depths greater than about 8 to 15 km, depending on the geothermal gradient, temperatures may be high enough for metamorphic reactions to begin, and low-grade nonfoliated metamorphic rocks form. Metamorphism due only to the consequences of very deep burial is called **burial metamorphism**. Note that burial metamorphism causes the organic molecules of oil to break up; for this reason, oil drillers stop drilling when the bottom of the hole reaches depths at which burial metamorphism has begun.

Dynamic Metamorphism: Metamorphism along Faults

Faults are surfaces on which one piece of crust slides, or shears, past another. Near the Earth's surface (in the upper 10–15 km) this movement can fracture rock, breaking it into angular fragments or even crushing it to a powder. But at greater depths rock is so warm that it behaves like soft plastic as shear along the fault takes place. During this process, the minerals in the rock recrystallize. We call this process **dynamic metamorphism**, because it occurs as a consequence of shearing alone, under metamorphic conditions, without requiring a change in temperature or pressure. The resulting rock, a **mylonite**, has a foliation that roughly parallels the fault (▶Fig. 8.25a–c). (Recrystallization processes during the formation of mylonite are a bit different from those we mentioned earlier; for reasons

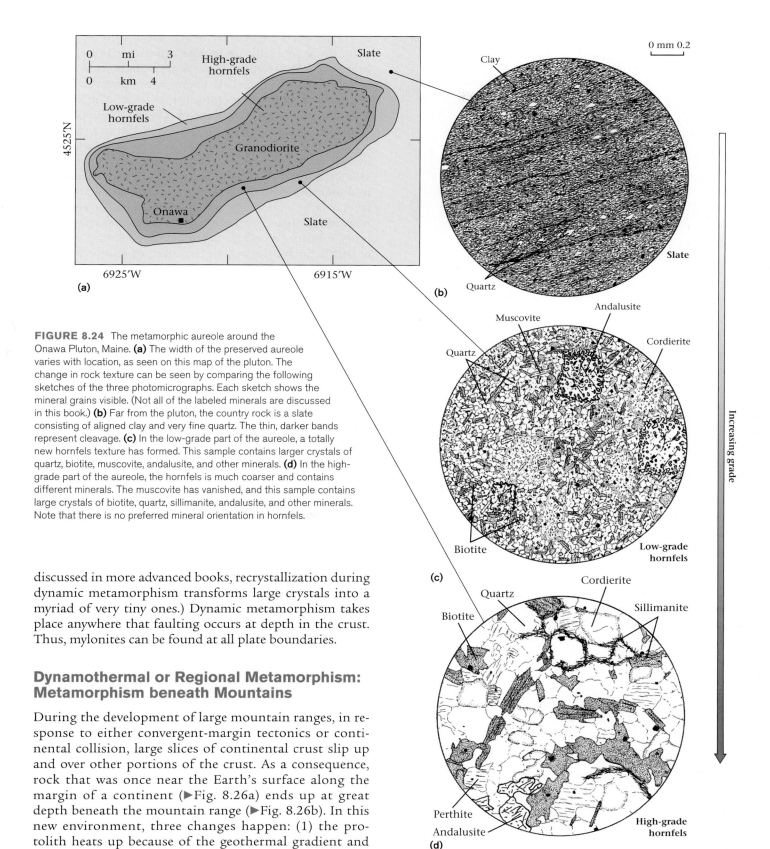

FIGURE 8.24 The metamorphic aureole around the Onawa Pluton, Maine. **(a)** The width of the preserved aureole varies with location, as seen on this map of the pluton. The change in rock texture can be seen by comparing the following sketches of the three photomicrographs. Each sketch shows the mineral grains visible. (Not all of the labeled minerals are discussed in this book.) **(b)** Far from the pluton, the country rock is a slate consisting of aligned clay and very fine quartz. The thin, darker bands represent cleavage. **(c)** In the low-grade part of the aureole, a totally new hornfels texture has formed. This sample contains larger crystals of quartz, biotite, muscovite, andalusite, and other minerals. **(d)** In the high-grade part of the aureole, the hornfels is much coarser and contains different minerals. The muscovite has vanished, and this sample contains large crystals of biotite, quartz, sillimanite, andalusite, and other minerals. Note that there is no preferred mineral orientation in hornfels.

discussed in more advanced books, recrystallization during dynamic metamorphism transforms large crystals into a myriad of very tiny ones.) Dynamic metamorphism takes place anywhere that faulting occurs at depth in the crust. Thus, mylonites can be found at all plate boundaries.

Dynamothermal or Regional Metamorphism: Metamorphism beneath Mountains

During the development of large mountain ranges, in response to either convergent-margin tectonics or continental collision, large slices of continental crust slip up and over other portions of the crust. As a consequence, rock that was once near the Earth's surface along the margin of a continent (▶Fig. 8.26a) ends up at great depth beneath the mountain range (▶Fig. 8.26b). In this new environment, three changes happen: (1) the protolith heats up because of the geothermal gradient and because of igneous activity, (2) the protolith is subjected

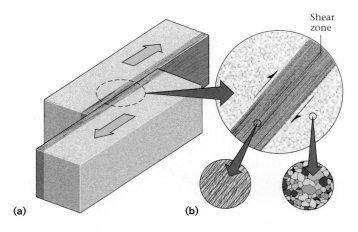

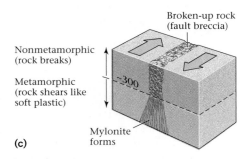

(a)

(b)

(c)

FIGURE 8.25 Dynamic metamorphism along a fault zone. **(a)** Note the band of sheared rock on either side of the slip surface. **(b)** The rock outside the shear zone has a different texture from that of the rock inside. **(c)** The block formed in (a) must have developed at a depth where metamorphic conditions exist, so that mylonite forms; otherwise, it would break up during movement.

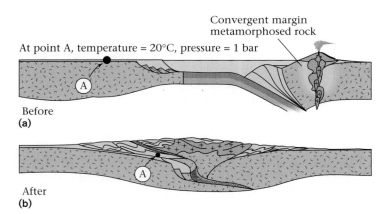

At point A, temperature = 20°C, pressure = 1 bar

Before
(a)

After
(b)

FIGURE 8.26 **(a)** Metamorphism occurs where there is plutonic activity along a convergent boundary. Some metamorphism may be thermal, but because of compression and shearing along convergent boundaries, some may be dynamothermal. **(b)** The sedimentary rock that lay at the top of a passive margin (point A) gets carried to great depth in a continental collision that leads to mountain building. As a result, it undergoes dynamothermal metamorphism. A broad region beneath a collision will lie in the field of metamorphism.

to greater pressure because of the weight of overburden, and (3) the protolith undergoes squashing and shearing because of the differential stress generated by plate interaction. As a result of these changes, the protolith transforms into foliated metamorphic rock. The type of foliated rock that forms depends on the grade of metamorphism—slate forms at shallower depths, whereas schist and gneiss form at greater depths. Since the metamorphism we've just described involves not only heat but also shearing and squashing, we can call it **dynamothermal metamorphism**. Typically, such metamorphism affects a large region, so geologists also call it **regional metamorphism**. Erosion eventually removes the mountains, exposing a belt of metamorphic rock that once lay at depth. Such belts may be hundreds of kilometers wide and thousands of kilometers long.

Hydrothermal Metamorphism of the Ocean Floor

Hot magma rises beneath the axis of mid-ocean ridges, so when cold seawater sinks through cracks down into the oceanic crust along ridges, it heats up and transforms into hydrothermal fluid. This fluid then rises through the crust, near the ridge, causing **hydrothermal metamorphism** of ocean-floor basalt; this metamorphism produces chlorite, giving the rock a greenish hue (▶Fig. 8.27a, b). This fluid eventually escapes through vents back into the sea; these vents are called black smokers (see Chapter 4).

Metamorphism in Subduction Zones: The Blueschist Puzzle

Blueschist is a relatively rare rock that contains an unusual blue-colored amphibole called glaucophane. Laboratory experiments indicate that glaucophane requires very high pressure but relatively low temperature to form. Such conditions do not develop in continental crust—usually, at the high pressure needed to generate glaucophane, temperature in continental crust is also high (see Box 8.1). So to figure out where blueschist forms, we must determine where high pressure can develop at relatively low temperature.

Plate tectonics theory provides the answer to this puzzle. Researchers found that blueschist occurs only in the accretionary prisms that form at subduction zones (see art, pp. 250–251). They realized that because prisms grow to be over 20 km thick, rock at the base of a prism feels high pressure (due to the weight of overburden). But because the subducted oceanic lithosphere beneath the prism is cool, temperatures at the base of the prism remain relatively low. Under these conditions, glaucophane can form. Because of shear between the subducting plate and the overriding plate, blueschist develops a foliation.

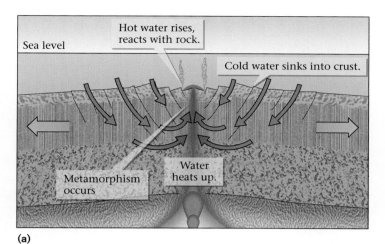

(a)

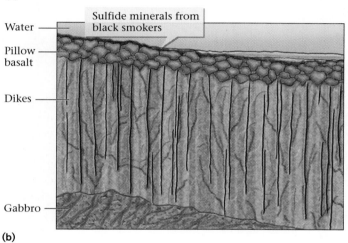

(b)

FIGURE 8.27 **(a)** Along a mid-ocean ridge, the circulation of hydrothermal fluids in response to igneous activity along the ridge causes metamorphism of basalt in the oceanic crust. **(b)** Hydrothermal metamorphism is concentrated along cracks and pores where fluid had access.

Shock Metamorphism

When large meteorites slam into the Earth, a vast amount of kinetic energy (see Appendix A) transforms into heat, and a pulse of extreme compression (a shock wave) propagates into the Earth. The heat may be sufficient to melt or even vaporize rock at the impact site, and the extreme compression of the shock wave causes the crystal structure of quartz grains in rocks below the impact site to suddenly undergo a phase change to a more compact mineral called coesite. Such changes are called **shock metamorphism** to emphasize their relationship to a large impact. The discovery of shock metamorphism at what is now known as Meteor Crater, Arizona, proved that the structure indeed resulted from impact, and when astronauts sampled the Moon, they discovered that the regolith covering the lunar surface contains the products of shock metamorphism produced by countless impacts.

Metamorphism in the Mantle

Discussions of metamorphic rocks rarely mention the mantle, because mantle rocks rarely crop out on Earth. But if you think about the process of convection in the mantle, it becomes clear that rocks in the mantle must have undergone metamorphic change several times during Earth history. Specifically, as peridotite (ultramafic rock) from shallower depths in the asthenosphere slowly cools, it becomes denser and sinks. As it sinks, pressure acting on it increases until, at a depth of about 440 km, olivine in the peridotite undergoes a phase change and transforms into different minerals that are stable at higher pressure. The process happens again when the rock sinks below 660 km, and it occurs at other depths as well. Later in Earth history, when the rock reaches great depth in the mantle, it heats up and becomes relatively buoyant. Therefore, it slowly rises back toward the base of the lithosphere and, in the process, undergoes phase changes again, but this time the phase changes produce minerals that are stable at lower pressures.

> ### Take-Home Message
>
> Contact metamorphism develops around igneous intrusions. Regional metamorphism occurs beneath mountain ranges, where realms of high temperature, pressure, and differential stress develop. Other phenomena (e.g., impact shock and fluid circulation) also cause metamorphism.

8.7 WHERE DO YOU FIND METAMORPHIC ROCKS?

When you stand on an outcrop of metamorphic rock, you are standing on material that once lay many kilometers beneath the surface of the Earth. In areas of regional metamorphism, high-grade rocks rose from a greater depth than did low-grade rocks—some high-grade rocks were once tens of kilometers below the surface. Where do exposures of metamorphic rock occur, and how does metamorphic rock return to the Earth's surface?

Occurences of Metamorphic Rock

If you want to study metamorphic rocks, you can start by taking a hike in a collisional or convergent mountain range (e.g., Fig. 8.21). During the formation of such ranges, rocks undergo both contact metamorphism and regional metamorphism; because of **exhumation**, the process by which overlying rock is removed and deeper rock rises, these rocks are exposed as towering cliffs of gneiss and schist. Exhumation can occur relatively quickly, geologically speaking— some metamorphic rocks now visible in actively growing mountain belts formed only a few million years ago. Where ancient mountain ranges once existed, we still find belts of

Environments of Metamorphism

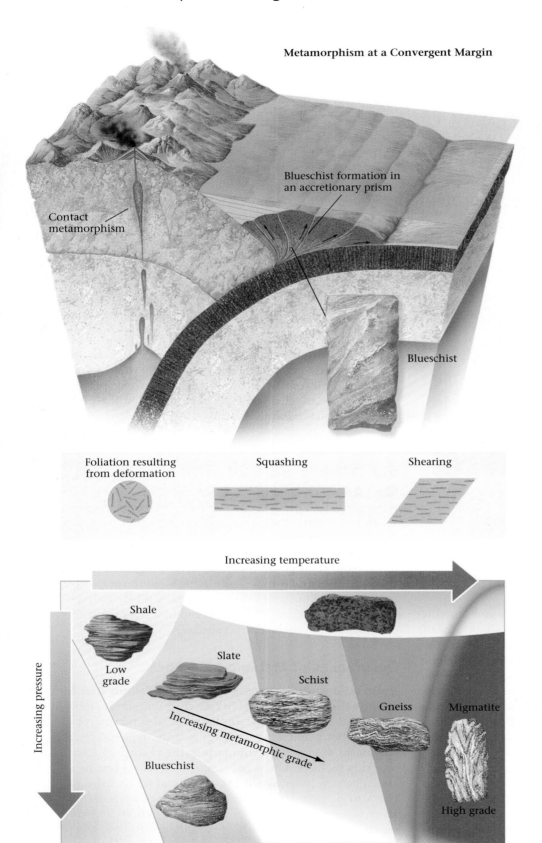

Metamorphism at a Convergent Margin

Contact metamorphism

Blueschist formation in an accretionary prism

Blueschist

Foliation resulting from deformation

Squashing

Shearing

Increasing temperature

Increasing pressure

Shale

Low grade

Slate

Schist

Increasing metamorphic grade

Gneiss

Migmatite

Blueschist

High grade

Mylonite in a shear zone

Hornfels formation

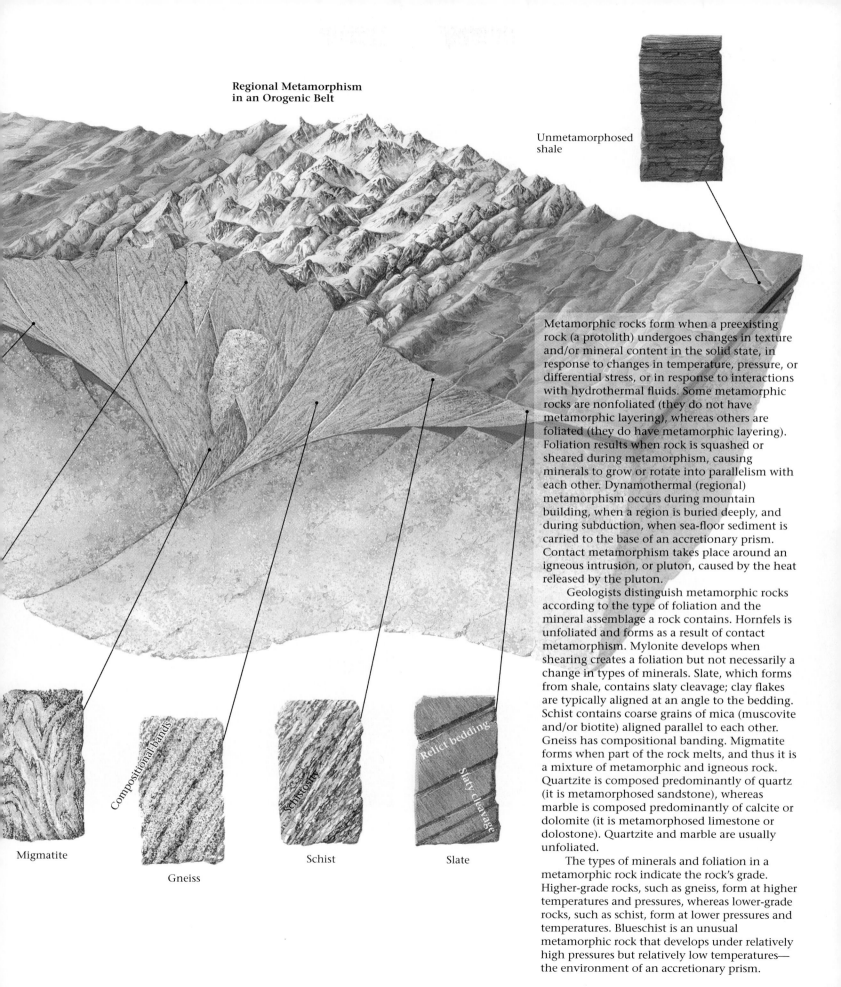

Regional Metamorphism
in an Orogenic Belt

Unmetamorphosed
shale

Metamorphic rocks form when a preexisting
rock (a protolith) undergoes changes in texture
and/or mineral content in the solid state, in
response to changes in temperature, pressure, or
differential stress, or in response to interactions
with hydrothermal fluids. Some metamorphic
rocks are nonfoliated (they do not have
metamorphic layering), whereas others are
foliated (they do have metamorphic layering).
Foliation results when rock is squashed or
sheared during metamorphism, causing
minerals to grow or rotate into parallelism with
each other. Dynamothermal (regional)
metamorphism occurs during mountain
building, when a region is buried deeply, and
during subduction, when sea-floor sediment is
carried to the base of an accretionary prism.
Contact metamorphism takes place around an
igneous intrusion, or pluton, caused by the heat
released by the pluton.

Geologists distinguish metamorphic rocks
according to the type of foliation and the
mineral assemblage a rock contains. Hornfels is
unfoliated and forms as a result of contact
metamorphism. Mylonite develops when
shearing creates a foliation but not necessarily a
change in types of minerals. Slate, which forms
from shale, contains slaty cleavage; clay flakes
are typically aligned at an angle to the bedding.
Schist contains coarse grains of mica (muscovite
and/or biotite) aligned parallel to each other.
Gneiss has compositional banding. Migmatite
forms when part of the rock melts, and thus it is
a mixture of metamorphic and igneous rock.
Quartzite is composed predominantly of quartz
(it is metamorphosed sandstone), whereas
marble is composed predominantly of calcite or
dolomite (it is metamorphosed limestone or
dolostone). Quartzite and marble are usually
unfoliated.

The types of minerals and foliation in a
metamorphic rock indicate the rock's grade.
Higher-grade rocks, such as gneiss, form at higher
temperatures and pressures, whereas lower-grade
rocks, such as schist, form at lower pressures and
temperatures. Blueschist is an unusual
metamorphic rock that develops under relatively
high pressures but relatively low temperatures—
the environment of an accretionary prism.

Migmatite

Compositional bands

Schistosity

Relict bedding

Slaty cleavage

Gneiss

Schist

Slate

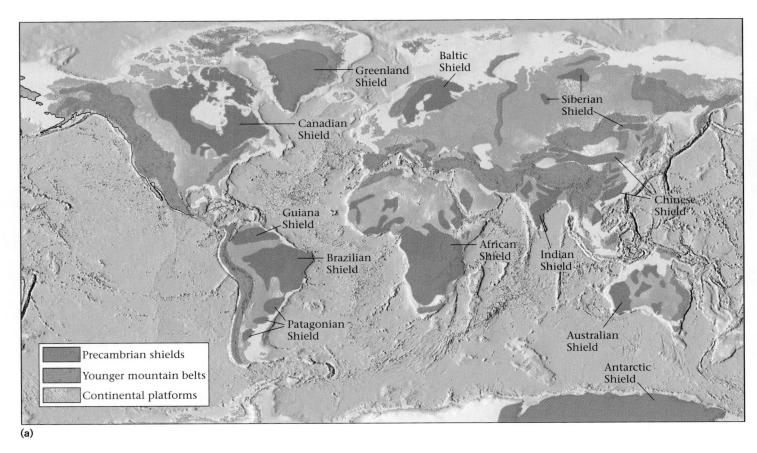

(a)

(b)　　　　　　　　　　　　　　　　　　　　　(c)

FIGURE 8.28 **(a)** The distribution of shield areas (exposed Precambrian metamorphic and igneous rock) on the Earth. **(b)** The Canadian Shield as viewed from the air. **(c)** The walls of the Black Canyon of the Gunnison River in Colorado display high-grade metamorphic rocks. These rocks underlie the Rocky Mountains of Colorado. The stripes are pegmatite dikes that intruded the dark rock.

metamorphic rocks cropping out at the ground surface even though the high peaks of the range have long since eroded away.

Huge expanses of metamorphic rock crop out in continental shields. A **shield** is the older portion of a continent, where extensive areas of Precambrian rock crop out at the ground surface, because overlying younger rock has eroded away (▶Fig. 8.28a). The shield of North America is called the Canadian Shield because it encompasses about half the land area of Canada (▶Fig. 8.28b). Large shields also occur in South America, northern Europe, Africa, India, and Siberia. Rocks in shields were metamorphosed during a succession of Precambrian mountain-building events that were responsible for building the continent in the first place—the oldest rocks on Earth occur in shields.

In the United States and much of Europe, a veneer of Paleozoic and Mesozoic sedimentary rocks covers most of the Precambrian metamorphic rocks, so you can see these metamorphic rocks only where they were uplifted and exposed by erosion in younger mountain belts, or where rivers have cut down deeply enough to expose basement (▶Fig. 8.28c; see Chapter 7). For example, at the base of the Grand Canyon, erosion by the Colorado River exposes dark cliffs of the 1.8-billion-year-old Vishnu Schist (Fig. 7.2).

Exhumation

As we noted earlier, geologists refer to the overall process by which deeply buried rocks end up back at the surface as **exhumation**. Exhumation results from several processes in the Earth System that happen simultaneously. Let's look at the specific processes that contribute to bringing high-grade metamorphic rocks from below a collisional mountain range back to the surface. First, as two continents progressively push together, the rock caught between them squeezes upward, or is uplifted, much like a ball of dough pressed in a vise (▶Fig. 8.29a); the upward movement takes place by slip on faults and by the plastic-like flow of rock. Second, as the mountain range grows, the crust at depth beneath it warms up and gets weak. Eventually, the range starts to collapse under its own weight, much like a block of soft cheese placed in the hot sun, in a process called extensional collapse (▶Fig. 8.29b; see Chapter 11). As a result of this collapse, the upper crust spreads out laterally. This movement stretches the upper part of crust in the horizontal direction and causes it to become thinner in the vertical direction. As the upper part of the crust becomes thinner, the deeper crust ends up closer to the surface. Third, erosion takes place at the surface (▶Fig 8.29c); weathering, landslides, river flow, and glacial flow together play the role of a giant rasp, stripping away rock at the surface and exposing rock that was once below the surface. As the weight of the overlying rock is removed, the underlying rock rises isostatically, like the deck of a cargo ship from which cargo has been removed.

> ### Take-Home Message
> Metamorphic rocks can be found in mountain ranges and in shields. Their exposure at the Earth's surface requires exhumation, which involves uplift, erosion, and in some cases, the thinning of upper crust by faulting.

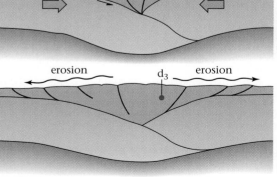

(a)

Block of cheese   Hot sun

(b)

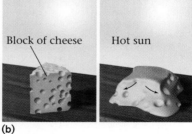

Rough wood surface   Rasp

(c)

FIGURE 8.29 Three geologic phenomena together contribute to exhumation in a collisional mountain belt. Because of exhumation, the vertical distance between a point at depth in the belt and the surface decreases with time. **(a)** Collision squeezes rock in the mountain belt upward, like dough pressed in a vise. **(b)** The crust beneath the mountain range becomes warm and weak, so the mountain belt collapses, like a block of soft cheese placed in the hot sun. **(c)** Throughout the history of the mountain belt, erosion grinds rock off the surface and removes it, much like a giant rasp. Removal of overlying weight causes the surface of the crust to rise in order to maintain isostatic compensation.

See for yourself . . .

Precambrian Metamorphic Terranes

A metamorphic terrane is a region of crust composed of metamorphic rock. In unvegetated exposures you can see regional grain (the map trend of foliations), nonconformities between basement (metamorphic rock) and cover (overlying sedimentary strata), and contacts between different rock types. Explore these examples! The thumbnail images provided on this page are only to help identify tour sites. Go to wwnorton.com/studyspace to experience this flyover tour.

Wind River Mountains, Wyoming
(Lat 42°51'30.49"N, Long 109°02'40.81"W)

At these coordinates, zoom to 750 km (466 miles), and you can see all of Wyoming. The NW-SE-trending Wind River Mountains lie in west-central Wyoming (Image G8.1). Zoom to 250 km (155 miles). A large fault bounds the SW side of the range. During mountain building, between 80 and 40 Ma, movement on the fault pushed up Precambrian metamorphic basement and warped overlying beds of Paleozoic and Mesozoic sedimentary rock. Erosion of sedimentary rock exposed the basement. Zoom to 40 km (25 miles) to see the contrast between tilted bedding of the sedimentary strata and knobby Precambrian gneiss. Tilt your field of view and look NW; the contrast becomes clearer (Image G8.2).

G8.1

G8.2

Canadian Shield, East of Hudson Bay
(Lat 61°09'50.15"N, Long 76°42'43.88"W)

At these coordinates, zoom to 230 kilometers. You can see the east coast of Hudson Bay and a 150 km-wide swath of the Canadian shield, here covered only by sparse tundra vegetation (Image G8.3). The land surface displays the trend of metamorphic foliation. Note that a belt of E-W trending grain truncates a region of N-S-trending grain—the contact between these two provinces represents the contact between Archean rocks (to the south) and Proterozoic rocks (which form the E-W-trending band). This contact originally formed deep below the Earth's surface and was subsequently exhumed.

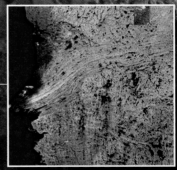

G8.3

Pilbara Craton, Western Australia
(Lat 21°14'10.00"S, Long 119°09'39.15"E)

Archean rocks underlie much of northwestern Australia, a crustal province called the Pilbara craton. Here, felsic intrusives and high-grade gneisses occur in light-colored dome-shaped bodies (circular to elliptical regions in map view). Low- to intermediate-grade "greenstone" (mafic and ultramafic metavolcanic rocks, interlayered locally with metasedimentary rocks) comprise curving belts that surround the domes. Zoom to an altitude of about 65 km (40 miles) at the coordinates given, and you can see domes surrounded by greenstone belts (Image G8.4). Zoom closer, and you can detect folded layering within the greenstone belts.

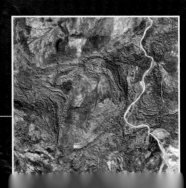

Chapter Summary

- Metamorphism refers to changes in a rock that result in the formation of a metamorphic mineral assemblage and/or a metamorphic foliation, in response to change in temperature and/or pressure, to the application of differential stress, and to interaction with hydrothermal fluids.

- Metamorphism involves recrystallization, metamorphic reactions (neocrystallization), phase changes, pressure solution, and/or plastic deformation. If hot-water solutions bring in or remove elements, we say that metasomatism has occurred.

- Metamorphic foliation can be defined either by preferred mineral orientation (aligned inequant crystals) or by compositional banding. Preferred mineral orientation develops where differential stress causes the squashing and shearing of a rock, so that its inequant grains align parallel with each other.

- Geologists separate metamorphic rocks into two classes, foliated rocks and nonfoliated rocks, depending on whether the rocks contain foliation.

- The class of foliated rocks includes slate, metaconglomerate, phyllite, schist, amphibolite, and gneiss. The class of nonfoliated rocks includes hornfels, quartzite, marble, and amphibolite, though the latter three can have a foliation. Migmatite, a mixture of igneous and metamorphic rock, forms under conditions where partial melting begins.

- Rocks formed under relatively low temperatures are known as low-grade rocks, whereas those formed under high temperatures are known as high-grade rocks. Intermediate-grade rocks develop between these two extremes. Different assemblages of minerals form at different grades.

- Geologists track the distribution of different grades of rock by looking for index minerals. Isograds indicate the locations at which index minerals first appear. A metamorphic zone is the region between two isograds.

- A metamorphic facies is a group of metamorphic mineral assemblages that develop under a specified range of temperature and pressure conditions. The assemblage in a given rock depends on the composition of the protolith, as well as on the metamorphic conditions.

- Thermal metamorphism (also called contact metamorphism) occurs in an aureole surrounding an igneous intrusion. Dynamically metamorphosed rocks form along faults, where rocks are only sheared, under metamorphic conditions. Dynamothermal metamorphism (also called regional metamorphism) results when rocks are buried deeply during mountain building.

- Metamorphism occurs because of plate interactions: the process of mountain building in either convergent or collisional zones causes dynamothermal metamorphism; shearing along plate boundaries causes dynamic metamorphism; and igneous plutons in rifts cause thermal metamorphism. The circulation of hot water causes hydrothermal metamorphism of oceanic crust at mid-ocean ridges. Unusual metamorphic rocks called blueschists form at the base of accretionary prisms. Metamorphism also results from the shock of meteorite impact and from convection in the mantle.

- We find extensive areas of metamorphic rocks in mountain ranges. Vast regions of continents known as shields expose ancient (Precambrian) metamorphic rocks. Metamorphic rocks return to the Earth's surface due to exhumation.

Geopuzzle Revisited

Marble is one of many different types of metamorphic rocks, formed due to changes in mineral content and/or rock texture, which happen when a rock is subjected to changes in temperature, pressure, and/or differential stress. The marble in Michelangelo's sculptures started as fossiliferous limestone (in the Jurassic). It was buried deeply, heated, and sheared during the formation of the Apennine Mountains. Exhumation, due partly to extensional collapse and partly to erosion, has returned the rock to the surface.

Key Terms

burial metamorphism (p. 246)
contact metamorphism (p. 245)
differential stress (p. 232)
dynamic metamorphism (p. 246)
dynamothermal (regional) metamorphism (p. 248)
exhumation (pp. 249, 253)
foliation (p. 235)
gneiss (p. 236)
hornfels (p. 238)

hydrothermal metamorphism (p. 248)
metamorphic aureole (p. 245)
metamorphic facies (p. 242)
metamorphic grade (p. 241)
metamorphic mineral (p. 229)
metamorphic rock (p. 229)
metamorphic texture (p. 229)
metamorphic zones (p. 244)

metamorphism (p. 229)

metasomatism (p. 234)

mylonite (p. 246)

phyllite (p. 235)

preferred mineral orientation
    (p. 232)

protolith (p. 229)

schist (p. 236)

schistosity (p. 236)

shield (p. 253)

shock metamorphism
    (p. 249)

slate (p. 235)

thermal metamorphism
    (p. 245)

vein (p. 234)

Review Questions

1. How are metamorphic rocks different from igneous and sedimentary rocks?

2. What two features characterize most metamorphic rocks?

3. What phenomena cause metamorphism?

4. What is metamorphic foliation, and how does it form?

5. How is a slate different from a phyllite? How does a phyllite differ from a schist? How does a schist differ from a gneiss?

6. Why are hornfels nonfoliated?

7. What is a metamorphic grade, and how can it be determined? How does grade differ from "facies"?

8. How does prograde metamorphism differ from retrograde metamorphism?

9. Describe the geologic settings where thermal, dynamic, and dynamothermal metamorphism take place.

10. Why does metamorphism happen at the sites of meteor impacts, along mid-ocean ridges, and deep in the mantle?

11. How does plate tectonics explain the peculiar combination of low-temperature but high-pressure minerals found in a blueschist?

12. Where would you go if you wanted to find exposed metamorphic rocks? How did such rocks return to the surface of the Earth after being at depth in the crust?

On Further Thought

1. Do you think that you would be likely to find a broad region (hundreds of km across and hundreds of km long) in which the outcrop consists of high-grade hornfels? Why or why not? (*Hint*: Think about the causes of metamorphism and the conditions under which a hornfels forms).

2. Would we likely find broad regions of gneiss and schist on the Moon? Why or why not?

3. Imagine that you take a field trip across a mountain range, starting at the front of the range (where rocks are non-metamorphic) and moving toward its interior. Your trip progresses from low-grade rocks to high-grade rocks, with your last stop at the highest-grade outcrop. Below are some of your rock descriptions. In the spaces provided, write the stop designation, with A being the lowest-grade rock and D being the highest grade. About how much overburden has been removed during exhumation to provide exposure of the rocks at stop D?

Stop _____: Coarse-grained schist. The rock contains quartz, feldspar, muscovite, and biotite. In a few horizons, the rock contained small crystals of kyanite.

Stop _____: Fine-grained metabasalt, containing chlorite and zeolite, giving the rock a greenish color.

Stop _____: A dark-black, coarse-grained rock—I'm not sure of an appropriate name yet. Fresh samples contain aluminum-rich amphibole, some garnet, and pyroxene.

Stop _____: This outcrop contains thick layers of quartzite, locally with thin interbeds of phyllite.

Suggested Reading

Barker, A.J. 2004. *Introduction to Metamorphic Textures and Microstructures*. London: Routledge.

Bucher, K., and Frey, M. 2002. *Petrogenesis of Metamorphic Rocks*, 7th. ed. New York: Springer.

Fry, N. 1991. *The Field Description of Metamorphic Rocks*. New York: John Wiley & Sons.

Winter, J.D. 2001. *Introduction to Igneous and Metamorphic Petrology*. Upper Saddle River, N.J.: Prentice-Hall.

Yardley, B.W.D. 1996. *An Introduction to Metamorphic Petrology*. Upper Saddle River, N.J.: Prentice Hall.

The Rock Cycle

The three rock types of the Earth System. (a) Igneous rock, here formed by the cooling of lava at a volcano. (b) Sedimentary rock, here eroding to form sediment. (c) Metamorphic rock, here exposed in a mountain belt. Over time, materials composing one rock type may be incorporated in another.

C.1 INTRODUCTION

"Stable as a rock." This familiar expression implies that rock is permanent, unchanging over time. But it isn't. In the time frame of Earth history, a span of over 4.5 billion years, atoms making up one rock type may be rearranged or moved elsewhere, eventually becoming part of another rock type. Later, the atoms may move again to form a third rock type and so on. Geologists call the progressive transformation of Earth materials from one rock type to another the **rock cycle** (▶Fig. C.1), one of many examples of cycles acting in or on the Earth. (James Hutton, the eighteenth-century Scottish geologist, was the first person to visualize and describe the rock cycle.) We focus on the rock cycle here because it illustrates the relationships among the three rock types described in the previous three chapters.

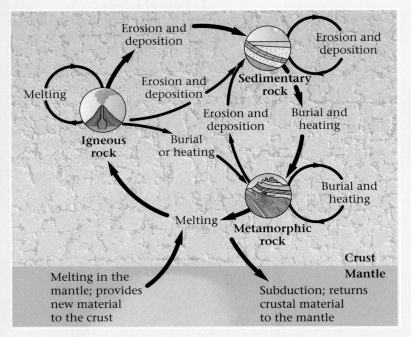

FIGURE C.1 The stages of the rock cycle, showing various alternative pathways.

A **cycle** (from the Greek word for circle or wheel) is a series of interrelated events or steps that occur in succession and can be repeated, perhaps indefinitely. During **temporal cycles**, such as the phases of the Moon or the seasons of the year, events happen according to a timetable, but the materials involved do not necessarily change. The rock cycle, in contrast, is an example of a geologic **mass-transfer cycle**, one that involves the transfer or movement of materials (mass) to different parts of the Earth System. (The hydrologic cycle, which we will learn about in Interlude F, is another mass-transfer cycle.)

There are many paths around or through the rock cycle. For example, igneous rock may weather and erode to produce sediment, which lithifies to form sedimentary rock. The new sedimentary rock may become buried and form metamorphic rock, which then could partially melt to create magma. This magma later solidifies to form new igneous rock. We can symbolize this path as igneous → sedimentary → metamorphic → igneous. But alternatively, the metamorphic rock could be uplifted and eroded to form new sediment and then new sedimentary rock without melting, taking a shortcut through the cycle that we can symbolize as igneous → sedimentary → metamorphic → sedimentary. Likewise, the igneous rock could be metamorphosed directly, without first turning to sediment. This metamorphic rock could again be turned into sedimentary rock, defining another shortcut: igneous →

metamorphic → sedimentary. To get a clearer sense of how the rock cycle works, we'll look at one example.

C.2 THE ROCK CYCLE IN THE CONTEXT OF THE THEORY OF PLATE TECTONICS

Material can enter the rock cycle when basaltic magma rises from the mantle. Suppose the magma erupts and forms basalt (an igneous rock) at a continental hot-spot volcano (▶Fig. C.2a). Interaction with wind, rain, and vegetation gradually weathers the basalt, physically breaking it into smaller fragments and chemically altering it to create clay. As water washes over the newly formed clay, it carries the clay away and transports it downstream—if you've ever seen a brown-colored river, you've seen clay en route to a site of deposition. Eventually the river reaches the sea, where the water slows down and the clay settles out.

Let's imagine, for this example, that the clay settles out along the margin of continent X and forms a deposit of mud. Gradually, through time, the mud becomes progressively buried and the clay flakes pack tightly together, resulting in a new sedimentary rock, shale. The shale resides 6 km below the continental shelf for millions of years, until the adjacent oceanic plate subducts and a neighboring continent, Y, collides with X. The shale gets buried *very* deeply when the edge of the encroaching continent pushes over it. As the mountains grow, the shale that had once been 6 km below the surface now ends up 20 km below the surface, and under the pressure and temperature conditions present at this depth, it metamorphoses into schist (▶Fig. C.2b).

The story's not over. Once mountain building stops, erosion grinds away the mountain range, and exhumation brings some of the schist to the ground surface. This schist erodes to form sediment, which is carried off and deposited elsewhere to form new sedimentary rock—this material takes a shortcut through the rock cycle. But other schist remains preserved below the surface. Eventually, continental rifting takes place at the site of the former mountain range, and the crust containing the schist begins to split apart. When this happens, some of the schist partially melts and a new felsic magma forms. This felsic magma rises to the surface of the crust and freezes to create rhyolite, a new igneous rock (▶Fig. C.2c). In terms of the rock cycle, we're back at the beginning, having once again made igneous rock (▶Fig. C.2d).

Note that atoms, as they pass through the rock cycle, do not always stay within the same mineral. In our example, a silicon atom in a pyroxene crystal of the basalt may become part of a clay crystal in the shale, part of a muscovite crystal in the schist, and part of a feldspar crystal in the rhyolite.

TIME 1

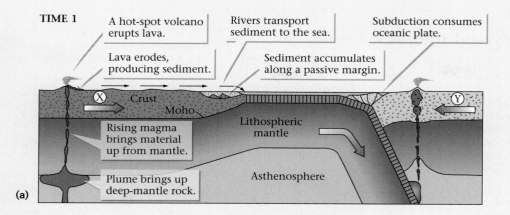

A hot-spot volcano erupts lava.

Rivers transport sediment to the sea.

Subduction consumes oceanic plate.

Lava erodes, producing sediment.

Sediment accumulates along a passive margin.

X Crust
Moho

Rising magma brings material up from mantle.

Lithospheric mantle

Y

Plume brings up deep-mantle rock.

Asthenosphere

(a)

TIME 2

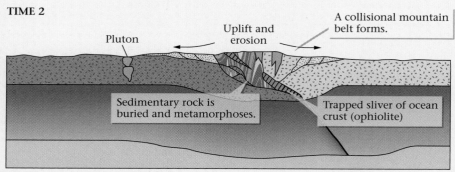

Pluton

Uplift and erosion

A collisional mountain belt forms.

Sedimentary rock is buried and metamorphoses.

Trapped sliver of ocean crust (ophiolite)

(b) ▨ Metamorphic rock ☐ Sediment eroded from mountains

TIME 3

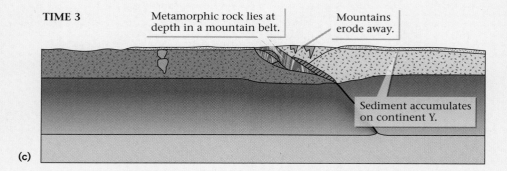

Metamorphic rock lies at depth in a mountain belt.

Mountains erode away.

Sediment accumulates on continent Y.

(c)

TIME 4

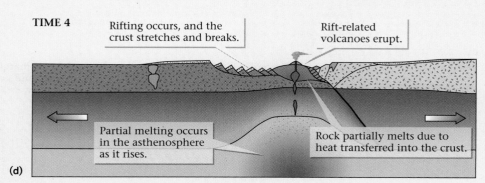

Rifting occurs, and the crust stretches and breaks.

Rift-related volcanoes erupt.

Partial melting occurs in the asthenosphere as it rises.

Rock partially melts due to heat transferred into the crust.

(d)

FIGURE C.2 **(a)** At the beginning of the rock cycle (time 1), atoms, originally making up peridotite in the mantle, rise in a mantle plume. The peridotite partially melts at the base of the lithosphere, and the atoms become part of a basaltic magma that rises through the lithosphere of continent X and erupts at a volcano. At this time, the atoms become part of a lava flow, that is, an igneous rock. Weathering breaks the lava down, and the resulting clay is transported to a passive-margin basin. After the clay is buried, the atoms become part of a shale—a sedimentary rock. Note that the ocean floor to the east of the passive-margin basin is being consumed beneath continent Y. **(b)** At time 2, continents X and Y collide, and the shale is buried deeply beneath the resulting mountain range (at the dot). Now the atoms become part of a schist—a metamorphic rock. **(c)** At time 3, the mountain range erodes away. The schist rises but does not reach the surface. **(d)** At time 4, rifting begins to split the continents apart, and igneous activity occurs again. At this time, the atoms of the schist become part of a new melt, which eventually freezes to form a rhyolite, another igneous rock.

Rock-Forming Environments and the Rock Cycle

Rocks form in many different environments. Igneous rocks develop where melt rises from depth and cools. Intrusive igneous rocks form where magma cools underground. Extrusive igneous rocks form where lava and ash erupt at the surface.

Weathering and erosion break up existing rock and produce sediment. Different kinds of sediments develop in different places, reflecting both the composition of the source and the setting in which the sediment is deposited. We distinguish among sediment that accumulates in alluvial fans, desert dunes, river channels and floodplains, deltas, coral reefs, coastlines, the continental shelf, the deep sea, and the toe of a glacier. When this sediment eventually gets buried and undergoes lithification, new sedimentary rocks form.

Drainage networks collect surface water that can transport sediment to the ocean.

Sand dunes form from grains carried by the wind.

In a desert environment, rock weathers and fragments. Debris falls in landslides.

Flash floods carry sediment out of canyons to form an alluvial fan.

Volcanic eruptions emit lava and ash, which form new igneous rock at Earth's surface.

Sedimentary rocks make a cover on the surface of continents.

The crust and lithospheric mantle stretch and thin in a rift.

Magma rises from the mantle. Heat from this magma causes contact metamorphism.

Deep levels of continents consist of ancient metamorphic and igneous rocks. This is the basement of the continents.

Continental margins slowly sink and are buried by new sediment.

Partial melting occurs in the asthenosphere to produce new magma.

km
0
10
20
30
40
50
60
70
80
90
100

Glaciers erode rock and
can transport sediment of all sizes.

In a region of continental
collision, rocks that were near
the surface are deeply buried
and metamorphosed.

In humid climates,
thick soils develop.

Magma that cools and solidifies
underground forms igneous
intrusions.

Along coastal plains, rivers meander. Sediment
collects in the channel and floodplain.

Where a river enters the sea, sediment
settles out to form a delta.

Reefs grow from calcite-secreting organisms.
These will eventually turn into limestone.

Many different kinds of sediment
accumulate along coastlines, building
out a continental shelf.

Underwater avalanches carry a cloud of sediment
that settles to form a submarine fan.

Fine clay and plankton shells settle on
the oceanic crust.

The oceanic crust consists
of igneous rocks formed at
a mid-ocean ridge.

Under certain conditions, preexisting rocks can undergo change in the solid state—
metamorphism—which produces metamorphic rocks. Contact metamorphism is due to heat
released by an intrusion of magma. Regional metamorphism occurs where tectonic processes
cause rocks from the surface to be buried very deeply.

Because the Earth is dynamic, environments change through time. Tectonic processes cause
new igneous rocks to form. When exposed at the surface, these rocks weather to make sediment.
The slow sinking of some regions creates sedimentary basins in which sediment accumulates and
new sedimentary rocks form. Later, these rocks may be buried deeply and metamorphosed. Uplift
as a result of mountain building exposes the rocks to the surface, where they may once again be
transformed into sediment. This progressive transformation is called the rock cycle.

C.3 RATES OF MOVEMENT THROUGH THE ROCK CYCLE

We have seen that not all atoms pass through the rock cycle in the same way. Similarly, not all atoms pass through the rock cycle at the same rate, and for that reason, we find rocks of many different ages at the surface of the Earth. Some rocks remain in one form for less than a few million years, whereas others stay unchanged for most of Earth history. Rocks exposed on Precambrian shields have remained unchanged for billions of years—the Canadian Shield of North America includes rocks as old as 3.9 billion years. In contrast, a rock with an Appalachian Mountain address has passed through stages of the rock cycle many times in the past few billion years, because the eastern margin of North America has been subjected to multiple events of basin formation, mountain building, and rifting since the shield to the west developed.

Studies during the past two decades suggest that most of the rock now making up the Earth's continental crust contains atoms that were extracted from the mantle over 2.5 billion years ago. Yet we see rocks of many different ages in the continents today. That is because geologic processes recyle these atoms again and again, similar to the way people recycle the metal of old cars to make new ones. And just as the number of late-model cars on the road today exceeds the number of vintage cars, younger rocks are more common than ancient rocks. At the surface of continents, sedimentary rocks created during the last several hundred million years are the most widespread type, whereas rocks recording the early history of the Earth are quite rare. But even though most continental crustal rocks are recycled, some new ones continue to be freshly extracted from the mantle each year, adding to the continent at volcanic arcs or hot spots.

Do the atoms in continental rocks ever get a chance to start the rock cycle all over, by returning to the mantle? Yes. Some sediment that erodes off a continent ends up in deep-ocean trenches, and some of this is dragged back into the mantle by subduction. In fact, recent research suggests that metamorphic and igneous rocks at the base of the continental crust may be removed and carried back down into the mantle at subduction zones.

Our tour of the rock cycle has focused on continental rocks. What about the oceans? Oceanic crust consists of igneous rock (basalt and gabbro) overlaid with sediment. Because a layer of water blankets the crust, erosion does not

The atoms in these rocks have passed through all stages of the rock cycle. They started in igneous rocks, then were eroded to form sediment that lithified to become sedimentary rock. This rock was buried, metamorphosed, and exhumed. It is now in the process of turning into soil, aided by the root wedging of a hardy cactus.

affect it, so oceanic crustal rock does not follow the path into the sedimentary loop of the rock cycle. But sooner or later, oceanic crust subducts. When this happens, the rock of the crust first undergoes metamorphism, for as it sinks, it is subjected to progressively higher temperatures and pressures. And eventually, a little of the rock may melt and become new magma, which then rises at a volcanic arc.

C.4 WHAT DRIVES THE ROCK CYCLE IN THE EARTH SYSTEM?

The rock cycle occurs because the Earth is a dynamic planet. The planet's internal heat and gravitational field drive plate movements. Plate interactions cause the uplift of mountain ranges, a process that exposes rock to weathering, erosion, and sediment production. Plate interactions also generate the geologic settings in which metamorphism occurs, where rock melts to provide magma, and where sedimentary basins develop.

At the surface of the Earth, the gases released by volcanism collect to form the ocean and atmosphere. Heat (from the Sun) and gravity drive convection in the atmosphere and oceans, leading to wind, rain, ice, and currents—the agents of weathering and erosion. Weathering and erosion grind away at the surface of the Earth and send material into the sedimentary loop of the cycle. In sum, external energy (solar heat), internal energy (Earth's internal heat), and gravity all play roles in driving the rock cycle, by keeping the mantle, crust, atmosphere, and oceans in constant motion.

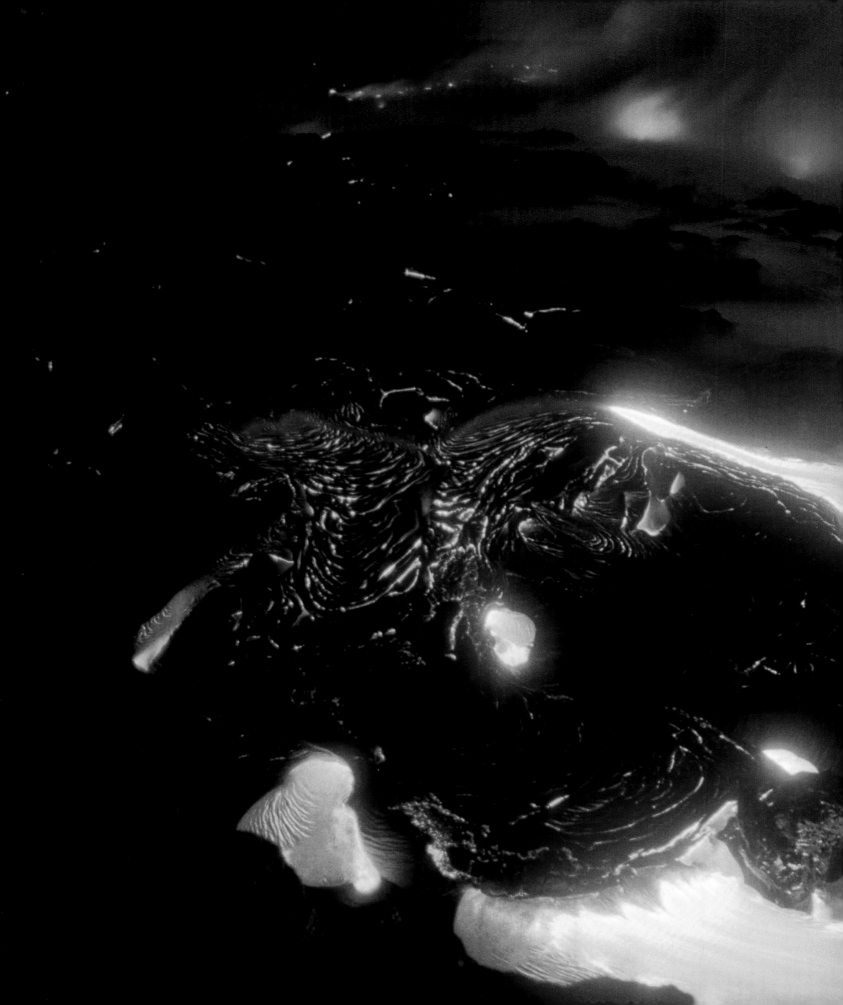

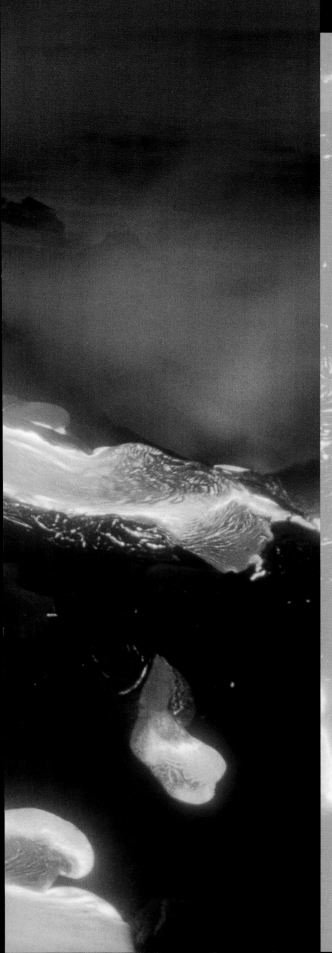

Tectonic Activity of a Dynamic Planet

Earlier in this book, we learned that the map of the Earth constantly, but ever so slowly, changes in response to plate movements and interactions. We now turn our attention to the dramatic consequences of such tectonic activity in the Earth System: volcanoes (Chapter 9), earthquakes (Chapter 10), and mountains (Chapter 11).

Why does molten rock rise like a fountain out of the ground, or explode into the sky at a volcano? Why does the ground shake and heave, in some cases so violently that whole cities topple, during an earthquake? How can the energy released by an earthquake tell us about the insides of the Earth, thousands of kilometers below the surface? What processes cause the land surface to rise several kilometers above sea level to form mountain belts? How do rocks bend, squash, stretch, and break in response to forces caused by plate interactions? Read on, and you will not only be able to answer these questions, but you will also see how the answers help people deal with some of the deadliest natural hazards that threaten society.

The glowing red stream that lights the night sky of Hawaii consists of molten lava flowing into the sea. When hot lava touches the water, the water flashes into steam. The eruption of volcanoes, the shaking of earthquakes, and the uplift of mountains provide dramatic proof that the Earth remains a dynamic planet.

The Wrath of Vulcan: Volcanic Eruptions

Geopuzzle

Why do volcanoes exist? Why do they occur where they do? Are all eruptions the same?

This 1989–1990 eruption of Redoubt Volcano, Alaska, produced immense clouds of ash. A jumbo jet flew through the ash and lost power in all four engines. Fortunately, after the plane lost about 2.5 km (8,000 feet) of altitude, the engines restarted and the plane was able to land. Volcanoes are hazards, and volcanoes are drama!

> *"Glowing waves rise and flow, burning all life on their way, and freeze into black, crusty rock which adds to the height of the mountain and builds the land, thereby adding another day to the geologic past . . . I became a geologist forever, by seeing with my own eyes: the Earth is alive!"*
>
> *–Hans Cloos (1944) on seeing an eruption of Mt. Vesuvius*

9.1 INTRODUCTION

Every few hundred years, one of the hills on Vulcano, an island in the Mediterranean Sea off the western coast of Italy, rumbles and spews out molten rock, glassy cinders, and dense "smoke" (actually a mixture of various gases, fine ash, and very tiny liquid droplets). Ancient Romans thought that such eruptions happened when Vulcan, the god of fire, fueled his forges beneath the island to manufacture weapons for the other gods. Geologic study suggests, instead, that eruptions take place when hot magma, formed by melting inside the Earth, rises through the crust and emerges at the surface. No one believes the myth anymore, but the island's name evolved into the English word **volcano**, which geologists use to designate either an erupting vent through which molten rock reaches the Earth's surface, or a mountain built from the products of eruption.

On the main peninsula of Italy, not far from Vulcano, another volcano, Mt. Vesuvius, towers over the nearby Bay of Naples. Two thousand years ago, Pompeii was a prosperous Roman resort and trading town of 20,000 inhabitants, sprawled at the foot of Vesuvius (▶Fig. 9.1a, b). Then, one morning in 79 C.E., earthquakes signaled the mountain's awakening. At 1:00 P.M. on August 24, a dark, mottled cloud boiled up above Mt. Vesuvius's summit to a height of 27 km. As lightning sparked in its crown, the cloud drifted over Pompeii, turning day into night. Blocks and pellets of rock fell like hail, while fine ash and choking fumes enveloped the town. Frantic people rushed to escape, but for many it was too late. As the growing weight of volcanic debris began to crush buildings, an avalanche of ash swept over Pompeii, and by the next day the town had vanished beneath a 6-m-thick gray-black blanket. The ruins of Pompeii were protected so well by their covering that when archaeologists excavated the town 1,800 years later, they found an amazingly complete record of Roman daily life. In addition, they discovered open spaces in the debris. Out of curiosity, the archaeologists filled the spaces with plaster, and realized that the spaces were fossil casts of Pompeii's unfortunate inhabitants, their bodies twisted in agony or huddled in despair (▶Fig. 9.1c).

Clearly, volcanoes are unpredictable and dangerous. Volcanic activity can build a towering, snow-crested mountain or can blast one apart. It can provide the fertile soil that enables agriculture to thrive, or it can snuff out a civilization in a matter of minutes. Because of the diversity of volcanic activity and its consequences, this chapter sets out ambitious goals. We first review the products of volcanic eruptions and the basic characteristics of volcanoes. Then we look again at the different kinds of volcanic eruptions on Earth. Volcanoes are not randomly distributed around the globe—their positions reflect the locations of plate boundaries, rifts, and hot spots. Finally, we examine the hazards posed by volcanoes, efforts by geoscientists to predict eruptions and help minimize the damage they cause, and the possible influence of eruptions on climate and civilization.

9.2 THE PRODUCTS OF VOLCANIC ERUPTIONS

The drama of a volcanic eruption transfers materials from inside the Earth to our planet's surface. Products of an eruption come in three forms—lava flows, pyroclastic debris, and gas (see art, pp. 278–279).

Lava Flows

Sometimes it races down the side of a volcano like a fast-moving, incandescent stream, sometimes it builds into a rubble-covered mound at a volcano's summit, and sometimes it oozes like a sticky but scalding paste. Clearly, not all **lava** (molten rock that has extruded onto the Earth's surface) behaves in the same way when it rises out of a volcano. Therefore, not all **lava flows** (moving masses of molten lava, or sheets of rock formed when lava solidifies) look or behave the same. Why? The character of a lava flow primarily reflects its viscosity, or resistance to flow.[1] Not all lavas have the same viscosity. Differences in viscosity depend on a variety of factors. For example, lava containing less SiO_2 (silica) is less viscous than lava containing more silica, because silica molecules tend to link together in long chains that tangle and cannot move past each other. Hot lava is less viscous than cool lava because thermal vibrations break up the bonds holding molecules together, so they can move past each other more easily and crystal-poor lava is less viscous than crystal-rich lava, because solid crystals inhibit flow.

---

[1]Recall that a material that can flow easily is said to be *less* viscous than a material that cannot flow easily. For example, water is less viscous than honey.

(a)

What a geologist imagines
(b)

(c)

FIGURE 9.1 **(a)** Pompeii, once buried by 6 m of volcanic debris from Mt. Vesuvius, was excavated by archaeologists in the late nineteenth century. Vesuvius rises in the distance. **(b)** What a geologist imagines: When Mt. Vesuvius erupted in 79 C.E., it was probably much larger, as depicted in this sketch. The pellets are hot volcanic bombs and lapilli. **(c)** A plaster cast of an unfortunate inhabitant of Pompeii, found buried by ash in the corner of a room, where the person crouched for protection. The flesh rotted away, leaving only an open hole that could be filled by plaster.

To illustrate the different ways in which lava behaves, we now examine flows of different compositions (▶Fig. 9.2a–c). Geologists give names to different lava compositions by specifying the silica content (SiO_2) relative to the sum of the iron oxide (FeO) and magnesium oxide (MgO) content. Lavas high in silica are called silicic, felsic, or rhyolitic; lavas with an intermediate silica content are called intermediate or andesitic; and lavas low in silica are called mafic or basaltic. Note that we used the same terms when discussing igneous rocks (Chapter 6).

Basaltic lava flows. Basaltic (mafic) lava has very low viscosity when it first emerges from a volcano because it contains relatively little silica, and is hot. Thus, on the steep slopes near the summit of a volcano, it flows very quickly (▶Fig. 9.3a). The fastest known flow raced down the side of Nyirangongo

Volcano in the Democratic Republic of Congo (Africa) in 2002 at an initial speed of 100 km/hour. Typically, however, flows travel at under 30 km/hour and slow down to a less-than-walking pace after they have traveled several kilometers and have started to cool. Most flows measure less than 10 km long, but larger flows on Hawaii have moved 50 km, and in the Columbia Plateau region of Oregon and Washington the ends of some flows are as far as 500 km from the source. How can lava travel so far? Although all the lava in a flow moves when it first emerges, rapid cooling causes the surface of the flow to crust over after the flow has moved a few kilometers from the source. The solid crust serves as insulation, allowing the hot interior of the flow to remain liquid. An insulated, tunnel-like conduit through which lava moves within a flow is called a **lava tube**. In some cases, lava tubes drain and become empty tunnels (▶Fig. 9.3e). Once a lava

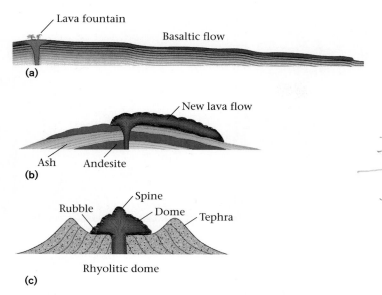

(a)

Lava fountain

Basaltic flow

(b)

New lava flow

Ash   Andesite

(c)

Rubble   Spine   Dome   Tephra

Rhyolitic dome

FIGURE 9.2 The character of a lava flow depends on the viscosity of the lava. Hotter lavas are less viscous than cooler lavas, and mafic lavas are less viscous than silicic lavas. **(a)** A basaltic lava flow is very fluid-like and can travel a great distance, forming a thin sheet. **(b)** An andesitic flow is too viscous to travel far, and it tends to break up as it flows. **(c)** Rhyolitic lava is so viscous that it piles up at the vent in the shape of a dome. In some cases, a tower-like spine pushes up from the center of the dome.

flow has covered the land and freezes, it becomes a solid blanket of rock (▶Fig. 9.3b).

The surface texture of a basaltic lava flow when it finally freezes reflects the timing of freezing relative to its movement. Flows with warm, pasty surfaces wrinkle into smooth, glassy, rope-like ridges; geologists have adopted the Hawaiian word **pahoehoe** (pronounced "pa-hoy-hoy") for such flows (▶Fig. 9.3c). If the surface layer of the lava freezes and then breaks up due to the continued movement of lava underneath, it becomes a jumble of sharp, angular fragments, yielding a rubbly flow also called by its Hawaiian name, **a'a'** (pronounced "ah-ah") (▶Fig. 9.3d). Footpaths made by people living in volcanic regions follow the smooth surface of pahoehoe flows rather than the rough, foot-slashing surface of a'a' flows.

During the final stages of cooling, lava flows contract and may fracture into roughly hexagonal columns. This type of fracturing is called **columnar jointing** (▶Fig. 9.4a, b).

Basaltic flows that erupt underwater look different from those that erupt on land, because the lava cools so much more quickly. In fact, water removes heat thirty times faster than air does—that's why you soon get hypothermia when immersed in cold water. Because of rapid cooling, submarine basaltic lava forms a glass-encrusted blob, or pillow, on freezing (▶Fig. 9.4c). The rind of a pillow momentarily stops the flow's advance, but within minutes the pressure of the lava squeezing into the pillow breaks the

rind, and a new blob of lava squirts out, perhaps moving 0.5 to 2 m before itself freezing into a pillow. The process repeats until the lava pillows freeze through and through, and the lava at the vent pushes up and makes another layer of pillows above the first (see Fig. 6.23a). As a result, a mound of pillow lava develops.

Andesitic lava flows. Because of its higher silica content and thus its greater viscosity, andesitic lava cannot flow as easily as basaltic lava. When erupted, andesitic lava first forms a large mound above the vent. This mound then advances slowly down the volcano's flank at only about 1 to 5 m a day, in a lumpy flow with a bulbous snout (Fig. 9.2b). Typically, andesitic flows are less than 10 km long. Because the lava moves so slowly, the outside of the flow has time to solidify; so as it moves, the surface breaks up into angular blocks, and the whole flow looks like a jumble of rubble. In places where submarine eruptions yield andesitic lava, hyaloclastite, a mass of splintery, glassy volcanic fragments, builds up around the vent.

Rhyolitic lava flows. Rhyolitic lava is the most viscous of all lavas because it is the most silicic and the coolest. Therefore, it tends to accumulate either in a dome-like mass, called a **lava dome**, above the vent or in short and bulbous flows rarely more than 1 to 2 km long (Fig. 9.2c). Sometimes rhyolitic lava freezes while still in the vent, and then pushes upward as a column-like spire or spine up to 100 m above the vent. Rhyolitic flows, where they do form, have broken and blocky surfaces, because the rind of the flow shatters as the inner part fills with lava and expands.

Volcaniclastic Deposits

Not all of the material that erupts at a volcano ends up as part of a lava flow. In some cases, bubble-filled lava begins to solidify in the vent of a volcano, forming scoria or pumice (see Chapter 6) that may eventually shatter and be ejected from the vent when an eruption begins. During basaltic eruptions, lava may fountain from a vent, producing clots of lava that cool and become solid or nearly solid clasts by the time they land (▶Fig. 9.5a). During andesitic or rhyolitic eruptions, powerful explosions spray lava into the air, forming droplets that instantly freeze into **volcanic ash**, composed of tiny glass shards. The blast of a volcanic explosion can also shatter the preexisting solid rock that makes up the volcano itself, forming chunks of varied sizes (▶Fig. 9.5b). Geologists refer to all fragmental material (ash, pumice or scoria fragments, and clots of frozen lava) erupted from a volcano as **pyroclastic debris** (from the Latin word *pyro*, meaning fire). A more general term, volcaniclastic debris, includes not only pyroclastic debris but also fragments of preexisting rock that break up and become dispersed during an eruption.

(a)

(b)

(c)

(d)

(e)

FIGURE 9.3 Characteristics of basaltic (mafic) lava flows. **(a)** A fast-moving 1984 lava flow just after emerging from a vent on Mauna Loa Volcano, Hawaii. The flow is about 20 m wide. **(b)** A basaltic lava flow covering a highway in Hawaii. **(c)** A pahoehoe lava flow forms on Hawaii. Note the characteristic smooth, ropy surface. The field of view is 2 m. **(d)** An a'a' flow has a rough and rubbly surface as illustrated by this example near Sunset Crater, Arizona. **(e)** Lava travels to the end of a flow through a tunnel called a lava tube. Some tubes drain, as did this one in Hawaii, now exposed in a road cut.

(a)   (b)   (c)

FIGURE 9.4 **(a)** Lava flows contract when they cool, and some crack to form columnar joints. These columns crop out in Yellowstone Park. **(b)** In cross section, the columns are somewhat hexagonal. **(c)** Lava pillows form as a consequence of subaqueous basalt eruption on the floor of the South Pacific Ocean. The pillow in the foreground is approximately 1 m across.

The finest pyroclastic debris, volcanic ash, (▶Fig. 9.5c) consists of powder-sized glass shards and pulverized rock generated during explosions. **Lapilli** (from the Latin for little stones) are pea- to plum-sized fragments. In many cases, lapilli consist of pumice or scoria fragments (see Chapter 6); lapilli composed of scoria are also known as cinders. If lapilli forms from low-viscosity lava, it may become streamlined while flying through the air, yielding teardrop-shaped glassy beads known as "Pelé's tears," after the Hawaiian goddess of volcanoes. When low-viscosity droplets rise from a pool of lava at the volcano's vent, they trail thin strands of lava behind them, and these strands freeze into filaments of brown glass known as "Pelé's hair." Not all lapilli are chunks of coherent rock. A type known as accretionary lapilli forms when a volcano erupts ash into rain or snow—under such conditions, the ash clumps together into small balls as it falls (▶Fig. 9.5d).

Coarser pyroclastic debris includes **blocks** and **bombs**, which are apple- to refrigerator-sized fragments (▶Fig. 9.5e, f). Specifically, blocks are chunks of preexisting igneous rock torn from the walls of the vent, whereas bombs form when large lava blobs enter the air in a molten state and then solidify. Because they are soft while traveling, bombs become streamlined before landing. In fact, if they are still soft on impact, they flatten like droppings of soft cow manure.

During some eruptions, bombs and lapilli accumulate around or within the volcanic vent, building up a deposit known as volcanic agglomerate. Ash, or ash mixed with lapilli, becomes **tuff** when transformed into coherent rock. Geologists refer to tuff formed from debris that settles from the air, like falling snow, as air-fall tuff. Unconsoli-dated deposits of pyroclastic grains, regardless of size, that have been erupted from a volcano constitute pyroclastic deposits, or **tephra**.

Not all tuff starts as an air-fall deposit. Some tuff forms from fast-moving, turbulent avalanches of hot ash and lapilli that rushed down the flank of the volcano (▶Fig. 9.6a). Such an avalanche, called a **pyroclastic flow** (or *nuée ardente*, French for glowing cloud), can form when gravity overcomes the upward force and buoyancy of a rising ash column, so that the column collapses and ash surges downward. A sheet of tuff formed from a pyroclastic flow is an **ignimbrite**. Thick ignimbrite sheets are so hot, immediately after deposition, that hot ash in the sheet—squeezed by the weight of overlying debris—may fuse together to produce hard welded tuff.

A devastating pyroclastic flow erupted from Mt. Pelée, a volcano on the otherwise quiet tropical West Indies island of Martinique. In April 1902, a small eruption shed fine, white air-fall ash over the town of St. Pierre, at the foot of the volcano. The air began to reek of sulfur, so inhabitants walked around with handkerchiefs covering their noses. Officials did not fully comprehend the threat and did not order an evacuation. Like a cork, frozen lava blocked the volcano's vent, but the pressure in the gas-rich magma beneath continued to build. On the morning of May 8, the cork popped, and like the froth that streams down the side of a champagne bottle, a pyroclastic flow swept down Pelée's flank. The cloud of burning ash, blocks, gas, and debris, at a temperature of 200° to 450°C, rode a cushion of air and may have reached a velocity of over 300 km per hour before it slammed into St. Pierre two minutes later. One breath of the super-hot ash meant instant

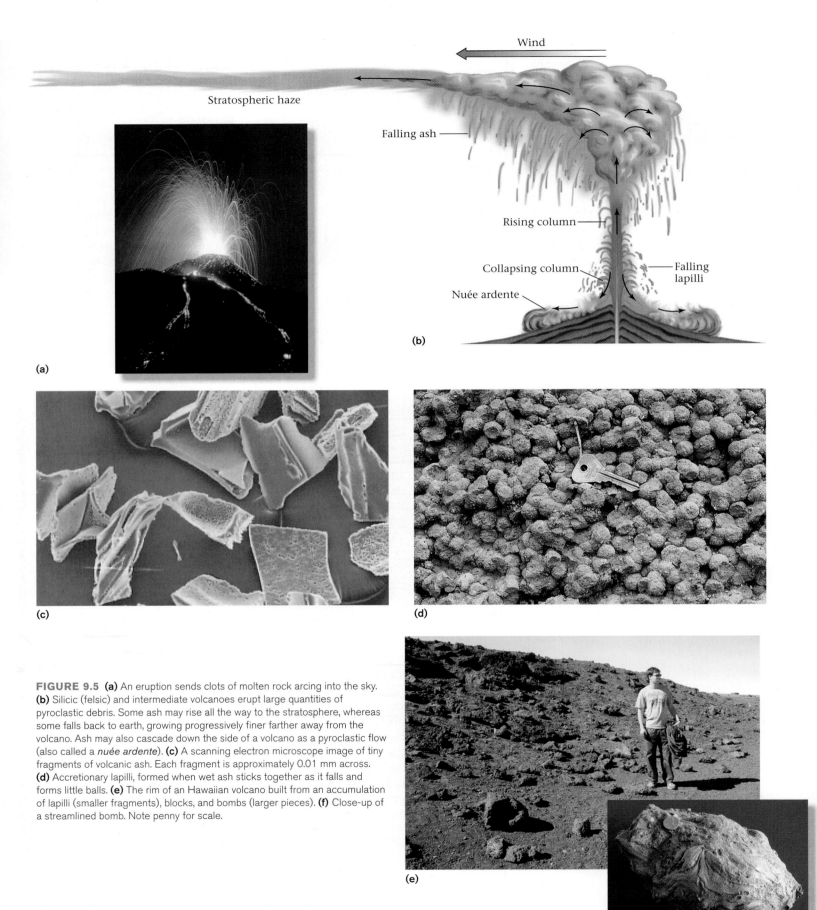

Wind

Stratospheric haze

Falling ash

Rising column

Collapsing column

Falling lapilli

Nuée ardente

(a)

(b)

(c)

(d)

(e)

(f)

FIGURE 9.5 **(a)** An eruption sends clots of molten rock arcing into the sky.
(b) Silicic (felsic) and intermediate volcanoes erupt large quantities of
pyroclastic debris. Some ash may rise all the way to the stratosphere, whereas
some falls back to earth, growing progressively finer farther away from the
volcano. Ash may also cascade down the side of a volcano as a pyroclastic flow
(also called a *nuée ardente*). **(c)** A scanning electron microscope image of tiny
fragments of volcanic ash. Each fragment is approximately 0.01 mm across.
(d) Accretionary lapilli, formed when wet ash sticks together as it falls and
forms little balls. **(e)** The rim of an Hawaiian volcano built from an accumulation
of lapilli (smaller fragments), blocks, and bombs (larger pieces). **(f)** Close-up of
a streamlined bomb. Note penny for scale.

death, and within moments 28,000 people lay dead of asphyxiation or incineration. Buildings toppled into a chaos of rubble and twisted metal, stockpiles of rum barrels exploded and sent flaming liquor into the streets, and ships at anchor in the harbor capsized. Only two people survived; one was a prisoner who, though burned, was protected from the brunt of the cataclysm by the stout walls of his underground cell. Similar eruptions began in the late 1990s on the island of Montserrat, another volcano in the eastern Caribbean (▶Fig. 9.6b, c). Numerous pyroclastic flows buried once-lush fields and forests and destroyed the towns, but fortunately this occurred after the island's 12,000 inhabitants had been evacuated from the danger zone.

Once it has accumulated on the flank of a volcano, tephra and other volcaniclastic debris may not yet have reached the end of its journey. Because such material is fairly weak, gravity may eventually cause masses of this material to slip downslope in semicoherent bodies called slumps. Also, water from rain and melting snow or ice may mix with the material to produce a chaotic, mobile slurry called a volcanic debris flow. Within a debris flow, viscous ashy mud buoys cobbles and boulders of all sizes. Particularly wet debris flows, called **lahars**, can rush down channels at high speeds and can travel tens of kilometers from the volcano (▶Fig. 9.7a). When this wet debris finally stops moving and drains, the resulting deposits consist of large clasts suspended in ashy mud. Debris flows and slumps of volcaniclastic debris occur not only subaerially but also underwater, on the flanks of islands. Further, not all volcaniclastic debris moves downslope in slumps or debris flows. A stream may erode and transport the debris, as it would any clastic sediment, eventually sorting the debris into size fractions before deposition finally occurs. Geologists use the general term **volcaniclastic deposits** for accumulations of volcaniclastic debris, regardless of whether it consists of tephra or of material that was first transported in debris flows or streams before it finally accumulated (▶Fig. 9.7b).

FIGURE 9.6 (a) A pyroclastic flow rushes down the side of a volcano in Japan. **(b)** Pyroclastic flows have left a swath of devastation on the flank of the Montserrat volcano in the Caribbean. The ash has accumulated to make a delta along the shore. **(c)** Some of the ash erupted from Montserrat has blanketed the town of Plymouth.

(a)

(c)

(b)

(a)

(b)

FIGURE 9.7 (a) A lahar flowed down a river on the flanks of Mt. St. Helens following the volcano's 1980 eruption, destroying homes and property along the banks of the river. (b) Close-up photo of volcaniclastic deposits formed from a debris flow in Puerto Rico. The different colored fragments are small blocks of lava. Note the lens cap for scale.

Volcanic Gas

Most magma contains dissolved gases, including water, carbon dioxide, sulfur dioxide, and hydrogen sulfide (H_2O, CO_2, SO_2, and H_2S). In fact, up to 9% of a magma may consist of gaseous components. Generally, lavas with more silica contain a greater proportion of gas. Volcanic gases come out of solution when the magma approaches the Earth's surface and pressure decreases, just as bubbles come out of solution in a soda or in champagne when you pop the bottle top off. Because of the sulfur in volcanic gas, the cloud above a volcano typically smells like rotten eggs. The SO_2 reacts with water in the air to create an aerosol of corrosive sulfuric acid. **Aerosols** are very tiny liquid droplets or solid particles that can remain suspended in air.

In low-viscosity magma (basalt), gas bubbles can rise faster than the magma moves, and thus most reach the surface of the magma and enter the atmosphere. Nevertheless, some bubbles freeze into the lava to create holes called **vesicles** (▶Fig. 9.8). In high-viscosity magmas (andesite and rhyolite), the gas has trouble escaping because bubbles can't push through the sticky lava. As these magmas approach the Earth's surface, and the weight of overlying lava decreases, the gas expands so that in some cases bubbles may account for as much as 50 to 75% of the volume of the magma. The gas can cause explosive pressures to build inside or beneath the volcano.

> ### Take-Home Message
>
> Volcanoes erupt lava (ranging in composition from mafic to felsic), pyroclastic debris (ranging in size from tiny ash shards to large blocks), and gas. Some ash falls like snow, whereas some rushes down the flanks of volcanoes in glowing avalanches.

9.3 THE ARCHITECTURE AND SHAPE OF VOLCANOES

As we saw in Chapter 6, melting in the upper mantle and lower crust produces magma, which rises into the upper crust. Typically, this magma accumulates underground in a **magma chamber**, an open space or a zone of highly fractured rock that can contain a large quantity of magma. Some of the magma freezes in the magma chamber and transforms into intrusive igneous rock, but some rises through an opening or conduit to the Earth's surface and erupts to form a volcano. In some volcanoes, the conduit has the shape of a vertical pipe, whereas in others the con-

FIGURE 9.8 Vesicles are the holes made by gas bubbles trapped in a freezing lava. This boulder of basalt, from Sunset Crater National Monument in Arizona, contains vesicles of various sizes.

duit is a vertical crack called a **fissure** (▶Fig. 9.9a–c). Initially, a fissure may erupt a curtain of lava. But as the eruption wanes, lava only comes out of discrete vents along a fissure (▶Fig. 9.9d).

With time, the solid products of eruption (lava and/or pyroclastic debris) accumulate around a conduit to form a mound or cone. A row of small cones may form along a fissure. At the top of the mound, a circular depression called a **crater** (shaped like a bowl, up to 500 m across and 200 m deep) develops, either during eruption as material accumulates around the summit vent or just after eruption as the summit collapses into the drained conduit. Eruptions that happen in the summit crater are summit eruptions. In some volcanoes, a secondary conduit or fissure breaks through along the sides, or flanks, of the volcano, causing a flank eruption (▶Fig. 9.10a).

After major eruptions, the center of the volcano may collapse into the large, partly drained magma chamber below, creating a **caldera**, a big circular depression (up to thousands of meters across and up to several hundred meters deep) with steep walls and a fairly flat floor (▶Fig. 9.10a–d). If new magma flows into the magma chamber, it may push up the floor of the caldera to form a resurgent dome. If the new magma erupts, a lava dome may form in the crater (see art, pp. 278–279). Note that calderas differ from craters in terms of size, shape, and mode of formation (▶Fig. 9.10e).

Geologists distinguish among three different shapes of subaerial volcano. **Shield volcanoes**, so named because they resemble a soldier's shield lying on the ground, are broad, gentle domes (▶Fig. 9.11a, b). Most shields form from low-viscosity basaltic lava flows. **Cinder cones** consist of cone-shaped piles of tephra. The slope of the cone approaches the angle of repose of tephra, meaning the steepest slope that the pile can attain without collapsing from the pull of gravity (between 30 and 35°, like a sandpile) (▶Fig. 9.11c). Typically, cinder cones are symmetrical and have deep craters at their summits. **Stratovolcanoes**, also called composite volcanoes, are large and cone shaped, and consist of alternating layers of lava and tephra (▶Fig. 9.11d, e). Their shape, exemplified by Japan's Mt. Fuji, supplies the classic image most people have of a volcano, although this shape may be disrupted by explosions or landslides. Stratovolcanoes tend to be steeper near the summit.

The hills or mountains resulting from volcanic eruptions come in a great range of sizes (▶Fig. 9.12). Shield volcanoes tend to be the largest, followed by stratovolcanoes.

> ### Take-Home Message
>
> Volcanoes erupt from chimney-like conduits or from fissures. Basaltic eruptions build shield volcanoes, fountaining lava builds cinder cones, and alternating ash and lava eruptions produce stratovolcanoes. Explosions may produce large calderas.

Cinder cones tend to be relatively small and are often found on the surface of larger volcanoes. Submarine volcanoes don't fit any of these categories, because they usually grow as irregularly shaped mounds, modified by huge landslides along their margins.

FIGURE 9.9 **(a)** Some volcanoes erupt out of a circular vent above a tube-shaped conduit. **(b)** Other volcanoes erupt out of a long crack, called a fissure, and produce a curtain of lava. **(c)** A "curtain of fire" formed as lava erupts from a fissure. **(d)** A row of small cones, formed by eruption of lava at discrete vents along a fissure.

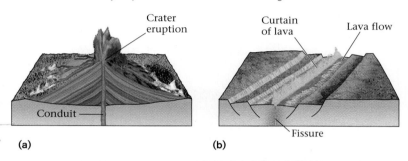

(a) (b)

(c)

(d)

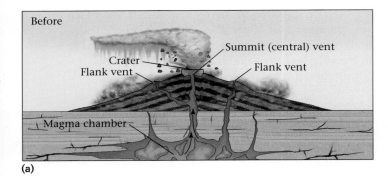

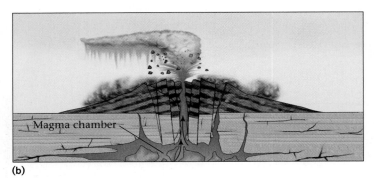

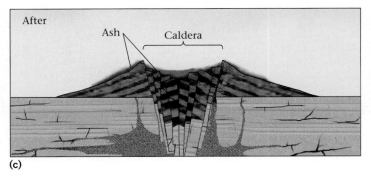

(d)

(e)

FIGURE 9.10 **(a)** The plumbing beneath a volcano can be complex. A central vent may lie directly above the magma chamber, but some of the lava may erupt at flank vents. **(b)** During an eruption, the magma chamber beneath a volcano is inflated with magma. **(c)** If the eruption drains the magma chamber, the volcano collapses downward to form a circular depression called a caldera. **(d)** A moderate-sized caldera has formed at the summit of Mt. Kilimanjaro, in the East African Rift. **(e)** Crater Lake, Oregon, is a caldera that subsequently filled with water to form a deep lake. An episode of renewed eruption produced the small cone of Wizard Island.

9.4 ERUPTIVE STYLES: WILL IT FLOW, OR WILL IT BLOW?

The 1983 eruption of Kilauea in Hawaii produced lakes and rivers of lava that cascaded down the volcano's flanks. In contrast, the 1980 eruption of Mt. St. Helens in Washington climaxed with a tremendous explosion that blanketed the surrounding countryside with tephra. Clearly, different volcanoes erupt in different ways; in fact, successive eruptions from the same volcano may differ from each other. Geologists refer to the character of an eruption as the eruptive style, and make the following distinctions.

• **Effusive eruptions:** These eruptions produce mainly lava flows (▶Fig. 9.13a). Most yield low-viscosity basaltic lavas, which can stream tens to hundreds of kilometers. In some effusive eruptions, lava lakes develop around the vent, whereas in others, lava sprays up in fountains that produce a cinder cone around the vent. To understand why fountaining occurs, watch the droplets of liquid ejected into the air above a frothing glass of soda. The bursting bubbles of gas eject liquid into the air. It's the rise of gas that propels lava upward in fountains.

• **Explosive (pyroclastic) eruptions:** These eruptions produce clouds and avalanches of pyroclastic debris

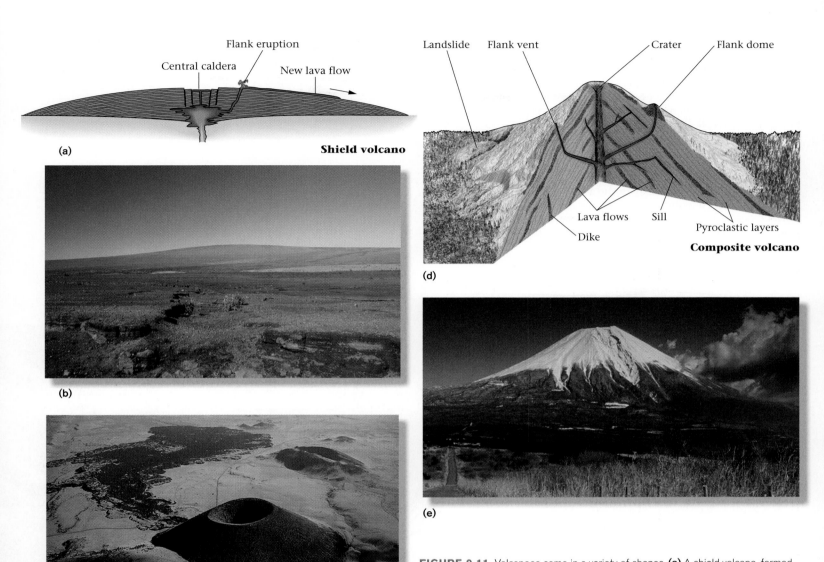

(a) Shield volcano

(b)

(c) Cinder cone

(d) Composite volcano

Labels in (a): Flank eruption, Central caldera, New lava flow

Labels in (d): Landslide, Flank vent, Crater, Flank dome, Lava flows, Sill, Pyroclastic layers, Dike

(e)

FIGURE 9.11 Volcanoes come in a variety of shapes. **(a)** A shield volcano, formed from basaltic lavas with low viscosity, has very gentle slopes. **(b)** A shield volcano on Hawaii. **(c)** A cinder cone is a pile of ash whose sides assume the angle of repose. A lava flow came from this example in Arizona. **(d)** A stratovolcano consists of alternating tephra and lava. **(e)** Mt. Fuji, in Japan, is a composite volcano. It last erupted in 1707 but recently has been showing signs of renewed activity.

FIGURE 9.12 These profiles emphasize that volcanoes come in different sizes. Large shield volcanoes, like Hawaii, are many times larger than cinder cones.

0 5 10
Km

Sea level

Large shield (Hawaii)

Small shield (Kilimanjaro)

Medium stratovolcano (Fuji)

Large stratovolcano (Shasta)

Small stratovolcano (Vesuvius)

Large cinder cone (Sunset Crater)

Volcano

Volcanic eruptions are a sight to behold and, in some cases, a hazard to fear. Beneath a volcano, magma formed in the upper mantle or the lower crust rises to fill a magma chamber near the Earth's surface. When the pressure in this magma chamber becomes great enough, magma is forced upward through a conduit, or crack, to the ground surface and erupts.

Once molten rock has erupted at the surface, it is called lava. Some lava flows down the side of the volcano in a lava flow. Lava flows eventually cool, forming solid rock. In some cases, lava spatters or fountains out of the volcanic vent in little blobs or drops that cool quickly in the air to create fragmental igneous rock called tephra, or cinders. Larger blobs ejected by a volcano become volcanic bombs, which attain a streamlined shape as they fall. Cinders may accumulate in a cone-shaped pile called a cinder cone.

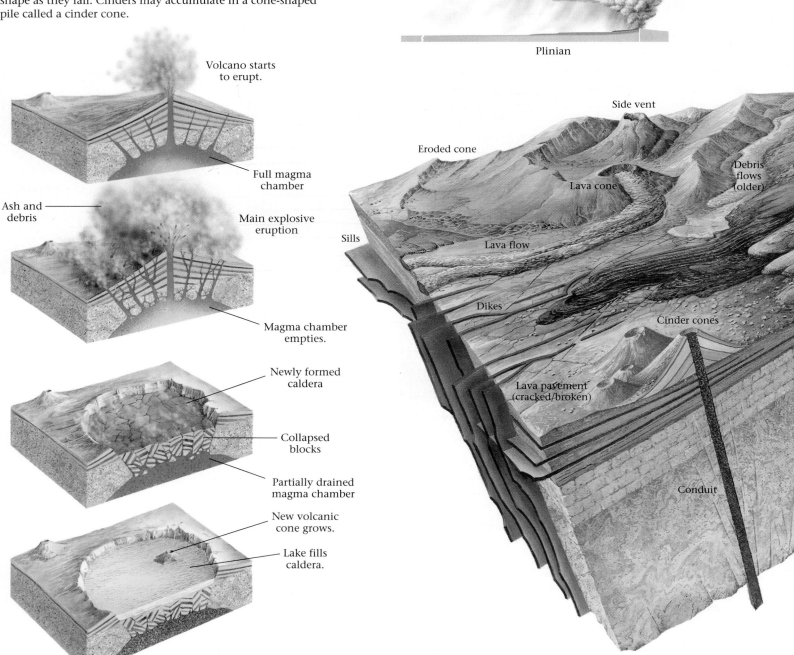

Styles of eruption

Vulcanian

Hawaiian

Hydrovolcanic

Strombolian

Plinian

Volcano starts to erupt.

Full magma chamber

Ash and debris

Main explosive eruption

Magma chamber empties.

Newly formed caldera

Collapsed blocks

Partially drained magma chamber

New volcanic cone grows.

Lake fills caldera.

Caldera formation (e.g., Crater Lake, Oregon)

Side vent

Eroded cone

Debris flows (older)

Lava cone

Sills

Lava flow

Dikes

Cinder cones

Lava pavement (cracked/broken)

Conduit

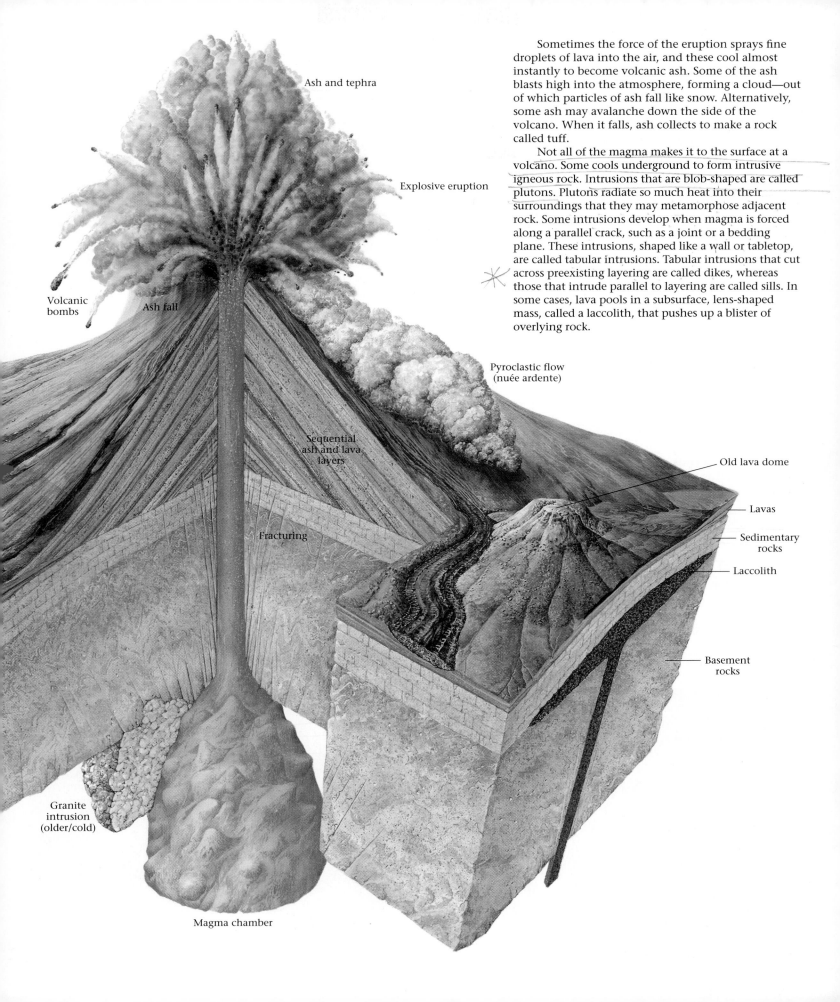

Ash and tephra

Explosive eruption

Volcanic bombs

Ash fall

Sequential ash and lava layers

Pyroclastic flow (nuée ardente)

Fracturing

Old lava dome

Lavas

Sedimentary rocks

Laccolith

Basement rocks

Granite intrusion (older/cold)

Magma chamber

Sometimes the force of the eruption sprays fine droplets of lava into the air, and these cool almost instantly to become volcanic ash. Some of the ash blasts high into the atmosphere, forming a cloud—out of which particles of ash fall like snow. Alternatively, some ash may avalanche down the side of the volcano. When it falls, ash collects to make a rock called tuff.

Not all of the magma makes it to the surface at a volcano. Some cools underground to form intrusive igneous rock. Intrusions that are blob-shaped are called plutons. Plutons radiate so much heat into their surroundings that they may metamorphose adjacent rock. Some intrusions develop when magma is forced along a parallel crack, such as a joint or a bedding plane. These intrusions, shaped like a wall or tabletop, are called tabular intrusions. Tabular intrusions that cut across preexisting layering are called dikes, whereas those that intrude parallel to layering are called sills. In some cases, lava pools in a subsurface, lens-shaped mass, called a laccolith, that pushes up a blister of overlying rock.

(►Box 9.1). Pyroclastic eruptions happen when gas expands in the rising magma but cannot escape. Eventually, the pressure becomes so great that it blasts the lava, along with previously solidified volcanic rock, out of the volcano. The process is similar to the way the rapidly expanding gas accompanying the explosion of gunpowder in a cartridge shoots a bullet out of a gun.

In some cases, an explosive eruption blasts the volcano apart and leaves behind a large caldera. Such explosions, awesome in their power and catastrophic in their consequences, eject cubic kilometers of igneous particles upward at initial speeds of up to 90 m per second. Convection in the cloud can carry ash up through the entire troposphere and into the stratosphere. The resulting plume of debris resembles the mushroom cloud above a nuclear explosion. Coarse-grained ash and lapilli settle from the cloud close to the volcano, while finer ash settles farther away.

Some explosive eruptions take place when water gains access to the hot rock around the magma chamber and suddenly transforms into steam—the steam pressure blasts the volcano apart and energetically expels debris. Geologists refer to pyroclastic eruptions involving the reaction of water with magma as **phreatomagmatic eruptions** (►Fig. 9.13b).

The type of volcano (shield, cinder cone, or stratovolcano) depends on its eruptive style. Volcanoes that have only effusive eruptions become shield volcanoes, those that generate small pyroclastic eruptions yield cinder cones, and those that alternate between effusive and large pyroclastic eruptions become stratovolcanoes. Large explosions yield calderas and blanket the surrounding countryside with sheets of ignimbrite.

Why are there such contrasts in eruptive style and therefore in volcano shape? Eruptive style depends on the viscosity and gas pressure of the magma in the volcano. These characteristics, in turn, depend on the composition and temperature of the magma and on the environment (subaerial or submarine) in which the eruption occurs. Let's look at these controls in detail.

- *The effect of viscosity on eruptive style:* Low-viscosity (basaltic) lava flows out of a volcano easily, whereas high-viscosity (andesitic and rhyolitic) lava can clog up a volcano's plumbing and lead to a buildup of pressure. Thus, basaltic eruptions are typically effusive and produce shield volcanoes, whereas rhyolitic eruptions are explosive.

- *The effect of gas pressure on eruptive style:* The injection of magma into the magma chamber and conduit generates an outward push or pressure inside the volcano. The presence of gas within the magma increases this pressure, because gas expands greatly as it rises toward the Earth's surface. In runny (basaltic) magma, gas bubbles can rise to the surface of the magma and pop, causing the lava to fountain into the sky; a small cinder cone of bombs and cinders results. In a viscous (andesitic or rhyolitic) magma, however, the gas bubbles cannot escape and thus move with the magma toward the Earth's surface. As pressure on the magma from overlying rock decreases, the gas bubbles expand and create a tremendous outward pressure. Eventually, the gas pressure shatters the partially solidified magma and sends a large cloud of pyroclastic debris into the sky or down the flank of the volcano, causing a pyroclastic eruption. Rhyolitic and andesitic magmas contain more gas, and thus eruptions of these magmas are more explosive than are eruptions of basaltic magmas.

FIGURE 9.13 Contrasting eruptive styles. **(a)** This effusive eruption on Hawaii, though relatively quiet, has produced a large lake of molten lava. The surface of the lake has frozen to form a black crust, but convection within the lake cracks the crust and allows us to see the red molten rock below. **(b)** This eruption of Surtsey, off the coast of Iceland, is a phreatomagmatic eruption, caused by steam explosions that result when seawater enters the magma chamber.

(a)

(b)

- *The effect of the environment on eruptive style:* The ability of a lava to flow depends on where it erupts. Lava flowing on dry land cools more slowly than lava erupting underwater. Thus, a basaltic lava that could flow easily down the flank of a subaerial volcano will pile up in a mound of pillows around the vent of a submarine volcano.

Traditionally, geologists have classified volcanoes according to their eruptive style, each style named after a well-known example (Hawaiian, Vulcanian, etc.) as described in specialized books on volcanoes (see art, pp. 278–279). Below, we focus on relating eruptive styles to the geologic setting in which the volcano forms, in the context of plate tectonics theory (▶Fig. 9.14).

Take-Home Message

Effusive eruptions produce lava, which can flow for tens of kilometers, sometimes through lava tunnels. Explosive eruptions occur when pressure builds up in the volcano or if water enters the volcano and turns to steam. Eruptive style reflects lava composition.

9.5 HOT-SPOT ERUPTIONS

A hot spot is a point on the surface of the Earth where volcanism that is not a direct consequence of the normal relative motion between two plates takes place (see Chapter 6). This means that hot-spot volcanoes are not a direct consequence of standard subduction or of sea-floor spreading. Although some of Earth's more than fifty hot-spot volcanoes do straddle plate boundaries, many have erupted in the interior of plates—both oceanic and continental—far

FIGURE 9.14 A map showing the distribution of volcanoes around the world, and the basic geologic settings in which volcanoes form, in the context of plate-tectonics theory.

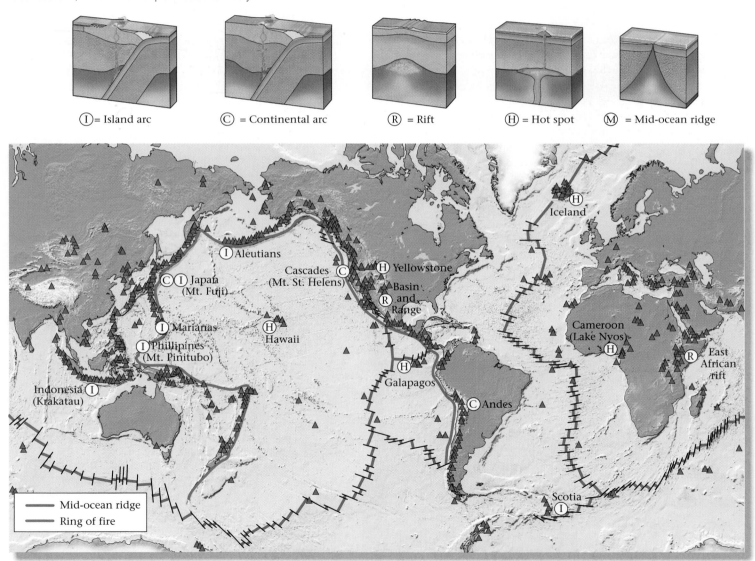

Ⓘ = Island arc   Ⓒ = Continental arc   Ⓡ = Rift   Ⓗ = Hot spot   Ⓜ = Mid-ocean ridge

BOX 9.1 THE HUMAN ANGLE

Volcanic Explosions to Remember

Explosions of volcanic-arc volcanoes generate enduring images of destruction. Let's look at two notable historic cases (see ▶Fig. 9.15).

Mt. St. Helens, a snow-crested stratovolcano in the Cascade mountain chain, had not erupted since 1857. However, geologic evidence suggested that the mountain had had a violent past, punctuated by many explosive eruptions. On March 20, 1980, an earthquake announced that the volcano was awakening once again. A week later, a crater 80 m in diameter burst open at the summit and began emitting gas and pyroclastic debris. Geologists, who set up monitoring stations to observe the volcano, noted that its north side was beginning to bulge markedly, suggesting that the volcano was filling with magma and that the magma was making the volcano expand like a balloon. Their concern that an eruption was imminent led local authorities to evacuate people in the area.

The climactic eruption came suddenly. At 8:32 A.M. on May 18, the geologist, David Johnston, monitoring the volcano from a distance of 10 km, shouted over his two-way radio, "Vancouver, Vancouver, this is it!" An earthquake had triggered a huge landslide that caused 3 cubic km of the volcano's weakened north side to slide away. The sudden landslide released pressure on the magma in the volcano, causing a sudden and violent expansion of gases that blasted through the side of the volcano (▶Fig. 9.16a–c). Rock, steam, and ash screamed north at the speed of sound and flattened a forest and everything in it over an area of 600 square km (▶Fig. 9.16d, e). Tragically, Johnston, along with sixty others, vanished forever. Water-saturated ash flooded river valleys, carrying away everything in its path. Seconds after the sideways blast, a vertical column carried about 540 million tons of ash (about 1 cubic km) 25 km into the sky, where

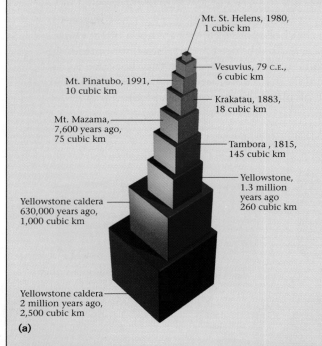

- Mt. St. Helens, 1980, 1 cubic km
- Vesuvius, 79 C.E., 6 cubic km
- Mt. Pinatubo, 1991, 10 cubic km
- Krakatau, 1883, 18 cubic km
- Mt. Mazama, 7,600 years ago, 75 cubic km
- Tambora , 1815, 145 cubic km
- Yellowstone, 1.3 million years ago, 260 cubic km
- Yellowstone caldera 630,000 years ago, 1,000 cubic km
- Yellowstone caldera 2 million years ago, 2,500 cubic km

(a)

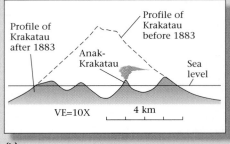

Profile of Krakatau after 1883

Profile of Krakatau before 1883

Anak-Krakatau

Sea level

VE=10X   4 km

(b)

FIGURE 9.15 **(a)** The chart shows the relative amounts of pyroclastic debris (in cubic km) ejected during major historic eruptions of the past two centuries. Notice that the 1815 Tambora eruption was over five times bigger than the 1883 Krakatau eruption, which in turn was over five times larger than the 1980 Mt. St. Helens eruption. **(b)** Profile of Krakatau, before and after the eruption. Note that a new resurgent dome (Anak Krakatau) has formed. **(c)** A map illustrating the dimensions of the region destroyed by the 1980 eruption of Mt. St. Helens. The arrows indicate the blast direction.

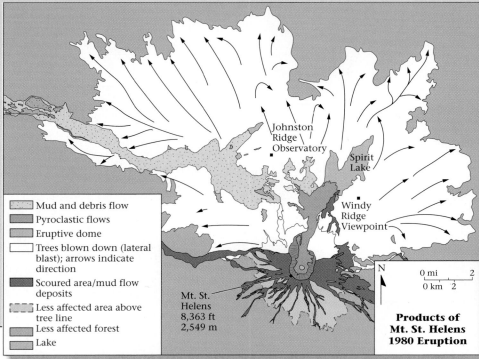

Johnston Ridge Observatory

Spirit Lake

Windy Ridge Viewpoint

- Mud and debris flow
- Pyroclastic flows
- Eruptive dome
- Trees blown down (lateral blast); arrows indicate direction
- Scoured area/mud flow deposits
- Less affected area above tree line
- Less affected forest
- Lake

Mt. St. Helens 8,363 ft 2,549 m

N

0 mi   2
0 km   2

Products of Mt. St. Helens 1980 Eruption

(c)

the jet stream carried it away so that it was able to circle the globe. In towns near the volcano, a blizzard of ash choked roads and buried fields. Measurable quantities of ash settled over an area of 60,000 square km. When the eruption was over, the once cone-like peak of Mt. St. Helens had disappeared—the summit now lay 440 m lower, and the once snow-covered mountain was a gray mound with a large gouge in one side. The volcano came alive again in 2004, but did not explode.

An even greater explosion happened in 1883 (see Fig. 9.15b). Krakatau, a volcano in the sea between Indonesia and Sumatra, where the Indian Ocean floor subducts beneath Southeast Asia, had grown to become a 9-km-long island rising 800 m (2,600 feet) above the sea. Then, on May 20, the island began to erupt with a series of large explosions, yielding ash that settled as far as 500 km away. Smaller explosions continued through June and July, and steam and ash rose from the island, forming a huge black cloud that rained ash into the surrounding straits. Ships sailing by couldn't see where they were going, and their crews had to shovel ash off the decks.

The climax came at 10 A.M. on August 27, perhaps when the volcano cracked and the magma chamber flooded with seawater. The resulting blast, five thousand times greater than the Hiroshima atomic-bomb explosion, could be heard as far as 4,800 km away, and subaudible sound waves traveled around the globe seven times. Giant waves pushed out by the explosion slammed into coastal towns, killing over 36,000 people. Near the volcano, a layer of pumice up to 40 m thick fell from the sky. When the air finally cleared, Krakatau was gone, replaced by a submarine caldera some 300 m deep. All told, the eruption shot 20 cubic km of rock into the sky. Some ash reached elevations of 27 km. Because of this ash, the world was treated to spectacular sunsets for the next few years.

FIGURE 9.16 The eruption of Mt. St. Helens, 1980. **(a)** Before the eruption, the magma chamber is empty. **(b)** The magma chamber fills, and the side of the volcano bulges outward. **(c)** The weakened north flank suddenly slipped, releasing the pressure on the magma chamber. The sudden decrease in pressure caused dissolved gases in the magma to expand and blast laterally out of the volcano. **(d)** The eruptive cloud. **(e)** The neighboring forest, flattened by a blast of rock, steam, and ash.

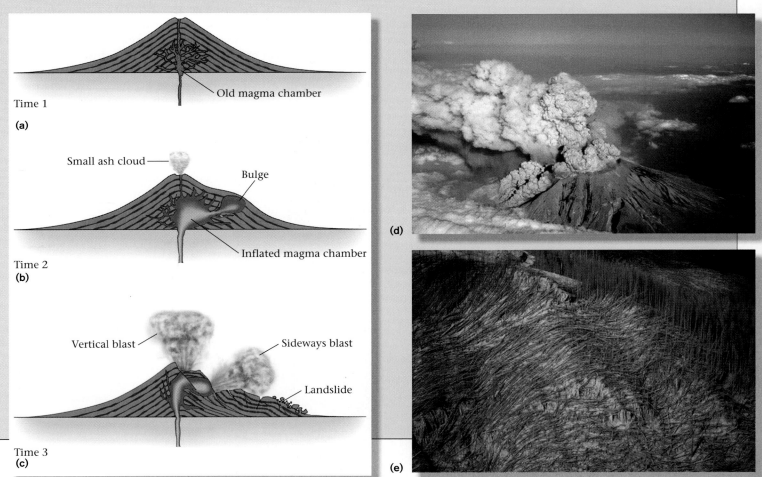

from plate boundaries. Hot-spot volcanoes possibly form as a consequence of melting at the top of column-like mantle plumes, for a plume brings hot rock from the base of the mantle up to the base of the lithosphere, where decompression causes the rock to partially melt. However, this model remains controversial, and geologists are testing alternatives. Let's look at two well-known examples of hot-spot volcanism.

Oceanic Hot-Spot Volcanoes (Hawaii)

An oceanic hot-spot eruption initially produces an irregular mound of pillow lava. With time, the volcano grows up above the sea surface and becomes an island. But when the volcano emerges from the sea, the basalt lava that erupts no longer freezes so quickly, and thus flows as a thin sheet over a great distance. Thousands of thin basalt flows pile up, layer upon layer, to build a broad, dome-shaped shield volcano with gentle slopes (Fig. 9.11a, b). Note that such shield volcanoes develop their distinctive shape because the low-viscosity, hot basaltic lava that constitutes them spreads out like pancake syrup and cannot build up into a steep cone. As the volcano grows, portions of it can't resist the pull of gravity and slip seaward, creating large slumps and debris flows that collect along the base of island. Thus, in cross section, hot-spot volcanoes are quite complex (▶Fig. 9.17).

The big island of Hawaii, the largest oceanic hot-spot volcano on Earth today, currently consists of five shield volcanoes, each built around a different vent. The island now towers over 9 km above the adjacent ocean floor (about 4.2 km above sea level), the greatest relief from base to top of any mountain on Earth; by comparison, Mt. Everest rises 8.85 km above the plains of India. Calderas up to 3 km wide have formed at the summit, and basaltic lava has extruded from both conduits and fissures. During some eruptions, the lava fountains into the air, or fills deep lava lakes in craters (Fig. 9.13a). The lakes gradually drain to feed streams of lava that cascade down the flanks of the volcano. Lava tubes within flows carry lava all the way to the sea, where the glowing molten rock drips into the water and instantly disappears in a cloud of steam.

Continental Hot-Spot Volcanoes (Yellowstone National Park)

Yellowstone National Park lies over a continental hot spot (▶Fig 9.18a, b). Though volcanoes are not erupting in the park today, they have done so in the fairly recent geologic past. In fact, during the past 2 million years, three immense explosive eruptions ripped open the land that is now the park. The last of these happened about 640,000 years ago. During this event, 1,000 cubic km of volcaniclastic debris blasted into the atmosphere or rushed across the countryside as immense ash flows. The 0.64 Ma eruption produced an immense caldera, up to 70 km across, which overlaps earlier calderas (▶Fig. 9.18c, d, e). When the debris settled, it blanketed an area of 2,500 square km with tuffs that, in the park, reached a thickness of 400 m (▶Fig. 9.18f, g). The park's name reflects the brilliant color of volcaniclastic debris exposures in the park's canyons. Magma remains in the crust beneath the park today; energy radiating from this magma heats groundwater that rises to fill hot springs and spurt out of steaming geysers. Eruptive activity, producing basalt flows and more pyroclastic debris, continued until about 70,000 years ago, and will likely happen in the future.

Why do eruptions at Yellowstone differ in style from those of Hawaii? To arrive at an answer, recall that Hawaii formed on mafic-composition oceanic crust, whereas Yellowstone formed on felsic- to intermediate-composition continental crust. In Hawaii, rising mafic magma from the mantle could not melt the surrounding mafic crust, because the crust has a melting temperature that is very close to that of the magma. In Yellowstone, however, though some of the rising mafic magma made it to the surface, some remained trapped in the continental crust. There it did provide sufficient heat to cause partial melting, for

FIGURE 9.17 The inside of an oceanic hot-spot volcano is a mound of pillow basalt built on the surface of the oceanic crust. When the mound emerges above sea level, a shield volcano forms on top. Volcanic debris accumulates along the margin of the volcano. The weak material occasionally slumps seaward on sliding surfaces (indicated with arrows).

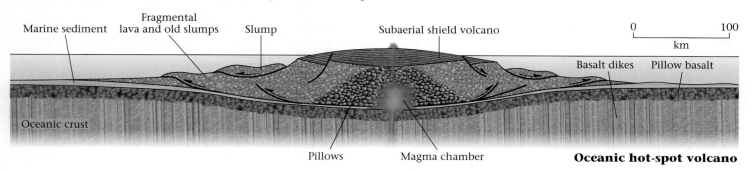

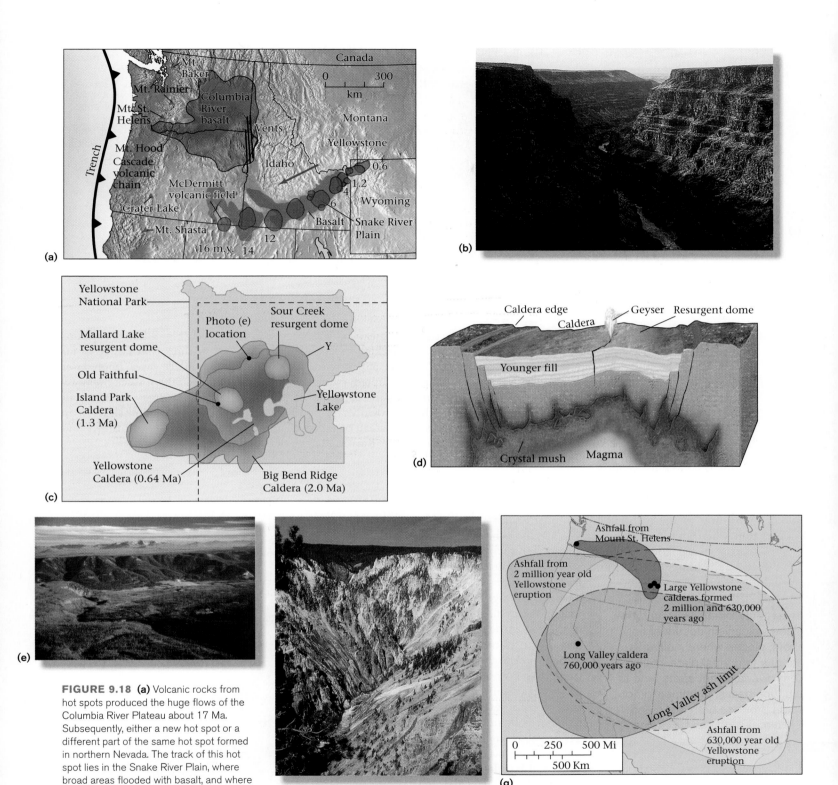

FIGURE 9.18 **(a)** Volcanic rocks from hot spots produced the huge flows of the Columbia River Plateau about 17 Ma. Subsequently, either a new hot spot or a different part of the same hot spot formed in northern Nevada. The track of this hot spot lies in the Snake River Plain, where broad areas flooded with basalt, and where several calderas formed in succession, beginning 16 Ma. **(b)** Basalt flows exposed along the walls of a canyon in the Snake River Plain. **(c)** A map of the Yellowstone Park area shows the location of the three most recent caldera, and the two present resurgent domes. **(d)** A speculative block diagram displays the subsurface geometry of the caldera. **(e)** A photograph of the edge of the caldera. The land in the foreground that has dropped down is within the caldera. **(f)** The "Grand Canyon of Yellowstone" shows exposures of yellow tuff. **(g)** A map showing the distribution of ash from the Yellowstone eruption. Note that the Long Valley and at Mt. Saint Helens eruptions produced much less ash.

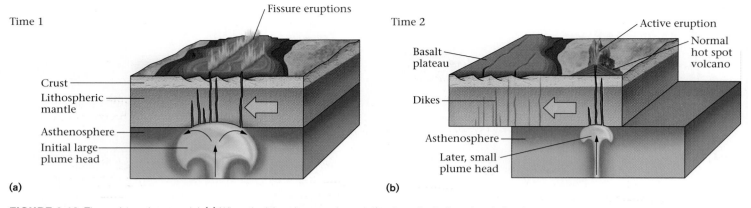

Time 1

Time 2

Fissure eruptions

Active eruption

Crust

Lithospheric mantle

Asthenosphere

Initial large plume head

Basalt plateau

Normal hot spot volcano

Dikes

Asthenosphere

Later, small plume head

(a)

(b)

FIGURE 9.19 The evolving plume model. **(a)** When the lithosphere cracks and rifts above the bulbous head of a plume, huge amounts of magma rise and erupt through fissures, producing sheets of basalt that pile up to form a plateau. **(b)** Later, the bulbous plume head no longer exists, leaving only a narrower plume stalk. The lithosphere has moved relative to the plume, so a track of hot-spot volcanoes begin to form.

continental crust has a lower melting temperature. Partial melting of continental crust yields rhyolitic magma.

The Yellowstone caldera lies at the end of a long chain of calderas and stacks of lava flows whose remnants crop out in the Snake River Plain of Idaho (►Fig. 9.18c). This chain marks the track of the Yellowstone hot spot. Though many geologists argue that the hot spot indicates the presence of a mantle plume below, other explanations for this volcanism have been suggested in recent years.

Flood-Basalt Eruptions

According to the mantle-plume model of hot-spot formation, when a plume first rises, it has a bulbous head in which there is a huge amount of partially molten rock. If the crust above the plume stretches and rifts, voluminous amounts of lava erupt along fissures. A particularly large amount of magma is available because of the size of the plume head and because the very hot asthenosphere of the plume undergoes a greater amount of partial melting than does the cooler asthenosphere that normally underlies rifts. The low-viscosity lava spreads out in sheets over vast areas. Geologists refer to these sheets as **flood basalt**. Over time, eruption of basalts builds a broad plateau (►Fig. 9.19a, b). Geologists refer to broad areas covered by flood basalt as large igneous provinces (LIPs; see Chapter 6). Once the plume head has drained, the volume of eruption decreases, and normal hot-spot eruptions take place.

About 15 million years ago, rifting above a plume created the region that now constitutes the Columbia River Plateau of Washington and Oregon (Fig. 9.18a). Eruptions yielded sheets of basalt up to 30 m thick that flowed as far as 550 km from the source. Gradually, layer upon layer erupted, creating a pile of basalt up to 500 m thick over a region of 220,000 square km. Even larger flood-basalt provinces formed elsewhere in the world, notably the Deccan Plateau of India (►Fig. 9.20), the Paraná Basin of Brazil, and the Karroo Plateau of South Africa.

> ### Take-Home Message
>
> At over fifty locations around the globe, volcanism occurs at hot spots, places where melting is not due to normal plate interactions. Oceanic hot spots tend to produce basalt. Continental hot spots produce both mafic and felsic rock, and can be explosive.

FIGURE 9.20 The flood basalts of western India, known as the "Deccan traps," are exposed in a canyon near the village of Ajanta. Between about 100 B.C.E. and 700 C.E., Buddhists carved a series of monasteries and meeting halls into the solid basalt. These are decorated by huge statues, carved in place, as well as spectacular frescoes, painted on cow-dung plaster.

9.6 ERUPTIONS ALONG PLATE BOUNDARIES AND RIFTS

Most volcanic activity on Earth occurs along divergent or convergent plate boundaries, or in rifts. We now examine characteristic eruptions in these different settings.

Mid-Ocean Ridge Volcanism

Eruptions of lava occur along the entire length of mid-ocean ridges, plate boundaries at which new sea-floor crust forms. In fact, products of mid-ocean ridge volcanism cover 70% of our planet's surface. We don't generally see this volcanic activity, however, because the ocean hides most of it beneath a blanket of water. Mid-ocean ridge volcanoes, which develop along fissures parallel to the ridge axis, are not all continuously active. Each one turns on and off in a time scale measured in tens to hundreds of years. They erupt basalt, which, because it's underwater, forms pillow-lava mounds. Water that heats up as it circulates through the crust near the magma chamber bursts out of hydrothermal (hot-water) vents (see Chapter 4).

Iceland is one of the few places on Earth where mid-ocean ridge volcanism has built a mound of basalt that protrudes above the sea. The island formed where a mantle plume lies beneath the Mid-Atlantic Ridge—the presence of this plume means that far more magma erupted here than beneath other places along mid-ocean ridges. Because Iceland straddles a divergent plate boundary, it is being stretched apart, with faults forming as a consequence. Indeed, the central part of the island is a narrow rift, in which the youngest volcanic rocks of the island have erupted (▶Fig. 9.21a, b); this rift is the trace of the Mid-Atlantic Ridge. Faulting cracks the crust and so provides a conduit to a magma chamber. Thus, eruptions on Iceland tend to be fissure eruptions, yielding either curtains of lava that are many kilometers long or linear chains of small cinder cones (Fig. 9.9b).

Not all volcanic activity on Iceland occurs subaerially. Some eruptions take place under glaciers. During 1996, for example, an eruption at the base of a 600-m-thick glacier melted the ice and produced a column of steam that rose several kilometers into the air. Meltwater accumulated under the ice for six days, until it burst through the edge of the glacier and became a flood that lasted two days and destroyed roads, bridges, and telephone lines. Some of Iceland's volcanic activity occurs under the sea. Continuing eruptions off the coast yielded the island of Surtsey, whose birth was first signaled by huge quantities of steam bubbling up from the ocean. Eventually, steam pressure explosively ejected ash as high as 5 km into the atmosphere. Surtsey finally emerged from the sea on November 14, 1963, building up a cone of ash and lapilli that rose almost 200 m above sea level in just three months (Fig. 9.13b). Waves could easily have eroded the cinder cone away, but the island has survived because lava erupted from the vent and flowed over the cinders, effectively encasing them in an armor-like blanket of solid rock.

Eruptions along Convergent Boundaries

Most of the subaerial volcanoes on Earth lie along convergent plate boundaries (subduction zones). The volcanoes form when volatile compounds such as water and carbon dioxide are released from the subducting plate and rise into the overlying hot mantle, causing melting and producing magma that then rises through the lithosphere and erupts. Some of these volcanoes start out as submarine volcanoes and later grow into volcanic island arcs, such as the

FIGURE 9.21 (a) Iceland consists of volcanic rocks that erupted from a hot spot along the Mid-Atlantic Ridge. Because the island straddles a divergent boundary, it gradually stretches, leading to the formation of faults. The central part of the island is an irregular northeast-trending rift, where we find the youngest rocks of the island. **(b)** The surface of Iceland has dropped down along the faults that bound the central rift. This low-altitude aerial photo shows an escarpment formed where slip occurred on a fault.

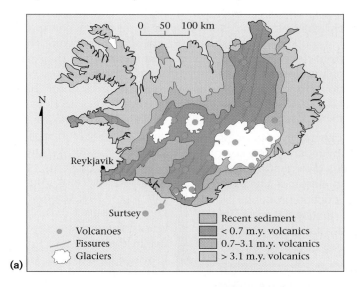

(a)

(b)

Marianas of the western Pacific. Others grow on continental crust, building continental volcanic arcs, such as the Cascade mountain chain of Washington and Oregon (Geotour 9). Typically, individual volcanoes in volcanic arcs lie about 50 to 100 km apart. Subduction zones border more than 60% of the Pacific Ocean, creating a 20,000-km-long chain of volcanoes known as the **Ring of Fire**. See Chapters 4 and 6 for illustrations.

Many different kinds of magma form at volcanic arcs. As a result, these volcanoes sometimes have effusive eruptions and sometimes pyroclastic eruptions—and occasionally they explode. Such eruptions yield composite volcanoes such as the elegant symmetrical cone of Mt. Fuji (Fig. 9.11e) and the blasted-apart hulk of Mt. St. Helens (Fig. 9.15; Geotour 9).

Volcanoes of island arcs initially erupt underwater. Thus, their foundation consists of volcanic material that froze in contact with water, or of volcanic debris that was deposited underwater. The layers that make up the foundation include pillow basalts, hyaloclastites, and submarine debris flows.

Eruptions in Continental Rifts

The rifting of continental crust yields a wide array of different types of volcanoes, because (as in the case of continental hot spots) the magma that feeds these volcanoes comes both from the partial melting of the mantle and from the partial melting of the crust. Thus, rifts host basaltic fissure eruptions, in which curtains of lava fountain up or linear chains of cinder cones develop. But they also host explosive rhyolitic volcanoes and, in some places, even stratovolcanoes.

Rift volcanoes are active today in the East African Rift (Fig. 9.10d; Geotour 9). During the past 25 million years, rift volcanoes were active in the Basin and Range Province of Nevada, Utah, and Arizona. About 1 billion years ago, a narrow but deep rift formed in the middle of the United States and filled with over 15 km of basalt; this Mid-Continent Rift runs from the tip of Lake Superior to central Kansas.

> **Take-Home Message**
>
> Most volcanic activity takes place on plate boundaries. The sea generally hides divergent-boundary volcanism—Iceland is an exception. Subduction produces island arcs and continental arcs. The latter include stratovolcanoes. Volcanism also occurs along rifts.

9.7 BEWARE: VOLCANOES ARE HAZARDS!

Like earthquakes, volcanoes are natural hazards that have the potential to cause great destruction to humanity, in both the short term and the long term. According to one estimate, volcanic eruptions in the last two thousand years have caused about a quarter of a million deaths—much fewer than those caused by earthquakes, but nevertheless a sizable number. With the rapid expansion of cities, far more people live in dangerous proximity to volcanoes today than ever before, so if anything, the hazard posed by volcanoes has gotten worse—imagine if a Krakatau-like explosion were to occur next to a major city today. Let's now look at the different kinds of threats posed by volcanic eruptions.

Hazards Due to Eruptive Materials

Threat of lava flows. When you think of an eruption, perhaps the first threat that comes to mind is the lava that flows from a volcano. Indeed, on many occasions lava has overwhelmed towns. Basaltic lava from effusive eruptions is the greatest threat, because it can flow quickly and spread over a broad area. In Hawaii, recent lava flows have buried roads, housing developments, and cars (Fig. 9.3b). In one place, basalt almost completely submerged a parked (and empty) school bus (▶Fig. 9.22a). Usually people have time to get out of the way of such flows, but not necessarily with their possessions. All they can do is watch helplessly from a distance as an advancing flow engulfs their home (▶Fig. 9.22b). Before the lava even touches it, the building may burst into flames from the intense heat. Similarly, forests, orchards, and sugarcane fields are burned and then buried by rock, their verdure replaced by blackness. The most disastrous lava flow in recent time came from the eruption in 2002 of Mt. Nyiragongo, a 3.7-km-high volcano in the Democratic Republic of Congo (▶Fig. 9.22c; Geotour 9). Lava flows traveled almost 50 km and flooded the streets of Goma, encasing the streets with a 2-m-thick layer of basalt. The flows destroyed almost half the city and turned 300,000 people into refugees.

Threat of ash and lapilli. During a pyroclastic eruption, large quantities of ash erupt into the air, later to fall back to Earth. Close to a volcano, pumice and lapilli tumble out of the sky, smashing through or crushing roofs of nearby buildings (for this reason, Japanese citizens living near volcanoes keep hard hats handy), and can accumulate into a blanket up to several meters thick. Winds can carry fine ash over a broad region. In the Philippines, for example, a typhoon spread heavy airfall ash from the 1991 eruption of Mt. Pinautubo so that it covered a 4,000-square-km area (▶Fig. 9.22d). Because of heavy rains, the ash became soggy and heavy, and it was particularly damaging to roofs. Ash buries crops, may spread toxic chemicals that poison the soil, and insidiously infiltrates machinery, causing moving parts to wear out.

Fine ash from an eruption can also present a hazard to airplanes. Ash clouds rise so fast that they may be at airplane heights (11 km) long before the volcanic eruption has been reported, especially if the eruption occurs in a remote locality; and at high elevations, the ash cloud may be too dilute for a pilot to see. Like a sandblaster, the sharp, angular ash abrades turbine blades, greatly reducing engine efficiency, and the ash, along with sulfuric acid formed from the volcanic gas, scores windows and damages the fuselage. Also, when heated inside a jet engine, the ash melts, creating a liquid that sprays around the turbine and freezes; the resulting glassy coating restricts the air flow and causes the engine to flame out.

For example, in 1982, a British Airways 747 flew through the ash cloud over a volcano on Java. Corrosion turned the windshield opaque, and ingested ash caused all four engines to fail. For thirteen minutes, the plane glided earthward, dropping from 11.5 km (37,000 feet) to 3.7 km (12,000 feet) above the black ocean below. As passengers assumed a brace position for ditching at sea, the pilots tried repeatedly to restart the engines. Suddenly, in the oxygen-rich air of lower elevations, the engines roared back to life. The plane swooped back into the sky and headed for an emergency landing in Jakarta, where, without functioning instruments and with an opaque windshield, the pilot brought his 263 passengers and crew back to the ground safely. To land, he had to squint out an open side window, with only his toes touching the controls. In 1989, the same fate befell a KLM 747 en route to Anchorage. The plane encountered ash from the Redoubt Volcano (see Chapter Opener), lost power in all four engines and all instruments, and sank about 2.6 km (8,000 feet) before the pilot could restart the engines and bring the plane in for a landing. During the month after the 1991 eruption of Mt. Pinautubo, fourteen jets flew through the resulting ash cloud and, of these, nine had to make emergency landings because of engine failure.

Threat of pyroclastic flows. Pyroclastic flows race down the flanks of a volcano at speeds of 100 to 300 km per hour (▶Fig. 9.22e). The largest can travel tens to hundreds of kilometers. The volume of ash contained in such glowing avalanches is not necessarily great—St. Pierre on Martinique was covered only by a thin layer of dust after the pyroclastic flow from Mt. Pelée had passed (see Chapter 6)—but the cloud can be so hot and poisonous that it means instant death to anyone caught in its path, and because it moves so fast, the force of its impact can flatten buildings and forests (Fig. 9.6a–c).

Other Hazards Related to Eruptions

Threat of the blast. Most exploding volcanoes direct their fury upward. But some, such as Mt. St. Helens, explode sideways. The forcefully ejected gas and ash, like the blast of a bomb, flattens everything in its path. In the case of Mt. St. Helens, the region around the volcano had been a beautiful pine forest; but after the eruption, the once-towering trees were stripped of bark and needles and lay scattered over the hill slopes like matchsticks (Fig. 9.16e).

Threat of landslides and floods. Eruptions commonly trigger large landslides along the volcano's flanks. The debris, composed of ash and solidified lava that erupted earlier, can move quite fast (250 km per hour) and far. During the eruption of Mt. St. Helens, 8 billion tons of debris took off down the mountainside, careened over a 360-m-high ridge, and tumbled down a river valley, until the last of it finally came to rest over 20 km from the volcano. In Iceland, a unique hazard develops. Some eruptions occur under ice, producing pools of meltwater that eventually burst out at the end of the glacier and destroy areas downstream. These abrupt floods of water and volcaniclastic debris are called *jokulhlaupt.*

Threat of lahars. When volcanic ash and other debris mix with water, the result is a slurry that resembles freshly mixed concrete. This slurry, known as a lahar, can flow downslope at speeds of over 50 km per hour. Because lahars are denser and more viscous than water, they pack more force than flowing water and can literally carry away everything in their path. The lahars of Mt. St. Helens traveled more than 40 km from the volcano, following existing drainages. When they had passed, they left a gray and barren wake of mud, boulders, broken bridges, and crumpled houses, as if a giant knife had scraped across the landscape. Widespread lahars also swept down the flanks of Mt. Pinautubo in 1991, the water provided by typhoonal and monsoonal rains.

Lahars may develop in regions where snow and ice cover an erupting volcano, for the eruption melts the snow and ice, thereby creating an instant supply of water. Perhaps the most destructive lahar of recent times accompanied the eruption of the snow-crested Nevado del Ruiz in Colombia on the night of November 13, 1985. The lahar surged down a valley of the Rio Lagunillas like a 40-m-high wave, hitting the sleeping town of Armero, 60 km from the volcano. Other pulses of lahar followed. When they had passed, 90% of the buildings in the town were gone, replaced by a 5-m-thick layer of mud (▶Fig. 9.22f), which now entombs the bodies of 25,000 people.

Threat of earthquakes. Earthquakes accompany almost all major volcanic eruptions, for the movement of magma breaks rocks underground. Such earthquakes may trigger landslides on the volcano's flanks, and can cause buildings to collapse and dams to rupture, even before the eruption itself begins.

Threat of tsunamis (giant waves). Where explosive eruptions occur in the sea, the blast and the underwater collapse of a caldera generate huge sea waves, tens of meters (in rare cases, over 100 meters) high. Most of the 36,000 deaths attributed to the 1883 eruption of Krakatau were due not to

FIGURE 9.22 **(a)** This empty school bus was engulfed by a basalt flow in Hawaii. **(b)** When lava at over 1,000°C comes close to a house, the house erupts in flame. **(c)** Residents of Goma, in west-central Africa, walking over lava-filled streets after a 2002 eruption of a nearby volcano. **(d)** A blizzard of ash falling from the eruption of Mt. Pinautubo, in the Philippines, blankets a nearby town in ghostly white. **(e)** A pyroclastic flow rushes toward fleeing firefighters in Japan, during the eruption of Mt. Unzen. **(f)** A devastating lahar buried the town of Armero, Colombia.

ash or lava, but rather to tsunamis that slammed into nearby coastal towns (see Box 9.1).

Threat of gas. We have already seen that volcanoes erupt not only solid material, but also large quantities of gases such as water vapor, carbon dioxide, sulfur dioxide, and hydrogen sulfide. Usually the gas eruption accompanies the lava and ash eruption, with the gas contributing only a minor part of the calamity. But occasionally, gas erupts alone and snuffs out life in its path without causing any other damage. Such an event occurred in 1986 near Lake Nyos in Cameroon, western Africa.

Lake Nyos is a small, but deep lake filling the crater of an active hot-spot volcano in Cameroon. Though only 1 km across, the lake reaches a depth of over 200 m. Because of its depth, the cool bottom water of the lake does not mix with warm surface water, and for many years the bottom water remains separate from the surface water. During this time, carbon dioxide gas slowly bubbles out of cracks in the floor of the crater and dissolves in the cool bottom water. Apparently, by August 21, 1986, the bottom water had become supersaturated in carbon dioxide. On that day, perhaps triggered by a landslide or wind, the lake "burped" and, like an exploding seltzer bottle, expelled a forceful froth of CO_2 bubbles (together constituting 1 cubic km of gas). Because it is denser than air, this invisible gas flowed down the flank of the volcano and spread out over the countryside for about 23 km before dispersing. Though not toxic, carbon dioxide cannot provide oxygen for metabolism or oxidation. (For this reason, it is the principal component of dry fire extinguishers.) When the gas cloud engulfed the village of Nyos, it quietly put out the cooking fires and suffocated the sleeping inhabitants, most of whom died where they lay. The next morning, the landscape looked exactly as it had the day before, except for the lifeless bodies of 1,742 people and about 6,000 head of cattle (▶Fig. 9.23). Recently, engineers have

FIGURE 9.23 Cattle near Lake Nyos, Cameroon, fell where they stood, victims of a cloud of carbon dioxide.

been testing methods to de-gas the lake gradually, to avoid a similar disaster in the future.

The threat of gas is more common than widely recognized. For example, on windless days CO_2 collects in depressions and gullies along the flanks of volcanoes. Children and animals wandering into these areas quickly collapse. In Swahili, these danger spots are called *mazukus* (evil winds).

Which Threats are Most Dangerous?

It's hard to compile statistics on how fatalities occur as a consequence of volcanic eruption, but a recent study has done just that. Lava flows, though dramatic, actually cause only a small percentage of the fatalities, because the flows generally move slowly enough that people can get out of their way. The greatest number of fatalities (almost 30%) result from pyroclastic flows, because these can strike so fast that people cannot escape. Other leading causes of death are mudflows (about 15%), tsunamis (about 20%), and indirect causes (almost 25%). The last item in the list recognizes the fact that when eruptions blanket and kill crops, disrupt transportation, and destroy communities, they cause starvation and illness that can lead to death. All other effects of volcanoes (earthquakes, floods, gas, ash falls, lava) together account for about 10% of fatalities.

> ### Take-Home Message
> Volcanoes can be dangerous! The lava flows, pyroclastic debris, explosions, mud flows (lahars), landslides, earthquakes, and gas clouds produced during eruptions can destroy cities and farmland. Ash flows move very fast and incinerate everything in their path.

9.8 PROTECTION FROM VULCAN'S WRATH

Active, Dormant, and Extinct Volcanoes

In the geologic record, volcanoes come and go. For example, while a particular convergent plate boundary exists, a volcanic arc exists; but if subduction ceases, the volcanoes in the arc die and erode away. Even when alive, individual volcanoes erupt only intermittently. In fact, the average time between successive eruptions (the repose time) ranges from a few years to a few centuries, and in some cases millennia.

Geologists refer to volcanoes that are erupting, have erupted recently, or are likely to erupt soon as **active volcanoes**. They distinguish these from **dormant volcanoes**, which have not erupted for 10,000 years but do have the potential to erupt again in the future. Volcanoes that were active in the past but have shut off entirely and will never erupt in the future are called **extinct volcanoes**. For example,

See for yourself...
Volcanic Features

There are more than 1,500 active volcanoes on Earth and thousands more volcanic landscapes. You visited some of these in Geotours 4 and 6. Here, we continue the journey. The thumbnail images provided on this page are only to help identify tour sites. Go to wwnorton.com/studyspace to experience this flyover tour.

Yellowstone Falls
(Lat 44°43'4.96"N, Long 110°29'45.44"W)

At these coordinates, hover above the Grand Canyon of the Yellowstone from an elevation of 5 km (3 miles). The canyon walls expose bright yellow tuffs deposited during cataclysmic eruptions that occurred during the past 2 million years. Drop to an elevation of 3 km (2 miles), look downstream, and tilt to get a better view (Image G9.1).

G9.1

Mt. Saint Helens, Washington
(Lat 46°12'1.24"N, Long 122°11'20.73"W)

Box 9.1 discussed the explosion of Mt. Saint Helens. To see the damage for yourself, fly to the coordinates and zoom to 30 km (18.5 miles). The breached crater, the blowdown zone, the slumps, and the lahars are all visible (Image G9.2). To get a better sense of the devastation, descend to 13 km (8 miles), tilt your image, and fly around the mountain (Image G9.3).

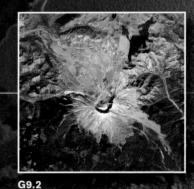

G9.2

G9.3

Smoking Volcano, Ecuador
(Lat 1°27'58.96"S, Long 78°26'58.24"W)

Fly to this locality and zoom to 75 km (46.5 miles). You are looking at the Andean Volcanic Arc in Ecuador, a consequence of subduction of Pacific Ocean floor beneath South America. Here, you see four volcanoes (Image G9.4). When this image was taken, Tungurahua was erupting, producing a plume of ash that winds blew to the southwest. The largest volcano in view is Chimborazo, the snow-covered peak at the western edge of the image.

G9.4

G9.5

Mt. Etna, Sicily
(Lat 37°45'5.49"N, Long 14°59'38.41"E)

At these coordinates, from an elevation of 50 km (31 miles), you can see Mt. Etna, which dominates the landscape of eastern Sicily (Image G9.5). This volcano, erupts fairly frequently—in this image small clouds of volcanic smoke rise from the 3200 m-high summit. Note the numerous lava flows on its flanks. Zoom to 5 km (3 miles), rotate the image, and tilt it so you are looking south. You can see calderas on the summit (Image G9.6).

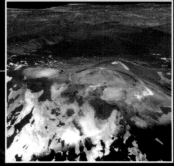

G9.6

G9.7

Mt. Vesuvius, Italy
(Lat 40°49'18.24"N, Long 14°25'32.04"E)

Eruption of Mt. Vesuvius in 79 C.E. destroyed Herculaneum and Pompeii. Fly to the coordinates provided and you'll be hovering over the volcano's crater. Zoom to 50 km (31 miles), and you can see the entire bay of Naples—several volcanic calderas lie on the west side of Naples. In fact, the entire bay is a caldera. Zoom to 15 km (9 miles), tilt the view, and fly around Vesuvius (Image G9.7). The central peak of the volcano lies within a larger caldera (Somma) that formed 17,000 years ago.

G9.8

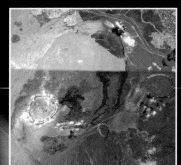

G9.9

Hawaiian Volcanoes
(Lat 19°28'20.47"N, Long 155°35'32.82"W)

At these coordinates, you'll be above the caldera of Mauna Loa, one of the shield volcanoes on Hawaii. From 25 km (15.5 miles) , you can see that the caldera is elliptical, and that other, smaller calderas occur to the SW. Tilt the image, and you can see a large fissure cutting the frozen lava lake in the Caldera. Distinctive lava flows spilled out of the ends of the caldera (Image G9.8). Fly about 33 km (20.5 miles) ESE to find the caldera of Kilauea (Image G9.9). From here, fly 41 km (25.5 miles) to the NNE to cross Mauna Kea, home to an observatory (white buildings).

Mt. Shishildan, Alaska
(Lat 54°45'36.59"N, Long 163°58'28.65"W)

Unimak Island, in the Aleutian chain, hosts a large stratovolcano, Mt. Shishildan. From an elevation of 35 km (22 miles), you'll note that the snowline starts halfway up the mountain, but that the peak itself is black (Image G9.10). That's because the volcano is active, and recent eruptions have buried and melted snow at the peak. From an elevation of 10 km (6 miles), you can see a red glow at the peak. Note that another, smaller volcano lies to the east.

G9.10

G9.11

East African Rift (Two Locations)

To understand volcanism in the East African rift, it is necessary to visit several locations. Here we provide two: (1) Mt. Kilimanjaro (Lat 3°3'53.63"S, Long 37°21'31.02"E): From a height of 10 km, you can see the caldera at the top of Africa's highest volcano (Image G9.11). The glaciers on the summit have been shrinking rapidly and may be gone in 20 years. Slumping has produced steep cliffs. Fly 45 km NE to find a long chain of cinder cones marking eruptions along a fault in the rift. (2) Goma region (Lat 1°39'27.40"S Long 29°14'15.27"E): At these coordinates, zoom to 80 km (50 miles), and tilt your view to look north (Image G9.12). The chain of lakes marks the western arm of the East African Rift. Active volcanoes lie just north of the city of Goma. Zoom to 3 km (2 miles) and tilt the image to look north—note the lava flow covering a portion of the airport runway and the city (Image G9.13).

G9.12

G9.13

Images provided by *Google Earth*™ mapping service/DigitalGlobe, TerraMetrics, NASA, Europa Technologies—copyright 2008.

geologists consider Hawaii's Kilauea to be active, for it currently erupts and has erupted frequently during recorded history. Similarly, Pompeii's Mt. Vesuvius, which has erupted over fifty times in the last two millennia, most recently in 1944, is an active volcano even though it currently emits no cloud of gas and ash. By this definition, Mt. Rainier in the Cascades last erupted centuries to millennia ago, but since subduction continues along the western edge of Oregon and Washington, the volcano could erupt in the future, and so it is considered active. Yellowstone Park, in contrast, is dormant, since the last eruptions were more than 10,000 years ago. Devil's Tower, in Wyoming, is the remnant of a volcano that was active millions of years ago but is now extinct, for the geologic cause of volcanism in the area no longer exists.

Clearly, when we know whether a volcano is active, dormant, or extinct, we have a basis for determining the hazard that the volcano represents. Active volcanoes are clearly an immediate threat. Dormant volcanoes are also likely to be a threat, but on a longer time scale. Extinct volcanoes are no threat at all. Commonly, the amount of erosion that affects a volcano provides a key to its classification. Active volcanoes display eruptive landforms and typically have a coating of recently erupted lava or pyroclastic debris that is free of vegetation or weathering. Dormant volcanoes have been dissected by erosion, and may be covered by lush vegetation, but nevertheless they still look like volcanoes. Extinct volcanoes have been eroded so much that they no longer look like volcanoes.

Predicting Eruptions

Little can be done to predict an eruption at a given volcano beyond a few months or years, except to define the repose time. But short-term (weeks to months) predictions of impending volcanic activity, unlike short-term predictions of earthquakes, are actually feasible. Some volcanoes send out distinct warning signals announcing that an eruption may take place very soon, for as magma squeezes into the magma chamber, it causes a number of changes that geologists can measure.

- *Earthquake activity:* Movement of magma generates vibrations in the Earth. And when magma flows into a volcano, rocks surrounding the magma chamber crack, and blocks slip with respect to each other. Such cracking and shifting also causes earthquakes. Thus, in the days or weeks preceding an earthquake, the region between 1 and 7 km beneath a volcano becomes seismically active. Earthquakes are the most reliable indicator of an impending eruption.

- *Changes in heat flow:* The presence of hot magma increases the local heat flow, the amount of heat passing through rock. In some cases, the increase in the heat flow melts snow or ice on the volcano, triggering floods and lahars even before an eruption occurs.

- *Changes in shape:* As magma fills the magma chamber inside a volcano, it pushes outward and can cause the surface of the volcano to bulge; the same effect happens when you blow into a balloon. Geologists now use laser sighting, accurate tiltmeters, surveys using global positioning satellites (GPS), and a technique called satellite interferometry (which uses radar beams from satellites to measure distance) to detect changes in a volcano's shape as magma rises.

- *Increases in gas emission and steam:* Even though magma remains below the surface, gases bubbling out of the magma, or steam formed by the heating of groundwater by the volcano, percolate upward through cracks in the Earth and rise from the volcanic vent. So an increase in the volume of gas emission, or of new hot springs, indicates that magma has entered the ground below.

Because geologists can determine when magma has moved into the magma chamber of a volcano, government agencies now send monitoring teams to a volcano at the first sign of activity. These teams set up instruments to record earthquakes, measure the heat flow, determine changes in the volcano's shape, and analyze emissions. In the case of Mt. St. Helens, the results are posted daily on the web. Sometimes, the monitoring comes to naught because the magma freezes in the magma chamber without ever erupting. But in other cases, the work becomes dangerous, and over the years volcanologists have been killed by the eruptions they were trying to observe. Such tragedies happen because although monitoring can yield a prediction that an eruption is imminent, it usually cannot pinpoint the exact time or eruptive style.

Controlling Volcanic Hazards

Danger assessment maps. Let's say that a given volcano has the potential to erupt in the near future. What can we do to prevent the loss of life and property? Since we can't prevent the eruption, the first and most effective precaution is to define the regions that can be directly affected by the eruption—to compile a volcanic hazard-assessment map (▶Fig. 9.24). These maps delineate areas that lie in the path of potential lava flows, lahars, debris flows, or pyroclastic flows.

River valleys initiating on the flanks of a volcano are particularly dangerous areas, because lahars may flow down them. Before the 1991 eruption of Mt. Pinautubo in the Philippines, geologists had defined areas potentially in the path of pyroclastic flows, and had predicted which river valleys were likely hosts for lahars. Athough the predicted pyroclastic-flow paths proved to be accurate, the region actually affected by lahars was much

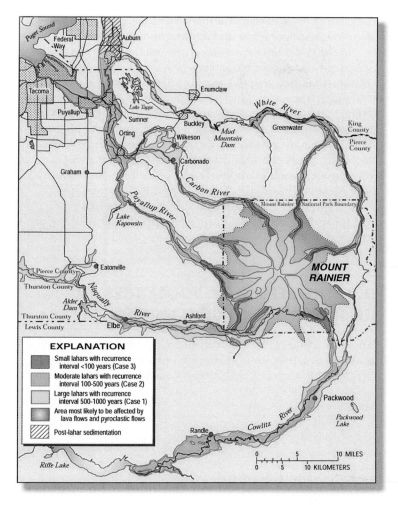

FIGURE 9.24 A volcanic hazard-assessment map for the Mt. Rainier area in Washington (courtesy of the U.S. Geological Survey). The different colors on the map indicate different kinds of hazards. Note that lahars can travel long distances down river valleys—some may threaten the city of Tacoma.

greater. Nevertheless, many lives were saved by evacuating people in areas thought to be under threat.

Evacuation. Unfortunately, because of the uncertainty of prediction, the decision about whether or not to evacuate is a hard one. In the case of Mt. St. Helens, hundreds of lives were saved in 1980 by timely evacuation, but in the case of Mt. Pelée, thousands of lives were lost because warning signs were ignored. In 1976, debate over the need for an evacuation around a volcano on Guadeloupe, in the French West Indies, was fierce. Eventually, the population of a threatened town was evacuated, but as months passed, the volcano did not erupt. Instead, tempers did, and anger at the cost of the evacuation translated into lawsuits.

Diverting flows. In traditional cultures, people believed that gods or goddesses controlled volcanic eruptions, so when a

volcano rumbled, they provided offerings to appease the deity. Sometimes, people have used direct force to change the direction of a flow or even to stop it. For example, during a 1669 eruption of Mt. Etna, a lively volcano on the Italian island of Sicily, basaltic lava formed a glowing orange river that began to spill down the side of the mountain. When the flow approached the town of Catania, 16 km from the summit, fifty townspeople protected by wet cowhides boldly hacked through the chilled side of the flow to create an opening through which the lava could exit. They hoped thereby to cut off the supply of lava feeding the end of the flow, near their homes. Their strategy worked, and the flow began to ooze through the new hole in its side. But unfortunately, the diverted flow began to move toward the neighboring town of Paterno. Five hundred men of Paterno then chased away the Catanians so that the hole would not be kept open, and eventually the flow swallowed part of Catania.

More recently, people have used high explosives to blast breaches in the flanks of flows, and have built dams and channels to divert flows. Major efforts to divert flows from a 1983 eruption of Mt. Etna, and again in 1992, were successful. Inhabitants of Iceland used a particularly creative approach in 1973 to stop a flow before it overran a town: they sprayed cold seawater onto the flow to freeze it in its tracks (▶Fig. 9.25). The flow did stop short of the town, but whether this was a consequence of the cold shower it received remains unknown.

Volcanoes in the Landscape

Why do volcanoes look the way they do? First of all, the shape of a volcano depends on whether it has been erupting recently or ceased erupting long ago. For erupting volcanoes,

FIGURE 9.25 As a basalt flow encroached on this town in Iceland, firefighters used forty-three pumps to dump over 6 million cubic m of seawater on the lava to freeze it and stop the flow.

the shape (shield, stratovolcano, or cinder cone) depends primarily on the eruptive style, because at an erupting volcano, the process of construction happens faster than the process of erosion. For example, in southern Mexico, the volcano Paracutín began to spatter out of a cornfield on February 20, 1943. Its eruption continued for nine years, and by the end, 2 cubic km of tephra had piled up into a cone almost half a kilometer high. But once a volcano stops erupting, erosion attacks. The rate at which a volcano erodes depends on whether it's composed of pyroclastic debris or lava. Cinder cones and ash piles can wash away quickly. For example, in the summer of 1831, a cinder cone grew 60 m above the surface of the Mediterranean Sea. As soon as the island appeared, Italy, Britain, and Spain laid claim to it, and shortly the island had at least seven different names. But the volcano stopped erupting, and within six months it was gone, fortunately before a battle for its ownership had begun. In contrast, stratovolcanoes or shield volcanoes, which have been armor-plated by lava flows, can withstand the attack of water and ice for quite some time.

In the end, however, erosion wins out, and you can tell a dormant volcano that has not erupted for a long time from a volcano that has erupted recently by the extent to which landslides, rivers, or glaciers have carved into its flanks (▶Fig. 9.26a–b). When a volcano becomes extinct, its softer exterior completely erodes away, leaving behind the plug of harder frozen magma that once lay just beneath the volcano, as well as the network of dikes that radiate from this plug. You can see a good example of such a landform at Shiprock, New Mexico (Fig. 6.11b). Devil's Tower in Wyoming (▶Fig. 9.26c) represents the consequence of even more erosion—its rock probably formed as an intrusion in the crust below a volcano.

Take-Home Message

Volcanoes don't erupt continuously and don't last forever, so we distinguish among active, dormant, and extinct volcanoes. It is possible to predict eruptions and take precautions. Once a volcano ceases to erupt, erosion destroys its eruptive shape.

9.9 THE EFFECT OF VOLCANOES ON CLIMATE AND CIVILIZATION

The consequences of volcanism go far beyond the mere building of a mountain. Eruptions may effect climate and perhaps the course of civilization. Let's see how.

FIGURE 9.26 **(a)** The shape of an active volcano is defined by the surface of the most recent lava flow or ash fall. Little erosion affects the surface. **(b)** An inactive volcano that has been around long enough for the surface to be modified by erosion. In humid climates, these volcanoes have gullies carved into their flanks and may be partially covered with forest. **(c)** Devil's Tower, Wyoming, rises 260 m above the surrounding land surface. It formed when a mass of magma solidified in the crust beneath a volcano, about 40 million years ago. Huge columnar joints, 2.5 m wide at the base, developed when the magma cooled. In Native American legend, the ribbed surface of Devil's Tower represents the claw marks of a giant bear, trying to reach a woman who sought refuge on the Tower's summit.

Time 1

(a)

Time 2

(b)

(c)

Volcanoes and Climate

In 1783, Benjamin Franklin was living in Europe, serving as the American ambassador to France. The summer of that year seemed to be unusually cool and hazy. Franklin, who was an accomplished scientist as well as a statesman, couldn't resist seeking an explanation for this phenomenon, and soon learned that, in June of 1783, a huge volcanic eruption had taken place in Iceland. He wondered if the "smoke" from the eruption had prevented sunlight from reaching the Earth, thus causing the cooler temperatures. Franklin reported this idea at a meeting, and by doing so, may well have been the first scientist ever to suggest a link between eruptions and climate.

Franklin's idea seemed to be confirmed in 1815, when Mt. Tambora in Indonesia exploded. Tambora's explosion ejected over 100 cubic km of ash and pumice into the air (compared with 1 cubic km from Mt. St. Helens). Ten thousand people were killed by the eruption and the associated tsunami. Another 82,000 died of starvation. The sky became so hazy that stars dimmed by a full magnitude. Temperatures dipped so low in the Northern Hemisphere that 1816 became known as "the year without a summer." The unusual weather of that year left a permanent impact on Western culture. Memories of fabulous sunsets and the hazy glow of the sky may have inspired the luminous and atmospheric quality that made the landscape paintings of the English artist J. M. W. Turner so famous (▶Fig. 9.27). Two English writers also documented the phenomenon. Lord Byron's 1816 poem "Darkness" contains the gloomy lines "The bright Sun was extinguish'd, and the stars / Did wander darkling in the eternal space . . . Morn came and went—and came, and brought no day." Two years later, Mary Shelley, trapped in her house by bad weather, wrote *Frankenstein*, with its numerous scenes of gloom and doom.

Geoscientists have witnessed other examples of eruption-triggered coolness more recently. In the months following the 1883 eruption of Krakatau and the 1991 eruption of Pinautubo, global temperatures dipped.

Classical literature provides more evidence of the volcanic impact on climate. For example, Plutarch wrote around 100 C.E., "Among events of divine ordering there was . . . after Caesar's murder . . . the obscuration of the Sun's rays. For during all the year its orb rose pale and without radiance . . . and the fruits, imperfect and half ripe, withered away." Similar conditions appear to have occurred in China the same year, as described in records from the Han dynasty, and may have been a consequence of volcanic eruption.

To study the effect of volcanic activity on climate even further in the past, geologists have studied ice from the glaciers of Greenland and Antarctica. Glacial ice has layers, each of which represents the snow that fell in a single year. Some layers contain concentrations of sulfuric acid, formed when SO_2 from volcanic gas dissolves in the water from

FIGURE 9.27 The glowing sunset depicted in this 1840 painting by the English artist J. M. W. Turner was typical in the years following the 1815 eruption of Mt. Tambora in Indonesia.

which snow forms. These layers indicate years in which major eruptions occurred. Years in which ice contains acid correspond to years during which the thinness of tree rings elsewhere in the world indicates a cool growing season.

How can a volcanic eruption create these cooling effects? When a large explosive eruption takes place, fine ash and aerosols enter the stratosphere. It takes only about two weeks for the ash and aerosols to circle the planet. They stay suspended in the stratosphere for many months to years, because they are above the weather and do not get washed away by rainfall. The haze they produce causes cooler average temperatures, because it absorbs incoming visible solar radiation during the day but does not absorb the infrared radiation that rises from the Earth's surface at night. A Krakatau-scale eruption can lead to a drop in global average temperature of about 0.3° to 1°C. According to some calculations, a series of large eruptions over a short period of time could cause a global average temperature drop of 6°C.

The observed effect of volcanic eruptions on the climate provides a model with which to predict the consequences of a nuclear war. Researchers have speculated that so much dust and gas would be blown into the sky in the mushroom clouds of nuclear explosions that a "nuclear winter" would ensue.

Volcanoes and Civilization

Not all volcanic activity is bad. Over time, volcanic activity has played a major role in making the Earth a habitable planet. Eruptions and underlying igneous intrusions have produced the rock making up the Earth's crust, and gases emitted by volcanoes provided the raw materials from which the atmosphere and oceans formed. The black smokers surrounding vents along mid-ocean ridges may have served as a birthplace for life, and volcanic islands in the oceans have hosted populations whose evolution adds to the diversity of life on the planet. Volcanic activity continues to bring nutrients (potassium, sulfur, calcium, and phosphorus) from Earth's interior to the surface, and provides fertile soils that nurture plant growth. And in more recent times, people have exploited the mineral and energy resources generated by volcanic eruptions.

Volcanoes and people have lived in close association since the first human-like ancestors walked the Earth 3 million years ago. In fact, one of the earliest relics of human ancestors consists of footprints fossilized in a volcanic ash layer in East Africa. Since volcanic ash contains abundant nutrients that make crops prosper, people tend to populate volcanic regions. It's amazing how soon after an eruption a volcanic soil in a humid climate sprouts a cloak of green. Only twenty years after the eruption of Mt. St. Helens, new plants covered much of the affected area.

But as we have seen, volcanic eruptions also pose a hazard. Eruptions may even lead to the demise of civilizations. The history of the Minoan people, who inhabited several islands in the eastern Mediterranean during the Bronze Age, illustrates this possibility. Beginning around 3000 B.C.E., the Minoans built elaborate cities and prospered. Then their civilization waned and disappeared (▶Fig. 9.28a). Geologists have discovered that the disappearance of the Minoans came within 150 years of a series of explosive eruptions of the Santorini volcano in 1645 B.C.E. Remnants of the volcano now constitute Thera, one of the islands of Greece. After a huge eruption, the center of the volcano collapsed into the sea, leaving only a steep-walled caldera (▶Fig. 9.28b). Archaeologists speculate that pyroclastic debris from the eruptions periodically darkened the sky, burying Minoan settlements and destroying crops. In addition, related earthquakes crumbled homes, and large tsunamis generated by the eruptions washed away Minoan seaports. Perhaps the Minoans took these calamities as a sign of the gods' displeasure, became demoralized, and left the region. Or perhaps trade was disrupted, and bad times led to political unrest. Eventually the Mycenaeans moved in, bringing the culture that evolved into that of classical Greece. The Minoans, though, were not completely forgotten. Plato, in his Dialogues, refers to a lost city, home of an advanced civilization that bore many similarities to that of the Minoans. According to Plato, this city, which he named Atlantis, disappeared beneath the waves of the sea. Perhaps this legend evolved from the true history of the Minoans, as modified by Egyptian scholars who passed it on to Plato.

Numerous cultures living along the Pacific Ring of Fire have evolved religious practices that are based on volcanic

> ### Take-Home Message
>
> The ash, gases, and aerosols produced by an explosive eruption can be blown around the globe, and this material can cause significant global cooling. Climatic effects, as well as other consequences of eruptions, may have hastened the end of some civilizations.

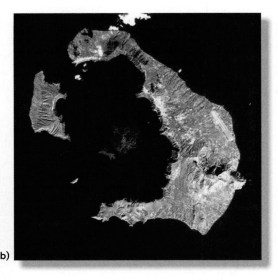

(b)

FIGURE 9.28 **(a)** Archaeologists have uncovered Minoan cities in the eastern Mediterranean, remnants of a culture that disappeared before the rise of classical Greece. **(b)** The cataclysmic eruption of the Santorini volcano in about 1645 B.C.E. may have contributed to the demise of the Minoan culture. All that is left of Santorini is a huge caldera whose rim still lies above sea level, forming the island of Thera. The ring of islands is about 20 km in diameter, as can be seen in this view from space.

(a)

activity—no surprise, considering the awesome might of a volcanic eruption in comparison with the power of humans. In some cultures, this reverence took the form of sacrifice in hopes of preventing an eruption that could destroy villages and bury food supplies. In traditional Hawaiian culture, Pelé, goddess of the volcano, created all the major landforms of the Hawaiian Islands. She gouged out the craters that top the volcano mountains, her fits and moods bring about the eruptions, and her tears are the smooth, glassy lapilli ejected from the lava fountains.

9.10 VOLCANOES ON OTHER PLANETS

We conclude this chapter by looking beyond the Earth, for our planet is not the only one in the Solar System to have hosted volcanic eruptions. We can see the effects of volcanic activity on our nearest neighbor, the Moon, just by looking up on a clear night. The broad, darker areas of the Moon, the **maria** (singular "mare," after the Latin word for seas), consist of flood basalts that erupted over 3 billion years ago (▶Fig. 9.29). They cover 17% of the lunar surface and occur only on the near side.

On Venus, about 22,000 volcanic edifices have been identified (▶Fig. 9.30a). Some of these even have caldera structures at their crests. Though no volcanoes currently erupt on Mars, the planet's surface displays a record of a spectacular volcanic past. The largest known mountain in the Solar System, Olympus Mons (▶Fig. 9.30b), is an extinct shield volcano on Mars. The base of Olympus Mons is 600 km across, and its peak rises 25 km above the surrounding plains, making it three times as high as Mt. Everest. A caldera 65 km in diameter developed on its summit.

Active volcanism currently occurs on Io, one of the many moons of Jupiter. Cameras in the *Galileo* spacecraft have recorded huge volcanoes on Io in the act of spraying plumes of sulfur gas into space (▶Fig. 9.30c) and have tracked immense, moving lava flows. Different colors of erupted material make the surface of this moon resemble a pizza. What causes the heat that produces all the melt? Researchers have proposed that the volcanic activity is due to tidal power: Io moves in an elliptical orbit around the huge mass of Jupiter and near Jupiter's other, larger moons. The gravitational pull exerted by these objects alternately stretches and then squeezes Io, creating sufficient friction to keep Io's mantle hot.

> ### Take-Home Message
>
> Space exploration reveals that volcanism not only occurs on Earth, but has also left its mark on other terrestrial planets. Satellites have detected active eruptions on the icy moons of Jupiter and Saturn, but these do not produce silicate lava.

FIGURE 9.29 The maria of the Moon, the broad dark areas, are composed of flood basalts.

The *Cassini* space craft has detected geysers of water vapor mixed with ice particles and other gases spewing from large cracks at the south pole of Enceladus, a 500-km-diameter moon of Saturn. These eruptions deposit ice along the edges of cracks (▶Fig. 9.30d).

Chapter Summary

- Volcanoes are vents at which molten rock (lava), pyroclastic debris (ash, pumice, and fragments of volcanic rock), gas, and aerosols erupt at the Earth's surface. A hill or mountain created from the products of an eruption is also called a volcano.

- The characteristics of a lava flow depend on its viscosity, which in turn depends on its temperature and composition. Rhyolitic lavas tend to be more viscous than basaltic lavas.

- Basaltic lavas can flow great distances. Pahoehoe flows have smooth, ropy surfaces, whereas a'a' flows have rough, rubbly surfaces. Andesitic and rhyolitic lava flows tend to pile into mounds at the vent.

- Pyroclastic debris includes powder-sized ash, marble-sized lapilli, and apple- to refrigerator-sized blocks. Some debris falls from the air, whereas some forms glowing avalanches that rush down the side of the volcano.

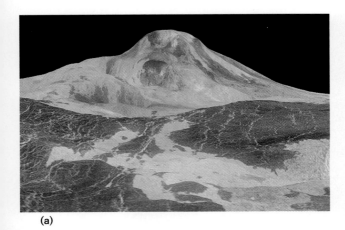

(a)

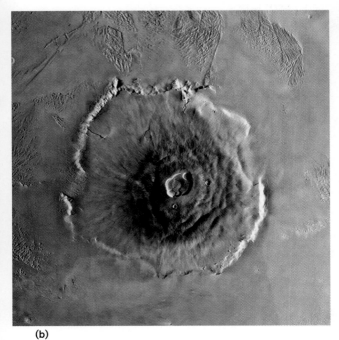

(b)

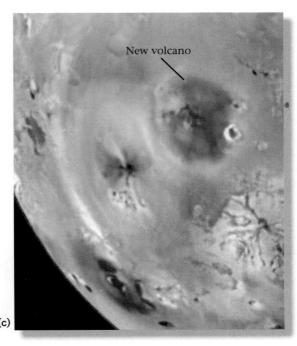

New volcano

(c)

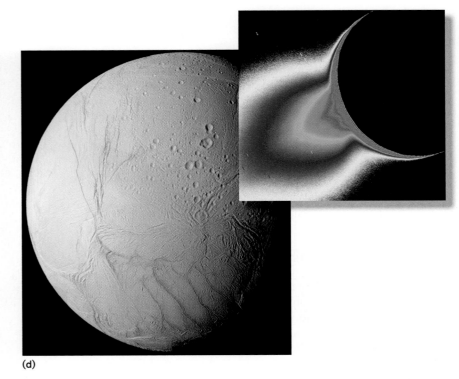

(d)

FIGURE 9.30 **(a)** This is a 3-D radar image, produced by the *Magellan* spacecraft, of Maat Mons, an 8-km-high volcano on Venus. The volcano produced lava flows hundreds of kilometers long. **(b)** Satellites orbiting Mars have provided this digital image of Olympus Mons, an immense shield volcano. Notice the caldera at the summit. **(c)** A satellite image caught a volcano on Io, one of the moons of Jupiter, in the act of erupting. The bluish bubble is a cloud of erupting gas. **(d)** Enceladus erupts water vapor and other gases. These deposit blue ice along the edges of cracks. From the side, false-color imagery shows the eruptions of Enceladus (inset).

- Eruptions may occur at a volcano's summit or from fissures on its flanks. The summit of an erupting volcano may collapse to form a bowl-shaped depression called a caldera.

- A volcano's shape depends on the type of eruption. Shield volcanoes are broad, gentle domes. Cinder cones are steep-sided, symmetrical hills composed of tephra. Stratovolcanoes can become quite large, and consist of alternating layers of pyroclastic debris and lava.

- The type of eruption depends on several factors, including the lava's viscosity and gas content. Effusive eruptions produce only flows of lava, whereas explosive eruptions produce clouds and flows of pyroclastic debris.

- Different kinds of volcanoes form in different plate-tectonic settings.

- Volcanic eruptions pose many hazards: lava flows overrun roads and towns, ash falls blanket the landscape, pyroclastic flows incinerate towns and fields, landslides and lahars bury the land surface, earthquakes topple structures and rupture dams, tsunamis wash away coastal towns, and invisible gases suffocate nearby people and animals.

- Eruptions can be predicted through changes in heat flow, changes in the shape of the volcano, earthquake activity, and the emission of gas and steam.

- We can minimize the consequences of an eruption by avoiding construction in danger zones and by drawing up evacuation plans. In a few cases, it may be possible to divert flows.

- Volcanic gases and ash, erupted into the stratosphere, may keep the Earth from receiving solar radiation and thus may affect climate. Eruptions bring nutrients from inside the Earth to the surface, and eruptive products may evolve into fertile soils.

- Immense maria of flood basalts cover portions of the Moon. The largest known volcano in the Solar System, Olympus Mons, towers over the surface of Mars. Satellites have documented eruptions on Io, a moon of Jupiter, and on Enceladus, a moon of Saturn.

Geopuzzle Revisited

Earth's volcanoes develop because there are places in the upper mantle and crust where partial melting takes place and magma forms. Though some magma freezes underground, some rises through conduits to the surface and erupts as lava or pyroclastic debris. The special places where melting takes place can be understood in the context of plate tectonics theory. Volcanism occurs at divergent and convergent boundaries, in rifts, and at hot spots. Not all eruptions are the same, in part because not all lava has the same composition, and in part because of local circumstances (for example, the presence of water). Some eruptions spew out lava that flows in fast-moving streams, whereas others end in a cataclysmic explosion that blankets the countryside in ash.

Key Terms

a'a' (p. 269)
active volcanoes (p. 291)
aerosols (p. 274)
blocks (p. 271)
bombs (p. 271)
caldera (p. 275)
cinder cone (p. 275)
columnar jointing (p. 269)
crater (p. 275)
dormant volcanoes (p. 291)
effusive eruption (p. 276)
explosive (pyroclastic) eruption (p. 276)
extinct volcanoes (p. 291)
fissure (p. 275)
flood basalt (p. 286)
ignimbrite (p. 271)
lahar (p. 273)
lapilli (p. 271)
lava (p. 267)
lava dome (p. 269)
lava flows (p. 267)
lava tube (p. 268)
magma chamber (p. 274)
maria (p. 299)
pahoehoe (p. 269)
phreatomagmatic eruptions (p. 280)
pyroclastic debris (p. 269)
pyroclastic flow (p. 271)
Ring of Fire (p. 288)
shield volcano (p. 275)
stratovolcano (p. 275)
tephra (p. 271)
tuff (p. 271)
vesicles (p. 274)
volcanic ash (p. 269)
volcano (p. 267)
volcaniclastic deposits (p. 273)

Review Questions

1. Describe the three different kinds of material that can erupt from a volcano.

2. Describe different types of lava flows.

3. Describe the differences between a pyroclastic flow and a lahar.

4. How is a crater different from a caldera?

5. Describe the differences among shield volcanoes, stratovolcanoes, and cinder cones. How are these differences explained by the composition of their lavas and other factors?

6. Why do some volcanic eruptions consist mostly of lava flows, whereas others are explosive and have no flow?

7. Explain how viscosity, gas pressure, and the environment affect the eruptive style of a volcano.

8. Describe the activity in the mantle that leads to hot-spot eruptions.

9. How do continental rift eruptions form flood basalts?

10. Contrast an island volcanic arc with a continental volcanic arc.

11. Identify some of the major volcanic hazards, and explain how they develop.

12. How do scientists predict volcanic eruptions?

13. Explain how steps can be taken to protect people from the effects of eruptions.

14. How have volcanoes affected civilization?

15. Describe the nature of volcanism on the other planets and moons in the Solar System.

On Further Thought

1. Mt. Fuji is a 3.6-km-high stratovolcano in Japan formed as a consequence of subduction (see below). With *Google*

Earth™ you can reach the volcano at Lat 35° 21' 46.72'' N Long 138° 43' 49.38'' E. It contains volcanic rocks with a range of compositions, including some andesitic rocks. Why do andesites erupt at Mt. Fuji? Very little andesite occurs on the Marianas Islands, which are also subduction-related volcanoes. Why?

2. The Long Valley Caldera, near the Sierra Nevada Mountains, exploded about 700,000 years ago and produced an immense ash fall called the Bishop Tuff. About 30 km to the northwest lies Mono Lake, with an island in the middle and a string of craters extending south from its south shore (see below). Hot springs and tufa deposits can be found along the lake. You can see the lake on *Google Earth*™ at Lat 37° 59' 56.58'' N Long 119° 2' 18.20'' W. Explain the origin of Mono Lake. Do you think that it represents a volcanic hazard?

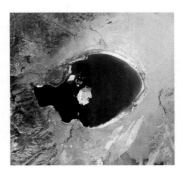

3. The city of Albuquerque lies along the Rio Grande River in New Mexico. Within the valley, numerous volcanic features crop out. Using *Google Earth*™ you can fly to Albuquerque and then along the river to find many examples. Many of the volcanoes are basaltic, but in places you will see huge caldera remnants. In fact, the city of Los Alamos lies atop thick ash deposits. What causes the volcanism in the Rio Grande Valley, and why are there different kinds of volcanism? Look at the photo of the volcanic cluster (see below).

These occur north of Santa Fe, at Lat 36° 45' 27.51'' N Long 105° 47' 24.85'' W—use *Google Earth*™ to get a closer view. Judging from the character of these volcanoes, would you say they are active? Why? How would you evaluate the volcanic hazard of this region? (*Hint*: Use the Web to find a map of seismicity for the Rio Grande Valley region, and think about its implications.)

Suggested Reading

Cattermole, P. 1996. *Planetary Volcanism*, 2nd ed. Chichester, England: John Wiley & Sons.

Chester, D. 1993. *Volcanoes and Society*. London: Edward Arnold.

De Boer, J. Z., and D. T. Sanders. 2001. *Volcanoes in Human History: The Far-Reaching Effects of Major Eruptions*. Princeton: Princeton University Press.

Decker, R. W., and B. B. Decker. 1997. *Volcanoes*. New York: W. H. Freeman.

Fisher, R. V., G. Heiken, and J. B. Hulen. 1997. *Volcanoes: Crucibles of Change*. Princeton: Princeton University Press.

McGuire, B., C. Kilburn, and J. Murray. 1995. *Monitoring Active Volcanoes*: London: UCL Press.

Pinna, M. 2002. Etna ignites. *National Geographic* (February): 68–87.

Scarth, A., 2004. *La Catastrophe: The Eruption of Mt. Pelée, the Worst Volcanic Disaster of the 20th Century*. Oxford: Oxford University Press.

Schminck, H. U. 2004. *Volcanism*. Berlin: Springer.

Sigurdsson, H., et al., eds. 2000. *Encyclopedia of Volcanoes*. San Diego: Academic Press.

Smith, R. B., and L. J. Siegel. 2000. *Windows into the Earth: The Geologic Story of Yellowstone and Grand Teton National Parks*: Oxford: Oxford University Press.

Winchester, S. 2003. *Krakatoa: The Day the World Exploded*. New York: Harper Collins.

A Violent Pulse: Earthquakes

In the tragic aftermath of an earthquake that rocked India, Pakistan, and Afghanistan in October of 2005, rescue workers struggle to free victims trapped beneath the rubble of this apartment building in Islamabad, Pakistan. When the ground shakes, walls may tumble.

Geopuzzle

Why do earthquakes happen where they do? Can people predict when an earthquake will occur?

We learn geology the morning after the earthquake.
—*Ralph Waldo Emerson (American poet, 1803–1882)*

10.1 INTRODUCTION

As the morning of January 17, 1994, approached, residents of Northridge, a suburb near Los Angeles, slept peacefully in anticipation of the Martin Luther King Day holiday. But beneath the quiet landscape, a disaster was in the making. For many years, the slow movement of the Pacific Plate relative to the North American Plate had been bending the rocks making up the California crust. But like a stick that you flex with your hands, rock can bend only so far before it snaps (▶Fig. 10.1a, b). Under California, the "snap" hap-

pened at 4:31 A.M., 10 km down. It sent pulses of energy racing through the crust at an average speed of 11,000 km (7,000 miles) per hour, 10 times the speed of sound.

When energy pulses, or shocks, reached the Earth's surface, the ground bucked up and down and swayed from side to side. Sleepers bounced off their beds, homes slipped off their foundations, and freeway bridges disconnected from their supports. As more and more shocks arrived, walls swayed and toppled, roofs collapsed, and rail lines buckled (▶Fig. 10.2). Early risers brewing coffee in their kitchens tumbled to the floor, under attack by dishes and cans catapulting out of cupboards. Trains careened off their tracks, and steep hill slopes bordering the coast gave way, dumping heaps of rock, mud, and broken houses onto the beach below. Ruptured gas lines fed fires that had been ignited in the rubble by water heaters and sparking wires. Then, forty seconds after it started, the motion stopped, and the shouts and sirens of rescuers replaced the crash and clatter of breaking masonry and glass. A strong **earthquake**—an episode of ground shaking—had occurred.

Earthquakes have affected the Earth since the formation of its solid crust. Most are a consequence of lithosphere-plate movement; they punctuate each step in the growth of mountains, the drift of continents, and the opening and closing of ocean basins. And, perhaps of more relevance to us, earthquakes have afflicted human civilization since the construction of the first village as they have directly caused the deaths of over 3.5 million people during the past two millennia (▶ Table 10.1). Ground shaking, giant waves, landslides, and fires associated with earthquakes turn cities to rubble. The destruction caused by some earthquakes may even have changed the course of civilization.

What does an earthquake feel like? When you're in one, time seems to stand still, so even though most earthquakes

FIGURE 10.1 Most earthquakes happen when rock in the ground first bends slightly and then suddenly snaps and breaks, like a stick you flex in your hands. **(a)** Before an earthquake, the crust bends (the amount of bending is greatly exaggerated here). **(b)** When the crust breaks, sliding suddenly occurs on a fault, generating vibrations.

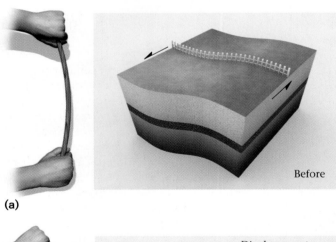

(a)

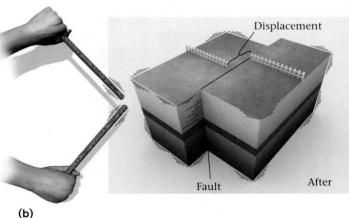

(b)

FIGURE 10.2 In the 1994 Northridge, California, earthquake, this building facade tore free of its supports and collapsed.

TABLE 10.1 Some Notable Earthquakes

Year	Location	Number of Deaths
2005	Pakistan	80,000
2004	Sumatra	230,000
2003	Bam, Iran	41,000
2001	Bhuj, India	20,000
1999	Calaraca/Armenia, Colombia	2,000
1999	Izmit, Turkey	17,000
1995	Kobe, Japan	5,500
1994	Northridge, California	51
1990	Western Iran	50,000
1989	Loma Prieta, California	65
1988	Spitak, Armenia	24,000
1985	Mexico City	9,500
1983	Turkey	1,300
1978	Iran	15,000
1976	T'ang-shan, China	255,000
1976	Caldiran, Turkey	8,000
1976	Guatemala	23,000
1972	Nicaragua	12,000
1971	Los Angeles	50
1970	Peru	66,000
1968	Iran	12,000
1964	Anchorage, Alaska	131
1963	Skopje, Yugoslavia	1,000
1962	Iran	12,000
1960	Agadir, Morocco	12,000
1960	Southern Chile	6,000
1948	Turkmenistan, USSR	110,000
1939	Erzincan, Turkey	40,000
1939	Chillan, Chile	30,000
1935	Quetta, Pakistan	60,000
1932	Gansu, China	70,000
1927	Tsinghai, China	200,000
1923	Tokyo, Japan	143,000
1920	Gansu, China	180,000
1915	Avezzano, Italy	30,000
1908	Messina, Italy	160,000
1906	San Francisco	500
1896	Japan	22,000
1886	Charleston, South Carolina	60
1866	Peru and Ecuador	25,000
1811–12	New Madrid, Missouri (3 events)	few
1783	Calabria, Italy	50,000
1755	Lisbon, Portugal	70,000
1556	Shen-shu, China	830,000

take less than a minute, they seem much longer. Because of the lurching, bouncing, and swaying of the ground and buildings, people become disoriented, panicked, and even seasick. Some people recall hearing a dull rumbling or a series of dull thumps, as well as crashing and clanging. Earthquakes may even shake dust into the air, creating a fine, fog-like mist.

Earthquakes are a fact of life on planet Earth: almost 1 million detectable earthquakes happen every year. Fortunately, most cause no damage or casualties, because they are too small or they occur in unpopulated areas. But a few hundred earthquakes per year rattle the ground sufficiently to damage buildings and injure their occupants, and every five to twenty years, on average, a great earthquake triggers a horrific calamity. What geologic phenomena generate earthquakes? Why do earthquakes take place where they do? How do they cause damage? Can we predict when earthquakes will happen, or even prevent them from happening? These questions have puzzled **seismologists** (from *seismos*, Greek for shock or earthquake), geoscientists who study earthquakes, for decades. In this chapter, we seek some of the answers, answers that can help those living in earthquake-prone regions to cope.

10.2 WHAT CAUSES EARTHQUAKES TO HAPPEN?

Ancient cultures offered a variety of explanations for **seismicity** (earthquake activity), most of which involved the action or mood of a giant animal or god. For example, in Japanese folklore, a giant catfish, Namazu, is said to have lived in the mud below the surface of the ground (▶Fig. 10.3). If the gods did not restrain him, he would thrash about and shake the ground. In Indian folklore, earthquakes happened when one of eight elephants holding up the Earth shook its head, and in Siberia, earthquakes were thought to take place when a dog hauling the Earth in a sled stopped to scratch. Native American cultures of the West Coast thought earthquakes were caused by arguments among the turtles holding up the Earth. Scientific studies conducted during the past 150 years show that seismicity can occur for several reasons, including

- the sudden formation of a new **fault** (a fracture on which sliding occurs),
- a sudden slip on an existing fault,
- a sudden change in the arrangement of atoms (i.e., a phase change) in the minerals comprising rock,
- movement of magma in a volcano,
- the explosion of a volcano,

FIGURE 10.3 This painting depicts the Japanese legend of Namazu, the giant catfish whose thrashings were thought to cause earthquakes.

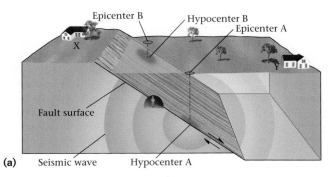

(a)

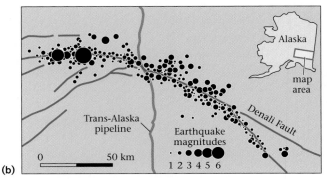

(b)

FIGURE 10.4 **(a)** The energy of an earthquake radiates from the hypocenter (focus), the place underground where rock has suddenly broken. The point on the ground surface (i.e., on a map) directly above the hypocenter is the epicenter. During a single earthquake, only part of a fault may slip. House X is on the footwall, and house Y is on the hanging wall. The miner excavating a tunnel along the fault has the hanging wall over his head and the footwall under his feet. **(b)** A simplified map of the Denali National Park region, Alaska, shows the trace of the Denali fault and epicenters of earthquake events along the fault during a period in late 2002. The diameter of a circle represents the size (magnitude) of the shock it represents.

- a giant landslide,
- a meteorite impact, or
- underground nuclear-bomb tests.

As we learned in Chapter 2, the place in the Earth where rock ruptures and slips, or the place where an explosion occurs, is the **hypocenter** (or **focus**) of the earthquake. Energy radiates from the hypocenter. The point on the surface of the Earth that lies directly above the hypocenter is the **epicenter** (▶Fig. 10.4).

The formation and movement of faults cause the vast majority of destructive earthquakes, so typically the hypocenter of an earthquake lies on a fault plane (the surface of the fault). Thus, we'll begin our investigation with a look at how faults develop and why their movement generates earthquakes.

Faults in the Crust

Faults are fractures on which slip or sliding occurs (see Chapter 11). They can be pictured as planes that cut through the crust. Some faults are vertical, but most slope at an angle. The nineteenth-century miners who encountered faults in mine tunnels referred to the rock mass above a sloping fault plane as the hanging wall, because it hung over their heads, and the rock mass below the fault plane as the footwall, because it lay beneath their feet (Fig. 10.4). The miners described the direction in which rock masses slipped on a fault by specifying the direction that the hanging wall moved in relation to the footwall, and we still use these terms today. When the hanging wall slips down the slope of the fault, it's a normal fault, and when the hang-

ing wall slips up the slope, it's a reverse fault if steep and a thrust fault if shallowly sloping (▶Fig. 10.5a–d). Strike-slip faults have near-vertical planes on which slip occurs parallel to an imaginary horizontal line, called a strike line, on the fault plane—no up or down motion takes place here (▶Fig. 10.5e). In Chapter 11, we will discuss these further and introduce other types of faults.

Normal faults form in response to stretching or extension of the crust, reverse or thrust faults develop in response to squeezing (compression) and shortening of the crust, and strike-slip faults form where one block of crust slides past another laterally. By measuring the distance between the two ends of a marker, such as distinctive sedimentary bed or igneous dike that's been offset by a fault, geologists define the **displacement**, the amount of slip, on the fault (▶Fig. 10.6a, b).

Faults are found almost everywhere—but don't panic! Not all of them are likely to be the source of earthquakes. Faults that have moved recently or are likely to move in the

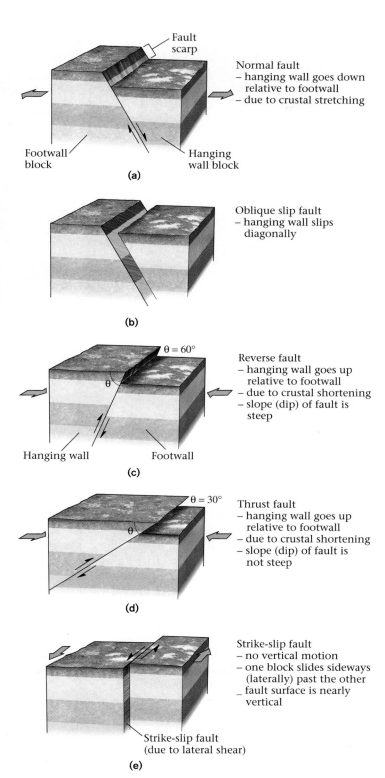

Fault scarp

Normal fault
– hanging wall goes down relative to footwall
– due to crustal stretching

Footwall block

Hanging wall block

(a)

Oblique slip fault
– hanging wall slips diagonally

(b)

θ = 60°

θ

Reverse fault
– hanging wall goes up relative to footwall
– due to crustal shortening
– slope (dip) of fault is steep

Hanging wall   Footwall

(c)

θ = 30°

θ

Thrust fault
– hanging wall goes up relative to footwall
– due to crustal shortening
– slope (dip) of fault is not steep

(d)

Strike-slip fault
– no vertical motion
– one block slides sideways (laterally) past the other
– fault surface is nearly vertical

Strike-slip fault (due to lateral shear)

(e)

FIGURE 10.5 The basic types of faults. Note that faults are distinguished from each other by the nature of the slip. **(a)** Normal fault. **(b)** Oblique-slip fault. **(c)** Reverse fault. **(d)** Thrust fault. **(e)** Strike-slip fault.

near future are called active faults (and if they generate earthquakes, news media sometimes refer to them as "earthquake faults"); faults that last moved in the distant past and probably won't move again in the near future (but are still recognizably faults because of the displacement across them) are called inactive faults. Some faults have been inactive for billions of years.

The intersection between a fault and the ground surface is a line we call the **fault trace**, or fault line (Fig. 10.6b). In places where an active normal or reverse fault intersects the ground, one side of the fault moves vertically with respect to the other side, creating a small step called a **fault scarp** (Fig. 10.5a). Active strike-slip faults tend to form narrow bands of low ridges and narrow depressions, because they break up the ground when they move (▶Fig. 10.6c; see art, pp. 310–311). Not all active faults, though, intersect the ground surface. Those that don't are called blind faults or hidden faults (▶Fig. 10.7). Many fault traces that we see on maps represent inactive faults. The portion of the fault that we now see may once have been far below the surface of the Earth, becoming visible only because overlying material has been eroded away.

Formation of Faults, Friction, and Stick-Slip

As we learned in Chapter 8, stress is the push, pull, or shear that a material feels when subjected to a force. In a drawing, we can represent stress with arrows showing the direction in which the stress acts. Objects can change shape in response to the application of a stress; this change in shape is referred to as a strain. (We will learn more about stress and strain in Chapter 11.)

What does stress have to do with fault formation? Stress causes faulting. To see why, imagine that you grip each side of a brick-shaped rock with a clamp (▶Fig. 10.8a–c). Now, suppose you apply a stress to the rock by moving one side upward and the other side downward. As soon as the movement begins, the rock begins to change shape (a line traced across the middle of the brick bends into an S-like curve), but it doesn't break, and if you were to remove the stress at this stage, the rock would return to its original shape, just as a stretched rubber band returns to its original shape when you let go. A change in the shape of an object that disappears when stress is removed is called an elastic strain. If you apply a larger stress, so that the sides of the rock shift farther, the rock starts to crack. First, a series of small cracks form, but as movement continues, the cracks grow and connect to each other to create a fracture that cuts across the entire block of rock. The instant this thoroughgoing fracture forms, the rock on one side of the fracture slides past the rock on the other side, and the fracture becomes a fault. And as soon as the fault forms, the once-bent line across the middle separates into

(a)

What a geologist sees
(b)

(c)

FIGURE 10.6 **(a)** This wooden fence was built across the San Andreas Fault. During the 1906 San Francisco earthquake, slip on the fault broke and offset the fence; the displacement of the fence indicates that the fault is strike-slip, as we see no evidence of up or down motion. The rancher quickly connected the two ends of the fence so no cattle could escape. **(b)** The amount the fence was offset indicates the displacement on the fault. **(c)** A photo taken looking down from a helicopter, showing the trace of the strike-slip fault that ruptured the ground surface during the Hector Mine earthquake in the southern California desert in 1999. Note the cracks in the ground and the small ridges and depressions, and how the fault offsets the dirt road.

two segments that no longer align with each other, and the stress in the rock decreases (i.e., there is a stress drop). Also, the elastic strain disappears.

If you slide a book across a tabletop, it eventually slows down and stops because of friction. Similarly, once a fault forms and rock starts to slip, it doesn't slip forever because of friction. Friction, the resistance to sliding on a surface, regulates movement. Friction occurs because, in reality, no surface can be perfectly smooth—rather, all surfaces con-

FIGURE 10.7 A hidden (or "blind") fault does not intersect the ground surface. Rather, it dies out at the fault tip. In this example, a fold (curving beds) has developed in response to slip on the fault in the region above the tip (end) of the fault.

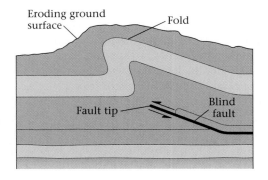

tain little bumps or protrusions. As one surface moves against another, the protrusions on one surface snag the protrusions of the opposing surface, acting like little anchors that slow down and eventually stop the movement (▶Fig. 10.9a, b). Once friction stops movement on a fault, the stress begins to increase again. Eventually, the magnitude of stress becomes so great that friction can no longer prevent movement. The instant this happens, the fault slips once again and the stress drops once again. The stress required to overcome friction and reactivate an existing fault tends to be less than the stress that's needed to fracture intact rock and form a new fault. In effect, once a fault has formed, it's like a permanent scar that is weaker than the surrounding crust. Thus, existing faults may reactivate many times.

To summarize, between faulting events, stress builds up. In some cases, the stress causes intact rock to rupture and a new fault to form. In other cases, stress overcomes friction on an existing fault and the fault slips again. In either case, after a faulting event, stress drops and elastic strain stored in rock decreases—the rock rebounds so layers near the fault are no longer bent. Friction stops the movement and the fault locks, until stress builds up enough again to cause slip. This overall image of how earthquakes occur is now known as **elastic-rebound theory**, and the start-stop movement on a fault as **stick-slip**

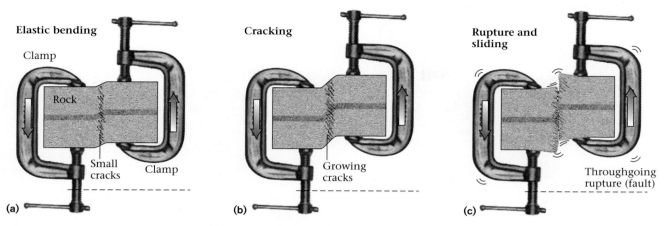

Elastic bending

Clamp

Rock

Small
cracks   Clamp

(a)

Cracking

Growing
cracks

(b)

**Rupture and
sliding**

Throughgoing
rupture (fault)

(c)

FIGURE 10.8 The stages in the development of a fault can be illustrated by breaking a rock block gripped at each end by a clamp. **(a)** As one clamp moves relative to the other (indicated by arrows), the rock feels a stress and begins to bend elastically and develop strain. **(b)** If bending continues, the rock begins to crack, and then the cracks grow and start to connect. **(c)** When the cracks connect sufficiently to make a throughgoing fracture, the rock ruptures into two pieces. The instant one piece slides past another, the rupture becomes a fault.

behavior (▶Fig. 10.10a–f). When a fault slips, the whole fault does not move at once; rather, the slipped area starts at a certain point and then grows outward (▶Fig. 10.11a). In some cases, the slip grows symmetrically from the epicenter. But sometimes the slip grows dominantly in one direction.

How Faulting Generates Earthquakes

How does the formation of a fault generate an earthquake? The moment a fault slips, the rock around the fault is subject to a sudden push or pull. Like a hammer blow, this movement sends pulses of energy (shocks) into the sur-

rounding rock. As the energy pulses pass, the rock moves back and forth like a vibrating bell. The series of shocks generated by the sudden slip on the fault and the subsequent vibration create the shaking we feel as an earthquake. The greater the amount of slip and the larger the amount of rock that moves, the greater the size of the vibrations and, therefore, the larger the earthquake.

A major earthquake may be preceded by smaller ones, called **foreshocks**, which possibly reflect the development of the smaller cracks that will eventually link up to form a major rupture. Small earthquakes that follow a major earthquake, called **aftershocks**, may occur for days to several weeks. The largest aftershock tends to be 10 times smaller than the main shock; most are much smaller. Aftershocks happen because the movement of rock during the main earthquake produces new points of contact. At these points, stresses may be large enough to reactivate small portions of the main fault or to activate small, nearby faults.

The Amount of Slip on Faults, and Ground Distortion due to Earthquakes

How much of a fault slips during an earthquake? The answer depends on the size of the earthquake: the larger the earthquake, the larger the slipped area. For example, the major earthquake that hit San Francisco, California, in 1906 ruptured a 430-km-long (measured parallel to the Earth's surface) by 15-km-high (measured perpendicular to the Earth's surface) segment of the San Andreas Fault. The maximum observed displacement was 7 m, in a strike-slip sense. Slip on a thrust fault caused the 1964 Good Friday earthquake in southern Alaska: at depth in the Earth, slip reached

FIGURE 10.9 On a microscopic scale, real surfaces have bumps and protrusions. **(a)** Before movement, the protrusions in rock lock together, causing friction that prevents sliding. One block is "anchored" to the other. **(b)** Like a boat whose anchor cable snaps, when the protrusions break off, the blocks can slide.

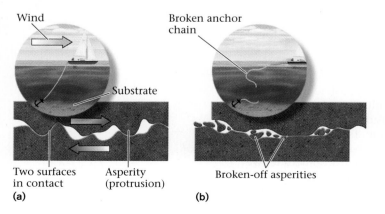

Wind

Broken anchor
chain

Substrate

Two surfaces
in contact   Asperity
(protrusion)

(a)

Broken-off asperities

(b)

Faulting in the Crust

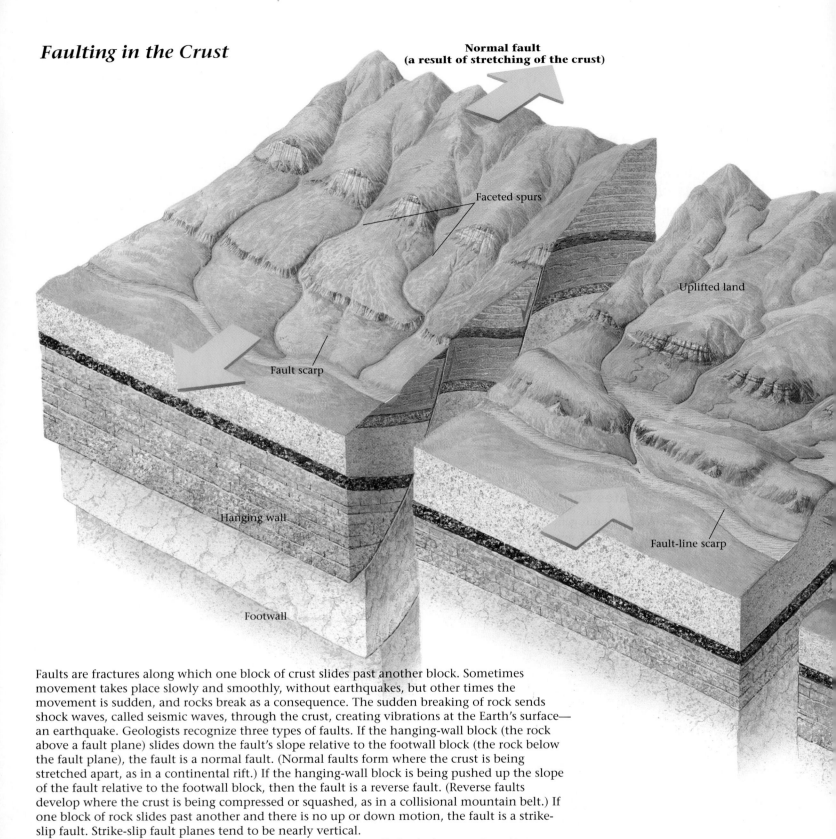

Normal fault (a result of stretching of the crust)

Faceted spurs

Uplifted land

Fault scarp

Hanging wall

Fault-line scarp

Footwall

Faults are fractures along which one block of crust slides past another block. Sometimes movement takes place slowly and smoothly, without earthquakes, but other times the movement is sudden, and rocks break as a consequence. The sudden breaking of rock sends shock waves, called seismic waves, through the crust, creating vibrations at the Earth's surface—an earthquake. Geologists recognize three types of faults. If the hanging-wall block (the rock above a fault plane) slides down the fault's slope relative to the footwall block (the rock below the fault plane), the fault is a normal fault. (Normal faults form where the crust is being stretched apart, as in a continental rift.) If the hanging-wall block is being pushed up the slope of the fault relative to the footwall block, then the fault is a reverse fault. (Reverse faults develop where the crust is being compressed or squashed, as in a collisional mountain belt.) If one block of rock slides past another and there is no up or down motion, the fault is a strike-slip fault. Strike-slip fault planes tend to be nearly vertical.

   If a fault displaces the ground surface, it creates a ledge called a fault scarp. Sometimes we can identify the trace (or line) of a fault on the land surface because the rock of the hanging wall has a different resistance to erosion than the rock of the footwall; a ledge formed along this line due to erosion is a fault-line scarp. Where fault scarps cut a system of rivers and valleys, the ridges are truncated to make triangular facets. Strike-slip faults may offset ridges, streams, and orchards sideways. If there is a slight extension along the fault, the land surface sinks, and a sag pond develops.

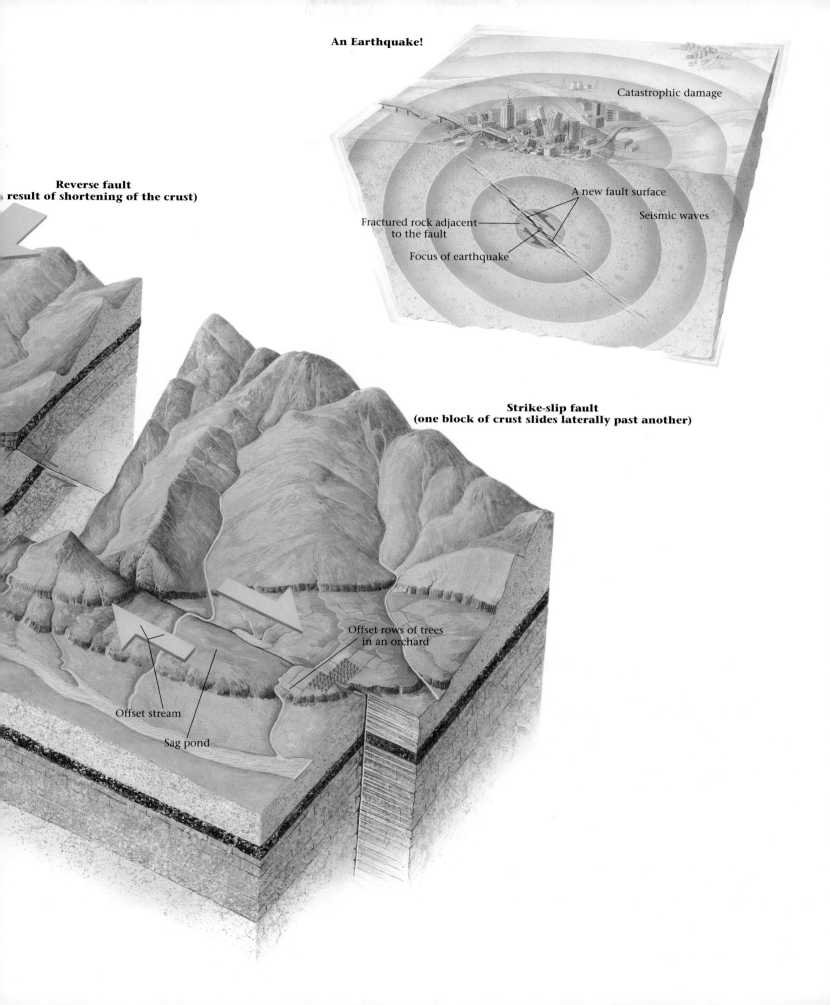

An Earthquake!

Catastrophic damage

A new fault surface

Fractured rock adjacent
to the fault

Seismic waves

Focus of earthquake

**Reverse fault
(a result of shortening of the crust)**

**Strike-slip fault
(one block of crust slides laterally past another)**

Offset rows of trees
in an orchard

Offset stream

Sag pond

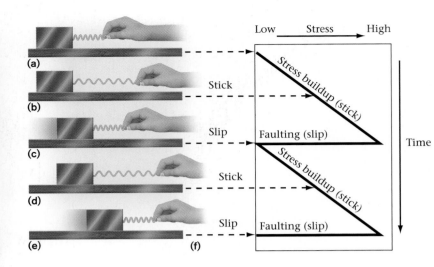

FIGURE 10.10 The concept of stick-slip behavior can be illustrated by means of a model consisting of a heavy block with a spring attached. **(a)** The block starts at rest. **(b)** We pull on the spring and the spring stretches, but friction keeps the block in place. Stretching the spring builds up stress. **(c)** Suddenly, the block slides along the tabletop (analogous to the slip on a fault), and the spring relaxes (analogous to a drop in stress). **(d)** We pull on the spring again, and the stress builds up once more. **(e)** The block slides again, and the stress drops. **(f)** The graph shows how stress gradually builds up and then suddenly drops during stick-slip behavior.

a maximum of 12 m, and at the Earth's surface, the faulting uplifted the ground over a 500,000-square-km area by as much as 2 m. Smaller earthquakes, like the one that hit Northridge in 1994, resulted in about 0.5 m slip on a break that was about 5 km long and 5 km wide. The smallest-felt earthquakes (which rattle the dishes but not much more) reflect displacements measured in millimeters to centimeters. Note that the greatest displacement is not necessarily the epicenter.

Because of the displacement that takes place on faults during an earthquake, the ground surface over the fault may undergo a change in shape. For example, slip on a thrust fault may cause a region to warp upward, even if the fault plane itself does not cut the ground surface. Recently, researchers have developed a new technique, called Interferometric Synthetic Aperture Radar (abbreviated

InSAR), to help detect subtle ground-surface distortion associated with earthquakes. To create an InSAR image, a satellite uses radar to make a precise map of ground elevation of a region at two different times (weeks to years apart). A computer compares the two images and detects differences in elevation as small as the wavelength of radar energy. A printout of the result portrays these differences as color bands that indicate the change in elevation between the time the first image was taken and the time the second image was taken (▶Fig. 10.11b, c). Each band represents a certain amount of change in elevation.

Although the cumulative movement on a fault during a human life span may not amount to much, over geologic time the cumulative movement becomes significant. For example, if earthquakes occurring on a reverse fault cause 1 cm of uplift over 10 years, on average, the fault's

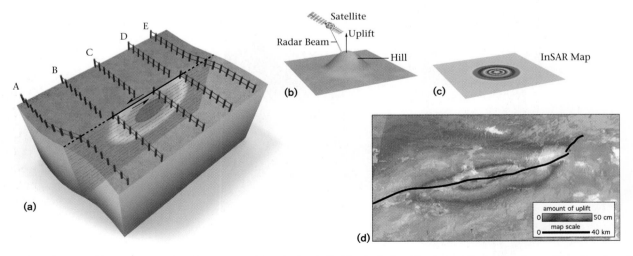

FIGURE 10.11 **(a)** During an earthquake on a preexisting fault, not all of the fault slips. Slip starts at the hypocenter, and then the slipped area grows outward; on a large fault, this growth takes tens of seconds. In this example, the slipped area on a strike-slip fault intersects the ground surface; fences beyond the end of this intersection have not been offset. **(b)** A satellite can use radar to map uplift of the ground related to faulting. **(c)** An InSAR map of the hill. Color bands can be thought of as contour lines. **(d)** An InSAR map of the area bordering a fault in Tibet.

movement will yield 1 km of uplift after 1 million years. Thus, earthquakes mark the incremental movements that create mountains. To take another example, movement on the San Andreas Fault in California averages around 6 cm per year. As a result, Los Angeles, which is to the west of the fault, will move northward by 6,000 km in 100 million years.

Can Faults Slip without Earthquakes?

When a material is subjected to stress cracks and fractures, we say that it has undergone brittle deformation. For example, a glass plate shattering on the floor is a type of brittle deformation. Similarly, faulting that generates earthquakes represents brittle deformation. Generally, rock must be fairly cool and must be stressed fairly quickly to behave in a brittle way. If a rock is warm or weak, or if stress builds slowly, it can bend and flow without breaking. We call such behavior ductile deformation. If the glass plate were heated to a high temperature, it could be bent in a ductile manner, like chewing gum. Ductile deformation does not cause earthquakes.

Because the temperature of the Earth increases with depth, most brittle deformation and, therefore, earthquake-generating faulting in continental crust occurs in the upper 15 to 20 km of the crust. At greater depths, shear and movement can take place, but they are caused by ductile deformation. In oceanic plates, earthquake faulting happens even in plates that have been subducted to depths of 670 km.

In some cases, movement on faults in the upper 15 to 20 km of the continental crust takes place slowly and steadily, without generating earthquakes. When movement on a fault happens without generating earthquakes, we call the movement **fault creep**. Seismologists do not completely understand fault creep, but speculate that it occurs in particularly weak rock, which can change shape without breaking or can slip smoothly without creating shock waves.

> **Take-Home Message**
>
> Most earthquakes happen due to the sudden rupture of rock accompanying the formation or reactivation of a fault. A hypocenter is the location in the Earth where an earthquake occurs, and an epicenter is the point on the surface of the Earth directly above. Friction stops slip on a fault. But stress then builds until it can trigger another slip event.

10.3 HOW DOES EARTHQUAKE ENERGY TRAVEL?

How does the energy emitted at the hypocenter of an earthquake travel to the surface or even pass through the entire Earth? Like other kinds of energy, earthquake energy travels in the form of waves. We call these waves **seismic waves** (or earthquake waves). You feel them if you hold one end of a brick and bang the other end with a hammer—the energy of the hammer blows travels to your fingertip in the form of waves.

Seismologists distinguish among different types of seismic waves on the basis of where and how the waves move. **Body waves** pass through the interior of the Earth (meaning within the body of the Earth), whereas **surface waves** travel along the Earth's surface. Waves in which particles of material move back and forth parallel to the direction in which the wave itself moves are called **compressional waves**. As a compressional wave passes, the material first compresses (or squeezes) together, then dilates (or expands). To see this kind of motion in action, push on the end of a spring and watch as the little pulse of compression moves along the length of the spring. Waves in which particles of material move back and forth perpendicular to the direction in which the wave itself moves are called **shear waves**. To see shear-wave motion, jerk the end of a rope up and down and watch how the up-and-down motion travels along the rope. With these concepts in mind, we can define four basic types of seismic waves (▶Fig. 10.12a–f):

- **P-waves** (P stands for primary) are compressional body waves.
- **S-waves** (S stands for secondary) are shear body waves.
- **R-waves** (R stands for Rayleigh, the name of a physicist) are surface waves that cause the ground to ripple up and down.
- **L-waves** (L stands for Love, the name of a seismologist) are surface waves that cause the ground to ripple back and forth in a snake-like movement.

P-waves travel the fastest and thus arrive first. S-waves travel more slowly, at about 60% of the speed of P-waves, so they arrive later. Surface waves (R- and L-waves) are the slowest of all.

Friction absorbs energy as waves pass through a material, and waves bounce off layers and obstacles in the Earth, so the amount of energy carried by seismic waves decreases the farther they travel. People near the epicenter of a large earthquake may be thrown off their feet, but those far away barely feel it. Similarly, an earthquake caused by slip on a fault deep in the crust causes less damage than one caused by slip on a fault near the surface.

> **Take-Home Message**
>
> Earthquake energy travels as seismic waves. Body waves (P-waves and S-waves) travel through the interior of the Earth, whereas surface waves travel along the surface. Ground shaking, due to arrival of waves, generally decreases with distance from the hypocenter.

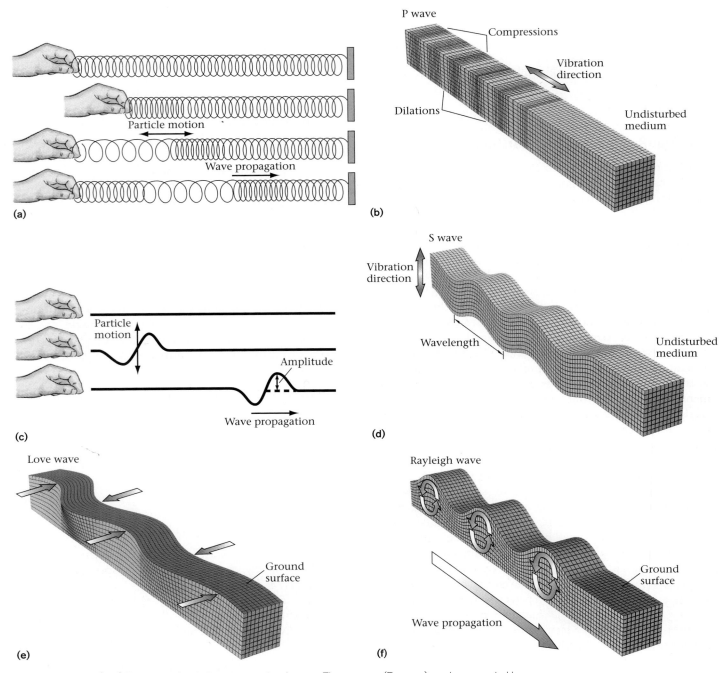

FIGURE 10.12 (a, b) Two ways of picturing compressional waves. These waves (P-waves) can be generated by pushing on the end of a spring. The pulse of energy compresses in sequence down the length of spring. Note that the back-and-forth motion of the coils occurs in the same direction the wave travels. The wavelength of P-waves is defined by the distance between successive pulses of compression. **(c, d)** Two ways of picturing shear waves. These waves (S-waves) resemble the waves in a rope. Note that the back-and-forth motion occurs in a direction perpendicular to the direction the wave travels. The wavelength of S-waves is defined as the distance between successive peaks (or troughs). **(e)** L-waves cause the surface of the ground to shear sideways, like a moving snake. **(f)** R-waves make the surface of the ground go up and down. They differ from water waves, whose particles follow a clockwise circular path as the wave passes (see Chapter 18). R-wave particles, instead, follow an elliptical counterclockwise orbit, as shown.

10.4 HOW DO WE MEASURE AND LOCATE EARTHQUAKES?

Most news reports about earthquakes provide information on the size and location of an earthquake. What does this information mean, and how do we obtain it? What's the difference between a great earthquake and a minor one? How do seismologists locate an epicenter? Understanding how a seismograph works and how to read the information it provides will allow you to answer these questions.

Seismographs and the Record of an Earthquake

When, in 1889, a German physicist realized that a pendulum in his lab had moved in reaction to a deadly earthquake that had occurred in Japan, his observation confirmed speculations that earthquake energy can pass through the planet. On reading of this discovery, other researchers saw a way to construct an instrument, called a **seismograph** (or seismometer), that can systematically record the ground motion from an earthquake happening anywhere on Earth. Seismologists use two basic configurations of seismographs—one for measuring vertical (up-and-down) ground motion and the other for measuring horizontal (back-and-forth) ground motion (▶Fig. 10.13a, b). Ideally, the instruments are placed on bedrock in sheltered areas, away from traffic and other urban noise (▶Fig. 10.13c).

A mechanical vertical-motion seismograph consists of a heavy weight suspended like a pendulum from a spring (▶Fig. 10.14a–c). The spring connects to a sturdy frame that has been bolted to the ground. A pen extends sideways from the weight and touches a vertical revolving cylinder of paper; the axle around which the cylinder rotates has been connected to the seismograph frame. When an earthquake wave arrives and causes the ground surface to move up and down, it also makes the seismograph frame move up and down. The weight, however, because of its inertia (the tendency of an object at rest to remain at rest) remains fixed in space. As a consequence, the revolving paper roll moves up and down under the pen, which traces out the waves representing the up-and-down movement. (If the paper cylinder did not revolve, the pen would simply move back and forth in the same place on the paper.) A mechanical horizontal-motion seismograph works on the same principle, except that the paper cylinder is horizontal and the weight is suspended from a wire. Sideways back-and-forth movement of this seismograph causes the pen to trace out waves (Fig. 10.13b). In sum, the key to a seismograph is the presence of a weight that stays fixed in space while everything else moves around it.

The waves traced by the pen on a seismograph provide a record of the earthquake called a **seismogram** (▶Fig. 10.14d, e). In order for scientists to determine when a particular

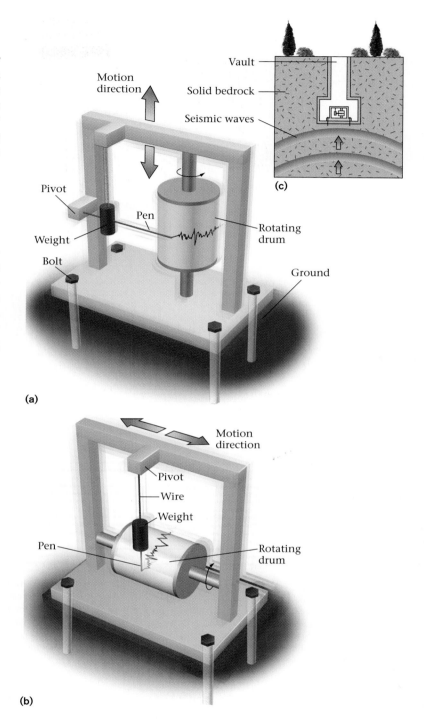

(a)

(b)

(c)

FIGURE 10.13 **(a)** A vertical-motion seismograph records up-and-down ground motion. **(b)** A horizontal-motion seismograph records back-and-forth ground motion. **(c)** Seismographs are bolted to the bedrock in a protected shelter or vault.

earthquake wave arrives, the record also displays lines representing time. At first glance, a typical seismogram looks like a messy squiggle of lines, but to a seismologist it contains a wealth of information. The horizontal axis represents time,

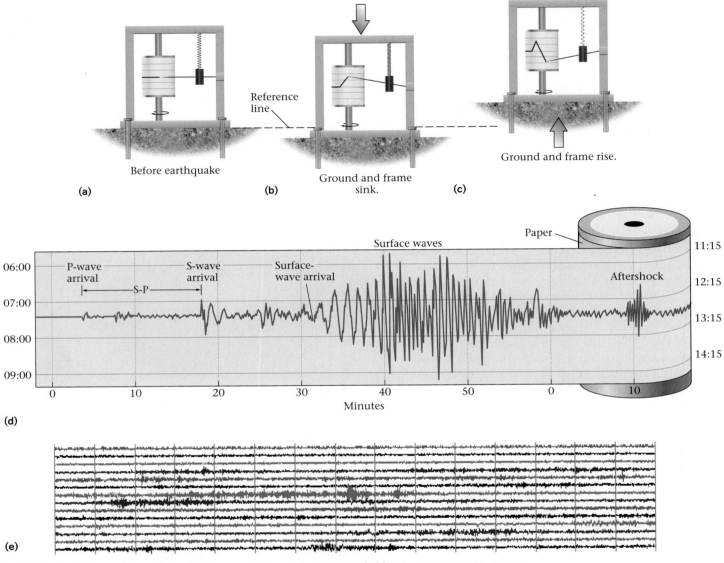

FIGURE 10.14 How a seismograph works (here, a vertical-motion seismograph). **(a)** Before an earthquake, the pen traces a straight line. **(b)** During an earthquake, when the ground and the frame of a seismograph go down, the weight stays in place because of inertia, so the pen rises relative to the paper roll. **(c)** When the ground and the seismograph frame rise, the pen goes down. **(d)** This closeup of the record (seismogram) for a single earthquake shows the signals generated by different kinds of seismic waves. **(e)** Digital seismic record from a seismograph station in Arkansas. The vertical lines represent minutes. Colors have no meaning, they just make the figure more readable. All of the earthquakes shown are small. Each horizontal line is the record for fifteen minutes.

and the vertical axis represents the amplitude (the size) of the seismic waves. The instant at which an earthquake wave appears at a seismograph station is the **arrival time** of the wave. The first squiggles on the record represent P-waves, because P-waves travel the fastest. They typically cause the ground to lurch up and down. Next come the S-waves, causing back-and-forth motion. And finally the surface waves (Rayleigh and Love) arrive, causing rolling motions of the ground. Typically, the surface waves have the largest amplitude, and arrive over a relatively long period of time.

Seismologists the world over have agreed to certain standards for measuring earthquakes, and they calibrate their seismographs to a precise atomic-clock time signal so that they can compare seismograms from different parts of the world. Today, seismologists work with digital records produced by modern electronic seismographs. In these instruments, the simple weight has been replaced by a heavy cylindrical magnet surrounding a coil of wire (▶Fig. 10.15). The coil connects to a solid frame anchored to the ground. During an earthquake, the magnet stays in

BOX 10.1 THE REST OF THE STORY

Quakes on Other Planets?

When the *Apollo* astronauts landed on the Moon in the 1960s and 1970s, they left behind seismographs that could measure moonquakes, shaking events on the Moon. The instruments found that moonquakes happen far less often than earthquakes (only about 3,000 a year) and are very small. Geologists were not surprised, because plate movement does not occur on the Moon, so there's no volcanism, rifting, subduction, or collisions to generate the forces that cause earthquakes.

But if plates don't move on the Moon, then what causes moonquakes? Undoubtedly some are caused by the impacts of meteorites, which hit the Moon's surface like large hammers. Most moonquakes, though, seem to occur when the Moon reaches its closest distance to the Earth, as it travels along its elliptical orbit, suggesting that gravitational attraction between the Earth and the Moon creates tidal-like motions that crack the Moon's lithosphere or make the faults in the Moon slip.

Not much is known yet about seismicity on other terrestrial planets. Since plate tectonics does not occur on Mercury, Venus, or Mars, it's not likely that these planets have many strong quakes, except due to meteorite impacts. But stress that develops due to thermal effects (cooling of the planet's interior or heating of the planet's surface), or due to the gravitational pull of other objects may cause some cracking and shaking. In the 1970s, the Viking lander placed a seismometer on Mars that survived only briefly, but did record one or two marsquakes during about 1,200 Martian days. Recently, investigators have found Martian landscape features that resemble features caused by tremors on Earth. New spacecraft to be launched in the near future will attempt to place new seismometers on the planet to listen for seismic activity once again. No seismic instruments have yet been placed on Venus, but imagery of the planet's surface also suggests landscape features (escarpments, etc.) similar to those triggered by earthquakes.

---

place while the coil and the frame move. This movement generates an electrical current, whose voltage indicates the amount of movement. A computer precisely records variations in voltage indicative of motion. Modern seismographs are so sensitive that they can record ground movements of a millionth of a millimeter (only 10 times the diameter of an atom)—movements that people can't feel. Seismologists can now access data from thousands of stations constituting the worldwide seismic network. Governments have supported this network because in addition to recording natural earthquakes, it can also detect nuclear-bomb tests.

FIGURE 10.15 In a modern electronic seismograph, the weight has been replaced by a heavy magnet. Movement of the wire coil inside the magnet, in response to ground motion, generates an electric current. The voltage of the current represents the size of the motion.

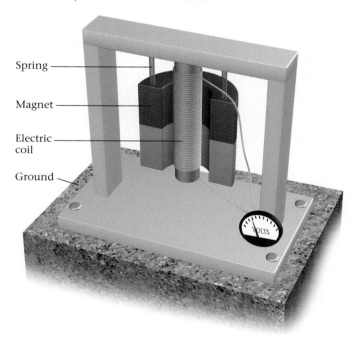

Spring

Magnet

Electric coil

Ground

VOLTS

Finding the Epicenter

If an earthquake happens in or near a populated area, seismologists can approximate the location of the epicenter by noting where there's the most damage. But how do we find the epicenters of earthquakes that occur in uninhabited areas or in the ocean lithosphere, thousands of kilometers from land?

The difference in velocity, and therefore arrival time, of the different kinds of earthquake waves provides the key. P-waves and S-waves pass through the interior of the Earth at different velocities. The farther the waves travel, the greater the distance between them. For an analogy, consider an automobile race. The first car travels at 100 mph and the second at 80 mph (▶Fig. 10.16a). At the starting gate (representing the earthquake's epicenter), the cars line up next to each other. But a half hour into the race, the faster car (P-wave) is 10 miles ahead of the second car (S-wave), and an hour into the race, the faster car is 20 miles ahead of the second car. Note that P- and S-waves travel along curved paths through the Earth (▶Fig. 10.16b). Interlude D explains why. Seismologists use the time delay between the arrivals at a seismograph station of the P-wave and the S-wave for the calculation. The delay between P-wave and S-wave arrival times increases as the distance from the epicenter increases (▶Fig. 10.16c). We can see this time delay on a graph, with a line called a **travel-time curve**, which plots the time since the earthquake began on the vertical axis and the distance to the epicenter on the horizontal axis (▶Figure 10.16d).

To use a travel-time curve to determine the distance of an epicenter, start by measuring the time difference between the P- and S-waves on your seismograph (Fig. 10.16c). Then draw a line segment on a piece of paper to represent this amount of time, at the scale used for the vertical axis of the graph. Move the line segment back and forth until one end lies on the P-wave curve and the other end lies on the S-wave curve. You have now determined the distance at which the time difference between the two waves equals the time difference you observed. Extend the line down to the horizontal axis, and simply read off the distance to the epicenter (Fig. 10.16d).

The analysis of one seismogram tells you only the distance between the epicenter and the seismograph station; it does not tell you in which direction the epicenter lies. To determine the map position of the epicenter, we use a method called triangulation, by plotting the distance between the epi-

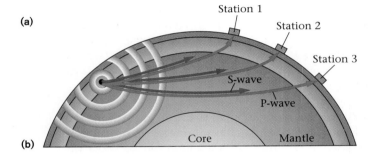

(a)

(b)

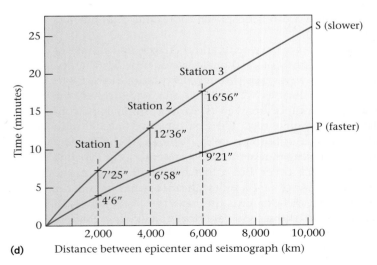

(c)

(d)

Distance between epicenter and seismograph (km)

FIGURE 10.16 (a) Different seismic waves travel at different velocities, like cars racing at different speeds. **(b)** Thus, different waves arrive at different times at seismograph stations. P-waves arrive first, then S-waves. **(c)** The greater the distance between the epicenter and the seismograph station, the greater the time delay between the P-wave and S-wave arrival times. In this example, station 1 is closest to the epicenter, and station 3 is farthest away from it. Note that the P-wave arrives later at station 3 than at station 1, and that the time interval between P- and S-wave arrivals is greater at station 3 than at station 1. Arrivals at station 2 are in between. **(d)** We can represent the contrasting arrival times of P-waves and S-waves on a travel-time curve. **(e)** If an earthquake epicenter lies 2,000 km from station 1, we draw a circle with a radius of 2,000 km around the station. Following the same procedure for stations 2 and 3, we can locate the epicenter: it lies at the intersection point of the three circles.

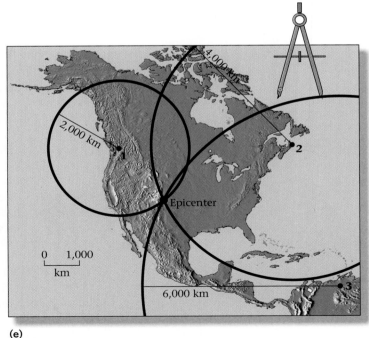

(e)

center and three stations. For example, say you know that the epicenter lies 2,000 km from station 1, 4,000 km from station 2, and 6,000 km from station 3. On a map, draw a circle around each station. Each circle's radius is the distance between the station and the epicenter, at the scale of the map. The epicenter lies at the intersection of the three circles, for this is the only point at which the epicenter has the appropriate measured distance from all three stations (▶Fig. 10.16e).

Note that different surface waves travel at different speeds. Thus, close to the epicenter, the surface waves haven't had time to spread out, so they arrive at once, like a clump of runners near the starting line, and an earthquake is over in a matter of tens of seconds. On a seismograph of the same earthquake recorded thousands of kilometers from the epicenter, the surface waves may arrive over an interval of tens of minutes.

Defining the Size of an Earthquake

Some earthquakes shake the ground violently and cause extensive damage, while others can barely be felt. Seismologists have developed means to define size in a uniform way, so that they can make systematic comparisons among earthquakes. Seismologists use two distinct approaches for characterizing the relative sizes of earthquakes—the first yields a number called the intensity and the second yields a number called magnitude.

Mercalli intensity scale. In 1902, the Italian scientist Giuseppe Mercalli developed the first widely used scale for characterizing earthquake size. This scale, called the **Mercalli intensity scale**, defines the intensity of an earthquake by the amount of damage it causes—that is, by its destructiveness. We denote different Mercalli intensities (M) with Roman numerals, as shown in ▶Table 10.2. Note that the specification of the intensity of an earthquake depends on a subjective assessment of the damage, not on a particular measurement with an instrument. Note also that the Mercalli intensity varies with distance from the epicenter, because earthquake energy dies out as the waves travel farther through the Earth. Thus, there is not one single intensity value for an earthquake. Seismologists draw lines, called contours, around the epicenter, delimiting zones in which the earthquake has a specific intensity (▶Fig. 10.17). A larger earthquake has a large intensity value over the epicenter; also, its intensity contours surround a wider area. Maps showing the regional variation in intensity expected for earthquakes in a given location provide useful guidelines for urban planners trying to specify building codes.

The distance between intensity contours depends on the nature of the crust when the earthquake occurs. In the eastern United States, where the crust is strong, earthquake energy travels farther, so a broader region will feel the earthquake and contours are far apart. In the western United

TABLE 10.2 Mercalli Intensity Scale

M	Destructiveness (Perceptions of the Extent of Damage)
I	Detected only by seismic instruments; causes no damage.
II	Felt by a few stationary people, especially in upper floors of buildings; suspended objects, like lamps, may swing.
III	Felt indoors; standing automobiles sway on their suspensions; it seems as though a heavy truck is passing.
IV	Shaking awakens some sleepers; dishes and windows rattle.
V	Most people awaken; some dishes and windows break, unstable objects tip over; trees and poles sway.
VI	Shaking frightens some people; plaster walls crack, heavy furniture moves slightly, and a few chimneys crack, but overall little damage occurs.
VII	Most people are frightened and run outside; a lot of plaster cracks, windows break, some chimneys topple, and unstable furniture overturns; poorly built buildings sustain considerable damage.
VIII	Many chimneys and factory smokestacks topple; heavy furniture overturns; substantial buildings sustain some damage, and poorly built buildings suffer severe damage.
IX	Frame buildings separate from their foundations; most buildings sustain damage, and some buildings collapse; the ground cracks, underground pipes break, and rails bend; some landslides occur.
X	Most masonry structures and some well-built wooden structures are destroyed; the ground severely cracks in places; many landslides occur along steep slopes; some bridges collapse; some sediment liquifies; concrete dams may crack; facades on many buildings collapse; railways and roads suffer severe damage.
XI	Few masonry buildings remain standing; many bridges collapse; broad fissures form in the ground; most pipelines break; severe liquefaction of sediment occurs; some dams collapse; facades on most buildings collapse or are severely damaged.
XII	Earthquake waves cause visible undulations of the ground surface; objects are thrown up off the ground; there is complete destruction of buildings and bridges of all types.

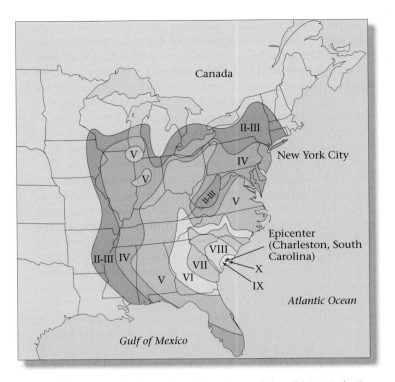

FIGURE 10.17 This map shows the contours of Mercalli intensity for the 1886 Charleston, South Carolina, earthquake. Note that near the epicenter, ground shaking reached M = X, and in New York City, ground shaking reached M = II to III.

States, where the crust is warmer and weaker, earthquake energy travels only a short distance, so contours are close.

Earthquake magnitude scales. When you read a report of an earthquake disaster in the news, you will likely come across a phrase that reads something like, "An earthquake with a magnitude of 7.2 struck the city yesterday at noon." What does this mean? The **magnitude** of an earthquake is a number that indicates its relative size, as determined by measuring the maximum amplitude of ground motion recorded by a seismograph at a given distance from the epicenter. By "amplitude of ground motion," we mean the amount of up-and-down or sideways motion of the ground. The larger the ground motion, the greater the deflection of a seismograph pen or needle as it traces out a seismogram. As an example, a magnitude 7.2 earthquake causes more ground movement, and thus greater deflections of a seismograph pen or needle, than does a magnitude 5.8 earthquake. Similarly, as you will see, a magnitude 7.2 earthquake releases more energy than does a magnitude 5.8 earthquake.

Earthquake intensity and earthquake magnitude have very different meanings. Remember that *intensity* is based on the amount of damage caused by an earthquake, so for a given earthquake, the intensity decreases with increasing distance from the epicenter. In contrast, *magnitude* is based

on ground motion recorded by a seismogram, and for a calculation of magnitude, a seismologist must accommodate for the distance between the epicenter and the seismograph. Thus, magnitude does not depend on this distance, and a calculation based on data from any seismograph anywhere in the world will yield the same result if the calculation correctly accommodates for distance.

In 1935, the American seismologist Charles Richter proposed the concept of defining and measuring earthquake magnitude. The scale he proposed came to be known as the **Richter scale**. To ensure that a magnitude on this scale has the same meaning, regardless of who measures it, Richter prescribed detailed guidelines for determining magnitude. Specifically, you can determine a Richter magnitude by measuring the amplitude of the largest deflection, on a seismograph, generated in response to seismic waves that have a period of one second, as recorded by a seismic station 100 km from the epicenter. (The period for a set of earthquake waves is the time interval between the arrival of successive waves; period = 1/frequency.) Because the amount of deflection depends on the distance between the seismograph and the epicenter, and since most seismograph stations do not happen to lie exactly 100 km from the epicenter, seismologists use a chart to adjust for distance from the epicenter (▶Fig. 10.18). Richter realized that different kinds of seismographs record earthquakes differently, so he required that magnitude measurements use data produced by a particular design of seismograph.

Richter's scale became so widely used that it became commonplace for news reports to include wording such as, "The earthquake registered a 7.2 on the Richter scale." These days, however, seismologists actually use several different magnitude scales, not just the Richter scale, and not all yield exactly the same number for a particular earthquake. We need to use alternate scales because the original Richter scale works well only for shallow, nearby earthquakes (earthquakes whose epicenters are less than 600 km from the seismograph, and whose hypocenters are less than 15 km below the surface). Because of the distance limitation, a number on the original Richter scale is now called a local magnitude (M_L).

To apply Richter's concept to the description of distant earthquakes, seismologists developed a new scale based on measuring the amplitudes of certain R-waves. A number on this scale is called a surface-wave magnitude (M_s). The surface-wave magnitude scale, however, is not suitable for an earthquake whose hypocenter is more than 50 km below the surface, because such earthquakes do not create large surface waves. So to describe the size of deeper earthquakes, seismologists determine a body-wave magnitude (m_b), which is based on measurement of P-waves. Note that by an unfortunately confusing convention, some magnitudes use upper-case M and some use lower-case m.

The M_L, m_b, and M_s scales have limitations—they cannot accurately define the sizes of very large earthquakes, be-

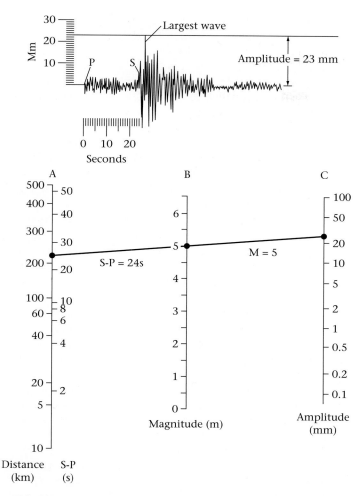

FIGURE 10.18 To calculate the Richter magnitude from a seismogram, first measure the S-minus-P time to determine the distance to the epicenter; then measure the height, or amplitude (in mm), of the largest wave recorded by the seismograph. Draw a line from the point on column A representing the S-minus-P time to the point on column C representing the wave amplitude, and read the Richter magnitude (m) off column B. Note that if the earthquake were much closer, then the same amplitude in the seismogram would yield a smaller-magnitude earthquake. We must take the distance to the epicenter into account because seismic waves grow smaller in amplitude with increasing distance from the epicenter.

cause for an earthquake above a given size, the scales give roughly the same magnitude regardless of how large the earthquake really is. For example, an earthquake with an M_L, m_b, or M_s of 8.3 could actually be much larger than a real magnitude 8.3 earthquake. Because of this problem, seismologists developed the **moment magnitude** (M_w) scale. To calculate the moment magnitude, it is necessary to measure the amplitude of a number of different seismic waves, determine the area of the slipped portion of the fault that moved, determine how much slip occurred, and define physical characteristics of the rock that broke during faulting. Typically, the larger the area that slips, and the larger the amount of slip, the greater the earthquake. And,

for that same slipped area and amount of slip, the rupture of stronger rock produces a greater earthquake than does the rupture of weaker rock. Moment magnitude numbers may be different from other magnitude numbers for the same earthquake. For example, the great 1964 earthquake in Alaska had an M_s of 8.4 but an M_w of 9.2, whereas the 1906 San Francisco earthquake had an M_s of 8.3 and an M_w of 7.9. The largest recorded earthquake in history hit Chile in 1960 and had an M_s of 8.5 but an M_w of 9.5. The Sumatra earthquake of 2004 registered an M_w of 9.0.

What magnitude is given in modern reports of earthquakes in the newspaper? It depends on how soon the article appears after an earthquake has taken place. For early reports, seismologists report a preliminary magnitude, which is an M_L, m_b, or M_s, because these magnitudes can be calculated fairly quickly. Later on, after they have had the chance to collect the necessary data, seismologists report an M_w, which becomes the number now generally used for archival records.

All magnitude scales are logarithmic, meaning that an increase of one unit of magnitude represents a tenfold increase in the maximum ground motion. Thus, a magnitude 8 earthquake results in ground motion that is *10 times* greater than that of a magnitude 7 earthquake, and a thousand times greater than that of a magnitude 5 earthquake. To make life easier, seismologists use familiar adjectives to describe the size of an earthquake, as listed in ▶Table 10.3. The intensity of an earthquake at its epicenter depends, approximately, on the magnitude of the earthquake.

Note that an earthquake magnitude scale is not a 10-point scale, as is sometimes erroneously stated in the news. For example, it is possible for an event with a magnitude of,

TABLE 10.3 Adjectives for Describing Earthquakes

Adjective	Magnitude	Approximate maximum intensity at epicenter	Effects
great	>8.0	X to XII	major to total destruction
major	7.0 to 7.9	IX to X	great damage
strong	6.0 to 6.9	VII to VIII	moderate to serious damage
moderate	5.0 to 5.9	VI to VII	slight to moderate damage
light	4.0 to 4.9	IV to V	felt by most; slight damage
minor	<3.9	III or smaller	felt by some; hardly any damage

say, −1 to occur, although it would be too weak to feel. (The smallest earthquake that you can feel at the epicenter has a magnitude of about 2.) Nevertheless, an earthquake larger than magnitude 10 probably won't occur, because it would require an almost impossibly large fault rupture to develop.

As we pointed out earlier, earthquakes release energy. Seismologists can calculate the energy release from the moment magnitude. Not all versions of this calculation yield the same result, so energy estimates must be taken as an approximation. According to some researchers, a magnitude 6 earthquake releases about as much energy as the atomic bomb that was dropped on Hiroshima in 1945, and the 1964 Alaska earthquake released significantly more energy than the largest hydrogen bomb ever detonated. Note that an increase in magnitude by one integer represents approximately a thirty-two-fold increase in energy. Thus, a magnitude 8 earthquake releases about 1 million times more energy than a magnitude 4 earthquake (▶Fig. 10.19a). In fact, a single magnitude 8.9 earthquake releases as much energy as the entire *average* global annual release of seismic energy coming from all other earthquakes combined! Fortunately, such large earthquakes occur much less frequently than small earthquakes; there are about 100,000 magnitude 3 earthquakes every year, but a magnitude 8 earthquake happens on average only about once every 3 to 5 years (▶Fig. 10.19b). Though earthquake energy seems immense to us because of the damage it can do, the combined average annual production of earthquake energy actually represents less than 1/1,000 of the heat energy received by the Earth from the Sun in a given year.

> **Take-Home Message**
>
> Seismographs measure earthquake vibrations and allow seismologists to determine the location and size of earthquakes. We specify earthquake size by an intensity value (damage caused) or a magnitude (the amplitude of ground motion). Magnitudes are logarithmic, meaning that a magnitude 8 event is 10 times larger than a magnitude 7 event.

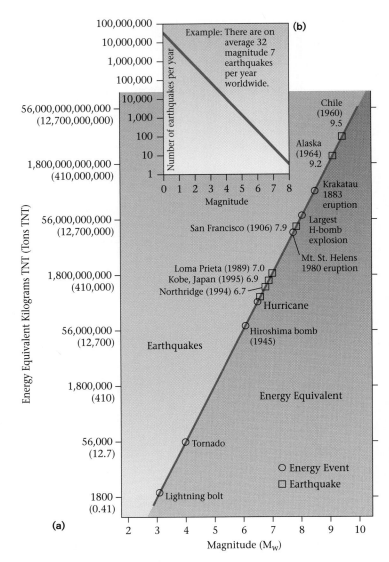

FIGURE 10.19 (a) This graph emphasizes that the energy released by an earthquake increases dramatically with magnitude. **(b)** This graph illustrates how the number of earthquakes of a given magnitude decreases with increasing magnitude. According to the graph, there are over 1 million m = 1 earthquakes per year, but only about 10,000 m = 4 earthquakes.

10.5 WHERE AND WHY DO EARTHQUAKES OCCUR?

Earthquakes do not take place everywhere on the globe. By plotting the distribution of earthquake epicenters on a map, seismologists find that most, but not all, earthquakes occur in fairly narrow **seismic belts**, or seismic zones. Most seismic belts follow plate boundaries. Earthquakes within these belts are plate-boundary earthquakes. Those that occur away from plate boundaries are intraplate earthquakes (the prefix *intra* means within) (▶Fig. 10.20). Note that 80% of the earthquake energy released on Earth comes from plate-boundary earthquakes in the belts surrounding the Pacific Ocean. Most of the remainder comes from earthquakes in the collision zone on the northern sides of the African and Indo-Australian Plates. Though numerous earthquakes happen along mid-ocean ridges and within plates, together they account for relatively little energy.

Seismologists distinguish three classes of earthquakes on the basis of hypocenter (focus) depth: shallow-focus earthquakes occur in the top 20 km of the Earth, intermediate-focus earthquakes take place between 20 and 300 km, and deep-focus earthquakes occur down to a depth of about 670 km. Most earthquakes in continental crust are shallow focus, because at depths greater than 15 to 20 km, the rock

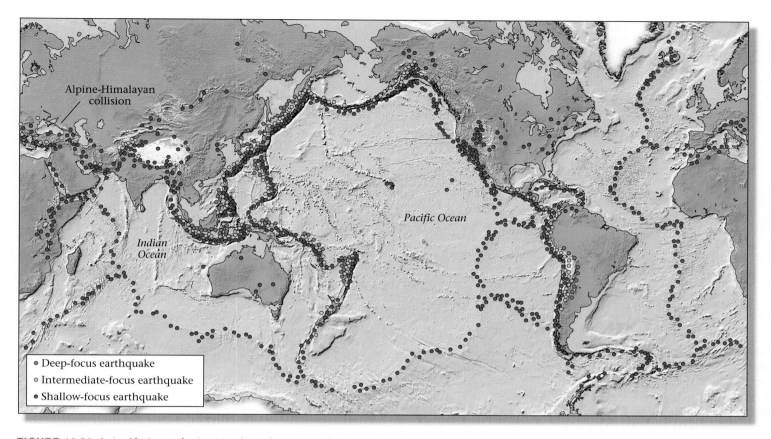

- Deep-focus earthquake
- Intermediate-focus earthquake
- Shallow-focus earthquake

FIGURE 10.20 A simplified map of epicenters shows that most earthquakes occur in distinct belts that define plate boundaries, and most take place in the highly populated Pacific Rim region. Epicenters that do not lie along plate boundaries are the result of rifting or collision, or are in intraplate settings. Intermediate- and deep-focus earthquakes occur only along convergent plate boundaries, except for a few that happen in collisional zones. For more detailed versions of this map, see the inside front cover of this book and Appendix B.

of continental crust is too ductile, so it flows instead of breaks when subjected to stress. As we will see, intermediate- and deep-focus earthquakes occur only in subducting oceanic plates. Shallow-focus earthquakes cause by far the most damage, because the earthquake waves they produce do not lose much energy before vibrating the Earth's surface. In this section, we look at the characteristics of earthquakes in various geologic settings and learn why earthquakes take place where they do.

Earthquakes at Plate Boundaries

The majority of earthquakes happen at plate boundaries, because the relative motion between plates takes place primarily by slip at their boundaries. Away from plate boundaries, lithosphere plates are fairly rigid and generally too strong to break in response to the forces present, so much less movement takes place in plate interiors. We find different kinds of faulting at different types of plate boundaries.

Divergent plate-boundary seismicity. At divergent plate boundaries (mid-ocean ridges), two oceanic plates form and move apart. Divergent boundaries are segmented, and spreading segments are linked by transform faults. Therefore, two kinds of faults develop at divergent boundaries. Along spreading segments, newly formed crust at or near the mid-ocean ridge stretches and ruptures, generating normal faults, whereas along the transform faults that link spreading segments, strike-slip faulting occurs (▶Fig. 10.21). Seismicity along mid-ocean ridges takes place at shallow depths (less than 10 km), but because most ridges are out in the ocean, far away from settled areas, these earthquakes don't cause damage. Only in a few localities (such as Iceland) that lie astride a mid-ocean ridge can divergent plate boundary earthquakes be of concern.

Convergent plate-boundary seismicity. Convergent plate boundaries are complicated regions at which several different kinds of earthquakes take place. Shallow-focus earthquakes occur in both the subducting plate and the overriding plate.

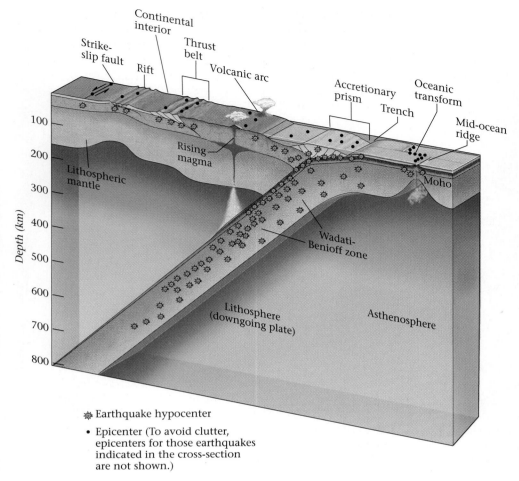

Depth (km): 100, 200, 300, 400, 500, 600, 700, 800

Strike-slip fault
Rift
Continental interior
Thrust belt
Volcanic arc
Rising magma
Lithospheric mantle
Accretionary prism
Trench
Oceanic transform
Mid-ocean ridge
Moho
Wadati-Benioff zone
Lithosphere (downgoing plate)
Asthenosphere

✳ Earthquake hypocenter
• Epicenter (To avoid clutter, epicenters for those earthquakes indicated in the cross-section are not shown.)

FIGURE 10.21 Earthquakes along oceanic divergent boundaries and transform boundaries are all shallow focus. Shallow-focus earthquakes also occur in the overriding plate of a convergent boundary. Earthquakes in the downgoing plate of a convergent boundary define a Wadati-Benioff zone and include intermediate- and deep-focus earthquakes.

caused by shear between the downgoing plate and the mantle, and by the slab pull of the deeper part of the plate on the shallower part.

Considering the depth at which they occur, intermediate- and deep-focus earthquakes in Wadati-Benioff zones present a problem: Shouldn't the Earth be too warm to undergo brittle deformation at such depths? Seismologists suggest that intermediate-focus earthquakes can happen because the downgoing plate takes a long time to heat up, and the rock in the plate's interior remains cool and brittle enough to break even down to about 300 km. Deep-focus earthquakes, however, remain a mystery. Some researchers suggest that they occur when minerals in the rock suddenly "collapse" (undergo a phase change; see Chapter 8) as a result of great pressure, and the atoms rearrange into new minerals that take up less space. The sudden change in volume would generate a shock. The fate of subducting lithosphere below a depth of 670 km remains unclear. Recent evidence suggests that some plate material accumulates at 670 km, whereas some sinks still deeper but no longer generates earthquakes. Evidently, earthquakes can't form in subducted plates at depths greater than 670 km because all minerals have completed the transition to more compact forms, and the rock is too warm for brittle breakage.

Specifically, as the downgoing plate begins to subduct, it bends and shears along the base of the overriding plate. The bending stretches the downgoing plate and generates small normal faults in the plate. Of more importance, large thrust faults develop along the contact between the downgoing and overriding plates. This shear between the two plates generates great earthquakes. In some cases, the push applied by the downgoing plate compresses, or squeezes, the overriding plate and causes shallow-focus reverse faulting in the overriding plate. Most of the Earth's seismic energy is released at convergent plate boundaries.

In contrast to other types of plate boundaries, convergent plate boundaries also host intermediate- and deep-focus earthquakes. These occur in the downgoing slab as it sinks into the mantle, defining the sloping band of seismicity called a **Wadati-Benioff zone**, after the seismologists who first recognized it (Fig. 10.21). Intermediate- and deep-focus earthquakes happen partly in response to stresses

Earthquakes in southern Alaska, Japan, the western coast of South America, Mexico, and other places along the Pacific Rim are examples of convergent boundary earthquakes. One such event happened on Good Friday of 1964, near Anchorage, Alaska. This earthquake had a seismic-moment magnitude of 9.2; the earthquake wrecked substantial parts of towns all along the coast (▶Fig. 10.22). In 1995, an earthquake with a magnitude (M_w) of 6.9 struck Kobe, Japan, causing immense devastation and killing over 5,000 people (▶Fig. 10.23a, b). This earthquake, like most large ones, noticeably changed the shape of the ground surface, as indicated by InSAR measurements (see Fig. 10.11b, c).

Recently, geologists have recognized that, in 1700, a huge earthquake, with a magnitude (M_w) of 8.7–9.2, accompanied slip on the Cascadia subduction zone off the now-densely populated coast of Oregon and Washington. An earthquake of similar size today would be disastrous.

FIGURE 10.22 A portion of 4th Avenue in Anchorage, Alaska, collapsed during the 1964 earthquake due to slumping of the substrate.

Transform plate-boundary seismicity. At transform plate boundaries, where one plate slides past another without the creation or consumption of oceanic lithosphere, most faulting results in strike-slip motion. The majority of transform faults in the world link segments of oceanic ridges (Fig. 10.21), but a few, such as the San Andreas Fault of California and the Alpine Fault of New Zealand, cut across continental lithosphere (▶Fig. 10.24). All transform-fault earthquakes are shallow focus, so the larger ones on land can cause disaster.

As an example of a transform-fault earthquake, consider the slip of the San Andreas Fault near San Francisco in 1906 (▶Fig. 10.25a, b). In the wake of the gold rush, San Francisco was a booming city with broad streets and numerous large buildings. But it was built on the transform boundary along which the Pacific Plate moves north at an average of 6 cm per year relative to North America. Because of the stick-slip behavior of the fault, this movement doesn't occur smoothly but happens rather in sudden jerks, each of which causes an earthquake. At 5:12 A.M. on April 18, the fault slipped by as much as 7 m, and earthquake waves slammed into the city. Witnesses watched in horror as the streets undulated like ocean waves. Buildings swayed and banged together, laundry lines stretched and snapped, church bells rang, and then towers, facades, and houses toppled. Two main shocks hit the city, the first lasting about 40 seconds and the other arriving 10 seconds later and lasting about 25 seconds—judging from the damage, seismologists estimate that the largest shock would have registered a seismic moment magnitude of 7.9. Most building collapse took place downtown (▶Fig. 10.25c). Fire followed soon after (perhaps started by overturned cooking stoves), consuming huge areas of the city, for most buildings were made of wood. In the end, about five hundred people died, and a quarter of a million were left homeless. Telegraphs instantly sent pleas for help. Officials responded rapidly and, within hours, relief trains began to arrive. Within days, tent cities housed all survivors.

FIGURE 10.23 **(a)** Simplified map of earthquake epicenters (black dots) and plate boundaries in Japan. The sizes of the dots indicate magnitude. Note the location of Kobe. **(b)** A collapsed building, after the Kobe earthquake.

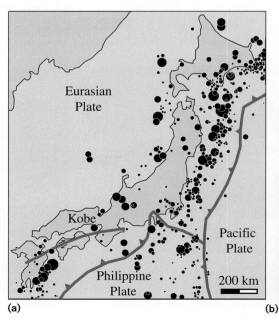

(a)

(b)

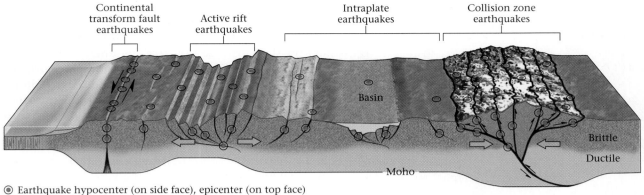

● Earthquake hypocenter (on side face), epicenter (on top face)

FIGURE 10.24 Earthquakes within continents: earthquakes occur in continental transform faults (such as the San Andreas), in continental rifts (the East African Rift, the Basin and Range Province), in intraplate settings (usually by reactivating old faults), and in collision zones (mountain ranges). Continental earthquakes mostly happen in the brittle crust, at depths of less than about 15 km.

FIGURE 10.25 **(a)** Map showing the portion of the San Andreas Fault that ruptured during the 1906 earthquake (red). A large portion of the fault in southern California ruptured during the 1857 Fort Tejon earthquake (yellow), with an $M_w = 8.0$. **(b)** A computer sketch showing the traces of active faults in the San Francisco Bay area. **(c)** A street in San Francisco after the 1906 earthquake. **(d)** Photo showing the collapse of a bridge in San Francisco due to the 1989 earthquake.

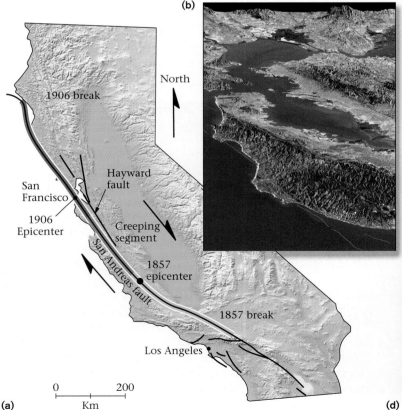

The San Francisco earthquake has not been the only one to strike along the San Andreas and nearby related faults. Over a dozen major earthquakes have happened on these faults during the past two centuries, including an 1857 magnitude 8.7 earthquake just east of Los Angeles, and the 1989 magnitude 7.1 Loma Prieta earthquake, which occurred 100 km south of San Francisco but nevertheless shut down a World Series game and caused the collapse of a double-decker freeway (▶Fig. 10.25d). Note that the 1994 magnitude 6.7 earthquake at Northridge, described earlier, occurred not on a strike-slip fault but rather on a blind thrust fault formed where a bend in the San Andreas Fault caused local compression of the crust.

Earthquakes at Continental Rifts and Collision Zones

Continental rifts. The stretching of continental crust at continental rifts generates normal faults (Fig. 10.24). Active rifts today include the East African Rift, the Basin and Range Province (mostly in Nevada, Utah, and Arizona), and the Rio Grande Rift (in New Mexico). In all these places, shallow-focus earthquakes occur, similar in nature to the earthquakes at mid-ocean ridges. But in contrast to mid-ocean ridges, these seismic zones can be located under populated areas, (such as Albuquerque and Salt Lake City), and thus cause major damage.

Collision zones. Zones where two continents collide after the oceanic lithosphere that once separated them has been completely subducted yield great mountain ranges such as the Alpine-Himalayan chain. Though a variety of earthquakes happen in these zones, the most common are due to movement on thrust faults, which formed when the crust was compressed, or squeezed (Fig. 10.24).

The complex collision between Africa and Europe generated the stress that triggered an earthquake near Lisbon, Portugal, on All Saints' Day (November 1), 1755. According to an eyewitness account,

> The first alarm was a rumbling noise that sounded like exceptionally heavy traffic. Then there was a brief pause and a devastating shock followed, lasting over two minutes, that brought down roofs, walls, facades of churches, palaces and houses and shops in a dreadful deafening roar of destruction. Close on this came a third trembling to complete the disaster, and then a dark cloud of suffocating dust settled fog-like on the ruins of the city.[1]

About fifteen minutes later, fires began to spread and consume the rubble, and then the waters of the harbor "rocked

---

[1]James R. Newman, "The Lisbon Earthquake," in Frank H. T. Rhodes and Richard O. Stone, eds., *Language of the Earth* (New York: Pergamon Press, 1981), p. 59.

and rose menacingly, and then poured in three great towering waves over its banks."[1] Aftershocks lasted for days. Innumerable artworks by great Renaissance masters were destroyed, along with a library that housed all the records of Portuguese exploration. The effects of the Lisbon earthquake went beyond the immediate destruction. Voltaire (1694–1778), the great French writer, immortalized the earthquake in his novel *Candide*. How, pondered Voltaire, could this world be the best of all possible worlds if such a calamity could happen and kill good people? The earthquake, and Voltaire's writing about it, brought to an end the "era of optimism" and ushered in the "era of realism" in Europe. The 2005 earthquake in Pakistan is a more recent example. Here, the stresses causing the earthquake were generated by the collision of India with Asia.

Intraplate Earthquakes

Some earthquakes occur in the interiors of plates and are not associated with plate boundaries, active rifts, or collision zones (Fig. 10.24). These **intraplate earthquakes** account for only about 5% of the earthquake energy released in a year. Almost all have a shallow hypocenter. Seismologists are still trying to understand the causes of intraplate earthquakes. Most favor the idea that force applied to the boundary of a plate can cause the interior of the plate to break at weak, preexisting fault zones, some of which date back to the Precambrian. Alternatively, the activity may be due to forces resulting from shear between the lithosphere and the underlying asthenosphere, or when the upper crust readjusts to loads from glaciers or sedimentary piles.

Because the crust in an intraplate setting tends to be composed of stronger rock, overall, than the crust along a plate boundary, it tends to transmit seismic waves more efficiently than does plate-boundary crust. In an analogy, if you were to bang a hammer on a block of solid rock and on a block of styrofoam, you would feel the impact on the other side of the rock, but not on the other side of the styrofoam because the impact would be absorbed by the styrofoam. Thus, intraplate earthquakes can be felt over a very broad region.

In Europe, a number of intraplate earthquakes have been recorded. For example, an earthquake with a moment magnitude of 5 to 5.4 hit northwestern England in 1990, and a magnitude 4.8 earthquake hit central England, near Birmingham, in 2002. In North America, intraplate earthquakes occur in distinct clusters, most notably in the vicinity of New Madrid, Missouri; Charleston, South Carolina; eastern Tennessee; Montréal (Quebec); and the Adirondack Mountains (New York). A magnitude 7.3 earthquake occurred near Charleston in 1886, ringing church bells up and down the coast and vibrating buildings as far away as Chicago (Fig. 10.17). In Charleston itself, over 90% of the buildings were damaged, and sixty people died. In the early nineteenth century, the region of New Madrid, which lies near the Mississippi River in southernmost Missouri, was inhabited by a

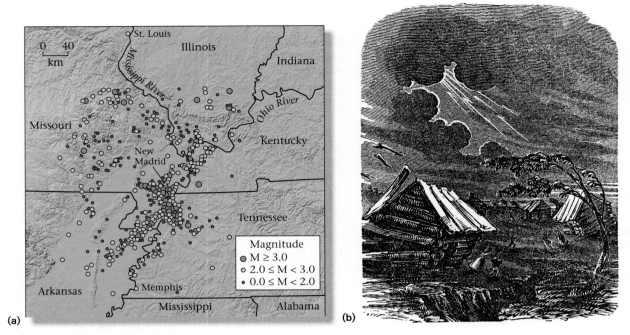

(a)

(b)

FIGURE 10.26 **(a)** The epicenters of earthquakes in the New Madrid, Missouri, area, recorded by modern seismic instruments. The earthquakes, most of which are minor, appear to occur along two sets of ancient, weak faults that break in response to stress in the interior of North America. The New Madrid zone is not a plate boundary. **(b)** The New Madrid earthquake destroyed cabins and disrupted the Mississippi River.

small population of Native Americans and an even smaller population of European descent (▶Fig. 10.26a, b). During the winter of 1811–1812, three magnitude 8 to 8.5 earthquakes struck the region. In the words of a settler:

> The Mississippi first seemed to recede from its banks, and its waters gathered up like a mountain, leaving for a moment many boats on the bare sand. . . . Then, rising 15 to 20 feet perpendicularly . . . the banks overflowed with a retrograde current rapid as a torrent. The boats were now torn from their moorings . . . the river took with it whole groves of young cottonwood trees.[2]

The earthquakes resulted from slip on faults in a Precambrian rift that underlies the Mississippi Valley. Both St. Louis, Missouri, and Memphis, Tennessee, grew close to the epicenter, so earthquakes in the area could be disastrous if they occurred today.

Induced Seismicity

Most earthquakes happen independent of human activity. But the timing of some earthquakes relative to human-

[2]John Gribbin, *This Shaking Earth: Earthquakes, Volcanoes, and Their Impact on Our World* (New York: Putnam, 1978), p. 15.

caused events suggests that, in certain cases, people *can* influence seismicity. **Induced seismicity**, or seismic events caused by actions of people, generally occurs in response to changes in groundwater pressure. In the Earth, the pressure of water can slightly push apart the opposing surfaces of faults and thereby decrease the friction that resists motion on them. So when people increase groundwater pressure, a fault may suddenly slip.

Seismologists observed such a relationship near Denver, Colorado, when engineers pumped waste water from a military installation down a well; as soon as the pumping began, earthquakes started. Induced seismicity is a particular danger when people build dams and create large reservoirs in valleys overlying faults. Faulting generally breaks up rock, making it more erodible, so it is no surprise that deep valleys commonly form over large faults. When a reservoir fills over a fault, water seeps down into the fault and, under the pressure caused by the water column above, triggers earthquakes. A big earthquake could destroy the dam, causing calamity downstream.

> **Take-Home Message**
>
> Most, but not all, earthquakes happen along plate boundaries. Events also occur in rifts and collision zones, and along weak faults within plate interiors. Most catastrophic earthquakes happen at convergent boundaries and along continental transforms.

10.6 HOW DO EARTHQUAKES CAUSE DAMAGE?

An area ravaged by a major earthquake is a heartbreaking sight. The sorrow etched on the faces of survivors mirrors the inconceivable destruction. This destruction comes as a result of many processes.

Ground Shaking and Displacement

An earthquake starts suddenly, with the shock of a P-wave arrival. Generally, P-waves are almost perpendicular to the ground surface when they arrive and cause the ground to buck up and down (►Fig. 10.27a). Next come the S-waves, which also reach the surface at a steep angle (►Fig. 10.27b). These waves are more complicated but tend to cause back-and-forth motion, parallel to the ground surface. Almost immediately afterward L-waves, the first surface waves, arrive and cause snake-like side-to-side motion (►Fig. 10.27c). Finally, the R-waves arrive and cause a rolling motion as particles near the surface of the ground follow elliptical movements (►Fig. 10.27d). Rayleigh waves last longer than other waves and may cause the most damage. Interference among the different kinds of waves causes motion, like choppiness on the surface of a pond, to be anything but regular. Feeling the ground move is a terrifying and disorienting experience that throws off your sense of balance. If the motions are large enough, they may knock you down or even throw you in the air. Ground accelerations caused by moderate earthquakes are in the range of 10 to 20% of g (where g is the acceleration resulting from gravity), and during a great earthquake accelerations may approach 1 g. For great earthquakes, the ground's movement can have an amplitude of as much as 1 m, but for moderate earthquakes, motions fall in the range of a few centimeters or less.

The nature and severity of the shaking at a given location depend on four factors: (1) the magnitude of the earthquake, because larger-magnitude events release more energy; (2) the distance from the hypocenter, because earthquake energy decreases as waves pass through the Earth (►Box 10.2); (3) the nature of the substrate at the location (i.e., the character and thickness of different materials beneath the ground surface); in places underlain by unconsolidated sediment or landfill, earthquake waves tend to be amplified and thus cause more ground motion than they would in hard, coherent bedrock; and (4) the frequency of the earthquake waves (where frequency equals the number of oscillations that pass a point in a specified interval of time). Box 10.2 illustrates how wave frequency affects the intensity of an earthquake.

People can withstand a few g's of acceleration, so if you're out in an open field during an earthquake, ground motion alone won't kill you—you may be knocked off your feet and bounced around a bit, but your body is too flexible to break. Buildings aren't so lucky (►Fig. 10.28a–g). When

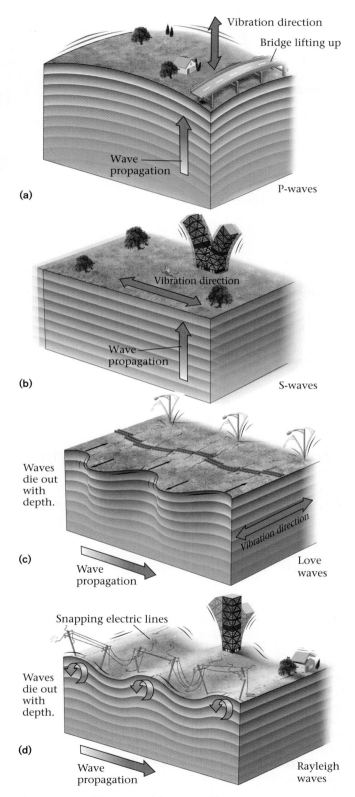

FIGURE 10.27 **(a)** P-waves. **(b)** S-waves. **(c)** Love waves cause the ground to undulate laterally. **(d)** Rayleigh waves make the ground undulate in a rolling motion, like the sea surface.

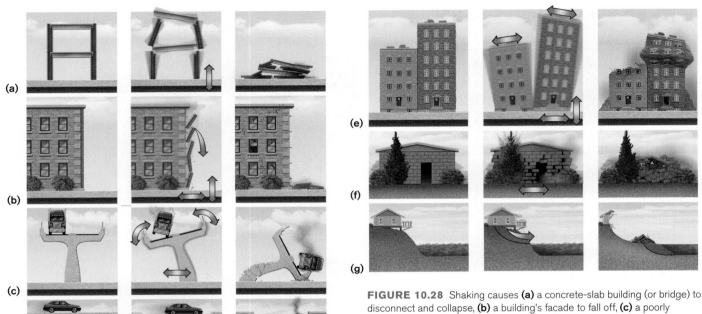

FIGURE 10.28 Shaking causes **(a)** a concrete-slab building (or bridge) to disconnect and collapse, **(b)** a building's facade to fall off, **(c)** a poorly supported bridge to topple, **(d)** a bridge span to disconnect and collapse, **(e)** neighboring buildings to collide and shatter (floors inside a tall building may collapse), **(f)** a concrete-block, brick, or adobe building to crack apart and collapse, **(g)** a steep cliff to collapse, carrying buildings with it.

BOX 10.2 GEOLOGIC CASE STUDY

When Earthquake Waves Resonate—Beware!

When you shine a flashlight, you produce white light. If you aim that beam at a prism, the light spreads into a spectrum of different colors with each color representing light waves of a different frequency. Thus, the original beam of white light contained all these frequencies. Similarly, when an earthquake occurs, it produces seismic waves with a variety of different frequencies. As the waves travel away from the hypocenter, however, the Earth acts like a filter in that high-frequency waves (waves with short wavelengths) lose energy more rapidly than low-frequency waves (waves with long wavelengths). You've experienced this phenomenon if you've ever heard a car stereo playing loud rock music—when you stand near the car, you hear all the sound frequencies and can make out soprano voices and high-frequency guitar notes, but if you are far away, all you can hear are the low-frequency thump-thump-thumps of the bass guitar and bass drum. Because of this effect, the frequency content (the variety of different wave frequencies) of the earthquake changes with distance from the hypocenter.

Why is frequency content important? Different frequencies of waves cause different amounts of ground acceleration. In particular, high-frequency waves cause more acceleration than low-frequency waves. The frequency of waves is also important because of a phenomenon called resonance. **Resonance** happens when each new wave arrives at just the right time to add more energy to a system. To understand resonance, picture a boy on a swing. If the boy pumps his legs at just the right time, he swings higher, but if not he slows down. The same phenomenon occurs if you slide a block of Jell-O that is resting on a plate back and forth on a table. If you move the plate too fast, the Jell-O merely trembles, but if your motion is at just the right frequency, resonance begins and the block sways wildly.

Resonance played a major role in accentuating the damage during an earthquake in Mexico City. On September 19, 1985, a magnitude 8.1 earthquake occurred 350 km away, on the convergent plate boundary along which the Cocos Plate grinds beneath North America. Though the epicenter was far away, ground movements in Mexico City were very intense. That's because Mexico City sat on a thick sequence of unconsolidated lake-bed sediments, exposed when the Spanish conquistadors drained Lake Texcoco. The sedimentary basin, somewhat like a lens, focused earthquake energy on the city. Because of the distance from the epicenter, high-frequency waves had weakened, but because of the nature of the sedimentary basin, low-frequency waves were amplified. These waves had just the right frequency to make buildings between eight and eighteen stories high begin to resonate. In some cases, neighboring buildings slammed together like clapping hands. Engineers had not designed the buildings to accommodate such motion, so many buildings collapsed. Between 8,000 and 30,000 people died, and another 250,000 were left homeless.

earthquake waves pass, buildings sway, twist back and forth, or lurch up and down, depending on the type of wave motion. As a result, connectors between the frame and the facade of a building may separate, so the facade crashes to the ground. The flexing of walls shatters windows and makes roofs collapse. Floors may rise up and slam down on the columns that support them. This motion can crush the columns, so some buildings collapse with their floors piling on top of one another like pancakes in a stack (▶Fig. 10.29a). The majority of earthquake-related deaths and injuries result when people are hit by flying debris or are crushed beneath falling walls or roofs. Aftershocks worsen the problem, because they may be the final straw that breaks already weakened buildings, trapping rescuers (▶Fig. 10.29b).

Roads, rail lines, and bridges suffer the same fate as buildings: ground motion breaks and buckles them (▶Fig. 10.29b–e). In the case of bridges, the span may rise off its supports, then slam down and buckle them so the bridge collapses, or the span may slip off the supports. For unlucky drivers on the bridge, the road literally disappears in front of them.

Ground motion can cause water in lakes, bays, reservoirs, and pools to slosh back and forth, in some cases thousands of kilometers from the epicenter. The water's rhythmic movement, known as a **seiche**, can build up waves almost 10 m high and can last for hours. Seiches capsize small boats and flood shoreline homes. And if they occur in reservoirs, seiches may wash over and weaken retaining dams.

If the ruptured portion of a fault intersects the ground surface, it displaces the ground surface. Strike-slip faults cause lateral displacements, while normal or reverse faults cause vertical displacements and fault scarps. If a building, fence, road, pipeline, or rail line straddles a fault trace, slippage on the fault can crack the structure and separate it into two pieces. Note that the amount of displacement on a fault varies along the length of the fault, and is not necessarily greatest at the epicenter.

Landslides and Avalanches

The shaking of an earthquake can cause ground on steep slopes or ground underlain by weak sediment to give way. This movement results in a landslide, the tumbling and/or flow of soil and rock downslope. A landslide destroys everything that it carries—trees, buildings, roads—and most of what's in its path, and may completely bury roads and buildings at the base of the slope. Avalanches of snow may also be triggered by earthquakes and, like landslides, cause havoc in their path.

Landslides commonly occur along the coast of California, for movement on faults has rapidly uplifted this coastline in the past few million years, resulting in the development of steep cliffs. When earthquakes take place,

the cliffs collapse, locally carrying expensive homes down to the beach below (▶Fig. 10.30a). Such events lead to the misperception that "California will someday fall into the sea." Although small portions of the coastline do collapse, the state as a whole remains firmly attached to the continent, despite what Hollywood scriptwriters say.

Landslides into water may cause huge waves. For example, in 1958, a magnitude 8.3 earthquake in southeastern Alaska triggered a landslide at the head of Lituya Bay. The splash from the displaced water washed the forest off the opposite wall of the bay up to an elevation of 516 m.

Sediment Liquefaction

Places where the substrate consists of wet sediment can be particularly hazardous during an earthquake. In beds of wet sand or silt, ground shaking causes the sediment grains to try to settle together. But because the open spaces (pores) between grains contain water, water pressure in the pores increases. When this happens, the friction that held the grains in place suddenly decreases, and the wet silt or sand becomes a fluid-like slurry—it becomes quicksand. Similarly, in certain types of damp clay, the clay flakes stick together only because of weak hydrogen bonds (see Appendix A). When still, the clay behaves like a solid gel, but when it is shaken, the hydrogen bonds break and the clay transforms into a viscous liquid. Clay that displays this behavior is called thixotropic clay, or quickclay. The behavior of thixotropic clay resembles that of ketchup—if you slowly turn a bottle of ketchup over, the ketchup doesn't flow, but if you shake the bottle first, the ketchup flows easily. The abrupt loss of strength of a wet sediment (either sand or clay) in response to ground shaking is called **liquefaction**, and it can cause major damage during an earthquake. Buildings whose foundations lie in liquefied material may sink or even tip over (▶Fig. 10.30b).

If liquefaction occurs in the sediment beneath a slope, the ground above may give way and slide down the slope. Such an event happened during the 1964 Alaska earthquake, when the sediment beneath a housing development in the coastal suburb of Turnagain Heights liquefied (▶Fig. 10.31a–c). The land broke into a series of blocks that slumped seaward on the liquefied sediment, carrying with it dozens of houses that transformed into a jumble of splintered wood and shattered windows. Along 4th Avenue, in the city of Anchorage, the ground level dropped by as much as 3 m due to slumping.

In some cases, the liquefaction of sand layers below the ground surface makes the sand erupt through holes or cracks in overlying sediment, producing small mounds of sand called sand volcanoes, or sand boils. Sometimes fountains of wet sand from underground spurt 10 m into the sky during an earthquake. Liquefaction may also cause bedding in unconsolidated sequences of sediment to break

FIGURE 10.29 (a) During a 1999 earthquake in Turkey, this concrete building collapsed when the supports gave way and piled onto one another like pancakes. (b) New apartment buildings are commonly placed on columns, making room for parking below. When the columns collapse, the building crushes parked cars. (c) A neighborhood of houses was flattened during a 1999 magnitude 6.0 earthquake in Armenia. (d) Reinforced concrete bridge supports were crushed during the 1994 Northridge earthquake. (e) An elevated bridge, balanced on a single row of columns, simply tipped over during the 1995 Kobe, Japan, earthquake.

FIGURE 10.30 **(a)** A landslide along the California coast, triggered by the Northridge earthquake, carried part of a luxury home down to the beach below. **(b)** The foundation of this apartment complex in Hsingchung, Taiwan, liquefied during a 1999 earthquake, and the buildings tipped over, intact.

(a)

(b)

FIGURE 10.31 The Turnagain Heights (Alaska) disaster. **(a)** Before the earthquake, wet clay packed together in a subsurface layer of compacted but wet mud. **(b)** Ground shaking caused liquefaction of the wet clay layer in the sediment beneath a housing development. As a consequence, the land slumped and slid seaward on an array of sliding surfaces, carrying the houses with it. The ground broke into many slices. **(c)** The housing development after the slide.

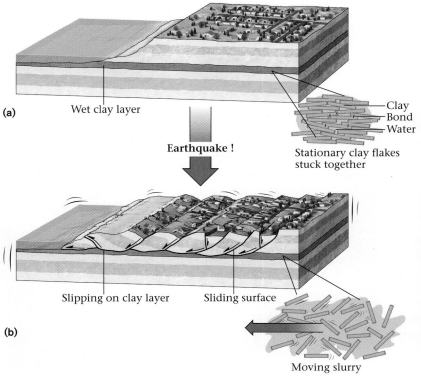

(a) Wet clay layer

Earthquake !

Clay
Bond
Water

Stationary clay flakes stuck together

(b) Slipping on clay layer   Sliding surface

Moving slurry

(c)

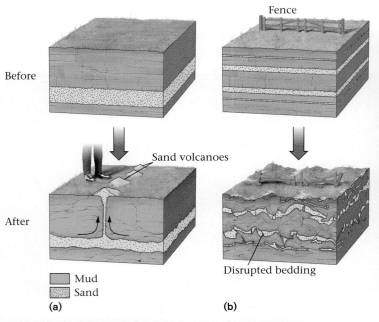

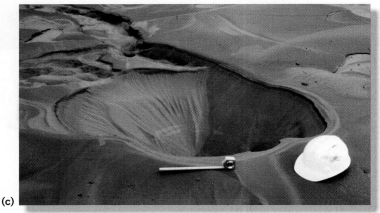

FIGURE 10.32 (a) Ground shaking may cause wet sand buried below the surface to squirt up between cracks in overlying compacted mud. The result is a row of sand volcanoes. Generally, these are very small, but those formed after the 1811 New Madrid (Missouri) quake were tens of meters across. (b) Thin layers of sand interlayered with mud may get disrupted, folded, and broken up as a result of ground shaking. (c) A sand volcano. The tape measure indicates scale. (d) Photo of ground cracking due to liquefaction of sediment underground.

FIGURE 10.33 Broken gas mains erupt fountains of flame.

up and become contorted (▶Fig. 10.32a–d). Ground settling due to underlying liquefaction may cause large cracks to develop in the overlying sediments, and these may gape open at the ground surface.

Fire

The shaking during an earthquake can make lamps, stoves, or candles with open flames tip over, and it may break wires or topple power lines, producing sparks. As a consequence, areas already turned to rubble, and even areas not so badly damaged, may be consumed by fire. Ruptured gas pipelines and oil tanks feed the flames, sending columns of fire erupting skyward (▶Fig. 10.33). Firefighters might not even be able to reach the fires, because the doors to the firehouse won't open or rubble blocks the streets. Moreover, firefighters may find themselves without water, for ground shaking can damage water lines.

If the fire gets a good start, it can become an unstoppable inferno. Most of the destruction of the 1906 San Francisco earthquake, in fact, resulted from a fire that ignited after the shaking had stopped. For 3 days, the blaze spread through the city until firefighters contained it by blasting a firebreak. But by then, 500 blocks of structures had turned to ash, causing 20 times as much financial loss as the shaking itself. When a large earthquake hit Tokyo in September 1923, fires set by cooking stoves spread quickly through the wood-and-paper buildings, creating an inferno that heated the air above the city. The hot air rose like a balloon, and when cool air rushed in, creating wind gusts of over 100 mph, the wind stoked the blaze and incinerated 120,000 people.

Tsunamis

The azure waters and palm-fringed islands of the Indian Ocean's east coast hide one of the most seismically active plate boundaries on Earth—the Sunda trench. Along this convergent boundary, the Indian Ocean floor subducts at about 6 cm per year, leading to slip on large thrust faults. Just before 8:00 A.M. on December 26, 2004, the crust above a 1,300-km-long by 100-km-wide portion of one of these faults lurched up and westward by as much as 15 m. The result was a magnitude 9.3 earthquake—a true giant that released as much energy as a 250-megaton hydrogen bomb, more than 4 times larger than the largest bomb ever ex-

ploded, and more powerful than all the earthquakes of the previous 10 years combined. The rupture propagated northward from the epicenter at a rate of 0.5 to 3 km per second, ripping the Earth's surface as it passed. The length of the fault that eventually ruptured was so huge—the largest ever recorded—that the earthquake lasted 9 minutes. This slip pushed the sea floor up by tens of centimeters. The rise of the sea floor, in turn, shoved up the overlying water. Because the area that rose was so broad, the volume of displaced water was immense. As a consequence, a tragedy of unimaginable extent was about to unfold.

Water from above the upthrust sea floor began moving outward from above the fault zone, a process that generated a series of giant waves, or **tsunamis**, traveling at speeds of about 800 km/hour (500 mph)—almost the speed of a jet plane (▶Fig. 10.34a–d). Fifteen minutes later, the first tsunami struck Banda Aceh, a city at the north end of the island of Sumatra, as it was waking to a beautiful, cloudless day. First, the sea receded much farther than anyone had ever seen, exposing large areas of reefs that normally remained submerged even at low tide. People walked out onto the exposed reefs in wonder. But then, with a rumble that grew to a roar, a wall of frothing water began to build in the distance and approach land. Puzzled bathers first watched, and then ran inland in panic when the threat became clear (▶Fig. 10.35a–c). As the tsunami approached shore, friction with the seafloor had slowed it to less than 30 km/hour, but it still moved faster than people can run (▶Fig. 10.35d). In places, the wave front reached heights of 15 to 30 m (45 to 100 feet) as it slammed into Banda Aceh (▶Fig. 10.35e).

FIGURE 10.34 Producing tsunamis. **(a)** Before a tsunami forms. **(b)** Sinking of the sea floor above a normal fault creates a void, and water rushes to fill it. Water mounds up over the fault. **(c)** A similar process happens in response to a reverse or thrust fault. **(d)** This time, the rising sea floor shoves up the water surface.

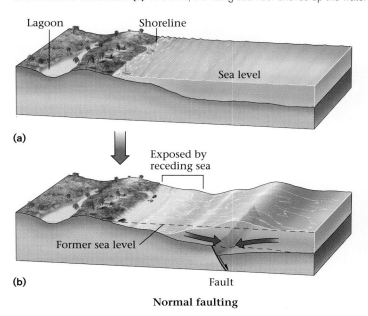

Normal faulting

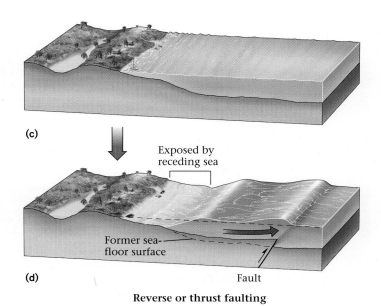

Reverse or thrust faulting

(a) Before the wave.

(b) Water recedes—the wave appears on the horizon.

(c) The wave reaches the shore and engulfs boats.

(d)

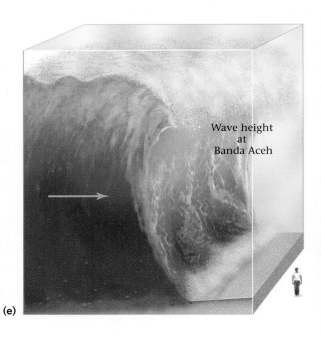

Wave height at Banda Aceh

(e)

FIGURE 10.35 The Indian Ocean tsunami disaster of 2004. **(a)** The tranquil coast of Sumatra, before the wave. **(b)** Water recedes and exposes a coral reef, while the wave builds on the horizon. **(c)** The wave inundates the coast of Sumatra, submerging sailboats anchored off shore. **(d)** The wave blasts through a grove of palm trees as it strikes the coast of Thailand. **(e)** At its highest, the tsunami's front was over 15 m high.

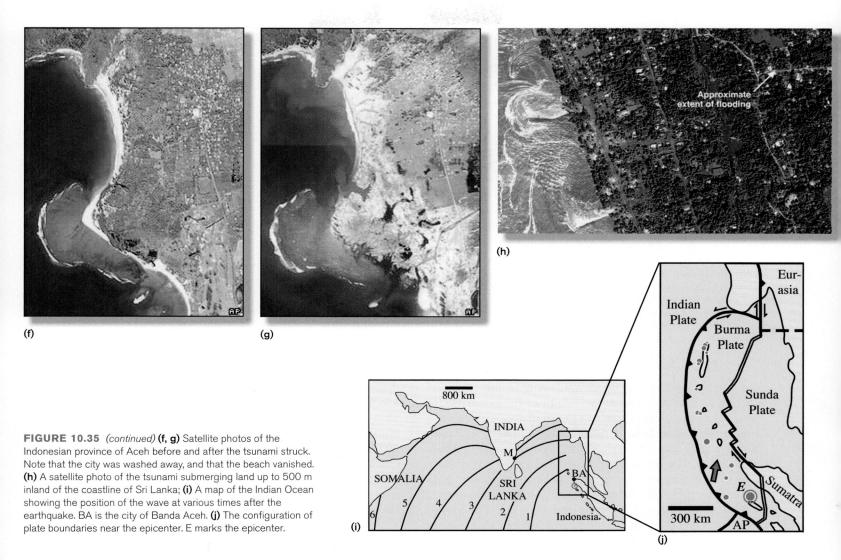

FIGURE 10.35 *(continued)* **(f, g)** Satellite photos of the Indonesian province of Aceh before and after the tsunami struck. Note that the city was washed away, and that the beach vanished. **(h)** A satellite photo of the tsunami submerging land up to 500 m inland of the coastline of Sri Lanka; **(i)** A map of the Indian Ocean showing the position of the wave at various times after the earthquake. BA is the city of Banda Aceh. **(j)** The configuration of plate boundaries near the epicenter. E marks the epicenter.

The impact of the water ripped boats from their moorings, snapped trees, battered buildings into rubble, and tossed cars and trucks like toys. And the water just kept coming, eventually flooding low-lying land as far as 7 km inland. It drenched forests and fields with salt water (deadly to plants) and buried fields and streets with up to a meter of sand and mud (►Fig. 10.35f–h). Eventually the water slowed and then began to rush back to the shore, but at Banda Aceh, at least two more tsunamis struck before the first one had entirely receded, so the water remained high for some time. When the water level finally returned to normal, a jumble of flotsam, as well as the bodies of unfortunate victims, floated out to sea and drifted away forever.

Geologists refer to the tsunami that struck Banda Aceh as a near-field (or local) tsunami, because of its proximity to the earthquake. But the horror of Banda Aceh was merely a preamble to the devastation that would soon visit other stretches of Indian Ocean coast. Far-field (or distant)

tsunamis crossed the ocean and struck Sri Lanka 2.5 hours after the earthquake, the coast of India half an hour later, and the coast of Africa, on the west side of the Indian Ocean, 5.5 hours after the earthquake (►Fig. 10.35i, j). Coastal towns vanished, fishing fleets sank, and beach resorts collapsed into rubble. In the end, more than 220,000 people died that day.

Tsunami is a Japanese word that translates literally as "harbor wave," an apt name because tsunamis can be particularly damaging to harbor towns. This name replaces the older term, "tidal wave." Tsunamis have no relationship to tides, and calling them tidal waves can be misleading. Tsunamis can be caused by undersea earthquakes but, as we will see in Chapter 16, they can also be set off by large undersea landslides.

Regardless of cause, tsunamis are very different from familiar, wind-driven storm waves. Large, wind-driven waves can reach heights of 10 to 30 meters in the open ocean. But

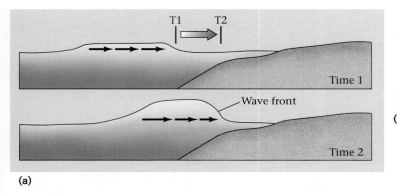

(a)

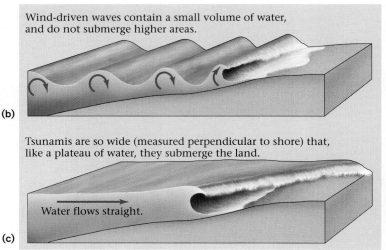

Wind-driven waves contain a small volume of water, and do not submerge higher areas.

(b)

Tsunamis are so wide (measured perpendicular to shore) that, like a plateau of water, they submerge the land.

Water flows straight.

(c)

FIGURE 10.36 (a) As a tsunami approaches the shore, friction between water and the sea floor slows the wave, allowing the back part of the wave to catch up with the front part. As a consequence, the wave becomes much higher than it was in the open ocean as the front moves from its position at Time 1 to its position at Time 2. **(b)** Wind-driven waves may be quite high, but they are so narrow that they contain relatively little water, and they only go partway up the beach until the water stops moving landward. The water in such a wave has a circular motion. **(c)** When a tsunami washes ashore, the water moves landward in a straight line. A tsunami is like a plateau of water—it contains so much water that the water keeps moving inland and submerges a broad area.

even such monsters have wavelengths of only tens of meters, and thus affect only the surface realm of the ocean. (Water beneath a wave moves down to a depth of about half a wavelength, so a submarine cruising at 100 m depth will move through calm water, even if a storm rages above.) In contrast, although a tsunami in deep water may cause a rise in sea level of at most only a few tens of centimeters—a ship crossing one wouldn't even notice—tsunamis have wavelengths of tens to hundreds of kilometers and thus affect the entire depth of the ocean. In simpler terms, we can think of the width of a tsunami, in map view, as being more than 100 times the width of a wind-driven wave. Because of this difference, a storm wave and a tsunami have very different effects when they strike the shore.

When a water wave approaches the shore, friction between the base of the wave and the sea floor slows the bottom of the wave, so the back of the wave catches up to the front, and the added volume of water builds the wave higher (▶Fig. 10.36a). The top of the wave may fall over the front of the wave and cause a breaker. In the case of a wind-driven wave, the breaker may be tall when it washes onto the beach, but the wave is so narrow that it doesn't contain much water. Thus, the wave makes it only partway up the beach before it runs out of water, friction slows it to a stop, and gravity causes the water to spill back seaward. In the case of a tsunami, the wave is so wide that, as friction slows the wave, it builds into a plateau that can be tens of meters high and many kilometers wide. Thus, when a tsunami reaches shore, it contains so much water that it crosses the beach and just keeps on going (▶Fig. 10.36b, c).

The 2004 Indian Ocean event remains etched in people's minds because of the immense death toll and the nonstop news coverage. But it is not unique. Tsunamis generated by the great Chilean earthquake of 1960 ($M_w = 9.5$) destroyed coastal towns of South America and crossed the Pacific, causing a 10.7-m-high wall of water to strike Hawaii 15 hours later. When, 21 hours after the earthquake, the tsunami reached Japan, it flattened coastal villages and left 50,000 people homeless. The tsunami following the 1964 Good Friday earthquake in Alaska destroyed ports at Valdez and Kodiak (▶Fig. 10.37a, b). Typically, an earthquake generates several tsunamis that arrive on distant shores as much as an hour apart (▶Fig. 10.37c).

Because of the danger of tsunamis, predicting their arrival can save thousands of lives. A tsunami warning center in Hawaii keeps track of earthquakes around the Pacific and uses data relayed from tide gauges and sea-floor pressure gauges to determine whether a particular earthquake has generated a tsunami (▶Fig. 10.38a). If observers detect a tsunami, they flash warnings to authorities around the Pacific. In recent years, to help in the effort, researchers have used computer models that predict how tsunamis propagate (▶Fig. 10.38b). Unfortunately, no warning system existed in the Indian Ocean, and even though Hawaiian observers detected the earthquake and realized that a tsunami was likely, they did not have the means to contact many local authorities. Even if they had, affected towns had no evacuation plans in place. A multinational team has begun work to remedy this situation.

> **Take-Home Message**
>
> Earthquakes cause devastation because ground shaking topples buildings. But destruction can also be caused by landslides, tsunamis, and sediment liquefaction triggered by the shaking. After an earthquake, survivors face the threats of fire and disease.

(a)

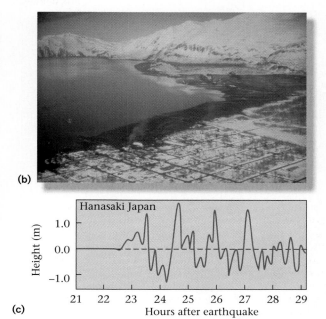

(b)

(c)

FIGURE 10.37 (a) In the aftermath of a tsunami that hit Kodiak, Alaska, after the 1964 earthquake, boats rest amid the rubble of a coastal town. (b) During the 1964 earthquake, the tsunami also struck Valdez. The wave washed away the snow near the shore. (c) Tidal gauge at Hanasaki, Japan, 12 hours after the great 1960 earthquake in Chile.

Disease

Once the ground shaking and fires have stopped, disease may still threaten lives in an earthquake-damaged region. Earthquakes cut water and sewer lines, destroying clean water supplies and exposing the public to bacteria, and they cut transportation lines, preventing food and medicine from reaching the city. The severity of such problems depends on the ability of emergency services to cope. During the January 1999 magnitude 6.3 earthquake in Colombia, over 2,000 people died and tens of thousands were left homeless. Cut off from aid by impassable mountain roads, survivors faced the threat of cholera and starvation until an airlift could bring in supplies. After the 2005 Pakistan earthquake, the inability to reach people in the quake-ravaged area contributed to the deaths of an additional 200,000 people.

10.7 CAN WE PREDICT THE "BIG ONE"?

We have seen that large earthquakes occurring near population centers cause catastrophe. Needless to say, many lives could be saved if only it were possible to know where an earthquake will happen, so people could build stronger structures and develop plans for dealing with the consequences, or when an earthquake will happen, so people could evacuate dangerous buildings, and turn off gas and electricity.

So . . . can seismologists predict where and when earthquakes will occur? The answer depends on the time frame of the prediction. With our present understanding of the distribution of seismic zones and the frequency at which

(a)

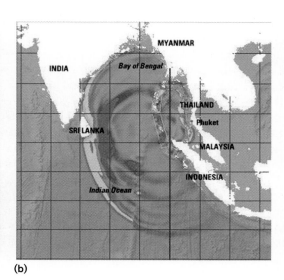

(b)

FIGURE 10.38 (a) Tsunami buoy. (b) A computer model showing the 2004 Indian Ocean tsunami about 2 hours after the earthquake. The yellow areas have the greatest wave height.

See for yourself . . .
Seismically Active Faults

In many locations, faulting during an earthquake ruptures the ground. Below, we visit examples of fault zones that have left a mark on the land surface. The journey also illustrates potential hazards associated with seismic activity. The thumbnail images provided on this page are only to help identify tour sites. Go to wwnorton.com/studyspace to experience this flyover tour.

San Andreas Fault near Palmdale, California
(Lat 34°30'35.20"N, Long 118°1'8.12"W)

If you fly to the above coordinates and zoom up to an elevation of 350 km (220 miles), you will see the coast of California, the San Gabriel Mountains, and a triangular area of high, flatter land that comprises the western Mojave Desert (Image G10.1). The San Andreas fault is the NW-trending line at the bottom of the triangle and the Garlock fault is the NE-trending line at the top. Zoom to an elevation of 9 km (5 miles), and you'll see a stream flowing north from the San Gabriels, near Palmdale. Where the stream crosses the fault, the channel steps to the right by over 1 km (0.6 miles), due to the displacement on the fault (Image G10.2). Narrow pressure ridges and sag ponds delineate the trace of the fault. Tilt the image and reorient so you are looking northwest, along the trace of the fault. You can see canals, roads, and reservoirs that cross the fault (Image G10.3). Fly along the fault and you will get a sense of how the fault disrupts the landscape.

San Andreas Fault, San Francisco
(Lat 37°34'41.53"N, Long 122°24'41.88"W)

Here, from an altitude of 20 km (13 miles), the trace of the San Andreas fault runs parallel to Highway 280, along the peninsula between San Francisco Bay on the east and the Pacific Ocean on the west. Erosion along the fault trace has produced a narrow NW-trending valley (Image G10.4). The dam straddling the fault holds back the San Andreas Reservoir. Zoom down to an elevation of 2 km (1.3 miles), tilt your view, and orient the image so that you are looking along the fault. Now, slowly fly NW, along the trace of the fault. You'll reach the end of the San Andreas Reservoir, cross housing developments, and eventually reach the ocean (Image G10.5). Fly a short distance up the coast to Lat 37°40'35.70"N, Long 122°29'35.51"W, head slightly out to sea, turn your view so you are looking east, and look back at the coast. You'll see a cliff, with waves lapping at its base (Image G10.6). This cliff formed because of tectonic uplift associated with faulting. Along the cliff, there are several spoon-shaped indentations—these are slump scarps, places where the cliff face slid down to the sea relatively recently. Places along the cliff face that are devoid of vegetation represent the most recent slumps. Slumping may be triggered by earthquakes. Note that developers have built homes up to the cliff edge, and onto the protrusions between existing slumps.

G10.1

G10.2

G10.3

G10.4

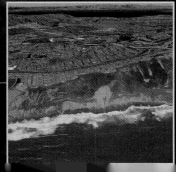

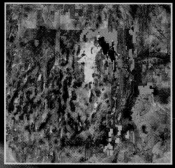

G10.7

G10.8

G10.9

Salt Lake City, Utah
(Lat 40°55'22.81"N, Long 111°51'52.48"W)

Fly to these coordinates just north of Salt Lake City and rise to an elevation of about 1500 km (930 miles). You can see the Basin and Range Province, an active rift that has stretched the crust between Salt Lake City and Reno (Image G10.7). The Colorado Plateau lies to the east, and the Sierra Nevada to the west. The Wasatch Mountains form a high ridge of uplifted land along the western edge of the Colorado Plateau. Zoom down to an elevation of 8 km (5 miles) and tilt the image (Image G10.8). You are looking at the Wasatch Front, an escarpment that delineates the western face of the Wasatch Mountains, just east of I-15. The Wasatch Front is a somewhat eroded fault scarp, formed when normal faulting dropped the Great Salt Lake basin down, relative to the Wasatch Mountains. Note how the scarp truncates the ridges between the streams flowing out of the mountains. Fly south along the range front to Lat 40°35'36.44"N, Long 111°47'27.01"W, and you will get another view of the Wasatch Front. Zoom to 7 km (4 miles), tilt the image to see the horizon, and look east (Image G10.9). This image also emphasizes how mountain building can be a long-term consequence of continued faulting.

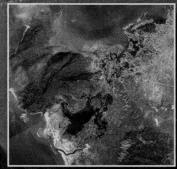

G10.10

G10.11

G10.12

Tsunami Damage, Banda Aceh
(Lat 5°30'46.75"N, Long 95°16'14.68"E)

Fly to these coordinates and zoom up to an elevation of 16 km. You will see an isthmus near the city of Banda Aceh, at the north end of Sumatra, that was attacked on both sides by the December, 2004 tsunami (Image G10.10). The land that was submerged has turned brown, because the saltwater waves washed away most vegetation and killed the remainder. Now, zoom down to the northern part of the image (Lat 5°33'25.50"N, Long 95°17'12.52"E), drop to an elevation of 1 km (0.6 miles), and tilt the image to look north (Image G10.11). The seaside part of the city was totally destroyed. Rotate the image so you are looking south, and fly to about Lat 5°31'54.31"N, Long 95°17'38.25"E. Here you see the boundary between the land that was submerged and the land that was not—the unscathed land is much greener (Image G10.12). Note that the highway in the center of the image nearly disappears as it crosses the boundary.

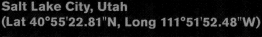

earthquakes occur, we can make long-term predictions (on the time scale of decades to centuries). For example, with some certainty, we can say that a major earthquake will rattle western California during the next 100 years, and that a major earthquake probably won't strike central Canada during the next ten years because western California straddles a plate boundary and central Canada does not. But despite extensive research, seismologists cannot make accurate short-term predictions (on the time scale of hours to weeks). Thus, we *cannot* say, for example, that an earthquake will happen in New Madrid, Missouri at 2:43 P.M., on January 17, 2020.

In this section, we look at the scientific basis of both long- and short-term predictions, and consider the consequences of a prediction. Seismologists refer to studies leading to predictions as **seismic-risk**, or seismic-hazard **assessment**. On the basis of this work, they produce maps that assign levels of seismic risk to different regions.

Long-Term Predictions

When making a long-term prediction, we use the word *probability*, because a prediction only gives the likelihood of an event. For example, a seismologist may say, "The probability of a major earthquake occurring in the next 20 years in this state is 20%." This sentence implies that there's a one-in-five chance of the earthquake happening during a 20-year period. Urban planners and civil engineers can use long-term predictions to help develop building codes for a region—stronger buildings make sense for regions with greater seismic risk. They may also use predictions to determine whether or not to build vulnerable structures such as nuclear power plants, hospitals, or dams in potentially seismic areas. Seismologists base long-term earthquake predictions on two pieces of information: the identification of seismic zones and the **recurrence interval** (the *average* time between successive events).

To identify a seismic zone, seismologists produce a map showing the epicenters of earthquakes that have happened during a set period of time (say, 30 years). Clusters or belts of epicenters define the seismic zone (see Figs. 10.26 and 10.23a). The basic premise of long-term earthquake prediction can be stated as follows: a region in which there have been many earthquakes in the past will likely experience more earthquakes in the near future. Seismic zones, therefore, are regions of greater seismic risk. This doesn't mean that a disastrous earthquake can't happen far from a seismic zone—they can and do—but the risk is less.

Epicenter maps can be produced using data from only the past 50 years or so, because before that time seismologists did not have enough seismographs to locate epicenters accurately. Fortunately, geologic mapping can contribute information about faulting back through longer periods of time, because seismic zones by definition contain active faults. Thus, geologists can assess seismic risk by examining landforms for evidence of recent faulting. For example, they can look for young fault scarps, determine whether recent sedimentary deposits have been cut by faults, and decide whether faulting has offset stream channels or has created small ponds or ridges (▶Fig. 10.39a–c).

To determine the recurrence interval for large earthquakes in a seismic zone, seismologists must determine when large earthquakes have happened in the past. Since the historical record does not provide information far enough back in time, we must study geologic evidence for great earthquakes. For example, we can examine sedimentary strata near a fault to find layers of sand volcanoes and disrupted bedding in the stratigraphic record. Each layer, whose age can be determined by using radiocarbon dating of plant fragments, records the time of an earthquake. We can obtain additional information by dating offset soil horizons now buried beneath other sediment (▶Fig. 10.39d).

As an example, imagine that disrupted layers formed 260, 820, 1,200, 2,100, and 2,300 years ago. By calculating the number of years between successive events and taking the average, we can say that the recurrence interval between events is about 510 years. The **annual probability** of an earthquake is 1/recurrence interval. This means that the probability of an earthquake happening in any given year is 1 in 510, or about 0.2%. (See caption 10.40c for a different example.) Note that a recurrence interval does not specify the exact number of years between events, only the average number. Since the public tends to misinterpret the recurrence interval, and thinks that it indicates the *exact* number of years between events, a statement of annual probability may give a better sense of seismic risk. With such information, seismologists are able to produce regional maps illustrating seismic risk (▶Fig. 10.40a–c).

Seismologists suspect that places, called **seismic gaps**, where a known active fault has not slipped for a long time may be particularly dangerous (▶Fig. 10.40d). In a seismic gap, either the fault must be moving by nonseismic fault creep, or the stress is building up to be released by a major earthquake in the future. Many seismic gaps have been identified along the seismic belts bordering the Pacific Ocean.

Short-Term Predictions

Short-term prediction, which could lead to such precautions as evacuating dangerous buildings, shutting off gas and electricity, and readying emergency services, is not reliable and may never be. Chinese seismologists claimed to have predicted that a major earthquake would happen near Haicheng during the first week in February 1975, and authorities evacuated the city before a major quake did strike

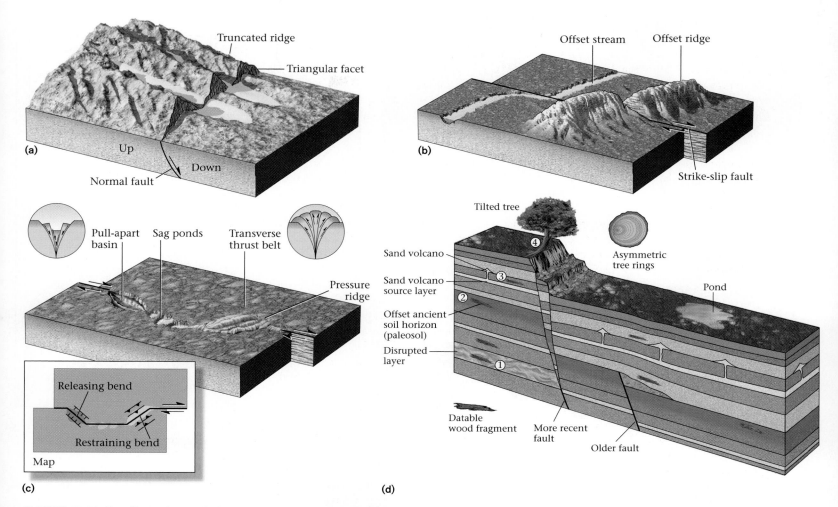

FIGURE 10.39 The effects of recent fault movement on the land surface. **(a)** A normal fault truncates ridges, creating triangular facets. **(b)** A strike-slip fault offsets stream channels and ridges that cross it. **(c)** A strike-slip fault may bend along its length. Some bends restrain motion and therefore feel compression, whereas some bends do not and thus open up during fault motion. Where there is compression, you find pressure ridges or transverse thrust belts. Where there is extension, you find sag ponds or pull-apart basins. The inset map shows the orientation of the restraining bend and the releasing bend, relative to the slip direction on the fault. **(d)** Here, earthquake events are represented by disrupted bedding (1); an offset ancient soil horizon, or paleosol (2); a layer of sand volcanoes (3); and a bent tree caused by the formation of a fault scarp (4). We can date these events from the wood fragments in the strata. If the disrupted bedding formed in 100 B.C.E., the offset paleosol in 800 C.E., the sand volcanoes in 1340, and the tilted tree in 1950, then the time gaps between earthquakes are 900, 540, and 610 years. The recurrence interval is 683 years (900 + 540 + 610 divided by 3), with a large degree of uncertainty.

on February 4. But no one predicted the T'ang-shan earthquake of 1976, which killed over a quarter-million people, the largest death toll of any earthquake in the twentieth century. This failure hints that the earlier prediction may have been a coincidence. More recently, seismologists predicted that an earthquake would take place in 1988 on the San Andreas fault in Parkfield, California, because an earthquake has occurred there every 20 to 22 years since 1857. But the prediction proved false—an earthquake did not happen in Parkfield until 2004. In fact, many seismologists feel that seismicity is a somewhat random process that can't be accurately predicted.

Nevertheless, there are clues to imminent earthquakes, and seismologists have been working hard to understand them. The first clue comes from the detection of foreshocks (▶Fig. 10.41). A swarm (a cluster of events during a short period of time) of foreshocks may indicate the cracking that precedes a major rupture along a fault zone. In this regard, foreshocks are analogous to the cracking noises you hear just before a tree limb breaks off and falls down. But foreshocks do not always occur, and even if they do, they usually can be identified only in hindsight, because they may be indistinguishable from other small earthquakes in a seismic zone.

Another possible data source for short-term predictions comes from precise laser surveying of the ground. Before an earthquake, a region of crust may undergo distortion, either in response to the buildup of elastic

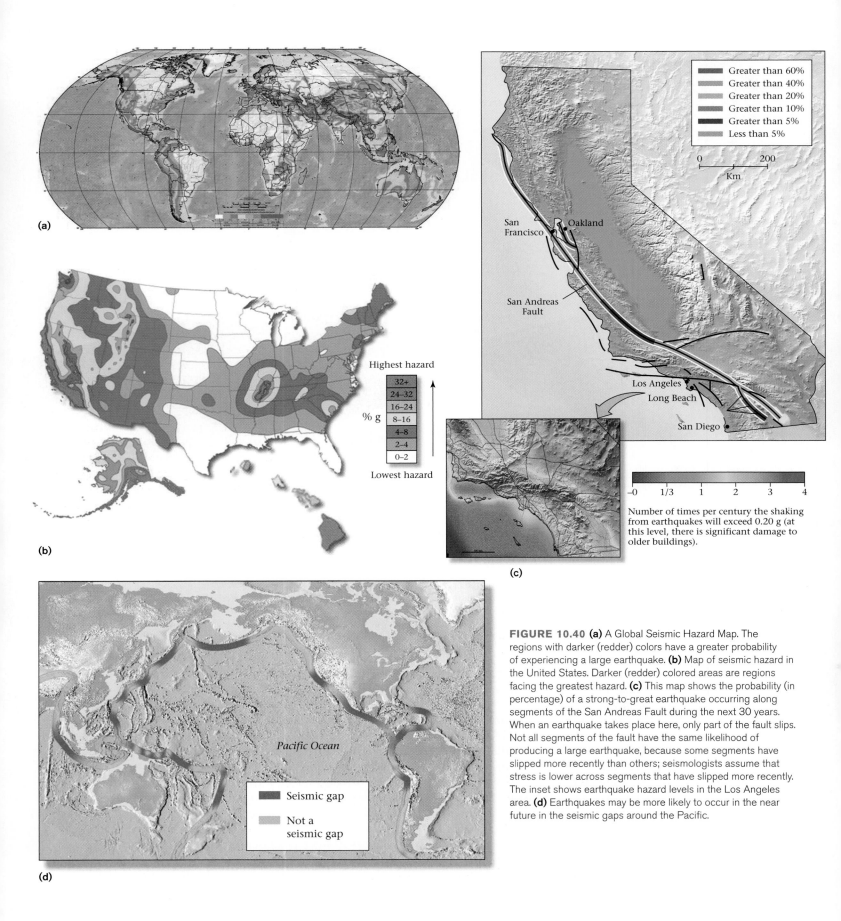

(a)

(b)

Highest hazard

32+
24–32
16–24
% g 8–16
4–8
2–4
0–2

Lowest hazard

(c)

Greater than 60%
Greater than 40%
Greater than 20%
Greater than 10%
Greater than 5%
Less than 5%

0   200
Km

San Francisco
Oakland

San Andreas Fault

Los Angeles
Long Beach

San Diego

–0   1/3   1   2   3   4

Number of times per century the shaking from earthquakes will exceed 0.20 g (at this level, there is significant damage to older buildings).

(d)

Pacific Ocean

Seismic gap

Not a seismic gap

FIGURE 10.40 (a) A Global Seismic Hazard Map. The regions with darker (redder) colors have a greater probability of experiencing a large earthquake. **(b)** Map of seismic hazard in the United States. Darker (redder) colored areas are regions facing the greatest hazard. **(c)** This map shows the probability (in percentage) of a strong-to-great earthquake occurring along segments of the San Andreas Fault during the next 30 years. When an earthquake takes place here, only part of the fault slips. Not all segments of the fault have the same likelihood of producing a large earthquake, because some segments have slipped more recently than others; seismologists assume that stress is lower across segments that have slipped more recently. The inset shows earthquake hazard levels in the Los Angeles area. **(d)** Earthquakes may be more likely to occur in the near future in the seismic gaps around the Pacific.

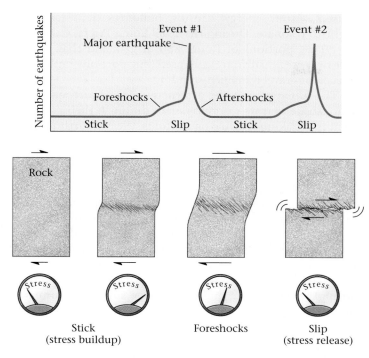

FIGURE 10.41 A sudden increase in the number of small earthquakes along a fault segment may be a possible precursor to a large earthquake along that segment; that is, the small earthquakes may be foreshocks. The pattern may represent stick-slip behavior.

strain in the rock, or because of the development of small, open cracks that cause the crust to increase in volume. The land surface may bulge or sink, or straight lines on the ground may become bent. The detection of such movements by laser surveying can hint at an upcoming earth-quake. More recently, researchers have been able to use InSAR data (Fig. 10.11b, c) to detect such distortion of the land surface.

Geologists have also begun to use computer models of stress to predict where stress buildups may lead to earth-quakes. Use of these stress-triggering models has provided substantial insight into seismic activity along the North Anatolian Fault, in Turkey (▶Fig. 10.42a). This fault is a large strike-slip fault along which Turkey slips westward. (Turkey is essentially being squeezed like a watermelon seed out of the way of Arabia, which is moving northward and is colliding with Asia.) Earthquakes happen again and again along the fault—in fact, ruins of several ancient cities lie along the trace of the fault. Since 1939, 11 major earth-quakes have occurred along the fault—each ruptured a dif-ferent portion of the fault (▶Fig. 10.42b). Overall, there has been a westward progression of faulting. By modeling the stress changes resulting from each earthquake, geologists identified places where stress accumulates, and predicted that the next big earthquake would happen at the western end of the fault. It did. In August 1999, a magnitude (M_w) 7.4 earthquake devastated Izmit, a city just south of Istan-bul, killing more than 17,000 people and leaving more than 600,000 homeless.

Recently, seismologists have begun to explore the possi-bility that shock waves from a large earthquake may trigger other earthquakes far away. Other possible warning signs that have been explored but have *not* yet been scientifically proven to be precursors of earthquakes include the follow-ing: changes in the water level in wells; appearance of gases, such as radon or helium, in wells; changes in the electrical conductivity of rock underground; and unusual animal be-havior (e.g., dogs howling). Believers in these proposed clues

FIGURE 10.42 (a) A map of Turkey, showing the Anatolian Fault. (b) A graph representing regions of the fault that slipped during various earthquakes. The horizontal axis represents location along the fault, and the vertical axis represents the amount of slip.

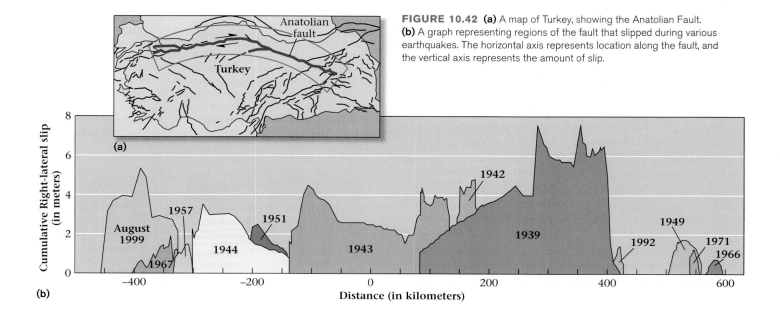

suggest that they all reflect the occurrence of cracking in the crust prior to an earthquake. But most investigators remain skeptical.

As long as short-term predictions remain questionable, emergency service planners must ask: What if a prediction is wrong? Should schools and offices be shut because of a prediction? Should millions be spent to evacuate people? Should a city be deserted, allowing for the possibility of looters? Should the public be notified, or should only officials be notified, creating a potential for rumor? If the prediction proves wrong, can seismologists be sued? No one really knows the answers to these questions.

Take-Home Message

Researchers can determine regions where earthquakes are more likely, but not exactly when and where an event will occur. Seismic hazard is greater where seismicity has happened more frequently in the past and therefore has a shorter recurrence interval.

10.8 EARTHQUAKE ENGINEERING AND ZONING

If we can't avoid earthquakes, can we prepare for them? A glance at Table 10.1 illustrates that the loss of human life from earthquakes varies widely. The loss depends on a number of factors, most notably the proximity of an epicenter to a population center, the depth of the hypocenter, the style of construction in the epicentral region, whether or not the earthquake occurred in a region of steep slopes or along the coast, whether building foundations are on solid bedrock or on weak substrate, whether the earthquake happened when people were outside or inside, and whether the government was able to provide emergency services promptly.

For example, the 1988 earthquake in Armenia was not much bigger than a 1971 earthquake in southern California, but it caused almost 500 times as many deaths (24,000 versus 50). The difference in death toll reflects differences in the style and quality of construction, and the characteristics of the substrate. The unreinforced concrete-slab buildings and masonry houses of Armenia collapsed, whereas the structures in California had, by and large, been erected according to building codes that take into account stresses caused by earthquakes. Most flexed and twisted but did not fall down and crush people. The terrible 1976 earthquake in T'ang-shan, China, was such a calamity because the ground beneath the epicenter had been weakened by coal mining and collapsed, and because buildings were poorly constructed. Mexico City's 1985 earthquake, as we have seen, proved disastrous because the city lies over a sedimentary basin whose composition and bowl-like shape

focused seismic energy, causing buildings of a certain height to resonate. During the 1989 Loma Prieta, California, quake, portions of Route 880 in Oakland that were built on a weak substrate collapsed, but portions built on bedrock or gravel remained standing.

The differences in the destructiveness of earthquakes demonstrate that we can mitigate or diminish their consequences by taking sensible precautions. Clearly, earthquake engineering, the designing of buildings that can withstand shaking, and earthquake zoning, the determination of where land is stable and where it might collapse, can help save lives and property.

In regions prone to large earthquakes, buildings should be constructed so they are able to withstand vibrations without collapsing (►Fig. 10.43a–c). They should be somewhat flexible so that ground motions can't crack them, and supports should be strong enough to maintain loads far in excess of the loads caused by the static (nonmoving) weight of the building. Bridge support columns should also be constructed with earthquakes in mind. Wrapping steel cables around the columns makes them many times stronger. Bolting the bridge spans to the top of a column prevents the spans from bouncing off. Concrete-block buildings, unreinforced concrete, and unreinforced brick buildings crack and tumble under conditions where wood-frame, steel-girder, or reinforced concrete buildings remain standing. Traditional heavy, brittle tile roofs shatter and bury the inhabitants inside, but sheet-metal or asphalt shingle roofs do not. Loose decorative stone and huge open-span roofs do not fare well when vibrated, and should be avoided.

Similarly, developers should avoid construction on land underlain by weak, sedimentary mud that could liquefy. They should not build on top of, on, or at the base of steep escarpments (which could fail and produce landslides), and they should avoid locating large population centers down-

FIGURE 10.43 How to prevent damage and injury during an earthquake. **(a)** Wrapping a bridge's support columns in cable (preventing buckling of the columns) and bolting the span to the columns (preventing the span from separating from the columns) will prevent the bridge from collapsing so easily. **(b)** Buildings will be stronger if they are wider at the base and if cross beams are added inside. **(c)** Placing buildings on rollers (or shock absorbers) will lessen the severity of the vibrations.

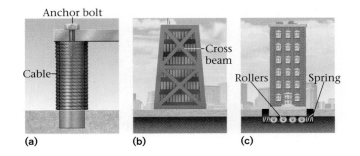

Anchor bolt

Cable

Cross beam

Rollers   Spring

(a)   (b)   (c)

FIGURE 10.44 If an earthquake strikes, take cover under a sturdy table near a wall.

stream of dams (which could fail, causing a flood). And they should avoid constructing vulnerable buildings directly over active faults, whose movement could crack and destroy the structure. Cities in seismic zones need to draw up emergency plans to deal with disaster. Communication centers should be located in safe localities, and strategies need to be implemented for providing supplies under circumstances where roads may be impassable.

Finally, communities and individuals should learn to protect themselves during an earthquake. In your home, keep emergency supplies, bolt bookshelves to walls, strap the water heater in place, install locking latches on cabinets, know how to shut off the gas and electricity, know how to find the exit, have a fire extinguisher handy, and know where to go to find family members. Schools and offices should have earthquake-preparedness drills (▶Fig. 10.44). If an earthquake strikes, stay outdoors and away from buildings if you can. If you're inside, stand near a wall or in a doorway near the center of the building, or crouch under a heavy table. And if you're on the road, stay away from bridges. As long as plates continue to move, earthquakes will continue to shake. But we can learn to live with them.

> **Take-Home Message**
>
> Earthquakes are a fact of life on this dynamic planet. People in regions facing high seismic risk should build on stable ground, avoid unstable slopes, and design construction that can survive shaking. Evacuation planning saves lives after an event.

Chapter Summary

- Earthquakes are episodes of ground shaking, caused when earthquake waves reach the ground surface. Earthquake activity is called seismicity.

- Most earthquakes happen when rock breaks during faulting. A fault is a fracture on which sliding occurs. The place where rock breaks and earthquake energy is released is called the hypocenter (focus), and the point on the ground directly above the hypocenter is the epicenter.

- We can distinguish among normal, reverse, thrust, and strike-slip faults, based on the relative motion of rock across the fault. The amount of movement is called the displacement.

- Active faults are faults on which movement is currently taking place. Inactive faults ceased being active long ago, but can still be recognized because of the displacement across them. Displacement on active faults that intersect the ground surface may yield a fault scarp. The intersection of the fault with the ground is the fault trace.

- According to elastic-rebound theory, during fault formation, rock elastically strains, then cracks. Eventually, cracks link to form a thoroughgoing rupture on which sliding occurs. When this happens, the elastically strained rock breaks and vibrates, and this generates an earthquake.

- Faults exhibit stick-slip behavior in that they move in sudden increments.

- Earthquakes in the continental crust can only happen in the brittle, upper part of the crust. At depth, where rocks become ductile, earthquakes don't occur.

- Earthquake energy travels in the form of seismic waves. Body waves, which pass through the interior of the Earth, include P-waves (compressional waves) and S-waves (shear waves). Surface waves, which pass along the surface of the Earth, include R-waves (Rayleigh waves) and L-waves (Love waves).

- We can detect earthquake waves by using a seismograph.

- Seismograms demonstrate that different earthquake waves arrive at different times, because they travel at different velocities. Using the difference between P-wave and S-wave arrival times, seismologists can pinpoint the epicenter location.

- The Mercalli intensity scale is based on documenting the damage caused by an earthquake. Magnitude scales, such as the Richter scale, are based on measuring the amount of ground motion, as indicated by traces of waves on a seismogram. The seismic-moment magnitude scale takes into account the amount of slip, the length and depth of the rupture, and the strength of the ruptured rock.

- A magnitude 8 earthquake yields about 10 times as much ground motion as a magnitude 7 earthquake, and releases about 32 times as much energy.

- Most earthquakes occur in seismic belts, or zones, of which the majority lie along plate boundaries. Intraplate earthquakes happen in the interior of plates. Different kinds of earthquakes happen at different kinds of plate boundaries. Shallow-focus earthquakes associated with normal faults occur at divergent plate boundaries and in rifts. Earthquakes associated with thrust and reverse faulting occur at convergent and collisional boundaries. At convergent plate boundaries, we also observe intermediate- and deep-focus earthquakes, which define Wadati-Benioff zones. Shallow-focus strike-slip earthquakes occur along transform boundaries.

- Earthquake damage results from ground shaking (which can topple buildings), landslides (set loose by vibration), sediment liquefication (the transformation of compacted clay into a muddy slurry), fire, and tsunamis (giant waves).

- Seismologists predict that earthquakes are more likely in seismic zones than elsewhere, and can determine the recurrence interval (the average time between successive events) for great earthquakes. But it may never be possible to pinpoint the exact time and place at which an earthquake will take place.

- Earthquake hazards and casualties can be reduced with better construction practices and zoning, and by knowing what to do during an earthquake.

Geopuzzle Revisited

Most earthquakes happen when rock abruptly breaks during the formation of a new fault, or during renewed slip on an existing fault. Stress drives the process, building up slowly until it exceeds the rock's strength. Major seismic belts (regions of frequent earthquake activity) coincide with plate boundaries, collision zones, and rifts, and thus delineate regions where relative plate motion is being accommodated. A few seismic belts, however, do occur along ancient, weak faults within plates. Seismic risk is clearly greater in seismic belts, for stress builds up more rapidly in these regions. But it is not possible to predict exactly where or when an earthquake will occur within a zone.

Key Terms

aftershock (p. 309)
annual probability (p. 342)
arrival time (p. 316)
body waves (p. 313)
compressional waves (p. 313)
displacement (p. 306)
earthquake (p. 304)
elastic-rebound theory (p. 308)
epicenter (p. 306)

fault (p. 305)
fault creep (p. 313)
fault scarp (p. 307)
fault trace (p. 307)
focus (p. 306)
foreshock (p. 309)
hypocenter (p. 306)
induced seismicity (p. 328)
intraplate earthquakes (p. 327)
L-waves (Love waves) (p. 313)
liquefaction (p. 331)
magnitude (p. 320)
Mercalli intensity scale (p. 319)
moment magnitude (p. 321)
P-waves (p. 313)
R-waves (Rayleigh waves) (p. 313)
recurrence interval (p. 342)
resonance (p. 330)

Richter scale (p. 320)
S-waves (p. 313)
seiche (p. 331)
seismic belts (zones) (p. 322)
seismic gaps (p. 342)
seismicity (p. 305)
seismic-risk assessment (p. 342)
seismic waves (p. 313)
seismogram (p. 315)
seismograph (seismometer) (p. 315)
seismologist (p. 305)
shear waves (p. 313)
stick-slip behavior (p. 308)
surface waves (p. 313)
travel-time curve (p. 318)
tsunami (p. 335)
Wadati-Benioff zone (p. 324)

Review Questions

1. Compare normal, reverse, and strike-slip faults.

2. Describe elastic rebound theory and the concept of stick-slip behavior.

3. Compare brittle and ductile deformation.

4. Describe the motions of the four types of seismic waves. Which are body waves, and which are surface waves?

5. Explain how the vertical and horizontal components of an earthquake are detected on a seismograph.

6. Explain the contrasts among the different scales used to describe the size of an earthquake.

7. How does seismicity on mid-ocean ridges compare with seismicity at convergent or transform boundaries? Do all earthquakes occur at plate boundaries?

8. What is a Wadati-Benioff zone, and why is it important in understanding plate tectonics?

9. Describe the types of damage caused by earthquakes.

10. What is a tsunami, and why does it form? What is a seiche?

11. Explain how liquefaction occurs in an earthquake, and how it can cause damage.

12. How are long-term and short-term earthquake predictions made? What is the basis for determining a recurrence interval, and what does a recurrence interval mean?

13. Why is it difficult to make accurate short-term predictions? What clues might suggest an earthquake may happen fairly soon?

14. What types of structures are most prone to collapse in an earthquake? What types are most resistant to collapse?

15. What should you do if you feel an earthquake starting?

On Further Thought

1. Is seismic risk greater in a town on the west coast of South America or in one on the east coast? Explain your answer.

2. The northeast-trending Ramapo Fault crops out north of New York City near the east coast of the United States. Precambrian gneiss forms the hills to the northwest of the fault, and Mesozoic sedimentary rock underlies the lowlands to the southeast. (You can see the fault on Google Earth by going to Lat 41° 10' 21.12'' N Long 74° 5' 12.36'' W. Once you're there, tilt the image and fly northeast along the fault.) Where the fault crosses the Hudson River, there is an abrupt bend in the river. A nuclear power plant was built near this bend. There are numerous bogs, containing sediments deposited over the past several thousand years, on the surface of the basin. Geologic studies suggest that the Ramapo Fault first formed during the Precambrian, was reactivated during the Paleozoic, and was the site of major displacement during the Mesozoic rifting that separated North America from Africa. Imagine that you are a geologist with the task of determining the seismic risk of the fault. What evidence of present-day or past seismic activity could you look for? What does the long-term geologic history of the fault imply about its strength?

3. On the seismogram of an earthquake recorded at a seismic station in Paris, France, the S-wave arrives 6 minutes after the P-wave. On the seismogram obtained by a station in Mumbai, India, for the same earthquake, the difference between the P-wave and S-wave arrival times is 4 minutes. Which station is closer to the epicenter? From the information provided, can you pinpoint the location of the epicenter? Explain.

Suggested Reading

Bolt, B. A. 2003. *Earthquakes*, 5th ed. New York: Freeman.

Bryant, E. 2001. *Tsunami: The Underrated Hazard*. Cambridge: Cambridge University Press.

Fradkin, P. L. 1999. *Magnitude 8*. Berkeley: University of California Press.

Geschwind, C. H. 2001. *California Earthquakes: Science, Risk and the Politics of Hazard Mitigation*. Baltimore: Johns Hopkins University Press.

Hough, S. E. 2004. *Earthshaking Science: What We Know (and Don't Know) about Earthquakes*. Princeton, N.J.: Princeton University Press.

Ritchie, D., and A. E. Gates. 2001. *Encyclopedia of Earthquakes and Volcanoes*. New York: Facts on File.

Shearer, P. 1999. *Introduction to Seismology*. Cambridge: Cambridge University Press.

Stein, S., and M. Wysession. 2002. *An Introduction to Seismology, Earthquakes and Earth Structure*. Boston: Blackwell Science.

Scholz, C. H. 2002. *The Mechanics of Earthquakes and Faults*, 2nd ed. Cambridge: Cambridge University Press.

Yeats, R. S. 2001. *Living with Earthquakes in California: A Survivor's Guide*. Corvallis: Oregon State University Press.

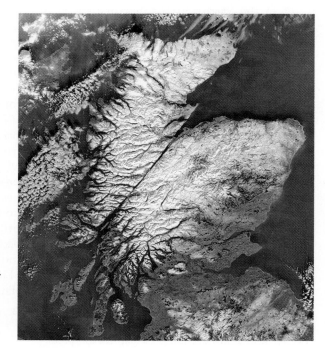

THE VIEW FROM SPACE A view of Scotland in winter shows the rugged highlands blanketed by snow. A narrow, northeast-southwest-trending valley slashes across the landscape. This valley delineates the trace of the Great Glen Fault, a structure that first formed over 200 million years ago. Fracturing made the rock along the fault more susceptible to weathering and erosion, leading to formation of the valley. In places, the valley has filled with water to form deep, elongate lakes. One of these is the infamous Loch Ness, supposedly home to the fictitious "Loch Ness monster." The Great Glen fault is not a present-day plate boundary, so it is not very active, seismically. Nevertheless, it is a crustal weakness, so occasionally small to moderate earthquakes happen along it.

Seeing Inside the Earth

D.1 INTRODUCTION

We live on the Earth's skin, and can see light years into space just by looking up. But when we look down—that's another story! We can't use our eyes to look through rock, because it is opaque. So how do we learn about what's inside this planet? Tunneling and drilling aren't much help because they do little more than prick the surface—the deepest mine (a gold mine in South Africa that reaches a depth of 3.6 km below the surface) represents less than 0.06% of the Earth's radius, and the deepest drill hole (drilled to a depth of 12.3 km in Precambrian rock of the Kola Peninsula in northwest Russia) represents less than 0.19% of the Earth's radius. Fortunately, as discussed in Chapter 2, nineteenth-century geologists realized that measurements of the Earth's mass and shape provide indirect clues to the mystery of what's inside. From these clues they determined that the Earth is not homogeneous but rather consists of three concentric layers: a crust (of low density), a mantle (of intermediate density), and a core (of high density). Further study showed that the crust beneath continents differs from the crust beneath oceans. Continental crust consists of a variety of felsic, intermediate, and mafic rocks, whereas oceanic crust consists almost entirely of mafic-composition rocks (▶Fig. D.1). However, the determination of the depths of the boundaries between Earth's layers and the division of the layers into sublayers with distinct properties could not be made until the twentieth century, when studies of seismic waves became available. By studying the speed of seismic waves and the direction in which these waves travel through the Earth, seismologists can effectively "see" details of the planet's internal layers.

In this interlude, we look at the behavior of seismic waves as they pass through our planet, and we learn how this behavior characterizes Earth's interior. We begin by reviewing a few key points about seismic waves, then move on to the phenomena of wave reflection and refraction. Then, we witness the discoveries of the different layer boundaries in the Earth. Finally, we briefly explore other data sources that provide clues to the interior. This interlude, incorporating the information about earthquakes and seismic waves provided in Chapter 10, completes the journey to the center of the Earth that we began in Chapter 2.

FIGURE D.1 The nineteenth-century three-layer image of the Earth, showing the crust, mantle, and core. The inset shows the contrast between continental crust (average composition is felsic to intermediate) and oceanic crust (average composition is mafic).

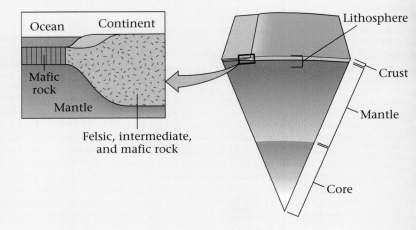

D.2 MOVEMENT OF SEISMIC WAVES THROUGH THE EARTH

Energy travels from one location to another in the form of waves. For example, the impact of a pebble on the surface of a pond produces water waves that eventually cause a stick meters away to bob up and down; the takeoff of a jet produces sound waves that rattle the windows of nearby houses; the Sun's nuclear furnace produces light waves that, when focused by a magnifying glass, can set a leaf on fire. Let's now consider in greater depth how seismic energy travels from the rupture on a fault to a seismograph on the other side of our planet.

Wave Fronts and Travel Times

A sudden rupture of intact rock or the frictional slip of rock on a fault produces seismic waves. These waves move outward from the point of rupture—the earthquake hypocenter—in all directions at once. A single earthquake produces many kinds of waves, distinguished from each other by where and how they move. Surface waves (R-waves and L-waves) propagate along the planet's surface, whereas body waves (P-waves and S-waves) pass through the interior. P-waves are compressional, and resemble the waves generated when you push a spring back and forth in a direction parallel to the length of the spring. S-waves are shear, and resemble the waves generated when you wiggle a rope back and forth perpendicular to the length of the rope (see Fig. 10.12).

The boundary between the rock through which a wave has passed and the rock through which it has not yet passed is called a wave front. A wave front expands outward from the earthquake focus like a growing bubble. We can represent a succession of waves in a drawing by a series of concentric wave fronts. The changing position of an imaginary point on a wave front as the front moves through rock is called a **seismic ray**. Seismic rays are perpendicular to wave fronts, so that each point on the wave front follows a slightly different ray (▶Fig. D.2). The time it takes for a wave to travel from the focus to a seismograph station along a given ray is the **travel time** along that ray.

The ability of a seismic wave to travel through a certain material and the velocity at which it travels depend on the character of the material. Factors such as density (mass per unit volume), rigidity (how stiff or resistant to twisting a material is), and compressibility (how much a material's volume changes in response to squashing) all affect seismic-wave movement. Studies of seismic waves reveal the following:

- Seismic waves travel at different velocities in different rock types (▶Fig. D.3a). For example, P-waves travel at 8 km per second in peridotite (an ultramafic rock), but at only 3.5 km per second in sandstone. Therefore, waves accelerate or slow down if they pass from one rock into another. P-waves in rock travel about 10 to 25 times faster than sound waves in air. But even at this rate, they take about 20 minutes to pass entirely through the Earth along a diameter.

- In general, seismic waves travel faster in a solid than in a liquid. Specifically, seismic waves travel more slowly in magma than in solid rock, and more slowly in molten iron alloy than in solid peridotite (▶Fig. D.3b).

- Both P-waves and S-waves can travel through a solid, but only P-waves can travel through a liquid (▶Fig. D.3c). To see why, picture what happens if you push down on the water surface in a pool—you send a pulse of compression (a P-wave) to the bottom of the pool. Now move your hand sideways through the water (shear). The water in front of your hand simply slides or flows past the water deeper down—your shearing motion has no effect on the water at the bottom of the pool (▶Fig. D.4a, b).

Reflection and Refraction of Wave Energy

Shine a flashlight into a container of water so that the light ray hits the boundary (or interface) between water and air at an angle. Some of the light bounces off the water surface and heads back up into the air, while some enters the water (▶Fig. D.5a). The light ray that enters the water bends at the air-water boundary, so that the angle between the ray and the boundary in the air is different from the angle between the ray and the boundary in the water. Physicists refer to the light ray that bounces off the air-water boundary and heads back into the air as the reflected ray, and the ray that bends

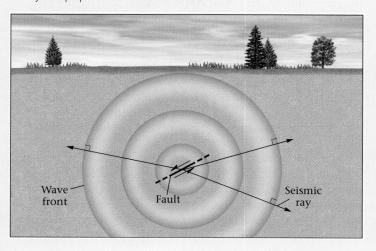

FIGURE D.2 An earthquake sends out waves in all directions. Seismic rays are perpendicular to the wave fronts.

Wave front — Fault — Seismic ray

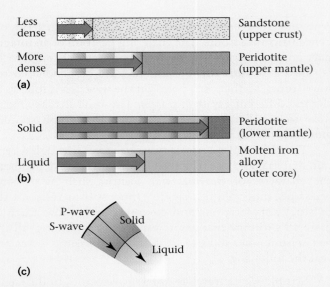

FIGURE D.3 **(a)** Seismic waves travel at different velocities in different rock types. For example, they travel faster in peridotite than in sandstone. **(b)** Seismic waves travel faster in solid peridotite than in a liquid such as molten iron alloy. **(c)** Both P-waves and S-waves can travel through a solid, but only P-waves can travel through a liquid.

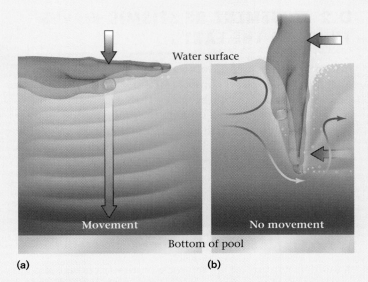

FIGURE D.4 **(a)** Pushing down on a liquid creates a compressive pulse (P-wave) that can travel through a liquid. **(b)** Moving your hand sideways through water does not generate a shear wave; the moving water simply flows past the water deeper down.

at the boundary as the refracted ray. The phenomenon of bouncing off is **reflection**, and the phenomenon of bending is **refraction**. Wave reflection and refraction take place at the interface between two materials, if the wave travels at different velocities in the two materials.

The amount and direction of refraction at a boundary depend on the contrast in wave velocity across the boundary, and on the angle at which a wave hits the interface. As a rule, if waves enter a material through which they will travel more slowly, the rays representing the waves bend away from the interface. If the waves enter a material through which they will travel faster, the rays bend toward the interface. For example, the light ray in ▶Figure D.5b bends down when hitting the air-water boundary, because light travels more slowly in water. This relation makes sense if you picture a car driving from a paved surface diagonally onto a sandy beach—the wheel that rolls onto the sand first slows down relative to the wheel still on the pavement, causing the car to turn. If the ray were to pass from a material in which it travels slowly into one in which it travels more rapidly, it would bend up (▶Fig. D.5c).

Since seismic energy travels in the form of waves, seismic waves, like rays of light in water, reflect and/or refract when reaching the interface between two rock layers if the waves travel at different velocities in the two layers. For example, imagine a layer of sandstone overlying a layer of peridotite. Seismic velocities in sandstone are slower than in peridotite, so as seismic waves reach the boundary, some reflect, and some refract.

D.3 SEISMIC STUDY OF EARTH'S INTERIOR

Let's now utilize your knowledge of seismic velocity, refraction, and reflection to see how each of the major layer boundaries inside the Earth was discovered.

Discovering the Crust-Mantle Boundary

The concept that seismic waves refract at boundaries between different layers led to the first documentation of the core-mantle boundary. In 1909, Andrija Mohorovičić, a Croatian seismologist, noted that P-waves arriving at seismograph stations less than 200 km from the epicenter traveled at an average speed of 6 km per second, whereas P-waves arriving at seismographs more than 200 km from the epicenter traveled at an average speed of 8 km per second. To explain this observation, he suggested that P-waves reaching nearby seismographs followed a shallow path through the crust, in which they traveled more slowly, whereas P-waves reaching distant seismographs followed a deeper path through the mantle, in which they traveled more rapidly.

To understand Mohorovičić's proposal, examine ▶Figure D.6a, which shows P-waves, depicted as rays, generated by an earthquake in the crust. Ray C, the shallower wave, travels through the crust directly to a seismograph. Ray M, the deeper wave, heads downward, refracts at the crust-mantle boundary, curves through the mantle,

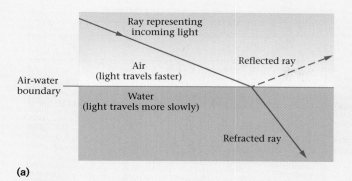
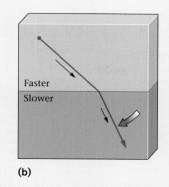
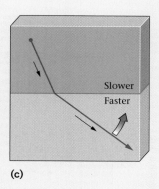

(a)

Ray representing incoming light

Reflected ray

Air-water boundary

Air (light travels faster)

Water (light travels more slowly)

Refracted ray

(b) Faster / Slower

(c) Slower / Faster

FIGURE D.5 (a) A ray of light, when it reaches the boundary between water and air, partly reflects and partly refracts. The refracted ray bends down as it enters the water. **(b)** A ray that enters a slower medium bends away from the boundary (like light reaching water from air). **(c)** A ray that enters a faster medium bends toward the boundary.

refracts again at the boundary, and then proceeds through the crust up to the seismograph. At stations less than 200 km from the epicenter, ray C arrives first, because it has a shorter distance to travel. But at stations more than 200 km from the epicenter (▶Fig. D.6b), ray M arrives first, even though it has farther to go, because it travels faster for much of its length. Calculations based on this observation require the crust-mantle boundary beneath continents to be at a depth of about 35 to 40 km. As we learned in Chapter 2, this boundary is now called the **Moho**, in honor of Mohorovičić.

Defining the Structure of the Mantle

After studying materials erupted from volcanoes, geologists concluded that the entire mantle has roughly the chemical composition of the ultramafic igneous rock called peridotite. If the density, rigidity, and compressibility of peridotite were exactly the same at all depths, seismic velocities would be the same everywhere in the mantle, and seismic rays would be

straight lines. But by studying travel times, seismologists have determined that seismic waves travel at different velocities at different depths. Let's now look at variations in seismic velocity that depend on mantle depth, and consider how these variations affect the shape of seismic rays.

Between about 100 and 200 km deep in the mantle beneath oceanic lithosphere, seismic velocities are slower than in the overlying lithospheric mantle (▶Fig. D.7). In this **low-velocity zone**, the prevailing temperature and pressure conditions cause peridotite to melt partially, by up to 2%. The melt, a liquid, coats solid grains and fills voids between grains. Because seismic waves travel more slowly through liquids than through solids, the coatings of melt slow seismic waves down. In the context of plate tectonics theory, the low-velocity zone is the weak layer on which oceanic lithosphere plates move. Below the low-velocity zone, the mantle does not contain melt. Geologists do not find a well-developed low-velocity zone beneath continents.

Below about 200 km, seismic-wave velocities everywhere in the mantle increase with depth. Seismologists

FIGURE D.6 (a) Seismic waves traveling only in the crust reach a nearby seismograph first, because they have a shorter distance to travel. Here, the waves following ray C arrive before those following ray M. **(b)** Seismic waves traveling for most of their path in the mantle reach a distant seismograph first, because they travel faster in the mantle than in the crust. Here, waves following ray M arrive before those following ray C. This observation led to the discovery of the Moho, the boundary surface between the crust and mantle.

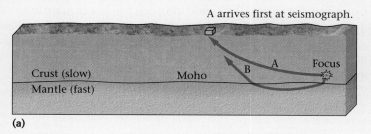

A arrives first at seismograph.

Crust (slow)

Mantle (fast)

Moho

B   A   Focus

(a)

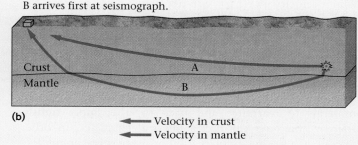

B arrives first at seismograph.

Crust

Mantle

A

B

(b)

← Velocity in crust
← Velocity in mantle

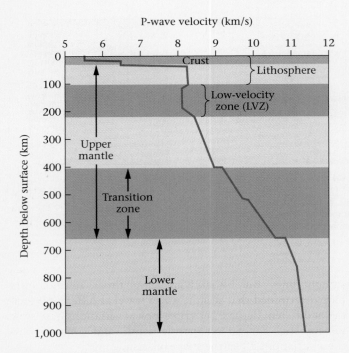

FIGURE D.7 The velocity of P-waves in the mantle changes with depth. Note the low-velocity zone between 100 and 200 km, and the sudden jumps in velocity defining the transition zone between 410 and 660 km.

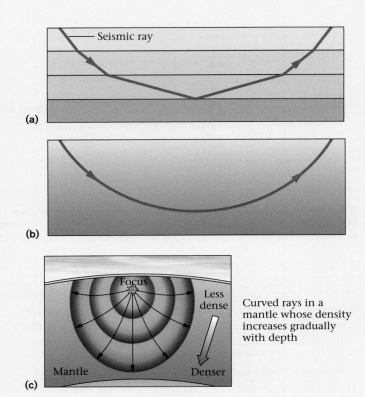

FIGURE D.8 **(a)** In a stack of layers in which seismic waves travel at different velocities (fastest in the lowest layer), a seismic ray gradually bends around and heads back to the surface. The curve consists of several distinct segments. **(b)** If the mantle's density increased gradually with depth, the ray would define a smooth curve. **(c)** Since the velocity of seismic waves increases with depth, wave fronts are oblong and seismic rays curve.

interpret this increase to mean that mantle peridotite becomes progressively less compressible and more rigid with depth. This proposal makes sense, considering that the weight of overlying rock increases with depth, and as pressure increases, the atoms making up rock squeeze together more tightly and bonds are not so free to move. Because of refraction, the progressive increase in seismic velocity with depth causes seismic rays to curve in the mantle. To understand the shape of a curved ray, look at ▶Figure D.8a, which represents a portion of the mantle as a series of layers, each permitting a slightly greater seismic-wave velocity than the layer above. Every time a seismic ray crosses the boundary between adjacent layers, it refracts a little toward the boundary. After the ray has crossed several layers, it has bent so much that it begins to head back up toward the top of the stack. Now if we replace the stack of distinct layers with a single layer in which velocity increases with depth at a constant rate, the wave follows a smoothly curving path (▶Fig. D.8b, c).

Between 410 km and 660 km deep (see Fig. D.7), seismic velocity increases in a series of abrupt steps, so the stack of layers in Figure D.8a is actually a somewhat realistic image. Experiments suggest that these **seismic-velocity discontinuities** occur at depths where pressure causes atoms in minerals to rearrange and pack together more tightly, thereby changing the rock's physical character. In other words, each seismic-velocity discontinuity corresponds to a

phase change (see Chapter 8). For example, at 410 km, the mineral olivine, a major constituent of peridotite, becomes unstable and collapses to form a different mineral called magnesium spinel. Magnesium spinel has the same chemical composition as olivine but a different internal arrangement of atoms and stays stable to a depth of about 660 km. At this depth, atoms rearrange to form an even denser crystal structure called perovskite structure. Because of these seismic-velocity discontinuities, as we learned in Chapter 2 but now can see more clearly, seismologists subdivide the mantle into the **upper mantle** (above 410 km), the **transition zone** (between 410 and 660 km), and the **lower mantle** (below 660 km) (see Fig. 2.13).

At this point, you may be wondering, Is there any way to test ideas about the identity of minerals that occur deep in the mantle? The answer is yes. Researchers can use a device called a diamond anvil to simulate deep-mantle pressure and temperature conditions. In a diamond-anvil experiment, tiny samples of minerals *thought to occur* in the mantle are squeezed between two diamonds. Simultaneously, the samples are subjected to the light of an intense laser beam (▶Fig. D.9). Diamond is the hardest mineral known, and the pressure between the two diamonds can reach values comparable to those found in the deep mantle. Laser light passes through clear diamonds, and thus can heat the sample to mantle temperatures. Using sophisticated instruments, researchers measure seismic velocities in the sample while it remains in the diamond anvil at high pressures and temperatures. By comparing the velocities observed in the laboratory with those observed for the real mantle, researchers gain insight into whether or not the sample could be a mantle mineral.

Discovering the Core-Mantle Boundary

During the first decade of the twentieth century, seismologists installed seismographs at many stations around the world, expecting to be able to record waves produced by a large earthquake anywhere on Earth. In 1914, one of these seismologists, Beno Gutenberg, discovered that P-waves from an earthquake do not arrive at seismographs lying in a band between 103° and 143° from the earthquake epicenter, as measured along the circumference of the Earth. This band is now called the **P-wave shadow zone** (▶Fig. D.10). If the density of the Earth increased gradually with depth all the way to the center, the shadow zone would not exist, because rays passing into the interior would curve up and reach every point on the surface. Thus, the presence of a shadow zone means that deep in the Earth a major interface exists where seismic waves *abruptly* refract down (implying that the velocity of seismic waves suddenly decreases). This

interface, now called the **core-mantle boundary**, lies at a depth of about 2,900 km. The density contrast across this boundary is greater than the density contrast between the crust and water.

To see why the P-wave shadow zone exists, follow the two seismic rays labeled A and B in Figure D.10. Ray A curves smoothly in the mantle (we are ignoring seismic-velocity discontinuities in the mantle) and passes just above the core-mantle boundary before returning to the surface. It reaches the surface 103° from the epicenter. In contrast, ray B just penetrates the boundary and refracts down into the core. Ray B then curves through the core and refracts again when it crosses back into the mantle. As a consequence, ray B intersects the surface at more than 143° from the epicenter.

The downward bending of seismic waves when they pass from the mantle down into the core indicates that seismic velocities in the core are slower than in the mantle. Thus, even though the core is deeper and denser than the mantle, the outer part of the core must be less rigid than the mantle.

FIGURE D.10 P-waves do not arrive in the interval between 103° and 143° from an earthquake's epicenter, defining the P-wave shadow zone. The wave that arrives at 103° passed just above the core-mantle boundary. The next wave bent down so far that it arrives at about 160°. And the wave that arrives at 143° bent only slightly as it passed through the core. The inset shows the shadow zone on a globe. Note that P-waves bend down at the core-mantle boundary because velocities are slower in the core than in the mantle. (The core is less rigid.)

FIGURE D.9 A laboratory apparatus for studying the characteristics of minerals under very high pressures and temperatures. The green laser beam is heating up a microscopic sample being squeezed between two diamonds hidden at the center of the metal cylinder.

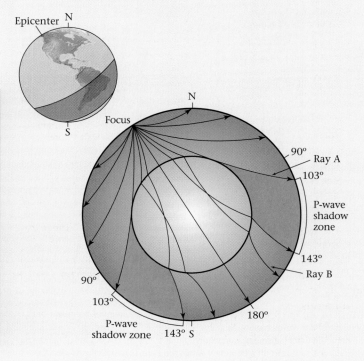

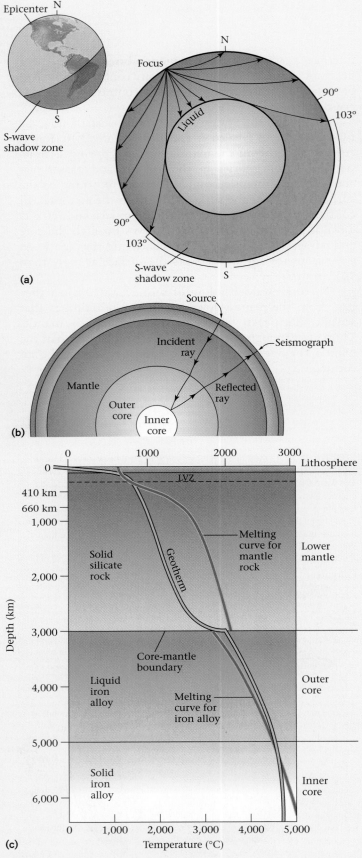

(a)

(b)

(c)

Discovering the Nature of the Core

Based on density calculations and on the study of meteorites thought to be fragments of another planet's interior, seismologists concluded that the core consists of iron alloy, which is less rigid than peridotite. But is the core the same through and through, or is it too divided into layers? A study of S-waves gave seismologists the answer. They found that S-waves do not arrive at stations located between 103° and 180° from the epicenter (a band called the **S-wave shadow zone**). This means that S-waves cannot pass through the core at all—otherwise, an S-wave headed straight down through the Earth would appear on the other side. Remember that S-waves are shear waves, which by their nature can travel only through solids. Thus, the fact that S-waves do not pass through the core means that the core, or at least part of it, consists of liquid (▶Fig. D.11a).

At first, seismologists thought that the entire core might be liquid iron alloy. But in 1936, a Danish seismologist, Inge Lehmann, discovered that P-waves passing through the core reflected off a boundary within the core. She then proposed that the core is made up of two parts: an **outer core** consisting of liquid iron alloy and an **inner core** consisting of solid iron alloy. Lehmann's work defined the existence of the inner core but could not locate the depth at which the inner core–outer core interface occurs. This depth was located by measuring the exact time it took for seismic waves generated by nuclear explosions to penetrate the Earth, bounce off the inner core–outer core boundary, and return to the surface (▶Fig. D.11b). The measurements showed that the inner core–outer core boundary occurs at a depth of about 5,155 km.

Why is the outer core molten, while the inner core is solid? An examination of ▶Figure D.11c provides some insight. This graph shows two curves: (1) the geotherm, which indicates the change in temperature as depth increases in the Earth; and (2) the melting curve, which indicates the temperature at which materials melt as depth increases in the Earth. As the graph shows, the geotherm lies to the left of the melting curve through most of the mantle and in the inner core. This means that temperatures in most of the mantle and in the inner core are not high enough to cause melting, under the high pressures found in these regions, so these regions are solid. But the geotherm lies to the right of the melting curve in the low-velocity zone of the mantle and in the outer core, so these regions contain molten material.

Before we leave the core, let's look again at one important characteristic of the liquid outer core, its convection. As noted earlier in the book, this convection is important because it generates Earth's magnetic field. Because of the Earth's rotation, convective cells takes the form of spirals that rise in a direction parallel to the Earth's rotation axis, and thus lead to a dipolar field that roughly parallels the rotation axis. Geologists realized that convection in the

FIGURE D.11 (a) The S-wave shadow zone covers about a third of the globe, and exists because shear waves cannot pass through the liquid outer core. **(b)** The solid inner core was detected when seismologists observed that some seismic waves generated by nuclear explosions reflected off a boundary within the core. **(c)** A graph showing the geotherm and melting curve for the Earth. Note that the melting occurs where the geotherm lies to the right of the melting curve.

outer core cannot be explained by thermal contrasts between the top and bottom of the core, for the outer core consists of metal, which is a very good heat conductor. Recent research suggests that the density differences leading to convection are largely due to differences in chemical composition between the top and bottom of the outer core. These compositional differences arise because the solid inner core is slowly growing as the Earth, overall, cools. The new crystals of solid iron that form along the surface of the inner core do not have room in their crystal structure for low-density elements such as silicon, sulfur, or oxygen. So these elements are expelled into the base of the molten outer core. Thus, at any given time, the base of the outer core has a lower density than the upper parts, and it starts to rise convectively.

The Velocity-versus-Depth Profile

By the time of World War II, seismologists had identified the crust-mantle boundary (the Moho) and the core-mantle boundary, and had recognized that the core is divided into two parts. In other words, the prevailing image of the Earth's interior being an onion-like sequence of concentric zones had been established. After World War II, seismologists set to the task of refining this image, a task made easier by the Cold War: because of the need to detect nuclear explosions, the Western nuclear powers built a vast array of precise seismograph stations scattered around the world. Through painstaking effort, seismologists used data from this array to develop a graph known as a **velocity-versus-depth curve**. We've already presented the upper part of the curve (Fig. D.7), and ►Figure D.12 presents the rest of it. This curve shows the depths at which seismic velocity suddenly changes and thus helped to establish the principal layers and sublayers in the Earth.

Fine-Tuning the Image: Seismic Tomography

In recent years, seismologists have developed a technique, called **seismic tomography**, to produce three-dimensional images of the Earth's interior. (This technique resembles that used to produce three-dimensional CAT scans of the human body. *CAT*, in the medical context, stands for computerized axial tomography.) In seismic tomography studies, researchers compare the *observed* travel time of seismic waves following a specific ray path with the *predicted* travel time that waves following the same path would have if the average velocity-versus-depth model of Figures D.7 and D.12 were completely correct. They found that waves following some paths take more time than predicted, whereas waves following other paths take less time than predicted. By repeating the measurements for many different wave paths in many different directions, researchers can outline three-dimensional regions

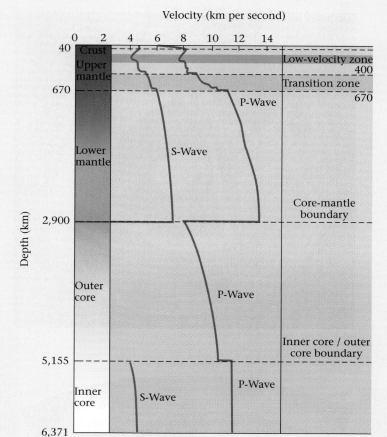

FIGURE D.12 The velocity-versus-depth profile for the Earth.

of the mantle in which waves travel unexpectedly fast or unexpectedly slow. Fast regions are probably cooler and more rigid than their surroundings, and slow regions are probably warmer and less rigid than their surroundings. Tomographic studies emphasize that the simple onion-like layered image of the Earth, with velocities increasing with depth at the same rate everywhere, is an oversimplification. In reality, the velocities of seismic waves vary significantly with location at a given depth. Results of tomographic studies can be displayed by three-dimensional models, cross sections, or maps (►Fig. D.13a–d). Generally, warmer colors (reds) on these images indicate slower and presumably warmer regions, whereas cooler colors (blues and purples) indicate faster and presumably cooler regions.

Seismic tomography studies provide new insight into the Earth's interior. For example: (1) A region of fast (cool) mantle lies in the asthenosphere beneath North America (see Fig D.13b). This region may represent the remnants of ocean lithosphere that was subducted during the Mesozoic. The fact that the fast region extends into the lower mantle suggests that some subducted plates

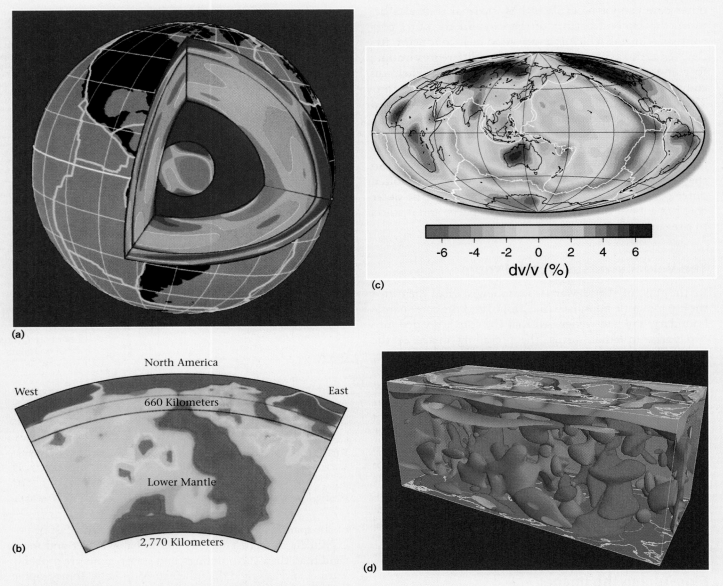

(a)

(b)

West
North America
East
660 Kilometers
Lower Mantle
2,770 Kilometers

(c)

dv/v (%)
-6 -4 -2 0 2 4 6

(d)

FIGURE D.13 Tomographic images show regions of faster and slower velocities. Redder colors indicate regions where seismic waves travel more slowly than expected. Blue and purple colors show regions where seismic waves travel faster than expected. (The mantle is cooler and more rigid.) In the slowest regions, waves travel 1.5% more slowly, whereas in the fastest regions, waves travel 1.5% faster than expected. **(a)** Tomographic image of the whole Earth. The brown region is the outer core, a liquid. **(b)** Closeup of the mantle beneath North America. The high-velocity area may be the remnant of a cold subducted slab. **(c)** A map showing seismic velocity variations at 70-km depth. Note that slow regions underlie mid-ocean ridges, where hot asthenosphere is rising. The scale indicates the % difference in velocity, relative to the expected velocity. **(d)** A 3-D tomographic model of the mantle. The top of the block is a map of the Earth's surface. Continents are outlined in white.

may sink into the lower mantle. In fact, some researchers suggest that there is a "plate graveyard" at the base of the mantle, where denser portions of subducted plates sink and accumulate. But others maintain that most subducted plates accumulate at the base of the upper mantle. (2) A region of slow (warm) mantle occurs in the region beneath mid-ocean ridges (see Fig. D.13c). This result supports the concept that warm asthenosphere rises be-

neath ridges. (3) A 200-km-thick layer of slow (warm) mantle lies just above the core-mantle boundary. This layer, called the D″ layer (pronounced "dee double prime") may represent the region in which the mantle has absorbed heat radiating from the core. Some researchers speculate that it may be the source of mantle plumes, but this proposal remains a subject for debate. (4) Tomographic studies of the inner core have shown that the inner core also has a pattern of fast and slow regions. Significantly, the orientation of the pattern changes over time, relative to the orientation of the pattern in the mantle. Researchers interpret this to mean that the inner core rotates slightly faster than the rest of the Earth—it makes an extra rotation about once every 20 to 25 years. (5) Taken together, tomographic studies have begun to give us an image of mantle convection. This image has inspired researchers to develop supercomputer models of what this convection looks like (▶Fig. D.14a, b).

Even though many important questions remain, tomography has led geologists to picture the Earth's insides as a dynamic place, not just a region of static, concentric shells (▶Fig. D.15). This image should become even clearer, for a major new research initiative, called EarthScope, has begun. This initiative involves placing hundreds of seismographs in an array across the United States. Just as digital photographs have higher resolution when taken by a camera with a 10-megapixel sensor than one with a 2-megapixel sensor, the greater number of seismographs in the EarthScope array will provide a higher resolution image of the Earth's interior. Researchers are also using other measurements, such as variation in the pull of Earth's gravity, to characterize the interior (▶Box. D.1).

FIGURE D.14 **(a)** This image is a computer-generated 3-D model depicting convective flow in the mantle. The upwelling regions, depicted in yellow, consist of rising hot mantle. The downwelling regions, depicted in blue, consist of sinking cooler mantle. The red sphere inside is the surface of the core. **(b)** Another computer model showing possible development of convective plumes inside the Earth. Here, the mantle is shown as transparent, so only the proposed hot source of the plumes—the D" layer—and the plumes themselves are visible.

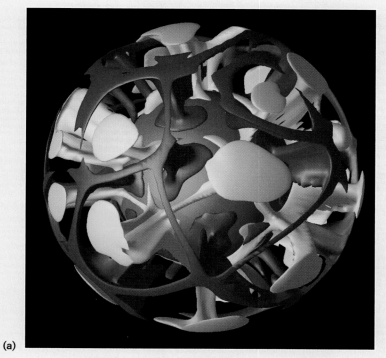

(a)

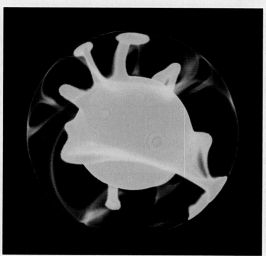

(b)

D.4 SEISMIC-REFLECTION PROFILING

Seismic techniques are also letting us fine-tune our image of the crust. During the past half century, geologists have found that by using dynamite, by banging large weights against the Earth's surface, or by releasing bursts of compressed air into the water, they can create artificial seismic waves that propagate down into the Earth and reflect off the boundaries between different layers of rock in the crust. By recording the time at which these reflected waves

FIGURE D.15 The modern view of a complex and dynamic Earth interior. Note the convecting cells, the mantle plumes, and the subducted-plate graveyards.

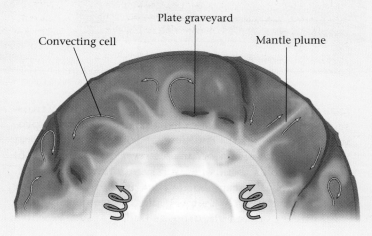

Convecting cell   Plate graveyard   Mantle plume

Is the Earth Really Round? Introducing the Geoid

In Chapter 1, we noted that as planetesimals grow into planets they eventually get large enough, warm enough, and soft enough for gravitational force to transform them from irregular shapes into spheres. On a perfect sphere, the pull of gravity is exactly the same at all points on the surface. Using the jargon of physics, we can say that the surface of a perfectly spherical planet is an equipotential surface, meaning that all points on its surface have the same gravitational potential energy.

But is the Earth really a *perfect* sphere? More precisely, is the Earth's equipotential surface perfectly spherical? The answer is . . . no. To start with, Earth's rotation produces centrifugal force, which flattens the

planet and causes the radius from the equator to the center (6,378 km, or 3,963 miles) to exceed the radius from the pole to the center (6,357 km, or 3,950 miles). But flattening due to centrifugal force isn't the whole story. Satellite measurements show that the Earth's equipotential surface actually has broad bumps and dimples. If these irregularities are greatly exaggerated, our planet would look somewhat like a warped pear (▶Fig. D.16a, b). This shape, a more accurate representation of the Earth's equipotential surface, is called the **geoid**. You can visualize the geoid as the shape that sea level would have if the Earth were covered by a global ocean and sea level depended on

the pull of gravity. Where gravity is stronger, the surface would be lower (for gravity pulls the surface down), and where gravity is weaker, the surface would be higher. The difference between the highest point on the geoid and the lowest point is about 210 m.

A comparison of a geoid map to a seismic tomography map (such as Figure D.13d) suggests that the broad bumps and dimples of the geoid are due to upwelling and downwelling associated with convection in the mantle. Warmer mantle rocks are less dense, so they exert less gravitational pull, so geoid highs are over seismically fast (warm) regions, whereas geoid lows are over seismic slow (cool) regions of the mantle.

FIGURE D.16 **(a)** An exaggerated 3-D representation of the geoid showing how the Earth's surface is distorted by highs and lows in the gravity field. **(b)** A map of the Earth's gravity field. Reds are gravity highs and dark blues are gravity lows.

(a)

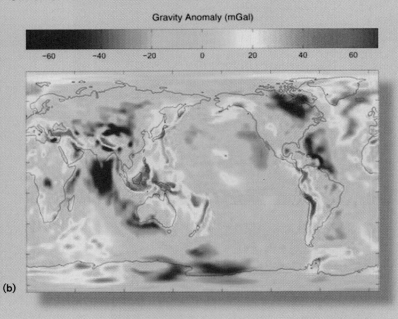

Gravity Anomaly (mGal)

−60   −40   −20   0   20   40   60

(b)

return to the surface, geologists effectively create a cross-sectional view of the crust called a **seismic-reflection profile** (▶Fig. D.17a–d). This image defines the depths at which layers of strata occur and reveals the presence of subsurface folds (bends in layers) and faults. Oil companies must obtain seismic-reflection profiles, despite their high cost, because they allow geologists to identify likely locations for oil and gas deposits underground. Research

geologists at universities have used the technique to obtain images of the Moho and even the upper mantle.

In the past decade, computers have become so sophisticated that geologists can now produce three-dimensional seismic-reflection images of the crust. These provide so much detail that geologists can trace out a ribbon of sand representing the channel of an ancient stream even where the sand lies buried kilometers below the Earth's surface (▶ Fig. D.17e).

(a)

(b)

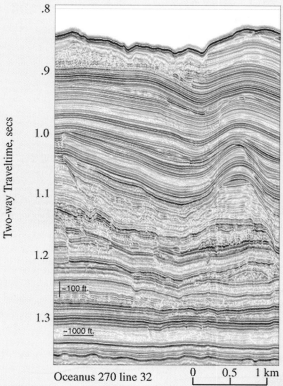

~100 ft.

~1000 ft.

Oceanus 270 line 32

0   0.5   1 km

(c)

(d)

FIGURE D.17 (a) Trucks thumping on the ground to generate the signal needed for making a seismic-reflection profile. (b) Analyzing data with a computer. (c) A seismic-reflection profile. The colored stripes are layers of strata. (d) A ship collecting seismic data at sea. (e) This image shows layers of subsurface strata in 3-D. Computers can expose different cross-section and map-view slices of the image. From such data, important features such as faults (indicated by colored surfaces) can be located.

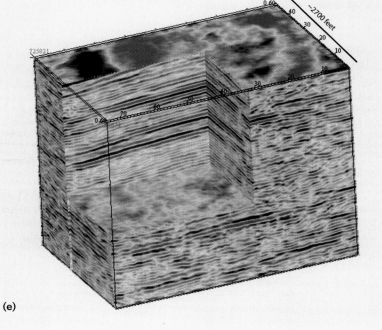

~2700 feet

(e)

Crags, Cracks, and Crumples: Crustal Deformation and Mountain Building

Geopuzzle

Mountain belts include some of the most spectacular scenery on Earth. Why do mountain ranges form? Why are the rocks within these ranges bent and broken? And how long can a mountain range survive?

Some of the world's most beautiful scenery can be found in mountainous regions. Here, gazing on Mt. Cook in the Southern Alps of New Zealand, we see evidence of the many processes that contribute to the development of mountain scenery. Compression between two plates uplifted rock to an elevation of over 3.7 km above sea level. Landslides, along with erosion by glaciers and rivers, create jagged peaks towering above debris-choked valleys.

Innumerable peaks, black and sharp, rose grandly into the dark blue sky, their bases set in solid white, their sides streaked and splashed with snow, like ocean rocks with foam. . . . [Mountains] are nature's poems carved on tables of stone. . . . How quickly these old monuments excite and hold the imagination!

—John Muir, from **Wilderness Essays**

11.1 INTRODUCTION

Geographers call the peak of Mt. Everest "the top of the world," for this mountain, which lies in the Himalayas of south Asia, rises higher than any other on Earth (▶Fig. 11.1). The cluster of flags on Mt. Everest's summit flap at 8.85 km (29,029 feet) above sea level—almost at the cruising height of modern jets. No one can live very long at the top, for the air there is too thin to breathe. In fact, even after spending weeks acclimating to high-altitude conditions at a base camp a couple of kilometers below the summit, most climbers need to use bottled oxygen during their summit attempt. In 1953, the British explorer Sir Edmund Hillary and Tenzing Norgay, a Nepalese guide, were the first to reach the summit. By 2004, about 2,200 more people had also succeeded—but 185 died trying. One in eight climbs ends in death because success depends not just on the skill of the climber, but also on the path of the jet stream, a 200-km-per-hour current of air that flows at high elevations (see Chapter 20). If the jet stream crosses the summit, it engulfs climbers in heat-robbing winds that can freeze a person's face, hands, and feet even if they're swaddled in high-tech clothing.

FIGURE 11.1 Mt. Everest and the surrounding Himalayas.

Mountains draw nonclimbers as well, for everyone loves a vista of snow-crested peaks. Their stark cliffs, clear air, meadows, forests, streams, and glaciers provide a refuge from the mundane. For millennia, mountain beauty has inspired the work of artists and poets, and in some cultures mountains have served as a home to the gods. Geologists feel a special fascination with mountains, for they provide one of the most obvious indications of dynamic activity on Earth. To make a mountain, Earth forces lift cubic kilometers of rock skyward against the pull of gravity. This uplift then provides the fodder for erosion, which, over time, grinds away at a mountain to make sediment, and in the process sculpts jagged topography.

The process of forming a mountain not only uplifts the surface of the crust, but also causes rocks to undergo **deformation**, a process by which rocks change shape, tilt, or break in response to squeezing, stretching, or shearing. Deformation produces **geologic structures,** including **joints** (cracks), **faults** (fractures along which one body of rock slides past another), **folds** (bends or wrinkles), and **foliation** (layering resulting from the alignment of mineral grains or the development of compositional bands). Mountain building may also involve metamorphism and melting. In this chapter, we learn about the phenomena that happen during mountain building—deformation, igneous activity, sedimentation, metamorphism, uplift, and erosion—and discover why they occur, in the context of plate tectonics theory.

11.2 MOUNTAIN BELTS AND THE CONCEPT OF OROGENY

With the exception of the large volcanoes formed over hot spots, mountains do not occur in isolation, but rather as part of linear ranges variously called **mountain belts**, orogenic belts, or **orogens** (from the Greek words oros, meaning mountain, and *genesis*, meaning formation). Geographers define about a dozen major mountain belts and numerous smaller ones worldwide (▶Fig. 11.2). Some large orogens contain smaller ranges within.

A mountain-building event, or **orogeny**, has a limited lifetime. The process begins, lasts for tens of millions of years, and then ceases. After an orogeny ceases, erosion may eventually bevel the land surface almost back to sea level, sometimes in as little as 50 million years. Thus, the mountain ranges we see today are comparatively young; most of Earth's present-day mountainous topography didn't exist before the Cretaceous Period. But even long after erosion

> **Take-Home Message**
>
> Mountains are not isolated points, but rather occur in belts or ranges. They are produced in response to orogeny (a mountain-building event).

FIGURE 11.2 Digital map of world topography, showing the locations of major mountain ranges.

has eliminated its peaks, a belt of "deformed" (contorted or broken) and metamorphosed rocks remains to define the location of an ancient orogen. The rock record indicates that mountain-building processes contributed to the formation of the first continental crust.

Why do orogens form? Scientific attempts to answer this question date back to the birth of geology, but explanations of the origin and distribution of mountains became possible only with the discovery of plate tectonics theory: orogens develop because of subduction at convergent plate boundaries, rifting, continental collisions, and, locally, because of motion on transform faults.

11.3 ROCK DEFORMATION IN THE EARTH'S CRUST

Deformation and Strain

As noted above, orogeny causes deformation (bending, breaking, tilting, squashing, stretching, or shearing), which in turn yields geologic structures. To get a visual sense of

deformation, let's compare a road cut along a highway in the central Great Plains of North America, a region that has not undergone orogeny, with a cliff in the Alps.

The road cut, which lies at an elevation of only about 100 m above sea level, exposes nearly horizontal beds of sandstone and shale. These beds have the same orientation that they had when first deposited (▶Fig. 11.3). Sand grains in sandstone beds of this outcrop have a nearly spherical shape (the same shape they had when deposited), and clay flakes in the shale lie roughly parallel to the bedding, because of compaction. Rock of this outcrop is **undeformed**, meaning that it contains no geologic structures other than a few joints.

In the Alpine cliff, exposed at an elevation of 3 km, rocks look very different. Here we find layers of quartzite and slate (the metamorphic equivalents of sandstone and shale) in contorted beds whose shapes resemble the wrinkles in a rug that has been pushed across the floor. These wrinkles are folds. Quartz grains in the quartzite are not spheres, but resemble flattened eggs, and the clay flakes in slate are aligned parallel to each other and tilt at a steep angle to the bedding. In fact, the rock splits on planes called slaty cleavage that parallel the flattened

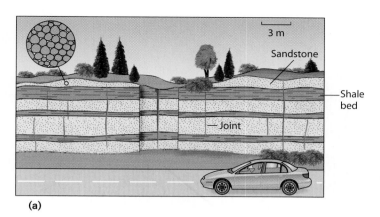

(a)

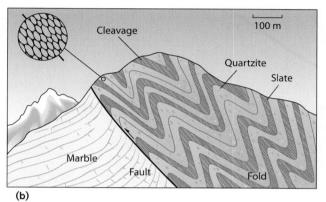

(b)

FIGURE 11.3 **(a)** This road cut exposes flat-lying beds of Paleozoic shale and sandstone along a highway in the interior region of North America. The region has *not* been involved in orogeny subsequent to the deposition of the beds. A few vertical joints cut the beds. Inset: An enlargement shows that the undeformed sandstone has spherical grains. **(b)** In this diagram of an Alpine mountain cliff, note the folded layers of quartzite and slate and the fault. Inset: Grains of sand in the quartzite have become flattened and are aligned parallel to each other. The slate has slaty cleavage.

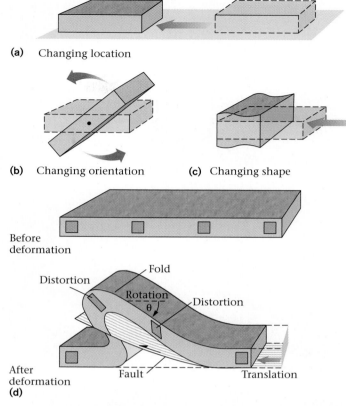

(a) Changing location

(b) Changing orientation

(c) Changing shape

Before deformation

After deformation
(d)

FIGURE 11.4 The components of deformation. **(a)** A block of rock changes location when it moves from one place to another. **(b)** It changes orientation when it tilts or rotates around an axis. **(c)** It changes shape when its dimensions change, or once planar surfaces become curved. **(d)** Folds and faults result from deformation, because they involve changes in location (e.g., sliding has occurred on a fault), orientation (a layer has tilted to form a fold), and shape (the squares in the undeformed layer have become rectangles or parallelograms in the deformed layer).

sand grains and clay flakes, and thus cut across the bedding at a steep angle. (As you recall from Chapter 6, slaty cleavage is a type of foliation, or metamorphic layering.) Finally, if we try tracing the quartzite and slate layers along the outcrop face, we find that they abruptly terminate at a sloping surface marked by shattered rock. This surface is a fault. In this example, thick beds of marble lie below this surface, so the fault juxtaposes two different rock units—the quartzite and slate must have moved along the fault from where they formed to get to their present location.

Clearly, the beds in the Alpine cliff have been deformed, and as a result the cliff exposes a variety of geologic structures. Beds no longer have the same shape and position that they had when first formed, and the shape and orientation of grains has changed. In sum, deforma-

tion includes one or more of the following (▶Fig. 11.4a–d): (1) a change in location (translation), (2) a change in orientation (rotation), (3) a change in shape (distortion). Deformation can be fairly obvious when observed in an outcrop (▶Fig. 11.5a–c). In a broad sense, geologic structures produced by deformation can be thought of as geometric shapes—planes, lines, and/or curves. ▶Box 11.1 explains how geologists specify the orientation of such shapes.

Geologists refer to the change in shape that deformation causes as **strain**. We distinguish among different kinds of strain: if a layer of rock becomes longer, it has undergone stretching, but if the layer becomes shorter, it has undergone shortening, and if a change in shape involves the movement of one part of a rock body past another so that angles between features in the rock change, the result is shear strain (▶Figs. 11.7a–d and 11.8).

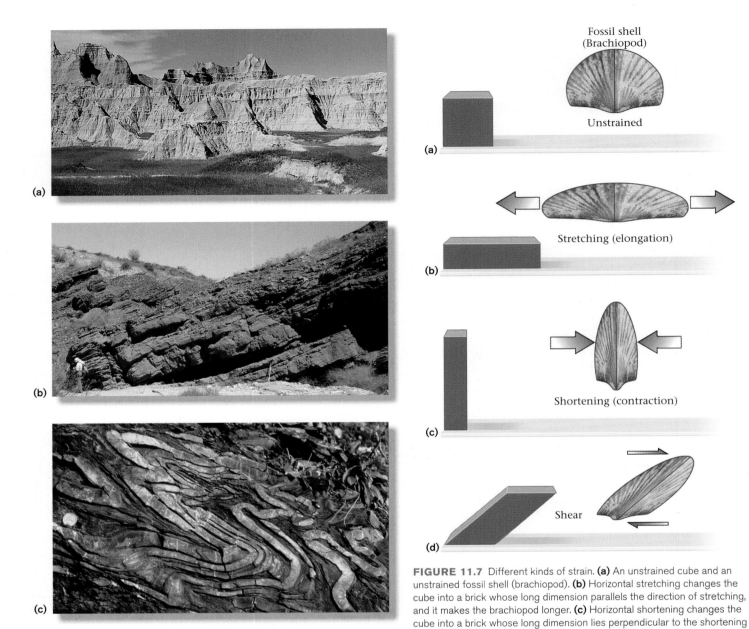

FIGURE 11.5 **(a)** Undeformed, flat-lying beds of sediment in Badlands National Monument, South Dakota. **(b)** Tilted beds of strata in Arizona. The tilting is a manifestation of deformation. **(c)** Folded layers of quartzite in Vermont. The folding is also a manifestation of deformation. Note the coin for scale.

Fossil shell (Brachiopod)

Unstrained

Stretching (elongation)

Shortening (contraction)

Shear

FIGURE 11.7 Different kinds of strain. **(a)** An unstrained cube and an unstrained fossil shell (brachiopod). **(b)** Horizontal stretching changes the cube into a brick whose long dimension parallels the direction of stretching, and it makes the brachiopod longer. **(c)** Horizontal shortening changes the cube into a brick whose long dimension lies perpendicular to the shortening direction, and it makes the brachiopod taller. **(d)** Shear strain tilts the cube over and transforms it into a parallelogram, and it changes the angular relationships in the brachiopod.

Kinds of Deformation: Brittle and Ductile Behavior

We saw in Chapter 10 that rocks can temporarily change shape when subjected to force (push, pull, or shear), developing an elastic strain, and then change back when the force that caused the strain is removed. But rocks can also develop a permanent strain in two fundamentally different ways. During **brittle deformation**, a material breaks into two or more pieces, like a plate shattering on the floor, whereas during **ductile deformation**, a material changes shape without breaking, like a ball of dough squeezed beneath a book (▶Fig. 11.9a–d). Joints and faults are brittle structures, whereas folds and foliations are ductile structures. What actually happens in a rock during the different kinds of deformation? Recall that rocks are solids in which chemical bonds, like little springs, link atoms together. During elastic deformation, these bonds

BOX 11.1 THE REST OF THE STORY

Describing the Orientation of Structures

When we are discussing geologic structures, it's important to be able to communicate information about their orientation. For example, does a fault exposed in an outcrop at the edge of town continue beneath the nuclear power plant 3 km to the north, or does it go beneath the hospital 2 km to the east? If we know the fault's orientation, we might be able to answer this question. To describe the orientation of a geologic structure, geologists picture the structure as a simple geometric shape, then specify the angles that the shape makes with respect to a horizontal plane (a flat surface parallel to sea level), a vertical plane (a flat surface perpendicular to sea level), and the north direction (a line of longitude).

Let's start by observing *planar* structures such as faults, beds, and joints. We call these structures planar because they resemble a geometric plane. A planar structure's orientation can be specified by its strike and dip. The **strike** is the angle between an imaginary horizontal line (the strike line) on the plane and the direction to true north (▶Fig. 11.6a, b). We measure the strike with a magnetic compass (▶Fig. 11.6d). The **dip** is the angle of the plane's slope (more precisely, the angle between a horizontal plane and the dip line, an imaginary line parallel to the steepest slope on the plane, as measured in a vertical plane perpendicular to the strike). We measure the dip angle with a clinometer, a type of protractor that measures slope angles. A

horizontal plane has a dip of 0°, and a vertical plane has a dip of 90°. We represent strike and dip on a geologic map using the symbol shown in Figure 11.6b.

A linear structure resembles a line rather than a plane; examples of linear structures include scratches or grooves on a rock surface. Geologists specify the orientation of linear structures by giving their plunge and bearing (▶Fig. 11.6c). The **plunge** is the angle between a line and horizontal, as measured with a clinometer, in the vertical plane that contains the line. A horizontal line has a plunge of 0°, and a vertical line has a plunge of 90°. The **bearing** is the compass heading of the line (more precisely, the angle between the projection of the line on the horizontal plane and the direction to true north).

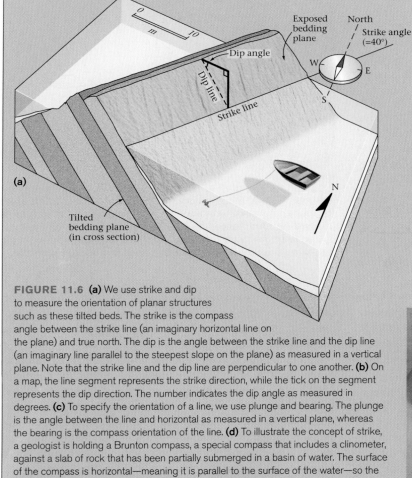

(a)

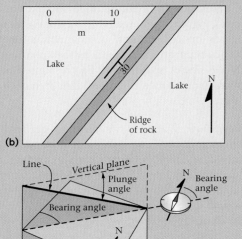

(b)

(c)

(d)

FIGURE 11.6 (a) We use strike and dip to measure the orientation of planar structures such as these tilted beds. The strike is the compass angle between the strike line (an imaginary horizontal line on the plane) and true north. The dip is the angle between the strike line and the dip line (an imaginary line parallel to the steepest slope on the plane) as measured in a vertical plane. Note that the strike line and the dip line are perpendicular to one another. **(b)** On a map, the line segment represents the strike direction, while the tick on the segment represents the dip direction. The number indicates the dip angle as measured in degrees. **(c)** To specify the orientation of a line, we use plunge and bearing. The plunge is the angle between the line and horizontal as measured in a vertical plane, whereas the bearing is the compass orientation of the line. **(d)** To illustrate the concept of strike, a geologist is holding a Brunton compass, a special compass that includes a clinometer, against a slab of rock that has been partially submerged in a basin of water. The surface of the compass is horizontal—meaning it is parallel to the surface of the water—so the edge of the compass in contact with the slab is a strike line.

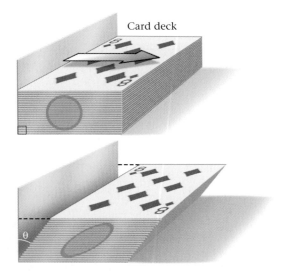

Card deck

Before

Dough

After

GEOLOGY

(a) Brittle deformation

(b) Ductile deformation

FIGURE 11.8 You can simulate shear strain by moving a deck of cards so that each card slides a little with respect to the one below. Note how a circle drawn on the side of the deck changes shape to become an ellipse, and that the angle between the bottom of the deck and the back side of the deck has changed from a right angle to an acute angle.

FIGURE 11.9 (a) Brittle deformation occurs when you drop a plate and it shatters. (b) Ductile deformation takes place when you squash a soft ball of dough beneath a book and the dough flattens into a pancake without breaking. (c) Cracks (joints) in an outcrop result from brittle deformation. (d) Folds, like these in the marble of a quarry wall, form without breaking rock, and thus result from ductile deformation.

stretch and bend, but do not break. During brittle deformation, many bonds break at once so that rocks can no longer hold together, whereas during ductile deformation, some bonds break but new ones quickly form, so that rocks do not separate into pieces as they change shape.

Why do rocks inside the Earth sometimes deform brittlely and sometimes ductilely? The behavior of a rock depends on:

- *Temperature:* Warm rocks tend to deform ductilely, whereas cold rocks tend to deform brittlely. To see this contrast, try an experiment with a candle. Chill a candle in a freezer, then press its middle against the edge of a table—the candle will brittlely snap in two. But if you first warm the candle in an oven, it will ductilely bend without breaking when pressed against the table.

- *Pressure:* Under great pressures deep in the Earth, rock behaves more ductilely than it does under low pressures near the surface. Pressure effectively prevents rock from separating into fragments.

- *Deformation rate:* A sudden change in shape causes brittle deformation, whereas a slow change in shape causes ductile deformation. For example, if you hit a marble bench with a hammer, it shatters, but if you leave it alone for a century, it gradually sags without breaking.

- *Composition:* Some rock types are softer than others; for example, halite (rock salt) can deform ductilely under conditions in which granite behaves brittlely.

Considering that pressure and temperature both increase with depth in the Earth, geologists find that in typical continental crust, rocks behave brittlely above about 10

to 15 km, whereas they behave ductilely below. We call this depth the brittle-ductile transition. Earthquakes in continental crust happen only above this depth because these earthquakes involve brittle breaking.

In some cases, both brittle and ductile structures occur in the same outcrop. For example, in our Alpine cliff (Fig. 11.3b), you can see both faulting (brittle deformation) and folding (ductile deformation). Such an occurrence may seem like a paradox at first. But the juxtaposition of styles happens simply because of changes in the deformation rate during orogeny. Slow deformation yielded the folds, whereas a pulse of rapid deformation caused the fault to form.

Force, Stress, and the Causes of Deformation

Up to this point, we've focused on picturing the *consequences* of deformation. Describing the *causes* of deformation is a bit more challenging, in the context of an introductory geology book. In captions for displays about mountains, museums and national parks typically dispense with the issue by using the phrase "The mountains were caused by forces deep within the Earth." But what does this mean? Isaac Newton defined force using the equation: force = mass × acceleration. According to this equation, if you apply a force to an object, the object speeds up, slows down, or changes direction. Applying this concept to geology, we see that phenomena such as plate interactions (for example, continent-continent collisions) apply forces to rock and thus cause rock to change location, orientation, or shape. In other words, the application of forces in the Earth indeed causes deformation.

However, geologists use the word *stress* instead of *force* when talking about the cause of deformation. We define the **stress** acting on a plane as the force applied *per unit area* of the plane. Written as an equation this becomes: stress = force/area. The need to distinguish between stress and force arises because the actual consequences of applying a force depend not just on the magnitude of force but also on the area over which the force acts. A pair of simple experiments shows why. Experiment 1: Stand on a single, empty aluminum can (▶Fig. 11.10a). All of your weight—a force—focuses entirely on the can, and the can crushes. Experiment 2: Place a board over 100 cans, and stand on the board (▶Fig. 11.10b). In this case, your weight is distributed across 100 cans, so the force acting on any one can is not enough to crush it. In both experiments, the force caused by the weight of your body was the same, but in experiment 1 the force was applied over a small area so the single can felt a *large* stress, whereas in experiment 2 the same force was applied over a large area so only a *small* stress developed. How does this concept apply to geology? During mountain building, the force of one plate interacting with another is distributed across the area of contact between the two plates, so the deformation resulting at any specific location actually depends on the stress developed at that location, not on the total force involved in the plate interaction.

Different kinds of stress occur in rock bodies. As we learned in Chapter 10, **compression** develops when a rock is squeezed, **tension** occurs when a rock is pulled apart, and **shear stress** develops when one side of a rock body moves sideways past the other side (▶Fig. 11.11a–d). **Pressure** refers to a special stress condition in which the same push acts on all sides of an object. Note that "stress" and "strain" have different meanings to geologists (though we tend to use them interchangeably in everyday English): stress refers to the amount of force per unit area of a rock, whereas strain refers to the change in shape of a rock. Thus, stress *causes* strain. Specifically, compression causes shortening, tension leads to stretching, and shear stress generates shear strain. Pressure can cause an object to become smaller (push in on all sides equally). With our knowledge of stress and strain, we can now look at the nature and origin of various classes of geologic structures.

> **Take-Home Message**
>
> Mountain building produces stress (compression, tension, shear), which in turn causes deformation. Stress breaks rock under brittle conditions or bends and distorts rock, without breaking it, under ductile conditions. Strain is a measure of distortion.

11.4 WHAT STRUCTURES FORM DURING BRITTLE DEFORMATION?

Joints: Natural Cracks in Rocks

If you look at the photographs of rock outcrops in this book, you'll notice thin dark lines that cross the rock faces (Fig. 11.9c). These lines represent traces of natural cracks, along which the rock brittlely broke and separated into two pieces. Geologists refer to such natural cracks as joints. Rock bodies do *not* slide past each other on joints. Since joints are roughly planar structures, we define their orientation by their strike and dip; see Box 11.1.

Joints develop in response to tensional stress in brittle rock: a rock splits open because it has been pulled slightly apart. They may form for a variety of geologic reasons. For example, some joints form when a rock cools and contracts, because contraction makes one part of a rock pull away from the adjacent part. Others develop when rock layers

FIGURE 11.10 **(a)** When you stand on a single can, you apply enough force to the can to crush it, for the can feels a large stress. **(b)** When you stand on a board resting on 100 cans, you apply the same force to the board, but now it is spread out over 100 cans. Therefore, each can feels only a small stress and does not crumple.

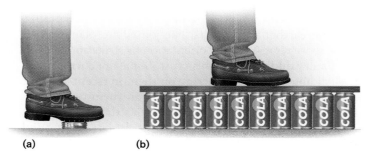

(a) (b)

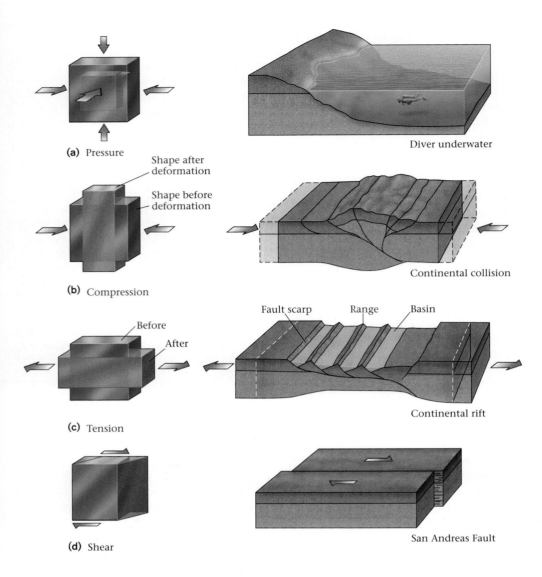

(a) Pressure

Shape after deformation

Shape before deformation

(b) Compression

Diver underwater

Continental collision

Before

After

(c) Tension

Fault scarp   Range   Basin

Continental rift

(d) Shear

San Andreas Fault

FIGURE 11.11 We represent the direction and magnitude of stress acting on each face of an object by arrows; the lengths of the arrows represent the magnitude of the stress. **(a)** Pressure occurs when an object feels the same stress on all sides. A diver feels pressure when submerged. **(b)** Compression takes place when an object is squeezed. Compression occurs during growth of a collisional mountain range. **(c)** Tension is created when the opposite ends of an object are pulled in opposite directions. Tension occurs during growth of a continental rift. **(d)** Shear stress occurs when one surface of an object slides relative to the other surface (we depict the shear direction with half arrows). Shear stress parallel to the Earth's surface causes slip on the San Andreas Fault.

veins contain small quantities of valuable metals, such as gold.

Geotechnical engineers, people who study the geologic characteristics of construction sites, pay close attention to jointing when recommending where to put roads, dams, and buildings. Water flows much more easily through joints than it does through solid rock, so it would be a bad investment to situate a water reservoir over rock with closely spaced joints—the water would leak down into the joints. Also, building a road on a steep cliff composed of jointed rock could be risky, for joint-bounded blocks separate easily from bedrock, and the cliff might collapse.

formerly at depth undergo a decrease in pressure as overlying rock erodes away, and thus change shape slightly. Still others form when rock layers bend.

Rock bodies may contain two categories of joints. Systematic joints are long planar cracks that occur fairly regularly through a rock body, whereas nonsystematic joints are short cracks that occur in a range of orientations and are randomly spaced. A group of systematic joints constitutes a joint set, a spectacular example of which can be seen in sandstone beds of Arches National Park, in Utah (▶Fig. 11.12a). Erosion has created narrow gullies along the joints. In sedimentary rocks, systematic joints typically are vertical planes (▶Fig. 11.12b).

If groundwater seeps through joints for a long period of time, minerals such as quartz or calcite can precipitate out of the groundwater and fill the joint. Such mineral-filled joints are called **veins** and look like white stripes cutting across a body of rock (▶Fig. 11.12c). Some

Introducing Faults

After the San Francisco earthquake of 1906, geologists found a rupture that ripped across the landscape near the city. Where this rupture crossed orchards, it offset rows of trees, and where it crossed a fence, it broke the fence in two; the western side of the fence moved northward by about 2 m (see Fig. 10.6a). The rupture represented the trace of the San Andreas Fault (▶Fig. 11.13a, b). As we have seen, a **fault** is a fracture on which sliding occurs. Slip events, or faulting, generate earthquakes. Faults, like joints, are planar structures, so we represent their orientation by strike and dip.

Faults riddle the Earth's crust. Some are currently active (sliding has been occurring on them in recent geologic time), but most are inactive (sliding on them ceased millions of years ago). Some faults, such as the San Andreas, intersect the ground surface and thus displace the ground when they move. Others accommodate the sliding of rocks

(a)

(b)

(c)

FIGURE 11.12 (a) This bedding plane in sandstone of Arches National Park, Utah, contains many systematic joints. **(b)** These vertical joints exposed on a cliff face near Ithaca, New York, run from the surface down into bedrock. **(c)** The veins in this outcrop, composed of milky white quartz, fill fractures in gray shale.

in the crust at depth, and remain invisible at the surface unless they are later exposed by erosion (▶Fig. 11.13c, d).

Geologists study faults not only because the movement on some faults causes earthquakes, but also because they juxtapose bodies of rock that did not originally lie adjacent to each other and thus complicate the arrangement of rocks at the Earth's surface. For example, in our Alpine cliff, Fig. 11.3b, movement on a fault placed quartzite and slate beds against marble beds. We must understand these rearrangements in order to predict where resources lie underground.

Fault Classification

Geologists have developed terminology to classify faults and describe movement on them. (We introduced this terminology in Chapter 10, and add to it here.) The fault plane can be vertical, horizontal, or at some angle in between, and we can describe its orientation by a strike and dip. In the case of nonvertical faults (those that slope at an angle), we define the **hanging-wall block** as the rock above the fault plane, and the **footwall block** as the rock below the fault plane (▶Fig. 11.14a). If you stand in a tunnel along a fault plane, the hanging-wall block looms over your head, and the footwall block lies under your feet. We distinguish several types of faults.

- *Dip-slip vs. strike-slip vs. oblique-slip faults:* On **dip-slip faults**, sliding occurs up or down the slope of the fault (therefore, up or down the dip); on **strike-slip faults**, one block slides past another (therefore, parallel to the strike line); and on **oblique-slip faults**, sliding occurs diagonally on the fault plane (▶Fig. 11.14b–d).

- *Types of dip-slip faults:* We subdivide dip-slip faults into two kinds, depending on which way the hanging-wall block moves relative to the footwall block. On **thrust faults** and **reverse faults**, the hanging-wall block moves up the slope of the fault. Thrust faults differ from reverse faults only in terms of the fault-plane's slope (or dip)—thrust faults have a slope (or dip) of less than about 35°, whereas reverse faults have a dip of greater than 35° (▶Fig. 11.15). On **normal faults**, the hanging-wall block moves down the slope of the fault. "Normal" and "reverse" are relicts of nineteenth-century miners' jargon. Normal faults were simply more common in the mines where faults were first recognized. But globally, normal faults aren't any more common or typical than reverse faults.

- *Types of strike-slip faults:* Geologists distinguish between two types of strike-slip faults on the basis of the relative movement of one side of the fault with respect to the other. If you stand facing the fault, you can say that it is a left-lateral strike-slip fault if the block on the far side slipped to your left, and that it is a right-lateral strike-slip fault if the block on the far side slipped to your right. Note that strike-slip faults commonly have a vertical dip, so we generally cannot define the hanging-wall or footwall block on such faults.

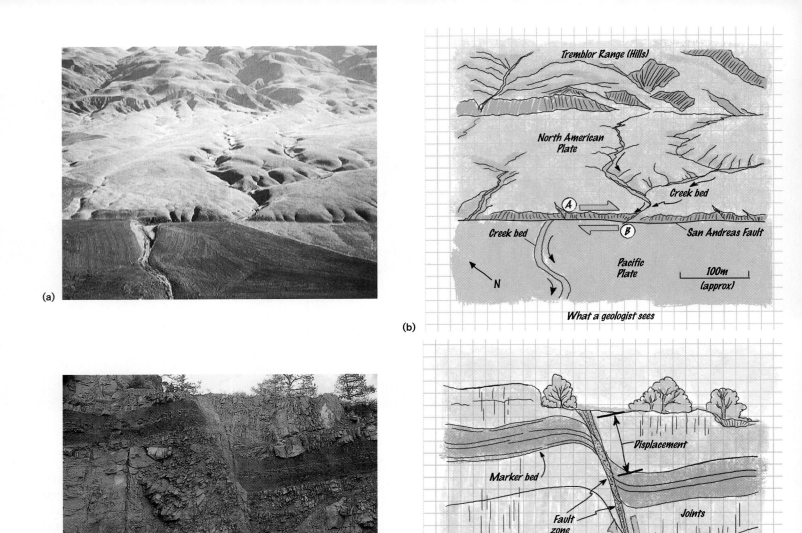

FIGURE 11.13 **(a)** An oblique air photo showing the San Andreas Fault displacing a creek flowing from the Tremblor Range (background) into the Carizzo Plain, California. **(b)** What a geologist sees in the previous photo. **(c)** A road cut in the Rocky Mountains of Colorado, showing a fault offsetting strata in cross section. Note that the fault is actually a band of broken rock about 50 cm wide. **(d)** What a geologist sees looking at the Rocky Mountain road cut.

Recognizing Faults

How do you recognize a fault when you see one? Some faults offset distinctive beds (Fig. 11.13b; ►Fig. 11.16a, b). In such cases, the fault is the plane on which offset occurs. Geologists refer to the amount of offset as the **displacement** on the fault: the displacement on the fault shown in Figure 11.13c, d is about 2 m. In some cases, faults juxtapose two different rock units so the fault is the surface of contact between the two units (for example, Fig. 11.3b). Note that in the example of Figure 11.3, we can't see distinctive markers, so without additional information we can't measure the displacement directly. Typically, thrust

or reverse faults cutting sedimentary beds place older beds on younger ones, whereas normal faults place younger beds on older. In some cases, layers of rock cut by a fault undergo folding during or just before slip; the resulting folds are informally called drag folds (Fig. 11.4d).

Faults may also leave their mark on the landscape. Those that intersect the ground surface while they are active can displace natural landscape features, such as stream valleys or glacial moraines (Fig. 11.13a), and also human-made features, such as highways, fences, or rows of trees in orchards. Displacement on a dip-slip or oblique-slip fault makes a small step on the ground surface; this step is called a **fault scarp**

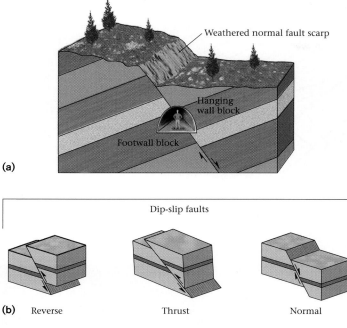

(a)

Weathered normal fault scarp

Hanging wall block

Footwall block

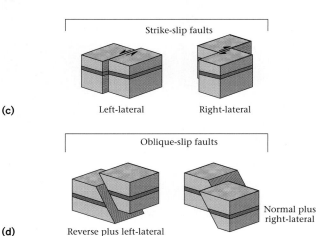

Dip-slip faults

(b) Reverse — Thrust — Normal

Strike-slip faults

(c) Left-lateral — Right-lateral

Oblique-slip faults

(d) Reverse plus left-lateral — Normal plus right-lateral

FIGURE 11.14 **(a)** A hanging-wall block and footwall block, relative to a sloping fault surface. The weathered fault scarp is an exposure of the fault at the ground surface. **(b)** Three types of dip-slip faults, on which sliding parallels the dip line. **(c)** Two types of strike-slip faults, on which sliding parallels the strike line. **(d)** Two examples of oblique-slip faults, on which sliding takes place diagonally along the surface.

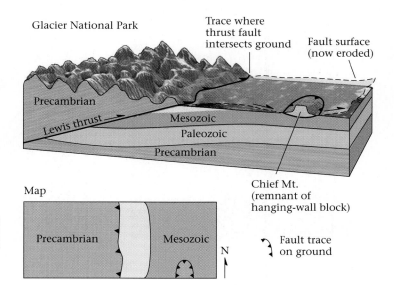

Glacier National Park

Trace where thrust fault intersects ground

Fault surface (now eroded)

Precambrian

Lewis thrust

Mesozoic

Paleozoic

Precambrian

Chief Mt. (remnant of hanging-wall block)

Map

Precambrian — Mesozoic

N

Fault trace on ground

FIGURE 11.15 This large thrust fault (the Lewis thrust) puts older rock (Precambrian) over younger rock (Mesozoic). Erosion has removed much of the hanging-wall block, but a small remnant still lies to the east of the mountains. On the geologic map of the region, the triangular barbs point to the hanging-wall block. The hanging wall has moved about 100 km relative to the footwall.

(►Fig. 11.16c). And because faults tend to break up rock, the fault may be preferentially eroded. If this happens, the fault trace (the line of intersection between the fault and the ground surface) will be marked by a linear valley. Even if a fault did not intersect the ground surface while active, it may influence the landscape later in geologic history. For example, if fault movement juxtaposed strong rock with weak rock, long-term erosion produces a step in the landscape, with stronger rock above the step and weaker rock below.

Fault surfaces and their borders typically look different from bedding planes. For example, faulting under brittle conditions may crush or break adjacent rock. If this shattered rock consists of visible angular fragments, then it is called fault breccia (►Fig. 11.16d), but if it consists of a fine powder, then it is called fault gouge. Some fault surfaces are polished and grooved by the movement of the hanging wall past the footwall. Polished fault surfaces are called slickensides, and linear grooves on fault surfaces are slip lineations (►Fig. 11.16e). We specify the orientation of a slip lineation by giving its plunge and bearing (see Box 11.1).

The shear on some faults takes place under ductile conditions at depth in the crust. Where this happens, rock does not break up into breccia or gouge along the fault zone, but rather shears ductilely to form a band of fine-grained foliated rock called mylonite. The fine grain size of mylonite results not from brittle fracturing during shear, but rather from a type of metamorphic recrystallization that subdivides large grains into small ones. Geologists refer to faults in which movement occurred ductilely as **shear zones** (►Fig. 11.17).

Fault Systems and the Resulting Strain

Numerous related faults are often found in groups called fault systems. Individual faults in a system may merge at depth with a nearly horizontal **detachment fault**. The faulting in a thrust-fault system shortens the crust; slip makes

(a)

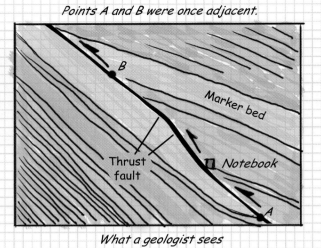

Points A and B were once adjacent.

Marker bed

B

Thrust
fault

Notebook

A

What a geologist sees

(b)

FIGURE 11.16 (a) A thrust fault on which a distinct layer has been offset. **(b)** A geologist's sketch emphasizes the offset. Point B was originally adjacent to point A. **(c)** A fault scarp formed after an earthquake in Nevada. **(d)** This fault breccia along a fault consists of broken-up rock. **(e)** Slip lineations on a fault surface.

(c)

(d)

(e)

FIGURE 11.17 Photograph of a shear zone in Precambrian rock of northwestern Scotland. On the left is grainy, unsheared granite. It grades to the right into pinkish mylonite with a strong foliation. This rock is in contact with black mylonite formed by shear of metamorphosed basalt. This rock grades to the right into unsheared metamorphosed basalt. The white splotches are lichen.

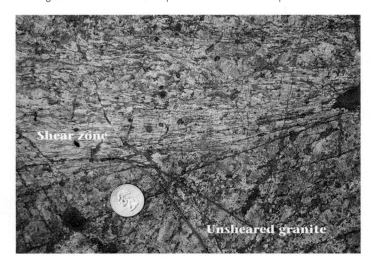

Shear zone

Unsheared granite

slices of the crust overlap like shingles (▶Fig. 11.18a). Faulting in normal-fault systems, on the other hand, stretches the crust (▶Fig. 11.18b). In such systems, elongate basins separate high blocks of rock. Individual basins may be bounded on each side by a normal fault, with each fault dipping toward the basin. Such a fault-bounded basin is called a graben; the high block between two grabens is a horst (▶Fig. 11.19a, b). Typically, normal faults become shallower with depth and merge with a detachment fault. As the hanging-wall block above such a curved fault slides down the fault, a triangular basin called a half-graben develops between the fault and the top surface of the rotated block (see Fig. 11.18b).

> **Take-Home Message**
>
> Brittle deformation produces joints (cracks along which slip *has not* occurred) and faults (fractures along which slip *has* occurred). Faulting offsets rock layers and, locally, the ground surface. Thrust faults form during crustal shortening, and normal faults during stretching.

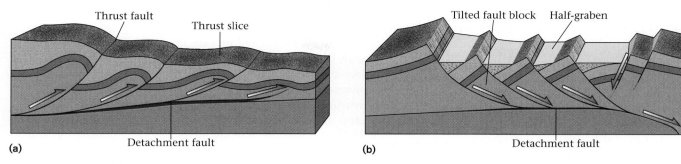

(a)   (b)

FIGURE 11.18 **(a)** In this thrust-fault system, several related thrust faults merge at depth with a detachment fault. Note that displacement on the thrusts shortens the layers of rock above the detachment. **(b)** A normal-fault system consists of several related normal faults. In this example, several faults dip in the same direction and merge with a detachment fault at depth. Displacement on the faults stretches the layers of rock above the detachment, and the rock layers have tilted as a consequence of slip on the faults. The wedge-shaped depressions bounded by a fault on one side and the tilted rock layer on the other are half-grabens.

11.5 WHAT STRUCTURES FORM DUE TO DUCTILE DEFORMATION?

Introducing Folds

Imagine a carpet lying flat on the floor. Push on one end of the carpet, and it will wrinkle or contort into a series of wave-like curves. Layers of rock can do the same: planar layers can be contorted into curves, which geologists call folds. In other words, a **fold** consists of curving rock layers. The layers can be beds of sedimentary rock, flows of extrusive igneous rock, or foliation of metamorphic rock. Folds may also be defined by curving dikes, sills, or veins.

FIGURE 11.19 **(a)** Horsts and grabens cutting through marble exposed in a quarry wall in Brazil. **(b)** A geologist's sketch of the quarry wall, indicating the positions of the faults.

Fold Terminology and Identification

Not all folds look the same—some look like an arch, some like a trough, and some have other shapes. To describe these shapes, we first label the parts of a fold. The **hinge** refers to the portion of the fold where curvature is greatest, and the **limbs** are the sides of the fold that show less curvature. The axial plane is an imaginary surface that encompasses the hinges of successive layers. With these terms in hand, we can now describe types of folds.

- *Anticlines, synclines, and monoclines:* A fold that has an archlike shape in which the limbs dip away from the hinge is called an **anticline**, whereas a fold with a

(a)

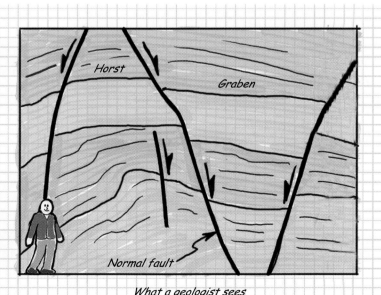

(b)   *What a geologist sees*

trough-like shape in which the limbs dip toward the hinge is called a **syncline** (▶Fig. 11.20a, b). A **monocline** has the shape of a carpet draped over a stairstep (▶Fig. 11.20c, d).

- *Open and tight folds:* If the angle between the limbs is large, then the fold is an open fold, but if the angle between the limbs is small, then the fold is a tight fold (▶Fig. 11.20e, f).

- *Nonplunging and plunging folds:* If the hinge is horizontal, the fold is called a nonplunging fold, but if the hinge is tilted, the fold is called a plunging fold (▶Fig. 11.20g).

- *Domes and basins:* A fold in the shape of an overturned bowl is called a **dome**, whereas a fold shaped like a right-side-up bowl is called a **basin** (▶Fig. 11.20h). Some small domes and basins are found at outcrops, but others measure hundreds of kilometers across.

Now see if you can use this terminology to identify the various folds shown in ▶Figure 11.21a–e.

You can recognize folds not only in cross section, but also by the pattern of rock layers on the ground surface (▶Fig. 11.22a). For example, a nonplunging anticline involving sedimentary layers appears as a series of parallel stripes, with the oldest layer in the center and progressively younger layers away from the center; the stripes are symmetrically positioned around the hinge (▶Fig. 11.22b). In a nonplunging syncline involving sedimentary layers, the youngest layers crop out in the center and the oldest at the margins (Fig. 11.22b). Layers in a plunging fold have a U-shape on the ground surface (▶Fig. 11.22c, d, e). We can represent the hinge of the fold by a heavy line bordered by outward-pointing arrows for an anticline and inward-pointing arrows for a syncline. Domes and basins both show circular outcrop patterns that look like a bull's-eye. The oldest layer occurs in the center of a dome, while the youngest layer is located in the center of a basin (▶Fig. 11.23a, b). Note that because some layers erode more easily than others, the shapes of folds may be indicated by the pattern of ridges and valleys.

FIGURE 11.20 **(a)** An anticline (archlike fold) and **(b)** a syncline (troughlike fold), showing the hinge, limbs, and axial plane. **(c)** A monocline. Note how it resembles a stair step. **(d)** Photo of a monocline involving sedimentary strata of the Colorado Plateau, in Arizona. **(e)** An open fold has a large angle, ø, between the limbs. **(f)** A tight fold has a small angle, ø, between the limbs. **(g)** If the hinge of a fold is tilted, it's a plunging fold; if the hinge is horizontal, it's a nonplunging fold. **(h)** Domes (dome-shaped folds) and basins (bowl-shaped folds).

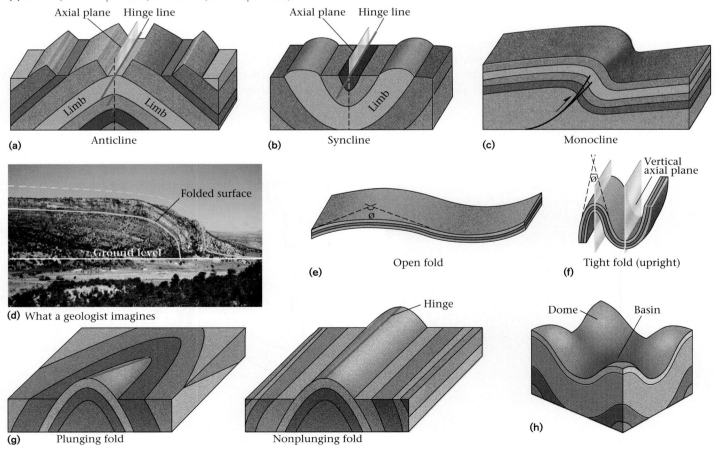

FIGURE 11.21 (a) An open anticline, exposed in a road cut near Kingston, New York. **(b)** An open syncline, exposed in a road cut near Sideling Hill, in Maryland. **(c)** A tight fold exposed along the coast of Brazil. **(d)** A train of folds exposed in sea cliffs in eastern Ireland. Note that the axial planes of these folds are not vertical. **(e)** What a geologist sees in the previous photo.

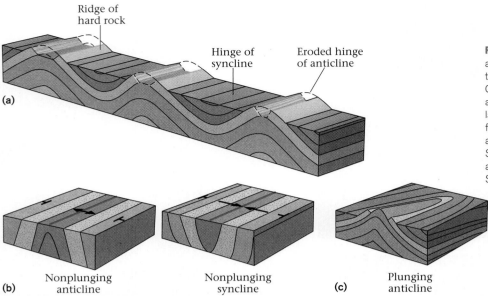

Ridge of
hard rock

Hinge of
syncline

Eroded hinge
of anticline

(a)

FIGURE 11.22 **(a)** After erosion, folds involving alternating strong and weak rock layers may control the positions of valleys and ridges on the land surface. Commonly, valleys form over synclines and in the hinge area of anticlines. **(b)** On the ground surface, the same layers appear on either side of the hinge of a nonplunging fold. **(c)** On the ground surface, the layers of a plunging anticline curve around the hinge. **(d)** Aerial photo of Sheep Mountain anticline, in Wyoming, a plunging anticline. **(e)** What a geologist sees looking at the Sheep Mountain anticline.

(b) Nonplunging anticline

Nonplunging syncline

(c) Plunging anticline

(d)

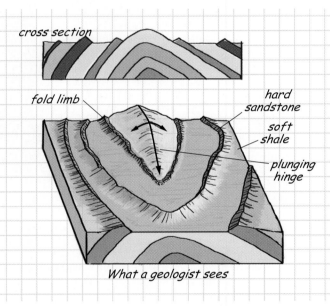

cross section

fold limb

hard sandstone

soft shale

plunging hinge

What a geologist sees

(e)

FIGURE 11.23 **(a)** A dome. **(b)** A basin.

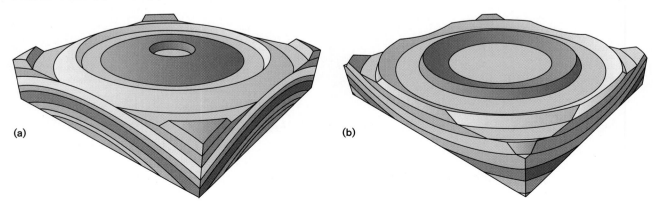

(a)

(b)

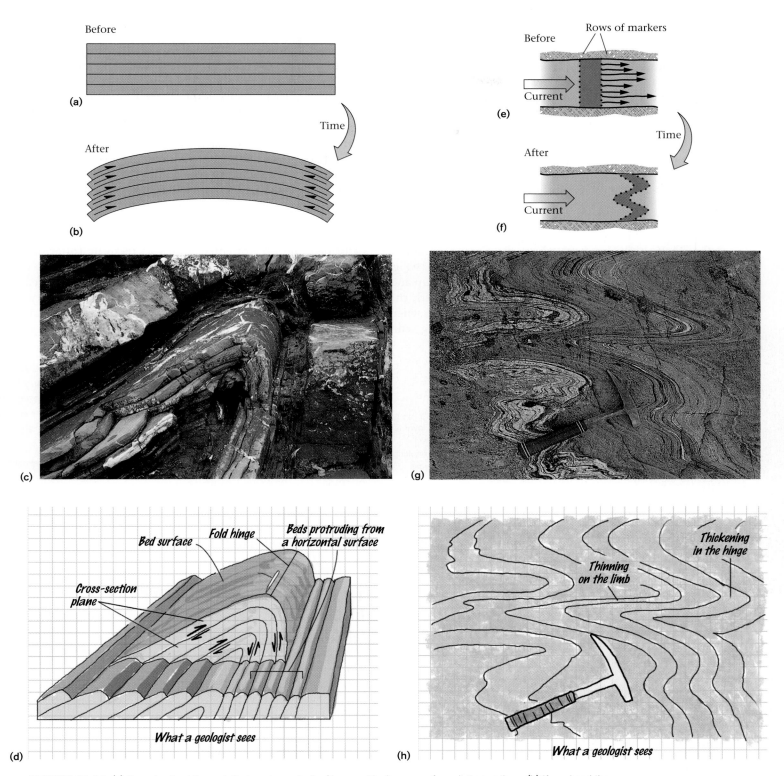

FIGURE 11.24 **(a)** To understand flexural slip, imagine a stack of layers with slippery surfaces between them. **(b)** If you bend the layers, but keep surfaces in contact, flexural slip occurs. **(c)** Photo of a small fold formed by flexural slip. Note that the sedimentary beds maintain the same thickness around the fold. **(d)** What a geologist sees looking at the previous photo. **(e)** To understand the concept of flow folding, imagine two rows of buoys on a river. This illustration shows the buoys before flow. **(f)** Not all the buoys flow at the same rate, so after flow, a fold shape has formed. **(g)** Photo of small flow folds exposed in northern Scotland. **(h)** What a geologist sees looking at the previous photo. Note how rock flowed into the hinge zone of the fold.

Forming Folds

Folds develop in two principal ways. During formation of flexural folds, a stack of layers bends, and slip occurs between the layers. The same phenomenon happens when you bend a deck of cards—to accommodate the change in shape, the cards slide with respect to each other (▶Fig. 11.24a–d). Flow folds form when the rock overall is so soft that it behaves like weak plastic; these folds develop simply because different parts of the rock body move at different rates (▶Fig. 11.24e–h). You can see the same process happen when scum or bubbles on the surface of a slow-moving river twist into complex curves.

Folds develop for a variety of reasons. Some layers wrinkle up, or buckle, in response to end-on compression (▶Fig. 11.25a). Others form where shear stress gradually moves one part of a layer up and over another part (▶Fig. 11.25b). Still others develop where rock layers move up and over bends in a fault and must curve to conform with the fault's shape (▶Fig. 11.25c). Finally, some folds form when a block of basement moves and bends the overlying sedimentary layers (▶Fig. 11.25d).

FIGURE 11.25 Different causes of folding. **(a)** If a layer becomes shortened along its length, it buckles (wrinkles up like a rug). **(b)** If a layer is sheared, it gradually bends over on itself to form a fold. **(c)** When layers move up and over step-shaped faults, they must bend into folds. **(d)** Faulting at depth may fold a layer closer to the ground surface. The folded layers drape over the uplifted fault block to form a monocline.

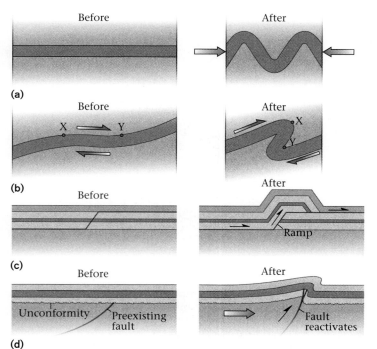

Tectonic Foliation

In an undeformed sandstone, the grains of quartz are roughly spherical, and in an undeformed shale, clay flakes press together into the plane of bedding so that shales tend to split parallel to the bedding. During ductile deformation, however, internal changes take place in a rock that gradually modify the original shape and arrangement of grains. For example, quartz grains may transform into cigar shapes, elongate ribbons, or tiny pancakes, and clay flakes may recrystallize or reorient so that they lie at an angle to the bedding. Overall, deformation can produce inequant grains and can cause them to align parallel with one another, thereby generating metamorphic foliation in the rock. We refer to layering created by the alignment of deformed and/or reoriented grains as **tectonic foliation** (▶Fig. 11.26a).

Chapter 8 introduced foliations such as slaty cleavage, schistosity, and gneissic layering. Here we add to the story by noting that they form in response to flattening and shearing in ductilely deforming rocks. In other words, foliations indicate that the rock has developed a strain. In rocks with slaty cleavage, the cleavage planes lie perpendicular to the direction of the shortening strain. Thus, the cleavage tends to be parallel to the axial plane of folds that developed in association with the cleavage. In schists and gneisses, the foliation commonly lies parallel to or at a slight angle to the direction of shear, because shear smears grains out into the plane of shearing (▶Fig. 11.26b–f).

> **Take-Home Message**
>
> Folds are curves defined by the shape of rock layers. They form when a region undergoes shortening or shearing. Under certain conditions, deformation also yields tectonic foliation.

11.6 IGNEOUS, SEDIMENTARY, AND METAMORPHIC PROCESSES IN OROGENIC BELTS

The process of orogeny establishes geologic conditions appropriate for the formation of a great variety of rocks. We'll consider examples from all three rock categories (▶Fig. 11.27):

- *Igneous activity during orogeny:* In mountain belts formed along convergent plate boundaries, magma forms as a consequence of subduction. In rifts, stretching and thinning of lithosphere causes decompression melting of the underlying mantle. And during continental collision, melting may take place where deep portions of the crust undergo heating.

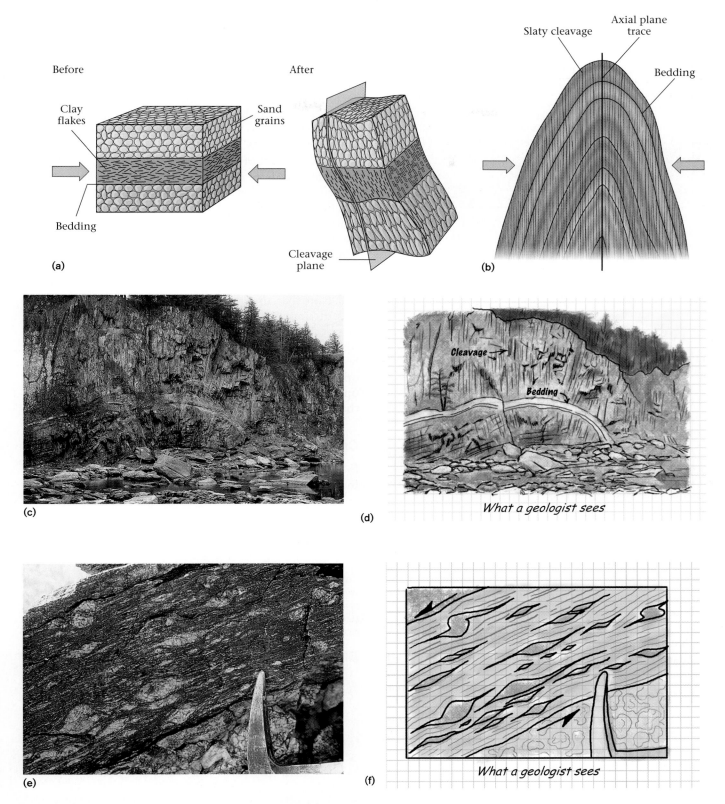

FIGURE 11.26 The development of tectonic foliation in rock. **(a)** Compression of a layer. Shortening occurs in one direction and lengthening in the other. Quartz grains flatten, and clay grains reorient. As a result, the rock develops cleavage. **(b)** Slaty cleavage oriented parallel to the axial plane of a fold and perpendicular to the direction of shortening. **(c)** A stream cut showing axial-planar cleavage. **(d)** A geologist's sketch of the stream cut. **(e)** Schistosity oriented at a low angle to the direction of shear. Note how large grains are all parallel to each other. **(f)** A geologist's sketch of the outcrop in **(e)**, showing shear movement.

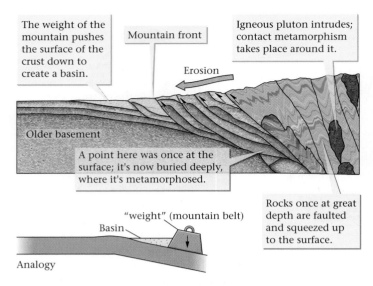

The weight of the mountain pushes the surface of the crust down to create a basin.

Mountain front

Igneous pluton intrudes; contact metamorphism takes place around it.

Erosion

Older basement

A point here was once at the surface; it's now buried deeply, where it's metamorphosed.

Rocks once at great depth are faulted and squeezed up to the surface.

"weight" (mountain belt)

Basin

Analogy

FIGURE 11.27 Various rocks form during orogeny. Sediment eroded from the orogen fills a sedimentary basin next to the mountain; rocks buried deeply in the orogen become squeezed, sheared, and heated to form metamorphic rocks; and igneous rocks intrude from below.

- *Sedimentation during orogeny:* Weathering and erosion in mountain belts generate vast quantities of sediment. This sediment tumbles down slopes and gets carried away by glaciers or streams that transport it to low areas where it accumulates in large fans or deltas. The weight of collisional mountain belts pushes down the surface of the lithosphere, thereby producing a deep sedimentary basin at the border of the range.

- *Metamorphism during orogeny:* Contact metamorphic aureoles (see Chapter 8) form adjacent to igneous intrusions in orogens. And regional metamorphism occurs where compression during mountain building produces faults on which one part of the crust thrusts over another. Rock of the footwall can end up at great depth and thus can be subjected to high temperature and pressure. Because deformation accompanies this process, the resulting metamorphic rocks contain tectonic foliation.

> **Take-Home Message**
>
> Mountain belts are regions where igneous, sedimentary, and metamorphic processes produce new rocks. Large sedimentary basins form adjacent to collisional orogens.

11.7 UPLIFT AND THE FORMATION OF MOUNTAIN TOPOGRAPHY

Leonardo da Vinci, the Renaissance artist and scientist, enjoyed walking in the mountains, sketching rock ledges and examining the rocks he found there. In the process, he dis-

covered marine shells (fossils) in limestone beds cropping out a kilometer above sea level, and suggested that the rock containing the fossils had risen from below sea level up to its present elevation. Contemporary geologists agree with Leonardo, and now refer to the process by which the surface of the Earth moves vertically from a lower to a higher elevation as **uplift**. Mountain building produces substantial uplift of the Earth's surface (▶Fig. 11.28).

What kinds of distances are we talking about when referring to uplift in mountain ranges? As noted earlier, Mt. Everest rises 8.85 km above sea level. Although this distance may seem monumental—that's equal to 5,000 people standing one on top of another—it represents only about 0.06% of the Earth's diameter. In fact, if the Earth were shrunk to the size of a billiard ball, its surface (mountains and all) would feel smoother than that of an actual billiard ball. Note that, in general, the individual peaks that you see in a mountain range represent only a fraction of its total height, for the plain at the base of the mountains may be significantly higher than sea level. Nevertheless, mountain heights are spectacular, and in this section we look at why uplift occurs, how erosion carves rugged landscapes into uplifted crust, and why mountains can't get much higher than Mt. Everest.

Crustal Roots and Mountain Heights

In the mid-1800s, Sir George Everest undertook a survey of India, which at the time formed part of the British Empire. Sir George, aside from contributing his name to Earth's highest mountain, discovered that the mass of the Himalaya Mountains was great enough to deflect the plumb bob (a

FIGURE 11.28 There are over 2 km of vertical relief between the base of Wyoming's Grand Teton Mountains and the peak. And the base already lies at a high elevation. The exposed rocks once lay 12 km or more beneath the surface of the Earth. Clearly, mountain building results in large vertical displacements of the surface of the crust.

lead weight at the end of a string) he was using to determine the vertical direction when setting up surveying equipment. But he was surprised to note that the amount of deflection caused by the mountains was actually *less* than it should have been, given their size. Why? Keeping in mind that Earth's crustal rocks are less dense than its mantle rocks, a nineteenth-century British scientist, George Airy, came up with an explanation. Airy suggested that the crust of the Earth is thicker beneath the Himalayas than elsewhere, and thus that a low-density crustal root protrudes downward into the dense mantle beneath the range. A mountain range with a low-density crustal root has less mass overall than a mountain range underlain by dense mantle, and so it exerts less pull on a plumb bob.

Work in the twentieth century confirmed that collisional and convergent-boundary mountain belts do sit above crustal roots. Whereas typical continental crust has a thickness of about 35 to 40 km (measured from the surface to the Moho), the crust beneath some mountain belts may reach a thickness of 50 to 70 km (about double its normal thickness) (►Fig. 11.29a). Mountain building in these belts shortens the crust horizontally and thickens it vertically.

FIGURE 11.29 (a) Mountain belts have crustal roots, meaning that where the land surface rises to a higher elevation, the crust underneath is thicker. (b) In general, the lithosphere obeys Archimedes' principle of buoyancy. The surface of a thicker (longer) buoyant block rises to a higher elevation than the surface of a thinner (shorter) buoyant block. Also, the base of a thicker block extends down to a greater depth.

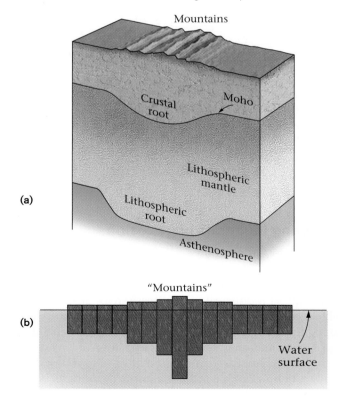

Crustal roots are important because without their buoyancy, mountain ranges would not be so high. To see why, try an experiment in a bathtub.

Prepare four 6-cm-thick wooden blocks, two 8-cm-thick blocks, and one 12-cm-thick block. Line the blocks up in a row, with the thinner (shorter) blocks on the outside, in a tub full of water (►Fig. 11.29b). You'll find that the tops of the thinner blocks lie at a lower elevation than the tops of the thicker blocks. That's because when you place a block of wood in water, it experiences a buoyancy force that pushes the block up until, as stated in Archimedes' principle, the mass of the water displaced by the block equals the mass of the whole block (see Box 4.1). Since wood (0.8 gm/cm³) is 20% less dense than water (1.0 gm/cm³), a certain volume of water has the same mass as a larger volume of wood. Thus, wood floats and juts out of the water by 20%. The thicker the original block, the higher its top surface will be.

The row of floating blocks resembles the cross section of a collisional mountain range, with the highest part of the range lying over the thickest crustal root. Such mountain belts are like icebergs, with only a small part of their mass above sea level and the bulk below. The extra thickness of continental crust beneath mountain ranges is the reason for their high elevations.

Keep in mind that in the Earth System, it's not just the continental crust that floats on the mantle. Rather, it's the continental lithosphere, including the crust plus the lithospheric mantle, as a whole that floats on the asthenosphere. The soft, but solid, asthenosphere can slowly flow out of the way. The crust acts like a buoy that holds the lithosphere up; the bigger the buoy (the thicker the crust), the higher the lithosphere floats. The dense lithospheric root, in some cases, may eventually drop off and cause the mountain range to rise a bit higher, like a hot-air balloon that has dropped some ballast.

The Concept of Isostasy

The condition that exists when the buoyancy force pushing lithosphere up equals the gravitational force pulling lithosphere down is called **isostasy**, or isostatic equilibrium. In most places, isostatic equilibrium exists at the surface of the crust, so that the surface elevation of the crust reflects the level at which the lithosphere naturally floats. If a geologic event happens that changes the density or thickness of the lithosphere, then the surface of the crust slowly rises or falls to reestablish isostatic equilibrium, a process called isostatic compensation.

To picture the process of isostatic compensation, let's return to the bathtub. If you lay a 4-cm-thick block of wood on top of a 6-cm-thick block, the base of the lower block sinks to compensate for the load until isostasy has been reestablished. Because wood is less dense than water, the surface of the resulting 10-cm-thick block will lie at a

BOX 11.2 THE REST OF THE STORY

Gravity Anomalies

Geologists use precise instruments called gravimeters, consisting of a weight suspended from a spring, to measure the pull of gravity at specific points on the Earth's surface. A **gravity anomaly** occurs wherever the measured value of gravity does not match the expected value (as determined by a calculation) at a location. A positive anomaly, or gravity high, indicates that gravitational pull is stronger than expected, whereas a negative anomaly, or gravity low, means that gravity is weaker than expected. Maps of gravity anomalies define locations where isostatic compensation has not been achieved, and thus can help delineate variations in the density of the crust. For example, a gravity anomaly map of the midcontinent region of North America shows a pronounced northeast-southwest-trending gravity high that can be traced from Minnesota, across Iowa, and into Kansas (▶Fig. 11.30). This anomaly coincides with the Midcontinent Rift, a large graben that formed about 1.1 Ga and filled with dense basalt. The excess mass of the basalt produces extra-strong gravitational pull.

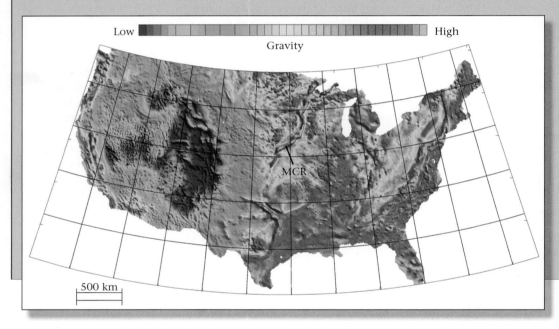

FIGURE 11.30 A gravity map of the United States, as prepared by Raed Aldouri. Redder colors are "gravity highs," meaning places where the pull of gravity is anomalously high. Bluer areas are "gravity lows," meaning places where the pull of gravity is anomalously low. The pronounced linear gravity high in the north central part of the country represents the Midcontinent rift (MCR).

higher elevation than did the surface of the 6-cm-thick block. If you lift the 4-cm-block off, the surface of the 6-cm-block rises. We see this same phenomenon happen at the surface of the Earth. For example, when glaciers grow during ice ages, their weight depresses the surface of the crust, and when they melt away after ice ages, the surface of the crust rises. Such isostatic compensation on Earth takes a long time to occur, for asthenosphere must flow out of the way when lithosphere sinks and must flow back under the lithosphere when lithosphere rises, and asthenosphere can flow only very slowly.

The process of mountain uplift is commonly a manifestation of isostatic compensation. For example, uplift associated with collision or convergence occurs because rock deformation substantially thickens the crust, creating a crustal root. The thicker crust rises and floats higher, as the base of the crust sinks. Not all regions are isostatically compensated (▶Box 11.2).

Sculpting Mountains by Erosion

The image we most often associate with a mountain range is that of rugged topography with spire-like peaks, knife-edge ridges, precipitous cliffs, and deep valleys. This type of landscape is a consequence of erosion by ice and water, which, over time, sculpts uplifted land. The specific style of topography found within a mountain range depends on the climate and on various geologic factors.

The climate determines whether glaciers or rivers erode a mountain. If conditions in the mountains become

(a)

(b)

FIGURE 11.31 Evidence of erosion in mountain ranges. **(a)** Glacially carved peaks in Switzerland; **(b)** Deep valleys cut by rivers into the weathered bedrock of eastern Brazil.

cool enough, glaciers form and, as they flow, carve pointed peaks and steep-sided valleys, as we will see in Chapter 22. But in many ranges, we can see glacially carved landscapes even though there are no glaciers. Such landscapes formed during the last ice age and stopped forming when the ice melted away, less than 14,000 years ago (▶Fig. 11.31a). At lower elevations or in warmer climates, mountain landscapes reflect the consequences of river erosion (▶Fig. 11.31b). Soil formation and vegetation growth may blunt escarpments, creating rounded hills. In desert regions, mountains are typically covered with rubble and expose steep rock escarpments, for thick soil does not develop, and rock debris falling off cliffs litters their slopes. Geologic factors such as bedrock composition also affect topography, for resistant rock units (e.g., quartzite or granite) typically stand up as high ridges, whereas weak rock units (e.g., shale) tend to erode.

In some cases, the geometry of geologic structures affects the shape of mountains. Mountain ridges formed by tilted layers of rock typically have an asymmetric shape, with a steep face on one side (where a cliff cuts across a layer) and a gentle slope on the other side (parallel to the layering, where the top layer becomes the surface of a hill). Such an asymmetric ridge is a **cuesta** (▶Fig. 11.32). The shape of cuestas to the east of Paris helped France defend itself against Germany during World War I; gentle slopes dipped toward Paris, whereas steep cliffs faced Germany, making it hard for the German army to advance. If beds dip steeply, they form a narrow, more symmetrical ridge called a **hogback**.

What Goes Up Must Come Down: Uplift Has Its Limits

Mountains much higher than Mt. Everest cannot exist on the Earth, for two reasons. First, erosion attacks mountains as soon as they rise, and in some cases it can tear mountains down as fast as uplift occurs. Second, rock does not have infinite strength. As mountains rise, the weight of overlying rock presses down on the rock that is now buried at depth in the crust. While this is happening, the rock buried at depth is gradually growing warmer and softer because of heat rising from the Earth's interior. Eventually, the weight of the mountain range causes the warm, soft rock at depth to flow slowly. Effectively, the mountains begin to collapse under their own weight and spread laterally like soft cheese that has been left out in the summer sun (▶Fig. 11.33a, b). Geologists call this

FIGURE 11.32 A cuesta is an asymmetric ridge underlain by dipping strata. Note the contrast between the shallow dip slope and the cliff face.

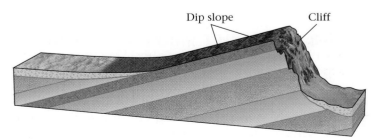

Dip slope   Cliff

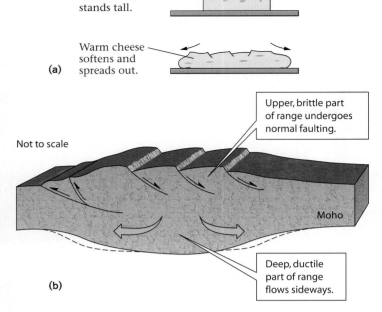

Cold cheese
stands tall.

Warm cheese
softens and
spreads out.

(a)

Not to scale

Upper, brittle part
of range undergoes
normal faulting.

Moho

Deep, ductile
part of range
flows sideways.

(b)

FIGURE 11.33 **(a)** Cheese spreads out sideways as it warms up and softens. **(b)** Similarly, mountain belts spread out sideways once they reach a certain thickness. The ductile crust at depth flows, whereas the upper (brittle) crust is broken by normal faults.

process orogenic collapse. Together, uplift, erosion, and orogenic collapse bring metamorphic and plutonic rocks that were once deep in the crust up toward the Earth's surface. This overall process is called exhumation.

A mountain range can exist only if the rate of uplift exceeds the rate of erosion. Once the rate of erosion becomes faster, the mountain range gradually wears away, perhaps eventually down to sea level. As each kilometer erodes from the top of the range, the base of the range rises by about one-third of a kilometer to maintain isostasy. Thus, with time, the excess crustal thickness in the orogen slowly diminishes.

> **Take-Home Message**
>
> Mountain building uplifts rock several kilometers above sea level. In some belts, crust thickens substantially. Eventually, such crust may collapse under its own weight. This process, along with erosion, limits the height of mountain belts.

11.8 CAUSES OF MOUNTAIN BUILDING

Before plate tectonics theory was established, geologists were just plain confused about how mountains formed. For example, some geologists thought that mountains rose in response to the cooling of the Earth; they argued that as it cools, the Earth shrinks, so its crust wrinkles like the skin of an old apple. But this idea had to be discarded when the discovery of natural radioactivity led to the realization that the Earth is not cooling rapidly, for radioactive decay produces new heat. Plate tectonics theory provided a new approach to interpreting mountains: as noted earlier, mountains form in response to convergent boundary–continental margin interactions, continental collisions, and rifting. Here we look at these different settings and the types of mountains and geologic structures that develop in each one.

Mountains Related to Convergence

Along the margins of continental convergent plate boundaries, where oceanic lithosphere subducts beneath a continent, a continental volcanic arc forms, and compression between the two plates causes a mountain range to rise (▶Fig. 11.34a). Some convergent-margin orogens last as long as 200 million years. During this time, offshore island volcanic arcs, oceanic plateaus, and small fragments of continental crust may drift into the convergent margin. These blocks are too buoyant to subduct, so they collide with the convergent margin and accrete, or attach, to the continent. Geologists refer to such blocks before they attach as **exotic terranes** and after they have attached as accreted terranes. Once an exotic terrane accretes to a continent, a new subduction zone typically forms on the ocean side of the terrane, and a new volcanic arc develops on top of the terrane. In some convergent-margin orogens, numerous exotic terranes attach over time, making the continent grow laterally (to the side). The western half of the North American Cordillera consists largely of accreted terranes (▶Fig. 11.34b). Orogens that grow laterally by the attachment of exotic terranes have come to be known as **accretionary orogens**.

If plate movements push the continent tightly against the subduction zone, compression on the continent side of the volcanic arc generates a **fold-thrust belt**. In such belts, a thrust-fault system has developed above a nearly horizontal detachment fault, and folds form as strata are pushed up the fault. Convergence along the western coast of South America, for example, has generated a fold-thrust belt on the eastern side of the Andes. A similar belt formed on the eastern side of the North American Cordillera during the Mesozoic and Cenozoic.

Mountains Related to Continental Collision

Once the oceanic lithosphere between two continents completely subducts, the continents themselves collide with each other. Continental collision results in the creation of large mountain ranges such as the present-day Himalayas or the Alps (▶Fig. 11.35a, b) and the Paleozoic Appalachian Mountains. The final stage in the growth of the Appalachians happened when Africa and North America collided.

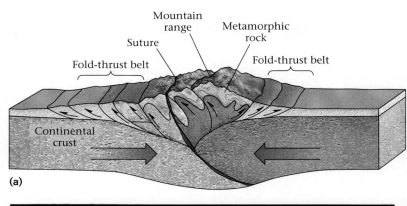

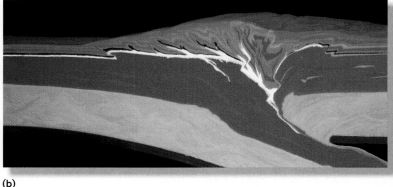

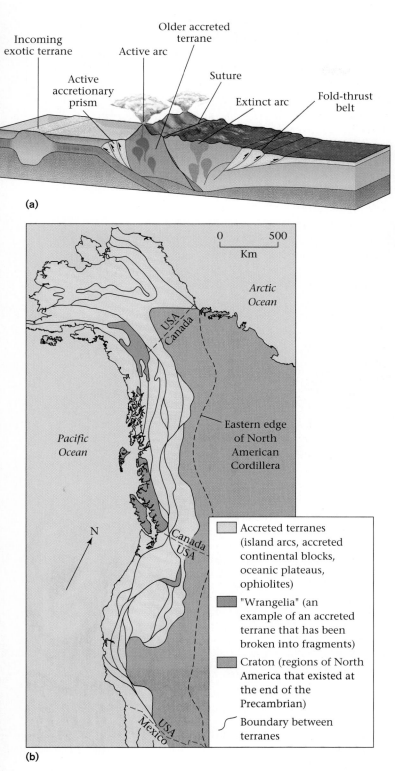

(a)

(b)

FIGURE 11.34 (a) In a convergent-margin orogen, volcanic arcs form, and there may be compression. Where this occurs, a large mountain range develops. Exotic terranes may collide with the convergent margin and accrete to the orogen. **(b)** Much of the western portion of the North American Cordillera consists of accreted terranes.

FIGURE 11.35 (a) In a collisional orogen, two continents collide. The compression that results from the collision shortens and thickens the continental crust so that a large mountain range develops. Fold-thrust belts form along the margins of the orogen. **(b)** A museum sand model showing development of a collisional mountain belt. The colored layers of sand represent layers of rock. As the left side of the model was dragged under the right, the sand was scraped off, folded, and faulted, building a high ridge.

During collision, intense compression generates fold-thrust belts on the margins of the orogen. In the interior of the orogen, where one continent overrides the edge of the other, high-grade metamorphism occurs, accompanied by formation of flow folds and tectonic foliation. During this process, the crust below the orogen grows to as much as twice its normal thickness. Gradually, rocks squeeze up-ward in the hanging walls of large thrust faults and later become exposed by exhumation. Finally, as noted earlier, rock at depth in the orogen heats up and becomes so weak that the mountain belt may collapse and spread out side-ways. The broad Tibetan Plateau may have formed in part when crust, thickened during the collision of India with Asia, spread to the northeast (see art, pp. 392–393).

Mountains Related to Continental Rifting

Continental rifts are places where continents are splitting in two. When rifts first form, there is generally significant uplift, and this uplift contributes to creating mountainous topography (▶Fig. 11.36). Uplift occurs, in part, because as

See for yourself...
Mountains and Structures

Some of the best examples of geologic structures, and some of the most spectacular examples of mountain ranges, are in fairly remote areas. Your computer can now take you there. Note that structures control erosion, and that you can trace them out more easily in dry regions. The thumbnail images provided on this page are only to help identify tour sites. Go to wwnorton.com/studyspace to experience this flyover tour.

Erosion in the Alps
(Lat 46°12'56.50"N, Long 10°33'59.43"E)

Fly to the coordinates provided and zoom to an elevation of about 1,100 km (680 miles). You are looking down on the northern end of the Italian Peninsula. The Alps, a Cenozoic collisional mountain belt, is the snowy arc of peaks that encompasses northern Italy and parts of France and Switzerland (Image G11.1). Drop to an elevation of 47 km (29 miles), tilt the image, and look north. You'll see a typical mountain landscape—its ruggedness a consequence of erosion by glaciers and rivers (Image G11.2).

G11.1

G11.2

Mt. Everest—the Top of the World
(Lat 27°59'17.84"N, Long 86°55'18.01"E)

Ever wonder what the view looks like from the top of Mt. Everest, the highest point on Earth? Fly to the coordinates provided, zoom to 1,300 km (800 miles) and you'll be looking down on the summit of Mt. Everest, in the Himalayas, a range that formed due to the collision of India with Asia. Tilt your view so you are looking north (Image G11.3). In the foreground is the Ganges Plain, and in the distance, the dry Tibet Plateau. Zoom to 15 km (9 miles) and rotate to look southeast. You can see the surrounding peaks and glaciers and the plains in the far distance (Image G11.4). Fly around the summit to see the view in all directions.

G11.3

G11.4

Joints in Arches National Park
(Lat 38°47'52.13"N, Long 109°35'26.37"W)

Zoom to an elevation of 5 km (3 miles) at the specified coordinates in southeastern Utah and you'll see a prominent set of NW-SE-trending joints cutting horizontal white sandstone beds (Image G11.5). Weathering along the joints made them into narrow troughs. A second set of smaller NE-SW trending joints also cut the rocks. The major joints are about 20–50 m (65–165 feet) apart. Now, fly SSE for about 5 km (3 miles) and you'll find a set of joints cutting red sandstone. Zoom to 3.3 km (2 miles), tilt the image and look NNE (Image G11.6). Arches of the park develop where holes weather through the walls between joint-controlled troughs.

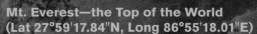

G11.5

G11.6

G11.7

Horsts and Grabens, Canyonlands (Lat 38°6'1.40"N, Long 109°54'40.14"W)

From an elevation of 20 km (12 miles) at these coordinates in Canyonlands Park, SE Utah, you can see a set of NE-trending grabens—troughs bordered on each side by a normal fault. Fly down to 7 km (4 miles), tilt, and look NE (Image G11.7). You'll be gazing down the axis of grabens which formed when the region southeast of the river valley stretched in a NW-SE direction. This stretching occurred because river erosion removed lateral support from a sequence of strata that lie above a weak salt layer.

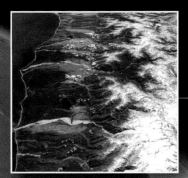
G11.8

Alpine Fault, New Zealand (Lat 43°21'15.86"S, Long 170°15'23.18"E)

At these coordinates near the west coast of South Island, New Zealand, zoom to 50 km (30 miles). You can see a high, ice-capped mountain range, called the Southern Alps, formed by motion on the Alpine Fault. Though a significant component of displacement on the fault is strike-slip, its motion also accommodates thrusting—eastern Zealand is being pushed up to the west. Tilt your image and look NE and you'll be looking along the fault (Image G11.8). The fault runs along the abrupt break in slope between the mountains to the SE, and the plains to the NW.

Faults and Folds, Makran Range (Lat 25°54'0.31"N, Long 64°14'8.63"E)

Fly to the coordinates provided and zoom to 10 km (6 miles) (Image G11.9). You can see a series of ENE-trending ridges that represent beds of strata tilted by folding and faulting in the Makran Range of southern Pakistan, about 63 km (40 miles) north of the coast. Note that ridges are offset where they are cut by a series of NE-SW trending valleys. The valleys are eroded left-lateral strike-slip faults. Zoom to 2 km (1 mile), and tilt the view so you are looking up the valley of the most prominent fault. A dry stream follows the valley—faulting broke up the rock making it easier to erode.

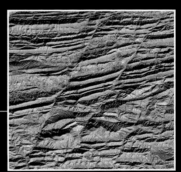
G11.9

G11.10

Appalachian Fold Belt, Pennsylvania (Lat 40°53'24.28"N, Long 77°4'28.11"W)

From an elevation of 114 km (70 miles), at these coordinates, you can see the curving traces of large folds in the Appalachian Mountains. These folds formed about 280 million years ago when compression wrinkled layers of Paleozoic sedimentary rock. The hinge zones of anticlines have been eroded; resistant ledges of sandstone define the fold limbs. Zoom to 20 km (12 miles), tilt the view, and rotate so you are looking west. You will be peering down the hinge of the folds (Image G11.10). This region is the Pennsylvania Valley and Ridge Province. Note that ridges remained forested.

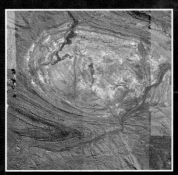
G11.11

Dome and Syncline, Western Australia (Lat 22°52'9.85"S, Long 117°18'1.58"E)

Here we see an exposure of Precambrian crust, part of the Pilbara craton. Zoom to 50 km (30 miles) and you'll see a 30 km (18 miles)-wide circular region of light rock—a dome of granite and gneiss—surrounded by folded metasedimentary and metavolcanic rock (Image G11.11). The layering within the sedimentary and volcanic rocks stands out—resistant layers form ridges, and outlines the shape of folds. Now fly to the fold on the SW side of the dome. Reorient the view so you are looking west, and tilt the view. You are looking along the plunging hinge of a syncline.

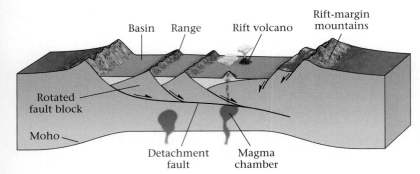

FIGURE 11.36 When the crust stretches in a continental rift, rift-related mountains form, as do normal faults. Displacement on the faults leads to the tilting of crustal blocks and the formation of half-grabens. The half-grabens fill with sediment eroded from the adjacent mountains. Later, the exposed ends on the tilted blocks create long, narrow ranges, and the half-grabens become flat basins. In the United States, the region containing such a structure is called the Basin and Range Province (located in Nevada, Utah, and Arizona).

the lithosphere thins, hot asthenosphere rises, making the remaining lithosphere less dense. Because the lithosphere is less dense, it becomes more buoyant and thus rises to reestablish isostatic equilibrium.

As heating in a rift takes place, stretching causes normal faulting in the brittle crust above, creating a normal-fault system. Movement on the normal faults drops down blocks of crust, creating deep basins separated by narrow, elongate mountain ranges that contain tilted rocks. These ranges are sometimes called fault-block mountains. In addition, the rising asthenosphere beneath the rift partially melts, generating magmas that rise to form volcanoes within the rift. Today, the East African Rift clearly shows the configuration of rift-related mountains and volcanoes. And in North America, rifting yielded the broad Basin and Range Province of Utah, Nevada, and Arizona. If you drive across the province from east to west, you'll pass over two dozen fault-block mountain ranges, separated from each other by sediment-filled basins.

> **Take-Home Message**
>
> Mountain belts form in association with subduction, collision, and rifting. At convergent margins, slivers of crust may attach to a continent over time. Collision tends to generate folds and cause regional metamorphism. Rifting produces tilt-block mountains.

11.9 CRATONS AND THE DEFORMATION WITHIN THEM

A **craton** consists of crust that has not been affected by orogeny for at least 1 billion years. Because orogeny happened so long ago in cratons, their crust has become quite

cool, and therefore relatively strong and stable. We can divide cratons into two provinces: shields, in which Precambrian metamorphic and igneous rocks crop out at the ground surface, and the cratonic platform, where a relatively thin layer of Phanerozoic sediment covers the Precambrian rocks (▶Fig. 11.37; see Fig. 13.9).

In shield areas, we find intensively deformed metamorphic rocks—abundant examples of shear zones, flow folds, and tectonic foliation. That's because the crust making the cratons was deformed during a succession of orogenies in the Precambrian. Recent studies of the Canadian Shield, which occupies much of the eastern two-thirds of Canada, for example, reveal the traces of Himalaya-like collision zones, Andean-like convergent boundaries, and East African–like rifts, all formed more than 1 billion years ago (some over 3 billion years ago). These orogens are so old that erosion has worn away the original topography, in the process exposing deep crustal rocks at the Earth's surface.

In the cratonic platform, we can't see the Precambrian rocks and structures, except where they are exposed by deep erosion. Younger strata do display deformation features, but in contrast to the deformation of orogens, cratonic-platform deformation is less intense. The cratonic platform of the U.S. Midwest region includes two classes of structures: regional basins and domes, and local zones of folds and faults.

Regional basins and regional domes are broad areas that gradually sank or rose, respectively (▶Fig. 11.38a, b). They are illustrated by a slice of the upper crust running across

FIGURE 11.37 Digital elevation map of North America showing the platform and shield areas. CP=Colorado Plateau.

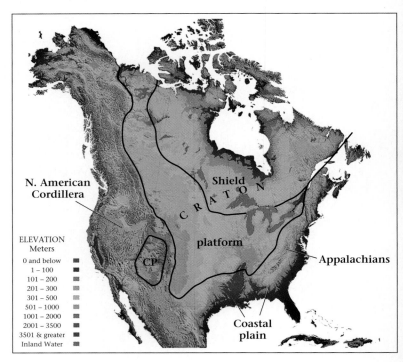

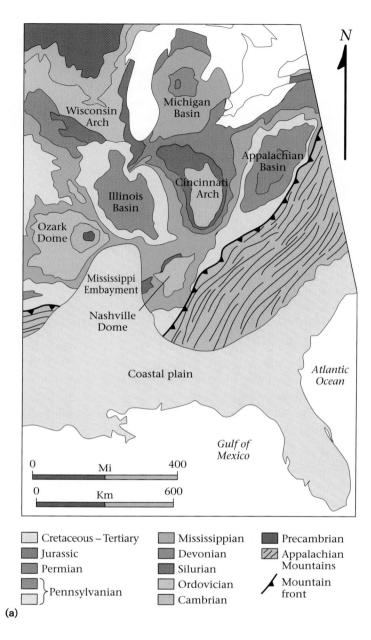

(a)

Cretaceous – Tertiary
Jurassic
Permian
Pennsylvanian

Mississippian
Devonian
Silurian
Ordovician
Cambrian

Precambrian
Appalachian Mountains
Mountain front

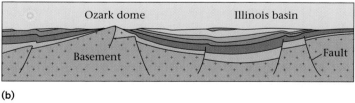

(b)

FIGURE 11.38 (a) Geologic map of the mid-continent region of the United States, showing the basins and domes and the faults that cut the region. **(b)** This cross section illustrates the geometry of the regional basins and domes. Note how layers of strata thin toward the crest of the Ozark Dome and thicken toward the center of the Illinois Basin. Some of the faults originated as normal faults and later moved again as reverse faults.

Missouri and Illinois. In Missouri, strata arch across a broad dome, the Ozark Dome, whose diameter is 300 km. Individual sedimentary layers thin toward the top of the dome, because less sediment accumulated on the dome than in adjacent basins. Erosion during more recent geologic history has produced the characteristic bull's-eye pattern of a dome, with the oldest rocks (Precambrian granite) exposed near the center. In the Illinois Basin, strata appear to warp downward into a huge bowl that is also about 300 km across. Strata get thicker toward the center, indicating that the floor of the basin sank so there was more room for sediment to accumulate. The Illinois Basin also has a bull's-eye shape, but here the youngest strata are in the center. Geologists refer to the broad vertical movements that generate huge, but gentle, mid-continent domes and basins as **epeirogeny**.

Folds and faults are hard to find in the cratonic platform, because most do not cut the ground surface. But subsurface studies indicate that faults do occur at depth. Monoclines, step-shaped folds, develop over these faults; the folds form as a block of basement pushes up. Most of these zones were likely active when major orogenies happened along the continental margin. This relation suggests that the orogenies created enough stress in the craton to cause faults to move, but not enough to generate large mountains or to create foliation.

> **Take-Home Message**
>
> Cratons are portions of continents that consist of old (Precambrian) and relatively stable crust. Parts of cratons may be covered by Paleozoic sedimentary rocks. Variations in the dip and thickness of these strata define regional basins, arches, and domes.

11.10 LIFE STORY OF A MOUNTAIN RANGE: A CASE STUDY

Perhaps the easiest way to bring together all the information in this chapter is to look at the life story of one particular mountain range—let's take the Appalachian Mountains of North America as our example. (We've simplified the story a bit, for ease of reading.) Geologists have constructed the range's life story by studying its structures, by determining the ages of igneous and metamorphic rocks, and by searching for strata formed from sediment eroded from the range.

About 1 billion years ago, the Appalachian region was involved in a massive collision with another continent (►Fig. 11.39). This event, called the Grenville orogeny, yielded a belt of deformed and metamorphosed rocks that underlie the eastern fifth of the continent. For a while after the Grenville event, the Appalachian region lay in the middle of a supercontinent. But this supercontinent rifted apart around 600 million years ago. Eventually, new ocean formed to the east,

The Collision of India with Asia

The Himalaya Mountains and other important highlands of southern Asia are a consequence of the collision of India, a small but very old and strong block of continental lithosphere, with Asia about 55 million years ago. At the time of the collision, the southern margin of Asia consisted of several smaller crustal blocks that had become stitched together by recent collisions, and thus was composed of younger, warmer, and softer lithosphere. Since then, the strong lithosphere of India has continued to push slowly into the weaker lithosphere of Asia.

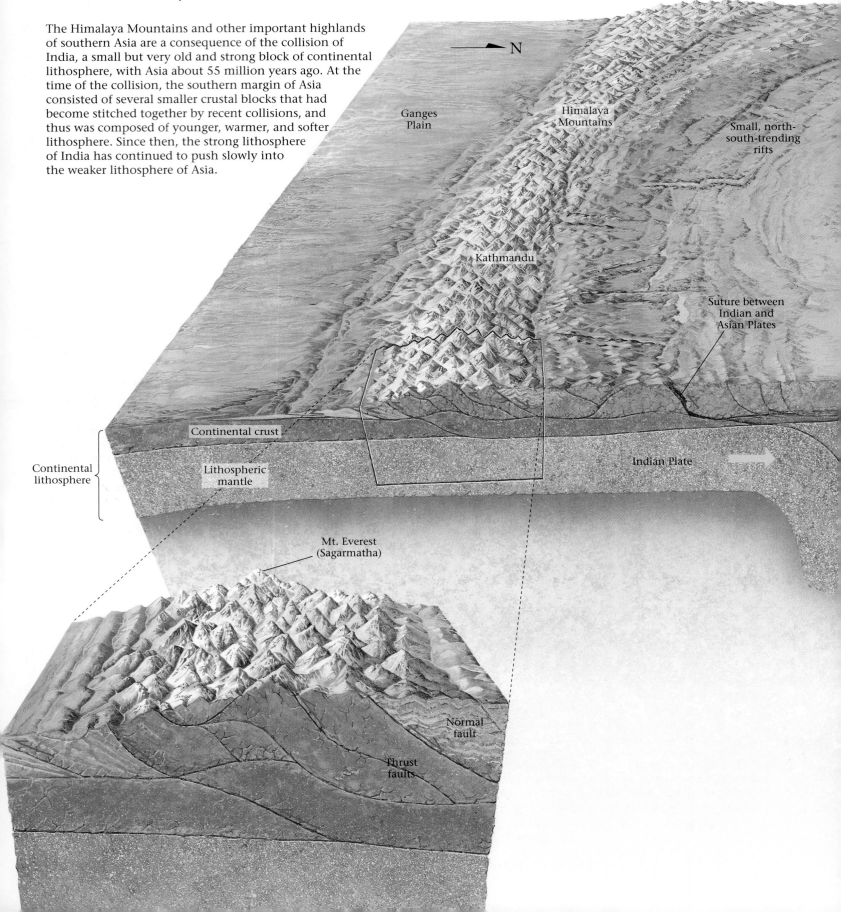

N

Ganges Plain

Himalaya Mountains

Small, north-south-trending rifts

Kathmandu

Suture between Indian and Asian Plates

Continental crust

Continental lithosphere

Lithospheric mantle

Indian Plate

Mt. Everest (Sagarmatha)

Normal fault

Thrust faults

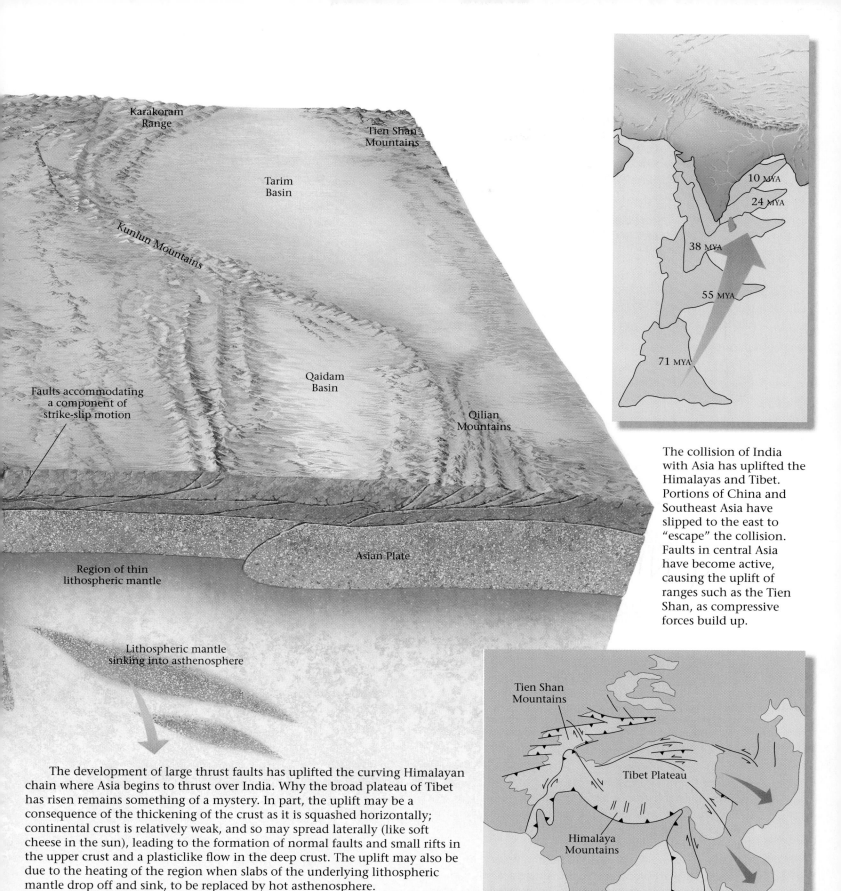

Karakoram Range

Tien Shan Mountains

Tarim Basin

Kunlun Mountains

Faults accommodating a component of strike-slip motion

Qaidam Basin

Qilian Mountains

Asian Plate

Region of thin lithospheric mantle

Lithospheric mantle sinking into asthenosphere

10 MYA

24 MYA

38 MYA

55 MYA

71 MYA

The collision of India with Asia has uplifted the Himalayas and Tibet. Portions of China and Southeast Asia have slipped to the east to "escape" the collision. Faults in central Asia have become active, causing the uplift of ranges such as the Tien Shan, as compressive forces build up.

Tien Shan Mountains

Tibet Plateau

Himalaya Mountains

The development of large thrust faults has uplifted the curving Himalayan chain where Asia begins to thrust over India. Why the broad plateau of Tibet has risen remains something of a mystery. In part, the uplift may be a consequence of the thickening of the crust as it is squashed horizontally; continental crust is relatively weak, and so may spread laterally (like soft cheese in the sun), leading to the formation of normal faults and small rifts in the upper crust and a plasticlike flow in the deep crust. The uplift may also be due to the heating of the region when slabs of the underlying lithospheric mantle drop off and sink, to be replaced by hot asthenosphere.

As India has pushed into Asia, it may have squeezed blocks of China and Southeast Asia sideways, toward the east; this motion is accommodated by slip on strike-slip faults. The collision may also have caused reverse faults in the interior of Asia to become active, uplifting a succession of small mountain ranges, such as the Tien Shan.

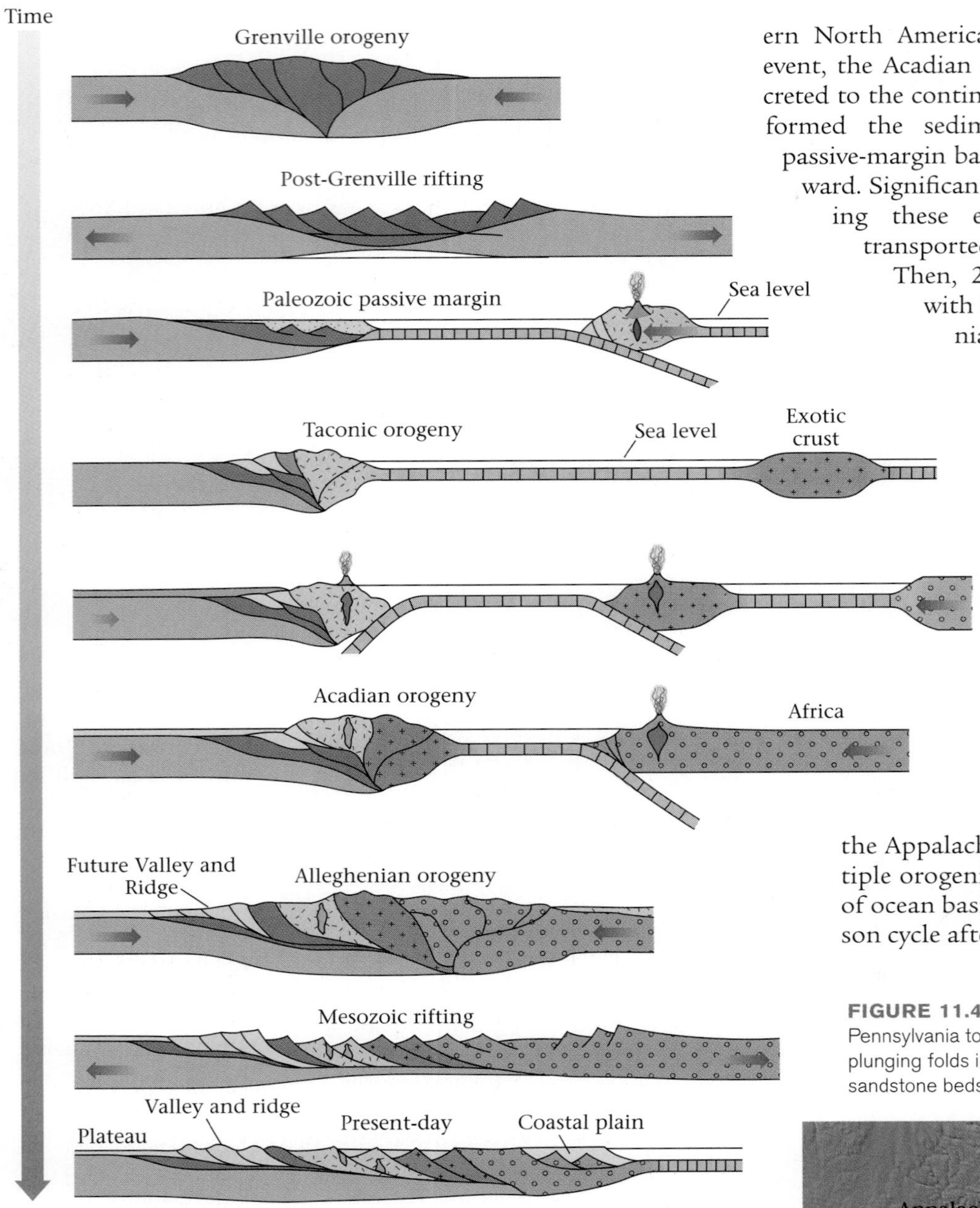

Time

Grenville orogeny

Post-Grenville rifting

Paleozoic passive margin — Sea level

Taconic orogeny — Sea level — Exotic crust

Acadian orogeny — Africa

Future Valley and Ridge — Alleghenian orogeny

Mesozoic rifting

Plateau — Valley and ridge — Present-day — Coastal plain

FIGURE 11.39 These idealized stages show the tectonic evolution of the Appalachian Mountains. Note that although mountains do form during rifting events, geologists traditionally assign names only to the collisional or convergent events.

and the former rifted margin of eastern North America cooled, sank, and evolved into a passive-margin sedimentary basin. From 600 to about 420 million years ago, this basin filled with a thick sequence of sediment.

Between around 420 and 370 million years ago, two collisions took place between North America and exotic terranes. During the first convergent event, called the Taconic orogeny, a crustal block and volcanic arc collided with east-

ern North America, and during the second convergent event, the Acadian orogeny, continental crustal slivers accreted to the continent. The accretion of these terranes deformed the sediment that had accumulated in the passive-margin basin, and made the continent grow eastward. Significant strike-slip displacement occurred during these events; thus, slivers of crust were transported along the margin of the continent. Then, 270 million years ago, Africa collided with North America. This event, the Alleghenian orogeny, yielded a huge mountain range resembling the present-day Himalayas and created a wide fold-thrust belt along the mountains' western margin. Eroded folds of this belt make up the topography of the present Valley and Ridge Province in Pennsylvania (▶Fig. 11.40).

When the Alleghenian orogeny ceased, the Appalachian region once again lay in the interior of a supercontinent (Pangaea), where it remained until about 180 million years ago. At that time, rifting split the region open again, creating the Atlantic Ocean. As you can see from this example, major ranges such as the Appalachians incorporate the products of multiple orogenies and reflect the opening and closing of ocean basins (a sequence of events called the Wilson cycle after J. Tuzo Wilson).

FIGURE 11.40 Relief map of the Valley and Ridge Province, Pennsylvania to Virginia. The ridges, which outline the shapes of plunging folds in the fold-thrust belt, are composed of resistant sandstone beds.

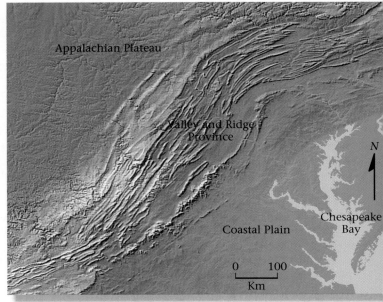

Appalachian Plateau

Valley and Ridge Province

Coastal Plain

Chesapeake Bay

N

0   100
Km

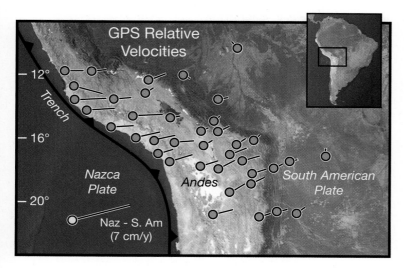

FIGURE 11.41 Convergence between the Nazca Plate and South America creates the broad Andean orogen along the western coast. Here, the line at the yellow dot indicates the rate of motion of Nazca relative to the interior of South America. The lines at the red dots indicate the rate of movement of places in the Andes relative to the interior of South America. Note that the rate generally decreases to the east.

11.11 MEASURING MOUNTAIN BUILDING IN PROGRESS

Mountains are not just "old monuments," as John Muir mused. The rumblings of earthquakes and the eruptions of volcanoes attest to present-day movements in mountains. Geologists can measure the rates of these movements through field studies and satellite technology. For example, geologists can determine where coastal areas have been rising relative to the sea level by locating ancient beaches that now lie high above the water. And they can tell where the land surface has risen relative to a river by identifying places where a river has recently carved a new valley down into sediments that it had previously deposited. In addition, geologists now can use the satellite **global positioning system (GPS)** to measure rates of uplift and horizontal shortening in orogens. With present technology, we can "see" the Andes shorten horizontally at a rate of a couple of centimeters per year, and we can "watch" as mountains along this convergent boundary rise by a couple of millimeters per year (▶Fig. 11.41).

Chapter Summary

- Mountains occur in linear ranges called mountain belts, orogenic belts, or orogens. An orogen forms during an orogeny, or mountain-building event. Orogenies, which last for millions of years, are a consequence of continental collision, subduction at a convergent plate boundary, or rifting.

- Mountain building causes rocks to bend, break, squash, stretch, and shear. Because of such deformation, rocks change their location, orientation, and shape.

- During brittle deformation, rocks crack and break into two or more pieces. During ductile deformation, rocks change shape without breaking.

- Rocks undergo three kinds of stress: compression, tension, and shear.

- Strain refers to the way rocks change shape when subjected to a stress. Compression causes shortening, tension causes stretching, and shear stress leads to shear strain.

- Deformation results in the development of geologic structures. Brittle structures include joints and faults. Ductile structures include folds and foliation.

- Structures can be visualized as geometric lines or planes. We can define the orientation of a plane by giving its strike and dip and the orientation of a line by giving its plunge and bearing.

- Joints are natural cracks in rock, formed in response to tension under brittle conditions. Some joints develop when rock cools and contracts; others form when erosion decreases the pressure on rocks buried at depth.

- Veins develop when minerals precipitate out of water passing through joints.

- Faults are fractures on which there has been shearing. In the case of nonvertical faults, the rock above the fault plane is the hanging-wall block, and the rock below the fault plane is the footwall block.

- On normal faults, the hanging-wall block slides down the surface; on reverse faults, the hanging-wall block slides up the surface; on strike-slip faults, rock on one side of the fault slides horizontally past the other; and on oblique-slip faults, rock slides diagonally across the surface.

- Faults can be recognized by the presence of broken rock (breccia) or fine powder (gouge). Scratches or grooves on fault surfaces are called slip lineations.

- Folds are curved layers of rock. Anticlines are arch-like folds, synclines are trough-like, monoclines resemble the shape of a carpet draped over a stair step, basins are shaped like a bowl, and domes are shaped like an overturned bowl.

- Tectonic foliation forms when grains are flattened or rotated so that they align parallel with each other, or when new platy grains grow parallel to each other.

- The process of orogeny may yield new igneous, metamorphic, and sedimentary rocks.

- Large mountain ranges are underlain by relatively buoyant roots. The height of such mountains is controlled by isostasy.

- Once uplifted, mountains are sculpted by the erosive forces of glaciers and rivers. Also, when the crust thickens during mountain building, the lower part eventually becomes warm and weak and begins to flow, leading to orogenic collapse.

- Mountain belts formed by convergent margin tectonism may incorporate accreted terranes.

- Continental collision, which resulted in the Alps, the Himalayas, and the Appalachians, generates metamorphic rocks and tectonic foliation. Fold-thrust belts form on the continental edge of collisional and convergent-margin orogens.

- Tilted blocks of crust in rifts become narrow, elongate mountain ranges, called fault-block mountains.

- Cratons are the old, relatively stable parts of continental crust. They include shields, where Precambrian rocks are exposed at the surface, and platforms, where Precambrian rocks are buried by a thin layer of sedimentary rock. Broad regional domes and basins form in platform areas because of epeirogeny.

- With modern satellite technology, it is now possible to measure the slow movements of mountains.

Key Terms

accretionary orogens (p. 386)
anticline (p. 375)
basin (p. 376)
bearing (p. 367)
brittle deformation (p. 366)
compression (p. 369)
craton (p. 390)
cuesta (p. 385)
deformation (p. 363)
detachment fault (p. 373)
dip (p. 367)
dip-slip fault (p. 371)
displacement (p. 372)
dome (p. 376)
ductile deformation (p. 366)
epeirogeny (p. 391)
exotic terranes (p. 386)
fault (pp. 363, 370)
fault scarp (p. 372)
fold (pp. 363, 375)
fold-thrust belt (p. 386)
foliation (p. 363)
footwall block (p. 371)
geologic structures (p. 363)
global positioning system (GPS) (p. 395)
gravity anomaly (p. 384)
hanging-wall block (p. 371)
hinge (p. 375)

hogback (p. 385)
isostasy (isostatic equilibrium) (p. 383)
joint (p. 363)
limbs (of fold) (p. 375)
monocline (p. 376)
mountain belts (p. 363)
normal fault (p. 371)
oblique-slip fault (p. 371)
orogenic belts (orogens) (p. 363)
orogeny (p. 363)
plunge (p. 367)
pressure (p. 369)
reverse fault (p. 371)
shear stress (p. 369)
shear zone (p. 373)
strain (stretching, shortening) (p. 365)
stress (p. 369)
strike (p. 367)
strike-slip fault (p. 371)
syncline (p. 376)
tectonic foliation (p. 380)
tension (p. 369)
thrust fault (p. 371)
undeformed (p. 364)
uplift (p. 382)
veins (p. 370)

Geopuzzle Revisited

Mountain ranges form in response to the interaction of lithosphere plates. Some ranges form along convergent plate boundaries, some in association with rifting, and some where two continents collide after the sea floor between them has been subducted. Stress (compression, tension, or shear) that develops during mountain building causes deformation, producing such geologic structures as folds, faults, and foliations. Mountain building involves uplift, and in many cases igneous activity and metamorphism. As soon as a range has risen above sea level, erosion attacks it. In fact, if the geologic processes causing uplift slow, erosion can eventually bevel a range nearly back to sea level. Thus, mountain ranges do not last forever.

Review Questions

1. What changes do rocks undergo during formation of an orogenic belt such as the Alps?

2. What is the difference between brittle and ductile deformation?

3. What factors influence whether a rock will behave in brittle or ductile fashion?

4. How are stress and strain different?

5. How is a fault different from a joint?

6. Compare the motion of normal, reverse, and strike-slip faults.

7. How do you recognize faults in the field?

8. Describe the differences among an anticline, a syncline, and a monocline.

9. Discuss the relationship between foliation and deformation.

10. Explain how certain kinds of igneous, sedimentary, and metamorphic rocks are formed during orogeny.

11. Describe the principle of isostasy.

12. What happens to the isostatic equilibrium of a mountain range as it is eroded away?

13. What happens to a mountain range when its uplift rate slows down?

14. Discuss the processes by which mountain belts are formed in convergent margins, in continental collisions, and in continental rifts.

15. How are the structures of a craton different from those of an orogenic belt?

On Further Thought

1. Nanga Parbat is an 8 km-high mountain at the northwest end of the Himalayan Range (Lat 35° 14′ 15″N, Long 74° 35′ 21″E). Geologists have found high-grade metamorphic rocks and migmatite at the mountain's peak (see Chapter 8). Their work suggests that these rocks were at a pressure of 8 kbar about 10 million years ago.

   (a) The graph in Box 8.1 of Chapter 8 shows change in temperature with depth (the "geotherm") for a collisional mountain belt. According to this graph, how hot were the rocks now exposed on Nanga Parbat when they formed 10 million years ago? If the rocks were *dry*, what metamorphic facies formed at these conditions? What does the presence of migmatite at Nanga Parbat mean?

   (b) Based on Box 8.1, what thickness of crust lay above the Nanga Parbat rocks 10 million years ago? At what average rate did the Nanga Parbat rocks move from this depth to the present-day ground surface? (Express your answer in millimeters per year). Note: This is one of the fastest exhumation rates yet discovered.

   (c) Using *Google Earth*™, go to the coordinates of Nanga Parbat, zoom to an elevation of about 15 km, tilt your view, and slowly fly around the mountain. What geologic phenomena are contributing to the rapid erosion of Nanga Parbat today?

2. Imagine that a geologist sees two outcrops of resistant sandstone, as depicted in the cross section sketch below. The region between the outcrops is covered by soil. A distinctive bed of cross-bedded sandstone occurs in both outcrops, so the geologist correlated the western outcrop (on the left) with the eastern outcrop. The curving lines in the bed indicate the shape of the cross beds.

   (a) Keeping in mind how cross beds form (see Chapter 7), sketch how the cross bedded bed connected from one outcrop to the other, before erosion? What geologic structure have you drawn?

   (b) Is the bedrock directly beneath the geologist older than or younger than the sandstone bed of the outcrop?

West

East

Suggested Reading

Condie, K. 1997. *Plate Tectonics and Crustal Evolution*. 4th ed. Woburn, Mass.: Butterworth-Heinemann.

Davis, G. H., and S. J. Reynolds. 1996. *Structural Geology of Rocks and Regions*. 2nd ed. New York: Wiley.

McPhee, J. 1998. *Annals of the Former World*. New York: Farrar, Straus & Giroux.

Moores, E. M., and R. J. Twiss. 1995. *Tectonics*. New York: Freeman.

Van der Pluijm, B. A., and S. Marshak, 2004. *Earth Structure: An Introduction to Structural Geology and Tectonics*, 2nd ed. New York: W. W. Norton.

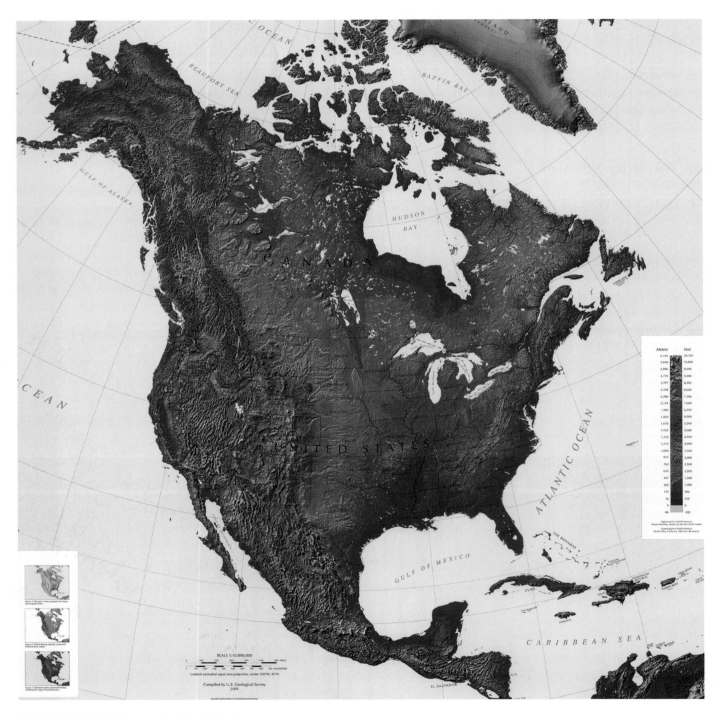

REFERENCE MAP Topography of North America.

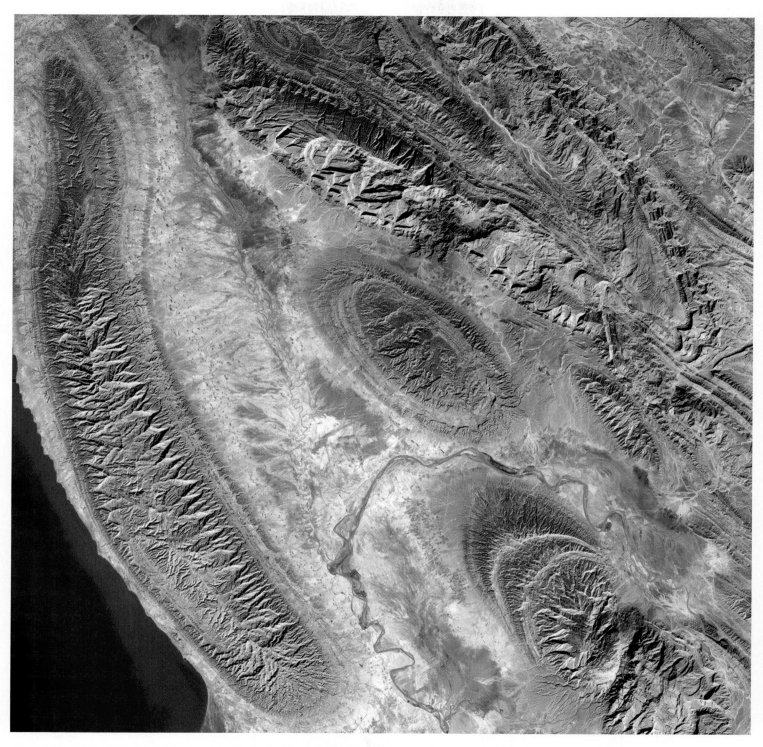

THE VIEW FROM SPACE Subduction of the Persian Gulf beneath the Asian continent, in the vicinity of Konari, Iran, produced doubly plunging folds. Dipping strata outline the ellipsoidal shape of the folds. The large one, near the coast, is about 10 km wide. Some of the folds are cored by salt. This is a tectonically active landscape, so deformation is warping the land surface.

History before History

Perhaps the most important contribution that the science of geology has made to humanity's understanding of the Earth is the demonstration that our planet existed long, long before humans took their first footsteps. In Part IV, we peer back into this history. First we look at fossils, remnants of ancient life that allow geologists to correlate life's evolution with that of Earth. Then, in Chapter 12, we learn how geologists gaze into "deep time"— geologic time, or the time since Earth formed—first by determining the relative ages of geologic features (whether one feature is older or younger than another), and then by learning how to calculate numerical ages (ages in years) on the basis of the ratios of radioactive elements to their daughter products in minerals. With the background provided in Chapter 12, we're ready for Chapter 13's brief synopsis of Earth history, from the birth of the planet to the present. We see how plate tectonics has redistributed continents and built mountains, how sea level has risen and fallen, and how Earth's climate has changed.

Every rock exposure has a tale to tell about Earth history. On the coast of France at Etretat, a site that inspired several famous paintings of the impressionist school in the nineteenth century, the layers of chalk record a time before the cliffs rose and the entire region in view was submerged by the sea.

Memories of Past Life:
Fossils and Evolution

E.1 THE DISCOVERY OF FOSSILS

If you pick up a piece of sedimentary rock, you may find shapes that look like shells, bones, leaves, or footprints (▶Fig. E.1). The origin of these shapes mystified early thinkers. Some thought that the shapes had simply grown underground, in solid rock. Today, geologists consider such **fossils** (from the Latin word *fossilis*, which means "dug up") to be remnants or traces of ancient living organisms now preserved in rock, and that they form when organisms become buried by sediment. This viewpoint, though first proposed by the Greek historian Herodotus in 450 B.C.E. and revived by Leonardo da Vinci in 1500 C.E., did not become widely accepted until the publication in 1669 of a book on the subject by a Danish physician named Nicholaus Steno (1638–1686). Steno noted the similarity be-

FIGURE E.1 This bedding surface in limestone contains fossils of organisms that lived about 420 million years ago. These particular species no longer exist on Earth.

tween fossils known as "dragon's tongues" and the teeth of modern sharks, and concluded that fossil-containing rocks originated as loose sediment that had incorporated the remains of organisms, and that when the sediment hardened into rock, the organisms became part of the rock. The understanding of fossils increased greatly thanks to the efforts of a British scientist, Robert Hooke (1635–1703), who described and sketched fossils in detail and who realized that most represent **extinct species**, meaning species that lived in the past but no longer exist. During the following two centuries, geologists described thousands of fossils and established museum collections (▶Fig. E.2).

The nineteenth century saw **paleontology**, the study of fossils, ripen into a science. Work with fossils went beyond description alone when William Smith, a British engineer who supervised canal construction in England during the 1830s, noted that different fossils occur in different layers of strata within a sequence of sedimentary rocks. In fact, Smith realized that he could define a distinctive succession of fossils, and that a given species would be present only for a specific interval of strata. This discovery made it possible to use fossils as a basis for determining the age of one sedimentary rock layer. Fossils, therefore, became an indispensable tool for studying geologic history and the evolution of life. In this interlude, we introduce fossils, an understanding of which serves as essential background for the next two chapters.

E.2 FOSSILIZATION

What Kinds of Rocks Contain Fossils?

Most fossils are found in sediments or sedimentary rocks. The fossils form when organisms die and become buried by sediment, or when organisms travel over or through

(a)

cases shearing, that occur during intermediate- and high-grade metamorphism. Similarly, fossils generally do not occur in igneous rocks that crystallize directly from melt, for organisms can't live in molten rock, and if engulfed by molten rock, they will be incinerated. Occasionally, however, lava flows preserve the shapes of tree trunks, because lava may surround a tree and freeze before the tree completely burns up—the resulting hole in the lava is, strictly speaking, a fossil. Also, fossils can occur in deposits formed from air-fall ash, for the ash settles just like sediment and can bury an organism or a footprint. In fact, ash preserved the footprints of 3.6-million-year-old human ancestors in ash deposits now exposed in Olduvai Gorge, in the East African Rift (▶Fig. E.3).

FIGURE E.3 The famous fossil footprints at a site called Laetoli, in Olduvai Gorge, Tanzania. They were left when *Australopithecus* walked— on two feet—over ash that had recently been erupted by a nearby volcano, and then had been dampened by rain. A second ash eruption buried, and thereby preserved, the footprints.

(b)

FIGURE E.2 (a) Robert Hooke published these sketches of fossil ammonites (an organism with a chambered shell) in 1703. They were among the first such sketches to be published. **(b)** A drawer of labeled fossils in a museum. Paleontologists from around the world study such collections to help identify unknown specimens.

sediment and leave their mark. Rocks formed from sediments deposited under oxygen-free conditions in quiet water, such as lake beds or lagoons, preserve particularly fine specimens. Rocks made from sediments deposited in high-energy environments, on the other hand—where strong currents tumble shells and bones and break them up—contain at best only small fragments of fossils mixed with other clastic grains.

Fossils can survive low grades of metamorphism, but not the recrystallization and new mineral growth, and in some

Forming a Fossil

Paleontologists refer to the process of forming a fossil as **fossilization**. To see how a typical fossil develops in sedimentary rock, let's follow the fate of an old dinosaur as it searches for food along a riverbank (▶Fig. E.4). On a scalding summer day, the hungry dinosaur, plodding through the muddy ground, succumbs to the heat and collapses, dead, into the mud. Over the coming days, scavengers strip the skeleton of meat and scatter the bones among the dinosaur footprints. But before the bones have time to weather away, the river floods and buries the bones, along with the footprints, under a layer of silt. More silt from succeeding floods buries the bones and prints still deeper in a chemically stable environment, below the depth that can be reworked by currents or disrupted by burrowing organisms. Later, sea level rises, and a thick sequence of marine sediment buries the fluvial sediment. The weight of overlying sediment flattens the bones somewhat.

Eventually, the sediment containing the bones turns to rock (siltstone). The footprints remain outlined by the contact between the siltstone and mud, while the bones reside within the siltstone. Minerals precipitating from groundwater passing through the siltstone gradually replace some of the chemicals constituting the bones, until the bones themselves have become rock-like. The buried bones and footprints are now fossils. One hundred million years later, uplift and erosion exposes the dinosaur's grave. Part of a fossil bone protrudes from a rock outcrop. A lucky paleontologist observes the fragment and starts excavating, gradually uncovering enough of the bones to permit reconstruction of the beast's skeleton. Further digging uncovers footprints. The dinosaur rises again,

FIGURE E.4 How a dinosaur eventually becomes a fossil.

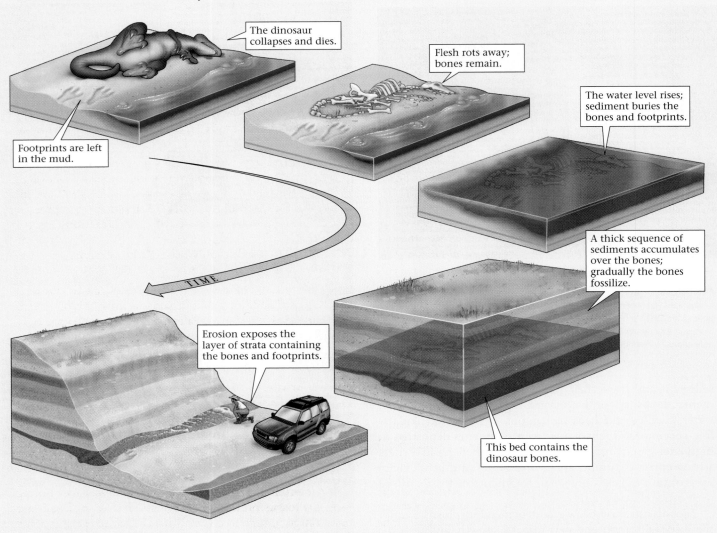

The dinosaur collapses and dies.

Footprints are left in the mud.

Flesh rots away; bones remain.

The water level rises; sediment buries the bones and footprints.

A thick sequence of sediments accumulates over the bones; gradually the bones fossilize.

TIME

Erosion exposes the layer of strata containing the bones and footprints.

This bed contains the dinosaur bones.

but this time in a museum. In recent years, bidding wars have made some fossil finds extremely valuable. For example, a skeleton of a *Tyrannosaurus rex*, a 67-million-year-old dinosaur found in South Dakota, sold at auction for $7.6 million in 1997. The specimen, named Sue after its discoverer, now stands in the Field Museum of Chicago.

Similar tales can be told for fossil seashells buried by sediment settling in the sea, for insects trapped in hardened tree sap (amber), and for mammoths drowned in the muck of a tar pit. In all cases, fossilization involves the burial and preservation of an organism or the trace of an organism. Once buried, the organism may be altered to varying degrees by pressure from overlying rock and chemical interaction with groundwater.

The Many Different Kinds of Fossils

Perhaps when you think of a fossil, you picture either a dinosaur bone or the imprint of a seashell in rock. But there are many types of fossils: **body fossils** are whole bodies or pieces of bodies; **trace fossils** are features left by an organism as it passed by; and **chemical fossils** are chemicals first formed by organisms and now preserved in rock. Let's look at examples of these categories.

- *Frozen or dried body fossils:* In a few environments, whole bodies of organisms may be preserved. Most of these fossils are fairly young, by geologic standards—their ages can be measured in thousands, not millions, of years. Examples include woolly mammoths that became incorporated in the permafrost (permanently frozen ground) of Siberia (►Fig. E.5a). In desert climates, organisms become desiccated (dried out) and can last for a long time.

- *Body fossils preserved in amber or tar:* Insects landing on the bark of trees may become trapped in the sticky sap or resin the trees produce. This golden syrup envelops the insects and over time hardens into **amber**, the semiprecious "stone" used for jewelry. Amber can preserve insects, as well as other delicate organic material such as feathers, for 40 million years or more (►Fig. E.5b).

  Tar similarly acts as a preservative. In isolated regions where oil has seeped to the surface, the more volatile components of the oil evaporate away and bacteria degrade what remains, leaving behind a sticky residue. At one such locality, the La Brea Tar Pits in Los Angeles, tar accumulated in a swampy area. While grazing or drinking at the swamp, or while hunting at the swamp, animals became mired in the tar and sank into it. Their bones have been remarkably well preserved for over 40,000 years.

- *Preserved or replaced bones, teeth, and shells:* Bones (the internal skeletons of vertebrate animals) and shells (the external skeletons of invertebrate animals) consist of durable minerals that may survive in rock. Some bone

or shell minerals are not stable, and they recrystallize (►Fig. E.5c). But even when this happens, the shape of the bone or shell remains in the rock.

- *Permineralized organisms:* **Permineralization** refers to the process by which minerals precipitate in porous material, such as wood or bone, from groundwater solutions that have seeped into the pores. **Petrified wood**, for example, forms by permineralization of wood, so that cell interiors are replaced with silica, causing the wood to become chert. In fact, the word *petrified* literally means "turned to stone." The more resistant cellulose of the wood transforms into an organic film that remains after permineralization, so that the fine detail of the wood's cell structure can be seen in a petrified log (►Fig. E.5d). The colorful bands in a petrified log come from impurities such as iron or carbon. Petrified wood typically forms when a volcanic eruption rapidly buries a forest in silica-containing ash.

- *Molds and casts of bodies:* As sediment compacts around a shell, it conforms to the shape of the shell or body. If the shell or body later disappears because of weathering and dissolution, a cavity called a **mold** remains (►Fig. E.5e). (Sculptors use the same term to refer to the receptacle into which they pour bronze or plaster.) A mold preserves the delicate shape of the organism's surface; it looks like an indentation on a rock bed. The sediment that had filled the mold also preserves the organism's shape; this **cast** protrudes from the surface of the adjacent bed.

- *Carbonized impressions of bodies:* Impressions are simply flattened molds created when soft or semisoft organisms or parts of organisms (leaves, insects, shell-less invertebrates, sponges, feathers, jellyfish) get pressed between layers of sediment. Chemical reactions eventually remove the organic chemicals that composed the organism, leaving only a thin film of carbon on the surface of the impression (►Fig. E.5f).

- *Trace fossils:* These include footprints, feeding traces, burrows, and dung that organisms leave behind in sediment (►Fig. E.5g, h). Paleontologists refer to pieces of fossilized dung as coprolites.

- *Chemical fossils:* Organisms consist of complex organic chemicals. When buried with sediment and subjected to diagenesis, some of these chemicals are destroyed, but some either remain intact or break down to form different, but still distinctive, chemicals. A distinctive chemical derived from an organism and preserved in rock is called a chemical fossil. (Such chemicals may also be called molecular fossils or biomarkers.)

Paleontologists also find it useful to distinguish among different fossils on the basis of their size. Macrofossils are fossils large enough to be seen with the naked eye. But some

FIGURE E.5 **(a)** A frozen mammoth, found in the permafrost (permanently frozen ground) of Siberia about 100,000 years after it died. It still had flesh and fur. **(b)** A piece of amber containing two fossil insects. **(c)** Fossil dinosaur bones exposed on a tilted bed of sandstone in Dinosaur National Monument, Utah. **(d)** Petrified wood from the Petrified National Forest, Arizona. Petrified wood is much harder than the surrounding tuff and thus remains after the tuff has eroded away. **(e)** Molds and casts of organisms. **(f)** The carbonized impression of fern fronds. **(g)** Dinosaur footprints in mudstone. **(h)** Worm burrows on a block of siltstone (lens cap for scale).

rocks and sediments also contain abundant microfossils, which can be seen only with a microscope (▶Fig. E.6). Microfossils include remnants of plankton, algae, bacteria, and pollen. Some deep-sea sediments consist almost entirely of microfossils derived from plankton. Pollen proves to be a particularly valuable microfossil for studying ancient climates (paleoclimates); pollen can be used to identify the types of plants that lived in a certain area in the past.

Fossil Preservation

Not all living organisms become fossils when they die. In fact, only a small percentage do, for it takes special circumstances—namely, one or more of the following four—to create a fossil.

- *Death in an oxygen-poor environment:* A dead squirrel by the side of the road won't become a fossil. As time passes, birds, dogs, or other scavengers may come along and eat the carcass. And if that doesn't happen, maggots, bacteria, and fungi will infest the carcass and gradually digest it. Flesh that has not been eaten or does not rot reacts with oxygen, or oxidizes, in the atmosphere and is transformed into carbon dioxide gas. The remaining skeleton weathers in the open air and turns to dust. Thus, before the dead squirrel can become incorporated in sediment, it has vanished. In order for fossilization to occur, a carcass must settle into an oxygen-poor environment, where oxidation reactions happen slowly, where scavenging organisms aren't so abundant, and where bacterial metabolism takes place very slowly. In such environments the organism won't rot away before it has a chance to be buried and preserved.

- *Rapid burial:* If an organism dies in a depositional environment where sediment accumulates rapidly, it may be buried before it has time to rot, oxidize, be eaten, be completely broken up, or be consumed by burrowing organisms. For example, if a storm suddenly buries an oyster bed with a thick layer of silt, the oysters die and become part of the sedimentary rock derived from the sediment.

- *The presence of hard parts:* Organisms without durable shells or skeletons, collectively called hard parts, commonly won't be fossilized, for soft flesh decays long before hard parts do in most depositional conditions. For this reason, paleontologists know much more about the fossil record of bivalves (a class of organisms, including clams and oysters, with strong shells) than they do about the fossil record of jellyfish (which have no shells) or spiders (which have very fragile shells).

- *Lack of diagenesis or metamorphism:* Metamorphism may destroy fossils, either by dissolving them away, or by causing recrystallization or neocrystallization sufficient to obscure the fossil's shape.

FIGURE E.6 Fossil plankton from deep-marine sediment. Because of their small size, these specimens are considered to be microfossils. This fossil is about 0.001 mm in diameter—this is about 1/10 the diameter of a pinpoint.

By carefully studying modern organisms, paleontologists can provide rough estimates of the **preservation potential** of organisms, meaning the likelihood that an organism will be buried and eventually transformed into a fossil. For example, in a typical modern-day shallow-marine environment, such as the mud-and-sand sea floor close to a beach, about 30% of the organisms have sturdy shells and thus a high preservation potential, 40% have fragile shells and a low preservation potential, and the remaining 30% have no hard parts at all and are not likely to be fossilized except in special circumstances. Of the 30% with sturdy shells, though, few happen to die in a depositional setting where they actually *can* become fossilized.

Extraordinary Fossils: A Special Window to the Past

Though only hard parts survive in most fossilization environments, paleontologists have discovered a few special locations where rock contains relicts of soft parts as well; such fossils are known as **extraordinary fossils**. We've already seen how extraordinary fossils such as insects and even feathers can be preserved in amber, and how complete skeletons have been found in tar pits. Extraordinary fossils can also be preserved on the oxygen-poor floors of lakes or lagoons or the deep ocean. Here, organic-rich mud accumulates, oxidation cannot occur, and flesh does not rot before burial. Carcasses of animals that settle into the mud gradually become fossils, but because they were buried before the destruction of their soft parts, fossil impressions of their soft parts surround the fossils of their bones.

A small quarry near Messel, in western Germany, for example, has revealed extraordinary fossils of 49-million-year-old mammals, birds, fish, and amphibians that died in a shallow-water lake (▶Fig. E.7a). Bird fossils from the quarry include the delicate imprints of feathers; bat fossils come complete with impressions of ears and wings, and other mammal fossils have an aura of carbonized fur. In southern Germany, exposures of the Solenhofen Limestone, an approximately 150-million-year-old rock derived from lime mud deposited in a stagnant lagoon, contain extraordinary fossils of six hundred species, including *Archaeopteryx*, one of the earliest birds (▶Fig. E.7b). And exposures of the Burgess Shale in the Canadian Rockies of British Columbia have yielded a plentitude of fossils showing what shell-less invertebrates that inhabited the deep-sea floor about 510 million years ago looked like (▶Fig. E.7c). The Burgess Shale fauna (animal life) is so strange—for example, it includes organisms with circular jaws—that it has been hard to determine how these organisms are related to present-day ones. Indeed, many of the forms of life represented by Burgess Shale fossils became extinct *without* leaving any descendants that evolved into modern species.

In a few cases, extraordinary fossils include actual tissue, a discovery that has led to a research race to find the oldest preserved DNA. (DNA, short for deoxyribonucleic acid, is the complex molecule, shaped like a double helix, that contains the code that guides the growth and development of an organism. Individual components of this code are called genes, and the study of genes and how they transmit information is called genetics.) Paleontologists have isolated small segments of DNA, from amber-encased insects, that is over 40 million years old. The amounts are not enough, however, to clone extinct species, as suggested in the popular movie *Jurassic Park*.

E.3 CLASSIFYING LIFE

The classification of fossils follows the same principles used for classifying living organisms, so before we learn how paleontologists classify fossils, we must first examine how biologists classify life forms. The principles of classification were first proposed in the eighteenth century by Carolus Linnaeus, a Swedish biologist. The study of how to classify organisms is now referred to as **taxonomy**. Linnaeus's scheme has a hierarchy of divisions. First, all life is divided into kingdoms, and each kingdom consists of one or more phyla. A phylum, in turn, consists of several classes; a class, of several orders; an order, of several families; a family, of several genera; and a genus, of one or more species (▶Fig. E.8). Kingdoms, then, are the broadest category and species the narrowest. When it is necessary, biologists and paleontologists separate out subsets of individual

(a)

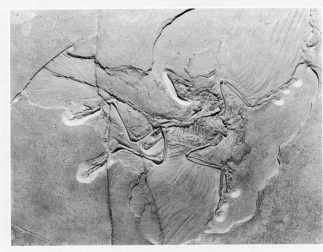

(b)

(c)

FIGURE E.7 Extraordinary fossils: **(a)** a mammal from Messel, Germany; **(b)** *Archaeopteryx* from the Solenhofen Limestone; **(c)** reproduction of a creature from the Burgess Shale, in Canada.

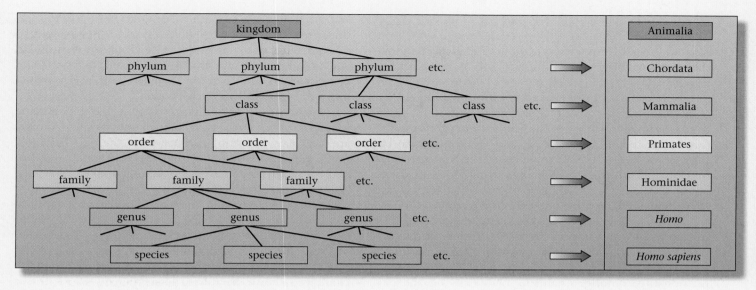

FIGURE E.8 The taxonomic subdivisions. The right-hand column shows how the names apply to human beings.

categories—thus, there are subphyla and subclasses. Biologists now recognize six kingdoms (▶Fig. E.9):

- *Archaea:* microorganisms typically found in extreme environments such as hot springs and in very acidic or very alkaline water;
- *Eubacteria:* "true bacteria," including the kind that cause infections;
- *Protista:* various unicellular and simple multicellular organisms; these include algae (such as diatoms) and forams, two of the major plankton types in the oceans;
- *Fungi:* mushrooms, yeast, and so on;
- *Plantae:* trees, grasses, ferns, and so on; and
- *Animalia:* sponges, corals, snails, dinosaurs, ants, people, and so on.

More recently, biologists have realized that Archaea are as different from Eubacteria as both are from the other kingdoms. Specifically, the last four kingdoms share enough common characteristics, at the cellular level, to be lumped together as a group called **Eukarya**. Thus, biologists now divide life into three domains: Archaea, Eubacteria, and Eukarya. The cells of Eukaryotic organisms have distinct nuclei and internal membranes, whereas those of Eubacteria and Archaea do not. (The latter two domains have prokaryotic cells.)

Figure E.8 shows how humans are classified according to Linnaeus's scheme. Humans, together with monkeys, apes, and lemurs, are primates because they have opposable thumbs and similar facial characteristics. Primates, together with dogs, cats, horses, cows, elephants, whales, and many other animals, are mammals because they nurse their young with milk and most of them have fur or hair.

Mammals, along with reptiles, birds, and fish, are chordates because they have an internal skeleton with a spinal canal. And chordates, along with clams, insects, lobsters, and worms, are animals because they share a similar cell structure. Note that genus and species names are always given in Latin and are italicized.

FIGURE E.9 The basic kingdoms of life on Earth. Archaea and Eubacteria are both prokaryotes, because among other characteristics they don't have nuclei. All other life forms are eukaryotes, because they consist of cells with a nucleus.

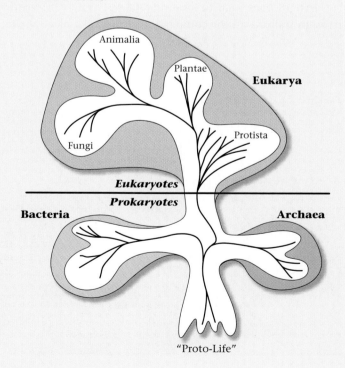

E.4 CLASSIFYING FOSSILS

As you'll recall from Chapter 5, mineralogists recognize thousands of different types of minerals. A given mineral has a specific crystal structure and a chemical composition that varies within a definable range. In contrast, paleontologists traditionally distinguish different fossil species from each other according to **morphology** (the form or shape) alone. A fossil clam is a fossil clam because it looks like one. In some cases, the characteristics that distinguish one fossil species from another may be pretty subtle (e.g., the number of ridges on the surface of a shell, or the relative length of different leg bones), but even beginners can distinguish the major groups of fossils from each other on sight. Though presently morphology provides the primary basis for fossil identification, paleontologists may someday be able to determine relationships among organisms by specifying the percentages of shared protein sequences constituting the fossil's DNA.

When it came to naming fossils, paleontologists began by comparing fossil forms with known organisms. They looked at the nature of the skeleton (was it internal or external?), the symmetry of the organism (was it bilaterally symmetric like a mammal, or did it have five-fold symmetry like a starfish?), the design of the shell (in the case of invertebrates), and the design of the jaws or feet (in the case of vertebrates). For example, a fossil organism with a spiral shell that does not contain internal chambers is classified a member of the class Gastropoda (the snails) within the phylum Mollusca (▶Fig. E.10).

Not all fossil species resemble living families of organisms. In such cases, the comparisons must take place at the level of orders or even higher. For example, a group of extinct organisms called trilobites have no close living relatives. But they were clearly segmented invertebrate animals, and as such they resemble arthropods such as insects and crustaceans. Thus, they are considered to be an extinct member of the phylum Arthropoda.

Sometimes there's nothing magical about classifying fossils; major characteristics at the level of class or higher are fairly easy to distinguish just from the way they look. For instance, skeletons of birds (class Aves) are hard to mistake for skeletons of mammals (class Mammalia), and snail shells (class Gastropoda) are hard to mistake for clam shells (class Pelecypoda). But classification can also be difficult, especially if specimens are incomplete, and in many cases classification can be controversial.

▶Figure E.11 shows examples of some of the major classes and subphyla of invertebrate fossils. With this chart, you should be able to identify many of the fossils you'll find in a common bed of limestone. Particularly common invertebrate fossils include the following.

- *Trilobites:* These have a segmented shell that is divided lengthwise into three parts. They are a type of arthropod.
- *Gastropods* (snails): Most fossil specimens of gastropods have a shell that does not contain internal chambers. Slugs are gastropods without shells.
- *Bivalves* (clams and oysters): These have a shell that can be divided into two similar halves. The plane of symmetry is parallel to the plane of the shell.
- *Brachiopods* (lamp shells): The top and bottom parts of these shells have different shapes, and the plane of symmetry is perpendicular to the plane of the shell. Shells typically have ridges radiating out from the hinge.
- *Bryozoans:* These are colonial animals. Their fossils resemble a screen-like grid of cells. Each cell is the shell of a single animal.
- *Crinoids* (sea lilies): These organisms look like a flower but actually were animals. Their shells have a stalk consisting of numerous circular plates stacked on top of each other.
- *Graptolites:* These look like tiny carbon-saw blades in a rock. They are remnants of colonial animals that floated in the sea.
- *Cephalopods:* These include ammonites, with a spiral shell, and nautiloids, with a straight shell. Their shells contain internal chambers and have ridged surfaces. These organisms were squid-like.
- *Corals:* These include colonial organisms that form distinctive mounds or columns. Paleozoic examples are solitary.

FIGURE E.10 Examples of the diversity of gastropods (snails). Note that although all these shells have a spiral shape, they differ in detail from each other. Some gastropods have no shells at all.

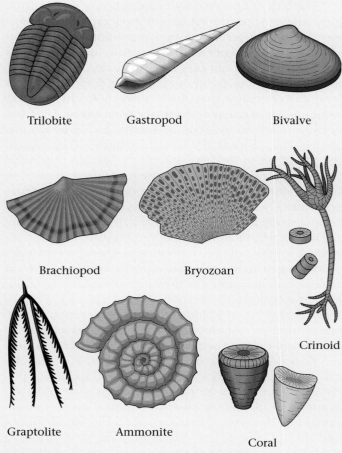

Trilobite

Gastropod

Bivalve

Brachiopod

Bryozoan

Crinoid

Graptolite

Ammonite

Coral

FIGURE E.11 Common types of invertebrate fossils.

E.5 THE FOSSIL RECORD

A Brief History of Life

As we will see in Chapter 12, the fossil record defines the long-term evolution of life on planet Earth. Archaea and bacteria fossils are found in rocks as old as about 3.7 billion years, indicating that these organisms are the earliest forms of life on Earth. Researchers once thought that archaea somehow developed by chemical reactions in concentrated "soups" that formed when seawater trapped in shallow pools evaporated. Growing evidence suggests, however, that they may instead have developed in warm groundwater beneath the Earth's surface or at hydrothermal vents on the sea floor.

For the first billion years or so of life history, archaea and bacteria were the only types of life on Earth. Then, about 2.5 Ga (billion years ago), organisms of the protist kingdom first appeared. Early multicellular organisms—shell-less invertebrates of the animal kingdom, and fungi of the plant kingdom—came into existence perhaps 1.0 to 1.5 Ga. Within each kingdom, life radiated (divided) into different phyla, and within each phylum, life radiated into different classes. The great variety of shelly invertebrate classes appeared a little over 540 million years ago, and many organisms that persist today appeared at different times since then: first came invertebrates, then fish, land plants, amphibians, reptiles, and finally mammals and birds. Researchers have been working hard to understand the phylogeny (the evolutionary relationships) among organisms, using both the morphology of organisms and, more recently, the study of genetic material. Ideas about which groups radiated from which ancestors is shown in a chart called the tree of life, or, more formally, the **phylogenetic tree**. Study of the DNA of different organisms is beginning to enable researchers to understand the relationship between molecular processes and evolutionary change.

Is the Fossil Record Complete?

By some estimates, more than 250,000 species of fossils have been collected and identified to date, by thousands of geologists working on all continents during the past two centuries. These fossils define the framework of life evolution on planet Earth. But the record is not complete—not every intermediate step in the evolution of every organism can be accounted for by known fossils. Considering that as many as 5 million species may be living on Earth today (not counting bacteria), over the billions of years that life has existed there may have been 5 billion to 50 billion species. Clearly, known fossils represent at most a tiny percentage of these species. Why is the record so incomplete?

First, despite all the fossil-collecting efforts of the past two centuries, paleontologists have not even come close to sampling every cubic centimeter of sedimentary rock exposed on Earth. Just as biologists have not yet identified every living species of insect, paleontologists have not yet identified every species of fossil. New species and even genera of fossils continue to be discovered every year.

Second, not all organisms are represented in the rock record, because not all organisms have a high preservation potential. As noted earlier, fossilization occurs only under special conditions, and thus only a minuscule fraction of the organisms that have lived on Earth become fossilized. There may be few, if any, fossils of a vast number of extinct species, so we may have no way of knowing that they ever existed.

Finally, as we will learn in Chapter 12, the sequence of sedimentary strata that exists on Earth does not account for every minute of time since the formation of our planet. Sediments accumulate only in environments where conditions are appropriate for deposition and not for erosion—sediments do not accumulate, for example, on the dry great plains or on mountain peaks, but do accumulate in the sea and in the floodplains and deltas of rivers. Because

Earth's climate changes through time and because sea level rises and falls, certain locations on continents are sometimes sites of deposition and sometimes aren't, and on occasion are sites of erosion. Therefore, the sequence of strata at a location records only part of geologic time.

In sum, a rock sequence provides an incomplete record of Earth history, organisms have a low probability of being preserved, and paleontologists have found only a small percentage of the fossils preserved in rock. So the incompleteness of the fossil record should come as no surprise.

E.6 EVOLUTION AND EXTINCTION

Darwin's Grand Idea

As a young man in England in the early nineteenth century, Charles Darwin had been unable to settle on a career but had developed a strong interest in natural history. As a consequence, he jumped at the opportunity to serve as a naturalist aboard the H.M.S. *Beagle* on an around-the-world surveying cruise. During the five years of the cruise, from 1831 to 1836, Darwin made detailed observations of plants, animals, and geology in the field and amassed an immense specimen collection from South America, Australia, and Africa. Just before departing on the voyage, a friend gave him a copy of Charles Lyell's 1830 textbook *Principles of Geology*, which argued in favor of James Hutton's proposal that the Earth had a long history and that geologic time extended much farther into the past than did human civilization. A visit to the Galápagos Islands, off the coast of Peru, was a turning point in Darwin's thinking. The naturalist was most impressed with the variability of Galápagos finches. He marveled not only at the fact that different varieties of the bird occurred on different islands, but at how each variety was adapted to utilize a particular food supply. With Lyell's writings in mind, Darwin developed a hypothesis that the finches had begun as a single species but had branched into several different species when isolated on different islands. This proposal implied that a species could change, or undergo **evolution**, throughout long periods of time and that new species could appear.

On his return to England, Darwin discussed his hypothesis with fellow biologists and geologists, and over succeeding years he gathered supporting evidence. When Alfred Russell Wallace, a naturalist working in Indonesia, wrote to Darwin (in 1858) outlining almost identical thoughts about evolution, Darwin realized that he needed to publish. Darwin and Wallace jointly presented the concept of evolution at a scientific meeting, so as to share credit, and in 1859 Darwin published the revolutionary book *On the Origin of Species by Means of Natural Selection*, in which he outlined evolution and proposed how it occurred.

The crux of Darwin's argument is simply this: because populations of organisms cannot grow exponentially forever, they must be limited by competition for scarce resources in the environment. In nature, only organisms capable of survival can pass on their characteristics to the next generation. In each new generation, some individuals have characteristics that make them more fit, whereas some have characteristics that make them less fit. The fitter organisms are more likely to survive and produce offspring. Thus, beneficial characteristics that they possess get passed on to the next generation. Darwin called this process **natural selection**, because it occurs on its own in nature. (He noted that the same process occurs when farmers artificially select animals or plants for breeding to develop new kinds of domestic livestock or crops.) According to Darwin, when natural selection takes place over long periods of time—geologic time—natural selection eventually produces new organisms that differ so significantly from their distant ancestors that new organisms can be considered to constitute a new species. If environmental conditions change, or if competitors enter the environment, species that do not evolve and become better adapted to survive and compete eventually die off and become extinct.

Darwin's view of evolution was not just a philosophical speculation—it was a scientific proposal that made predictions that could be tested by observations. Darwin himself provided many observations that support evolution. And in succeeding decades paleontologists, biologists, and anthropologists have made countless observations at countless sites around the world that support evolution. Thus, the idea has gained the status of a theory—it is an idea that has been successfully supported by many observations, that can be used to make testable predictions, and that so far has not been definitively disproved by any observation or experiment. Thus, we now refer to Darwin's idea as the **theory of evolution**.

In the century and a half since Darwin published his work, the science of genetics has developed and has provided insight into how evolution works. Progress began in the late nineteenth century when an Austrian monk, Gregor Mendel, studied peas in the garden of his monastery and showed that genetic mutations led to new traits that could be passed on to offspring. Traits that make an organism less likely to survive are not passed on, either because the organism dies before it has offspring or because the offspring themselves cannot survive, but traits that make an organism better suited to survival are passed on to succeeding generations. With the discovery of DNA in 1953, biologists began to understand the molecular nature of genes and mutations, and thus of evolution. And with the genome projects of the twenty-first century, which define the detailed architecture of DNA molecules for a given species, it is now possible to pinpoint the exact arrangement of genes responsible for specific traits.

The theory of evolution provides a conceptual framework in which to understand paleontology. By studying fossils in sequences of strata, paleontologists are able to observe progressive changes in species through time, and can determine when some species go extinct and other species appear. But because of the incompleteness of the fossil record, questions remain as to the rates at which evolution takes place during the course of geologic time, and new proposals of specific phylogenies often are met with skepticism. Traditionally, it was assumed that evolution happened at a constant, slow rate—this concept is called **gradualism**. More recently, however, researchers have suggested that evolution takes place in fits and starts: evolution occurs very slowly for quite a while (species are in equilibrium), and then during a relatively short period it takes place very rapidly. This concept is called **punctuated equilibrium**. Factors that could cause sudden pulses of evolution include (1) a sudden **mass extinction event**, during which many organisms disappear, leaving ecological niches open for new species to colonize; (2) a sudden change in the Earth's climate that puts stress on organisms—organisms that evolve to survive the new stress survive, while others become extinct; (3) the sudden formation of new environments, as may happen when rifting splits apart a continent and generates a new ocean with new coastlines; and (4) the isolation of a breeding population.

Extinction: When Species Vanish

Extinction occurs when the last members of a species die, so there are no parents to pass on their genetic traits to offspring. Some species become extinct as they evolve into new species, whereas others just vanish, leaving no hereditary offspring. These days, we take for granted that species become extinct, because a great number have, unfortunately, vanished from the Earth during human history. Before the 1770s, however, few geologists thought that extinction occurred; they thought that fossils that didn't resemble known species must have living relatives somewhere on Earth. Considering that large parts of the Earth remained unexplored, this idea wasn't so far-fetched. But by the end of the eighteenth century, it became clear that numerous fossil organisms did not have modern-day counterparts. The bones of mastodons and woolly mammoths, for example, were too different from those of elephants to be of the same species, and the animals were too big to hide.

What causes extinction? Initially, paleontologists assumed that Noah's flood (recounted in the Old Testament) was the cause. But as the fossil record began to accumulate, and the principle of fossil succession was documented, they realized that not all extinction could have happened in a single event. Some authors proposed that extinction was the consequence of a series of global catastrophic floods, but continued geologic research demonstrated that such floods have not taken place.

Twentieth-century studies concluded that many different phenomena can contribute to extinction. Some extinctions may happen suddenly, when all members of a species die off in a short time, whereas others may occur over longer periods, when the replacement rate of a population simply becomes lower than the mortality rate. By examining the number of species on Earth through time (i.e., by studying variations in the diversity of life, or biodiversity), paleontologists have found that there is a varying rate of extinction. Generally, the rate is fairly slow, but on occasion a **mass extinction event** occurs, during which a large number of species worldwide disappear. At least five major mass extinction events have happened during the past half billion years (▶Fig. E.12). These events define the boundaries between some of the major intervals into which geologists divide time. For example, a major extinction event marks the end of the Cretaceous Period, 65 million years ago. During this event, dinosaur species (with the exception of their modified descendants, the birds) vanished, along with many marine invertebrate species. Some researchers have suggested that extinction events are periodic, but this idea remains controversial.

Following are some of the geologic factors that may cause extinction.

- *Global climate change:* At times, the Earth's mean temperature has been significantly colder than today's,

FIGURE E.12 This graph illustrates the variation in diversity of life with time. Steep dips in the curve mean that the number of species on Earth suddenly decreased substantially. The largest dips represent major mass extinctions.

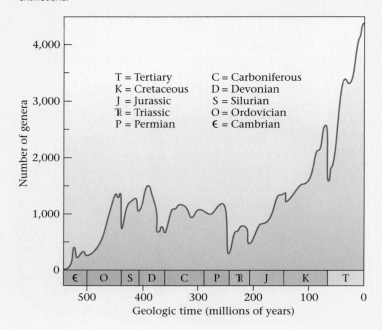

T = Tertiary
K = Cretaceous
J = Jurassic
Ŧ = Triassic
P = Permian

C = Carboniferous
D = Devonian
S = Silurian
O = Ordovician
Є = Cambrian

Number of genera

Geologic time (millions of years)

whereas at other times it has been much hotter. These shifts affect ocean currents and sea level, and may trigger ice ages or droughts. Because of a change in climate, an individual species may lose its habitat, and if it cannot adapt to the new habitat or migrate to stay with its old one, the species will disappear.

- *Tectonic activity:* Tectonic activity causes mountain building, the gradual vertical movement of the crust over broad regions, changes in sea-floor spreading rates, and changes in the amount of volcanism. These phenomena can make sea level rise or fall, or can bring about changes in elevation, thus modifying the distribution and area of habitats. Species that cannot adapt die off.

- *Asteroid or comet impact:* Many geologists have concluded that impacts of large meteorites or asteroids with the Earth have been catastrophic for life (▶Fig. E.13). An impact would send so much dust and debris into the atmosphere that it would blot out the Sun and plunge the Earth into darkness and cold. Such a change, though relatively short-lived, could interrupt the food chain.

- *The appearance of a predator or competitor:* Some extinctions may happen simply because a new predator appears on the scene. Some researchers suggest that this phenomenon explains the mass extinction that occurred during the past 20,000 years, when a vast number of large mammal species vanished from North America. The timing of these extinctions appears to coincide with the appearance of the first humans (fierce predators) on the continent. If a more efficient competitor appears, the competitor steals an ecological niche from the weaker species, whose members can't obtain enough food and thus die out.

FIGURE E.13 Painting of an asteroid impact in Yucatán that possibly eliminated most if not all dinosaur species.

Suggested Reading

Benton, M. J. 2003. *When Life Nearly Died: The Greatest Mass Extinction of All Time.* London: Thames & Hudson.

Clark, J. A., and J. O. Farlow. 2002. *Gaining Ground: The Origin and Early Evolution of Tetrapods.* Bloomington: Indiana University Press.

Cutle, A. 2003. *The Seashell on the Mountaintop: A Story of Science, Sainthood, and the Humble Genius Who Discovered a New History of the Earth.* New York: E. P. Dutton.

Knoll, A. H. 2003. *Life on a Young Planet: The First Three Billion Years of Evolution on Earth.* Princeton, N.J.: Princeton University Press.

Prothero, D. R. 2003. *Bringing Fossils to Life: An Introduction to Paleontology, 2nd edition.* Boston: WCB/McGraw-Hill.

ANOTHER LOOK Crack open a sedimentary rock, and you may find fossils. The cobble of siltstone on the left, collected near Catskill, New York, contains Middle Devonian brachiopods; the block of sandstone on the right, collected in Brazil, contains the well preserved scales and jaw bones of a Cretaceous fish.

CHAPTER 12

Deep Time: How Old Is Old?

How do geologists determine the age of rocks? This cliff face in Missouri shows two rock units. The sedimentary layers on the left were deposited on the rhyolite to the right. From this relationship, we can determine the relative age of the two units—the rhyolite is older. But to determine the age of the rhyolite in years—its numerical age—we must date it radiometrically.

Geopuzzle

Geologists state that the Earth is very, very old—4.57 billion years old, to be exact. How was this age determined? Is there any place on Earth where we can see a succession of rocks representing the entirety of Earth history?

If the Eiffel Tower were now representing the world's age, the skin of paint on the pinnacle-knob at its summit would represent man's share of that age; and anybody would perceive that that skin was what the tower was built for. I reckon they would, I dunno.

—Mark Twain (1835–1910)

12.1 INTRODUCTION

In May of 1869, a one-armed Civil War veteran named John Wesley Powell set out with a team of nine geologists and scouts to explore the previously unmapped expanse of the Grand Canyon, the greatest gorge on Earth. Though Powell and his companions battled fearsome rapids and the pangs of starvation, most managed to emerge from the mouth of the canyon 3 months later (▶Fig. 12.1). During their voyage, seemingly insurmountable walls of rock both imprisoned and amazed the explorers, and led them to pose important questions about the Earth and its history, questions that even casual tourists to the canyon ponder today: Did the Colorado River sculpt this marvel? If so, how long did it take? When did the rocks making up the walls of the canyon form? Was there a time before the colorful layers accumulated? These questions pertain to **geologic time**, the span of time since Earth's formation.

In this chapter, we first learn how geologists developed the concept of geologic time and thus a frame of reference for describing the ages of rocks, fossils, structures, and landscapes. Then we look at the tools geologists use to determine the age of the Earth and its features. With the concept of geologic time in hand, a hike down a trail into the Grand Canyon becomes a trip into the distant past, into what authors call *deep time*. The geological discovery that our planet's history extends billions of years into the past changed humanity's perception of the Universe as profoundly as did the astronomical discovery that the limit of space extends billions of light years beyond the edge of our Solar System.

12.2 TIME: A HUMAN OBSESSION

When you plan your daily schedule, you have to know not only where you need to be, but when you need to be there. Because time assumes such significance in human consciousness today, we have developed elaborate tools to measure it and formal scales to record it. We use a second

FIGURE 12.1 Woodcut illustration of the "noonday rest in Marble Canyon," from J. W. Powell, *The Exploration of the Colorado River and Its Canyons* (1895). The explorers have just entered a quiet stretch of the river: "We pass many side canyons today that are dark, gloomy passages back into the heart of the rocks."

as the basic unit of time measurement. What exactly is a second? From 1900 to 1968, we defined the second as 1/31,556,925.9747 of the year 1900, but now we define it as the duration of time that it takes for a cesium atom to change back and forth between two energy states 9,192,631,770 times. This change is measured with a device called an atomic clock. An atomic clock has an accuracy of about 1 second per million years. We sum 60 seconds into 1 minute, 60 minutes into 1 hour, and 24 hours into 1 day, about the time it takes for Earth to spin once on its axis.

In the preindustrial era, each locality kept its own time, setting noon as the moment when the Sun reached the highest point in the sky. But with the advent of train travel and telegraphs, people needed to calibrate schedules from place to place. So in 1883, countries around the globe agreed to divide the world into 15°-wide bands of longitude called time zones—in each time zone, all clocks keep the same standard time. The times in each zone are set in relation to Greenwich Mean Time (GMT), the time at the astronomical observatory in Greenwich, England. Today, the world standard for time is determined by a group of about 200 atomic clocks that together define Coordinated Universal Time (CUT). CUT is the basis for the global positioning system (GPS), used for precise navigation.

12.3 THE CONCEPT OF GEOLOGIC TIME

The Birth of Geologic Time

Most people develop a sense for the duration of time by remembering when events took place in terms of human lifetimes. So if an average generation spans 20 years, then recorded history began about 200 to 400 generations ago. Many cultures viewed geologic time to be about the same duration as historical time. For example, in 1654, the Irish archbishop James Ussher added up the generations of patriarchs described in the Judeo-Christian Old Testament and concluded that the Earth came into being on October 23, 4004 B.C.E. Scholars at the time assumed that the Earth formed essentially just as we see it.

Beginning in the Renaissance, however, scientists studying geological features began to speculate that geological time might far exceed historical time. The discovery of seashell-like shapes in rocks triggered this revolution in thinking. Although most people dismissed these shapes, which came to be known as fossils (see Interlude E), as a coincidence or a supernatural trick, some concluded that they were the relicts of ancient sea creatures. This idea did not gain wide acceptance, though, until the work of Nicolaus Steno (1638–1686).

Steno, born Niels Stenson in Denmark, served as a court physician in Florence, Italy. During walks in the nearby Apennine Mountains, Steno frequently came upon fossils and wondered how they ended up in solid rock hundreds of meters above sea level. One day, local fishermen gave Steno a shark's head, which he dissected out of curiosity. Inside its mouth, he found rows of distinctive triangular teeth, identical to so-called tongue stones, fossils thought to be the petrified tongues of dragons (▶Fig. 12.2). Steno became convinced that these fossils, and by implication others as well, did not form supernaturally but were the relicts of ancient life, and introduced this idea in his 1669 book *Forerunner to a Dissertation on a Solid Naturally Occurring within a Solid*. The title may seem peculiar until you think about the puzzle Steno was trying to solve: How did a solid object (a fossil) get into another solid object (a sedimentary rock)? Steno concluded that the rock must have once been loose sand that incorporated the teeth and shells while soft and only later hardened into rock. His discovery implied that geologic features developed not all at once, but rather by a series of events that took a long time, longer than human history.

FIGURE 12.2 A fossilized shark's tooth. Before Steno explained the origin of such fossils, they were known as dragons' tongues.

The next major step in the development of the idea that geologic time exceeded historical time came from the work of a Scottish doctor and gentleman farmer named James Hutton (1726–1797). He based his work on relationships that he observed in the rocky crags of his native land. Hutton lived during the Age of Enlightenment, when philosophers such as Voltaire, Kant, Hume, and Locke encouraged people to cast aside the constraints of dogma and think for themselves, and when the discovery of physical laws by Sir Isaac Newton made people look to natural, not supernatural, processes to explain the Universe. As Hutton wandered around Scotland, he came up with an idea that provides the foundation of geology and led to his title: "father of geology." This idea, the **principle of uniformitarianism**, states simply that physical processes we observe today also operated in the past and were responsible for the formation of the geologic features we see in outcrops. More concisely, the principle of uniformitarianism implies that *the present is the key to the past*. If this principle is correct, Hutton reasoned, the Earth must be much older than human history allowed, for observed geologic processes work very slowly. In his 1785 book *Theory of the Earth with Proofs and Illustrations*, Hutton mused on the issue of geologic time, and suggested that in the case of Earth, "there is no vestige of a beginning, no prospect of an end."

> **Take-Home Message**
>
> The scientific concept that the Earth is very old began to develop during the eighteenth century. Geologists distinguish between relative age (the age of one feature with respect to another) and numerical age (the age of a feature in years).

RELATIVE AGE

Iraq War

Gulf War (Persian Gulf)

Falkland War (Argentina-England)

Six-Day War (Arab-Israeli)

Vietnam War

Bay of Pigs Invasion (Cuba-U.S.)

Korean War

World War II

Spanish Civil War

World War I

Russian Revolution

Russo-Japanese War

Boer War (South Africa)

(a)

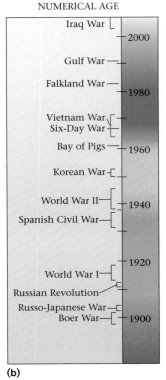

NUMERICAL AGE

Iraq War

— 2000

Gulf War

Falkland War
— 1980
Vietnam War
Six-Day War
Bay of Pigs — 1960

Korean War

World War II
— 1940
Spanish Civil War

— 1920
World War I
Russian Revolution
Russo-Japanese War
Boer War — 1900

(b)

FIGURE 12.3 The difference between relative and numerical age. **(a)** The relative ages of selected wars in the last 100 or so years. **(b)** The numerical ages of these same wars. Clearly, this chart provides more information, for it displays the duration of events and indicates the amount of time between events.

Relative versus Numerical Age

In the early nineteenth century, geologists struggled to develop ways to divide and describe geologic time. Like historians, geologists want to establish both the sequence of events that created an array of geologic features (such as rocks, structures, and landscapes) and the exact dates on which the events happened. We specify the age of one feature with respect to another as its **relative age** and the age of a feature given in years as the **numerical age** (or, in older literature, the absolute age) (▶Fig. 12.3a, b). Geologists developed ways of defining relative age long before they did so for numerical age, so we will look at relative-age determination first.

12.4 PRINCIPLES FOR DEFINING RELATIVE AGE

Nineteenth-century geologists such as Charles Lyell recast the ideas of Steno and Hutton into formal, usable geological principles. These principles, defined below, continue to

provide the basic framework within which geologists read the record of Earth history.

Physical Principles

- **The principle of uniformitarianism:** As noted earlier, physical processes we observe operating today also operated in the past, at roughly comparable rates (▶Fig. 12.4a). Geologists emphasize that physical processes do not occur at *exactly* the same rate through time. Factors such as climate change affect erosion rates, and occasional catastrophic events, such as a large meteor impact, may trigger abrupt global extinctions.

- **The principle of superposition:** In a sequence of sedimentary rock layers, each layer must be younger than the one below, for a layer of sediment cannot accumulate unless there is already a substrate on which it can collect. Thus, the layer at the bottom of a sequence is the oldest, and the layer at the top is the youngest (▶Fig. 12.4b).

- **The principle of original horizontality:** Sediments on Earth settle out of a fluid in a gravitational field. Typically, the surfaces on which sediments accumulate (such as a floodplain or the bed of a lake or sea) are fairly horizontal. If sediments were deposited on a steep slope, they would likely slide downslope before lithification, and so would not be preserved as sedimentary layers. Therefore, when layers of sediment are originally deposited, they are fairly horizontal. With this principle in mind, we know that when we see folds and tilted beds, we are seeing the consequences of deformation that postdates deposition (▶Fig. 12.4c).

- **The principle of original continuity:** Sediments generally accumulate in continuous sheets. If today you find a sedimentary layer cut by a canyon, then you can assume that the layer once spanned the canyon but was later eroded by the river that formed the canyon (▶Fig. 12.4d).

- **The principle of cross-cutting relations:** If one geologic feature cuts across another, the feature that has been cut is older. For example, if an igneous dike cuts across a sequence of sedimentary beds, the beds must be older than the dike (▶Fig. 12.4e). If a fault cuts across and displaces layers of sedimentary rock, then the fault must be younger than the layers. But if a layer of sediment buries a fault, the sediment must be younger than the fault.

- **The principle of inclusions:** If an igneous intrusion contains fragments of another rock, the fragments must be older than the intrusion. If a layer of sediment deposited on an igneous layer includes pebbles of the igneous rock, then the sedimentary layer must be younger. The fragments (xenoliths) in an igneous body and the pebbles in the sedimentary layer are inclusions, or pieces of one material incorporated in another. The rock containing the inclusion must be younger than the inclusion (▶Fig. 12.4f).

- **The principle of baked or chilled contacts:** An igneous intrusion "bakes" (metamorphoses) surrounding rocks. The rock that has been baked must be older than the intrusion (▶Fig. 12.4g). Similarly, since an intrusion injects into cooler rocks, the margin of the intrusion cools rapidly and is finer grained than the interior. Such chilled margins are visible in outcrops (▶Fig. 12.4h).

Now let's use these principles to determine the relative ages of the features shown in ▶Figure 12.5a (p. 422). In so doing, we develop a **geologic history** of the region, defining the relative ages of events that took place there.

First, note what kinds of rocks and structures the figure contains. Most of the rocks constitute a sequence of folded sedimentary beds, but we also see a granite pluton, a basalt dike, and a fault. Let's start our analysis by looking at the sedimentary sequence. By the principle of superposition, we conclude that the oldest layer is the limestone labeled 1, for it occurs at the bottom. Progressively younger beds lie above the limestone. We can confirm that layer 1 predates layer 2 by applying the principle of inclusions, for layer 2 contains pebbles (inclusions) of layer 1. Thus, the sedimentary beds from oldest to youngest are 1, 2, 3, 4, 5, 6, 7. (There are other rocks below layer 1, but we do not see them in our cross section.) Considering the principle of original horizontality, we conclude that the layers were folded sometime after deposition.

Now let's look at the relationships between the igneous rocks and the sedimentary rocks. The granite pluton cuts across the folded sedimentary rocks, so the intrusion of the pluton occurred after the deposition of the sedimentary beds and after they were folded. The layer of igneous rock that parallels the sedimentary beds could be either a sill that intruded between the sedimentary layers or a flow that spread out over the sandstone and solidified before the shale was deposited. From the principle of inclusions, we deduce that the layer is a sill, because it contains xenoliths of both the underlying sandstone and the overlying shale. Since the sill is folded, it intruded before folding took place. By applying the principle of baked contacts, we also can tell that the sill intruded before the pluton did, because the baked zones (metamorphic aureole) surrounding the pluton affected the sill. The dike cuts across both the pluton and the sill, as well as all the sedimentary layers, and thus formed later.

Finally, let's consider the fault and the land surface. Because the fault offsets the granite pluton and the sedimentary beds, by the principle of cross-cutting relations the fault must be younger than those rocks. But the fault itself has been cut by the dike, and so must be older than the dike. The present land surface erodes all rock units and the fault, and thus must be younger. We can now propose the following geologic history for this region (▶Fig. 12.5b–h): (b) deposition of the sedimentary sequence, in order from layers 1 to 8; (c) intrusion of the sill; (d) folding of the sedimentary layers and the sill; (e) intrusion of the granite pluton; (f) faulting; (g) intrusion of the dike; (h) formation of the land surface.

Adding Fossils to the Story: Fossil Succession

As England entered the Industrial Revolution in the late eighteenth and early nineteenth centuries, new factories demanded coal to fire their steam engines. The government decided to build a network of canals to transport coal and iron, and hired an engineer named William Smith (1769–1839) to survey the excavations. These excavations provided fresh exposures of bedrock that previously had been covered by vegetation. Smith learned to recognize distinctive layers of sedimentary rock and to identify the **fossil assemblage** (the group of fossil species) that the layers contain (▶Fig. 12.6). He also discovered that a particular fossil assemblage occurs only in a limited interval of strata: the assemblage cannot be found in strata above or below this interval. Thus, assemblages succeed one another in a definable, consistent order. This observation implies that fossil species are not randomly distributed in a sequence of beds, and that once a fossil species disappears in the sequence it does not reappear higher in the sequence—that is, extinction is forever. The predictability of fossil distribution, which allows geologists to arrange fossil species in a progression from older at the bottom to younger at the top, has been found at countless locations around the world and has been codified as the **principle of fossil succession**.

To see how this principle works, examine ▶Figure 12.7a, which depicts a sequence of strata. Bed 1 at the base contains fossil species A, bed 2 contains fossil species A and B, bed 3 contains B and C, bed 4 contains C, and so on. From these data, we can define the **range** of specific fossils in the sequence, meaning the interval in the sequence in which the fossils occur. Note that the sequence contains a definable succession of fossils (A, B, C, D, E, F), that the range in which a particular species occurs may overlap with the range of other species, and that once a species vanishes (becomes extinct), it does not reappear higher in the sequence.

(a)

Time 1   Time 2   Time 3

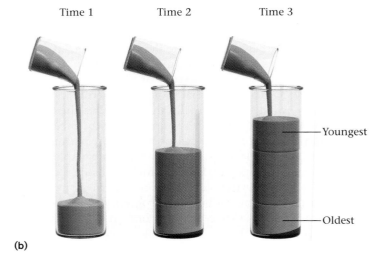

Youngest

Oldest

(b)

FIGURE 12.4 Geologic principles: **(a)** Uniformitarianism: Geologists assume that the physical processes that formed the mud cracks of the dried-up mud puddle on the left also formed the mud cracks preserved on the surface of the Paleozoic bed shown on the right. Note that the Paleozoic mud cracks were inside solid rock and are visible now only because the overlying bed has been removed by erosion. **(b)** Superposition: If you fill a glass cylinder with different colors of sand, the oldest sand (white) must be on the bottom. **(c)** Original horizontality: When sediment is originally deposited, like this silt on the tidal flat near Mont St. Michel, France (shown on the left), it forms a flat layer. If layers do not become deformed, they can remain horizontal for hundreds of millions of years, as illustrated by these beds of Paleozoic sandstone in Wisconsin (on the right).

(c)

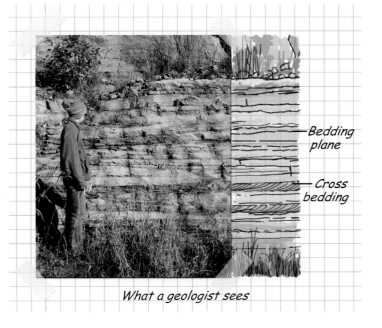

Bedding plane

Cross bedding

What a geologist sees

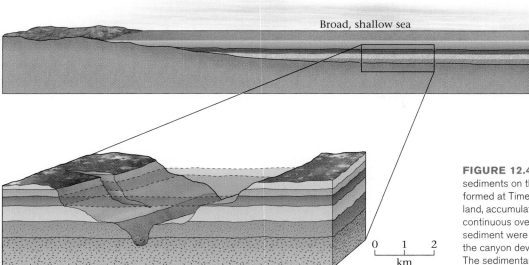

Sea-level rise

Broad, shallow sea

0 50 100
km

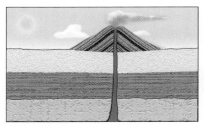

(d)

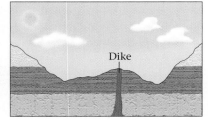

Time 1

(e)

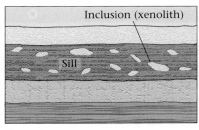

Inclusion (xenolith)

Sill

(f)

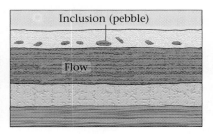

Inclusion (pebble)

Flow

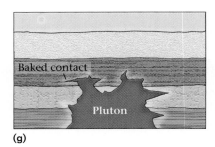

Baked contact

Pluton

(g)

0 1 2
km

Dike

Time 2

FIGURE 12.4 *(continued.)* **(d)** Original continuity: The sediments on the floor of this broad, shallow sea, which formed at Time 1 when the sea level rose and flooded the land, accumulated in nearly horizontal layers that are continuous over a broad region. We assume that the layers of sediment were originally continuous across the canyon before the canyon developed at Time 2. **(e)** Cross-cutting relations: The sedimentary beds existed first. Then they were cut by igneous magma rising to a volcano, so today we see a dike cutting across the bedding. Since the dike does the cutting, it must be younger. **(f)** Inclusions: In the left-hand example, a sill intrudes between a limestone layer and a sandstone. The sill incorporates fragments (inclusions) of limestone and sandstone, so it must be younger than the inclusions. In the right-hand example, the igneous rock is a lava flow that existed before the sandstone was deposited; the sandstone contains pebbles (inclusions) of lava. **(g)** Baked contacts: The intrusion of the pluton creates a metamorphic aureole (baked contact) in the surrounding rock, so the pluton must be younger. Note that the pluton also crosscuts bedding, confirming this interpretation. **(h)** Chilled margin: The basalt on the left is part of a dike that intruded the basalt on the right. The left basalt is younger because it contains a chilled margin (the darker, finer-grained rock).

Chilled Margin

Younger   Older

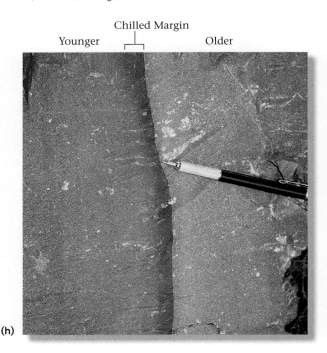

(h)

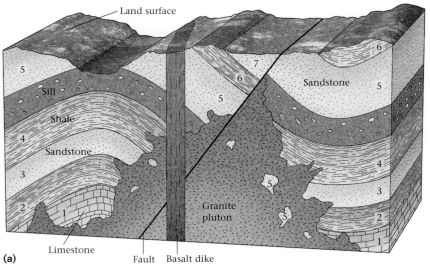

Land surface

5

Sill

Shale

4

Sandstone

3

2

1

Limestone

Fault   Basalt dike

6

7

6

5

Sandstone   5

4

3

5   2

5   1

Granite
pluton

(a)

FIGURE 12.5 (a) Geologic principles allow us to interpret the sequence of events leading to the development of the features shown here. Beds 1–7 were deposited first. Intrusion of the sill came next, followed by folding, intrusion of the granite, faulting, intrusion of the dike, and erosion to yield the present land surface. **(b–h)** The sequence of geologic events leading to the geology shown in **(a)**.

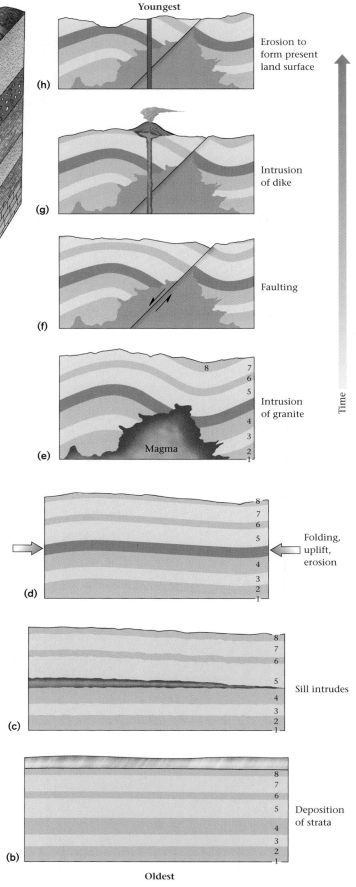

Youngest

(h) Erosion to form present land surface

(g) Intrusion of dike

(f) Faulting

(e) Intrusion of granite

Magma

8   7
6
5
4
3
2
1

(d) Folding, uplift, erosion

8
7
6
5
4
3
2
1

(c) Sill intrudes

8
7
6
5
4
3
2
1

(b) Deposition of strata

8
7
6
5
4
3
2
1

Oldest

Time

FIGURE 12.6 A bedding surface containing many fossils.

1 cm

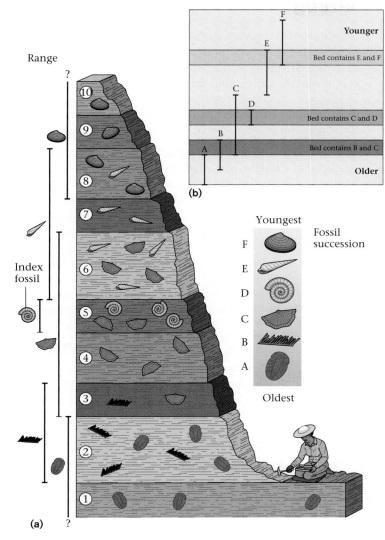

FIGURE 12.7 (a) The principle of fossil succession. Note that each species has only a limited range in a succession of strata, and ranges of different fossils may overlap. Widespread fossils with a short range are index fossils. (b) Overlapping fossil ranges can be used to limit the relative age of a given bed and to determine the relative ages of beds. For example, a bed containing fossils E and F must be younger than a bed containing B and C. Note that a bed containing C alone could be older than or younger than a bed containing D alone.

Take-Home Message

Relative-age determination is based on geologic principles, including: uniformitarianism (the present is the key to the past), superposition (younger strata overlie older strata), cross-cutting relations (younger features cut older ones), and fossil succession (fossil species occur in a predictable order).

Because of the principle of fossil succession, we can define the relative ages of strata by looking at fossils. For example, if we find a bed containing fossil A, we can say that the bed is older than a bed containing, say, fossil F (▶Fig. 12.7b). Geologists have now identified and determined the relative ages of about 250,000 fossil species.

12.5 UNCONFORMITIES: GAPS IN THE RECORD

James Hutton often strolled along the coast of Scotland because the shore cliffs provided good exposures of rock, stripped of soil and shrubbery. He was particularly puzzled by an outcrop along the shore at Siccar Point. One sequence of rock exposed there consisted of alternating beds of gray sandstone and shale, whereas another consisted of red sandstone and conglomerate (▶Fig. 12.8a, b). The beds of gray sandstone and shale were nearly vertical, but the beds of red sandstone and conglomerate were

FIGURE 12.8 (a) The Siccar Point unconformity in Scotland, on the coast about 60 km east of Edinburgh. (b) A geologic interpretation of the unconformity.

(a)

(b)   What a geologist sees

nearly horizontal. Further, the red layers seemed to lie across the truncated ends of the vertical gray layers, like a handkerchief lying across a row of books. We can imagine that as Hutton examined this odd geometric relationship, the tide came in and deposited a new layer of sand on top of the rocky shore. With the principle of uniformitarianism in mind, Hutton suddenly realized the significance of what he saw. Clearly, the gray sandstone–shale sequence had been deposited, then tilted, and then truncated by erosion before deposition of the red sandstone–conglomerate beds. At some time in the past, the gray sandstone–shale sequence had formed a substrate, and the red sandstone and conglomerate had been deposited above.

Hutton deduced that the surface between the gray and red rock sequences represented a long interval of time during which new strata had not been deposited and older strata had been eroded away. We now call a surface, representing a period of nondeposition and possibly erosion, an **unconformity**. The interval of time between deposition of the youngest rock below an unconformity and deposition of the oldest rock above is called a hiatus. Essentially, an unconformity forms wherever the land surface does not receive and accumulate sediment. Geologists recognize three kinds of unconformity:

- *Angular unconformity:* Rocks below an angular unconformity were tilted or folded before the unconformity developed (▶Fig. 12.9a). Thus, an angular unconformity cuts across the underlying layers; the layers below have a different orientation from the layers above. (We can see an angular unconformity in the outcrop at Siccar Point.) Angular unconformities form where rocks were either folded or tilted by faulting before being uplifted and eroded.

- *Nonconformity:* A nonconformity is a type of unconformity at which sedimentary rocks overlie intrusive igneous rocks and/or metamorphic rocks (▶Fig. 12.9b). The photo in Figure 7.2 (Chapter 7) illustrates a nonconformity. The igneous or metamorphic rocks must have cooled, been uplifted, and been exposed by erosion to form the substrate on which new sedimentary rocks were deposited. At a nonconformity, you typically find pebbles of the igneous or metamorphic rock in the lowest bed of the sedimentary sequence.

- *Disconformity:* Imagine that a sequence of sedimentary beds has been deposited beneath a shallow sea. Then sea level drops, and the recently deposited beds become exposed for some time. During this time, no new sediment accumulates, and some of the preexisting sediment gets eroded away. Later, sea level rises, and a new sequence of sediment accumulates over the old. The boundary between the two sequences is a disconformity (▶Fig. 12.9c). Even though the beds above and below the disconformity are parallel, the contact between them represents an

interruption in deposition. Disconformities may be hard to recognize unless you notice evidence of erosion (such as stream channels) or of a paleosol (an ancient soil horizon, formed at the ground surface but then later buried by younger strata and preserved in the rock record) at the disconformity surface, or can identify a distinct gap in the succession of fossils (▶Fig. 12.9d).

The succession of strata at a particular location provides a record of Earth history there (▶see art, pp. 426–427). But because of unconformities, the record preserved in the rock layers is incomplete (▶Fig. 12.10a, b). It's as if geologic history is being chronicled by a tape recorder that turns on only intermittently—when it's on (times of deposition), the rock record accumulates, but when it's off (times of nondeposition and possibly erosion), an unconformity develops. Because of unconformities, no single location on Earth contains a complete record of Earth history.

> **Take-Home Message**
>
> At a given location, sediments do not accumulate continuously. Surfaces representing intervals of nondeposition and possible erosion are unconformities. Because of unconformities, the geologic record at any given location is incomplete.

12.6 STRATIGRAPHIC FORMATIONS AND THEIR CORRELATION

Geologists summarize information about the sequence of strata at a location by drawing a **stratigraphic column** (Fig. 12.10a). Typically, we draw columns to scale, so that the relative thicknesses of layers portrayed on the column represent the thicknesses of layers in the outcrop. Then, for ease of reference, geologists divide the sequence of strata represented on a column into **stratigraphic formations** (formations for short), recognizable intervals of a specific rock type or group of rock types deposited during a specific time interval, that can be traced over a fairly broad region. The boundary surface between two formations is a type of geologic **contact**.

Let's see how the concept of a stratigraphic formation applies to the Grand Canyon. The walls of the canyon look striped, because they expose a variety of rock types that differ in color and in resistance to erosion. Geologists identify major contrasts and use them as a basis to divide the strata into formations, each of which may consist of many beds (▶Fig. 12.11). Note that some formations include a single rock type, while others include interlayered beds of two or more rock types. Also, note that not all formations have the same thickness. Typically, geologists name a formation after a locality where it was first identified. If the formation consists of only one rock type, we

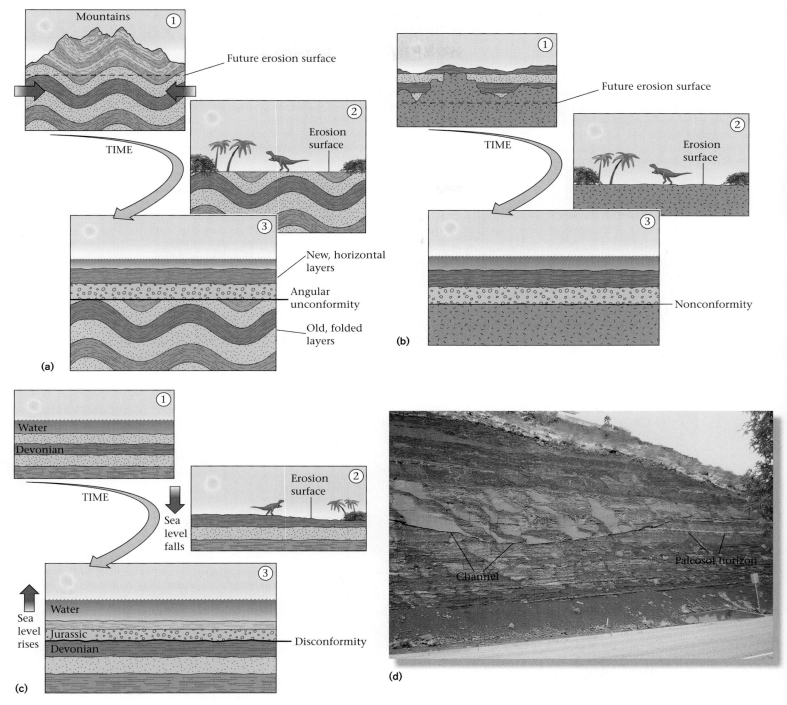

FIGURE 12.9 **(a)** The stages during the development of an angular unconformity: (1) mountains form and layers are folded; (2) erosion removes the mountains, creating an erosion surface; (3) sea level rises and new horizontal layers of sediment are deposited. **(b)** The stages during the development of a nonconformity: (1) a pluton intrudes sedimentary rocks; (2) erosion removes all the sedimentary layers and cuts down into the crystalline rock, forming an erosion surface; (3) sea level rises and new sedimentary layers are deposited above the erosion surface. **(c)** The stages during the development of a disconformity: (1) layers of sediment are deposited; (2) sea level drops and an erosion surface forms; (3) sea level rises and new sedimentary layers accumulate. Note that regardless of the details, an unconformity represents a surface of erosion and/or a period of nondeposition. **(d)** A road cut in Utah shows two disconformities; an older one marked by a paleosol (the greenish horizon) and a younger one marked by a channel.

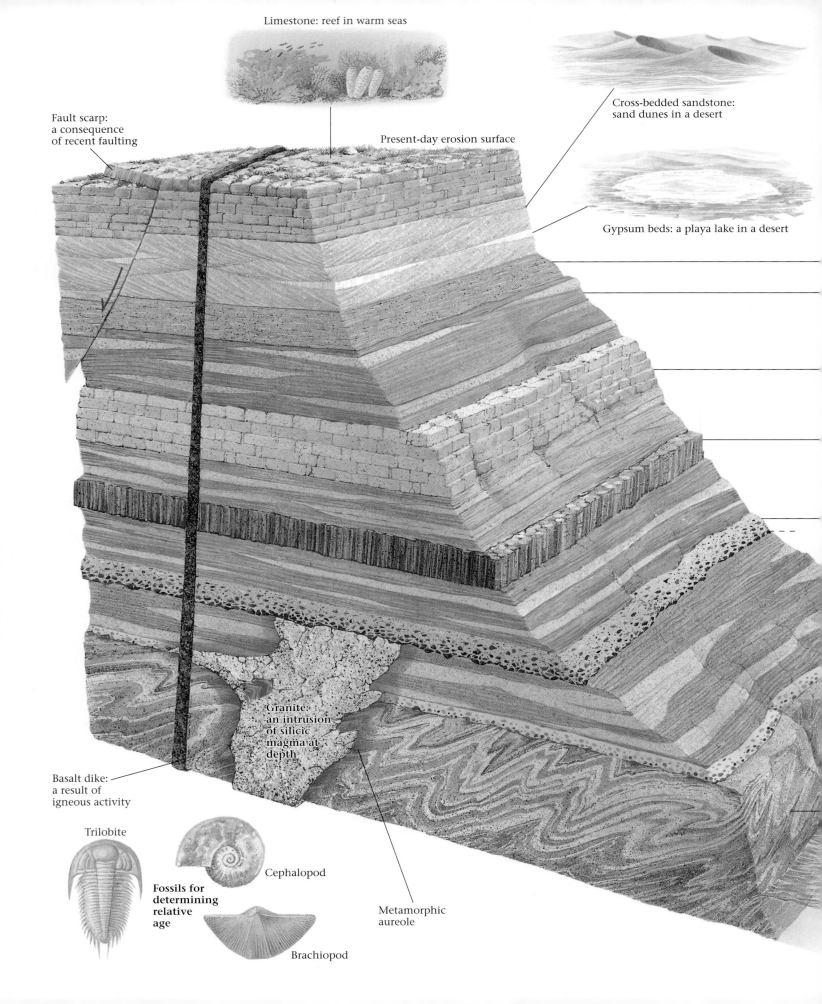

Limestone: reef in warm seas

Fault scarp: a consequence of recent faulting

Present-day erosion surface

Cross-bedded sandstone: sand dunes in a desert

Gypsum beds: a playa lake in a desert

Granite: an intrusion of silicic magma at depth

Basalt dike: a result of igneous activity

Trilobite

Cephalopod

Fossils for determining relative age

Metamorphic aureole

Brachiopod

The Record in Rocks: Reconstructing Geologic History

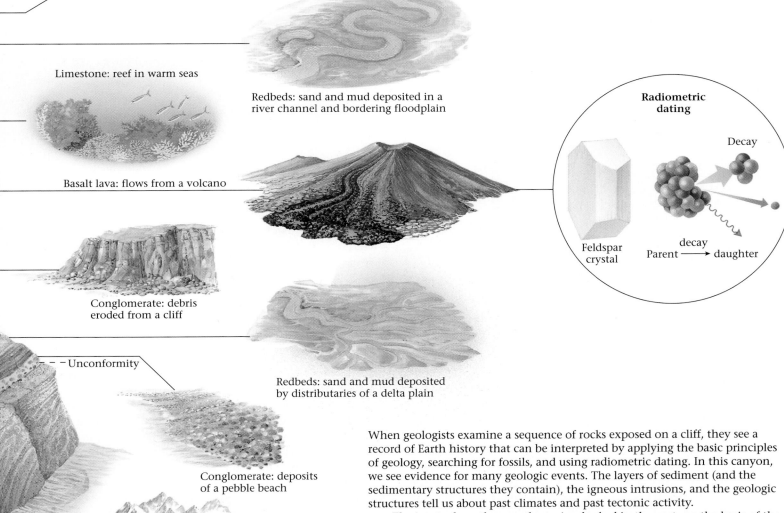

Ignimbrite (welded tuff): an explosive volcanic eruption

Limestone: reef in warm seas

Redbeds: sand and mud deposited in a river channel and bordering floodplain

Basalt lava: flows from a volcano

Radiometric dating

Decay

Feldspar crystal

decay
Parent → daughter

Conglomerate: debris eroded from a cliff

Redbeds: sand and mud deposited by distributaries of a delta plain

Unconformity

Conglomerate: deposits of a pebble beach

Gneiss: metamorphism at depth beneath a mountain belt

When geologists examine a sequence of rocks exposed on a cliff, they see a record of Earth history that can be interpreted by applying the basic principles of geology, searching for fossils, and using radiometric dating. In this canyon, we see evidence for many geologic events. The layers of sediment (and the sedimentary structures they contain), the igneous intrusions, and the geologic structures tell us about past climates and past tectonic activity.

The insets show the way the region looked in the past, on the basis of the record in the rocks. For example, the presence of gneiss at the base of the canyon indicates that at one time the region was a mountain belt, for the protoliths of the gneiss were buried deeply. Unconformities indicate that the region underwent uplift and erosion. Sedimentary successions record transgressions and regressions of the sea, igneous rocks are evidence of volcanic and intrusive activity, and faults indicate deformation.

Clearly, the land surface portrayed in this painting was sometimes a river floodplain or a delta (indicated by redbeds), sometimes a shallow sea (limestone), and sometimes a desert dune field (cross-bedded sandstones). And at several times in the past, volcanic activity occurred in the region. We can gain insight into the age of the sedimentary rocks by studying the fossils they contain, and the age of the igneous and metamorphic rocks by using radiometric dating methods.

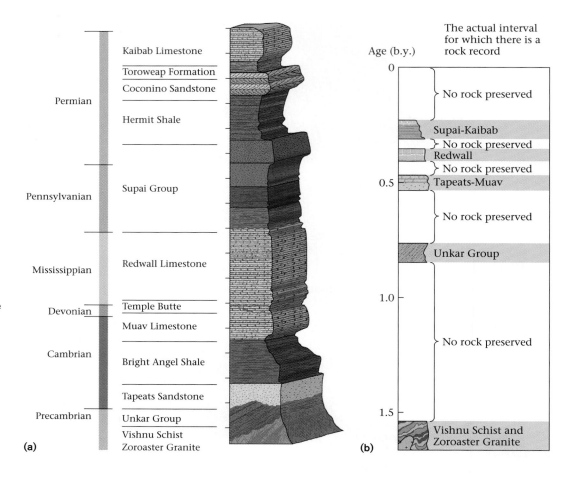

FIGURE 12.10 **(a)** The sequence of strata in the Grand Canyon (shown beneath the arrow in Fig. 12.11a) can be represented on a stratigraphic column. The vertical scale gives relative thicknesses. The right-hand edge of the column represents resistance to erosion (e.g., Coconino Sandstone is more resistant than Hermit Shale). **(b)** Because of unconformities, the stack of strata exposed in the Grand Canyon represents only bits and pieces of geologic history. If the strata are projected on a numerical time scale, you can see that large intervals of time are not accounted for.

Permian — Kaibab Limestone / Toroweap Formation / Coconino Sandstone / Hermit Shale

Pennsylvanian — Supai Group

Mississippian — Redwall Limestone

Devonian — Temple Butte

Cambrian — Muav Limestone / Bright Angel Shale / Tapeats Sandstone

Precambrian — Unkar Group / Vishnu Schist / Zoroaster Granite

(a)

The actual interval for which there is a rock record

Age (b.y.)

0 — No rock preserved — Supai-Kaibab — No rock preserved — Redwall — No rock preserved

0.5 — Tapeats-Muav — No rock preserved — Unkar Group

1.0 — No rock preserved

1.5 — Vishnu Schist and Zoroaster Granite

(b)

may incorporate that rock type in the name (e.g., Kaibab Limestone), but if the formation contains more than one rock type, we use the word formation in the name (e.g., Toroweap Formation; note that both words are capitalized). Several related formations in a succession may be lumped together as a group.

While he was excavating canals in England, William Smith discovered that formations cropping out at one locality resembled formations cropping out at another, in that their beds looked similar and contained similar fossil assemblages. In other words, Smith was able to define the age relationship between the strata at one locality and the strata at another, a process called **correlation**.

How does correlation work? Typically, geologists correlate formations between *nearby* regions based on similarities in rock type. We call this method of correlation lithologic correlation (▶Fig. 12.12a–c). For example, the sequence of strata on the southern rim of the Grand Canyon clearly correlates with the sequence on the northern rim, because they contain the same rock types in the same order. In some cases, a sequence contains a key bed, or marker bed, a particularly unique bed that

provides a definitive basis for correlation. But to correlate units over *broad* areas, we must rely on fossils to define the relative ages of sedimentary units. We call this method fossil correlation. Geologists must rely on fossil correlation for studies of broad areas because sources of sediments and depositional environments may change from one location to another. The beds deposited at one location during a given time interval may look quite different from the beds deposited at another location during the same time interval.

Now let's look at an example of fossil correlation by tracing the individual formations exposed in the Grand Canyon into the mountains just north of Las Vegas, 150 km to the west (▶Fig. 12.13a, b). Near Las Vegas, we find a sequence of sedimentary rocks that includes a limestone formation called the Monte Cristo Limestone. The Monte Cristo Limestone contains fossils of the same age as occur in the Redwall Limestone of the Grand Canyon, but it is much thicker. Because the formations contain fossils of the same age, we conclude that they were deposited during the same time interval, and thus we say that they correlate with one another. Note also that not

Location of cross section

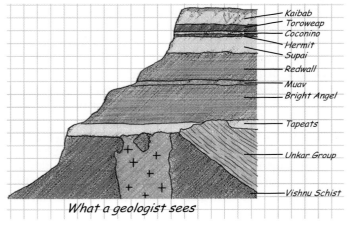

What a geologist sees

Kaibab
Toroweap
Coconino
Hermit
Supai
Redwall
Muav
Bright Angel
Tapeats
Unkar Group
Vishnu Schist

FIGURE 12.11 The succession of rocks in the Grand Canyon can be divided into formations, based on notable changes in rock type and changes in fossil assemblages. The photo shows the ridge represented by the cross section.

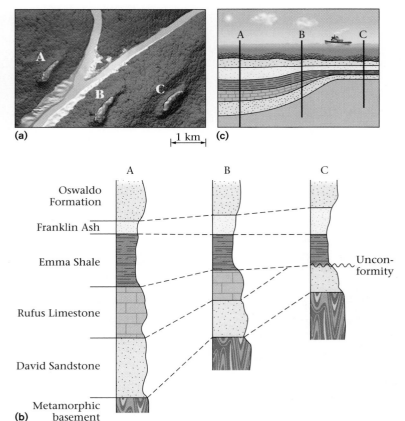

(a)

|← 1 km →|

(c)

A   B   C

Oswaldo Formation

Franklin Ash

Emma Shale

Rufus Limestone

David Sandstone

Metamorphic basement

(b)

Unconformity

FIGURE 12.12 The principle of lithologic correlation. (a) This map shows three outcrops of rock are a few kilometers apart. (b) The stratigraphic sections at each location are somewhat different, but the columns can be correlated with one another by matching rock types. Note that the Franklin Ash is a key bed, a distinctive layer that can be correlated with certainty. And the Rufus Limestone "pinches out" (thins and disappears along its length), so the contact between the Emma Shale and the David Sandstone in column C is an unconformity. (c) Geologists reconstruct a sedimentary basin using correlation. The eastward thinning of sedimentary layers suggests that the basin tapered to the east, as shown by this cross section.

only are the units thicker in the Las Vegas area than in the Grand Canyon area, but there are more of them. This discovery indicates that during part of the time when thick sediments were deposited near Las Vegas, none accumulated near the Grand Canyon. Thus, the contact beneath the Grand Canyon's Redwall Limestone is an unconformity.

The contrast between the stratigraphic columns of Las Vegas and the Grand Canyon represents a contrast in depositional setting. The Las Vegas column accumulated on the thinner crust of a passive-margin basin that sank (subsided) rapidly and remained submerged below the sea almost continuously. The Grand Canyon column, on the other hand, accumulated on the thicker crust of a craton that periodically emerged above sea level.

By correlating strata at many locations, William Smith realized that he could trace individual formations of strata over fairly broad regions. In 1815, he plotted the distribution of formations and created the first modern **geologic map**, which portrays the spatial distribution of rock units at the Earth's surface (▶Fig. 12.14a–c). We can use the information provided by a geologic map to identify geologic structures. Appendix B provides a geologic map of part of North America.

Take-Home Message

A recognizable sequence of beds that can be mapped across a broad region is called a stratigraphic formation. The boundary surface between two formations is a type of contact. Geologists correlate formations regionally on the basis of rock type and fossil content.

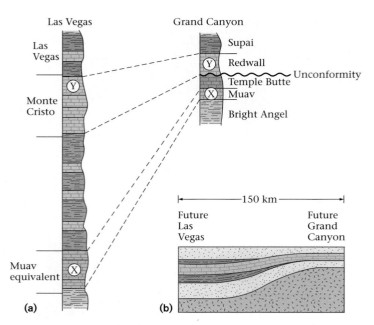

FIGURE 12.13 The principle of fossil correlation: **(a)** Because they both contain "P-age" fossils, we can say that the Redwall Limestone of the Grand Canyon correlates with the Monte Cristo Limestone near Las Vegas. But the fossils in the Muav Limestone, which lies directly beneath the Redwall at the Grand Canyon, are "N-age" fossils and thus correlate with those of a unit 1,000 m below the Monte Cristo Limestone. The sequence of strata between the two limestones at Las Vegas is not represented by any rock at the Grand Canyon; thus, the Redwall-Muav contact is an unconformity. **(b)** A sedimentary basin thinned radically between Las Vegas and the Grand Canyon. Thus, the sequence of strata deposited at the Grand Canyon was thinner and less complete than the section deposited at Las Vegas.

12.7 THE GEOLOGIC COLUMN

As stated earlier, no one locality on Earth provides a complete record of our planet's history, because stratigraphic columns can contain unconformities. But by correlating rocks from locality to locality at millions of places around the world, geologists have pieced together a *composite* stratigraphic column, called the **geologic column**, that represents the entirety of Earth history (▶Fig. 12.15a, b). The column is divided into segments, each of which represents a specific interval of time. The largest subdivisions break Earth history into the Hadean (not shown in Figure 12.15), Archean, Proterozoic, and Phanerozoic **Eons**. (The first three together constitute the **Precambrian**.) The suffix "-zoic" means "life," so Phanerozoic means "visible life," and Proterozoic means "earlier life." The earliest life, bacteria and archaea, appeared during the Archean Eon. In the Phanerozoic Eon, organisms with hard parts (shells and, later, skeletons) became widespread, so there are abundant fossils from this eon, whereas in Precambrian time, only small organisms with no shells existed, so Precambrian fossils are rare and hard to find.

The Phanerozoic Eon is subdivided into **eras**. In order from oldest to youngest, they are the Paleozoic ("ancient life"), Mesozoic ("middle life"), and Cenozoic ("recent life") eras. We can further divide each era into **periods** and each period into **epochs**.

Where do the names of the periods come from? They refer either to localities where a fairly complete stratigraphic column representing that time interval was identified (e.g., rocks representing the Devonian Period crop out near Devon, England) or to a characteristic of the time (rocks from the Carboniferous Period contain a lot of coal), or they come from Latin roots. The terminology was not set up in a planned fashion that would make it easy to learn. Instead, it grew haphazardly in the years between 1760 and 1845, as geologists began to refine their understanding of geologic history and fossil succession.

Note that the way geologists divide time into named intervals resembles the way historians divide human history. For example, intervals of Chinese history have names such as Han dynasty and Chin dynasty, while intervals of British history are called Elizabethan, Victorian, or Edwardian, each named after the reigning monarch of the time. Geologists know that Jurassic rocks are younger than Permian rocks, just as interior designers know that Victorian furniture is younger than Elizabethan furniture.

The succession of fossils in the geologic column defines the course of life's evolution throughout Earth history (▶Fig. 12.16). In fact, geologists now give the age range of fossil species in terms of the periods and epochs of the geologic column, and can use fossils alone to determine the relative age of a sedimentary rock. Simple bacteria appeared in the Archean Eon, whereas complex shell-less invertebrates did not evolve until the late Proterozoic. The appearance of invertebrates with shells defines the Precambrian-Cambrian boundary. At this time there was a sudden diversification (development of many different species) in life, with many new genera appearing over a relatively short interval—this event is called the **Cambrian explosion**. The first vertebrates, fish, appeared in the Ordovician Period. Before the Silurian Period, the land surface was barren of multicellular life; in the Silurian, land plants spread over the landscape, followed in the Devonian by the first amphibians. Though reptiles evolved during the Pennsylvanian Period, the first dinosaurs did not pound across the land until the Triassic. Dinosaurs continued to inhabit the Earth until their sudden extinction at the end of the Cretaceous Period. For this reason, we refer to the Mesozoic Era as the Age of Dinosaurs. Small mammals also appeared in the Triassic Period, but the diversification of mammals to fill a wide range of ecological niches did not happen until the beginning of the Cenozoic Era, so we now call the Cenozoic the Age of Mammals. Birds also appeared

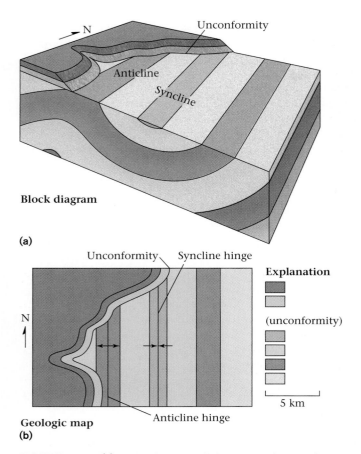

Block diagram

(a)

Geologic map
(b)

FIGURE 12.14 **(a)** The block diagram depicts an angular unconformity with horizontal strata above and folded strata below. **(b)** The geologic map shows what this ground surface would look like if viewed from above. On the map, the contacts have been projected onto a flat piece of paper. **(c)** A simplified geologic map of California.

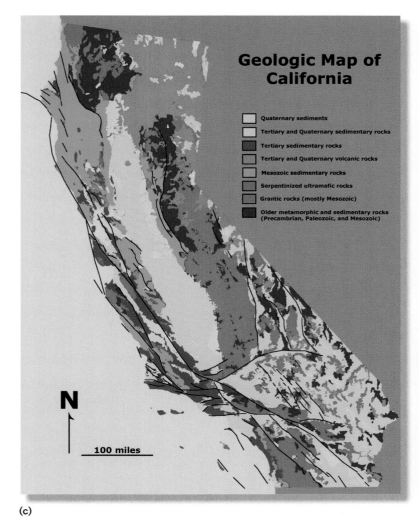

(c)

during the Mesozoic (specifically, at the beginning of the Cretaceous Period), but they underwent great diversification in the Cenozoic Era.

To conclude our discussion, let's examine a case study and see how the geologic column comes into play when correlating strata across a region. We return to the Colorado Plateau of Arizona and Utah, in the southwestern United States (▶Fig. 12.17 a–g); because of the lack of vegetation in the region, you can easily see bedrock exposures on the walls of cliffs and canyons. Some of these exposures are so beautiful that they have become national parks. Geologists can specify the relative ages of the units by using the names of periods in the geological column. The oldest sedimentary rock of the region crops out at the base of the Grand Canyon. This rock, the Unkar Group, rests unconformably on schist and granite, does not contain shelly fossils, and underlies Cambrian sandstone. Thus, the Unkar Group is Precambrian, as is the schist and granite below it. The walls of the Grand Canyon expose strata up

through the Permian age—the light-colored rock that crops out at the top of the canyon is the Permian Kaibab Limestone. Rocks younger than the Kaibab no longer remain at the Grand Canyon, due to erosion, so to see younger strata, we must travel to Zion Canyon. Here, many layers of red sandstone and shale overlie the Kaibab Limestone. Fossils indicate that these strata were deposited in Triassic and Jurassic times. To see what lay above the sandstone strata at Zion, we must move further west to Bryce Canyon or the cliffs of Cedar Breaks. The oldest strata exposed here correlate with the youngest strata at Zion. But at Bryce and Cedar Breaks, these rocks are overlain with Jurassic, Cretaceous, and Tertiary strata. Note that by using the names of the geologic column, geologists can efficiently define relative-age relationships among the strata of the different parks. Also, as spectacular as the exposures of the Colorado Plateau may be, they contain only a partial record of geologic time, for the sequence includes many unconformities.

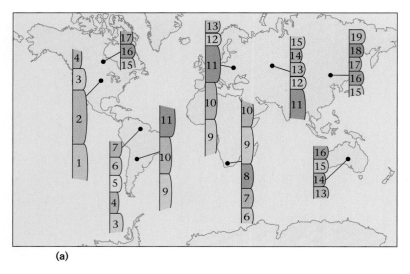

(a)

FIGURE 12.15 (a) The geologic column was constructed by determining the relative ages of stratigraphic columns from around the world. Each of these little columns represents the stratigraphy at a given location. **(b)** By correlation, the strata in the columns can be stacked in a sequence representing most of geologic time. This is the geologic column. Note that the column was first built without knowledge of numerical ages, so even though we depict the sequence of periods, we cannot indicate their relative duration here.

	Eon	Era	Period	Epoch
19			Quaternary	Holocene
18				Pleistocene
17				Pliocene
16		Cenozoic		Miocene
15			Tertiary	Oligocene
14				Eocene
13				Paleocene
12	Phanerozoic			
11			Cretaceous	
10		Mesozoic	Jurassic	
9			Triassic	
8			Permian	
7			Carboniferous (Pennsylvanian, Mississippian)	
6		Paleozoic	Devonian	
5			Silurian	
4			Ordovician	
3			Cambrian	
2	Precambrian	Proterozoic		
1		Archean		

(b) Geologic Column

Walking through these parks is thus like walking through time. Each rock layer gives an indication of the climate and topography of the past. Like the pages of a book, the layers of rock contain a record of Earth's history. For example, when the Precambrian metamorphic and igneous rocks exposed in the inner gorge of the Grand Canyon formed, the region was a high mountain range, perhaps as dramatic as the Himalayas today. When the fossiliferous beds of the Kaibab Limestone at the rim of the canyon accumulated, the region was a Bahama-like carbonate reef and platform, bathed in a warm, shallow sea. And when the rocks making up the towering red cliffs of the Navajo Sandstone in Zion Canyon were deposited, the region was a Sahara-like desert, blanketed with huge sand dunes.

> **Take-Home Message**
>
> Correlation of stratigraphic sequences from around the world allowed production of a chart, the geologic column, that represents the entirety of Earth history. The column, developed using only relative age relations, is subdivided into eons, periods, and epochs.

12.8 HOW DO WE DETERMINE NUMERICAL AGE? THE RADIOMETRIC CLOCK

By saying that World War II began after World War I, a historian is specifying the relative ages of the two cataclysms. By saying that World War II began in 1939 and that World War I began in 1914, a historian not only indicates relative ages but also places events on a scaled time line—calibrated in years—of human history. When we specify the date of an event in years, we are providing the event's numerical age.

It's fairly straightforward to determine the numerical age of a historical event that happened subsequent to the invention of writing. Historians consult books and journals keyed to a calendar. Most of the Earth's geologic history, however, occurred before the dawn of civilization, so geologists have no written record from which to determine numerical ages of rocks. In fact, the development of a method for determining numerical ages of rocks has

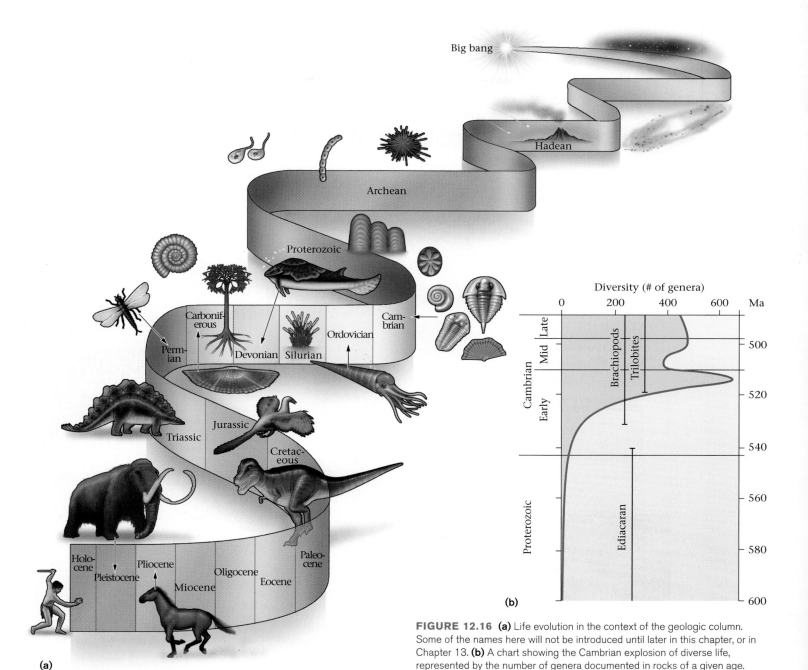

FIGURE 12.16 (a) Life evolution in the context of the geologic column. Some of the names here will not be introduced until later in this chapter, or in Chapter 13. **(b)** A chart showing the Cambrian explosion of diverse life, represented by the number of genera documented in rocks of a given age.

proven to be elusive. It was not until the 1950s that geologists developed techniques that provide insight into the numerical ages of rocks. These techniques are based on the measurement of radioactive elements in rocks, so the science of dating geological materials in years is now called **radiometric dating**. In this section, we introduce radiometric dating, or geochronology. To set the stage for this discussion, we must begin by describing the process of radioactive decay.

Isotopes, Radioactive Decay and the Concept of a Half-Life

All atoms of a given element have the same number of protons in their nucleus—we call this number the atomic number (see Appendix A). However, not all atoms have the same number of neutrons in their nucleus. Therefore, not all atoms of a given element have the same atomic weight—the approximate number of protons plus neutrons. Different

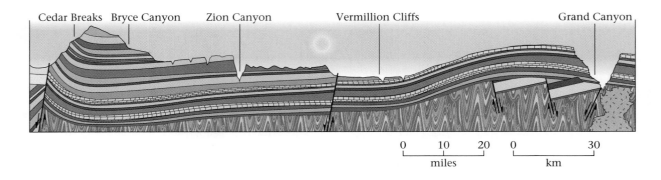

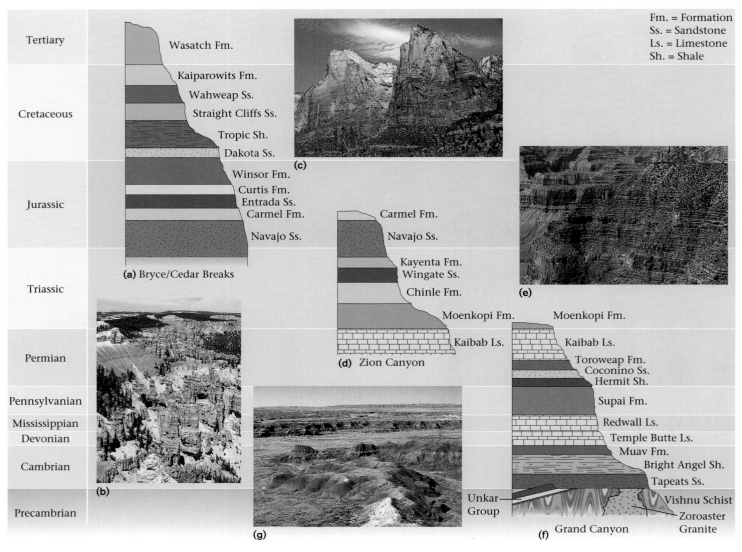

FIGURE 12.17 The correlation of strata among the various national parks of Arizona and Utah: **(a, b)** Bryce Canyon/Cedar Breaks, **(c, d)** Zion Canyon, **(e, f)** the Grand Canyon, **(g)** the Painted Desert. The inset at the top shows a cross section of the region.

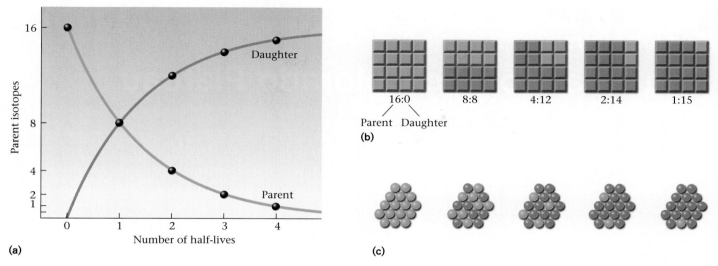

(a)

(b)

Parent   Daughter

(c)

FIGURE 12.18 The concept of a half-life. **(a)** Graph showing the decrease in the number of parent isotopes as time passes; the curve is exponential, meaning that the rate of change in the parent-daughter ratio decreases with time. **(b)** The ratio of parent-to-daughter isotopes changes with the passage of each successive half-life. **(c)** A cluster of isotopes undergoing decay. There's no way to predict which parent will decay in a given time interval.

versions of an element, called isotopes of the element, have the same atomic number but different atomic weights. For example, all uranium atoms have 92 protons, but the uranium-238 isotope (abbreviated ^{238}U) has an atomic weight of 238 and thus has 146 neutrons, whereas the ^{235}U isotope has an atomic weight of 235 and thus has 143 neutrons.

Some isotopes of an element are stable, meaning that they last essentially forever. Radioactive isotopes are unstable: after a given time, they undergo a change called **radioactive decay**, which converts them into a different element. Radioactive decay can take place by a variety of reactions.[1] All these reactions change the atomic number of the nucleus and thus form a different element. In these reactions, the isotope that undergoes decay is the **parent isotope,** while the decay product is the **daughter isotope**. For example, rubidium-87 (^{87}Rb) decays to strontium-87 (^{87}Sr), potassium-40 (^{40}K) decays to argon-40 (^{40}Ar), and uranium-238 (^{238}U) decays to form lead-206 (^{206}Pb).

Physicists cannot specify how long an individual radioactive isotope will survive before it decays, but they can measure how long it takes for half of a group of isotopes to

decay. This time is called the **half-life**. ▶Figure 12.18 can help you visualize the concept of a half-life. Imagine a crystal containing 16 radioactive parent isotopes. (In real crystals, the number of atoms would be much larger.) After one half-life, 8 isotopes have decayed, so the crystal now contains 8 parent and 8 daughter isotopes. After a second half-life, 4 of the remaining parent isotopes have decayed, so the crystal contains 4 parent and 12 daughter isotopes. And after a third half-life, 2 more parent isotopes have decayed, so the crystal contains 2 parent and 14 daughter isotopes. Note again that we cannot predict which specific isotopes decay at which time, only that during a half-life, half the parent isotopes decay to form daughter isotopes. For a given decay reaction, the half life is a constant.

Radiometric Dating Technique

Like the ticktock of a clock, radioactive decay proceeds at a known rate and thus provides a basis for telling time. In other words, because an element's half-life is a constant, we can calculate the age of a mineral by measuring the ratio of parent to daughter isotopes in the mineral.

How do geologists actually obtain a radiometric date (or radiometric age)? First, we must find the right kind of elements to work with. Although there are many different pairs of parent and daughter isotopes among the known elements, only a few have long enough half-lives and occur in sufficient abundance to be useful for radiometric dating.

---

[1] Radioactive reactions include: alpha decay (ejection of two protons and two neutrons from the nucleus); beta decay (transformation of a neutron of the nucleus into a proton by ejection of an electron); and electron capture (joining of a proton with an electron to form a new neutron). Some decay reactions take several steps. Specifically, an original parent may decay to form another, different radioactive isotope that, in turn, decays to form a stable daughter.

See for yourself...
The Strata of the Colorado Plateau

The Colorado Plateau encompasses portions of Arizona, Utah, New Mexico, and Colorado. Uplift of the region during the past 20 million years raised the land surface to elevations of 2 to 3 km. Erosion then cut down into the plateau to reveal portions of a 2 km (1.2 miles)-thick sequence of strata. These rocks, which are well exposed because of the plateau's present dryness, tell a story of changing climates and changing sea levels over the past half billion years. *Google Earth*™ allows you to fly through the canyons and along the cliffs of this magical landscape to see the record in the rocks for yourself. The thumbnail images provided on this page are only to help identify tour sites. Go to wwnorton.com/studyspace to experience this flyover tour.

Grand Canyon National Park, Arizona
(Lat 36°5'36.05"N, Long 112°7'0.88"W)

These coordinates take you to Plateau Point, on the edge of the inner gorge. You can reach the Point by trail from Park Headquarters on the south rim of the canyon. Zoom to an elevation of 4 km (2.5 miles), then tilt your view so you are looking due north (Image G12.1). You can see a stratigraphic section from Precambrian basement, through the Paleozoic to the Kaibab Limestone at the top. Try to spot the formations shown on Figure 12.17. Now, fly to the river, pivot your view, and proceed downstream (Image G12.2). You'll be able to see continuous exposure of the unconformity at the base of the Tapeats Sandstone. Note that the narrow inner gorge cuts down into hard Precambrian gneiss—the wider outer gorge cuts into sedimentary rock.

G12.1

G12.2

Vermillion Cliffs, Arizona
(Lat 36°49'4.81"N, Long 111°37'56.59"W)

Highway 89 crosses Marble Canyon at the coordinates provided. Just upstream of the bridge, rafters enter the Colorado River for a journey downstream into the Grand Canyon. Zoom to 5 km (3 miles), tilt your view, and look SSW. You can see the gash of Marble Canyon (Image G12.3) bounded by cliffs of Permian Kaibab Limestone. In the right foreground, you can see small hills surrounded by subtle rings that trace out beds of horizontal strata from the Lower Triassic Moenkopi Formation. Rotate your view to look NW and you'll see the steep escarpment of the Vermillion Cliffs, which expose a complete section of the Moenkopi, a unit

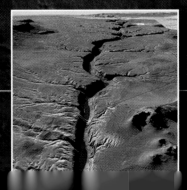

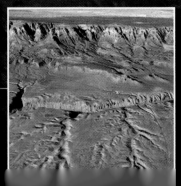

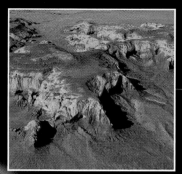

G12.5

Monument Valley, Arizona
(Lat 36°55'25.00"N, Long 110°7'8.98"W)

The towering cliffs at these coordinates, along the northern border of Arizona, expose brilliant red strata deposited at the end of the Permian and beginning of the Triassic. Tilt your view and pivot to look due west (Image G12.5). Weak beds (the Organ Rock Shale) comprise a moderate slope of stairstep-like ledges at the base of the cliff. The vertical cliff consists of massive de Chelly Sandstone, and the thin-bedded unit at top is Moenkopi Shale, capped locally by beds of the Chinle Formation. Note that the Permian strata here differ from the same-age strata of Marble Canyon—this difference demonstrates that the environment of deposition can change with location at a given time.

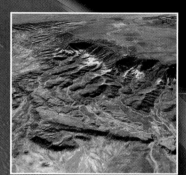

G12.6

Painted Desert, Arizona
(Lat 35°5'8.13"N, 109°47'14.78"W)

This portion of the Colorado Plateau exposes a colorful sequence of Late Triassic fluvial deposits, including a variety of varicolored sandstones and shales. Fly to the coordinates given, zoom to an elevation of 3 km (1.9 miles), tilt your view, and pivot to see the gentle escarpment (Image G12.6). The sequence of strata was deposited in a terrestrial setting. Numerous soil horizons formed as the strata were exposed prior to each new phase of deposition. The colors represent the paleosols—ancient soil horizons—preserved in the stratigraphic record.

G12.7

G12.8

Upper Zion Canyon Park, Utah
(Lat 37°13'34.59"N, Long 112°53'12.03"W)

Fly to these coordinates and zoom to an elevation of 10 km (6 miles). You're looking down on Route 9, traversing the upper part of Zion Park (Image G12.7). Note that the erosional pattern is strongly controlled by N-S-trending joints. Now zoom down to 3.5 km (2 miles), then tilt and pivot your view so you are looking west along Route 9 (Image G12.8). From this vantage, you see a thick sequence of cross-bedded Jurassic sandstone recording a time when much of the Colorado Plateau region was a Sahara-like desert covered by large sand dunes.

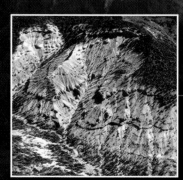

G12.9

Cedar Breaks, Utah
(Lat 37°40'3.94"N, Long 112°52'7.89"W)

At Cedar Breaks National Monument, you're at the west edge of the Colorado Plateau. A view to the west looks out over the Basin and Range Rift of western Utah. For our purpose, fly to an elevation of 2 km (1.2 miles) at the coordinates given. Tilt and pivot your view so you are gazing east, and you will see the multi-colored Tertiary strata eroded along an escarpment (Image G12.9). Note that the section includes beds of resistant sandstone, standing out as ridges. But much of the sequence consists of weak shale, the deposits of lakes and floodplains. You would see similar rocks at Bryce Canyon, about 40 km (25 miles) to the west, but as of this writing, the imagery available for Bryce is very poor.

Particularly useful elements are listed in ▶Table 12.1. Each radioactive element has its own half-life. Note that carbon-dating is *not* used for dating rocks (▶Box 12.1).

Second, we must identify the right kind of minerals to work with. Not all minerals contain radioactive elements, but fortunately some common minerals do. For example, feldspar, mica, and hornblende contain potassium and rubidium, zircon contains uranium, and garnet contains samarium. Once we have identified appropriate minerals containing appropriate elements, we can set to work. Radiometric dating consists of the following steps.

- *Collecting the rocks:* Geologists collect fresh (unweathered) rocks for dating. The chemical reactions that happen during weathering may allow isotopes to leak out of minerals, in which case a date from the rock has no valid meaning.

- *Separating the minerals:* The fresh rocks are crushed, and the appropriate minerals are separated from the debris.

- *Extracting parent and daughter isotopes:* To separate out the parent and daughter isotopes from minerals, geologists either dissolve the minerals in acid or evaporate portions of them with a laser. This work must take place in a very clean lab, to avoid contaminating samples with parent or daughter isotopes from the atmosphere (▶Fig. 12.19a).

- *Analyzing the parent-daughter ratio:* Geologists pass the dissolved or evaporated atoms through a mass spectrometer, a complex instrument that uses a magnet to separate isotopes from one another according to their respective weight, and then measures the ratio of parent to daughter isotopes (▶Fig. 12.19b).

At the end of the laboratory process, geologists can define the ratio of parent to daughter isotopes in a mineral, and from this ratio, calculate the age of the mineral. Needless to say, the description of the procedure here has been simplified—in reality, obtaining a radiometric date is time consuming and expensive and requires complex calculations. When they report radiometric dates, geologists report the uncertainty of the measurement. Uncertainty, which defines the range of values in which the true measurement probably lies, arises because no instrument can count atoms perfectly. Uncertainties for radiometric dates may be on the order of 1% or less. For example, a date may be reported as 200 ± (plus or minus) 2 million years. Newer methods of dating produce results with uncertainties as small as 0.1%.

FIGURE 12.19 (a) A lab where samples are prepared for radiometric dating. The air must be exceedingly clean so that stray parent or daughter isotopes don't contaminate the samples. **(b)** The heart of this mass spectrograph, used for measuring isotope ratios, is a large magnet (the yellow coils).

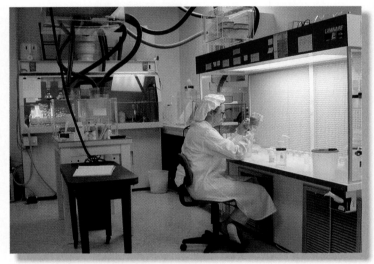

(a)

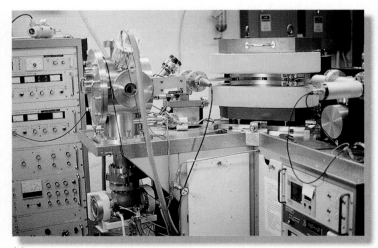

(b)

TABLE 12.1   Isotopes Used in the Radiometric Dating of Rocks

Parent → Daughter	Half-Life (years)	Minerals in which the Isotopes Occur
^{147}Sm → ^{143}Nd	106 billion	Garnets, micas
^{87}Rb → ^{87}Sr	48.8 billion	Potassium-bearing minerals (mica, feldspar, hornblende)
^{238}U → ^{206}Pb	4.5 billion	Uranium-bearing minerals (zircon, apatite, uraninite)
^{40}K → ^{40}Ar	1.3 billion	Potassium-bearing minerals (mica, feldspar, hornblende)
^{235}U → ^{207}Pb	713 million	Uranium-bearing minerals (zircon, uraninite, apatite)

Sm = samarium, Nd = neodymium, Rb = rubidium, Sr = strontium, U = uranium, Pb = lead, K = potassium, Ar = argon

BOX 12.1 THE REST OF THE STORY

Carbon-14 Dating

Many people who have heard of **carbon-14** (**^{14}C**) **dating** assume that it can be used to define the numerical age of rocks. But this is not the case. Rather, ^{14}C dating tells us the ages of *organic materials*—such as wood, cotton fibers, charcoal, flesh, bones, and shells—that contain carbon originally extracted from the atmosphere by photosynthesis in plants. ^{14}C, a radioactive isotope of carbon, forms naturally in the atmosphere when cosmic rays (charged particles from space) bombard atmospheric nitrogen-14 (^{14}N) atoms. When plants consume carbon dioxide during photosynthesis, or when animals consume plants, they ingest a tiny amount of ^{14}C along with ^{12}C, the more common isotope of carbon. After an organism dies and can no longer exchange carbon with the atmosphere, the ^{14}C in its body begins to decay back to ^{14}N. Thus, the ratio of ^{14}C to ^{12}C changes at a rate determined by the half-life of ^{14}C.

We can use ^{14}C dating to determine the age of prehistoric fire pits or of organic debris in sediment. ^{14}C has a short half-life—only 5,730 years. Thus, the method cannot be used to date anything older than about 70,000 years, for after that time essentially no ^{14}C remains in the material. But this range makes it a useful tool for geologists studying sediments of the last ice age and for archaeologists studying ancient cultures or prehistoric peoples. Again, since rocks do not contain organic carbon, and may be significantly older than 70,000 years, we cannot determine the age of rocks by using the ^{14}C dating method.

What Does a Radiometric Date Mean?

At high temperatures, isotopes in a crystal lattice vibrate so rapidly that chemical bonds can break and reattach relatively easily. As a consequence, parent and daughter isotopes escape from or move into crystals, so parent-daughter ratios are meaningless. Because radiometric dating is based on the parent-daughter ratio, the "radiometric clock" starts only when crystals become cool enough for both parent and daughter isotopes to be locked into the lattice. The temperature below which isotopes are no longer free to move is called the **closure temperature** of a mineral. The closure temperature is typically significantly cooler than the melting temperature of a mineral. Not all minerals have the same closure temperature; for example, the closure temperature of hornblende (an amphibole) is higher than that of biotite (a mica). When we specify a radiometric date for a rock, we are defining the time at which a specific mineral in the rock cooled below its closure temperature.

With the concept of closure temperature in mind, we can interpret the meaning of radiometric dates. In the case of igneous rocks, radiometric dating tells you when a magma or lava cooled to form a solid, cool igneous rock. In the case of metamorphic rocks, a radiometric date tells you when a rock cooled from the high temperature of metamorphism down to a low temperature. If a rock cools quickly (as when a lava flow freezes), then all minerals yield roughly the same age, but if a rock cools slowly (as when a pluton cools slowly at depth in the Earth), minerals with high closure temperatures give older ages than minerals with low closure temperatures.

Can we radiometrically date a sedimentary rock directly? No. If we date the minerals in a sedimentary rock, we determine only when the minerals making up the sedimentary rock first crystallized as part of an igneous or metamorphic rock, not when the minerals were deposited as sediment or when the sediment lithified to form a sedimentary rock. For example, if we date the feldspar grains contained in a granite pebble in a conglomerate, we're dating the time when the granite cooled below feldspar's closure temperature, not when the pebble was deposited by a stream. The age of mineral grains in sediment, however, can be useful. In recent years, geologists have undertaken studies to determine the ages of detrital (clastic) grains; by doing so, they can learn the age of the rocks in the region where the sediment originated.

Other Methods of Determining Numerical Age

Counting rings in trees or layers in sediment. The changes in seasons affect a wide variety of phenomena, including the following.

- *The growth rate of trees:* Trees grow seasonally, with rapid growth during the spring and no growth during the winter.
- *The organic productivity of lakes and seas:* During the winter, less light reaches the Earth, water temperatures decrease, and the supply of nutrients decreases.
- *The sediment supply carried by rivers:* During the rainy season, the volume and velocity of water in rivers increase, and thus rivers carry more sediment.
- *The growth rate of chemically precipitated sedimentary rocks (such as travertine):* Factors that control precipitation

rates, such as the rate of groundwater flow and temperature, change seasonally in some locations.

- *The growth rate of shell-secreting organisms:* Organisms grow and produce new shell material on a seasonal basis.
- *The layering in glaciers:* During the snowy season, more snow falls relative to dust than during the dry season; annual snowfall includes a dusty layer and a clean layer. These contrasts are preserved as layers in sediment.

As a consequence of these seasonal changes, **growth rings** develop in trees, travertine deposits, and shelly organisms, and **rhythmic layering** develops in sedimentary accumulations and glacier ice (▶Fig. 12.20a–c). By counting rings or layers, geologists can determine how long an organism survived, or how long a sedimentary accumulation took to form. If rings or layers have developed right up to the pres-

ent, we can count backward and determine how long ago they began to form.

Dendrochronologists, scientists who study and date tree rings, have found that not only do tree rings count time, they also preserve a record of the climate: in warm, rainy years, trees grow much faster than they do in drought years. Thus, by studying tree rings, dendrochronologists can provide a dated record of the climate in prehistoric times. The oldest living trees, bristlecone pines, are almost 4,000 years old. By correlating the older rings of these trees with rings in logs preserved in sediment accumulations, dendrochronologists have extended the tree-ring record back for many thousands of years (▶Fig. 12.20d).

Similarly, geologists have found that glacier ice preserves a valuable record of past climate, for the ratio of different oxygen isotopes in the water molecules making up the ice reflects the global temperature at the time the snow fell to create the ice. Ice cores drilled through the

(a)

(b)

(c)

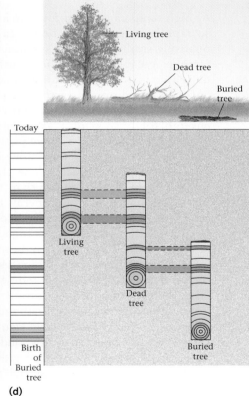

(d)

FIGURE 12.20 (a) A clamshell with growth rings; **(b)** tree rings; **(c)** rhythmic layers in the ice of a glacier. **(d)** Dendrochronology is based on the correlation of tree rings. Each of the columns in the diagram represents a core drilled out of a tree. Distinctive clusters of closely spaced rings indicate dry seasons. By correlation, researchers extend the climate record back in time before the oldest living tree started to grow.

Greenland ice cap contain a continuous record back through 750,000 years.

Magnetostratigraphy. As was discussed in Chapter 3, the polarity of Earth's magnetic field flips every now and then through geological time. Geologists have determined when reversals took place, and have constructed a reference column showing the succession of reversals through time (▶Fig. 12.21). This reference column resembles the bar code on a supermarket product. By comparing the pattern of the reversals in a sequence of strata with the bar code of the reference column, a study known as **magnetostratigraphy**, geologists can determine the age of the sequence.

Fission tracks. In certain minerals, the ejection of an atomic particle during the decay of a radioactive isotope damages the nearby crystal lattice, creating a line called a **fission track**. This track resembles the line of crushed grass left behind when you roll a tire across a lawn. As time passes, more atoms undergo fission, so the number of fission tracks in the crystal increases (▶Fig. 12.22a, b). Therefore, the number of fission tracks in a given volume of a crystal represents the age of the crystal. Geologists have been able to measure the rate at which fission tracks are produced, and thus can determine the age of a mineral grain by counting the fission tracks within it.

> **Take-Home Message**
>
> Radiometric dating provides numerical ages (in years). To obtain a radiometric date, geologists measure the ratio of parent radioactive isotopes to stable daughter products. The date gives the time at which a mineral cooled below its closure temperature.

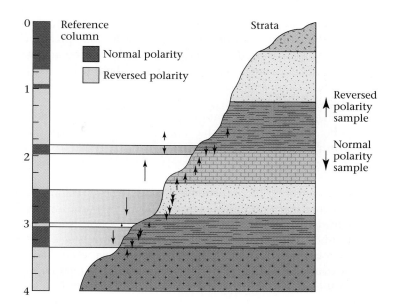

FIGURE 12.21 Magnetostratigraphy involves comparing the sequence of polarity reversals in strata with the sequence of polarity reversals in a global reference column to determine the age of the strata.

FIGURE 12.22 (a) Photomicrograph of fission tracks in a crystal. Though only a few atoms wide, the tracks are visible because the sample has been treated with a solvent that preferentially dissolves the crystal along the tracks. **(b)** How a fission track is formed. The red atom disrupts the crystal structure when it hits.

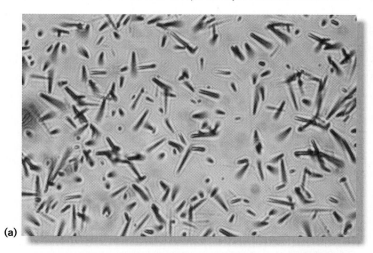

(a)

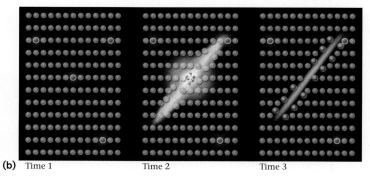

(b) Time 1   Time 2   Time 3

12.9 HOW DO WE ADD NUMERICAL AGES TO THE GEOLOGIC COLUMN?

We have seen that radiometric dating can be used to date the time when igneous rocks formed and when metamorphic rocks metamorphosed, but not when sedimentary rocks were deposited. So how do we determine the numerical age of a sedimentary rock? We must answer this question if we want to add numerical ages to the geologic column. Remember, the column was originally constructed by studying only the *relative* ages of fossil-bearing sedimentary rocks.

Geologists obtain dates for sedimentary rocks by studying cross-cutting relationships between sedimentary rocks and datable igneous or metamorphic rocks. For example, if we find a sedimentary rock layer deposited unconformably on a datable igneous or metamorphic rock, we know that the sedimentary rock must be younger. If we find a datable igneous dike or pluton that cuts across beds of sedimentary rock, then the datable rock must be younger. And if a datable lava flow or ash layer spread out over a layer of sediment and then was buried by another layer of sediment, then the datable rock or ash must be younger than the underlying sediment and older than the overlying sediment.

Let's look at an example of dating sedimentary rocks (►Fig. 12.23). Imagine that a layer of sandstone contains fossilized dinosaur bones. Geologists assign the fossils to the Cretaceous Period by correlating them with fossils in known Cretaceous strata from elsewhere. The sandstone was deposited unconformably over an eroded granite pluton that has a numerical age, determined by uranium-lead dating, of 125 million years. This same layer of sandstone has been cut by a basalt dike whose numerical age, based on potassium-argon dating, is 80 million years. These measurements mean that this Cretaceous sandstone bed was deposited sometime between 125 million and 80 million years ago. Note that the data provide an age range, not an exact age. Thus, we can only conclude that the Cretaceous Period includes the time interval of *at least* 125 to 80 million years ago.

Geologists have searched the world for localities where cross-cutting relations between datable igneous rocks and sedimentary rocks can be recognized. By radiometrically dating the igneous rocks, they have been able to provide numerical ages for the boundaries between all the geological periods. For example, work from around the world shows that the Cretaceous Period actually began about 145 million years ago and ended 65 million years ago. So the bed in Figure 12.23 was deposited during the middle part of the Cretaceous, not at the beginning or end.

The discovery of new data requires changing the numbers defining the boundaries of periods, which is why the term *numerical age* is preferred to *absolute age*. In fact, around 1995, new research showed that the Cambrian-Precambrian boundary occurred at about 542 million years ago, in contrast to previous, less definitive studies that had placed the boundary at 570 million years ago. ►Figure 12.24 shows the currently favored numerical ages of periods and eras in the geologic column. This dated column is commonly called the **geologic time scale**. Because of the numerical constraints provided by the geologic time scale, when geologists say that the first dinosaurs appeared during the Triassic Period, they mean that dinosaurs appeared between 251 and 200 million years ago.

> **Take-Home Message**
>
> Sedimentary rocks cannot be dated directly. So attaching numerical ages to the periods of the geologic column required geologists to radiometrically date igneous or metamorphic rocks in known cross-cutting relations with fossiliferous sedimentary rocks.

12.10 WHAT IS THE AGE OF THE EARTH?

The mind grows giddy gazing so far back into the abyss of time.

—John Playfair (1747–1819), British geologist who popularized the works of Hutton

Determining Earth's Numerical Age

In the eighteenth and nineteenth centuries, before the discovery of radiometric dating, scientists came up with a great variety of clever solutions to the question "How old is the Earth?"—all of which have since been proven wrong. One approach assumed that sediments have been deposited at a constant rate of about 1 mm per year. Dividing the thickest known pile of sediments (about 20 km) by this number yields an age of about 20 million years. But this approach doesn't take unconformities into account, and it neglects the fact that some sediments eventually transform into metamorphic rocks. Another approach assumed that the world's oceans started out as freshwater that then slowly accumulated salt. By measuring the rate at which rivers bring the ions making salt into the sea, we arrive at an age of about 90 million years. It turns out, however, that the salt content of the sea remains fairly constant over time, because excess

FIGURE 12.23 The Cretaceous sandstone bed was deposited unconformably on a 125-million-year-old granite pluton, so it must be younger than the granite. The 80-million-year-old dike cuts across the sandstone, so the sandstone must be older than the dike. Thus, this Cretaceous sandstone bed must be between 125 and 80 million years old. Similarly, the Paleocene bed was unconformably deposited over the dike and lies beneath a 50-million-year-old ash. Thus, the Paleocene bed must be between 80 and 50 million years old. Note that the data give only an age range.

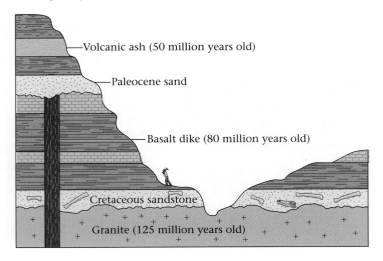

Volcanic ash (50 million years old)

Paleocene sand

Basalt dike (80 million years old)

Cretaceous sandstone

Granite (125 million years old)

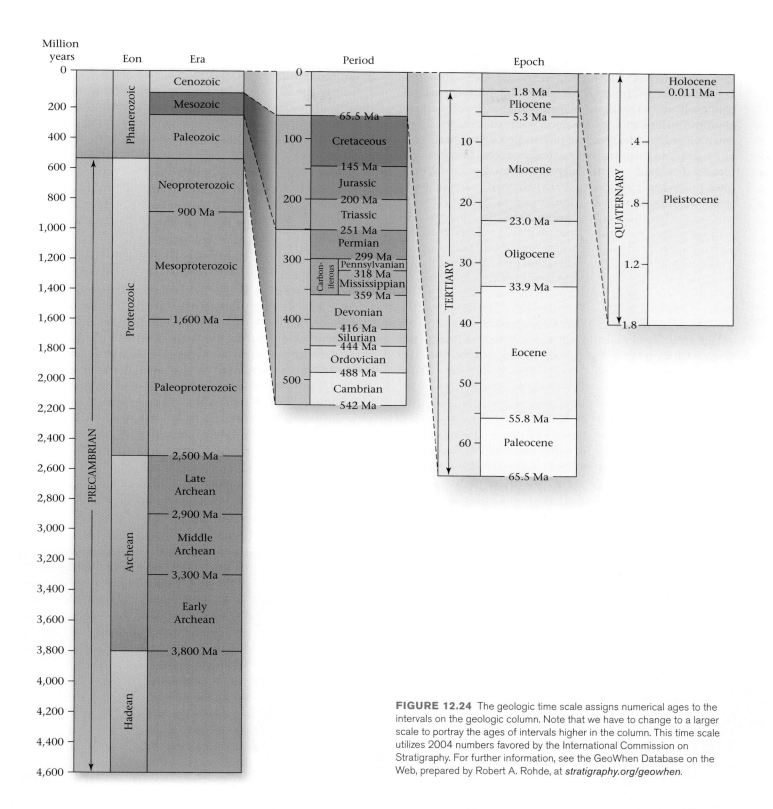

FIGURE 12.24 The geologic time scale assigns numerical ages to the intervals on the geologic column. Note that we have to change to a larger scale to portray the ages of intervals higher in the column. This time scale utilizes 2004 numbers favored by the International Commission on Stratigraphy. For further information, see the GeoWhen Database on the Web, prepared by Robert A. Rohde, at *stratigraphy.org/geowhen*.

salt precipitates out of seawater to form evaporite minerals; so salt concentration is not a measure of the Earth's age. Lord William Kelvin, a nineteenth-century physicist renowned for his discoveries in thermodynamics, made the most influential scientific estimate of the Earth's age of his time. Kelvin calculated how long it would take for the Earth to cool down from a temperature as hot as the Sun's, and concluded that the Earth is about 20 million years old.

In sum, estimates of the Earth's age proposed before the twentieth century implied that the planet was not much more than 100 million years old. This view contrasted with the idea being promoted by followers of Hutton, Lyell, and Darwin, who argued that if the concepts of uniformitarianism and evolution were correct, the Earth must be much older, because physical processes that shape the Earth and form its rocks and the process of natural selection that yields the diversity of species take a long time. Geologists and physicists debated the age issue for many years. The route to a solution came in 1896, when the physicist Henri Becquerel announced the discovery of radioactivity. Geologists immediately realized that the Earth's interior was producing heat from the decay of radioactive material. This realization uncovered the flaw in Kelvin's argument: Kelvin had assumed that no heat was added to the planet after it first formed. Because radioactivity constantly generates new heat in the Earth, the planet has cooled down much more slowly than Kelvin had calculated. In 1904, the British physicist Ernest Rutherford presented this discovery to an audience that included Kelvin, as he later recounted:

> I came into the room, which was half dark, and presently spotted Lord Kelvin in the audience and realized that I was in trouble at the last part of the speech dealing with the age of the Earth, where my views conflicted with his. To my relief, Kelvin fell fast asleep, but as I came to the important point, I saw the old bird sit up, open an eye and cock a baleful glance at me!

The discovery of radioactivity not only invalidated Kelvin's estimate of the Earth's age, it also led to the development of radiometric dating.

Since the 1950s, geologists have scoured the planet to identify its oldest rocks. Samples from several localities (Wyoming, Canada, Greenland, and China) have yielded dates as old as 3.96 billion years. And sandstones found in Australia contain clastic grains of zircon that yielded dates of 4.1 to 4.2 billion years, indicating that rock as old as 4.2 billion years did once exist (▶Fig. 12.25a). Models of the Earth's formation assume that all objects in the Solar System developed at roughly the same time from the same nebula. Radiometric dating of meteors and Moon rocks have yielded ages as old as 4.57 billion years (▶Fig. 12.25b); geologists take this to be the approximate age of the Earth, leaving more than enough time for the rocks and life forms of the Earth to have formed and evolved.

We don't find 4.57 billion-year-old (Ga) rocks in the crust because during the first half-billion years of Earth history, rocks in the crust remained too hot for the radiometric clock to start (their temperature stayed above the closure temperature), and/or crust that has formed before about 4.0 Ga was destroyed by an intense meteorite bombardment at about 4.0 Ga. Geologists have named the time interval between the birth of the Earth and the formation of the oldest dated rock the Hadean Eon (see Fig. 12.24).

FIGURE 12.25 (a) One of the oldest dated rocks on Earth, this 3.96-billion-year-old gneiss comes from the Northwest Territories of Canada. **(b)** A Moon rock.

(a)

(b)

Picturing Geologic Time

The number 4.57 billion is so staggeringly large that we can't begin to comprehend it. If you lined up this many pennies in a row, they would make an 87,400-km-long line that would wrap around the Earth's equator more than twice (▶Fig. 12.26). At the scale of our penny chain, human history is only about 100 city blocks long.

Another way to grasp the immensity of geologic time is to make a scale model, which we do by equating the entire 4.57 billion years to a single calendar year. On this scale, the oldest rocks preserved on Earth date from early February, and the first bacteria appear in the ocean on February 21. The first shelly invertebrates appear on October 25, and the first amphibians crawl out onto land on November 20. On December 7, the continents coalesce into the supercontinent of Pangaea. The first mammals and birds appear about December 15, along with the dinosaurs, and the Age of Dinosaurs ends on December 25. The last week of December represents the last 65 million years of Earth history, covering the entire Age of Mammals. The first human-like ancestor appears on December 31 (New Year's Eve) at 3 P.M., and our species, *Homo sapiens*, shows up only about an hour before midnight. The last ice age ends a minute before midnight, and all of recorded human history takes place in the last 30 seconds. Put another way, human history occupies the last 0.000001% of Earth history.

Take-Home Message

Because of heat-generating processes during early Earth history, the oldest rocks on Earth cooled below their closure temperature a half billion years after the Earth formed. Thus, the age of the Earth, 4.57 billion years, comes from dating meteorites.

FIGURE 12.26 We can use the analogy of distance to represent the duration of geologic time.

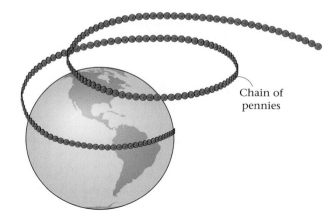

Chain of pennies

Chapter Summary

- The concept of geologic time, the span of time since the Earth's formation, developed when early geologists suggested that the Earth must be very old if geologic features formed by same natural processes we see today.

- Relative age specifies whether one geologic feature is older or younger than another; numerical age provides the age of a geologic feature in years.

- Using such principles as uniformitarianism, superposition, original horizontality, original continuity, cross-cutting relations, inclusions, and baked and chilled contacts, we can construct the geologic history of a region.

- The principle of fossil succession states that the assemblage of fossils in a sequence of strata changes from base to top. Once a fossil species becomes extinct, it never reappears.

- Strata are not deposited continuously at a location. An interval of nondeposition and/or erosion is called an unconformity. Geologists recognize three kinds: angular unconformity, nonconformity, and disconformity.

- A stratigraphic column shows the succession of formations in a region. A given succession of strata that can be traced over a fairly broad region is called a stratigraphic formation. The boundary surface between two formations is a contact. The process of determining the relationship between strata at one location and strata at another is called correlation. A geologic map shows the distribution of formations.

- A composite stratigraphic column that represents the entirety of geologic time is called the geologic column. The geologic column's largest subdivisions, each of which represents a specific interval of time, are eons. Eons are further subdivided into eras, eras into periods, and periods into epochs.

- The numerical age of rocks can be determined by radiometric dating. This is because radioactive elements decay at a constant rate. During radioactive decay, parent isotopes transform into daughter isotopes. The decay rate for a given element is known as its half-life, the time it takes for half of the parent isotopes to decay. The ratio of parent to daughter isotopes in a mineral grain indicates the mineral's age.

- The radiometric date of a mineral specifies the time at which the mineral cooled below a certain temperature. We can use radiometric dating to determine when an igneous rock solidified and when a metamorphic rock cooled from high temperatures. To date sedimentary strata, we must examine cross-cutting relations with dated igneous or metamorphic rock.

- Other methods for dating materials include counting rings in trees, layers in shells and glaciers, and fission tracks in mineral grains. We can also study the sequence of magnetic reversals in strata.

- From the radiometric dating of meteors and Moon rocks, geologists conclude that the Earth formed about 4.57 billion years ago. Our species, *Homo sapiens*, has been around for only 0.000001% of geologic time.

Geopuzzle Revisited

The age of the Earth comes from radiometric dating of meteorites thought to have formed at the same time as the Earth. This technique, based on measuring the ratio of parent radioactive elements to stable daughter products, tells us when minerals drop below a certain temperature. The Earth was too hot for the first half billion years of its existence for the radiometric clocks of crustal rocks to start ticking, so the oldest known rocks on Earth are only about 4 billion years old. Layers of strata record Earth's history. But as geologic environments change, a given locality that accumulates sediment during part of this history may be exposed and subjected to erosion later on. Thus, there is no single place on Earth where a sequence of strata records all of Earth history. But correlation of strata from around the world has let geologists produces a composite chart, the geologic column, that represents all of geologic time.

Key Terms

baked contacts (p. 419)
Cambrian explosion (p. 430)
carbon-14 (^{14}C) dating (p. 439)
chilled contacts (p. 419)
closure temperature (p. 439)
contact (p. 424)
correlation (p. 428)
cross-cutting relations
    (p. 418)
daughter isotope (p. 435)

eon (p. 430)
epoch (p. 430)
era (p. 430)
fission track (p. 441)
formation (stratigraphic
    formation) (p. 424)
fossil assemblage (p. 419)
fossil succession (p. 419)
geologic column (p. 430)
geologic history (p. 419)

geologic map (p. 429)
geologic time (p. 416)
geologic time scale (p. 442)
growth rings (p. 440)
half-life (p. 435)
inclusions (p. 418)
magnetostratigraphy
    (p. 441)
numerical age (absolute age)
    (p. 418)
original continuity (p. 418)
original horizonality
    (p. 418)
parent isotope (p. 435)
period (p. 430)

Precambrian (p. 430)
radioactive decay (p. 435)
radiometric dating
    (geochronology) (p. 433)
range (p. 419)
relative age (p. 418)
rhythmic layering (p. 440)
stratigraphic column
    (p. 424)
stratigraphic formations
    (p. 424)
superposition (p. 418)
unconformity (p. 424)
uniformitarianism (pp. 417,
    418)

Review Questions

1. Compare numerical age and relative age.

2. Describe the principles that allow us to determine the relative ages of geologic events.

3. How does the principle of fossil succession allow us to determine the relative ages of strata?

4. How does an unconformity develop?

5. Describe the differences among the three kinds of unconformities.

6. Describe two different methods of correlating rock units. How was correlation used to develop the geologic column? What is a stratigraphic formation?

7. What does the process of radioactive decay entail?

8. How do geologists obtain a radiometric date? What are some of the pitfalls in obtaining a reliable one?

9. Why can't we date sedimentary rocks directly?

10. Why is carbon-14 dating useful in archaeology, but useless for dating dinosaur fossils?

11. How are growth rings and ice cores useful in determining the ages of geologic events?

12. How are the reversals of the Earth's magnetic field useful in dating strata?

13. Why did early scientists think the Earth was less than 100 million years old?

14. How did the discovery of radioactivity invalidate Kelvin's assumptions about the Earth's age and also provide a method for obtaining its true age?

15. What is the age of the oldest rocks on Earth? What is the age of the oldest rocks known? Why is there a difference?

On Further Thought

1. Where on Earth might you find the most complete stratigraphic record of the past 100 million years? (*Hint*: Think about environments of deposition that have remained submerged for most of the past 100 million years). Would you expect to see a complete stratigraphic record of the past 100 million years in the central United States, or in central France? Explain your answer.

2. Imagine an outcrop exposing a succession of alternating sandstone and conglomerate beds. A geologist studying the outcrop notes the following:

   - The sandstone beds contain fragments of land plants, but the fragments are too small to permit identification of species.
   - A layer of volcanic ash overlies the sandstone bed. Radiometric dating indicates that this ash is 300 Ma.
   - A paleosol occurs at the base of the ash layer.
   - A basalt dike, dated at 100 Ma, cuts both the ash and the sandstone-conglomerate sequence.
   - Pebbles of granite in the conglomerate yield radiometric dates of 400 Ma.

   On the basis of these observations, how old is the sandstone conglomerate? (Specify both the numerical age range and the period or periods of the geologic column during which it formed.) If the igneous rocks were not present, could you still specify the maximum or minimum ages of the sedimentary beds? Explain.

3. Fossil correlation indicates that the beds below an unconformity were deposited at the end of the Carboniferous, and that the beds above the unconformity were deposited at the beginning of the Cretaceous. How much missing geologic time (in years) does the unconformity represent?

Suggested Reading

Berry, W. B. N. 1987. *Growth of a Prehistoric Time Scale*. 2nd ed. Palo Alto, Calif.: Blackwell Scientific Publications.

Brookfield, M.E. 2004. *Principles of Stratigraphy*. Palo Alto: Blackwell.

Burchfield, J. D. 1975. *Lord Kelvin and the Age of the Earth*. New York: Science History Publications.

Dalrymple, G. B. 1991. *The Age of the Earth*. Palo Alto, Calif.: Stanford University Press.

Dickin, A. P. 2005. *Radiogenic Isotope Geology*. 2nd ed. Cambridge: Cambridge University Press.

Faure, G., and T. Mensing. 2004. I*sotopes: Principles and Applications*. 3rd ed. New York: Wiley.

Gould, S. J. 1987. *Time's Arrow, Time's Cycle*. Cambridge, Mass.: Harvard University Press.

Gradstein, F. J., and A. Smith. 2005. *A Geologic Time Scale 2004*. Cambridge: Cambridge University Press.

Prothero, D. R., Jr., and R. H. Dott. 2003. *Evolution of the Earth*. 7th ed. New York: McGraw-Hill.

Repcheck, J. 2003. *The Man Who Found Time: James Hutton and the Discovery of the Earth's Antiquity*. Cambridge, Mass.: Perseus.

THE VIEW FROM SPACE The pages of Earth history stand on end in Namibia, southwestern Africa. Here, in a false-color image, the Ugab River cuts across layer upon layer of strata that were tilted to near-vertical by a Precambrian mountain-building event. Subsequent erosion exposed the strata, and the desert climate keeps it clear of vegetation. The field of view is 45 km.

A Biography of Earth

A group of geology students examining an outcrop near Baraboo, Wisconsin. Field observations allow geologists to piece together a history of the Earth.

I weigh my words well when I assert that the man who should know the true history of the bit of chalk which every carpenter carries about in his breeches pocket, though ignorant of all other history, is likely, if he will think his knowledge out to its ultimate results, to have a truer and therefore a better conception of this wonderful Universe and of man's relation to it than the most learned student who [has] deep-read the records of humanity [but is] ignorant of those of nature.

—*Thomas Henry Huxley, from*
On a Piece of Chalk *(1868)*

FIGURE 13.1 Horizontal chalk beds, created from layers of deep-sea sediment, exposed along the coast of England. The inset shows the Seven Sisters cliffs in Sussex, United Kingdom.

13.1 INTRODUCTION

In 1868, a well-known British scientist, Thomas Henry Huxley, presented a public lecture on geology to an audience in Norwich, England. Seeking a way to convey his fascination with Earth history to people with no previous geological knowledge, he focused his audience's attention on the piece of chalk he had been writing with (see epigraph above). And what a tale the chalk had to tell! Chalk, a type of limestone, consists of microscopic marine algae shells and shrimp feces. The specific chalk that Huxley held came from beds deposited in Cretaceous time (the name *Cretaceous*, in fact, derives from the Latin word for chalk). These beds are now exposed along the White Cliffs of Dover (▶Fig. 13.1). Geologists in Huxley's day knew of similar chalk beds in outcrops throughout much of Europe, and had discovered that the chalk contains not only plankton shells but also fossil species of swimming reptiles, fish, and invertebrates that do not inhabit the seas of today. Clearly, when the chalk was deposited, warm seas holding unfamiliar creatures covered much of what is now dry land.

Clues in his humble piece of chalk allowed Huxley to demonstrate to his audience that the geography and inhabitants of the Earth in the past differed markedly from those today, and thus that *the Earth has a history*. In the many decades since Huxley's lecture, field and laboratory studies worldwide have allowed geologists to develop an overall image of Earth's history. With every new geologic discovery, this image, once a blur, comes progressively more into focus. We now see a complex, evolving Earth System in which physical and biological components interact pervasively in ways that have transformed a formerly barren, crater-pocked surface into one of countless environments supporting a diversity of life. Fossil finds of the past two decades demonstrate that life began when the Earth was still quite young and that life has affected surface and near-surface processes ever since. The map of our planet constantly changes as continents rift, drift, and collide, and as ocean basins open and close.

We will discuss the nature of overall global change further in Chapter 23. In this chapter, we offer a concise geological biography of the planet, from its birth 4.57 billion years ago to the present. After learning the methods geologists use to interpret the past, we see how continents came into existence and have waltzed across the globe ever since. We also learn when mountain-building events took place and how Earth's climate and sea level have changed through time—in fact, the climate alternates from being relatively warm (greenhouse conditions) to being relatively cool (icehouse conditions). And while these physical changes have taken place, life has evolved. To simplify the discussion, we will use the following abbreviations: Ga (for billion years ago), Ma (million years ago), and Ka (thousand years ago).

13.2 METHODS FOR STUDYING THE PAST

When historians outline human history, they describe daily life, wars, economics, governments, leaders, inventions, and explorations. When geologists outline Earth history, we describe the distribution of depositional environments, mountain-building events (orogenies), past climates, life evolution, the changing positions of continents, the past configuration of plate boundaries, and changes in the composition of the atmosphere and oceans. Historians collect data by reading written accounts, examining relics and monuments, and, for more recent events, listening to recordings or watching videos. Geologists collect data by examining rocks, geologic structures, and fossils and, for more recent events, by studying sediments, ice cores, and tree rings.

Figuring out Earth's past hasn't been an easy task for geologists, because the available record isn't complete. The Earth materials that hold the record of the past don't form continuously through time, and erosion may destroy some materials. Also, it is no surprise that the record of early Earth history is vaguer than the record for more recent history, for older rocks are more likely to have been eroded away, and because the uncertainty attached to age measurement may become larger as the rocks get older. Nevertheless, we can find enough of the record at least to outline major geologic events of the past. Following are a few examples of how we use observational data to study Earth history.

- *Identifying ancient orogens:* We identify present-day orogens (mountain belts) by looking for regions of high, rugged topography. However, since it takes as little as 50 million years to erode the peaks of a mountain range entirely away, we cannot identify orogens of the past simply by studying topography. Rather, we must look for the rock record they leave behind. Orogeny causes igneous activity, deformation (folds, faults, and foliation), and metamorphism. Thus, we can recognize an ancient orogen by looking for a belt of crust containing these features (▶Fig. 13.2), and we can determine the ages of rocks formed during orogeny by using radiometric dating.

  Orogeny also leads to the development of unconformities, for uplift exposes rocks to erosion, and to subsidence of foreland sedimentary basins (basins located on the continent adjacent to the mountain front), for the weight of the mountain belt pushes crust down and produces a depression that traps sedi-

ment. This sediment provides a record of the erosion of the mountains.

- *Recognizing the growth of continents:* Not all continental crust formed at the same time. In order to determine how a continent grew, geologists find the ages of different regions of the crust by using modern radiometric dating techniques. They can figure out not only when rocks originally formed from magmas rising out of the mantle, but also when the rocks were metamorphosed during a subsequent orogeny. The identities of the rock types making up the crust indicate the tectonic environment in which the crust formed.

- *Recognizing past depositional environments:* The environment at a particular location changes through time. To learn about these changes, we study successions of sedimentary rocks, for the environment controls both the type of sediment deposited at a location and the type of organisms that live there.

- *Recognizing past changes in relative sea level:* We can determine whether sea level has gone up or down by looking for changes in the depositional environment. For example, occurrence of a marine limestone above an alluvial-fan conglomerate indicates a rise in sea level at that site.

- *Recognizing positions of continents in the past:* To help us find out where a continent was located in the past, we have three sources of information. First, the study of paleomagnetism can tell us the latitude of a continent in the past (see Chapter 3; ▶Fig. 13.3a, b). Second, we can study marine magnetic anomalies to reconstruct the change over time in the width of an ocean basin between continents (▶Fig. 13.3c). Third, we can compare rocks and/or fossils from different continents to see if there are correlations that could indicate that the continents were adjacent.

- *Recognizing past climates:* We can gain insight into past climates by looking at fossils and rock types that formed at given latitudes. For example, if organisms requiring semitropical conditions lived near the poles during a given time period, then the atmosphere overall

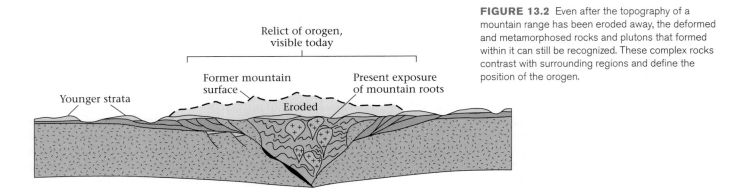

FIGURE 13.2 Even after the topography of a mountain range has been eroded away, the deformed and metamorphosed rocks and plutons that formed within it can still be recognized. These complex rocks contrast with surrounding regions and define the position of the orogen.

Relict of orogen, visible today

Former mountain surface

Present exposure of mountain roots

Younger strata

Eroded

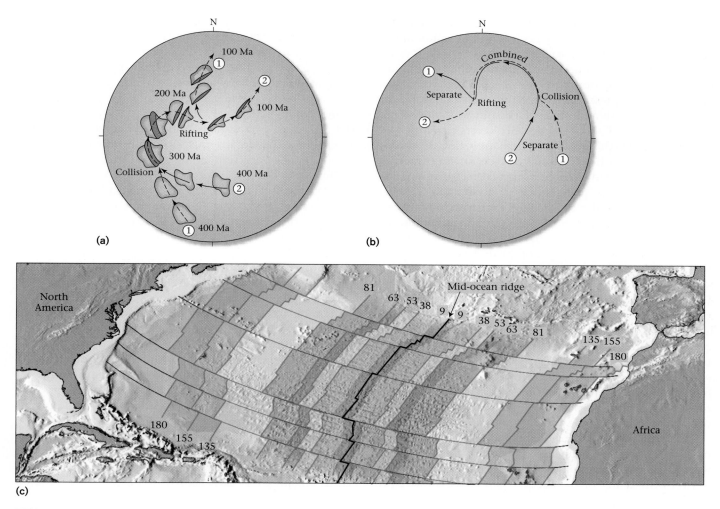

FIGURE 13.3 Geologists use apparent polar-wander paths as clues to the past positions of continents. **(a)** This hypothetical example shows the movement of two continents (1 and 2). At 400 Ma (million years ago), the continents are separate. They collide at 300 Ma and move together as a supercontinent until 200 Ma. Then they rift apart and drift away from each other. **(b)** The apparent polar-wander paths for the two continents. Note that the paths coincide when the continents move together as a supercontinent. **(c)** Geologists use marine magnetic anomalies to define relative motions of continents for the past 200 million years. This map shows the sea floor at various times in the past (in Ma) corresponding with specific marine magnetic anomalies. If you remove the strip of sea floor between the coast and the 81-Ma anomaly (or between the coast and the 63-Ma anomaly, etc.), you will see the relative positions of the continents at 81 Ma (or 63 Ma, etc.).

must have been warmer. Geologists have also learned how to use the ratios of isotopes for certain elements in fossil shells as a measure of past temperatures.

- *Recognizing life evolution:* Progressive changes in the assemblage of fossils in a sequence of strata represent changes in the assemblage of organisms inhabiting Earth through time, and thus indicate the occurrence of evolution.

Take-Home Message

Geologists use a variety of approaches to study Earth's past. Stratigraphic and paleomagnetic analyses, along with radiometric dating, for example, provide constraints on how climate, the position of continents, life, and sea level have changed over time.

13.3 THE HADEAN EON: HELL ON EARTH?

A viscid pitch boiled in the fosse below
and coated all the bank with gluey mire.
I saw the pitch; but I saw nothing in it
except the enormous bubbles of its boiling
which swelled and sank, like breathing, through all the pit.

–Dante, The Inferno, Canto XXI

James Hutton, the eighteenth-century Scottish geologist who was the first to provide convincing evidence that the Earth was vastly older than human civilization, could not measure Earth's age directly, and indeed speculated that there may be "no vestige of a beginning." But radiometric dating studies of recent decades have shown that it is possible, in fact, to assign a numerical age to Earth's formation. Specifically, dates obtained for a class of meteorites thought to be representative of the planetesimal cloud out of which the Earth and other planets formed consistently yield an age of 4.57 billion years. Geologists currently take this age to be Earth's birth date, perhaps within 30 million years of the time when the Sun's nuclear furnace first fired up. But a clear record of Earth history, as recorded in continental crustal rocks, does not begin until about 3.80 Ga, for crustal rocks older than 3.80 billion years are exceedingly rare. Geologists refer to the mysterious time interval between the birth of Earth and 3.80 Ga as the Hadean Eon (from the Greek *Hades*, the god of the underworld). A number of clues provide the basic framework of Hadean history.

The Hadean Eon began with the formation of the Earth by the accretion of planetesimals (see Chapter 1). As the planet grew, collection and compression of matter into a dense ball generated substantial heat. Each time another meteorite collided with the Earth, its kinetic energy added more heat, and radioactive decay produced still more heat within the newborn planet. During Earth's first 150 million years of existence, many short-lived radioactive isotopes still existed, so physicists estimate that radioactive decay produced 5 times as much heat then as it does today.

Eventually, the Earth became hot enough to partially melt, and when this happened, by about 4.5 Ga, it underwent **internal differentiation**. Gravity pulled molten iron down to the center of the Earth, where it accumulated to form the core. A mantle, composed of ultramafic rock, remained as a thick shell surrounding the core (see Chapter 2). Internal differentiation, which may have occurred relatively quickly, generated sufficient heat to make the Earth even hotter. Researchers suggest that soon after—or perhaps during—differentiation, a Mars-sized protoplanet collided with the Earth. The energy of this collision blasted away a significant fraction of Earth's mantle, which mixed with the shattered fragments of the colliding planet's mantle to form a ring of silicate-rock debris orbiting the Earth. Heat generated by the collision substantially melted Earth's remaining mantle, making it so weak that the iron core of the colliding body sank through it and merged with the Earth's core. Meanwhile, the ring of debris surrounding the Earth coalesced to form the Moon, which, when first formed, was less than 20,000 km away. (By comparison, the Moon is 384,000 km from Earth today.)

In the wake of differentiation and Moon formation, the Earth was so hot that its surface was likely an ocean of seething magma, supplied by intense eruption of melts rising from the mantle—it resembled Dante's Inferno (▶Fig. 13.4). Here and there,

rafts of solid rock formed temporarily on the surface of the magma ocean, but these eventually sank and remelted. This stage lasted at least until about 4.4 Ga. After that time, because of the decrease in radioactive heat generation, Earth might have become cool enough for solid rocks to form at its surface. The evidence for this statement comes from western Australia, where geologists have found 4.4 Ga grains of a durable mineral called zircon in sandstone beds. The zircons must originally have formed in igneous rock, but were eroded and later deposited in sedimentary beds.

During the hot early stages of the Hadean Eon, rapid "outgassing" of the Earth's mantle took place. This means that volatile elements or compounds originally incorporated in mantle minerals were released and erupted at the Earth's surface, along with lava. The gases accumulated to constitute an unbreathable atmosphere of water vapor (H_2O), methane (CH_4), ammonia (NH_3), hydrogen (H_2), nitrogen (N_2), carbon dioxide (CO_2), sulfur dioxide (SO_2), and other gases. Some researchers have speculated that gases from comets colliding with Earth during the Hadean may have contributed components of the early atmosphere. The early atmosphere was probably much denser—perhaps 250 times denser—than our present atmosphere.

If the Earth's surface was sufficiently cool for solid crust to form at 4.4 Ga, then it's fair to ask: Could oceans have formed at that time? Detailed analyses of oxygen isotopes in the 4.4 Ga zircons of Australia suggest that the grains formed at a time when cool, liquid water existed at the Earth's surface. From this result, some researchers have concluded that very early oceans did exist. But the question remains a subject of intense debate, for there are alternative explanations for the zircon data.

FIGURE 13.4 A painting of the early Hadean Earth. Note the magma ocean, with small crusts of solid basalt floating about. If the sky had been clear, streaks of meteors would have filled it, and the Moon would have appeared much larger than it does today, because it was closer. An observer on Earth probably could not have seen the sky, however, because the atmosphere contained so much water vapor and volcanic ash.

Though mineral grains as old as 4.4 Ga exist, the oldest *whole rock* yet found on Earth has an age of only 4.03 Ga. This rock, the Acasta Gneiss, crops out in the Northwest Territories of Canada. What destroyed the pre–4.03 Ga crust (and oceans, if they existed) of the Earth? The answer comes from studies of cratering on the Moon. These studies suggest that the Moon—and, therefore, all inner planets of the Solar System—underwent intense meteor bombardment between 4.0 and 3.9 Ga. Researchers speculate that this bombardment would have pulverized and/or melted any crust that had existed on Earth at the time, and would have vaporized any existing water. Only after the bombardment ceased could long-lasting crust and oceans begin to form. The discovery of 3.85 Ga marine sedimentary rocks in Greenland suggests that the appearance of land and sea happened quite soon after bombardment ceased. What did the Earth's surface look like at this time? An observer probably would detect small, barren land masses and abundant volcanoes poking up above an acidic sea. But both land and sea might have been partially covered by ice, and both would have been obscured by murky, dense air.

> **Take-Home Message**
>
> Little is known about the Hadean, the first 700 million years of Earth history. For part of this time, the surface may have been molten. Crust and oceans may have formed at about 4 Ga, but the planet's surface was destroyed around 3.9 Ga by meteorite bombardment.

13.4 THE ARCHEAN EON: THE BIRTH OF THE CONTINENTS AND THE APPEARANCE OF LIFE

The date that marks the boundary between the **Archean** (from the Greek word for "ancient") **Eon** and the Hadean has been placed at about 3.8 Ga, the time at which the record of crustal rocks starts to become progressively more complete.

Geologists still argue about whether plate tectonics operated in the early Archean. Some researchers picture an early Archean Earth with rapidly moving small plates, numerous volcanic island arcs, and abundant hot-spot volcanoes. Others propose that early Archean lithosphere was too warm and buoyant to subduct, and that plate tectonics could not have operated until the later part of the Archean. These authors argue that plume-related volcanism was the main source of new crust until the late Archean. Regardless of which model ultimately proves better, it is clear that the Archean was a time of significant change in the map of the Earth. During that time, the volume of continental crust increased significantly.

How did the continents come to be? According to one model, relatively buoyant (felsic and intermediate) crustal rocks formed both at subduction zones and at hot-spot volcanoes. Frequent collisions sutured volcanic arcs and hot-spot volcanoes together, creating progressively larger blocks called **protocontinents** (▶Fig. 13.5a, b). Some of these protocontinents developed rifts that filled with basalt. As the Earth gradually cooled, protocontinents became cooler and stronger, and by 2.7 Ga the first **cratons**, long-lived blocks of durable continental crust, had developed. By the end of the Archean Eon, about 80% of continental crust had formed (▶Fig. 13.6). Volcanic activity during this time continued to supply gas to the atmosphere.

Archean cratons contain five principal rock types: *gneiss* (relics of Archean metamorphism in collisional zones), *greenstone* (metamorphosed relics of ocean crust trapped between colliding blocks of basalts that had filled early continental rifts, or of flood basalts produced at hot spots), *granite* (formed from magmas generated by the partial melting of the crust in continental volcanic arcs or above hot spots), *graywacke* (a mixture of sand and clay eroded from the volcanic areas and dumped into the ocean), and *chert* (formed by the precipitation of silica in the deep sea). Archean shallow-water sediments are rare, either because continents were so small that depositional environments in which such sediments could accumulate didn't exist, or because any that were present have since eroded away.

Once land areas had formed, rivers flowed over their stark, unvegetated surfaces. Geologists reached this conclusion because sedimentary beds from this time contain clastic grains that were clearly rounded by transport in liquid water. Salts weathered out of rock and transported to the sea in rivers made the oceans salty.

Clearly, the Archean Eon saw many firsts in Earth history—the first continents, and probably the first life. Geologists use three sources of evidence to identify early life:

- *Chemical (molecular) fossils, or biomarkers:* These are durable chemicals that are produced only by the metabolism of living organisms.
- *Isotopic signatures:* Carbon can occur as ^{12}C, ^{13}C, or ^{14}C. Organisms preferentially incorporate ^{12}C instead of ^{13}C, by a slight amount. Thus, by analyzing the ratio of ^{12}C to ^{13}C in carbon-rich sediment, geologists can determine if the sediment once contained the bodies of organisms.
- *Fossil forms:* Given appropriate depositional conditions, shapes representing bacteria or archaea cells can be preserved in rock. However, identification of such fossil forms can be controversial—similar shapes can result from inorganic crystal growth.

The search for the earliest evidence of life continues to make headlines in the popular media. It is a story of high hopes and intense frustrations. For example, isotopic signatures of life were discovered in 3.8-billion-year-old

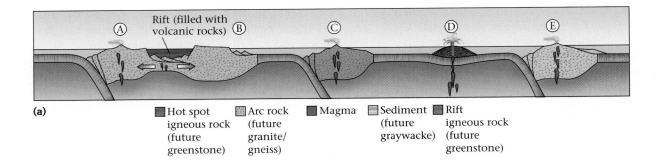

(a)

Legend:
- ■ Hot spot igneous rock (future greenstone)
- ▨ Arc rock (future granite/gneiss)
- ■ Magma
- ▨ Sediment (future graywacke)
- ■ Rift igneous rock (future greenstone)

(b)

- ▨ Granite   ▨ Ocean crust   ▨ Gneissic fabric

FIGURE 13.5 (a) In the Archean Eon, rocks that would eventually make up the continental crust began to form. At early subduction zones, volcanic island arcs, composed of relatively buoyant rocks, developed, and large shield volcanoes formed over hot spots. Larger crustal blocks underwent rifting, creating rift basins that filled with volcanic rocks, and the erosion of crustal blocks deposited graywacke on the sea floor. **(b)** Successive collisions brought all the buoyant fragments together to form a protocontinent. Melting at the base of the protocontinent may have caused the formation of plutons.

metamorphosed sedimentary rocks of western Greenland, suggesting that life started right at the beginning of the Archean Eon. But this proposal has been disputed by researchers who argue that the signatures are the result of later interaction between the rock and groundwater. As another example, fossil forms resembling filamentous bacteria were found in 3.4 to 3.5 Ga chert beds of western

FIGURE 13.6 As time progressed, the area of the Earth covered by continental crust increased, though not all geologists agree about the rate of increase. This model shows growth beginning about 3.9 Ga and continuing rapidly until the end of the Archean Eon. The rate slowed substantially in the Proterozoic Eon.

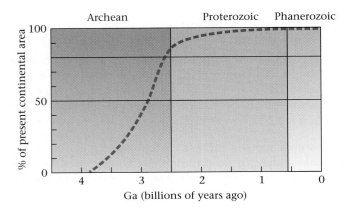

Australia, but this proposal has been disputed by researchers who argue that these fossils are simply inorganic mineral crystals.

Nevertheless, most geologists currently conclude that life has existed on Earth since at least 3.5 Ga, for rocks of this age contain clear isotopic signatures of organisms. The oldest undisputed fossils of bacteria and archaea occur in 3.2 Ga rocks. Some rocks of this age contain **stromatolites**, distinctive mounds of sediment produced by mats of cyanobacteria. Stromatolites form because cyanobacteria secrete a mucuslike substance to which sediment settling from water sticks. As the mat gets buried, new cyanobacteria colonize the top of the sediment, building a mound upward (▶Fig. 13.7a–c). The occurrence of stromatolites in sedimentary beds after 3.2 Ga means that when the Earth was one-quarter of its present age, its surface hosted prokaryotic life (simple cells without a nucleus; see Interlude E). Biomarkers in Archean sediments indicate that photosynthetic organisms appeared by 2.7 Ga.

What specific environment on the Archean Earth served as the actual cradle of life? In other words, where did the first DNA-containing cells begin to grow, extract energy from the chemical bonds of molecules that they ingested, and reproduce? In 1953, Stanley Miller tried to answer this question by placing methane, hydrogen, water, and ammonia in a glass vial; over a period of a few days, he subjected the mixture to electrical sparks. When he analyzed the contents of

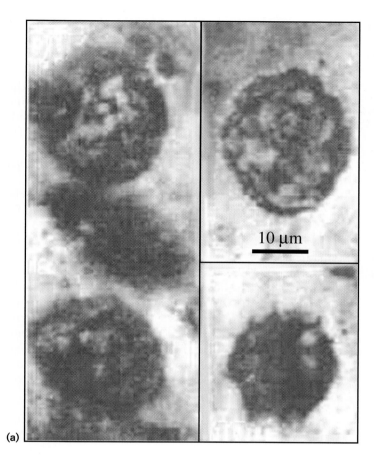

(a)

(b)

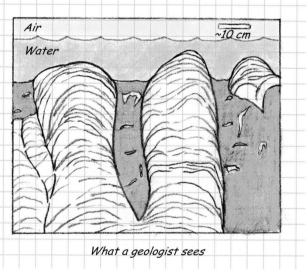

What a geologist sees

(c)

FIGURE 13.7 **(a)** Some of the first fossils, found in 3.2-million-year-old rocks, are spherical bacteria cells from chert of South America. **(b)** This photo shows a polished vertical slice through a stromatolite deposit, as displayed in the American Museum of Natural History. This specimen is actually Proterozoic, not Archean, but is shown here because the internal structure is so clear. It was collected in Mauritania in west Africa. **(c)** A geologist's interpretation of what these mounds looked like at the time of their formation. Note that pockets of debris collect between the stromatolite mounds. Only the top surface of the mound is alive.

the vial a few days later, he found a mixture of chemicals called amino acids, from which protein—an essential ingredient of life—forms. This experiment led many researchers to conclude that life began in warm pools of water, beneath a methane- and ammonia-rich atmosphere streaked by bolts of lightning. The only problem with this hypothesis is that more recent evidence suggests that the early atmosphere consisted mostly of CO_2 and N_2, with relatively little methane and ammonia. More recent studies point instead to submarine hot-water vents, so-called black smokers, as the host of the first organisms. These vents emit clouds of ion-charged solutions from which sulfide minerals precipitate and build chimneys. The earliest life in the Archean Eon may well have been thermophilic (heat-loving) bacteria or archaea that dined on pyrite at dark depths in the ocean along-side these vents or in pores below the surface near the vents. Only when photosynthesis, the ability to extract energy from sunlight, evolved could organisms populate shallow marine environments. And only after ozone appeared in the atmosphere could organisms rise to the shallowest depths, for ozone filters out dangerous ultraviolet light.

As the Archean Eon came to a close, the first continents existed, and life had colonized not only the depths of the sea but also the shallow marine realm. Plate tectonics had commenced, and thus

> **Take-Home Message**
>
> During the Archean, the first continents formed, the oceans were established, and life appeared (in the form of archaea and bacteria). Continents may have formed by collisions among volcanic arcs and hot-spot volcanoes.

continental drift was taking place, collisional mountain belts were forming, and erosion was occurring. The atmosphere was gradually accumulating oxygen, although probably this gas still accounted for only a very small percentage of the air. The stage was set for another major change in the Earth System.

13.5 THE PROTEROZOIC EON: TRANSITION TO THE MODERN WORLD

The Proterozoic Eon (from the Greek words meaning "first life") spans roughly 2 billion years, from about 2.5 billion years ago to the beginning of the Cambrian Period, 542 million years ago—thus, it encompasses almost half of Earth's known history. The eon received its now misleading name before fossil bacteria and archaea were discovered in Archean rock. During Proterozoic time, Earth's surface environment changed from the unfamiliar world of possibly small, fast-moving plates, small continents, and an oxygen-poor atmosphere to the more familiar world of mostly large plates, large continents, and an oxygen-rich atmosphere.

First, let's look at changes to the continents. New continental crust continued to form during the Proterozoic Eon at progressively slower rates. Even though crust formed by the middle of the eon, over 90% of the Earth's continental crust had formed. Also, collisions between the Archean continents, as well as the accretion of new volcanic island arcs and hot-spot volcanoes, resulted in the assembly of large continents whose interiors were located far from orogenic activity. These continents cooled and strengthened to become cratons. Most of the large cratons that exist today had formed by about 1.8 Ga (▶Fig. 13.8).

Let's look at the geology of a large craton more closely, by examining the interior of North America (▶Fig. 13.9). The Precambrian rocks constituting this craton crop out extensively in Canada. Geologists refer to this region as the Canadian Shield. A **shield** consists of a broad, low-lying region of exposed Precambrian rocks. In the United States, Phanerozoic strata bury most of the Precambrian rocks, forming a province called a **cratonic platform**, or **continental platform**. Here we see Precambrian rocks only where they have been exposed by erosion of the sedimentary cover. The Canadian Shield includes several Archean blocks that are sutured together along huge collisional belts. Of these belts, the longest is called the Trans-Hudson

FIGURE 13.8 The distribution of Precambrian crust. Note that only small remnants of Archean crust remain.

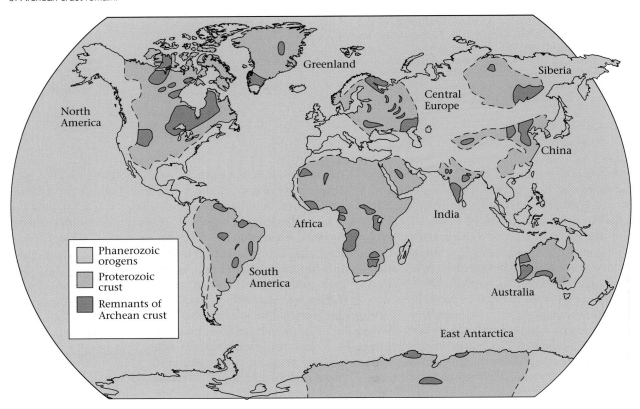

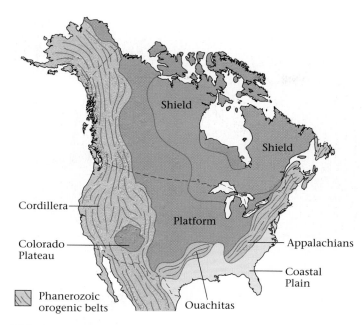

FIGURE 13.9 On this map of North America, we see four different geological provinces: shield areas, where Precambrian rocks of the craton crop out extensively at the ground surface; platform areas, where Phanerozoic sedimentary rocks have buried Precambrian rocks of the craton; Phanerozoic orogenic belts, composed of rocks that formed or were deformed in mountain belts during the past half-billion years; and the coastal plain, low land buried by Cretaceous and Tertiary sediment.

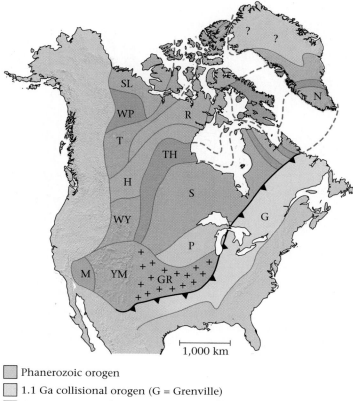

1,000 km

- Phanerozoic orogen
- 1.1 Ga collisional orogen (G = Grenville)
- 1.6–1.7 Ga accreted crust covered by 1.3–1.5 Ga granite and rhyolite, where patterned (GR = granite-rhyolite province)
- 1.6–1.7 Ga accreted crust (YM = Yavapai and Mazatzal)
- 1.8 Ga accreted crust (P = Penokean)
- 1.8 Ga collisional orogen (TH = Trans-Hudson; WP = Wopmay)
- 1.9 Ga collisional orogen (T = Thelon)
- Archean rocks, later deformed and metamorphosed in the Proterozoic (H = Hearn; R = Rae)
- Relics of Archean crust (WY = Wyoming; M = Mojave; S = Superior; N = Nain; SL = Slave)

FIGURE 13.10 The North American craton consists of a collage of different fragments, stitched together during orogenies of Precambrian time. (The explanation names these fragments and orogens.) We distinguish between collisional orogens, where large blocks joined, and accretionary orogens, composed of volcanic island arcs, hot-spot volcanoes, and continental slivers plastered onto the margin of a larger continent at a convergent margin.

orogen (▶Fig. 13.10). In the United States, Archean rocks underlie part of the Midwest and much of Wyoming, but most of the U.S. craton consists of crust that formed when a series of volcanic island arcs and continental slivers accreted (attached) to the southern margin of the Canadian Shield between 1.8 and 1.6 Ga. Such orogens, formed by the attachment of numerous buoyant slivers to an older, larger block, are called **accretionary orogens**. In the Midwest, huge granite plutons intruded much of this accreted region, and rhyolite covered it, between 1.5 and 1.3 Ga.

Successive collisions ultimately brought together most continental crust on Earth into a single supercontinent, named **Rodinia**, by around 1 Ga. The last major collision during the formation of Rodinia produced a large collisional orogen called the **Grenville orogen**. Rocks that underwent metamorphism and deformation during this event crop out in eastern Canada and along the crest of the Appalachian Mountains in the eastern United States. If you look at a popular (though not totally proven) reconstruction of Rodinia, you can identify the crustal provinces that would eventually become the familiar continents of today (▶Fig. 13.11a). In this reconstruction, the future Antarctica and Australia lay somewhere along the western coast of North America, while fragments of the future South America lay to the east of Greenland and Canada.

Several studies suggest that sometime between 800 and 600 Ma, Rodinia "turned inside out," in that Antarctica, India, and Australia broke away from western North America and swung around and collided with the future South America, possibly forming a short-lived supercontinent that some geologists refer to as **Pannotia** (▶Fig. 13.11b). The breakup of Rodinia resulted in the formation of a passive-margin basin (a thick accumulation of sediment along a tectonically inactive coast) along the western edge of North America.

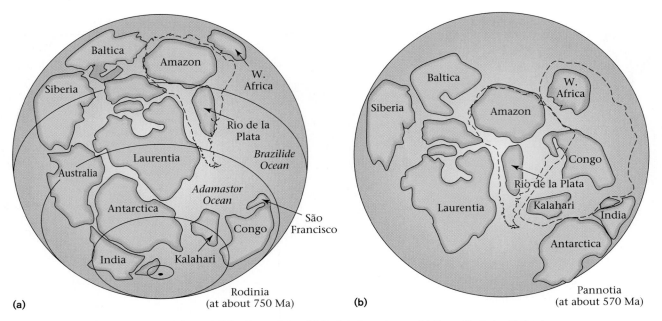

(a)

Rodinia
(at about 750 Ma)

(b)

Pannotia
(at about 570 Ma)

FIGURE 13.11 Supercontinents at the end of the Precambrian. **(a)** Rodinia formed around 1 Ga and lasted until about 700 Ma. **(b)** According to one model, by about 570 Ma, Antarctica, India, and Australia had broken off the western margin of Rodinia and had swung around and collided with the eastern margin of the future South America, to create a new, short-lived supercontinent called Pannotia. Pannotia broke apart at about 550 Ma.

Much of the transformation of the Earth's atmosphere from the oxygen-poor volcanic gas mix of the Archean Eon to the oxygen-rich mix we breathe today occurred during the Proterozoic Eon (▶Fig. 13.12; note that the present concentration of oxygen, 21%, was not attained until the Phanerozoic Eon). We discuss the evidence for this claim in ▶Box 13.1. This new oxygen-rich atmosphere had a profound effect on the Earth, for it permitted a great diversification of life and the eventual conquest of the land by living organisms. Life could become more complex because oxygen-dependent (aerobic) metabolism produces energy much more efficiently than does oxygen-free (anaerobic) metabolism—eating sulfide minerals may sustain bacteria, but it can't keep multicellular organisms on the move! The addition of substantial oxygen to the atmosphere also set the stage for the eventual move of life onto the land, hundreds of millions of years after the Proterozoic. Oxygen provides the raw material from which ozone (O_3) forms. As we have noted, ozone protects the land surface from harmful ultraviolet rays.

So far, we've talked only of *prokaryotic* organisms, those which consist of single cells that do not contain a nucleus. How and when did *eukaryotic* cells, the type of cells that make up more complex organisms (protozoans, true algae, fungi, plants, and animals) appear on Earth? Biologists have been struggling with the "how" question for decades, and have concluded that the first eukaryotic cells originated when one cell incorporated another to form a more complex organism. Geologists have been

struggling with the "when" question for decades as well, and the answer hasn't come easily. Chemical fossils hint that eukaryotic organisms existed as early as 2.7 Ga, and controversial fossil forms of eukaryotic organisms have been found in rocks ranging from 2.1 to 1.2 billion years of age. Undeniable eukaryotic fossil organisms occur in 1.0-billion-year-old rocks.

FIGURE 13.12 The graph shows the change in the oxygen content of Earth's atmosphere through time. The vertical axis represents the percentage of the atmosphere that consists of oxygen. Today, oxygen accounts for about 21% of the atmosphere, but in Archean time it constituted only about 0.000000001%.

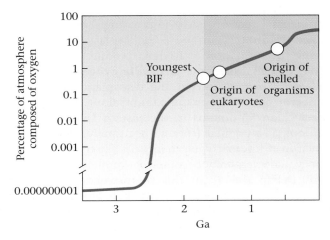

BOX 13.1 THE REST OF THE STORY

The Mystery of Atmospheric Oxygen

Without oxygen, the great variety of life that exists on Earth could not survive. Presently, the atmosphere contains 21% oxygen, but this has not always been the case. Throughout most of the Archean Eon, and into the beginning of the Proterozoic Eon, the atmosphere contained less than 1% oxygen. Several lines of evidence lead geologists to conclude that a transformation from an oxygen-poor to an oxygen-rich atmosphere occurred about 2.4 to 2.2 Ga, during the early part of the Proterozoic.

One line of evidence comes from examining clastic grains in sandstones. In sediments deposited *before* 2.2 Ga, pyrite (iron sulfide) occurs as clasts in sediment. This could only be possible if the atmosphere before 2.2 Ga contained very little oxygen, for in an oxygen-rich atmosphere, pyrite undergoes chemical weathering (oxidation) and doesn't survive long enough at the Earth's surface to become a sedimentary clast.

Another line of evidence comes from studying the age of redbeds, clastic sedimentary rocks colored by the presence of bright red hematite (iron oxide). Redbeds form when oxygen-rich groundwater flows through sediment during lithification, and such rocks appear in the geologic record only *after* 2.2 Ga.

A third line of evidence may come from studying **banded-iron formation (BIF)** deposits, humanity's main source of iron ore. BIF is a marine sedimentary rock composed of alternating layers of gray iron-rich minerals (hematite or magnetite) and bright red chert (▶Fig. 13.13). Sediment forming this rock was deposited *only* during the Archean and the early part of the Proterozoic; BIF deposits did not form after 1.88 Ga. This observation suggests that after 1.88 Ga, the ocean no longer contained abundant quantities of dissolved iron from which the iron minerals in BIF formed. The decrease of dissolved iron in seawater may reflect an increase in the amount of dissolved oxygen in seawater, caused by an increase in the amount of oxygen in the air, for iron cannot dissolve in oxygen-rich water. Alternatively, the decrease in dissolved iron may reflect an increase in chemical weathering of the land by oxidation; such weathering would have dumped various ions into the ocean, and these ions could react with dissolved iron and remove it.

Why did the atmosphere change so radically during the Paleoproterozoic? At one time, geologists assumed that the change marked the first appearance of photosynthetic organisms. But this can't be the whole story, for cyanobacteria had already been in existence for over a half-billion years at the time the atmosphere became oxygen rich. It now appears that the enrichment of oxygen probably reflects, in part, the appearance of new organisms or an increase in abundance of environments where photosynthetic organisms could live, and partly a variety of other chemical changes in the Earth System. The complete story has not yet been worked out.

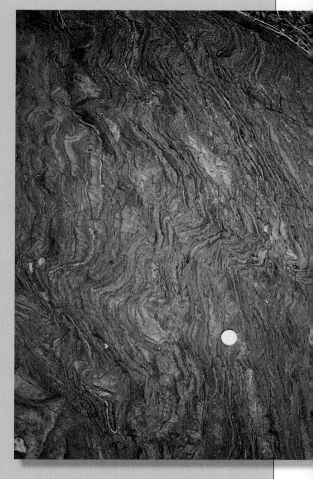

FIGURE 13.13 An outcrop of BIF in the Iron Ranges of Michigan's Upper Peninsula. The red stripes are jasper (red chert) and the gray stripes are hematite. The rock was folded during a mountain-building event long after deposition. The coin indicates scale.

The last half-billion years of the Proterozoic Eon saw the remarkable transition from simple organisms into complex ones. Ciliate protozoans (single-celled organisms covered with fibers that give them mobility) appear at about 750 Ma. A great leap forward in complexity of organisms occurred during the next 150 million years, for sediments deposited perhaps as early as 620 Ma and certainly by 565 Ma contain several types of multicellular organisms that together constitute the **Ediacaran fauna**, named for a region in southern Australia. Ediacaran species survived into the beginning of the Cambrian before becoming extinct. Though their nature is enigmatic, fossil forms suggest that some of these invertebrate (shell-less) organisms resembled jellyfish, while others resembled worms (▶Fig. 13.14a).

What triggered the appearance of radically new forms of life, as represented by the Ediacaran fauna? Perhaps it reflects the complex interplay of life and geology in the Earth System. A radical change in the Earth System may have stressed and/or extinguished existing life, making room for newer forms to evolve. The geologic record *does* show two major changes in the Earth System happening at

(a)

(b)

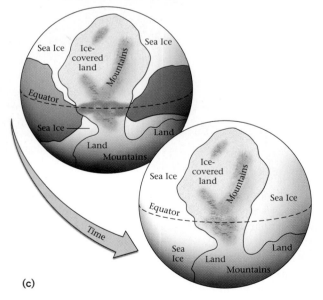

(c)

FIGURE 13.14 (a) "Dickinsonia," a fossil of the Ediacaran fauna, an assemblage of soft-bodied organisms that appeared in late Proterozoic seas. **(b)** Investigators examining Proterozoic glacial till in Africa. The existence of such till indicates that low-latitude land masses were glaciated during the Proterozoic. **(c)** Two stages in the development of snowball Earth. First, glaciers start growing on land, then the sea freezes over.

the end of the Proterozoic Eon, either or both of which may have influenced the course of evolution.

First, the appearance of the Ediacaran fauna corresponds with the formation and subsequent dispersal of Pannotia. Such major changes in the distribution of continents, in the prominence of mountain belts, and in the nature of plate boundaries likely affected the chemistry and temperature of the sea and the atmosphere, and thus the variety of ecological niches.

Second, radical climate shifts occurred on Earth at the end of the Proterozoic Eon. Specifically, accumulations of glacial sediments occur in virtually all sequences of the latest Proterozoic. What's strange about the occurrence of these sediments is that they can be found even in continental coastal regions that were located at the equator during these times. This observation implies that the *entire* planet was cold enough for glaciers to form at the end of the Proterozoic. Geologists still are debating the history of these global "ice ages" (see Chapter 22), but in one model glaciers covered the land and perhaps the entire ocean froze, creating what geologists refer to as **snowball Earth** (▶Fig. 13.14b, c). The shell of ice cut off the oceans from the atmosphere, causing oxygen levels in the sea to drop drastically, so many life forms died off. The icy sheath also prevented atmospheric CO_2 from dissolving in seawater, but it did not prevent volcanic activity from continuing to add CO_2 to the atmosphere. According to this model, Earth would have remained a snowball forever, were it not for volcanic CO_2. CO_2 is a greenhouse gas, meaning that it traps heat in the atmosphere much as glass planes trap heat in a greenhouse (see Chapter 23); thus, as the CO_2 concentration increased, Earth warmed up and eventually the glaciers rapidly melted. Life may have survived snowball Earth conditions only near submarine black smokers and hot springs. When the ice vanished, life rapidly expanded into new environments, and new species, such as the Ediacaran fauna, could evolve.

> **Take-Home Message**
>
> During the Proterozoic, stable crustal blocks formed, and then sutured together to form large continents, and toward the end of the Proterozoic, supercontinents. Multicellular organisms appeared, and the atmosphere began to be rich in oxygen.

13.6 THE PHANEROZOIC EON: LIFE DIVERSIFIES, AND TODAY'S CONTINENTS FORM

The Phanerozoic (Greek for visible life) Eon encompasses the last 542 million years of Earth history. Its name reflects the appearance of diverse organisms with hard shells or skeletons that became the well-preserved fossils you can easily find in rock outcrops. Geologists divide the Phanerozoic into three eras—the Paleozoic (Greek for "ancient life"), the Mesozoic ("middle life"), and the Cenozoic ("recent life")—to emphasize the changes in Earth's living population throughout the eon. The solid Earth also changed during this time. Continental blocks rearranged, and new supercontinents formed, only to break apart again. Numerous orogenies built mountain ranges, the most recent of which persist today. Shell-secreting organisms, burrowing organisms, and plants invaded the sedimentary environment and changed the style of deposition. In the following sections, we look at the changes in the planet's map (its paleogeography) and in its life forms during the three eras of the Phanerozoic Eon.

13.7 THE PALEOZOIC ERA: FROM RODINIA TO PANGAEA

The Early Paleozoic Era (Cambrian-Ordovician Periods, 542–444 Ma)

Paleogeography. As the Proterozoic Eon drew to a close, most continental crust had accumulated in a supercontinent called Pannotia, formed when Rodinia turned inside out. Recall that during this reorganization, Antarctica, India, and Australia rifted away from the future North America, and a passive-margin basin formed along North America's western margin. At the beginning of the Paleozoic Era (Cambrian Period), Pannotia broke up, yielding smaller continents including Laurentia (composed of

North America and Greenland), Gondwana (South America, Africa, Antarctica, India, and Australia), Baltica (Europe), and Siberia (▶Fig. 13.15). As these continents drifted apart, Laurentia, Baltica, and Siberia stayed at low latitudes, but Gondwana drifted toward the South Pole, and for a brief interval in the Late Ordovician Period, much of it became ice covered.

Following the breakup of Pannotia, several more passive-margin basins formed around the globe, including one on the eastern coast of Laurentia (what is now eastern North America). In addition, sea level rose, so that vast areas of continents were flooded with shallow seas called **epicontinental seas**. By the end of the Cambrian Period, the only dry land in Laurentia was an oval region centered on Hudson Bay, so most of what is now the United States lay beneath water (▶Fig. 13.16). In many places, water depths in epicontinental seas reached only a few meters, creating a well-lit environment in which life abounded. Deposition in the seas yielded a layer of fossiliferous sediment that you can see today near the floor of the Grand Canyon. A relatively thin layer of sediment formed in Laurentia's interior (the region that would become the continental platform), but a much thicker layer accumulated in the passive-margin basins that fringed the continent. Sea level, however, did not stay high for the entire early Paleozoic Era; regressions and transgressions occurred, the former

FIGURE 13.16 A paleogeographic map of North America, showing the regions of dry land and shallow sea in the Early Cambrian Period. Note the passive margins, which define the borders of the continent. At this time, the regions between the edge of the passive margin and the coast did not yet exist. The transcontinental arch was a ridge of dry land.

FIGURE 13.15 The distribution of continents in the Cambrian Period (about 510 Ma). Note that Gondwana lay near the equator, and Laurentia straddled the equator. The map shows how the Earth might have looked as viewed from the South Pole. Note that Florida had not yet connected to North America (Laurentia).

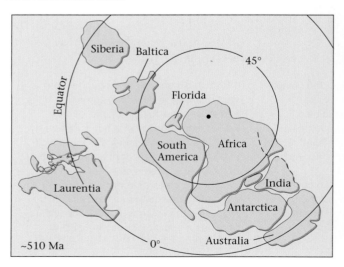

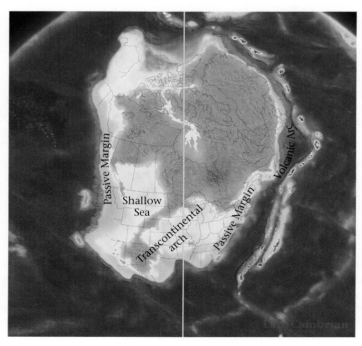

BOX 13.2 SCIENCE TOOLBOX

Stratigraphic Sequences and Sea-Level Change

As we have seen in this chapter, there have been intervals in geologic history when the interior of North America was submerged beneath a shallow sea, and times when it was high and dry. This observation implies that sea level, relative to the surface of the continent, rises and falls through geologic time. When the continental surface was dry and exposed to the atmosphere, weathering and erosion ground away at previously deposited rock and, as a result, created a continent-wide unconformity. In the early 1960s, an American stratigrapher named Larry Sloss introduced the term **stratigraphic sequence** to refer to the strata deposited on the continent during periods when continents were submerged. Such sequences are bounded above and below by regional unconformities.

We can picture the deposition of an idealized sequence as follows. As the starting condition, much of the surface of a continent lies above sea level, and an unconformity develops. Then, as a transgression takes place, the margin of the continent floods first, then the interior. Thus, the preexisting unconformity gets progressively buried toward the interior, and the bottom layers of the newly deposited sequence tend to be older near the margin of the continent than toward the interior. The base of a sequence consists of terrestrial strata (river alluvium). As transgression progresses, this sediment is buried by nearshore sediment, then by deeper-water sediment (▶Fig. 13.17a). Eventually, nearly the entire width of the continent, with the exception of highlands and mountain belts, lies below sea level. When regression begins, the interior of the continent is exposed first and then the margins. So a new unconformity forms first in the interior and then later along the margins. Many shorter-duration transgressions and regressions may happen during a single long-duration rise or fall.

Sloss recognized six major stratigraphic sequences in North America, and he named each after a nation of Native Americans (▶Fig. 13.17b). Each sequence represents deposition during an interval of time lasting tens of millions of years. The transgressions and regressions could have been caused by global ("eustatic") sea level change, or they could reflect mountain-building

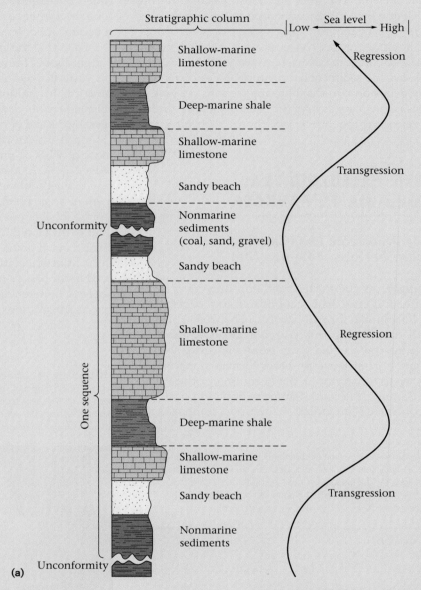

Stratigraphic column   Low ◄—— Sea level ——► High

Shallow-marine limestone

Deep-marine shale

Shallow-marine limestone

Sandy beach

Unconformity

Nonmarine sediments (coal, sand, gravel)

Sandy beach

One sequence

Shallow-marine limestone

Deep-marine shale

Shallow-marine limestone

Sandy beach

Transgression

Nonmarine sediments

(a) Unconformity

Regression

Transgression

Regression

Transgression

processes, or they could reflect the tilting of continents.

In more recent decades, studies of strata along continental passive margins have provided an even more detailed record of sequences, because there is less erosion in these regions.

Taken together, this information may provide insight into the history of the global rise and fall of sea level, though this interpretation remains controversial (▶Fig.

13.17c). Some sea-level changes may reflect changes in sea-floor-spreading rates and thus in the volume of mid-ocean ridges; some may reflect hot-spot activity; some may reflect variations in Earth's climate that caused the formation or melting of ice sheets; others may represent changes in areas of continents accompanying continental collisions; and still others may reflect the warping of continental surfaces as a result of mountain building.

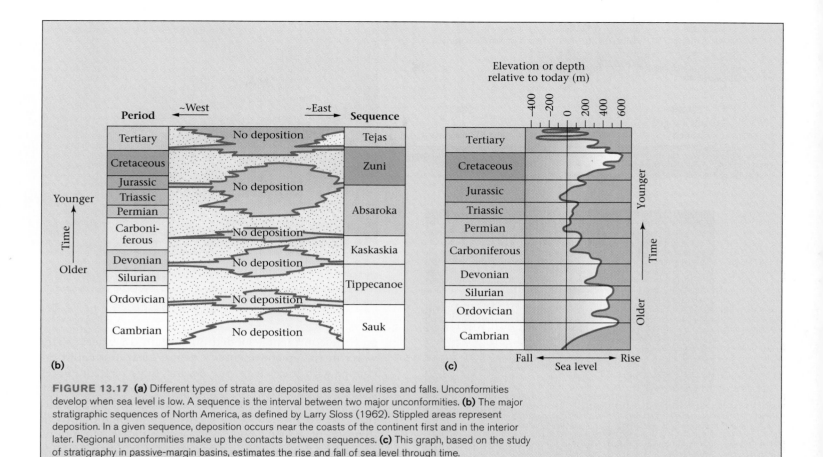

FIGURE 13.17 (a) Different types of strata are deposited as sea level rises and falls. Unconformities develop when sea level is low. A sequence is the interval between two major unconformities. **(b)** The major stratigraphic sequences of North America, as defined by Larry Sloss (1962). Stippled areas represent deposition. In a given sequence, deposition occurs near the coasts of the continent first and in the interior later. Regional unconformities make up the contacts between sequences. **(c)** This graph, based on the study of stratigraphy in passive-margin basins, estimates the rise and fall of sea level through time.

marked by unconformities and the latter by accumulations of sediment (▶Box 13.2).

The geologically peaceful world of the early Paleozoic Era in Laurentia abruptly came to a close in the Middle Ordovician Period, for at this time Laurentia's eastern margin rammed into a volcanic island arc and other crustal fragments. The resulting collision, called the **Taconic orogeny**, deformed and metamorphosed strata of the former passive-margin basin and created a mountain range (▶Fig. 13.18a, b). This event was the first stage in the development of the Appalachian orogen.

Life evolution. Beginning with the Cambrian Period, life on Earth left a clear record of evolution, because so many organisms developed shells of durable minerals. The fossil record indicates that soon after the Cambrian began, life underwent remarkable diversification. This event, which paleontologists refer to as the **Cambrian explosion of life**, took about 20 million years. What caused this explosion? No one can say for sure, but considering that it occurred roughly at the time a supercontinent broke up, it may have had something to do with the new ecological niches and

the isolation of populations that resulted when small continents formed and drifted apart.

The first animals to appear in the Cambrian Period had simple tube- or cone-shaped shells, but soon thereafter, the shells became more complex (▶Fig. 13.19). Small fossils called conodonts, which resemble teeth, are found in Cambrian strata. Their presence suggests that creatures with jaws had appeared at this time, hinting that shells evolved as a means of protection against predators. By the end of the Early Cambrian, trilobites were grazing the sea floor. Trilobites shared the environment with mollusks, brachiopods, and echinoderms (see Interlude E). Thus, a complex food chain arose, which included plankton, bottom feeders, and at the top, giant predators such as the 2-m-long anomalocaris. Mounds of sponges with mineral skeletons formed the first reefs, and graptolites (floating organisms that contained a serrated mineral band), coral, and gastropods appeared. The Ordovician saw the appearance of crinoids and the first vertebrate animals, jawless fish. At the end of the Ordovician, mass extinction took place, perhaps because of the brief glaciation and associated sea-level lowering of the time.

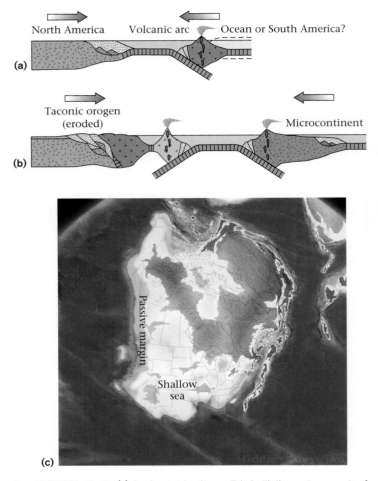

FIGURE 13.19 A museum diorama illustrates what marine organisms living during the early Paleozoic may have looked like. Here, we see such creatures as trilobites and nautiloids.

FIGURE 13.18 **(a)** A volcanic island arc collided with the eastern margin of North America to cause the Taconic orogeny. Some geologists have suggested that South America lay behind the volcanic arc, so that at the end of the Taconic orogeny eastern North America bordered western South America. **(b)** Sometime after the collision, the high land of the Taconic orogen eroded away, and other blocks, such as the Avalon microcontinent, approached from the east. Their eventual collision with North America caused the Acadian orogeny. **(c)** A geologists's interpretation of topography in the Middle Ordovician.

But although the sea teemed with organisms during the early Paleozoic Era, there were no land organisms for most of this time, so the land surface was a stark landscape of rock and sediment, subjected to rapid erosion rates. Our earliest record of primitive land plants and green algae comes from the Late Ordovician Period, but these plants were very small and occurred only along bodies of water.

The Middle Paleozoic Era (Silurian-Devonian Periods, 444–359 Ma)

Paleogeography. As the world entered the Silurian Period, the global climate warmed, leading to so-called **greenhouse conditions**. Because of this warming, sea level rose,

and the continents flooded once again. In places where water in the epicontinental seas was clear and could exchange with water from the oceans, huge reef complexes grew, forming a layer of fossiliferous limestone on the land. In other places, where the seas were stagnant and cut off from the oceans, evaporation rates were so high that thick deposits of gypsum and halite accumulated. Mountains formed by the Taconic orogeny in eastern North America eroded away, and their roots were buried beneath a thin layer of shallow-water limestone.

More orogeny took place during the middle Paleozoic Era. For example, in eastern North America, the Appalachian region that had been affected by the Taconic orogeny underwent a collisional orogeny once again in the Devonian Period, this time with small continental masses that drifted in from the east (Fig. 13.18b; ►Fig. 13.20). Geologists refer to this event as the **Acadian orogeny**. (The Acadian orogeny corresponds to the **Caledonian orogeny** of northern Europe.) Note that because of the Taconic and Acadian orogenies, the easternmost United States consists of crust that was attached to North America during the Paleozoic. Large mountains rose during the Acadian orogeny, providing a source for the Catskill Delta, an enormous fan of conglomerates, sandstones, and shales that spread across western New York, Pennsylvania, Ohio, Kentucky, and Tennessee (►Fig. 13.21a, b). Similar rocks, called the Old Red Sandstone, crop out in the United Kingdom.

Throughout the middle Paleozoic, the western margin of North America continued to be a passive-margin basin, filling with thick accumulations of sediment that can be seen today in Nevada or eastern California. Finally, in the Late Devonian, the quiet environment of the west-coast passive margin ceased, possibly because of a collision with an island arc. This event, known as the

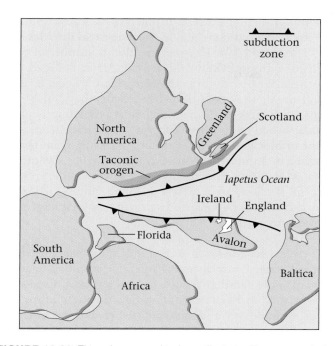

FIGURE 13.20 This paleogeographical map illustrates the movement of the Avalon microcontinent (including England and southern Ireland) toward its ultimate collision with the eastern margin of North America. During the process, the intervening Iapetus Ocean was subducted. The remnant of the Taconic orogen fringed the eastern margin of North America.

FIGURE 13.21 (a) This paleogeographical map of North America depicts the distribution of land and sea and the position of mountain belts in the Late Devonian Period. Note that the Acadian orogeny occurred in the east and, somewhat later, the Antler orogeny in the west. The Acadian orogen shed a vast delta, called the Catskill Delta, of sand and conglomerate into the eastern interior of North America. (b) Redbeds (conglomerate and sandstone) of the Catskill Delta. (c) Skeleton of Tiktaalik, perhaps the first animal to walk on land.

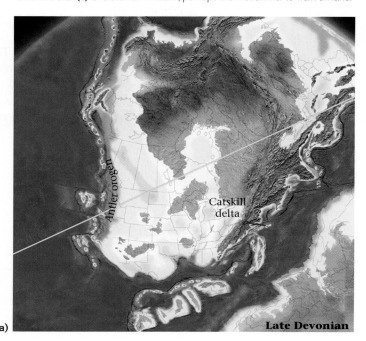

(a)

(b)

(c)

Antler orogeny, pushed slices of deep-marine strata eastward, and up and over shallow-water strata. It was the first of many orogenies to affect the western margin of the continent.

Life evolution. Life on Earth underwent radical changes in the middle Paleozoic Era. In the sea, new species of trilobites, eurypterids, gastropods, crinoids, and bivalves replaced species that had disappeared during the mass extinction at the end of the Ordovician Period. On land, vascular plants, with veins for transporting water and food, rooted in the soil for the first time. These plants produced the first wood and seeds. With the evolution of veins and wood, plants could grow much larger, and by the Late Devonian Period the land surface hosted swampy forests with tree-sized relatives of club mosses and ferns. Also at this time, spiders, scorpions, insects, and crustaceans came to exploit both dry-land and freshwater habitats, and jawed fish such as sharks and bony fish began to cruise the oceans. Finally, at the very end of the Devonian Period, the first amphibians crawled out onto land and breathed with lungs. In 2004, researchers found a new species, called *Tiktaalik*, that appears to be transitional between fish and amphibians—it has both gills and lungs (▶Fig. 13.21c).

The Late Paleozoic Era (Carboniferous-Permian Periods, 359–251 Ma)

Paleogeography. After the peak of the greenhouse conditions and high sea levels in the middle Paleozoic Era, the climate cooled significantly and Earth entered the late Paleozoic icehouse. Seas gradually retreated from the continents, so that during the Carboniferous Period regions that had hosted the limestone-forming reefs of epicontinental seas now became coastal swamps and river deltas in which sand, shale, and organic debris accumulated. During the Carboniferous Period, Laurentia lay near the equator (▶Fig. 13.22a, b), so it retained tropical and semitropical conditions that favored lush growth in coal swamps even though the overall global climate had cooled; this growth left thick piles of woody debris that transformed into coal on burial. Much of Gondwana and Siberia, in contrast, lay at high latitudes. By the Permian Period, ice sheets covered high-latitude regions.

The late Paleozoic Era also saw a succession of continental collisions, culminating in the formation of a single supercontinent, **Pangaea**, by the Middle Permian Period. The largest collision occurred in Carboniferous and Permian time, when Laurentia collided with Gondwana, causing the **Alleghenian orogeny** of North America. (The Alleghenian corresponds to the **Hercynian orogeny** of Europe.) During this event, the final stage in the development of the Appalachians, eastern North America squashed against northwestern Africa, and what is now the Gulf Coast region of North America squashed against the northern margin of South America. The collision was oblique, not head-on, so in addition to the development of thrust faults and folds, significant strike-slip faulting took place. A vast mountain belt grew, in which deformation generated huge faults and folds. We now see the eroded remnants of rocks deformed during this event in the Appalachian and Ouachita mountains.

Along the continental side of the Alleghenian orogen, a wide band of deformation called the Appalachian fold-thrust belt formed. In this province, you can see a distinctive style of deformation, called thin-skinned deformation, characterized by displacement on thrust faults in sedimentary strata above the Precambrian basement (▶Fig. 13.23). At depth, the thrust system merges with a near-horizontal sliding surface, or detachment. Movement on the faults generates folds in overlying beds.

Forces generated during the Alleghenian orogeny were so strong that faults in the continental crust clear across North America moved, creating uplifts (local high areas) and sediment-filled basins in the Midwest and even in the present-day Rocky Mountains (Colorado, New Mexico, and Wyoming). Geologists refer to the late Paleozoic uplifts of the Rocky Mountain region as the **Ancestral Rockies**.

The assembly of Pangaea involved a number of other collisions around the world as well. Africa collided with southern Europe to form the Hercynian orogen, a rift or small ocean in Russia closed to create the Ural Mountains, and parts of China along with other fragments of Asia attached to southern Siberia. Note that the arrangement of blocks in Pangaea differs markedly from that of Rodinia or Pannotia. Because of the immense size of Pangaea, its interior climate did not feel the moderating effect of the sea, so a desert of sand dunes, red mud, debris-choked channels, and salt pans covered large areas of the land surface.

Life evolution. The fossil record indicates that during the late Paleozoic Era, plants and animals continued to evolve toward more familiar forms. In coal swamps, fixed-wing insects such as huge dragonflies flew through a tangle of ferns, club mosses, and scouring rushes, and by the end of the Carboniferous Period insects such as the cockroach, with foldable wings, appeared (▶Fig. 13.24). Forests containing gymnosperms ("naked seed" plants such as conifers) and cycads (trees with a palmlike stalk peaked by a fan of fern-like fronds) became widespread in the Permian Period. Amphibians and, later, reptiles populated the land. The appearance of reptiles marked the evolution of a radically new component in animal reproduction: eggs with a protective covering. Such eggs permitted reptiles to reproduce without returning to the water, and thus allowed the group to populate previously uninhabitable environments. Early members of the reptile group included the carnivorous, fin-backed *pelycosaurs* and *therapsids*, the forerunners of mammals. The late Paleozoic Era came to a close with major mass-extinction events, during which over 90% of marine species disappeared. Why these particular events occurred remains a controversy; but there is evidence that the terminal Permian mass extinction occurred as a result of a huge meteor impact that, like the terminal Mesozoic event we will discuss shortly, disrupted the food chain by clouding the atmosphere with debris from the collision and with smoke from burning forests.

> **Take-Home Message**
>
> At the beginning of the Paleozoic, the late Precambrian supercontinent broke apart. North America was then bordered by passive margins, until collisions brought together all land to form Pangaea. The interiors of continents flooded during sea-level rise. The planet also saw life diversify into complex forms, first in the sea and then later on land.

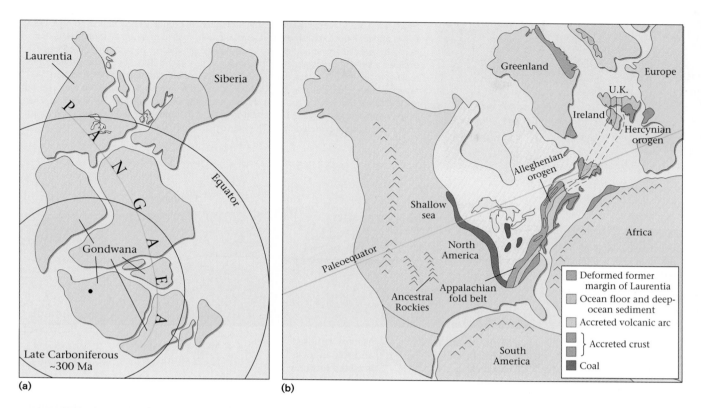

(a)

(b)

Legend for (b):
- Deformed former margin of Laurentia
- Ocean floor and deep-ocean sediment
- Accreted volcanic arc
- } Accreted crust
- Coal

Labels in (a): Laurentia, Siberia, PANGAEA, Gondwana, Equator, Late Carboniferous ~300 Ma

Labels in (b): Greenland, Europe, U.K., Ireland, Hercynian orogen, Alleghenian orogen, Shallow sea, North America, Africa, Appalachian fold belt, Ancestral Rockies, South America, Paleoequator

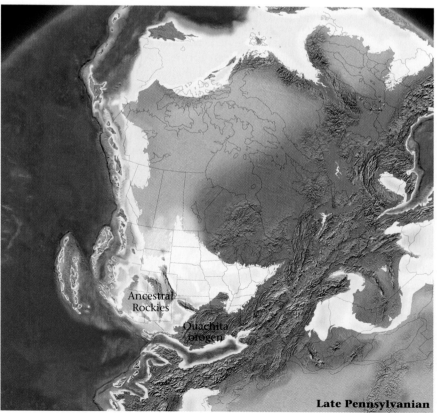

Ancestral Rockies

Ouachita orogen

Late Pennsylvanian

(c)

FIGURE 13.22 (a) This map shows the distribution of the continents of Pangaea as viewed from the South Pole (black dot). **(b)** This paleogeographical map shows the distribution of land and sea and the location of the Alleghenian and Hercynian orogens. Extensive coal swamps lay along the interior coast. The Ancestral Rockies (late Paleozoic uplifts of the Rocky Mountain region) also developed at this time, as did smaller uplifts in the Midwest. **(c)** A geologist's representation of topography at the end of the Paleozoic.

FIGURE 13.23 A cross section of the Alleghenian orogen. Note that the northwestern half is a thin-skinned fold-thrust belt, involving deformation of strata by thrust faults and folds above a sliding surface at the top of the Precambrian basement. The Blue Ridge exposes a slice of Precambrian North American basement. But in the metamorphic regions to the east, the crust consists of slices (accreted terranes) added on to North America during the Paleozoic Era. In the Piedmont, the accreted terranes are exposed, but in the Coastal Plain, these rocks have been buried by Cretaceous and Tertiary strata. The inset shows a satellite image of the fold-thrust belt in Pennsylvania. Resistant beds form ridges that trace out eroded folds. The field of view is 80 km.

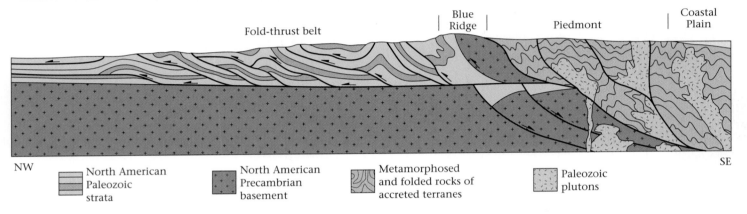

Fold-thrust belt   Blue Ridge   Piedmont   Coastal Plain

NW   SE

- North American Paleozoic strata
- North American Precambrian basement
- Metamorphosed and folded rocks of accreted terranes
- Paleozoic plutons

13.8 THE MESOZOIC ERA: WHEN DINOSAURS RULED

The Early and Middle Mesozoic Era (Triassic–Jurassic Periods, 251–145 Ma)

Paleogeography. Pangaea, the supercontinent formed in the late Paleozoic Era, lasted for about 100 million years. During the Late Triassic and Early Jurassic Periods, rifts formed

FIGURE 13.24 A museum diaorama illustrating life in a Carboniferous coal swamp. The wingspan of the giant dragonfly was about 1 m (3 feet).

and the supercontinent began to break up. Rifting started along the boundary between North America and Africa, and yielded deep rift basins. By the end of the Jurassic Period, rifting had succeeded, and the Mid-Atlantic Ridge of the North Atlantic Ocean formed (▶Fig. 13.25a). At first, the Atlantic was narrow and shallow, and evaporation made its water so salty that thick evaporite deposits, which now underlie much of the Gulf Coast region, accumulated.

According to the record of sedimentary rocks, Earth in the Triassic and Early Jurassic had greenhouse conditions, but during the Late Jurassic and Early Cretaceous, it had cool conditions. In the early Mesozoic, the interior of Pangaea remained a nonmarine environment in which red sandstones and shales, now exposed in the spectacular cliffs of Zion National Park, were deposited (▶Fig. 13.25b). The stratigraphic record indicates that by the Middle Jurassic Period, sea level began to rise. During the resulting transgression, a shallow sea submerged much of the Rocky Mountain region.

On the western margin of North America, subduction became the order of the day, and a convergent-margin orogen formed. Beginning with Late Permian and continuing through Mesozoic time, subduction created volcanic island arcs and caused them, along with microcontinents and hot-spot volcanoes, to collide with North America. The resulting orogenies included the Sonoma orogeny (in the Late Permian–Early Triassic periods) and the Nevadan orogeny (in the Late Jurassic), both of which created deformed and metamorphosed rocks. Thus, North America grew in land area by the addition (accretion) of crustal fragments and island arcs—exotic terranes—on its western mar-

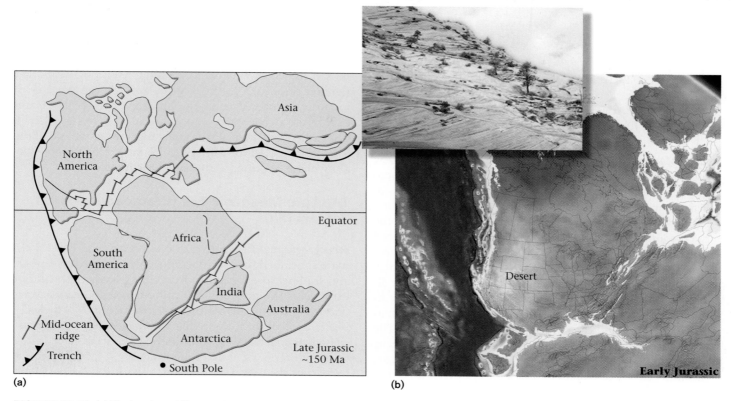

(a)

(b)

FIGURE 13.25 **(a)** The breakup of Pangaea began in the Late Triassic Period. By the Late Jurassic the North Atlantic had formed. Note that at this time, the South Atlantic remained closed, and southern Asia assembled out of a number of smaller continental fragments and volcanic island arcs. **(b)** Detail showing North America in the Early Jurassic. Rifts occur on the east, and subduction on the west. Note the sand sea in what is now Utah. The inset shows the sand-dune deposits that now make up the Navajo Formation of Utah.

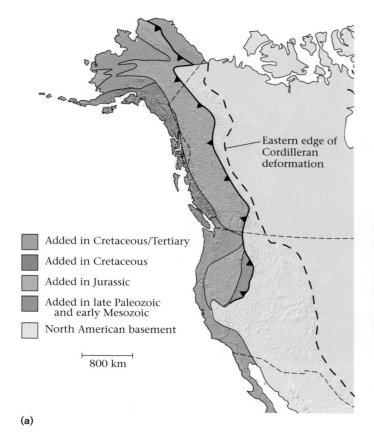

(a)

Eastern edge of Cordilleran deformation

Added in Cretaceous/Tertiary

Added in Cretaceous

Added in Jurassic

Added in late Paleozoic and early Mesozoic

North American basement

800 km

FIGURE 13.26 **(a)** During the Mesozoic Era, convergent-boundary tectonics took place along the western margin of North America. Volcanic island arcs, hot-spot volcanoes, and continental slivers accreted to the margin; the map shows the location of these blocks (accreted terranes) today. **(b)** During the Jurassic, giant dinosaurs inhabited the land.

(b)

gin (►Fig. 13.26a). From the end of the Jurassic through the Cretaceous Period, a major continental volcanic arc, the Sierran arc, formed along the western margin of North America; we'll learn more about this arc later.

Life evolution. During the early Mesozoic Era, a variety of new species of already established plant and animal groups appeared, filling the ecological niches left vacant by the Late Permian mass extinction. New creatures such as swimming reptiles (e.g., *plesiosaurs*) appeared in the sea. Corals became the predominant reef builders of the day, and have remained so ever since. On land, gymnosperms and reptiles diversified, and the Earth saw its first turtles and flying reptiles (*pterosaurs*). At the end of the Triassic Period, the first true dinosaurs appeared. Dinosaurs differed from other reptiles in that their legs were positioned under their bodies rather than off to the sides, and they were possibly warm-blooded. (The warm-blooded hypothesis has been supported by the recent discovery of a dinosaur heart; according to paleontologists, this heart closely resembles that of a bird.) By the end of the Jurassic Period, gigantic sauropod dinosaurs (such as seis-

mosaurus, which weighed up to 100 tons), along with other familiar ones such as stegosaurus and the carnivorous allosaurus, thundered across the landscape, and the first feathered birds (archaeopteryx) took to the skies (►Fig. 13.26b). The earliest ancestors of mammals appeared at the end of the Triassic Period, in the form of small, rat-like creatures. Flowering plants appeared at the end of the Jurassic.

The Late Mesozoic Era (Cretaceous Period, 145–65 Ma)

Paleogeography. In the Cretaceous Period, the Earth's climate continued to shift to warmer, greenhouse conditions, and sea level rose significantly, reaching levels that had not existed for the previous 200 million years. Great seaways flooded most of the continents, producing the thick deposits of chalk in Europe that Thomas Henry Huxley discussed, as well as layers of limestone and sandstone in the western interior of North America (►Fig. 13.27). In the latter part of the Cretaceous Period, a shark could have swum from the Gulf of Mexico to the Arctic Ocean via the West-

FIGURE 13.27 **(a)** Paleogeographical map of North America in the Early Cretaceous Period, showing the distribution of land and sea and the convergent orogen of the Cordillera. Note the position of the Sierran arc and the Sevier fold-thrust belt (inset). An interior seaway stretched from the Gulf of Mexico to the Arctic Ocean. **(b)** A geologist's representation of topography in the Late Cretaceous.

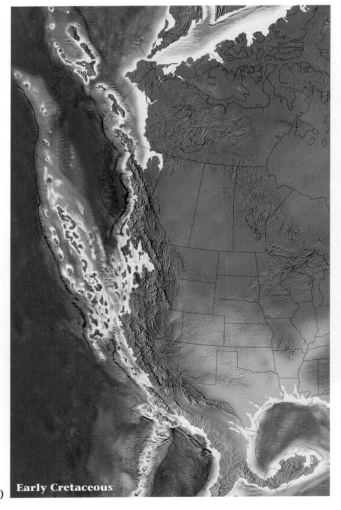

(a) Early Cretaceous

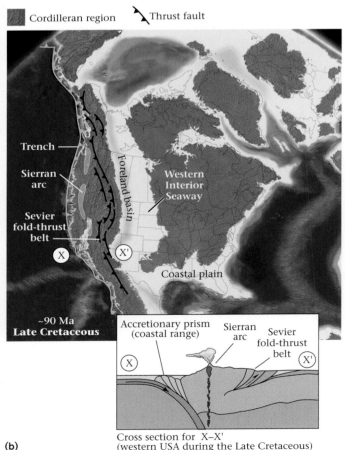

(b)

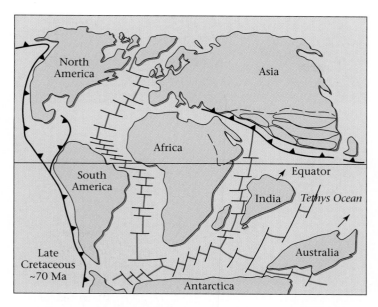

FIGURE 13.28 By the Late Cretaceous Period, the Atlantic Ocean had formed, and India was moving rapidly northward, ultimately colliding with Asia.

ern Interior Seaway of North America. The eastern and southern margins of North America were also submerged, creating an environment in which the sediments that now underlie the coastal plain collected. During this warm period, glaciers that once covered high-latitude regions of Gondwana and Siberia vanished.

The breakup of Pangaea continued through the Cretaceous Period, with the opening of the South Atlantic Ocean and the separation of South America and Africa from Antarctica and Australia. India broke away from Gondwana and headed rapidly northward toward Asia (▶Fig. 13.28). Along the continental margins of the newly formed Mesozoic oceans, new passive-margin basins developed that, like their predecessors of the early Paleozoic Era, filled with great thicknesses of sediments. For example, along the Gulf Coast, a wedge of sediment over 15 km thick has accumulated.

Along western North America, the **Sierran arc**, a large continental volcanic arc that had initiated at the end of the Jurassic Period, continued to be active. This arc resembled the one that currently lies along the Andes on the western edge of South America. Though the volcanoes of the arc have long since eroded away, we can see their roots in the form of the plutons that now constitute the granitic batholith of the Sierra Nevada. A thick accretionary prism, formed from sediments and debris scraped off the subducting oceanic plate, was present to the west of the Sierran arc and now crops out in the Coast Ranges of California. Compressional forces along the western North American convergent boundary activated large thrust faults east of the arc, an event geologists refer to as the **Se-**

vier orogeny**. This orogeny produced a fold-thrust belt whose remnants you can see today in the Canadian Rockies and in western Wyoming. The weight of the orogen helped push the surface of the continent down, forming a wide foreland basin that filled with sediment. Formation of the basin may also reflect warping down of the continental margin in response to the downward pull of the subducting plate beneath it. This foreland basin constituted the western part of the Western Interior Seaway.

At the end of the Cretaceous Period, continued compression along the convergent boundary of western North America caused large faults in the region of Wyoming, Colorado, eastern Utah, and northern Arizona to slip. In contrast to the faults of fold-thrust belts, these faults penetrated deep into the Precambrian rocks of the continent, and thus movement on them generated **basement uplifts**: overlying layers of Paleozoic strata warped into large monoclines, folds whose shape resembles the drape of a carpet over a step (▶Fig. 13.29b, c). This event, which geologists call the **Laramide orogeny**, formed the structure of the present Rocky Mountains (▶Fig. 13.29a, d). Some geologists have suggested that the contrast in the location of faulting between the Sevier and Laramide orogenies may reflect contrasts in the dip of the subducting plate. During the Laramide orogeny, the subducting plate entered the mantle at a shallower angle, and therefore scraped along and applied stress to the base of the continent farther inland. This change in dip angle did not occur in Canada, so during the Laramide orogeny the Canadian Rockies simply continued to grow eastward as a fold-thrust belt (Fig. 13.29a).

Fascinating new research suggests possible relationships among tectonic events, global climate, and sea-level changes in the Cretaceous Period. For example, the breakup of Pangaea led to the formation of many mid-ocean ridges, and geologists have determined (by studying marine magnetic anomalies) that sea-floor spreading took place about three times faster on these ridges during the Cretaceous Period than it does today. Thus, more of the oceanic crust was younger and warmer then than it is today. And since young sea floor lies at a shallower depth than older crust, mid-ocean ridges occupied more space than they do today. The ridges therefore displaced seawater, causing sea level to rise and continents to flood.

By mapping the sea floor, geologists have also discovered that huge submarine plateaus, composed of hot-spot volcanic rocks, formed during the Cretaceous Period. These plateaus might also have contributed to the displacement of seawater. The presence of huge hot spots implies that the Cretaceous may have been a time when particularly large mantle plumes, called **superplumes**, reached the base of the lithosphere. Hot-spot and mid-ocean-ridge activity may have influenced the climate, because the volcanoes they produced released huge quantities of carbon dioxide (CO_2) into the atmosphere, perhaps as

(a)

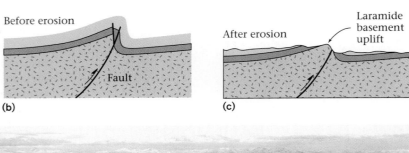

Before erosion

Fault

(b)

After erosion

Laramide basement uplift

(c)

(d)

FIGURE 13.29 **(a)** During the Laramide orogeny, deformation shifted eastward, from the Sevier fold-thrust belt to the belt of Laramide uplifts. **(b)** The geometry of a basement uplift before erosion. Note that the fault penetrates the basement, and that the overlying sediment warps into a fold that resembles a carpet draped over a stair step. **(c)** After erosion, the Precambrian rocks in the core of the uplift crop out at the ground surface. **(d)** The Rocky Mountain Front as seen looking west over Denver, Colorado.

much as eight times the amount found today. An increase in the concentration of CO_2 in the atmosphere leads to an increase in atmospheric temperature, for CO_2 in the air traps infrared radiation rising from the Earth and causes the air to warm up. (In this way, CO_2 acts as a greenhouse gas, in that it plays the same role as glass panes in a greenhouse; see Chapter 23.) This increase would melt ice sheets, and thus contribute to the rise of sea level.

Life evolution. In the seas of the late Mesozoic world, modern fish, called teleost fish, appeared and became dominant. What set them apart from earlier fish were their short jaws, rounded scales, symmetrical tails, and specialized fins. Huge swimming reptiles and gigantic turtles (with shells up to 4 m across) preyed on the fish. On land, cycads largely vanished, and **angiosperms** (flowering plants), including hardwood trees, began to compete successfully with conifers for dominance of the forest. Angiosperms can produce seeds much more rapidly than conifers and can attract insects to help with pollination. Dinosaurs reached their peak of success at this time, inhabiting almost all environments on Earth. Social herds of grazing dinosaurs roamed the plains, preyed on by the fearsome *Tyrannosaurus rex* (a Cretaceous, not Jurassic, dinosaur, despite what Hollywood

says!). Pterosaurs, with wingspans of up to 11 m, soared overhead, and birds began to diversify. Mammals also diversified and developed larger brains and more specialized teeth, but they remained small and rat-like.

The "K-T boundary event." Geologists first recognized the K-T boundary (K stands for Cretaceous and T for Tertiary) from eighteenth-century studies that identified an abrupt global change in fossil assemblages. Until the 1980s, most geologists assumed the faunal turnover took millions of years. But modern dating techniques indicate that this change happened almost instantaneously and that it indicated the sudden mass extinction of most species on Earth. The dinosaurs, which had ruled the planet for over 150 million years, simply vanished, along with 90% of some plankton species in the ocean and up to 75% of plant species. What kind of catastrophe could cause such a sudden and extensive mass extinction? From data collected in the 1970s and 1980s, most geologists have concluded that the Cretaceous Period came to a close, at least in part, as a result of the impact of a 10-km-wide meteorite at the site of the present-day Yucatán peninsula in Mexico (▶Fig. 13.30).

The discoveries leading up to this conclusion provide fascinating insight into how science works. The story

FIGURE 13.30 This painting illustrates the collision of a huge bolide with the Earth at the end of the Cretaceous Period.

photosynthesis to all but cease, and thus would break the food chain and probably trigger extinctions.

Geologists suggest that the meteorite landed on the northwestern coast of the Yucatán peninsula, where a 100-km-wide by 16-km-deep scar called the Chicxulub crater lies buried beneath younger sediment. A layer of glass spherules up to 1 m thick occurs at the K-T boundary in strata near the site. And radiometric dating indicates that igneous melts in the crater formed at 65 ± 0.4 Ma, exactly the time of the K-T boundary event. The discovery of this event has led geologists to speculate that other such collisions may have punctuated the path of life evolution throughout Earth history (see art, pp. 476–477).

Take-Home Message

The Mesozoic was the age of dinosaurs, for these beasts roamed all continents during this eon. In North America, a major convergent plate boundary formed along the west coast, while rifting of Pangaea led to the formation of the Atlantic Ocean on the east coast.

began when Walter Alvarez, a geologist studying strata in Italy, noted that a thin layer of clay interrupted the deposition of deep-sea limestone precisely at the K-T boundary. Cretaceous plankton shells constituted the limestone below the clay layer, whereas Tertiary plankton shells made up the limestone just above the clay. Apparently, for a short interval of time at the K-T boundary, all the plankton died, so that only clay settled out of the sea. When Alvarez and his father, Luis (a physicist), and other colleagues analyzed the clay, they learned that it contained iridium, a very heavy element found only in extraterrestrial objects. Soon, geologists were finding similar iridium-bearing clay layers at the K-T boundary all over the world. Further study showed that the clay layer contained other unusual materials, such as tiny glass spheres (formed from the flash freezing of molten rock), wood ash, and shocked quartz (grains of quartz that had been subjected to intense pressure). Only an immense impact could explain all these features. The glass spherules formed when melt sprayed in the air from the impact site, the ash resulted when forests were set ablaze by the impact, the iridium came from fragments of the colliding object, and the unusual quartz grains were created by the shock of the impact.

The impact caused so much destruction because it not only formed a crater, blasting huge quantities of debris into the sky, but probably also generated 2-km-high tsunamis that inundated the shores of continents and generated a blast of hot air that set forests on fire. The blast and the blaze together would have ejected so much debris into the atmosphere that for months there would have been perpetual night and winter-like cold. In addition, chemicals ejected into the air would have combined with water to produce acid rain. These conditions would cause

13.9 THE CENOZOIC ERA: THE FINAL STRETCH TO THE PRESENT

Paleogeography. During the last 65 million years, the map of the Earth has continued to change, gradually producing the configuration of continents we see today. The final stages of the Pangaea breakup separated Australia from Antarctica and Greenland from North America, and formed the North Sea between Britain and continental Europe. The Atlantic Ocean continued to grow because of sea-floor spreading on the Mid-Atlantic Ridge, and thus the Americas moved westward, away from Europe and Africa. Meanwhile, the continents that once constituted Gondwana drifted northward as the intervening Tethys Ocean was consumed by subduction. Collisions of the former Gondwana continents with the southern margins of Europe and Asia resulted in the formation of the largest orogenic belt on the continents today, the **Alpine-Himalayan chain** (▶Fig. 13.31). India and a series of intervening volcanic island arcs and microcontinents collided with Asia to form the Himalayas and the Tibetan Plateau to the north. Africa, along with some volcanic island arcs and microcontinents, collided with Europe to produce the Alps in the west and the Zagros Mountains of Iran in the east. Finally, collision between Australia and New Guinea led to orogeny in Papua New Guinea.

As the Americas moved westward, convergent plate boundaries evolved along their western margins. In South America, convergent-boundary activity has built the Andes mountain belt, which remains an active orogen to the present

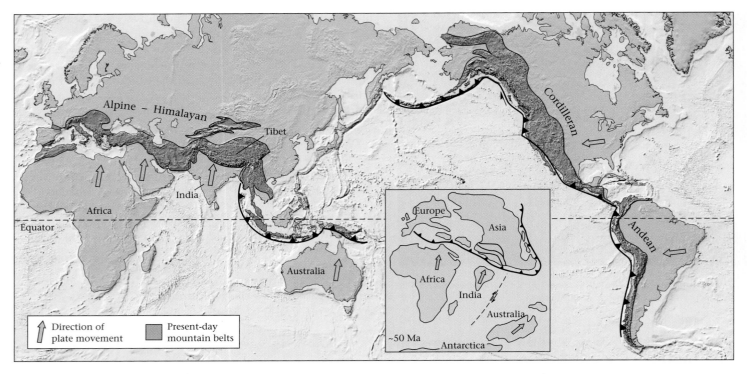

FIGURE 13.31 The two main continental orogenic systems on the Earth today. The Alpine-Himalayan system formed when pieces of Gondwana (Africa, India, and Australia) migrated north and collided with Asia (inset). The Cordilleran and Andean systems are the consequence of convergent-boundary tectonism along the eastern Pacific.

day. In North America, convergent-boundary activity continued without interruption until the Eocene Epoch, yielding, as we have seen, the Laramide orogen. Then, because of the rearrangement of plates off the western shore of North America, beginning about 40 Ma, a transform boundary replaced the convergent boundary in the western part of the continent (▶Fig. 13.32a–c). When this happened, volcanism and compression ceased in western North America, the San Andreas fault system formed along the coast of the United States, and the Queen Charlotte fault system developed off the coast of Canada. Along the San Andreas and Queen Charlotte Faults today, the Pacific Plate moves northward with respect to North America at a rate of about 6 cm per year. The Queen Charlotte Fault links to the Aleutian subduction zone along southern Alaska, where the Pacific Ocean floor undergoes subduction. In the western United States, convergent-boundary tectonics continues only in Washington, Oregon, and northern California where subduction of the Juan de Fuca Plate produces the Cascade volcanic chain.

As convergent tectonics ceased in the western United States, south of the Cascades, the region began to undergo rifting (stretching) in roughly an east-west direction. The result was the formation of the **Basin and Range Province**, a broad continental rift that has caused the region to stretch to twice its original width (▶Fig. 13.33). The Basin and Range gained its name from its topography—the province contains

long, narrow mountain ranges separated from each other by flat, sediment-filled basins. This geometry reflects the normal faulting resulting from stretching: crust of the region was broken up by faults, and movement on the faults created elongate depressions. The Basin and Range Province terminates just north of the Snake River Plain, the track of the hot spot that now lies beneath Yellowstone National Park.

Recall that in the Cretaceous Period, the world experienced greenhouse conditions and sea level rose so that extensive areas of continents were submerged. During the Cenozoic Era, however, the global climate rapidly shifted to icehouse conditions, and by the early Oligocene Epoch, Antarctic glaciers reappeared for the first time since the Triassic. The climate continued to grow colder through the Late Miocene Epoch, leading to the formation of grasslands in temperate climates. About 2.5 Ma, the Isthmus of Panama formed, separating the Atlantic completely from the Pacific, changing the configuration of oceanic currents, and allowing the Arctic Ocean to freeze over.

In the overall cold climate of the last 2 million years, the Quaternary Period, continental glaciers have expanded and retreated across northern continents at least 20 times, resulting in the **Pleistocene ice age** (▶Fig. 13.34). Each time the edge of the glacier advanced, sea level fell so much that the continental shelf became exposed to air, and a land bridge formed across the Bering Strait, west of Alaska,

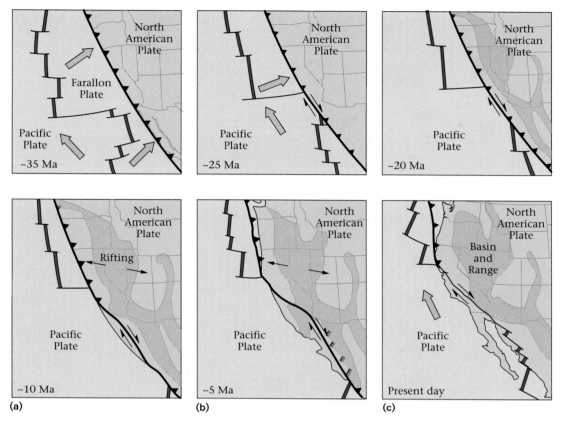

FIGURE 13.32 The western margin of North America changed from a convergent-plate boundary into a transform-plate boundary when the Farallon-Pacific Ridge was subducted. **(a, b)** The Farallon Plate was moving toward North America, while the Pacific Plate was moving parallel to the western margin of North America. **(c)** The Basin and Range Province opened as the San Andreas Fault developed. Subduction along the West Coast today occurs only where the Juan de Fuca Plate, a remnant of the Farallon Plate, continues to subduct.

providing migration routes for animals and people from Asia into North America; a partial land bridge also formed from southeast Asia to Australia, making human migration to Australia easier. Erosion and deposition by the glaciers created much of the landscape we see today in northern temperate regions. About 11,000 years ago, the climate warmed, and we entered the interglacial time interval we are still experiencing today (see Chapter 22). This most recent interval if time is the Holocene.

Life evolution. When the skies finally cleared in the wake of the K-T boundary catastrophe, plant life recovered, and soon forests of both angiosperms and gymnosperms reappeared. A new group of plants, the grasses, sprang up and began to dominate the plains in temperate and subtropical climates by the middle of the Cenozoic Era. The dinosaurs, however, were gone for good, for no examples of such great beasts survived; their descendants are the birds of today. Mammals rapidly diversified into a variety of forms to take their place. In fact, most of the modern groups of mammals that exist today originated at the beginning of the

Cenozoic Era, giving this time the nickname "Age of Mammals." During the latter part of the era, remarkably huge mammals appeared (such as mammoths, giant beavers, giant bears, and giant sloths), but these became extinct over the past 10,000 years, perhaps because of hunting by humans.

It was during the Cenozoic that our own ancestors first appeared. Ape-like primates diversified in the Miocene Epoch (about 20 Ma), and the first human-like primate appeared at about 4 Ma, followed by the first members of the human genus, *Homo*, at about 2.4 Ma. (Note that people and dinosaurs did *not* inhabit Earth at the same time!) Perhaps the evolution of *Homo*, with its significantly larger brain, accompanied climate changes that led to the spread of grasslands, allowing primates to

> **Take-Home Message**
>
> The Cenozoic began with a bang—literally—as the age of dinosaurs came to a close, probably in the wake of a huge meteorite impact. During the Cenozoic, the modern mountain belts and plate boundaries of the planet became established, and mammals diversified.

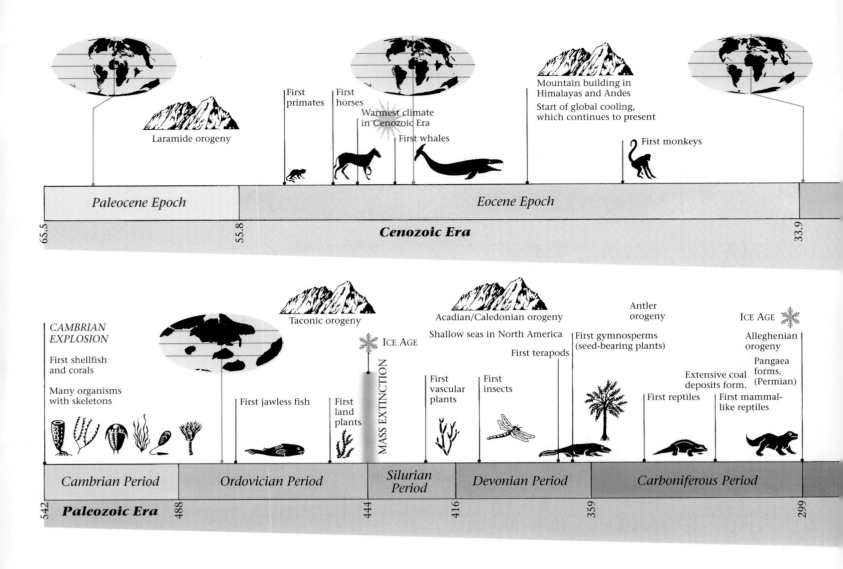

Cenozoic Era

First primates · First horses · Warmest climate in Cenozoic Era · First whales

Laramide orogeny

Mountain building in Himalayas and Andes

Start of global cooling, which continues to present

First monkeys

Paleocene Epoch	Eocene Epoch

65.5 · 55.8 · 33.9

CAMBRIAN EXPLOSION

First shellfish and corals

Many organisms with skeletons

First jawless fish

First land plants

Taconic orogeny

ICE AGE

MASS EXTINCTION

First vascular plants

First insects

Shallow seas in North America

Acadian/Caledonian orogeny

First terapods

Antler orogeny

First gymnosperms (seed-bearing plants)

ICE AGE

Alleghenian orogeny

Extensive coal deposits form.

Pangaea forms. (Permian)

First reptiles

First mammal-like reptiles

Cambrian Period	Ordovician Period	Silurian Period	Devonian Period	Carboniferous Period

542 · 488 · 444 · 416 · 359 · 299

Paleozoic Era

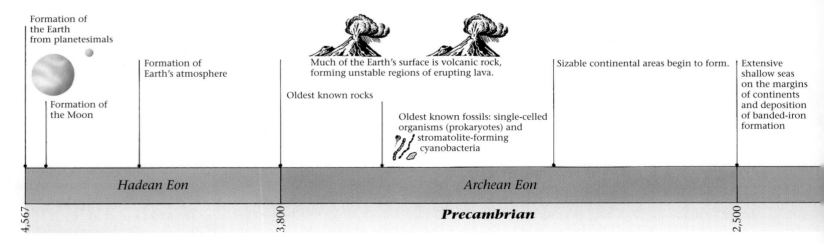

Formation of the Earth from planetesimals

Formation of the Moon

Formation of Earth's atmosphere

Oldest known rocks

Much of the Earth's surface is volcanic rock, forming unstable regions of erupting lava.

Oldest known fossils: single-celled organisms (prokaryotes) and stromatolite-forming cyanobacteria

Sizable continental areas begin to form.

Extensive shallow seas on the margins of continents and deposition of banded-iron formation

Hadean Eon	Archean Eon

4,567 · 3,800 · 2,500

Precambrian

The Evolution of Earth

Earth has not had a static history; because of plate tectonics and its consequences (continental drift, sea-floor spreading, volcanism, etc.), the map of the Earth constantly changes. Distinct mountain-building events, or orogenies, have taken place during this process. The fossil record suggests that life appeared within the first few hundred million years of our planet's existence, soon after a liquid-water

ocean had accumulated, and like the planet itself, has constantly changed ever since. The progressive change of the assemblage of species of life on Earth is called evolution.

The earliest life forms were microscopic. By the end of the Precambrian, complex multicellular organisms had formed, and a burst of evolution at the Precambrian-Cambrian boundary yielded a

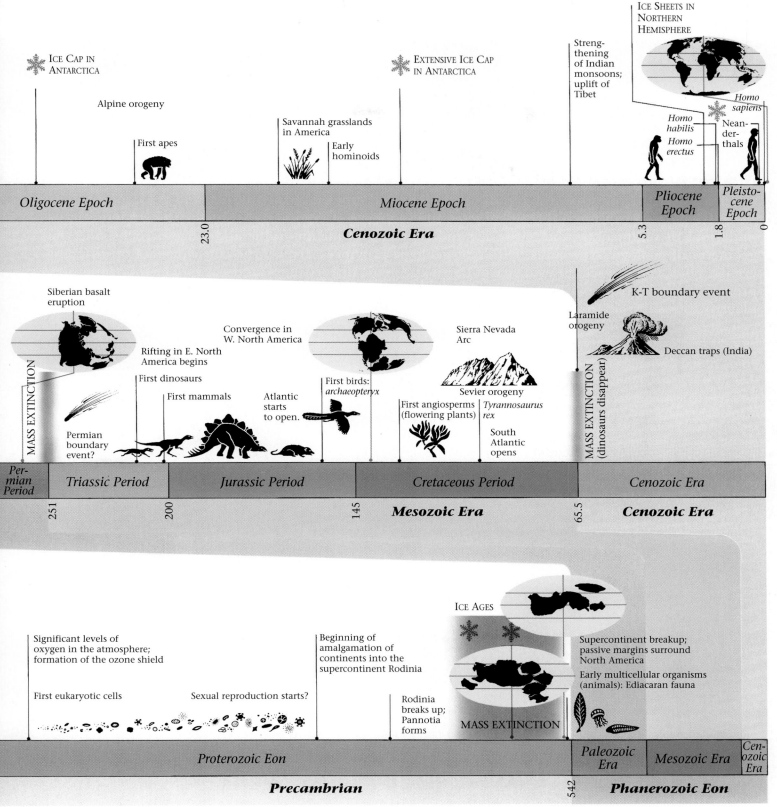

ICE CAP IN ANTARCTICA

Alpine orogeny

First apes

ICE SHEETS IN NORTHERN HEMISPHERE

EXTENSIVE ICE CAP IN ANTARCTICA

Savannah grasslands in America

Early hominoids

Strengthening of Indian monsoons; uplift of Tibet

Homo sapiens

Homo habilis

Homo erectus

Neanderthals

| Oligocene Epoch | Miocene Epoch | Pliocene Epoch | Pleistocene Epoch |

23.0

Cenozoic Era

5.3

1.8

0

Siberian basalt eruption

K-T boundary event

Laramide orogeny

Deccan traps (India)

Convergence in W. North America

Rifting in E. North America begins

Sierra Nevada Arc

First dinosaurs

MASS EXTINCTION

First mammals

First birds: *archaeopteryx*

Atlantic starts to open.

Sevier orogeny

First angiosperms (flowering plants)

Tyrannosaurus rex

South Atlantic opens

Permian boundary event?

MASS EXTINCTION (dinosaurs disappear)

| Permian Period | Triassic Period | Jurassic Period | Cretaceous Period | Cenozoic Era |

251

200

145

Mesozoic Era

65.5

Cenozoic Era

ICE AGES

Significant levels of oxygen in the atmosphere; formation of the ozone shield

Beginning of amalgamation of continents into the supercontinent Rodinia

Supercontinent breakup; passive margins surround North America

First eukaryotic cells

Sexual reproduction starts?

Rodinia breaks up; Pannotia forms

MASS EXTINCTION

Early multicellular organisms (animals): Ediacaran fauna

| Proterozoic Eon | Paleozoic Era | Mesozoic Era | Cenozoic Era |

Precambrian

542

Phanerozoic Eon

Geologic time (million years ago)

diversity of invertebrates with shells. During the past half-billion years, several new major groups of organisms have appeared, and countless species have become extinct. Evidence suggests that evolution is not a continuous, gradual process, but occurs in pulses, separated by intervals of time during which the assemblage of species is fairly stable.

The last few hundred million years of Earth history have seen life leave the ocean and spread across the land. During the Mesozoic Era, dinosaurs roamed the Earth—then vanished abruptly 65 million years ago, perhaps as a result of a meteorite colliding with the Earth. Since then, mammals have diversified into a great variety of species. And the last 100,000 years or so have witnessed the evolution of our own species, *Homo sapiens*. Considering the changes that people have brought about to the Earth System, the appearance of our species clearly was a major event in the history of the planet. The time scale is the 2004 time scale (International Committee on Stratigraphy).

See for yourself . . .
Earth Has a History

The Earth displays the record of a long and complex geologic history, marked by the collision of continents, the rifting of crust, the building and eventual erosion of mountains, and the deposition of strata. Geologists can unravel the character and sequence of events in this history by studying accumulations of sedimentary rocks, the age and character of metamorphic and igneous rocks, and the geometry of geologic structures. Different geologic provinces display the consequences of different events. Thus, you can see the evidence of Earth's geologic history in its landscapes. The following sites provide examples from North America. The thumbnail images provided on this page are only to help identify tour sites. Go to wwnorton.com/studyspace to experience this flyover tour.

Folds of the Canadian Shield
(Lat 58°4'44.52"N, Long 69°36'31.14"W)

Let's start by looking at some of the oldest crust of this continent, which is exposed in the Canadian Shield. This region is part of the craton, a tectonically inactive region of continental crust that formed in the Precambrian. Though it had a complex geologic past, it has not endured mountain building during the last billion years—instead, it has been subjected to erosion. As a result, rocks that were formed at depths of over 10 km (6 miles) now crop out. Fly to the coordinates provided and you'll find yourself in northeastern Canada, west of Ungava Bay (Quebec). Zoom to an elevation of 70 km (45 miles) (Image G13.1). You can see a subdued landscape—none of the hills are very high, and none of the valleys are very deep. Nevertheless, the ups and downs define swirling patterns, some emphasized by dark lakes. These patterns are controlled by the folded foliation of Precambrian gneiss.

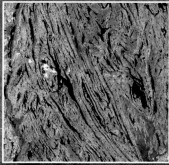
G13.1

Appalachian Mountains, Tennessee
(Lat 36°28'33.93"N, Long 83°45'28.63"W)

Fly to this locality and hover at an elevation of 100 km (62 miles). You can see a portion of the fold belt (known geographically as the Appalachian Valley and Ridge province) in northern Tennessee and southern Kentucky (Image G13.2). If you zoom down to an elevation of 20 km (12 miles) in the lower right corner of the image, tilt to see the horizon, and pivot the image so you are looking northeast, the Valley and Ridge topography is clear (Image G13.3). Resistant stratigraphic formations hold up ridges, and weaker ones erode away to form valleys. The deformation here involves Cambrian through Pennsylvania strata, so geologists deduce that the deformation is a consequence of a late Paleozoic event. This event is the Alleghanian orogeny, due to the collision of Africa with North America. This collision compressed the crust and caused it to wrinkle along its eastern margin, somewhat like a rug on a slippery floor.

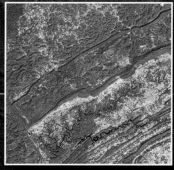
G13.2

G13.3

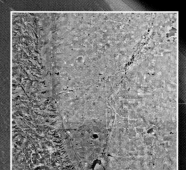
G13.4

Maroon Bells, Colorado
(Lat 39°4'17.65"N, Long 106°59'19.42"W)

Fly to the coordinates given and you'll find a set of peaks known as the Maroon Bells, south of Aspen, Colorado. Zoom to an elevation of 7.5 km (4.5 miles), tilt so you just see the horizon, and rotate your view so you are looking due east (Image G13.4). You can now see layering of the reddish strata that make up these mountains—fly around the mountains to get a broader perspective. Here, we have the record of two major orogenic events: (1) The late Paleozoic Ancestral Rockies event, which formed deep basins and uplifts. The reddish strata of the mountains were deposited in one of the basins. Because the event happened at roughly the same time as the Alleghanian orogeny, it might also be due to compression caused by the collision of Africa and South America; and (2) The Mesozoic/Cenozoic Laramide orogeny, during which the region was uplifted by several kilometers. Erosion subsequent to the Laramide orogeny produced the present landscape. Some valleys were carved by glaciers during the last ice age.

G13.5

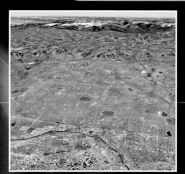
G13.6

The Rocky Mountain Front, Colorado
(Lat 39°44'55.71"N, Long 104°59'44.74"W)

Fly to this location and zoom to an elevation of 70 km (45 miles) and you are hovering over Denver, Colorado. The field of view (Image G13.5) encompasses the Rocky Mountain front which marks the boundary between the Rocky Mountains to the west and the plains to the east. Now, zoom down to an elevation of 14 km (8.7 miles), tilt your view so you just see the horizon, and rotate so you are looking southwest. You can now see the abrupt face of the mountain front (Image G13.6). The snow-capped peak on the horizon is Mt. Evans, one of several 4.3 km (14,000 feet)-high peaks in Colorado. Fly about 12 km (7.5 miles) to the mountain front, just where highway I-70 crosses the front (Lat 39° 38'3.75"N Long 105° 10'14.28"W), drop down to an elevation of 3.5 km (2 miles), and tilt your view so you are looking north. Here you see the hogbacks at the front of the range (Image G13.7). These are asymmetric ridges composed of strata dipping toward the east at about 35°. We see two components of orogeny—uplift and deformation. These developed during the Laramide orogeny.

G13.7

G13.8

G13.9

Basin and Range Rift
(Lat 38°50'30.06"N, Long 114°46'41.99"W)

Fly to these coordinates and zoom to an elevation of 150 km (93 miles). Here, you see an approximately 120 km (75 miles)-wide block of the Basin and Range Province, in Nevada, just west of the Utah border (Image G13.8). This region is a continental rift, formed during the Cenozoic, after the Laramide orogeny ceased. The tilted fault blocks now form N-S trending narrow ranges. Because of their higher elevation, they are lightly forested. The basins between these blocks have filled with sediment. Their flat surfaces are at lower elevation and have desert climates. Streams occasionally flow along the axes of the basins. If you zoom down to 14 km (8.7 miles), tilt the image to just see the horizon, and look due east at the easternmost range in the previous view, then you can see how a range rises between two basins (Image G13.9). The floor of the basin in this view occasionally fills with water, forming a desert feature called a playa lake.

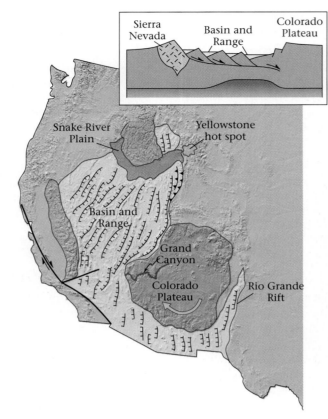

FIGURE 13.33 The Basin and Range Province is a rift (inset). The northern part has opened more than the southern part, causing the Sierran arc to swing westward and rotate. The Rio Grande Rift is a small rift that links to the Basin and Range. The Colorado Plateau is a block of craton bounded by the Rio Grande Rift to the east and the Basin and Range to the west.

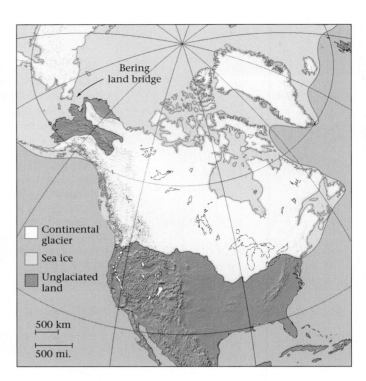

FIGURE 13.34 North America during the maximum advance of the Pleistocene ice sheet. Because sea level was so low (water was stored in the ice sheet on land), a land bridge formed across the Bering Strait, allowing people and animals to migrate from Asia to America.

leave the trees. Life on the ground provides a longer time for infant development and the growth of a large brain. Fossil evidence, primarily from Africa, indicates that *Homo erectus*, capable of making stone axes, appeared about 1.6 Ma, and the line leading to *Homo sapiens* (our species) diverged from *Homo neanderthalensis* (Neanderthal man) about 500,000 years ago. According to the fossil record, modern people appeared about 150,000 years ago. Thus, much of human evolution took place during the radically shifting climatic conditions of the Pleistocene Epoch.

Chapter Summary

- Earth formed about 4.57 billion years ago. For part of the first 600 million years, the Hadean Eon, the planet was so hot that its surface was a magma ocean. We have hardly any rock record of this time interval, but we can gain insight into it by studying the Moon and meteorites.

- The Archean Eon began about 3.8 Ga, when permanent continental crust formed. The crust assembled out of volcanic arcs and hot-spot volcanoes that were too buoyant to subduct. Oceans and stable continental blocks, called cratons, also formed. The atmosphere contained little oxygen, but the first life forms—bacteria and archaea—appeared.

- In the Proterozoic Eon, which began at 2.5 Ga, Archean cratons collided and sutured together along orogenic belts and large Proterozoic cratons. Photosynthesis by organisms added oxygen to the atmosphere, and iron precipitated out of the ocean to deposit banded-iron formations. By the end of the Proterozoic, complex but shell-less marine invertebrates populated the planet. Most continental crust accumulated to form a supercontinent called Rodinia at about 1 Ga. Rodinia broke apart and reorganized to form another supercontinent, Pannotia, at the end of the Proterozoic Eon.

- At the beginning of the Paleozoic Era, Pannotia broke apart, yielding several smaller continents. Sea level rose and fell a number of times, creating sequences of strata in continental interiors. Continents began to collide and coalesce again, leading to new orogenies and, by the end of the era, another supercontinent, Pangaea. Early Paleozoic evolution produced many invertebrates with shells, and jawless fish. Land plants and insects appeared in the middle Paleozoic. And by the end of the eon, there were land reptiles and gymnosperm trees.

- In the Mesozoic Era, Pangaea broke apart and the Atlantic Ocean formed. Convergent-boundary tectonics dominated along the western margin of North America, causing mountain-building events such as the Sevier and Laramide orogenies. Dinosaurs appeared in Late Triassic time and became the dominant land animal through the Mesozoic Era. During the Cretaceous Period, sea level was very high, and the continents flooded. Angiosperms (flowering plants) appeared at this time, along with modern fish. A huge mass extinction event (the K-T boundary event), which wiped out the dinosaurs, occurred at the end of the Cretaceous Period, probably because of the impact of a large bolide with the Earth.

- In the Cenozoic Era, continental fragments of Pangaea began to collide again. The collision of Africa and India with Asia and Europe formed the Alpine-Himalayan orogen. Convergent tectonics has persisted along the margin of South America, creating the Andes, but ceased in North America when the San Andreas Fault formed. Rifting in the western United States during the Cenozoic Era produced the Basin and Range Province. Various kinds of mammals filled niches left vacant by the dinosaurs, and the human genus, *Homo*, appeared and evolved throughout the radically shifting climate of the Pleistocene Epoch.

Geopuzzle Revisited

The Earth has a long and complex history. According to geologic studies, our planet formed at about 4.57 Ga. The first land and water may have appeared as early as 4.04 Ga, but the record of the earliest crust was destroyed by meteorite bombardment about 3.9 Ga. The first long-lived continental crust appeared about 3.8 Ga. Oceans also appeared about 3.8 Ga and have lasted ever since. Subtle evidence suggests that life appeared by around 3.5 Ga. Thus, life did not become established until the Earth was a billion years old. Most continental crust formed by about 2.5 Ga. Some has survived intact, but most has been recycled by tectonic processes. The amount of dry land, however, changes over time due to the rise and fall of sea level. As continents drift, collide, and break apart, mountain belts form and then later erode away. The present ranges on Earth have been around only since the beginning of the Cenozoic.

Key Terms

Acadian orogeny (p. 464)
accretionary orogens (p. 457)
Alleghenian orogeny (p. 466)
Alpine-Himalayan chain (p. 473)
Ancestral Rockies (p. 466)
angiosperms (p. 472)
Antler orogeny (p. 465)
Archean Eon (p. 453)
banded-iron formation (BIF) (p. 459)
basement uplifts (p. 471)
Basin and Range Province (p. 474)
Caledonian orogeny (p. 464)
Cambrian explosion (p. 463)
continental platform (p. 456)
craton (p. 453)
cratonic (continental) platform (p. 456)
Ediacaran fauna (p. 459)
epicontinental seas (p. 461)

greenhouse conditions (p. 464)
Grenville orogen (p. 457)
Hercynian orogen (p. 466)
internal differentiation (p. 452)
Laramide orogeny (p. 471)
Pangaea (p. 466)
Pannotia (p. 457)
Pleistocene ice age (p. 474)
protocontinent (p. 453)
Rodinia (p. 457)
Sevier orogeny (p. 471)
shield (p. 456)
Sierran arc (p. 471)
snowball Earth (p. 460)
stratigraphic sequence (p. 462)
stromatolite (p. 454)
superplume (p. 471)
Taconic orogeny (p. 463)

Review Questions

1. List some methods by which geologists study the past.

2. Why are there no rocks on Earth that yield radiometric dates older than 4 billion years?

3. Describe the condition of the crust, atmosphere, and oceans during the Hadean Eon.

4. Describe the five principal rock types found in Archean protocontinents. Under what kind of environmental conditions did each type form?

5. What are stromatolites? How do they form?

6. How did the atmosphere and tectonic conditions change during the Proterozoic Eon?

7. How do banded-iron formations tell us about atmospheric oxygen levels?

8. What evidence do we have that the Earth nearly froze over twice during the Proterozoic Eon?

9. How did the Cambrian explosion of life change the nature of the living world?

10. How did the Taconic and Acadian orogenies affect the eastern coast of North America?

11. How did the Alleghenian and Ancestral Rocky orogenies affect North America?

12. What are the major classes of organisms to appear in the Paleozoic?

13. Compare the typical sedimentary deposits of the early Paleozoic greenhouse Earth with those of the late Paleozoic–early Mesozoic icehouse Earth.

14. Describe the plate tectonic conditions that led to the formation of the Sierran arc and the Sevier thrust belt.

15. How did the plate tectonic conditions of the Laramide orogeny differ from more typical subduction zones?

16. What life forms appeared during the Mesozoic?

17. What caused the flooding of the continents during the Cretaceous Period?

18. What could have caused the K-T extinctions?

19. What continents formed as a result of the breakup of Pangaea?

20. What are the causes of the uplift of the Himalayas and the Alps?

21. What events led to the end of the Mesozoic greenhouse and the development of glaciers on the Arctic and Antarctic during the Cenozoic Era?

On Further Thought

1. During intervals of the Paleozoic, large areas of continents were submerged by shallow seas. Using *Google Earth*™ or a comparable program, tour North America from space. Do any present-day regions *within* North America consist of continental crust that was submerged by seawater? What about regions offshore? (*Hint*: Look at the region just east of Florida.)

2. Describe the field evidence that geologists could have used to determine that the "transcontinental arch" in North America was above sea level during intervals of the Paleozoic when the surrounding area of the continent was submerged.

3. Geologist have concluded that 80% to 90% of Earth's continental crust had formed by 2.5 Ga. But if you look at a geological map of the world, you find that only about 10% of the Earth's continental crustal surface is labeled "Precambrian." Why?

Suggested Reading

Clarkson, E. K. 1998. *Invertebrate Palaeontology and Evolution*. 4th ed. Boston: Blackwell.

Cloud, P. 1988. *Oasis in Space: Earth History from the Beginning*. New York: Norton.

Condie, K. C. 1989. *Plate Tectonics and Crustal Evolution*. New York: Pergamon.

Cutler, A. 2004. *The Seashell on the Mountaintop: How Nicolaus Steno Solved an Ancient Mystery and Created the Science of the Earth*. New York: Plume Paperback.

Fastovsky, D. E., and D. B. Weishampel. 2005. *The Evolution and Extinction of the Dinosaurs*. 2nd ed. New York: Cambridge University Press.

Gould, S. J., et al. 1993. *The Book of Life*. New York: W. W. Norton.

Gradstein, F., J. Ogg, and A. Smith. 2004. *A Geologic Time Scale 2004*. Cambridge: Cambridge University Press.

Knoll, A. H. 2003. *Life on a Young Planet*. Princeton, N.J.: Princeton University Press.

Nisbet, E. G. 1987. *The Young Earth: An Introduction to Archean Geology*. Boston: Allen and Unwin.

Prothero, D. R., and R. H. Dott, Jr. *Evolution of the Earth*. 7th ed. New York: McGraw-Hill.

Rodgers, J. J. W. 1994. *A History of the Earth*. Cambridge: Cambridge University Press.

Rollinson, H. 2007. *Earth Systems*. Boston: Blackwell.

Smith, R. B., and L. J. Siegel. 2000. *Windows into the Earth: The Geologic Story of Yellowstone and Grand Teton National Parks*. New York: Oxford University Press.

Stanley, S. M. 2004. *Earth System History*. 2nd ed. New York: Freeman.

Windley, B. F. 1995. *The Evolving Continents*. 3rd ed. New York: Wiley.

THE VIEW FROM SPACE The present-day Bahamas serve as an example of what the interior of the United States might have looked like during intervals of the Paleozoic. Shallow land areas were submerged and became the site of shallow-marine sedimentation.

Earth Resources

Many of the materials we use in our daily lives come from geologic materials—the Earth itself is, essentially, a natural resource. In Chapter 14, we look at the energy resources that come from the Earth. These include fossil fuels, such as oil and coal, as well as nuclear fuel and moving water. Chapter 15 focuses on nonenergy resources, particularly the mineral deposits from which we obtain metals.

By the end of Part V, we'll realize that many natural resources are not renewable and thus must be conserved if we are to avoid shortages. Also, we'll see how our use of resources has had and continues to have an impact on the environment and on transnational politics.

An airplane view shows us the great pit of the Bingham Mine in Utah. This huge excavation, which can be seen by satellites 2000 km in space, provides ore that has yielded tons of copper, gold, and silver. Earth materials provide energy, metals, and other resources.

Squeezing Power from a Stone: Energy Resources

Geopuzzle

Most of the energy used today comes from the burning of oil, gas, and coal. How much longer will this pattern of energy usage last, and why?

Workers struggle to extinguish the inferno erupting from an oil well fire in Kuwait. Hundreds of wells were set ablaze at the end of the Gulf War in 1991. The flames are a dramatic display of the energy held in oil.

14.1 INTRODUCTION

In the extreme chill of a midwinter night in northwestern Canada, a pan of water freezes almost instantly. But the low temperature doesn't stop a wolf from stalking its prey—the wolf's legs move through the snow, its heart pumps, and its body radiates heat. These life processes require energy. **Energy** provides the capacity to do work, to cause something to happen, or to cause change in a system (see Appendix A). The wolf's energy comes from the metabolism of special chemicals such as sugar, protein, and carbohydrates in its body. These chemicals, in turn, came from the food the animal eats. In order to survive, a wolf must catch and eat mice and rabbits, so to a wolf, these animals are energy resources. In a general sense, we use the term **resource** for any item that can be employed for a useful purpose, and, more specifically, **energy resource** for something that can be used to produce heat, produce electricity, or move vehicles. Matter that stores energy in a readily usable form is also called a **fuel**.

The earliest humans needed about the same quantity of fuel per capita as a wolf, and thus could maintain themselves simply by hunting and gathering. But when people discovered how to use fire for cooking and heating, their need for energy resources began to exceed that of other animals, for they now needed fuel to feed their fires. Before the dawn of civilization, wood and dried dung provided adequate fuel. But as people began to congregate in towns, they also required energy for agriculture and transportation, and new resources such as animal power, wind, and flowing water came into use. By the end of the seventeenth century, energy-resource needs began to outpace the supplies available at the Earth's surface, for not only did the population continue to grow, but new industries such as iron smelting also became commonplace. In fact, to feed the smelting industry, woodcutters mowed down most European forests, so that when the Industrial Revolution began in the eighteenth century, workers had to mine supplies of underground coal to feed the new steam engines that were driving factories.

Since the Industrial Revolution, society's hunger for energy has increased almost unabated (▶Fig. 14.1). In the United States today, for example, the average person uses more than 110 times the amount of energy used by a prehistoric hunter-gatherer. Most energy for human consumption in the industrial world now comes from oil and natural gas. To a lesser extent, we also continue to use coal, wind, and flowing water, and in the last half century we've added nuclear energy, geothermal energy, and solar energy to the list of energy resources. As the twenty-first century dawns, we have started to work toward developing new ways of obtaining energy resources, such as extraction of natural gas from coal beds and from gas hydrates (strange, icelike substances on the sea floor),

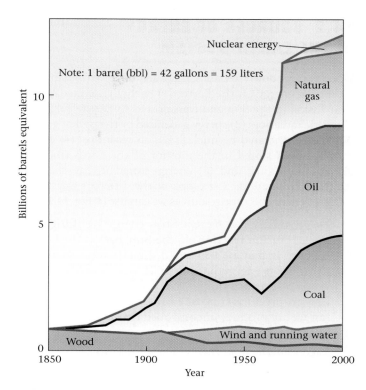

FIGURE 14.1 The graph demonstrates how energy needs have increased in the past 150 years, and how different energy resources have been used to fill those needs. Oil and natural gas together now account for more than half the world's energy usage.

gasification of coal (turning coal into various burnable gases), production of hydrogen fuel cells, and production of alcohol from plant crops.

Why is a chapter in a geology book devoted to such energy resources? Simply because most of these resources originate in geologic materials or processes. Oil, gas, coal, and the fuel for atomic power plants come from rocks; geothermal energy is a product of Earth's internal heat; and the movement of wind and water involves cycles in the Earth System. To understand the source and limitations of energy resources and to find new resources, we must understand their geology. That's why the multibillion-dollar-a-year energy industry employs tens of thousands of geologists—these are the people who find new fuel supplies.

In this chapter, we begin by looking at various types of energy resources on Earth. Then we focus on **fossil fuels** (oil, gas, and coal)—combustible materials derived from organisms that lived in the past—in detail and other types of energy sources more briefly. The chapter concludes by outlining the dilemmas we will face as energy resources begin to run out, and as the products of energy consumption enter our environment in dangerous amounts.

14.2 SOURCES OF ENERGY IN THE EARTH SYSTEM

When you get down to basics, there are only five fundamental sources of energy on the Earth: (1) energy generated by nuclear fusion in the Sun and transported to Earth via electromagnetic radiation; (2) energy generated by the pull of gravity; (3) energy generated by nuclear fission reactions; (4) energy that has been stored in the interior of the Earth since the planet's beginning; and (5) energy stored in the chemical bonds of compounds. Let's look at the different ways these forms of energy become resources we can use (▶Fig. 14.2).

- *Energy directly from the Sun:* Solar energy, resulting from nuclear fusion reactions in the Sun, bathes the Earth's surface. It may be converted directly into electricity, using solar-energy panels, or it may be used to heat water or warm a house. (No one has yet figured out how to produce controlled nuclear fusion on Earth, but we *can* produce uncontrolled fusion by exploding a hydrogen bomb.)

- *Energy directly from gravity:* The gravitational attraction of the Moon, and to a lesser extent the Sun, causes ocean tides, the daily up-and-down movement of the sea surface. The flow of water in and out of channels during tidal changes can drive turbines.

- *Energy involving both solar energy and gravity:* Solar radiation heats the air, which becomes buoyant and rises. As this happens, gravity causes cooler air to sink. The resulting air movement, wind, powers sails and windmills. Solar energy also evaporates water, which enters the atmosphere. When the water condenses, it rains and falls on the land, where it accumulates in streams that flow downhill in response to gravity. This moving water powers waterwheels and turbines.

- *Energy via photosynthesis:* Green plants absorb some of the solar energy that reaches the Earth's surface. Their green color comes from a pigment called chlorophyll. With the aid of chlorophyll, plants produce sugar through a chemical reaction called **photosynthesis**. In chemist's shorthand, we can write this reaction as

$$6CO_2 + 12H_2O + light \rightarrow 6O_2 + C_6H_{12}O_6 + 6H_2O.$$
$$\text{carbon} \quad \text{water} \qquad\quad \text{oxygen} \quad \text{sugar} \quad\quad \text{water}$$
$$\text{dioxide}$$

Plants use the sugar produced by photosynthesis to manufacture more complex chemicals, or they metabolize it to provide themselves with energy.

Burning plant matter in a fire releases potential energy stored in the chemical bonds of organic chemicals. During burning, the molecules react with oxygen and break apart to produce carbon dioxide, water, and carbon (soot):

$$\text{plant} + O_2 \xrightarrow{\text{burning}} CO_2 + H_2O + C \text{ (soot)} \\ + \text{ other gases} + \text{heat energy}$$

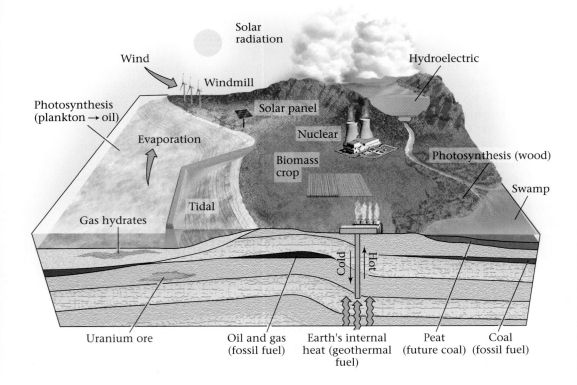

FIGURE 14.2 The diverse sources of energy on Earth. Solar energy can be used directly, to drive wind or to cause water to evaporate and make rain and ultimately running water. It also provides energy for photosynthesis, which produces wood and plankton; these ultimately become fossil fuels (coal and oil, respectively). Radioactive material from the Earth powers nuclear reactors, and the Earth's internal heat creates geothermal resources. Gravity plays a role in producing wind and water power.

The flames you see in fire consist of glowing gases released and heated by this reaction.

People have burned wood to produce energy for centuries. More recently, plant material (biomass) from crops such as corn and sugar cane has been used to produce ethanol, a flammable alcohol.

- *Energy from chemical reactions:* A number of inorganic chemicals can burn to produce light and energy. The energy results from exothermic (heat-producing) chemical reactions. A dynamite explosion is an extreme example of such energy production. Recently, researchers have been studying electrochemical devices, such as hydrogen fuel cells, that produce electricity directly from chemical reactions.

- *Energy from fossil fuels:* Oil, gas, and coal come from organisms that lived long ago, and thus store solar energy that reached the Earth long ago. We refer to these substances as "fossil fuels," to emphasize that they were derived from ancient organisms and have been preserved in rocks for geologic time. Burning fossil fuels produces energy in the same way that burning plant matter does.

- *Energy from nuclear fission:* Atoms of radioactive elements can split into smaller pieces, a process called nuclear fission (see Appendix A). During fission, a tiny amount of mass is transformed into a large amount of energy, called nuclear energy. This type of energy runs nuclear power plants and nuclear submarines.

> **Take-Home Message**
>
> Energy comes from several sources—solar radiation, gravity, chemical reactions, radioactive decay, and internal heat. Solar radiation and gravity together drive wind and water movement. Products of photosynthesis can be preserved in rocks as fossil fuels.

- *Energy from Earth's internal heat:* Some of Earth's internal energy dates from the birth of the planet, while some is produced by radioactive decay in minerals. This internal energy heats water underground.

The resulting hot water, when transformed to steam, provides **geothermal energy** that can drive turbines.

14.3 OIL AND GAS

What Are Oil and Gas?

For reasons of economics and convenience, industrialized societies today rely primarily on oil (petroleum) and natural gas for their energy needs. Oil and natural gas consist of **hydrocarbons**, chainlike or ringlike molecules made of carbon and hydrogen atoms. For example, bottled gas (propane) has the chemical formula C_3H_9. Chemists consider hydrocarbons to be a type of organic chemical, so named because similar chemicals make up living organisms.

Some hydrocarbons are gaseous and invisible, some resemble watery liquids, some appear syrupy, and some are solid (►Fig. 14.3). The viscosity (ability to flow) and the volatility (ability to evaporate) of a hydrocarbon product depend on the size of its molecules. Hydrocarbon products composed of short chains of molecules tend to be less viscous (they can flow more easily) and more volatile (they evaporate more easily) than products composed of long chains, simply because the long chains tend to tangle up with each other. Thus, short-chain molecules occur in gaseous form at room temperature, moderate-length-chain molecules occur in liquid form, and long-chain molecules occur in solid form as tar.

FIGURE 14.3 The diversity of hydrocarbon products we use: natural gas piped to houses for heating and cooking, bottled gas (propane), gasoline for cars, kerosene for heating or illumination, diesel fuel for trucks, lubricating oil for motors, and solid tar, which melts when heated and can be used to make asphalt. Note that these products are listed in order of increasing viscosity.

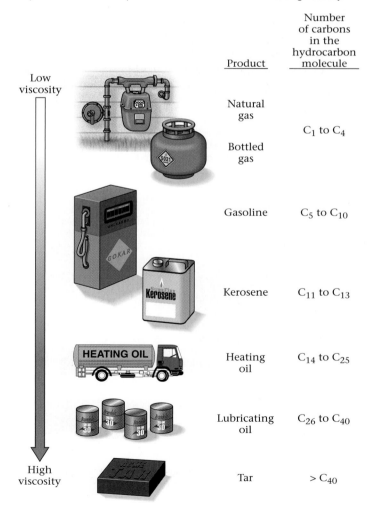

	Product	Number of carbons in the hydrocarbon molecule
Low viscosity	Natural gas	
	Bottled gas	C_1 to C_4
	Gasoline	C_5 to C_{10}
	Kerosene	C_{11} to C_{13}
	Heating oil	C_{14} to C_{25}
	Lubricating oil	C_{26} to C_{40}
High viscosity	Tar	$> C_{40}$

Why can we use hydrocarbons as fuel? Simply because hydrocarbons, like wood, burn—they react with oxygen to form carbon dioxide, water, and heat. As an example, we can describe the burning of gasoline by the reaction

$$2C_8H_{18} + 25O_2 \rightarrow 16CO_2 + 18H_2O + \text{heat energy}$$

During such reactions, the potential energy stored in the chemical bonds of the hydrocarbon molecules converts into usable heat energy.

Where Do Oil and Gas Form?

Many people incorrectly believe that hydrocarbons come from buried trees or the carcasses of dinosaurs. In fact, the primary sources of the organic chemicals in oil and gas are dead algae and plankton. (Plankton, as we have seen, are the tiny plants and animals—typically around 0.5 mm in diameter—that float in sea or lake water.) When algae and plankton die, they settle to the bottom of a lake or sea. Because their cells are so tiny, they can be deposited only in quiet-water environments in which clay also settles, so typically the cells mix with clay to create an organic-rich, muddy ooze. For this ooze to be preserved, it must be deposited in oxygen-poor water. Otherwise, the organic chemicals in the ooze would react with oxygen or be eaten by bacteria, and thus would decompose quickly and disappear. In some quiet-water environments (oceans, lagoons, or lakes), dead algae and plankton can get buried by still more sediment before being destroyed. Eventually, the resulting ooze lithifies and becomes black *organic* shale (in contrast to ordinary shale that consists only of clay), which contains the raw materials from which hydrocarbons eventually form. Thus, we refer to organic shale as a **source rock**.

If organic shale is buried deeply enough (2 to 4 km), it becomes warmer, since temperature increases with depth in the Earth. Chemical reactions slowly transform the organic material in the shale into waxy molecules called **kerogen** (▶Fig. 14.4). Shale containing kerogen is called **oil shale**. If the oil shale warms to temperatures of greater than about 90°C, the kerogen molecules break down to form oil and natural gas molecules. At temperatures over about 160°C, any remaining oil breaks down to form natural gas, and at temperatures over 225°–250°C, organic matter transforms into graphite. Thus, oil itself forms only in a relatively narrow range of temperatures, called the **oil window** (▶Fig. 14.5). For regions with a geothermal gradient of 25°C/km, the

> ### Take-Home Message
>
> Oil and gas are hydrocarbons, molecules composed of carbon-hydrogen chains. Gaseous hydrocarbons consist of short chains, liquids of medium chains, and solids (tar) of long chains. Burning a hydrocarbon is simply a rapid oxidation reaction.

FIGURE 14.4 Plankton, algae, and clay settle out of water and become progressively buried and compacted, gradually being transformed into black organic shale. When heated for a long time, the organic matter in black shale is transformed into oil shale, which contains kerogen. Eventually, the kerogen transforms into oil and gas. The oil may then start to seep upward out of the shale. The red arrow indicates pressure, which increases as more sediment accumulates above.

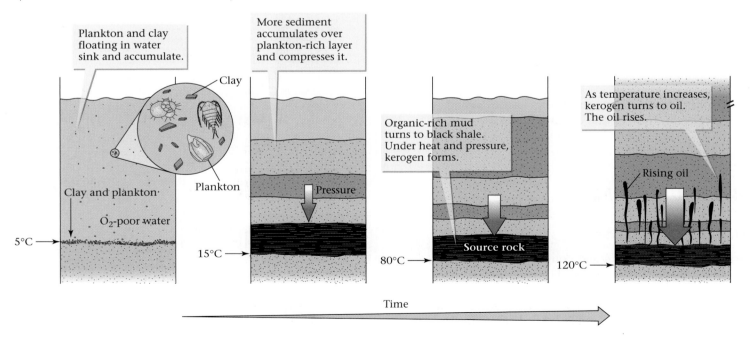

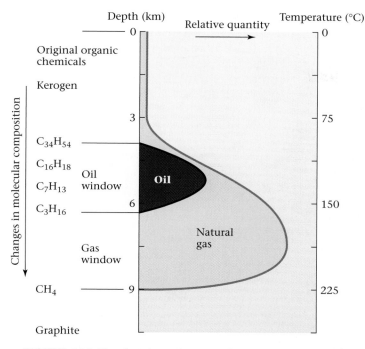

Depth (km) Relative quantity Temperature (°C)

Changes in molecular composition

Original organic chemicals

Kerogen

$C_{34}H_{54}$

$C_{16}H_{18}$

C_7H_{13}

C_3H_{16}

Oil window

Gas window

CH_4

Graphite

Oil

Natural gas

0 — 0
3 — 75
6 — 150
9 — 225

FIGURE 14.5 The oil window is the range of temperature conditions (i.e., depth) at which hydrocarbons form. In regions with a geothermal gradient of 25°C per km, oil occurs only at depths of less than about 6.5 km, the gradient portrayed on this graph, Note that gas can be found at greater depths. The length of hydrocarbon chains decreases with increasing depth, because at higher temperatures longer chains break to form smaller ones.

oil window lies at depths of about 3.5 to 6.5 km, whereas gas can survive down to 9 km. If the geothermal gradient is low (15°C/km), oil survives only below about 11 km. Thus, hydrocarbon reserves can only exist in the topmost 15 to 25% of the crust.

14.4 HYDROCARBON SYSTEMS: THE MAKING OF A RESERVE

Oil and gas do not occur in all rocks at all locations. That's why the desire to control oil fields, regions that contain significant amounts of accessible oil underground, has sparked bitter wars. A known supply of oil and gas held underground is a **hydrocarbon reserve**; if the reserve consists dominantly of oil, it is an "oil reserve." Reserves are not randomly distributed around the Earth (▶Fig. 14.6). Currently, countries bordering the Persian Gulf contain the world's largest reserves. In this section, we learn that the development of a reserve requires the existence of four geologic features: a source rock, a reservoir rock, a migratory pathway, and a trap. A particular association of all of these components—along with the processes of hydrocar-

bon generation, migration, and accumulation that ultimately produce a reserve from a given source—is called a **hydrocarbon system**.

Source Rocks and Hydrocarbon Generation

We saw above that the chemicals that become oil and gas start out in algae and plankton bodies. Their cells accumulate along with clay to form an organic ooze, which when lithified becomes black organic shale. Geologists refer to organic-rich shale as a source rock because it is the source for the organic chemicals that ultimately become oil and gas. If black shale resides in the oil window, the organic material within transforms into kerogen, and then into oil and gas. This process is **hydrocarbon generation**.

Reservoir Rocks and Hydrocarbon Migration

Wells drilled into source rocks do not yield much oil because kerogen can't flow easily from the rock into the well. Thus, any organic matter in an oil shale remains trapped among the grains and can't move easily. So to obtain oil, companies drill instead into **reservoir rocks**, rocks that contain (or could contain) an abundant amount of *easily accessible* oil and gas, meaning hydrocarbons that can be extracted out of the ground.

To be a reservoir rock, a body of rock must have space in which the oil or gas can reside, and must have channels through which the oil or gas can move. The space can be in the form of open spaces, or **pores**, between clastic grains (which exist because the grains didn't fit together tightly and because cement didn't fill all the spaces during cementation) or in the form of cracks and fractures that developed after the rock formed. In some cases, groundwater passing through rock dissolves minerals and creates new space. **Porosity** refers to the amount of open space in a rock (▶Fig. 14.7). Oil or gas can fill porosity just as water fills the holes in a sponge. Not all rocks have the same porosity. For example, shale typically has a porosity of 10%, while sandstone has a porosity of 35%. That means that about a third of a block of sandstone actually consists of open space.

Permeability refers to the degree to which pore spaces connect to each other. Even if a rock has high porosity, it is not necessarily permeable (Fig. 14.7). In a permeable rock, the holes and cracks (pores) are linked, so a fluid is able to flow slowly through the rock, following a tortuous pathway.

Keeping the concepts of porosity and permeability in mind, we can see that a poorly cemented sandstone makes a good reservoir rock, because it is both porous and permeable. A highly fractured rock can be porous and permeable, even if there is no pore space between individual grains. The greater the porosity, the greater the capacity of a reservoir rock to hold oil; and the greater the rock's permeability, the easier it is for the oil to be extracted.

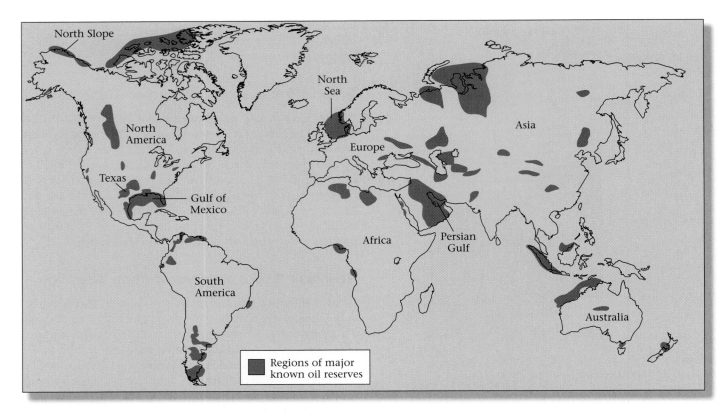

FIGURE 14.6 The distribution of oil reserves around the world. The largest fields occur in the region surrounding the Persian Gulf.

In an oil well, which is simply a hole drilled into the ground to where it penetrates reservoir rock, oil flows from the permeable reservoir rock into the well and then up to the ground surface. If the oil in the rock is under natural pressure, it may move by itself, but usually producers must set up a pump literally to suck the oil up and out of the hole.

To fill the pores of a reservoir rock, oil and gas must first migrate (move) from the source rock into a reservoir rock, which they will do over millions of years of geologic time (▶Fig. 14.8). Why do hydrocarbons migrate? Oil and gas are less dense than water, so they try to rise toward the Earth's surface to get above groundwater, just as salad oil rises above the vinegar in a bottle of salad dressing. Natural gas, being less dense, ends up floating above oil. In other words, buoyancy drives oil and gas upward. Typically, a hydrocarbon system must have a good **migration pathway**, such as a set of permeable fractures, in order for large volumes of hydrocarbons to move.

Traps and Seals

The existence of reservoir rock alone does not create a reserve, because if hydrocarbons can flow easily into a reservoir rock, they can also flow out. If oil or gas escapes from the reservoir rock and ultimately reaches the Earth's surface, where it leaks away at an **oil seep**, there will be none left underground to pump. Thus, for an oil reserve to exist, oil and gas must be *trapped* underground in the reservoir rock, by means of a geologic configuration called a **trap**. A field contains one or more traps.

There are two components to an oil or gas trap. First, a **seal rock**, a relatively impermeable rock such as shale, salt, or unfractured limestone, must lie above the reservoir rock and stop the hydrocarbons from rising further. Second, the seal and reservoir rock bodies must be arranged in a geometry that collects the hydrocarbons in a restricted area. Geologists recognize several types of hydrocarbons trap geometries, four of which are described in ▶Box 14.1.

Note that when we talk about trapping hydrocarbons underground, we are talking about a temporary process in the context of geologic time. Oil and gas may be trapped for millions of years, but eventually they will manage to pass through a seal rock, because no rock is absolutely impermeable. Also, in some cases, microbes eat hydrocarbons in the

> **Take-Home Message**
>
> Oil and gas reserves develop only where there is a source (organic shale), a migration pathway, a reservoir (porous and permeable rock), and a trap. For oil to form and survive, temperatures must remain within the "oil window."

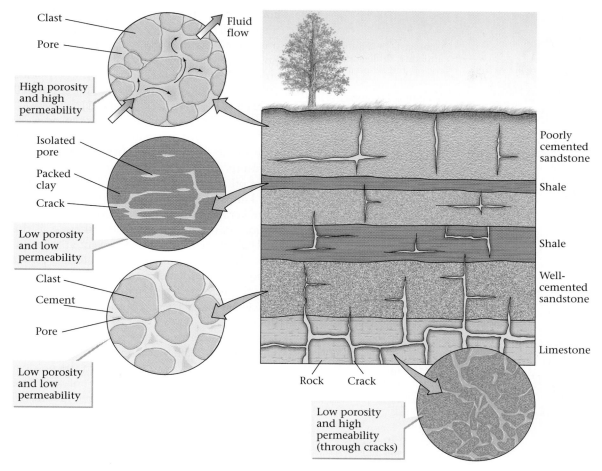

FIGURE 14.7 The porosity and permeability of a sedimentary rock depend on the character of the rock. For example, poorly cemented sandstone can have high porosity and permeability, whereas well-cemented sandstone does not. Shale tends to be impermeable and to have low porosity, and the porosity in limestone is commonly due to the presence of cracks. Rocks with high porosity and permeability make the best reservoir rocks. Not only can such rocks hold a lot of oil, but the oil can also flow relatively easily and thus can be pumped out efficiently.

subsurface. Thus, innumerable oil fields that existed in the past have vanished, and the oil fields we find today, if left alone, will disappear millions of years in the future.

14.5 OIL EXPLORATION AND PRODUCTION

Birth of the Oil Industry

People have used oil since the dawn of civilization—as a cement, as a waterproof sealant, and even as a preservative to embalm mummies. But early on, the only available oil came from natural seeps, places where oil-filled reservoir rock intersects the Earth's surface or where fractures connect a reservoir rock to the Earth's surface, so that oil flows out on the ground on its own.

In the United States, during the first half of the nineteenth century, people collected "rock oil" (later called petroleum, from the Latin words *petra*, meaning rock, and *oleum*, meaning oil) at seeps and used it to grease wagon axles and to make patent medicines. But such oil was rare and expensive. In 1854, George Bissel, a New York lawyer, came to the realization that oil might have broader uses, particularly as fuel for lamps (to replace whale oil). Bissel and a group of investors contracted Edwin Drake, a colorful character who had drifted among many professions, to find a way to drill for oil in rocks beneath a hill near Titusville, Pennsylvania, where an oily film floated on the water of springs. Using the phony title "Colonel" to add respectability, Drake hired drillers and obtained a steam-powered drill. Work was slow and the investors became discouraged, but the very day that a letter arrived ordering Drake to stop drilling, his drillers found that the hole, which had reached a depth of 21.2 m, had filled with oil. They set up a pump,

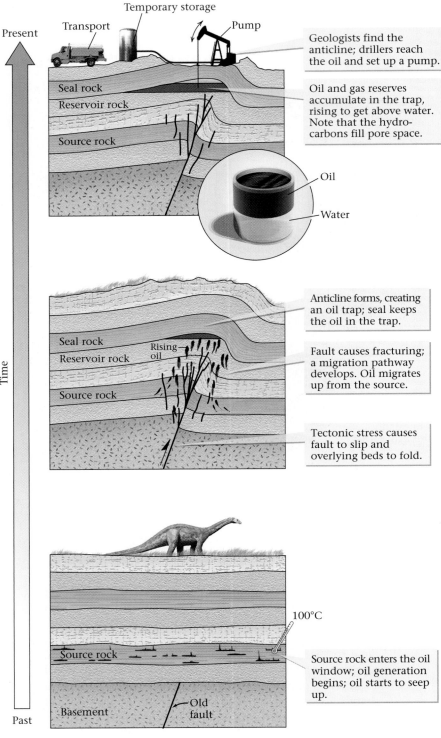

Transport

Temporary storage

Pump

Present

Geologists find the anticline; drillers reach the oil and set up a pump.

Seal rock
Reservoir rock

Oil and gas reserves accumulate in the trap, rising to get above water. Note that the hydro-carbons fill pore space.

Source rock

Oil

Water

Anticline forms, creating an oil trap; seal keeps the oil in the trap.

Seal rock
Reservoir rock

Rising oil

Fault causes fracturing; a migration pathway develops. Oil migrates up from the source.

Source rock

Tectonic stress causes fault to slip and overlying beds to fold.

Time

100°C

Source rock

Source rock enters the oil window; oil generation begins; oil starts to seep up.

Basement

Old fault

Past

FIGURE 14.8 Initially, oil resides in the source rock. Oil gradually migrates out of the source rock and rises into the overlying water-saturated reservoir rock. Oil rises because it is buoyant relative to water—it tries to float on water. The oil is trapped beneath a seal rock. If natural gas exists, it floats to the top of the oil.

and on August 27, 1859, for the first time in history, oil was pumped out of the ground. No one had given much thought to the question of how to store the oil, so workers dumped it into empty whisky barrels. This first oil well yielded 10 to 35 barrels a day, which sold for about $20 a barrel (1 barrel equals 42 gallons).

Within a few years, thousands of oil wells had been drilled in many states, and by the turn of the twentieth century civilization had begun its addiction to oil (▶Box 14.2). Initially, most oil went into the production of kerosene for lamps. Later, when electricity took over from kerosene as the primary source for illumination, gasoline derived from oil became the fuel of choice for the newly invented automobile. Oil was also used to fuel electric power plants. In its early years, the oil industry was in perpetual chaos. When "wildcatters" discovered a new oil field, there would be a short-lived boom during which the price of oil could drop to pennies a barrel. In the midst of this chaos, John D. Rockefeller established the Standard Oil Company, which monopolized the production, transport, and marketing of oil. The Supreme Court eventually broke Standard Oil down into several companies (including Exxon, Chevron, Mobil, Sohio, Amoco, Arco, Conoco, and Marathon). Oil became a global industry governed by the complex interplay of politics, profits, supply, and demand. The business has evolved, and as the twenty-first century began, many of the companies have merged.

The Modern Search for Oil

Wildcatters discovered the earliest oil fields either by blind luck or by searching for surface seeps. But in the twentieth century, when all known seeps had been drilled and blind luck became too risky, oil companies realized that finding new oil fields would require systematic exploration. The modern-day search for oil is a complex, sometimes dangerous, and often exciting procedure with many steps.

Source rocks are always sedimentary, as are most reservoir and seal rocks, so geologists begin their exploration by looking for a region containing appropriate sedimentary rocks. Then they compile a geologic map of the area, showing the distribution of rock units. From this information, it may be possible to construct a preliminary cross section depicting the geometry of the sedimentary layers underground as they would appear on an imaginary vertical slice through the Earth.

BOX 14.1 THE REST OF THE STORY

Types of Oil and Gas Traps

Geologists who work for oil companies spend much of their time trying to identify underground traps. No two traps are exactly alike, but we can classify most into the following four categories.

- *Anticline trap:* In some places, sedimentary beds are not horizontal, as they are when originally deposited, but have been bent by the forces involved in mountain building. These bends, as we have seen, are called folds. An anticline is a type of fold with an arch-like shape (►Fig. 14.9a; see Chapter 11). If the layers in the anticline include a source rock overlain by a reservoir rock, overlain by a seal rock, then we have the recipe for an oil reserve. The oil and gas rise from the source rock, enter the reservoir rock, and rise to the crest of the anticline, where they are trapped by the seal rock.

- *Fault trap:* A fault is a fracture on which there has been sliding. If the slip on the fault crushes and grinds the adjacent rock to make an impermeable layer along the fault, then oil and gas may migrate upward along bedding in the reservoir rock until they stop at the fault surface (►Fig. 14.9b). Alternatively, a fault trap develops if the slip on the fault juxtaposes an impermeable rock layer against a reservoir rock.

- *Salt-dome trap:* In some sedimentary basins, the sequence of strata contains a thick layer of salt, deposited when the basin was first formed and seawater covering the basin was shallow and very salty. Sandstone, shale, and limestone overlie the salt. The salt layer is not so dense as sandstone or shale, so it is buoyant and tends to rise up slowly through the overlying strata. Once the salt starts to rise, the weight of surrounding strata squeezes the salt out of the layer and up into a growing, bulbous **salt dome**. As the dome rises, it bends up the adjacent layers of sedimentary rock. Oil and gas in reservoir rock layers migrate upward until they are trapped against the boundary of the salt dome, for salt is not permeable (►Fig 14.9c).

- *Stratigraphic trap:* In a stratigraphic trap, a tilted reservoir rock bed "pinches out" (thins and disappears along its length) between two impermeable layers. Oil and gas migrating upward along the bed accumulate at the pinch-out (►Fig. 14.9d).

FIGURE 14.9 **(a)** Anticline trap. The oil and gas rise to the crest of the fold. **(b)** Fault trap. The oil and gas collect in tilted strata adjacent to the fault. **(c)** Salt-dome trap. The oil and gas collect in the tilted strata on the flanks of the dome. **(d)** Stratigraphic trap. The oil and gas collect where the reservoir layer pinches out.

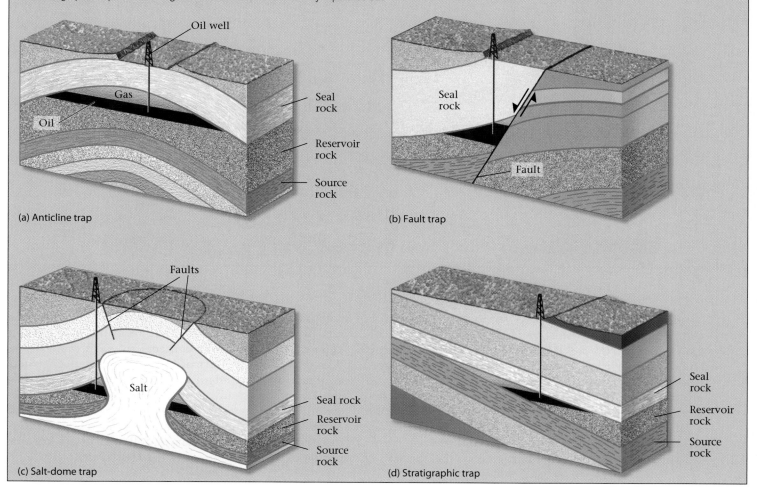

(a) Anticline trap

(b) Fault trap

(c) Salt-dome trap

(d) Stratigraphic trap

BOX 14.2 THE HUMAN ANGLE

Spindletop

As demand for oil increased in the last decade of the nineteenth century, oil became "black gold," worth a fortune to anyone lucky enough to find it. One Texas mechanic, Patillo Higgins, not only dreamed of oil riches, but thought he knew a place to find them. Higgins had noticed a strange gas bubbling slowly up through little water springs on Spindletop Hill, where he took his Sunday school class for walks. When he lit the bubbles with a match, they burned. Did oil lie beneath the hill? Higgins struggled to convince investors to help him find out. Finally, in 1899, his arguments convinced Anthony Lucas, who brought a drilling crew to Spindletop.

Drilling began in late 1900, and soon after, Lucas and his crew reached a small oil pocket, which to their delight could yield 50 barrels a day, more than the average wells of Pennsylvania. But as long as drilling equipment was in place, Lucas decided to keep drilling. The operation proceeded routinely until January 10, 1901, when suddenly, without warning, the drilling platform began to vibrate violently. Workers fled for cover as the drilling pipe shot upward and out of the ground. As tons of equipment rained from the sky, a black fountain of oil erupted from the ground and blasted skyward to a height of almost 50 m (▶Fig. 14.10a). Oil from the Spindletop "gusher" rained over the pasture at the astounding rate of 75,000 barrels per day! No one had ever seen so much oil before.

News of the gusher chattered over telegraph lines, and pandemonium ensued. Tens of thousands of people flocked to Spindletop Hill, creating a boomtown of mud streets, plank shacks, and saloons. Land prices exploded from $10 per acre to almost $1 million per acre, and soon the hill looked like a pin cushion, punctured by over 200 wells (▶Fig. 14.10b). Oil hysteria had moved to Texas.

FIGURE 14.10 Photos of oil dericks at the Spindletop Oil Field, in the early 1900s. **(a)** A gusher of "black gold" erupting from a well. **(b)** Without regulations to guide well spacing, wells were drilled one next to the other.

(a)

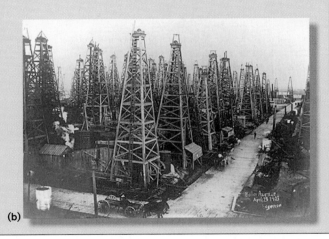
(b)

To add detail to the cross section, an exploration company makes a **seismic-reflection profile** of the region. To construct a seismic profile, a special vibrating truck or a dynamite explosion sends seismic waves (shock waves that move through the Earth) into the ground (▶Fig. 14.11a). The seismic waves reflect off contacts between rock layers, just as sonar waves sent out by a submarine reflect off the bottom of the sea. Reflected seismic waves then return to the ground surface, where sensitive instruments (geophones) record their arrival. A computer measures the time between the generation of a seismic wave and its return, and from this information defines the depth to the contacts at which the wave reflected. With such information, the computer constructs an image of the configuration of underground rock layers. Technological advances now enable geologists to create 3-D seismic-reflection profiles of the subsurface both under land and under water (▶Fig. 14.11b, c). Unfortunately, it may cost millions of dollars to create just one profile.

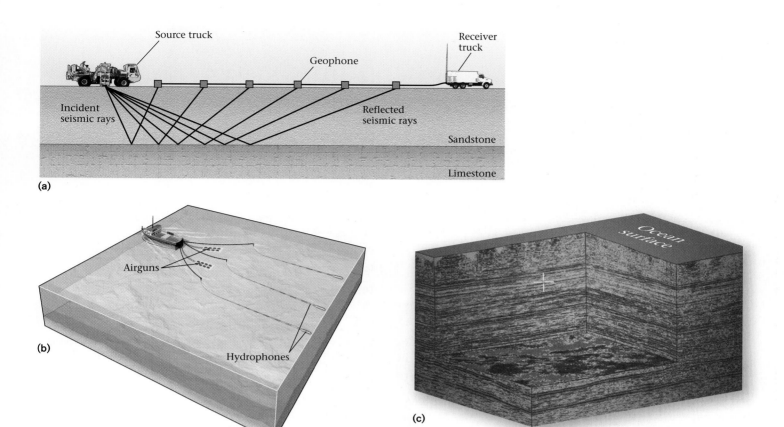

FIGURE 14.11 (a) The basic concept of seismic-reflection profiling. A signal sent by a source truck (e.g., a "Vibroseis" truck, which sends energy into the crust via a heavy plate placed on the ground beneath the middle of the truck) reflects off a bedding plane in the subsurface. Seismic receivers detect the reflections of different rays, and computers in a recording truck record the information. The computers process the data to determine the location of the bedding plane. **(b)** Seismic-reflection studies can be obtained in submarine strata by use of a special ship that tows airguns (devices that send sound pulses into the water) and strings of detectors. When a ship tows several rows of detectors at once, a 3-D image of the strata can be produced. **(c)** A modern 3-D seismic-reflection image of the subsurface. Each vertical face of this block is a seismic-reflection profile. The horizontal surface shows a map-like image at a specified depth beneath the surface. The colored bands represent beds of sedimentary rock.

Geologists continue to discover new tools that help in oil exploration. For example, tiny seeps of oil host communities of hydrocarbon-eating bacteria. Mapping the distribution of bacteria species in the soil may, therefore, indicate the presence of an oil reserve below.

If geological studies identify a trap, and if the geologic history of the region indicates the presence of good source rocks and reservoir rocks, geologists make a recommendation to drill. (They do not make such recommendations lightly, as drilling a deep well may cost over $10 million.) Once the decision has been made, drillers go to work. These days, drillers use rotary drills to grind a hole down through rock. A rotary drill consists of a rotating pipe tipped by a bit, a bulb of metal studded with industrial diamonds or hard metal prongs (▶Fig. 14.12a, b). As the bit rotates, it scratches and gouges the rock, turning it into powder and chips. Drillers pump "drilling mud," a slurry of water mixed with clay, down the center of the pipe. The

mud squirts out of holes in the drill bit, cooling the bit and flushing rock cuttings up and out of the hole. The weight of the mud also keeps oil down in the hole and prevents "gushers," fountains of oil formed when underground pressure causes the oil to rise out of the hole on its own.

Drillers use derricks (towers) to hoist the heavy drill pipe. To drill in an offshore oil reserve, one that occurs in strata of the continental shelf, the derrick must be constructed on an offshore-drilling facility. Offshore-drilling facilities can be immense (▶Fig. 14.12c); some tower more than 100 stories from base (below sea level) to top. They serve as artificial towns for housing the drilling crew. In shallower water, facilities anchor on the sea floor, but in deeper water, they float on huge submerged pontoons. Drill holes can be aimed in any direction (not just vertical), so drillers can reach many traps from the same facility.

On completion of a hole, workers remove the drilling rig and set up a pump. Some pumps resemble a bird pecking for

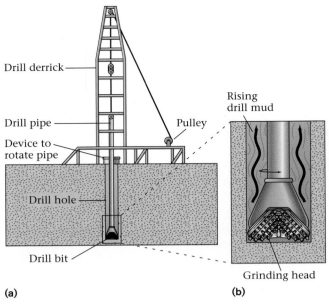

Drill derrick

Drill pipe

Pulley

Device to rotate pipe

Drill hole

Drill bit

(a)

Rising drill mud

Grinding head

(b)

(c)

FIGURE 14.12 **(a)** The derrick in this drilling platform is used to hoist the drill pipe, which comes in segments. **(b)** A drill bit, typically consisting of three diamond-studded parts, connects to the bottom of the pipe. The diameter of the bit is greater than the diameter of the pipe. Drilling mud is pumped down through the pipe, comes out through holes in the bit, and then rises between the pipe and the walls of the hole, thereby flushing cuttings out of the hole. **(c)** An offshore-drilling facility in southern California. **(d)** A row of pumps in an Illinois oil field.

(d)

TABLE 14.1 **Statistics of Oil Reserves, Oil Production, and Oil Consumption[1]**

Country	Reserves[2] (billion bbl)	Production (millions bbl/day)	Consumption[3] (millions bbl/day)
Saudia Arabia	265	8.5	1.4
Iraq	115	2.4	0.5
Kuwait	99	1.8	0.3
Iran	96	3.8	1.1
United Arab Emirates	63	2.6	0.3
Russia	54	7.0	2.5
Venezuela	48	3.1	0.5
China	31	3.3	4.9
Libya	30	1.4	0.2
Mexico	27	3.6	1.9
Nigeria	24	2.2	0.3
United States	22	8.1	20.0
Qatar	15	0.6	0.03
Norway	10	3.4	0.2
Algeria	9	1.5	0.2
Brazil	8	1.6	2.1
World	1032	75.2	76.0

[1]Data comes from various sources. The countries listed are the top 15 in terms of total reserves; [2]Taking population into account: UAE has 39,359 bbl/person, Saudia Arabia has 10,875 bbl/person, Russia has 336 bbl/person, and the U.S. has 77 bbl/person. [3]The 6 largest consumers worldwide are: U.S. (20), Japan (5.4), China (4.8), Germany (2.8), Russia (2.5), South Korea (2.1). Note that the largest consumers are not countries with large reserves.

grain; their heads move up and down to pull up oil that has seeped into the drill hole (►Fig. 14.12d). You may be surprised to learn that simple pumping gets only about 30% of the oil in a reservoir rock out of the ground. Thus, oil companies use secondary recovery techniques to coax out up to 50% more oil. For example, a company may drive oil toward a drill hole by forcing steam into the ground nearby. The steam heats the oil in the ground, making it less viscous, and pushes it along. In some cases, drillers create artificial fractures in rock around the hole so that the oil has easy routes to follow from rock to well.

Presently, the vast majority of known oil reserves are distributed among only twenty-five fields, known as super-giant fields (Fig. 14.6). The largest occur around the Persian Gulf (in Saudi Arabia, Kuwait, Iraq, and neighboring countries; ►Table 14.1). The United States is the largest consumer of oil (at a rate of 7 million barrels per day, about 25% of world consumption), but lost its position as the largest producer in the 1970s. Oil reserves in the United States now account for only about 4% of the world total. Thus, today the United States must import more than half of the oil it uses. The industrialized countries use vastly more oil per capita than do the developing countries. For example, 1,000 people in the United States use 70 barrels per day. In China, 1,000 people use 4 barrels per day, and in Ethiopia, 0.3 barrels per day.

Once extracted directly from the ground, "crude oil" flows first into storage tanks and then into a pipeline or tanker, which transports it to a refinery (▶Fig. 14.13a). At a refinery, workers distill crude oil into several separate components by heating it gently in a vertical pipe called a **distillation column** (▶Fig. 14.13b). Lighter molecules rise to the top of the column, while heavier molecules stay at the bottom. The heat may also "crack" larger molecules to make smaller ones. Chemical factories buy the largest molecules left at the bottom and transform them into plastics.

Take-Home Message

Searching for oil is a complex and expensive process. Geologists use seismic-reflection profiles to locate possible traps. Drilling taps reserves and pumping brings crude oil to the surface, where it is processed at refineries that crack hydrocarbon molecules.

14.6 ALTERNATIVE RESERVES OF HYDROCARBONS

Tar Sands (Oil Sands)

So far, we've focused our discussion on hydrocarbon reserves that can be pumped from the subsurface in the form of a liquid or a gas. But in several locations around the world, most notably Alberta (in western Canada) and Venezuela, vast reserves of very viscous, tarlike "heavy oil" exist. This heavy oil, known also known as bitumen, has the consistency of heavy molasses, and thus cannot be pumped directly from the ground; it can fill the pore spaces of sand or of poorly cemented sandstone, constituting up to 12% of the sediment or rock volume. Sand or sandstone containing such high concentrations of bitumen is known as **tar sand** or oil sand (▶Fig. 14.14a).

The hydrocarbon system that leads to the generation of tar sands begins with the production and burial of a source rock in a large sedimentary basin (specifically, a type of basin called a foreland basin, which forms along the continent side of a mountain belt; see Chapter 7). When subjected to temperatures of the oil window, the source rock yields oil and gas, which migrate into sandstone layers and then up the dip of tilted layers to the edge of the basin, where they become caught in stratigraphic traps (see Box 14.1). Initially, these hydrocarbons have relatively low viscosity; in the geologic past, they could have been pumped easily. But over time, microbes (bacteria) attacked the oil reserve underground, digested lighter, smaller hydrocarbon molecules, and left behind only the larger molecules, whose presence makes the remaining oil so viscous. Geologists refer to such a transformation process as **biodegradation**. Generation of tar sand by biodegradation is yet another example of the interaction between physical and biological components of the Earth System.

Production of usable oil from tar sand is difficult and expensive, but not impossible. It takes about 2 tons of tar sand to produce one barrel of oil. Oil companies mine near-surface deposits in open-pit mines, then heat the tar sand in a furnace to extract the oil. Producers then crack the heavy oil molecules to produce smaller, more usable molecules. Trucks dump the drained sand back into the mine pit. To extract oil from deeper deposits of tar sand, oil companies drill wells and pump steam or solvents down into the sand to liquefy the oil enough so that it can be pumped out.

Oil Shale

Vast reserves of organic-rich shale have not been subjected to temperatures of the oil window, or if they were, they did not stay within the oil window long enough to complete the transformation to

(a)

FIGURE 14.13 (a) The Alaska pipeline. This pipeline transports oil 1,300 km (800 miles) from the North Slope oil fields on the Arctic coast of Alaska to a tanker port in Valdez, on the southern coast. The pipeline (1.5 m, or 48 inches, in diameter) was built between 1975 and 1977 at a cost of $8 billion. **(b)** An oil refinery; note the distilling towers (or columns).

(b)

oil, and thus still contain a high proportion of kerogen. Shale that contains at least 15 to 30% kerogen is called oil shale. Oil shale is not the same as coal, because the organic matter within it exists in the form of hydrocarbon molecules, not as elemental carbon.

Lumps of oil shale can be burned directly, and thus have been used as a fuel since ancient times. In the 1850s, researchers developed techniques to produce liquid oil from oil shale. The process involves heating the oil shale to a temperature of 500°C; at this temperature, the shale decomposes and the kerogen transforms into liquid hydrocarbon and gas. Large supplies of oil shale occur in Estonia, Scotland, China, and Russia, and in the Green River Basin of Wyoming in the United States. As is the case with tar sand, production of oil from oil shale is possible, but very expensive. In addition to the expense of mining and environmental reclamation, producers must pay for the energy needed to heat the shale. It takes about 40% of the energy yielded by a volume of oil shale to produce the oil.

Natural Gas

Natural gas consists of volatile short-chain hydrocarbons, including methane, ethane, propane, and butane. It occurs in the pores of reservoir rock above oil, because it "floats" over the oil. Where temperatures in the subsurface are so high that oil molecules break apart to form gas, gas-only fields develop.

Gas burns more cleanly than oil (burning gas produces primarily carbon dioxide and water, whereas burning oil also produces complex organic pollutants), and thus has become the preferred fuel for home cooking and heating. But gas transportation requires expensive high-pressure pipelines or container ships, so even though gas is much more abundant than oil, the world's population still consumes more oil than gas. However, in recent years many industries and electrical generation plants have switched to natural gas, so demand for gas has increased.

Gas Hydrate

Until now, we have discussed natural gas as if it occurs only in association with petroleum deposits, and as if it forms only as a result of the breakdown of hydrocarbons at high temperatures in deeply buried sedimentary rocks. However, researchers increasingly recognize that the Earth System holds vast quantities of hydrocarbons in other forms. We will discuss coalbed methane later in this chapter, after we discuss coal. Here, we introduce gas hydrate.

Gas hydrate is a chemical compound consisting of a methane (CH_4) molecule surrounded by a cage-like arrangement of water molecules. An accumulation of gas hydrate produces a whitish solid that resembles ordinary water ice (▶Fig. 14.14b). Gas hydrate forms when anaero-bic bacteria (bacteria that live in the absence of oxygen) eat organic matter such as dead plankton that have been incorporated into the sea floor. When the bacteria digest organic matter, they produce methane as a byproduct, and the methane bubbles into the cold seawater that fills pore spaces in sediments. Under pressures found at water depths greater than 300 m, the methane dissolves in water and produces gas hydrate molecules; the reaction can occur at shallower depths at colder, polar latitudes. Exploration tests suggest that gas hydrate occurs as layers interbedded with sediment, and/or as a cement holding together the sediment, at depths of between 90 and 900 m beneath the sea floor. Geol-ogists estimate that an immense amount of methane lies trapped in gas hydrate layers. In fact, worldwide there may be more organic carbon stored in gas hydrate than in all other reservoirs combined (▶Fig. 14.14c)! So far, however, techniques for safely recovering gas hydrate from the sea floor have not been devised.

> **Take-Home Message**
>
> Large reserves of hydrocarbons occur in tar sands and oil shale as natural gas and as gas hydrate. But these alternative sources are difficult to process and/or transport, and thus cannot be exploited until new technologies evolve and the price of oil increases.

14.7 COAL: ENERGY FROM THE SWAMPS OF THE PAST

Coal, a black, brittle, sedimentary rock that burns, consists of elemental carbon mixed with minor amounts of organic chemicals, quartz, and clay (▶Fig. 14.15). Note that coal and oil do not have the same composition or origin. In contrast to oil, coal forms from plant material (wood, stems, leaves) that once grew in **coal swamps**, regions that resembled the wetlands and rain forests of modern tropical to semitropical coastal areas (▶Fig. 14.16a). Like oil and gas, coal is a fossil fuel because it stores solar energy that reached Earth long ago. The United States burns about 1 billion tons of coal per year, mostly at electrical-generating stations, yielding about 23% of the country's energy supply.

Significant coal deposits could not form until vascular land plants appeared in the late Silurian Period, about 420 million years ago. The most extensive deposits of coal in the world occur in Carboniferous-age strata (deposited between 286 and 354 million years ago; ▶Fig 14.16b). In fact, geologists coined the name *Carboniferous* because strata representing this interval of the geologic column contain so much coal. The abundance of Carboniferous coal reflects (1) the past position of the continents (during the Carboniferous Period, North America, Europe, and northern Asia straddled the equator, and thus had warm climates

(a)

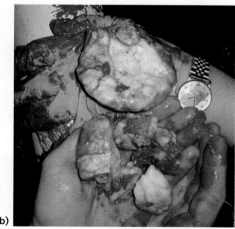

(b)

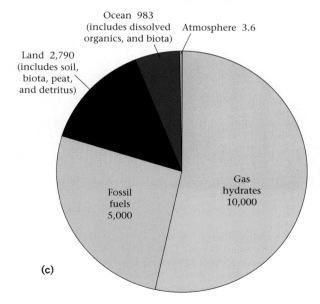

(c)

Ocean 983
(includes dissolved
organics, and biota)

Atmosphere 3.6

Land 2,790
(includes soil,
biota, peat,
and detritus)

Gas
hydrates
10,000

Fossil
fuels
5,000

FIGURE 14.14 **(a)** A large tar-sand mine in Canada. Giant shovels dig up the sand, and trucks haul it to the processing plant. **(b)** Gas hydrate samples (the white, icy material) brought up from the sea floor. **(c)** Graph showing the proportion of organic matter in different materials. Note that gas hydrate may contain the most organic carbon. (Numbers are × 10^5 tons of carbon.)

FIGURE 14.15 Chunks of freshly mined coal await processing at a mine in Indiana.

in which vegetation flourished) and (2) the height of sea level (at this time, shallow seas bordered by coal swamps covered vast parts of continental interiors). Because of their antiquity, Carboniferous coal deposits contain fossils of long-extinct species such as giant tree ferns, primitive conifers, and giant horsetails (▶Fig. 14.16c). Extensive coal deposits can also be found in strata of Cretaceous age (about 65 to 144 million years ago).

The Formation of Coal

How do the remains of plants transform into coal? The vegetation of an ancient swamp must fall and be buried in an *oxygen-poor* environment, such as stagnant water, so that it can be incorporated in a sedimentary sequence without first reacting with oxygen or being eaten. Compaction and partial decay of the vegetation transforms it into **peat**. Peat, which contains about 50% carbon, itself serves as a fuel in many parts of the world, where deposits formed from moss and grasses in bogs during the last several thousand years; it can easily be cut out of the ground, and once dried, it will burn (▶Fig. 14.17).

To transform peat into coal, the peat must be buried deeply (4–10 km) by sediment. Such deep burial can happen where the surface of the continent gradually sinks, creating a depression, or sedimentary basin, that can collect sediment. At any given time, the type of sediment deposited at a location depends on sea level, which rises and falls over geologic time. As sea level rises, the shoreline migrates inland (transgression takes place), and the location of a coal swamp becomes submerged beneath deeper water. Marine silt and mud then bury the swamp. When sea level later falls relative to the land, the shoreline migrates seaward (regression occurs) (▶Fig. 14.18).

(a)

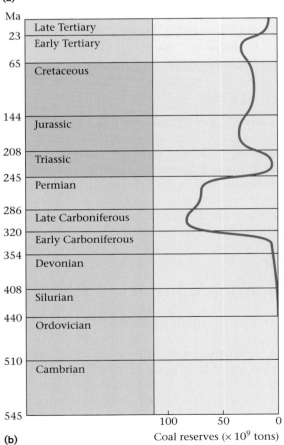

(b)

(c)

FIGURE 14.16 **(a)** This museum diorama depicts a Carboniferous coal swamp. **(b)** The graph shows the distribution of coal reserves in rocks of different ages. Notice that most reserves occur in late Paleozoic (Carboniferous) strata, a time when the continents were locked together to form Pangaea, a supercontinent that straddled the equator. **(c)** Fossils of fern leaves from a Carboniferous coal bed.

FIGURE 14.17 Digging out peat from a peat bog for fuel.

remove additional hydrogen, nitrogen, and sulfur, yielding progressively higher concentrations of carbon.

The Classification of Coal

Geologists classify coal according to the concentration of carbon. With increasing burial, peat transforms into a soft, dark-brown coal called **lignite**. At higher temperatures (about 100°–200°C), lignite, in turn, becomes dull, black **bituminous coal**. At still higher temperatures (about 200°–300°C), bituminous coal is transformed into shiny, black **anthracite coal** (also called hard coal). The progressive transformation of peat to anthracite coal, which occurs as the coal layer is buried more deeply and becomes warmer, reflects the completeness of chemical reactions that remove water, hydrogen, nitrogen, and sulfur from the organic chemicals of the peat and leave behind carbon (▶Fig. 14.19a–d). As the carbon content of coal increases, we say the **coal rank** increases.

The formation of anthracite coal requires high temperatures that develop only on the borders of mountain belts, where mountain-building processes can push thick sheets of rock along thrust faults and over the coal-bearing sediment, so the sediment ends up at depths of 8 to 10 km, where temperatures reach 300°C. In the interiors of mountain belts, temperatures become even higher, and rock begins to undergo metamorphism, at which point almost all elements except carbon leave the coal. The remaining carbon atoms rearrange to form graphite, the gray mineral used to make pencils; thus, coal cannot exist in metamorphic rocks.

The burning of coal is a chemical reaction: C (carbon) + O_2 (oxygen) → CO_2 (carbon dioxide). Therefore, the different ranks of coal produce different amounts of energy when burned (▶Table 14.2). Anthracite contains more carbon per kilogram than lignite does, so it produces more energy.

During successive transgressions and regressions in a sinking sedimentary basin (see Chapter 7), many kilometers of sediment containing numerous peat layers eventually accumulate. At depth in the pile, the weight of overlying sediment compacts the peat and squeezes out any remaining water. Then, because temperature increases with depth in the Earth, deeply buried peat gradually heats up. Heat accelerates chemical reactions that gradually destroy plant fiber and release elements such as hydrogen, nitrogen, and sulfur in the form of gas. These gases seep out of the reacting peat layer, leaving behind a residue concentrated with carbon. Once the proportion of carbon in the residue exceeds about 70%, we have coal. With further burial and higher temperatures, chemical reactions

Finding and Mining Coal

Because the vegetation that eventually becomes coal was initially deposited in a sequence of sediment, coal occurs as sedimentary beds ("seams," in mining parlance) interlayered with other sedimentary rocks (▶Fig. 14.20). To find coal, geologists search for sequences of strata that were deposited in tropical to semitropical, shallow-marine to terrestrial (fluvial or deltaic) environments—the environments in which a swamp could exist. The sedimentary strata of continents contain huge quantities of discovered coal, or **coal reserves**. For example, economic seams (beds of coal about 1–3 m thick, thick enough to be worth mining) of Cretaceous age occur in the U.S. and Canadian Rocky Mountain region, while economic seams of Carboniferous

FIGURE 14.18 Sea level transgresses and regresses over time, with the result that a coal swamp along the coast migrates inland, and the swamp's deposits eventually get buried by other strata. Notice that the floor of the basin gradually sinks, so there is room for all the strata. (See Fig. 7.37 for the consequences of progressive transgression and regression.)

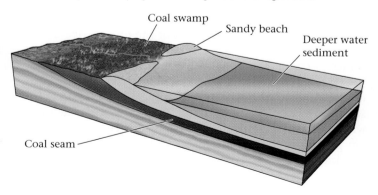

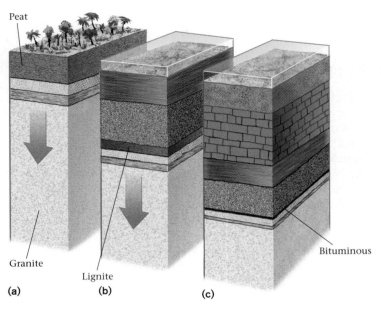

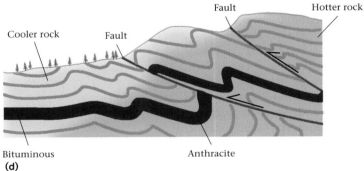

FIGURE 14.19 The evolution of coal from peat. **(a)** A layer of peat accumulates beneath a coal swamp. **(b)** Later, after the peat has been substantially buried, it compacts, loses water, releases hydrogen, nitrogen, and sulfur, and becomes lignite. **(c)** Still later, after even more burial, the lignite compacts and alters still further to form bituminous coal. **(d)** In this mountain range, the coal bed has been folded, and faults have transported warmer rock from greater depth over the coal. As a result, the coal becomes even warmer and turns into anthracite coal.

TABLE 14.2 Types of Coal

Material	% Carbon	Energy Content[1]	Rank
Peat	50	1,500 kcal/kg	
Lignite	70	3,500 kcal/kg	Low-rank coal
Bituminous coal	85	6,500 kcal/kg	Mid-rank coal
Anthracite coal	95	7,500 kcal/kg	High-rank coal

[1]1 kcal (kilocalorie) = 1,000 calories. A calorie is the heat needed to raise the temperature of 1 gram of water by 1°C. 1 kg (kilogram) = 2.2 pounds.

FIGURE 14.20 Coal is found in sedimentary beds interlayered with other strata (sandstone, shale, and limestone).

age are found throughout the midwestern United States (►Fig. 14.21a–c).

The way in which companies mine coal depends on the depth of the coal seam. If the coal seam lies within 100 m of the ground surface, strip mining proves to be most economical. In strip mines, miners use a giant shovel called a drag line to scrape off soil and layers of sedimentary rock above the coal seam (►Fig. 14.22a, b). Drag lines are so big that they could swallow a two-car garage without a trace. Once the drag line has exposed the seam, it then scrapes out the coal and dumps it into trucks or onto a conveyor belt. Before modern environmental awareness took hold, strip mining left huge scars on the landscape. Without topsoil, the rubble and exposed rock of the mining operation remained barren of vegetation. In many contemporary mines, however, the drag-line operator separates out and preserves topsoil. Then, when the coal has been scraped out, the operator fills the hole with the rock that had been stripped to expose the coal and covers the rock back up with the saved topsoil, on which grass or trees may eventually grow. Within years to decades, the former mine site can become a pasture or a forest.

Deep coal can be obtained only by underground mining. Miners dig a shaft down to the depth of the coal seam and then create a maze of tunnels, using huge

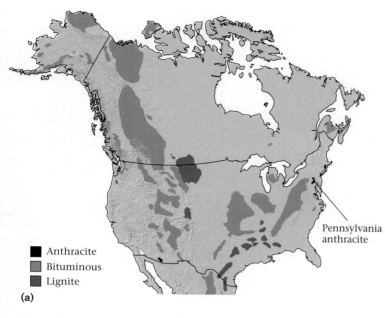

Anthracite
Bituminous
Lignite

Pennsylvania anthracite

(a)

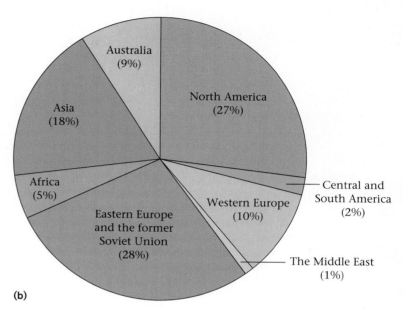

Australia (9%)

Asia (18%)

North America (27%)

Africa (5%)

Central and South America (2%)

Western Europe (10%)

Eastern Europe and the former Soviet Union (28%)

The Middle East (1%)

(b)

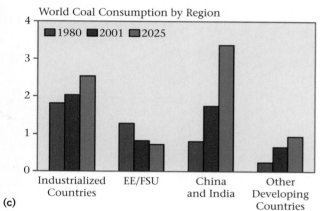

World Coal Consumption by Region

1980 2001 2025

Industrialized Countries   EE/FSU   China and India   Other Developing Countries

(c)

FIGURE 14.21 **(a)** The distribution of coal reserves in North America. **(b)** The distribution of coal reserves by region. **(c)** World coal consumption by region. The vertical axis is "Billion Short Tons," where 1 short ton = 2,000 lbs. = 907 kg.

FIGURE 14.22 **(a)** The configuration of a coal strip mine. **(b)** A drag line.

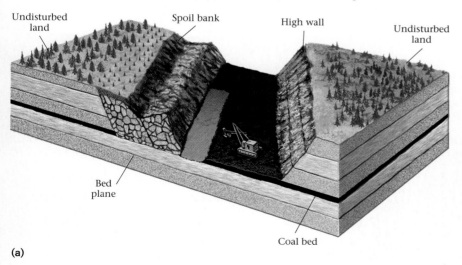

Undisturbed land

Spoil bank

High wall

Undisturbed land

Bed plane

Coal bed

(a)

(b)

See for yourself . . .
Sources of Energy

The geologic features associated with energy resources are not all visible at the ground surface, but you can see evidence of where resources are being extracted, and of how they are being transported. The thumbnail images provided on this page are only to help identify tour sites. Go to wwnorton.com/studyspace to experience this flyover tour.

Ackerly Oil Field near Lamesa, Texas
(Lat 32°33'18.42"N, Long 101°46'55.39"W)

Fly to this locality and zoom to 8 km (5 miles). You can see a grid of farm fields (1.6 km, or 1 mile, wide) within which small roads lead to patches of dirt (Image G14.1). Each patch hosts or hosted a single pump for extracting oil from underground. In effect, an oil field viewed from the air looks like a pattern of spots connected by a web of roads. These wells are in the Ackerly Oil field of the Permian Basin in west Texas. Drilling here began in the 1940s to tap oil reserves in Permian sandstone beds that now lie at depths of between 1,500 and 2,500 m (5,000 and 8,200 feet). The beds tilt slightly and grade laterally into impermeable shales, so producers are exploiting stratigraphic traps. To date, 50 million barrels of oil have been extracted from this field.

G14.1

Ghawar Oil Field, Saudi Arabia
(Lat 26°24'12.91"N, Long 49°13'54.89"E)

Sandstone beds beneath the desert landscape of Saudi Arabia contain vast quantities of oil. The Ghawar field—200 km (125 miles) long and 40 km (25 miles) wide—is one of the largest oil fields. At the coordinates given, zoom to 1 km and tilt your image so you are looking north. Here, you can see an abandoned drilling site (Image G14.2). Now fly to Lat 24°47'48.02"N, Long 49°10'42.04"E, zoom to 4 km, and you'll find an active portion of the field where wells feed a processing station linked to pipelines (Image G14.3). Oil of the Ghawar field resides in an anticline trap whose reservoir rock consists of sandstone, 2,100 m (7,000 feet) below the surface. Oil companies have drilled 3,500 wells into the anticline. So far, the field has yielded 55 *billion* barrels of oil.

G14.2

G14.3

Offshore Well, Gulf Coast USA
(Lat 29°41'10.94"N, Long 93°43'46.32"W)

Much of the active oil production in the United States takes place along the Gulf Coast. Zoom to 1,500 km (930 miles) at this location, and you can see the bathymetry beneath this region (Image G14.4). Note the broad continental shelf which bulges southward as a broad lobe bounded by an abrupt slope (the Sigsbee Escarpment). This shelf is underlain by a very thick pile of sediment that has slowly slipped southward on weak salt horizons deep in the section. The salt has also risen to form domes, making the sea floor quite bumpy. Many of the traps in this region border salt domes. Zoom to 500 m (1,640 feet) to see a near-shore oil-drilling platform (Image G14.5).

G14.4

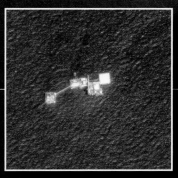

G14.5

G14.6

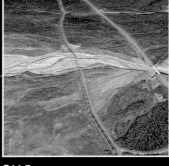

G14.7

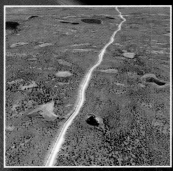

G14.8

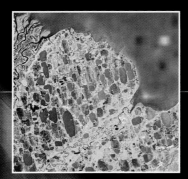

G14.9

Alaska Pipeline
(Lat 64°55'38.64"N, Long 147°37'23.71"W)

The oil of the isolated North Slope oil fields near Prudhoe Bay moves through the Alaska pipeline across the entire state, crossing permafrost (permanently frozen ground) and seismic belts, to reach ports on the south coast where it is loaded into supertankers. Fly to the coordinates provided, zoom to 650 m (0.4 miles), tilt your image, and look northwest. You can see the Alaska pipeline just north of Fairbanks (Image G14.6). At Lat 63°24'16.92"N, Long 145°44'31.72"W, zoom to 1.5 km (0.9 miles) and tilt the image so you are looking north. Here, you can see a section of the fault near the active Denali fault (Image G14.7). The pipeline sits on sliding pedestals that allow it to move without breaking during an earthquake. At Lat 63°45'15.94"N, Long 145°50'17.23"W, the pipeline crosses permafrost. Zoom to 1.5 km (0.9 miles), look north, and you can see how melting of the permafrost produces the shallow circular to elliptical depressions on the landscape (Image G14.8). From an altitude of 30 km (18 miles) at Lat 70°17'31.41"N, Long 148°31'24.30"W, you can see roads and pipelines crossing the stark tundra landscape of Prudhoe Bay (Image G14.9).

G14.10

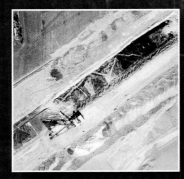

G14.11

Coal Mine, Farmersburg Indiana
(Lat 39°14'57.59"N, Long 87°20'53.20"W)

Fly to this locality and zoom to 8 km (5 miles). You are looking at an active coal strip mine exploiting Pennsylvanian-age coal seams. Note several points about this mine (Image G14.10). First, the active part of the mine consists of a 3.4 km (2 mile)-long trench. To produce the coal, giant bulldozers first scrape off the soil layers to expose bedrock. Then workers drill a series of holes, fill them with explosives, and blast the bedrock into fragments. The dragline scrapes out the rock and sets it aside. The dragline then piles the coal into giant trucks which can carry as much as 150 tons in a load. Once a dragline completes a trench, bulldozers fill it back in and resurface it with topsoil. The reclaimed land can be cultivated or reforested. Zoom down to 900 m (2,950 feet) and, in the image online at the time of this writing, you can see the giant drag haul, with a tiny pickup truck nearby (Image G14.11).

grinding machines that chew their way into the coal (▶Fig. 14.23). Underground coal mining can be very dangerous, not only because the sedimentary rocks forming the roof of the mine are weak and can collapse, but also because methane gas released by chemical reactions in coal can accumulate in the mine, leading to the danger of a small spark triggering a deadly mine explosion. Unless they breathe through filters, underground miners also risk contracting black-lung disease from the inhalation of coal dust. The dust particles wedge into tiny cavities of the lungs and gradually cut off the oxygen supply or cause pneumonia.

Coalbed Methane

The natural process by which coal forms underground yields large quantities of methane, a type of natural gas. Over time, some of the gas escapes to the atmosphere, but vast amounts remain within the coal. Such **coalbed methane**, trapped in strata too deep to be reached by mining, is a valuable energy resource that has become a target for exploration in many regions of the world.

Obtaining coalbed methane from deep layers of strata involves drilling, rather than mining. Drillers penetrate a coal bed with a hole, and then start pumping out groundwater. As a result of pumping out water, the pressure in the vicinity of the drillhole decreases relative to the surrounding bed. Methane bubbles into the hole and then up to the ground surface, where condensers force it into tanks for storage. Disposal of the water produced by coalbed methane extraction can be a major problem. If the water is pure, it can be used for irrigation, but in deep coalbeds the water may be saline and thus can't be used for crops. Producers either pump this water back underground or evaporate it in large ponds, so that they can extract and collect the salt.

Coal Gasification

Traditional burning of coal produces clouds of smoke, containing fly ash (solid residue left after the carbon in coal has been burned) and noxious gases (including sulfur dioxide, SO_2, which forms because coal contains sulfur-bearing minerals such as pyrite). Today, smoke can be partially cleaned by expensive scrubbers, but pollution is still a problem. Alternatively, coal can be transformed into various gases, as well as solid byproducts, *before* burning. The gases burn quite cleanly. The process of producing clean-burning gases from solid coal is called **coal gasification**; the process was invented in the late eighteenth and early nineteenth centuries and was used extensively to produce fuels during World War II.

Coal gasification involves the following steps. First, pulverized coal is placed in a large container. Then, a mixture of steam and oxygen passes through the coal at high pressure. As a result, the coal heats up to a high temperature but does not ignite; under these conditions, chemical reactions break down the carbon molecules in coal and produce hydrogen and other gases such as carbon monoxide. Solid ash, as well as sulfur and mercury, concentrate at the bottom of the container and can be removed before the gases are burned, so that the contaminants do not go up the chimney and into the atmosphere. The hydrogen can also be concentrated to produce fuel cells (which we will discuss later). A large research effort, called FutureGen, is currently beginning in the United States to test technologies that not only produce gas from coal but also pump greenhouse gases (CO_2) back underground. The goal is to produce a zero-emissions fossil-fuel plant.

FIGURE 14.23 Machinery used in underground coal mining. The rotating head grinds away at the coal.

Underground Coalbed Fires

Coal will burn not only in furnaces but also in surface and subsurface mines, as long as the fire has access to oxygen. Coal mining of the past two centuries has exposed much more coal to the air and has provided many more opportunities for fires to begin; once started, a coalbed fire that progresses underground (sucking in oxygen from joints in overlying rock) may be very difficult to extinguish. Some fires begin as a result of lightning strikes, some from spontaneous combustion (when coal reacts with air, it heats up), some in the aftermath of methane explosions, and some when people intentionally set trash fires in mines.

Major coalbed fires (▶Fig. 14.24a–b) can be truly disastrous. The most notorious of these fires in North America began as a result of trash burning in a mine near the town of Centralia, Pennsylvania. For the past forty years the fire has progressed underground, eventually burning coal beneath

(a)

(b)

FIGURE 14.24 **(a)** A burning coalbed glows red on the wall of a mine in China. **(b)** A coalbed fire beneath Centralia, Pennsylvania, produces noxious gas.

the town itself. The fire produces toxic fumes that rise through the ground and make the overlying landscape uninhabitable, and it also causes the land surface to collapse and sink (Fig. 14.24a). Eventually, inhabitants had to abandon many neighborhoods of the town. Much longer-lived fires occur elsewhere—one in Australia may have been burning for 2,000 years. And today, over 100 major coalbed fires are burning in northern China; these can be detected by satellite imagery, which shows warm spots on the overlying ground surface (Fig. 14.24b). Recent estimates suggest that 200 million tons of coal burn in China every year, an amount equal to approximately 20% of the annual national production of coal in China. These fires produce as much CO_2 gas as all the cars and light trucks now on the road in the United States.

> **Take-Home Message**
>
> Coal forms from the remains of plant material. When buried deeply, this organic material undergoes reactions that concentrate carbon; higher-rank coal contains more carbon. Coal occurs in sedimentary successions and must be mined underground or in open pits.

14.8 NUCLEAR POWER

How Does a Nuclear Power Plant Work?

So far, we have looked at fuels (oil, gas, and coal) that release energy when burned. During burning, a chemical reaction between the fuel and oxygen releases the potential energy stored in the *chemical bonds* of the materials. Nuclear power, however, comes from a different process: the fission, or breaking, of the *nuclear bonds* that hold protons and neutrons together in the nucleus provides the energy. Fission splits an atom into smaller pieces.

Nuclear power plants were first built to produce electricity during the 1950s. A **nuclear reactor**, the heart of the plant, commonly lies within a dome-shaped shell (containment building) made of reinforced concrete (▶Fig. 14.25a). The reactor contains nuclear fuel—pellets of concentrated uranium oxide or a comparable radioactive material—packed into metal tubes called fuel rods. Fission occurs when a neutron strikes a radioactive atom, causing it to split; for example, uranium-235 splits into barium-141 plus krypton-92 plus three neutrons plus energy (▶Fig. 14.25b). During fission, a tiny fraction of the matter composing the original atom is transformed into a large amount of thermal and electromagnetic energy. Just one gram of nuclear-reactor fuel yields as much energy as 2.7 barrels of oil. The neutrons released during the fission of one atom strike other atoms, thereby triggering more fission in a self-perpetuating process called a **chain reaction**. If enough radioactive atoms cram into a small enough space, a "critical mass" forms: the chain reaction happens so fast that the mass explodes, making an atomic bomb. Such an explosion cannot occur in a reactor, because the fuel pile does not contain a critical mass.

In a nuclear power plant, pipes carry water close to the heat-generating fuel. The heat transforms the water into high-pressure steam. The pipes then carry this steam to a turbine, where it rotates fan blades; the rotation drives a dynamo that generates electricity. Eventually the steam goes into cooling towers, where it condenses back into water that can be reused in the plant or returned to the environment (▶Fig. 14.26a, b).

The Geology of Uranium

^{235}U, an isotope of uranium containing 143 neutrons, is the most common fuel for conventional nuclear power plants. ^{235}U accounts for only about 0.7% of naturally occurring uranium; most uranium consists of ^{238}U, an isotope with 146 neutrons. Thus, to make a fuel suitable for use in a power plant, the ^{235}U concentration in a mass of natural uranium must be increased by a factor of 2 or 3, an expensive process called enrichment.

(a)

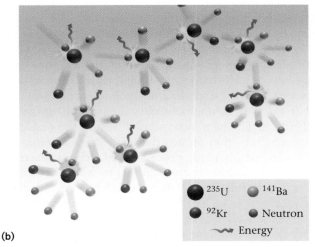

235U

141Ba

92Kr

Neutron

Energy

(b)

FIGURE 14.25 **(a)** This nuclear power plant has two reactors, each in its own containment building. **(b)** A chain reaction involving the radioactive decay of uranium.

Where does uranium come from? The Earth's radioactive elements, including uranium, probably developed during the explosion of a supernova before the existence of the solar system. Uranium atoms from this explosion became part of the nebula out of which the Earth formed and thus were incorporated into the planet. They gradually rose into the upper crust in granitic magma.

Even though granite contains uranium, it does not contain very much. But nature has a way of concentrating uranium. Hot water circulating through a pluton after intrusion dissolves the uranium and precipitates it, along with other elements, in cracks. Uranium from veins typically occurs in the mineral pitchblende (UO_2). Uranium may be further concentrated once plutons, and the associ-

ated uranium-rich veins, weather and erode at the ground surface. Sand derived from a weathered pluton washes down a stream, and as it does so, uranium-rich grains stay behind because they are much heavier than quartz and feldspar grains. The world's richest uranium deposits, in fact, occurs in ancient stream-bed deposits. Uranium deposits may also form when groundwater percolates through uranium-rich sedimentary rocks: the uranium dissolves in the water and moves with the water to another location, where it precipitates out of solution and fills the pores of the host sedimentary rock.

To find uranium deposits, prospectors use a Geiger counter, an instrument that detects radioactivity. These days, mining companies can quickly explore large regions using an instrument like a Geiger counter towed behind an airplane. If the instrument detects an unusually high concentration of radioactivity, land-based geologists explore the site further.

Nuclear Problems

Textbooks written in the 1950s touted nuclear power as a magical source that could meet our limitless energy needs cleanly for centuries to come. Currently, there are over 100 nuclear power plants in the United States, and these provide about 20% of the country's electrical power. Other countries rely much more on nuclear energy. For example, nuclear power provides 22% of the electrical power in the United Kingdom, 34% in Japan, 47% in Ukraine, and 76% in France.

But in recent decades, many people have become concerned about plant safety and the disposal of radioactive waste. These concerns, along with vast cost overruns involved in the construction of nuclear plants, have slowed the growth of the nuclear power industry substantially. To date, nuclear plants on the whole have been remarkably safe. But maintaining safety requires work. Operators must constantly cool the nuclear fuel with circulating water, and the rate of nuclear fission must be regulated by the insertion of boron-steel control rods, which absorb neutrons and thus decrease the number of collisions between neutrons and radioactive atoms. If left uncontrolled, so many neutrons begin to dash about in the fuel rods that the rate of fission becomes too fast and the fuel rods become too hot, in which case they melt, causing a **meltdown**. Were a meltdown to occur, the fuel could melt through pipes in the containment building and cause a steam explosion that would scatter radioactive debris into the air.

Only one near-meltdown has occurred since the beginning of the nuclear age—at the power plant in Chernobyl, Ukraine, in April 1986. Operators of the Chernobyl plant, while testing a procedure for cooling the reactor, shut down the safety devices. During the test, the fuel pile became too hot, triggering a steam explosion that raised the roof of the reactor and spread fragments of the reactor and

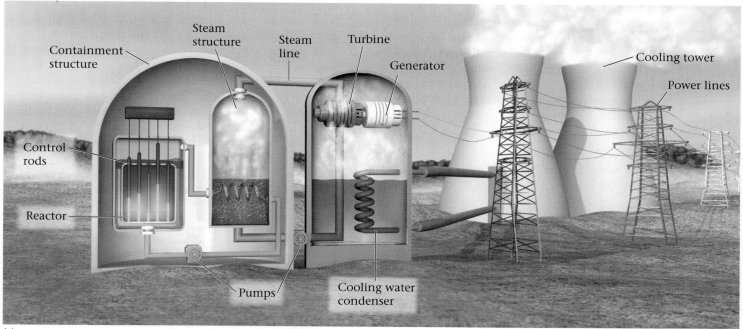

(a)

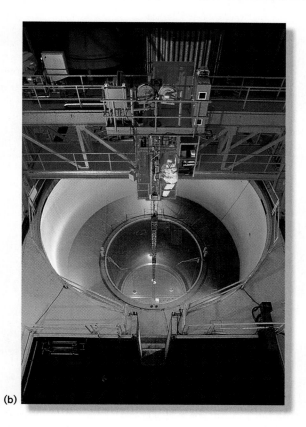

(b)

FIGURE 14.26 (a) A nuclear reactor can be used to generate electricity. Water, heated by the reactor, flows into a steam generator that produces high-pressure steam. The steam, in turn, drives a turbine. A condenser transforms the steam back into water. **(b)** Nuclear fuel rods in a reactor.

its fuel around the plant grounds. Within 6 weeks, twenty people had died from radiation sickness. Further, about 10% of the radioactive material in the reactor entered the atmosphere and dispersed over Eastern Europe and Scandinavia; no one yet knows the full effects of this fallout on the health of exposed populations. In the United States, the worst nuclear accident happened in the Three Mile Island plant near Harrisburg, Pennsylvania, in 1979. Because of faulty pumps, the reactor was not properly cooled for several hours and suffered damage. However, hardly any radioactivity entered the environment, and no one appears to have been directly injured. These days, there is concern that even a safely operated plant could pose a hazard if the target of a terrorist attack.

Nuclear waste is the radioactive material produced in a nuclear plant. It includes spent fuel, which contains radioactive daughter products, as well as water and equipment that have come in contact with radioactive materials. Radioactive elements emit gamma rays and X-rays that can damage living organisms and cause cancer. Some radioactive material decays quickly (in decades to centuries), but some remains dangerous for thousands of years or more. High-level waste contains greater than 1 million times the safe level of radioactivity, intermediate-level waste contains between 1 thousand and 1 million times the safe level, and low-level waste contains less than 1 thousand times the safe level.

Nuclear waste cannot just be stashed in a warehouse or buried in a town landfill. If the waste were simply buried, groundwater passing through the dump site might transport radioactive elements into municipal water supplies or nearby lakes or streams. Ideally, waste should be sealed in containers that will last for thousands of years (the time needed for the short-lived radioactive atoms to undergo decay) and stored in a place where it will not come in contact with the environment. Finding an appropriate place is not easy. Geoscientists have suggested the following possibilities.

- Underground tunnels drilled into a mountain composed of solid, dry rock in a region safe from damage by earthquakes or volcanoes.
- The interiors of impermeable salt domes.
- Landfills surrounded by clay, for clay can absorb and trap radioactive atoms.
- Landfills in regions where the groundwater composition can react with radioactive atoms to form nonmovable minerals.
- The deep-ocean floor at a location where it will soon be subducted.

So far, experts disagree about which is the best way to dispose of nuclear waste. The U.S. government favors storing waste at Yucca Mountain, in the Nevada desert. The mountain consists of fairly dry rhyolite, the area is far from a population center, and the water table (the top surface of groundwater) lies 330 m (1,000 feet) below the proposed repository. But because of continuing disagreement, much waste has been stored in temporary facilities.

> ### Take-Home Message
>
> Controlled fission in reactors produces nuclear power. The fuel consists of uranium or other elements obtained by mining. Reactors run the risk of meltdown, though this has not yet happened. Reactors also produce radioactive waste that is difficult to store.

14.9 OTHER ENERGY SOURCES

Geothermal Energy

As the name suggests, *geothermal energy* refers to heat and electricity produced by using the internal heat of the Earth. Geothermal energy exists because the Earth grows progressively hotter with depth; recall that the geothermal gradient, the rate of temperature change, varies between 15° and 50°C per km in the upper part of the crust. Where the geothermal gradient is high, we find high temperatures at relatively shallow depths. Groundwater in such areas absorbs heat from the rock and becomes hot.

We use geothermal energy to produce heat and electricity in two ways. In some places, we simply pump hot groundwater out of the ground and run it through pipes to heat houses and spas. Elsewhere, the groundwater is so hot that when it rises to the Earth's surface and decompresses, it turns to steam. This steam is then used to drive turbines and generate electricity (▶Fig. 14.27a, b).

In volcanic areas such as Iceland or New Zealand, geothermal energy provides a major portion of energy needs. But on a global basis, it meets only a small proportion of

FIGURE 14.27 (a) Surface water sinks down into the ground as groundwater until it becomes heated at depth (possibly by magma below). The hot water rises and, when it reaches shallow depths, turns to steam (as a result of decompression), and runs turbines. **(b)** A geothermal energy plant.

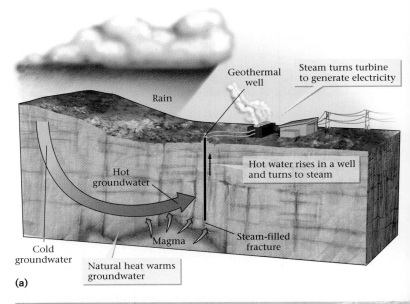

(a)

(b)

energy needs because few cities lie near geothermal resources. Furthermore, this energy supply can be destroyed, for if people pump groundwater out of the ground faster than it can be replenished, the hot-water supply diminishes.

Hydroelectric and Wind Power

As water flows downslope, its potential energy converts into kinetic energy. In a modern **hydroelectric power plant**, the water flow drives turbines, which in turn drive generators that produce electricity. In order to increase the rate and volume of water flow, engineers build dams to create reservoirs that retain water and raise it to a higher elevation—the water flows through pipes down to turbines at the foot of the dam (▶Fig. 14.28a).

At first glance, hydroelectric power seems ideal, because it produces no smoke or radioactive waste, and because reservoirs can also be used for flood control, irrigation, and recreation. But unfortunately, reservoirs may also bring unwanted changes to a region's landscape and ecology. Damming a fast-moving river may flood a spectacular canyon, eliminate exciting rapids, and destroy a river's ecosystem. Further, the reservoir traps sediment, so floodplains downstream lose their sediment and nutrient supply. Reservoirs have not always remained full of water, because planners can underestimate water supplies.

Not all hydroelectric power plants utilize river water. In a few places, engineers have employed the potential energy stored in ocean water at high tide (▶Fig. 14.28b, c). To do this, they build a floodgate dam across an inlet. Water flows into the inlet when the tide rises, only to be trapped when the gate is closed. After the tide has dropped outside the floodgate, the water retained by the floodgate flows back to sea via a pipe that carries it through a power-generating turbine.

Wind power has been used for millennia as a way to provide power for mills and water pumps. In recent decades, numerous wind farms have been established around the world to generate electricity (▶Fig. 14.29a). The electricity is clean, but wind production has a serious drawback: it requires construction of large, somewhat noisy towers. Not everyone wants a giant wind farm in their "backyard"—because of the visual effect, because of the noise, and because the towers pose a hazard for birds.

Solar Power

The Sun drenches the Earth with energy in quantities that dwarf the amounts stored in fossil fuels. Were it possible to harness this energy directly, humanity would have a reliable and totally clean solution for powering modern technology. But using solar energy is not quite so simple, for the energy is diffuse—on the sunniest days, each square meter of the Earth's surface receives about 1,000 watts of

(a)

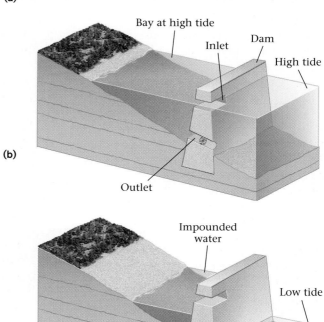

(b)

(c)

FIGURE 14.28 **(a)** A hydroelectric dam. This example is the Grand Coulee Dam, located on the Columbia River in central Washington State; it is the third-largest producer of hydroelectric power in the world. **(b, c)** Oceanic tides can be used to generate electricity. At high tide, water fills the volume behind the dam. At low tide, water flows out via a turbine.

energy. How can this energy be concentrated sufficiently to produce heat and electricity? At present, energy consumers have two options for the direct use of solar energy.

Solar collectors constitute the first option. A **solar collector** is a device that collects energy to produce heat. One class of solar collectors includes mirrors or lenses that

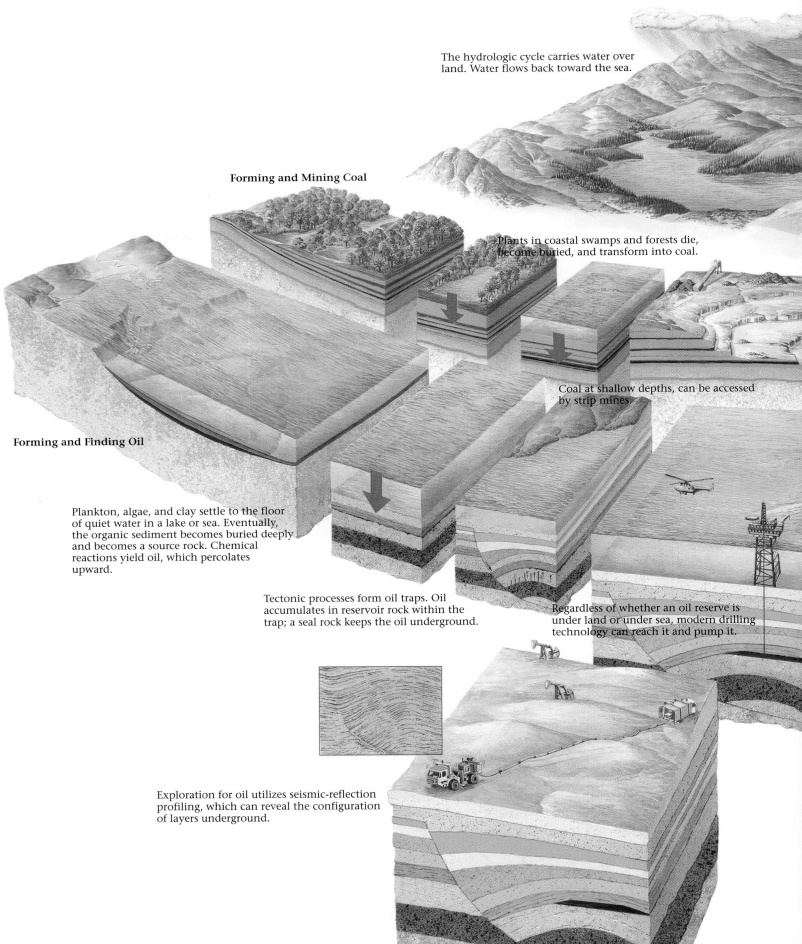

Water, Wind, and Tides

The hydrologic cycle carries water over land. Water flows back toward the sea.

Forming and Mining Coal

Plants in coastal swamps and forests die, become buried, and transform into coal.

Coal at shallow depths, can be accessed by strip mines.

Forming and Finding Oil

Plankton, algae, and clay settle to the floor of quiet water in a lake or sea. Eventually, the organic sediment becomes buried deeply and becomes a source rock. Chemical reactions yield oil, which percolates upward.

Tectonic processes form oil traps. Oil accumulates in reservoir rock within the trap; a seal rock keeps the oil underground.

Regardless of whether an oil reserve is under land or under sea, modern drilling technology can reach it and pump it.

Exploration for oil utilizes seismic-reflection profiling, which can reveal the configuration of layers underground.

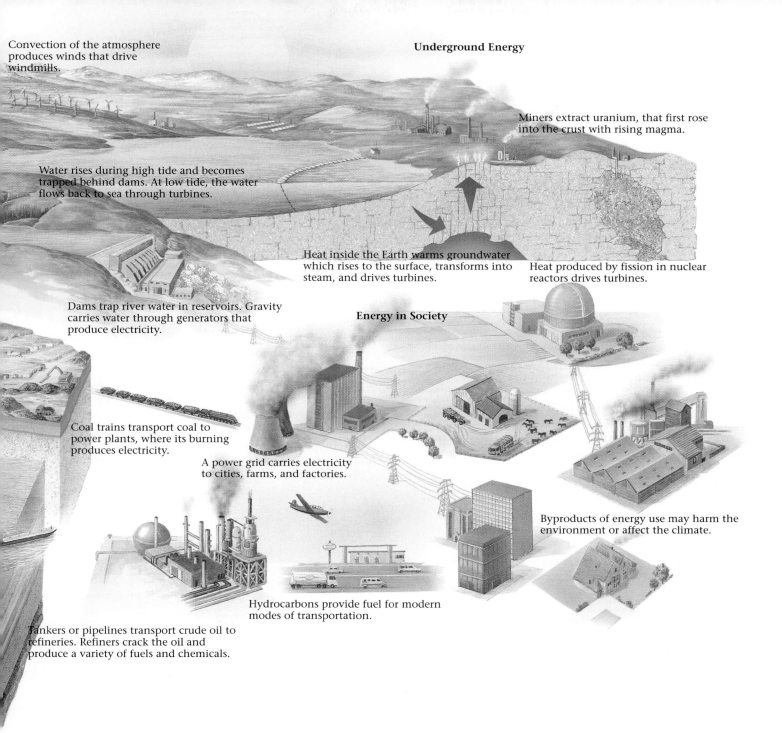

Convection of the atmosphere produces winds that drive windmills.

Underground Energy

Miners extract uranium, that first rose into the crust with rising magma.

Water rises during high tide and becomes trapped behind dams. At low tide, the water flows back to sea through turbines.

Heat inside the Earth warms groundwater which rises to the surface, transforms into steam, and drives turbines.

Heat produced by fission in nuclear reactors drives turbines.

Dams trap river water in reservoirs. Gravity carries water through generators that produce electricity.

Energy in Society

Coal trains transport coal to power plants, where its burning produces electricity.

A power grid carries electricity to cities, farms, and factories.

Byproducts of energy use may harm the environment or affect the climate.

Hydrocarbons provide fuel for modern modes of transportation.

Tankers or pipelines transport crude oil to refineries. Refiners crack the oil and produce a variety of fuels and chemicals.

Power from the Earth

Modern society, for better or worse, uses vast amounts of energy to produce heat, to drive modes of transportation, and to produce electricity (see Appendix B). This energy comes either from geologic materials stored in the Earth, or from geologic processes happening at our planet's surface. For example, oil and gas fill the pores of reservoir rocks at depth below the surface, coal occurs in sedimentary beds, and uranium concentrates in ore deposits. A hydroelectric power plant taps into the hydrologic cycle, windmills operate because of atmospheric convection, and geothermal energy comes from hot groundwater. Ultimately, the energy in the sources we've

just listed comes from the Sun, from gravity, from Earth's internal heat, and/or from nuclear reactions. Oil, gas, and coal are fossil fuels because the energy they store first came to Earth as sunlight, long ago.

As energy usage grows, easily obtainable energy resources dwindle, the environment can be degraded, and the composition of the atmosphere has changed. The pattern of energy use that forms the backbone of society today may have to change radically in the not-so-distant future, if we wish to avoid a decline in living standards.

focus light striking a broad area into a smaller area. On a small scale, such devices can be used for cooking. Another class of solar collectors consists of a black surface placed beneath a glass plate. The black surface absorbs light that has passed through the glass plate and heats up. The glass does not let the heat escape. When a consumer runs water between the glass and the black surface, the water heats up; the hot water then can be stored in an insulated tank.

Photovoltaic cells constitute the second option. The use of **photovoltaic cells** (solar cells) allows light energy to convert directly into electricity (▶Fig. 14.29b). Most photovoltaic (PV) cells consist of two wafers of silicon pressed together. Silicon is a semiconductor, meaning that it can conduct electricity only when doped with impurities; "doping" means that manufacturers intentionally add atoms other than silicon to the wafer. One wafer is doped with atoms of arsenic, which have extra electrons and can serve as electron donors. The other wafer is doped with boron, an element that lacks electrons and can serve as an electron acceptor. Electrons would like to flow from the donor wafer to the acceptor wafer, but their way is blocked by neutral atoms at the contact surface between the two wafers. When light strikes the cell, the electrons gain enough energy to cross the boundary. If a wire loop connects the back side of one wafer to the back side of the other, an electrical current flows when light strikes the cell (▶Fig. 14.29c). Although PV cells sound like a great way to produce electricity, they are fairly inefficient. A typical cell converts only about 10% to 15% of the energy that it receives into electricity. Thus, until the technology improves significantly, huge areas of land would have to be covered with sheets of PV cells if the cells were to provide a significant proportion of global energy supplies.

Biofuels

Until the nineteenth century, the burning of biomass—wood, charcoal, or other plant and animal materials—provided most of the energy needed to drive civilization. But, as noted earlier, wood cannot regenerate quickly enough to provide a sufficient energy source for modern needs. In recent years, farmers have begun to produce rapidly growing crops specifically for the purpose of creating biomass for fuel production. The resulting liquids are called **biofuels**.

The most commonly used biomass fuel, ethanol (a type of alcohol), is currently produced from the fermentation of sugar derived from corn or sugar cane. Ethanol can substitute for gasoline in car engines. The production of ethanol includes the following steps. (1) Producers grind grain into a fine powder, mix it with water, and cook it to produce a mash of starch. (2) They add an enzyme to the mash, which converts the mash into sugar. (3) They mix in yeast and allow it to ferment. Fermentation produces ethanol and CO_2. (4) The fermented mash is then distilled to concentrate the ethanol. In recent years, researchers have begun to develop methods to produce ethanol from cellulose. Such methods would allow perennial grasses to become a renewable source of liquid fuel.

Another product, called biodiesel, has also entered the market as an alternative to petroleum products. Biodiesel is produced by chemical modification of fats or vegetable oils. It can be mixed with petroleum-derived diesel to provide a fuel for trucks and buses.

Debate over the use of biofuels has political overtones, because even though its use provides a market for agricultural goods, it may take almost as much energy to produce biofuels as the fuels provide. A switch to cellulosic ethanol may improve this situation greatly.

Fuel Cells

In a fuel cell, chemical reactions produce electricity directly. Let's consider a hydrogen fuel cell. Hydrogen gas flows through a tube across an anode (a strip of platinum) that has been placed in a water solution containing an electrolyte (a substance that enables the solution to conduct electricity). At the same time, a stream of oxygen gas flows onto a separate platinum cathode that also has been placed into the solution. A wire connects the anode and the cathode to create an electrical circuit (▶Fig. 14.29d). In this configuration, hydrogen reacts with oxygen to produce water. About 40% to 80% of the chemical energy released by this reaction produces electricity that flows through the wire, and the remainder becomes heat.

Fuel cells are efficient and clean. Their limitation lies in the need for a design that will prevent the cells from damage by impact and that will enable them to store hydrogen, an explosive gas, in a safe way. Also, it takes energy to produce the hydrogen used in fuel cells. Hydrogen could be manufactured by the electrolysis of water but the energy needed to drive the reaction must come from another source. Hydrogen may also be a byproduct of the gasification of coal.

> **Take-Home Message**
>
> Geothermal energy utilizes hot groundwater that has been warmed by heat from Earth's interior. Flowing water (in rivers or tides), flowing air, solar panels, and fuel cells can also produce energy. Biomass can be transformed into burnable alcohol and biodiesel.

14.10 ENERGY CHOICES, ENERGY PROBLEMS

The Crisis of the 1970s

Energy usage in industrialized countries grew with dizzying speed through the mid–twentieth century. Of the many possible sources (see art, pp. 514–515), people came to rely mostly on oil, which burns more efficiently than coal and, being a liquid, can be transported easily even though oil is not the

FIGURE 14.29 (a) A wind farm in southwestern England. The towers are about 50 m high. **(b)** An example of a photovoltaic cell; **(c)** Diagram illustrating the way in which a photovoltaic cell produces electricity. **(d)** Diagram illustrating the way in which a hydrogen fuel cell works.

FIGURE 14.30 (a) Estimated world reserves of various energy resources. The numbers at the ends of each column are in "Quads," where 1 Quad = 1 quadrillion BTU. (1 BTU [British Thermal Unit] = 1,055 Joules = 252 calories; 1 barrel of oil produces 5.8 million BTU). **(b)** The proportion of world energy supply provided by different energy resources in 2003.

most abundant energy resource (▶Fig. 14.30a, b). Eventually, oil supplies within the borders of industrialized countries could no longer match the demand, and these countries began to import more oil than they produced themselves. Through the 1960s, oil prices remained low (about $1.80 a barrel), so this was not a problem.

In 1973, however, a complex tangle of politics and war led the Organization of Petroleum-Exporting Countries (OPEC) to limit its oil exports. In the United States, fear of an oil shortage turned to panic, and motorists began lining up at gas stations, in many cases waiting for hours to fill their tanks. The price of oil rose to $18 a barrel, and newspaper headlines proclaimed an "Energy Crisis!" Governments in industrialized countries instituted new rules to encourage oil conservation—speed limits dropped, manu-

facturers produced smaller cars, power companies switched back to coal, home insulation became tax deductible, and conservation slogans appeared next to light switches.

During the last two decades of the twentieth century, the oil market stabilized, though political events occasionally led to price jumps and short-term shortages—a shortage in 1979, for example, pushed the price of oil temporarily to over $30 a barrel (≈$60/bbl in present dollars). During most of the 1980s and 1990s, oil hovered at about $25/bbl. But since 2004, oil prices have been rising steadily, passing the $135/bbl mark in mid-2008. Will there come a day when shortages arise not because of an embargo or high prices but because there is no more oil to produce? To answer these questions, we need to examine the world oil supply. What we find there may be unnerving.

The Oil Crunch

To understand the issues involved in predicting the future of energy supplies, we must first classify energy resources. We call a particular resource "renewable" if nature can replace it within a short time relative to a human life span (in months or, at most, decades). We call a resource "nonrenewable" if nature takes a very long time (hundreds to perhaps millions of years) to replenish it. Oil, gas, and coal are nonrenewable resources, in that the rate at which humans consume these materials far exceeds the rate at which nature replenishes them, so we will inevitably run out of oil. The question is, When?

Historians in the future may refer to our time as the **Oil Age**, because so much of our economy depends on oil. How long will this Oil Age last? Geologists estimate that as of 2000, there were about 850 to 1,000 billion barrels of proven conventional reserves (oil that had been found). There may be an additional 100 to 2,000 billion barrels not yet found. Thus, the world probably holds between 1,100 and 3,000 billion barrels of obtainable oil (not counting oil in tar sands or oil shales). Presently, humanity guzzles oil at a rate of about 28 billion barrels per year. The United States alone consumes more than 20 million barrels a day. At this rate of consumption, oil supplies will last until sometime between 2050 and 2150.

Unfortunately, problems may begin long before the supply runs out completely, because as oil becomes scarcer, it becomes harder to produce. Only the first third of the oil in a field flows easily into wells; another third may be urged into wells using expensive techniques, but the remaining third stays put, trapped in isolated pores. Already, geologists see the beginning of the end of the Oil Age, for consumption now exceeds the rate of discovery of new oil by a factor of 3. Further, burning oil has environmental consequences. In the end, the Oil Age will probably have lasted less than three centuries. On a time line representing the 4,000 years since the construction of the Egyptian pyramids, this looks like a very small blip (▶Fig. 14.31a, b). We may indeed be living during a unique interval of human history.

Can Other Energy Sources Meet the Need?

Are there alternatives to conventional oil? Perhaps. Certainly the world contains vast fossil-fuel supplies in other forms, enough to last for centuries if they can be exploited in an economical and environmentally sound way. These supplies include tar sand, oil shale, natural gas, and coal.

All told, perhaps 1.5 trillion barrels of hydrocarbons are trapped in tar sands and 3 trillion barrels trapped in oil shale. In addition, there are coal deposits that could meet our energy needs for the next few centuries. But extracting tar sand, oil shale, and coal requires the construction of

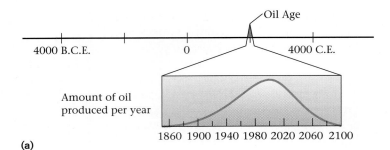

(a)

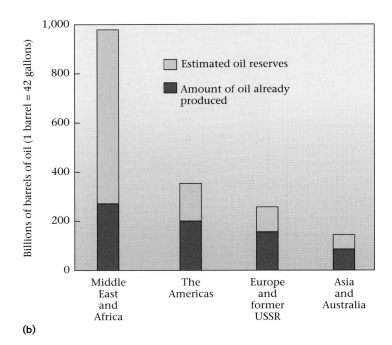

(b)

FIGURE 14.31 (a) The predicted history of the Oil Age, which may be a blip in human history like the Bronze Age. **(b)** The graph shows the amount of oil and gas reserves used and the estimated amounts still remaining.

huge mines, which might scar the landscape. Natural gas and coalbed methane are probably our best alternatives in the near future, but they remain very expensive to transport.

Can nuclear power or hydroelectric power fill the need? Vast supplies of uranium, the fuel of traditional nuclear plants, remain untapped (Fig. 14.30a), and in addition, nuclear engineers have designed alternative plants, powered by "breeder reactors," that essentially produce their own fuel. But many people view nuclear plants with suspicion, because of concerns about radiation, accidents, terrorism, and waste storage, and these concerns have stalled the industry. A substantial increase in hydroelectric power production is not likely, as most appropriate rivers have already been dammed, and industrialized countries have little appetite for taming any more. Similarly, the growth of geothermal-energy output seems limited.

Because of the problems that would result from relying more on coal, hydroelectric, and nuclear energy, researchers have been increasingly exploring "clean energy" options. One possibility is solar power. Unfortunately, affordable technologies for large-scale solar energy do not yet exist, and the idea of covering the landscape with solar cells is an unattractive one. Similarly, we can turn to wind power for small-scale energy production in areas where strong gusts blow, but covering the landscape with windmills is no more appealing. Fusion power may be possible, but physicists and engineers have not yet figured out a way to harness it. Despite their inherent problems, natural gas, coal, and nuclear power will probably step in for oil when the Oil Age comes to a close, but clearly society will be facing difficult choices in the not-so-distant future about where to obtain energy, and we will need to invest in the research needed to discover new alternatives.

Environmental Issues

Environmental concerns about energy resources begin right at the source. Oil drilling requires substantial equipment, the use of which can damage the land. Oil spills from pipelines or trucks sink into the subsurface and contaminate groundwater, and oil spills from ships create slicks that spread over the sea surface and foul the shoreline (▶Fig. 14.32). Coal and uranium mining also scar the land and can lead to the production of **acid mine runoff**, a dilute solution of sulfuric acid produced when sulfur-bearing minerals such as pyrite (FeS_2) in mines react with rainwater. The runoff enters streams and kills fish and plants. Exten-

sive coal mining may also cause the ground surface to subside. This happens when an underground mine collapses, so that the ground sinks. As a result, building foundations may crack. Coal-seam fires may also pose a problem. If ignited, a seam may smolder underground for years, releasing noxious gases that seep through cracks and pores to the ground surface, making the region above uninhabitable.

Numerous air-pollution issues also arise from the burning of fossil fuels, which sends soot, carbon monoxide, sulfur dioxide, nitrous oxide, and unburned hydrocarbons into the air, all pollutants that cause smog. Coal, for example, commonly contains sulfur, primarily in the form of pyrite, which enters the air as sulfur dioxide (SO_2) when coal is burned. This gas combines with rainwater to form sulfuric acid (H_2SO_4), or **acid rain**. For this reason, many countries now regulate the amount of sulfur that coal can contain when it is burned. Because of these regulations, low-sulfur coal, such as the coal that occurs in Cretaceous seams of the western United States, commands a higher price per ton than the higher-sulfur coals of the Midwest.

But even if pollutants can be decreased, the burning of fossil fuels still releases carbon dioxide (CO_2) into the atmosphere. CO_2 is important because it traps heat in the Earth's atmosphere much as the glass traps heat in a greenhouse—this is the "greenhouse effect." Too much CO_2 may lead to a global increase in atmospheric temperature (global warming), which in turn may alter the distribution of climatic belts. We'll learn more about this issue in Chapter 23.

> ### Take-Home Message
> Society faces difficult choices, for energy demand increases while conventional resources decrease. Oil is a nonrenewable resource and may run out in 50 to 100 years, bringing the Oil Age to a close. Use of most energy resources has negative environmental consequences. For example, burning fossl fuel produces greenhouse gases.

FIGURE 14.32 A marine oil spill. Volatile components evaporate, leaving a tar-like residue.

Chapter Summary

- Energy resources come in a variety of forms: energy directly from the Sun; energy from tides, flowing water, or wind; energy stored by photosynthesis (either in contemporary plants or in fossil fuels); energy from inorganic chemical reactions; energy from nuclear fission; and geothermal energy (from Earth's internal heat).

- Oil and gas are hydrocarbons, a type of organic chemical. The viscosity and volatility of a hydrocarbon depend on the length of its molecules.

- Oil and gas originally develop from the bodies of plankton and algae, which settle out in a quiet-water, oxygen-poor depositional environment and form black organic shale. Later, chemical reactions at elevated temperatures convert the dead plankton into kerogen, and then oil.

- For an oil reserve to be usable, oil must migrate from a source rock into a porous and permeable rock called a reservoir rock. Unless the reservoir rock is overlain by an impermeable seal rock, the oil will escape to the ground surface. The subsurface configuration of strata that ensures the entrapment of oil in a good reservoir rock is called an oil trap.

- Substantial volumes of hydrocarbons exist in tar sand, oil shale, and gas hydrate.

- For coal to form, the plant material must be deposited in an oxygen-poor environment so that it does not completely decompose. Compaction near the ground surface creates peat, which, when buried deeply and heated, transforms into coal. Coal has a high concentration of carbon.

- Coal is classified into ranks, based on the amount of carbon it contains: lignite (low rank), bituminous (higher rank), and anthracite (still higher rank).

- Coal occurs in beds, interlayered with other sedimentary rocks. Coal beds can be mined by either strip mining or underground mining.

- Coalbed methane and coal gasification provide additional sources of energy.

- Nuclear power plants generate energy by using the heat released from the nuclear fission of radioactive elements. The heat turns water into steam, and the steam drives turbines.

- Some economic uranium deposits occur as veins in igneous rock bodies; some are found in sedimentary beds.

- Nuclear reactors must be carefully controlled to avoid overheating or meltdown. The disposal of radioactive nuclear waste can create environmental problems.

- Geothermal energy uses Earth's internal heat to transform groundwater into steam that drives turbines; hydroelectric power uses the potential energy of water; and solar energy uses solar cells to convert sunlight to electricity.

- We now live in the Oil Age, but oil supplies may last only for another century. Natural gas may become our major energy supply in the near future.

- Most energy resources have environmental consequences. Oil spills pollute the landscape, and the sulfur associated with some fuel deposits causes acid mine runoff. The burning of coal can produce acid rain, and the burning of coal and hydrocarbons produces smog and may cause global warming.

Geopuzzle Revisited

Hydrocarbons (oil and gas) and coal are fossil fuels, meaning that they store the energy that was produced by photosynthesis millions to hundreds of millions of years ago. As such, they are nonrenewable resources, because geologic processes cannot replace them as fast as humans consume them. Conventional hydrocarbon supplies may last for only another 50 to 100 years. There are other sources of hydrocarbons, but they are expensive to utilize. Coal supplies can last longer, but regardless of supply, the use of fossil fuels has significant environmental consequences ranging from landscape destruction to global warming.

Key Terms

acid mine runoff (p. 519)
acid rain (p. 519)
anthracite coal (p. 503)
biodegradation (p. 499)
biofuel (p. 516)
bituminous coal (p. 503)
chain reaction (p. 509)
coal (p. 500)
coal gasification (p. 508)
coal rank (p. 503)
coal reserves (p. 503)
coal swamps (p. 500)
coalbed methane (p. 508)
distillation column (p. 499)
energy (p. 487)
energy resource (p. 487)
fossil fuels (p. 487)
fuel (p. 487)
gas hydrate (p. 500)
geothermal energy (p. 489)
hydrocarbon generation (p. 491)
hydrocarbon reserve (p. 491)
hydrocarbons (p. 489)
hydrocarbon system (p. 491)
hydroelectric power plant (p. 513)

kerogen (p. 490)
lignite (p. 503)
meltdown (p. 510)
migration pathway (p. 492)
nuclear reactor (p. 509)
nuclear waste (p. 511)
Oil Age (p. 518)
oil seep (p. 492)
oil shale (pp. 490)
oil window (p. 490)
peat (p. 501)
permeability (p. 491)
photosynthesis (p. 488)
photovoltaic cells (p. 516)
pores (p. 491)
porosity (p. 491)
reservoir rock (p. 491)
resource (p. 487)
salt dome (p. 495)
seal rock (p. 492)
seismic-reflection profile (p. 496)
solar collector (p. 513)
source rock (p. 490)
tar sand (p. 499)
trap (p. 492)

Review Questions

1. What are the fundamental sources of energy?

2. How does the length of a hydrocarbon chain affect its viscosity and volatility?

3. What is the source of the organic material in oil?

4. What is the oil window, and why does oil form only there?

5. How is organic matter trapped and transformed to create an oil reserve?

6. What are the different kinds of oil traps?

7. What are tar sand and oil shale, and how can oil be extracted from them?

8. What are gas hydrates, and where do they occur?

9. How do porosity and permeability affect the oil-bearing potential of a rock?

10. Where is most of the world's oil found? At present rates of consumption, how long will it last?

11. How is coal formed?

12. Explain how coal is transformed in rank from peat to anthracite.

13. What are some of the environmental drawbacks of mining and burning coal?

14. What is coalbed methane, and how is it extracted?

15. Describe how a nuclear reaction is initiated and controlled in a nuclear reactor.

16. Where does uranium form in the Earth's crust? Where does it usually accumulate in minable quantities?

17. What are some of the drawbacks of nuclear energy?

18. Discuss the pros and cons of various alternative energy sources.

19. What is geothermal energy? Why is it not more widely used?

20. What is the difference between renewable and nonrenewable resources?

21. What is the likely future of hydrocarbon production and use in the twenty-first century?

On Further Thought

1. Much of the oil production in the United States takes place at offshore platforms along the coast of the Gulf of Mexico. Many of the traps within this province are associated with salt domes. Consider the geologic setting of the Gulf Coast, in the context of the theory of plate tectonics, and explain why an immensely thick sequence of sediment accumulated in this region, and why so many salt-dome traps formed.

2. Coal in West Virginia occurs in flat-lying (horizontal) seams interbedded with sandstone and shale. The region where the coal occurs is very hilly. In recent years, coal companies have gained access to the coal by removing the tops of the hills and exposing the seams. Miners dump the debris of sandstone and shale into adjacent valleys. How does this approach change the relief of the landscape? What are the potential environmental consequences of using this approach?

3. Ethanol can potentially be used as an alternative to petroleum as a liquid fuel. Ethanol can be produced by processing corn, sugar cane, or certain perennial grasses (e.g., switchgrass, or *Miscanthus*). What factors should be considered in determining which of these crops would be the most appropriate for use as a source of ethanol in North America?

Suggested Reading

Aubrecht, G. J. 2005. *Energy: Physical, Environmental, and Social Impact*. 3rd ed. Upper Saddle River, N.J.: Benjamin Cummings.

Benka, S. G., et al. 2002 (April). The Energy Challenge. *Physics Today* (special issue).

Cassedy, E. S., Jr. 2000. *Prospects for Sustainable Energy: A Critical Assessment*. Cambridge: Cambridge University Press.

Cassedy, E. S., and P. Z. Grossman. 1998. *Introduction to Energy: Resources, Technology, and Society*. 2nd ed. Cambridge: Cambridge University Press.

Conaway, C. F. 1999. *The Petroleum Industry: A Nontechnical Guide*. Tulsa, Okla.: PennWell.

Deffeyes, K. S. 2003. *Hubbert's Peak: The Impending World Oil Shortage*. Princeton, N.J.: Princeton University Press.

Energy Information Administration. 2002. *Annual Energy Review*. Washington, D.C.: U.S. Department of Energy.

Freese, B. 2004. *Coal: A Human History*. New York: Penguin USA.

Gluyas, J. G., and R. Swarbrick. 2003. *Petroleum Geoscience*. Malden, Mass.: Blackwell Science.

Hoffmann, P., and T. Harkin. 2001. *Tomorrow's Energy: Hydrogen, Fuel Cells, and the Prospects for a Cleaner Planet*. Cambridge, Mass.: MIT Press.

Hunt, J. M. 1996. *Petroleum Geochemistry and Geology*. 2nd ed. New York: Freeman.

Hyne, N. J. 2001. *Nontechnical Guide to Petroleum Geology, Exploration, Drilling and Production*. 2nd ed. Tulsa, Okla.: PennWell.

Schobert, H. H. 2001. *Energy and Society: An Introduction*. London: Taylor & Francis.

Selley, R. C. 1997. *Elements of Petroleum Geology*. 2nd ed. New York: Academic Press.

Thomas, L., and A. Kellerman. 2002. *Coal Geology*. New York: Wiley.

Williams, A., M. Pourkashanian, and N. Skorupska. 2000. *Combustion and Gasification of Coal*. London: Taylor & Francis.

Yergin, D. 1993. *The Prize: The Epic Quest for Oil, Money and Power*. New York: Free Press.

CHAPTER
15

Riches in Rock:
Mineral Resources

Geopuzzle

Your lifestyle depends on a great
variety of materials—bricks, copper, concrete,
steel, gravel, and aluminum are all essential for
the construction of buildings, roads, cars, and
computers. Where do these materials come from?

Gold bracelets by the dozen are on display in a marketplace in Kuwait. For each ounce of
gold, miners must process 30 tons of rock. Riches do come from the Earth—but at a cost.

Truth, like gold, is to be obtained not by its growth, but by washing away from it all that is not gold.

—*Leo Tolstoy (Russian author, 1828–1910)*

15.1 INTRODUCTION

In June 1845, over a year after leaving Missouri, James Marshall arrived by horse at Sutter's Fort, in central California, to make a new life. Having just finished a stint as a rancher and a few months as a soldier, Marshall decided to go into the lumber trade and convinced Captain John Sutter, a former army officer, to finance the construction of a sawmill in the foothills of the Sierra Nevada Mountains. Marshall's scruffy crew finished the mill by the beginning of 1848. As Marshall stood admiring the new building, he noticed a glimmer of metal in the gravel that littered the bed of the adjacent stream. He picked it up, banged it between two rocks to test its hardness, and shouted, "Boys, by God, I believe I have found a gold mine!" For a short while, Marshall and Sutter managed to keep the discovery secret. But word of the gold soon spread, and within weeks the workers at Marshall's mill had disappeared into the mountains to seek their own fortunes.

All through that year, gold fever spread throughout California and eventually reached the East Coast. As a result, 1849 brought 40,000 prospectors to California. These "forty-niners," as they were called, had abandoned their friends and relatives on the gamble that they could strike it rich. Many traveled for months by sea, sailing through stormy waters around the southern tip of South America, to reach San Francisco, a town of mud streets and plank buildings. In many cases, crews abandoned their ships in the harbor to join the scramble to the gold fields (▶Fig. 15.1). Although $20 million worth of gold was mined in 1849, few of the forty-niners became rich. Most lost whatever wealth they found in the saloons and gambling halls that sprouted around the gold fields.

Gold is but one of the many **mineral resources**—minerals extracted from the Earth's upper crust that are of use to civilization. Without these resources, industrialized societies could not function. Geologists divide mineral resources into two categories: metallic mineral resources (rocks containing gold, copper, aluminum, iron, etc.) and nonmetallic mineral resources (building stone, gravel, sand, gypsum, phosphate, salt, etc.). In this chapter, we look at the nature of mineral resources, the geologic phenomena responsible for their formation, and the ways people mine them. We conclude by considering limits to mineral reserves.

FIGURE 15.1 Forty-niners mining in the Sierra Nevada Mountains.

15.2 METALS AND THEIR DISCOVERY

What Is a Metal?

Metals are opaque, shiny, smooth solids that can conduct electricity and can be bent, drawn into wire, or hammered into thin sheets. They look and behave quite differently from wood, plastic, meat, or rock. This is because, unlike in other substances, the atoms that make up metals are held together by **metallic bonds**, meaning that the outer electrons flow from atom to atom fairly easily (see Appendix A). Despite the mobility of their electrons, metals are solids, so their atoms lie fixed in a regular lattice defining a crystal structure. For example, the atoms in pure copper are arranged in wafer-like layers (▶Fig. 15.2a).

Not all metals behave the same way. Some are noted for their strength, others for their hardness, and others for their malleability (how easily they can be bent or molded). The behavior of a metal depends on the strength of bonds between atoms and on its crystalline structure. When you bend or stretch a strip of copper, for example, the wafer-like layers slip past each other quite readily (▶Fig. 15.2b). The shape and dimensions of crystals in a piece of metal are determined by how fast the molten metal cools and solidifies when removed from a furnace, by tempering (alternate heating and cooling), and by cold working (manipulating the shape of the metal after it is cooled). Before the development of the modern science of metallurgy, metalworkers learned how to obtain the qualities they wanted in a metal

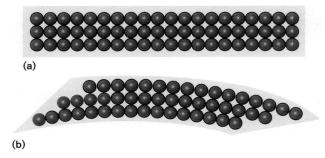

FIGURE 15.2 (a) A copper crystal has a waferlike appearance (the circles represent copper atoms). **(b)** When the crystal is deformed, one sheet of atoms slides past its neighbor.

object simply by trial and error. Skilled metalworkers passed trade secrets from master to apprentice for generations. But now engineers with a detailed knowledge of chemistry can control the properties of a metal product.

The Discovery of Metals

Certain metals—namely, copper, silver, gold, and mercury—can occur in rock as **native metals** (▶Fig. 15.3a–c). Native metals consist only of metal atoms, and thus look and behave like metal. Prehistoric hunters collected nuggets of native metal from stream beds and pounded them together with stone hammers to make tools. Eventually, because native metals are rare and durable, people began to use them

as money. Gold, in particular, attained popularity as a currency because of its unique warm yellow glow and its resistance to tarnish or rust. Though iron does not occur as a native metal in rocks, chunks of iron fall to Earth from space in the form of meteorites. Ancient peoples used meteorite iron for tool making when they could find it.

If we had to rely solely on native metals as our source of metal, we would have access to only a tiny fraction of our current metal supply. Most of the metal atoms we use today originated as ions bonded to nonmetallic elements in a great variety of minerals that themselves look nothing like metal. Only because of the chance discovery by some prehistoric genius that certain rocks, when heated to high temperatures in fire (a process now called **smelting**), decompose to yield metal plus a nonmetallic residue called slag do we now have the ability to produce sufficient metal for the needs of industrialized society.

Of the three principal metals in use today—copper, iron, and aluminum—copper began to be used first, because copper smelting from sulfide minerals is relatively easy. Copper implements appeared as early as 4000 B.C.E., although pure copper has limited value because it is too soft to retain a sharp edge. Around 2800 B.C.E., Sumerian craftsmen discovered that copper could be mixed with tin to produce bronze, an **alloy** (a compound containing two or more metals) whose strength exceeds that of either metal alone, and warriors came to rely on bronze for their swords.

FIGURE 15.3 (a) Gold nuggets come from quartz veins. At an early stage, much of the quartz remains. **(b)** With time, most of the quartz breaks away, leaving only gold. The largest single mass of gold on record, composed of small pieces of gold that grew together, weighed in at about 93 kg (3,000 oz) before being melted down. The largest single nugget weighed 71 kg (2,284 oz) and also came from Australia. It was also melted down. The largest nugget in existence today weighs 27.7 kg (approximately 891 oz, or 61 lb) and resides in a display case at a Las Vegas casino. **(c)** Native copper from northern Michigan.

(a)

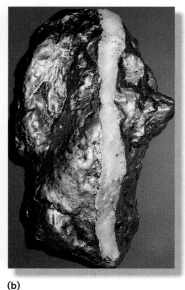

(b)

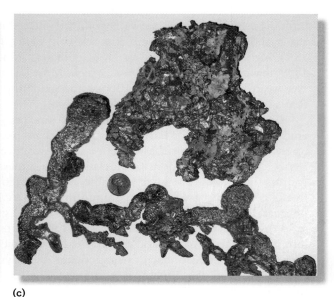

(c)

Iron proves superior to copper or bronze for many purposes, because of its strength, hardness, and abundance. Still, people didn't start using iron widely until 1,500 years after they had begun to use bronze. The delay was due, in part, to the fact that iron has a very high melting temperature that can be difficult to reach. But, in addition, the metal generally occurs in iron-oxide minerals (such as hematite, Fe_2O_3), and the liberation of iron metal from oxide minerals requires a chemical reaction, not just simple heating. The widespread use of iron became possible only after someone discovered that metallic iron can be produced by heating iron-oxide minerals in the presence of carbon monoxide gas, which comes from burning charcoal (▶Fig. 15.4a, b). Chemists describe this reaction by the formula

$$Fe_2O_3 + 3CO \rightarrow 2\ Fe + 3CO_2.$$

More recently, people learned to make steel, an alloy of iron and carbon, and stainless steel, an alloy of iron and chromium that resists corrosion.

Aluminum is abundant in rocks of the crust and, in many ways, is preferable to iron because it weighs less. But it does not occur in native form, and the extraction of aluminum from minerals requires complex methods that have become economically practical only since the 1880s.

These days, in addition to iron, copper, tin, and aluminum, we use a vast array of different metals. Some are known as **precious metals** (gold, silver, and platinum) and others as **base metals** (copper, lead, zinc, and tin) because of the difference in their price. Of the sixty-three or so metals in use today, people knew of only nine (gold, copper, silver, mercury, lead, tin, antimony, iron, and arsenic) before the year 1700.

> ### Take-Home Message
>
> Metals can be bent and drawn into wire. Some occur in native form, but most are bonded to other atoms in minerals. People discovered how to smelt (extract) metals from minerals long ago. Copper came into use first, then bronze, iron, and finally aluminum.

15.3 ORES, ORE MINERALS, AND ORE DEPOSITS

What Is an Ore?

As we have seen, metals occur in two forms—as native elements (copper, silver, or gold) or as ions mixed with non-metallic elements in a great variety of minerals. If you pick up a chunk of common granite and analyze its mineral content, you will find that it consists mostly of quartz and feldspar, with minor proportions of biotite (mica), hornblende, and garnet (see Chapter 5). Some of these minerals contain metals: potassium feldspar ($KAlSi_3O_8$), for example, contains about 8% aluminum. But even though common granite includes metal ions, we don't mine granite to produce these metals. Why? Simply because the minerals in granite contain relatively little metal, so it would cost more to separate the metals from granite than you would get if you were to sell them. To obtain the metals needed for industrialized society, we mine **ore**, rocks containing a concentrated accumulation of native metals or ore minerals.

Ore minerals (or economic minerals) are minerals that contain metal in high concentrations and in a form that can be easily extracted. Galena (PbS), for example, is about 50% lead, so we consider it to be an ore mineral of lead (▶Fig. 15.5a). We obtain most of our iron from the oxide minerals, hematite and magnetite (▶Fig. 15.5b).

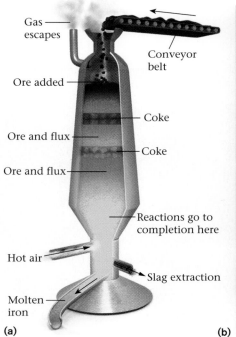

FIGURE 15.4 (a) During smelting, workers use very hot air to heat iron ore and coke (carbon) placed in a furnace to temperatures of 1,250° to 1,400°C. The iron ore separates into molten iron, which stays at the base of the furnace, and slag, which floats to the top. Slag consists of various silicates and metal oxides, and can be used for making roads and fertilizer. (Flux is a chemical added to make the ore melt at a lower temperature.) **(b)** Molten iron being poured.

Gas escapes

Conveyor belt

Ore added

Coke

Ore and flux

Coke

Ore and flux

Reactions go to completion here

Hot air

Slag extraction

Molten iron

(a)

(b)

(a)

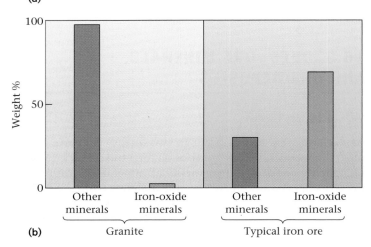

(b)

FIGURE 15.5 (a) This block of limestone, containing galena, can be considered lead ore. **(b)** Less than 2% of granite consists of iron-oxide minerals, whereas iron ore may include over 80%.

TABLE 15.1   Some Common Ore Minerals

Metal	Mineral Name	Chemical Formula
Copper	Chalcocite	Cu_2S
	Chalcopyrite	$CuFeS_2$
	Bornite	Cu_5FeS_4
	Azurite	$Cu_3(CO_3)_2(OH)_2$
	Malachite	$Cu_2(CO_3)(OH)_2$
Iron	Hematite	Fe_2O_3
	Magnetite	Fe_3O_4
Tin	Cassiterite	SnO_2
Lead	Galena	PbS
Mercury	Cinnabar	HgS
Zinc	Sphalerite	ZnS
Aluminum	Kaolinite	$Al_2Si_2O_5(OH)_4$
	Corundum	Al_2O_3
Chrome	Chromite	$(Fe,Mg)(Cr,Al,Fe)2O_4$
Nickel	Pentlandite	$(Ni,Fe)_9S_8$
Titanium	Rutile	TiO_2
	Ilmenite	$FeTiO_3$
Tungsten	Sheelite	$CaWO_4$
Molybdenum	Molybdenite	MoS_2
Magnesium	Magnesite	$MgCO_3$
	Dolomite	$CaMg(CO_3)_2$
Manganese	Pyrolusite	MnO_2
	Rhodochrosite	$MnCO_3$

Geologists have identified many different kinds of ore minerals (▶Table 15.1). As you can see from the chemical formulas, many ore minerals are sulfides, in which the metal occurs in combination with sulfur (S), or oxides, in which the metal occurs in combination with oxygen (O). Numerous ore minerals are colorful and come in interesting shapes, and some have a metallic luster (▶Fig. 15.6).

To be an ore, a rock must not only contain ore minerals, it must also have a sufficient amount to make it worth mining. Iron constitutes about 6.2% of the continental crust's weight, whereas it makes up 30 to 60% of iron ore. The concentration of a useful metal in an ore determines the **grade** of the ore—the higher the concentration, the higher the grade. Whether or not an ore is worth mining depends on the price of metal in the market. For example, in 1880, copper-bearing rocks needed to contain at least 3% copper to be considered economic ore, but as of 1970, rock containing only about 0.3% copper can be considered

FIGURE 15.6 Colorful ore minerals containing copper. The blue mineral is azurite and the green is malachite.

economic. This change reflects new technology for mining and processing the ore.

How Do Ore Deposits Form?

Ore minerals do not occur uniformly through rocks of the crust. If they did, we would not be able to extract them economically. Fortunately for humanity, geologic processes concentrate these minerals in ore deposits. Simply put, an **ore deposit** is an economically significant occurrence of ore. The various kinds of ore deposits differ from each other in terms of which ore minerals they contain and which kind of rock body they occur in. Below, we introduce a few examples.

Magmatic deposits. When a magma cools, sulfide ore minerals crystallize early and then, because sulfides tend to be dense, sink to the bottom of the magma chamber, where they accumulate; this accumulation is a **magmatic deposit**. When the magma freezes solid, the resulting igneous body may contain a solid mass of sulfide minerals at its base. Because of their composition, we consider these masses to be a type of **massive-sulfide deposit** (▶Fig. 15.7).

Hydrothermal deposits. Hydrothermal activity involves the circulation of hot-water solutions through a magma or through the rocks surrounding an igneous intrusion. These fluids dissolve metal ions. When a solution enters a region of lower pressure, lower temperature, different acidity, and/or different availability of oxygen, the metals come out of solution and form ore minerals that precipitate in fractures and pores, creating a **hydrothermal deposit** (▶Fig. 15.8). Such deposits may form within an igneous intrusion or in surrounding country rock. If the resulting ore minerals disperse through

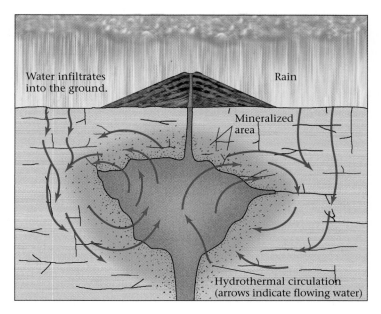

FIGURE 15.8 Water circulating through a granite pluton dissolves and redistributes metals, leading to formation of hydrothermal deposits.

the intrusion, we call the deposit a disseminated deposit, but if they precipitate to fill cracks in preexisting rock, we call the deposit a vein deposit (veins are mineral-filled cracks; ▶Fig. 15.9). Hydrothermal copper deposits commonly occur in porphyritic igneous intrusions; these are known as porphyry copper deposits. Typically, vein deposits include quartz in addition to the ore minerals. For example, native gold commonly appears as flakes in milky-white quartz veins.

In recent years, geologists have discovered that hydrothermal activity at the submarine volcanoes along

FIGURE 15.7 Heavy, metal crystals can sink to the bottom of a magma chamber to form a massive-sulfide deposit. The sulfide concentrate may become an ore body in the future.

FIGURE 15.9 The difference between a vein deposit and a disseminated deposit; vein ore occurs as crystals filling cracks, whereas disseminated ore occurs as crystals scattered through the rock.

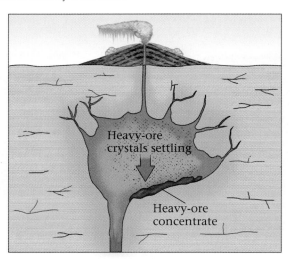

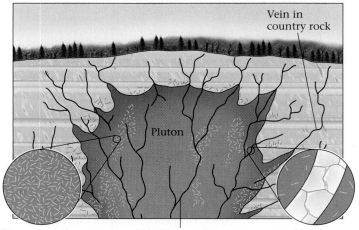

mid-ocean ridges leads to the eruption of hot water, containing high concentrations of dissolved metal and sulfur, from a vent. When this hot water comes in contact with cold seawater, the dissolved components instantly precipitate as tiny crystals of metal sulfide minerals (▶Fig. 15.10). The erupting water, therefore, looks like a black cloud, so the vents are called black smokers. The minerals in the cloud eventually sink and form a pile of nearly pure ore minerals around the vent. Since the ore minerals typically are sulfides, the resulting hydrothermal deposits constitutes another type of massive-sulfide deposit.

Secondary-enrichment deposits. Sometimes groundwater passes through ore-bearing rock long after the rock first formed. This groundwater dissolves some of the ore minerals and carries the dissolved ions away. When the water eventually flows into a different chemical environment (e.g., one with a different amount of oxygen or acid), it precipitates new ore minerals, commonly in concentrations that exceed that of the original deposit. A new ore deposit formed from metals that were dissolved and carried away from a preexisting ore deposit is called a **secondary-enrichment deposit** (▶Fig. 15.11a, b). Some of these deposits contain spectacu-

larly beautiful copper-bearing carbonate minerals, such as azurite and malachite (Fig. 15.6).

MVT ores. Geologists have discovered that some groundwater beneath mountain belts sinks several kilometers down into the crust, and follows a curving flow path that eventually brings it back to the surface perhaps hundreds of kilometers away from the mountain range. At the depths reached by the groundwater, temperatures become high enough that the water dissolves metals. When the water returns to the surface and enters cooler rock, these metals precipitate in ore minerals. Ore deposits formed in this way, containing lead- and zinc-bearing minerals, appear in dolomite beds of the Mississippi Valley region, and thus have come to be known as **Mississippi Valley–type (MVT) ores**.

FIGURE 15.11 (a) In the process of secondary enrichment, water passing down through the ore body oxidizes the ore minerals (i.e., the minerals react when dissolved in the water), dissolves them, and carries the dissolved ions out of the rock. **(b)** The ore then reprecipitates just below the water table, which is the level at which pores in the rock are filled with water, where chemical conditions are less oxidizing than in the situation described in (a) above.

FIGURE 15.10 Aprons of sulfide ore precipitate around "black smokers," along a mid-ocean ridge.

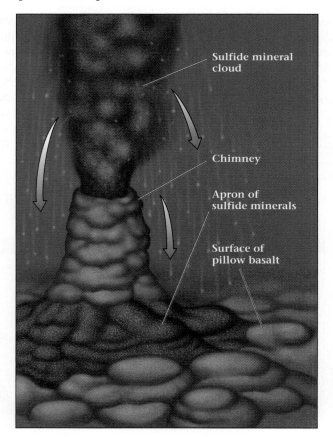

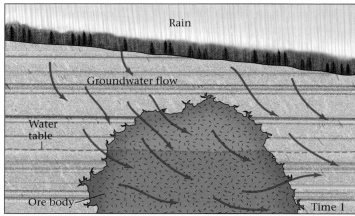

(a)

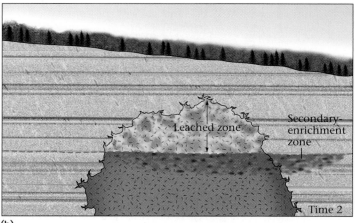

(b)

(a)

(b)

FIGURE 15.12 (a) Precambrian banded-iron formation near Ishpiming, Michigan. (b) Manganese nodules on the sea floor. Each nodule is a few centimeters in diameter.

Sedimentary deposits of metals. Some ore minerals accumulate in sedimentary environments under special circumstances. For example, between 2 and 2.5 billion years ago, the atmosphere, which previously had contained little oxygen, evolved into the oxygen-rich atmosphere we breathe today. This change affected the chemistry of seawater and led to the precipitation of iron-oxide minerals that settled as sediment on the sea floor. As we learned in Chapter 13, the resulting iron-rich sedimentary layers are known as **banded-iron formations (BIF)** (▶Fig. 15.12a), because after lithification they consist of alternating beds of gray iron oxide (magnetite or hematite) and red beds of jasper (iron-rich chert). Microbes may have participated in the precipitation process.

The chemistry of seawater in some parts of the ocean today leads to the deposition of manganese-oxide minerals on the sea floor. These minerals grow into lumpy accumu-

lations known as **manganese nodules** (▶Fig. 15.12b). Mining companies have begun to explore technologies for vacuuming up these nodules; geoscientists estimate that the worldwide supply of nodules contains 720 years' worth of copper and 60,000 years' worth of manganese, at current rates of consumption.

Residual mineral deposits. Recall from Chapter 7 that as rainwater sinks into the Earth, it leaches (dissolves) certain elements and leaves behind others, as part of the process of forming soil. In rainy, tropical environments, the residuum left behind in soils after leaching includes concentrations of iron or aluminum. Locally, these metals become so concentrated that the soil itself becomes an ore deposit (▶Fig. 15.13a, b). We refer to such deposits as **residual mineral deposits**. Most of the aluminum ore

FIGURE 15.13 (a) When rainwater sinks through the soil, it dissolves and removes many elements. (b) A thick soil forms, containing a residuum of iron or aluminum.

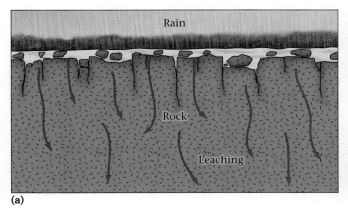

(a)

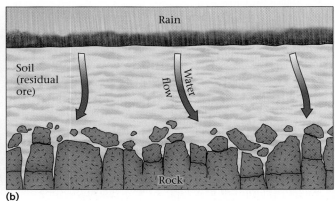

(b)

See for yourself . . .
Large Open-Pit Mines

When ore bodies lie close to the ground surface, it generally proves to be cheaper to extract the ore by excavating an open pit than by boring underground tunnels. Large open-pit mines can be seen from space. The thumbnail images provided on this page are only to help identify tour sites. Go to wwnorton.com/studyspace to experience this flyover tour.

Bingham Copper Mine, Utah
(Lat 40°31'14.66"N, Long 112°9'1.97"W)

Fly to the coordinates provided and zoom to 15 km (9 miles). You can see the east flank of the Oquirrh Mountains, one of the ranges of the Basin and Range rift. The broad treeless area at this site, 25 km (16 miles) SSE of the center of Salt Lake City (Image G15.1), is the Bingham Copper Mine, the largest open-pit mine in the world—it's visible from at least an elevation of 2,000 km (1,200 miles) in space! At the time of this writing, only the east half of the mine can be seen at good resolution. Zoom down to 3 km (2 miles) and tilt the image so you are looking north (Image G15.2). You can see the terraces that the miners cut to maintain a stable slope as they follow the ore body into the Earth. Ramps provide access for giant trucks to haul ore out of the bottom of the mine. The largest trucks can carry over 240 tons in a single load. The mine is now 4 km (2.5 miles) across and 1.2 km (0.75 miles) deep. Since digging began in 1906, mining operations have produced over 17 million tons of copper, in addition to vast quantities of gold and silver. The ore minerals occur in a porphyritic igneous rock, intruded during the Cenozoic.

G15.1

G15.2

Iron Mine near Ouro Preto, Brazil
(Lat 20°9'54.45"S, Long 43°30'7.86"W)

Fly to these coordinates and zoom to 15 km (9.3 miles) (Image G15.3). You are hovering over an active mine in one of the world's largest iron-mining districts, the Quadrilátero Ferrífero (iron quadrangle). Here, in the hilly highlands of eastern Brazil, immense deposits of banded iron formation (BIF) accumulated about 2.2 billion years ago. Subsequently, these strata were deformed and metamorphosed, and now they occur as bands of magnetite and hematite. Zoom closer, to an elevation of 3 km (1.8 miles), tilt the image, and look north (Image G15.4). You can see the terracing as well as pools filled with iron-colored water. These are settling ponds to let solids sink out of the waters used in processing ore. The ore removed from this mine is crushed and then transported by train, truck, and conveyor either to factories in Brazil or to Atlantic ports from which ships carry it overseas.

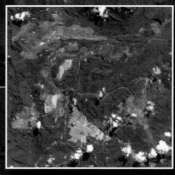

G15.3

G15.4

Images provided by *Google Earth*™ mapping service/DigitalGlobe, TerraMetrics, NASA, Europa Technologies—copyright 2008.

mined today comes from bauxite, a residual mineral deposit created by the extreme leaching of rocks (e.g., granite) containing aluminum-bearing minerals.

Placer deposits. Ore deposits may develop when rocks containing native metals erode and create a mixture of sand grains and metal flakes or nuggets (pebble-sized fragments). The heavy metal grains (e.g., gold) accumulate in sand or gravel bars along the course of rivers, for the moving water carries away lighter mineral grains but can't move the metal grains so easily. Concentrations of metal grains in stream sediments are a type of **placer deposit** (▶Fig. 15.14). (The term is also used for concentrations of diamonds.) Panning further concentrates gold flakes or nuggets—swirling water in a pan causes the lighter sand grains to wash away, leaving the gold behind. Placer deposits may eventually be buried and lithify to become part of a new sedimentary rock.

Where Are Ore Deposits Found?

The Inca empire of fifteenth-century Peru boasted elaborate cities and temples, decorated with fantastic masks, jewelry, and sculpture made of gold. Then, around 1532, Spanish conquistadors arrived in ships, led by commanders who quipped, "We Spaniards suffer from a disease that only gold can cure." The Incas, already weakened by civil war, were no match for the armor-clad Spaniards with their guns and horses. Within six years, the Inca empire had vanished, and Spanish ships were transporting Inca treasure back to Spain. Why did the Incas possess so much gold? Or to ask the broader question, what geologic factors control the distribution of ore? Once again, we can find the answer by considering the consequences of plate tectonics.

FIGURE 15.14 The formation of a placer deposit. Ore-bearing rock is eroded, and clasts containing native metals fall into a stream. Sorting by the stream concentrates the metals.

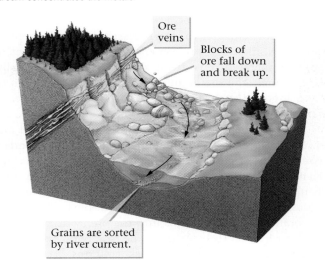

Ore veins

Blocks of ore fall down and break up.

Grains are sorted by river current.

Several of the ore-deposit types mentioned above occur in association with igneous rocks. As we learned in Chapter 6, igneous activity does not happen randomly around the Earth, but rather concentrates at convergent plate boundaries (in the overriding plate of a subduction zone), at divergent plate boundaries (along mid-ocean ridges), at continental rifts, or at hot spots. Thus, magmatic and hydrothermal deposits (and secondary enrichment deposits derived from these) occur along plate boundaries, along rifts, or at hot spots. Placer deposits are typically found in the sediments eroded from magmatic or hydrothermal deposits.

Consider the Inca gold. The Inca empire was situated in the Andes Mountains, which had formed as a result of compression and volcanic activity where the Pacific Ocean floor subducts beneath the South American Plate. As the mountains rose, erosion stripped away surface rocks to expose the large granite plutons that had intruded into the continental crust beneath. The magma that froze to make the granite brought gold, copper, and silver atoms with it. Some of the gold precipitated along with quartz to form veins in the plutons. Inca miners quarried these veins and separated the gold, or panned for gold in the streams choked with sediment eroded from the plutons. Plutons that contain similar ore deposits developed in the western United States during the Mesozoic and Cenozoic Eras.

As noted earlier, some massive sulfide deposits accumulate at the base of a magma chamber, and some precipitate from black smokers along a mid-ocean ridge system. Thus, some massive sulfides form in plutons at convergent margins, while some are interlayered with sea-floor basalt. Miners can gain access to sea-floor deposits only in places where the collision of continents traps a sliver of sea floor and slides it up and onto continental crust. During the rifting of continents, large magma dikes form by the partial melting of the mantle along the axis of the rift. These bring many valuable metals with them, which may accumulate in magmatic or hydrothermal deposits.

Some ore deposits are not a direct result of plate tectonics activity, and thus are not directly associated with plate boundaries. For example, banded iron formed along passive continental margins during the Precambrian. Its occurrence today reflects the present distribution of Precambrian rocks, and thus most major exposures occur in shield areas of continents. Bauxite forms where that aluminum-rich bedrock occurs, thick soils form, and extreme leaching takes place during soil formation. Thus, many bauxite deposits form on granite bedrock in stable continental areas that now lie in tropical regions.

> **Take-Home Message**
>
> Ores can be processed economically to produce metals. Ores form by: settling from melt, precipitating from hot water, deposition from currents, alteration by groundwater, and extreme weathering. The distribution of ores can be explained by plate tectonics.

15.4 ORE-MINERAL EXPLORATION AND PRODUCTION

Imagine an old prospector clanking through the desert with a broken-down donkey, eyeing the hillsides for "shows" of ore (evidence of ore minerals at the ground surface). If he finds a show of minerals, he pries out chunks of the rock with a pick, and the poor donkey hauls the rock back to town for an assay, a test to determine how much extractable metal the rock contains. Mining laws permitted a prospector to, literally, "stake a claim" by marking off an area of ground with stakes. The prospector would then have the exclusive right to dig up ore at that spot and sell it. Old claims still litter the desert in Arizona and Nevada.

What does a show look like? Typically, prospectors looked for milky-white quartz veins and/or exposures in which rocks were stained by the oxidation of metal-containing minerals (▶Fig. 15.15). Some prospectors panned streams in search of gold flakes in the stream gravel. When a prospector did find ore, word usually spread fast and others rushed to stake neighboring claims.

These days, large mining companies employ geologists to survey ore-bearing regions systematically. The geologists focus their studies on rocks that developed in settings appropriate for ore formation. Once such a region has been identified, they measure the local strength of Earth's magnetic field and the local pull of gravity. These measurements lead them to ore bodies, because ore minerals tend to be denser and more magnetic than average rocks (▶Fig. 15.16a, b). Geologists also sample rocks and soils to test for metal content, and may even analyze plants in the area to detect traces of metals, for plants absorb metals through their roots. Once geologists have identified a possible ore deposit, they drill holes to sample subsurface rock and to determine the ore deposit's shape and extent. Ore-mineral exploration takes geologists into jungles, deserts, and tundras worldwide.

If calculations show that the mining of an ore deposit will yield a profit, and if environmental concerns can be accommodated, a company builds a mine. Mines can be below or above ground, depending on how close the ore deposit is to the surface. To make an **open-pit mine** (▶Fig. 15.16c), workers first drill a series of holes into the solid bedrock and then fill the holes with high explosives. They must space the holes carefully and must set off the charges in a precise sequence, so that the bedrock shatters into appropriate-sized blocks for handling. When the dust settles, large front-end loaders dump the ore into giant ore trucks, which can carry as much as 200 tons of ore in a single load. (In comparison, a loaded cement mixer weighs about 70 tons.) The tires on these trucks are so huge that a tall person comes up only to the base of the hub. The trucks transport waste rock (rock that doesn't contain ore) to a tailings pile and the ore to a crusher, a giant set of moving steel jaws that smash the ore into small fragments. Workers then separate ore minerals from other minerals and send the ore-mineral concentrate to a processing plant, where it undergoes smelting or treatment with acidic solutions to separate metal atoms from other atoms. Eventually, workers melt the metal and then pour it into molds to make ingots (brick-shaped blocks) for transport to a manufacturing facility.

If the ore deposit lies more than about a hundred meters below the Earth's surface, miners must make an **underground mine**. To do so, they first dig a tunnel into the side of a mountain. (The entrance to the tunnel is called an adit.) Then the miners sink a vertical shaft in which they install an elevator. At the level in the crust where the ore body appears, they build a maze of tunnels into the ore by drilling holes into the rock and then blasting. The rock removed must be carried back to the surface. Rock columns between the

FIGURE 15.15 Stained rock is an indicator of ore. The stain comes when ore reacts with air and water. The photo encompasses about 1 m of section.

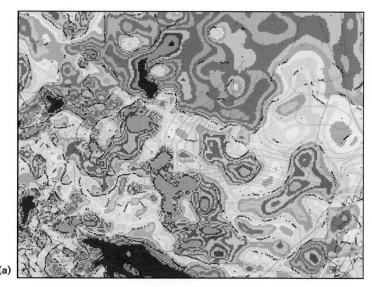

(a)

(c)

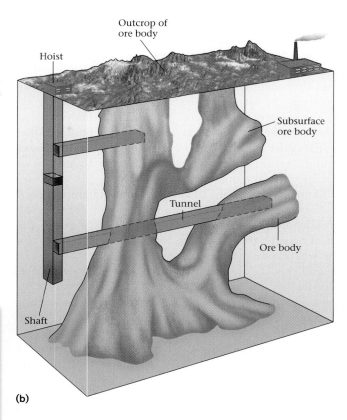

(b)

FIGURE 15.16 (a) On this magnetic-anomaly map, yellowish areas are regions with normal intensity, reddish areas are positive anomalies (greater-than-magnetic intensity), and purplish areas are negative anomalies (less-than-expected magnetic intensity). In some cases, ore-bearing rocks occur where there are positive anomalies. (b) The three-dimensional shape of an ore body underground, and the workings that miners dig to gain access to the ore body. Note that shafts are vertical and tunnels are horizontal. (c) An open-pit mine in Montana.

15.5 NONMETALLIC MINERAL RESOURCES

So far, this chapter has focused on resources that contain metal. But society uses many other geological materials, commonly known as industrial minerals, as well. From the ground we get the stone used to make roadbeds and buildings, the chemicals for fertilizers, the gypsum in drywall, the salt filling salt shakers, and the sand used to make glass—the list is endless. This section looks at a few of these geological materials and explains where they come from.

Dimension Stone

The Parthenon, a colossal stone temple rimmed by forty-six carved columns, has stood atop a hill overlooking the city of Athens for almost 2,500 years. No wonder—"stone,"

tunnels hold up the ceiling of the mine. The deepest mine on the planet, located in South Africa, currently reaches a depth of 3.5 km, where temperatures exceed 55°C, making mining there a very uncomfortable occupation. Miners face danger from mine collapse and rock falls. Some miners have been killed or injured by "rock bursts," sudden explosions of rock off the ceiling or walls of a tunnel. These explosions happen because the rock surrounding the adit is under such great pressure that it sometimes spontaneously fractures.

Take-Home Message

Geologists study outcrops for ore shows and drill holes, and make maps of rock magnetism in order to discover ore deposits. Mining takes place either in open-pit mines or in shafts and tunnels underground.

an architect's word for rock, outlasts nearly all other construction materials. We use stone to make facades, roofs, curbs and steps, and countertops and floors. We value stone for its visual appeal as well as its durability. The names that architects give to various types of stone may differ from the formal names that geologists use. For example, architects refer to any polished carbonate rock as marble, whether or not it has been metamorphosed. Likewise, they refer to any rocks containing feldspar and quartz as granite, regardless of whether the rock has an igneous or a metamorphic texture.

To obtain intact slabs and blocks of rock (granite or marble)—known as **dimension stone** in the trade—for architectural purposes, workers must carefully cut rock out of the walls of quarries (▶Fig. 15.17a). Note that a quarry provides stone, whereas a mine supplies ore. To cut stone slabs, quarry operators split rock blocks from bedrock by hammering a series of wedges into the rock. Or they cut it off bedrock by using a wireline saw, a thermal lance, or a water jet. A wireline saw consists of a loop of braided wire moving between two pulleys. In some cases, as the wire moves along the rock surface, the quarry operator spills abrasive (sand or garnet grains) and water onto the wire. The movement of the wire drags the abrasive along the rock and grinds a slice into it. Alternatively, the quarry operator may use a diamond-coated wire, cooled with pure water. A thermal lance looks like a long blowtorch: a flame of burning diesel fuel stoked by high-pressure air pulverizes rock, and thereby cuts a slot. More recently, quarry operators have begun to use an abrasive water jet, which squirts out water and abrasives at very high pressure, to cut rock.

Crushed Stone and Concrete

Crushed stone forms the substrate of highways and railroads and is the raw material for manufacturing cement, concrete, and asphalt. In crushed-stone quarries (▶Fig. 15.17b), operators use high explosives to break up bedrock into rubble that they then transport by truck to a jaw crusher, which reduces the rubble into usable-size fractions.

Much modern construction utilizes mortar and concrete (see ▶Box 15.1), human-made rock-like materials formed when a slurry composed of sand and/or gravel mixed with cement and water is allowed to harden. The hardening takes place when a complex assemblage of minerals grows by chemical reactions in the slurry; these minerals bind together the grains of sand or gravel in mortar or concrete. (Note that the word **mortar** refers to the substance that holds bricks or stone blocks together, whereas **concrete** refers to the substance that workers shape into roads or walls by spreading it out into a layer or by pouring it into a form.) The **cement** in mortar or concrete starts out as a powder composed of lime (CaO), quartz (SiO_2), aluminum oxide (Al_2O_3), and iron oxide (Fe_2O_3). Typically,

lime accounts for 66% of cement, silica for 25%, and the remaining chemicals for about 9%.

It appears that the ancient Romans were the first to use cement—they made it from a mixture of volcanic glass and limestone. During recent centuries, most cement has been produced by heating specific types of limestone (which happened to contain calcite, clay, and quartz in the correct proportions) in a kiln up to a temperature of about 1,450°C; the heating releases CO_2 gas and produces "clinker," chunks consisting of lime and other oxide compounds. Manufacturers crush the clinker into cement powder and pack it in bags for transport. But natural limestone with the exact composition of cement is fairly rare, so most cement used today is **Portland cement**, made by mixing limestone, sandstone, and shale in just the correct proportions to provide the correct chemical makeup. Isaac Johnson, an English engineer, came up with the recipe for Portland cement in 1844, and he named it after the town of Portland, England, because he thought it resembled rock exposed there.

FIGURE 15.17 (a) An active quarrying operation, showing large blocks of cut stone; **(b)** a crushed-limestone quarry.

(a)

(b)

BOX 15.1 THE HUMAN ANGLE

The Sidewalks of New York

Untold tons of concrete have gone into the construction of New York City. In fact, with the exception of a few city parks, most of the walking space in the city consists of concrete (▶Fig. 15.18). And concrete skyscrapers tower above the concrete plain. Where does all this concrete come from?

Much of the sand used in New York concrete was deposited during the last ice age. As vast glaciers moved southward over 14,000 years ago, they ground away the igneous and metamorphic rocks that constituted central and eastern Canada. These ancient rocks contained abundant quartz, and since quartz lasts a long time (it does not undergo chemical weathering easily), the sediment transported by the glaciers retained a large amount of quartz. Glaciers deposited this sediment in huge piles called moraines (see Chapter 22). As the glaciers melted, fast-moving rivers of meltwater washed the sediment, sorting sand from mud and pebbles. The sand was deposited in bars in the meltwater rivers, and these relict bars now provide thick lenses of sand that can be economically excavated.

What about the cement? Cement contains a mixture of lime, derived from limestone, and other elements (such as silica)

derived from shale and sandstone. The bedrock of New York, though, consists largely of schist and gneiss, not sedimentary rocks. Fortunately, a source of rocks appropriate for making cement lies up the Hudson River. A rock unit called the Rosendale Formation, which naturally contains exactly the

FIGURE 15.18 A sidewalk in New York City.

right mixture of lime and silica needed to make a durable cement, crops out in low ridges just to the west of the river. Beginning in the late 1820s, workers began quarrying the Rosendale Formation for cement, creating a network of underground caverns. Quarry operators followed the Rosendale beds closely, making horizontal mine tunnels where the beds were horizontal, tilted mine tunnels where the beds tilted, and vertical mine tunnels where the beds were vertical. They then dumped the excavated rock into nearby kilns and roasted it to produce lime mixed with other oxides. The resulting powder was packed into barrels, loaded onto barges, and shipped downriver to New York. As demand for cement increased, operators eventually dug open-pit quarries from which they excavated other limestone and shale units, mixing them together in the correct proportion to make Portland cement.

The rocks making up the Rosendale Formation consist of cemented-together shell fragments and small, reef-like colonies of organisms. In other words, the lime in the concrete of New York sidewalks was originally extracted from seawater by living organisms—brachiopods, crinoids, and bryozoans—over 350 million years ago.

Nonmetallic Minerals in Your Home

We use an astounding variety of nonmetallic geologic resources (▶Table 15.2) without ever realizing where they come from. Consider the materials in a house or apartment. The concrete foundation consists of cement, made from limestone mixed with sand or gravel. The bricks in the exterior walls originated as clay, formed from the chemical weathering of silicate rocks and perhaps dug from the floodplain of a stream. To make *bricks*, workers mold wet clay into blocks, which they then bake. Baking drives out water and causes metamorphic reactions that recrystallize the clay. Clay is also the raw material from which pottery, porcelain, and other ceramic materials are made.

The glass used to glaze windows consists largely of silica, formed by first melting and then freezing pure quartz sand from a beach deposit or a sandstone formation. Quartz may also be used in the construction of photovoltaic cells for solar panels. Gypsum board (drywall), used

to construct interior walls, comes from a slurry of water and the mineral gypsum sandwiched between sheets of paper. Gypsum ($CaSO_4 \cdot 2H_2O$) occurs in evaporite strata precipitated from seawater or saline lake water. Evaporites provide other useful minerals as well, such as halite.

The asbestos that was once used to make roof shingles is derived from serpentine, a rock created by the reaction of olivine with water. The olivine may come from oceanic lithosphere, thrust onto continental crust during continental collisions. The copper in the house's electrical wiring most likely comes from the ores in a porphyry copper deposit; the iron in the nails probably comes from

> **Take-Home Message**
>
> Society uses a great variety of nonmetallic geologic materials. These include dimension stone, crushed stone, concrete (made from roasted limestone), evaporites (including gypsum), and clay (to make bricks).

TABLE 15.2 Common Nonmetallic Resources

Limestone	Sedimentary rock made of calcite; used for gravel or cement.
Crushed stone	Any variety of coherent rock (limestone, quartzite, granite, gneiss).
Siltstone	Beds of sedimentary rock; used to make flagstone.
Granite	Coarse igneous rock; used for dimension stone.
Marble	Metamorphosed limestone; used for dimension stone.
Slate	Metamorphosed shale; used for roofing shingles.
Gypsum	A sulfate salt precipitated from saltwater; used for wallboard.
Phosphate	From the mineral apatite; used for fertilizer.
Pumice	Frothy volcanic rock; used to decorate gardens and paths.
Clay	Very fine mica-like mineral in sediment; used to make bricks or pottery.
Sand	From sandstone, beaches, or riverbeds; quartz sand is used for construction and for making glass.
Salt	From the mineral halite, formed by evaporating saltwater; used for food, melting ice on roads.
Sulfur	Occurs either as native sulfur, typically above salt domes, or in sulfide minerals; used for fertilizer and chemicals.

banded-iron formations; and the plastic used in everything from countertops to light fixtures comes from oil formed from the bodies of plankton and algae that died millions of years ago. As you can see, geologic processes acting over millions to billions of years have been at work to produce the materials making your home.

Chemicals employed for agricultural purposes also come from the ground. For example, potash (K_2CO_3) comes from the minerals in evaporite deposits. Phosphate (PO_4^{-3}) comes from the mineral apatite, formed by diagenesis of mud on the ocean floor. Truly, without the geologic resources of the Earth, modern society would grind to a halt.

15.6 GLOBAL MINERAL NEEDS

How Long Will Resources Last?

The average citizen of an industrialized country uses 25 kilograms (kg) of aluminum, 10 kg of copper, and 550 kg of iron and steel in a year's time (▶Fig. 15.19; ▶Table 15.3). If you combine these figures with the quantities of energy resources and nonmetallic geologic resources a person uses, you get a total of about 15,000 kg (15 metric tons) of resources used per capita each year. Thus, the population of the United States consumes about 4 billion metric tons of geologic material per year. To create this supply, workers must mine, quarry, or pump 18 billion metric tons (see art, pp. 538–539). By comparison, the Mississippi River transports 190 million metric tons of sediment per year.

Mineral resources, like oil and coal, are nonrenewable resources. Once mined, an ore deposit or a limestone hill disappears forever. Natural geologic processes do not happen fast enough to replace the deposits as quickly as we use them. Geologists have calculated **reserves** (measured quantities of a commodity) for various mineral deposits just as they have for oil. Based on current definitions of reserves (which depend

TABLE 15.3 Yearly Per Capita Usage of Geologic Materials in the United States

4,100 kg	Stone
3,860 kg	Sand and gravel
3,050 kg	Petroleum
2,650 kg	Coal
1,900 kg	Natural gas
550 kg	Iron and steel
360 kg	Cement
220 kg	Clay
200 kg	Salt
140 kg	Phosphate
25 kg	Aluminum
10 kg	Copper
6 kg	Lead
5 kg	Zinc

1 kg = 2.205 pounds

FIGURE 15.19 We consume vast quantities of mineral resources in a year, as the diagram indicates. The numbers indicate the weight of the material used per person per year.

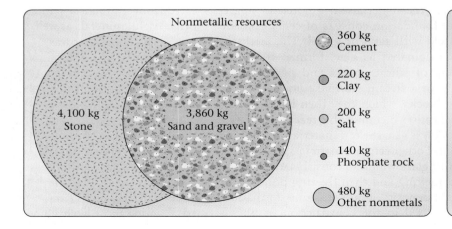

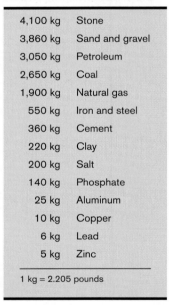

Nonmetallic resources

4,100 kg Stone

3,860 kg Sand and gravel

360 kg Cement

220 kg Clay

200 kg Salt

140 kg Phosphate rock

480 kg Other nonmetals

Metallic resources

550 kg Iron and steel

25 kg Aluminum

10 kg Copper

6 kg Lead

5 kg Zinc

6 kg Manganese

9 kg Other metals

TABLE 15.4 **Lifetimes (in Years) of Currently Known Ore Resources**

Metal	World Resources	U.S. Resources
Iron	120	40
Aluminum	330	2
Copper	65	40
Lead	20	40
Zinc	30	25
Gold	30	20
Platinum	45	1
Nickel	75	less than 1
Cobalt	50	less than 1
Manganese	70	0
Chromium	75	0

new ways of mining become available (e.g., providing access to nodules on the sea floor or to deeper parts of the crust). Further, increased efforts at conservation and recycling can cause a dramatic decrease in rates of consumption, and thereby stretch the lifetime of existing reserves.

Ore deposits do not occur everywhere, because their formation requires special geological conditions. As a result, some countries possess vast supplies, whereas others have none. In fact, no single country owns all the mineral resources it needs, so nations must trade with each other to maintain supplies, and global politics inevitably affects prices. Many wars have their roots in competition for mineral reserves, and it is no surprise that the outcomes of some wars have hinged on who controls these reserves.

The United States worries in particular about supplies of so-called **strategic metals**, which include manganese, platinum, chromium, and cobalt—metals alloyed with iron to make the special-purpose steels needed in the aerospace industry. At present, the country must import 100% of the manganese, 95% of the cobalt, 73% of the chromium, and 92% of the platinum it consumes. Principal reserves of these metals lie in the crust of countries that have not always practiced open trade with the United States. As a defense precaution, the United States stockpiles these metals in case supplies are cut off.

on today's prices) and rates of consumption, supplies of some metals may run out in only decades to centuries (▶Table 15.4). But these estimates may change as supplies become depleted and prices rise (making previously uneconomical deposits worth mining). And supplies could increase if geologists discover new reserves or if

FIGURE 15.20 **(a)** The orange color in this acid mine runoff is from dissolved iron in the water. **(b)** This vegetation-free zone near Sudbury, Ontario, developed in response to acidic smelter smoke. A large tailings pile can be seen in the distance. Once the tall smokestack came into use, the smoke blew farther away, and the land in the foreground became revegetated.

(a)

(b)

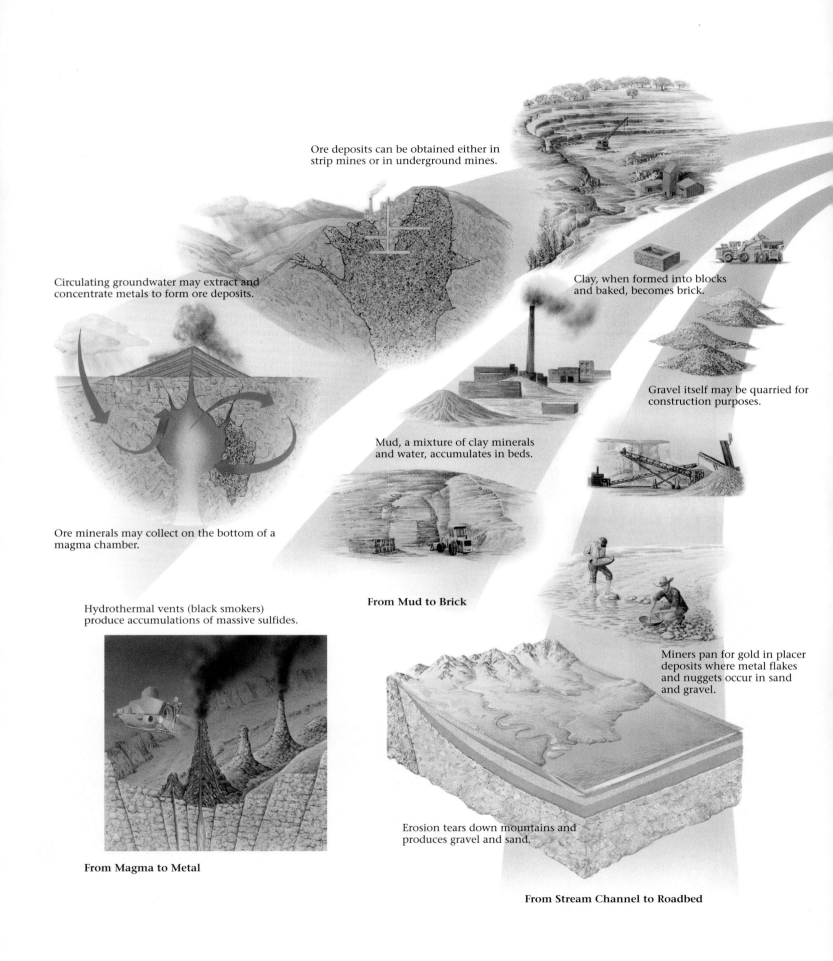

Ore deposits can be obtained either in strip mines or in underground mines.

Circulating groundwater may extract and concentrate metals to form ore deposits.

Ore minerals may collect on the bottom of a magma chamber.

Hydrothermal vents (black smokers) produce accumulations of massive sulfides.

Clay, when formed into blocks and baked, becomes brick.

Gravel itself may be quarried for construction purposes.

Mud, a mixture of clay minerals and water, accumulates in beds.

From Mud to Brick

Miners pan for gold in placer deposits where metal flakes and nuggets occur in sand and gravel.

From Magma to Metal

Erosion tears down mountains and produces gravel and sand.

From Stream Channel to Roadbed

Mining and processing ore has environmental consequences, including acid runoff, acid rain, and groundwater contamination.

Geologic materials are the substance from which cities grow.

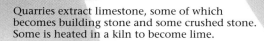

A mixture of lime, other elements, sand, and water, when allowed to harden, becomes concrete.

Mixed with water, spread into sheets, and wrapped in paper, gypsum makes drywall.

In quarries, operators dig up gypsum, crush it to powder, and ship it to factories.

Gypsum is a salt that precipitates when saline lakes evaporate. It grows as white or clear crystals.

Quarries extract limestone, some of which becomes building stone and some crushed stone. Some is heated in a kiln to become lime.

From Lake Bed to Drywall

Over millions of years, shells and shell fragments collect and eventually form beds of limestone.

Organisms extract ions from water and construct shells.

From Sea Floor to Sidewalk

Forming and Processing Earth's Mineral Resources

The raw materials from which we manufacture the buildings, roads, wires, and coins of modern society were produced by geologic processes. For example, ore deposits—the concentrations of minerals that are a source of metal—formed during a variety of magmatic or sedimentary processes. Limestone, a rock used for buildings and for making concrete, began as an accumulation of seashells. Brick began as clay, a byproduct of chemical weathering. And the gypsum of drywall began as an accumulation of salt along a desert lake. Metal, gravel, lime, and gypsum are all examples of Earth's mineral resources. We can use some mineral resources right from the Earth, simply by digging them out. But most become usable only after expensive processing.

Mining and the Environment

Mining leaves a big footprint in the Earth System. Some of the gaping holes that open-pit mining creates in the landscape have become so big that astronauts can see them from space. Both open-pit and underground mining yield immense quantities of waste rock, which miners dump in tailings piles. Some tailings piles grow into artificial hills 200 meters high and many kilometers long. Lacking soil, tailings piles tend to remain unvegetated for a long time. Mining also exposes ore-bearing rock to the atmosphere, and since many ore minerals are sulfides, they react with rainwater to produce **acid mine runoff**, which can severely damage vegetation downstream (▶Fig. 15.20a). Ore processing tends to release noxious chemicals that can mix with rain and spread over the countryside, damaging life. Before the installation of modern environmental controls, smoke from ore-processing plants caused severe air pollution; plumes of smoke from the old smelters in Sudbury, Ontario, for example, created a wasteland for many kilometers downwind (▶Fig. 15.20b). Recent years have seen efforts to reclaim mining spoils, and new technologies have been developed to extract metals in ways that are less deleterious to the environment and that treat waste more efficiently. Clearly, a mine has the potential to become a scar on the landscape, the size of which depends on the efforts of miners to minimize damage.

Take-Home Message

Mineral resources are nonrenewable, so minerals have limited reserves. Also, reserves are not distributed uniformly around the planet, so supplies are not necessarily accessible to consumers. Mineral utilization has significant environmental consequences.

Chapter Summary

- Industrial societies use many types of minerals, all of which must be extracted from the upper crust. We distinguish two general categories: metallic resources and nonmetallic resources.

- Metals are materials in which atoms are held together by metallic bonds. They are malleable and make good conductors.

- Metals come from ore. An ore is a rock containing native metals or ore minerals (sulfide, oxide, or carbonate minerals with a high proportion of metal) in sufficient quantities to be worth mining. An ore deposit is an accumulation of ore.

- Magmatic deposits form when sulfide ore minerals settle to the floor of a magma chamber. In hydrothermal deposits, ore minerals precipitate from hot-water solu-

tions. Secondary-enrichment deposits form when groundwater carries metals away from a preexisting deposit. Mississippi Valley–type deposits precipitate from groundwater that has passed long distances through the crust. Sedimentary deposits precipitate out of the ocean. Residual mineral deposits in soil are the result of severe leaching in tropical climates. Placer deposits develop when heavy metal grains accumulate in sediment along a stream.

- Many ore deposits are associated with igneous activity in subduction zones, along mid-ocean ridges, along continental rifts, or at hot spots.

- Nonmetallic resources include dimension stone for decorative purposes, crushed stone for cement and asphalt production, clay for brick making, sand for glass production, and many other materials. A large proportion of materials in your home have a geological ancestry.

- Mineral resources are nonrenewable. Many are now or may soon become in short supply.

Geopuzzle Revisited

The stuff of everyday life comes from geologic materials. For example: baking mud makes brick, smelting copper-bearing minerals yields metallic copper, and mixing crushed and baked limestone with sand and water forms concrete. Geologists play a key role in identifying sources of such mineral resources, and in addressing the environmental issues that arise from using such resources.

Key Terms

acid mine runoff (p. 540)
alloy (p. 524)
banded-iron formation (BIF) (p. 529)
base metals (p. 525)
cement (p. 534)
concrete (p. 534)
dimension stone (p. 534)
grade (p. 526)

hydrothermal deposit (p. 527)
magmatic deposit (p. 527)
manganese nodules (p. 529)
massive-sulfide deposit (p. 527)
metallic bonds (p. 523)
metals (p. 523)
mineral resources (p. 523)

Review Questions

1. Describe how people have used copper, bronze, and iron throughout history.

2. Why don't we use an average granite as a source for useful metals?

3. What kinds of concentrations of a metal are required for it to be economically minable?

4. Describe various kinds of economic mineral deposits.

5. What procedures are used to locate and mine mineral resources today?

6. How is stone cut from a quarry?

7. What are the ingredients of cement? How is Portland cement made?

8. How many kilograms of mineral resources does the average person in an industrialized country use in a year?

9. Compare the estimated lifetimes of ore supplies (worldwide and in the United States) of iron, aluminum, copper, gold, and chromium.

10. What are some environmental hazards of large-scale mining?

On Further Thought

1. Imagine that an ore deposit of a certain metal contains 0.6% grade ore. This means that 0.6 percent by weight of a block of ore consists of the metal. The pure metal, on the open market, sells for $8,000/ton. It costs $15/ton to mine the ore, $15/ton to transport the ore to the processing plant, and $15/ton to process the ore and produce pure metal. Start-up costs (building the mine and building the processing factory) are about $100 million. How much profit does the company make when it sells a ton of metal? How much ore does the operation have to mine to pay back the start-up costs? Considering that a giant dump truck in a mine can carry 200 tons of ore at a time, how many dump-truck loads will have been transported at the break-even point? If the mine has 8 trucks that can each make 6 loads a day, about how many years will it take to break even?

2. An ore deposit at a location in Arizona has the following characteristics: one portion of the ore deposit is an intrusive igneous rock in which tiny grains of copper sulfide minerals are dispersed among the other minerals of the rock. Another nearby portion of the ore deposit consists of limestone in which malachite fills cavities and pores in the rock. What types of ores are these? Describe the geologic history that led to the formation of these deposits.

Suggested Reading

Brands, H. W. 2002. *The Age of Gold: The California Gold Rush and the New American Dream*. New York: Doubleday.

Carr, D. D., and N. Herz, eds. 1988. *Concise Encyclopedia of Mineral Resources*. Cambridge, Mass.: MIT Press.

Craddock, P., and J. Lang, eds. 2003. *Mining and Metal Production through the Ages*. London: British Museum Publications.

Craig, J. R., D. J. Vaughan, and B. J. Skinner. 1989. *Resources of the Earth*. Englewood Cliffs, N.J.: Prentice-Hall.

Dorr, A. 1987. *Minerals: Foundations of Society*. Alexandria, Va.: American Geological Institute.

Evans, A. M. 1992. *Ore Geology and Industrial Minerals: An Introduction*. 3rd ed. Malden, Mass.: Blackwell Science.

Evans, A. M. 1997. *An Introduction to Economic Geology and Its Environmental Impact*. Malden, Mass.: Blackwell Science.

Evans, A. M., M. Whateley, and C. Moon (eds.). 2006. *Introduction to Mineral Exploration*. Malden, Mass.: Blackwell.

Guilbert, J. M., and C. F. Park. 2007. *The Geology of Ore Deposits*. Long Grove, Ill.: Waveland Press.

Guilbert, J. M., and C. F. Park, Jr. 1986. *The Geology of Ore Deposits*. New York: Freeman.

Kesler, S. E. 1994. *Mineral Resources, Economics, and the Environment*. New York: Macmillan.

Manning, D. A. C. 1995. *Introduction to Industrial Minerals*. London: Chapman & Hall.

National Research Council. 1990. *Competitiveness of the U.S. Minerals and Metals Industry*. Washington, D.C.: National Academy Press.

Robb, L. J. 2004. *Introduction to Ore-Forming Processes*. Malden, Mass.: Blackwell.

Sawkins, F. J. 1984. *Metal Deposits in Relation to Plate Tectonics*. New York: Springer-Verlag.

Stone, I. 1956. *Men to Match My Mountains*. New York: Doubleday; reprinted 1982 by Berkley Books. See pp. 128–51.

Processes and Problems at the Earth's Surface

In the last part of this book, we focus on Earth's surface and near-surface realms. This portion of the Earth System, which encompasses the boundaries between the lithosphere, hydrosphere, and atmosphere, displays great variability, for the dynamic interplay between internal processes (driven by Earth's internal heat) and external processes (driven by the warmth of the Sun), under the influence of Earth's gravitational field, has resulted in a diverse array of landscapes. In Chapters 16 through 22, we examine five of these landscapes, plus groundwater and the atmosphere, and finally, in Chapter 23, we see how forces at work in the Earth System cause the planet to change constantly.

Snow-capped peaks of the Rocky Mountains provide a dramatic landscape. The existence and shape of these peaks reflects the constant battle between tectonic forces that uplift the land surface, and erosive forces that grind it away.

Ever-Changing Landscapes and the Hydrologic Cycle

This view of a windy inlet along the coast of Norway illustrates various reservoirs for water in the Earth System.

Talk of mysteries! Think of our life in nature—daily to be shown matter, to come in contact with it—rocks, trees, wind on our cheeks! the solid earth! the actual world! the common sense! Contact! Contact! Who are we? Where are we?

—Henry David Thoreau (1817–1862)

F.1 INTRODUCTION

The Earth's surface is at once a place of endless variety and intricate detail. Observe the height of its mountains, the expanse of its seas, the desolation of its deserts, and you may be inspired, frightened, or calmed. It's no wonder that artists and writers across the ages have sought inspiration from the **landscape**—the character and shape of the land surface in a region—for landscapes display the diversity of human emotion (▶Fig. F.1a–f). Geologists, like artists and writers, savor the impression of a dramatic landscape, but on seeing one, they can't help but ask, How did it come to be, and how will it change in the future?

The subject of landscape development and evolution, and the **landforms** (individual shapes such as mesas, valleys, cliffs, and dunes) that constitute it, dominate most of the chapters in Part VI. Geologists who study landscape development are known as geomorphologists. This interlude, a general introduction to the topic, explains the driving forces

FIGURE F.1 A great variety of landscapes on Earth. **(a)** The desert ranges of the Mojave Desert, in southeastern California. **(b)** The peaks of the Grand Tetons, in Wyoming. **(c)** As viewed from an airplane at high altitude, the texture of a stream-eroded landscape appears very complex. **(d)** A rock and sand seascape along the coast of Brazil. **(e)** Buttes at Monument Valley, Arizona. **(f)** Steep cliffs in Australia's Blue Mountains.

behind landscape development, identifies factors that control which landscape develops in a given locality, and describes the hydrologic cycle—the pathway water molecules follow as they move from ocean to air to land and back to ocean. We focus on the hydrologic cycle here because so many processes on or near Earth's surface involve water in its various forms. We conclude by introducing landscapes on other planets.

F.2 SHAPING THE EARTH'S SURFACE

If the Earth's surface were totally flat, the great diversity of landscapes that embellish our vistas would not exist. But the surface isn't flat, because a variety of geologic processes cause portions of the surface to move up or down relative to adjacent regions. We refer to the upward movement of the land surface as **uplift**, and the sinking or downward

TABLE F.1 Causes of Uplift and Subsidence

Causes of Uplift

- **Thickening of the crust due to deformation.** At convergent and collisional boundaries, compression causes the crust to shorten horizontally (by development of folds, faults, and foliations) and thicken in the vertical direction. Because of isostasy (see Chapter 11), lithosphere with thickened crust floats relatively higher on the asthenosphere, with the result that the surface of the crust in mountain belts rises.

- **Heating of the lithosphere.** Heating decreases the thickness and density of the lithosphere, so to maintain isostatic equilibrium, it floats higher. Intrusion or extrusion of igneous rocks thickens the crust or builds volcanoes on top of the surface; these phenomena, therefore, cause uplift.

- **Rebound due to unloading.** Removal of a heavy load (such as a glacier or mountain) causes the Earth's surface to rise, somewhat how the surface of a trampoline rises when you step off of it.

Causes of Subsidence

- **Thinning of the crust due to deformation.** In rifts, where the crust undergoes horizontal stretching, the axis of the rift drops down by slip on normal faults.

- **Cooling of the lithosphere.** Cooling thickens the lithosphere and makes it denser, so to maintain isostatic equilibrium, the lithosphere sinks down and its surface lies at a lower elevation.

- **Sinking due to loading.** Where a heavy load (such as a glacier or volcano) forms on the Earth's surface, the lithosphere warps downward, somewhat how the surface of a trampoline warps down when you stand on it.

movement of the land surface as **subsidence**. Both uplift and subsidence occur for a variety of reasons (▶Table F.1).

When uplift or subsidence takes place, the elevation difference does not remain the same forever, because other components of the Earth System kick into action. Material at higher elevations becomes unstable and susceptible to downslope movement (the tumbling or sliding of rock and sediment from higher elevations to lower ones); moving water, ice, and air cause **erosion** (the grinding away and removal of the Earth's surface); and where moving fluids slow down, *deposition* of sediment takes place. Downslope movement, erosion, and deposition redistribute rock and sediment, ultimately stripping it in from higher areas and collecting it in low areas. As a consequence, a great variety of both **erosional landforms** (those carved by erosion) and **depositional landforms** (those built from an accumulation of sediment) can form.

Because lithosphere floats on asthenosphere (see the discussion of isostasy in Chapter 11), the removal of 1 km of rock of the top of a mountain range causes the crust to rise by about 1/3 km, just as the removal of heavy containers from the deck of a cargo ship makes the ship rise. Because the crust rises as erosion takes place, *more than* 5 km of rock must erode

from a 5-km-high mountain range to return the land surface to sea level. Similarly, in a depositional setting, a 1-km-thick layer of sediment depresses the crust by about 1/3 km.

The energy that drives landscape evolution comes from three sources: **internal energy**, the heat within the Earth, which drives the plate motions and mantle plumes that cause displacement of the crust's surface; **external energy**, energy coming to the Earth from the Sun, which causes the atmosphere and ocean to flow; and **gravitational energy**, which pulls rock down slopes at the surface and, along with external energy, causes convection. Landscape evolution, in fact, reflects a "battle" between (1) tectonic processes such as collision, convergence, and rifting, which create **relief** (see ▶Fig. F.3a) (an elevation difference between two locations) in an area, and (2) processes such as downslope movement, erosion, and deposition, which destroy relief by removing material from high areas and depositing it in low ones. If, in a particular region, the rate of uplift exceeds the rate of erosion, the land surface rises; if the rate of subsidence exceeds the rate of deposition, the land surface sinks. Without uplift and subsidence, Earth's surface would long ago have been beveled to a flat plain, and without erosion and deposition, high and low areas would have lasted for the entirety of Earth history.

How rapidly do uplift, subsidence, erosion, and deposition take place? The Earth's surface can rise or sink by as much as 3 m during a single major earthquake. But averaged over time, the rates of uplift and subsidence range between 0.01 and 10 mm per year (▶Fig. F.2a). Similarly, erosion can carve out several meters of *substrate*, the material just below the ground surface, during a single flood, storm, or landslide (▶Fig. F.2b). And deposition during a single event can produce a layer of debris tens of meters thick in a matter of minutes to days. But, averaged over time, erosional and depositional rates also vary between 0.10 and 10 mm per year. Although these rates seem small, a change in surface elevation of just 0.5 mm (the thickness of a fingernail) per year can yield a net change of 5 km in 10 million years. Uplift can build a mountain range, and erosion can whittle one down to near sea level—it just takes time!

F.3 TOOLS OF THE TRADE: TOPOGRAPHIC MAPS AND PROFILES

We can distinguish one landform from another by its shape—for example, as you will see in succeeding chapters, a river-carved valley simply does not look like a glacially carved valley. Landform shapes are manifested by variations in elevation within a region. Geologists use the term **topography** to refer to such variations.

(a)

(b)

FIGURE F.2 **(a)** Uplifted beach terraces along a coast appear where the coast is rising relative to sea level. Wave erosion eats into the land, creating a terrace. Eventually, the terrace rises out of the reach of the waves, which then cut a new one. **(b)** So much erosion can take place during a single hurricane that the foundations of houses built along the beach become undermined.

How can we convey information about topography—a three-dimensional feature—on a two-dimensional sheet of paper? Geologists do this by means of a **topographic map**, which uses contour lines to represent variations in elevation (▶Fig. F.3a–d). A **contour line** is an imaginary line along which all points have the same elevation. For example, if you walk along the 200-m contour line on a hill slope, you stay at exactly the same elevation. As another example, the shoreline on a flat, calm body of water is a contour line. In other words, you can picture the contour line as the intersection between the land surface and an imaginary horizontal plane. Contour lines form a closed loop around a hill, and they form a V shape that points upstream where they cross a river valley.

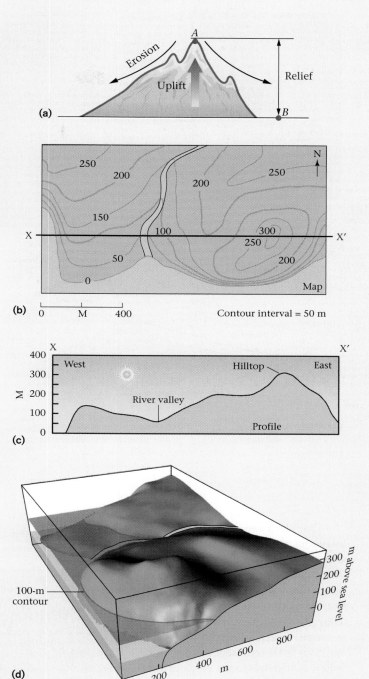

FIGURE F.3 **(a)** Relief is simply the vertical elevation difference between two points (A and B) at the surface of the Earth. **(b)** A topographic map depicts the 3-D shape of the land surface on a 2-D map through the use of contour lines. The difference in elevation between two adjacent lines is the contour interval. **(c)** A topographic profile (along section line X-X') shows the shape of the land surface as seen in a vertical slice. **(d)** You can picture a contour line as the intersection of a horizontal plane with the land surface. This block diagram shows the area of (a) in 3-D.

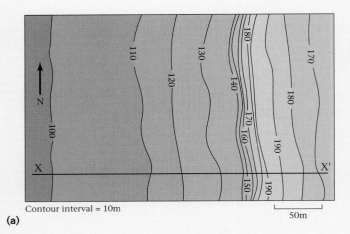

Contour interval = 10m

50m

(a)

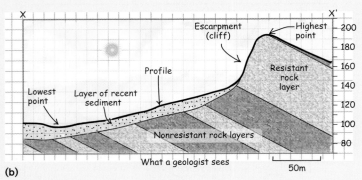

What a geologist sees

50m

(b)

FIGURE F.4 **(a)** Topographic map showing a north-south trending escarpment, indicated by the closely spaced contour lines. **(b)** Topographic profile showing a geologic interpretation of the subsurface.

are provided in an introductory geology laboratory manual.) If we add a representation of geologic features under the ground surface, then we have a geologic cross section. In some cases, geologists gain insight into subsurface geology simply by looking at the shape of a landform. For example, a steep cliff in a region of dipping sedimentary strata may indicate the presence of a resistant (difficult to erode) layer; low areas may be underlain with nonresistant (easy to erode) layers (▶Fig. F.4a, b).

In recent years, geologists have used satellites to produce highly detailed digital images of the Earth's surface. The satellite sends a radar beam down to the surface, and the beam bounces back to the satellite. The time it takes for the beam to make its round trip represents land elevation. The resulting maps, called **digital elevation maps**, provide images that can be analyzed with a computer. The maps can be produced with shading and color that gives the impression of a 3-D shape even on a 2-D sheet of paper (▶Fig. F.5).

F.4 FACTORS CONTROLLING LANDSCAPE DEVELOPMENT

Imagine traveling across a continent. On your journey, you pass plains, swamps, hills, valleys, mesas, and mountains. Some of these features are erosional landforms, in that they result from the breakdown and removal of rock or sediment and develop where agents of erosion such as water, wind, or ice carve into the substrate. Of these three

The elevation difference between two adjacent contour lines on a topographical map is called the **contour interval**. For a given topographic map, the contour interval is constant, so the spacing between contour lines represents the steepness of a slope. Specifically, closely spaced contour lines represent a steep slope, whereas widely spaced contour lines represent a gentle slope.

We can also represent variations in elevation by means of a topographic profile. A profile is the trace of the ground surface as it would appear on a vertical plane that sliced into the ground—put another way, it's the shape of the ground surface as viewed from the side. (The details of how to produce a topographic profile from a topographical map

FIGURE F.5 Digital elevation model of Oahu, Hawaii, as viewed obliquely from the south. The surface colors were taken from a satellite photograph and were laid down over the land surface.

agents, water has the greatest effect on a global basis. Other features are depositional landforms, in that they result from the deposition of sediment where the medium carrying the sediment evaporates, slows down, or melts. The specific landforms that develop at a given locality, and that together make up the landscape, reflect six factors.

- *Eroding or transporting agents:* Water, ice, and wind all can cause erosion and transport sediment, but landforms produced by glacial erosion or deposition differ from landforms produced by river erosion or deposition, and both differ from landforms carved by the wind. The differences reflect the varying abilities of water, ice, and wind to carve into the substrate and to carry debris.

- *Elevation:* In regions where the land surface has risen to a high elevation, rugged mountain ranges with deep valleys and steep cliffs can form, but in places where the land is a wide plain near sea level, valleys and cliffs do not exist. The elevation difference, or relief, between adjacent places in a landscape determines the height and steepness (angle) of slopes.

- *Climate:* The average mean temperature and the volume of precipitation in a region determines whether running water, flowing ice, or wind is the main agent of erosion or deposition.

- *Life activity:* Organisms such as bacteria, worms, trees, and elephants affect landscape development. Although some life activity weakens the substrate, some protects it. The ecology of an area depends on the climate. A landscape that would consist of gently rolling, forested hills in a temperate climate might consist of barren escarpments in a desert climate.

- *Substrate composition:* A substrate's composition determines how it responds to erosion. Regions underlain with hard rock erode less easily than regions underlain with unconsolidated sediment.

- *Time:* Landscapes change in response to continued erosion and/or deposition acting over time. In erosional landscapes, more erosion happens as time passes, whereas in depositional landscapes, more deposition takes place as time passes.

Although water, wind, and ice are responsible for the development of most landscapes, human activities have had an increasingly important impact on the Earth's surface. We have dug pits (mines) where once there were mountains, built hills (tailings piles and landfills) where once there were valleys, and made steep slopes gentle and gentle slopes steep (▶Fig. F.6a–c). By constructing concrete walls, we modify the shapes of coastlines, change the courses of rivers, and fill new lakes (reservoirs). In cities, buildings and pavements completely seal the ground and cause water that might once have seeped down into the

(a)

(b)

(c)

FIGURE F.6 Human workings on a geological scale. (a) The pyramids of Egypt are essentially human-made hills that rise above the desert sands. They have lasted for thousands of years. (b) People routinely cut deep valleys through high ridges, as in this highway road cut in Colorado, to make a level grade for a road. Think how long it would take a river to cut such a gorge. (c) This stone dam holds back a reservoir in Colorado. Think how long it would take a glacier to pile up so much sediment.

ground to spill into streams instead, increasing their flow. The area of land covered by pavement or buildings in the United States now exceeds the area of Ohio! And in the country, agriculture, grazing, water usage, and deforestation substantially alter the rates at which natural erosion and deposition take place. For example, agriculture greatly increases the rate of erosion, because for much of the year farm fields have no vegetation cover.

F.5 THE HYDROLOGIC CYCLE

Nothing that is can pause or stay—
The moon will wax, the moon will wane,
The mist and cloud will turn to rain,
The rain to mist and cloud again,
Tomorrow be today.

—Henry Wadsworth Longfellow (1807–1882)

As is evident from the discussion above, water in its various forms (liquid, gas, and solid) plays a major role in erosion and deposition on Earth's surface. Our planet's water is found in certain distinct reservoirs (containers), namely: the oceans, glacier ice (today, mostly in Antarctica and Greenland), groundwater, lakes, soil moisture, living organisms, the atmosphere, and rivers (▶Table F.2). Together, these constitute the hydrosphere. Water constantly flows from reservoir to reservoir, and this never-ending passage is called the **hydrologic cycle** (see art, pp. 552–553). Perhaps without realizing it, Longfellow, an American poet fascinated with reincarnation, provided an accurate if somewhat romantic image of the hydrologic cycle. Without this cycle, the erosive forces of running water (rivers and streams) and flowing ice (glaciers) would not exist.

The *average* length of time that water stays in a particular reservoir during the hydrologic cycle is called the **residence time** (▶Table F.3). Water in different reservoirs has different residence times. For example, a typical molecule of water remains in the oceans for 4,000 years or less, in lakes and ponds for 10 years or less, in rivers for 2 weeks or less, and in the atmosphere for 10 days or less. Groundwater residence times are highly variable and depend on how deep the groundwater flows. Water can stay underground for anywhere from 2 weeks to 10,000 years before it inevitably moves on to another reservoir.

To get a clearer sense of how the hydrologic cycle operates, let's follow the fate of seawater that has just reached the surface of the ocean. Solar radiation heats the water, and the increased thermal energy of the vibrating water molecules allows them to evaporate (break free from the liquid) and drift upward in a gaseous state to become part of the atmosphere. About 417,000 cubic km (102,000 cubic miles), or about 30% of the total ocean volume, evaporates every year. Atmospheric water vapor moves with the wind to higher elevations, where it cools, undergoes condensation (the molecules link together to form a liquid), and rains or snows. About 76% of this water precipitates (falls out of the air) directly back into the ocean. The remainder precipitates onto land; most of this becomes trapped temporarily in the soil, or in plants and animals, and soon re-

TABLE F.3 Estimated Residence Time of Water in Earth's Reservoirs

H₂O Reservoir	Average Residence Time
Ice caps	10,000 to 200,000 years
Deep groundwater	3,000 to 10,000 years
Oceans and Inland seas	3,000 to 3,500 years
Shallow groundwater	100 to 200 years
Valley glaciers	20 to 100 years
Fresh-water lakes	50 to 100 years
Winter snow	2 to 6 months
Rivers and streams	2 to 6 months
Soil moisture	1 to 2 months
Atmosphere	5 to 15 days
Living organisms	hours to days

TABLE F.2 Major Water Reservoirs of the Earth*

H₂O Reservoir	Volume (km³)	% of Total Water	% of Fresh Water
Oceans and seas	1,338,000,000	96.5	—
Glaciers, Icecaps, Snow	24,064,000	2.05	68.7
Saline groundwater	12,870,000	0.76	—
Fresh groundwater	10,500,000	0.94	30.1
Permafrost	300,000	0.022	0.86
Fresh-water lakes	91,000	0.007	0.26
Salt lakes	85,400	0.006	—
Soil moisture	16,500	0.001	0.05
Atmosphere	12,900	0.001	0.04
Swamps	11,470	0.0008	0.03
Rivers and streams	2,120	0.0002	00.006
Living organisms	1,120	0.0001	00.003

*SOURCE: Gleick, P. H. (1996) in *Encyclopedia of Climate and Weather.* New York: Oxford Univ. Press

turns directly to the atmosphere by what is called evapotranspiration: the sum of evaporation from bodies of water, evaporation from the ground surface, and transpiration (release as a metabolic byproduct) from plants and animals. Rainwater that did not become trapped in the soil or in living organisms either enters lakes or rivers and ultimately flows back to the sea as surface water, becomes trapped in glaciers, or sinks deeper into the ground to become groundwater. Groundwater also flows and ultimately returns to the Earth's surface reservoirs. In sum, during the hydrologic cycle, water moves among the ocean, the atmosphere, reservoirs on or below the land surface, and living organisms.

F.6 LANDSCAPES OF OTHER PLANETS

The dynamic, ever-changing landscape of Earth contrasts markedly with those of other terrestrial planets. Each of the terrestrial planets and moons has its own unique surface landscape features, reflecting the interplay between the object's particular tectonic and erosional processes. Let's look at a few examples: the Moon, Mars, and Venus.

Our Moon has a static, pockmarked landscape generated exclusively by meteorite impacts and volcanic activity. Because no plate tectonics occurs on the Moon, no new mountains form, and because no atmosphere or ocean exists, there is no hydrologic cycle and no erosion from rivers, glaciers, or winds. Therefore, the lunar surface has remained largely unchanged for over billions of years. The

landscape can be divided into the Lunar Highlands, the heavily cratered, light-colored regions of the moon exposing rocks over 4.0 billions of years old, and the mare, vast plains of flood basalt possibly formed in response to impacts over 3.8 billion years ago. These impacts were so huge that they caused melting in the Moon's mantle and extrusion of flood basalts (▶Fig. F.7a, b).

Landscapes on Mars differ from those of the Moon because Mars *does* have an atmosphere (though much less dense than that of Earth) whose winds generate huge dust storms, some of which obscure nearly the entire surface of the planet for months at a time. The landscapes of Mars also differ from the Moon's because Mars probably once had surface water (▶Box F.1 and ▶Fig. F.8a, b). Thus, the Martian surface appears to consist of four kinds of materials: volcanic flows and deposits (primarily of basalt), debris from impacts, wind-blown sediment, and water-laid sediment. There is even evidence that soil-forming processes affected surface materials. Martian winds not only deposit sediment, they also slowly erode impact craters and polish surface rocks.

Landscapes on Mars also differ from those on Earth, because Mars does not have plate tectonics. So, unlike Earth, Mars has no mountain belts or volcanic arcs. In fact, most landscape features on Mars, with the exception of wind-related ones, are over 3 billion years old. Long ago, a huge mantle plume formed. This plume caused the uplift of a 9-km-high bulge (the Tharsis Ridge) that covers an area comparable to that of North America (▶Fig. F.9). Thermal activity also led to the eruption of gargantuan hot-spot volcanoes, such as the 22-km-high Olympus Mons. Mars also

FIGURE F.7 (a) This lunar landscape is virtually the same landscape that would have been visible 2 billion years ago, because the Moon's surface is static. **(b)** A closeup view of the lunar landscape, with the lunar rover and an astronaut for scale.

(a)

(b)

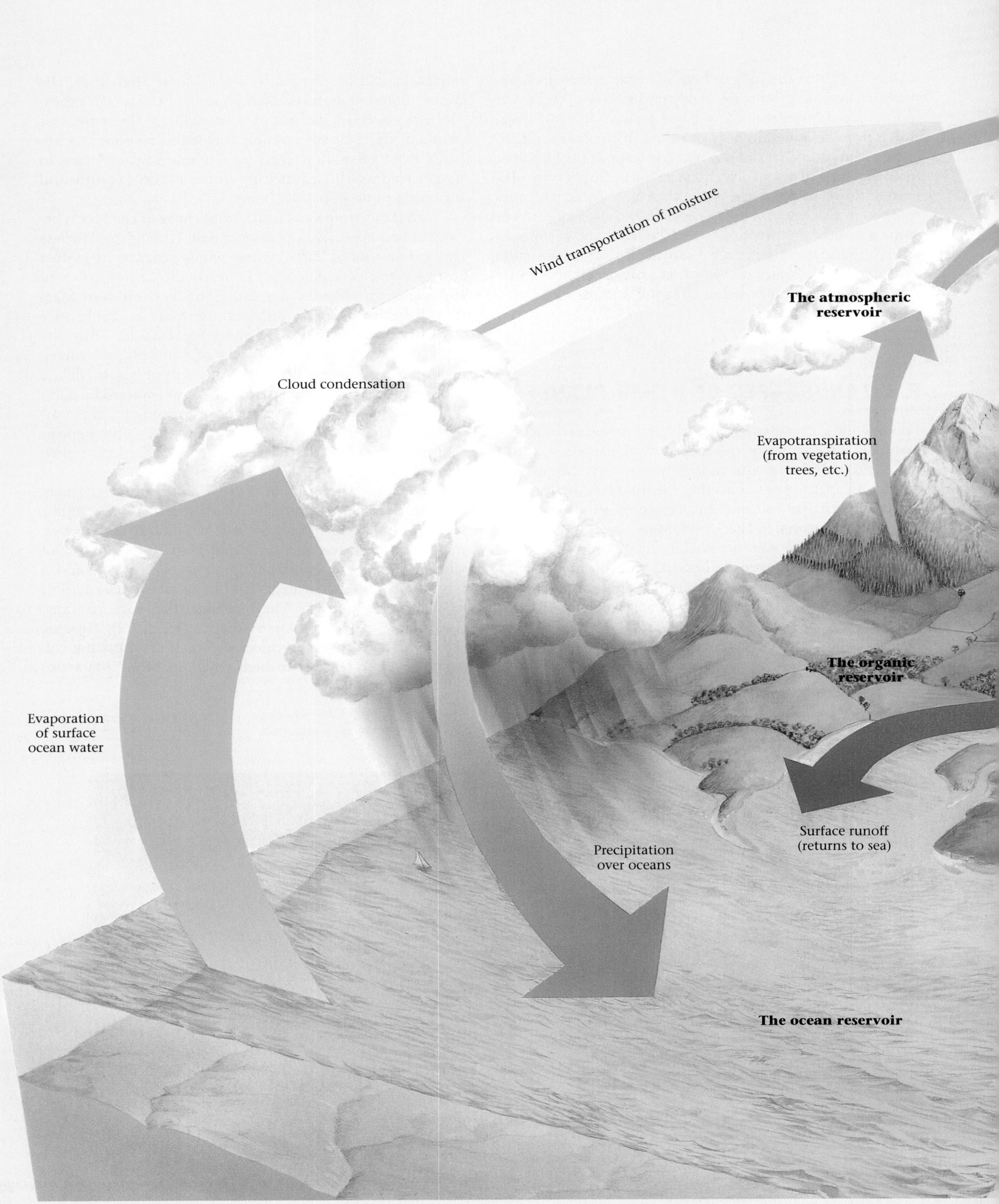

Wind transportation of moisture

The atmospheric reservoir

Cloud condensation

Evapotranspiration
(from vegetation,
trees, etc.)

Evaporation
of surface
ocean water

**The organic
reservoir**

Surface runoff
(returns to sea)

Precipitation
over oceans

The ocean reservoir

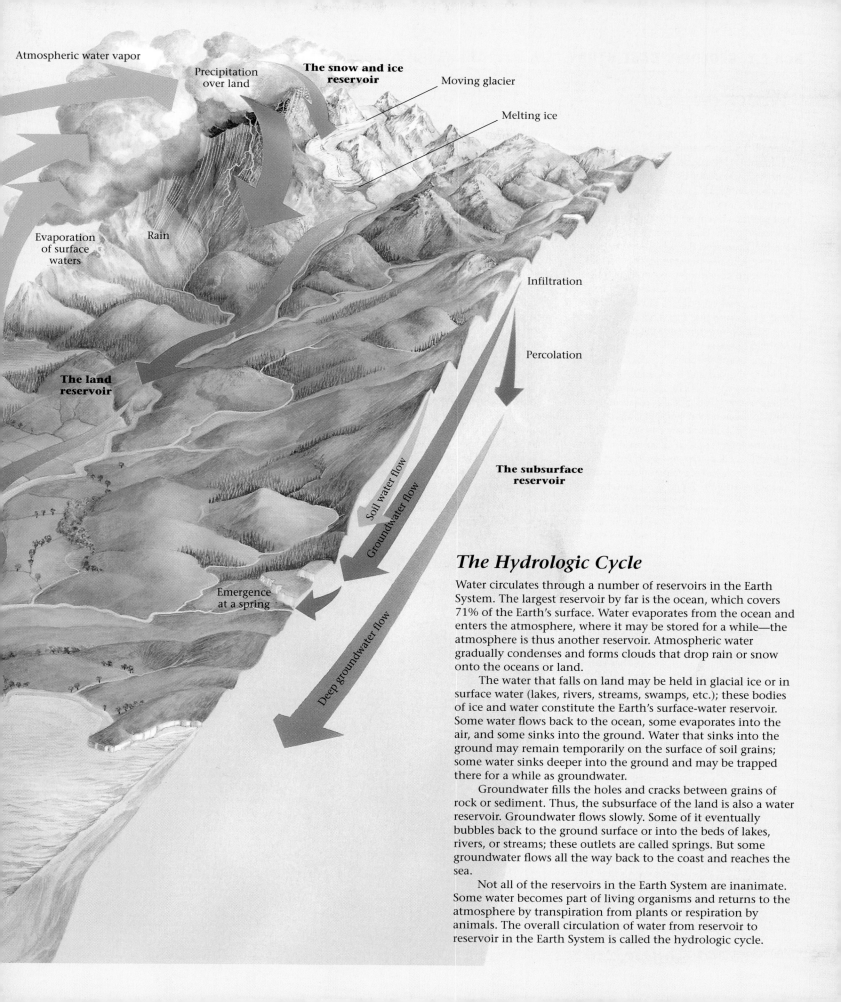

Atmospheric water vapor

Precipitation over land

The snow and ice reservoir

Moving glacier

Melting ice

Evaporation of surface waters

Rain

Infiltration

Percolation

The land reservoir

Soil water flow

Groundwater flow

The subsurface reservoir

Emergence at a spring

Deep groundwater flow

The Hydrologic Cycle

Water circulates through a number of reservoirs in the Earth System. The largest reservoir by far is the ocean, which covers 71% of the Earth's surface. Water evaporates from the ocean and enters the atmosphere, where it may be stored for a while—the atmosphere is thus another reservoir. Atmospheric water gradually condenses and forms clouds that drop rain or snow onto the oceans or land.

The water that falls on land may be held in glacial ice or in surface water (lakes, rivers, streams, swamps, etc.); these bodies of ice and water constitute the Earth's surface-water reservoir. Some water flows back to the ocean, some evaporates into the air, and some sinks into the ground. Water that sinks into the ground may remain temporarily on the surface of soil grains; some water sinks deeper into the ground and may be trapped there for a while as groundwater.

Groundwater fills the holes and cracks between grains of rock or sediment. Thus, the subsurface of the land is also a water reservoir. Groundwater flows slowly. Some of it eventually bubbles back to the ground surface or into the beds of lakes, rivers, or streams; these outlets are called springs. But some groundwater flows all the way back to the coast and reaches the sea.

Not all of the reservoirs in the Earth System are inanimate. Some water becomes part of living organisms and returns to the atmosphere by transpiration from plants or respiration by animals. The overall circulation of water from reservoir to reservoir in the Earth System is called the hydrologic cycle.

Water on Mars?

In 1877, an Italian astronomer named Giovanni Schiaparelli studied the surface of Mars with a telescope and announced that long, straight *canali* criss-crossed the planet's surface. *Canali* should have been translated into the English word "channel," but perhaps because of the recent construction of the Suez Canal, newspapers of the day translated the word into the English "canal," with the implication that the features were constructed by intelligent beings. An eminent American astronomer, Percival Lowell, began to study the "canals" and suggested that they had been built to carry water from the polar ice caps to the Martian deserts.

Late-twentieth-century satellite mapping of Mars showed that the "canals" do not exist—they were simply optical illusions. There are no lakes, oceans, rainstorms, or flowing rivers on the surface of Mars today. The atmosphere of Mars has such low density, and thus exerts so little pressure on the planet's surface, that any liquid water released at the surface in recent time would quickly evaporate. Thus, there is no hydrologic cycle on Mars, as there is on Earth. But three crucial questions remain: Does liquid water ever form, even for short periods of time, on the Martian surface today? Was there ever a significant amount of running water or standing water on Mars in the past? If there once was significant water on the planet, where is the water now? The question of the presence of water lies at the heart of the even more basic question: Even the simplest life as we know it requires water, so is there, or was there, life on Mars?

Many planetary geologists believe that the case for liquid water on Mars is quite strong. Much of the evidence comes from comparing landforms on the planet's surface with landforms of known origin on Earth. High-resolution images of Mars reveal a number of landforms that look as if they formed in response to the action of water. Examples include networks of channels resembling river networks on Earth (▶Fig. F.8a, b) scour features, deep gullies, and streamlined deposits of sediment. Dark streaks on the walls of craters and canyons look like the products of short-term floods emanating from springs in the crater walls. Some researchers speculate that the northern third of the planet was once a vast ocean.

Studies by the *Odyssey* satellite in 2003, and by Mars rovers (*Spirit* and *Opportunity*) that landed on the planet in 2004, have added intriguing new data to the debate. *Odyssey* detected hints that hydrogen, an element in water, exists beneath the surface of the planet over broad regions, and the Mars rovers have documented the existence of hematite and gypsum minerals that form in the presence of water. The rovers have also found sedimentary deposits that appear to have been deposited in water. Researchers speculate that Mars was much wetter in its past, perhaps billions of years ago, when it had volcanic activity and a denser atmosphere. But once the atmosphere became less dense, the water evaporated and now lies hidden underground or trapped in polar ice caps. Sudden local ruptures of the ice layer may produce short-term releases of water on the surface of Mars.

FIGURE F.8 Water-related landscape features on Mars, as photographed by satellites. **(a)** A streamlined island in the middle of what looks like a broad stream channel. Water flow could have carved this island. **(b)** A river-like network of channels. Small tributary channels join a large trunk channel.

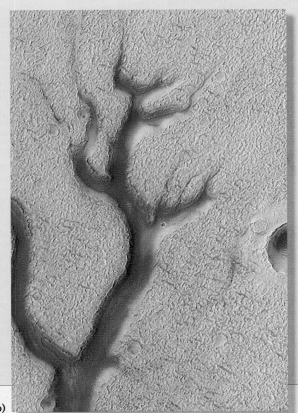

300 m

(a)　　　　(b)

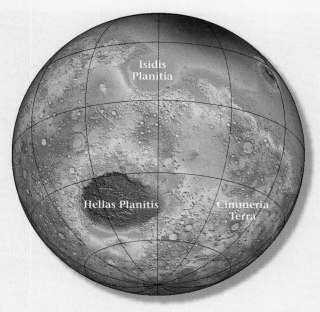

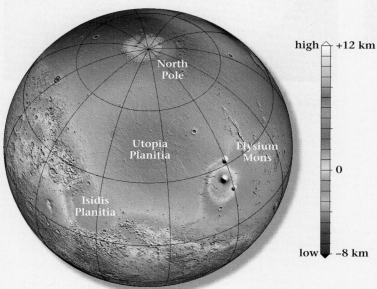

high △ +12 km

0

low ▽ −8 km

FIGURE F.9 Digital elevation models depicting the surface of Mars from different perspectives. Note the huge bulge of the Tharsis Ridge, the giant volcano of Olympic Mons, and the deep canyon of Valles Marineris. The northern end of the planet is low.

those of the Moon and Mars. Further, Venus has a dense atmosphere that protects it from impacts by smaller objects. Because there has been relatively little cratering since the resurfacing event, volcanic and tectonic features dominate the landscape of Venus (▶Fig. F.10). Satellites have used radar to reveal a variety of volcanic constructions (e.g., shield volcanoes, lava flows, and calderas). Rifting on Venus produced faults, some of which occur in association with volcanic features. Liquid water cannot survive the scalding temperatures of Venus's surface, so no hydrologic cycle operates there and no life exists. Because of the density of the atmosphere, winds are too slow to cause much erosion or deposition, thus leaving volcanic landforms virtually unchanged.

boasts the largest known canyon, the Valles Marineris, a gash over 3,000 km long and 8 km deep. No comparable feature exists on Earth. Because Mars has no vegetation and no longer has rain, its surface does not weather and erode like that of Earth, so it still bears the scars of impact by swarms of meteors earlier in the history of the solar system, scars that have long since disappeared on Earth. Mars does have a hydrologic cycle, of sorts, in that it has ice caps that grow and recede on an annual basis.

Venus is closer to the size of the Earth, and may still have operating mantle plumes. Virtually the entire surface of Venus was resurfaced by volcanic eruptions about 300 to 1,600 Ma, making the planet's surface much younger than

During the past two decades, spacecraft visiting the moons of the outer planets have sent home amazing images of surface features that differ markedly from any found on Earth. As an example, consider Enceladus, a 500-km-diameter moon of Saturn (▶Fig. F.11). Much of Enceladus's surface is cracked and wrinkled and largely crater free, suggesting that tectonic movements rifted and folded this moon's crust subsequent to the intense meteorite bombardment episodes of early solar system history. The still-cratered terrains may be older and stabler regions of the crust.

After this brief side trip to other planets, let's now return to Earth. Our planet has the greatest diversity of landscapes in the Solar System.

FIGURE F.10 A radar map of the surface of Venus as it would appear if the atmosphere were removed. All these features are invisible to Earth-bound observers because of cloud cover. The colors represent elevation. The light tan is high, and the dark blue is low.

FIGURE F.11 A photo of Enceladus taken by the Cassini spacecraft. Note the fractures and lack of cratering in the southern hemisphere.

Unsafe Ground: Landslides and Other Mass Movements

What goes up must come down. As a consequence, rock and regolith forming the substrate of hill slopes occasionally give way and slide downslope. The results can be disastrous, as happened when a mudslide buried homes in La Conchita, California.

Geopuzzle

Picture a mountain slope covered by an ancient forest. The next day, the forest along with tens of meters of its substrate (soil and bedrock) are gone, having tumbled downhill to form a massive pile of debris in the river valley below. What triggers such catastrophic landslides? Can they be prevented?

557

16.1 INTRODUCTION

It was Sunday, May 31, 1970, a market day, and thousands of people had crammed into the Andean town of Yungay, Peru, to shop. Suddenly they felt the jolt of an earthquake, strong enough to topple some masonry houses. But worse was to come. This earthquake also broke an 800-m-wide ice slab off the end of a glacier at the top of Nevado Huascarán, a nearby 6.6-km-high mountain peak. Gravity instantly pulled the ice slab down the mountain's steep slopes. As it tumbled down over 3.7 km, the ice disintegrated into a chaotic avalanche of chunks traveling at speeds of over 300 km per hour. Near the base of the mountain, most of the avalanche channeled into a valley and thickened into a moving sheet as high as a ten-story building that ripped up rocks and soil along the way. Friction transformed the ice into water, which when mixed with rock and dust created 50 million cubic meters of mud, a slurry viscous enough to carry boulders larger than houses. This mass, sometimes floating on a compressed air cushion that allowed it to pass without disturbing the grass below, traveled over 14.5 km in less than 4 minutes.

At the mouth of the valley, most of the mass overran the village of Ranrahica before coming to rest and creating a dam that blocked the Santa River. But part of it shot up the sides of the valley and became airborne for several seconds, flying over the ridge bordering Yungay. As the town's inhabitants and visitors stumbled out of earthquake-damaged buildings, they heard a deafening roar and looked up to see the churning mud cloud bursting above the nearby ridge. The town was completely buried under several meters of mud and rock. When the dust had settled, only the top of the church and a few palm trees remained visible to show where Yungay once lay (▶Fig. 16.1a, b)—18,000 people are forever entombed beneath the mass. Today, the site is a grassy meadow with a hummocky (irregular and lumpy) surface, spotted with crosses left by mourning relatives.

Could the Yungay tragedy have been prevented? Perhaps. A few years earlier, climbers had recognized the instability of the glacial ice on Nevado Huascarán, and Peruvian newspapers had published a warning, but alas, no one took notice. In the aftermath of the event, geologists discovered that Yungay had been built on ancient layers of debris, from past avalanches. The government has since prevented new towns from rising in the danger zone.

People often assume that the ground beneath them is *terra firma*, a solid foundation on which they can build their lives. But the catastrophe at Yungay says otherwise. Much of the Earth's surface is unstable and capable of moving downslope in a matter of seconds to weeks. Geologists refer to the gravitationally caused downslope transport of rock, regolith (soil, sediment, and debris), snow, and ice as **mass movement**, or mass wasting. Like earthquakes, volcanic eruptions, storms, and floods, mass movements are a type of **natural hazard**, meaning a natural feature of the environment that

(a) Before

(b) After

FIGURE 16.1 **(a)** Before the May 1970 earthquake, the town of Yungay, Peru, perched on a hill within view of the ice-covered mountain Nevado Huascarán. **(b)** Three months after the earthquake, the town lay buried beneath debris. A landslide scar remains visible on the mountain today.

can cause damage to living organisms and to buildings. Unfortunately, mass movement becomes more of a threat every year, because as the world's population grows, cities expand into areas of unsafe ground. In fact, by some estimates, mass movements may, on average, be *the most costly* natural hazard. But mass movement also plays a critical role in the rock cycle, for it's the first step in the transportation of sediment. And it plays a critical role in the evolution of landscapes: it's the most rapid means of modifying the shapes of slopes.

In this chapter, we look at the types, causes, and consequences of mass movement, and the precautions society can take to protect people and property from its dangers. You might want to consider this information when selecting a site for your home or when voting on land-use propositions for your community.

16.2 TYPES OF MASS MOVEMENT

Though in everyday language people commonly refer to all mass-movement events as landslides, geologists and civil engineers tend to distinguish among different types of mass movement on the basis of four factors: the type of material involved (rock, regolith, or snow and ice), the velocity of the movement (fast, intermediate, or slow), the character of the moving mass (chaotic cloud, slurry, or coherent body), and the environment in which the movement takes place (subaerial or submarine). Below, we look at mass movements that occur on land roughly in order from slow to very fast.

Creep, Solifluction, and Rock Glaciers

In temperate climates, the upper few centimeters of ground freeze during the winter, only to thaw again the following spring. Because water increases in volume by 9.2% when it freezes, the water-saturated soil and underlying fractured rock expand outward, and particles in the regolith move out perpendicular to the slope during the winter. During the spring thaw, water becomes liquid again, and gravity makes the particles sink vertically and thus migrate downslope slightly. This gradual downslope movement of regolith is called **creep**. You can't see creep by staring at a hill slope because it occurs too slowly, but over a period of years creep causes trees, fences, gravestones, walls, and foundations built on a hillside to tilt downslope. Trees that continue to grow after they have been tilted display a pronounced curvature at their base (▶Fig. 16.2a–d).

In Arctic or high-elevation regions, regolith freezes solid to great depth during the winter. In the brief summer thaw, only the uppermost 1 to 3 m of the ground thaws. Since meltwater cannot sink into permanently frozen ground, or permafrost, the melted layer becomes soggy and weak and flows slowly downslope in overlapping sheets.

FIGURE 16.2 **(a)** Creep on hill slopes accompanies the annual freeze-thaw cycle. The clast originally came from the marker bed. It rose perpendicular to the slope when the slope froze and dropped down when the slope thawed. After three years, it has migrated downslope to the position shown. **(b)** As rock layers weather and break up, the resulting debris creeps downslope. **(c)** Soil creep causes walls to bend and crack, building foundations to sink, trees to bend, and power poles and gravestones to tilt. **(d)** Trees that grow in creeping soil gradually develop pronounced curves.

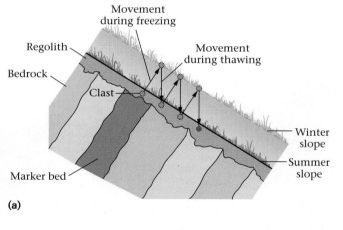

(a)

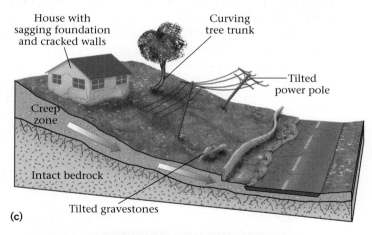

(c)

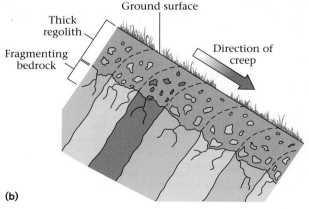

(b)

(d)

Geologists refer to this kind of creep, characteristic of tundra regions (cold, treeless regions), as **solifluction** (▶Fig. 16.3a). Another type of slow mass movement in cold regions takes place in rock glaciers, which consist of a mixture of rock fragments and ice, with the rock fragments making up the major proportion (▶Fig. 16.3b). Rock glaciers develop where the volume of debris falling into a valley equals or exceeds the volume of glacial ice forming from snow.

FIGURE 16.3 (a) Solifluction on a hill slope in the tundra. (b) A rock glacier in Alaska.

(a)

(b)

Slumping

Near Pacific Palisades, along the coast of southern California, Highway 1 runs between the beach and a 120-m-high cliff. Between March 31 and April 3, 1958, a 1-km-long section of the highway disappeared beneath a mass of regolith that had moved down the adjacent cliff. When the movement stopped, the face of the cliff lay 200 m farther inland than it had before. It took weeks for bulldozers to uncover the road. During such slumping, a mass of regolith detaches from its substrate along a spoon-shaped sliding surface and slips semicoherently downslope (▶Fig. 16.4a). We call the moving mass a slump, and the surface on which it slips a **failure surface** (▶Fig. 16.4b). On average, the sea cliffs of southern California retreat (move inland) by up to a few meters a year because of slumping.

The distinct, curving step at the upslope edge of a slump, where the regolith detached, is called a **head scarp**. Immediately below the head scarp, the land surface sinks below its previous elevation. Farther downslope, at the toe, or end, of the slump, the ground elevation rises as the

FIGURE 16.4 (a) A slump on a hillslope. (b) Note the curving failure surface in this slump. The dashed line indicates the slope's shape before slumping.

(a)

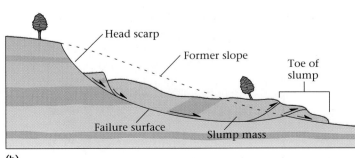

(b)

slump mass rides up and over the preexisting land surface. The toe may break into a series of slices that form curving ridges at the ground surface. Slumps come in all sizes, from only a few meters across to tens of kilometers across. Slumps move at speeds from millimeters per day to tens of meters per minute. They typically break up as they move, and structures (such as houses, patios, and swimming pools) built on them crack and fall apart.

Mudflows and Debris Flows

Rio de Janeiro, Brazil, originally occupied only the flatlands bordering beautiful crescent beaches that had formed between steep hills. But in recent decades, the population has grown so much that the city has expanded up the sides of the hills, and in many places densely populated communities of makeshift shacks cover the slopes. These communities, which have no storm drains, were built on the thick regolith that resulted from long-term weathering of bedrock in Brazil's tropical climate. In 1988, particularly heavy rains saturated the regolith, which turned into a viscous slurry of mud that flowed downslope. Whole communities disappeared overnight, replaced by a hummocky muddle of mud and debris. And at the base of the cliffs, the flowing mud caused high-rise buildings and homes to collapse (▶Fig. 16.5a).

In areas such as the hill-slope communities of Rio, where neither vegetation nor drainage systems protect the ground from rainfall, water mixes with regolith to create a slurry that moves downslope. If the slurry consists of just mud, it's a **mudflow**, but if the mud is mixed with larger rock fragments, it's a **debris flow**. The speed at which mud or debris moves depends on the slope angle and on the water content. Flows move faster if they are more liquid (i.e., less viscous), and if they move on steeper slopes. On a gentle slope, viscous mud flows like molasses, but on a steep slope, low-viscosity mud may move at over 100 km per hour. Because mud and debris flows have greater viscosity than clear water, they can carry large rock chunks, as well as houses and cars. They typically follow channels downslope, and at the base of the slope they spread out into a broad lobe.

Particularly devastating mudflows spill down the river valleys bordering volcanoes. These mudflows, known as **lahars**, consist of a mixture of volcanic ash from a currently erupting or previously erupted pyroclastic cloud, and water from the snow and ice that melts in a volcano's heat or from heavy rains (▶Fig. 16.5b; see Chapter 9). One of the most destructive recent lahars occurred on November 13, 1985, in the Andes Mountains in Colombia. That night, a major eruption melted a volcano's thick snowcap, creating hot water that mixed with ash. A scalding lahar rushed down river valleys and swept over the nearby town of Armero while most inhabitants were asleep (see Fig. 9.22e). Of the 25,000 residents, 20,000 perished.

(a)

(b)

FIGURE 16.5 **(a)** The aftermath of a mudflow in Rio de Janeiro. **(b)** The aftermath of a lahar that flowed down a stream valley after the 1980 Mt. St. Helens volcanic eruption.

Landslides (Rock and Debris Slides)

In the early 1960s, engineers built a huge new dam across a river on the northern side of Monte Toc, in the Italian Alps, to create a reservoir for generating electricity. This dam, the Vaiont Dam, was an engineering marvel, a concrete wall rising 260 m (as high as an 85-story skyscraper) above the valley floor (▶Fig. 16.6a). Unfortunately, the dam's builders did not recognize the hazard posed by nearby Monte Toc. The side of Monte Toc facing the reservoir was underlain with dipping limestone beds interlayered with weak shale beds. These beds dipped parallel to the surface of the mountain and curved under the reservoir (▶Fig. 16.6b). As the reservoir filled, the flank of the mountain cracked, shook, and rumbled. Local residents began to call Monte Toc *la montagna che cammina* ("the mountain that walks").

After several days of rain, Monte Toc began to rumble so much that on October 9, 1963, engineers lowered the water level in the reservoir. They thought the wet ground might slump a little into the reservoir, but no more than that, so no one ordered the evacuation of the town of Longarone, a few kilometers down the valley. Unfortunately, the engineers underestimated the problem. At 10:30 that evening, a huge chunk of Monte Toc—600 million tons of rock—detached from the mountain and slid downslope into the reservoir. Some debris rocketed up the opposite wall of the valley to a height of 260 m above the original reservoir level. The displaced water of the reservoir spilled over the top of the dam and rushed down into the valley below. When the flood had passed, nothing of Longarone and its 1,500 inhabitants remained. Though the dam itself still stands, it holds back only debris and has never provided any electricity.

Geologists refer to such a sudden movement of rock and debris down a nonvertical slope as a **landslide**. If the mass consists only of rock, it may also be called a rock slide (the case in the Vaiont Dam disaster), and if it consists mostly of regolith, it may also be called a debris slide. Once a landslide has taken place, it leaves a landslide scar on the slope and forms a debris pile at the base of the slope.

Slides happen when bedrock and/or regolith detaches from a slope and shoots downhill on a failure surface roughly parallel to the slope surface. Thus, landslides generally occur where a weak layer of rock or sediment at depth below the ground parallels the land surface. (At the Vaiont Dam, the plane of weakness that would become the failure surface was a weak shale bed.) Slides may move at speeds of up to 300 km per hour; they are particularly fast when a cushion of air gets trapped beneath, so there is virtually no friction between the slide and its substrate, and the mass moves like a hovercraft. Rock and debris slides sometimes have enough momentum to climb the opposite side of the valley into which they fall.

Landslides, like slumps, come at a variety of scales. Most are small, involving blocks up to a few meters across. Some, such as the Vaiont slide, are large enough to cause a catastrophe. ▶Figure 16.6c shows a large rock slide in the Uinta Mountains of Utah. A large section of forest that once surrounded the lake is now buried under the debris. Had this slide occurred in a populated area, it could have been devastating.

Avalanches

In the winter of 1999, an unusual weather system passed over the Austrian Alps. First it snowed; then the temperature warmed and the snow began to melt. But then the weather turned cold again, and the melted snow froze into a hard, icy

(a)

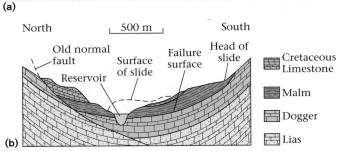

(b)

(c)

FIGURE 16.6 (a) Vaiont Dam and the debris now behind it. The exposed failure surface is visible in the distance. **(b)** A cross section parallel to the face of Vaiont Dam before the landslide. Monte Toc is to the right. Note the failure surface at the base of the weak Malm Shale. The landslide completely filled the reservoir. Its surface after movement is indicated by the dashed line. The names Malm, Dogger, and Lias refer to epochs in the Jurassic Period. **(c)** A large rock slide in the Uinta Mountains, Utah buried a section of forest that once surrounded this lake.

crust. This cold snap ushered in a blizzard that blanketed the ice crust with tens of centimeters (1–2 ft.) of snow. Wind built the snow into huge overhanging drifts, called cornices, on the leeward side of peaks and ridges. Skiers delighted in the bounty of white, but not for long. With the frozen snow layer underneath acting as a failure surface, the heavy layer of new snow began to slide down the mountain, accelerating as it moved and then disintegrating and mixing with air. It became a roaring cloud—an avalanche—traveling at hurricane speeds, and it flattened everything in its path. Trees and ski lodges toppled like toothpicks before the mass of snow finally reached the valley floor and slowed to a halt (▶Fig. 16.7a, b). Unfortunately, many of those who survived the impact succumbed to suffocation in the minutes that followed. The avalanche blocked roads into the region, tragically slowing rescue efforts.

Avalanches are turbulent clouds of debris mixed with air that rush down steep hill slopes at high velocity. If the debris consists of snow, like the Austrian avalanche, it's a snow avalanche. If it consists of fragments of rock and dust, it's a debris avalanche. The moving air-debris mass is denser than clear air. Thus it hugs the ground and acts like an extremely strong and viscous wind that can knock down and blow away anything in its path. As illustrated by the Austrian example, snow avalanches pose a particular threat when frozen snow layers get buried and thus can act as a failure surface for the overlying snow. Typically, avalanches happen again and again in the same area, creating pathways, called avalanche chutes, in which no mature trees grow. Not all snow avalanches are the same—some are clouds of light powder traveling at up to 250 km per hour, whereas some are much slower flows of wet snow. The snow in an avalanche can break off a cornice, or it can start as a huge slab that detaches from its substrate in response to little more than the weight of a single skier or a loud noise. In some cases, avalanches occur because of failure of a glacier. We've already mentioned the example of Yungay, Peru. A more recent example occurred in the Caucasus Mountains of Russia. In September 2002, a 3-million-ton portion of the Maili Glacier detached and charged 16 km (10 miles) down a gorge. It buried the village of Karmadon with up to 170 m (500 ft.) of ice blocks and other debris.

Rockfalls and Debris Falls

Rockfalls and **debris falls**, as their names suggest, occur when a mass free-falls from a cliff (▶Fig. 16.8a). Commonly, rockfalls happen when a rock separates from a cliff face along a joint. Friction and collision with other rocks may bring some blocks to a halt before they reach the bottom of the slope; these blocks pile up to form a **talus**, a sloping apron of rocks along the base of the cliff (Fig. 16.8a). Rock or debris that has fallen a long way can reach speeds of 300 km per hour, and may have so much

(a)

(b)

FIGURE 16.7 (a) Aftermath of the 1999 avalanches in the Austrian Alps. **(b)** Trees that were flattened by an avalanche, now exposed after the snow melted.

momentum that it keeps moving as a debris avalanche when it reaches the base of a cliff. Large, fast rockfalls push the air in front of them, creating a short blast of hurricane-like wind. For example, the wind alone from a 1996 rockfall in Yosemite National Park flattened over 2,000 trees.

Most rockfalls involve only a few blocks detaching from a cliff face and dropping into the talus. But some falls dislodge immense quantities of rock. In September 1881, a 600-m-high crag of slate, undermined by quarrying, suddenly collapsed onto the Swiss town of Elm in a valley of the Swiss Alps. Over 10 million cubic meters of rock fell to the valley floor, burying Elm and its 115 inhabitants to a depth of 10 to 20 m.

FIGURE 16.8 **(a)** Successive rock falls have littered the base of this sandstone cliff with boulders. Note the talus at the base of the cliff. **(b)** A 2005 rockfall along a highway in Oregon.

(a)

(b)

Rockfalls typically take place along steep highway road cuts, leading to the posting of "falling-rock zone" signs (▶Fig. 16.8b). Such rockfalls occur with increasing frequency as the road cut ages, because frost wedging and/or root wedging pries fragments loose, and water infiltrates the outcrop and weakens clay-rich layers.

Submarine Mass Movements

So far, we've focused on mass movements that occur subaerially, for these are the ones we can see and are affected by most. But mass wasting also happens underwater. The sedimentary record contains abundant evidence of submarine mass movements, because after they take place, they tend to be buried by younger sediments and are preserved.

Geologists distinguish three types of submarine mass movements, according to whether the mass remains coherent or disintegrates as it moves (▶Fig. 16.9a-c). The degree to which the mass comes apart during movement typically depends on the amount of water mixed with the sediment in the moving mass. In **submarine slumps**, semicoherent blocks (olistostromes) slip downslope on weak mud detachments (▶Fig. 16.10a). In some cases, the layers constituting the blocks become contorted as they

FIGURE 16.9 Submarine mass movements. **(a)** Slump blocks remain semicoherent as they move. **(b)** In a debris flow, the mass becomes a viscous slurry of chunks floating in a mud matrix. **(c)** In a turbidity current, sediment is suspended like a cloud in water. Submarine slumping and debris flow can generate tsunamis.

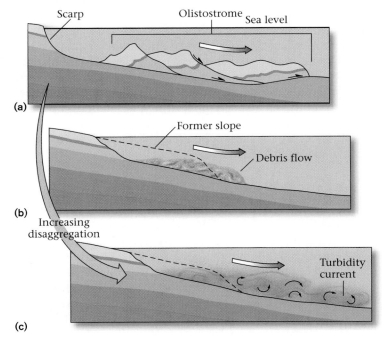

move, like a tablecloth that has slid off a table. In **submarine debris flows**, the moving mass breaks apart to form a slurry containing larger clasts (pebbles to boulders) suspended in a mud matrix. And in **turbidity currents**, sediment disperses in water to create a turbulent cloud of suspended sediment that avalanches downslope. As the turbidity current slows, sediment settles out in sequence from coarse to fine, creating graded beds (see Chapter 7).

In recent years, marine geologists have used a sensitive type of sonar, called GLORIA (geologic long-range inclined asdic [sonar]), to map the sea floor. The instrument "sees" sideways and can map a 60-km-wide swath of the ocean floor all at once. GLORIA maps have documented huge slumps along the margins of Hawaii. These have substantially modified the shape of the island and have created a broad apron of hummocky sea floor around them (▶Fig. 16.10b). Major slumping events, which seem to recur (on average) every 100,000 years, may generate catastrophic tsunamis in the Pacific basin. Marine geologists have mapped similar slumps along the edges of continental shelves, which suggests that tsunamis generated by submarine slumps and debris flows could be a major hazard worldwide (▶Box 16.1). A submarine slide set in motion by a 1998 earthquake in Papua New Guinea generated a tsunami that devastated 40 km of the coast and killed 2,100 people.

> ### Take-Home Message
>
> Geologists distinguish among different kinds of mass movement according to the speed and character of the flow. Slow movements include creep, solifluction, and slumping. Mudflows and debris flows move faster, and avalanches and rockfalls move the fastest. All mass movements are hazards. Submarine occurrences may generate tsunamis.

16.3 WHY DO MASS MOVEMENTS OCCUR?

We've seen that mass movements travel at a range of different velocities, from the slowest (creep) to the faster (slumps, mud and debris flows, and rock and debris slides), to the fastest (snow and debris avalanches, and rock and debris falls—see art, pp. 572–573). The velocity, in turn, depends on the steepness of the slope and the water or air content of the mass. In order for these movements to take place, the stage must be set by the following phenomena: fracturing and weathering, which weaken materials at Earth's surface so that they cannot hold up against the pull of gravity; and the development of relief, which provides slopes down which masses move.

FIGURE 16.10 (a) A digital bathymetric map showing a submarine slump along the coast, as viewed from an oblique angle (image courtesy MBARI). The red areas are higher, and the purple areas are lower. Note that the slump spreads out for several kilometers over the sea floor. **(b)** A map of Hawaii, showing the larger submarine landslides that have formed around the islands. Yellow areas delineate the underwater slides. Dark blue areas are deeper water, and light blue areas are shallower water.

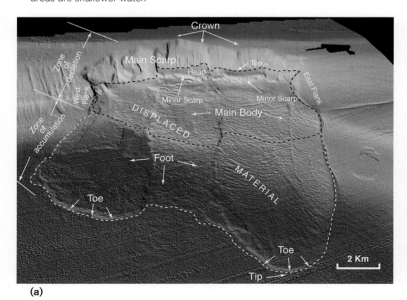

(a)

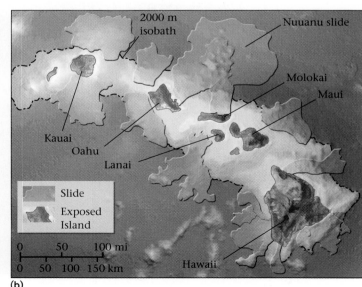

(b)

BOX 16.1 GEOLOGIC CASE STUDY

The Storegga Slide and the North Sea Tsunamis

The Firth of Forth, a long inlet of the North Sea, forms the waterfront of Edinburgh, Scotland. At its western end, it merges with a broad plain in which mud and peat have been accumulating since the last ice age. Around 1865, geologists investigating this sediment discovered an unusual layer of sand containing bashed seashells, marine plankton, and torn-up fragments of substrate. This sand layer lies sandwiched between mud layers at an elevation of up to 4 m above the high-tide limit and 80 km inland from the shore. How could the sand layer have been deposited? The mystery simmered for many decades. During this time, layers of shelly sand, similar to the one found in Scotland, were discovered at many other localities along both sides of the North Sea (▶Fig. 16.11). In some cases, the layer occurred 20 m above the high-tide limit. Geologists studying the coast also found locations where coastal cliffs appeared to have been eroded by wave action at elevations well out of reach of normal storm waves.

While land-based geologists puzzled about these unusual sand beds and coastal erosional features, marine geologists investigating the continental shelf off the western coast of Norway discovered a region of very irregular sea floor underlain with a jumble of chaotic blocks, some of which are 10 km by 30 km across and 200 m thick. When mapped out, they indicate that a 290-km-long sector of the continental shelf had collapsed in a series of at least three submarine slides that together constitute about 5,580 cubic km of debris—the overall feature is called the Storegga Slide. Further studies show that the Storegga Slide formed during three movement events: one occurred 30,000 years ago, the second about 7,950 years ago, and the third about 6,000 years ago. Now the pieces of the puzzle were in place. Slides the size of those constituting the Storegga Slide had displaced enough ocean water to create tsunamis (see Fig. 16.10).

Tsunamis produced by the second Storegga Slide may coincide with the disappearance of Stone Age tribes along the North Sea coast, suggesting the tribes simply washed away. If such a calamity happened in the past, could it happen again, with submarine-slide-generated tsunamis inundating coastal cities with large populations? Geologists now realize that submarine landslides can trigger tsunamis every bit as devastating as earthquake- and volcano-generated tsunamis. A slide in 1929 along the coast of Newfoundland, for example, not only created a turbidity current that broke the trans-Atlantic telephone cable, but also generated tsunamis that washed away houses and boats along the coast of Newfoundland at elevations of up to 27 m above sea level. With a bit of looking, geologists have found huge boulders flung by tsunamis onto the land, layers of sand and gravel deposited well above the high-tide limit, and erosional features high up on shoreline cliffs along many coastal areas, even in areas (such as the Bahamas and southeastern Australia) far from seismic or volcanic regions. And submarine mapping shows that many large slumps occur all along continental shelves. It's no wonder that Edward Bryant, in his recent book on tsunamis, calls them "The Underrated Hazard."

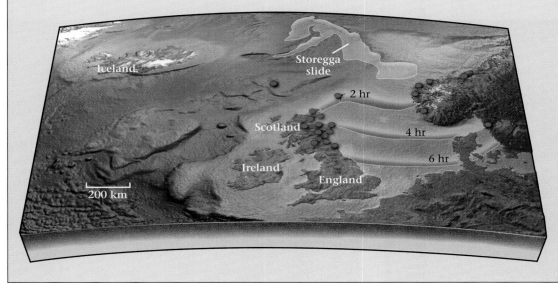

FIGURE 16.11 Map of the North Sea region showing the location of sites (red dots) where marine sand layers occur significantly above the high-tide limit. These were caused by tsunamis generated by movement of the Storegga Slide. The map shows the estimated position of the tsunamis at 2 hours, 4 hours, and 6 hours.

Weakening the Surface: Fragmentation and Weathering

If the Earth's surface were covered by intact (unbroken) rock, mass movements would be of little concern, for intact rock has great strength and could form stalwart mountain faces that would never tumble, even if they were vertical. But the rock of Earth's upper crust has been affected by jointing and faulting, and in many locations the surface has a cover of regolith resulting from the weathering of rock in Earth's corrosive atmosphere. Fragmented rock and regolith are much weaker than intact rock and can indeed collapse in response to Earth's gravitational pull (▶Fig. 16.12). Thus, jointing, faulting, and weathering make mass movements possible.

Why are regolith and fractured rock so much weaker than intact rock? Intact rock is held together by the strong chemical bonds within mineral crystals, by mineral cement, or by the interlocking of grains. In contrast, a joint-bounded or fault-bounded block is held in place only by friction between the block and its surroundings. Regolith is unconsolidated; that is, it consists of unattached grains. Dry regolith holds together because of friction between adjacent grains and/or because weak electrical charges cause grains to attract each other. Slightly wet regolith holds together because of water's surface tension. Surface tension, the phenomenon that makes water form drops, exists because water molecules have a positively charged side and a negatively charged side, so the molecules bond to mineral surfaces and attract each other. (Because of surface tension, damp sand holds together to form a sand castle, whereas dry sand collapses into a shapeless pile.)

Slope Stability: The Battle between Downslope Force and Resistance Force

Mass movements do not take place on all slopes, and even on slopes where such movements are possible, they occur only occasionally. Geologists distinguish between stable slopes, on which sliding is unlikely, and unstable slopes, on which sliding will likely happen. When material starts moving on an unstable slope, we say that slope failure has occurred. Whether a slope fails or not depends on the balance between two forces—the downslope force, caused by gravity, and the resistance force, which inhibits sliding. If the downslope force exceeds the resistance force, the slope fails and mass movement results.

Imagine a block sitting on a slope. We can represent the gravitational attraction between this block and the Earth by an arrow (a vector) that points straight down, toward the Earth's center of gravity. This arrow can be separated into two components—the downslope force parallel to the slope and the normal force perpendicular to the slope. We can represent the resistance force by an arrow pointing uphill. If the downslope force is larger than the resistance force, then the block moves; otherwise, it stays in place (▶Fig. 16.13a, b). Note that for a given mass, the magnitude of the downslope force increases as the slope angle increases, so downslope forces are greater on steeper slopes.

What causes the resistance force? As we saw above, chemical bonds in mineral crystals, cement, and the jigsaw-puzzle-like interlocking of crystals hold intact rock in place, friction holds an unattached block in place, electrical charges and friction hold dry regolith in place, and surface

FIGURE 16.12 Perfectly intact rock is rare at the surface of the Earth. Most outcrops, such as this one on the western coast of Ireland, are highly jointed.

FIGURE 16.13 (a) Gravity, represented by the black arrow, pulls a block toward the center of the Earth. The gravitational force has two components, the downslope force parallel to the slope and the normal force perpendicular to the slope. On gentle slopes, the normal force is larger than the downslope force. The resistance force, caused by friction and represented by an arrow pointing upslope, is larger in this example than the downslope force. **(b)** If the slope angle increases, the normal force becomes smaller than the downslope force. If the downslope force then becomes greater than the resistance force, the block starts to move.

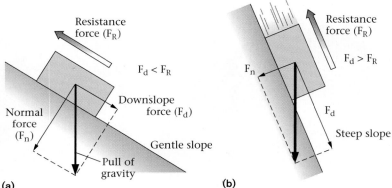

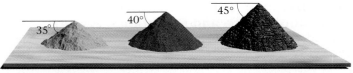

FIGURE 16.14 The angle of repose is the steepest slope that a pile of unconsolidated sediment can have and remain stable. Angles of repose depend on the size and shape of grains. **(a)** Fine, well-rounded sand has a small angle of repose. **(b)** Coarse, angular sand has a larger angle. **(c)** Large, irregularly shaped pebbles have a large angle of repose.

tension holds slightly wet regolith in place. Because of resistance force, granular debris tends to pile up and create the steepest slope it can without collapsing. The angle of this slope is called the **angle of repose**, and for most dry, unconsolidated materials (such as dry sand) it typically has a value of between 30° and 37°. The angle depends partly on the shape and size of grains, which determine the amount of friction across boundaries. For example, larger angles of repose (up to 45°) tend to form on slopes composed of large, irregularly shaped grains, for these grains interlock with each other (▶Fig. 16.14a–c).

In many locations, the resistance force is less than might be expected because a weak surface exists at some depth below ground level. This weak surface separates unstable rock and debris above from the substrate below. If downslope movement begins on the weak surface, we say that failure has occurred and that the weak surface has become a failure surface. Geologists recognize several different kinds of weak surfaces that are likely to become failure surfaces (▶Fig. 16.15a–c). These include: wet clay layers; wet, unconsolidated sand layers; surface-parallel joints (also known as exfoliation joints); weak bedding planes (shale beds and evaporite beds are particularly weak); and metamorphic foliation planes.

Failure surfaces that dip parallel to the slope are particularly likely to fail because the downslope force is parallel to the surface. For example, consider the 1959 landslide that occurred in Madison Canyon, in southwestern Montana. On August 17 of that year, shock waves from a strong earthquake jarred the region. The southern wall of the canyon is underlain with metamorphic rock with a strong foliation that provided a plane of weakness. When the ground vibrated, rock detached along a foliation plane and tumbled downslope. Unfortunately, twenty-eight campers lay sleeping on the valley floor. They were probably awakened by the hurricane-like winds blasting in front of the moving mass, but seconds later were buried under 45 m of rubble.

Fingers on the Trigger: What Causes Slope Failure?

What triggers an individual mass-wasting event? In other words, what causes the balance of forces to change so that the downslope force exceeds the resistance force, and a slope suddenly fails? Here we look at various phenomena—natural and human-made—that trigger slope failure.

Shocks, vibrations, and liquefaction. Earthquake tremors, the passing of large trucks, or blasting in construction sites may cause a mass that was on the verge of moving actually to start moving. For example, an earthquake-triggered slide dumped debris into Lituya Bay, in southeastern Alaska, in 1958. The debris displaced the water in the bay, creating a 300-m-high (1,200 feet) splash that washed the slope on the opposing side of the bay clean of their forest and carried fishing boats many kilometers out to sea. The vibrations of an earthquake break bonds that hold a mass in place and/or cause the mass and the slope to separate slightly, thereby decreasing friction. As a consequence, the resistance force decreases, and the downslope force sets the mass in motion.

FIGURE 16.15 Different kinds of surfaces become failure surfaces in different geologic settings. **(a)** In exfoliated massive granite, exfoliation joints become failure surfaces. **(b)** In sedimentary rock, bedding planes become failure surfaces. **(c)** In metamorphic rock, foliation planes, especially schistosity (the parallel alignment of mica flakes), become failure surfaces.

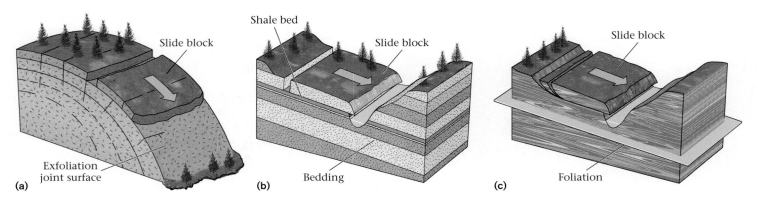

Shaking produces a unique effect in certain types of clay, called quick clay. **Quick clay**, which consists of damp clay flakes, behaves like a solid when still, for surface tension holds water-coated flakes together. But shaking separates the flakes from one another and suspends them in the water, thereby transforming the clay into a slurry that flows like a fluid (▶Fig. 16.16a, b).

Shaking can also cause **liquefaction** of wet sand. The shaking causes the sand grains to try to fit together more tightly, which, in turn, increases the water pressure in the pores between grains, destroying the cohesion between the grains. Without cohesion, the mixture of sand and water turns into a weak slurry.

Changing slope angles, slope loads, and slope support. Factors that make a slope steeper or heavier may cause the slope to fail (▶Fig. 16.17a–c). For example, when a river eats away at the base of the slope, or a contractor excavates at the base of a slope, the slope becomes steeper and the downslope force increases. But while this happens, the resistance force stays the same. Therefore, if the excavation continues, the downslope force will eventually exceed the resistance force and the slope will fail. The same phenomenon occurs when it rains heavily, for the addition of water to regolith not only makes the regolith heavier—thereby increasing the downslope force—but it also weakens failure surfaces, thereby decreasing resistance force. So, as trucks dump loads of waste on the side of a tailing pile, the pile gradually becomes steeper than the angle of repose, and when this happens, collapse becomes inevitable. The largest observed landslide in U.S. history, the Gros Ventre Slide, which took place in 1925 on the flank of Sheep Mountain, near Jackson Hole, Wyoming, illustrates this phenomenon (▶Fig. 16.18a–c). Almost 40 million cubic meters of rock, soil, and forest detached from the side of the mountain and slid 600 m down a slope, filling the valley and creating a 75-m-high natural dam across the Gros Ventre River, for the river itself had removed support.

In retrospect, the geology of the slide area made this landslide almost inevitable. The flank of Sheep Mountain is

a **dip slope**, meaning that bedding parallels the face of the mountain. The Tensleep Formation, the stratigraphic unit exposed at the surface, consists of interbedded sandstone and shale. In the past, a thick sandstone layer spanned the valley and propped up sandstone farther up the side of the mountain. But river erosion cut down through the sand layer to a weak shale (the Amsden Shale) beneath. Rainfall in the weeks before the landslide made the ground heavier than usual and increased the amount of water seeping into the shale layer, weakening it further. By June, the downslope force exceeded the restraining force, and a huge slab of rock upslope broke off and raced downhill, with the wet shale layer acting as a failure surface.

In some cases, excavation results in the formation of an overhang. When such **undercutting** has occurred, rock making up the overhang eventually breaks away from the slope and falls. Overhangs commonly develop above a weak horizontal layer that erodes back preferentially, or along seacoasts and rivers where the water cuts into a fairly strong slope (▶Fig. 16.19a, b).

Changing the slope strength. The stability of a slope depends on the strength of the material constituting it. If the material weakens with time, the slope becomes weaker and eventually collapses. Three factors influence the strength of slopes: weathering, vegetation cover, and water.

With time, chemical weathering produces weaker minerals, and physical weathering breaks rocks apart. Thus, a formerly intact rock composed of strong minerals is transformed into a weaker rock or into regolith.

We've seen that thin films of water create cohesion between grains. Water in larger quantities, though, decreases cohesion, because it fills pore spaces entirely and keeps grains apart. Though slightly damp sand makes a better sand castle than dry sand, a slurry of sand and water can't make a castle at all. Likewise, the saturation of regolith with water during a torrential rainstorm weakens the regolith so much that it may begin to move downslope as a slurry. If the water weakens a specific subsurface layer, then the layer becomes a failure surface. Similarly, if the water table (the top surface of the groundwater layer) rises above a weak failure surface after water has sunk into the ground, overlying rock or regolith may start to slide over the further weakened failure surface.

Water infiltration has a particularly notable effect in regions underlain by swelling clays. These clays possess a mineral structure that allows them to absorb water: water molecules form sheets between layers of silica tetrahedra, which cause clay flakes to swell to several times their original size. Such swelling pushes up the ground surface, making it crack, and weakens the upper layer of the ground, making it susceptible to creep or slip. When the clay dries, it shrinks, and the ground surface subsides. This up-and-down movement is enough to wrinkle road surfaces and crack foundations.

FIGURE 16.16 (a) In a quick clay, before shaking, the grains stick together. **(b)** During shaking, the grains become suspended in water, and the formerly solid mass becomes a movable slurry.

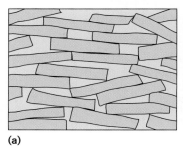

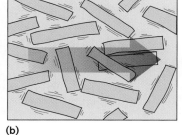

(a) (b)

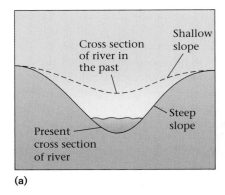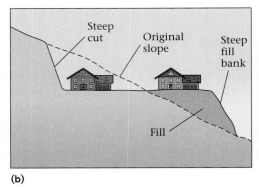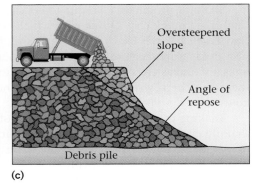

(a) **(b)** **(c)**

FIGURE 16.17 Slope angles may become steeper, making the slopes unstable. **(a)** A river can cut into the base of a slope, steepening the sides of the valley. **(b)** Cutting terraces in a hill slope creates a steeper slope. **(c)** Adding debris to the top of an unconsolidated sediment pile may cause the angle of repose to be exceeded.

FIGURE 16.18 **(a)** The huge Gros Ventre Slide took place after heavy rains had seeped into the ground, weakening the Amsden Shale and making the overlying Tensleep Formation heavier. The slope was already unstable because the Gros Ventre River had cut down to the shale, and the bedding planes dipped parallel to the slope. **(b)** The motion begins. **(c)** After the slide moved, it filled the river valley and dammed the river, creating a lake. A huge landslide scar formed on the hill slope. **(d)** Photo of the Gros Ventre Slide.

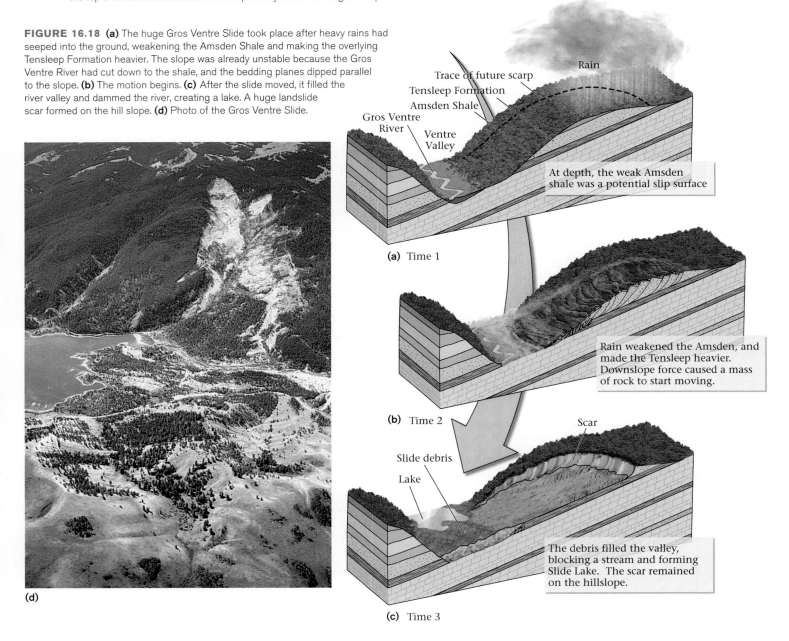

Rain

Trace of future scarp
Tensleep Formation
Amsden Shale
Gros Ventre River
Ventre Valley

At depth, the weak Amsden shale was a potential slip surface

(a) Time 1

Rain weakened the Amsden, and made the Tensleep heavier. Downslope force caused a mass of rock to start moving.

(b) Time 2

Slide debris
Lake
Scar

The debris filled the valley, blocking a stream and forming Slide Lake. The scar remained on the hillslope.

(c) Time 3

(d)

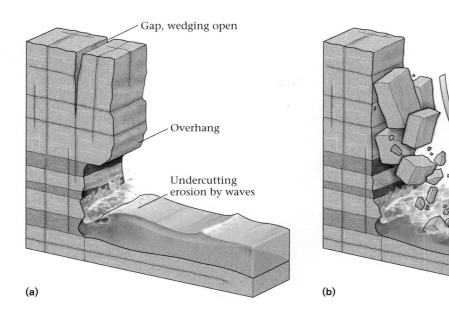

Gap, wedging open

Overhang

Undercutting
erosion by waves

(a)

Rock fall

(b)

FIGURE 16.19 (a) Undercutting by waves removes the support beneath an overhang. (b) Eventually, the overhang breaks off along joints, and a rockfall takes place.

In the case of slopes underlain with regolith, vegetation tends to strengthen the slope, because the roots hold otherwise unconsolidated grains together. Also, plants absorb water from the ground, thus keeping it from turning into slippery mud. The removal of vegetation therefore has the net result of making slopes more susceptible to downslope mass movement. In 2003, terrifying wildfires, stoked by strong winds, destroyed the ground-covering vegetation in many areas of California. When heavy rains followed, the barren ground of this hilly region became saturated with water and turned into mud, which then flowed downslope, damaging and destroying many homes and roads. Deforestation in tropical rain forests, similarly, leads to catastrophic mass wasting of the forest's substrate (▶Fig. 16.20).

> **Take-Home Message**
>
> Weathering and fragmentation weaken slope materials and make them more susceptible to mass movement. Failure occurs when downslope pull exceeds the resistance force. This may happen due to shocks, changing slope angles and strength, and changing slope support.

16.4 PLATE TECTONICS AND MASS MOVEMENTS

The Importance of the Tectonic Setting

Most unstable ground on Earth ultimately owes its existence to the activity of plate tectonics. As we've seen, plate tectonics causes uplift, generates relief, and causes

FIGURE 16.20 Deforestation makes slopes more susceptible to mass movement, as shown in this example from Puebla, Mexico. The slide destroyed the small village of Acalama, killing all but 30 of its 150—200 residents.

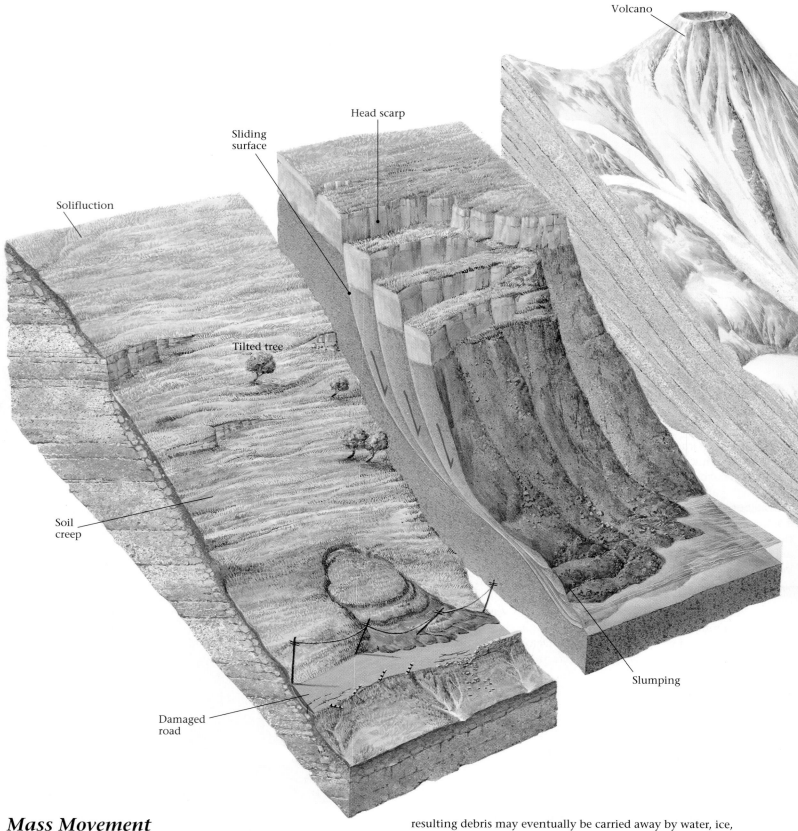

Solifluction

Sliding surface

Head scarp

Volcano

Tilted tree

Soil creep

Damaged road

Slumping

Mass Movement

In Earth's gravity field, what goes up must come down—sometimes with disastrous consequences. Rock and regolith are not infinitely strong, so every now and then slopes or cliffs give way in response to gravity, and materials slide, tumble, or career downslope. This downslope movement, called mass movement, or mass wasting, is the first step in the process of erosion and sediment formation. The resulting debris may eventually be carried away by water, ice, or wind.

The kind of mass wasting that takes place at a given location reflects the composition of the slope (is it composed of weak soil, loose rock, or hard rock containing joints?), the steepness of the slope, and the climate (is the slope wet or dry, frozen or unfrozen?). Stronger rocks can hold up steep cliffs, but with time, rock breaks free along joints and tumbles or slides

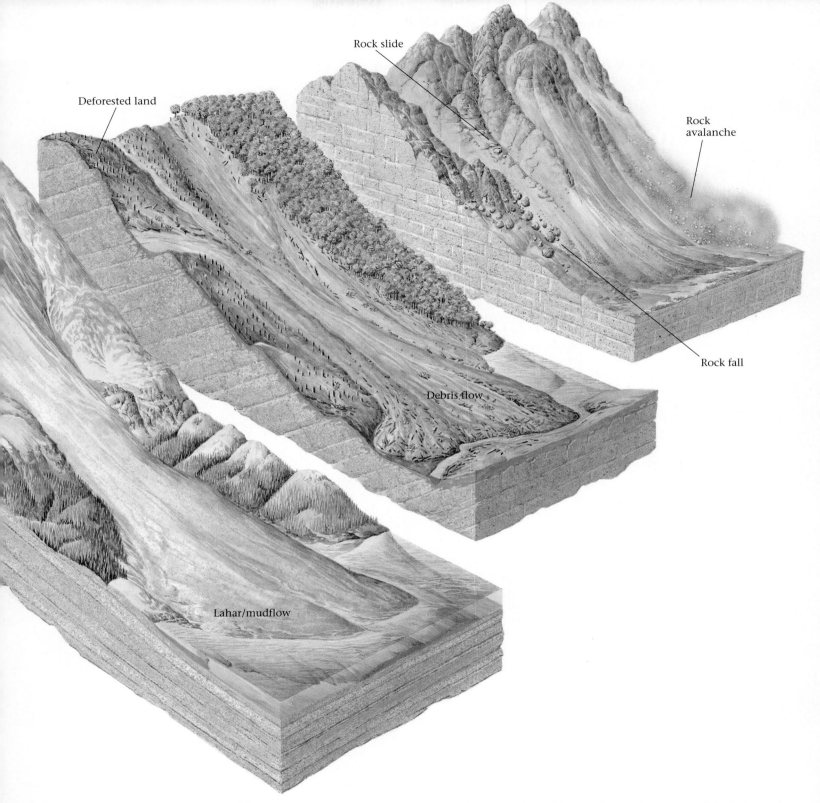

Rock slide

Deforested land

Rock avalanche

Rock fall

Debris flow

Lahar/mudflow

down weak surfaces. Coherent regolith may slowly slide down slopes, whereas water-saturated regolith may flow rapidly. Episodes of mass movement may be triggered by an oversteepened slope (when a river has cut away at the base of a cliff), a heavy rainfall that saturates the slope, an earthquake that shakes debris free, or a volcanic eruption, which not only shakes the ground but melts snow and ice to saturate regolith.

Geologists classify mass-wasting events by the rate and character of the movement. Soil creep accompanies seasonal freezing and thawing, which causes soil gradually to migrate downslope; if it creeps over a frozen substrate, it's called

solifluction. Slumping involves semicoherent slices of earth that move slowly down spoon-shaped sliding surfaces, leaving behind a head scarp. Mudflows and debris flows happen where regolith has become saturated with water and moves downslope as a slurry. When volcanoes erupt and melt ice and snow at their summit, or if heavy rains fall during an eruption, water mixes with ash, creating a fast-moving lahar. Steep, rocky cliffs may suddenly give way in rockfalls. If the rock breaks up into a cloud of debris that rushes downslope at high velocity, it is a rock avalanche. Snow avalanches are similar, but the debris consists only of snow.

See for yourself...
Examples of Landslides

Landslides cause distinctive scars on the Earth's surface, such as head scarps, hummocky ground, and/or disrupted vegetation. Examples of these features occur worldwide. The thumbnail images provided on this page are only to help identify tour sites. Go to wwnorton.com/studyspace to experience this flyover tour.

Slumping Hawaii
(Lat 19°18'21.37"N, Long 155°5'2.55"W)

These coordinates take you to Volcanoes National Park, along the south coast of the Big Island of Hawaii, where active lava flows reach the sea. From an altitude of 5 km (3 miles), you can see black lava flows that descended over the Holei Pali cliff (Image G16.1). The 250 m (820 foot)-high cliff is the head scarp of the Hilina slump. Here, we see only the upper 2.5 km (1.5 miles) of the slump on land. The rest is submerged and extends about 40 km (25 miles) offshore. Tilt your view to look east along the scarp (Image G16.2).

G16.1   G16.2

1964 Slumps, Anchorage, Alaska
(Lat 61°12'52.06"N, Long 149°54'31.39"W)

During the 1964 earthquake, several coastal slumps moved. Fly to the coordinates given and zoom to 2 km (1.2 miles)—you are hovering over downtown Anchorage (Image G16.3). Note the curving head scarps near the shore marking slumps that sank by about 7 m (21 feet) during the 1964 quake. Fly SW along the coast until you see the airport. At Lat 61°11'51.51"N, Long 149°58'25.11"W you'll find a crescent-shaped band of woods along the shore (Image G16.4). This is Earthquake Park, what's left of the Turnagain Heights neighborhood, destroyed by 1964 slumping.

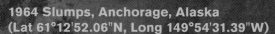

G16.3   G16.4

Roadside Landslide, Pacific Coast Highway, California
(Lat 34°3'31.44"N, Long 118°58'15.85"W)

Portions of the Pacific Coast Highway lie at the base of unstable slopes—landslides down these slopes block the road. At the coordinates provided, from an elevation of 800 m (2,600 feet), you can see the subtle scar left by one of these landslides, and the remnants of debris cleared by bulldozers along the base of the cliff (Image G16.5).

G16.5

Portuguese Bend Landslide, California
(Lat 33°44'46.94"N, Long 118°22'7.83"W)

The land here started to slump about 37,000 years ago. At the coordinates given, from an altitude of 5 km (3 miles), you can see the head scarp of the 3 km (1.8 mile)-wide slump (Image G16.6). In the 1950s, developers built a housing project on the hummocky land of the slump. The southeastern portion of the slump began moving again, ultimately destroying 150 homes.

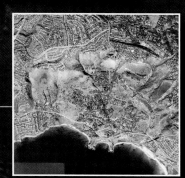

G16.6

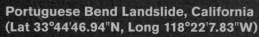

G16.7

G16.8

La Conchita Mudslide, California (Lat 34°21'50.29"N, Long 119°26'46.85"W)

At La Conchita, developers built a housing project in the narrow strip of land between the cliffs and the shore. When heavy rains drench this landscape, the slope becomes unstable. In 1995 and again in 2005, mudflows buried homes. From an elevation of 2 km (1.2 miles), you can see the development (Image G16.7). Note that the cliff bounds a broad, uplifted terrace. This terrace was a wave-cut platform formed at sea level. Tectonic movements caused the terrace to rise. In addition to the mudslide, which also destroyed a road traversing the hill, you can see gullies and canyons that have been incised by successive floods.

Mt. Saint Helens Lahars, Washington (Lat 46°10'36.32"N, Long 122°10'4.44"W)

Fly to these coordinates, and you are over the southeast flank of Mt. Saint Helens. Zoom to an elevation of 12 km (7.5 miles), tilt your view, and pivot to look NW (Image G16.8). The southeast flank was not destroyed by the 1980 explosion, but rather was the site of numerous lahars. The gray lahars look like spills of gravy down the side of the mountain. In the stream valleys, the lahars traveled much farther.

Gros Ventre Slide, Wyoming (Lat 43°37'29.78"N, Long 110°33'3.49"W)

Fly to the coordinates provided. From an elevation of 10 km (6 miles), you can see Lower Slide Lake, northeast of Jackson, Wyoming (Image G16.9). This lake formed in 1925, when the Gros Ventre slide tumbled down Sheep Mountain and blocked the Gros Ventre River. Zoom to 6 km (3.7 miles), tilt your view, and look southeast (Image G16.10). The scarp left by the landslide is obvious, as is the hummocky ground on top of the debris.

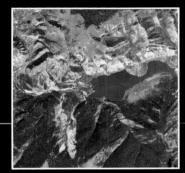

G16.9

G16.10

Debris Fall, Yungay, Peru (Lat 9°8'53.88"S, Long 77°43'20.35"W)

Remnants of the debris fall of 1970 that buried parts of Yungay, Peru, can still be seen decades later. Fly to the latitude and longitude provided, zoom to 15 km (9 miles), tilt to see the horizon, and look NE (Image G16.11). Note the towering, ice-covered peak of Nevado Huascarán in the far distance, and the valley down which the debris flow traveled in the middle distance. In the foreground, the broad apron of debris is now farmed, but it does not host many homes.

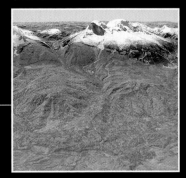

G16.11

Rockfalls, Canyonlands National Park, Utah (Lat 38°29'52.65"N, Long 110°0'50.20"W)

In southeastern Utah, at the coordinates provided, you can see evidence of rockfalls along the banks of the Green River. Zoom to an elevation of 3 km (1.8 miles) and look down (Image G16.12). Note that NW-trending joints cut the whitish, resistant Permian sandstone layer that forms the top of the mesa. When the cliff breaks away along a joint, blocks topple down the slope and break up. Zoom down to 2 km (1.2 miles), tilt to see the horizon, and pivot so you are looking east (Image G16.13). You can see a large rockfall that sent debris almost down to the bottom of the slope.

G16.12

G16.13

Images provided by *Google Earth*™ mapping service/DigitalGlobe, TerraMetrics, NASA, Europa Technologies—copyright 2008.

faulting, which fragments the crust. And, of course, earthquakes on plate boundaries trigger devastating landslides. Spend a day along the steep slopes of the Alpine Fault, a plate boundary that transects New Zealand, and you can *hear* mass movement in progress: during heavy rains, rockfalls and landslides clatter with astounding frequency, as if the mountains were falling down around you. To see the interplay of plate tectonics and other factors, let's consider the example of mass movements in southern California.

A Case Study: Slumping in Southern California

Southern Californians pay immense prices for the privilege of building homes on cliffs overlooking the Pacific. The sunset views from their backyard patios are spectacular. But the landscape is not ideal from the standpoint of stability. Slumps and mudflows on coastal cliffs have consumed many homes over the years, with a cost to their owners (or insurance companies) of untold millions of dollars (see Fig. 10.30a). What is special about southern California that makes it so susceptible to mass wasting?

First, California lies along an active plate boundary. The coast borders the San Andreas fault, a transform fault accommodating the northward movement of the Pacific Plate with respect to North America. Faulting has shattered the rock of California's crust. The fractures not only act as planes of weakness, they also provide paths for water to seep into bedrock and cause chemical weathering. The resulting clay and other slippery minerals further weaken the rock. Also, the rocks in many areas are weak to begin with, because they formed as part of an accretionary prism, a chaotic mass of clay-rich sediment that was scraped off subducting oceanic lithosphere during the Mesozoic Era. Though most of the movement between the North American and Pacific plates involves strike-slip displacement on the San Andreas fault, there is a component of compression across the fault. This compression leads to uplift and slope formation. Since the uplifted region borders the coast, wave erosion steepens and in some places undercuts cliffs. And, because it is a plate boundary, numerous earthquakes rock the region, thus shaking regolith loose.

California is also susceptible to mass movements because of its climate. In general, the region is hot and dry and thus supports only semidesert flora. Brush fires remove much of this cover, leaving large areas with no dense vegetation. But since the region lies on the West Coast, it endures occasional heavy winter rains. The water sinks quickly into the sparsely vegetated ground, adds weight to the mass on the slope, and weakens failure surfaces.

Development of cities and suburbs is the final factor that triggers mass movements in southern California. Development has oversteepened and overloaded slopes, and has caused the water content of regolith to change. The consequences can be seen in an event called the Portuguese Bend Slide in the Palos Verdes area near Los Angeles. The Portuguese Bend region borders the Pacific coast and is underlain with a thick, seaward-dipping layer of weak volcanic ash (now altered to weak clay) resting on the shale. The land slopes down to the sea and, as a result, the weak ash acts as a failure plane. Downslope movement initiated on this glide plane 37,000 years ago. Movement began again in 1956 in response to development. To provide a foundation for homes and roads, developers deposited a 23-m-thick layer of fill over the ground surface. Residents began to water their lawns and to use septic tanks that were susceptible to leaking. The water seeped into the ground and decreased the strength of the ash layer. Because of the decrease in strength, the added weight, and the erosion of the toe of the hill by the sea, the upper 30 m of land began to move. Between 1956 and 1985, the Portuguese Bend Slide moved at rates of up to 2.5 cm per day. Eventually, portions of a 260-acre region slid by over 200 m, and, in the process, over 150 homes were destroyed (▶Fig. 16.21).

> **Take-Home Message**
>
> Were it not for plate tectonics, Earth's surface would show far less relief. Tectonic movements result in the formation of the slopes down which mass movements occur. Plate motions also set the stage for earthquakes, which, in turn, trigger mass-movement events.

FIGURE 16.21 The Portuguese Bend Slide viewed from the air.

16.5 HOW CAN WE PROTECT AGAINST MASS-MOVEMENT DISASTERS?

Identifying Regions at Risk

Clearly, landslides, mudflows, and slumps are natural hazards we cannot ignore. Too many of us live in areas where mass wasting has the potential to kill people and destroy property. In many cases, the best solution is avoidance: don't build, live, or work in an area where mass movement will take place. But avoidance is only possible if we know where the hazards are.

To pinpoint dangerous regions, geologists look for landforms known to result from mass movements, for where these movements have happened in the past, they might happen again in the future. For example, the Portuguese Bend Slide occurred on top of at least two other slides that had happened in the past several thousand years. Features such as slump head scarps, swaths of forest in which trees have been flattened and point downslope, piles of loose debris at the base of hills, and hummocky land surfaces all indicate recent mass wasting.

Geologists may also be able to detect regions that are beginning to move (▶Fig. 16.22). For example, roads, buildings, and pipes begin to crack over unstable ground. Power lines may be too tight or too loose because the poles to which they are attached move together or apart. Visible cracks form on the ground at the potential head of a slump, while the ground may bulge up at the toe of the slump. Subsurface cracks may drain the water from an area and kill off vegetation; another area may sink and form a swamp. Slow movements cause trees to develop pronounced curves at their base.

In some cases, the activity of land masses moving too slowly to be perceptible to people can be documented with sensitive surveying techniques that can detect a subtle tilt

FIGURE 16.22 The features shown here indicate that a large slump is beginning to develop. Note the cracks at the site of the growing head scarp, which drain water and kill trees. Power-line poles crossing the unstable ground bend, and the lines become overtight. Fences and roads that straddle the scarp begin to break up. Houses that straddle the scarp begin to crack, and their foundations sink.

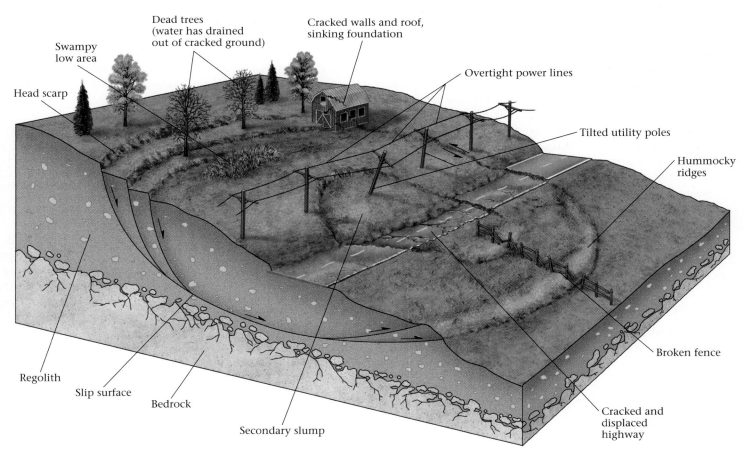

of the ground or changes in distance between nearby points. Specifically, measurements with satellite data (GPS and LIDAR) permit geologists to rapidly identify movements of just a few millimeters that may indicate the reactivation of a slump. (GPS, which stands for "global positioning satellite," allows accurate surveying and can indicate whether a point on the surface of the Earth has moved during the time between two measurements. LIDAR, which stands for "light detection and ranging," uses laser beams to measure the shape of the land and produces very accurate digital elevation maps. Changes in the shape of the land surface shown on such maps provides evidence of movement.)

If various clues indicate that a land mass is beginning to move, and if conditions make accelerating movement likely (e.g., persistent rain, rising floodwaters, or continuing earthquake aftershocks), then officials may order an evacuation. Evacuations have saved lives, and ignored warnings have cost lives. But unfortunately, some mass movements happen without any warning, and some evacuations prove costly but unnecessary.

Even if there is no evidence of recent movement, a danger may still exist: just because a steep slope hasn't collapsed in the recent past doesn't mean it won't in the future. In recent years, geologists have begun to identify such potential hazards (by using computer programs that evaluate factors that trigger mass wasting) and create maps that portray the degree of risk for a certain location. These factors include the following: slope steepness; strength of substrate; degree of water saturation; orientation of bedding, joints, or foliation relative to the slope; nature of vegetation cover; potential for heavy rains; potential for undercutting to occur; and likelihood of earthquakes. From such hazard-assessment studies, geologists compile **landslide-potential maps**, which rank regions according to the likelihood that a mass movement will occur. In any case, common sense suggests that you should avoid building on or below particularly dangerous slide-prone slopes. In Japan, regulations on where to build in regions susceptible to mass wasting, careful monitoring of ground movements, and well-designed evacuation plans have drastically reduced property damage and the number of fatalities.

Preventing Mass Movements

In areas where a hazard exists, people can take certain steps to remedy the problem and stabilize the slope (▶Fig. 16.23a–h).

- *Revegetation:* Since bare ground is much more vulnerable to downslope movement than vegetated ground, stability in deforested areas will be greatly enhanced if owners replant the region with vegetation that sends down deep roots.

- *Regrading:* An oversteepened slope can be regraded or terraced so that it does not exceed the angle of repose.

- *Reducing subsurface water:* Because water weakens material beneath a slope and adds weight to the slope, an unstable situation may be remedied either by improving drainage so that water does not enter the subsurface in the first place, or by removing water from the ground.

- *Preventing undercutting:* In places where a river undercuts a cliff face, engineers can divert the river. Similarly, along coastal regions they may build an offshore breakwater or pile **riprap** (loose boulders or concrete) along the beach to absorb wave energy before it strikes the cliff face.

- *Constructing safety structures:* In some cases, the best way to prevent mass wasting is to build a structure that stabilizes a potentially unstable slope or protects a region downslope from debris if a mass movement does occur. For example, civil engineers can build retaining walls or bolt loose slabs of rock to more coherent masses in the substrate in order to stabilize highway embankments. The danger from rock falls can be decreased by covering a road cut with chainlink fencing or by spraying road cuts with "shotcrete," a cement that coats the wall and prevents water infiltration and consequent freezing and thawing. Highways at the base of an avalanche chute can be covered by an avalanche shed, whose roof keeps debris off the road.

- *Controlled blasting of unstable slopes:* When it is clear that unstable ground threatens a particular region, the best solution may be to blast the unstable ground or snow loose at a time when its movement can do no harm.

Clearly, the cost of preventing mass-wasting calamities is high, and people might not always be willing to pay the price. In such cases, they have a choice of avoiding the risky area, taking the chance that a calamity will not happen while they are around, buying appropriate insurance, or counting on relief agencies to help if disaster does strike. Once again, geology and society cross paths.

> **Take-Home Message**
>
> Various features of the landscape, as well as detailed measurements by satellites, may help geologists to identify unstable slopes and send out warnings. Systematic study, in fact, allows production of landslide-potential maps. Engineers may use a variety of techniques to stabilize slopes physically.

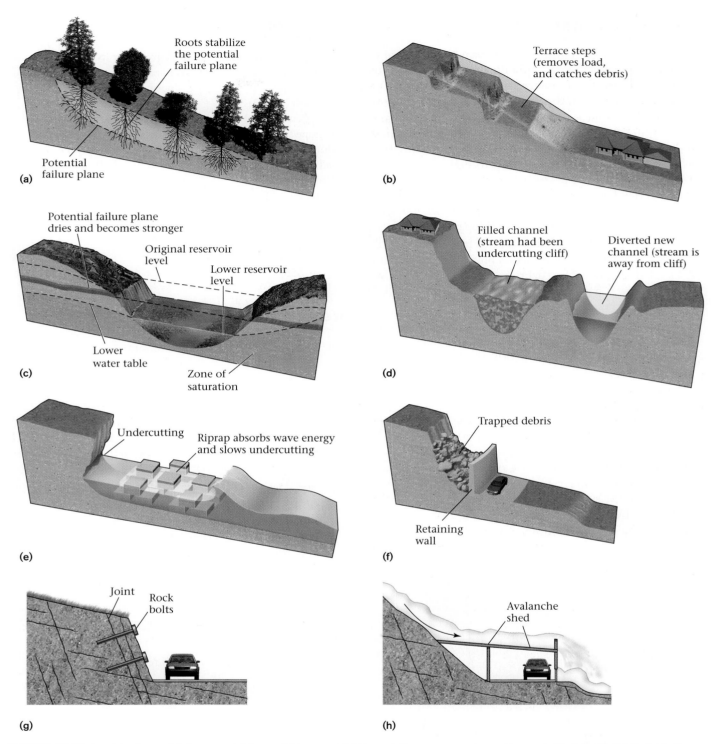

FIGURE 16.23 A variety of remedial steps can stabilize unstable ground. **(a)** Revegetation removes water, and tree roots bind regolith. **(b)** Redistributing the mass on a slope can stabilize it. Terracing can help catch debris. **(c)** Lowering the level of the water table may strengthen a potential failure surface. **(d)** Relocating a river channel can prevent undercutting. **(e)** Adding riprap can slow undercutting of coastal cliffs; **(f)** A retaining wall can trap falling rock. **(g)** Bolting or screening a cliff face can hold loose rocks in place. **(h)** An avalanche shed diverts debris or snow over a roadway.

Chapter Summary

- Rock or regolith on unstable slopes has the potential to move downslope under the influence of gravity. This process, called mass movement, or mass wasting, plays an important role in the erosion of hills and mountains.

- Slow mass movement, caused by the freezing and thawing of regolith, is called creep. In places where slopes are underlain with permafrost, solufluction causes a melted layer of regolith to flow down slopes. During slumping, a semicoherent mass of material moves down a spoon-shaped failure surface. Mudflows and debris flows occur where regolith has become saturated with water and moves downslope as a slurry.

- Landslides (rock and debris slides) move very rapidly down a slope; the rock or debris breaks apart and tumbles. During avalanches, debris mixes with air and moves downslope as a turbulent cloud. And in a debris fall or rock fall, the material free-falls down a vertical cliff.

- Intact, fresh rock is too strong to undergo mass movement. Thus, for mass movement to be possible, rock must be weakened by fracturing (joint formation) or weathering.

- Unstable slopes start to move when the downslope force exceeds the resistance force that holds material in place. The steepest angle at which a slope of unconsolidated material can remain without collapsing is the angle of repose.

- Downslope movement can be triggered by shocks and vibrations, a change in the steepness of a slope, a change in the strength of a slope, deforestation, weathering, or heavy rain.

- Geologists produce landslide-potential maps to identify areas susceptible to mass movement. Engineers can help prevent mass movements using a variety of techniques.

Geopuzzle Revisited

Gravity constantly applies a downslope force. For a time, the strength of material making up the substrate of a slope may be strong enough to resist this relentless pull. But heavy rains, ground shaking, undercutting, deforestation, and/or a change in the water table depth can destabilize a slope until it finally gives way and a landslide takes place. Roots, retaining walls, and other natural or human-built features may delay a landslide, but in the context of geologic time, gravity always wins. Mass wasting events, such as landslides, contribute to the erosion of uplifted land on islands and continents. They can also take place under the sea.

Key Terms

angle of repose (p. 568)	mudflow (p. 561)
avalanche (p. 563)	natural hazard (p. 558)
creep (p. 559)	quick clay (p. 569)
debris fall (p. 563)	riprap (p. 578)
debris flow (p. 561)	rockfall (p. 563)
dip slope (p. 569)	solifluction (p. 560)
failure surface (p. 560)	submarine debris flow
head scarp (p. 560)	(p. 564)
lahar (p. 561)	submarine slump (p. 564)
landslide (p. 562)	talus (p. 563)
landslide-potential maps	turbidity current (p. 565)
(p. 578)	undercutting (p. 569)
liquefaction (p. 569)	
mass movement (wasting)	
(p. 558)	

Review Questions

1. What factors distinguish the various types of mass movement?

2. How does a slump differ from creep? How does it differ from a mudflow or debris flow?

3. How does a rock or debris slide differ from a slump? What conditions trigger a snow avalanche?

4. How are submarine slumps similar to those above water? How might they be related to tsunamis?

5. How does a small amount of water between grains hold material together? How does this change when the sediment is oversaturated?

6. What force is responsible for downslope movement? What force helps resist that movement?

7. How does the angle of repose change with grain size? How does it change with water content?

8. What factors trigger downslope movement?

9. How do geologists predict whether an area is susceptible to mass wasting?

10. What steps can people take to reduce the risk of mass wasting?

On Further Thought

1. Imagine that you have been asked by the World Bank to determine whether or not it makes sense to build a dam in a steep-sided, east-west-trending valley in a small central Asian nation. The local government has lobbied for the dam, because the climate of the country has gradually been getting drier, and the farms of the area are running out of

water. The World Bank is considering making a loan to finance construction of the dam, a process that would employ thousands of now-jobless people. Initial investigation shows that the rock of the valley floor consists of schist containing a strong foliation that dips south. Outcrop studies reveal that abundant fractures occur in the schist along the valley floor; the surface of most fractures are coated with slickensides. Moderate earthquakes have rattled the region. What would you advise the bank? Explain the hazards and what might happen if the reservoir were filled.

2. Marine geologists have been studying the nature and distribution of submarine slumps around the world in order to understand the tsunami threat that submarine slumping poses. Initial results indicate that submarine slumps occur both along active margins (continental margins that coincide with convergent plate boundaries) and along passive margins. Slumps occur much more frequently, but are smaller, along convergent plate boundaries than along passive margins. Suggest an explanation for this observation. Is there a significant tsunami threat in the Atlantic Ocean?

Suggested Reading

Brabb, E. E., and B. L. Harrod, eds. 1989. *Landslides: Extent and Economic Significance*. Brookfield, Va.: Balkema.

Cornforth, D. 2005. *Landslides in Practice: Investigation, Analysis, and Remedial/Preventative Options in Soils*. New York: Wiley.

Costa, J. E., and G. F. Wieczorek. 1987. *Reviews in Engineering Geology*. Vol. 7, *Debris Flows, Avalanches: Process, Recognition, and Mitigation*. Boulder, Colo.: Geological Society of America.

Crozier, M. J. 1986. *Landslides: Causes, Consequences, and Environment*. Dover, N.H.: Croom Helm.

Dikau, R., D. Brunsden, and L. Schrott, eds. 1996. *Landslide Recognition: Identification, Movement, and Causes*. Chichester, England: Wiley.

Evans, S. G., and J. V. Degraff, eds. 2003. *Catastrophic Landslides: Effects, Occurrence, and Mechanisms*. Boulder, Colo.: Geological Society of America.

Glade, T., and Crozier, M. J., 2005. *Landslide Hazard and Risk*. New York: Wiley.

Slosson, J. E., A. G. Keene, and J. A. Johnson, eds. 1993. *Reviews in Engineering Geology*. Vol. 9, *Landslides/Landslide Mitigation*. Boulder, Colo.: Geological Society of America.

Voight, B., ed. 1978. *Rockslides and Avalanches*. Vol. 1, *Natural Processes*. New York: Elsevier.

Zaruba, Q., and V. Mencl. 1969. *Landslides and Their Control*. New York: Elsevier.

THE VIEW FROM SPACE The Earth is not the only home to landslides. Detailed imagery from NASA's Viking Orbiter satellite has revealed the consequences of huge landslides on the surface of Mars. In this view, the *Ganges Chasma* landslide has taken a bite out of the plateau that borders the south wall of *Valles Marineris*, and has spread debris in a huge fan. Note how the landslide took part of a meteorite crater with it. This view is 60 km across.

Streams and Floods: The Geology of Running Water

Geopuzzle

Why do stream networks develop, and why do streams sometimes overflow their channels and flood the surrounding landscape?

Water that falls on land drains back to the sea via rivers. Here, the Niagara River drops dramatically over Niagara Falls, giving a visible display of the power of running water.

As many fresh streams meet in one salt sea, as many lines close in the dial's center, so may a thousand actions, once afoot, end in one purpose.

—William Shakespeare (1564–1616)

17.1 INTRODUCTION

By the 1880s, Johnstown, built along the Conemaugh River in scenic western Pennsylvania, had become a significant industrial town with numerous steel-making factories. Recognizing the attraction of the region as a summer retreat from the heat and pollution of nearby Pittsburgh, speculators built a mud and gravel dam across the river, upstream of Johnstown, to trap a pleasant reservoir of cool water. A group of industrialists and bankers bought the reservoir and established the exclusive South Fork Hunting and Fishing Club, a cluster of lavish fifteen-room "cottages" on the shore. Unfortunately, the dam had been poorly designed, and debris blocked its spillway (a passageway for surplus water), setting the stage for a monumental tragedy. On May 31, 1889, torrential rain drenched Pennsylvania, and the reservoir surface rose until water flowed over the dam and down its face. Despite frantic attempts to strengthen the dam, the soggy structure abruptly collapsed, and the reservoir emptied into the Conemaugh River Valley. A 20-m-high wall of water roared downstream and slammed into Johnstown, transforming bridges and buildings into twisted wreckage (▶Fig. 17.1). When the water subsided, 2,300 people lay dead, and Johnstown became the focus of national sympathy. Clara Barton mobilized the recently founded Red Cross, which set to work building dormitories, and citizens nationwide donated everything from clothes to beds. Nevertheless, it took years for the town to recover, and many residents simply picked up and left. Despite several lawsuits, no one payed a penny of restitution, but the South Fork Hunting and Fishing Club abandoned its property.

The unlucky inhabitants of Johnstown had experienced the immense power of **running water**, water that flows down the surface of sloping land in response to the pull of gravity. Geologists use the term **stream** for any channelized body of running water, meaning water that flows along a **channel**, an elongate depression or trough. In everyday English, however, we also refer to large streams as rivers and small ones as creeks. Streams drain water from the landscape and carry it into lakes or to the sea, much as culverts drain water from a parking lot. In the process, streams relentlessly erode the landscape and trans-

FIGURE 17.1 During the disastrous 1889 flood in Johnstown, Pennsylvania, the force of the water was able to move and tumble sturdy buildings.

port sediment and debris to sites of deposition. Generally, a stream stays within the confines of its channel, but when the supply of water entering a stream exceeds the channel's capacity, water spills out and covers the surrounding land, thereby causing a **flood**, such as the one that washed away Johnstown.

Earth is the only planet in the Solar System that currently hosts flowing streams; Mars probably did in the past, but most if not all surface water on the red planet dried up long ago (see Interlude F). Streams are of great importance to human society, not only because of how they modify the landscape—during normal flow but especially during floods—but also because they provide avenues for travel and commerce, nutrients and sediments for agriculture, water for irrigation and consumption, and sources for power. In this chapter, we examine how streams operate in the Earth System. First we learn about the origin of running water and about the architecture of streams and stream networks. Then we look at the process of stream erosion and deposition and at the landscapes that form in response to these processes. Finally, we consider the nature and consequences of flooding.

17.2 DRAINING THE LAND

Where does the Water in Streams Come From?

Stand by a river and watch—you'll see volumes of water traveling past you, from somewhere upstream, and on toward the river's mouth. Where does this water come

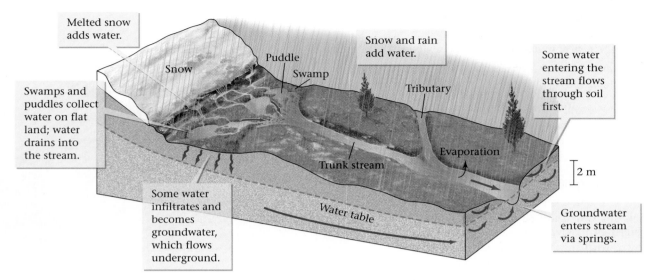

FIGURE 17.2 Excess surface water comes from rain, melting ice or snow, and groundwater springs. Where the ground is flat, the water accumulates in puddles or swamps, but on sloping ground, it flows downslope, collecting in natural troughs called streams.

from? The answer may seem obvious at first: it must originally have fallen from the sky as meteoric water (rain or snow). But on close examination, the story becomes complex, for **runoff**—the portion of meteoric water that eventually ends up in streams—includes water that has passed through a variety of surface and subsurface reservoirs in the hydrologic cycle (see Interlude F). Specifically, meteoric water can follow one of several pathways to a stream (►Fig. 17.2):

- Some water falls directly from the sky onto the surface of a stream. Generally, this water makes only a small contribution to a stream's volume.

- In places where the ground surface is flat or forms a depression, water accumulates in a standing body (puddle, swamp, pond, or lake). When the water level in the standing body becomes higher than the lowest point along the body's bank, an outlet forms, through which water spills into a stream.

- On slopes, water can move as **sheetwash**—a thin film up to a few millimeters thick—down the ground surface to a lower elevation. This water either enters a stream directly or first enters a standing body and later flows through an outlet into a stream.

- Where the ground surface is permeable, water can infiltrate down and become subsurface water (see Interlude F and Chapter 19). Subsurface water includes soil moisture (water adhering to particles in soil), vadose-zone water (water that partially fills cracks and pores in rock or regolith below the soil but above the water table), and groundwater (water that completely fills cracks and pores in the region below the water table).

Subsurface water enters streams either where heavy rains push existing soil water and vadose-zone water back to the surface, or where the stream channel lies below the water table so that groundwater bubbles up through the channel walls.

Note that if it's cold enough, water remains in solid form (snow or ice), until melting takes place. Meltwater can move to a stream via the surface or subsurface pathways listed above.

Water flowing in streams may pause temporarily in downstream lakes or reservoirs. But all *fresh* standing bodies of water have an outlet through which water escapes and continues to the sea. On a global basis, 36,000 cubic km of water becomes runoff every year, about 10% of the total volume that passes through the hydrologic cycle.

Forming Streams and Drainage Networks

How does a stream channel form in the first place? To find an answer to this question, remember that any flowing fluid can cause **downcutting**, the process of eroding or digging into substrate. The efficiency of downcutting depends on several factors, including (1) the velocity of the flow, for faster flow erodes more rapidly than slower flow; (2) the strength of the substrate, for weaker substrate can be eroded more rapidly than stronger substrate; and (3) the amount of vegetation cover, for unvegetated ground can be eroded more rapidly than land held together by plant roots. With these controls in mind, we can now complete our answer. Overland flow initiates as sheetwash—if you've ever looked down while standing on a smooth concrete of a sidewalk in a heavy rain, you've witnessed sheetwash (▶Fig. 17.3a). But natural ground surfaces, unlike sidewalks, do not have uniform slope, uniform resistance to erosion, or uniform vegetation cover. Thus, the velocity of natural sheetwash and the ease with which it can dig into its substrate varies with location. Where sheetwash is faster, or the substrate weaker and less protected, downcutting proceeds more rapidly and lowers the land relative to its surroundings (▶Fig. 17.3b). More sheetwash immediately diverts into the newly lowered land surface in which, therefore, flow velocity increases and erosion takes place even more rapidly. If this process continues for enough time, it produces a distinct channel. Note that downcutting is, in effect, a "positive feedback" process. Once downcutting begins to produce a channel, water flow focuses into the channel, so the channel deepening accelerates. Localized mass wasting on slopes may jump-start the downcutting process by rapidly producing the depression that focuses water flow.

As its flow increases, a stream channel also begins to lengthen up its slope, a process called **headward erosion** (▶Fig. 17.3c, d). Headward erosion occurs because the flow is more intense at the entry to the channel (upslope) than in the surrounding sheetwashed areas. At the same time, new channels form nearby; these merge with the main channel, because once a channel forms, the surrounding land slopes into it. An array of linked streams evolves, with the smaller streams, or **tributaries**, flowing into a single larger stream, or **trunk stream**. The array of interconnecting streams together constitute a **drainage network**.

Like transportation networks, drainage networks reach into all corners of a region, providing conduits for the removal of runoff. The configuration of tributaries and trunk streams defines the map pattern of a drainage network. This pattern depends on the shape of the landscape and the composition of the substrate. Geologists recognize several types of networks on the basis of their map pattern:

- *Dendritic:* When rivers flow over a fairly uniform substrate with a fairly uniform initial slope, they develop a dendritic network, which looks like the pattern of branches connecting to the trunk of a deciduous tree (▶Fig. 17.4a). In fact, the word *dendritic* comes from the Greek *dendros,* meaning tree.

- *Radial:* Drainage networks forming on the surface of a cone-shaped mountain flow outward from the mountain peak, like spokes on a wheel. Such a pattern defines a radial network (▶Fig. 17.4b).

- *Rectangular:* In places where a rectangular grid of fractures (vertical joints) breaks up the ground, channels form along the preexisting fractures, and streams join each other at right angles, creating a rectangular network (▶Fig. 17.4c).

- *Trellis:* In places where a drainage network develops across a landscape of parallel valleys and ridges, major tributaries flow down a valley and join a trunk stream

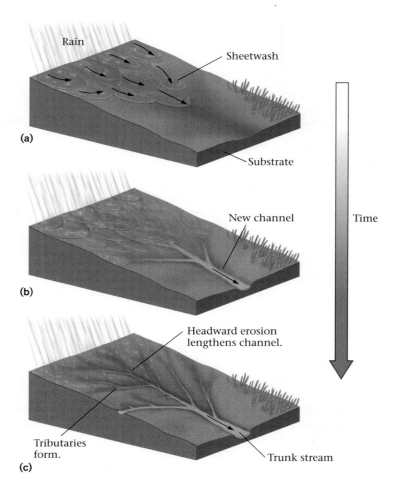

FIGURE 17.3 (a) Drainage on a slope first occurs when sheetwash, overlapping films or sheets of water, moves downslope. (b) Where the sheetwash happens to move a little faster, it scours a channel. (c) The channel grows upslope, a process called headward erosion, and new tributary channels form. The interconnecting streams make up a drainage network. (d) Headward erosion in Canyonlands National Park, Utah, where the main canyon and its tributaries are slowly cutting upstream.

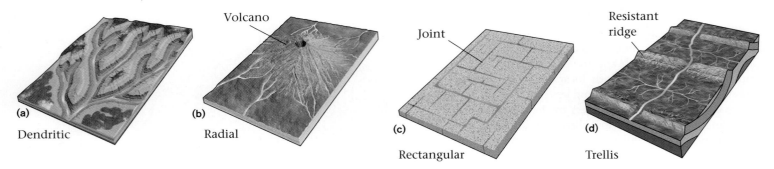

FIGURE 17.4 **(a)** Dendritic patterns resemble the branches of a tree and form on land with a uniform substrate. **(b)** A radial network drains a conical mountain, in this case, a volcano. **(c)** Rectangular patterns develop where a gridlike array of vertical joints controls drainage. **(d)** Trellis patterns (resembling a garden trellis) form where drainage networks cross a landscape in which ridges of hard rock separate valleys of soft rock. In this example, the alternation is due to folding of the rock layers.

that cuts across the ridges; the place where a trunk stream cuts across a resistant ridge is a water gap. The resulting map pattern resembles a garden trellis, so the arrangement of streams constitutes a trellis network (►Fig. 17.4d).

Drainage Basins and Divides

A drainage network collects water from a broad region, variously called a drainage basin, catchment, or **watershed**, and feeds it into the trunk stream, which carries the water away. The highland, or ridge, that separates one watershed from another is a **drainage divide** (►Fig. 17.5). A continental divide separates drainage that flows into one ocean from drainage that flows into another. For example, if you straddle the North American continental divide and pour a cup of water out of each hand, the water in one hand flows

to the Atlantic, and the water in the other flows to the Pacific. This continental divide is not, however, the only important divide in North America. A divide runs along the crest of the Appalachians, separating Atlantic Ocean drainage from Gulf of Mexico drainage, and another one runs just south of the border between Canada and the United States, separating Gulf of Mexico drainage from Hudson Bay (Arctic Ocean) drainage. These three divides bound the Mississippi drainage basin, which drains the interior of the United States (►Fig. 17.6).

FIGURE 17.6 The Mississippi drainage basin is one of several drainage basins in North America. The continental divide separates basins that drain into the Atlantic (and waters connected to the Atlantic) from basins that drain into the Pacific.

FIGURE 17.5 A drainage divide is a relatively high ridge that separates one drainage basin from another.

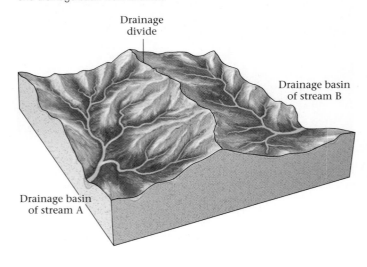

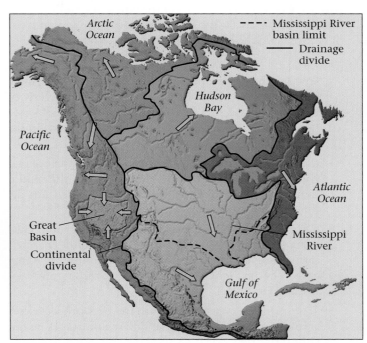

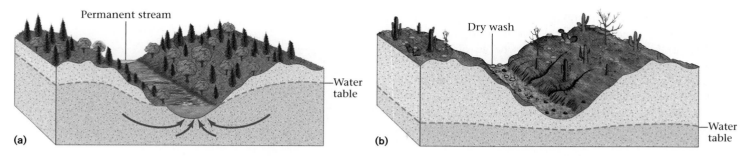

FIGURE 17.7 **(a)** If the bed of a stream channel lies below the water table, then springs add water to the stream, and the stream contains water even during periods when there is no rainfall. Such streams are permanent. **(b)** If the stream bed lies above the water table, then the stream flows only during rainfall or spring thaws, when water enters the stream faster than it can infiltrate into the ground. Such streams are ephemeral. In desert regions, a dry stream bed is called a dry wash, wadi, or arroyo.

Streams That Last, Streams That Don't: Permanent and Ephemeral Streams

Some streams flow all year long, whereas others flow for only part of the year; in fact, some flow only for a brief time after a heavy rain. The character of a stream depends on the depth of the water table. If the bed, or floor, of a stream lies below the water table, then the stream flows year-round (▶Fig. 17.7a). In such **permanent streams**, found in humid or temperate climates, water comes not only from upstream or from surface runoff, but also from springs through which groundwater seeps. But if the bed of a stream lies above the water table, then water flows only when the rate at which water enters the stream channel exceeds the rate at which water infiltrates the ground below the channel (▶Fig. 17.7b). Such streams can be permanent only if supplied by abundant water from upstream. In dry climates with intermittent rainfall and high evaporation rates, water entirely sinks into the ground, and the stream dries up when the supply of water stops. Streams that do not flow all year are called **ephemeral streams**. Ephemeral streams only flow during rainstorms or after spring thaws. A dry ephemeral stream bed (channel floor) is called a dry wash, wadi, or arroyo.

> **Take-Home Message**
>
> Water in streams comes from standing bodies spilling through outlets, sheetwash on the surface, and groundwater. Stream channels form by downcutting and lengthen by headward erosion. Eventually, networks of tributaries flow into a trunk stream and drain the land.

17.3 DISCHARGE AND TURBULENCE

Geologists and engineers describe the amount of water a stream carries by its **discharge**, the volume of water passing through an imaginary cross section drawn across the stream perpendicular to the bank, in a unit of time. We can specify stream discharge either in cubic feet per second (ft^3/s) or in cubic meters per second (m^3/s). Stream discharge depends on two factors: the cross-sectional area of the stream (A_c; the area measured in a vertical plane perpendicular to the flow direction) and the average velocity at which water moves in the downstream direction (v_a). Thus, we can calculate stream discharge by using the simple formula $D = A_c \times v_a$. Stream discharge can be determined at a stream-gauging station, where instruments measure the velocity and depth of the water (▶Fig. 17.8).

Different streams have different average discharges. Fundamentally, discharge depends on the size of the watershed and on the amount of meteoric water falling in the watershed. The Amazon River has the largest average discharge

FIGURE 17.8 Geologists obtain information needed to calculate discharge at a stream-gauging station. First, they make a survey of the channel so they know its shape and can calculate the area (A_c). Then they measure its depth, using a well, and its velocity, using a current meter (either a propeller that spins in response to moving water as shown or, more recently, Doppler radar). The current meter takes measurements at various points in the stream, for velocity changes with location, and provides data for calculation of average velocity (v_a).

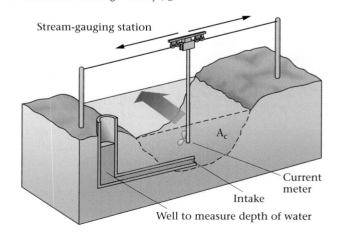

in the world—about 200,000 m³/s, or 15% of the total amount of runoff on Earth. The next-largest stream, the Congo River, has an average discharge of 40,000 m³/s, whereas the "mighty" Mississippi's is only 17,000 m³/s. Also, the discharge of a given stream varies along its length. For example, the discharge in a temperate region *increases* in the downstream direction, because each tributary that enters the stream adds more water, whereas the discharge in an arid region may *decrease* downstream, as progressively more water seeps into the ground or evaporates. Discharge can also be affected by human activity. If people divert the river's water for irrigation, the river's discharge decreases downstream. Finally, the discharge at a given location can vary with time: in a temperate climate, a stream's discharge during the spring may be double or triple the amount during a hot summer, and a flood may increase the discharge to more than 100 times normal.

The average velocity of stream water (va) can be difficult to calculate, because the water doesn't all travel at the same velocity for two reasons. First, friction along the sides and floor of the stream slows the flow. Thus, water near the channel walls or the stream bed (the floor of the stream) moves more slowly than water in the middle of the flow, and the fastest-moving part of the stream flow lies near the surface in the center of the channel. In a curved channel, the fastest flow shifts toward the outside curve, somewhat like a car swerves to the outer edge of a curve on a highway. Therefore, the deepest part of a channel, its **thalweg**, lies near the outside curve. In fact, as the water flows toward the outside wall of a curving channel, it begins a spiral motion; because water near the surface can flow faster toward the outer bank, water deeper down must flow toward the inner bank to replace the surface water. The amount by which friction slows the flow depends both on the roughness of the walls and bed and on the channel shape. A wide, shallow stream channel has a larger wetted perimeter (the area in which water touches the channel walls) than does a semicircular channel, so water flows more slowly in the former than in the latter (▶Fig. 17.9a–c). Second, **turbulence**, or turbulent flow, is a twisting, swirling motion that, on a large scale, can create eddies (whirlpools) in which water curves and actually flows upstream or circles in place (▶Fig. 17.10). Turbulence develops in part because the shearing motion of one volume against its neighbor causes the neighbor to spin, and in part because obstacles such as boulders deflect volumes, forcing them to move in a different direction.

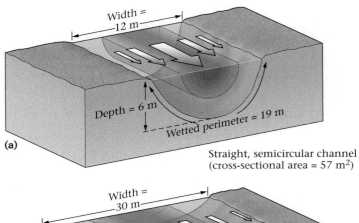

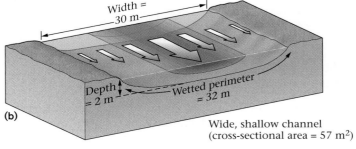

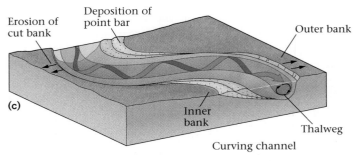

FIGURE 17.9 (a) In a straight, semicircular channel, the maximum velocity occurs near the surface in the center of the stream. The deepest part of the channel is the thalweg. **(b)** The maximum velocity also occurs in the center of a wide, shallow channel, but the maximum velocity of a shallow channel is less than that of a semicircular channel, for a given cross-sectional area. That's because its wetted perimeter (where water touches the channel walls) is greater than that of the semicircular channel (even though its cross-sectional area is the same), so there is more friction between the water and the channel walls, which slows down the flow. **(c)** In a curved channel, the fastest flow shifts toward the outer edge of the stream, over the thalweg. Also, the water follows a spiral path.

17.4 THE WORK OF RUNNING WATER

How does a Stream Erode?

The energy that makes running water move comes from gravity. As water flows downslope from a higher to a lower elevation, the gravitational potential energy stored in water transforms into kinetic energy. About 3% of this energy goes into the work of eroding the walls and beds of stream channels. Running water causes erosion in four ways.

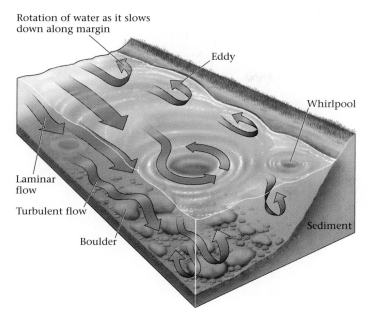

Rotation of water as it slows down along margin

Eddy

Whirlpool

Laminar flow

Turbulent flow

Boulder

Sediment

FIGURE 17.10 In a turbulent flow, because of shearing between volumes of water and movement over boulders, the water swirls in curving paths and becomes caught in eddies.

- *Scouring:* Running water can remove loose fragments of sediment, a process called **scouring**.
- *Breaking and lifting:* In some cases, the push of flowing water can break chunks of solid rock off the channel floor or walls. In addition, the flow of a current over a clast can cause the clast to rise, or lift off the substrate.
- *Abrasion:* Clean water has little erosive effect, but sand-laden water acts like sandpaper and grinds or rasps away at the channel floor and walls, a process called **abrasion**. In places where turbulence produces long-lived whirlpools, abrasion by sand or gravel carves a bowl-shaped depression, called a **pothole**, into the floor of the stream (▶Fig. 17.11a, b).
- *Dissolution:* Running water dissolves soluble minerals as it passes, and carries the minerals away in solution.

The efficiency of erosion depends on the velocity and volume of water and on its sediment content. A large volume of fast-moving, turbulent, sandy water causes more erosion than a trickle of quiet, clear water. Thus, most erosion takes place during floods, which supply streams with large volumes of fast-moving, sediment-laden water.

How Do Streams Transport Sediment?

The Mississippi River received the nickname "Big Muddy" for a reason—its water can become chocolate brown because of all the clay and silt it carries. All streams carry sediment, though not the same amount.

Geologists refer to the total volume of sediment carried by a stream as its sediment load. The sediment load consists of three components:

- *Dissolved load:* Running water dissolves soluble minerals in the sediment or rock of its substrate, and groundwater seeping into a stream through the channel walls brings dissolved minerals with it. These ions constitute a stream's **dissolved load**.
- *Suspended load:* The **suspended load** of a stream consists of tiny solid grains (silt or clay size) that swirl along with the water without settling to the floor of the channel; this sediment makes the water brown (▶Fig. 17.11c).
- *Bed load:* The **bed load** of a stream consists of large particles (such as sand, pebbles, or cobbles) that bounce or roll along the stream floor (▶Fig. 17.11d). Typically, bed-load movement involves **saltation**, a process during which grains on the channel floor get knocked into the water column momentarily, follow a curved trajectory downstream, and gradually sink to the bed again, where they knock other grains into the water column.

When describing a stream's ability to carry sediment, geologists specify its competence and capacity. The **competence** of a stream refers to the maximum particle size it carries; a stream with high competence can carry large particles, whereas one with low competence can carry only small particles. A fast-moving, turbulent stream has greater competence (it can carry bigger particles) than a slow-moving stream, and a stream in flood has greater competence than a stream with normal flow. In fact, the huge boulders that litter the bed of a mountain creek move *only* during floods. The **capacity** of a stream refers to the total quantity of sediment it *can carry*. A stream's capacity depends on its competence and discharge.

How Do Streams Deposit Sediment?

A raging torrent of water can carry coarse and fine sediment— the finer clasts rush along with the water as suspended load, whereas the coarser clasts may bounce and tumble as bed load. If the flow velocity decreases, either because the gradient (downstream slope) of the stream bed becomes shallower or because the channel broadens out and friction between the bed and the water increases, then the competence of the stream decreases and sediment settles out. The size of the clasts that settle at a particular locality depends on the decrease in flow velocity. For example, if the stream slows by a small amount, only large clasts settle; if the stream slows by a greater amount, medium-sized clasts settle; and if the stream slows to almost a standstill, the fine grains settle. Thus, coarser sediment tend to settle out further upstream, where

(a)

(b)

(c)

FIGURE 17.11 **(a)** A polished, red sandstone canyon wall in northern Arizona. The canyon formed by the linkage of potholes. **(b)** Potholes formed in the bed of a creek in Ithaca, New York, by the grinding power of swirling gravel. **(c)** The Colorado River, at the west end of the Grand Canyon has a brownish color due to its sediment load. Note the debris falling into the river from the canyon walls. **(d)** Streams transport sediment in many forms. The dissolved sediment load consists of ions in solution. A suspended load consists of tiny grains distributed through the water. A bed load consists of grains that undergo saltation (they bounce along the bed) or grains that roll along the bed; part of the bed load stays in place during times of normal flow, but begins to move during a flood.

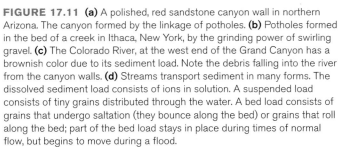

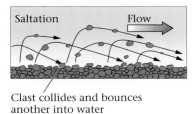

(d)

the gradient of the stream is steeper and water flows faster, whereas finer grains settle out further downstream, where the water flows more slowly. Because of this process of sediment sorting, stream deposits tend to be segregated by size—gravel accumulates in one location, and mud in another. Geologists refer to sediments transported by a stream as fluvial deposits (from the Latin fluvius, meaning river) or **alluvium**. Fluvial deposits may accumulate along the stream bed in elongate mounds, called **bars** (▶Fig. 17.12a, b). Some stream channels make broad curves. Water slows along the inner edge of a

(a)

curve, so crescent-shaped point bars bordering the shoreline develop. During floods, a stream may over-top the banks of its chan-nel and spread out over its floodplain, a broad flat area bordering the stream. Friction slows the water on the floodplain, so a sheet of silt and mud set-tles out. Where a stream empties at its mouth, into a standing body of water, the water slows. A wedge of sedi-ment, called a **delta**, accumulates. We discuss deltas in more detail later in the chapter.

> ### Take-Home Message
>
> Streams erode into the substrate by scouring, breaking and lifting, abrasion, and dissolution. They carry sediment as dissolved, sus-pended, or bed loads. Compe-tence, the ability to carry sediment, depends on velocity. Where the velocity of flow de-creases, sediment settles out.

17.5 HOW DO STREAMS CHANGE ALONG THEIR LENGTH?

Longitudinal Profiles

In 1803, under President Thomas Jefferson's leadership, the United States bought the Louisiana Territory, a vast tract of land encompassing the western half of the Missis-sippi drainage basin. At the time, the geography of the ter-ritory was a mystery. To fill the blank on the map, Jefferson asked Meriwether Lewis and William Clark to lead a voyage of exploration across the Louisiana Territory to the Pacific.

Lewis and Clark, along with about forty men, began their expedition at the mouth of the Missouri River, where it joins the Mississippi. At this juncture, the Missouri is a wide, languid stream of muddy water. The group found the Missouri's downstream reaches (intervals along the length of a stream), where the river's channel is deep and the water smooth, to be easy going. But the farther up-stream they went, the more difficult their voyage became, for the **stream gradient** (the slope of the stream channel) became progressively steeper, and the stream's discharge became less. When Lewis and Clark reached the site of

(b)

FIGURE 17.12 (a) Recently deposited gravel in a streambed of a steep mountain stream at Denali National Park, Alaska. The large cobbles and boulders were deposited during ferocious floods. **(b)** Point bars of mud deposited along a gentle, slowly moving stream in Brazil.

what is now Bismarck, North Dakota, they had to aban-don their original boats and haul smaller vessels up rapids where turbulent water plunges over a steep, bouldery bed, and around waterfalls, where water drops over an escarp-ment. When they reached what is now southwestern Montana, they abandoned these boats as well and trudged along the stream valley on foot or on horseback, strug-gling up steep gradients until they reached the continental divide.

If Lewis and Clark had been able to plot a graph show-ing their elevation above sea level relative to their distance along the Missouri, they would have found that the **longi-tudinal profile** of the Missouri, a cross-sectional image showing the variation in the river's elevation along its

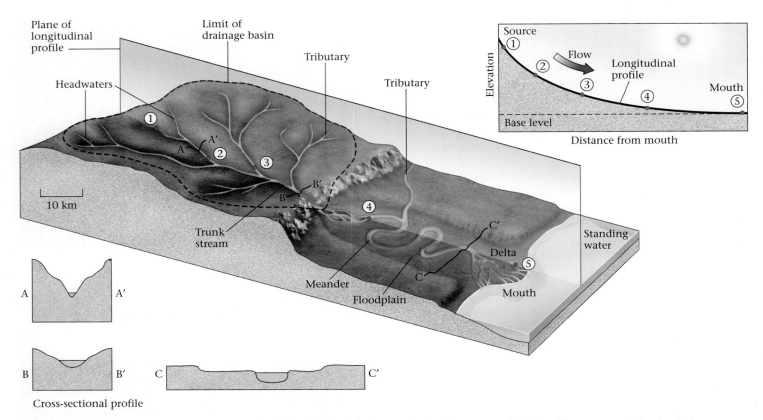

FIGURE 17.13 A drainage network collects water from a broad drainage basin, or watershed, via numerous tributaries. These carry water to a trunk stream and eventually to a standing body of water. Points 1–5 refer to locations along the longitudinal profile (inset). The cross-sectional profiles show how river valley shapes change along the length of a river.

length, is roughly a concave-up curve (►Fig. 17.13). This curve illustrates that stream gradient is steeper near its headwaters (source) than near its mouth. Real longitudinal profiles are not perfectly smooth curves, but rather display little plateaus and steps, representing interruptions by lakes or waterfalls. Near its headwaters, an idealized stream flows down deep valleys or canyons, whereas near its mouth, it flows over nearly horizontal plains.

The Base Level

Streams progressively deepen their channels by downcutting, but there is a depth below which a stream cannot downcut any further. The lowest elevation a stream channel's floor can reach at a locality is the **base level** of the stream. A local base level occurs upstream of a drainage network's mouth, and the ultimate base level (i.e., the lowest possible elevation along the stream's longitudinal profile) is determined by sea level. The trunk stream cannot downcut deeper than sea level, for if it did, it would have to flow upslope to enter the sea.

Lakes or reservoirs can act as local base levels along a stream, for where the stream enters such standing bodies of water, it slows almost to a halt and cannot downcut further (►Fig. 17.14a). A ledge of resistant rock can also act as a local base level, for the stream level cannot drop below the ledge until the ledge erodes away (►Fig. 17.14b). Finally, where a tributary joins a larger stream, the channel of the larger stream acts as the base level for the tributary; thus, the mouth of a tributary lies at the same elevation as the stream that it joins at the point of intersection. Local base levels do not last forever, because running water eventually removes the obstructions that create them.

Streams attain a concave-up longitudinal profile gradually. During this process, steps or ledges defining local base levels along

> **Take-Home Message**
>
> Streams have steeper regional gradients toward their sources, and gentler gradients near their mouths, so longitudinal profiles tend to be concave up. A stream cannot downcut any lower than its base level. Sea level is the ultimate base level for drainage networks.

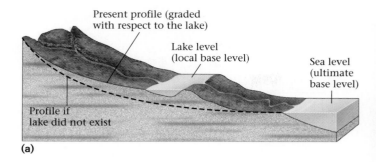

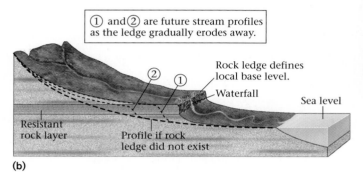

FIGURE 17.14 (a) A lake acts as a local base level. The longitudinal profile of the stream upstream of the lake lies above the profile for a graded stream (one that deposits as much sediment as it removes). Eventually, headward erosion of the stream below the lake will cause the lake to drain. (b) A resistant rock ledge also acts as a local base level. With time, the ledge will erode, and the waterfall will migrate upstream until the stream achieves grade. Sea level is the ultimate base level for a drainage network.

FIGURE 17.15 Erosion by the Colorado River and its tributaries have cut deep canyons into the rock of the Colorado Plateau.

the stream erode away, and low areas fill in, until any point along the stream approaches a condition such that there is no net erosion or deposition. The stream can carry all the sediment that has been supplied to it, and it deposits as much sediment as it removes. A stream that has reached this condition is called a graded stream.

17.6 STREAMS AND THEIR DEPOSITS IN THE LANDSCAPE

Valleys and Canyons

About 10 million years ago, a large block of crust, the region now known as the Colorado Plateau (located in Arizona, Utah, Colorado, and New Mexico), began to rise. Before the rise, the Colorado River had been flowing over a plain not far above sea level, causing little erosion. But as the land uplifted, the river began to downcut steadily. Eventually, its channel lay as much as 1.6 km below the surface of the

plateau at the base of a steep-walled gash now known as the Grand Canyon (▶Fig. 17.15). The formation of the Grand Canyon illustrates a general phenomenon. In regions where the land surface lies well above the base level, a stream can carve a deep trough, much deeper than the channel itself. If the walls of the trough slope gently, the landform is a valley. If they slope steeply, the landform is a canyon.

Whether stream erosion produces a valley or a canyon depends on the rate at which downcutting takes place relative to the rate at which mass wasting causes the walls on either side of the stream to collapse. In places where a stream downcuts through its substrate faster than the walls of the stream collapse, erosion creates a slot (steep-walled) canyon. Such canyons typically form in hard rock, which can hold up steep cliffs for a long time (▶Fig. 17.16a). In places where the walls collapse as fast as the stream downcuts, landslides and slumps gradually cause the slope of the walls to approach the angle of repose. When this happens, the stream channel lies at the floor of a valley whose cross-sectional shape resembles the letter V (▶Fig. 17.16b); this landform is called a **V-shaped valley**. Where the walls of the stream consist of alternating layers of hard and soft rock, the walls develop a stair-step shape such as that of the Grand Canyon (▶Fig. 17.16c).

In places where active downcutting occurs, the valley floor remains relatively clear of sediment, for the stream—especially when it floods—carries away sediment that has fallen or slumped into the channel from the stream walls. But if the stream's base level rises or its discharge decreases, the valley floor fills with sediment, creating an alluvium-filled

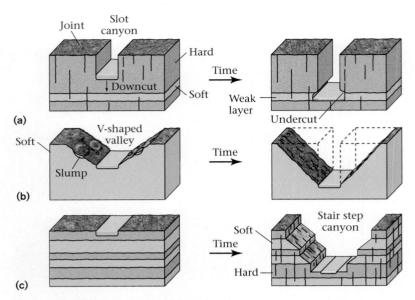

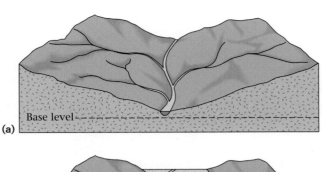

(a)

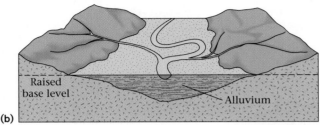

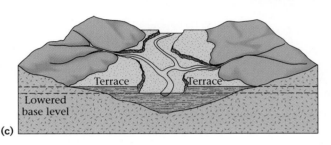

(a)

(b)

(c)

FIGURE 17.16 **(a)** If downcutting by a stream happens faster than mass wasting alongside the stream, as is typical when streams erode through hard rock, a slot canyon forms. The canyon widens with time as the stream undercuts the walls. **(b)** If mass wasting takes place as fast as downcutting occurs, a V-shaped valley develops. **(c)** In regions where the stream downcuts through alternating hard and soft layers, a stair-step canyon forms.

valley (▶Fig. 17.17a, b). The surface of the alluvium becomes a broad floodplain. If the stream's base level then drops again and/or the discharge increases, the stream will start to cut down into its own alluvium, a process that generates **stream terraces** bordering the present floodplain (▶Fig. 17.17c, d).

Rapids and Waterfalls

When Lewis and Clark forged a path up the Missouri River, they came to reaches that could not be navigated by boat because of **rapids**, particularly turbulent water with a rough surface (▶Fig. 17.18a). Rapids form where water flows over steps or large clasts in the channel floor, where the channel abruptly narrows, or where its gradient abruptly changes. The turbulence in rapids produces eddies, waves, and whirlpools that roil and churn the water surface, in the process creating whitewater, a mixture of bubbles and water. Modern-day whitewater rafters thrill to the unpredictable movement of rapids (▶Fig. 17.18b).

A **waterfall** forms where the gradient of a stream becomes so steep that the water literally free-falls down the stream bed (▶Fig. 17.19a, b). The energy of falling water may scour a depression, called a plunge pool, at the base of the waterfall. Some waterfalls develop where a stream crosses a resistant ledge of rock, some develop as a result of faulting, because displacement produces an escarpment.

FIGURE 17.17 The evolution of alluvium-filled valleys. **(a)** Stream erosion creates a valley. **(b)** Later, a rise in the base level or a decrease in discharge allows the valley to fill with alluvium. **(c)** Later, if the base level falls or discharge increases, the stream downcuts through the alluvium, and a new, lower floodplain develops. The remnants of the original alluvial plain remain as a pair of terraces, one on each side of the new floodplain. **(d, e)** Two episodes of renewed downcutting into an alluvium-filled valley produced these two terraces at the junction of two streams in Utah. The present floodplain is wetter and greener than the terraces.

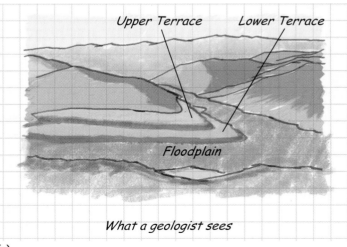

(d)

(e)

Class V rapids

(a)

(b)

FIGURE 17.18 **(a)** Rapids in the Grand Canyon (raft for scale). These rapids formed where a flood from a side canyon dumped debris into the channel of the Colorado River. **(b)** Whitewater boaters distinguish among different classes of rapids (I to V) by the velocity of water, the steepness of the stream, the size of standing waves, the size of obstacles, and the difficulty of navigating past obstacles. Here we see Class V rapids, which should be navigated only by experts!

(a)

(b)

FIGURE 17.19 **(a)** Iguaçu Falls, along the Brazil–Argentina border. **(b)** A waterfall emerging from a hanging valley, Milford Sound, New Zealand.

Waterfalls also occur where glacial erosion has deepened a trunk valley relative to tributary valleys (see Chapter 22).

Though a waterfall may appear to be a permanent feature of the landscape, all waterfalls eventually disappear as headward erosion slowly eats back the resistant ledge until the stream reaches grade. We can see a classic example of headward erosion at Niagara Falls. As water flows from Lake Erie to Lake Ontario, it drops over a 55-m-high ledge of hard Silurian dolostone, which overlies a weak shale. Erosion of the shale leads to undercutting of the dolostone. Gradually, the overhang of dolostone becomes unstable and collapses, with the result that the waterfall migrates upstream. Before the industrial age, the edge of Niagara Falls cut upstream at an average rate of 1 m per year, but since then, the diversion of water from the Niagara River into a hydroelectric power station has decreased the rate of

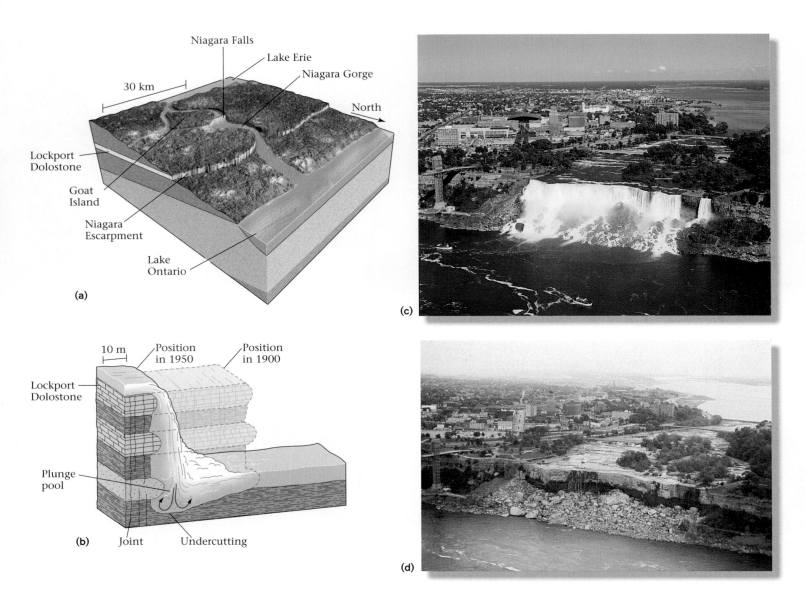

FIGURE 17.20 **(a)** Niagara Falls exists because Lake Erie lies at a higher elevation than Lake Ontario. The Lockport Dolostone, a resistant layer, is the local base level for Lake Erie. The Niagara Escarpment is located along the outcrop belt of the dolostone, and Niagara Falls first formed where the outlet of Lake Erie flowed over the escarpment. With time, the falls have cut upstream, at about 1 m per year or less, creating the Niagara Gorge. When the falls reach Lake Erie, the lake will drain. **(b)** This cross section of the falls shows how undercutting of the soft shale layers eventually causes the resistant layers of dolostone to break off at joints. **(c)** The American Falls, part of Niagara Falls. **(d)** The American Falls with no water, showing the escarpment and the jumble of dolostone blocks that have broken off. The water was diverted from the falls so that geologists could investigate the erosion rate.

headward erosion to half that. Nevertheless, at this rate, Niagara Falls will cut all the way back to Lake Erie in about 60,000 years (►Fig. 17.20a–d).

Alluvial Fans and Braided Streams

Where a fast-moving stream abruptly emerges from a mountain canyon into an open plain at the range front, the water that was once confined to a narrow channel spreads out over a broad surface. As a consequence, the water slows and abruptly drops its sedimentary load, forming a gently sloping apron of sediment (sand, gravel, and cobbles) called an **alluvial fan** (►Fig. 17.21a). The stream then divides into a series of small channels that spread out over the fan. During particularly strong floods, debris flows spread over and smooth out the fan's surface.

In some localities, streams carry abundant coarse sediment during floods but cannot carry this sediment during

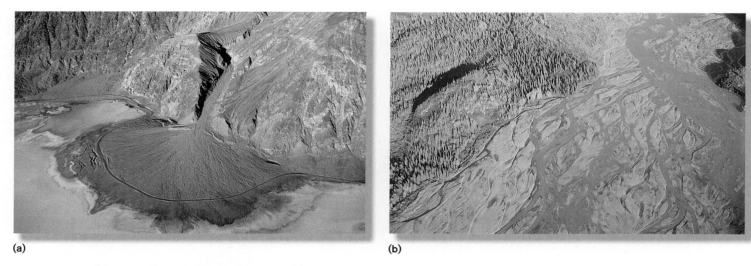

(a) **(b)**

FIGURE 17.21 **(a)** An alluvial fan in Death Valley, California. **(b)** A braided stream, carrying meltwater from a glacier, near Denali, Alaska.

normal flow. Thus, during normal flow, the sediment settles out and chokes the channel. As a consequence, the stream divides into numerous strands weaving back and forth between elongate bars of gravel and sand. The result is a **braided stream**—the name emphasizes that the streams entwine like strands of hair in a braid (▶Fig. 17.21b). Strands of the stream branch out at the upstream ends of bars and merge at the downstream end of bars. Because the gravelly sediment of a braided stream can't stick together, the stream cannot cut a deep channel with steep banks—the channel walls simply collapse, so the stream spreads out over a broad area. Braided streams commonly form in landscapes where streams fill with sediment-choked glacier meltwater.

Meandering Streams and Their Floodplains

A riverboat cruising along the lower reaches of the Mississippi River cannot sail in a straight line, for the river channel winds back and forth in a series of snake-like curves called **meanders** (▶Fig. 17.22a, b). In fact, the boat has to go 500 km along the river channel to travel 100 km as the crow flies. Meandering streams have many meanders and form where running water travels over a broad floodplain underlain by a soft substrate, in a region where the river has a very gentle gradient. The development of meanders increases the volume of the stream by increasing its length.

How do meanders form and evolve? Even if a stream starts out with a straight channel, natural variations in the water depth and associated friction (see Fig. 17.10) cause the fastest-moving current to swing back and forth. The water erodes the side of the stream more effectively where it flows faster, so it begins to cut away faster on the outer arc

of the curve. Thus, each curve begins to migrate sideways and grow more pronounced until it becomes a meander. On the outside edge of a meander, erosion continues to eat away at the channel wall, forming a cut bank, whereas on the inside edge, water slows down so that its competence decreases and sediment accumulates, creating a wedge-shaped deposit called a point bar. (Mark Twain, who worked as a riverboat pilot on the Mississippi River before writing such books as *Huckleberry Finn*, took his pen name from the signals the mate of a paddle-wheel steamer called out to the skipper to indicate water depth. "Mark twain" means two fathoms, or about 4 m deep.) With continued erosion, a meander may curve through more than 180°, so that the cut bank at the meander's entrance approaches the cut bank at its end, leaving a meander neck, a narrow isthmus of land separating the portions of the meander. Meandering streams only develop where the banks have sufficient strength to hold up cut banks. This may require plant roots to bind the sediment of the cut bank together.

People building communities along a riverbank may assume that the shape of a meander remains fixed for a long time. It doesn't. In a natural meandering river system, the river channel migrates back and forth across the floodplain. When erosion eats through a meander neck, a straight reach called a cutoff develops. The meander that has been cut off is called an **oxbow lake** if it remains filled with water, or an abandoned meander if it dries out (▶Fig. 17.22c).

Most meandering stream channels cover only a relatively small portion of a broad **floodplain** (▶Fig. 17.22d). In many cases, a floodplain terminates at its sides along a bluff, or escarpment. During a flood, water spills out from the stream channel onto the floodplain, and large floods may cover the entire region from bluff to bluff. As the water leaves the channel, friction between the ground and

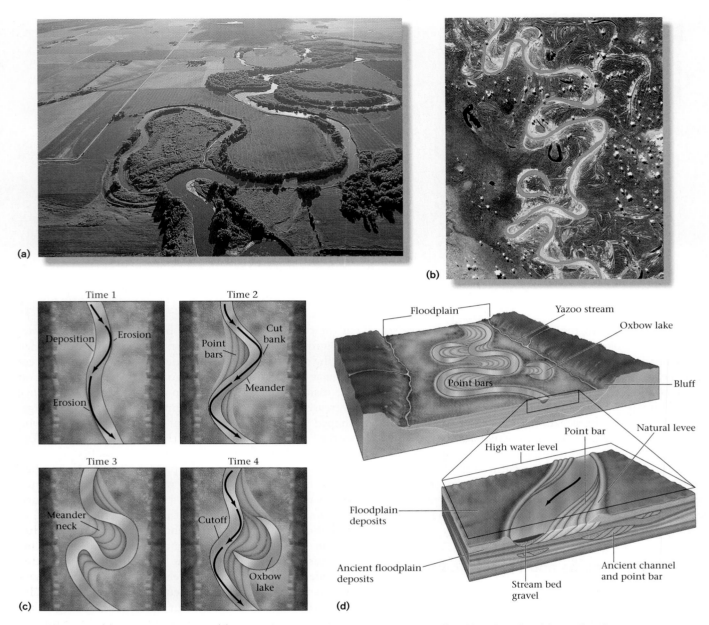

FIGURE 17.22 **(a)** A meandering stream. **(b)** A photo from space of a meandering stream in Peru. Note the oxbow lakes at the ridges representing older point bars. The field of view is 50 km wide. **(c)** Erosion occurs faster on the outer bank of a stream's curve, whereas deposition takes place on the inner curve. The meander becomes a progressively tighter curve until the stream cuts through the meander neck and forms a cutoff, thereby isolating an oxbow lake. **(d)** The landforms of a meandering stream. The detail of a stream channel shows a natural levee, the structure of a point bar, and floodplain deposits. Note that alluvium below the present-day channel includes ancient channels and point bars, surrounded by fine-grained floodplain deposits.

the thin sheet of water moving over the floodplain slows down the flow. This slowdown decreases the competence of the running water, so sediment settles out along the edge of the channel. Over time, the accumulation of this sediment creates a pair of low ridges, called **natural levees**, on either side of the stream. Natural levees may grow so large that the floor of the channel may become higher than the surface of the floodplain. In fact, the higher areas of New

Orleans, which have remained fairly dry during floods that submerged the rest of the city, are the parts built on the natural levees (▶Fig. 17.23). In places where large natural levees exist, the region between the bluffs and the levees becomes a low, marshy swamp. Also because of the levees, small tributaries may be blocked from joining the trunk stream; the tributaries, called yazoo streams, run in the floodplain parallel to the main river.

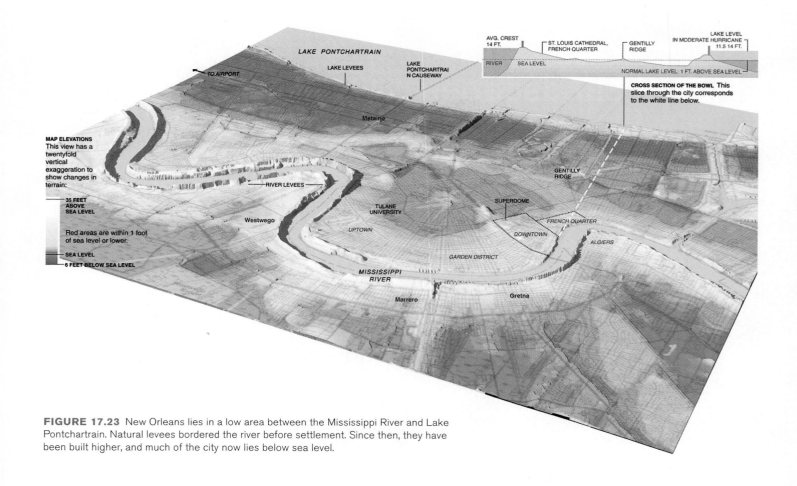

FIGURE 17.23 New Orleans lies in a low area between the Mississippi River and Lake Pontchartrain. Natural levees bordered the river before settlement. Since then, they have been built higher, and much of the city now lies below sea level.

Deltas: Deposition at the Mouth of a Stream

Along most of its length, only a narrow floodplain—green, irrigated farm fields—borders the Nile River in Egypt. But at its mouth, the trunk stream of the Nile divides into a fan of smaller streams, called **distributaries**, and the area of green agricultural lands broadens into a triangular patch. The Greek historian Herodotus noted that this triangular patch resembles the shape of the Greek letter delta (Δ), and so the region became known as the Nile Delta.

Deltas develop where the running water of a stream enters standing water, the current slows, the stream loses competence, and sediment settles out. Geologists refer to any wedge of sediment formed at a river mouth as a delta, even though relatively few have the triangular shape of the Nile Delta (▶Fig. 17.24a–d). Some deltas curve out into the sea, whereas others consist of many elongate lobes; the latter are called bird's-foot deltas, because they resemble the scrawny toes of a bird. The existence of several toes indicates that the main course of the river in the delta has

shifted on several occasions. These shifts occur when a toe builds so far out into the sea that the slope of the stream becomes too gentle to allow the river to flow. At this point, the river overflows a natural levee upstream and begins to flow in a new direction, an event called an avulsion. The distinct lobes of the Mississippi Delta, a bird's-foot delta, suggest that avulsion has happened several times during the past 9,000 years (▶Fig. 17.25a). New Orleans, built along one of the Mississippi's distributaries, may eventually lose its riverfront, for a break in a levee upstream of the city could divert the Mississippi into the Atchafalaya River channel.

The shape of a delta depends on many factors. Deltas that form where the strength of the river current exceeds that of ocean currents have a bird's-foot shape, since the sediment can be carried far offshore. In contrast, deltas that form where the ocean currents are strong have a Δ shape, for the ocean currents redistribute sediment in bars running parallel to the shore. And in places where waves and currents are strong enough to remove sediment as fast as it arrives, a river has no delta at all.

See for yourself . . .
Fluvial Landscapes

Streams stand out in the landscape, for they carve intricate shapes as their waters flow from high areas to low. In this Geotour, we visit a variety of landscapes whose features are a consequence of deposition and erosion in a fluvial setting. The thumbnail images provided on this page are only to help identify tour sites. Go to wwnorton.com/studyspace to experience this flyover tour.

Deep Gorge in the Himalayas (Lat 28°9'40.37"N, Long 85°25'53.21"E)

Fly to these coordinates, zoom to 10 km (6 miles), and tilt your image so you are looking up the valley to the east (Image G17.1). You are 52 km (42 miles) NNE of Katmandu, Nepal. This image illustrates how the upper reaches of a mountain stream have steep gradients and flow down deep gorges. If the rock walls of the gorge are relatively weak, the stream valley attains a V-shape.

G17.1

Headward Erosion, Canyonlands, Utah (Lat 38°17'58.16"N, Long 109°50'18.76"W)

At these coordinates, in Canyonlands National Park, fly to an altitude of 8 km (5 miles). You see a side canyon carved into horizontal strata by intermittent tributaries that flow into the Colorado River (Image G17.2). An abrupt scarp marks the upstream limit of each tributary. As the streams erode, they cut into the land by headward erosion, so the scarp migrates upstream. Zoom down to 4 km (2.5 miles), tilt the image, and look southwest to better visualize the concept of headward erosion (Image G17.3).

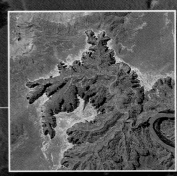

G17.2

G17.3

Meanders Along Rio Ucayali, Peru (Lat 7°27'1.80"S, Long 75°2'48.92"W)

At this locality, 560 km (347 miles) NNE of Lima, Peru, you will find a nearly horizontal landscape on the east side of the Andes. Because of the low gradient here, the Rio Ucayali has become a meandering stream. This view, from an elevation of 100 km (62 miles), shows numerous meanders within a flood plain (Image G17.4). You also see abandoned meanders, point bars, and oxbow lakes. You can find very similar meanders along the Mississippi River (USA) at Lat 31°29'29.67"N, Long 91°38'0.23"W.

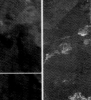

G17.4

Incised Meanders, Canyonlands, Utah (Lat 38°13'56.26"N, Long 109°49'27.64"W)

Return to Canyonlands National Park, Utah, zoom to 6.5 km (4 miles), and tilt the image to look northeast. You can see a meander of the Colorado River that has almost cut through the meander neck (Image G17.5). The meanders incised down through bedrock, and now lie at the floor of a steep-walled canyon.

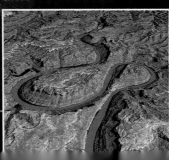

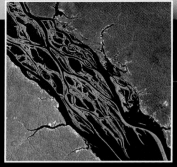

G17.6

Mid-Stream Bars, Rio Negro, Brazil
(Lat 2°43'31.82"S, Long 60°42'50.50"W)

At this locality, along the Rio Negro (a major tributary of the Amazon), 105 km (65 miles) upstream of Manaus, the river is about 15 km (9 miles) wide (Image G17.6). From an altitude of 60 km (37 miles), you can see numerous mid-stream bars of sediment that emerge from the river. These bars will be submerged during a flood, and may be shifted by the force of flowing water.

G17.7

G17.8

Point Bars, Trinity River, Texas
(Lat 30°10'2.73"N, Long 94°48'58.24"W)

Fly to this locality and zoom to an elevation of 14 km (8.7 miles). You are looking at a reach of the Trinity River in the vicinity of Dayton Lakes, 70 km (43 miles) to the northeast of Houston (Image G17.7). Along the inner curve of each meander, a bright tan point bar of sand has formed. In the green landscape surrounding the river, you will also be able to spot a number of abandoned meanders and oxbow lakes. Zoom down to 3 km (1.9 miles) and tilt so you see the horizon (Image G17.8). The bars and abandoned meanders stand out in the landscape.

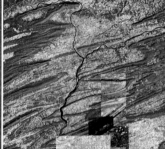

G17.9

Trellis Drainage Pattern, Pennsylvania
(Lat 40°45'31.75"N, Long 76°50'45.41"W)

Fly to these coordinates and zoom to 150 km (93 miles). You can see the Valley and Ridge Province of Pennsylvania, produced by the erosion of folded strata. Here, we see a trellis drainage pattern—tributaries flow down valleys and intersect the Susquehanna at right angles (Image G17.9). The Susquehanna itself cuts across ridges.

G17.10

Radial Drainage, Mt. Shasta
(Lat 41°24'17.85"N, Long 122°11'42.87"W)

Looking straight down from an elevation of 25 km (15.5 miles), you can see the peak of Mt. Shasta, a volcano in California. The streams that flow down its slopes, away from the peak, define a radial network (Image G17.10). This pattern resembles the spokes of a wheel.

Dendritic Drainage, Pennsylvania (Lat 41°30'29.73"N, Long 78°14'18.04"W)

Fly to these coordinates, zoom to an elevation of 30 km (18.6 miles), and you are looking down on a region of the Appalachian Plateau near the town of Emporium, in north-central Pennsylvania. The strata here consist of flat-lying beds of shale. Streams cutting down into this fairly homogeneous and soft substrate carve a dendritic network (Image G17.11). This means that the pattern of streams and tributaries resembles the pattern of veins on a leaf.

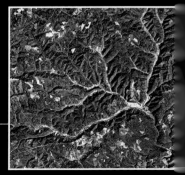

G17.11

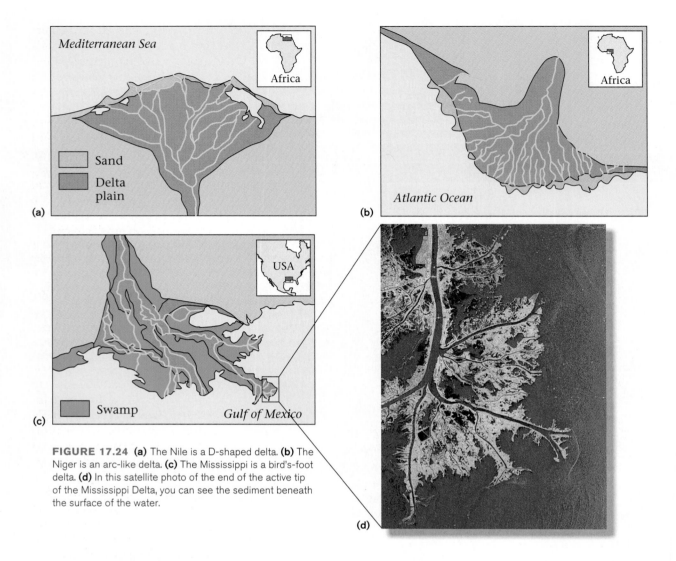

(a) Mediterranean Sea — Africa
Sand
Delta plain

(b) Atlantic Ocean — Africa

(c) USA — Swamp — Gulf of Mexico

(d)

FIGURE 17.24 **(a)** The Nile is a D-shaped delta. **(b)** The Niger is an arc-like delta. **(c)** The Mississippi is a bird's-foot delta. **(d)** In this satellite photo of the end of the active tip of the Mississippi Delta, you can see the sediment beneath the surface of the water.

With time, the sediment of a delta compacts and the land beneath the delta sinks, so the delta becomes a low swampland called a delta plain, across which distributaries (known as bayous in Louisiana) sluggishly flow. In tropical climates, mangrove trees, which can grow in shallow tidal flats, cover the seaward edge of the delta plain.

Why do rivers divide into distributaries at their mouths? When a river reaches standing water, its velocity slows down. The sediment settles out at the mouth to form a midstream bar. The presence of the bar causes the stream to split into two channels. Similar bars created at the mouths of these two sub-

sidiary channels cause each to separate in turn, until eventually numerous distributary channels have formed (►Fig. 17.25b).

17.7 THE EVOLUTION OF DRAINAGE

Beveling Topography

Imagine a place where continental collision uplifts a region (►Fig. 17.26a). At first, rivers have steep gradients, flow over many rapids and waterfalls, and cut deep valleys (►Fig. 17.26b). But with time, rugged mountains become low, rounded hills; once-deep, narrow valleys broaden into wide floodplains, with more gradual gradients (►Fig. 17.26c). As more time passes, even the low hills are beveled down, becoming small mounds or even disappearing altogether (►Fig. 17.26d). (Some geologists have referred to the resulting landscape as a peneplain, from the Latin *paene,* which means al-

Take-Home Message

Erosion carves valleys and canyons, with shapes that depend on the balance between slope-collapse and downcutting rates. Streams choked with sediment become braided; those following snakelike paths are meandering; those emptying into standing water build deltas.

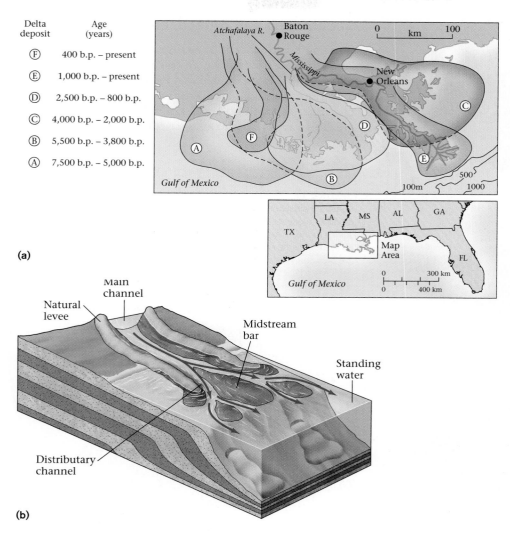

Delta deposit	Age (years)
F	400 b.p. – present
E	1,000 b.p. – present
D	2,500 b.p. – 800 b.p.
C	4,000 b.p. – 2,000 b.p.
B	5,500 b.p. – 3,800 b.p.
A	7,500 b.p. – 5,000 b.p.

(a)

(b)

FIGURE 17.25 (a) The map shows the different, dated lobes of the Mississippi Delta and the different channels that were their sources. The inset shows a current view of the delta, relative to Louisiana. A major flood could divert the main flow of the Mississippi into the channel of the Atchafalaya River, in which case a new delta would form to the west of the Mississippi's present mouth. **(b)** When a stream enters standing water, it deposits more sediment in the center of the channel than along the margins because the formerly fast-moving water at the center carried more sediment. The deposit builds a midstream bar, separating the stream into two distributaries. The same process happens at the mouth of each distributary, leading to further subdivisions.

stream drops, when land rises beneath a stream, or when the discharge of a stream increases. What is the evidence that rejuvenation took place at a given locality? In the case of a stream flowing in an alluvium-filled valley, renewed downcutting allows the stream to create a new floodplain at a lower elevation than the original one (see Fig. 17.17c). As we have seen, the younger floodplain tends to be narrower than the older, and the surface of the older floodplain becomes a terrace on either side of the new floodplain. In the case of a stream flowing on bedrock, a drop of the base level causes the stream to incise into the bedrock.

Stream Piracy and Drainage Reversal

Stream piracy sounds like pretty violent stuff. In reality, **stream piracy**, or stream capture, simply refers to a situation in which headward erosion causes one stream to intersect the course of another stream. When this happens, the pirate stream "captures" the water in the stream it intersects, so the captured stream starts flowing into the pirate stream (▶Fig. 17.27a, b). The piracy of a stream that had been flowing through a water gap transforms the water gap into a wind gap, a dry pathway through a high ridge. In 1775, the pioneer Daniel Boone blazed the "wilderness road" through the Cumberland Gap, a wind gap in the Appalachian Mountains at the border of Kentucky and Virginia, to provide other settlers with access to the Kentucky wilderness.

In some cases, plate tectonics can change even the course of mighty rivers. For example, in the early Mesozoic Era, when South America linked to Africa on its eastern coast, a "proto-Amazon" River flowed westward and drained the interior of Gondwana into the Pacific Ocean. Later, when South America separated from Africa, the Andes rose on South America's western coast. This event caused a **drainage reversal**—flow in the Amazon changed direction and began carrying water eastward to the newly formed Atlantic (▶Fig. 17.28a, b).

most; it lies at an elevation close to that of a stream's base level.) Through these stages, a fluvial landscape changes or evolves through time, and extensive denudation (the removal of rock and regolith from the Earth's surface) occurs.

Though the above model makes intuitive sense, it is an oversimplification. Plate tectonics can uplift the land again, and/or global sea-level change can lower the base level, so in reality peneplains rarely develop before downcutting begins again. In fact, geologists debate about whether they ever really form at all.

Where streams cut down into landscape that was originally near the stream's base level, **stream rejuvenation** has occurred. Rejuvenation happens when the base level of a

Time 1:
Swampy, low-relief land
(a)

Time 2:
Well-drained land
(b)

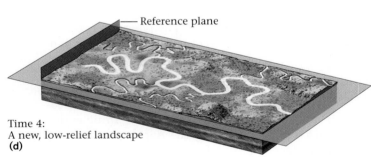

Time 3:
Valleys become broader.
(c)

Time 4:
A new, low-relief landscape
(d)

FIGURE 17.26 (a) A fluvial landscape is first uplifted, so that the base level lies at a lower elevation than does the stream channel. **(b)** Then the stream cuts down into the plain, leaving remnants of the plain as flat-topped mesas between valleys. **(c)** Later, the landscape consists of rounded hills dissected by tributaries that feed a trunk stream flowing on a floodplain near the base level. Valleys are V-shaped. **(d)** Still later, only a few remnant hills are left, for most of the landscape has been denuded to form a new peneplain, nearly at sea level. The decreased height of the land surface above the reference plane indicates the thickness of land that has eroded away.

Superposed and Antecedent Streams

In some locations, the structure and topography of the landscape does *not* appear to control the path, or course, of a stream. For example, imagine a stream that carves a deep canyon straight across a strong mountain ridge—why didn't the stream find a way around the ridge? We distinguish two types of streams that cut across resistant topographic highs:

FIGURE 17.27 (a) A drainage divide separates the Hades River drainage from the Persephone River drainage. Headward erosion by the Hades River gradually breaches the drainage divide, creating a water gap. **(b)** When the source of the Hades River reaches the channel of the Persephone River, Hades (the pirate stream) captures Persephone and carries off its water to the Styx Sea. As a result, the former channel of the Persephone River becomes a dry canyon or channel.

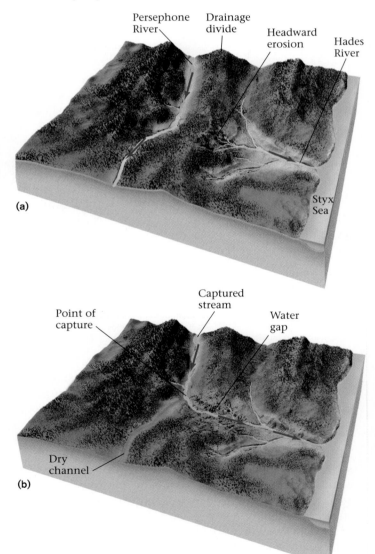

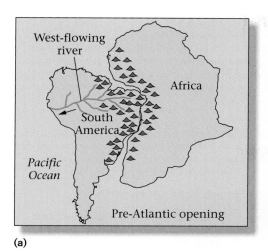

(a)

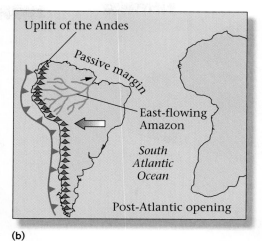

(b)

FIGURE 17.28 (a) In the early Mesozoic Era, highlands existed along the boundary between Africa and South America. A river drained westward across South America to the Pacific. (b) In the late Mesozoic Era, South America began to drift westward. Subduction caused uplift of the Andes, and the Amazon River reversed course and drained water to the Atlantic.

- *Superposed streams:* Imagine a region in which drainage initially forms on a layer of soft, flat strata that unconformably overlies folded strata. Streams carve channels into the flat strata; when they eventually erode down through the unconformity and start to downcut into the folded strata, they maintain their earlier course, ignoring the structure of the folded strata. Geologists call such streams **superposed streams**, because their preexisting geometry has been laid down on the rock structure (▶Fig. 17.29a, b).

- *Antecedent streams:* In some cases, tectonic activity (such as subduction or collision) causes a mountain range to rise up beneath an already established stream. If the stream downcuts as fast as the range rises, it can maintain its course and will cut right across the range. Geologists call such streams **antecedent streams** (from the Greek *ante,* meaning be-

fore), to emphasize that they existed before the range uplifted. Note that if the range rises faster than the stream downcuts, the new highlands divert (change) the stream's course so that it flows along the range face (▶Fig. 17.30a–d).

In some locations, antecedent streams display incised meanders that lie at the bottom of a steep-walled canyon; the canyon was carved out by the stream when uplift occurred in the region. The "goosenecks" of the San Juan River, in southern Utah, illustrate this geometry (▶Fig. 17.31a–c).

Take-Home Message
Stream-carved landscapes evolve over time as gradients diminish and the ridges and hills between valleys erode away. Superposed streams attain their shape before cutting down into rock structure, whereas antecedent streams cut while the land beneath them uplifts.

FIGURE 17.29 (a) A superposed stream first establishes its geometry while flowing over uniform, flat layers above an unconformity. (b) The stream gradually erodes away the layers and exposes underlying rock with a different structure. (In this example, the older strata are folded.) The drainage is superposed (let down) on the folded rocks, and appears to ignore structural control.

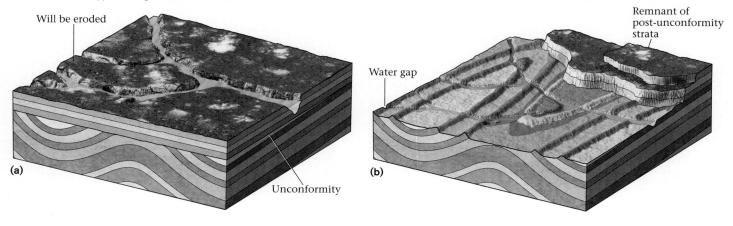

(a)

(b)

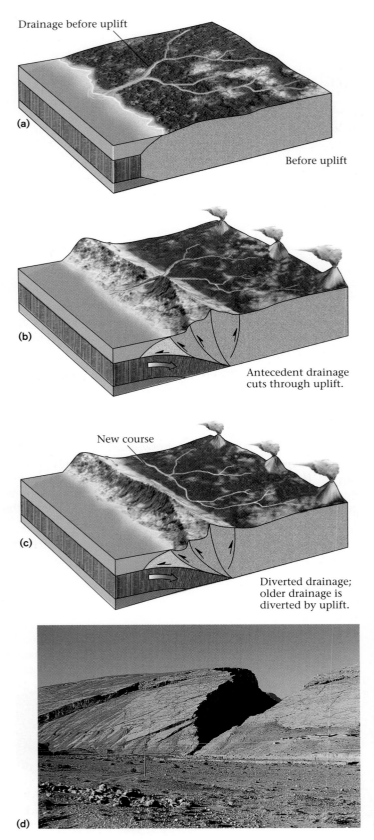

Drainage before uplift

(a) Before uplift

(b) Antecedent drainage cuts through uplift.

New course

(c) Diverted drainage; older drainage is diverted by uplift.

(d)

FIGURE 17.30 (*left*) **(a)** An antecedent stream flows across the land to the sea. **(b)** A mountain range develops across the path of the stream. If the stream erosion keeps pace with the rate of uplift, the stream cuts across the mountain range and is an antecedent stream. **(c)** If uplift happens faster than erosion, the stream is diverted and flows along the front of the range. This stream is not antecedent. **(d)** An antecedent stream cutting through a rapidly uplifting ridge in Pakistan.

FIGURE 17.31 (*below*) **(a)** A stream forms meanders while it flows across a plain. **(b)** Uplift of the land over which the stream flows causes the meanders to cut down and carve out canyons that meander like the stream. **(c)** The "goosenecks" of the San Juan River, in Utah, illustrate incised meanders.

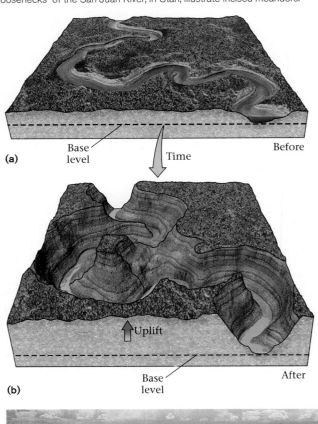

(a) Base level — Before

Time

Uplift

(b) Base level — After

(c)

17.8 RAGING WATERS

[And Enhil, the ruler of the gods, said,] "The earth bellows like a herd of wild oxen. The clamor of human beings disturbs my sleep. Therefore, I want Adad [god of the skies] to cause heavy rains to pour down upon the Earth, both day and night. I want a great flood to come like a thief upon the Earth, steal the food of these people and destroy their lives."

–from the **Epic of Gilgamesh**
(written in Sumeria, c. 2100 b.c.e.)

The Inevitable Catastrophe

Up to now, this chapter has focused on the variety of features and processes of a river system (see art, pp. 608–609). Now we turn our attention to the havoc that a stream can cause when it starts to flood. Floods can be catastrophic—they can strip land of forests and buildings, they can bury land in mud and silt, and they can submerge cities. A flood occurs when the volume of water flowing down a stream exceeds the volume of the stream channel, so water rises out of the normal channel and spreads out over the floodplain or delta plain (by breaking through levees), or it fills a canyon to a greater depth than normal. The news media may report that a river "crested at 9 feet [3 m] above flood stage at 10 P.M." This means that at 10, the water surface in the stream was 3 m higher than the top of the normal channel, and after that time it became lower. (The flood crest is the highest level that the stream reaches.) Because of its increased discharge, a stream in flood flows faster than it normally does, so it's more turbulent, has greater competence, and exerts more pressure on structures in its path. Muddy, fast-moving floodwater is denser than clear water, and pushes harder on objects in its path. It can buoy and transport sediment, as well as cars, buildings, and people.

Floods happen (1) during abrupt, heavy rains, when water falls on the ground faster than it can infiltrate and thus immediately becomes surface runoff; (2) after a long period of continuous rain, when the ground has become saturated with water and can hold no more; (3) when heavy snows from the previous winter melt rapidly in response to a sudden hot spell; or (4) when a dam holding back a lake or reservoir, or a levee holding back a river or canal suddenly collapses and releases the water that it held back.

Geologists find it convenient to divide floods into two general categories. Floods that occur regularly when rainfall is particularly heavy or when winter snows start to melt are called **seasonal floods**. Severe floods of this type take place in tropical regions that are drenched by monsoons. During the 1990 monsoon season in Bangladesh, for example, rain fell almost continuously for weeks. The delta plain became inundated; the resulting flood killed 100,000 people. The floods in Bangladesh can also be called delta-plain floods because the water submerges the delta plain. Similarly, floodplain floods submerge a floodplain (▶Fig. 17.32a–e).

Typically, seasonal floods take time—hours or days—to develop. Thus, in many cases, authorities can evacuate potential victims and organize efforts to protect property. Nevertheless, so many people live on deltas and floodplains that these floods can cause a staggering loss of life and property. A 1931 flood of the Yangtze River in China led to a famine that killed 3.7 million people, and an 1887 flood of China's Hwang (Yellow) River, so named because of the yellow silt it carries, killed as many as 2.5 million. More recently, the flooding of Italy's Arno River submerged the art treasures of Florence, and the flooding of North Dakota's Red River submerged the town of Grand Forks and threatened Winnipeg, Manitoba. Floods in Europe in September 2000 caused havoc in several cities. Seasonal floods struck Indonesia in 2007, killing dozens of people and displacing almost half a million, nearly half of whom became sick with diarrhea and skin infections from contact with filthy water and mud that submerged 60% of the capital and hundreds of square kilometers of farm land (Fig. 17.32e).

Events during which the floodwaters rise so fast that it may be impossible to escape from the path of the water are called **flash floods**. These happen during unusually intense rainfall or as a result of a dam collapse (as in the 1889 Johnstown flood). During a flash flood, a wall of water may slam downstream with great force, leaving devastation in its wake, but the floodwaters subside after a short time. Flash floods can be particularly unexpected in arid or semi-arid climates, where isolated thundershowers may suddenly fill the channel of an otherwise dry wash, whose unvegetated ground can absorb little water. Such a flood may even affect areas downstream that had not received a drop of rain.

Case Study: A Seasonal Flood (Midwestern United States)

In the spring of 1993, the jet stream, the high-altitude (10–15 km high) wind current that controls weather systems, drifted southward (see Chapter 20). For weeks, the jet stream's cool, dry air formed an invisible wall that trapped warm, moist air from the Gulf of Mexico over the central United States. When this air rose to higher elevations, it cooled, and the water it held condensed and fell as rain, rain, and more rain. In fact, almost a whole year's supply of rain fell in just that spring—some regions received 400% more than usual. Because the rain fell over such a short period, the ground became saturated and could no longer absorb additional water, so the excess entered the region's streams, which carried it into the Missouri and Mississippi rivers.

River Systems

Rivers, or streams, drain the landscape of surface runoff. Typically, an array of connected streams called a drainage network develops, consisting of a trunk stream into which numerous tributaries flow. The land drained is the network's watershed. A stream starts from a source, or headwaters, in the mountains, perhaps collecting water from rainfall or from melting ice and snow. In the mountains, streams carve deep, V-shaped valleys, and tend to have steep gradients. For part of its course, a river may flow over a steep, bouldery bed, forming rapids. It may drop off an escarpment, creating a waterfall. Rivers gradually erode landscapes and carry away debris, so after a while, if there is no renewed uplift, mountains are eroded into gentle hills. Over time, rivers can bevel once-rugged mountain ranges into nearly flat plains.

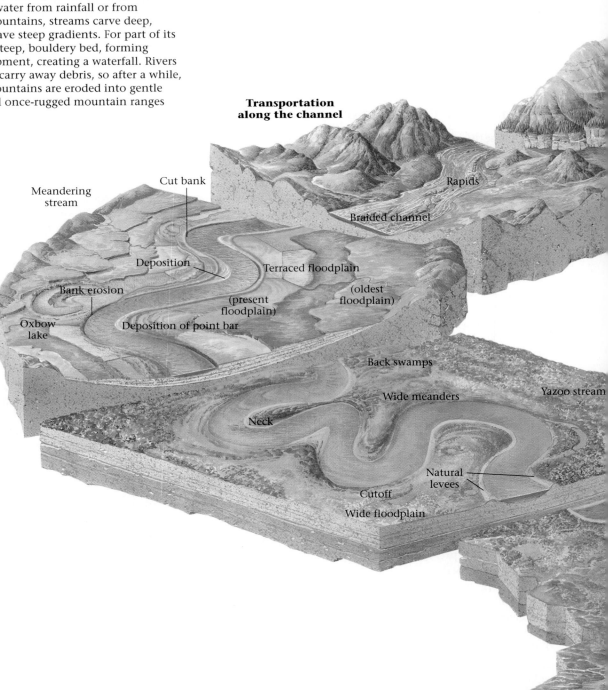

Transportation along the channel

Meandering stream

Cut bank

Deposition

Bank erosion

Oxbow lake

Deposition of point bar

Terraced floodplain

(present floodplain)

(oldest floodplain)

Rapids

Braided channel

Back swamps

Wide meanders

Yazoo stream

Neck

Cutoff

Wide floodplain

Natural levees

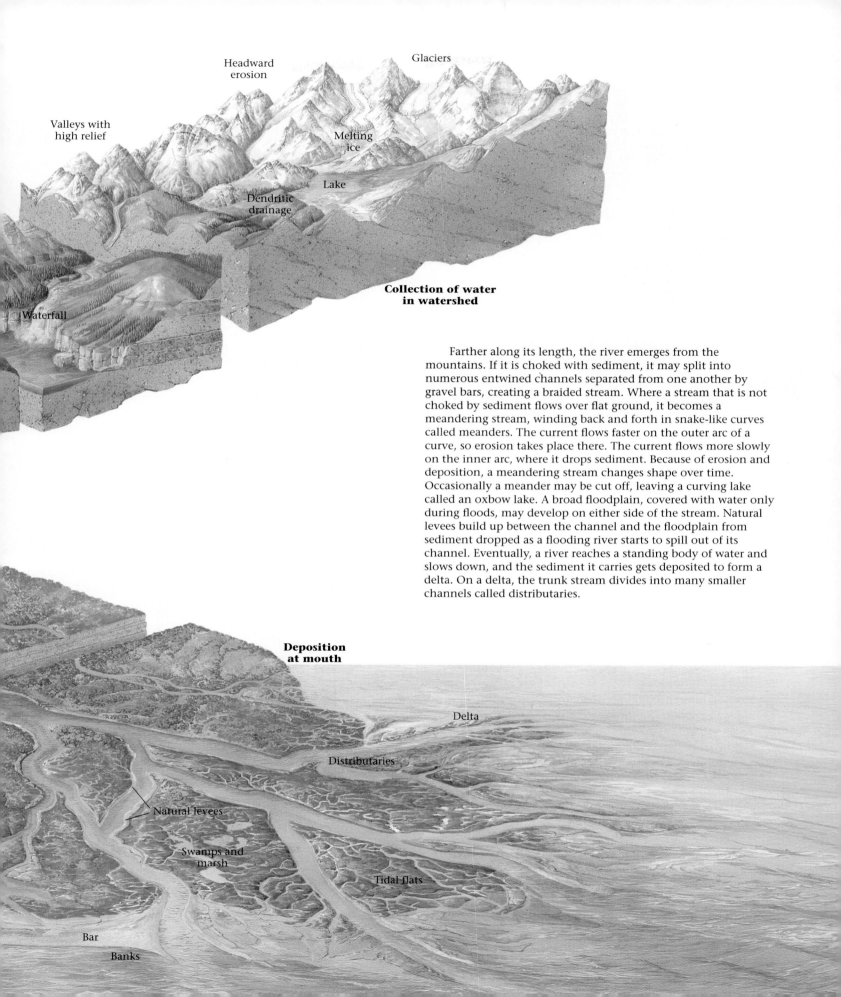

Headward
erosion

Glaciers

Valleys with
high relief

Melting
ice

Lake

Dendritic
drainage

Waterfall

**Collection of water
in watershed**

Farther along its length, the river emerges from the mountains. If it is choked with sediment, it may split into numerous entwined channels separated from one another by gravel bars, creating a braided stream. Where a stream that is not choked by sediment flows over flat ground, it becomes a meandering stream, winding back and forth in snake-like curves called meanders. The current flows faster on the outer arc of a curve, so erosion takes place there. The current flows more slowly on the inner arc, where it drops sediment. Because of erosion and deposition, a meandering stream changes shape over time. Occasionally a meander may be cut off, leaving a curving lake called an oxbow lake. A broad floodplain, covered with water only during floods, may develop on either side of the stream. Natural levees build up between the channel and the floodplain from sediment dropped as a flooding river starts to spill out of its channel. Eventually, a river reaches a standing body of water and slows down, and the sediment it carries gets deposited to form a delta. On a delta, the trunk stream divides into many smaller channels called distributaries.

**Deposition
at mouth**

Delta

Distributaries

Natural levees

Swamps and
marsh

Tidal flats

Bar

Banks

(a)

(b)

(c)

(d)

(e)

FIGURE 17.32 **(a)** Satellite photo showing the Mississippi and Missouri Rivers during a time of drought as seen from high elevation. **(b)** The same area during the 1993 flood. Notice how the floodplains of the rivers are totally submerged. **(c)** Great Falls, Montana, submerged by floodwaters in 1975. **(d)** After floodwaters receded (from an orchard in Arizona), they left a layer of mud and silt. **(e)** The devastating floods that hit Indonesia in 2007 brought not only severe property damage, but disease and infections as well, as the people had to wade through filthy water.

Eventually, the water in these rivers rose above the height of levees and spread out over the floodplain. By July, parts of nine states were under water (Fig. 17.32a, b).

The roiling, muddy flood uprooted trees, cars, and even coffins (which floated up from inundated graveyards). All barge traffic along the Mississippi came to a halt, bridges and roads were undermined and washed away, and towns along the river were submerged in muddy water. For example, in Davenport, Iowa, the riverfront district and baseball stadium were covered with 4 m (14 feet) of water. In Des Moines, Iowa, 250,000 residents lost their supply of drinking water when floodwaters contaminated the municipal water supply with raw sewage and chemical fertilizers. Rowboats replaced cars as the favored mode of transportation in towns where only the rooftops remained visible. In St. Louis, Missouri, the river crested 14 m (47 feet) above flood stage. When the water finally subsided, it left behind a thick layer of silt and mud, filling living rooms and kitchens in floodplain towns and burying crops in floodplain fields.

For 79 days, the flooding continued. In the end, more than 40,000 square km of the floodplain had been submerged, fifty people died, at least 55,000 homes were destroyed, and countless acres of crops were buried. Officials estimated that the flood caused over $12 billion in damage.

Case Study: A Flash Flood (Big Thompson Canyon)

On a typical sunny day in the Front Ranges of the Rocky Mountains, north of Denver, Colorado, the Big Thompson River seems quite harmless. Clear water, dripping from melting ice and snow higher in the mountains, flows down its course through a narrow canyon, frothing over and around boulders. In places, vacation cabins, campgrounds, and motels line the river, for the pleasure of tourists. The landscape seems immutable, but, as is the case with so many geologic features, permanence is an illusion.

On July 31, 1976, easterly winds blew warm, moist air from the Great Plains toward the Rocky Mountain front. As this air rose over the mountains, towering thunderheads built up, and at 7:00 P.M. rain began to fall. It poured, in quantities that even old-timers couldn't recall. In a little over an hour, 19 cm (7.5 inches) of rain drenched the watershed of the Big Thompson River. The river's discharge grew to more than 4 times the maximum recorded at any time during the previous century. The river rose quickly, in places reaching depths several meters above normal. Turbulent water swirled down the canyon at up to 8 m per second and churned up so much sand and mud that it became a viscous slurry. Slides of rock and soil tumbled down the steep slopes bordering the river and fed the torrent with even more sediment. The water undercut house foundations and washed the houses away, along with their inhabitants. Roads and bridges disappeared (►Fig. 17.33). Boulders that had stood like landmarks for generations bounced along in the torrent like beachballs, striking and shattering other rocks along the way; the largest rock known to be moved by the flood weighed 275 tons. Cars drifted downstream until they finally wrapped like foil around obstacles. When the flood subsided, the canyon had changed forever, and 144 people had lost their lives.

FIGURE 17.33 During the 1976 Big Thompson River flood, this house was carried off its foundation and dropped on a bridge.

Ice-Age Megafloods

Perhaps the greatest floods chronicled in the geologic record happen when natural ice dams burst. The Great Missoula Floods of about 11,000 years ago illustrate this phenomenon (see Box 22.2). This flood occurred at the end of the last ice age, when a glacier acted like a dam, holding back a large lake called Glacial Lake Missoula. When the glacier melted and the dam suddenly broke, the lake abruptly drained, and water roared over what is now eastern Washington, eventually entering the Columbia River Valley and flowing on out to the Pacific Ocean. If the glacier then grew again, the dam reformed, trapping a new lake, which could drain during a subsequent failure. These floods formed the channeled scablands of eastern Washington, a region where the soil and regolith have been stripped off the land surface, leaving barren, craggy rock (►Fig. 17.34a–c).

The hypothesis that the channeled scablands formed as a consequence of catastrophic flooding was first proposed by J. Harlan Bretz, who studied the landscape of the region in the 1920s. Initially, other geologists ridiculed Bretz because his idea seemed to violate the well-accepted principle of uniformitarianism (see Chapter 12). But Bretz steadily fought back, demonstrating that the scablands, an unglaciated region, are littered with boulders too large to have been carried by normal rivers, that hills in the region were giant ripples a thousand times larger than the ripples typically found in a stream bed, that the now-dry Grand Coulee was once a giant waterfall hundreds of times larger than Niagara Falls, and that deep pits in the region are actually huge potholes scoured by whirlpools. Ultimately the geologic community accepted the reality of the Great Missoula Floods and other such catastrophic events during Earth history.

Living with Floods

Mark Twain once wrote of the Mississippi that we "cannot tame that lawless stream, cannot curb it or confine it, cannot say to it, 'go here or go there,' and make it obey." Was Twain right? Since ancient times, people have attempted to confine rivers to set courses so as to prevent undesired flooding. In the twentieth century, flood-control efforts intensified as the population living along rivers increased. For example, since the passage of the 1927 Mississippi River Flood Control Act (drafted after a disastrous flood took place that year), the U.S. Army Corps of Engineers has labored to control the Mississippi. First, engineers built about 300 dams along the river's tributaries so that excess runoff could be stored in the reservoirs and later be released slowly. Second, they built sand and mud levees and concrete flood walls to increase the channel's volume. Such artificial levees isolate a discrete area of the floodplain (►Fig. 17.35a, b).

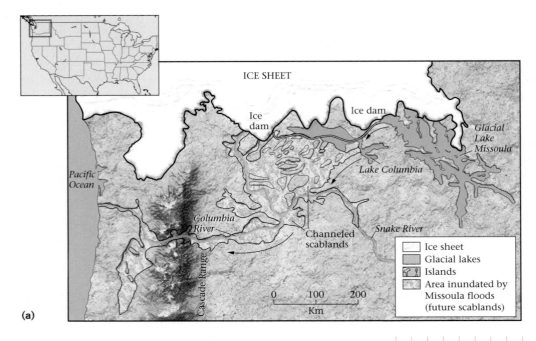

ICE SHEET

Ice dam

Ice dam

Glacial Lake Missoula

Pacific Ocean

Lake Columbia

Columbia River

Channeled scablands

Snake River

Cascade Range

☐	Ice sheet
▨	Glacial lakes
🏝	Islands
▨	Area inundated by Missoula floods (future scablands)

0   100   200
Km

(a)

FIGURE 17.34 (a) The map illustrates Glacial Lake Missoula, blocked by an ice dam. When the dam melted away and finally broke, the lake drained westward, eroding away all sediment cover and leaving behind scoured basalt—a region now known as the channeled scablands. **(b)** The channeled scablands, in Washington State, as viewed from the air. **(c)** A geologist's sketch of the photo.

(b)

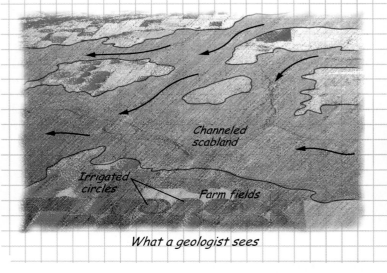

Channeled scabland

Irrigated circles

Farm fields

What a geologist sees

(c)

But although the corps' strategy worked for floods up to a certain size, it was insufficient to handle the 1993 flood. Because of the volume of water drenching the Midwest during the spring and early summer of that year, the reservoirs filled to capacity, and additional runoff headed downstream. The river rose until it spilled over the tops of some levees and undermined others. Undermining occurs when rising floodwaters increase the water pressure on the river side of the levee, forcing water through sand under the levee. In susceptible areas, water begins to spurt out of the ground on the dry side of the levee, thereby washing away the levee's support. The levee finally becomes so weak that it collapses, and water fills in the area behind it.

Sooner or later a flood comes along that can breach a river's levees, allowing water to spread out over the floodplain. Defensive efforts merely delay the inevitable, for it is unfeasible and too expensive to build levees high enough to handle all conceivable floods. And in some cases, building levees may be counterproductive, since they constrain water to a smaller area and thus make floodwaters rise to a higher level than they would if they were free to spread over a wide floodplain. Those who build on floodplains must face this reality and consider alternative ways to use the region that can accommodate occasional flooding. The cost of flood damage has quadrupled in recent years, despite the billions of dollars

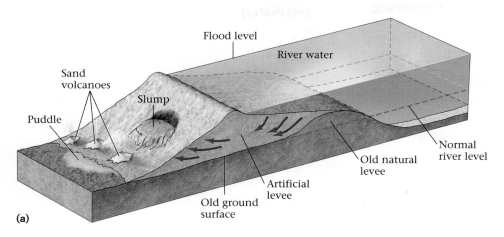

(a)

When making decisions about investing in flood-control measures, mortgages, or insurance, planners need a basis for defining the hazard or risk posed by flooding. If floodwaters submerge a locality every year, a bank officer would be ill advised to approve a loan that would promote building there. But if floodwaters submerge the locality very rarely, then the loan may be worth the risk. Geologists characterize the risk of flooding in two ways. The **annual probability** (more formally known as the "annual exceedance probability") of flooding indicates the likelihood that a flood of a given size or larger will happen at a specified locality during any given year. For example, if we say that a flood of a given size has an annual probability of 1%, then we mean that there is a 1 in 100 chance that a flood of at least this size will happen in any given year. The **recurrence interval** of a flood of a given size is defined as the *average* number of years between successive floods of at least this size. For example, if a flood of a given size happens once in 100 years, *on average,* then it is assigned a recurrence interval of 100 years and is called a 100-year-flood. Note that annual probability and recurrence interval are related:

$$\text{annual probability} = \frac{1}{\text{recurrence interval}}.$$

(b)

FIGURE 17.35 (a) When the water level on the river side of the levee is much higher than on the dry floodplain, pressure causes water to infiltrate the ground and flow through this artificial levee. The water spurts out of the ground on the dry side of the levee, generating sand volcanoes. Water saturates the levee, so the face of the levee slumps. The levee eventually collapses. **(b)** A concrete floodwall on Cape Girardeau, Missouri. When floods threaten, a crane drops a gate into the slot to hold out the Mississippi River. High-water marks are indicated by black lines.

FIGURE 17.36 Concept of a floodway. **(a)** Building artificial levees directly on natural levees creates a larger channel for a river. **(b)** Building artificial levees at a distance from the river creates a floodway on either side of the river, an even larger channel than in (a), and a surface of wetland that can absorb floodwaters.

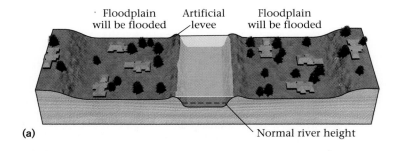

(a)

that have been spent on flood "control," because more people have settled in floodplains.

There are other ways to prevent floods besides building levees and reservoirs. For example, transforming portions of floodplains back into natural wetlands helps prevent floods, for wetlands absorb water like a sponge. A solution to flooding in some cases may lie in the removal, rather than the construction, of levees. Property may also be kept safe by defining **floodways**, regions likely to be flooded, and then by moving or abandoning buildings located there. Even the simple act of moving levees farther away from the river and creating natural habitats in the resulting floodways would decrease flooding damage immensely (►Fig. 17.36a, b).

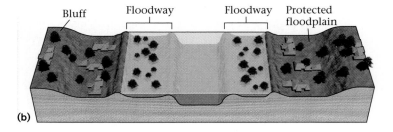

(b)

For example, the annual probability of a 50-year-flood is ⅟₅₀, which can also be written as 0.02 or 2%. To learn how to calculate annual probabilities and recurrence intervals of floods in more detail, see ▶Box 17.1.

Unfortunately, some people are misled by the meaning of recurrence interval, and think that they do not face a flooding hazard if they buy a home built within an area submerged by 100-year floods just after such a flood has occurred. Their confidence comes from making the incorrect assumption that because such flooding just happened, it can't happen again until "long after I'm gone." They may regret their decision because two 100-year floods can occur in consecutive years or even in the same year (alternatively, the interval between such floods could be, say, 210 years). Because the term *recurrence interval* can lead to confusion, it may be better to report risk in terms of annual probability.

Knowing the discharge during a flood of a specified annual probability, and knowing the shape of the river channel and the elevation of the land bordering the river, hydrologists can predict the extent of land that will be submerged by such a flood (▶Fig. 17.37). Such data, in turn, permit hydrologists to produce flood-hazard maps. In the United States, the Federal Emergency Management Agency (FEMA) produces Flood Insurance Rate Maps that show the 1% annual probability (100-year) flood area and the 0.2% annual probability (500-year) flood risk zones (▶Fig. 17.38).

> **Take-Home Message**
>
> Seasonal floods submerge floodplains and delta plains at certain times of the year. Flash floods are sudden and short lived. We can specify the probability that a certain-size flood will happen in a given year, but flood-control efforts meet with mixed success.

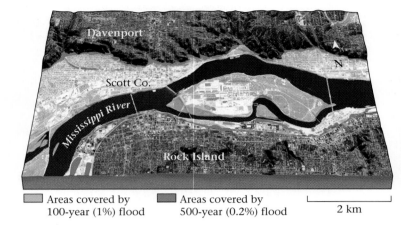

Areas covered by 100-year (1%) flood   Areas covered by 500-year (0.2%) flood   2 km

FIGURE 17.38 A flood hazard map for a region near Davenport, Iowa, as prepared by FEMA. It shows areas likely to be flooded.

17.9 RIVERS: A VANISHING RESOURCE?

As *Homo sapiens* evolved from hunter-gatherers into farmers, areas along rivers became attractive places to settle. Rivers serve as avenues for transportation and are sources of food, irrigation water, drinking water, power, recreation, and (unfortunately) waste disposal. Further, their floodplains provide particularly fertile soil for fields, replenished annually by seasonal floods. Considering the multitudinous resources that rivers provide, it's no coincidence that early civilizations developed in river valleys and on floodplains: Mesopotamia arose around the Tigris and Euphrates Rivers, Egypt around the Nile, India in the Indus Valley, and China along the Hwang (Yellow) River. Over the millennia, rivers have killed millions of people in floods, but they have been the lifeblood for hundreds of millions more. Nevertheless, over time, humans have increasingly tended to abuse or overuse the Earth's rivers. Here we note four pressing environmental issues.

Pollution. The capacity of some rivers to carry pollutants has long been exceeded, transforming them into deadly cesspools. Pollutants include raw sewage and storm drainage from urban areas, spilled oil, toxic chemicals from industrial sites, and excess fertilizer and animal waste from agricultural fields. Some pollutants directly poison aquatic life, some feed algae blooms that strip water of its oxygen, and some settle out to be buried along with sediments. River pollution has become overwhelming in developing countries, where there are few waste-treatment facilities.

Dam construction. In 1950, there were about 5,000 large (over 15 m high) dams worldwide, but today there are over 38,000. Damming rivers has both positive and negative results.

FIGURE 17.37 A 100-year flood covers a larger area than a 2-year flood.

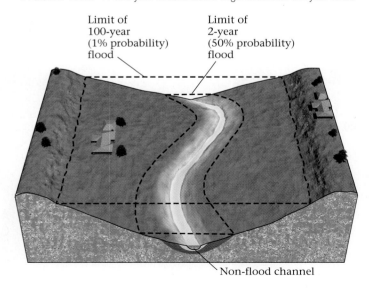

Limit of 100-year (1% probability) flood

Limit of 2-year (50% probability) flood

Non-flood channel

Reservoirs provide irrigation water and hydroelectric power, and they trap some floodwaters and create popular recreation areas. But sometimes their construction destroys "wild rivers" (the whitewater streams of hilly and mountainous areas) and alters the ecosystem of a drainage network by forming barriers to migrating fish, decreasing the nutrient supply to organisms downstream, by removing the source of sediment for the delta, and by eliminating seasonal floods that replenish nutrients in the landscape.

Overuse of water. Because of growing populations, our thirst for river water continues to increase, but the supply of water does not. The use of water has grown especially in response to the "Green Revolution" of the 1960s, during which huge new tracts of land came under irrigation. Today, 65% of the water taken out of rivers is used for agriculture, 25% for industry, and 9% for drinking and sewage transport. Civilization needed 3 times as much river water in 1995 as it did in 1950.

BOX 17.1 THE REST OF THE STORY

Calculating the Threat Posed by Flooding

How do we calculate the probability that a flood of a given size at a locality along a stream will happen in a given year? (Note that "size" in this context is indicated by the stream's discharge, as measured in cubic feet per second or cubic meters per second). First, researchers collect data on the stream's discharge at the locality for at least 10 to 30 years to get a sense of how the discharge varies during a year. Then, they pick the largest, or peak, discharge for each year and make a table listing the peak discharges. The largest peak discharge is given a *rank* of 1, the second-largest discharge is given a rank of 2, and so on. Researchers can then calculate the recurrence interval for each different discharge by using a simple equation:

$$R = (n + 1) \div m$$

R is the recurrence interval in years, n is the total number of years for which there is a record, and m is the rank.

Once the recurrence interval for each peak discharge has been calculated, the researchers plot a graph: the vertical axis represents peak discharge, and the horizontal axis represents recurrence interval. In order for all the data to fit on a reasonable-size graph, the horizontal axis must be logarithmic. Typically, the data for a stream plots roughly along a straight line (▶Fig. 17.39a). In the example shown in Figure 17.39a, a flood with a recurrence interval of 10 years (meaning an annual probability of 10%) has a peak discharge of about 460 cubic feet per second. We can extend the line beyond the data points (the dashed line on Fig. 17.39a) to make predictions about the re-

currence interval and, therefore, the annual probability, of floods with discharges larger than the ones that have been measured. As more data become available, the graph may need to be modified. In this example, note that the graph predicts that a 1% probability flood (a 100-year flood) will have a discharge of about 650 cubic feet per second.

The peak annual discharge of the Mississippi River at St. Louis has been measured almost continuously since 1850. A bar graph of these data shows that floods characterized as 100-year floods (meaning 1% probability floods) or larger happened in 1844, 1903, and 1993 (▶Fig. 17.39b). Note that the time between "100-year floods" is not exactly 100 years.

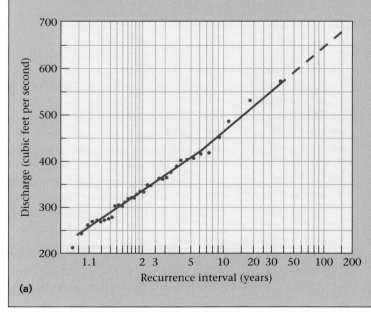

(a)

FIGURE 17.39 (a) A flood-frequency graph shows the relationship between the recurrence interval and the discharge for an idealized river. (b) The peak discharge of the Mississippi River as measured at St. Louis, Missouri. Each bar represents the largest discharge of a given year. The horizontal line represents the discharge of a 100-year flood.

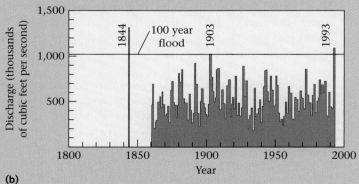

(b)

FIGURE 17.40 The Central Arizona Project canal, shunting water from the Colorado River to Phoenix.

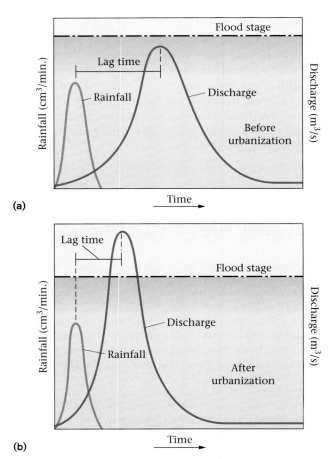

FIGURE 17.41 Hydrographs. **(a)** Before urbanization, rain infiltrated the ground, and vegetation slowed sheetwash. As a consequence, the peak discharge resulting from a heavy rain was smaller, and the peak runoff occurred after some time (the lag time) had passed. **(b)** With urbanization, the peak discharge happens after a shorter lag time, and is great enough to reach flood stage.

As a result, in some places human activity consumes the entire volume of a river's water, so that the channel contains little more than a saline trickle, if that, at its mouth. For example, except during unusually wet years, the Colorado River contains almost no water where it crosses the Mexican border, for huge pipes and canals carry the water instead to Phoenix and Los Angeles (▶Fig. 17.40). In the case of the Colorado River, states along its banks have established legal agreements that divide up the river's water. Unfortunately, the agreements were written during wet years when the river had unusually large discharge. Thus, the amount of water specified in the agreements actually exceeds the amount of water the river carries in most years.

Perhaps the most dramatic consequence of river over-consumption can be seen in central Asia, where almost all the water in the Amu Darya and Syr Darya rivers has been diverted into irrigation. These rivers once fed the Aral Sea. Now the sea has shrunk so much that "coastal" towns lie many kilometers from the coast, fishing trawlers rot in the desert (see Fig. 21.29d), and the catch of fish has dropped from 44,000 to 0 tons per year.

The effects of urbanization and agriculture on discharge. Although the consumption of water for agricultural and industrial purposes decreases the overall supply of river water, urbanization may actually increase the short-term supply. Cities cover the ground with impermeable concrete or blacktop, so rainfall does not soak into the ground but rather runs into storm sewers and then into streams, caus-ing local flooding. Stream discharge during a rainfall thus increases much more rapidly than it would without urban-ization. This change can be illustrated by dia-grams, called hydro-graphs, that show how discharge varies with time (▶Fig. 17.41a, b). Similarly, although dam-ming rivers decreases the amount of silt a river car-ries downstream, agri-culture may increase the sediment supply: agriculture decreases the vegetative cover on the land, so that when it rains, soil washes into streams.

> **Take-Home Message**
> People have greatly modified streams by constructing dams and levees, by adding pollutants, and by modifying the amount of water that enters streams. These changes have created significant problems in drainage networks.

Chapter Summary

- Streams are bodies of water that flow down channels and drain the land surface. Channels develop when sheetwash cuts into the substrate and concentrates the water flow; they grow by headward erosion. Streams carry water out of a drainage basin. A drainage divide separates two adjacent catchments.

- Drainage networks consist of many tributaries that flow into a trunk stream.

- Permanent streams exist where the water table lies above the bed of the channel. Where the water table lies below the channel bed, streams are ephemeral and dry up between rainfalls to form dry washes.

- The discharge of a stream is the volume of water passing a point in a second. Most streams are turbulent, meaning that their water swirls in complex patterns.

- Streams erode the landscape by scouring, lifting, abrading, and dissolving. The resulting sediment is divided among dissolved, suspended, and bed loads. The total quantity of sediment carried by a stream is its capacity. Capacity differs from competence, the maximum particle size a stream can carry. When stream water slows, it deposits alluvium.

- Longitudinal profiles (images of the shape of a stream bed in cross section from its source to its mouth) of streams are concave up. Typically, a stream has steeper gradients at its headwaters than near its mouth. Streams cannot cut below the base level.

- Streams cut valleys or canyons, depending on the rate of downcutting relative to the rate at which the slopes on either side of the stream undergo mass wasting. Where a stream flows down steep gradients and has a bed littered with large rocks, rapids develop, and where a stream plunges off a vertical face, a waterfall forms.

- Meandering streams wander back and forth across a floodplain. Such a stream erodes its outer bank and builds out sediment into a point bar on the inner bank. Eventually, a meander may be cut off and turn into an oxbow lake. Natural levees form on either side of the river channel. Braided streams consist of many entwined channels.

- Where streams or rivers flow into standing water, they deposit deltas. The shape of a delta depends on the balance between the amount of sediment supplied by the river and the amount of sediment redistributed or carried away by wave activity along the coast.

- With time, fluvial erosion can bevel landscapes to a nearly flat plain. If the base level drops or the land surface rises, stream rejuvenation causes the stream to start downcutting into the peneplain. The headward erosion of one stream may capture the flow of another.

- If an increase in rainfall or spring melting causes more water to enter a stream than the channel can hold, a flood results. Some floods are seasonal, in that they accompany monsoonal rains. Some floods submerge broad floodplains or delta plains. Flash floods happen very rapidly. Officials try to prevent floods by building reservoirs and levees.

- Rivers are becoming a vanishing resource because of pollution, damming, and overuse.

Geopuzzle Revisited

As soon as the land surface of a region rises above the ultimate base level (sea level), water starts flowing toward lower elevations. Eventually, the faster flow carves channels, with tributary channels flowing into a trunk channel. Streams eventually cut down to the base level. The channel volume reflects the usual discharge of the stream. If heavy rain, melting, or a dam rupture provides more water than the channel can hold, water spills over the channel walls and floods the surrounding landscape.

Key Terms

Review Questions

1. What role do streams have during the hydrologic cycle? Indicate various sources of water in streams.

2. Describe the four different types of drainage networks. What factors are responsible for the formation of each?

3. What factors determine whether a stream is permanent or ephemeral?

4. How does discharge vary according to the stream's length, climate, and position along the stream course?

5. Why is average downstream velocity always less than maximum downstream velocity?

6. How does a turbulent flow differ from a laminar flow?

7. Describe how streams and running water erode the Earth.

8. What are three components of sediment load in a stream?

9. Distinguish between a stream's competence and its capacity.

10. Describe how a drainage network changes, along its length, from head waters to mouth.

11. What factors determine the position of the base level?

12. What do lakes, rapids, waterfalls, and terraces indicate about the stream gradient and base level? Why do canyons form in some places, and valleys in others?

13. How does a braided stream differ from a meandering stream?

14. Describe how meanders form, develop, are cut off, and then are abandoned.

15. Describe how deltas grow and develop. How do they differ from alluvial fans?

16. How does a stream-eroded landscape evolve as time passes?

17. What is stream piracy? What causes drainage reversal?

18. How are superposed and antecedent drainages similar? How are they different?

19. What human activities tend to increase flood risk and damage?

20. What is the recurrence interval of a flood? Why can't someone say, "The hundred-year flood happened last year, so I'm safe for another hundred years"?

21. How have humans abused and overused the resource of running water?

On Further Thought

1. The northeastern two-thirds of Illinois, in the midwestern United States, was last covered by glaciers only 14,000 years ago. The rest of the state was last covered by glaciers over 100,000 years ago. Until the advent of modern agriculture, the recently glaciated area was a broad, grassy swamp, cut by very few stream channels. In contrast, the area that was glaciated over 100,000 years ago is not swampy and has been cut by numerous stream valleys. Why?

2. Records indicate that flood crests for a given amount of discharge along the Mississippi River have been getting *higher* since 1927, when a system of levees began to block off portions of the floodplain. Why?

3. The Ganges River carries an immense amount of sediment load, which has been building a huge delta in the Bay of Bengal. Look at the region using an atlas or *Google Earth*™, think about the nature of the watershed supplying water to the drainage network that feeds the Ganges, and explain why this river carries so much sediment.

4. Fly to Lat 43°16′20.55″S Long 170°24′8.52″E using *Google Earth*™, zoom to an elevation of 40 km, and look straight down. You will see a portion of the South Island in New Zealand. In this area, the Whataroa River flows northwest from the Southern Alps, a mountain range formed by movement on a plate boundary called the Alpine fault. The fault trace forms the abrupt boundary between the mountain front and the plains to the northwest. Describe the changes in the nature of the river as it crosses the fault—how is the downstream reach of the stream different from the upstream reach? To help you answer this question, zoom in and out, and change the tilt angle of your view, and compare your view to the drawing of Figure 17.13. Find the drainage divide of the Southern Alps and try to map it along the length of the range. You might want to print an image of your screen to provide a base on which to draw the divide.

5. Look closely at the graph of Figure 17.39 in Box 17.1. What is the recurrence interval of a flood with a discharge of 650 cubic feet per second? In a given year, how much more likely is a flood with a discharge of 200 cubic feet per second than a flood of 400 cubic feet per second? For the sake of discussion, imagine that the floodplain of the river is completely covered when a flood with an annual probabil-

ity 1/300 occurs. Would you build a new home in the flood-plain? Would it make a difference to you if the last flood with this probability happened 1 year ago? 100 years ago?

Suggested Reading

Barry, J.M. 1998. *Rising Tide: The Great Mississippi Flood of 1927 and How It Changed America.* Parsippany, N.J.: Simon and Schuster.

Bridge, J.S. 2003. *Rivers and Floodplains: Forms, Processes and Sedimentary Record.* Malden, Mass.: Blackwell Publishing Ltd.

Brierley, G.J., and K.A. Fryirs. 2005. *Geomorphology and River Management.* Malden, Mass.: Blackwell Publishing Ltd.

Changnon, S. A., 1996. *The Great Flood of 1993: Causes, Impacts, and Responses.* Boulder, CO; Westview Press.

Coleman, J. M., H. H. Roberts, and G. W. Stone. 1998. Mississippi River Delta: An overview. *Journal of Coastal Research* 14: 698–716.

Julien, P.Y. 2006. *River Mechanics.* Cambridge: Cambridge University Press.

Knighton, D. 1998. *Fluvial Forms and Processes: A New Perspective.* New York: A Hodder Arnold Publication, Oxford University Press.

Leopold, L. B. 1994. *A View of the River.* Cambridge, Mass.: Harvard University Press.

Leopold, L. B., M. G. Wolman, and J. P. Miller. 1995. *Fluvial Processes in Geomorphology.* Garden City, N.Y.: Dover.

Leopold, L. B. 1997. *Water, Rivers, and Creeks.* Sausalito, Calif.: University Science Books.

Mathur, A., and D.D. Cunha. 2001. *Mississippi Floods: Designing a Shifting Landscape.* New Haven: Yale University Press.

McCullough, D: 1987. *The Johnstown Flood.* Parsippany, N.J.: Simon and Schuster.

Miall, A.D. 2007. *The Geology of Fluvial Deposits.* New York: Springer.

Ro, C. 2007. *Fundamentals of Fluvial Geomorphology.* New York: Routledge.

Robert, A. 2003. *River Processes: An Introduction to Fluvial Dynamics.* London: Hodder Arnold.

Schumm, S.A. 2003. *The Fluvial System.* Caldwell, N.J.: Blackburn Press.

Schumm, S.A. 2005. *River Variability and Complexity.* Cambridge: Cambridge University Press.

Smith, K., and R. Ward. 1998. *Floods: Physical Processes and Human Impacts.* New York: Wiley.

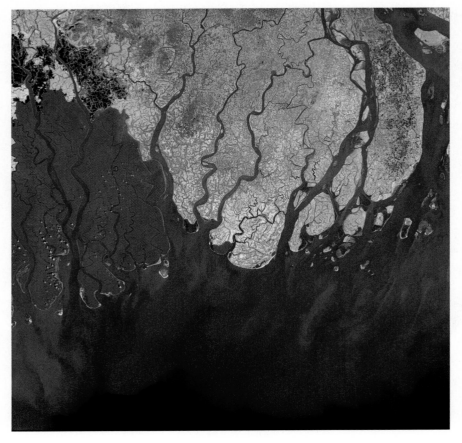

THE VIEW FROM SPACE The Ganges River drains the base of the Himalaya Mountains and carries sediment into the Indian Ocean. The river divides into many meandering channels and has built out a huge delta. These low-lying lands are home to millions of people, but many of these areas flood during the monsoon season, or during typhoons.

Restless Realm:
Oceans and Coasts

Geopuzzle

In 1992, 29,000 plastic bath toys fell off a cargo ship and spilled into the middle of the Pacific Ocean. Some of the toys are about to be washed onto the Atlantic Ocean beaches of the UK. Why?

Surfers off Maui ride the power of waves where the sea meets the shore.

On this wondrous sea
Sailing silently,
Ho! Pilot, ho!
Knowest thou the shore
Where no breakers roar —
Where the storm is o'er?

> *— Emily Dickinson (American poet, 1830–1886)*

18.1 INTRODUCTION

A thousand kilometers from the nearest shore, two scientists and a pilot wriggle through the entry hatch of the research submersible *Alvin*, ready for a cruise to the floor of the ocean and, hopefully, back. *Alvin* consists of a super-strong metal sphere embedded in a cigar-shaped tube (▶Fig. 18.1a). The sphere protects its crew from the immense water pressures of the deep ocean, and the tube holds motors and oxygen tanks. When the hatch seals, *Alvin* sinks at a rate of 1.8 km per hour, mostly through utter darkness, for light penetrates only the top few hundred meters of ocean water. On reaching the bottom, at a depth of 4.5 km, the cramped explorers turn on outside lights to reveal a stark vista of loose sediment, black rock, and the occasional sea creature. For the next 5 hours they take photographs and use a robotic arm to collect samples. When finished, they release ballast and rise like a bubble, reaching the surface about 2 hours later.

Alvin dives began in the 1970s, but humans have explored the ocean for tens of centuries. In fact, Phoenician traders circumnavigated Africa by 590 B.C.E., and Polynesian sailors used outrigger canoes to travel among South Pacific islands beginning around 700 C.E. Chinese naval ships may have circled the globe in the fifteenth century. European mapmakers have known that the ocean spanned the entire globe since Ferdinand Magellan's round-the-world voyage of 1519–22, but they could not systematically map the ocean until the late eighteenth century, when it became possible to determine longitude accurately. Subsequently, naval officers gathered data on water depths in the ocean (using a plumb line, a lead weight on the end of a cable), and by 1839 had determined that the greatest ocean depths could swallow the highest mountains without a trace.

A converted British navy ship, the H.M.S *Challenger*, made the first true ocean research cruise (▶Fig. 18.1b). Beginning in 1872, onboard scientists spent 4 years dredging rocks from the sea floor, analyzing water composition, collecting specimens of marine organisms, and measuring water depths and currents. But still our knowledge of the ocean remained spotty. In fact, we knew less about the ocean floor than we did about the surface of the Moon, for at least we could see the Moon with a telescope.

The fields of oceanography (the study of ocean water and its movements), marine geology (the study of the ocean floor), and marine biology (the study of life forms in the sea) expanded rapidly in the latter half of the twentieth century, as new technology became available and a fleet of oceanographic research ships invaded the seas. These ships can tow instrument-laden sleds just above the sea floor; the sleds use sonar to generate detailed bathymetric maps revealing the shape of the floor (▶Fig. 18.2a). Some ships send pulses of sound into the sea floor that reflect off layers in the subsurface and return to the ship to provide an image, called a seismic-reflection profile, of the layering in the oceanic crust (▶Fig. 18.2b; see Interlude D). Other ships, such as the *JOIDES Resolution*, drill holes as deep as 4 km into the sea floor and bring up samples of the oceanic

FIGURE 18.1 **(a)** The *Alvin*, a submersible used to explore the ocean floor. **(b)** The H.M.S *Challenger*, the first ship to undertake a cruise dedicated to ocean research.

(a)

(b)

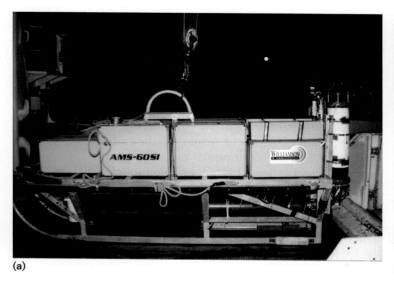

(a)

Depth Migration

(b)

FIGURE 18.2 (a) Researchers lowering sampling containers over the side of a modern oceanographic research vessel. **(b)** In this traditional seismic-reflection profile of the sea floor, the darker lines represent the boundaries between sedimentary layers in the oceanic crust.

crust. In addition, satellites circling the globe produce high-resolution maps of the ocean and its coasts. And while ships and satellites research the open ocean, land-based geologists study its margins.

When seen from space, Earth glows blue, for the oceans cover 70.8% of its surface. The sea provides the basis for life, tempers Earth's climate, and spawns its storms. It is a vast reservoir for water and chemicals that cycle into the atmosphere and crust, and for sediment washed off the continents. In this chapter, we first learn the fundamental characteristics of ocean basins and seawater, and the role they play in the Earth System. Then we focus on the landforms that develop along the **coast**, the region where the land meets the sea, and where over 60% of the global population lives today. Finally, we consider how to cope with the hazards of living on the coast.

> **Take-Home Message**
>
> Oceanic lithosphere differs from continental lithosphere and lies at an average depth of about −5 km. Trenches, ridges, and fracture zones represent plate boundaries, and abyssal plains are plate interiors. Wide continental shelves are submerged passive margins.

18.2 LANDSCAPES BENEATH THE SEA

The oceans exist because oceanic lithosphere and continental lithosphere differ markedly from one another in terms of composition and thickness (▶Fig. 18.3; see Chapter 2). The surface of the denser and thinner oceanic lithosphere

lies deeper than the surface of the relatively buoyant, thicker continental lithosphere, creating oceanic *basins* (low areas) that fill with water. On the present-day map of the world, the ocean encircles the globe. For purposes of reference, however, cartographers divide the ocean into several major parts, with somewhat arbitrary boundaries and significantly different volumes (▶Fig. 18.4a). Most continental crust (81%) lies in the Northern Hemisphere today (▶Fig. 18.4b); but because of plate tectonics, the map of Earth's surface was different in the past. In fact, the oldest oceanic crust visible today is only about 200 million years old, for subduction has consumed all older oceanic crust.

Have you ever wondered what the ocean floor would look like if all the water evaporated? Marine geologists can now provide a clear image of the ocean's **bathymetry**, or variation in depth, based originally on sonar measurements and more recently on measurements made by satellites. Such studies indicate that the ocean contains broad bathymetric provinces, distinguished from each other by their water depth.

Continental Shelves, Slopes, and Rises

Imagine you're in a submersible cruising just above the floor of the western half of the North Atlantic. If you start at the shoreline of North America and head east, you will find that extending from the shoreline for about 200 to 500 km, a **continental shelf**, a relatively shallow portion of the ocean in which water depth does not exceed 500 m, fringes the continent. Across the width of the shelf, the

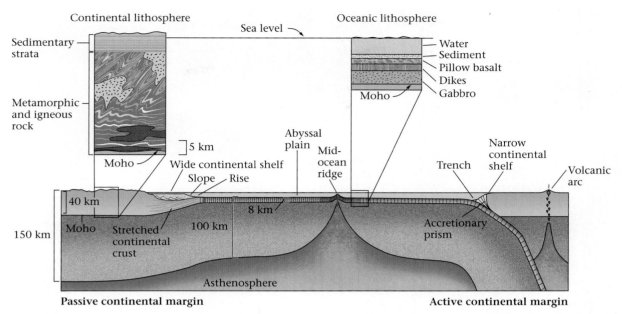

FIGURE 18.3 The bathymetric provinces of the sea floor. At a passive continental margin, a thick wedge of sediment accumulates in an ocean basin over continental lithosphere that had been stretched and thinned during the rifting that formed the basin. The flat surface of this sedimentary wedge creates a wide continental shelf. At an active continental margin, here a convergent plate boundary, a narrow continental shelf forms over an accretionary prism. Insets: These vertical slices through continental and oceanic lithosphere illustrate that oceanic lithosphere is thinner, and that the two kinds differ in composition.

ocean floor slopes seaward at only about 0.3°, an almost imperceptible amount. At its eastern edge, the continental shelf merges with the continental slope, which descends to depths of nearly 4 km at an angle of about 2°. From about 4 km down to about 4.5 km, a province called the continental rise, the angle decreases until at 4.5 km deep, you find yourself above a vast, nearly horizontal plain: the **abyssal plain**.

Broad continental shelves, like that of eastern North America, form along **passive continental margins**, margins that are not plate boundaries and thus lack seismicity (▶Fig. 18.5a; see Chapter 4). Passive margins originate after rifting breaks a continent in two; when rifting stops and sea-floor spreading begins, the stretched lithosphere at the boundary between the ocean and continent gradually cools and sinks. Sediment washed off the continent, as well as the shells of marine creatures, buries the sinking crust, slowly producing a pile of sediment up to 20 km thick. The top surface of this sedimentary pile constitutes the continental shelf. As discussed in Chapter 16, recent surveys show that large submarine slumps form along the continental slope. The movement of some may have generated tsunamis.

If you were to take your submersible to the western coast of South America and cruise out into the Pacific,

you would find a very different continental margin. After crossing a narrow continental shelf, the sea floor falls off at the relatively steep angle of 3.5° down to a depth of over 8 km. South America does not have a broad continental shelf because it is an **active continental margin**, a margin that coincides with a plate boundary and thus hosts many earthquakes (▶Fig. 18.5b). Off South America, the edge of the Pacific Ocean is a convergent plate boundary. The narrow shelf along a convergent plate boundary forms where an apron of sediment spreads out over the top of an accretionary prism, the pile of material scraped off the downgoing subducting plate. Here, the continental slope corresponds to the face of the accretionary prism.

At many locations, relatively narrow and deep valleys called **submarine canyons** dissect continental shelves and slopes (▶Fig. 18.5c). The largest submarine canyons start offshore of major rivers, and for good reason: rivers cut into the continental shelf at times when sea level was low and the shelf was exposed. But river erosion cannot explain the total depth of these canyons. Some slice almost 1,000 m down into the continental margin, far deeper than the maximum sea-level change. Submarine exploration demonstrates that much of the erosion of submarine canyons results from the flow of turbidity currents, avalanches of

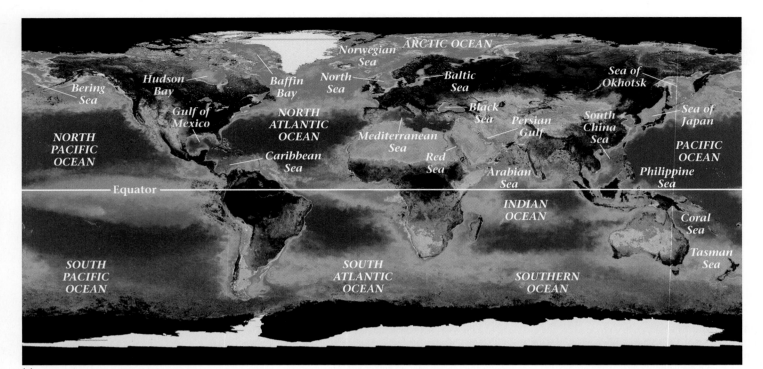

(a)

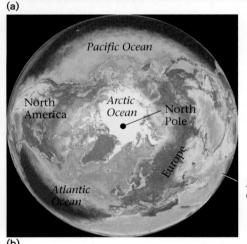

(b)

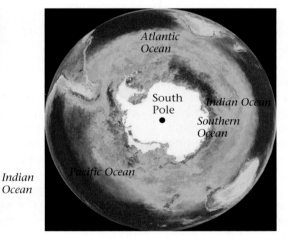

FIGURE 18.4 (a) A SeaWiFS satellite image showing the biologic productivity of the land and sea. In the sea, productivity is indicated by chlorophyll concentration, and on land, it is indicated by vegetation cover. The major oceans are labeled. **(b)** The Earth's oceans as viewed looking down on the poles. Note that the southern hemisphere (map on the right) is mostly ocean.

sediment mixed with water (see Chapter 7). When turbidity currents finally reach the base of the continental slope, turbidites (composed of graded beds) accumulate and build up into a submarine fan (▶Fig. 18.5d, e).

The Bathymetry of Oceanic Plate Boundaries

Can you see plate boundaries on the sea floor? Yes. As we noted in Chapter 4, each type of plate boundary is a distinctive bathymetric feature. For example, sea-floor spreading at a divergent plate boundary results in the formation of a mid-ocean ridge (▶Fig. 18.6a, b), along which the sea

floor rises 2 km above the depth of abyssal plains. Because of the normal faulting that accommodates stretching of the crust during sea-floor spreading, escarpments form the ridge axis. Strike-slip faulting along transform plate boundaries also breaks up the crust, so transform faulting forms fracture zones, narrow belts of ruptured and irregular sea floor. Transform faults are perpendicular to the ridge axis and link segments of ridge. Finally, subduction at a convergent plate boundary produces a deep trough called a trench, which borders a volcanic arc. Many trenches attain depths of over 8 km. In fact, the deepest point in the ocean, –11,035 m, lies in the Mariana trench of the western Pacific.

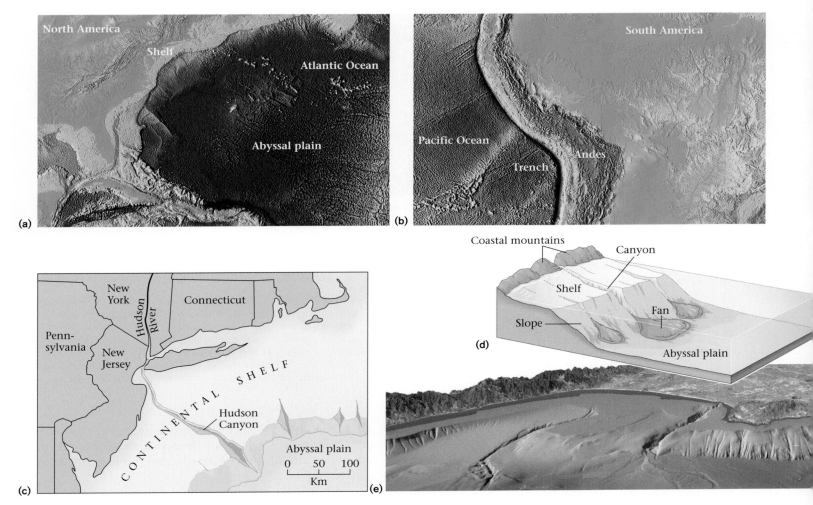

FIGURE 18.5 **(a)** Digital bathymetric map portraying the surface of the western Atlantic sea floor. Maps like this one are constructed by computer analysis of satellite data that measures variations in the height of the water surface and the pull of gravity, which depend on water depth. This map shows a passive continental margin, here a portion of the eastern coast of North America, with a broad continental shelf. Several seamounts protrude from the sea floor east of the continental shelf. **(b)** A digital bathymetric map of an active continental margin, here the subduction zone on the western coast of South America. **(c)** Submarine canyons along the east coast of the United States. A particularly large one starts at the mouth of the Hudson River. **(d)** Submarine fans accumulate at the base of submarine canyons. **(e)** A 3-D bathymetry of coastal California, off Los Angeles. Note the prominent submarine canyons (vertical exaggeration: 6x). The field of view is about 50 km.

Abyssal Plains and Seamounts

As oceanic crust ages and moves away from the axis of the mid-ocean ridge, two changes take place. First, the lithosphere cools, and as it does so, its surface sinks (to maintain isostatic compensation; see Chapter 11). Second, a blanket of **pelagic sediment** gradually accumulates and covers the basalt of the oceanic crust. This blanket consists mostly of microscopic plankton shells and fine flakes of clay, which slowly fall like snow from the ocean water and settle on the sea floor. Because the ocean crust gets progressively older away from the ridge axis, sediment thickness increases away from the ridge axis (▶Fig. 18.6c). Eventually,

over old sea floor, the sediment buries the escarpments that had formed at the mid-ocean ridge, resulting in a flat, featureless surface of the abyssal plain.

Numerous distinct, localized high areas rise above surrounding ocean depths (see Figs. 3.20a and 18.5a). These high areas result from hot-spot volcanic activity. If the product of this activity protrudes above sea level, it forms an oceanic island. Oceanic islands that lie over hot spots host active volcanoes, whereas those that have moved off the hot spot are extinct. In warm climates, coral reefs grow and surround oceanic islands. With time, oceanic islands erode and partially collapse due to slumping. Also, the seafloor beneath them ages and sinks. As a result, each island's peak

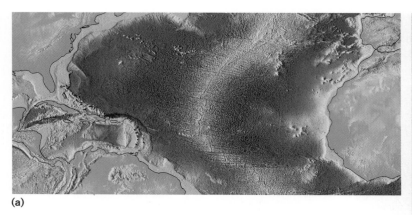

(a)

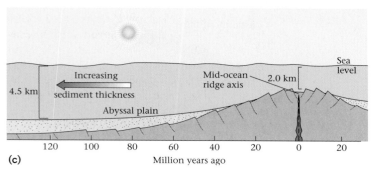

(c)

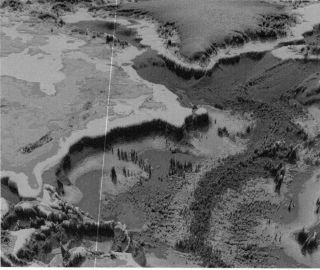

(b)

FIGURE 18.6 (a) A segment of a mid-ocean ridge, showing transform faults that link segments of the ridge. **(b)** A 3-D enlargement of the North Atlantic sea floor. Note how the Mid-Altantic Ridge intersects Iceland, an oceanic plateau. In this image, the ridge is red and purple. **(c)** The sea floor slopes away from a mid-ocean ridge and gradually flattens out to become an abyssal plain. Sediment increases in thickness away from the ridge axis, because the sea floor gets older as it moves away from the ridge axis.

eventually submerges, and what was once an island becomes a **seamount**. A seamount that submerges after being overgrown by a reef will have a flat top, and can be called a **guyot**. Oceanic islands and seamounts that developed above the same hot spot line up in a chain (a hot-spot track), with the oldest seamount at one end and the youngest seamount or island at the other (see Chapter 4). In places where hot-spot igneous activity was particularly voluminous, a broad oceanic plateau, underlain by flood basalt, forms.

> **Take-Home Message**
>
> Ocean water contains about 3.5% dissolved salt. Salinity and temperature vary with depth and location. The wind drives surface currents, whereas deep water circulates due to the density variations resulting from salinity and temperature variation.

18.3 OCEAN WATER AND CURRENTS

Composition

If you've ever had a chance to swim in the ocean, you may have noticed that you float much more easily in ocean water than you do in freshwater. That's because ocean water contains an average of 3.5% dissolved salt (▶Fig. 18.7); in contrast, typical freshwater contains only 0.02%

salt. The dissolved ions fit between water molecules without changing the volume of the water, so adding salt to water increases the water's density, and you float higher in a denser liquid.

Leonardo da Vinci, the famous Renaissance artist and scientist, speculated that sea salt came from rivers passing through salt mines, but modern studies demonstrate that most cations in sea salt—sodium (Na^+), potassium (K^+),

FIGURE 18.7 The composition of average seawater. The expanded part of the graph shows the proportions of ions in the salt of seawater.

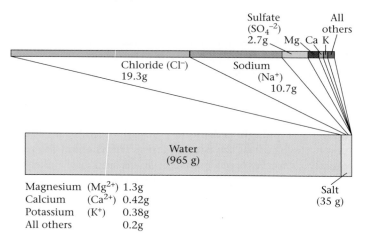

Magnesium (Mg^{2+}) 1.3g
Calcium   (Ca^{2+}) 0.42g
Potassium (K^+)   0.38g
All others   0.2g

calcium (Ca^{2+}), and magnesium (Mg^{2+})—come from the chemical weathering of rocks, and that the anions, chloride (Cl^-) and sulfate (SO_4^{-2}), from volcanic gases. Still, da Vinci was right in believing that dissolved ions get carried to the sea by flowing groundwater and river water: rivers deliver over 2.5 billion tons of salt every year.

There's so much salt in the ocean that if all the water suddenly evaporated, a 60-m-thick layer of salt would coat the ocean floor. This layer would consist of about 75% halite ($NaCl$), with lesser amounts of gypsum ($CaSO_4 \cdot H_2O$), anhydrite ($CaSO_4$), and other salts. Oceanographers refer to the concentration of salt in water as salinity. Although ocean salinity averages 3.5%, measurements from around the world demonstrate that salinity varies with location, ranging from about 1.0% to about 4.1% (▶Fig. 18.8a). Salinity reflects the balance between the addition of freshwater by rivers or rain and the removal of freshwater by evaporation, for when seawater evaporates, salt stays behind; salinity may also depend on water temperature, for warmer water can hold more salt in solution than can cold water.

The salinity of the ocean changes with depth. A graph of the variation in salinity with depth (▶Fig. 18.8b) indicates that such differences in salinity are found in seawater only down to a depth of about 1 km. Deeper water tends to be more homogenous. Oceanographers refer to the gradational boundary between surface-water salinities and deep-water salinities as the halocline.

Temperature

When the *Titanic* sank after striking an iceberg in the North Atlantic, most of the unlucky passengers and crew who jumped or fell into the sea died within minutes, because the seawater temperature at the site of the tragedy approached freezing, and cold water removes heat from a body very rapidly. Yet swimmers can play for hours in the Caribbean, where sea-surface temperatures reach 28°C (83°F). Though the *average* global sea-surface temperature hovers around 17°C, it ranges between freezing near the poles to almost 35°C in restricted tropical seas (▶Fig. 18.8c). The correlation of average temperature with latitude exists because the intensity of solar radiation varies with latitude.

The intensity of solar radiation also varies with the season, so surface seawater temperature varies with the season. But the difference is only around 2° in the tropics, 8° in the temperate latitudes, and 4° near the poles. (By contrast, the seasonal temperature change on land can be much greater—in central Illinois, for example, temperatures may reach 40°C [104°F] in the summer and drop to −32°C [−25°F] in the winter.) Seasonal seawater temperature changes remain in a narrow range because water can absorb or release large amounts of heat without changing temperature very much. Thus, the ocean regulates the temperatures of coastal regions; air temperature in Vancouver, on the Pacific coast of Canada, rarely drops below freezing even though it lies farther north than Illinois.

Water temperature in the ocean varies markedly with depth (▶Fig. 18.8d). Waters warmed by the Sun are less dense and tend to remain at the surface. An abrupt thermocline, below which water temperatures decrease sharply, reaching near freezing at the sea floor, appears at a depth of about 300 m, in the tropics. There is no pronounced thermocline in polar seas, since surface waters there are already so cold.

Currents: Rivers in the Sea

Since first setting sail on the open ocean, people have known that the water of the ocean does not stand still, but rather flows or circulates at velocities of up to several kilometers per hour in fairly well-defined streams called **currents**. Oceanographic studies made since the *Challenger* expedition demonstrate that circulation in the sea occurs at two levels: surface currents affect the upper hundred meters of water, and deep currents keep even water at the bottom of the sea in motion.

Surface Currents: A Consequence of the Wind

When the skippers of sailing ships planned their routes from Europe to North America, they paid close attention to the directions of surface currents, for sailing against a current slowed down the voyage substantially. If they headed due west at a high latitude, they would find themselves battling an eastward-flowing surface current, the Gulf Stream. Further, they found that the water moving in a surface current does not flow smoothly but displays some turbulence. Isolated swirls or ring-shape currents of water, called eddies, form along the margins of currents (▶Fig. 18.9).

Surface currents occcur in all the world's oceans (▶Fig. 18.10). They result from interaction between the sea surface and the wind—as moving air molecules shear across the surface of the water, the friction between air and water drags the water along. If we look at a map that shows global wind patterns along with oceanic currents, we can see this relationship (see Fig. 18.10 inset). But the movement of water resulting from wind shear does not exactly parallel the movement of the wind. This is a consequence of Earth's rotation, which generates the **Coriolis effect** (▶Box 18.1; ▶Fig. 18.11). This phenomenon causes surface currents in the Northern Hemisphere to veer toward the right and surface currents in the Southern Hemisphere to veer toward the left of the average wind direction (see Box 18.1).

Because of the geometry of ocean basins and the pattern of wind directions, surface currents in the oceans today trace out large circular flow patterns known as **gyres**, clockwise in the northern seas and counterclockwise in the

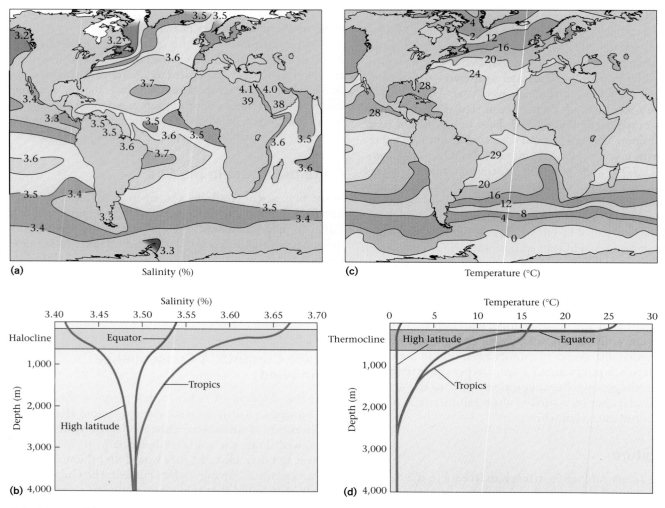

FIGURE 18.8 **(a)** The variations in salinity in the world ocean. The contour lines represent regions of different salinity (numbers are percentages). **(b)** The variation of salinity with depth in the ocean. **(c)** The variation in temperature with latitude. Contours are given in degrees Celsius. **(d)** The variation in temperature with depth.

southern seas. Individual currents within these gyres, have names (see Fig. 18.10). Northern- and southern-hemisphere gyres merge at the equator, creating an equatorial westward flow. Water in the center of a gyre becomes somewhat isolated from surface currents. Sailors refer to the center of the North Atlantic gyre as the Sargasso Sea, for sargassum, a tropical seaweed, accumulates in this sluggish water.

In the past, when continents were in different positions, the geometry of ocean currents was quite different. For example, the circum-Antarctic current that exists today did not appear until the Drake Passage, between South America and Antarctica, opened 25 million years ago. Currents that move from the poles to the equator bring cool water toward the equator, whereas currents that move from the equator toward the poles carry warm water poleward—this transport of heat moderates the global climate. Thus,

changes in the pattern of currents through geologic time affect the climate of the Earth System.

Upwelling, Downwelling, and Deep Currents

Surface currents are not the only means by which water flows in the ocean; it also circulates in the vertical direction. Oceanographers have now identified downwelling zones, places where near-surface water sinks, and upwelling zones, places where subsurface water rises.

What causes upwelling and downwelling? First, along coastal regions, these two phenomena exist because as the wind blows, it drags surface water along. If surface water moves toward the coast, then an oversupply of water develops along the shore and excess water must sink—that is,

FIGURE 18.9 A satellite image of the Gulf Stream, a current of warm surface water flowing northward along the east coast of North America. The colors represent different temperatures (red is warmest). Note the large eddies that form along the margin of the Gulf Stream.

downwelling occurs (▶Fig. 18.12a). Alternatively, if surface water moves away from the coast, then a deficit of water develops near the coast and water rises to fill in the gap—upwelling takes place (▶Fig. 18.12b).

Upwelling of subsurface water also occurs along the equator because the winds blow steadily from east to west. The Coriolis effect causes water to deflect to the right in the Northern Hemisphere and to the left in the Southern Hemisphere. Upwelling replaces this deficit and causes the surface water at the equator to be cooler and rich in nutrients. The nutrients foster an abundance of life in equatorial water.

Upwelling and downwelling can also be driven by contrasts in water density, caused by differences in temperature and salinity; we refer to the rising and sinking of water

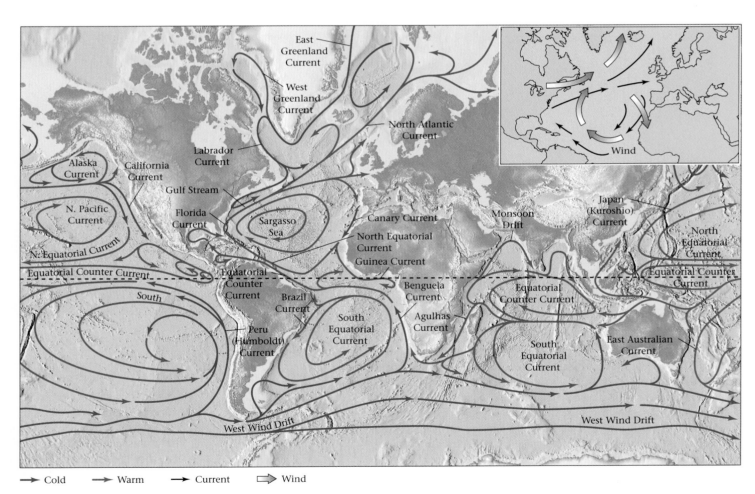

→ Cold → Warm → Current ⇨ Wind

FIGURE 18.10 The major surface currents of the world's oceans. Inset: The relationship between the prevailing wind direction and the North Atlantic Current.

BOX 18.1 SCIENCE TOOLBOX

The Coriolis Effect

Imagine you are spinning a playground merry-go-round counterclockwise around a vertical axis at a rate of 10 revolutions per minute. The circumference of the outer edge of the merry-go-round is 5 m. Thus, Emma, a child sitting at the outer edge, moves at a velocity of 50 m per minute, whereas David, a child sitting at the center, spins around an axis but moves at zero velocity. If Emma were to try throwing a ball to David by aiming directly along a radius, the ball would veer to the right of the radius and miss David, because the ball is not only moving in the direction parallel to a radius line, but is also moving in the direction par-

allel to the edge of the circle. If David were to throw a ball along a radius to Emma, this ball would miss Emma because the revolution of the merry-go-round moves her relative to the ball's trajectory (▶Fig. 18.11a, b).

The rotation of the Earth generates the same phenomenon. Earth spins counterclockwise around its axis, so a cannon shell fired along a line of longitude from the North Pole toward the equator veers to the right (west) because the Earth is moving faster to the east at the equator (▶Fig. 18.11c). A cannon shell fired parallel to a line of longitude from the equator to the North Pole veers to the right (east), because as it

moves north, it is traveling east faster than the land beneath it (▶Fig. 18.11d). Similarly, a cannon shell fired from the equator to the South Pole veers to the left (east).

In 1835, a French engineer named Gaspard Gustave de Coriolis (1792–1843) proposed that a similar effect would cause the deflection of winds and currents on the surface of the Earth. Because of this "Coriolis effect," north-flowing currents in the Northern Hemisphere deflect to the east, whereas south-flowing currents deflect to the west. The opposite is true in the Southern Hemisphere.

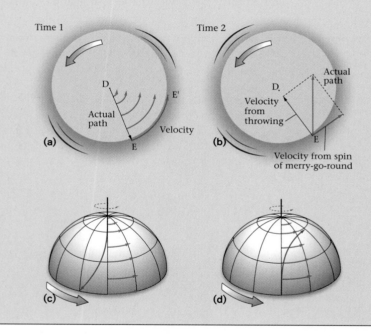

FIGURE 18.11 The Coriolis effect. **(a, b)** The velocity of a point on the rim of this spinning merry-go-round is greater than the velocity at the center. A ball thrown from point D to point E would follow a straight line, but while the ball is in the air, point E moves to point E′. Relative to the surface of the merry-go-round, the ball looks as if it follows a curved path—but remember, the ball goes straight; it's the surface that moves underneath the ball. A ball aimed from the rim to the center won't go straight to the center. Again, since the merry-go-round is moving under the ball, the ball appears to follow a curved path with respect to the surface of the merry-go-round. **(c, d)** The same phenomenon happens on Earth. A projectile shot from the pole to the equator in the Northern Hemisphere deflects to the west, whereas a projectile shot from the equator to the pole deflects to the east relative to the moving Earth.

driven by density contrasts as **thermohaline circulation**. During thermohaline circulation, denser water (cold and/or saltier) sinks, whereas water that is less dense (warm and/or less salty) rises. As a result, the water in polar regions sinks and flows back along the bottom of the ocean toward the equator. This process divides the ocean vertically into a number of distinct water masses, which mix only very slowly with one another. In the Atlantic Ocean, for example, the Antarctic Bottom Water sinks along the coast of Antarctica, and the North Atlantic Deep Water sinks in the north polar

region (▶Fig. 18.13). The combination of surface currents and thermohaline circulation, like a conveyor belt, moves water and heat among the various ocean basins (▶Fig. 18.14).

Take-Home Message

Gravitational attraction by the Moon and Sun, as well as centrifugal force, cause tidal bulges that move around the Earth. Because of the rise and fall of tides, intertidal regions are alternately submerged and exposed.

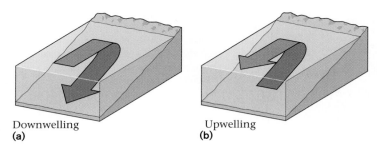

Downwelling
(a)

Upwelling
(b)

FIGURE 18.12 **(a)** Where surface water moves toward shore, it downwells to make room for more water. **(b)** Where surface water moves offshore, deep water upwells to replace the water that flowed away.

18.4 THE TIDES GO OUT . . . THE TIDES COME IN . . .

There is a tide in the affairs of men, which taken at the flood, leads on to fortune. Omitted, all the voyage of their life is bound in shallows and in miseries.

—William Shakespeare, **Julius Caesar**

A ship captain seeking to float a ship over reefs, a fisherman hoping to set sail from a shallow port, marines planning to attack a beach from the sea, a tourist eager to harvest shellfish from nearshore mud—all must pay attention to the rise and fall of sea level, a vertical movement called a **tide**, if they are to be successful. If the tide is too

low, ships run aground on reefs or stay trapped in harbors, and if the tide is too high, a beach may become too narrow to permit access. The tidal reach, meaning the difference between sea level at high tide and sea level at low tide, depends on location. The largest tidal reach on Earth is 16.8 m (54.6 feet). The intertidal zone, the region submerged at high tide and exposed at low tide, is a fascinating ecological niche.

During a rising tide, or flood tide, the shoreline (the boundary between water and land) moves inland, whereas during the falling tide, or ebb tide, the shoreline moves seaward. The horizontal distance over which the shoreline migrates between high and low tides depends on the slope of the shore surface (▶Fig. 18.15a,b). Where the slope is gentle and the tidal reach is large, the position of the shore can move a long way during a tidal cycle—at low tide, a broad tidal flat lies exposed to the air in the intertidal zone (Fig. 18.15b). Such settings can be hazards. For example, in February 2004, fifteen shellfish hunters lost their lives along the coast of northwestern England; they were far offshore, searching for cockles in the mud, when the flood tide came in. Arrival of a flood tide can create a visible wall of water, or tidal bore, ranging from a few centimeters to a couple of meters high, and moving at up to 35 km/h (i.e., faster than a person can run).

Tides are caused by a **tide-generating force**, which is due, in part, to the gravitational attraction of the Sun and Moon and, in part, to centrifugal force caused by the revolution of the Earth-Moon system around its center of mass. (To understand the meaning of this complex statement, see ▶Box 18.2.) Gravitational pull by the Moon contributes most of the gravitational part of the tide-generating force. The Sun, even though it is larger, is so far away that its contribution is only 46% that of the Moon. Tide-generating forces create two bulges in the global ocean, making this envelope of water more oval shaped than the solid Earth (▶Fig. 18.15c). One bulge, the sublunar bulge, lies on the side of the Earth closer to the Moon; it forms because the Moon's gravitational attraction is greatest at this point. The other, the secondary bulge, lies on the opposite (far) side of the Earth (12,000 km—the diameter of the Earth—farther from the Moon); it forms because the Moon's gravitational

FIGURE 18.13 Because of variations in density, primarily caused by variations in temperature, the oceans are vertically stratified into moving water masses. Each mass has a name. Note that Antarctic Bottom Water, which sinks down from the surface along the chilly shores of Antarctica, flows northward along the floor of the Atlantic at least as far as the equator.

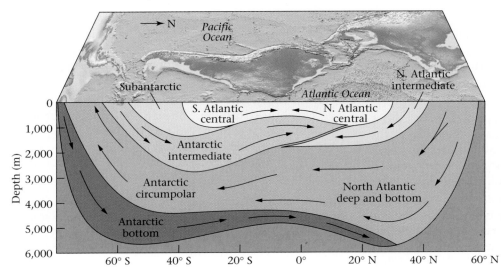

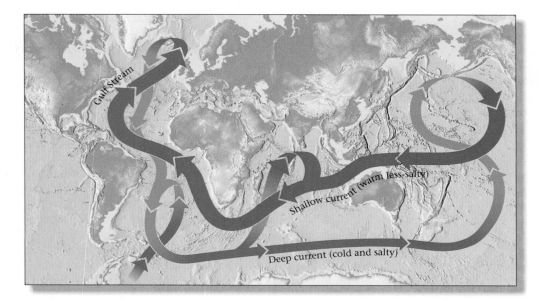

FIGURE 18.14 The exchange between upwelling deep water and downwelling surface water creates a global conveyor belt that circulates water throughout the entire ocean. A complete cycle takes hundreds of years to millennia.

Gulf Stream

Shallow current (warm less-salty)

Deep current (cold and salty)

attraction is weakest at this point. Here, centrifugal force can push water outward (▶Fig. 18.15d). A depression in the global ocean surface separates the two bulges. When a shore location lies under a tidal bulge, it experiences a high tide, and when it passes under a depression, it experiences low tide.

If the Earth's solid surface were smooth and completely submerged beneath the ocean, so that there were no continents or islands, the timing of tides would be fairly simple to understand. Because the Earth spins on its axis once a day, we would predict two high tides and two low tides at a given point per day. But the story isn't quite that simple—many other factors affect the timing and magnitude of tides. These include:

- *Tilt of the Earth's axis:* Because the spin axis of the Earth is not perpendicular to the plane of the Earth-Moon system, a given point passes between a high part of one bulge during one part of the day, and through a lower part of the other bulge during another part of the day, so the two high tides at the given point are not the same size (Fig. 18.15c).

- *The Moon's orbit:* The moon progresses in its 28-day orbit around the Earth in the same direction as the Earth rotates. High tides arrive 50 minutes later each day because of the difference between the time it takes for Earth to spin on its axis, and the time it takes for the Moon to orbit the Earth.

- *The Sun's gravity:* When the angle between the direction to the Moon and the direction to the Sun is 90°, we experience extra-low tides (neap tides) because the Sun's gravitational attraction counteracts the Moon's. When

the Sun is on the same side as the Moon, we experience extra-high tides (spring tides) because the Sun's attraction *adds* to the Moon's (▶Fig. 18.15e).

- *Focusing effect of bays:* In the open ocean, the maximum tidal reach is only a few meters. But in the Bay of Fundy, along the eastern coast of Canada, the tidal reach approaches 20 m. In a bay that narrows to a point, such as the Bay of Fundy, the flood tide brings a large volume of water into a small area, so the point experiences an especially large high tide.

- *Basin shape:* The shape of the basin containing a portion of the sea influences the sloshing of water back and forth within the basin as tides rise and fall. Depending on the timing and magnitude of this sloshing, this effect can locally add to the global tidal bulge or subtract from it, and thus can affect the rhythm of tides. In some locations, the net effect is to cancel one of the daily tides entirely, so that the locality experiences only one high tide and one low tide in a day.

- *Air pressure:* The effects of air pressure on tides can contribute to disaster. For example, during a hurricane the air pressure drops radically, so the sea surface rises; if the hurricane coincides with a high tide, the storm surge (water driven landward by the wind) can inundate the coast.

Because of the complexity of factors contributing to tides, the timing and magnitude of tides vary significantly along the coast. Nevertheless, at a given location the tides are periodic and can be predicted. Tides gave early civilizations a rudimentary way to tell time. In fact,

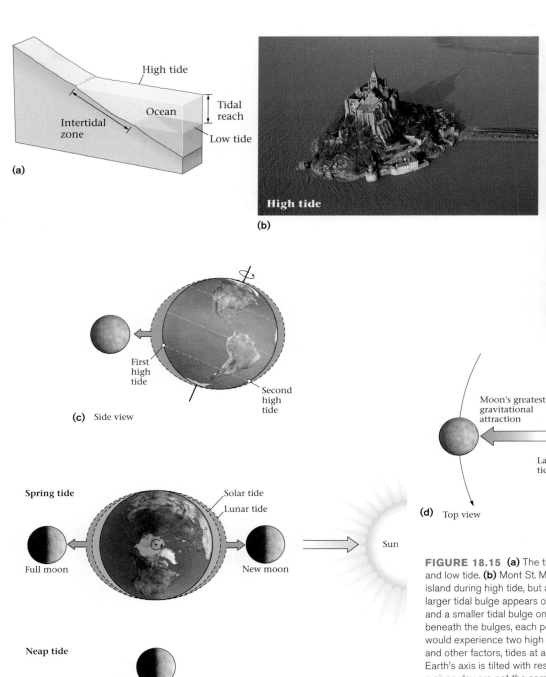

(a)

High tide

Tidal reach

Ocean

Intertidal zone

Low tide

(b) High tide

Low tide

(c) Side view

First high tide

Second high tide

Spring tide

Solar tide

Lunar tide

Full moon

New moon

Sun

Neap tide

Solar tide

Sun

Lunar tide

(e)

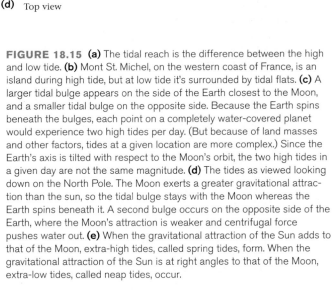

Moon's greatest gravitational attraction

North Pole

Largest tidal bulge

Smaller tidal bulge

(d) Top view

FIGURE 18.15 (a) The tidal reach is the difference between the high and low tide. **(b)** Mont St. Michel, on the western coast of France, is an island during high tide, but at low tide it's surrounded by tidal flats. **(c)** A larger tidal bulge appears on the side of the Earth closest to the Moon, and a smaller tidal bulge on the opposite side. Because the Earth spins beneath the bulges, each point on a completely water-covered planet would experience two high tides per day. (But because of land masses and other factors, tides at a given location are more complex.) Since the Earth's axis is tilted with respect to the Moon's orbit, the two high tides in a given day are not the same magnitude. **(d)** The tides as viewed looking down on the North Pole. The Moon exerts a greater gravitational attraction than the sun, so the tidal bulge stays with the Moon whereas the Earth spins beneath it. A second bulge occurs on the opposite side of the Earth, where the Moon's attraction is weaker and centrifugal force pushes water out. **(e)** When the gravitational attraction of the Sun adds to that of the Moon, extra-high tides, called spring tides, form. When the gravitational attraction of the Sun is at right angles to that of the Moon, extra-low tides, called neap tides, occur.

BOX 18.2 SCIENCE TOOLBOX

The Forces Causing Tides

Fundamentally, tides result from interaction between two forces: gravitational attraction exerted by the Moon and Sun on the Earth, and centrifugal force caused by the revolution of the Earth around the center of mass of the Earth-Moon system. To explain this statement, we must review some key terms from physics.

- **Gravitational pull** is the attractive force that one mass exerts on another. The magnitude of gravitational pull depends on the amount of mass in each object, and on the distance between the two masses.
- **Centrifugal force** is the apparent outward-directed ("center-fleeing") force that material on or in an object feels when the object moves in orbit around a point. Note that centrifugal force differs from *centripetal force*, the "center-seeking" force; this distinction can be confusing. To experience centripetal force, tie a ball to a string and swing it around your head. The string exerts an inward-directed centripetal force on the ball—if the string breaks, the centripetal force ceases to exist and the

ball heads off in a straight-line path. To picture centrifugal force, imagine that the ball is hollow and that you've placed a marble inside. As you twirl the ball around your head, the marble moves to the outer edge of the ball. The apparent force pushing the marble outward is the centrifugal force. But as such, centrifugal force is not a real force—it is simply a manifestation of inertia, and it exists *only* from the perspective, or reference frame, of the orbiting object. (A physics book explains this contrast in greater detail.)

- **Earth-Moon system** refers to this pair of objects viewed as a unit, as they move together through space.
- **Center of mass** is the point within an object, or a group of objects, about which mass is evenly distributed; put another way, it is the location of the average position, or the balance point, of the total mass in a single object or a group of objects. Because the Earth is 81 times more massive than the Moon, the center of mass of the Earth-Moon system actually lies 1,700 km below the surface of the Earth.

First, let's consider the origin of centrifugal force in the Earth-Moon system. To do this, we must consider the way in which the Earth-Moon system moves. The center of the Earth itself does not follow a simple orbit around the Sun. Rather, it is the *center of mass* of the Earth-Moon system that follows this trajectory; the Earth actually spirals around this trajectory as it speeds around the Sun. To picture this motion, imagine that the Earth-Moon system is a pair of dancers, one of whom is much heavier than the other. The dancers face each other, hold hands, and whirl in a circle as they drift across the dance floor (▶Fig. 18.16a). Each dancer's head orbits the center of mass.

Revolution of the Earth around the Earth-Moon system's center of mass generates centrifugal forces on both the Earth and the Moon that would cause the Earth and the Moon to fly away from one another, were it not for the gravitational attraction holding them together. We can see this by looking again at our dancer analogy (▶Fig. 18.16b)—the centrifugal force acting on each dancer points outward, away from his or her partner, and is the *same* for all points on each dancer. We can represent the direction and magnitude of this centrifugal force by arrows called vectors. (A vector is a number that has magnitude and direction.) In this case, the length of the arrow represents the magnitude of the force, and the orientation of the arrow indicates the direction of the force. If we think of the dancers as the Earth and the Moon, then centrifugal force vectors at all points on the surface of the Earth point away from the Moon (▶Fig. 18.17a). On the Earth, therefore, centrifugal force causes the surface of the ocean to bulge outward, away from the center of mass of the Earth-Moon system, on the far side of the Earth.

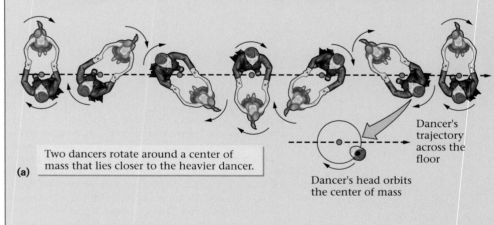

Two dancers rotate around a center of mass that lies closer to the heavier dancer.

(a)

Dancer's trajectory across the floor

Dancer's head orbits the center of mass

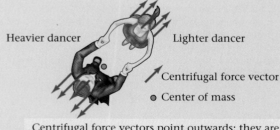

Heavier dancer   Lighter dancer

↗ Centrifugal force vector

● Center of mass

Centrifugal force vectors point outwards; they are the same magnitude for all points on a dancer.

(b)

FIGURE 18.16 **(a)** To picture the Earth-Moon system, imagine two dancers spinning around each other as they move along a straight-line trajectory. The center of mass of the two-dancer system lies closer to the heavier dancer. Each dancer orbits the center of mass. **(b)** Each point on each dancer feels a centrifugal force (represented by a vector) that points outward.

Now, let's consider how the force of gravity comes into play in causing tides. To simplify this discussion, we only examine the effect of the Moon's gravity on Earth. Vectors representing the magnitude and direction of the Moon's gravitational pull at any point on the surface of the Earth all point toward the center of the Moon. Because the magnitude of gravity depends on distance, the Moon exerts more attraction on the near side of the Earth than at the Earth's center, and less attraction on the far side of the Earth than at the Earth's center. Gravity, therefore, causes the surface of the ocean on the near side of the Earth to bulge toward the Moon.

In the Earth-Moon system, both centrifugal force and gravitational pull operate at the same time. How do they interact? If we draw vectors representing both centrifugal force and gravitational force at various points on or in the Earth, we see that the vectors representing centrifugal force do not have the same length as those representing gravitational attraction, except at the Earth's center. Moreover, the vectors representing centrifugal force do not point in the same direction as the vectors representing gravitational attraction. The force that the ocean water feels is the sum of the two forces acting on the water. You can determine the sum of two vectors by drawing the vectors so they touch head to tail—the sum is the vector that completes the triangle. This sum is the *tide-generating force*, and its magnitude and direction vary with location on the Earth.

Let's look at the tide-generating force a little more closely. On the side of the Earth closer to the Moon, gravitational vectors are larger than centrifugal force vectors, so adding the two gives a net tide-generating force that pulls the sea surface to bulge toward the Moon. On the side of the Earth further from the Moon, the centrifugal force vectors are larger, so centrifugal force caused by the orbiting of the Earth-Moon system around the center of mass causes the surface of the sea to bulge outward, away from the Moon (▶Fig. 18.17b). Thus, the ocean has two tidal bulges—one on the side close to the Moon, and one on the opposite side of the Earth. The bulge closer to the Moon is larger. Note that tidal bulges have nothing to do with the spinning of the Earth on its axis—this spin has no measurable effect on the sea surface.

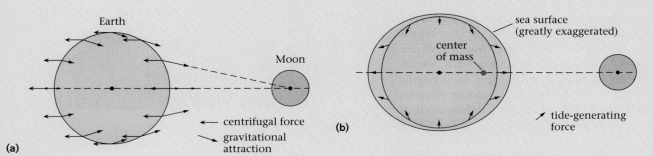

(a) centrifugal force
gravitational attraction

(b) tide-generating force

FIGURE 18.17 (a) Each point on the surface of the Earth feels the same centrifugal force (due to the spin of the Earth-Moon system around its center of mass), but feels a different gravitational attraction (due to the pull of the Moon). Centrifugal force vectors pointing away from the Moon are all the same magnitude, but gravitational force vectors pointing toward the center of the Moon, and their magnitude varies with distance from the Moon. **(b)** The tide-generating force is the sum of the centrifugal force vector and the gravitational force vector. Vectors on the side of the Earth close to the Moon are dominated by gravitational force and thus point toward the Moon, creating a tidal bulge. Vectors on the other side of the Earth are dominated by centrifugal force and point away from the Moon, creating a second tidal bulge.

in some languages the word for tide is the same as the word for time.

Friction between ocean water and the ocean floor causes the movement of the tidal bulge to lag slightly behind the movement of the Moon across the Earth. The Moon, therefore, exerts a slight pull on the side of the bulge. This pull acts like a brake and slows the Earth's spin, so that days grow longer at a rate of about 0.002 seconds per century. Over geologic time, the seconds add up; a day was only 21.9 hours long in the Middle Devonian Period (390 Ma). As the spinning Earth slows, the Moon moves farther away. During the Archean Eon (3.8 Ga), the Moon was 15,000 km closer, so the tidal reach on Earth was larger.

18.5 WAVE ACTION

Wind-driven waves make the ocean surface restless, an ever-changing vista. They develop because of the shear between the molecules of air in the wind and the molecules of water at the surface of the sea. It may seem surprising that so much friction can arise between two fluids, but it can, as Benjamin Franklin demonstrated. Franklin noted that oily waste spilled from ships on a windy day made the water

Take-Home Message

The friction of the wind against the sea surface causes waves to form. Within a wave, water moves in a circle; the amount of motion decreases with depth. Near shore, water piles up into breakers that refract when they approach the shore causing longshore drift.

surface smoother. He proposed that the oily coating on the water decreased frictional shear between the air and the water, and therefore prevented waves from forming.

When you watch a wave travel across the open ocean, you may get the impression that the whole mass of water constituting the wave moves with the wave. But drop a cork overboard and watch it bob up and down and back and forth; it does not move along with a wave. Within a wave, away from shore, a particle of water moves in a circular motion, as viewed in cross section. The diameter of the circle is greatest at the ocean's surface, where it equals the amplitude of the wave. With increasing depth, though, the diameter of the circle decreases until, at a depth equal to about half the "wavelength" (the horizontal distance between two wave troughs), there is no wave movement at all (▶Fig. 18.18a). Submarines traveling below this **wave base** cruise through smooth water, whereas ships toss about above.

The character of waves in the open ocean depends on the strength of the wind (how fast the air moves) and on the fetch of the wind (over how long a distance it blows). When the wind first begins to blow, it creates ripples in the water surface, pointed waves whose amplitude (the height from rest to crest or rest to trough) and wavelength are small. With continued blowing over a long fetch, swells, larger waves with amplitudes of 2 to 10 m and wavelengths of 40 to 500 m, begin to build. Hurricane wave amplitudes may grow to over 25 m. Swells may travel for thousands of kilometers across the ocean, well beyond the region where they formed.

How large can wind-driven waves in the open ocean get? It's not surprising that huge waves form during hurricanes. Oceanographers calculate that if hurricane-strength winds were to blow across the width of the Pacific for at least 24 hours, 15- to 20-m-high waves would develop. Particularly large waves may also form where two sets of wind-driven waves coming from different directions constructively interfere, so that wave crests add to each other. This happened in 1979, when waves generated by an east-blowing gale collided with waves generated by a west-blowing gale in the waters off Ireland during the Fastnet Yacht Race. Waves with amplitudes of over 15 m developed, and 23 of the 300 sailboats in the race capsized.

Wave interference, the interaction of wind-driven waves with strong currents, and focusing due to the shape of the coastline or sea floor can lead to the formation of **rogue waves**, defined as waves that are more than twice the size of most large waves passing a locality during a specified time interval. Long thought to exist only in the imagination of sailors, rogue waves now have been documented numerous times. For example, wave-measuring instruments on oil platforms in the North Sea recorded almost 500 encounters with rogue waves during a 10-year period. Some of the waves were 3 to 5 times higher than other large waves. The decks of large ships—including famous cruise ships—have been swamped by immense rogue waves in the open ocean (▶Fig. 18.18b). In 1995, for example, a 29-m-high wave struck the *Queen Elizabeth II*, and in 1933, a military ship encountered one at least 34 m (112 feet) high. Recent research proves that rogue waves are not so rare as once thought. Radar studies from satellites demonstrate that at any given time, there are about ten rogue waves around the world's oceans. If a rogue wave reaches the shore, it can wash unsuspecting bystanders off shore-side piers or beaches.

FIGURE 18.18 (a) Within a deep-ocean wave, water molecules follow a circular path. The diameter of the circle decreases with depth to the wave base, below which the wave has no effect. When a wave passes, the shape of the water surface changes, but water does not move as a mass. Note that the amplitude is one-half the wave height. **(b)** This photo shows water draining off the deck of a ship after a rogue wave passed. The deck normally sits 23 m (75 feet) above the water surface, so the wave—which is now in front of the ship—must be higher.

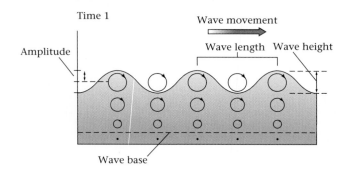

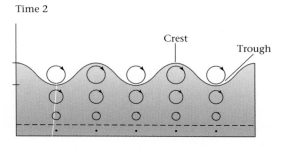

(a)

(b)

Waves have no effect on the ocean floor, as long as the floor lies below the wave base. However, near the shore, where the wave base just touches the floor, it causes a slight back-and-forth motion of sediment. Closer to shore, as the water gets shallower, friction between the wave and the sea floor slows the deeper part of the wave, and the motion in the wave becomes more elliptical. Eventually, water at the top of the wave curves over the base, and the wave becomes a breaker, ready for surfers to ride. Breakers crash onto the shore in the surf zone, sending a surge of water up the beach. This upward surge, or **swash**, continues until friction brings motion to a halt. Then gravity draws the water back down the beach as **backwash** (▶Fig. 18.19).

Waves may make a large angle with the shoreline as they're coming in, but they bend as they approach the shore, a phenomenon called **wave refraction;** right at the shore, their crests make no more than about a 5° angle with the shoreline (▶Fig. 18.20a). To understand why this happens, imagine a wave approaching the shore so that its crest makes an angle of 45° with the shoreline. The end of the wave closer to the shore touches bottom first and slows down because of friction, whereas the end farther offshore still continues to move at its original velocity, swinging the whole wave around so that it's more parallel with the shoreline.

Though refraction decreases the angle at which a wave rolls onto shore, the wave may still arrive at an angle. When the water returns seaward in the backwash, however, it must flow straight down the slope of the beach in response to gravity. Overall, this sawtooth-like flow results in a **longshore current**, which flows parallel to the beach (Fig. 18.20a). Also because of wave refraction, wave energy is focused on headlands (places where higher land protrudes into the sea), and is weaker in embayments (places set back from the sea). Thus, erosion happens at headlands, forming cliffs, whereas deposition takes place in embayments, forming a beach (▶Fig. 18.20b–d).

Waves pile water up on the shore incessantly. As the excess water moves back to the sea, it may create a strong, localized seaward flow perpendicular to the beach called a rip current (▶Fig. 18.21). Rip currents are the cause of many drownings every year along beaches, because they suddenly carry unsuspecting swimmers away from the beach.

> ### Take-Home Message
> Beaches form where there is abundant sediment; over time, sediment moves and builds spits and bars. Rocky coasts evolve due to wave erosion, fjords are submerged glacial valleys and estuaries are submerged river valleys. In warm climates, reefs grow offshore.

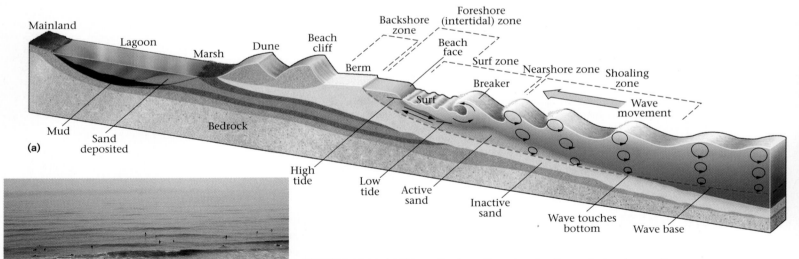

FIGURE 18.19 (a) This profile shows the various landforms of a beach, as well as a cross section of a barrier island. As a wave approaches the shore, it touches the bottom of the sea, at a depth of about half the wavelength. Due to friction, the wave slows down and the wavelength decreases, so the wave height must increase. Because the bottom of the wave moves more slowly than the top, the wave builds up into a breaker that carries water up onto the beach, with the top of the wave falling over the bottom. The water washing up on the beach is swash, and the water rushing back is backwash (indicated by arrows). **(b)** Small breakers forming along a California beach. Note how wave height builds toward the beach.

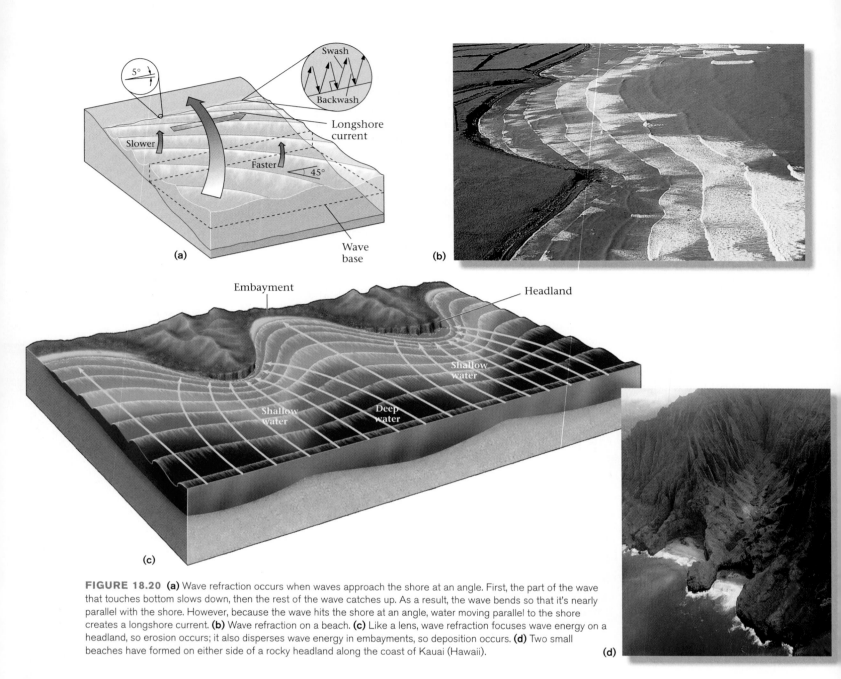

FIGURE 18.20 (a) Wave refraction occurs when waves approach the shore at an angle. First, the part of the wave that touches bottom slows down, then the rest of the wave catches up. As a result, the wave bends so that it's nearly parallel with the shore. However, because the wave hits the shore at an angle, water moving parallel to the shore creates a longshore current. (b) Wave refraction on a beach. (c) Like a lens, wave refraction focuses wave energy on a headland, so erosion occurs; it also disperses wave energy in embayments, so deposition occurs. (d) Two small beaches have formed on either side of a rocky headland along the coast of Kauai (Hawaii).

18.6 WHERE LAND MEETS SEA: COASTAL LANDFORMS

Tourists along the Amalfi coast of Italy thrill to the sound of waves crashing on rocky shores. But in the Virgin Islands, sunbathers can find seemingly endless white sand beaches next to calm seas. Large, dome-like mountains rise directly from the sea in Rio de Janeiro, Brazil, but a 100-m-high vertical cliff marks the boundary between the Nullarbor Plain of southern Australia and the Great Southern Ocean (▶Fig. 18.22a–d). And

New Orleans sprawls over a vast plain of former swamp. As these examples illustrate, coasts, the belts of land bordering the sea, vary dramatically in terms of topography and associated landforms (▶Fig. 18.23a–g).

Beaches and Tidal Flats

For millions of vacationers, the ideal holiday includes a trip to a **beach**, a gently sloping fringe of sediment along the shore. Some beaches consist of pebbles or boulders, whereas others consist of sand grains (▶Fig. 18.24a, b).

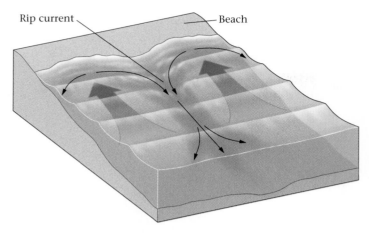

FIGURE 18.21 Waves bring water up on shore. The water may return to sea in a narrow rip current perpendicular to the shore.

This is no accident, for waves winnow out finer sediment like silt and mud and carry it to quieter water, where it settles. Storm waves, which can smash cobbles against one another with enough force to shatter them, have little effect on sand, for sand grains can't collide with enough energy to crack. Thus, cobble beaches exist only where nearby cliffs continuously supply large rock fragments.

The composition of sand itself varies from beach to beach, because different sands come from different sources. Sands derived from the weathering and erosion of silicic-to-intermediate rocks consist mainly of quartz; other minerals in these rocks chemically weather to form clay, which washes away in waves. Beaches made from the erosion of limestone or of recent corals and shell beds consist of carbonate sand, including masses of sand-sized chips of shells. And beaches derived by the recent erosion of basalt may have black sand, made of tiny basalt grains.

FIGURE 18.22 (a) A rocky shore in eastern Italy. (b) A sandy beach along the coast of St. John, U.S. Virgin Islands. (c) The sugar loafs (rounded mountains) rising out of the sea at Rio de Janeiro, Brazil. (d) The abrupt edge of the Nullarbor Plain in South Australia.

(a)

(b)

(c)

(d)

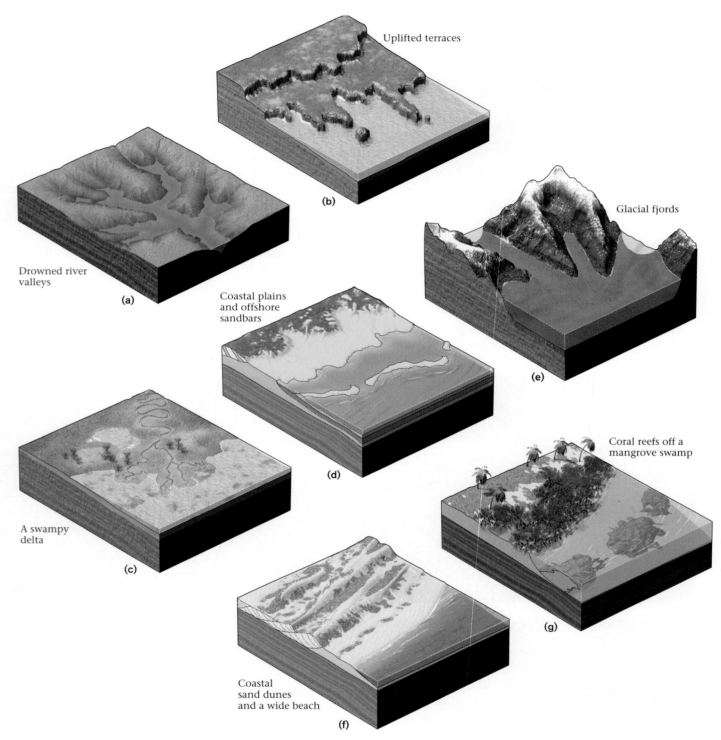

FIGURE 18.23 A wide variety of coastal landforms have developed on Earth. **(a)** Drowned river valleys, formed where sea level rises and floods valleys, create complex, irregular coastlines. **(b)** Uplifted terraces develop where the coastline rises relative to sea level and creates escarpments. **(c)** Swampy deltas form where a sediment-laden stream deposits sediment along the coast. **(d)** Along sandy coastal plains, large beaches and offshore bars appear. **(e)** Glacial fjords develop where sea level rises and floods a glacially carved valley. **(f)** Coastal dunes form where there is a large sand supply and strong wind. **(g)** In tropical environments, mangrove swamps grow along the shore, protected from wave action by offshore coral reefs.

(a)

(b)

FIGURE 18.24 (a) A pebble and cobble beach, Olympic Peninsula, Washington. The clasts were derived from nearby cliffs. (b) A sand beach on the western coast of Puerto Rico.

A beach profile, a cross section drawn perpendicular to the shore, illustrates the shape of a beach (Fig. 18.19a). Starting from the sea and moving landward, a beach consists of a foreshore zone, or intertidal zone, across which the tide rises and falls. The **beach face**, a steeper, concave part of the foreshore zone, forms where the swash of the waves actively scours the sand. The backshore zone extends from a small step, or escarpment, cut by high-tide swash to the front of the dunes or cliffs that lie farther inshore. The backshore zone includes one or more **berms**, horizontal to landward-sloping terraces that received sediment during a storm.

Geologists commonly refer to beaches as "rivers of sand," to emphasize that beach sand moves along the coast over time—it is not a permanent substrate. Wave action at the shore moves an active sand layer on the sea floor on a daily basis. Inactive sand, buried below this layer, moves only during severe storms or not at all. Where waves hit the beach at an angle, the swash of each successive wave moves active sand up the beach at an angle to the shoreline, but the backwash moves this sand down the beach parallel to the slope of the shore. This sawtooth motion causes sand gradually to migrate along the beaches, a process called **beach drift** (Fig. 18.20a). Beach drift, which happens in association with the longshore drift of water, can transport sand hundreds of kilometers along a coast in a matter of centuries. Where the coastline indents landward, beach drift stretches beaches out into open water to create a **sandspit**. Some sandspits grow across the opening of a bay, to form a baymouth bar (►Fig. 18.25).

The scouring action of waves piles sand up in a narrow ridge away from the shore called an offshore bar, which parallels the shoreline. In regions with an abundant sand supply, offshore bars rise above the mean high-water level and become **barrier islands** (►Fig. 18.26a). The water between a barrier island and the mainland becomes a quiet-water **lagoon**, a body of shallow seawater separated from the open ocean.

Though developers have covered some barrier islands with expensive resorts, in the time frame of centuries to millennia, barrier islands are temporary features. For example, wind and waves pick up sand from the ocean side of the barrier island and drop it on the lagoon side, causing the island to migrate landward. Storms may breach barrier islands and create an inlet (a narrow passage of water). Finally, beach drift gradually transports the sand of barrier islands and modifies their shape.

Tidal flats, regions of mud and silt exposed or nearly exposed at low tide but totally submerged at high tide, develop in regions protected from strong wave action (Fig. 18.15b; ►Fig. 18.26b). They are typically found along the

FIGURE 18.25 Beach drift can generate sandspits and baymouth bars. Sedimentation fills in the region behind a baymouth bar. As a result, the shoreline gets smoother with time.

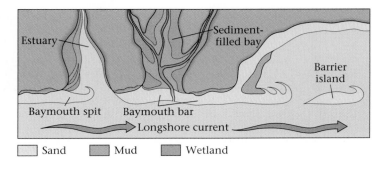

(a)

(b)

FIGURE 18.26 **(a)** The barrier islands of the Outer Banks off the coast of North Carolina, as viewed by *Apollo 9* astronauts. The white dots are clouds.
(b) Tidal flats are broad muddy areas submerged only at high tide. At low tide, boats at anchor rest in the mud of this tidal flat along the coast of Wales.

FIGURE 18.27 The sediment budget along a coast. Sediment is brought into the system by rivers, by the erosion of cliffs and moraines, and by wind. Sediment moves along the coast as a result of beach drift. And sediment leaves the system by being blown off the beach, by sinking into deeper water, or by being carried out by the longshore current.

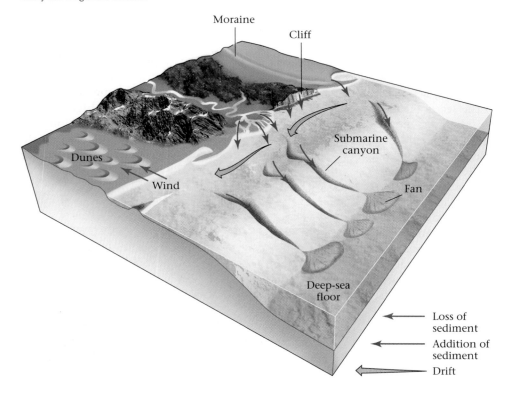

margins of lagoons or on shores protected by barrier islands. Here, mud and silt accumulate to form thick, sticky layers. In tidal flats that provide a home for burrowing organisms such as clams and worms, **bioturbation** ("stirring by life") mixes sediments together.

Because of the movement of sediment, the sediment budget (the difference between sand supplied and sand removed) plays an important role in determining the long-term evolution of a beach. Let's look at how the budget works for a small segment of beach (▶Fig. 18.27). Sand may be supplied to the segment from local rivers or by wind from nearby dune fields; it may also be brought from just offshore by waves or from far away by beach drift. (In fact, the large quantity of sand along beaches of the southeastern United States may have originated in Pleistocene glacial outwash far to the north.) Some of the sand from a stretch of beach may be removed by beach drift, whereas some gets carried offshore by waves, where it either settles locally or tumbles down a submarine canyon into the deep sea. If the lost sand cannot be replaced, the beach segment grows narrower, whereas if the supply of sand exceeds the amount that washes away, the beach becomes wider.

In temperate climates, winter storms tend to be stronger and more frequent than summer ones. The larger, shorter-wavelength waves of winter storms wash beach sand into deeper water and thus make the beach narrower, whereas the smaller, longer-wavelength summer waves bring sand in from offshore and deposit it on the beach (▶Fig. 18.28a, b).

Rocky Coasts

More than one ship has met its end smashed and splintered in the spray and thunderous surf of a rocky coast, where bedrock cliffs rise directly from the sea (Fig. 18.22a; ▶Fig. 18.29a). Lacking the protection of a beach, rocky coasts feel the full impact of ocean breakers. The water pressure generated during the impact of a breaker can pick up boulders and smash them together until they shatter, and it can squeeze air into cracks, creating enough force to

FIGURE 18.28 (a) In the winter, when waters are stormier, sand moves offshore, and the beach narrows and may become stonier. **(b)** During the summer, waves bring sand back to replenish the beach.

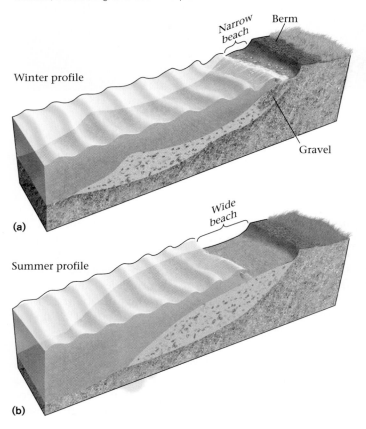

widen them. Further, because of its turbulence, the water hitting a cliff face carries suspended sand, and thus can abrade the cliff. The combined effects of shattering, wedging, and abrading, together called wave erosion, gradually undercut a cliff face and make a **wave-cut notch** (▶Fig. 18.29b, c). Undercutting continues until the overhang becomes unstable and breaks away at a joint, creating a pile of rubble at the base of the cliff that waves immediately attack and break up. In this process, wave erosion cuts away at a rocky coast, so that the cliff gradually migrates inland. Such cliff retreat leaves behind a **wave-cut bench**, or platform, which becomes visible at low tide (▶Fig. 18.29d).

Other processes besides wave erosion break up the rocks along coasts. For example, salt spray coats the cliff face above the waves and infiltrates into pores. When the water evaporates, salt crystals grow and push apart the grains, thereby weakening the rock. Biological processes also contribute to erosion, for plants and animals in the intertidal zone bore into the rocks and gradually break them up.

Many rocky coasts start out with an irregular coastline, with headlands protruding into the sea and embayments set back from the sea. Such irregular coastlines tend to be temporary features in the context of geologic time: wave energy focuses on headlands and disperses in embayments, a result of wave refraction. The resulting erosion removes debris at headlands, and sediment accumulates in embayments (Fig. 18.20c); thus, over time the shoreline becomes less irregular.

A headland erodes in stages (▶Fig. 18.30a–c). Because of refraction, waves curve and attack the sides of a headland, slowly eating through it to create a sea arch connected to the mainland by a narrow bridge (▶Fig. 18.31a). Eventually the arch collapses, leaving isolated sea stacks just offshore (▶Fig. 18.31b). Once formed, a sea stack protects the adjacent shore from waves. Therefore, sand collects in the lee of the stack, slowly building a tombolo, a narrow ridge of sand that links the sea stack to the mainland.

Coastal Wetlands

Let's move now from the crashing waves of rocky coasts to the gentlest type of shore, the **coastal wetland**, a vegetated, flat-lying stretch of coast that floods with shallow water but does not feel the impact of strong waves. In temperate climates, coastal wetlands include swamps (wetlands dominated by trees), marshes (wetlands dominated by grasses; ▶Fig. 18.32a), and bogs (wetlands dominated by moss and shrubs). So many marine species spawn in wetlands that despite their relatively small area when compared with the oceans as a whole, wetlands account for 10 to 30% of marine organic productivity.

In tropical or semitropical climates (between 30° north and 30° south of the equator), mangrove swamps thrive in wetlands (▶Fig. 18.32b). Mangrove tree roots can filter salt out of water, so the trees have evolved to survive in freshwater or saltwater. Some mangrove species form a

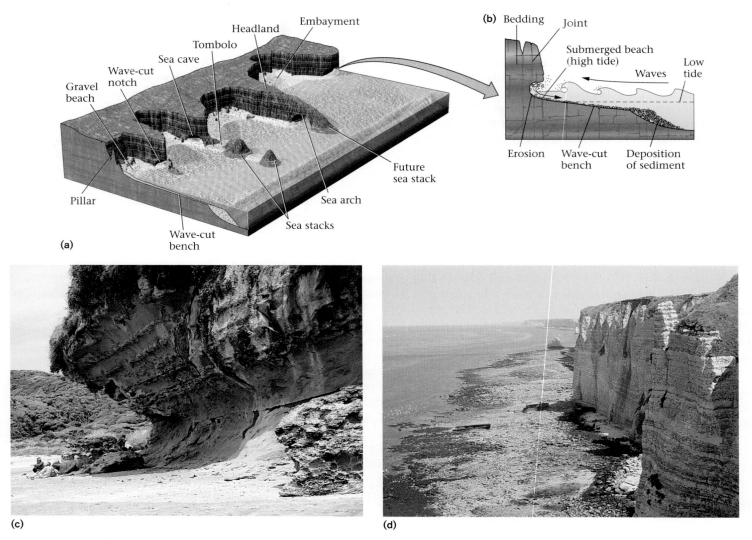

FIGURE 18.29 **(a)** The major landforms of a rocky shore include cliffs, sea caves, wave-cut notches, sea stacks, sea arches, wave-cut benches, and tombolos. Beaches tend to collect in embayments, whereas erosion happens at headlands. **(b)** Erosion by waves creates a wave-cut notch. Eventually, the overhanging rock collapses into the sea to form gravel on the wave-cut bench. **(c)** A wave-cut notch exposed along a rocky shore. **(d)** A wave-cut bench at the foot of the cliffs at Etrétat, France.

broad network of roots above the water surface, making the plant look like an octopus standing on its tentacles, and some send up small protrusions from roots that rise above the water and allow the plant to breathe. Dense stands of mangroves counter the effects of stormy weather and thus prevent coastal erosion.

Estuaries

Along some coastlines, a relative rise in sea level causes the sea to flood river valleys that merge with the coast, resulting in **estuaries**, where seawater and river water mix. You can recognize an estuary on a map by the dendritic pattern of its river-carved coastline (▶Fig. 18.33). Oceanic and fluvial waters interact in two ways within an estuary. In quiet estuaries, protected from wave action or river turbulence, the water becomes stratified, with denser oceanic saltwater flowing upstream as a wedge beneath less dense fluvial freshwater. Such saltwater wedges migrate about 100 km up the Hudson River in New York, and about 40 km up the Columbia River in Oregon. In turbulent estuaries, such as the Chesapeake Bay, oceanic and fluvial water combine to create nutrient-rich brackish water with a salinity between that of oceans and rivers. Estuaries are complex ecosystems inhabited by unique species of shrimp, clams, oysters, worms, and fish that can tolerate large changes in salinity.

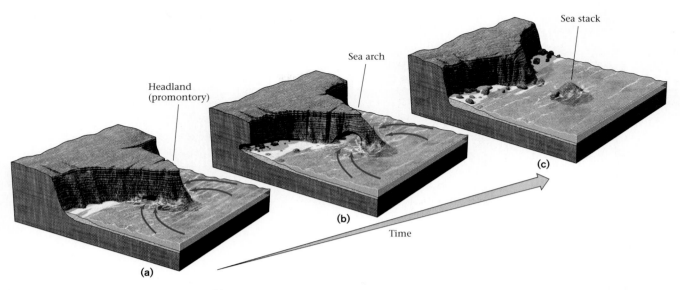

FIGURE 18.30 The erosion of a headland. **(a)** At first, wave refraction causes wave energy to attack the sides of a promontory, making a sea cave on either side. **(b)** Gradually erosion breaks through the promontory to create a sea arch. **(c)** The arch finally collapses, leaving a sea stack.

Fjords

During the last ice age, glaciers carved deep valleys in coastal mountain ranges. When the ice age came to a close, the glaciers melted away, leaving deep, U-shaped valleys (see Chapter 22). The water stored in the glaciers, along with the water within the vast ice sheets that covered continents during the ice age, flowed back into the sea and caused sea level to rise. The rising sea filled the deep valleys, creating **fjords**, or flooded glacial valleys. Coastal fjords are fingers of the sea surrounded by mountains; because of their deep-blue water and steep walls of polished rock, they are distinctively beautiful (▶Fig. 18.34a, b). Some of the world's most spectacular fjords decorate the western coasts of Norway, British Columbia, and New Zealand. Smaller examples appear along the coast of Maine and southeastern Canada.

FIGURE 18.31 **(a)** A sea arch exposed along a rocky coast of southern Australia. Another arch once bridged the gap to the cliff on the left, but it collapsed in 1990, stranding two tourists. **(b)** These limestone sea stacks along the southern coast of Australia, together with eight others, comprise a tourist attraction called the Twelve Apostles. One of the Apostles collapsed abruptly in 2005, right in front of the eyes of tourists.

(a)

(b)

See for yourself . . .
Landscapes of Oceans and Coasts

You can find a huge variety of different coastal and bathymetric features with *Google Earth*™ or comparable programs. In addition to visiting the stops identified in this Geotour, simply fly to a coast, tilt your view, and set the image in motion—you'll be amazed at what you can see! The thumbnail images provided on this page are only to help identify tour sites. Go to wwnorton.com/studyspace to experience this flyover tour.

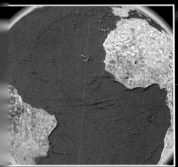
G18.1

Mid-Atlantic Ridge Bathymetry
(Lat 0°3'51.75"N, Long 24°47'35.66"W)

Fly to the coordinates provided and zoom to 8000 km (5000 miles) (Image G18.1). You can see the entire South Atlantic Ocean. Examine the bathymetry of the sea floor. Ridge segments and transforms are obvious.

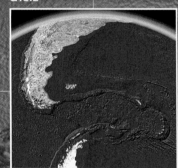
G18.2

Southern South America Bathymetry
(Lat 39°22'15.03"S, Long 65°9'1.31"W)

Zoom to 5000 km (3100 miles) at the coordinates provided. You are looking down on southern South America (Image G18.2). Compare the bathymetry of the west coast (an active convergent margin) to that of the east coast (a passive margin). Where does a broad continental shelf occur? Now, fly south to Lat 54°29'29.28"S, Long 52°1'4.39"W, zoom to 4500 km (2800 miles), and tilt to look north. You can see the Scotia Sea, between South America and Antarctica, and the Scotia volcanic arc (Image G18.3). Can you identify the pattern of plate boundaries?

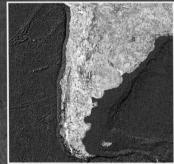
G18.3

Coral Reefs, Pacific (Lat 16°47'26.92"S, Long 150°58'1.27"W)

Fly to the coordinates given and zoom to 6 km (3.7 miles). You are looking down on the coral reef and lagoon of Huahine, one of the Society Islands in the South Pacific. Zoom down to 1.5 km (1 mile), tilt, and look north (Image G18.4). Note how the reef absorbs wave energy and protects the shore. Now, fly west to Lat 16°37'25.65"S, Long 151°29'32.80"W, and zoom to 25 km (15 miles). You are looking down on the island of Tahaa, an atoll surrounded by an offshore reef (Image G18.5).

G18.4

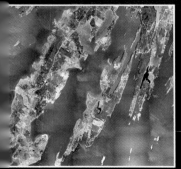
G18.6

Rocky Coast, Maine
(Lat 43°46'34.77"N, Long 69°58'28.58"W)

Fly to these coordinates and zoom to 10 km (6 miles). From this viewpoint, you can see the rocky coast of Maine (Image G18.6). Islands here were carved by glaciers during the last ice age. Post-ice age sea-level rise submerged the landscape.

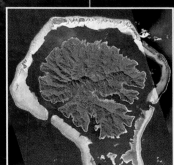

G18.7

Offshore Bar, near Cape Hatteras, North Carolina (Lat 35°8'24.05"N, Long 75°53'34.62"W)

At the coordinates provided, zoom to 28 km (17 miles) and tilt your view so that you are looking NW. You can see an offshore sand bar (Image G18.7). The bar formed due to redistribution of sand by waves and currents. Contrast this coast with the one you saw in Maine.

Chicago Shoreline (Lat 41°54'59.76"N, Long 87°37'36.41"W)

This locality, on the west shore of Lake Michigan, provides an excellent example of how the moving sand of a beach interacts with groins. Zoom to 2.5 km (1.5 miles) and look down on Chicago's beach front (Image G18.8). Note that sand has accumulated in asymmetric wedges due to a south-flowing current.

G18.8

Fjords of Norway (Lat 60°53'56.09"N, Long 5°12'31.84"E)

At the coordinates provided, zoom to 80 km (50 miles) and you see the intricate coast of Norway (Image G18.9). During the last ice age, glaciers carved deep valleys which filled with sea water when the ice melted and sea level rose. Zoom to 13 km (8 miles), tilt, and look north to see the steep cliffs bordering fjords. Now fly to Lat 61°12'34.22"N, Long 5°7'18.20"E. Here, the image resolution is better, and if you zoom to 5 km (3 miles), tilt, and look east, you can look inland, along the axis of a fjord (Image G18.10).

G18.9

G18.10

Organic Coast, Florida (Lat 25°7'48.94"N, Long 80°59'3.18"W)

Much of southern Florida, the Everglades, is a vast swamp, through which fresh water flows slowly south. Here, the transition from land to sea is gradual. Fly to the coordinates provided and look down from 40 km (25 miles) (Image G18.11). You see lagoons, swamps and bars, all colored by vegetation. Darker green areas are mangrove thickets. Zoom to 700 m (2300 feet) and tilt to look north (Image G18.12). You can see how the thickets stabilize the shore.

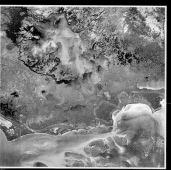

G18.11

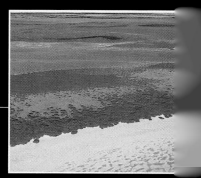

G18.12

Sandspit, Cape Cod (Lat 42°2'40.10"N, Long 70°11'31.46"W)

Cape Cod formed when a glacier deposited a 200 m (600 foot)-thick layer of sand and gravel to form a ridge called a moraine about 18,000 years ago. Eventually, sea level rose and currents began to transport sand along the shore. Fly to the coordinates provided and look down from an elevation of 20 km (12 miles) (Image G18.13). Note the large sand spit that protects Provincetown. On the north shore, you can see a beach and dunes.

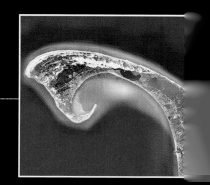

(a)

(b)

FIGURE 18.32 Examples of coastal wetlands: **(a)** a salt marsh; **(b)** a mangrove swamp.

FIGURE 18.34 **(a)** The subsurface shape of a fjord, a drowned U-shaped glacial valley. **(b)** Fjords in Norway have spectacular scenery.

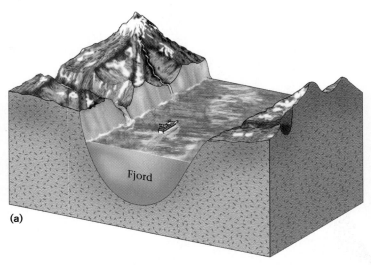

Fjord

(a)

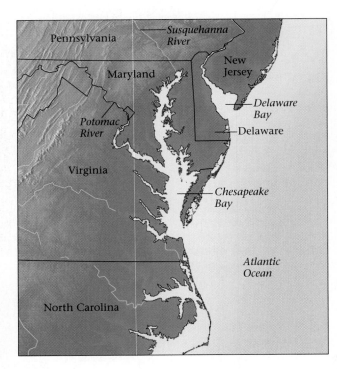

FIGURE 18.33 Chesapeake Bay, a large estuary along the East Coast of the United States, formed when sea level rose and flooded the Potomac and Susquehanna river valleys and the mouths of their tributaries.

Coral Reefs

In the Undersea National Park of the Virgin Islands, visitors swim through colorful growths of living coral. Some corals look like brains, others like elk antlers, still others like delicate fans (▶Fig. 18.35a). Sea anemones, sponges, and clams grow on and around the coral. Though at first glance coral looks like a plant, it is actually a colony of tiny invertebrates related to jellyfish. An individual coral animal, or polyp, has a tube-like body with a head of tentacles. Corals obtain part of their livelihood by filtering nutrients out of seawater; the remainder comes from algae that live on the corals' tissue. Corals

(b)

(a)

(b)

← Location of modern reef

(c)

FIGURE 18.35 (a) Corals and other organisms make up a reef as in this example from the Great Barrier Reef of northeast Australia. **(b)** A coral reef bordering Hawaii. **(c)** The distribution of coral reefs on Earth today.

have a symbiotic (mutually beneficial) relationship with the algae, in that the algae photosynthesize and provide nutrients and oxygen to the corals, while the corals provide carbon dioxide and nutrients for the algae.

Coral polyps secrete calcite shells, which gradually build into a mound of solid limestone whose top surface lies from just below the low-tide level down to a depth of about 60 m. At any given time, only the surface of the mound lives—the mound's interior consists of shells from previous generations of coral. The realm of shallow water underlain by coral mounds, associated organisms, and debris comprises a **coral reef** (▶Fig. 18.35b). Reefs absorb wave energy and thus serve as a living buffer zone that protects coasts from erosion. Corals need clear, well-lit, warm (18°–30°C) water with normal oceanic salinity, so coral reefs only grow along clean coasts at latitudes of less than about 30° (▶Fig. 18.35c).

Marine geologists distinguish three different kinds of coral reef, on the basis of their geometry (▶Fig. 18.36a–c).

> **Take-Home Message**
>
> Not all coasts look the same. Some are rimmed by beaches of moving sand. Some are dominated by wave-cut rocky cliffs, and some host living wetlands. Estuaries are submerged river valleys, and fjords are submerged glacial valleys.

A fringing reef forms directly along the coast, a barrier reef develops offshore (separated from the coast by a lagoon), and an atoll makes a circular ring surrounding a lagoon. As Charles Darwin first recognized back in 1859, coral reefs associated with islands in the Pacific start out as fringing reefs and then later become barrier reefs and finally atolls. Darwin suggested, correctly, that this progression reflects the continued growth of the reef as the island around which it formed gradually sinks. Eventually, the reef itself sinks too far below sea level to remain alive and becomes the cap of a guyot.

18.7 CAUSES OF COASTAL VARIABILITY

Plate Tectonic Setting

The tectonic setting of a coast plays a role in determining whether the coast has steep-sided mountain slopes or a broad plain that borders the sea (see art, pp. 652–653). Along an active margin, compression squeezes the crust and pushes it up, creating mountains like the Andes along the western coast of South America. Along a passive margin, the cooling and sinking of the lithosphere may create a

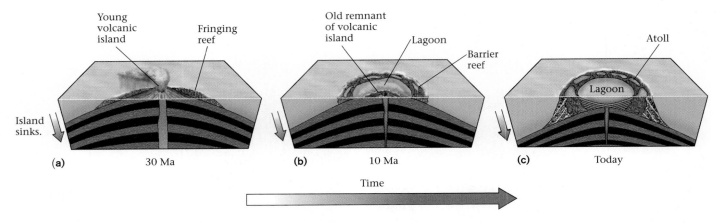

(a) 30 Ma   (b) 10 Ma   (c) Today

Time

FIGURE 18.36 The progressive change from a fringing reef around a young volcanic island to a ring-shaped atoll. **(a)** The reef begins to grow around the volcano. **(b)** The volcano subsides as the sea floor under it ages, so the reef is now a ring, separated from a small island (the peak of the volcano) by a lagoon. **(c)** The volcano has subsided completely, so that only an atoll surrounding a lagoon remains. When the lagoon fills with debris and together with the atoll finally sinks below sea level, the result is a guyot.

broad **coastal plain**, a flatland that merges with the continental shelf, as exists along the Gulf Coast and southeastern Atlantic coast of the United States. But not all passive margins have coastal plains. At some, the margin of the rift that gave birth to the passive margin remains at a high elevation, even tens of millions of years after rifting ceased. For example, highlands formed during recent rifting border the Red Sea, whereas highlands formed during Cretaceous rifting persist along portions of the Brazilian coast (Fig. 18.22c).

Relative Sea-Level Changes (Emergent and Submergent Coasts)

Sea level, relative to the land surface, changes during geologic time. Some changes develop due to vertical movement of the land. These may reflect plate-tectonic processes or the addition or removal of a load (such as a glacier) on the crust. Local changes in sea level reflect human activity. When people pump out groundwater, for example, the pores between grains in the sediment beneath the ground collapse, and the land surface sinks (see Chapter 19). Some relative sea-level changes, however, are due to a *global* rise or fall of the ocean surface. Such **eustatic sea-level changes** may reflect changes in the volume of mid-ocean ridges. An increase in the number or size of ridges, for example, displaces water and causes sea level to rise. Eustatic sea-level changes may also reflect changes in the volume of glaciers, for glaciers store water on land (▶Fig. 18.37).

Geologists refer to coasts where the land is rising or rose relative to sea level as **emergent coasts**. At emergent coasts, steep slopes typically border the shore. A series of step-like

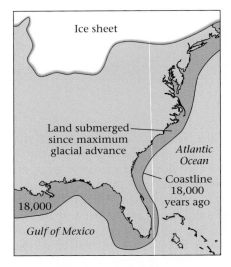

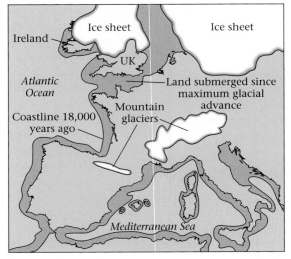

FIGURE 18.37
During the last ice age, North America's continental shelf lay exposed to the air, the United Kingdom and Ireland were not islands, and the Mediterranean Sea was cut off from the Atlantic Ocean.

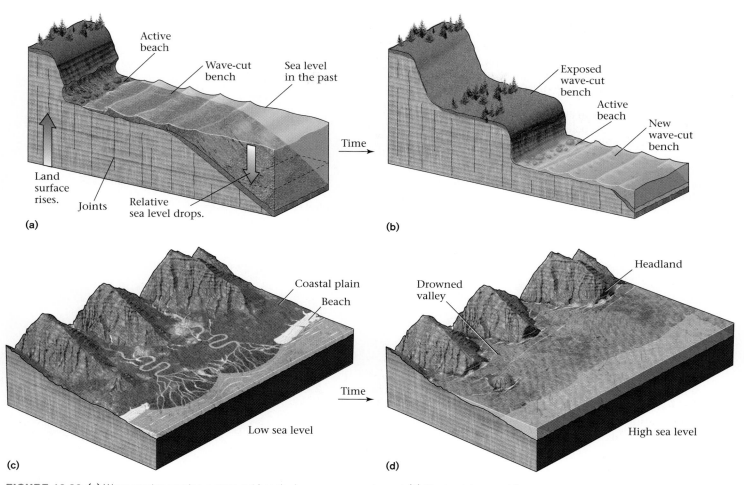

FIGURE 18.38 **(a)** Wave erosion creates a wave-cut bench along an emergent coast. **(b)** The land rises, and the bench becomes a terrace. **(c)** A coast before sea level rises. Rivers drain valleys onto a coastal plain. **(d)** As a submergent coast forms, sea level rises and floods the valleys, and waves erode the headlands.

terraces form along some emergent coasts (▶Fig. 18.38a, b). These terraces reflect episodic changes in relative sea level.

Those coasts at which the land sinks relative to sea level become **submergent coasts** (▶Fig. 18.38c, d). At submergent coasts, landforms include estuaries and fjords that developed when the sea flooded coastal valleys. Many of the coastal landforms of eastern North America are the consequences of submergence.

Sediment Supply and Climate

The quantity and character of sediment supplied to a shore affects its character. That is, coastlines where the sea washes sediment away faster than it can be supplied (erosional coasts) recede landward and may become rocky, whereas coastlines that receive more sediment than erodes away (accretionary coasts) grow seaward and develop broad beaches.

Climate also affects the character of a coast. Shores that enjoy generally calm weather erode less rapidly than those constantly subjected to ravaging storms. A sediment supply large enough to generate an accretionary coast in a calm environment may be insufficient to prevent the development of an erosional coast in a stormy environment. The climate also affects biological activity along coasts. For example, in the warm water of tropical climates, mangrove swamps flourish along the shore, and coral reefs form offshore. The reefs may build into a broad carbonate platform such as appears in the Bahamas today. In cooler climates, salt marshes develop, whereas in arctic regions, the coast may be a stark environment of lichen-covered rock and barren sediment.

> **Take-Home Message**
>
> The nature of coastal landscapes depends on three factors: Whether the sea level is rising or falling relative to the coast; whether the coast receives or loses sediment; and whether the climate is warm or cold.

Oceans and Coasts

The oceans of the world provide a diverse array of environments illustrating the full complexity of the Earth System. Tectonic processes and surface processes constantly battle with each other to produce submarine and subaerial landscapes.

Water in the ocean circulates in currents that transport heat from equator to pole. Interactions between the atmosphere and the ocean build waves that ripple the surface. Waves erode shorelines and transport sediment. Sea-floor features define the location of plate boundaries and hot spots. Coastal landforms depend on tectonic setting, climate, and sediment supply. Specifically, passive margins differ markedly from active convergent margins; equatorial coasts differ from sandy coasts. A great variety of organisms inhabit all these realms.

Wave erosion cuts notches at the base of cliffs and bevels wave-cut benches.

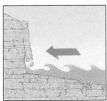

Along sandy shores, sand buil beaches, sand spits, and bars.

Turbidities flowing down submarine canyons produce submarine fans.

In tropical environments, mangroves live along the shore and coral reefs grow offshore.

Along rocky coasts, sea cliffs, sea arches, and sea stacks evolve.

At a passive margin, a broad continental shelf develops. Submarine slumping may occur along the shelf.

The ocean teems with l

At divergent plate boundaries, mid-ocean ridge rises. Transfo faults, marked by fracture zon link segments of the ridge.

The Global Conveyor

Surface winds drive surface currents in large gyres.

−0
−1
−2
−3 Km
−4
−5
−6

Cold water sinks at polar regions.

Tidewater glaciers produce icebergs.

At high latitudes, fjords form when the rising sea floods glacially carved valleys.

A river transports sediment to a delta.

Hot spots build chains of oceanic islands. Only the youngest island of the chain is active.

Bathymetry of the Sea Floor

Volcanic arcs form along convergent-margin coasts.

Seamounts and guyots are relics of hot spots.

At a convergent boundary, a trench bordered by an accretionary prism develops.

Waves and Beaches

The wind forms ocean waves. As a wave passes, water moves in a circular motion.

Near the shore, the top of the wave breaks over the base of the wave. Swash carries sand up the beach, and backwash carries sand back.

Sand may pile into dunes that build out over a lagoon, in which mud had accumulated.

18.8 COASTAL PROBLEMS AND SOLUTIONS

Contemporary Sea-Level Changes

People tend to view a shoreline as a permanent entity. But in fact, shorelines are ephemeral geologic features. On a time scale of hundreds to thousands of years, a shoreline moves inland or seaward depending on whether relative sea level rises or falls. In places where sea level is rising today, shoreline towns will eventually be submerged. For example, the Persian Gulf now covers about twice the area that it did 4,000 years ago. And if present rates of sea-level rise along the East Coast of the United States continue, major coastal cities such as Washington, New York, Miami, and Philadelphia may be inundated within the next millennium (▶Fig. 18.39).

Hurricanes and Coastal Floods

Hurricanes are immense storms that grow over the waters of equatorial oceans. Some are born and die at sea, but some move onto land. Winds in hurricanes can exceed 250 km/h (155 mph), and can generate waves in excess of 15 m. Because air rises beneath a hurricane, atmospheric pressure (effectively, the weight of the overlying atmosphere) decreases substantially in the region beneath a hurricane. Without the downward push of the air, the sea surface rises, so a bulge of high sea moves with the hurricane. When this combination of locally high sea level and powerful waves reach the shore, it causes a calamity. The sea inundates low areas along the shore, and the waves batter offshore reefs and the shore. In regions where the coast is a low-lying delta plain, the land can be submerged for days or more. Such catastrophic flooding has taken a dreadful toll on the Ganges delta in Bangladesh and, during Hurricane Katrina, on the New Orleans region on the Gulf Coast of the United States. We discuss hurricanes and their consequences more fully in Chapter 20.

Beach Destruction—Beach Protection?

In a matter of hours, a storm—especially a hurricane—can radically alter a landscape that took centuries or millenia to form. The backwash of storm waves sweeps vast quantities of sand seaward, leaving the beach a skeleton of its former self. The surf submerges barrier islands and shifts them toward the lagoon. Waves and wind together rip out mangrove swamps and salt marshes and fragment coral reefs, thereby destroying the organic buffer that normally protects the coast and leaving it vulnerable to erosion for years to come. Of course, major storms also destroy human constructions: erosion undermines shoreside buildings, causing them to collapse into the sea; wave impacts smash buildings to bits; and the storm surge—very high water lev-

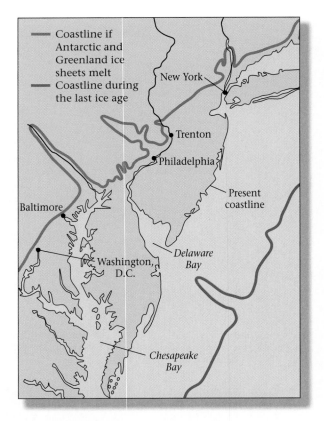

FIGURE 18.39 A possible sea-level rise in the future may flood major cities of the northeastern United States. The Washington–New York corridor would lie underwater.

els created when storm winds push water toward the shore—floats buildings off their foundations (▶Fig. 18.40a).

But even less dramatic events, such as the loss of river sediment, a gradual rise in sea level, a change in the shape of a shoreline, or the destruction of coastal vegetation, can alter the balance between sediment accumulation and sediment removal on a beach, leading to beach erosion (▶Fig. 18.40b). In some places, beaches retreat landward at rates of 1 to 2 m per year, forcing homeowners to pick up and move their houses. Even large lighthouses have been moved to keep them from washing away or tumbling down eroded headlands.

In many parts of the world, beachfront property has great value; but if a hotel loses its beach sand, it probably won't stay in business. Thus, property owners often construct artificial barriers to protect their stretch of coastline, or to shelter the mouth of a harbor from waves. These barriers alter the natural movement of sand in the beach system and thus change the shape of the beach, sometimes with undesirable results. For example, people may build groins, concrete or stone walls protruding perpendicular to the shore, to prevent beach drift from removing sand (▶Fig. 18.41a). Sand accumulates on the updrift side of the groin, forming a long triangular wedge, but sand erodes away on

(a)

(b)

FIGURE 18.40 **(a)** Damaged beachfront homes after a hurricane in Florida. **(b)** Wave erosion has completely removed the beach, and has started to erode a beach cliff along the coast of Cape Cod.

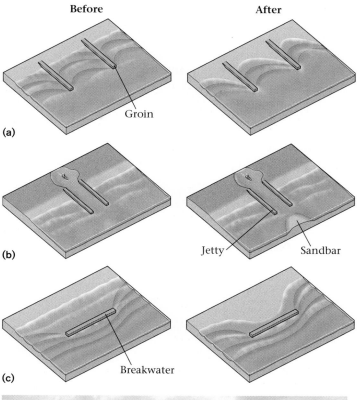

Before   After

(a) Groin

(b) Jetty   Sandbar

(c) Breakwater

(d)

FIGURE 18.41 **(a)** The construction of groins creates a sawtooth beach. **(b)** Jetties extend a river farther into the sea, but may result in the deposition of a sandbar at the end of the channel. **(c)** A breakwater causes the beach to build out in the lee. **(d)** Planners hope that riprap on the beach side of this parking lot in California will help slow erosion.

the downdrift side. Needless to say, the property owner on the downdrift side doesn't appreciate this process.

A pair of walls called jetties may protect the entrance to a harbor (▶Fig. 18.41b). But jetties erected at the mouth of a river channel effectively extend the river into deeper water, and thus may lead to the deposition of an offshore sandbar. Engineers may also build an offshore wall called a breakwater, parallel or at an angle to the beach, to prevent the full force of waves from reaching a harbor. With time,

however, sand builds up in the lee of the breakwater and the beach grows seaward, clogging the harbor (▶Fig. 18.41c). And to protect expensive shoreside homes, people build seawalls, out of riprap (large stone or concrete blocks) or reinforced concrete, on the landward side of the backshore zone (▶Fig. 18.41d). Seawalls reflect wave energy, which crosses the beach, back to sea. Unfortunately, this process increases the rate of erosion at the foot of the sea-wall, and thus during a large storm the seawall may be undermined so that it collapses (▶Fig. 18.42).

In some places, people have given up trying to decrease the rate of beach erosion, and instead have worked to increase the rate of sediment supply. To do this, they truck or ship in vast quantities of sand to replenish a beach. This procedure, called **beach nourishment**, can be hugely expensive and at best provides only a temporary fix, for the backwash and beach drift that removed the sand in the first place continue unabated as long as the wind blows

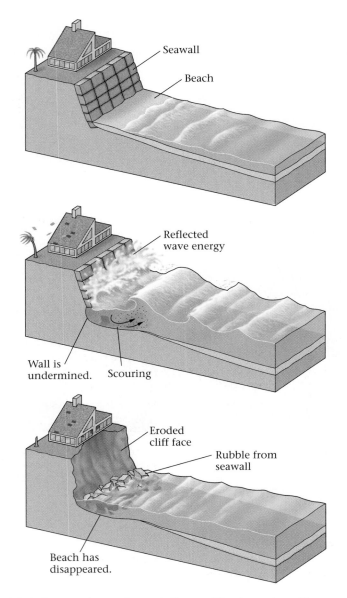

FIGURE 18.42 A seawall protects the sea cliff under most conditions, but during a severe storm the wave energy reflected by the seawall helps scour the beach. As a result, the wall may be undermined and collapse.

and the waves break. Clearly, beach management remains a controversial issue, for beachfront properties are expensive, but the shore is, geologically speaking, a temporary feature whose shape can change radically with the next storm.

Pollution and the Destruction of Organic Coasts

Bad cases of beach pollution create headlines. Because of beach drift, garbage dumped in the sea in an urban area may drift along the shore and be deposited on a tourist beach far from its point of introduction. For example, hospital waste from New York City has washed up on beaches tens of kilometers to the south. Oil spills—most commonly from ships that flush their bilges but also from tankers that have run aground or foundered in stormy seas—have contaminated shorelines at several places around the world.

Coasts in which living organisms control landforms along the shore are called **organic coasts**. These coasts, a manifestation of interaction between the physical and biological components of the Earth System, are particularly susceptible to changes in the environment. The loss of such landforms can increase a coast's vulnerability to erosion and, because they provide spawning grounds for marine organisms, can upset the food chain of the global ocean.

In wetlands and estuaries, sewage, chemical pollutants, and agricultural runoff cause havoc. Toxins settle along with clay and concentrate in the sediments, where they contaminate burrowing marine life and then move up the food chain. Fertilizers and sewage that enter the sea with runoff increase the nutrient content of water, creating algal blooms that absorb oxygen and therefore kill animal and plant life. Coastal wetlands face destruction by development—they have been filled or drained to be converted into farmland or suburbs, and have been used as garbage dumps. In most parts of the world, between 20 and 70% of coastal wetlands were destroyed in the last century.

Reefs, which depend on the health of delicate coral polyps, can be devastated by even slight changes in the environment. Pollutants and hydrocarbons, for example, will poison them. Organic sewage fosters algal blooms that rob water of dissolved oxygen and suffocate the coral. And agricultural runoff or suspended sediment introduced to coastal water during beach-nourishment projects reduces the light, killing the algae that live in the coral, and clogs the pores that coral polyps use to filter water. Changes in water temperature or salinity caused by dumping waste water from power plants into the sea or by global warming of the atmosphere also destroy reefs, for reef-building organisms are very sensitive to temperature changes. People can destroy reefs directly by dragging anchors across reef surfaces, by touching reef organisms, or by quarrying reefs to obtain construction materials. In the last decade, marine biologists have noticed that reefs around the world have lost their color and died. This process, called **reef bleaching**, may be due to the removal or death of symbiotic algae, in response to the warming of seawater triggered by El Niño, or may be a result of the dust carried by winds from desert or agricultural areas.

Chapter Summary

- The landscape of the sea floor depends on the character of the underlying crust. Wide continental shelves form over passive-margin basins, whereas narrow continental shelves form over accretionary prisms. Continental shelves are cut by submarine canyons. Abyssal plains develop on old, cool oceanic lithosphere. Seamounts and guyots form above hot spots.

- The salinity, temperature, and density of seawater vary with location and depth.

- Water in the oceans circulates in currents. Surface currents are driven by the wind and are deflected in their path by the Coriolis effect. The vertical upwelling and downwelling of water create deep currents. Some of this movement is thermohaline circulation, a consequence of variations in temperature and salinity.

- Tides—the daily rise and fall of sea level—are caused by a tide-generating force, mostly driven by the gravitational pull of the Moon.

- Waves are caused by friction where the wind shears across the surface of the ocean. Water particles follow a circular motion in a vertical plane as a wave passes. Waves refract (bend) when they approach the shore because of frictional drag with the sea floor.

- Sand on beaches moves with the swash and backwash of waves. If there is a longshore current, the sand gradually moves along the beach and may extend outward from headlands to form sand spits.

- At rocky coasts, waves grind away at rocks, yielding such features as wave-cut benches and sea stacks. Some shores are wetlands, where marshes or mangrove swamps grow. Coral reefs grow along coasts in warm, clear water.

- The differences in coasts reflect their tectonic setting, whether sea level is rising or falling, sediment supply, and climate.

- To protect beach property, people build groins, jetties, breakwaters, and seawalls.

- Human activities have led to the pollution of coasts. Reef bleaching has become dangerously widespread.

Geopuzzle Revisited

In the past 15 years, ocean currents have carried the armada of yellow duckies and blue turtles nearly around the world. Some of the toys have arrived on the shores of South America and Australia. Others floated through the Bering Strait and were carried by sea ice around the Arctic Ocean to the Atlantic. Their amazing voyage emphasizes how mobile ocean water is. Ocean currents serve as a major conveyor of heat in the Earth System. Also, as the toys have found out, wave action where the sea meets the land not only transports water, but also transports sediment and debris, making it a tool that can sculpt coastlines of amazing beauty.

Key Terms

abyssal plain (p. 623)
active continental margin (p. 623)
backwash (p. 637)
barrier island (p. 641)
bathymetry (p. 622)
beach (p. 638)
beach drift (p. 641)
beach face (p. 641)
beach nourishment (p. 656)
berm (p. 641)
bioturbation (p. 642)
center of mass (p. 634)
centrifugal force (p. 634)
coast (p. 622)
coastal plain (p. 650)
coastal wetland (p. 643)
continental shelf (p. 622)
coral reef (p. 649)
Coriolis effect (p. 627)
currents (p. 627)
Earth-Moon system (p. 634)
emergent coasts (p. 650)
estuary (p. 644)
eustatic sea-level change (p. 650)
fjord (p. 645)

gravitational pull (p. 634)
guyot (p. 626)
gyre (p. 627)
lagoon (p. 641)
longshore current (p. 637)
organic coasts (p. 656)
passive continental margin (p. 623)
pelagic sediment (p. 625)
reef bleaching (p. 657)
rogue wave (p. 636)
sandspit (p. 641)
seamount (p. 626)
submarine canyons (p. 623)
submergent coasts (p. 651)
swash (p. 637)
thermohaline circulation (p. 630)
tide (p. 631)
tide-generating force (p. 631)
wave base (p. 636)
wave refraction (p. 637)
wave-cut bench (platform) (p. 643)
wave-cut notch (p. 643)

Review Questions

1. How much of the Earth's surface is covered by oceans? How much of the world's population lives near a coast?

2. Describe the typical topography of a passive continental margin, from the shoreline to the abyssal plain.

3. How do the shelf and slope of an active continental margin differ from those of a passive margin?

4. Where does the salt in the ocean come from? How does the salinity in the ocean vary?

5. What factors control the direction of surface currents in the ocean? What is the Coriolis effect, and how does it affect oceanic circulation? Explain thermohaline circulation.

6. What causes the tides?

7. Describe the motion of water molecules in a wave. How does wave refraction cause longshore currents?

8. Describe the components of a beach profile.

9. How does beach sand migrate as a result of longshore currents? Explain the sediment budget of a coast.

10. Describe how waves affect a rocky coast, and how such coasts evolved.

11. What is an estuary? Why is it such a delicate ecosytem? What is the difference between an estuary and a fjord?

12. Discuss the different types of coastal wetlands. Describe the different kinds of reefs, and how a reef surrounding an oceanic island changes with time.

13. How do plate tectonics, sea-level changes, sediment supply, and climate change affect the shape of a coastline? Explain the difference between emergent and submergent coasts.

14. In what ways do people try to modify or "stabilize" coasts? How do the actions of people threaten the natural systems of coastal areas?

On Further Thought

1. In 1789, the crew of the H.M.S. *Bounty* mutinied. Near Tonga, in the Friendly Islands (approximately 20°S and 175°W), the crew, led by Fletcher Christian, forced the ship's commanding officer, Lieutenant Bligh, along with those crewmen who remained loyal to Bligh, into a rowboat and set them adrift in the Pacific Ocean. The castaways, amazingly survived, and 47 days later landed at Timor (near Sumatra), 6,700 km to the west. Why did they end up where they did? Considering how long the journey took them, how fast were they moving?

2. A hotel chain would like to build a new beach-front hotel along a north-south-trending stretch of beach where a strong long-shore current flows from south to north. The neighbor to the south has constructed an east-west-trending groin on the property line. Will this groin pose a problem? If so, what solutions could the hotel try?

3. Observations made during the last decade suggest that sea level is rising in response to global warming. This rise happens partly because water expands when it heats up, and partly because glacial ice sheets in Greenland and elsewhere are melting. Much of southern Florida lies at elevations of less than 6 m above sea level. What changes will take place to the region as sea level rises? To answer, keep in mind the concepts of emergent and submergent coasts, and assume that southern Florida will continue to lie in the subtropical realm.

Suggested Readings

Ballard, R. D., and W. Hively. 2002. *The Eternal Darkness: A Personal History of Deep-Sea Exploration*. Princeton, N.J.: Princeton University Press.

Bird, E. C. 2001. *Coastal Geomorphology: An Introduction*. New York: Wiley.

Davis, R. A. 1997. *The Evolving Coast*. New York: Holt.

Dean, R. G., and R. A. Dalrymple. 2001. *Coastal Processes with Engineering Applications*. Cambridge: Cambridge University Press.

Emanuel, K. 2005. *Divine Wind: The History and Science of Hurricanes*. Oxford: Oxford University Press.

Erickson, J. 2003. *Marine Geology*. London: Facts on File.

Garrison, T. 2002. *Oceanography: An Invitation to Marine Science*, 4th ed. Pacific Grove, Calif.: Wadsworth/Thomson.

Komar, P. D. 1997. *Beach Processes and Sedimentation*. 2nd ed. Upper Saddle River, N.J.: Pearson.

Kunzig, R. 1999. *The Restless Sea: Exploring the World beneath the Waves*. New York: Norton.

Seibold, E., and W. H. Berger. 1995. *The Sea Floor: An Introduction to Marine Geology*. 3rd ed. New York: Springer-Verlag.

Sverdrup, K. A., A. B. Duxbury, and A. C. Duxbury. 2002. *An Introduction to the World's Oceans*. 7th ed. New York: McGraw-Hill.

Viles, H., and T. Spencer. 1995. *Coastal Problems: Geomorphology, Ecology and Society at the Coast*. New York: Wiley.

Woodroffe, C. D. 2002. *Coasts: Form, Process and Evolution*. Cambridge: Cambridge University Press.

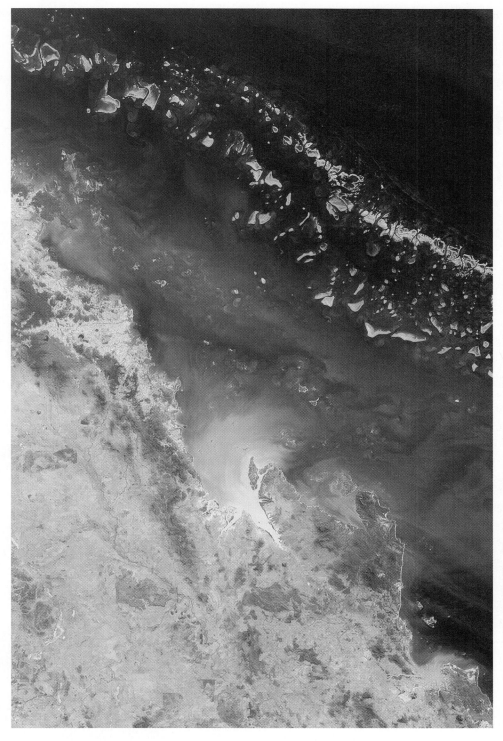

THE VIEW FROM SPACE The Great Barrier Reef fringes the northeast coast of Australia. It has been built from the shells of corals and other marine animals. This reef is the largest in the world—it forms the living coast of a continent.

A Hidden Reserve: Groundwater

Geopuzzle

In centuries past, the social center of a town would be the town's well, consisting of a hole dug down several meters into the ground. Usually, townspeople would line the well with stone, and to obtain water they would lower a bucket down the well, then lift up the water. Where did the water come from?

Groundwater can turn a desert green, as shown by these circular irrigated fields sprouting in the sands of Jordan. A water well lies at the center of each circle.

When the rain falls and enters the earth, when a pearl drops into the depth of the sea, you can dive in the sea and find the pearl, you can dig in the earth and find the water.

—Mei Yao-ch'en (Chinese Poet, 1002–1060)

19.1 INTRODUCTION

Imagine Rosa May Owen's surprise when, on May 8, 1981, she looked out her window and discovered that a large sycamore tree in the backyard of her Winter Park, Florida, home had suddenly disappeared. It wasn't a particularly windy day, so the tree hadn't blown over—it had just vanished! When Owen went outside to investigate, she found that more than the tree had disappeared. Her whole backyard had become a deep, gaping hole. The hole continued to grow for a few days until finally it swallowed Owen's house and six other buildings, as well as the municipal swimming pool, part of a road, and several expensive Porsches in a car dealer's lot (▶Fig. 19.1a).

What had happened in Winter Park? The bedrock beneath the town consists of limestone, a fairly soluble rock. Underground water had gradually dissolved the limestone, carving open rooms, or caverns, underground. On May 8, the roof of a cavern underneath Owen's backyard began to collapse, forming a circular depression called a sinkhole. The sycamore tree and the rest of the neighborhood simply dropped down into the sinkhole. It would have taken too much effort to fill in the hole with soil, so the community allowed it to fill with water, and now it's a circular lake, the cen-

terpiece of a pleasant municipal park. Similar lakes appear throughout central Florida (▶Fig. 19.1b).

The formation of the Winter Park sinkhole is one of the more dramatic reminders that significant quantities of water reside underground. In fact, at any given time, the volume of subsurface water is more than 200 times that held in all the lakes, swamps, and rivers at the Earth's surface combined (see Table F.2a). Where does the underground water come from? Some is ancient seawater that was trapped in pores when sediment was buried, and some is water rising from depth, released by metmorphic reactions or igneous activity. But most begins as rain or snow—meteoric water—that falls on the ground. Recall from Interlude F that some meteoric water evaporates directly back into the atmosphere, some becomes trapped in glaciers, and some becomes surface water that fills lakes or flows down streams. The remainder sinks, or infiltrates, into the ground. In effect, the upper part of the crust behaves like a giant sponge that can soak up meteoric water.

For millennia, subsurface freshwater (which accounts for about 30% of the freshwater on Earth) has been a major resource for homes, farms, and industry, so knowledge of this water has practical value. As a result, a large proportion of professional geologists specialize in hydrogeology, and they spend their careers either identifying usable sources of subsurface water or proposing strategies to clean contaminated supplies. In this chapter, we first examine where subsurface water resides and discuss how this water flows and interacts with rock and sediment. We then look into how subsurface water has been affected by human activities and conclude with a survey of landscape features that originate in response to interactions between water and rock in the subsurface realm.

FIGURE 19.1 **(a)** The Winter Park, Florida, sinkhole as seen from a helicopter; **(b)** numerous sinkhole lakes dot central Florida, as seen from high altitude.

(a)

(b)

19.2 WHERE DOES UNDERGROUND WATER RESIDE?

Porosity: Open Space in Rock and Regolith

The existence of water underground requires the existence of open space underground. What is the nature of this space? When asked this question, many people picture networks of caves containing spooky lakes and inky streams. Indeed, as we see later in this chapter, caves do provide room for water to drip and flow. But contrary to popular belief, only a small proportion of underground water actually occurs in caves. Most of this water resides in small open spaces or voids *within* regolith (sediment or soil) and within solid rock. Geologists use the term **pore** for any open space within a volume of re-

golith, or within a body of rock, and the term **porosity** for the total amount of open space within a material, specified as a percentage. If we say that a block of rock has 30% porosity, then 30% of the block consists of pores. To develop an intuitive image of what porosity looks like, take a sturdy glass jar and fill it with gravel composed of rounded pebbles. If you look closely at this gravel through the side of the jar, you will see air-filled pores between the pebbles, because the pebbles don't fit together perfectly. If you pour water into the jar, the water can trickle between grains down to the bottom of the jar, where it displaces air and fills the pores (▶Fig 19.2a).

How does porosity form? **Primary porosity** forms during sediment deposition and during rock formation. It includes the pores between clastic grains that exist because the grains don't fit together tightly during deposition. Such pores survive the process of lithification if cementation is incomplete (▶Fig. 19.2b). Primary porosity in clastic sediment and clastic sedimentary rock depends on the size

FIGURE 19.2 (a) A jar filled with pebbles, here shown in cross section, contains not only rock but also abundant open space, because the pebbles don't fit together tightly. If you pour some water in, the water trickles down the sides of the grains, partly filling the upper openings and completely filling the bottom ones. **(b)** A thin section of a rock composed of ooids (elliptical, snowball-like grains of calcite) that have been partially cemented together by blocky calcite crystals. The black areas are open spaces. **(c)** Various kinds of primary porosity in rock. Porosities are indicated as percentages.

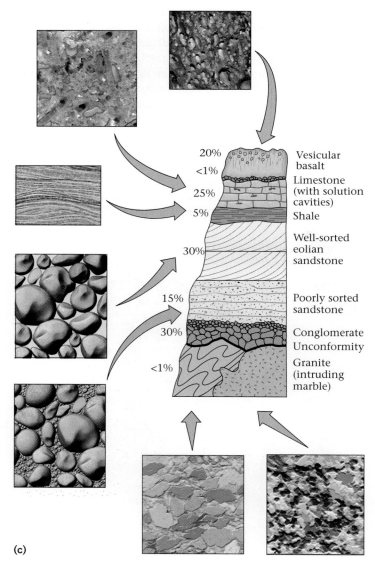

Glass beaker

Solid pebble

Open pore space

Saturated pore space

(a)

(b)

(c)

20% Vesicular basalt
<1%
25% Limestone (with solution cavities)
5% Shale
 Well-sorted eolian sandstone
30%
15% Poorly sorted sandstone
30% Conglomerate
 Unconformity
<1% Granite (intruding marble)

and sorting of the sediment and on the amount of compaction. For example, the overall porosity of a poorly sorted sediment is less than that of a well-sorted sediment because smaller clasts fill spaces between larger grains. Primary porosity decreases with increasing burial depth, because the weight of overburden pushes grains together. Primary porosity in chemical sedimentary rocks or biochemical sedimentary rocks develops because mineral crystals do not all grow snugly against their neighbors during precipitation. In crystalline igneous or metamorphic rock, primary porosity can form if grains do not interlock perfectly, but such porosity is rare. In fine-grained or glassy igneous rocks, primary porosity may consist of vesicles, air bubbles that were trapped during cooling. The amount of primary porosity ranges from less than 1% in crystalline igneous rocks and metamorphic rocks to over 30% in well sorted sands and poorly cemented sandstone (▶Fig. 19.2c).

Secondary porosity refers to new pore space in rocks, produced some time after the rock first formed. For example, when rocks fracture, the opposing walls of the fracture do not fit together tightly, so narrow spaces remain in between. Thus, joints and faults may provide secondary porosity for water (▶Fig. 19.3). Faulting may also produce breccia, a jumble of angular fragments, with the space between the fragments providing another type of secondary pore space. Finally, as groundwater passes through rock, it may dissolve and remove some minerals, creating solution cavities.

Permeability: The Ease of Flow

If solid rock completely surrounds a pore, the water in the pore cannot flow to another location. For groundwater to flow, therefore, pores must be linked by conduits (openings). The ability of a material to allow fluids to pass through an interconnected network of pores is a characteristic known as **permeability** (▶Fig. 19.4a, b). Water flows

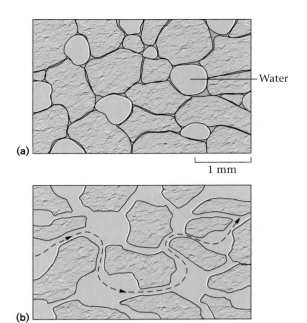

FIGURE 19.4 **(a)** Isolated, nonconnected pores in an impermeable material; **(b)** pores connected to each other by a network of conduits in permeable material.

easily through a permeable material. In contrast, water flows slowly or not at all through an impermeable material. The permeability of a material depends on several factors.

- *Number of available conduits:* As the number of conduits increases, permeability increases.
- *Size of the conduits:* More fluids can travel through wider conduits than through narrower ones.
- *Straightness of the conduits:* Water flows more rapidly through straight conduits than it does through crooked ones. In crooked channels, the distance a water molecule actually travels may be many times the straight-line distance between the two end points.

Note that the factors that control permeability in rock or sediment resemble those that control the ease with which traffic moves through a city. Traffic can flow quickly through cities with many straight, multilane boulevards, whereas it flows slowly through cities with only a few narrow, crooked streets.

Porosity and permeability are not the same. A material whose pores are isolated from each other can have high porosity but low permeability. For example, if you plug the bottom of a water-filled tube with a cork, the water remains in the tube even though cork is very porous; this is because woody cell walls separate adjacent pores in the cork and prevent communication between them, making cork impermeable. Vesicular basalt is an example of a rock that can have high porosity but low permeability.

FIGURE 19.3 Fractures in a rock provide secondary porosity.

Aquifers and Aquitards

With the concept of permeability in mind, hydrogeologists distinguish between **aquifers,** sediment or rocks that transmit water easily, and **aquitards,** sediment or rocks that do not transmit water easily and therefore retard the motion of water. Aquicludes do not transmit water at all. Aquifers that intersect the surface of the Earth are called unconfined aquifers, because water can percolate directly from the surface down into the aquifer, and water in the aquifer can rise to the surface. Aquifers that are separated from the surface

by an aquitard are called confined aquifers, because the water they contain is isolated from the surface (▶Fig. 19.5a).

Hydrogeologists distinguish among many diverse types of aquifers on the basis of the nature of the material making up the aquifer and on the configuration of the aquifer. To get a sense of this diversity, let's consider some examples: (1) *Mahomet aquifer:* This aquifer, named for a town in central Illinois, formed at the end of the Pleistocene ice age when sediment-choked meltwater streams filled pre-Pleistocene river valleys, which had been cut into bedrock, to the brim with over 50 m of permeable gravel (▶Fig. 19.5b). It provides water to almost 1 million people. (2) *Dakota aquifer:* During the Cretaceous, rivers deposited sand in floodplains and deltas along the western edge of the shallow sea that covered the interior of North America. When lithified, the deposits became the Dakota Sandstone. Mountain-building processes causing uplift of the Black Hills warped the Dakota Sandstone layer into an asymmetric trough. At the low point of the trough, the unit lies deep underground, beneath thick aquitards (▶Fig. 19.5c). Water infiltrates into the unit near the Black Hills and flows slowly eastward, filling the aquifer

FIGURE 19.5 **(a)** An aquifer is a high-porosity, high-permeability material. If it has access to the ground surface, it's an unconfined aquifer. If it's trapped below an aquitard, a rock with low permeability, it's a confined aquifer. **(b)** Cross-section sketch of the Mahomet aquifer, Illinois. **(c)** Cross-section sketch of the Dakota aquifer, South Dakota. **(d)** Map showing the distribution of the Ogallala Formation, which comprises the High Plains aquifer; **(e)** Cross-section sketch of aquifers in the basins of a rift.

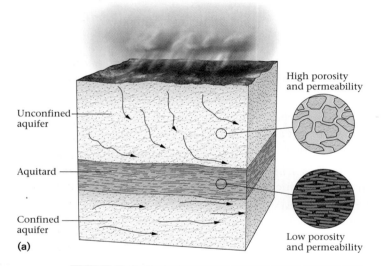

(a)

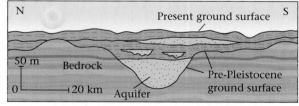

(b) Mahomet aquifer: buried gravel-filled valley

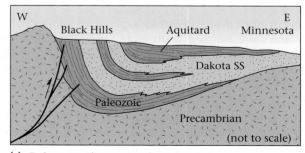

(c) Dakota aquifer: regional sandstone layer

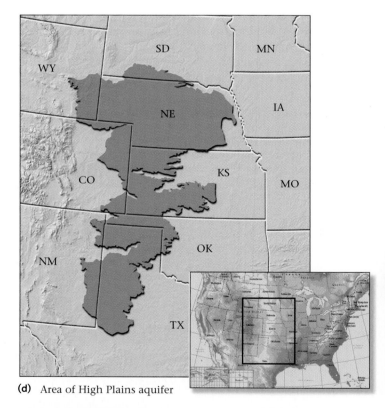

(d) Area of High Plains aquifer

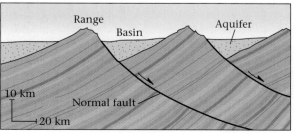

(e) Gravel-filled basins in the Basin and Range Province

over a 500-km-wide region. (3) *High Plains aquifer:* Erosion of the Rocky Mountains over 3 million years ago led to deposition of huge alluvial fans over the high plains region of the United States. The resulting deposits make up the Ogallala Formation, which still lies close to the surface. Water can infiltrate into the unit across a vast area (▶Fig. 19.5d). (4) *Phoenix aquifer:* Movement on normal faults in a continental rift, such as the Basin and Range Province of the western United States, develops half grabens that fill with sediment eroded off of adjacent fault-block ranges (▶Fig. 19.5e). The resulting deposits of coarse sediment are a source of groundwater for the communities, such as Phoenix, Arizona, and farms of the desert Southwest. (5) *Florida aquifer:* Areas underlain by limestone usually take advantage of solution-enhanced fractures and cavities. Florida, for example, is underlain by vast layers of limestone that formed when the region was a Bahama-like reef. Fractures in the limestone have been widened by dissolution and now provide space that holds large quantities of groundwater.

> **Take-Home Message**
>
> Most underground water fills pores and cracks in rock or sediment. Porosity refers to the total amount of open space within a material, whereas permeability indicates the degree to which pores connect. Aquifers have high porosity and permeability; aquitards don't.

19.3 GROUNDWATER AND THE WATER TABLE

In the last section, we used the very general term *underground water* for *all* water that occurs below the surface of the Earth. In detail, however, geologists distinguish among three categories of underground water: soil moisture, vadose-zone water, and groundwater. In places where soil exists, some infiltrating meteoric water adheres temporarily to the surfaces of clay, sand, or organic debris. This water comprises soil moisture—it either evaporates back into the atmosphere or is absorbed by plant roots. Not all land has a covering of soil, however, and even where soil exists, it grades downward into unaltered sediment or bedrock (see Chapter 7). Infiltrating water can enter permeable sediment and bedrock by percolating along cracks and through conduits connecting pores. Nearer the ground surface, water only partially fills pores, leaving some space that remains filled with air. In technical jargon, the region of the subsurface in which water only partially fills pores is called the **unsaturated zone**, or the vadose zone; the water that resides in this region is vadose-zone water. Deeper down, water completely fills, or saturates, the pores. This region is the **saturated zone**, or the phreatic zone. Geologists use the term **groundwater** specifically for subsurface water in the saturated zone, where water completely fills pores.

The term **water table** refers to the horizon that separates the undersaturated zone above from the saturated zone below (▶Fig. 19.6a). Material above the water table can be damp, but pores are not full. Typically, surface tension, the electrostatic attraction of water molecules to mineral surfaces, causes water to seep up from the water table (just as water rises in a thin straw), filling pores in the **capillary fringe**, a thin layer at the base of the unsaturated zone.

The depth of the water table in the subsurface varies greatly with location. In some places, the water table effectively lies above ground and defines the surface of a permanent stream, lake, or marsh (▶Fig. 19.6b, c). Elsewhere, the water table lies hidden below the ground surface: in humid regions, it typically lies within tons of meters of the ground surface, whereas in arid regions, it may lie deeper than hundreds of meters below the surface.

Rainfall affects the water-table depth—the level sinks during a dry season (▶Fig. 19.6d). If the water table drops below the floor of a river or lake, the river or lake dries up, because the water it contains infiltrates into the ground. In arid regions, there are no permanent streams, because the water table lies below the stream bed. Streams in such regions flow only during storms, when rain falls at a faster rate than the water can infiltrate.

We've defined the water table as the *top* of a groundwater reservoir. How do we define the *bottom* of a groundwater reservoir? For practical purposes, hydrogeologists specify the base of a reservoir in a given location as the subsurface horizon where an aquifer lies in contact with an underlying aquitard. For example, consider the Dakota aquifer. Hydrogeologists consider the unconformity between the base of the Dakota Sandstone (permeable sedimentary rock) and the underlying Precambrian rock (relatively impermeable metamorphic rock) to be the base of the aquifer (see Fig. 19.5c). However, research shows that liquid groundwater actually does circulate in basement igneous and metamorphic rock, even deeper in the crust, perhaps to depths of over 15 km. Most of this deep circulation takes place along fractures. In this regard, the ultimate base of groundwater in crust should perhaps be taken as the depth where high-grade metamorphism and plastic deformation currently take place. At this depth, plastic deformation closes up pore space, and water becomes involved in metamorphic reactions and no longer exists as a free liquid. The depth of this high-grade metamorphic horizon at a given location depends on the geothermal gradient.

> **Take-Home Message**
>
> Above the water table, pores are not saturated with water, but below the water table they are. Groundwater is the water in the saturated zone. Depth to the water table depends on climate and topography—the water table is higher beneath hills.

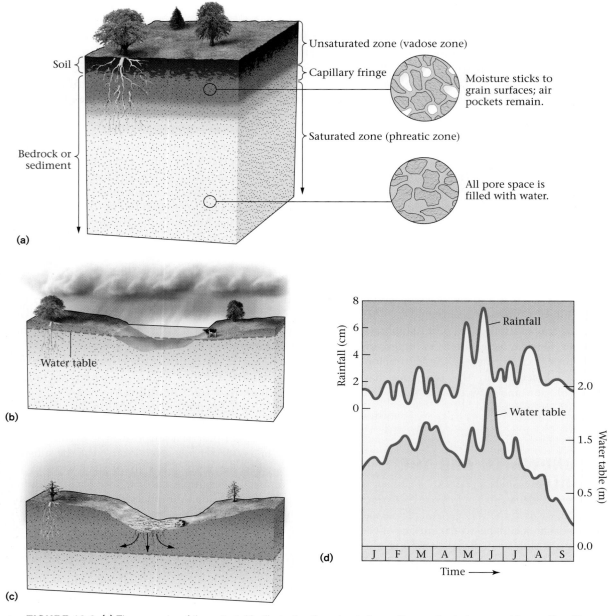

FIGURE 19.6 **(a)** The geometry of the water table, illustrating the saturated zone, the unsaturated zone, and the capillary fringe. **(b)** The surface of a permanent pond is the water table. **(c)** During the dry season, the water table can drop substantially, causing the pond to dry up. **(d)** The graph shows the relation between the height of the water table and rainfall between January and September, in a temperate region.

Topography of the Water Table

In hilly regions, if the ground has relatively low permeability, the water table is not a planar surface. Rather, its shape mimics, in a subdued way, the shape of the overlying topography (▶Fig. 19.7). This means that the water table lies at a higher elevation beneath hills than it does beneath valleys. But the relief (the vertical distance between the highest and lowest elevations) of the water table is not as great as that of the overlying land, so the surface of the water table tends to be smoother than that of the landscape.

At first thought, it may seem surprising that the elevation of the water table varies as a consequence of ground-surface topography. After all, when you pour a bucket of

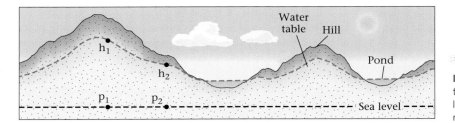

FIGURE 19.7 The shape of a water table beneath hilly topography. Note that the water table can be above sea level. Point h_1 on the water table is higher than point h_2, relative to a reference elevation (sea level).

water into a pond, the surface of the pond immediately adjusts to remain horizontal. The elevation of the water table varies because groundwater moves so slowly through rock and sediment that it cannot quickly assume a horizontal surface. When it rains on a hill and water infiltrates down to the water table, the water table rises a little. When it doesn't rain, the water table sinks slowly, but so slowly that when rain falls again, the water table rises before it has had time to sink very far.

Perched Water Tables

In some locations, layers of strata are discontinuous, meaning that they pinch out at their sides. As a result, lens-shaped layers of impermeable rock (such as shale) may lie within a thick aquifer. A mound of groundwater accumulates above this aquitard. The result is a **perched water table,** a quantity of groundwater that lies above the regional water table because an underlying lens of impermeable rock or sediment prevents the water from sinking down to the regional water table (▶Fig. 19.8).

FIGURE 19.8 The configuration of a perched water table. A lens of groundwater lies above, and the regional water table lies at greater depth.

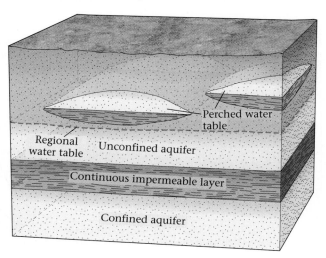

19.4 GROUNDWATER FLOW

Groundwater Flow Paths

What happens to water that has infiltrated down into the ground? Does it just sit, unmoving, like the water in a stagnant puddle, or does it flow and eventually find its way back to the surface? Countless measurements confirm that groundwater enjoys the latter fate—groundwater indeed flows, and in some cases it moves great distances underground. In this section we examine factors that drive groundwater flow and determine the path that this flow follows. Then, in the next section, we examine the rate (velocity) at which groundwater moves.

In the unsaturated zone—the region between the ground surface and the water table—water percolates straight down, like the water passing through a drip coffee maker, for this water moves only in response to the downward pull of gravity. But in the zone of saturation—the region below the water table—water flow is more complex, for in addition to the downward pull of gravity, water responds to differences in pressure. Pressure may cause groundwater to flow sideways, or even upward. (If you've ever watched water spray up from a fountain, you've seen pressure push water upward.) Thus, to understand the nature of groundwater flow, we must first understand the origin of pressure in groundwater. For simplicity, we'll only consider the case of groundwater in an unconfined aquifer.

Pressure in groundwater at a specific point underground is caused by the weight of all the overlying water from that point up to the water table. (The weight of overlying rock does not contribute to the pressure exerted on groundwater, for the contact points between mineral grains bear the rock's weight.) Thus, a point at a greater depth below the water table feels more pressure than does a point at lesser depth. If the water table is horizontal, the pressure acting on an imaginary horizontal reference plane at a specified depth below the water table is the same everywhere. But if the water table is not horizontal, as shown in Figure 19.7, the pressure at points on a horizontal reference plane at depth changes with location. For example, the pressure acting at point p_1, which lies below the hill in Figure 19.7, is greater than the pressure acting at point p_2, which lies below the valley, even though both p_1 and p_2 are at the same elevation (sea level, in this case).

Both the elevation of a volume of groundwater and the pressure within the water provide energy that, if given the chance, will cause the water to flow. Physicists refer to such stored energy as potential energy (see Appendix A). (To understand why elevation provides potential energy, imagine a bucket of water high on a hill; the water has potential energy, due to Earth's gravity, which will cause it to flow downslope if the bucket were suddenly to rupture. To understand why pressure provides potential energy, imagine a water-filled plastic bag sitting on a table; if you puncture the bag and then squeeze the bag to exert pressure, water spurts out.) The potential energy available to drive the flow of a given volume of groundwater at a location is called the **hydraulic head**. To measure the hydraulic head at a point in an aquifer, hydrogeologists drill a vertical hole down to the point and then insert a pipe in the hole. The height above a reference elevation (e.g., sea level) to which water rises in the pipe represents the hydraulic head—water rises higher in the pipe where the head is higher. As a rule, *groundwater flows from regions where it has higher hydraulic head to regions where it has lower hydraulic head.* This statement generally implies that groundwater flows from locations where the water table is higher to locations where the water table is lower.

Hydrogeologists have calculated how hydraulic head changes with location underground, by taking into account both the effect of gravity and the effect of pressure. They conclude that groundwater flows along concave-up curved paths, as illustrated in cross section (►Fig. 19.9). (Specialized books on hydrogeology provide the details of why flow paths have such a specific shape.) These curved paths eventually take groundwater from regions where the water table is high (under a hill) to regions where the water table is low (below a valley), but because of flow-path shape, some groundwater may flow deep down into the crust along the first part of its path, and then may flow back up, toward the ground surface, along the final part of its path. The location where water enters the ground (i.e., where the flow direction has a downward trajectory) is called the **recharge area**, and the location where groundwater flows back up to the surface is called the **discharge area** (Fig. 19.9). To apply this concept, look once again at the cross section of the Dakota aquifer shown in Figure 19.5c. You can see that water enters the aquifer at a recharge area along the flank of the Black Hills, sinks down into the ground, and then flows eastward, eventually returning to the surface at a discharge area in Minnesota.

Groundwater following short paths close to the Earth's surface travels tens of meters to a few kilometers before returning to the surface. Such movement constitutes local flow. Locally flowing groundwater has a residence time (i.e., stays underground) for only hours to weeks. Groundwater following paths of several kilometers to tens of kilometers constitutes intermediate flow and has a residence time of weeks to years. Groundwater following paths that carry it hundreds of kilometers across a sedimentary basin constitutes regional flow and stays underground for centuries to millennia (►Fig. 19.10).

Rates of Groundwater Flow: Darcy's Law

Flowing water in an ocean current moves at up to 3 km per hour (over 26,000,000 m per year), and water in a steep river channel can reach speeds of up to 30 km per hour (over 260,000,000 m per year). In contrast, groundwater moves at a snail's pace. Typical rates range between 0.01 and 1.4 m per day (about 4–500 m per year). Groundwater moves much more slowly than surface water for two reasons. First, groundwater moves by percolating through a complex, crooked network of tiny conduits; it must

FIGURE 19.9 The flow lines from the recharge area to the discharge area curve through the substrate. In fact, some groundwater descends to great depth and then rises back to the surface.

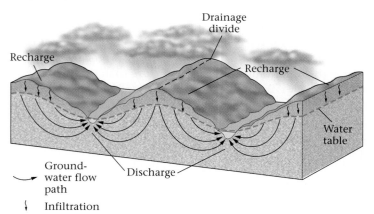

FIGURE 19.10 Cross section showing regional-scale groundwater flow in a sedimentary basin. The arrows indicate the flow paths of groundwater. Note that recharge occurs in highlands on one side of the basin, and discharge occurs on the far side of the basin. Some of the water flows through fractures deep in the basement.

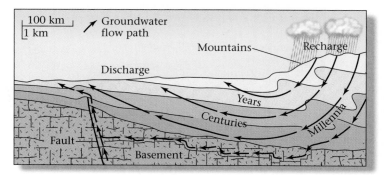

travel a much greater distance than it would if it could follow a straight path. Second, friction and/or electrostatic attraction between groundwater and conduit walls slows down the water flow. Hydrogeologists may measure the rate (velocity) of groundwater flow in a region by "tagging" water with dye or a radioactive element. They inject the tracer in one well and time how long it takes for the tracer to reach another well.

The rate at which groundwater flows at a given location depends on the permeability of the material containing the groundwater; groundwater flows faster in material with greater permeability than in less permeable material. Rate also depends on the **hydraulic gradient**, the change in hydraulic head per unit of distance between two locations as measured along the flow path. Recall that groundwater flows from a region where the hydraulic head is higher (due to higher elevation and/or greater pressure) to a region where the hydraulic head is lower. If there is a large difference in the hydraulic head over a given distance, then a greater amount of energy drives the flow, so the flow is faster.

To calculate the hydraulic gradient, we simply divide the difference in hydraulic head between two points by the distance between the two points as measured along the flow path. This can be written as a formula:

$$\text{hydraulic gradient} = \Delta h / j,$$

where Δh is the difference in head (given in meters or feet, because head can be represented as an elevation), and j is the distance between the two points (▶Fig. 19.11). A hydraulic gradient exists anywhere that the water table has a slope. In an unconfined aquifer, the hydraulic gradient is roughly equivalent to the slope of the water table, if the slope is gentle.

FIGURE 19.11 A hydraulic gradient (HG) is the change in hydraulic head (Δh) per unit of distance between two points (j) along the flow path.

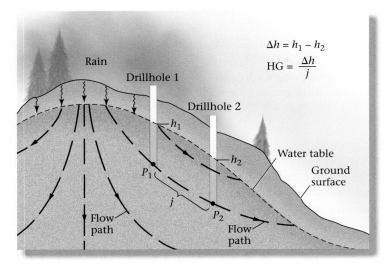

With an understanding of the controls on the rate of groundwater flow, we can consider the practical problem of determining the volume of water that will pass through an area underground during a specified time. In 1856, a French engineer named Henry Darcy addressed this problem because he wanted to find out whether the city of Dijon in central France could supply its water needs from groundwater. Darcy carried out a series of experiments in which he measured the rate of water flow through sediment-filled tubes tilted at varying angles. Darcy found that the volume of water that passed through a specified area in a given time, a quantity he called the *discharge* (Q), can be calculated from the equation

$$Q = K(\Delta h / j)A,$$

where $\Delta h / j$ is the hydraulic gradient, A is the area of porous rock or sediment through which the water is passing as measured in a plane perpendicular to flow direction (see Fig. 19.11), and K is a number called the hydraulic conductivity. The hydraulic conductivity takes into account permeability as well as gravity, the fluid's viscosity (stickiness), and the fluid's density. Hydrogeologists refer to this equation as **Darcy's law**.

To simplify discussion, geologists sometimes represent Darcy's law as follows:

$$\text{discharge} \propto (\text{water table slope}) \times (\text{permeability}),$$

where "$\propto$" means "is proportional to" or "depends on." Darcy's law states, therefore, that groundwater flows faster through very permeable rocks than through impermeable rocks, and that it flows faster where the water table has a steep slope than where the water table has a shallow slope. It is no surprise, therefore, that groundwater moves very slowly (0.5–3.0 cm per day) through a well-cemented bed of the Dakota Sandstone, but moves relatively quickly (over 15 cm per day) through a steeply sloping layer of unconsolidated gravel on the side of a mountain.

19.5 TAPPING GROUNDWATER SUPPLIES

Groundwater can be brought to the ground surface at wells or springs. **Wells** are holes that people dig or drill to obtain water. **Springs** are natural outlets from which groundwater flows. Wells and springs provide welcome sources of water but must be treated with care if they are to last.

> **Take-Home Message**
>
> Gravity and pressure cause groundwater to flow slowly from recharge to discharge areas. In essence, the rate of flow depends on the water table's slope and on permeability. Groundwater can follow curving flow paths that take it deep into the crust.

Wells

In an **ordinary well**, the base of the well penetrates an aquifer below the water table. Water from the pore space in the aquifer seeps into the well and fills it; the water surface in the well is the water table. Drilling into an aquitard, or into rock that lies above the water table, will not supply water, and thus yields a dry well. Some ordinary wells are seasonal and function only during the rainy season, when the water table rises. During the dry season, the water table lies below the base of the well, so the well is dry.

Ideally, we would like an ordinary well to be as shallow as possible (to decrease the cost of digging). So hydrogeologists search for particularly porous and permeable aquifers in which the water table lies near the surface. Contrary to legend, a dowser cannot find water simply by using a forked stick—when dowsers do strike water, either they have enjoyed dumb luck or they have studied published data about the water table in the area.

To obtain water from an ordinary well, you either pull water up in a bucket or pump the water out (▶Fig. 19.12a, b). As long as the rate at which groundwater fills the well exceeds the rate at which water is removed, the level of the water table near the well remains about the same. However, if users pump water out of the well too fast, then the water table sinks down around the well, a process called drawdown, so that the water table becomes a downward-pointing, cone-shaped surface called a **cone of depression** (▶Fig. 19.13). Drawdown may cause shallower wells that have been drilled nearby to run dry.

An **artesian well**, named for Artois Province in France, penetrates confined aquifers, in which water is under enough pressure to rise on its own to a level above the surface of the aquifer. If this level lies below the ground surface, the well is a nonflowing artesian well. But if the level lies above the ground surface, the well is a flowing artesian well, and water actively fountains out of the ground (▶Fig. 19.14a). Artesian wells occur in special situations where a confined aquifer lies beneath a sloping aquitard.

We can understand why artesian wells exist if we look first at the configuration of a city water supply (▶Fig. 19.14b). Water companies pump water into a high tank that has a significant hydraulic head relative to the surrounding areas. If the water were connected by a water main to a series of vertical pipes, pressure caused by the elevation of the water in the high tank would make the water rise in the pipes until it reached an imaginary surface, called a potentiometric surface, that lies above the ground. This pressure drives water through water mains to household water systems without requiring pumps. In an artesian system, water enters a tilted, confined aquifer that intersects the ground in the hills of a high-elevation recharge area (▶Fig. 19.14c). The confined groundwater flows down to the adjacent plains, which lie at a lower elevation. The

(a)

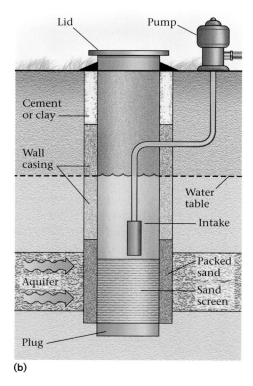

(b)

FIGURE 19.12 (a) In this ordinary well, the water has to be brought up by bucket. **(b)** Diagram showing how drillers set up a pump to extract water from an ordinary well. The packed sand filters the water.

potentiometric surface to which the water would rise were it not confined lies above this aquifer; in fact, it may lie above the ground surface of the plains. Pressure in the confined aquifer pushes water up a well. Where the potentiometric surface lies below ground, the well will be a nonflowing artesian well, but where the surface lies above the ground, the well will be a flowing artesian well.

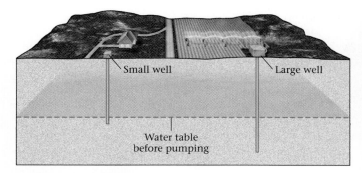

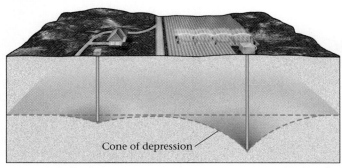

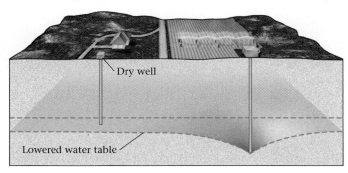

FIGURE 19.13 The base of an ordinary well penetrates below the water table. If groundwater is extracted faster than it can be replaced, a cone of depression forms around the well. Pumping by the big well, in this example, may lower the water table sufficiently to cause the small well to become dry.

Springs

More than one town has grown up around a spring, a place where groundwater naturally flows or seeps onto the Earth's surface, for springs provide fresh, clear groundwater for drinking or irrigation, without the expense of drilling or digging. Springs form under a variety of conditions:

- Where the ground surface intersects the water table in a discharge area (▶Fig. 19.15a): such springs typically occur in valley floors where they may add water to lakes or streams.
- Where downward-percolating water runs into an impermeable layer and migrates along the top surface of the layer to a hillslope (▶Fig. 19.15b).

(a)

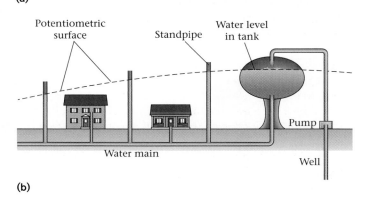

(b)

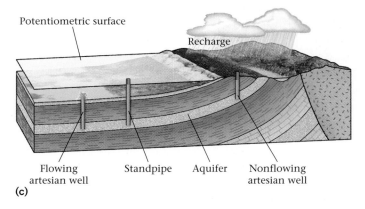

(c)

FIGURE 19.14 (a) A flowing artesian well in a Missouri field. **(b)** The configuration of a city water supply. Water will rise in vertical pipes up to the level of the potentiometric surface. **(c)** The configuration of an artesian system. Artesian wells flow if the potentiometric surface lies above the ground surface. Nonflowing artesian wells form where the potentiometric surface lies below the ground.

- Where a particularly permeable layer or zone intersects the surface of a hill (▶Fig. 19.15c): water percolates down through the hill and then migrates along the permeable layer to the hill face.

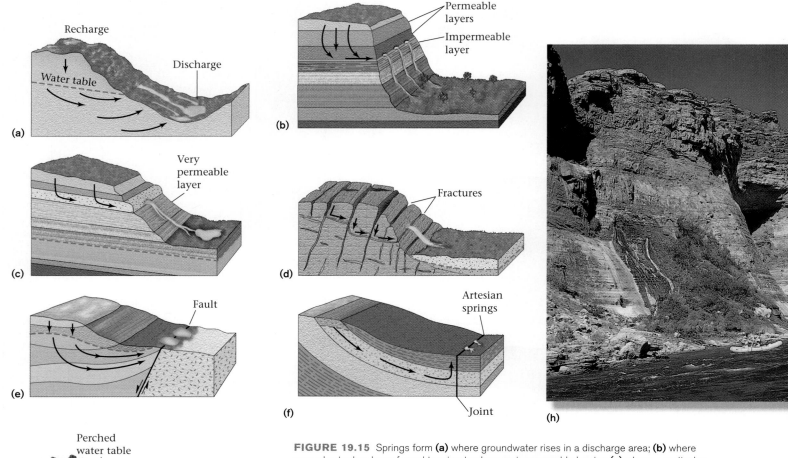

FIGURE 19.15 Springs form **(a)** where groundwater rises in a discharge area; **(b)** where groundwater has been forced to migrate along an impermeable barrier; **(c)** where a particular permeable layer transmits water to the surface of a hill; **(d)** where a network of interconnected fractures channels water to the hill face; **(e)** where groundwater collides with a steep impermeable barrier, and pressure pushes it up to the ground along the barrier. **(f)** An artesian spring forms where water from a confined aquifer migrates up a joint; **(g)** springs also form where a perched water table intersects the surface of a hill. **(h)** A spring arising from a perched water table intersecting a wall of the Grand Canyon.

- Where a network of interconnected fractures channels groundwater to the surface of a hill (▶Fig. 19.15d).
- Where flowing groundwater collides with a steep impermeable barrier, and pressure pushes it up to the ground along the barrier (▶Fig. 19.15e): faulting can create such barriers by juxtaposing impermeable rock against permeable rock.
- **Artesian springs** form if the ground surface intersects a natural fracture (joint) that taps a confined aquifer in which the pressure is sufficient to drive the water to the surface (▶Fig. 19.15f and ▶Box 19.1).
- Where a perched water table intersects the surface of a hill (▶Fig. 19.15g, h).

19.6 HOT SPRINGS AND GEYSERS

Hot springs, springs that emit water ranging in temperature from about 30° to 104°C, are found in two geologic settings. First, they occur where very deep groundwater, heated in warm bedrock at depth, flows up to the ground surface. This water brings heat with it as it rises. Such hot springs form in places where faults or fractures provide a high-permeability

> **Take-Home Message**
>
> Groundwater can be obtained at wells (built by people) and springs (natural outlets). In ordinary wells, water must be lifted to the surface, but in artesian wells and springs, it rises due to its hydraulic head. Pumping of groundwater lowers the water table.

BOX 19.1 THE HUMAN ANGLE

Oases

The Sahara of northern Africa is now one of the most barren and desolate places on Earth, for it lies in a climatic belt where rain seldom falls. But it wasn't always that way. During the last ice age, when glaciers covered parts of northern Europe on the other side of the Mediterranean, the Sahara enjoyed a more temperate climate, and the water table was high enough that permanent streams dissected the landscape. In fact, using ground-penetrating radar, geologists can detect these streams today—they look like ghostly valleys beneath the sand (▶Fig. 19.16a). When the climate warmed after the ice age, rainfall diminished, the water table sank below the floors of most stream beds, and the streams dried up.

Today, the water of the Sahara region lies locked in a vast underground aquifer composed of porous sandstone. Recharge into this aquifer comes from highlands bordering the desert, from occasional downpours, and from the Nile River (particularly Lake Nasser, created by the Aswan High Dam). In general, the water of the aquifer can be obtained only by drilling deep wells, but locally, water spills out at the surface—either because folding brings the aquifer particularly close to the ground so that valley floors intersect the water table, or because artesian pressure pushes groundwater up along joints of faults (▶Fig. 19.16b). In either case, the aquifer feeds springs that quench the thirst of desert and tropical plants and create an **oasis**, an island of green in the sand sea (▶Fig. 19.16c). Oases became important stopping points along caravan routes, allowing both people and camels to replenish water supplies.

In some oases, people settled and used the groundwater to irrigate date palms and other crops. For example, the Bahariya Oasis, about 400 km southwest of Cairo, Egypt, hosted a town of perhaps 30,000 between 300 B.C.E. and 300 C.E. During that time, the water table lay only 5 m below the ground and could easily be reached by shallow wells. Today, as a result of changing climates and centuries of use, the water table lies 1,500 m below the ground, almost out of reach. Bahariya's glorious past came to light in 1996, quite by accident. A man was riding his donkey in the desert near the oasis when the ground beneath the donkey suddenly caved in. The man had inadvertently opened the roof into a tomb filled with over 150 mummies, along with thousands of well-preserved artifacts. The site has since come to be known as the Valley of the Mummies.

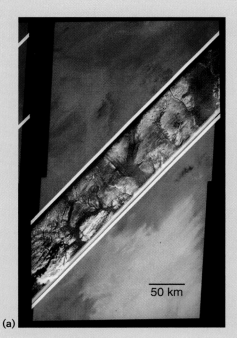

(a)

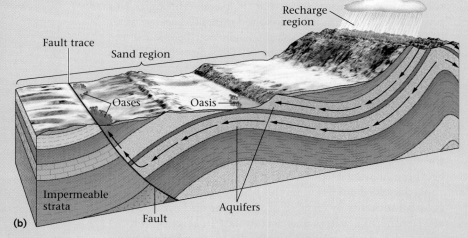

(b)

(c)

FIGURE 19.16 (a) A satellite image, taken using ground-penetrating radar, showing a long-abandoned drainage network now buried by Saharan sand. The gray strip shows darker channels and lighter high areas under the sand. The orange area is the sand surface that you see without radar. **(b)** This subsurface configuration of aquifers leads to the formation of an oasis, where groundwater reaches the surface. **(c)** An oasis in the Sahara.

FIGURE 19.17 In the city of Bath, England, the Romans built an elaborate spa around an artesian hot spring. The spring formed along a fault that tapped a supply of deep groundwater that still flows today. The baths of Bath have remained popular through the ages, and are currently undergoing renovation.

conduit for deep water, or where the water emitted in a discharge region followed a trajectory that first carried it deep into the crust. Second, hot springs develop in **geothermal regions**, places where volcanism currently takes place or has occurred recently, so that magma and/or very hot rock resides close to the Earth's surface. In hot springs, groundwater dissolves minerals from rock that it passes through. Hot groundwater contains more dissolved minerals because water becomes a more effective solvent when hot. People use the water emitted at hot springs to fill relaxing mineral baths (▶Fig. 19.17).

Numerous distinctive geologic features form in geothermal regions as a result of the eruption of hot water (▶Fig. 19.18a–d). In places where the hot water rises into soils rich in volcanic ash and clay, a viscous slurry forms and fills bubbling mud pots. Bubbles of steam rising through the slurry cause it to splatter about in goopy drops. Where geothermal waters spill out of natural springs and then cool, dissolved minerals in the water precipitate, forming colorful mounds or terraces of travertine and other minerals. Geothermal waters may accumulate in brightly colored pools—the gaudy greens, blues, and oranges of these pools come from thermophyllic (heat-loving) bacteria and archaea that thrive in hot water and metabolize the sulfur-containing minerals dissolved in the groundwater.

The most spectacular consequence of geothermal waters is a **geyser** (from the Icelandic word for gush), a fountain of steam and hot water that erupts episodically from a vent in the ground (▶Fig. 19.19a). To understand why a geyser erupts, we first need a picture of its underground plumbing. Beneath a geyser lies a network of irregular fractures in very hot rock; groundwater sinks and fills these fractures. Adjacent hot rock then superheats the water: it raises the

(a)

(b)

(c)

(d)

FIGURE 19.18 Features of the Yellowstone Park geothermal region. **(a)** Mud pots; **(b)** terraces of minerals precipitated at Mammoth Hot Springs; **(c)** colorful bacteria- and archaea-laden pools; **(d)** Old Faithful geyser. The geyser erupts somewhat predictably.

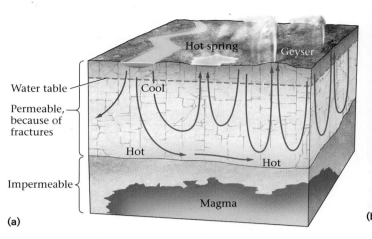

FIGURE 19.19 (a) Geysers and hot springs form where groundwater, heated at depth, rises to the surface. **(b)** Geysers in the geothermal region of Rotorua, New Zealand.

temperature above the temperature at which water at a pressure of 1 atm will boil. Eventually, this superhot water rises through a conduit to the surface. When some of this water transforms into steam, the resulting expansion causes water higher up to spill out of the conduit at the ground surface. When this spill happens, pressure in the conduit, from the weight of overlying water, suddenly decreases. A sudden drop in pressure causes the superhot water at depth instantly to turn to steam, and this steam quickly rises, ejecting all the water and steam above it out of the conduit in a geyser eruption. Once the conduit empties, the eruption ceases, and the conduit fills once again with water that gradually heats up, starting the eruptive cycle all over again.

Hot springs are found in many localities around the world: at Hot Springs, Arkansas, where deep groundwater rises to the surface; in Yellowstone National Park, above the magma chamber of a continental hot spot; around the Salton Sea in southern California, where the mid-ocean ridge of the Gulf of California merges with the San Andreas Fault; in the Geysers Geothermal Field of California, an area of significant geothermal power generation formed above a felsic magma intrusion; in Iceland, which has grown on top of an oceanic hot spot along the Mid-Atlantic Ridge; and in Rotorua, New Zealand, which lies in an active volcanic field above a subduction zone.

People live in some geothermal regions, though the areas have inherent natural hazards. In Rotorua, signs along the road warn of steam, which can obscure visibility, and steam indeed spills out of holes in backyards and parking lots (▶Fig. 19.19b). But all this hot water does offer a benefit: in Rotorua, waters circulate through pipes to provide home heating, and in geothermal areas worldwide, steam provides a relatively inexpensive means of generating electricity.

> **Take-Home Message**
>
> In geothermal regions, or in localities where discharged groundwater followed a flow path deep into the crust, hot springs appear. Geysers develop under special circumstances where pressure builds up sufficiently to eject water and steam forcefully.

19.7 GROUNDWATER USAGE PROBLEMS

Since prehistoric times, groundwater has been an important resource that people have relied on for drinking, irrigation, and industry. Groundwater feeds the lushness of desert oases in the Sahara, the amber grain in the North American high plains, and the growing cities of arid regions. Agricultural and industrial usage accounts for about 93% of all water usage, so as once-empty land comes under cultivation and countries become increasingly industrialized, demands on the groundwater supply soar. Groundwater provides only about 20% of the water we use worldwide, but this percentage has increased as surface-water resources decrease. And locally, groundwater is the sole water source.

Though groundwater accounts for about 95% of the liquid freshwater on the planet, accessible groundwater cannot be replenished quickly in important locations, and this leads to shortages. In the twentieth century, the problem was exacerbated by the contamination of existing groundwater. Such pollution, caused when toxic wastes and other impurities infiltrate down to the water table, may be invisible to us but may ruin a water supply for generations to come. In this section, we'll take a look at problems associated with the use of groundwater supplies.

Depletion of Groundwater Supplies

Is groundwater a renewable resource? In a time frame of 10,000 years, the answer is yes, for the hydrologic cycle will eventually resupply depleted reserves. But in a time frame of 100 to 1,000 years—the span of a human lifetime or a civilization—groundwater in many regions may be a nonrenewable resource. By pumping water out of the ground at a rate faster than nature replaces it, people are effectively "mining" the groundwater supply. In fact, in portions of the desert Sunbelt region of the United States, supplies of young groundwater have already been exhausted, and deep wells now extract 10,000-year-old groundwater. But such ancient water has been in rock so long that some of it has become too mineralized to be usable. A number of other problems accompany the depletion of groundwater.

- *Lowering the water table:* When we extract groundwater from wells at a rate faster than it can be resupplied, the water table drops. First, a cone of depression forms locally around the well; then the water table gradually

becomes lower in a broad region. As a consequence, existing wells, springs, and rivers dry up (▶Fig. 19.20a, b). To continue tapping into the water supply, we must drill progressively deeper. Pumping has lowered the water table in the High Plains aquifer (see Fig. 19.5d) by 15 m over broad areas, and by over 50 m locally.

The water table can also drop when people divert surface water from the recharge area. Such a problem has developed in the Everglades of southern Florida, a huge swamp where, before the expansion of Miami and the development of agriculture, the water table lay at the ground surface (▶Fig. 19.20c, d). Diversion of water from the Everglades' recharge area into canals has significantly lowered the water table, causing parts of the Everglades to dry up.

- *Reversing the flow direction of groundwater:* The cone of depression that develops around a well creates a local

FIGURE 19.20 **(a)** Before a water table is lowered, a large swamp exists. **(b)** Pumping by a nearby city causes the water table to sink, so the swamp dries up. **(c)** The Everglades, in Florida, before the advent of urban growth and intensive agriculture. Water flowed south from Lake Okeechobee, creating a vast swamp in which the water table lay just above the ground surface. In effect, the Everglades is a "river of grass." **(d)** Channelization and urbanization have removed water from the recharge area, disrupting the groundwater flow path in the Everglades. Since less water enters the Everglades, the water table has dropped, so locally the swamp has dried up. The decrease in the supply of freshwater has also led to saltwater (saline) intrusion along the coast.

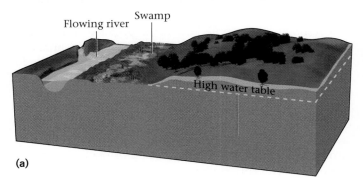

(a)

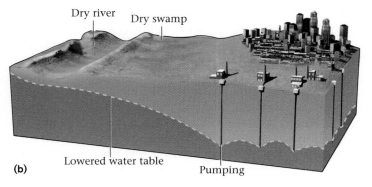

(b)

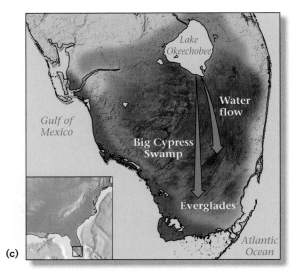

(c)

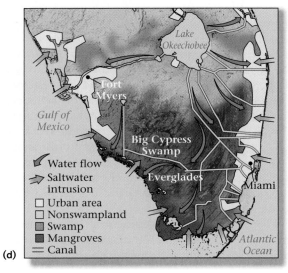

(d)

slope to the water table. The resulting hydraulic gradient may be large enough to reverse the flow direction of nearby groundwater (►Fig. 19.21a, b). Such reversals can lead to the contamination of a well, from pollutants seeping out of a septic tank.

- *Saline intrusion:* In coastal areas, fresh groundwater lies in a layer above saltwater that entered the aquifer from the adjacent ocean (►Fig. 19.21c, d). (Saltwater is denser than freshwater, so the fresh groundwater floats above it.) If people pump water out of a well too quickly, the boundary between the saline water and the fresh groundwater rises. And if this boundary rises above the base of the well, then the well will start to yield useless saline water.

- *Pore collapse and land subsidence:* When groundwater fills the pore space of a rock, it holds the grains of the rock or regolith apart, for water cannot be compressed. The extraction of water from a pore eliminates the support holding the grains apart, because the air that replaces the water *can* be compressed. As a result, the grains pack more closely together. Such pore collapse permanently decreases the porosity and permeability of a rock, and thus lessens its value as an aquifer (►Fig. 19.21e).

Pore collapse also decreases the volume of the aquifer, with the result that the ground above the aquifer sinks. Such land subsidence may cause fissures (►Fig. 19.21f) at the surface to develop and the ground to tilt. Buildings constructed over regions undergoing land subsidence may themselves tilt, or their foundations may crack. The Leaning Tower of Pisa, in Italy, tilts because the removal of groundwater caused its foundation to subside (►Fig. 19.21g). In the San Joaquin Valley of California, the land surface subsided by 9 m between 1925 and 1975, because water was removed to irrigate farm fields (►Fig. 19.21h). In coastal areas, land subsidence may even make the land surface sink below sea level. The flooding of Venice, Italy, is due to land subsidence accompanying the withdrawal of groundwater (►Fig. 19.21i).

To avoid such problems, communities have sought to prevent groundwater depletion either by directing surface water into recharge areas, or by pumping surface water back into the ground. For example, some communities have excavated to lower the land surface of a park, then configured storm sewers so that they drain onto its grassy surface. The park then acts as a catchment for storm water (►Fig. 19.22).

Groundwater Chemistry and Quality

As we've seen, groundwater is not distilled water, composed only of H_2O. Rather, it is a solution of ions obtained by reactions of water with the rock or sediment through which

it passes. The reactions between groundwater and rock or sediment are similar to the chemical weathering reactions that take place at the Earth's surface (see Chapter 7), and they include dissolution, oxidation, hydrolysis, and hydration. During these reactions, ions go into or drop out of solution, and new minerals—such as quartz, calcite, clay, feldspar, and various ore minerals—may form.

The concentration of dissolved ions in groundwater (i.e., the quantity of ions dissolved per unit volume of water) depends on factors such as the temperature, pressure, and acidity (pH) of the groundwater. For example, warmer groundwater can carry more ions in solution than can cooler groundwater. If groundwater has had a long time to react with the material in which it resides, an equilibrium may be reached so that the water contains just as many dissolved ions as possible, under the local environmental conditions.

Because groundwater flows, it eventually enters different environments. If groundwater enters an environment where it has the capacity to contain more ions, it may dissolve surrounding rock or sediment and create secondary porosity. Alternatively, if groundwater enters an environment in which it cannot hold as many dissolved ions, some of the ions in the water bond together and become solid mineral grains that precipitate to form cement or vein fill.

Much of the world's groundwater is crystal clear, and pure enough to drink right out of the ground. In fact, the commercial distribution of bottled groundwater ("spring water") has become a major business worldwide. But in some places, the chemicals contained in groundwater may make it unusable, or it must be treated before being used, even though it has not been contaminated by human-created materials. For example, groundwater that has passed through salt-containing strata may become salty and unsuitable for irrigation or drinking. Groundwater that has passed through limestone or dolomite contains dissolved calcium (Ca^{+2}) and magnesium (Mg^{+2}) ions; this water, called **hard water**, can be a problem because carbonate minerals precipitate from it to form "scale" that clogs pipes. Also, washing with hard water can be difficult because soap won't develop a lather. Groundwater that has passed through iron-bearing rocks may contain dissolved iron oxide that precipitates to form rusty stains. Some groundwater contains dissolved hydrogen sulfide, which comes out of solution when the groundwater rises to the surface; hydrogen sulfide is a poisonous gas that has a rotten-egg smell. In recent years, concern has grown about arsenic in groundwater. Arsenic, a highly toxic chemical, enters groundwater when arsenic-bearing minerals dissolve. In the United States, government standards require drinking water to contain less than 10 micrograms of arsenic per liter. Unfortunately, surveys suggest that about 5% of groundwater supplies contain 20 or more micrograms per liter.

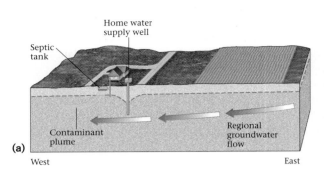

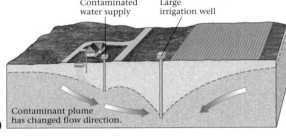

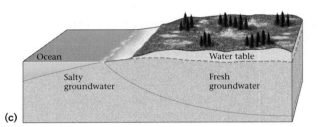

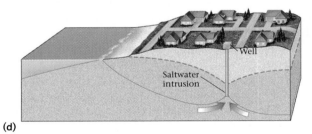

(a)

West   East

(b)

(c)

(d)

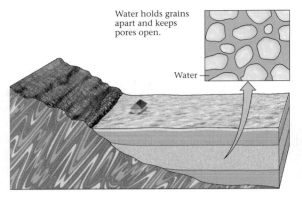

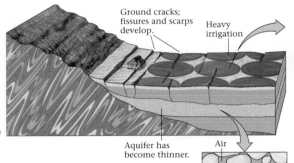

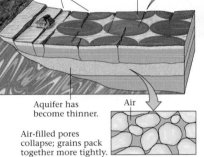

FIGURE 19.21 **(a)** Before pumping, effluent from a septic tank drifts west with the regional groundwater flow. **(b)** After pumping, it drifts east into the well, in response to the local slope of the water table. **(c)** Before pumping, fresh groundwater forms a large lens over salty groundwater. **(d)** Pumping too fast sucks saltwater from below into the well. **(e)** Pore space collapses when water is removed. The pore collapse makes the land subside, as indicated by fissures and cracked houses. **(f)** This photo shows a fissure near Picacho, Arizona. **(g)** The Leaning Tower of Pisa's foundation was destabilized by groundwater removal. **(h)** Evidence of land subsidence in the San Joaquin Valley, California. Former ground elevations are marked on the pole. **(i)** The canals of Venice provide transportation routes in a city slowly subsiding below sea level.

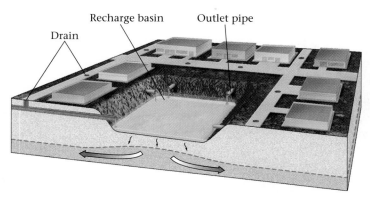

Drain   Recharge basin   Outlet pipe

FIGURE 19.22 Sketch of an enhanced recharge catchment in a city.

Groundwater Contamination

Rocks and sediment are natural filters capable of efficiently removing suspended solids (mud and solid waste) from groundwater, for these solids get trapped in the tiny pathways between pores. Clay flakes, because of their electrically charged surfaces, can remove certain ions from water. Thus, groundwater tends to be clear when it emerges from the ground in a spring or well. Nevertheless, dissolved, invisible organic and inorganic chemicals may be carried along with flowing groundwater; some dissolved chemicals are toxic (such as arsenic, mercury, and lead), whereas others are not (salt, iron, lime, and sulfur). Even nontoxic chemicals can make groundwater unusable—saltwater harms plants and animals, sulfur makes water smell like rotten eggs, iron stains surfaces, and lime makes water hard. In addition, liquids (called nonaqueous phase liquids, or NAPL) that do not dissolve or mix thoroughly with water can be pushed through the subsurface by flowing groundwater. In places where septic systems leak, sickness-causing coliform bacteria also may enter groundwater. The addition of such substances in quantities that make the groundwater dangerous to use is **groundwater contamination**. And once groundwater has been contaminated, it may stay so for as long as the water resides underground.

As we've noted, some contaminants in groundwater occur naturally. For example, sulfur, iron, calcium carbonate, methane, and salt can all be introduced to groundwater directly from the rock through which it is flowing. But in recent decades, contaminants have increasingly been introduced into aquifers through human activity (▶Fig. 19.23a). These contaminants include agricultural waste (pesticides, fertilizers, and animal sewage), industrial waste (dangerous organic and inorganic chemicals), effluent from "sanitary" landfills and septic tanks (including bacteria and viruses), petroleum products, radioactive waste (from weapons manufacture, power plants, and hospitals), and acids leached from sulfide minerals in coal and metal mines. Some of these contaminants seep into the ground from subsurface tanks, some in-

filtrate from the surface, and some are intentionally forced through injection wells (wells in which a liquid is pumped down into the ground under pressure so that it passes from the well back into the pore space of the rock or regolith). The cloud of contaminated groundwater that moves away from the source of contamination is called a **contaminant plume** (▶Fig. 19.23b–d). Staggering quantities of contaminating liquids (trillions of gallons in the United States alone) enter the groundwater system every year.

The best way to avoid contamination is to prevent contaminants from entering groundwater in the first place. This can be done by locating potential sources of contamination on impermeable bedrock so that they are isolated from the aquifer. If such a site is not available, the storage area should be lined with a thick layer of clay, for the clay not only acts as an aquitard, but it can hold on to contaminants. For this reason, environmental engineers commonly place landfills on top of a liner of packed clay, or place waste in durable, sealed containers. Government agencies have studied various options for safely storing the containers. One option involves stockpiling them in tunnels cut into salt domes, for salt is impermeable. Another option involves hiding waste containers in tunnels above the water table. For example, it may be that American nuclear waste will be stored in a network of tunnels 300 m beneath Yucca Mountain, Nevada. This mountain consists of dry, fairly impermeable tuff, below which the water table lies at great depth.

Fortunately, in some cases, natural processes can clean up groundwater contamination. Chemicals may be absorbed by clay, oxygen in the water may oxidize the chemicals, and bacteria in the water may metabolize the chemicals, thereby turning them into harmless substances. Where contaminants do make it into an aquifer, environmental engineers drill test wells to determine which way and how fast the contaminant plume is flowing; once they know the flow path, they can close wells in the path to prevent consumption of contaminated water. Alternatively, engineers attempt to clean the groundwater by drilling a series of extraction wells to pump it out of the ground. If the contaminated water does not rise fast enough, engineers drill injection wells to force clean water or steam into the ground beneath the contaminant plume (▶Fig. 19.24a, b). The injected fluids then push the contaminated water up into the extraction wells. More recently, environmental engineers have begun exploring techniques of **bioremediation**: injecting oxygen and nutrients into a contaminated aquifer to foster growth of bacteria that can react with and break down molecules of contaminants. They

> **Take-Home Message**
> Groundwater usage can cause problems. Overpumping lowers the water table, causes land subsidence, and causes saltwater intrusion. Contamination can ruin a groundwater supply for years to come. Remediation of groundwater is extremely expensive.

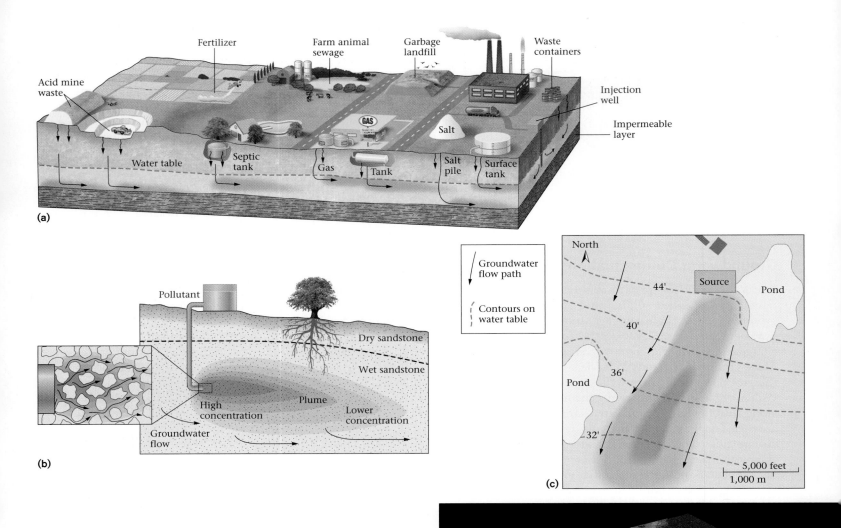

(a)

(b)

(c)

(d)

FIGURE 19.23 (a) The various sources of groundwater contamination. **(b)** A contaminant plume as seen in cross section. **(c)** A map of the contaminant plume emanating from a sewage-treatment basin at a military base on Cape Cod, Massachusetts. The darkness of the color represents the concentration of pollutant. The dashed contour lines indicate that the water table slopes to the south, so the plume moves south. Pollutants are no longer being added to the aquifer, so the greatest concentration of pollution is now south of the source. **(d)** Modern computer analysis can produce three-dimensional images of subsurface plumes. In this image, showing a coastal industrial facility, the ground has been removed to reveal the plume beneath. Red color indicates the greatest concentration of the contaminant.

have also been experimenting with "permeable reactive barriers," subsurface walls of materials such as iron filings which chemically react with contaminants to transform the contaminants into safer materials. To produce a barrier, engineers dig a trench in the path of the flow, fill it with the reactive material, and bury it. When the plume flows through the barrier, the reactions take place. Needless to say, cleaning techniques are expensive and generally only partially effective.

Unwanted Effects of Rising Water Tables

We've seen the negative consequences of sinking water tables, but what happens when the water table rises? Is that necessarily good? Sometimes, but not always. If the water table rises above the floor of a house's basement, water seeps through the foundation and floods the basement floor. Catastrophic damage occurs when a rising water

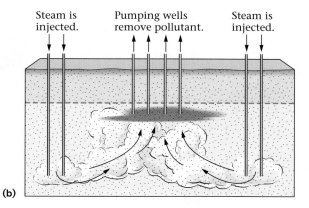

(a)

Waste leaks into ground.

Water table

Pollutant

Aquifer

Steam is injected.

Pumping wells remove pollutant.

Steam is injected.

(b)

FIGURE 19.24 **(a)** Leaky drums of chemicals introduce pollutants into the groundwater. **(b)** Steam injected beneath the contamination drives the contaminated water upward in the aquifer, where pumping wells remove it.

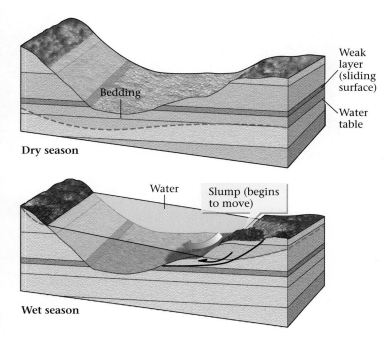

Weak layer (sliding surface)

Water table

Bedding

Dry season

Water

Slump (begins to move)

Wet season

FIGURE 19.25 When the water table rises, material above a weak sliding surface begins to slump, and a landslide may result.

table weakens the base of a hillslope. The addition of water into pore space buoys up the overlying rock or regolith and, therefore, makes the slope unstable and likely to slip. Thus, rising water tables can trigger landslides and slumps (▶Fig. 19.25).

19.8 CAVES AND KARST: A SPELUNKER'S PARADISE

Dissolution and the Development of Caves

In 1799, as legend has it, a hunter by the name of Hutchins was tracking a bear through the woods of Kentucky when the bear suddenly disappeared on a hill slope. Baffled, Hutchins plunged through the brambles trying to sight his prey. Suddenly he felt a draft of surprisingly cool air flowing down the slope from uphill. Now curious, Hutchins climbed up the hill and found a dark portal into the hill slope beneath a ledge of rocks. Bear tracks were all around—was the creature inside? Hutchins returned later with a lantern and cautiously stepped into the passageway. It led into a large, open room. Hutchins had discovered Mammoth Cave, an immense network—over 540 km in total length—of tunnel-like openings connecting underground rooms.

Large networks such as Mammoth Cave developed primarily in limestone bedrock, as a consequence of the dissolution of limestone by groundwater. Slightly acidic groundwater reacts with calcite to produce HCO_3^{-1} and Ca^{+2} ions, which readily dissolve and move away from the site of dissolution in flowing groundwater. Groundwater develops acidity because the water that ultimately becomes groundwater absorbed carbon dioxide (CO_2) partly from the atmosphere when it fell as rain, and more from organic-rich soil as it percolated down to the water table. When it absorbs CO_2, water transforms into carbonic acid.

Geologists have debated about the depth at which limestone caves form in the subsurface. Acidic rainwater clearly dissolves limestone near the ground surface, as indicated by the presence of extensive pitting on limestone bedrock surfaces and along joints in limestone bedrock. And some caves may also form at depth below the water table. It appears, however, that most dissolution takes place in limestone that lies just below the water table, for in this interval the acidity of the groundwater remains high, the mixture of groundwater and newly introduced rainwater is undersaturated (meaning it can dissolve more ions), and groundwater flow is fastest. The association between cave formation and the water table helps explain why openings in a cave system align along the same horizontal plane.

In recent years, geologists have discovered that not all caves form due to reactions with carbonic acid. An alternative type of cave-forming process, called sulfuric-acid

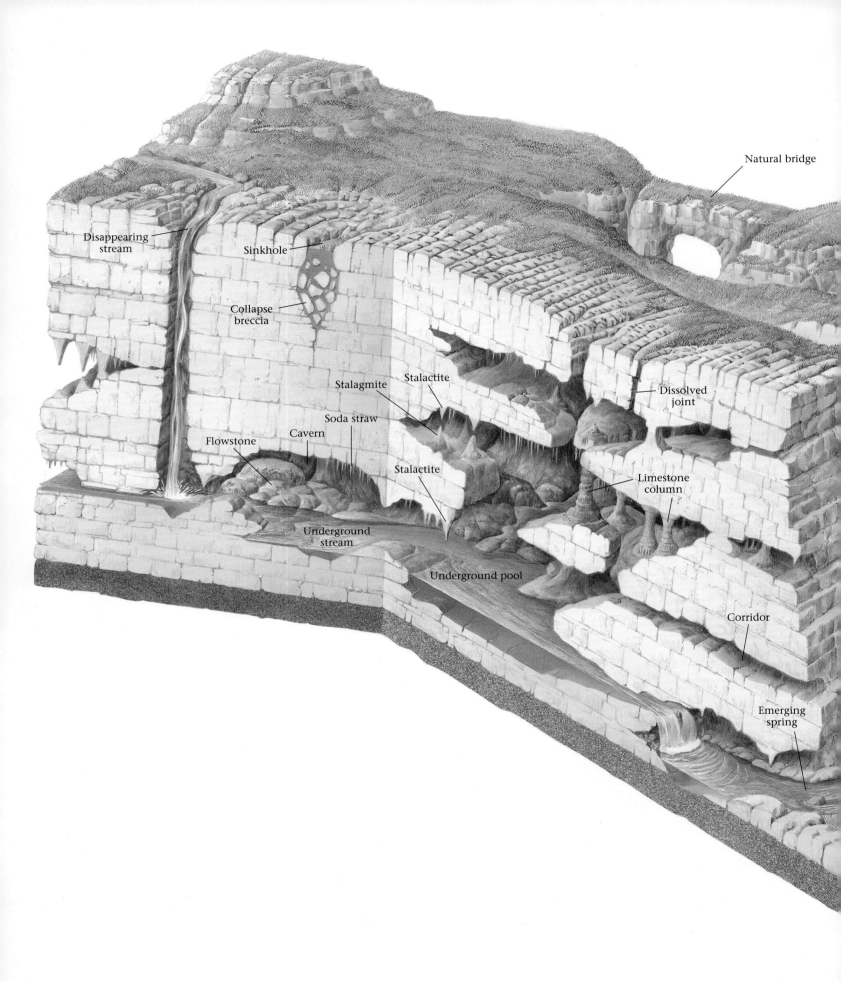

Natural bridge

Disappearing
stream

Sinkhole

Collapse
breccia

Stalagmite

Stalactite

Dissolved
joint

Soda straw

Flowstone

Cavern

Stalactite

Limestone
column

Underground
stream

Underground pool

Corridor

Emerging
spring

Caves and Karst Landscapes

Limestone, a sedimentary rock made of the mineral calcite, is soluble in acidic water. Much of the water that falls to the ground as rain, or seeps through the ground as groundwater, tends to be acidic, so in regions of the Earth where bedrock consists of limestone, there are signs of dissolution. Underground openings that develop by dissolution are called caves or caverns. Some of these may be large open rooms, whereas others are long, narrow passages. Underground lakes and streams may form on the floor. A cave's location depends on the orientation of bedding and joints, for these features localize the flow of groundwater.

Caves originally form at or near the water table (the subsurface boundary between rock or sediment in which pores contain air, and rock or sediment in which pores contain water). As the water table drops, caves empty of most water and become filled with air. In many locations, groundwater drips from the ceiling of a cave or flows along its walls. As the water evaporates and thus loses its acidity (because of the evaporation of dissolved carbon dioxide), new calcite precipitates. Over time, this calcite builds into cave formations, or speleothems, such as stalactites, stalagmites, columns, and flowstone.

Distinctive landscapes, called karst landscapes, develop at the Earth's surface over limestone bedrock. In such regions, the ground may be rough where rock has dissolved along joints, and where the roofs of caves collapse, sinkholes develop. If a surface stream flows through an open joint into a cave network, we say that the stream is "disappearing." The water from such streams may flow underground for a ways and then reemerge elsewhere as a spring. In some places, the collapse of subsurface openings leaves behind natural bridges.

speleogenesis, which takes place in regions where limestone overlies strata containing hydrocarbons, may be responsible for about 5 percent of caves. This process happens because microbes can convert organic sulfur within oil into hydrogen sulfide (H_2S) gas. The gas rises into the limestone, where it oxidizes either by reaction with air or by the action of microbes to produce sulfuric acid. Sulfuric acid is very strong and eats into calcite to produce gypsum and CO_2 gas. This process appears to be responsible for Carlsbad Caverns of New Mexico, an immense network that contains the Big Room, whose floor has an area 14 times that of a football field.

Cave Networks

Caves in limestone usually occur as part of a network. Networks include rooms, or chambers, which are large, open spaces sometimes with cathedral-like ceilings, and tunnel-shaped or slot-shaped passages. (See art, pp. 682–683.) The shape of the cave network reflects variations in permeability and in the composition of the rock from which the caves formed. Larger open spaces developed where the limestone was most soluble and where groundwater flow was fastest. Thus, in a sequence of strata, caves develop preferentially in the more soluble limestone beds. Passages in cave networks typically follow preexisting joints, for the joints provide secondary porosity along which groundwater can flow faster (▶Fig. 19.26). Because joints commonly occur in orthogonal systems (consisting of two sets of joints oriented at right angles to each other; see Chapter 11), passages form a grid.

Why do extensive cave networks, with large rooms and abundant corridors, develop in some locations but not in others? Most caves form in limestone, so without a thick layer of limestone in the subsurface, extensive networks can't form. The dissolution that forms caves occurs primarily in freshwater at or above the water table, so unless the water table lies above sea level and below the land surface, extensive networks can't form. For caves to develop, sufficient liquid water must be present. Thus, extensive networks form in temperate or tropical regions, drenched by rain. They do not form in polar regions, where ice covers the ground and water in the ground is permanently frozen; nor do they form in desert regions, where water is scarce. Organic matter in the overlying soil also plays a key role in cave development, because without organic matter in the overlying soil, groundwater will not develop sufficient acidity to dissolve limestone aggressively.

Groundwater temperature does not play a major role in determining whether an extensive cave network develops due to dissolution by carbonic acid, as long as the groundwater remains liquid. The solubility of CO_2 increases as the temperature of the water decreases, so it may seem at first that colder groundwater will dissolve limestone more rapidly than will warm groundwater. But it turns out that acidity, which reflects the amount of organic matter in overlying soil, plays a *more* important role than temperature in determining the ability of groundwater to dissolve limestone. Thus, caves develop more extensively in tropical regions, where jungle provides abundant organic litter, than in temperate regions, even though the groundwater of temperate regions is cooler than that of tropical regions.

Precipitation and the Formation of Speleothems

When the water table drops below the level of a cave, the cave becomes an open space filled with air. Acidic, calcite-containing water then emerges from the rock above the cave and drips from the ceiling or trickles down the walls. As soon as this water reenters the air, it evaporates a little and releases some of its dissolved carbon dioxide. As a result, calcium carbonate (limestone) precipitates out of the water. Rock formed by such precipitation is a type of travertine called dripstone, and the various intricately shaped formations that grow in caves by the accumulation of dripstone are called **speleothems**.

Cave explorers, or spelunkers, and geologists have developed a detailed nomenclature for different kinds of speleothems (▶Fig. 19.27). Where water drips from the ceiling of the cave, the precipitated limestone adds to the tip of an icicle-like cone called a **stalactite**. Initially, calcite precipitates around the outside of the drip, forming a delicate, hollow stalactite called a soda straw. But eventually, the soda straw fills up, and water migrates down the margin of the cone to form a more massive, solid stalactite. Where the drips hit the floor, the resulting precipitate builds an upward-pointing cone called a **stalagmite**. If the

FIGURE 19.26 Joints act as conduits for water in cave networks. Thus, caves and passageways lie along joints.

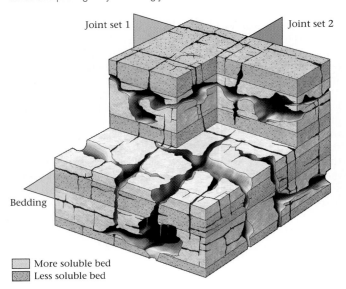

Joint set 1   Joint set 2

Bedding

☐ More soluble bed
▨ Less soluble bed

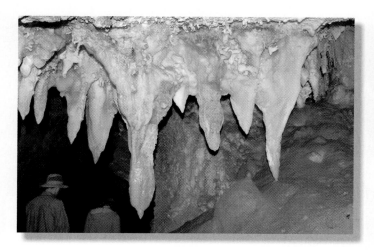

FIGURE 19.27 Large speleothems in a cave.

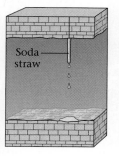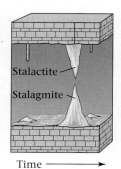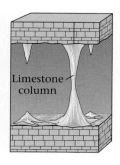

Time ⟶

FIGURE 19.28 The evolution of a soda straw stalactite into a limestone column.

cloth-like sheets of limestone called flowstone. Thin sheets of dripstone and flowstone tend to be translucent and, when lit from behind, glow with an eerie amber light.

The Formation of Karst Landscapes

Limestone bedrock underlies most of the Kras Plateau in Slovenia, along the eastern coast of the Adriatic Sea. The name *Kras,* which means bare, rocky ground, is apt because

process of dripstone formation in a cave continues long enough, stalagmites merge with overlying stalactites to create **limestone columns** (▶Fig. 19.28). If the groundwater flows along the surface of a wall, it drapes the wall with

FIGURE 19.29 **(a)** Numerous sinkholes of a karst landscape. **(b)** The Arecibo radio telescope in Puerto Rico was built in a sinkhole. **(c)** Natural Bridge, Virginia. **(d)** A disappearing stream.

(a)

(b)

(c)

(d)

See for yourself...
Evidence of Groundwater

You can see features at the surface of the Earth that result from the rise of groundwater at springs or due to pumping, or from underground erosion by groundwater. The thumbnail images provided on this page are only to help identify tour sites. Go to wwnorton.com/studyspace to experience this flyover tour.

Irrigation in the Saudi Desert
(Lat 25°18'18.36"N, Long 44°28'5.72"E)

Fly to the coordinates provided and zoom to 70 km (43 miles). You can see the desert of Saudi Arabia. Note the sand dunes in the upper right corner (Image G19.1). Porous rock holds large reserves of groundwater under the surface. Each circle in the image is a field, irrigated by groundwater pumped from a well at the center. Move to Lat 25° 17'53.71"N, Long 44°28'27.17"E and zoom to an elevation of 2 km (1.2 miles) to see a field close up (Image G19.2).

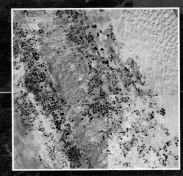

G19.1

G19.2

Water Table, Minnesota (Lat 46°34'49.28"N, Long 95°43'55.13"W)

Fly to this locality and look down from 12 km (7.5 miles) (Image G19.3). A large number of ponds fill kettles, depressions that formed at the end of the ice age when blocks of ice buried in glacial sediment melted away. The surface of the lakes represents the surface of the water table. Dry land exists where the ground surface lies slightly above the water table.

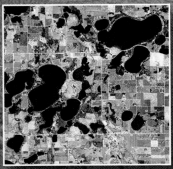

G19.3

Everglades, Florida (Lat 25°40'12.62"N, Long 80°42'1.11"W)

Zoom to 15 km (9 miles) at the coordinates provided, and you are hovering over the Everglades (Image G19.4). This swamp is actually a broad, slowly moving stream that flows from Lake Okeechobee to the tip of Florida. During the wet season, the water table effectively lies above ground, so only vegetated hammocks rise above the water table. But during the dry season, the water table lies below the surface. Why are the hammocks elongated? Some of the water has been diverted into canals, one of which occurs on the right side of this view.

G19.4

G19.5

Hot Springs, Yellowstone (Lat 44°27'44.65"N, Long 110°51'17.98"W)

Zoom to an elevation of 3 km (2 miles) at the coordinates provided and you're looking at Black Sand Basin, near Old Faithful geyser in Yellowstone National Park (Image G19.5). Magma heats the groundwater of Yellowstone. The hot water rises and either fills pools or erupts at geysers. This view shows examples of the pools. Bacteria grow in the pools and add color.

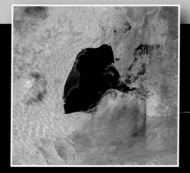
G19.6

Desert Oasis, Egypt (Lat 30°14'41.27"N, Long 28°55'58.95"E)

The Sahara Desert contains vast seas of sand. Yet here and there, ponds or vegetated areas fill depressions where springs bring groundwater to the surface. You can see such an oasis from 10 km (6 miles) at the locality specified above (Image G19.6).

Sinking Venice, Italy
(Lat 45°26'9.01"N, Long 12°20'0.02"E)

The city of Venice covers an island in the Venice Lagoon. From 20 km (12 miles) red roofs in the city stand out (Image G19.7). Zoom to 2.5 km (1.5 miles), and you can see that the main thoroughfares of Venice are canals; the city has subsided in part due to groundwater removal (Image G19.8).

G19.7

G19.8

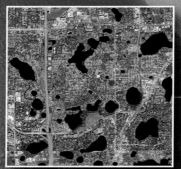
G19.9

Sinkholes in Florida (Lat 28°38'36.16"N, Long 81°22'18.69"W)

Fly to the coordinates provided and zoom to 6 km (4 miles). You are looking at the landscape around Interstate Highway 4. Here, the circular ponds fill sinkholes (Image G19.9). Superficially, they look like the kettles of Minnesota, but they form in a totally different manner—sinkholes develop when an underground cavern collapses. The water surface represents the water table.

Karst Landscape, Puerto Rico
(Lat 18°24'40.75"N, Long 66°25'7.08"W)

At this locality, a view looking down from 10 km (6 miles) reveals part of an extensive karst region (Image G19.10). 36 km (22.3 miles) WSW of this area (Lat 18°20'39.41"N, Long 66°45'10.10"W), a sinkhole was converted into the Arecibo Observatory radio telescope (Image G19.11).

G19.10

G19.11

Tower Karst of China (Lat 22°32'29.97"N, Long 107°26'18.17"E)

In southeastern China, at these coordinates, a view from 20 km (12 miles) illustrates a tower karst landscape (Image G19.12). The thick layer of limestone now at the surface was a cave network. When the roof of the network collapsed, rooms became sinkholes, and tunnels became valleys. Joints localized solution.

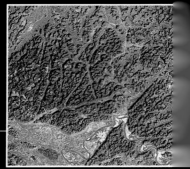

G19.12

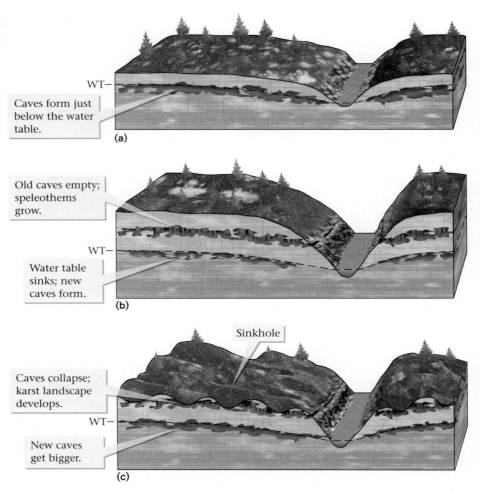

WT—

Caves form just below the water table.

(a)

Old caves empty; speleothems grow.

WT—

Water table sinks; new caves form.

(b)

Sinkhole

Caves collapse; karst landscape develops.

WT—

New caves get bigger.

(c)

FIGURE 19.30 The formation of caves and a karst landscape. **(a)** Dissolution takes place near the water table (WT) in an uplifted sequence of limestone. **(b)** Downcutting by an adjacent river lowers the water table, and the caves empty. Speleothems grow on the cave walls. **(c)** After roof collapse, the landscape becomes pockmarked with sinkholes.

FIGURE 19.31 (a) Tower karst in China; **(b)** painting of tower karst by an unknown Chinese artist.

of the rough, unvegetated land surface in the region. Throughout the area, the bedrock contains abundant caves, and the landscape is pockmarked by deep circular-to-elongate depressions, or sinkholes, which, as we have seen, can form where the roof of a cave collapses. Such sinkholes are called collapse sinkholes (▶Fig. 19.29a, b). Sinkholes can also form where acidic water accumulates in a pool at the surface, seeps down, and dissolves bedrock below. Eventually, a solution pit develops in the bedrock, and the overlying soil sinks downward, forming a depression called a solution sinkhole. The sinkholes of the Kras Plateau are separated from each other by hills or walls of bedrock. Where most of a cave collapsed, a **natural bridge** spans the cave remnant (▶19.29c). Where the water table rises above the floor of a sinkhole, the sinkhole fills to become a lake. Where surface streams intersect cracks or holes that link to the caves below, the water disappears into the subsurface and becomes an underground stream. Such **disappearing streams** reemerge from a cave entrance downstream (▶19.29d).

(a)

(b)

(a)

(b)

FIGURE 19.32 Life in caves: **(a)** blind fish and **(b)** snotites (gobs of bacteria).

Geologists refer to landscapes such as the Kras Plateau in which surface features reflect the dissolution of bed rock below as **karst landscapes**, from the Germanized version of "kras." Karst landscapes form in a series of stages (►Fig. 19.30a–c).

- *The establishment of a water table in limestone:* The story of a karst landscape begins after the formation of a thick interval of limestone. Limestone forms in seawater, and thus initially lies below sea level. If relative sea level drops, a water table can develop in the limestone below the ground surface. If, however, the limestone gets buried deeply, it must first be uplifted and exposed by erosion before it can contain the water table.

- *The formation of a cave network:* Once the water table has been established, dissolution begins and a cave network develops.

- *A drop in the water table:* If the water table later becomes lower, either because of a decrease in rainfall or because nearby rivers cut down through the landscape and drain the region, newly formed caves dry out. Downward-percolating groundwater emerges from the roofs of the caves; dripstone and flowstone precipitate.

- *Roof collapse:* If rocks fall off the roof of a cave for a long time, the roof eventually collapses. Such collapse creates sinkholes and troughs, leaving behind hills, ridges, and natural bridges of limestone.

Some karst landscapes contain many round sinkholes separated by hills. The giant Arecibo radio telescope in Puerto Rico, for example, consists of a dish formed by smoothing the surface of a 300-m-wide round sinkhole (Fig. 19.29b). In regions where vertical joints control roof collapse, steep-sided residual bedrock towers remain between sinkholes. A karst landscape with such spires is called tower karst. The surreal collection of pinnacles constituting the tower karst landscape in the Guilin region of China has inspired generations of artists to portray them on scroll paintings (►Fig. 19.31a, b).

Life in Caves

Despite their lack of light, caves are not sterile, lifeless environments. Caves that are open to the air provide a refuge for bats as well as for various insects and spiders. Similarly, fish and crustaceans enter caves where streams flow in or out. Species living in caves have evolved some unusual characteristics. For example, cave fish lose their pigment and in some cases their eyes (►Fig. 19.32a). Recently, explorers discovered caves in Mexico in which warm, mineral-rich groundwater currently flows. Colonies of bacteria metabolize sulfur-containing minerals in this water, and create thick mats of living ooze in the complete darkness of the cave. Long gobs of this bacteria slowly drip from the ceiling. Because of the mucus-like texture of these drips, they have come to be known as snotites (►Fig. 19.32b).

> **Take-Home Message**
>
> Reaction with natural acids dissolves limestone underground to form caverns. Most dissolution takes place near the water table. If the water table sinks, dripping of water in caves can produces speleothems. Collapse of a cavern network produces karst terrain.

Chapter Summary

- During the hydrologic cycle, water infiltrates the ground and fills the pores and cracks in rock and sediment. This subsurface water is called groundwater. The amount of open space in rock or sediment is its porosity, and the degree to which pores are interconnected, so that water can flow through, defines its permeability.

- Geologists classify rock and sediment according to their permeability. Aquifers are relatively permeable, and aquitards are relatively impermeable.

- The water table is the surface in the ground above which pores contain mostly air, and below which pores are filled with water. The shape of a water table is a subdued imitation of the shape of the overlying land surface.

- Groundwater flows wherever the water table has a hydraulic gradient, and moves slowly from recharge areas to discharge areas. Darcy's law shows that this rate depends on permeability and on the hydraulic gradient.

- Groundwater contains dissolved ions. These ions may come out of solution to form the cement of sedimentary rocks or to fill veins.

- Groundwater can be extracted in wells. An ordinary well penetrates below the water table, but in an artesian well, water rises on its own. Pumping water out of a well too fast causes drawdown, yielding a cone of depression. At a spring, groundwater exits the ground on its own.

- Hot springs and geysers release hot water to the Earth's surface. This water may have been heated by residing very deep in the crust, or by the proximity of a magma chamber or recently formed volcanic rock.

- Groundwater is a precious resource, used for municipal water supplies, industry, and agriculture. In recent years, some regions have lost their groundwater supply because of overuse or contamination.

- When limestone dissolves just below the water table, underground caves are the result. Soluble beds and joints determine the location and orientation of caves. If the water table drops, caves empty out. Limestone precipitates out of water dripping from cave roofs, and creates speleothems (such as stalagmites and stalactites). Regions where abundant caves have collapsed to form sinkholes are called karst landscapes.

Geopuzzle Revisited

Water resides underground in the pores and cracks of rock and regolith. Below the water table, this water completely fills open space. When it rains, some water infiltrates the ground and percolates downward until it reaches the water table—it is this water that becomes groundwater. The level of water in a well defines the water table.

Key Terms

aquifer (p. 664)	limestone column (p. 685)
aquitard (p. 664)	natural bridge (p. 688)
artesian springs (p. 672)	oasis (p. 673)
artesian well (p. 670)	ordinary well (p. 670)
bioremediation (p. 679)	perched water table
capillary fringe (p. 665)	(p. 667)
cone of depression (p. 670)	permeability (p. 663)
contaminant plume (p. 679)	pore (p. 662)
Darcy's law (p. 669)	porosity (p. 662)
disappearing streams	primary porosity (p. 662)
(p. 688)	recharge area (p. 668)
discharge area (p. 668)	saturated zone (phreatic
geothermal region (p. 674)	zone) (p. 665)
geyser (p. 674)	secondary porosity
groundwater (p. 665)	(p. 663)
groundwater contamination	speleothem (p. 684)
(p. 679)	springs (p. 669)
hard water (p. 677)	stalactite (p. 684)
hot springs (p. 672)	stalagmite (p. 684)
hydraulic gradient	unsaturated zone (vadose
(p. 669)	zone) (p. 665)
hydraulic head (p. 668)	water table (p. 665)
karst landscape (p. 689)	wells (p. 669)

Review Questions

1. How do porosity and permeability differ? Give examples of substances with high porosity but low permeability.

2. What factors affect the level of the water table? What factors affect the flow direction of the water below the water table?

3. How does the rate of groundwater flow compare with that of moving ocean water or river currents?

4. What does Darcy's law tell us about how the hydraulic gradient and permeability affect discharge?

5. How does the chemical composition of groundwater change with time? Why is "hard water" hard?

6. How does excessive pumping affect the local water table?

7. How is an artesian well different from an ordinary well?

8. Explain why hot springs form and what makes a geyser erupt.

9. Is groundwater a renewable or nonrenewable resource? Explain how the difference in time frame changes this answer.

10. Describe some of the ways in which human activities can adversely affect the water table.

11. What are some sources of groundwater contamination? How can it be prevented?

12. Describe the process leading to the formation of caves and the speleotherms within caves.

On Further Thought

1. The population of Desert Paradise (a fictitious town in the southwestern United States) has been doubling every seven years. Most of the new inhabitants are "snowbirds," people escaping the cold winters of more northerly latitudes. There are no permanent streams or lakes anywhere near DP. In fact, the only standing water in the town occurs in the ponds of the many golf courses that have been built recently. The water in these ponds needs to be replenished almost constantly, for without supplementing it, the water seeps into the ground quickly and dries up. The golf courses and yards of the suburban-style developments of DP all have lawns of green grass. DP has been growing on a flat, sediment-filled basin between two small mountain ranges. Where does the water supply of DP come from? What do you predict will happen to the water table of the area in coming years, and how might the land surface change as a consequence? Is there a policy that you might suggest to the residents of DP that could slow the process of change?

2. You are part of a cave-exploration team that is trying to map a cave network in a temperate region of flat-lying limestone beds that underlie a plateau. A set of NW-SE-trending sys-

tematic joints cuts the limestone beds. A river cuts through the region, and the entrance to the cave is along the valley wall. The surface of the river lies about 400 m below the surface of the plateau. What do you predict will be the trend of tunnels in the cave network, and how far below the surface of the plateau do you think the cave network extends?

Suggested Reading

Alley, W. M., et al. 1999. *Sustainability of Ground-Water Resources.* U.S. Geological Survey Circular 1186. Denver: U.S. Government Printing Office.

Deutsch, W. J. 1997. *Groundwater Geochemistry: Fundamentals and Applications to Contamination.* Boca Raton, Fla.: Lewis Publishers.

Domenico, P. A., and F. W. Schwartz. 1997. *Physical and Chemical Hydrogeology.* 2nd ed. New York: Wiley.

Freeze, R. A., and J. A. Cherry. 1979. *Groundwater.* Englewood Cliffs, N.J.: Prentice-Hall.

Glennon, R. J. 2002. *Water Follies: Groundwater Pumping and the Fate of America's Fresh Waters:* Washington, D.C.: Island Press.

Gunn, J. 2004. *Encyclopedia of Caves and Karst Science.* London: Fitzroy Dearborn.

Klimchouk, A. B. 2000. *Speleogenesis: Evolution of Karst Aquifers.* Huntsville, Ala.: National Speleological Society.

Schwartz, F. W., and H. Zhang. 2002. *Introduction to Groundwater Hydrology.* New York: Wiley.

White, W. B. 1997. *Geomorphology and Hydrology of Karst Terrains.* Oxford: Oxford University Press.

THE VIEW FROM ABOVE: A 3-D image of a karst landscape from a region of Jamaica called "Cockpit Country." The high areas are residual hills, and the low areas are collapsed caves and tunnels. The white patches are clouds.

An Envelope of Gas: Earth's Atmosphere and Climate

Geopuzzle

Where did the atmosphere come from, and why do different regions of the Earth host different climates? And what's the difference between "climate" and "weather"?

Fed by the warm waters of the Gulf of Mexico, Hurricane Katrina's spiraling winds flattened buildings, eroded coastal islands, and triggered catastrophic floods in 2005. This image shows the storm entering the Gulf. The colors represent the temperature of ocean water—orange is warmer and blue is colder.

Who has seen the wind?

Neither you nor I:

But when the trees bow down their heads,

The wind is passing by.

—*Christina Rossetti (British poet, 1830–1894)*

20.1 INTRODUCTION

On March 21, 1999, Bertrand Piccard, a Swiss psychiatrist, and Brian Jones, a British balloon instructor, became the first people to circle the globe nonstop in a balloon (▶Fig. 20.1). They began their flight on March 1, following more than twenty unsuccessful attempts by various balloonists during the previous two decades. Their airtight gondola, which could float in case they had to ditch in the sea, contained a heater, bottled air, food and water, and instruments for navigation and communication. The nature of their equipment hints at the challenges the balloonists faced, and why it took so many years before anyone finally succeeded in circling the globe.

Balloons can rise from the Earth only because an **atmosphere**, a layer consisting of a mixture of gases called **air**, surrounds our planet. Any object placed in such a fluid feels a buoyancy force, and if the object is less dense than the fluid, then the buoyancy force can lift it off the ground. Balloons rise because the gas (either helium or hot air) in a balloon is less dense than air. Balloonists control their vertical movements by changing either the buoyancy of the balloon or the weight of the payload, but they cannot directly control their horizontal motions—bal-

loons float with the **wind**, the flow of air from one place to another. In order to reach their destination before running out of supplies, long-distance balloonists must find a fast wind flowing in the correct direction. Thus, round-the-world balloonists study the **weather**, the physical conditions (the temperature, pressure, moisture content, and wind velocity and direction) of the atmosphere at a given time and location, in great detail to decide when and where to take off and how high to fly. Different winds blow at different elevations, so balloonists adjust their elevation to catch the best wind. On leaving their launching point in the Swiss Alps, Piccard and Jones entered a strong wind flowing from west to east. Despite a few problems (with fuel supplies and heaters), the balloonists circled the globe and touched down in Egypt.

In this chapter, we explore the envelope of air—the atmosphere—through which Piccard and Jones traveled. We begin by learning where the gases came from and how the atmosphere evolved in the context of the Earth System. Then we look at the structure of the atmosphere, and the global-scale and local-scale circulation of the lowermost layer; this circulation ultimately controls the weather, and can lead to the growth of storms. We conclude by exploring **climate**, the average weather conditions during the year.

20.2 THE FORMATION OF THE ATMOSPHERE

When the Earth formed about 4.57 Ga, it was initially surrounded by gas molecules gravitationally attracted to its surface from the gas and dust rings surrounding the newborn Sun (see Chapter 1). This primary atmosphere, which consisted mostly of hydrogen and helium and traces of other gases, survived only a short time. Heat from the Sun caused the light atoms (H and He) in it to move about so rapidly that they eventually escaped the attraction of Earth's gravity. Effectively, the primary atmosphere leaked into space and was blown away by the solar wind.

Even as the primary atmosphere was disappearing, volcanic activity on Earth released new gases that accumulated to form a secondary atmosphere after the Earth developed a magnetic field that deflected the solar wind. The elements in these gases had been bonded to minerals inside the Earth, but during melting, the elements separated from the minerals and bubbled out of volcanoes. Volcanic gas consists of about 70 to 90% water (H_2O), with smaller amounts of carbon dioxide (CO_2) and sulfur dioxide (SO_2), along with traces of other gases including nitrogen (N_2) and ammonia (NH_3). The secondary atmosphere consisted of these volcanic gases plus, according to some researchers, other gases brought to Earth by comets.

FIGURE 20.1 The balloon and gondola used by Piccard and Jones during their successful attempt to circle the globe in March 1999.

The original secondary atmosphere of the Earth underwent major evolutionary changes over geologic time. Specifically, when the Earth cooled sufficiently for water to condense, probably by 3.9 to 3.8 Ga (though possibly earlier), most of the water from the atmosphere fell as rain, accumulated on the surface, and either filled oceans, lakes, and streams or sank underground to become groundwater. As a consequence, the proportion of water in the atmosphere decreased. Once liquid water existed at Earth's surface, the concentration of CO_2 in the atmosphere began to decrease. This occurred because CO_2 dissolves in oceans and then combines with calcium to form solid carbonate minerals that precipitate to the sea floor. CO_2 also reacts with rocks exposed on the surface of continents to produce solid chemical-weathering products (see Chapter 7). Also over time, ultraviolet radiation from the Sun split apart molecules of NH_3 to produce nitrogen and hydrogen atoms. The light-weight hydrogen atoms escaped into space, but the nitrogen atoms combined to form N_2 molecules. Molecular nitrogen is a stable gas that does not chemically react with rocks, so once it has formed it remains in the air for a long time. Because of the accumulation of new N_2 molecules and the loss of atmospheric H_2O and CO_2, the proportion of nitrogen in the atmosphere gradually increased.

It is noteworthy that if the Earth's surface had been too hot for liquid water to exist on it, then CO_2 would not have been removed from the atmosphere, and Earth's atmosphere today would resemble the present atmosphere of Venus. Venus's atmosphere currently contains 96.5% CO_2, whereas Earth's contains only 0.033%. CO_2 is a greenhouse gas, meaning that it traps heat in the atmosphere. The high concentration of CO_2 makes Venus's atmosphere so hot that lead can melt at the planet's surface.

If you were suddenly to travel back through time and appear on Earth 3 billion years ago, you would instantly suffocate, for the atmosphere back then contained virtually no molecular oxygen (O_2). It took the appearance of life on Earth to add significant quantities of oxygen to the air, for O_2 is produced by photosynthesis. The first photosynthetic organisms, cyanobacteria (blue-green algae), appeared on Earth between 3.8 and 3.5 Ga and began to add O_2 to the atmosphere. By 2 Ga, Earth's atmosphere contained about 1% of its present oxygen level. Oxygen concentrations increased very slowly. At about 1.2 Ga, there may have been a boost in the production of O_2 with the appearance of photosynthetic algae. Only by about 600 Ma did oxygen levels in the air become substantial. How oxygen levels changed during the Phanerozoic remains the subject of debate, but the occurrence of charcoal in Silurian strata indicates that there must have been at least 13% oxygen in the air by about 425 Ma, when vascular plants appeared, for vegetation cannot burn without this concentration. According to one model, oxygen concentration reached a peak of about 35% during the late Paleozoic coal age, and have fluctuated since then, declining since the end of the Mesozoic to the present value of 21% (▶Fig. 20.2a, b).

Oxygen is important not only because it allows complex multicellular organisms to breathe, but also because it supplies the raw components for the production of **ozone** (O_3), a gas that absorbs harmful ultraviolet (short-wavelength) radiation from the Sun. Ozone, which accumulates primarily at an elevation of about 30 km, forms by a two-step reaction:

(1) O_2 + energy (from the Sun) $\rightarrow$ 2O
(2) $O_2 + O \rightarrow O_3$

Only when enough ozone had accumulated in the atmosphere could life leave the protective blanket of seawater, which also absorbs ultraviolet radiation. When this happened, terrestrial plants and animals could evolve. These organisms have themselves interacted with and modified the atmosphere ever since.

In sum, Earth's atmosphere today consists mostly of volcanic gas modified by interactions with the land and life. It must be maintained, though—if all life were to vanish and all volcanic activity to cease, lighter gases would leak into space in only a few million years.

20.3 THE ATMOSPHERE IN PERSPECTIVE

The Components of Air

Completely dry air consists of 78% nitrogen and 21% oxygen. The remaining 1% includes several gases in trace amounts. Trace gases are important in the Earth System. They include carbon dioxide (CO_2) and methane (CH_4), which are greenhouse gases that regulate Earth's atmospheric temperature; greenhouse gases allow solar radiation from the Sun to pass through, but trap infrared radiation rising from the Earth's surface. Trace gases also include ozone (O_3), which protects the surface from ultraviolet radiation. Not all trace gases are beneficial, though. Radon, produced by the natural decay of uranium in rocks, is radioactive and can cause cancer; it can seep from the Earth and collect in dangerous concentrations in the basements of houses.

> **Take-Home Message**
>
> Earth's original atmosphere was lost soon after the planet's formation, and was replaced by volcanic gas. Precipitation of water from this atmosphere formed the oceans, into which CO_2 dissolved. To the remaining N_2, photosynthesis added O_2.

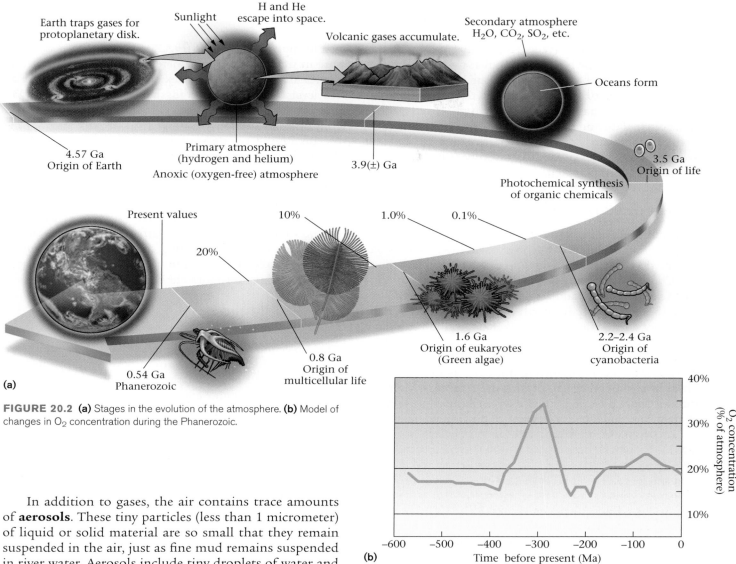

(a)

FIGURE 20.2 (a) Stages in the evolution of the atmosphere. **(b)** Model of changes in O_2 concentration during the Phanerozoic.

(b)

In addition to gases, the air contains trace amounts of **aerosols**. These tiny particles (less than 1 micrometer) of liquid or solid material are so small that they remain suspended in the air, just as fine mud remains suspended in river water. Aerosols include tiny droplets of water and acid and microscopic particles of sea salt, volcanic ash, clay, soot, and pollen (▶Fig. 20.3).

FIGURE 20.3 A recent forest fire produces smoke that adds aerosols (e.g., soot), as well as CO_2 gas, to the atmosphere.

Atmospheric Pollutants

During the past two centuries, human activity has added substantial amounts of pollutants (both gases and aerosols) to the air, primarily through the burning of fossil fuels and through industrial operations. Pollutants include sulfate (SO_4^{-2}) and nitrate (NO_3^-) molecules, which react with water to make a weak acid that then falls from the sky as **acid rain**. Where acid rain falls, lakes, streams, and the ground become more acidic and thus toxic to fish and vegetation (particularly coniferous trees). The burning of fossil fuels has significantly increased the amount of CO_2 in the atmosphere. As we discuss in Chapter 23, most researchers have concluded that this increase is a cause of global warming, the overall rise in atmospheric temperature that is now underway. In

1985, researchers discovered that certain pollutants, notably chlorofluorocarbons (CFCs), react with ultraviolet light from the Sun to release chlorine atoms, which, in turn, react with ozone and break it down. These reactions appear to happen mainly in high clouds above polar regions during certain times of the year, thus preferentially removing ozone from these regions to create an ozone hole. We'll examine human impact on the atmosphere further in Chapter 23.

Pressure and Density Variations

Air is not uniformly distributed in the atmosphere. In the Earth's gravity field, the weight of air at higher elevations presses down on and compresses air at lower elevations (see Chapter 2). **Air pressure**, the push that air can exert on its surroundings, and air density therefore increase toward the surface of the Earth (▶Fig. 20.4). Because the density of a gas reflects the number of gas molecules in a given volume, a gulp of air on the top of Mt. Everest, where the air pressure is about one-third that at sea level, contains about a third as many O_2 molecules as air at sea level. Therefore, most climbers seeking to reach Mt. Everest's summit must breathe bottled oxygen. Similarly, the cabin of an airliner must be pressurized to provide adequate oxy-

gen for normal breathing (Fig. 20.4). We measure air pressure in units called atmospheres (atm), where one atm is approximately the pressure exerted by the atmosphere at sea level (about 14.7 pounds per square inch, or 1,035 grams per square centimeter), or in bars, where 1 bar is about 0.986 atm.

Because of the decrease in air density with elevation, 50% of the atmosphere's molecules lie below an elevation of 5.6 km, 90% lie below 16 km, and 99.99997% lie below 100 km. Thus, even though the outer edge of the atmosphere, a vague boundary where the gas density becomes the same as that of interplanetary space, lies as far as 10,000 km from the Earth's surface, most of the atmosphere's molecules lie within a shell only 0.5% as wide as the solid Earth. Though thin, the atmospheric shell contains sufficient gas to turn the sky blue (▶Box 20.1).

Heat and Temperature

The molecules that constitute the atmosphere, or any gas, do not stand still but are constantly moving. We refer to the *total* kinetic energy (energy of motion; see Appendix A) resulting from the movement of molecules in a gas as its thermal energy, or heat. Note that heat and temperature are not the same—a gas's temperature is a measure of the *average* kinetic energy of its molecules. A volume of gas with a small number of rapidly moving molecules has a higher temperature but may contain less heat than a volume with a large number of slowly moving molecules. If we add heat to a gas, its molecules move faster and its temperature rises, and the gas will try to expand to occupy a larger volume.

Relations between Pressure and Temperature

When air moves from a region of higher pressure to a region of lower pressure, without adding or subtracting heat, it expands. When this happens, the air temperature decreases. Such a process is called **adiabatic cooling** (from the Greek *adiabatos*, meaning impassable); air cools at 6° to 10°C per kilometer that it rises. The reverse is also the case: if air moves from a region of lower pressure to a region of higher pressure, without adding or subtracting heat, it contracts, and the air temperature increases. Such a process is called **adiabatic heating**. Adiabatic cooling and heating are important processes in the atmosphere, because pressure changes with elevation. When air near the ground surface (where pressure is higher) flows up to higher elevations (where pressure is less), it undergoes adiabatic cooling, but when air from high elevations flows down and compresses, it undergoes adiabatic heating.

FIGURE 20.4 This graph shows air pressure versus elevation on the Earth. Note that climbers on top of Mt. Everest breathe an atmosphere that contains only about 33% of the atmospheric gases at sea level. A commercial airliner flies through air that is only 20% as dense as the air at sea level.

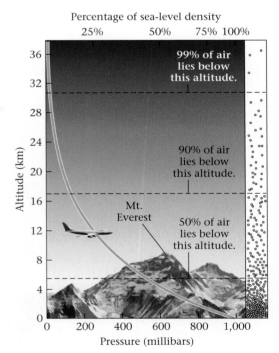

BOX 20.1 THE HUMAN ANGLE

Why Is the Sky Blue?

When astronauts standing on the Moon looked up, even during the day, they saw a black sky filled with stars and a nearly white Sun. On Earth, when we look up during the day, we see a blue sky when there are no clouds and a white sky when there are. The color we see in the sky is the result of the dispersal of energy that occurs when light interacts with particles in the atmosphere, a process called scattering. When light is scattered, a single ray divides into countless beams, each heading off in a different direction. It's similar to what happens when you shine a spotlight on a mirror ball over a dance floor.

Sunlight consists of a broad spectrum of electromagnetic radiation. Different components of the atmosphere scatter different wavelengths of this light, because the ability of a particle to scatter light depends on its size relative to the wavelength. Aerosols (soot, water droplets, dust, etc.) are larger and scatter all wavelengths of light, whereas gas molecules are smaller and scatter short-wavelength (blue) light but not long-wavelength (red) light. Since scattered light heads in all directions, some of it returns back to space; this light is called backscattered light. Because of backscattering, the intensity of light received on Earth's surface is less than it would be if Earth had no atmosphere. It's also because of scattering that shadows are not completely dark. Light is able to enter regions (beneath trees, for example) that are blocked from the Sun.

When you look at an object, the color you see is the color of light either emitted from the object or reflected off it. A plum appears violet because it reflects violet light and absorbs other wavelengths, and a traffic light appears green because it emits green light. On a clear day, with the Sun high in the sky, the gas in the atmosphere scatters primarily blue light. When we look up, we are seeing the blue light scattered off the gas molecules (▶Fig. 20.5a). (Other wavelengths of light pass through the atmosphere without scattering, so we don't see that light until it reflects off objects on the ground.) Because blue light is scattered, not all of it reaches the Earth (some has been backscattered to space), so the Sun appears yellow rather than white—if you subtract a little blue light from white light, you get yellow light. On a cloudy or hazy day, there are more aerosols (such as water droplets) in the atmosphere, and these scatter all wavelengths of light. Therefore, the sky appears white unless the clouds are so dense that they do not let light through, in which case they appear gray.

At the end of the day, when the Sun sinks to the horizon, light passes through a thicker amount of the atmosphere, and so much of the blue light scatters back to space that the sunlight reaching Earth contains mostly red wavelengths—if you subtract a lot of blue light from white light, you are left with red light. Thus, the Sun appears red, as does the light that reflects off clouds (▶Fig. 20.5b).

FIGURE 20.5 **(a)** Scattering of light by gas molecules leads to the brilliant blue of a clear sky at noon at the peak of a volcano, in Hawaii. **(b)** So much scattering happens when sunlight hits the atmosphere at a low angle that all that's left is red light, as during this sunrise on Cape Cod, Massachusetts.

Noon — Thin air layer

(a)

Sunrise — Thick air layer

(b)

Water in the Air

Earlier, we examined the percentages of gases in completely dry (water-free) air. In reality, however, air contains variable amounts of water—from 0.3% above a hot desert to 4% in a rainforest during a heavy downpour.

Meteorologists, scientists who study the weather, specify the water content of air by a number called the **relative humidity**, the ratio between the measured water content and the maximum possible amount of water the air could hold, expressed as a percentage.* The maximum possible amount varies with temperature. Warmer air can hold more water than colder air. When air contains as much water as possible, it is saturated, whereas air with less water is undersaturated. If we say that air at a given temperature has a relatively humidity of 20%, we mean the air contains only 20% of the water that it could hold at that temperature when saturated. Such air feels dry. Air with a relative humidity of 100% is saturated and feels very humid, or damp.

Because cold air can't hold as much water as warm air, air that is undersaturated when warm may become saturated when cooled, without the addition of any new water. The temperature at which the air becomes saturated is called the **dewpoint temperature**; dew forms when undersaturated air cools at night and becomes saturated, so that water condenses on surfaces. When the dewpoint temperature is below freezing, frost develops. And when moist air rises and adiabatically cools, its moisture condenses to form a cloud, a mist of tiny droplets (▶Fig. 20.6).

When water in the air changes from liquid to gas or vice versa (a process called a change of state), the temperature of the air also changes. That is, when water evaporates, it absorbs heat, so that molecules can break free from the liquid. The condensation of water reverses this process and therefore releases heat. The heat released during condensation was hidden, in the sense that it comes only from the change of state and does not require an external energy source. Thus, it is called the latent heat of condensation.

Atmospheric Layers

Temperature, in contrast to pressure, does not decrease continuously from the surface of the Earth to the outer edge of the atmosphere. In fact, starting from the surface and going up, temperature decreases, then increases, then decreases, then increases. Elevations where temperatures stop decreasing and start increasing or vice versa are called **pauses**. Atmospheric scientists divide the Earth's atmosphere into four layers, separated by pauses (▶Fig. 20.7).

---

*"Relative humidity" differs from "absolute humidity." The latter term refers to the mass of water in a volume of air. Absolute humidity is given in grams per cubic meter.

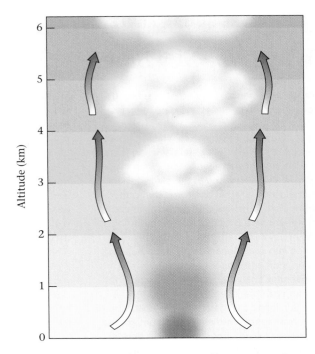

FIGURE 20.6 As air rises and enters regions of lower pressure, it expands and adiabatically cools. If the air contains moisture, the moisture condenses (it turns into water droplets or ice crystals) after the air has risen high enough. Here, we see moist air rising and becoming less dense; its moisture condenses at elevations above 3 km and produces a cloud.

- *Troposphere*: This layer starts at the surface of the Earth and rises to an elevation of 9 km at the poles and 12 km at the equator. Within this layer, the temperature decreases progressively from an average of 18°C at the surface to about −55°C at the top, a boundary called the tropopause. The name **troposphere** comes from the Greek *tropos*, which means turning. It is an appropriate name, because air in the troposphere constantly undergoes convection (see Appendix A). The heat that initiates movement in the troposphere comes primarily from infrared radiation rising from the Earth's surface. The radiation heats air at the base of the troposphere—in effect, the Earth bakes air from below. This air then rises, and cold air sinks to take its place. As we will see, this movement causes most weather phenomena, so the troposphere can also be thought of as the "weather layer."

- *Stratosphere*: Beginning at the tropopause and continuing up for about 10 km, the temperature stays about the same. Then it slowly rises, reaching a maximum of about 0°C at an elevation of about 47 km, a boundary called the stratopause. The layer between the tropopause and the stratopause is the **stratosphere**, so named because it doesn't convect and thus remains stable and stratified. The stratosphere doesn't mix with the underlying troposphere, because at the tropopause hotter (less dense) air already lies on top of cooler (denser) air. Most of the

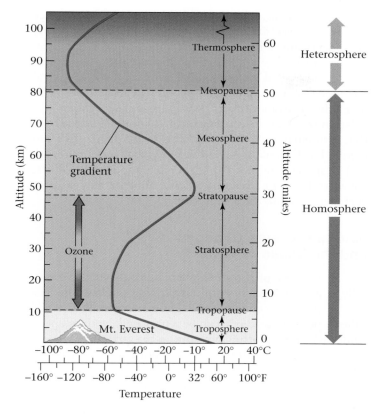

FIGURE 20.7 The principal layers of the atmosphere are separated from each other by pauses. At a pause, the temperature gradient in the atmosphere changes direction.

each other and shoot off in different directions. This constant, chaotic motion stirs the gases sufficiently to make a homogeneous mixture, so that the air in the lower three layers has essentially the same proportion of different gases regardless of location. For this reason, atmospheric scientists refer to the troposphere, stratosphere, and mesosphere together as the homosphere. In contrast, atoms and molecules in the low-density thermosphere collide so infrequently that this layer does not homogenize. Rather, gases separate into distinct layers based on composition, with the heaviest (nitrogen) on the bottom, followed in succession by oxygen, helium, and at the top, hydrogen, the lightest atom. To emphasize this composition, atmospheric scientists refer to the thermosphere as the heterosphere.

So far, we've distinguished atmospheric layers according to their thermal structure (troposphere, stratosphere, mesosphere, and thermosphere) and according to the degree their gases mix (homosphere and heterosphere). We need to add one more "sphere" to our discussion. The **ionosphere** is the interval between 60 and 400 km, and thus includes most of the mesosphere and the lower part of the thermosphere. It was given its name because in this layer, short-wavelength solar energy strips nitrogen molecules and oxygen atoms of their electrons and transforms them into positive ions. The ionosphere plays an important role in modern communication in that, like a mirror, it reflects radio transmissions from Earth so that they can be received over great distances.

The ionosphere also hosts a spectacular atmospheric phenomenon, the auroras (aurora borealis in the Northern Hemisphere and aurora australis in the Southern), which look like undulating, ghostly curtains of varicolored light in the night sky (▶Fig. 20.8). They appear when charged particles (protons and electrons) ejected from the Sun, especially when solar flares erupt, reach the Earth and interact with the ions in the ionosphere, making them release energy. Auroras occur primarily at high latitudes because Earth's magnetic field traps solar particles and carries them to the poles.

> **Take-Home Message**
>
> The atmosphere is a mixture of N_2 and O_2, with traces of other gases, including H_2O, the abundance of which (expressed as relative humidity) varies with time and location. The atmosphere is divided into several layers—we live in the troposphere.

ozone in Earth's atmosphere resides in the stratosphere. Heating in the stratosphere happens because ozone absorbs solar radiation.

- *Mesosphere*: The temperature decreases in the interval, called the **mesosphere**, from 47 to 82 km. At the mesopause, the top of the mesosphere, the temperature has dropped to −85°C. The mesosphere does not absorb much solar energy and thus cools with increasing distance from the hotter stratosphere below. Most meteors (shooting stars) begin burning here and have vaporized by the time they reach an altitude of 25 km.

- *Thermosphere*: The outermost layer of the atmosphere, the **thermosphere**, contains very little of the atmosphere's gas (less than 1%). The temperature increases with elevation in this layer, because gases of the thermosphere absorb short-wavelength solar energy. (The Sun broils this layer from above.) Because the thermosphere has so little gas, it contains very little heat, even though it registers a high temperature. Thus, an astronaut walking in space at an elevation of 200 km doesn't feel hot.

The densities of gases in the lower three layers of the atmosphere are great enough that moving atoms and molecules frequently collide. Like billiard balls, they bounce off

20.4 WIND AND GLOBAL CIRCULATION IN THE ATMOSPHERE

A gusty breeze on a summer day, the steady currents of air that once blew clipper ships across the oceans, and a fierce hurricane all are examples of the wind, the movement of air from one place to another. We can feel the wind, because of

FIGURE 20.8 The splendor of an aurora borealis lights up the night sky in Arctic Canada. The colors result when particles emitted from the Sun interact with atoms in the thermosphere of the Earth.

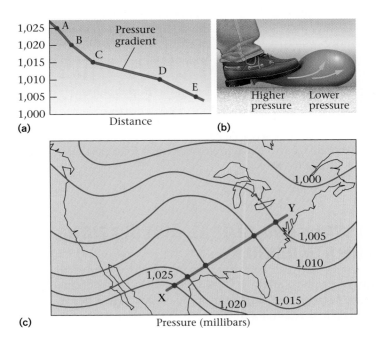

(a)

(b)

(c) Pressure (millibars)

the impacts of air molecules as they strike us. The existence of wind illustrates that the lower part of the atmosphere is in constant motion, swirling and overturning at rates between a fraction of a kilometer and a few hundred kilometers per hour. This circulation happens on two scales, local and global. Local circulation refers to the movement of air over a distance of tens to a thousand kilometers. Global circulation refers to the movement of volumes of air in paths that ultimately carry it around the entire planet. (We can picture local circulation as eddies in global-scale "rivers" of air.) To understand both kinds of circulation, we must first see what drives air from one place to another, and examine energy inputs into the atmosphere.

Lateral Pressure Changes and the Cause of Wind

The air pressure of the atmosphere not only changes vertically, it also changes horizontally at a given elevation. The rate of pressure change over a given horizontal distance, called a pressure gradient, can be represented by the slope of a line on a graph plotting pressure on the vertical axis (specified in bars) and distance on the horizontal axis (▶Fig. 20.9a). Winds form wherever a pressure gradient exists. Air always flows from a high-pressure region to a low-pressure region; in other words, it flows down a pressure gradient. To see why, step on one end of a balloon filled with air; you momentarily increase the pressure at that end, so the air flows toward the other end (▶Fig. 20.9b).

We can use a map to represent air pressure at a given elevation. A line on a map along which the air has a specified pressure is called an **isobar** (▶Fig. 20.9c). In other words, the pressure is the same all along an isobar. Isobars can never touch, because they represent different values of pressure. A difference in pressure exists between one isobar and the next, so the air starts flowing perpendicular to these lines. As we will see, however, the Coriolis effect modifies wind direction.

Energy Input into the Atmosphere: Convection

Ultimately, air circulation, like the movement of water in a heated pot, results from convection: heated air expands and becomes less dense, so it rises to be replaced by sinking

FIGURE 20.9 (a) This graph shows a profile from location X to location Y. Notice that the pressure gradient, the slope of the line, is greater between X and C than between C and D. **(b)** Air flows from a high-pressure region to a low-pressure region to cause wind. If you step on one end of a balloon, for example, you increase the pressure there, so that the air flows toward the unsqueezed end. **(c)** Isobars on a map are lines of equal pressure. Every place along the 1,005 isobar is experiencing a pressure of 1,005 millibars (1 millibar = 0.001 bar). By moving in any direction not parallel to the isobars, you will feel a pressure change. For example, if you walk from X to Y, you will experience a decrease in pressure.

cooler, denser air. In the case of Earth's atmosphere, the energy comes from solar radiation. Solar energy constantly bathes the Earth. Of this energy, 30% reflects back to space (off clouds, water, and land); air or clouds absorb 19%; and land and water absorb 51%. The energy absorbed by land and water later reradiates as infrared radiation and thus bakes the atmosphere from below. As noted earlier, greenhouse gases absorb part of this reradiated energy before it can return to space.

Because the Earth is a sphere, not all areas receive the same amount of incoming solar energy, or **insolation**: portions of the Earth's surface hit by direct rays of the Sun receive more energy per square meter than portions hit by oblique rays. We can simulate this contrast with a flashlight; if you point a flashlight beam straight down, you get a small but bright spot of light on the ground, but if you aim the beam so that it hits the ground at an angle of 45°, the spot covers a broader area but does not appear as bright (▶Fig. 20.10a). Higher latitudes thus receive less energy than lower latitudes (▶Fig. 20.10b). Because of the tilt of Earth's axis, the amount of solar radiation that any point on the surface receives changes during the year, which is why we have seasons (▶Box 20.2).

The contrast in the amount of solar radiation received by different latitudes means that polar regions are cooler at the surface than are equatorial regions. In 1735, George Hadley, a British mathematician, realized that this contrast could cause global air to circulate by convection. Specifically, he suggested that warm air at the equator would rise and flow toward the pole, to be replaced by cool polar air, which would flow to the equator at lower elevations (▶Fig. 20.11a). Hadley's proposal, however, did not take into account an important factor, namely the Earth's rotations and the resulting Coriolis effect. The Coriolis effect, as we learned in Chapter 18, refers to the deflection that happens to an object as it moves from the circumference to the center of a rotating disk or from the equator to the pole of a rotating sphere (and vice versa) (▶Fig. 20.11b).

Because of the Coriolis effect, northward-moving high-altitude air in the Northern Hemisphere deflects to the east, so by the latitude of 30°N, it basically moves due east and cannot make it the rest of the way to the pole. By the time it reaches this latitude, the air has also cooled significantly and thus starts to sink. This means that a global convection cell (a current that looks like a loop in cross section) develops that conveys warm air north from the equator to the subtropics (latitude 30°), where it cools and sinks. When the sinking air reaches low elevations, it divides, some moving back toward the equator near the surface and some moving north near the surface. A place where sinking air separates into two flows moving in opposite directions is a **divergence zone**.

Meanwhile, cool air from the polar region moves south near the surface and deflects to the west. At a latitude of about 60°, the near-surface polar air collides, or converges, with the northward-moving, near-surface mid-latitude air. This air must rise, because there is nowhere for the extra air to go but up. A place where two surface air flows meet so that air has to rise is called a **convergence zone**. The convergence zone at latitude 60° is called the **polar front**. The air that rises along the polar front divides at the top of the troposphere, with some air heading toward the equator at high altitude to complete a mid-latitude convection cell, and some heading toward the pole to create a polar convection cell. Because of seasonal changes on Earth, the exact position of the polar front changes during the year.

In sum, because of the Coriolis effect, circulating air in the troposphere splits into three globe-encircling convection cells in each hemisphere. The low-latitude cells, extending from the equator to a latitude of about 30°, are called **Hadley cells**, in honor of George Hadley. The mid-latitude cells are called **Ferrel cells**, in honor of the American meteorologist

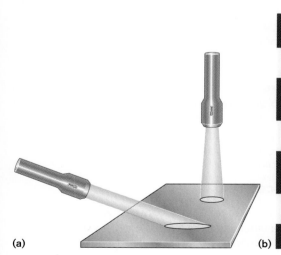

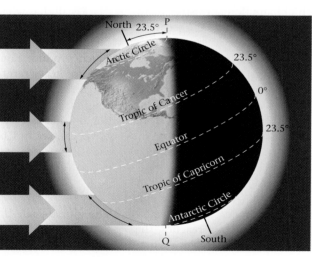

FIGURE 20.10 **(a)** A flashlight beam aimed straight down produces a narrower and more intense beam than a flashlight aimed obliquely. Thus, the area under the straight beam heats up more than the area under the oblique beam. **(b)** Sunlight hitting the Earth near the equator provides more heat per unit area of surface than sunlight hitting the Earth at a polar latitude. That is why the poles are colder.

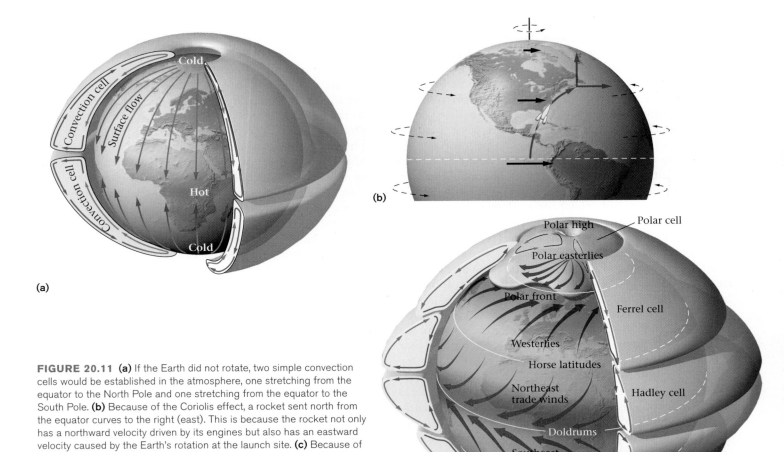

(a)

(b)

(c)

FIGURE 20.11 **(a)** If the Earth did not rotate, two simple convection cells would be established in the atmosphere, one stretching from the equator to the North Pole and one stretching from the equator to the South Pole. **(b)** Because of the Coriolis effect, a rocket sent north from the equator curves to the right (east). This is because the rocket not only has a northward velocity driven by its engines but also has an eastward velocity caused by the Earth's rotation at the launch site. **(c)** Because of the Coriolis effect, atmospheric circulation breaks into three convection cells within each hemisphere, named, from equator to pole, Hadley, Ferrel, and polar.

William Ferrel, who proposed them. The high-latitude cells are called simply **polar cells** (▶Fig. 20.11c).

In this three-cell-per-hemisphere model of atmospheric circulation, several belts of high and low pressure form around the Earth. The equatorial regions are marked by a belt of convergence called the intertropical convergence zone. In these regions, intense insolation makes air hot, so that it rises. The upward flow of hot, rising air leaves an area of low pressure in its wake; thus, the convergence zone at the equator is also called the equatorial low. The low-pressure area in equatorial regions fills with cooler air flowing in from higher latitudes at the base of the Hadley cells.

At the latitude of about 30° and at a high altitude, a Hadley cell converges with a Ferrel cell, causing cooler, dense air to sink. This downward movement forms a belt of high pressure called a subtropical high. (A subtropical high can also be called a subtropical divergence zone, because the downward-flowing air diverges when it reaches the Earth's surface.) Finally, the rise of air where the surface

flow of a polar cell converges with the surface flow of a Ferrel cell (at the latitude of 60°) creates a subpolar low, and the sinking of air at the pole itself creates a polar high.

Recall that where warm, moist air rises, it cools adiabatically. Cooler air can hold less moisture, so this air becomes oversaturated. The moisture condenses and forms clouds, which produce rain. Therefore, the equatorial lows are regions of heavy rainfall, which leads to the growth of tropical rain forests. In contrast, in high-pressure belts where cool, dry air sinks, air contracts and heats adiabatically. The resulting hot air can absorb moisture, so that it rains only rarely. Thus, as we will see in Chapter 21, the subtropical regions at latitude 30° include some of the Earth's major deserts.

Prevailing Surface Winds

The global-scale flow in the six major convective cells on Earth creates belts in which surface air generally flows in a consistent direction. Such airflows are called **prevailing**

BOX 20.2 THE REST OF THE STORY

The Earth's Tilt: The Cause of Seasons

If the Earth's axis were not tilted, we would not experience seasons, for the amount of solar energy received at a given latitude would stay the same all year. But at present, the axis tilts at 23.5° (this number changes with time, because of the axis's wobble; see Chapter 22), so the amount of radiation received at a given latitude varies during the year. As a result, the average daily temperature changes during the year, and this change, along with associated variations in amounts of precipitation, brings the seasons.

At any given time, half the Earth has daylight, and half experiences night. The boundary between these two hemispheres is called the terminator. Because of the Earth's tilt, the terminator does *not* always pass through the North and South Poles

(▶Fig. 20.12). On June 21, a special day called a solstice, the terminator lies 23.5° away from the poles, as measured along the Earth's surface. The line of latitude at this position is called the Arctic Circle in the Northern Hemisphere and the Antarctic Circle in the Southern Hemisphere. On June 21, a person standing on the Arctic Circle sees the midnight sun (the Sun is visible for an entire day), whereas anyone south of the Antarctic Circle has night for a full twenty-four hours. During the northern summer, regions north of the Arctic Circle see the midnight sun for more than a day, and the North Pole itself experiences the midnight sun for a full 6 months. Similarly, regions south of the Antarctic Circle experience 24-hour nights for more than a day, and the South Pole itself sees nothing but night for a

full 6 months. December 21 is the other solstice, but on this day all regions south of the Antarctic Circle see the midnight sun, whereas all regions north of the Arctic Circle have perpetual night. During the southern summer, the South Pole sees daylight for 6 months.

On the June 21 solstice, the Sun's rays are exactly perpendicular to the Earth (the Sun is directly overhead) at latitude 23.5°N, the Tropic of Cancer, and on December 21 the Sun's rays are exactly perpendicular to the Earth at 23.5°S, the Tropic of Capricorn. On September 22 and March 20, each known as an "equinox," the Sun is directly overhead at the equator. The two solstices and two equinoxes divide the year into four astronomical seasons.

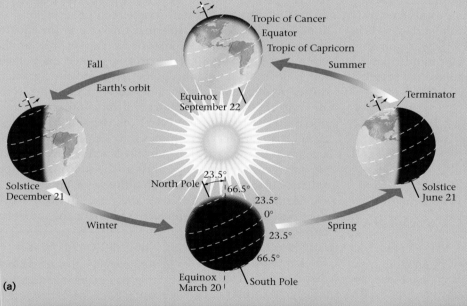

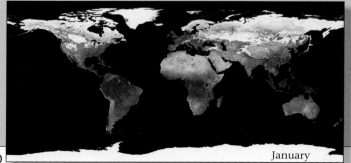

FIGURE 20.12 (a) Because of the tilt of the Earth's axis, we have seasons. At the Northern Hemisphere's summer solstice, the noonday sun is directly overhead at the Tropic of Cancer, and the region above the Arctic Circle sees the midnight sun; whereas at the winter solstice, the noonday sun is directly overhead at the Tropic of Capricorn and the region south of the Antarctic Circle sees the midnight sun. At the equinoxes, the noonday sun is directly overhead at the equator. **(b)** A satellite image of the Earth in January. Note the Northern Hemisphere snow cover. **(c)** A satellite image of the Earth in July.

winds. (Note that here we're examining *surface* winds, the base of the convective cell.) These simple patterns tend to be disrupted by local-scale winds caused by storms or affected by local topography. When describing winds, meteorologists label them according to the direction the air comes from. Thus, a westerly wind blows from west to east.

Let's start our tour of prevailing winds at the base of the Hadley cell in the Northern Hemisphere. Near-surface winds start to flow from 30°N to the south and are deflected west. Thus, between the equator and 30°N, surface winds come out of the northeast, and are called the northeast **trade winds**, so named because they once carried trading ships westward from Europe to the Americas. Trade winds in the Southern Hemisphere, which start flowing northward and then deflect to the west, end up flowing from southeast to northwest, and are called the southeast trade winds. Where the southeast and northeast trade winds merge at the equator, they are flowing almost due west (▶Fig. 20.13). But winds along the equator are very slow, because the air is mostly rising. Ships tended to be becalmed in this belt, which came to be called the **doldrums**.

At the base of a Ferrel cell, at mid-latitudes, surface air starts to move toward the north, but because of the Coriolis effect it curves to the east. Thus, throughout much of North America and Europe, the prevailing surface winds come out of the west or southwest and are known as the surface westerlies. In the subtropical high itself, where airflow is primarily downward, winds are weak and tend to shift in different directions. In the past, these conditions inhibited the progress

of sailing ships. Perhaps because so many horses being transported by ship died of heat exhaustion in the subtropical high, the region came to be known as the horse latitudes.

Finally, at the base of polar cells, surface air starts by flowing from the pole southward, but deflects to the west. The resulting prevailing winds, which are known as the "polar easterlies," flow from the polar high to the subpolar low, and converge with the westerlies of mid-latitudes at the polar front.

Note that prevailing winds, for much of their course, follow paths almost parallel to the high- and low-pressure belts that surround the Earth, and thus flow nearly parallel to isobars. This situation seems to contradict the statement made earlier that wind flows from regions of high pressure to regions of low pressure, and thus should flow at almost right angles to isobars. This apparent contradiction is a consequence of the Coriolis effect.

High-Altitude Winds in the Troposphere: The Jet Streams

In the upper atmosphere, a global-scale pressure gradient exists because of temperature differences between the equator and the pole. At the equator, air is warmer and thus expands. This causes the top of the troposphere there to rise relative to the top of the troposphere in polar regions, so, as noted earlier, the troposphere is thicker over the equator than over the poles. As a consequence, the air pressure at a given elevation above the equator is greater than at the same elevation above the poles. This pressure gradient causes overall high-altitude air to flow north; in fact, some air from the top of the Hadley cell spills over and moves north over the top of the Ferrel cell (▶Fig. 20.14a). Once again, the Coriolis effect comes into play and makes this air deflect to the east, and so in the Northern Hemisphere we have generally westerly winds at the top of the troposphere. These are called the high-altitude westerlies.

In two special places, over the polar front and over the horse latitudes, the pressure gradient at the top of the troposphere is particularly steep. Because of the steepness of the gradient, high-altitude westerlies flow particularly fast. These zones of rapid movement, where winds typically flow at speeds of between 200 and 400 km per hour, are called **jet streams** (▶Fig. 20.14b, c). The polar jet stream is faster and more important than the subtropical jet stream. Both can be viewed as fast-flow zones within an overall westerly high-altitude flow.

> **Take-Home Message**
> Air flows (i.e. the wind blows) due to pressure variations, which develop because of convection. Because of the Coriolis effect, global convection occurs in six ring-like cells. This pattern leads to prevailing winds, whose direction varies with latitude.

FIGURE 20.13 If the Earth had a uniform surface, distinct high- and low-pressure zones would form on its surface. But surface winds flowing from high-pressure zones to low-pressure zones are deflected by the Coriolis effect, so that for much of their course they flow almost parallel to pressure zones. Compare this figure with Figure 20.11c.

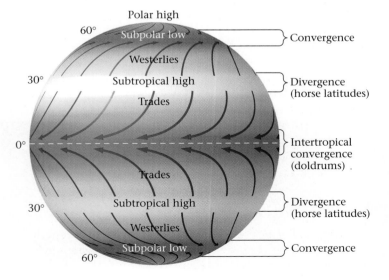

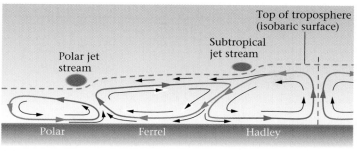

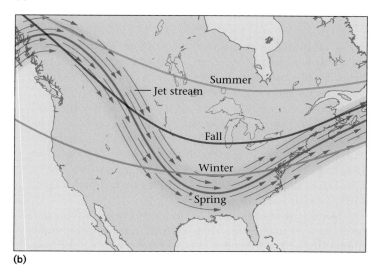

(a)

(b)

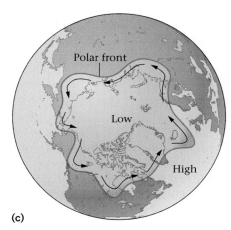
(c)

FIGURE 20.14 (a) In an equator-to-pole cross section of the atmosphere (simplified by removing the curvature of the Earth), we see that the troposphere varies in thickness. The atmosphere above the troposphere, however, is essentially uniform in thickness from equator to pole, so the tropopause is an isobaric surface. (All points on the surface have the same pressure.) Therefore, at a given elevation within the troposphere, there is a pressure gradient from equator to pole, causing a northward flow of air at the top of the troposphere. Because of the Coriolis effect, high-altitude westerly winds develop in the Northern Hemisphere. At places where the pressure gradient is particularly steep (at the boundary between convection cells), the air moves very fast, creating the jet streams. (b) The polar jet stream meanders over time and changes its overall position with the seasons; it tends to be farther south in the winter than in the summer. (c) Looking down on the Earth from the North Pole, we see the irregular shape of the polar front, along which the polar jet stream flows.

ing times than those flying west, because the former are helped along by strong tailwinds, whereas the latter battle head winds that slow them down.

20.5 WEATHER AND ITS CAUSES

If the surface of the Earth were a perfectly uniform sphere, airflow—driven by convection and modified by the Coriolis effect—might follow the global pattern just described. But, in fact, the Earth's surface is heterogeneous—parts are covered with water, parts with land. In the oceans, water currents affect the temperature of the sea surface, and on land, there are variations in altitude and vegetation cover. All these factors ensure that the heat that rises into the base of the atmosphere from the heated surface of the Earth varies with time and location in a complex way, and that moving air must interact with obstacles and either be forced to rise or be shoved to the side. Thus, it's not surprising that on a *local scale*, atmospheric flow can be quite turbulent.

The term *weather* refers to local-scale conditions as defined by temperature, air pressure, relative humidity, and wind speed (see ▶ Table 20.1 for the range of weather conditions that can exist on Earth). A specific set of weather conditions, reflecting the configuration of air movement in the atmosphere, that affects a region for a period of time is called a **weather system**. Weather systems can move across the surface of the Earth, carried by prevailing winds. In this section, we'll briefly review some of the more general aspects of weather and see how they result from the interaction of air masses along fronts.

Air Masses and Fronts

Air that remains or passes over a certain region for a length of time takes on characteristics that reflect both its interaction with the Earth's surface and the amount of insolation

The position of the polar front tends to undulate over time in wave-like motions. Thus, as viewed on a map, the positions of the polar front and therefore of the jet stream follow large, curving trajectories that change with time, sometimes bringing the jet stream down to southern latitudes of North America and sometimes running it across northern Canada. Further, the average position varies with the season. The jet stream, because of its high steady winds, affects airplane flights. Planes flying east have shorter fly-

TABLE 20.1 Weather Extremes

Lowest recorded air pressure at sea level	870 millibars (25.7 inches); during Typhoon Tip (1979), Pacific Ocean
Highest air pressure at sea level	1,084 millibars (32.02 inches); Agata, Siberia
Lowest air temperature (world)	–89°C (–129°F); Vostok, Antarctica
Lowest air temperature (North America)	–63°C (–81°F); Yukon, Canada
Highest air temperature (world)	58°C (136°F); Libya
Highest air temperature (North America)	57°C (134°F); Death Valley, California
Highest wind speed (world)	372 km per hour (231 mph); Mt. Washington, New Hampshire

it receives. For example, air that has hovered over a warm sea tends to become warm and moist, whereas air stuck over cold land becomes cold and dry. A body of air, at least 1,500 km across, that has recognizable physical characteristics is called an **air mass**. Air masses move within the overall global circulation of the atmosphere, and their paths are controlled by prevailing winds. Meteorologists refer to air masses that originate over land in polar regions as Arctic air masses, and those that originate over tropical or subtropical oceanic regions as maritime tropical air masses. (See ▶Fig. 20.15 for names of other air masses.) The weather of a region changes drastically when one air mass replaces another. For example, a summer day in a Midwestern state will be cool and dry under an Arctic air mass but hot and humid under a maritime tropical air mass.

The boundary between two air masses is called a **front**. Meteorologists recognize three kinds of fronts. At a "cold front," a cold air mass pushes underneath a warm air mass

(▶Fig. 20.16a). As a consequence, the warm, moist air flows up to higher elevations, where it expands and cools adiabatically. The moisture it contains then condenses to form large clouds from which heavy rains fall. As a "warm front" moves into a region, the warm air slowly rises over the cool air (▶Fig. 20.16b). Again, the air expands and cools, and the moisture condenses, so clouds develop over the boundaries between the two air masses. It's important to note that even though a front can be portrayed on a map as a line, it is in fact a sloping surface. Cold fronts have much steeper surfaces than warm fronts.

Not all air masses move at the same velocity. Typically, cold fronts move faster than warm fronts and overtake them. Where this happens, the cold front lifts up the base of the warm front, so that the warm front no longer intersects the ground surface. Meteorologists refer to the geometry that results when a cold front pushes underneath a warm front as an "occluded front" (▶Fig. 20.16c).

Cyclonic and Anticyclonic Flow

An air mass may be a region of distinctly high or low pressure. (Specifically, rising warm air produces a low-pressure mass, whereas sinking cold air produces a high pressure mass.) Thus, a pressure gradient exists along the margin of an air mass and, anywhere that a pressure gradient forms, winds begin to blow. Because air moves from high pressure to low pressure, winds tend to flow toward the center of a low-pressure mass, where they converge and create an excess that must flow up. Thus, air rises at the center of a low-pressure mass. In contrast, air spreads outward along the margins of a high-pressure mass, creating a deficit of air at the center that pulls air down from above. Because rising air expands and cools and its moisture condenses to form clouds, rain may fall as a low-pressure mass moves by. But sinking air gets compressed, warms up, and can absorb more moisture, so high-pressure masses tend to mean fair (clear and dry) weather.

If the Earth did not spin, airflow into a low-pressure mass and out of a high-pressure mass would be radial: that is, the air would move in a direction perpendicular to the

FIGURE 20.15 Air masses that form in different places have been assigned different names. The arrows indicate the average directions in which the air masses move.

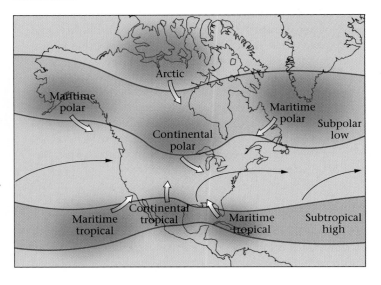

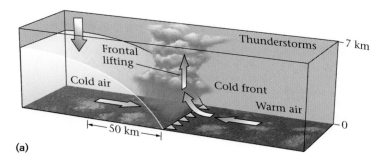

(a)

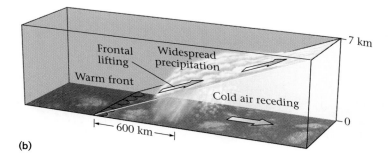

(b)

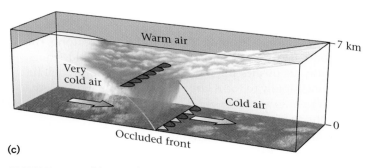

(c)

FIGURE 20.16 (a) A cold front develops where a cold air mass moves under a warm air mass. Clouds and thunderstorms form above the front. **(b)** A warm front develops where a warm air mass moves into a cold one. Note that the slope of a warm front is much gentler than that of a cold front. Clouds form over a broad area. **(c)** An occluded front develops where a fast-moving cold front overtakes a warm front and lifts the base of the warm front off the ground.

center of the air mass, like the spokes of a wheel. But because of the Coriolis effect, the wind deflects and moves in a spiral around the air mass, eventually flowing nearly parallel to isobars. In the Northern Hemisphere, the flow becomes counterclockwise (creating a **cyclone**) around a low-pressure mass, and clockwise (creating an **anticyclone**) around a high-pressure mass (▶Fig. 20.17). (Note that "cyclone" here does not refer to a tornado or hurricane.)

In the mid-latitudes that encompass the United States and much of Canada as well as much of Europe, the weather commonly reflects the movement of a large low-pressure air mass that moves from west to east. Air circulates counterclockwise around the mass, creating a mid-latitude cyclone, or **wave cyclone** (▶Fig. 20.18a). Wave

cyclones develop when air on one side of a cold front shears sideways past air on the other side (▶Fig. 20.18b). Just as the shear of air along the surface of the sea creates water waves, the shear of air along the cold front warps the face of the front into the shape of a wave. When this happens, warm air starts to move north, up and over the cold air mass, creating a warm front, whereas cold air circles around and starts to move south and downward, pushing the cold front forward. The two fronts meet at a V, the point of which lies near the center of the low-pressure mass. In a satellite image, a mid-latitude cyclone is a huge spiral mass of clouds, rotating counterclockwise; the clouds develop along both fronts and are centered on the low-pressure mass. The cold front of a wave cyclone tends to move faster than the warm front, so eventually the warm front becomes occluded, and the cyclone dies out.

Clouds and Precipitation

Let's now look at clouds a little more closely (▶Fig. 20.19). As noted earlier, a cloud is a region of the atmosphere where moisture has condensed to form tiny water droplets (about 20 micrometers across, less than a third of the diameter of a human hair; 1 micrometer = 0.001 mm), or has crystallized to form tiny grains of ice. Clouds that form at ground level make up **fog**. Because clouds reflect and scatter incoming sunlight, they keep the ground cooler during the day, but at night they prevent infrared radiation from escaping, and thus keep the ground warmer.

Droplets or ice grains in clouds form by condensation or precipitation, respectively, when the air becomes

FIGURE 20.17 Air spirals downward and clockwise (creating an anticyclone) at a high-pressure mass and upward and counterclockwise (creating a cyclone) at a low-pressure mass.

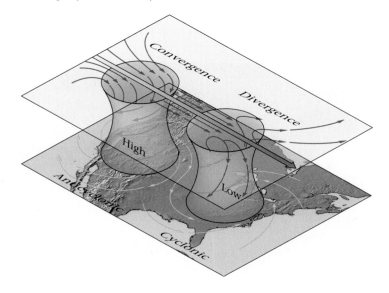

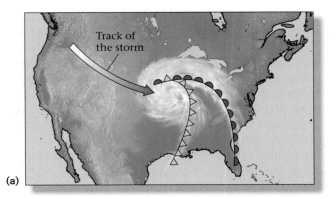

(a)

FIGURE 20.18 **(a)** A fully developed mid-latitude (wave) cyclone in the middle of its west-to-east trek across North America. Note that clouds form in association with the fronts and with the large low-pressure zone that develops where the two fronts merge. **(b)** The evolution of a mid-latitude cyclone. 1: Two air masses shear past each other. 2: An indentation, or "wave," develops along the front. 3: The indentation becomes more pronounced, and air starts circulating in a counterclockwise path (it becomes a cyclone). 4: When fully developed, the warm front is occluded. 5: When occlusion is complete, the cyclone dissipates.

Time ⟶

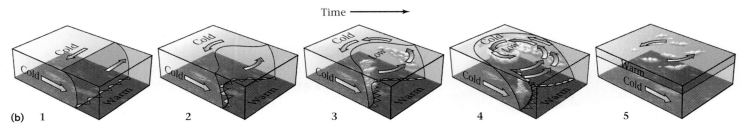

(b)   1   2   3   4   5

saturated with water vapor. During cloud formation, water condenses on **condensation nuclei**, preexisting solid or liquid aerosols. Air can become saturated (so that clouds develop) when evaporation provides additional water or when the air cools so that its capacity to hold water decreases. Cooling may take place at night, simply because of the loss of sunlight, or at any time that air rises and adiabatically cools. Meteorologists recognize several conditions that cause air to rise, and they refer to these as **lifting mechanisms**.

- *Convective lifting*: This process occurs where air on the ground warms, becomes buoyant, and rises. Clouds may form by convective lifting over an island that heats

FIGURE 20.19 Clouds can create a dramatic spectacle. Here, a towering anvil cloud forms, as viewed from an airplane.

up during the day, relative to the surrounding sea (▶Fig. 20.20a).

- *Frontal lifting*: Frontal lifting takes place along the fronts between air masses (Fig. 20.16). At cold fronts, warm air pushes up and over a steep wall of cold air, rising rapidly to form large clouds. At warm fronts, the advancing warm air rides up the gentle slope of the front and condenses.

- *Convergence lifting*: Where air converges or pushes together, as happens where air spirals up into a low-pressure zone or where two winds that have been deflected around an obstacle meet again, the air has nowhere to go but up, and thus rises and cools, forming clouds (▶Fig. 20.20b).

- *Orographic lifting*: This type of lifting happens where moisture-laden wind blows toward a mountain range, and on meeting the mountain range can go no farther and must rise. As a result, clouds form above the mountain range (▶Fig. 20.20c, d).

Rain, snow, and sleet precipitate from clouds in two ways, depending on the temperature of the cloud. In warm clouds, rain develops by **collision and coalescence**, during which the tiny droplets that compose the cloud collide and stick together to create a larger drop (▶Fig. 20.21a). Eventually, water drops that are too big to be held in suspension by circulating air start to fall, incorporating more droplets as they go. Depending on how far the drops have fallen and on how much moisture is available, drops reach different sizes before they hit ground. Typical raindrops have a diameter of 2 mm and fall at a velocity of about 20 km per hour. Any drops larger than 5 mm tend to break

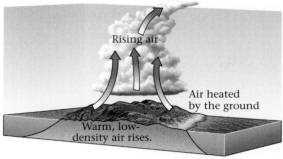

(a) Convective lifting

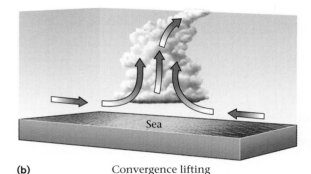

(b) Convergence lifting

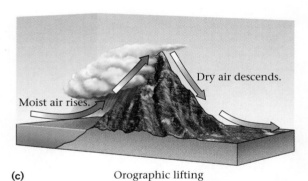

(c) Orographic lifting

(d)

FIGURE 20.20 **(a)** Convective lifting occurs where particularly warm air starts to rise. **(b)** Convergence lifting takes place where winds merge—the air has nowhere to go but up. **(c)** Orographic lifting occurs where winds off the sea run into a mountain range and are forced up. **(d)** Photo of clouds formed due to orographic lifting near St. George, Utah.

into smaller ones because of air resistance. If rain falls through colder air near the ground, it freezes to become sleet.

In cold clouds, the mist contains a mixture of very cold water droplets and tiny ice crystals. The water droplets evaporate faster than the ice (because water molecules are less tightly bound to water than to ice) and provide moisture that condenses onto preexisting ice crystals, leading to the growth of hexagonal snowflakes. If the air below the cloud is very cold, the snow falls as powder-like flakes; if the air is close to the melting temperature, large, wet clumps of snowflakes fall; and if the air is warmer than 0°C, the snow transforms to rain before it hits the ground. This kind of precipitation, involving the growth of ice crystals in a cloud at the expense of water droplets, is called the **Bergeron process**, after Tor Bergeron, the Swedish meteorologist who discovered it (▶Fig. 20.21b).

> **Take-Home Message**
>
> Weather, the local conditions of temperature, wind, and humidity, depends on the season, the interaction of air masses and fronts, and the cyclonic movements that occur where air rises or sinks. Water in rising air forms clouds, from which rain and snow may fall.

FIGURE 20.21 **(a)** Some rain forms when droplets collide and coalesce. Once a drop becomes large enough to fall, it incorporates more drops on the way down. Note that drops are flattened at their base because of air resistance (real raindrops are not teardrop-shaped). When a drop gets too big, it splits in two. **(b)** During the Bergeron process, water drops evaporate, releasing vapor that attaches to growing snowflakes, which then fall. If the air below a cloud is warm enough, the snowflakes melt and turn to rain before hitting the ground.

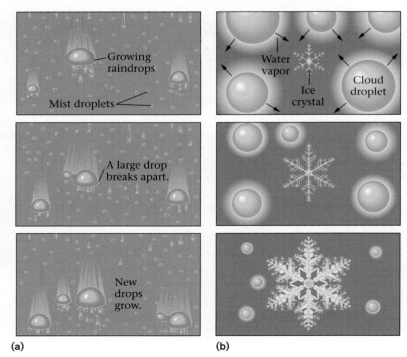

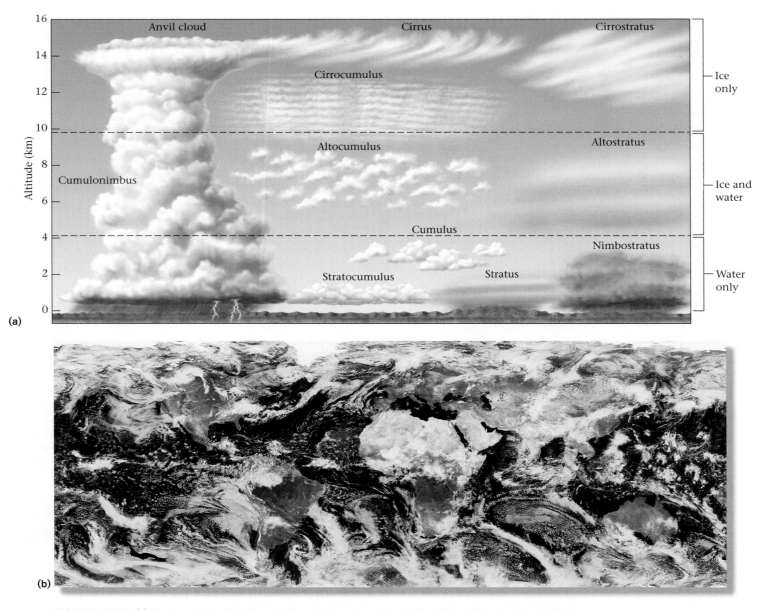

FIGURE 20.22 (a) The type of cloud that forms in the sky depends on the stability of the air, the elevation at which moisture condenses, and the wind speed. Note that cumulonimbus clouds develop vertically in that they grow across elevation boundaries. **(b)** A satellite photo of the Earth's surface displays the distribution of clouds.

Many kinds of clouds form in the troposphere. It wasn't until 1803, however, that Luke Howard, a British naturalist, proposed a simple terminology for describing clouds (▶Fig. 20.22). First, we divide clouds into types based on their shape: puffy, cotton-ball- or cauliflower-shaped clouds are **cumulus** (from the Latin word for "stacking"). Clouds that occur in relatively thin, stable layers and thus have a sheet-like or layered shape are called **stratus**. Clouds that have a wispy shape and taper into delicate, feather-like curls are called **cirrus**. We can then add a prefix to the name of a cloud to indicate its elevation: high-altitude clouds (above

about 7 km) take the prefix "cirro," mid-altitude clouds take the prefix "alto," and low-altitude clouds (below 2 km) do not have a prefix. Finally, we add the suffix "nimbus" or the prefix "nimbo" if the cloud produces rain.

Applying this cloud terminology, we see that a nimbostratus is a layered, sheet-like rain cloud, and a cumulonimbus is a rain-producing puffy cloud. Cumulonimbus clouds can be truly immense, with their bases lying at less than 1 km high and their tops butting up against the tropopause at an elevation of over 14 km. These clouds are vertically developed, because they grow across altitude divisions. Large cumulonimbus clouds spread laterally at the tropopause to form broad, flat-topped clouds called anvil clouds.

The differences in cloud types depend on whether the clouds develop in stable or unstable air. Stable air does not have a tendency to rise, because it is colder than its surroundings. Unstable air has a tendency to rise, because it is warmer than its surroundings. Cumulus clouds, which in time-lapse photography look as if they're boiling, form in unstable air. They billow because of updrafts (upward-moving air) and downdrafts (downward-moving air). Plane flights through these clouds will be rather bumpy. In contrast, stratus clouds indicate stable air.

20.6 STORMS: NATURE'S FURY

The rain from an overcast sky may be inconvenient for a picnic, but it won't threaten life or property and will be appreciated by farmers. In contrast, a storm poses a threat. A **storm** is an episode of severe weather, when winds, rainfall, snowfall, and, in some cases, lightning become strong enough to be bothersome and even dangerous (▶Fig. 20.23). Storms form where large pressure gradients develop (as may exist across a front), for pressure gradients produce strong winds. Storms also form where local conditions cause warm, moist air to rise. Rising moist air can trigger a storm because when the air reaches higher elevations, it condenses and, as we have seen, releases latent heat. This heat warms the air, makes it more buoyant, and thus causes it to rise still further, until it becomes cool enough to produce clouds. Meanwhile, at ground level, new moist air flows in beneath the clouds to replace the air that has already risen. This new air then immediately starts to rise and causes the clouds to

FIGURE 20.23 A thunderstorm can drench an area with rain and attack it with lightning.

build upwards. Warm, moist air effectively feeds the storm. Once the clouds become thick enough to start producing heavy rain, and/or the wind becomes strong enough to be troublesome, we can say that a storm has been born. We'll now look at various types of storms.

Thunderstorms

A thunderstorm is a local episode of intense rain accompanied by strong, gusty winds and by lightning. Over 2,000 are occurring at any given time somewhere on the Earth, and over 100,000 take place in the United States every year. Thunderstorms form where a cold front moves into a region of particularly warm, moist air (as happens, for example, in North America's mid-latitudes during the summer where cold polar air masses collide with warm Gulf air masses); where convective lifting is driven by solar radiation in a region with an immense supply of moisture (as occurs over tropical rain forests); or where orographic lifting causes clouds to form over mountains.

A typical thunderstorm has a relatively short life, lasting from under an hour to a few hours (▶Fig. 20.24). The storm begins when a cumulonimbus cloud, fed by a steady supply of warm, moist air, grows large. The rising hot air, kept warm by the addition of energy from the latent heat of condensation, creates updrafts that cause the cloud to stack, or billow upward, toward the tropopause. When this air adiabatically cools, precipitation begins.

If updrafts in the cloud are strong enough, ice crystallizes in the higher levels of the cloud, where temperatures are below freezing, building into ice balls known as **hail** (or hailstones). A discrete mass of hail may fall from a cloud over a few minutes to form a hail streak on the ground, typically 2 km by 10 km and elongated in the direction that the storm moves. Though most hailstones are pea-sized, the largest recorded hailstone reached a diameter of 14 cm and weighed 0.7 kilograms.

Once precipitation begins, a thunderstorm has reached its mature stage. Now, falling rain pulls air down with it, creating strong downdrafts. By this stage, the top of the cloud has reached the top of the troposphere and begins to spread laterally to form an anvil cloud. Because of the simultaneous occurrence of updrafts and downdrafts, a mature thunderstorm produces gusty winds and the greatest propensity for lightning. Eventually, downdrafts become the overwhelming wind; their cool air cuts off the supply of warm, moist air, so the thunderstorm dissipates.

Lightning accompanies thunderstorms because electrical charges separate in a storm cloud (▶Fig. 20.25a). Surprisingly, meteorologists still aren't sure why this happens, but they speculate that the rubbing of air and water molecules together in a cloud creates positively charged hydrogen ions and/or ice crystals that drift to the top of the cloud, and negatively charged OH⁻ ions and/or water

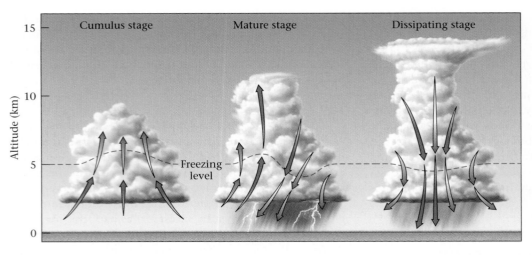

FIGURE 20.24 A thunderstorm evolves in three stages. First, warm, unstable, moist air rises and builds a cumulus cloud. Second, updrafts, fueled by more warm air at the surface and the heat released by condensation, cause the cloud to billow higher. Heavy rain falls, lightning flashes occur, and the downpour creates cold downdrafts. Finally, the cloud reaches the top of the troposphere, but downdrafts cut off the supply of warm air at the ground surface, and without this fuel, the thunderstorm dissipates.

droplets that sink to the base. The negative ions at the base of the cloud repel negative ions on the ground below, creating a zone of positive charge on the ground. Air is a good insulator, so the charge separation can become very large until a giant spark or pulse of current, a **lightning flash**, jumps across the gap. Essentially, lightning is like a giant short circuit across which a huge (30-million-volt) pulse of electricity flows.

Lightning flashes can jump from one part of a cloud to another, or from a cloud to the ground. In the case of cloud-to-ground lightning, the flash begins when electrons leak from the negatively charged base of the cloud incrementally downward across the insulating air gap, creating a conductive path, or leader. While this happens, positive charges flow upward to the cloud through conducting materials, such as trees or buildings (▶Fig. 20.25b). The instant that the charge flows connect, a strong current carries positive charges up into the cloud; this upward-flowing current is called the "return stroke" (▶Fig. 20.25c).

We hear **thunder**, the cracking or rumbling noise that accompanies lightning, because the immense energy of a flash almost instantaneously heats the surrounding air to a temperature of 8,000° to 33,000°C, and this abrupt expansion, like an explosion, creates sound waves that travel through the air to our ears. But because sound travels so much more slowly than light, we hear thunder after we see lightning. A 5-second time delay between the two means that the lightning flashed about 1.6 km (1 mile) away.

FIGURE 20.25 **(a)** Lightning flashes when a charge separation develops in a cloud, with a negative charge at the base and a positive charge at the top. The negative charge repels negative charges on the ground, so positive charges develop on the ground. A leader begins to descend from the cloud. **(b)** As the leader grows downward, positive charges begin to flow upward from an object on the ground. **(c)** When the connection is complete, the return stroke carries positive charges rapidly from the ground to the cloud, creating the main part of the flash.

(a)

(b)

(c)

(a)

(b)

FIGURE 20.26 **(a)** A tornado touches down near Mulvane, Kansas. **(b)** The intense winds of a tornado tore through Midwest City, Oklahoma, in April of 1999, completely destroying dozens of homes.

Tornadoes

Some thunderstorms grow to be quite violent and spawn one or more tornadoes. A **tornado** is a near-vertical, funnel-shaped cloud in which air rotates extremely rapidly around the axis (center line) of the funnel (▶Fig. 20.26a). In other words, a tornado is a vortex beneath a severe thunderstorm. The word probably comes either from the Spanish *tonar*, meaning to turn, or *tronar*, meaning thunder (perhaps referring to the loud noise generated by a tornado).

In the mid-latitudes of the Northern Hemisphere, where most tornadoes form, air in the funnel rotates counterclockwise around the center and spirals upward. Air in the fiercest tornadoes may move at speeds of up to 500 km per hour (about 300 mph). The diameter of the base of the funnel in a small tornado may be only 5 m across, while in the largest tornadoes it may be as wide as 1,500 m across. Because of the upward movement of air, air pressure near a tornado drops.

In North America, tornadoes drift with a thunderstorm from southwest to northeast, because of the prevailing wind direction, traveling at speeds of 0 to 100 km per hour. They tend to hopscotch across the landscape, touching down for a stretch, then rising up into the air for a while before touching down again. This characteristic leads to a bizarre incidence of damage—one house may be blasted off its foundation while its next-door neighbor remains virtually unscathed (▶Fig. 20.26b). Small tornadoes may cut a swath less than a kilometer long, but large tornadoes raze the ground for tens of kilometers, and the largest have left a path of destruction up to 500 km long (▶Fig. 20.27a, b). In 1925, for example, one of the most enormous tornadoes on record ripped across Missouri, Illinois, and Indiana, killing 689 people before it dissipated. In some cases, two or three tornadoes may erupt from a single thunderstorm. Massive thunderstorm fronts may produce a tornado swarm, dozens of tornadoes out of the same storm. In April 1974, a single thunderstorm system generated a swarm of at least 148 individual tornadoes, which killed 307 people over eleven states. On November 11, 2002, a chain of thunderstorms covering a belt from Ohio to Alabama spawned 66 tornadoes that left 66 people dead. The death toll would have

Over eighty people a year die from lightning strikes in the United States alone, and many more are seriously burned or shocked. Lightning that strikes trees heats the sap so quickly that the trees literally explode. Lightning can spark devastating forest fires and set buildings on fire. You can reduce the hazard to buildings by installing lightning rods, upward-pointing iron spikes that conduct electricity directly to the ground so that it doesn't pass through the building.

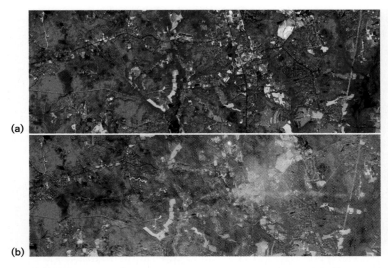

(a)

(b)

FIGURE 20.27 Two satellite images showing the swath cut by an F5 tornado that passed through Maryland on April 28, 2002. The field of view is 6 km × 17.8 km. **(a)** Before the tornado; **(b)** after the tornado. Red is vegetation. The turquoise strip shows where the tornado stripped vegetation away. The swath here is about 150 m wide and 5 km long.

damage: they have been known to drive straw through wood, lift cows and carry them unharmed for hundreds of meters, and raise railroad cars right off the ground.

Because of the range of damage a tornado can cause, T. T. Fujita of the University of Chicago proposed a scale that distinguishes among tornadoes on the basis of wind speed, path dimensions, and possible damage (▶Table 20.2). The wind speeds in the **Enhanced Fujita scale** are estimates, based on the damage assessment, because no one has yet succeeded in measuring winds directly in a tornado.

Tornadoes in North America form where strong westerlies exist at high altitudes while strong southeast winds develop near the ground surface (▶Fig. 20.28a–c). These opposing winds shear the air between them, so that the air begins to rotate in a horizontal cylinder. Tornado watchers search for the resulting funnel cloud, the precursor to a true tornado. As the associated thunderstorm matures, updrafts tilt one end of the cylinder up and downdrafts push the other end down, until the air in the cylinder starts spiraling inward and upward, gaining speed as do spinning figure skaters who pull their arms inward. At this stage, the cloud consists only of rotating moisture and may look grayish white. But if the process continues, the low end of the funnel touches ground, and at the instant of contact, dirt and debris get sucked into the tornado, giving it a dark color.

The special weather conditions that spawn tornadoes in the midwestern United States and Florida develop when cold polar air from Canada collides with warm tropical air from the Gulf of Mexico. These conditions happen most frequently during the months of March through September. So many tornadoes occur during the summer in a belt from Texas to Indiana that this region has the unwelcome nickname "tornado alley" (▶Fig. 20.29). During a 30-year span, the number of reported tornadoes per year ranged between about 420 and 1,100 in the United States

been higher were it not for warnings broadcast by the U.S. National Weather Service, which sent people scrambling for safety.

Tornadoes cause damage both because of the force of their rapidly moving wind and because of their low air pressure. The wind lifts trucks and tumbles them for hundreds of meters, uproots trees, and flattens buildings (Fig. 20.26b). Particularly large tornadoes can even rip asphalt off a highway. The low air pressure around a tornado can make windows pop out of buildings and may cause airtight buildings to explode, as the air inside suddenly expands relative to the air outside. In some cases, tornadoes cause strange kinds of

TABLE 20.2 Enhanced Fujita Scale for Tornadoes

Scale	Category	Wind Speed km/h (mph)	Average Path Length; Average Path Width	Typical Damage
EF0	Weak	104–137 (65–85)	0–1.6 km; 0–17 m	Branches and windows broken.
EF1	Moderate	138–177 (86–110)	1.6–5.0 km; 18–55 m	Trees broken; shingles peeled off; mobile homes moved off their foundations.
EF2	Strong	178–217 (111–135)	5–16 km; 56–175 m	Large trees broken; mobile homes destroyed; roofs torn off.
EF3	Severe	218–266 (136–165)	16–50 km; 176–556 m	Trees uprooted; cars overturned; well-constructed roofs and walls removed.
EF4	Devastating	267–322 (166–200)	50–160 km; .56–1.5 km	Strong houses destroyed; buildings torn off foundations; cars thrown; trees carried away.
EF5	Incredible	>322 (over 200)	160–500 km; 1.5–5.0 km	Cars and trucks carried more than 90 m; strong houses disintegrated; bark stripped off trees; asphalt peeled off roads.

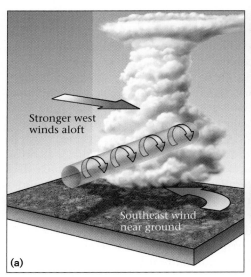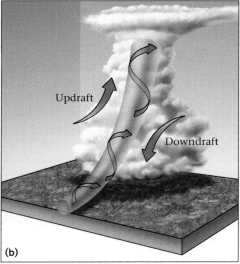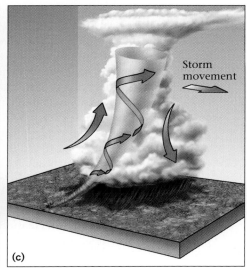

FIGURE 20.28 (a) Tornadoes initiate at intense fronts where high-altitude westerlies flow over low-altitude southeasterlies. The resulting shear creates a horizontal cylinder of rotating air in a cloud. **(b)** Updrafts and downdrafts in the cloud eventually tilt the cylinder, creating a tornado. **(c)** When the tornado touches down, destruction follows. Note that the spiraling winds of the funnel also circulate inside the main cloud; the tornado we see protruding from the base of the cloud is only the tip of a much larger flow.

(with an average of 770 per year; fewer than twenty strike Canada annually). On average, about eighty people a year die in tornadoes. But a single F5 event may kill hundreds.

Because of the threat tornadoes pose to life and property, meteorologists have worked hard to be able to forecast them. First they search for appropriate weather conditions. If these conditions exist, meteorologists issue a tornado watch. If observers spot an actual tornado forming, they

FIGURE 20.29 North American tornadoes are most common in "tornado alley," a band extending from Texas to Indiana, where the polar air mass collides with the Gulf Coast maritime tropical air mass. The storm systems are urged eastward by the jet stream and related high-altitude westerlies.

Number of tornadoes per year (per 26,000 sq. km, for a 27-year period)

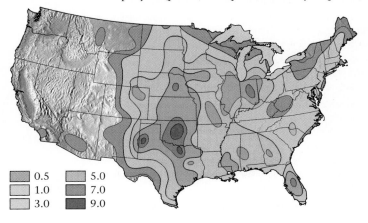

0.5	5.0
1.0	7.0
3.0	9.0

issue a tornado warning for the region in its general path (the exact path can't be predicted). If you hear a warning, it's best to take cover immediately in a basement, or at least in an interior room away from windows. With the invention of Doppler radar, which uses the Doppler effect (see Chapter 1) to identify rain moving in strong winds, meteorologists may detect tornadoes without even going outside.

Nor'easters

In some cases, large mid-latitude (wave) cyclones of North America affect the Atlantic coast. Because the cold, counterclockwise winds of these cyclones come out of the northeast, they are called **nor'easters**. Some nor'easters are truly phenomenal storms. One of the strongest occurred at the end of October 1991 (the "perfect storm" made famous by the book and movie of that name). During this storm, winds were not so strong as those of a hurricane, but they covered such a large area that waves in the open ocean built to a height of over 11 m, a disaster for ships. When the waves reached shore, they eroded huge stretches of beach. The rainfall from the storm caused extensive flooding inland.

Hurricanes

During the summer and early fall, cyclonic wind systems called "tropical disturbances" develop off the western coast of Africa, near the Cape Verde Islands (latitude 20°N). In these low-pressure regions, air converges and rises and, because of the Coriolis effect, begins to circulate counterclockwise. But because these storms form over the warm tropical waters of the central Atlantic, the air that rises within them is particularly warm and moist. As it rises, it cools, and moisture condenses, releasing latent heat of condensation. This latent heat is a fuel that provides energy to the storm—it causes the air to rise still higher, creating even lower pressure at the Earth's surface, which, in turn, sucks up even more warm, moist air. Thus, the storm grows in strength until it becomes a tropical depression whose winds may reach 61 km per hour.

As long as there continues to be a supply of warm, moist air, the storm continues to grow broader, and air within begins to rotate still faster. When the sustained wind speed exceeds 119 km per hour, a hurricane has been born. In other words, a **hurricane** (named for the Carib god of evil) is a huge, rotating storm, resembling a giant spiral in map view, in which sustained winds are greater than 119 km (74 miles) per hour (►Fig. 20.30a). Within a hurricane, air pressure becomes much lower because of the upward flow of air. Before the days of satellite forecasting, mariners watched their barometers closely, knowing that an extreme drop signaled the approach of a hurricane.

Once a storm reaches tropical-depression status, it is assigned a name. Each year, the first storm has a name beginning with "A," the second with "B," and so on in alphabetical order, with alternating male and female names. Names for less significant hurricanes may be reused, but the names of particularly notorious hurricanes are retired. Meteorologists classify hurricanes according to the **Saffir-Simpson scale** (►Table 20.3). Intense hurricanes have sustained winds of over 230 km per hour.

Traditionally, the designation "hurricane" applies to storms born in the east-central Atlantic that drift first westward and then northward with the prevailing winds. The path, or hurricane track, that they follow allows them to inflict damage to land regions in the Caribbean and the Gulf Coast and East Coast of North America (►Fig. 20.30b). Occasionally, hurricanes make it across Central America and then thrash the Pacific coast of Mexico and the western United States; some may even drift northwestward to Hawaii. Because hurricanes require warm water (warmer than 27°C), they develop only at latitudes south of about 20°. They do not form close to the equator because there is not enough atmospheric motion or Coriolis effect at that latitude. Similar storms that form between latitudes 20°N and 20°S in the western Pacific are called "typhoons"; those at latitude 20°N in the Indian Ocean are called "cyclones" (a second use of the word; ►Fig. 20.30c).

A typical hurricane consists of several spiral arms, called rain bands, extending inward to a central zone of relative calm known as the hurricane's eye. A rotating vertical cylinder of clouds, called the eye wall, surrounds the eye (►Fig. 20.31). The entire width of a hurricane ranges from 100 to 1,500 km, with the average around 600 km. Winds spiral toward the eye, and thus their angular momentum, and therefore speed, increases toward the center; as a result, the greatest rotary wind velocity occurs in the eye wall.

Hurricanes, in general, move along their track because of the prevailing motion of the atmosphere. Typically, a hurricane's velocity along its track, the storm-center velocity, ranges from 0 to 60 km per hour (on rare occasions, as fast as 100 km per hour). Because of the rotary motion of wind around the eye, winds on one side of the eye move in the opposite direction of those on the other side. On the side of the hurricane where winds move in the same direction as the storm's center, the

TABLE 20.3 Saffir-Simpson Scale for Hurricanes

Scale	Category	Wind Speed km/h (mph)	Air Pressure in Eye (millibars)	Damage
1	Minimal	119–153 (74–95)	980 or more	Branches broken; unanchored mobile homes damaged; some flooding of coastal areas; no damage to buildings; storm surge of 1.2 to 1.5 m.
2	Moderate	154–177 (96–110)	965–979	Some roofs, doors, and windows damaged; mobile homes seriously damaged; some trees blown down; small-boat moorings broken; storm surge of 1.6 to 2.4 m.
3	Extensive	178–209 (111–130)	945–964	Some structural damage to small buildings; large trees blown down; mobile homes destroyed; structures along coastal areas destroyed by flooding and battering; storm surge of 2.5 to 3.6 m.
4	Extreme	210–250 (131–155)	920–944	Some roofs completely destroyed; extensive window and door damage; major damage and flooding along coast; storm surge of 3.7–5.4 m. Widespread evacuation of regions within up to 10 km of the coast may be necessary.
5	Catastrophic	over 250 (over 155)	less than 920	Many roofs and buildings completely destroyed; extensive flooding; storm surge greater than 5.4 m. Widespread evacuation of regions within up to 16 km of the coast may be necessary.

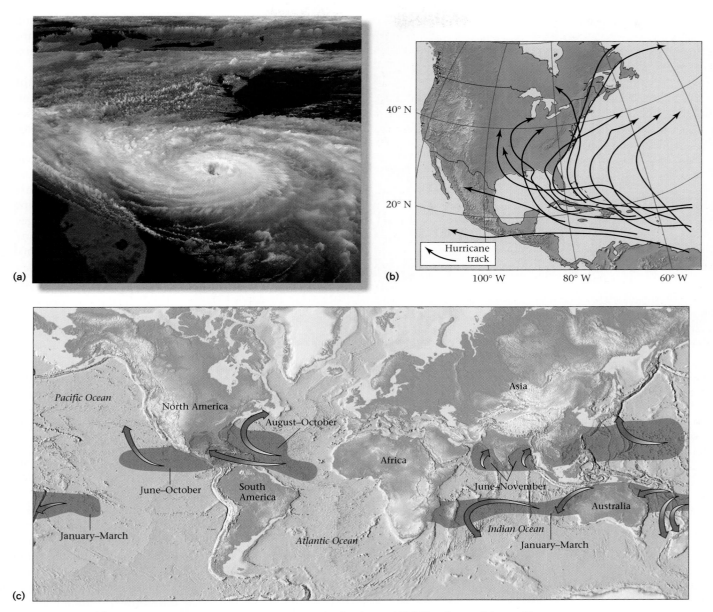

FIGURE 20.30 **(a)** An oblique view of Hurricane Hugo, approaching the Carolinas in 1989. Note the spiral shape of the storm and the relatively small eye. **(b)** The tracks of several important Atlantic hurricanes show how most begin at latitudes of 15° to 20° off the western coast of Africa, then drift westward and northward. **(c)** Some Atlantic hurricanes make it across Central America and lash the eastern Pacific. Similar storms occur in the western Pacific, where they are known as typhoons, and in the Indian Ocean, where they are known as cyclones; these storms also originate at latitudes of about 20°.

storm-center velocity adds to the rotary motion, making surface winds particularly fast. On the other side of the hurricane, the storm-center velocity subtracts from the rotary velocity, so the winds are slower.

Hurricanes cause damage in several ways (▶Fig. 20.32).

- *Wind*: Hurricane-force winds may reach such intensity that buildings cannot stand up to them. Hurricanes can tear off branches, uproot trees, rip off roofs, collapse walls, force vehicles off the road, knock trains

from their tracks, and blast mobile homes off their foundations.

- *Waves*: The force of hurricane winds shearing across the sea surface generates huge waves, sometimes tens of meters high. Out in the ocean, these waves can swamp or capsize even large ships. Near shore, they batter beachside property, rip up anchored boats and carry them inland, and strip beaches of their sand.

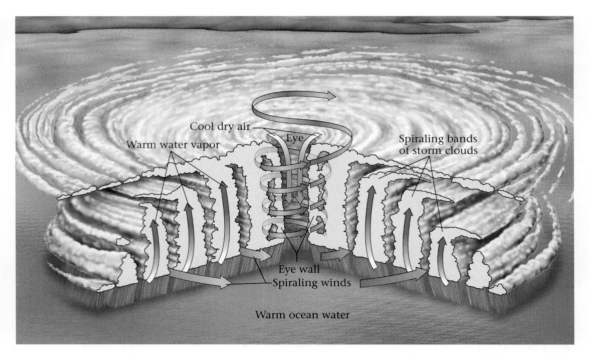

FIGURE 20.31 This cutaway diagram of a hurricane shows the spirals of clouds, the eye, and the eye wall. Dry air descends in the eye, creating a small region of calm weather.

- *Storm surge*: Coastal areas may be severely flooded by a **storm surge**, excess water that is carried landward by the hurricane. In the portion of a hurricane where winds blow onshore, water piles up and becomes deeper over a region of 60 to 80 km, allowing waves to break at a higher elevation than they otherwise would. Storm surges during hurricanes are exacerbated by the very low air pressure in a hurricane, which allows the sea surface to rise still higher—if a hurricane hits at high tide, the damage may be even worse. A storm surge during the 1900 hurricane that hit Galveston, Texas, flooded the city and killed 6,000 people. Storm surges in Bangladesh cause horrendous loss of life because the Indian Ocean submerges the low-lying delta plain of the Ganges. During one 1970 cyclone, the death toll reached 500,000.

- *Rainfall*: The intense rainfall during a hurricane (in some cases, half a meter of rain has fallen in a single day) causes streams far inland to flood. The flooding itself can submerge towns and cause death. Intense rains can also weaken the soil of steep slopes, especially in deforested areas, and thus can trigger mudslides.

On average, five Atlantic hurricanes happen every year. But in 1995, there were eleven—in fact, three or four tropical depressions and hurricanes existed at a single time, forming a deadly procession that marched across the Atlantic. Because hurricanes are nourished by warm ocean water, some atmospheric scientists fear that a warming of the climate may lead to more and fiercer hurricanes in the future. Many Atlantic hurricanes veer northward when they bump into high-pressure air masses over North America. This sends the storms up the eastern seaboard. Hurricanes die out when they lose their source of warm, moist air—when they cross onto land or move over the cold waters of high latitudes.

2005 was a particularly bad year for hurricanes in the North Atlantic. There were so many storms that meteorologists went through the entire list of names and had to add new ones. At one time, four hurricanes swirled over the Atlantic at once. Of these, Hurricane Katrina was by far the most devastating (▶Box 20.3).

FIGURE 20.32 In 1992, Hurricane Andrew caused immense damage in southern Florida before crossing the Gulf of Mexico and slamming into Louisiana.

Take-Home Message

Where large pressure gradients develop, or where warm moist air rises, storms develop. Thunderstorms produce lightning and heavy rain. Violent thunderstorms can produce tornadoes. Hurricanes develop over warm ocean water and can severely damage coasts.

20.7 GLOBAL CLIMATE

When we talk about the weather, we're referring to the atmospheric conditions at a certain location at a specified time. But when we speak of the characteristic weather conditions, the typical range of conditions, the nature of seasons, and the possible weather extremes of that region over a long time (say, 30 years), we use the term **climate**. For example, on a given summer day in Winnipeg, central Canada, it may be sunny, hot, and humid, and on a given winter day in southern Florida, it may drop below freezing. But averaged over a year, the weather in Winnipeg is more likely to be cooler and drier than the weather in southern Florida. The variables that characterize the climate of a region include its temperature (both the yearly average and the yearly range), its humidity, its precipitation (both the yearly amount and the distribution during the year), its wind conditions, and the character of its storms.

Climate Controls and Belts

Climatologists, scientists who study the Earth's climate, suggest that several distinct factors control the climate of a region.

- *Latitude*: This is perhaps the most significant factor, for the latitude determines the amount of solar energy a region receives, as well as the contrasts between seasons. Polar regions, which receive much less solar radiation over the year, have colder climates than equatorial regions. And the contrast between winter and summer is greater in mid-latitudes than at the poles or the equator. We can easily see the influence of latitude by examining the global distribution of temperature, represented on a map by **isotherms**, lines along which the temperature is exactly the same (▶Fig. 20.33). Because land and sea do not heat up at the same rate, because the distribution of clouds is not uniform, and because ocean currents transfer heat across latitudes, isotherms are not perfect circles.

- *Altitude*: Because temperature decreases with elevation, cold climates exist at high elevations, even at the equator. Hiking from the base of a high mountain in the Andes to its summit takes you through the same range of climate belts you would pass through on a hike from the equator to the pole.

- *Proximity of water*: Land and water have very different heat capacities (ability to absorb and hold heat). Land absorbs or loses heat quickly, whereas water absorbs or loses heat slowly. Also, water can absorb and hold on to more heat than land can because water is semitransparent; sunlight heats water down to a depth of up to 100 m, whereas sunlight heats land down to a depth of only a few centimeters. Thus, the proximity of the sea

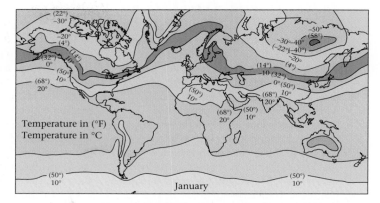

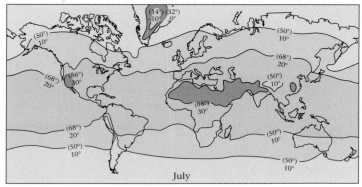

FIGURE 20.33 Isotherms of January roughly parallel lines of latitude, owing to contrasts in amounts of insolation, except where they are distorted by oceanic currents. The Gulf Stream in the Atlantic Ocean deflects isotherms northward, granting the United Kingdom, Ireland, and Scandinavia milder climates than they might otherwise have. The pattern of isotherms is different in July.

tempers the climate of a region: as a rule, locations in the interior of a continent experience a much greater range of weather conditions than regions along the coast.

- *Proximity to ocean currents*: Where a warm current flows, it may warm the overlying air, and where a cold current flows, it may cool down the overlying air. For example, the Gulf Stream brings warm water north from the Gulf of Mexico and keeps Ireland, the United Kingdom, and Scandinavia much warmer than they would be otherwise.

- *Proximity to orographic barriers*: An **orographic barrier** is a landform (such as a mountain range) that diverts airflow upward or laterally. This diversion affects the amount of precipitation and wind a region receives.

- *Proximity to high- or low-pressure zones*: Zones of high and low pressure, roughly parallel to the equator, encircle the planet (Fig. 20.13). Because land and sea have different heat capacities, they modify the zones, so that high-pressure zones tend to be narrower over land. Meteorologists refer to the resulting somewhat elliptical regions of high or low pressure as semipermanent pressure cells (▶Fig. 20.35). These influence prevailing wind direction and relative humidity.

BOX 20.3 THE HUMAN ANGLE

Hurricane Katrina!

Hurricane Katrina, in 2005, stands as the most expensive hurricane to strike the United States in historic time. Because of where Katrina passed, worst-case damage-prediction scenarios came to be reality, particularly in the historic city of New Orleans. Let's look at this storm's history.

Tropical storm Katrina came into existence over the Bahamas and headed west. Just before landfall in southeastern Florida, winds strengthened and the storm became Hurricane Katrina (►Fig. 20.34a). This hurricane sliced across the southern tip of Florida, causing several deaths and millions of dollars in damage. It then entered the Gulf of Mexico and passed directly over the Loop Current, an eddy (circular flow) of summer-heated water from the Caribbean that had entered the Gulf of Mexico. Water in the Loop Current reaches temperatures of 32°C (90°F), and thus stoked the storm, injecting it with a burst of energy sufficient for the storm to morph into a Category 5 monster whose swath of hurricane-force winds reached a width of 325 km. When it entered the central Gulf of Mexico, Katrina turned north and began to bear down on the Louisiana-Mississippi coast. The eye of the storm passed just east of New Orleans and then across the coast of Mississippi. Storm surges broke records, in places rising 7.5 m above sea level, and then washed coastal communities off the map along a broad stretch of the Gulf Coast (►20.34b, c). In addition to the devastating wind and surge damage, Katrina caused the drowning of New Orleans.

To understand what happened to New Orleans, we must consider the city's environmental history. New Orleans grew on the Mississippi Delta between the banks of the Mississippi River on the south and Lake Pontchartrain (actually a bay in the Gulf of Mexico) on the north (►Fig. 20.34d). The oldest part of the town, a neighborhood called the French Quarter, was built on the relatively high land of the Mississippi's natural levee. Newer parts of the city, however, spread out over the lower delta plain. As the decades passed, people modified the surrounding delta landscape by draining wetlands, by constructing artificial levees that confined the Mississippi River, and by extracting groundwater. Sediment beneath the delta compacted and the delta's surface has been starved of new sediment, so large areas of the delta sank below sea level. Today, most of New Orleans lies in a bowl-shaped depression as much as 2 m below sea level—the hazard implicit in this

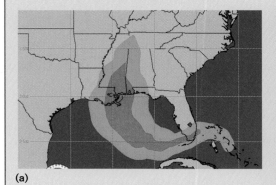

(a)

(b)

(c)

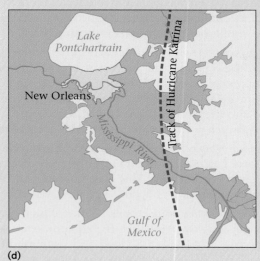

(d)

(e)

Broken Levees, Slow Repairs

FIGURE 20.34 (a) A wind-swath map of Hurricane Katrina. The red area was affected by hurricane-force winds, and the gold area was affected by tropical-storm force winds. Note how the hurricane grew after it entered the Gulf of Mexico. **(b)** An aerial photo, looking straight down at the coast of Mississippi, after Hurricane Katrina. The distribution of debris and the near-total destruction of buildings shows the extent to which the storm surge washed inland. **(c)** Officials observe storm damage. **(d)** A map of New Orleans and Lake Pontchartrain, showing the track of the Hurricane's eye. **(e)** Three breaches occurred in the levee system. This map shows the levee system and the region near downtown New Orleans that the rising waters submerged. The cross section inset shows the basin of New Orleans.

situation had been recognized for years (see Fig. 17.6)

The winds of Hurricane Katrina ripped off roofs, toppled trees, smashed windows, and triggered the collapse of weaker buildings, but their direct consequences were not catastrophic. However, when the winds blew storm surge into Lake Pontchartrain, its water level rose beyond most expectations and pressed against the system of artificial levees and flood walls that had been built to protect New Orleans (▶Fig. 20.34e). This system had been built to withstand a Category 3 storm, but when Katrina hit, it still had Category 4 strength, and upkeep of the levee system had been neglected for many years. Thus, hours after the hurricane eye had passed, the high water of Lake Pontchartrain finally found a weakness along

the flood walls bordering the 17th Street Canal and pushed out a 100-m-long section. The canal was one of four that had been built to carry water pumped out of New Orleans away to Lake Pontchartrain. There was no gate at its mouth to keep water out when the storm surge caused the lake to rise, and the flood walls were too weak to resist the push of the rising water. Preliminary studies suggest that the foundations of the walls along the canal had not been driven deeply enough, so that the bases of the walls were anchored in a very weak bed of plant debris, remnants of an ancient swamp, rather than in the firmer sand below—under pressure, the bases of the walls just gave way. Breaks eventually formed in other locations as well. A day after the hurricane was over, New Orleans began

to flood. As the water line climbed the walls of houses, brick by brick, residents fled first upstairs, then to their attics, then to their roofs. Water spread across the city until finally the bowl of New Orleans filled to the same level as Lake Pontchartrain and 80% of the city was submerged (▶Fig. 20.34f–h). The disaster grew to be of national significance, as the trapped population sweltered without food, drinking water, or adequate shelter. With no communications, no hospitals, and few police, the city almost descended into anarchy. It took days for outside relief to reach the city, and by then, many had died and much of New Orleans, a cultural landmark and major port, had become uninhabitable. Even two years later, many neighborhoods remained wastelands (▶Fig. 20.34i, j).

(f)

(i)

(j)

(g)

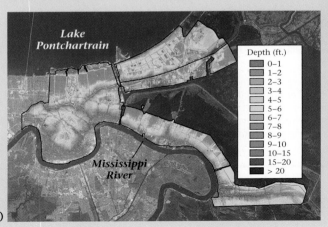
(h)

Depth (ft.)
0–1
1–2
2–3
3–4
4–5
5–6
6–7
7–8
8–9
9–10
10–15
15–20
> 20

Lake Pontchartrain

Mississippi River

FIGURE 20.34 (*continued*) **(f)** Water flowing across the levees bordering the 17th Street Canal. **(g)** Photo of a flooded New Orleans neighborhood. **(h)** NOAA map showing the depth of water in New Orleans during the flooding. **(i)** A photo of the exterior of a New Orleans home damaged by Hurricane Katrina. **(j)** A photo of the damaged interior of a New Orleans home.

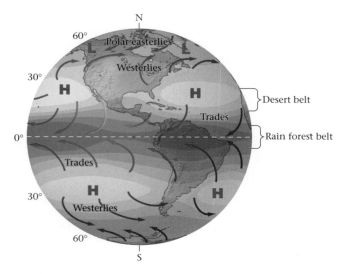

FIGURE 20.35 Because of land masses, atmospheric pressure belts vary in width to create lens-shaped, semipermanent high- and low-pressure cells. Compare this figure with Figure 20.13.

Climatologists, who have studied the distribution of climatic conditions around the globe, have developed a classification scheme for climates, based on such factors as the average monthly and annual temperatures and the total monthly and yearly amounts of precipitation. The vegetation of a region proves to be an excellent indicator of climate, because plants are sensitive to temperature and to the amount and distribution of rainfall. ▶Table 20.4 lists the principal types of climate belts (▶Fig. 20.36).

Climate Variability: Monsoons and El Niño

The climate at a certain location may change during the course of one or more years. Here, we look at two important examples of climate variability that affect human populations in notoriously significant ways.

A **monsoon** is a major reversal in the wind direction that causes a shift from a very dry season to a very rainy season. In southern Asia, home to about half the world's population, people depend on monsoonal rains to bring moisture for their crops.

The Asian monsoon develops primarily because Asia is so large that it includes vast tracts of land far from the sea. Further, a substantial part of this land, the Tibet Plateau, lies at a high elevation. During the winter, central Asia becomes very cold, much colder than coastal regions to the south. This coldness creates a stable high-pressure cell over central Asia. Dry air sinks and spreads outward from this cell and flows southward over southern Asia, pushing the intertropical convergence zone (ITCZ) out over the Indian Ocean, south of Asia (▶Fig. 20.37a). Thus, during the winter, southern Asia experiences a dry season. During the summer, central Asia warms up dramatically. As warm air

TABLE 20.4   Climate Types of the Earth

Climate Type	Regions and Characteristics
Tropical rainy	Tropical rain forests lie at equatorial latitudes and experience rain throughout the year. Rain commonly falls during afternoon thunderstorms. Tropical savanna (grasslands with brush and drought-resistant trees), which lie on either side of a rain forest, have a rainy season and a dry season. Rain forests and savannas may receive tropical monsoons.
Dry	Dry regions include deserts (regions with very little moisture or vegetation cover; vegetation that does exist has adapted to long periods without moisture) and steppes. Steppe regions (vast, grassy plains with no forest) border the desert and have somewhat more precipitation. Some steppe regions occur at high elevations, in latitudes where the climates would otherwise be more humid.
Humid mesothermal	This category includes humid subtropical climates, with moist air and warm temperatures for much of the year, in which mixed deciduous-coniferous forest thrives; Mediterranean climates, coastal regions with most rainfall in the winter, very hot summers, and scrub forests; and marine west-coast climates, where the sea tempers the climate and may create a coastal temperate rain forest.
Humid microthermal	These higher-latitude temperate climates, which occur only in the Northern Hemisphere, include humid continental regions, with long summers (as in the U.S. Midwest and mid-Atlantic states), in which deciduous forest thrives; humid continental regions with short summers, characterized by mixed deciduous-coniferous forest or coniferous-only forest; and subarctic climates, with very short, cool summers and coniferous forest that becomes lower and scrubbier at higher latitudes.
Polar	These cold climates include tundra and ice caps. Tundra are regions with no summer and an extremely cold winter, in which only low, cold-resistant plants (moss, lichen, and grass) can survive. Much of the ground in tundra is permafrost (permanently frozen ground). In ice-cap regions, near the poles, the climate is sub-freezing year-round, and any land not covered by ice has essentially no vegetation cover. Highlands are regions that lie at lower (nonpolar) latitudes but have such a high elevation that they have polarlike climates. When you enter a region above the treeline, you have entered a highland polar climate.

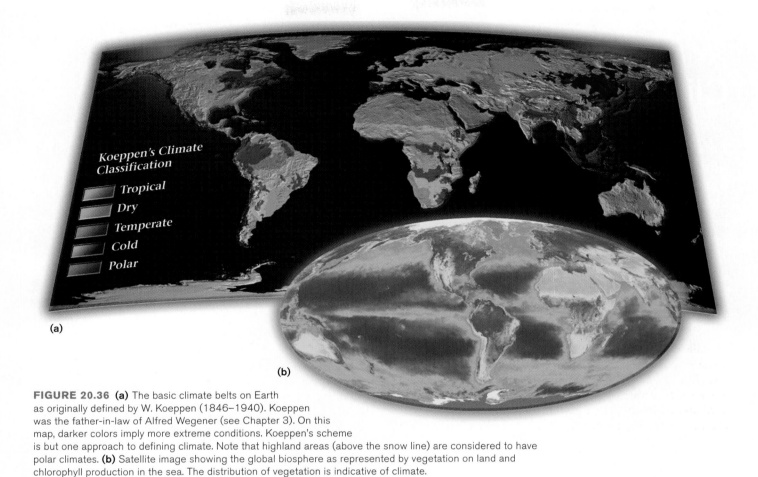

FIGURE 20.36 (a) The basic climate belts on Earth as originally defined by W. Koeppen (1846–1940). Koeppen was the father-in-law of Alfred Wegener (see Chapter 3). On this map, darker colors imply more extreme conditions. Koeppen's scheme is but one approach to defining climate. Note that highland areas (above the snow line) are considered to have polar climates. **(b)** Satellite image showing the global biosphere as represented by vegetation on land and chlorophyll production in the sea. The distribution of vegetation is indicative of climate.

rises over central Asia, a pronounced low-pressure cell develops, and the intertropical convergence zone moves north. When this happens, warm air flows northward from the Indian Ocean, bringing with it substantial moisture, and the summer rains begin. Rainfalls are especially heavy on the southern slope of the Himalayas, because orographic lifting leads to the production of huge cumulonimbus clouds (▶Fig. 20. 37b).

Long before the modern science of meteorology became established, fishermen from Peru and Ecuador who ventured into the coastal waters west of South America knew that in late December the fish population that provided their livelihood diminished. Because of the timing of this event, it came to be known as **El Niño**, Spanish for "the Christ child." Why did the fish vanish? Fish are near the top of a food chain that begins with plankton, which live off nutrients in the water. These nutrients increase when cold water upwells from the deep along the coast of South America. During El Niño, warm water currents flow eastward from the central Pacific, and the

cold, nutrient-rich water that supports the marine food chain remains at depth. With fewer nutrients, there are fewer plankton, and without the plankton, the fish migrate elsewhere.

To understand why El Niño occurs, we need to look at atmospheric flow and related surface ocean currents in the equatorial Pacific (▶Fig. 20.38a, b). When El Niño is not in progress, a major equatorial low-pressure cell exists in the western Pacific over Indonesia and Papua New Guinea, while a high-pressure cell forms over the eastern Pacific, along the coast of equatorial South America. This geometry means that air rises in the western Pacific, flows east, sinks in the eastern Pacific, and then flows west at the surface. The easterly surface winds blow warm surface water westward, so that it pools in the western Pacific. Cold water from the deep ocean rises along South America to replace the warm water that moved west. It is this rising cold water that brings nutrients to the surface. During El Niño, the low-pressure cell moves eastward over the central Pacific, and a high-pressure cell develops over

See for yourself . . .
Climate Belts of the Earth

This Geotour is different from others. Here, we focus on the overall character of vegetation as a function of latitude-controlled climate belts. We begin our tour at the south end of Africa, and jump north roughly along the 20°E longitude line. The thumbnail images provided on this page are only to help identify tour sites. Go to wwnorton.com/studyspace to experience this flyover tour.

30°S: Subtropical Desert, South Africa
(Lat 30°00'00.75"S, Long 19°47'37.67"E)

At these coordinates, from an elevation of 8 km (5 miles), you sense the dryness of the Kalahari Desert that occupies the southern end of Africa (Image G20.1).

G20.1

10°S: Transitional Tropical, Angola
(Lat 9°59'59.40"S, Long 20°00'23.46"E)

The view from 30 km (19 miles), at these coordinates, shows the widespread, but not dense, vegetation of regions that lie at the borders between the steppe and the tropical jungle (Image G20.2).

G20.2

0°: Tropical, Congo
(Lat 0°11'11.78"N, Long 14°23'59.54"E)

In equatorial Congo, you are in the heart of a rainforest. From 30 km (19 miles), you see nothing but trees (Image G20.3). You can see a dendritic drainage network in the jungle.

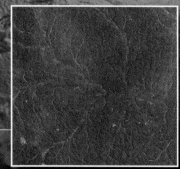
G20.3

10°N: Transitional Tropical, Southern Chad
(Lat 10°00'07.97"N, Long 20°00'21.10"E)

A journey north from the last stop leaves the jungle and enters the semi-arid steppe north of the equator. In southern Chad, from an altitude of 50 km (31 miles), you see moderate vegetation, and streams that do not fill their channels (Image G20.4).

G20.5

24°N: Subtropical Desert, Libya (Lat 24°22'06.32"N, Long 20°35'31.08"E)

At these coordinates, the view from an elevation of 80 km (50 miles) over south-eastern Libya shows the edge of a sand sea of the subtropical Sahara Desert (Image G20.5). You can see large sand dunes aligned with the wind direction. Note how rocky hills act as wind blocks and divert the sand.

40°N: Temperate, Albania (Lat 39°53'42.23"N, Long 20°0'40.18"E)

Your voyage north now jumps across the Mediterranean Sea, into Albania. At these coordinates, from an elevation of 10 km (6 miles), you can see a dry temperate region along the coast (Image G20.6). Farm fields make the landscape, in places, look like a checkerboard.

G20.6

G20.7

50°N: Temperate, Poland (Lat 49°59'45.66"N, Long 20°00'01.43"E)

Fly to the coordinates provided, and zoom to 10 km (6 miles). You are looking down on a temperate agricultural region of Poland (Image G20.7). Prior to civilization, this area was covered by forest.

70°N: Transitional Polar, Norway (Lat 69°51'20.44"N, Long 18°54'38.48"E)

At these coordinates, from an elevation of 15 km (9 miles), you see a landscape at the fringe of the polar climate belt (Image G20.8). Snow remains on hilltops down to fairly low elevations, even during the summer, and only sparse vegetation covers the landscape.

G20.8

80°N: Polar, Greenland (Lat 79°58'29.66"N, Long 19°20'50.08"W)

We have to shift westward to find land at this high latitude. So when you fly to the coordinates provided, and zoom to 50 km (31 miles), you are looking down on coastal Greenland (Image G20.9). Here, glaciers flow down to sea level, sea ice remains frozen, and hardly any vegetation survives.

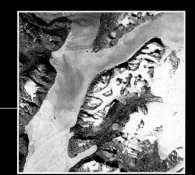

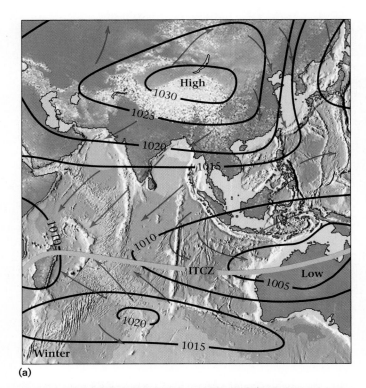

(a)

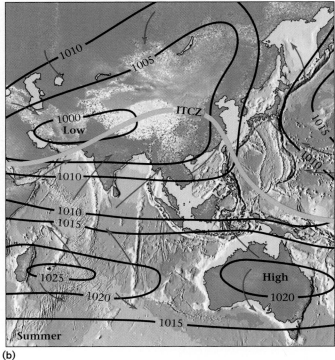

(b)

FIGURE 20.37 In the monsoonal climate of Asia, each year can be divided into a dry season and a wet season. **(a)** During winter, the dry season, a large high-pressure cell develops over central Asia, and the intertropical convergence zone lies well south of Asia. **(b)** During summer, the wet season, a low-pressure cell develops over Asia, so warm, moisture-laden air from the Indian Ocean flows landward. Orographic lifting along the Himalayas leads to cloud formation and intense rainfall.

Indonesia; so two convective cells develop. As a result, surface winds start to blow east in the western Pacific, driving warm surface water back to South America. This warm surface water prevents deep cold water from rising, sending the fish away. In effect, pressure cells oscillate back and forth across the Pacific, an event now called the **southern oscillation**.

El Niño gained world notoriety in late 1982 and early 1983 when a particularly large low-pressure cell developed in the eastern part of the Pacific. As a result, the jet streams stayed farther north than is typical. In effect, El Niño caused a temporary climate

> **Take-Home Message**
>
> Due to variations in solar heating, distance from oceans, currents, topography, and other reasons, Earth hosts many different climates (e.g., tropical, dry, temperate, cold, and polar). Certain regions of Earth have seasonal monsoons, or are affected by El Niño.

FIGURE 20.38 El Niño exists because of a change in winds and currents in the central Pacific. **(a)** During the times between El Niño, a low-pressure cell lies over the western Pacific, and surface trade winds blow to the west. These winds drive warm surface water westward, so cold water rises along the western coast of South America to replace it. **(b)** During El Niño, the low-pressure cell moves eastward, and the westward flow stops, so that cold water no longer upwells.

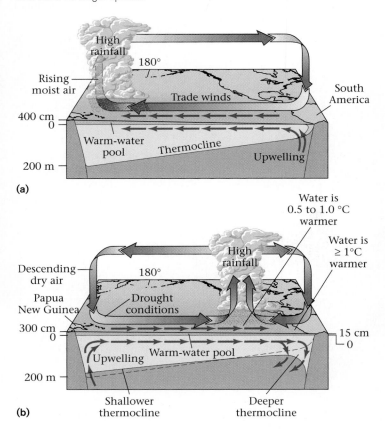

change worldwide. Drought conditions persisted in the normally rainy western Pacific, while unusually heavy rains drenched western South America. In North America, rains swamped the southern United States and storms battered California, winters were warmer than usual in Canada, and snowfalls were heavier in the Sierra Nevada, leading to spring floods.

Climatologists have been working intensely to understand the periodicity of El Niño. It is clear that strong El Niños take place around once every four years, with even stronger ones possibly happening at other intervals.

Chapter Summary

- The early atmosphere of the Earth contained high concentrations of water, carbon dioxide, and sulfur dioxide, gases erupted by volcanoes.

- After the oceans formed, much of the carbon dioxide was removed from the atmosphere. When photosynthetic organisms evolved, they produced oxygen, and the concentration of this gas gradually increased.

- Air consists mostly of nitrogen (78%) and oxygen (21%). Several other gases occur in trace amounts. The atmosphere also contains aerosols. Air pressure decreases with elevation. Thus, 90% of the air in the atmosphere occurs below an elevation of 16 km.

- When air rises, it expands and cools, a process called adiabatic cooling. If air is compressed, it heats up, a process called adiabatic heating.

- Air generally contains water. The ratio between the measured water content and the maximum possible amount of water that the air can hold is its relative humidity.

- The atmosphere is divided into layers, separated from each other by pauses. In the lowest layer, the troposphere, temperature decreases with elevation. The troposphere convects—its air movement causes weather. The other layers are the stratosphere, the mesosphere, and, at the top, the thermosphere.

- Air circulates on two scales, local and global. Winds blow because of pressure gradients: air moves from regions of higher pressure to regions of lower pressure.

- High latitudes receive less solar energy than low latitudes. This contrast initiates convection in the atmosphere. Because of the Coriolis effect, air moving north from the equator to the pole deflects to the east, and in each hemisphere, three convection cells develop (the Hadley, Ferrel, and polar cells).

- Prevailing surface winds—the northeast trade winds, the surface westerlies, and the polar easterlies—develop because of circulation in global convection cells.

- Air pressure at a given latitude decreases from the equator to the pole, causing a poleward flow of air. In the Northern Hemisphere, the Coriolis effect deflects this flow to generate high-altitude westerlies. These winds, where particularly strong, are known as jet streams.

- "Weather" refers to the temperature, air pressure, wind speed, and relative humidity at a given location and time. Weather reflects the interaction of air masses. The boundary between two air masses is a front.

- Air sinks in high-pressure air masses and rises in low-pressure air masses. Because of the Coriolis effect, the air begins to rotate around the center of the mass as a consequence, generating cyclones or anti-cyclones.

- Clouds, which consist of tiny droplets of water or tiny crystals of ice, form when the air is saturated with water and contains condensation nuclei on which water condenses.

- Thunderstorms begin when cumulonimbus clouds grow large. Friction between air and water molecules separates positive and negative charges. Lightning flashes when a giant spark jumps across the charge separation.

- Tornadoes, rapidly rotating funnel-shaped clouds, develop in violent thunderstorms. Nor'easters are large storms associated with wave cyclones. Hurricanes, huge rotating storms, originate over oceans where the water temperature exceeds 27 °C.

- "Climate" refers to the typical range of conditions, the nature of seasons, and the possible weather extremes of a region over a long time (say 30 years). Climate is controlled by latitude, altitude, proximity to water, ocean currents, orographic barriers, and high- or low-pressure zones. Climate classes can be recognized by the vegetation they support.

Geopuzzle Revisited

The atmosphere initially formed from gas emitted by volcanoes. The composition of the atmosphere evolved over time, due to the formation of the oceans and later, due to the evolution of photosynthesis. Climate is the overall condition of temperature and humidity, as well as their variability, averaged over time. Climate largely reflects variation in solar radiation, but also topography and distance from the ocean. Weather refers only to the condition in the atmosphere at a given time, and may change rapidly at a location.

- Monsoonal climates occur where there is a seasonal shift in the wind direction. El Niño is a temporary shift in weather conditions triggered by shifts in the position of high- and low-pressure cells in the Pacific.

Key Terms

acid rain (p. 695)
adiabatic cooling, heating (p. 696)
aerosols (p. 695)
air (p. 693)
air mass (p. 706)
air pressure (p. 696)
anticyclone (p. 707)
atmosphere (p. 693)
Bergeron process (p. 709)
cirrus (p. 710)
climate (pp. 693, 719)
collision and coalescence (p. 708)
condensation nuclei (p. 708)
convergence zone (p. 701)
cumulus (p. 710)
cyclone (p. 707)
dewpoint temperature (p. 698)
divergence zone (p. 701)
doldrums (p. 704)
El Niño (p. 723)
Enhanced Fujita scale (p. 714)
Ferrel cell (p. 701)
fog (p. 707)
front (p. 706)
Hadley cell (p. 701)
hail (hailstones) (p. 711)
hurricane (p. 716)
insolation (p. 701)
ionosphere (p. 699)

isobar (p. 700)
isotherm (p. 719)
jet stream (p. 704)
lifting mechanism (p. 708)
lightning flash (p. 712)
mesosphere (p. 699)
monsoon (p. 722)
nor'easter (p. 715)
orographic barrier (p. 719)
ozone (p. 694)
pauses (p. 698)
polar cell (p. 702)
polar front (p. 701)
prevailing winds (p. 702)
relative humidity (p. 698)
Saffir-Simpson scale (p. 716)
southern oscillation (p. 726)
storm (p. 711)
storm surge (p. 718)
stratosphere (p. 698)
stratus (p. 710)
thermosphere (p. 699)
thunder (p. 712)
tornado (p. 713)
trade winds (p. 704)
troposphere (p. 698)
wave cyclone (p. 707)
weather (p. 693)
weather system (p. 705)
wind (p. 693)

Review Questions

1. Describe the stages in the formulation and evolution of Earth's atmosphere. Where does the ozone in the atmosphere come from, and why is it important?

2. Describe the composition of air (considering both its gases and its aerosols). Why are trace gases important?

3. How does air pressure change with elevation? Does the density of the atmosphere also change with elevation? Explain why or why not.

4. Describe the atmosphere's structure from base to top. What characteristics define the boundaries between layers?

5. What is the relative humidity of the atmosphere? What is the latent heat of condensation, and what is its relevance to a thunderstorm or hurricane?

6. Explain the relation between the wind and variations in air pressure.

7. Why do changes in atmospheric temperature depend on latitude and the seasons? Why does global circulation break into three distinct convection belts in each hemisphere, separated by high-pressure or low-pressure belts?

8. Why do prevailing winds develop at the Earth's surface? Why do the jet streams form?

9. Explain the origin of cyclones and anticyclones, and note their relationship to high-pressure and low-pressure air masses. What is a mid-latitude cyclone?

10. How does a cold front differ from a warm front and from an occluded front?

11. Why do clouds form? (Include a discussion of lifting mechanisms.) What are the basic categories of clouds?

12. Under what conditions do thunderstorms develop? What provides the energy that drives clouds to the top of the troposphere? How do meteorologists explain lightning?

13. What conditions lead to the formation of a tornado? Where do most tornadoes appear?

14. Describe the stages in the development of a hurricane. Describe a hurricane's basic geometry.

15. What factors control the climate of a region? What special conditions cause monsoons? El Niño?

On Further Thought

1. When Columbus set sail from Spain, his route first took him southward to the Canary Islands, and then westward. His landfall in the new world, on an island in the Bahamas, was much further to the south than his point of departure. Why?

2. Explain why *large* rain forests occur in equatorial Africa (the Congo) and in equatorial South America (the Amazon). Why do *small* rain forests occur along the coastal regions of Washington State (northwestern North America) and along the coast of southern Chile (southwestern South America?)

3. Much more snow falls on the eastern and southern shores of Lake Michigan, one of the Great Lakes, than on the west-

ern shore. Why does this phenomenon, which weather reporters refer to as "lake-effect snow," occur?

4. A typhoon is approaching the east coast of Asia. Weather forecasters have predicted where the eye of the storm will make landfall on a narrow, north-south-trending island. People in the path of the storm have been told to evacuate, but cannot head inland because they are on a narrow island. To escape the highest winds, should they move to the north or to the south of the eye?

Suggested Reading

Barry, R. G., R. J. Chorley, and N. J. Yokoi. 2003. *Atmosphere, Weather and Climate*. 8th ed. New York: Routledge.

Bluestein, H. B. 1999. *Tornado Alley: Monster Storms of the Great Plains*. New York: Oxford University Press.

Burroughs, W. J. 1999. *The Climate Revealed*. Cambridge: Cambridge University Press.

Davies, P. 2000. *Inside the Hurricane*. New York: Holt.

Graedel, T. E., and P. J. Crutzen. 1993. *Atmospheric Change: An Earth System Perspective*. New York: Freeman.

Grazulis, T. P., and D. Flores. 2003. *The Tornado: Nature's Ultimate Windstorm*. Norman: University of Oklahoma Press.

Hartmann, D. 1994. *Global Physical Climatology*. New York: Academic Press.

Holton, J. 2004. *An Introduction to Dynamic Meteorology*. 4th ed. New York: Academic Press.

Larson, E. 2001. *Isaac's Storm: A Man, a Time, and the Deadliest Hurricane in History*. New York: Crown, Random House.

Lutgens, F. K., E. J. Tarbuck, and D. Tasa. 2003. *The Atmosphere: An Introduction to Meteorology*. 9th ed. Englewood Cliffs, N.J.: Prentice-Hall.

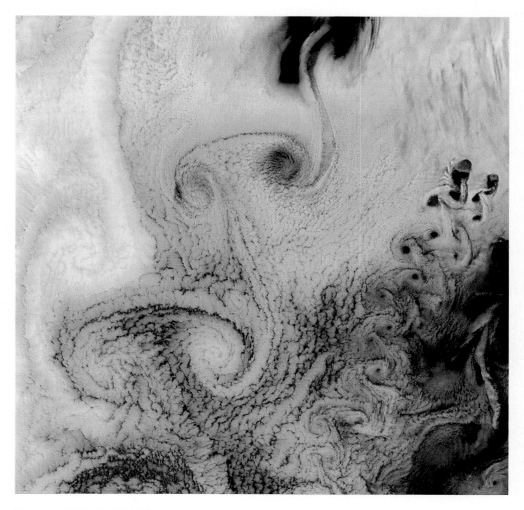

THE VIEW FROM SPACE A satellite view shows eddies in the atmosphere forming where the Aleutian Islands (Alaska) interfere with air flow.

Dry Regions:
The Geology of Deserts

Geopuzzle

Hollywood movies typically portray deserts as endless seas of hot sand, in which nothing at all grows. Is this portrayal accurate?

In a desert, very little vegetation grows because it rarely rains. But not all deserts are seas of sand. Here in the Sonoran Desert of northwestern Mexico, a saguaro cactus stands guard over a landscape of stony plains and barren rock cliffs.

The bare hills are cut out with sharp gorges, and over their stone skeletons scanty earth clings . . . A white light beat down, dispelling the last tract of shadow, and above hung the burnished shield of hard, pitiless sky.

— *Clarence King (1842–1901; First director of the U.S. Geological Society, describing a desert)*

21.1 INTRODUCTION

For generations, nomadic traders have saddled camels to traverse the Sahara Desert in northern Africa (▶Fig. 21.1) The Sahara, the world's largest desert, receives so little rainfall that it has hardly any surface water or vegetation. So camels must be able to walk for up to 3 weeks without drinking or eating. They can survive these journeys because they sweat relatively little, thereby conserving their internal water supply; they have the ability to metabolize their own body fat (up to 20 kg of fat make up the animal's hump alone) to produce new water; and they can withstand severe dehydration (loss of water). Most mammals die after losing only 10 to 15% of their body fluid, but camels can survive 30% dehydration with no ill effects. Camels do get thirsty, though. After a marathon trek across the desert, a camel may guzzle up to 100 liters of water in less than 10 minutes.

The survival challenges faced by a camel emphasize that deserts are lands of extremes—extreme dryness, heat, cold, and, in some places, beauty. Desert vistas include everything

FIGURE 21.1 Camels, as in this Sahara caravan in Mauritania, survive harsh desert conditions by storing water in fat reserves and by sweating little, if at all.

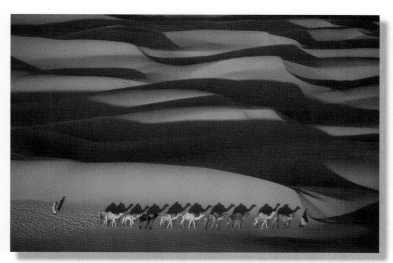

from sand seas to sagebrush plains, cactus-covered hills to endless stony pavements. Although less populated than other regions on Earth, deserts cover a significant percentage (about 25%) of the land surface, and thus constitute an important component of the Earth System. In this chapter, we take a look at the desert landscape. We learn why deserts occur where they do, and how erosion and deposition shape their surface. We conclude by exploring life in the desert and by examining the problem of desertification, the gradual transformation of temperate lands into desert.

21.2 WHAT IS A DESERT?

Formally defined, a **desert** is a region that is so arid (dry) that it contains no permanent streams, except for rivers that bring water in from temperate regions elsewhere, and supports vegetation on no more than 15% of its surface. In general, desert conditions exist where less than 25 cm of rain falls per year, on average. But rainfall alone does not determine the aridity of a region. Aridity also depends on rates of evaporation and on whether rainfall occurs only sporadically or more continuously during the year. If all the rain in a region drenches the land during isolated downpours only once every few years, the region becomes a desert, because the intervals of drought last so long that plants and permanent streams cannot survive. Similarly, if high temperatures and dry air cause evaporation rates from the ground to exceed the rate at which rainfall wets the ground, then the region becomes a desert even if it receives more than 25 cm per year of rain.

Note that the definition of a desert depends on a region's aridity, *not* on its temperature. Geologists distinguish between cold deserts, where temperatures generally stay below about 20°C for the year, and hot deserts, where summer daytime temperatures exceed 35°C. Cold deserts exist at high latitudes where the Sun's rays strike the Earth obliquely and thus don't provide much energy, at high elevations where the air is too thin to hold much heat, or in lands adjacent to cold oceans, where the cold water absorbs heat from the air above. Hot deserts develop at low latitudes where the Sun's rays strike the desert at a high angle, at low elevations where dense air can hold a lot of heat, and in regions distant from the cooling effect of cold ocean currents. The hottest recorded temperatures on Earth occur in low-latitude, low-elevation deserts—58°C (136°F) in Libya and 56°C (133°F) in Death Valley, California.

The ground surface absorbs so much heat in hot deserts that a layer of very hot air (up to 77°C, or 170°F) forms just above the ground. This layer refracts sunlight, creating a mirage, a wavering pool of light, on the ground. Mirages make the dry sand of a desert wasteland look like a shimmering lake and distant mountains look like islands

FIGURE 21.2 The shimmering in this desert mirage may look like water, but it's not. Mirages result from the interaction of light with a thin layer of hot air just above the ground surface.

heavy rains water accumulates into flash floods of immense power. Rocks and sediment do not undergo rapid chemical weathering, and humus (organic matter) does not collect on the ground surface. Thus, the desert land surface consists of any of the following: exposed bedrock, accumulations of clasts, relatively unweathered sediment, precipitated salt, or windblown sand. Overall, therefore, desert landscapes tend to be harsher and more rugged than temperate or tropical ones. If eastern North America were a desert, the Appalachians would not be gentle, forested hills but rather would consist of stark, rocky ridges.

> **Take-Home Message**
>
> Deserts are so dry that vegetation covers less than 15% of their surface. A desert is a desert because of aridity, not heat—some deserts are quite cold. Because of the lack of vegetation, deserts tend to have harsh, rugged landscapes.

(▶Fig. 21.2). Heat also contributes to aridity by increasing the rate of evaporation. In fact, evaporation rates in hot deserts may be so great that even when it rains, the ground stays dry because raindrops evaporate in mid-air. But even the hottest of hot deserts become cold at night: because of the dryness of the air, the lack of cloud cover, and the lack of foliage, deserts radiate their heat back into space at night. As a consequence, the air temperature at the ground surface in a desert may change by as much as 80°C in a single day.

The aridity of deserts causes weathering, erosion, and depositional processes to be different from those in temperate or tropical regions. Without plant cover, rain and wind batter and scour the ground, and during particularly

21.3 TYPES OF DESERTS

Each desert on Earth has unique characteristics of landscape and vegetation that distinguish it from others. Geologists group deserts into five different classes, based on the environment in which the desert forms (▶Fig. 21.3).

- *Subtropical deserts*: Subtropical deserts (e.g., the Sahara, Arabian, Kalahari, and Australian) form because of the pattern of convection cells in the atmosphere (▶Fig. 21.4; see also Chapter 20). At the equator, the air becomes warm and humid, for sunlight is intense and water rapidly evaporates from the ocean. The hot, moisture-laden air rises to great heights above the equator. As this air

FIGURE 21.3 The global distribution of deserts. Note that the largest deserts lie in the subtropical belts.

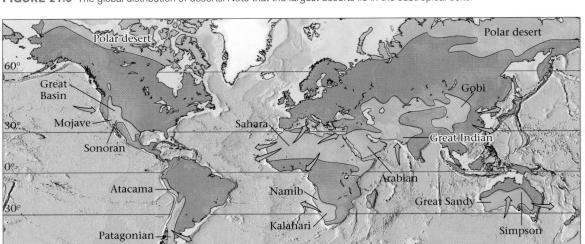

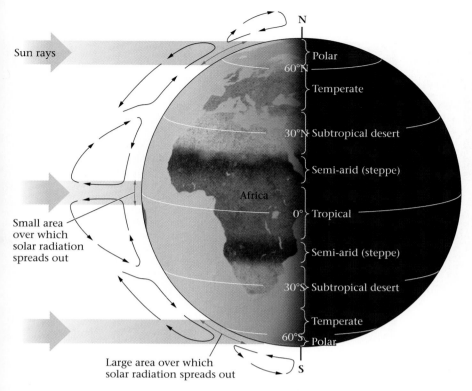

FIGURE 21.4 Rising air at the equator loses its moisture by raining over rain forests. When the air sinks over the subtropics, it warms and absorbs water. Thus, rainfall rarely occurs in the subtropics.

rises, it expands and cools, and can no longer hold so much moisture. Water condenses and falls in downpours that feed the lushness of the equatorial rain forest. The now-dry air high in the troposphere spreads laterally north or south. When this air reaches latitudes of 20° to 30°, a region called the subtropics, it has become cold and dense enough to sink. Because the air is dry, no clouds form, and intense solar radiation strikes the Earth's surface. The sinking, dry air condenses and heats up, soaking up any moisture present. In the regions swept by this hot air on its journey back to the equator, evaporation rates greatly exceed rainfall rates.

In subtropical deserts that border the sea, episodes of high sea level flood coastal areas. The eventual evaporation of stranded seawater leaves broad regions, called "sabkhas," where a salt crust covers a mire of organic-rich mud.

- *Deserts formed in rain shadows*: As air flows over the sea toward a coastal mountain range, the air must rise (▶Fig. 21.5). As the air rises, it expands and cools. The water it contains condenses and falls as rain on the seaward flank of the mountains, nourishing a coastal rain forest. When the air finally reaches the inland side of the mountains, it has lost all its moisture and can no longer provide rain. As a consequence, a **rain shadow** forms, and the land beneath the rain shadow becomes a desert. A rain-shadow desert can be found east of the Cascade Mountains in Washington.

- *Coastal deserts formed along cold ocean currents*: Cold ocean water cools the overlying air by absorbing heat, thereby decreasing the capacity of the air to hold moisture. For example, the cold Humboldt Current, which carries water northward from Antarctica to the western coast of South America, absorbs water from the breezes that blow east, over the coast. Thus, rain rarely falls on the coastal areas of Chile and Peru. As a result, this region hosts a desert landscape, including one of the driest deserts in the world, the Atacama (▶Fig. 21.6a–d). Portions of this narrow (less than 200-km-wide) desert, which lies between the Pacific coast on the west and the Andes on the east, received no rain at all between 1570 and 1971.

- *Deserts formed in the interiors of continents:* As air masses move across a continent, they lose moisture by dropping rain, even in the absence of a coastal mountain range. Thus, when an air mass reaches the interior of a

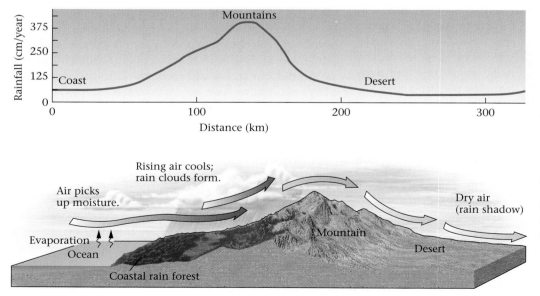

FIGURE 21.5 Moist air, when forced to rise by mountains, cools. As this happens, the moisture condenses and rain falls, nourishing coastal rain forests, so by the time the air reaches the inland side of the mountains, it no longer holds enough moisture to rain. Deserts form in the rain shadow of mountains. The graph shows the variation of rainfall from the coast, across a mountain range, and into the rain shadow.

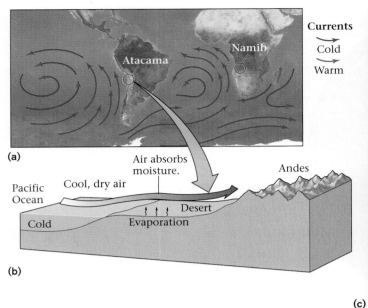

(a)

(c)

(d)

FIGURE 21.6 (a) Currents bringing cold water up from the Antarctic cool the air along the southwestern coasts of South America and Africa. (b) The cool, dry air absorbs moisture from the adjacent coastal land, keeping it dry, so coastal deserts form. (c) Satellite view of the west coast of South America. Note the Atacama (coastal) Desert (AD). (d) The Atacama Desert of South America is the driest place in the world.

pattern of air circulation means that the air flowing over these regions is dry) and, in part, for the same reason that coastal areas along cold currents are dry (cold air holds little moisture).

The distribution of deserts around the world through geologic time reflects the process of plate tectonics, for plate movements determine the latitude of land masses, the position of land masses relative to the coast, and the proximity of land masses to a mountain range. Because of continental drift, some regions that were deserts in the past are temperate regions now, and vice versa.

particularly large continent such as Asia, it has grown quite dry, so the land beneath becomes arid. The largest example of such a continental-interior desert, the Gobi, lies in central Asia, over 2,000 km away from the nearest ocean.

- *Deserts of the polar regions*: So little precipitation falls in Earth's polar regions (north of the Arctic Circle, at 66°30′N, and south of the Antarctic Circle, at 66°30′S) that these areas are, in fact, arid. Polar regions are dry, in part, for the same reason that the subtropics are dry (the global

Take-Home Message

Deserts form in many settings. The largest occur in subtropical or polar latitudes, due to global patterns of atmospheric circulation. Other examples form along coasts bordered by cold currents, in the rain shadows of mountains, or in continental interiors.

21.4 WEATHERING AND EROSIONAL PROCESSES IN DESERTS

Without the protection of foliage to catch rainfall and slow the wind, and without roots to hold regolith in place, rain and wind can attack and erode the land surface of deserts. The result, as we have noted, is that hillslopes are typically bare, and plains can be covered with stony debris or drifting sand.

Weathering and Soil Formation in Deserts

In the desert, as in temperate climates, physical weathering happens primarily when joints (natural fractures) split rock into pieces. Joint-bounded blocks eventually break free of bedrock and tumble down slopes, fragmenting into smaller pieces as they fall. In temperate climates, thick soil forms over bedrock, so it lies buried beneath the surface. In deserts, however, the jointed bedrock commonly remains exposed at the surface of hillslopes, creating rugged, rocky escarpments.

Chemical weathering happens more slowly in deserts than in temperate or tropical climates, because less water is available to react with rock. Still, rain or dew provides enough moisture for *some* weathering to occur. This water seeps into rock and leaches (dissolves and carries away) calcite, quartz, and various salts. Leaching effectively rots the rock by transforming it into a poorly cemented aggregate. Over time, the rock will crumble and form a pile of unconsolidated sediment, susceptible to transport by water or wind. If water seeping into the rock contains dissolved salts, salt crystals may grow in the rock when the water dries out. The growth of such crystals pushes neighboring crystals apart and can also weaken the rock.

Although enough rain falls in deserts to leach chemicals out of rock, there is insufficient water to flush the dissolved minerals entirely away. Thus, when the water percolates down and dries up, minerals precipitate in regolith below the surface. If calcite precipitates, it cements loose grains together, forming solid calcrete (▶Fig. 21.7a). The growth of hard masses such as calcrete can occur rapidly enough to incorporate tools abandoned by prospectors.

Shiny **desert varnish**, a dark, rusty brown coating of iron oxide, manganese oxide, and clay, covers the surface of many rock varieties in deserts (▶Fig. 21.7b). Desert varnish was once thought to form when water from rain or dew seeped into a rock, dissolved iron and magnesium ions, and carried the ions back to the surface of the rock by capillary action. More recent studies, however, have shown that desert varnish is not derived from the rock below, but actually forms when windborne dust settles on the surface of the rock; in the presence of moisture, microorganisms (bacteria) extract elements from the dust and transform it into iron or manganese oxide. The oxides bind together clay flakes. Such varnish won't form in humid climates, because rain washes the ions away too fast.

Desert varnish takes a long time to form. In fact, the thickness of a desert varnish layer provides an approximate estimate of how long a rock has been exposed at the ground surface. In past centuries, Native Americans used desert-varnished rock as a medium for art: by chipping away the varnish to reveal the underlying lighter-colored rock, they were able to create figures or symbols on a dark background. The resulting drawings are called **petroglyphs** (▶Fig. 21.7c).

Because of the lack of plant cover in deserts, variations in bedrock color stand out. Locally derived soils typically retain the color of the bedrock from which they were derived. Slight variations in the concentration of iron, or in the degree of iron oxidation, in adjacent beds result in spectacular color bands in rock layers and the thin soils derived from them. The Painted Desert of northern Arizona earned its name from the brilliant and varied hues of oxidized iron (▶Fig. 21.7d).

Water Erosion

Although rain rarely falls in deserts, when it does come, it can radically alter a landscape in a matter of minutes. Since deserts lack plant cover, rainfall, sheetwash, and stream flow all are extremely effective agents of erosion. It may seem surprising, but water generally causes more erosion than does wind in most deserts. (▶Fig. 21.8a).

Water erosion begins with the impact of raindrops, which eject sediment into the air. On a hill, the ejected sediment lands downslope, and thus during a rain, sediment gradually migrates to lower elevations. The ground quickly becomes saturated with water during a heavy rain, so water starts flowing across the surface, carrying the loose sediment with it. Within minutes after a heavy downpour begins, dry stream channels fill with a turbulent mixture of water and sediment, which rushes downstream as a flash flood. When the rain stops, the water sinks into the stream bed's gravel and disappears—such streams are called intermittent, or ephemeral streams (see Chapter 17). Because of the relatively high viscosity of the water (owing to its load of suspended sediment) and the velocity and turbulence of the flow, flash floods in deserts cause intense erosion—they undercut cliffs and transport huge boulders downstream. As rocks roll and tumble along, they strike each other and shatter, creating smaller pieces that can be carried still farther. Between floods, the stream floor consists of gravel littered with boulders.

Flash floods carve steep-sided channels into the ground. Scouring of bedrock walls by sand-laden water may polish the walls and create grooves. Dry stream channels in desert regions of the western United States are called **dry washes**, or **arroyos**, and in the Middle East and North Africa they are called wadis (▶Fig. 21.8b).

Wind Erosion

In temperate and humid regions, plant cover protects the ground surface from the wind, but in deserts, the wind has direct access to the ground. In hot deserts, gusts of hot air feel like blasts from a furnace. Wind, just like flowing water, can carry sediment both as suspended load and as bed load. **Suspended load** (fine-grained sediment such as dust and silt held in suspension) floats in the air and moves with it (▶Fig. 21.9). The suspended sediment can be carried so high into the atmosphere (up to several kilometers above the Earth's surface) and so far downwind (tens to hundreds of kilometers) that it may move completely out of its source region. In some cases, tiny vortices can churn up dust. In general, these vortices are very small (centimeters to meters high). But, in some cases, they become "dust devils" up to 100 m high, and look like miniature tornadoes. Cars driving down dirt roads in deserts have the same effect; they break dust free from the ground and generate plumes of suspended sediment.

FIGURE 21.7 **(a)** Calcrete forms where calcium carbonate cements gravel clasts together. This example occurs in western Arizona. **(b)** Desert varnish coats folded rock layers in Arizona, giving the outcrop a dark hue. **(c)** Desert varnish, composed of iron oxide, manganese oxide, and clay, is a dark coating on desert rock surfaces. Native American artists created petroglyphs, images on the rock surface, by chipping through the varnish to reveal the lighter rock beneath. **(d)** In the Painted Desert of Arizona, the different colors of the rock layers are due, in part, to the oxidation state of iron.

(a)

(b)

(c)

(d)

FIGURE 21.8 **(a)** These hills in the desert near Las Vegas, Nevada, are bone dry, but their shape reflects erosion by water. Note the numerous stream channels. **(b)** Gravel and sand are left behind on the floor of a dry wash after a flash flood. The wash has steep walls because downcutting happens so fast.

(a)

(b)

FIGURE 21.9 Dust clouds form in deserts when turbulent air carries very fine sediment in suspension.

Moderate to strong winds can roll and bounce sand grains along the ground, a process called **saltation** (▶Fig. 21.10). Saltating sand constitutes the wind's **surface load**. Saltation begins when turbulence caused by wind shearing along the ground surface lifts sand grains. The grains move downwind, following an asymmetric, arch-like trajectory. Eventually, they return to the ground, where they strike other sand grains, causing the new grains to bounce up and drift or roll downwind. The collisions between sand grains make the grains rounded and frosted. Saltating grains generally rise no more than 0.5 m. But where sand bounces on bedrock during a desert sandstorm, the grains may rise 2 m, and can strip the paint off a car.

The size of clasts that wind can carry depends on the wind velocity. Wind, therefore, does an effective job of sorting sediment, sending dust-sized particles skyward and sand-sized particles bouncing along the ground, while pebbles and larger grains remain behind. In some cases, wind carries away so much fine sediment that pebbles and cobbles become concentrated at the ground surface (▶Fig. 21.11). An

accumulation of coarser sediment left behind when fine-grained sediment blows away is called a **lag deposit**.

In many locations, the desert surface resembles a tile mosaic in that it consists of separate stones that fit together tightly, forming a fairly smooth surface layer above a soil composed of silt and clay. Such natural mosaics constitute **desert pavement** (▶Fig. 21.12a, b). Typically, desert varnish coats the top surfaces of the stones forming desert pavement. Geologists have come up with several explanations for the origin of desert pavements. Traditionally, pavements were thought to be lag deposits, formed when wind blows away the fine sediment between clasts, so that the clasts can settle down and fit together. More recently, researchers suggest that pavements form as wind-blown dust slowly sifts down onto the stones, and then washes down between the stones. In this model, the pavement is "born at the surface," meaning that the stones forming the pavement were never buried, but have been progressively lifted up as sediment collects and builds up beneath (▶Fig. 21.12c–e). Over time, the rocks at the surface crack, perhaps due to extreme heating by the desert sun. Sheetwash, during downpours, may wash away fine sediment between fragments, and when soils dry and shrink between storms, the clasts settle together, locking into a stable, jigsaw-like arrangement.

Desert pavements are remarkably durable and can last for hundreds of thousands of years if they are left alone. But like many features of the desert, they can be disrupted by human activity. For example, people driving vehicles across the pavement indent and crack its surface, making it susceptible to erosion. In parts of Arizona, vast desert pavements have become parking lots for campers who migrate to the desert in motor homes for the winter season, hoping to escape the snows of the north.

Just as sandblasting cleans the grime off the surface of a building, windblown sand and dust grind away at surfaces in the desert. Over long periods of time, such wind abrasion creates smooth faces, or facets, on pebbles, cobbles, and boulders. If a rock rolls or tips relative to the prevailing wind

FIGURE 21.11 The progressive development of a lag deposit. Pebbles in a desert are distributed through a matrix of finer sediment. With time, wind blows the finer sediment away, and the pebbles concentrate on the ground surface and may settle together to create a lag deposit. The deposit acts like armor, protecting the substrate from further erosion.

FIGURE 21.10 During saltation, sand grains roll and bounce along the ground surface. As they bounce, they follow parabolic paths.

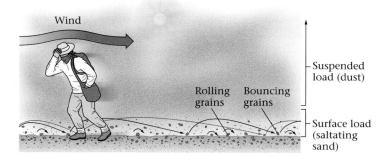

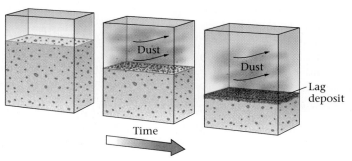

(a)

— Desert
pavement

— Soil

— Old
alluvium

(b)

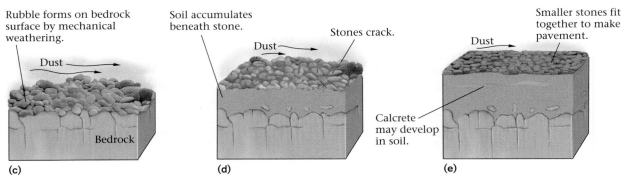

Rubble forms on bedrock
surface by mechanical
weathering.

Dust

Bedrock

(c)

Soil accumulates
beneath stone.

Dust

Stones crack.

(d)

Calcrete
may develop
in soil.

Smaller stones fit
together to make
pavement.

Dust

(e)

FIGURE 21.12 **(a)** A well-developed desert pavement in Arizona. **(b)** A trench dug in the Sonoran Desert of Arizona, showing how the desert pavement lies on top of a fine-grained soil. This pavement was lifted up from the surface of a now-buried alluvial fan. **(c)** Desert pavement forms in stages. First, loose pebbles and cobbles collect at the surface. These can be formed by mechanical weathering, as shown, or by deposition of alluvium. **(d)** Dust settles among stones and builds up a layer of soil beneath the stone layer. Stones crack into smaller pieces. **(e)** A durable, mosaic-like pavement has formed.

direction after it has been faceted on one side, or if the wind shifts direction, a new facet with a different orientation forms, and the two facets join at a sharp edge. Rocks whose surface has been faceted by the wind are faceted rocks, or **ventifacts** (▶Fig. 21.13a–d). Wind abrasion also gradually polishes and bevels down irregularities on a desert pavement and polishes the surfaces of desert-varnished outcrops, giving them a reflective sheen.

In places where a resistant layer of rock overlies a softer layer of rock, wind abrasion may create a formation consisting of a resistant block perched on an eroding

mushroom-like column of softer rock. These unusual features are called **yardangs** (▶Fig. 21.14). If a strong wind blows in only one direction, the yardangs become elongate, aligned with the wind direction.

Over time, in regions where the substrate consists of soft sediment, wind picks up and removes so much sediment that the land surface sinks. The process of lowering the land surface by wind erosion is called **deflation**. Shrubs can stabilize a small patch of sediment with their roots, so after deflation a forlorn shrub with its residual pedestal of soil stands isolated above a lowered ground surface (▶Fig. 21.15). In some places, the shape of the land surface twists the wind into a turbulent vortex that causes enough deflation to scour a deep, bowl-like depression called a blowout.

Satellite exploration in the last two decades shows that wind affects the desert-like surface of Mars just as it does

> **Take-Home Message**
>
> Desert soils don't contain much organic material, and may be rich in minerals that would be leached away in wetter climates. Water flows only rarely in deserts but nevertheless causes most of the erosion. Wind locally modifies landforms.

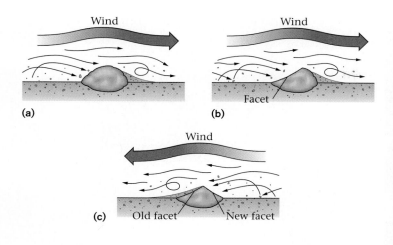

FIGURE 21.13 The progressive development of a ventifact. (a) Wind-blown sand and dust abrade the face of a rock. (b) Slowly, erosion carves a smooth surface called a facet. (c) Later, the wind shifts direction, and a new facet forms. The two facets join at a sharp edge. (d) Ventifacts formed by strong winds in Antarctica. The inset provides a closeup.

the deserts on Earth. In fact, during the Martian fall and spring, when differences in temperature between the ice-covered poles and the ice-free equator are greatest, strong winds blow from the poles to the equator, transporting so much dust that the planet's surface, as seen from Earth, visibly changes. At times, the entire planet becomes enveloped in a cloud of dust. Close-up images taken by spacecraft that have landed on Mars show that rocks have been abraded by saltating sand and that sand has accumulated on the lee side of the rocks (▶Fig. 21.16).

FIGURE 21.14 Yardangs are small landforms sculpted by the wind, where a resistant rock layer overlies a softer rock layer. They may be elongate, aligned with the wind direction. These yardangs formed in the Sahara of Egypt.

21.5 DEPOSITIONAL ENVIRONMENTS IN DESERTS

We've seen that erosion relentlessly eats away at bedrock and sediment in deserts. Where does the debris go? Below, we examine the various desert settings in which sediment accumulates.

Talus Aprons

Over time, joint-bounded blocks of rock break off ledges and cliffs on the sides of hills. Under the influence of gravity, the resulting debris tumbles downslope and accumulates as

FIGURE 21.15 Wind erosion in Death Valley has left bushes perched on little mounds, where the roots keep soil from blowing away.

FIGURE 21.16 This NASA photo shows the wind abrasion of clasts and small sand dunes on the surface of Mars. Clearly, the surface of the "red planet" now experiences desert conditions.

FIGURE 21.17 This talus apron along the base of a desert cliff formed from rocks that broke off and tumbled down the cliff.

a **talus apron** at the base of a hill. Talus aprons can survive for a long time in desert climates, so we typically see them fringing the bases of cliffs in deserts (▶Fig. 21.17). The angular clasts constituting talus aprons gradually become coated with desert varnish.

Alluvial Fans

Flash floods can carry sediment downstream in an ephemeral stream channel. When the turbulent water flows out into a plain at the mouth of a canyon, it spreads out over a broader surface and slows. As a consequence, sediment in the water settles out. In some cases, debris flows also emerge from the canyon and spread out. The resulting lenses of sediment cause the channel that has emerged from the mountains to subdivide into a number of sub-channels (distributaries) that diverge outward in a broad fan. The fan of distributaries spreads the sediment, or alluvium, out into a broad **alluvial fan**, a wedge- or apron-shaped pile of sediment (▶Fig. 21.18). Alluvial fans emerging from adjacent valleys may merge and overlap along the front of a mountain range, creating an elongate wedge of sediment called a **bajada**. Over long periods, the sediment of bajadas fills in adjacent valleys to depths of several kilometers.

Playas and Salt Lakes

Water from a flash flood may make it out to the center of an alluvium-filled basin, but if the supply of water is relatively small, it quickly sinks into the permeable alluvium without accumulating as a standing body of water. During a particularly large storm or an unusually wet spring, however, a temporary lake may develop over the low part of a basin. During drier times, such desert lakes evaporate entirely, leaving behind a dry, flat lake bed known as a **playa** (▶Fig. 21.19a). Over time, a smooth crust of clay and various salts (halite, gypsum, borax, and other minerals) accumulates on the surface of playas. Some of these minerals have industrial uses and thus have been mined. When it

FIGURE 21.18 The sediment constituting this alluvial fan in Death Valley, California, was carried to the mountain front during flash floods. Because the water slows down, it drops the sediment at the mountain front.

(a)

(b)

FIGURE 21.19 (a) A playa, a basin covered with clay and salt, develops where a shallow desert lake dries up. This one occurs on the floor of California's Death Valley. (b) Rocks slide along the slippery clay surface of California's Racetrack Playa when strong winds blow.

rains lightly, clay-covered playa surfaces become very slippery. Racetrack Playa, in California, becomes so slippery, in fact, that the wind sends stones sliding out across the surface, leaving grooves behind them to mark their path (►Fig. 21.19b).

Where sufficient water flows into a desert basin, it creates a permanent lake. If the basin is an **interior basin**, with no outlet to the sea, the lake becomes very

salty, because although its water escapes by evaporation in the desert sun, its salt cannot. The Great Salt Lake, in Utah, exemplifies this process. Even though the streams feeding the lake are fresh enough to drink, their water contains trace amounts of dissolved salt ions. Because the lake has no outlet, these ions have become concentrated in the lake over time, making it even saltier than the ocean.

Deposition from the Wind

As mentioned earlier, wind carries two kinds of sediment loads—a suspended load of dust-sized particles and a surface load of sand. Much of the dust is carried out of the desert and accumulates elsewhere, forming layers of fine-grained sediment called **loess**. Sand, however, cannot travel far, and accumulates within the desert in piles called **dunes**, ranging in size from less than a meter to over 300 m high. In favorable locations, dunes accumulate to form vast sand seas hundreds of meters thick. We'll look at dunes in more detail later in this chapter.

> **Take-Home Message**
>
> Sediment carried by water and wind in deserts accumulates in a variety of landforms. Alluvial fans form at the outlets of canyons, playas where water temporarily collects in basins, and dunes where large amounts of sand are available.

21.6 DESERT LANDSCAPES

The popular media commonly portray deserts as endless seas of sand, piled into dunes that hide the occasional palm-studded oasis. In reality, immense sand seas are merely one type of desert landscape. Some deserts are vast, rocky plains, others sport a stubble of cacti and other hardy desert plants, and still others contain intricate rock formations that look like medieval castles. Explorers of the Sahara, for example, traditionally distinguished among hamada (barren, rocky highlands), reg (vast, stony plains), and erg (sand seas in which large dunes form). In this section, we'll see how the erosional and depositional processes described above lead to the formation of such contrasting landscapes.

Rocky Cliffs and Mesas

In hilly desert regions, the lack of soil exposes rocky ridges and cliffs. As noted earlier, cliffs erode when rocks split away along vertical joints. When this happens, the cliff face retreats but retains roughly the same form. The process, commonly referred to as **cliff retreat**, or scarp retreat, occurs in fits and starts. A cliff may remain unchanged for

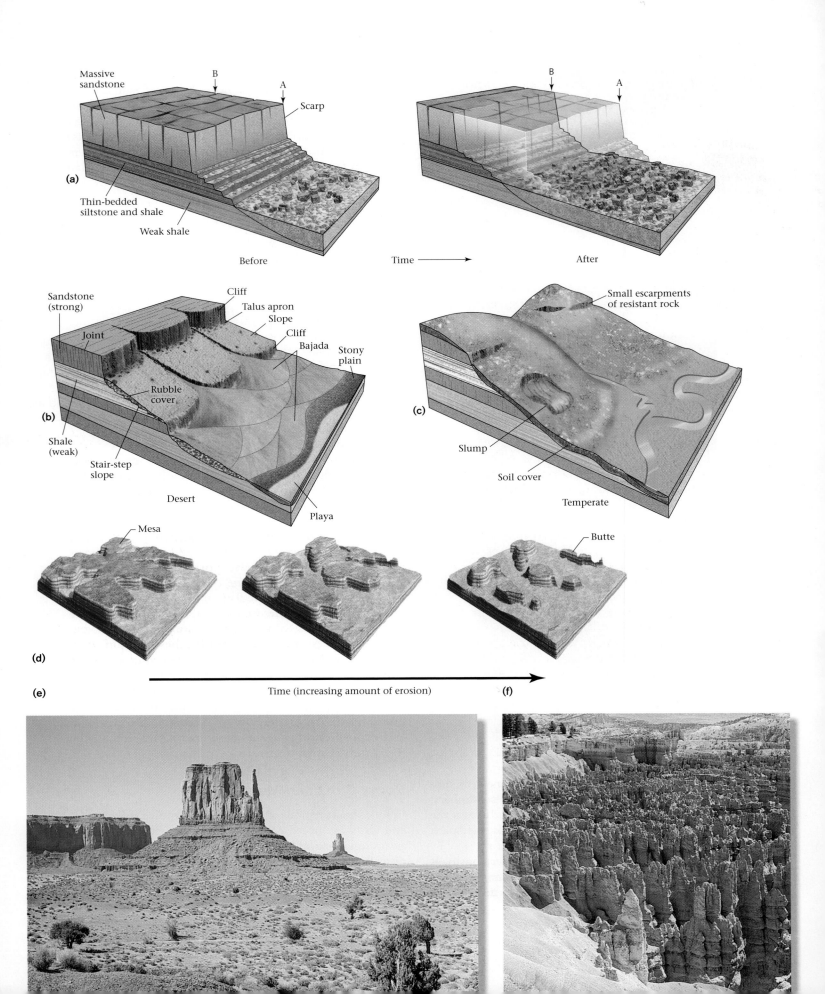

(a)
Massive sandstone
B
A
Scarp
Thin-bedded siltstone and shale
Weak shale
Before
Time
After

(b)
Sandstone (strong)
Joint
Cliff
Talus apron
Slope
Cliff
Bajada
Stony plain
Rubble cover
Shale (weak)
Stair-step slope
Desert
Playa

(c)
Small escarpments of resistant rock
Slump
Soil cover
Temperate

(d) Mesa

(e)

(f) Butte

Time (increasing amount of erosion)

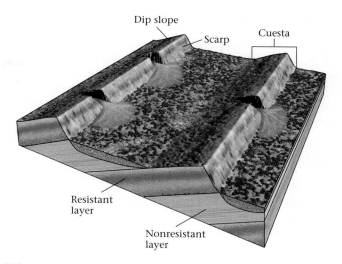

FIGURE 21.20 *(opposite)* **(a)** The process of cliff retreat in a desert. Rocks break off the cliff along joints that run parallel to the cliff face. The cliff face therefore maintains the same shape and orientation, even though erosion causes the cliff to retreat from point A to point B. **(b)** Stair-step cliffs appear where beds of strong rock are interlayered with thin beds of weak rock. Joints are more closely spaced in the thin layers. **(c)** If the sequence of rocks shown in (b) were to occur in a temperate or humid climate, the slope would instead be smooth and underlain by thick soil. **(d)** Because of cliff retreat, a once-continuous layer of rock evolves into a series of isolated remnants. **(e)** These buttes were carved from massive red sandstone layers in Monument Valley, Arizona. **(f)** These hoodoos, in Bryce Canyon, Utah, formed from the erosion of multicolored layers of sandstone, siltstone, and shale.

FIGURE 21.21 Asymmetric ridges called cuestas appear where the strata in a region are not horizontal. A joint-controlled cliff forms the steep side, while a dip slope makes up the gentle side. The surface of a dip slope, by definition, is parallel to the bedding of strata beneath.

decades or centuries, and then suddenly a block of rock falls off and crumbles into rubble at the foot of the cliff (▶Fig. 21.20a). Cliff height depends on bed thickness: in places where particularly thick, resistant beds crop out, tall cliffs develop. This is because large, widely spaced joints form in thick beds, so the collapse of a portion of the wall generates huge blocks. In thinly bedded shales, joints are small and closely spaced, so shale beds erode to make an overall gradual slope consisting of many tiny stair steps. Thus, cliffs formed from stratified rocks (such as beds of sandstone and shale) develop a step-like shape; strong layers (sandstone or limestone) become vertical cliffs, and weak layers (shale) become rubble-covered slopes (▶Fig. 21.20b). (This landscape contrasts with landscapes in humid climates, where thick soils form; ▶Fig. 21.20c.)

With continued erosion and cliff retreat, a plateau of rock slowly evolves into a cluster of isolated hills, ridges, or columns (▶Fig. 21.20d). Flat-lying strata or flat-lying layers of volcanic rocks erode to make flat-topped hills. These go by different names, depending on their size. Large examples (with a top surface area of several square km) are **mesas**, from the Spanish word for table. Medium-sized examples are **buttes** (▶Fig. 21.20e). Small examples, whose height greatly exceeds their top surface area, are **chimneys**. Erosion of strata has resulted in the skyscraper-like buttes of Monument Valley, Arizona, and the stark cliffs of Canyonlands National Park. Bryce Canyon National Park in Utah contains countless chimneys of brightly colored shale and sandstone—locally, these chimneys are called hoodoos (▶Fig. 21.20f). Natural arches, such as those of Arches National Monument, form when erosion along joints leaves narrow walls of rock. When the lower part of the wall erodes while the upper part remains, an arch results. (See art, pp. 748–749.)

In places where bedding dips at an angle to horizontal, flat-topped mesas and buttes don't form; rather, asymmetric ridges called **cuestas** develop. A joint-controlled cliff forms the steep front side of a cuesta, and the tilted top surface of a resistant bed forms the gradual slope on the backside (▶Fig. 21.21). Because the angle of the gradual slope is the same as the dip angle of the bed (the angle the bed surface makes with respect to horizontal), it is called a dip slope. If the bedding dip is steep to near vertical, a narrow symmetrical ridge, called a hogback, forms. If desert hills consist of homogeneous rock such as granite rather than stratified rock, they typically erode to make a pile of rounded blocks (see Fig. 7.10b).

With progressive cliff retreat on all sides of a hill, finally all that remains of the hill is a relatively small island of rock, surrounded by alluvium-filled basins. Geologists refer to such islands of rock by the German word **inselberg** (island mountain; ▶Fig. 21.22). Depending on the rock type or the orientation of stratification in the rock, and on rates of erosion, inselbergs may be sharp-crested, plateaulike, or loaf-shaped (steep sides and a rounded crest). Inselbergs with a loaf geometry, as exemplified by Uluru (Ayers Rock) in central Australia (▶Box 21.1), are also known as bornhardts.

FIGURE 21.22 Inselbergs are small islands of rock surrounded by pediments. Alluvium gradually covers the pediments.

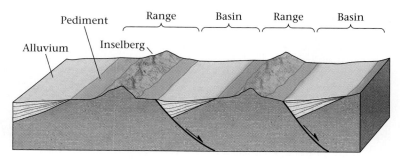

Stony Plains

The coarse sediment eroded from desert mountains and ridges washes into the lowlands and builds out to form gently sloping alluvial fans. The surfaces of these gravelly piles are strewn with pebbles, cobbles, and boulders, and are dissected by dry washes (wadis or arroyos). Portions of these stony plains evolve into desert pavements.

Pediments

When travelers began trudging through the desert of the southwestern United States during the nineteenth century, they found that in many locations the wheels of their wagons were rolling over flat or gently sloping bedrock surfaces. These bedrock surfaces extended outward like ramps from the steep cliffs of a mountain range on one side, to alluvium-filled valleys on the other (Fig. 21.22). Geologists now refer to such surfaces as **pediments**. A pediment is a consequence of erosion, left behind as a mountain front gradually retreats. Pediments develop when sheetwash during floods carries sediment away from the mountain front—as it moves, the sediment grinds away the bedrock that it tumbles over. Between erosional events, weathering weakens the surface of the pediment. Alluvium that is washed off pediments accumulates farther downslope, and may eventually build up sufficiently to bury the pediments.

Seas of Sand: The Nature of Dunes

A **sand dune** is a pile of sand deposited by a moving current. Dunes in deserts form because of the wind. They start to form where sand becomes trapped on the windward side of an obstacle, such as a rock or a shrub. (Dunes formed around shrubs are known as coppice dunes.) Gradually, the sand builds downwind into the lee of the obstacle. Once initiated, the dune itself affects the wind flow, and sand accumulates on the lee (downwind) side of the dune. Here, sand slides down the lee surface of the dune, so this surface is aptly named the slip face (▶Fig. 21.23).

In places where abundant sand accumulates, sand seas (ergs) bury the landscape. The wind builds the sand in these ergs into dunes that display a variety of shapes and sizes, depending on the character of the wind and the sand supply (▶Fig. 21.25). Where the sand is relatively scarce and the wind blows steadily in one direction, beautiful crescents called barchan dunes develop, with the tips of the crescents pointing downwind. If the wind shifts direction frequently, a group of crescents pointing in different directions overlap one another, creating a constantly changing star dune. Where enough sand accumulates to bury the ground surface completely, and only moderate winds blow, sand piles into simple, wave-like shapes called transverse dunes. The crests of transverse dunes lie perpendicular to the wind direction. Strong winds may break through transverse dunes and change them into parabolic dunes whose ends point in the upwind direction. Finally, if there is abundant sand and a strong, steady wind, the sand streams into longitudinal dunes (also called seif dunes after the Arabic word for sword) whose axis lies parallel to the wind direction. In the southern third of the Arabian Peninsula, a region called the Empty Quarter because of its total lack of population, a vast erg called the Rub al Khali contains seif dunes that stretch for almost 200 km and reach heights of over 300 m.

In a sand dune, sand saltates up the windward side of the dune, blows over the crest of the dune, and then settles on the steeper, lee face of the dune. The slope of this face attains the angle of repose, the slope angle of a freestanding pile of sand. As sand collects on this surface, it may become unstable and slide down the slope—as we've noted, geologists refer to the lee side of a dune as the slip face. As progressively more sand accumulates on the slip face, the crest of the dune migrates downwind, and former slip faces become preserved inside the dune. In cross section, these slip

> **Take-Home Message**
>
> Erosion of thick layers of horizontal sedimentary rock in deserts yields buttes and mesas; erosion of tilted layers forms cuestas. Stony plains develop if finer sediment blows or washes away. Seas of sand contain many types of dunes.

FIGURE 21.23 Progressive stages in the growth of a small sand dune.

Blowing sand   Obstacle

Time 1

Trapped sand

Time 2

Sand grows around obstacle.

Time 3

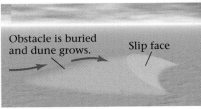

Obstacle is buried and dune grows.   Slip face

Time 4

BOX 21.1 THE REST OF THE STORY

Uluru (Ayers Rock)

Much of central Australia is an immense desert. Parts are barren and sandy, but much of the land is covered with scrub brush, which provides meager grazing for cattle and kangaroo. Uluru, also known by its English name, Ayers Rock, is a huge bornhardt that towers 360 m above the desert plain (▶Fig. 21.24a, b). This rock mass, 3.6 km long and 2 km wide, consists of nearly vertical dipping sandstone beds. It makes up one limb of a huge regional syncline. The other limb is also

a bornhardt, known locally as The Olgas, and the entire area in between is buried in alluvium. Uluru formed because its sandstone resisted erosion, whereas adjacent rock formations did not. Thus, over geologic time, alluvium buried the surrounding landscape, but Uluru remained high. Because its strata dip vertically, it has not developed the stair-step shape of mesas and buttes.

Because of its grandeur, Uluru plays a sacred role in Australian Aboriginal tradi-

tions. In the dreamtime (time of creation) legends of the Aboriginal people, erosional features on the surface of the rock are scars from a fierce battle between ancient clans. In recent years, the rock has attracted tourists from around the world, who risk life and limb to climb to its top. Plaques at the base of the rock record the names of those who slipped and fell from its steep side.

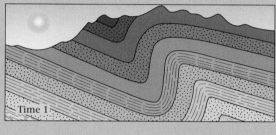

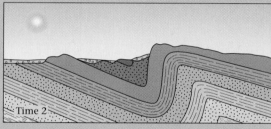

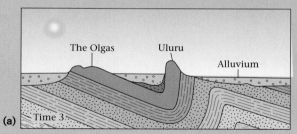

FIGURE 21.24 **(a)** Originally Uluru (Ayers Rock), in Australia, was part of a large syncline beneath a mountain. Erosion has beveled the mountain away, until all that's left now above a sediment-covered plain are the resistant sandstone limbs of the syncline; softer units are covered by sediment. Uluru is the vertical limb, and the Olgas are the gently dipping limb. **(b)** Uluru is a dramatic landform.

(a)   (b)

faces appear as cross beds (▶Fig. 21.26a–c). The surfaces of dunes are not, in general, smooth, but rather are covered with delicate ripples.

With the exception of star and longitudinal dunes, sand dunes migrate downwind as the wind continuously picks up sand from the gently dipping windward slope and drops it onto the leeward side, or slip face. Rates of migra-

tion can exceed 25 m per year. Because of moving sand on an active dune, vegetation can't grow there. If a change in climate brings more rain, however, plant cover may grow and stabilize the dunes. At the end of the last ice age, for example, the sand hills region of western Nebraska was a vast dune field, but in the past 11,000 years it has been covered by grasslands, and the dunes have become stabilized.

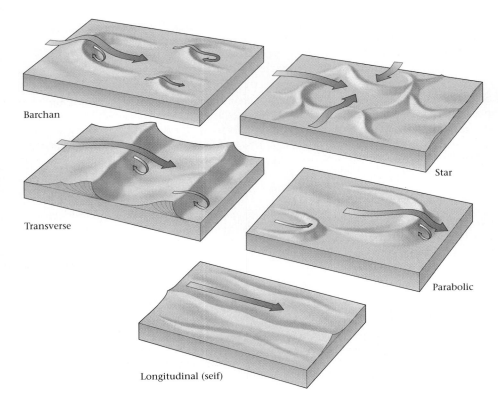

Barchan

Transverse

Star

Parabolic

Longitudinal (seif)

FIGURE 21.25 The various kinds of sand dunes.

21.7 LIFE IN THE DESERT

In the midst of a large erg, there seems to be nothing growing or moving at all. But most desert landscapes do include plants and animals. These organisms must possess special characteristics to enable them to survive in the desert: they must be able to withstand extremes of temperature—oppressive heat during the day and chilling cold at night—and to survive without abundant water.

Plants have evolved a number of different means to survive desert conditions. Some produce thick-skinned seeds that last until a heavy rainfall, then quickly germinate, grow, and generate new seeds only while water is available. The new generation of seeds then waits until the next rainfall to start the cycle over again. Other plants have evolved the ability to send roots down to find deep groundwater; these plants have very long taproot systems. Still others have shallow root systems that spread over a broad area so they can efficiently soak up water when it does rain.

FIGURE 21.26 **(a)** A sand dune with surface ripples. **(b)** Cross bedding inside a dune. **(c)** Cross beds exposed in the Mesozoic sandstone of Zion National Park, in Utah. (The photo has been reversed, so the cross beds have the same orientation as slip faces in the other parts of the figure.)

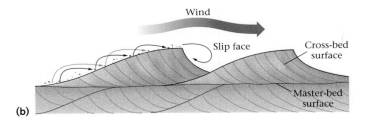

(b)

(a)

(c)

Many desert plants have thick, fleshy stems and leaves. These plants, known as **succulents**, can store water for long periods of time (▶Fig. 21.27). Because succulents may be the only source of water during a time of drought, they have developed threatening thorns or needles to keep away thirsty animals. Plant life is much more diverse in desert oases, the verdant islands that crop up where natural springs spill groundwater onto the surface (see Chapter 19). The nearly year-round supply of water in an oasis nourishes a variety of palms and other nonsucculent plants.

Animal life in the desert includes scavengers, hunters, and plant eaters. Animals face the same challenge as plants: to retain water and survive extreme temperatures. To accomplish these goals, desert animals have also evolved numerous strategies. Frogs, for example, burrow beneath the ground and remain dormant for months, waiting until the next rain. Reptiles escape the midday heat by crawling into dark cracks between rocks. Rodents forage for food only during the cool night. And kit foxes, jack rabbits, and mule deer have disproportionately large ears through which they efficiently lose body heat. Many desert mammals, such as camels, retain body water by not sweating.

Humans are not meant to live in the desert. The loss of body moisture in extreme heat can be so rapid that a person will die in less than 24 hours unless shaded from the Sun and supplied with at least 8 liters of water per day. Nevertheless, all but the most barren deserts are inhabited, though sparsely. Before civilization provided water wells, pipelines, and mechanized transportation, desert peoples lived in small nomadic groups, spaced far enough apart that they could live off the land; deserts have too little water to sustain agriculture or husbandry (▶Fig. 21.28a). Australian Aboriginal groups, for example, rarely included more than a dozen individuals. In the past, nomadic desert dwellers either built temporary shelters out of local materials or traveled with tents. Locally, people carved underground dwellings in sediment or soft rock, for rock is such a good insulator that a few meters below the ground surface it stays close to the region's mean temperature year round.

> **Take-Home Message**
>
> Plants and animals that live in the desert have special adaptations to survive in the absence of water. For example, some plants send roots deep into the ground to reach groundwater. Succulents have fleshy stems that can hold a lot of water.

(a)

(b)

FIGURE 21.27 These plants in the Sonoran Desert, Arizona, are well adapted for dry conditions. **(a)** Saguaro cactus. **(b)** Teddy-bear cholla.

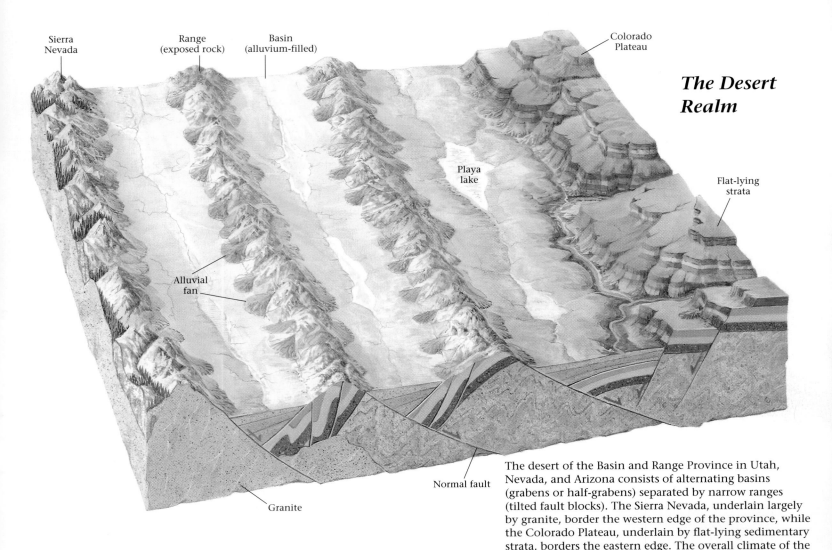

The Desert Realm

Sierra Nevada

Range (exposed rock)

Basin (alluvium-filled)

Colorado Plateau

Flat-lying strata

Playa lake

Alluvial fan

Normal fault

Granite

The desert of the Basin and Range Province in Utah, Nevada, and Arizona consists of alternating basins (grabens or half-grabens) separated by narrow ranges (tilted fault blocks). The Sierra Nevada, underlain largely by granite, border the western edge of the province, while the Colorado Plateau, underlain by flat-lying sedimentary strata, borders the eastern edge. The overall climate of the region is dry. Because of the great variety of elevations and rock types, the region hosts different desert landscapes.

Except in the case of large rivers, such as the Nile or the Colorado, which bring water into a desert region from a more temperate region, streams in deserts fill with water only after heavy rains. At other times, the stream channels are dry. These channels are called dry washes, arroyos, or wadis. When there is a heavy rain, water cannot be absorbed into the ground fast enough, so runoff enters dry washes and fills them very quickly, causing a flash flood. The turbulent, muddy water of a flash flood can transport even large boulders. This flash flood is rushing down a stream in the Sonoran Desert of Arizona.

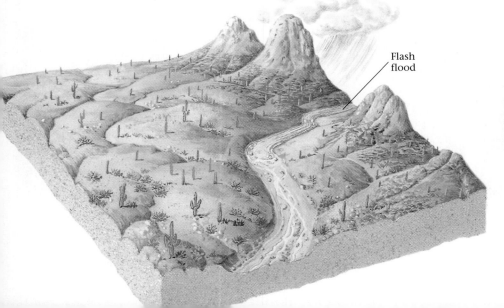

Flash flood

Barchan dune

Cross beds

Where there is a large supply of sand, a variety of sand dunes develop. The geometry of a particular sand dune (such as barchan, longitudinal, or star) depends on the sand supply and the wind. Inside sand dunes, we find cross beds.

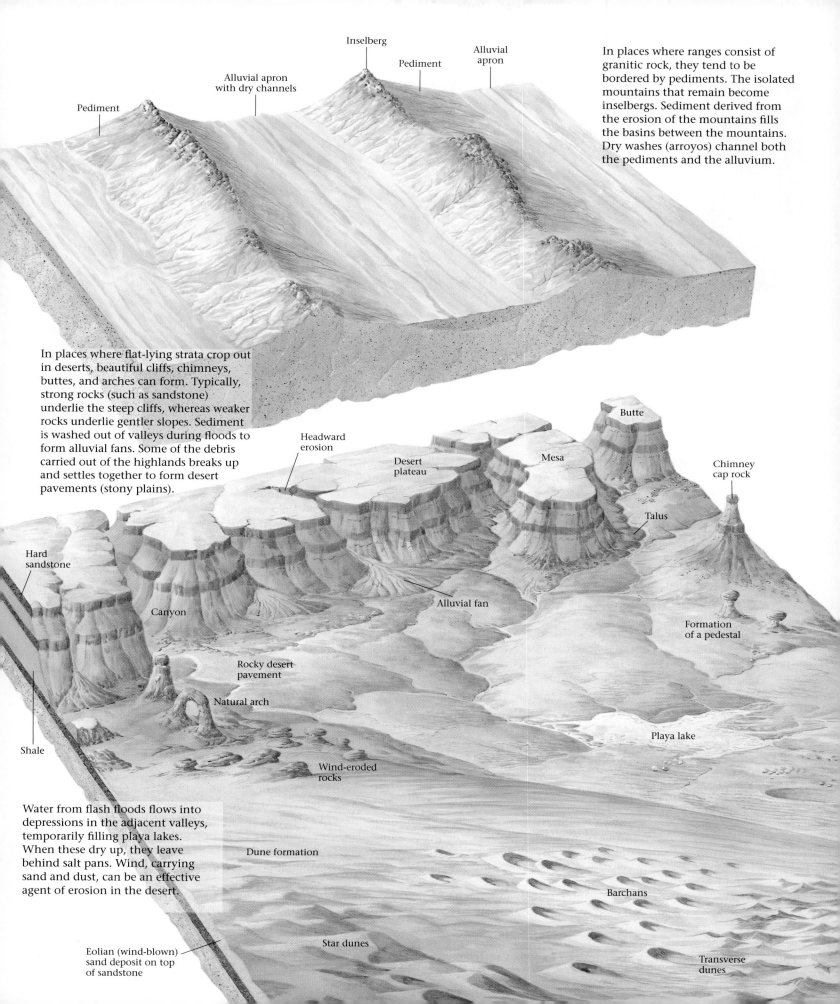

Pediment

Alluvial apron
with dry channels

Inselberg

Pediment

Alluvial
apron

In places where ranges consist of
granitic rock, they tend to be
bordered by pediments. The isolated
mountains that remain become
inselbergs. Sediment derived from
the erosion of the mountains fills
the basins between the mountains.
Dry washes (arroyos) channel both
the pediments and the alluvium.

In places where flat-lying strata crop out
in deserts, beautiful cliffs, chimneys,
buttes, and arches can form. Typically,
strong rocks (such as sandstone)
underlie the steep cliffs, whereas weaker
rocks underlie gentler slopes. Sediment
is washed out of valleys during floods to
form alluvial fans. Some of the debris
carried out of the highlands breaks up
and settles together to form desert
pavements (stony plains).

Headward
erosion

Desert
plateau

Mesa

Butte

Chimney
cap rock

Talus

Hard
sandstone

Canyon

Alluvial fan

Formation
of a pedestal

Rocky desert
pavement

Natural arch

Playa lake

Shale

Wind-eroded
rocks

Water from flash floods flows into
depressions in the adjacent valleys,
temporarily filling playa lakes.
When these dry up, they leave
behind salt pans. Wind, carrying
sand and dust, can be an effective
agent of erosion in the desert.

Dune formation

Barchans

Eolian (wind-blown)
sand deposit on top
of sandstone

Star dunes

Transverse
dunes

(a)

(b)

FIGURE 21.28 (a) Before the modern era, inhabitants of Australia's central desert (the red center, named for its red soil and rocks) lived in small nomadic groups. There was not enough water to supply permanent communities until the advent of technology. (b) Las Vegas is pushing into the desert even farther every year. People expect to cover the landscape with green, but water is a problem.

21.8 DESERT PROBLEMS

More and more people are moving into desert regions. In fact, the population of the desert in the southwestern United States is growing faster than in any other part of the country, and as desert cities grow, environmental problems soon follow (►Fig. 21.28b). As noted in Chapter 19, growing cities must either suck water out of the ground or bring in water via canals from rivers or reservoirs to meet their needs. So water tables in deserts are dropping, rivers are drying up, and the land surface is cracking. People also have imported exotic plants and animals that invade the countryside and upset the ecological balance.

The modern era has seen a remarkable change in desert margins. Natural droughts (periods of unusually low rainfall), aggravated by overpopulation, overgrazing, careless agriculture, and diversion of water supplies have transformed semi-arid grasslands into true deserts, leading to tragic famines that have killed millions of people. **Desertification**, the process of transforming nondesert areas to desert, has accelerated.

The consequences of desertification have devastated the Sahel, the belt of semi-arid land that fringes the southern margin of the Sahara Desert. In the past, the Sahel provided sufficient vegetation to support a small population of nomadic people and animals. But during the second half of the twentieth century, large numbers of people migrated into the Sahel to escape overcrowding in central Africa. The immigrants began farming and maintained large herds of cattle and goats. Plowing and overgrazing re-

moved soil-preserving grass and caused the soil to dry out. In addition, the trampling of animal hooves compacted the ground so it could no longer soak up water. In the 1960s and again in the 1980s, a series of natural droughts hit the region, bringing catastrophe (►Fig. 21.29a, b). Wind erosion stripped off the remaining topsoil. Without vegetation, the air grew drier, and the semiarid grassland of the Sahel became desert, with mass starvation as the result.

Other regions on Earth are developing the same problem. The Aral Sea in Kazakhstan, for example, has started to dry up. Diversion of the rivers that once fed the sea has so diminished its water supply that the area of the sea has shrunk dramatically. Boats that once plied its waters now lie as rusting hulks in a sea of dust (►Fig. 21.29c, d).

Desertification does not happen only in less industrialized nations. People in the western Great Plains of the United States and Canada suffered from the problem beginning in 1933, the fourth year of the Great Depression. Banks had failed, workers had lost their jobs, the stock market had crashed, and hardship burdened all. No one needed yet another disaster—but that year, even nature turned hostile. All through the fall, so little rain fell in the plains of Texas and Oklahoma that the region's grasslands and croplands browned and withered, and the topsoil turned to powdery dust. Then, on November 12 and 13, strong storms blew eastward across the plains. Without vegetation to protect the ground, the wind lapped at it, stripped off the topsoil, and sent it skyward to form rolling black clouds that literally blotted out the sun (►Fig. 21.29e). People caught in the resulting **dust storm** found themselves choking and gasping for

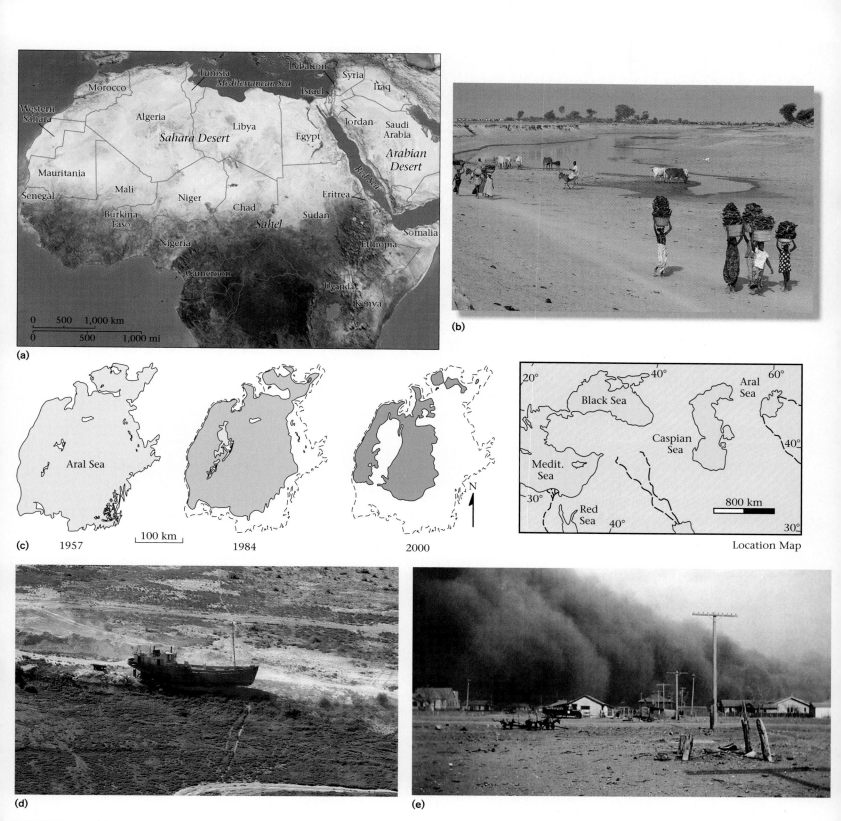

FIGURE 21.29 **(a)** The Sahel is the semi-arid land along the southern edge of the Sahara. As a result of grazing and agriculture, the vegetation in this region has vanished, and large parts have undergone desertification. **(b)** Drought in the Sahel has brought deadly consequences. Here, residents seek water from a dwindling pond. **(c)** The Aral Sea in central Asia has started to dry up because of diversion of rivers that once flowed into it. What was once the floor of the sea is now a parched land. The inset map shows the location. **(d)** Fishing vessels now lie stranded in the Aral Sea area. **(e)** During the Dust Bowl days in central Oklahoma (c. 1930s), dust storms stripped valuable topsoil off the land.

See for yourself...
Desert Landscapes

In the desert, the unvegetated land surface lies exposed, so landform shapes are clear and dramatic. High-resolution images allow you to see cliffs, dunes, fans, and even individual boulders. Here, we tour several deserts to see the variety of landscapes. Zoom out into space to remind yourself of where you are on the globe. The thumbnail images provided on this page are only to help identify tour sites. Go to wwnorton.com/studyspace to experience this flyover tour.

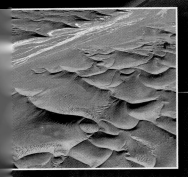

G21.1

Sand Dunes, Namib Desert, Africa (Lat 24°43'57.60"S, Long 15°26'59.90"E)

Fly to the coordinates provided, zoom to 20 km (12 miles), and you will see a reddish sand sea in the Namib Desert on the west coast of southern Africa. The color is due to oxidized iron. A dry wash cut through the sand and washed it away. Zoom to 8 km (5 miles), and tilt to look NNE to see star dunes, formed by shifting winds (Image G21.1).

G21.2

Uluru (Ayers Rock), Australia (Lat 25°20'52.42"S, Long 131°2'24.98"E)

At these coordinates, from 10 km (6 miles), you can see the famous inselberg, Uluru (Ayers Rock) in the "red center" of Australia (Image G21.2). Zoom to an elevation of 4 km (2.5 miles), tilt your view, and look southeast. Note that the beds of sandstone comprising Uluru have a vertical dip—they are one limb of a syncline. If you fly to Lat 24°10'33.30"S, Long 132°55'19.57"E, and zoom to 20 km (12 miles), you can see another fold whose shape, without vegetation cover, stands out in the landscape (Image G21.3).

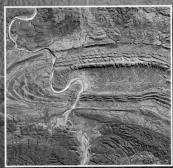

G21.3

Atacama Desert near Chala, Peru (Lat 15°50'40.47"S, Long 74°16'8.10"W)

The Atacama Desert, on the west coast of South America, exists because a cold current flows northwards just offshore. At the coordinates provided, from an elevation of 15 km (9 miles), you can see the stark dryness of the land, right up to the shore of the blue Pacific Ocean (Image G21.4).

G21.4

Tarim Basin of Western China (Lat 37°39'56.20"N, Long 82°19'7.20"E)

Fly to this location, zoom to an elevation of 60 km (37 miles), and tilt the view to look north (Image G21.5). The desert stretches as far as the eye can see. Sand dunes cover the land surface except in the green, irrigated oasis of the middle distance.

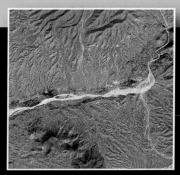

G21.6

Dry Wash in Arizona (Lat 33°39'49.26"N, Long 111°35'10.25"W)

Stream channels in deserts fill only intermittently. At other times, the water table lies below the surface of the ground so the channels are dry. At these coordinates, from 9 km (5.6 miles), you see the floor of a dry wash (Image G21.6). Adjacent slopes display a dendritic drainage pattern, because even though water flows only intermittently, water is still the dominant agent of erosion.

G21.7

Urbanizing a Desert, Tucson, Arizona (Lat 32°15'53.09"N, Long 110°51'19.87"W)

Much of southern Arizona lies in the Sonoran Desert. There's hardly any surface water, but the warm climate attracts a growing population, and the region's cities have ballooned. Zoom to 5 km (3 miles) and tilt your view to look north (Image G21.7). Suburbs of Tucson sprawl to the foot of the Catalina Mountains; intense watering keeps the golf courses green.

Playa in Death Valley, California (Lat 36°13'40.64"N, Long 116°45'55.28"W)

Here at Badwater, in Death Valley, you see the lowest point in North America (85 m or 280 feet below sea level). Zoom to an elevation of 3 km (2 miles), tilt your view, and look east (Image G21.8). The white surface is the salt crust of a dry playa lake. Note an alluvial fan building out onto the playa.

G21.8

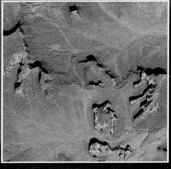

G21.9

Buttes of Monument Valley, Arizona (Lat 36°57'6.66"N, Long 110°5'19.66"W)

Fly to this locality and look down from an elevation of 10 km (6 miles) (Image G21.9). You are seeing the buttes and chimneys of Monument Valley. These landforms consist of massive sandstone eroding by cliff retreat when joint-bounded walls of rocks collapse.

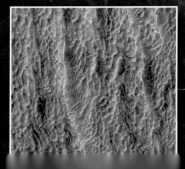

Sand Sea in the Sahara, Egypt (Lat 26°53'42.61"N, Long 26°5'5.02"E)

The Sahara is the largest continuous desert on Earth. At this locality, from an elevation of 10 km (6 miles), you see two scales of dunes (Image G21.10). Very large longitudinal dunes trend from NNW to SSE. On top of these, smaller dunes developed.

breath. When the dust finally settled, it had buried houses and roads under huge drifts, and dirtied every nook and cranny. The dust blew east as far as New England, where it turned the snow brown. What had once been a rich farmland in the southwestern plains turned into a wasteland that soon acquired a nickname, the Dust Bowl.

For several more years the drought persisted, leading to starvation and bankruptcy. Many of the region's residents were forced to move to wherever they could find work. Hundreds of Okies, as the natives of Oklahoma were then called, piled into jalopies and drove on old Route 66 out to California, looking for jobs in the state's still-green agricultural regions. Many were subjected to exploitation and persecution once they arrived in California. John Steinbeck dramatized this staggering human tragedy, which came to symbolize the Depression, in his novel *The Grapes of Wrath*.

Why did the fertile soils of the southern Great Plains suddenly dry up? The causes were complex; some were natural and some were human-induced. Drought episodically visits certain regions, but drought alone doesn't inevitably lead to dust-bowl conditions. They may result from human activity. Typically, the Great Plains region has a semi-arid climate in which only thin soil develops. But the plains were settled in the 1880s and 1890s, unusually wet years. Not realizing its true character, far more people moved into the region than it could sustain, and the land was farmed too intensively. When farmers used steel plows, they destroyed the fragile grassland root systems that held the thin soil in place. And when the drought of the 1930s came, it brought catastrophe.

Desertification can be reversed, but at a price. Planting and irrigation may transform desert into farm fields, orchards, forests, or lawns. But water to nourish the plants has to come from somewhere, and people obtain it by diverting rivers or by pumping groundwater, activities that create their own set of problems. River diversion robs regions downstream of their water supply, and pumping out too much groundwater lowers the water table so substantially that the pore space in aquifers collapses and the ground surface subsides. In short, people will need to rethink land-use policies in semi-arid lands to avoid catastrophe.

The Dust Bowl of the 1930s and the fate of the Sahel in Africa remind us of how fragile the Earth's green blanket of vegetation really is. Global climate changes can shift climatic belts sufficiently to transform agricultural regions into deserts. Some 5,000 years ago, the swath of land between the Nile Valley in Egypt and the Tigris-Euphrates Valley of Mesopotamia was known as the fertile crescent; here, people first abandoned their nomadic ways and settled in agricultural communities. Yet the original "land of milk and honey" has become a barren desert, in need of intensive irrigation to maintain any agriculture. The change in landscape reflects a change in climate—the beginning of Western civilization occurred during the warmest and wettest period of global climate since before the last ice age. So much water drenched the Middle East and North Africa that rivers flowed where Saharan sands now blow. Unfortunately, if the current trend in climate change continues, our present agricultural belts could someday become new Saharas.

Desertification has a potentially dangerous side effect—blowing dust. As desert areas expand in response to desertification, and as desert pavement gets disrupted, wind-blown dust becomes more of a problem. Not only do winds have larger areas of dry, dusty land to churn, but the dust generated from lands that were once agricultural and are now desert may contain harmful chemicals (e.g., residues of herbicides and pesticides) that can themselves become windborne. In recent years, satellite images have revealed that wind-blown dust from deserts can travel across oceans and affect regions on the other side. For example, dust blown off the Sahara can traverse the Atlantic and settle over the Caribbean (▶Fig. 21.30). Geologists are concerned that this dust, along with the fungi, toxic chemicals, and microbes that it carries, may infect corals with disease or in some other way inhibit their life processes. Thus, windblown dust can contribute to the destruction of coral reefs.

> ### Take-Home Message
> Desert populations have been burgeoning, so water supplies are problematic, and groundwater is being depleted. In lands bordering deserts, droughts and population pressures have led to desertification, the transformation of vegetated land into desert.

FIGURE 21.30 In this satellite image, we see a huge dust cloud blowing across the Atlantic. The dust originated in the Sahara.

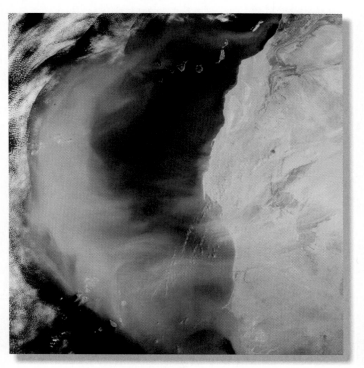

Chapter Summary

- Deserts generally receive less than 25 cm of rain per year. Vegetation covers no more than 15% of their surface.

- Subtropical deserts form between 20° and 30° latitude, rain-shadow deserts are found on the inland side of mountain ranges, coastal deserts are located on the land adjacent to cold ocean currents, continental-interior deserts exist in land-locked regions far from the ocean, and polar deserts form at high latitudes.

- In deserts, chemical weathering happens slowly, so rock bodies tend to erode primarily by physical weathering. Desert varnish forms on rock surfaces, and soils tend to accumulate soluble minerals.

- Water causes significant erosion in deserts, mostly during heavy downpours. Flash floods carry large quantities of sediments down ephemeral streams. When the rain stops, these streams dry up, leaving steep-sided washes.

- Wind causes significant erosion in deserts, because it picks up dust and silt and carries them as suspended load, and causes sand to saltate. Where wind blows away finer sediment, a lag deposit remains. Windblown sediment abrades the ground, creating a variety of features such as ventifacts and yardangs.

- Desert pavements are mosaics of varnished stones armoring the surface of the ground.

- Talus aprons form when rock fragments accumulate at the base of a slope. Alluvial fans form at a mountain front where water in ephemeral streams deposits sediment. When temporary desert lakes dry up, they leave playas.

- In some desert landscapes, erosion causes cliff retreat, eventually resulting in the formation of inselbergs. Pediments of nearly flat or gently sloping bedrock surround some inselbergs. The erosion of stratified rock yields such landforms as buttes.

- Where sand is abundant, the wind builds it into dunes. Common types include barchan, star, transverse, parabolic, and longitudinal (seif) dunes.

- Deserts contain a great variety of plant and animal life. All are adapted to survive extremes in temperature and without abundant water.

- Changing climates and land abuse may cause desertification, the transformation of semi-arid land into deserts. Windblown dust from deserts may waft across oceans.

Geopuzzle Revisited

Deserts are very arid regions that can sustain very little vegetation. They host only hardy plants and animals adapted to living with little water. Vast areas of sand dunes do cover some portions of some deserts, but desert landscapes can also host stony plains and rocky ridges. And, although many deserts are very hot, some occur at high elevations or at polar latitudes and can be very cold.

Key Terms

alluvial fan (p. 740)
arroyo (p. 735)
bajada (p. 740)
butte (p. 743)
chimney (p. 743)
cliff retreat (p. 741)
cuesta (p. 743)
deflation (p. 738)
desert (p. 731)
desert pavement (p. 735)
desert varnish (p. 737)
desertification (p. 750)
dry wash (p. 735)
dune (pp. 741, 744)
dust storm (p. 750)
inselberg (p. 743)

interior basin (p. 741)
lag deposit (p. 737)
loess (p. 741)
mesa (p. 743)
pediment (p. 744)
petroglyph (p. 735)
playa (p. 740)
rain shadow (p. 733)
saltation (p. 737)
succulents (p. 747)
surface load (p. 737)
suspended load (p. 735)
talus apron (p. 740)
ventifact (p. 738)
yardang (p. 738)

Review Questions

1. What factors determine whether a region can be classified as a desert?

2. Explain why deserts form.

3. Have today's deserts always been deserts? Keep in mind the consequences of plate tectonics.

4. How do weathering processes in deserts differ from those in temperate or humid climates?

5. Describe how water modifies the landscape of a desert. Be sure to discuss both erosional and depositional landforms.

6. Explain the ways in which desert winds transport sediment.

7. Explain how the following features form: (a) desert varnish; (b) desert pavement; (c) ventifacts; (d) yardangs.

8. Describe the process of formation of the different types of depositional landforms that develop in deserts.

9. Describe the process of cliff (scarp) retreat and the landforms that result from it.

10. What are the various types of sand dunes? What factors determine which type of dune develops?

11. Discuss various adaptations that life forms have evolved in order to be able to survive in desert climates.

12. What is the process of desertification, and what causes it? How can desertification in Africa affect the Caribbean?

On Further Thought

1. Death Valley, California, lies to the east of a high mountain range, and its floor lies below sea level. During the summer, Death Valley is very hot and dry. Explain why it has such weather.

2. You are working for an international non-governmental agency (NGO) and have been charged with the task of providing recommendations to an African nation that wishes to slow or halt the process of desertification within its borders. What are your recommendations?

3. The Namib Desert lies to the north and west of the Kala-hari Desert, in southern Africa. The reason that the former region is a desert is not the same as the reason that the latter is a desert. Explain this statement.

4. Lake Havasu, Arizona, is a resort town along the California/Arizona border, in the western part of the Sonoran Desert. Water sports are one of the attractions of the town. Where does the standing water of this location come from, and what factors can impact the supply in the future?

Suggested Reading

Abrahams, A. D., and A. J. Parsons, eds. 1994. *Geomorphology of Desert Environments.* London: Chapman and Hall.

Bagnold, R. A., 2005. *The Physics of Blown Sand and Desert Dunes.* Mineola, N.Y.: Dover Publications.

Cooke, R. U., A. Warren, A. Goudie, 1992. *Desert Geomorphology.* Boca Raton, FL: CRC Press.

Dregne, H. E. 1983. *Desertification of Arid Lands.* Chur, Switzerland: Harwood Academic.

Livingston, I., S. Stokes, and A. S. Goudie, eds. 2000. *Aeolian Environments, Sediment and Landforms.* New York: Wiley.

Tchakerian, V. P., ed. 1995. *Desert Aeolian Processes.* London: Chapman and Hall.

THE VIEW FROM SPACE Erosion of rock produces sand, which, under appropriate conditions in arid climates, can build into large dunes. These dunes, some of which are 300 m high, developed in the Namib Desert of Namibia, southwestern Africa.

Amazing Ice: Glaciers and Ice Ages

Glaciers are rivers or sheets of ice. They carve beautiful landscapes and deposit hills of sediment. During the ice ages, glaciers covered vast areas of continents. Here, we see the jagged, melting surface of a glacier in Alaska.

Geopuzzle

If modern society had existed about 12,000 years ago, would it have been possible to build New York City (USA) or Edinburgh (Scotland) at their present locations? Why?

I seemed to vow to myself that some day I would go to the region of ice and snow and go on and on till I came to one of the poles of the earth.

—*Ernest Shackleton (British polar explorer, 1874–1922)*

22.1 INTRODUCTION

There's nothing like a good mystery, and one of the most puzzling in the annals of geology came to light in northern Europe early in the nineteenth century. When farmers of the region prepared their land for spring planting, they occasionally broke their plows by running them into large boulders buried randomly through otherwise fine-grained sediment. Many of these boulders did not consist of local bedrock, but rather came from outcrops hundreds of kilometers away. Because the boulders had apparently traveled so far, they came to be known as **erratics** (from the Latin *errare*, to wander).

The mystery of the wandering boulders became a subject of great interest to early nineteenth-century geologists, who realized that deposits of *un*sorted sediment (containing a variety of different clast sizes) could not be examples of typical stream alluvium, for running water sorts sediment by size. Most attributed the deposits to a vast flood that they imagined had somehow spread a slurry of boulders, sand, and mud across the continent. In 1837, however, a young Swiss geologist named Louis Agassiz proposed a radically different interpretation. Agassiz often hiked among **glaciers** (slowly flowing masses of ice that survive the summer melt) in the Alps near his home. He observed that glacial ice could carry enormous boulders as well as sand and mud, because ice is solid and has the strength to support the weight of rock. Agassiz realized that because solid ice does not sort sediment as it flows, glaciers leave behind unsorted sediment when they melt. On the basis of these observations, he proposed that the mysterious sediment and erratics of Europe were deposits left by **ice sheets**, vast glaciers that had once covered much of the continent. In Agassiz's mind, Europe had once been in the grip of an **ice age**, a time when the climate was significantly colder and glaciers grew (▶Fig. 22.1).

Agassiz's radical proposal faced intense criticism for the next two decades. But he didn't back down, and instead challenged his opponents to visit the Alps and examine the sedimentary deposits that glaciers had left behind. By the late 1850s, most doubters had changed their minds, and the geological community concluded that the notion that Europe once had had Arctic-like climates was correct. Later in life, Agassiz traveled to the United States and documented many

FIGURE 22.1 During the ice age, glaciers covered much of North America and Europe, so that these continents would have resembled present-day Antarctica, where glacial ice extends from the foot of the Transantarctic Mountains (seen here on the horizon) to the far side of Antarctica, over 3,000 km away.

glacier-related features in North America's landscape, proving that an ice age had affected vast areas of the planet.

Glaciers, which have many forms, cover only about 10% of the land on Earth today (▶Fig. 22.2a, b), but during the most recent ice age, which ended only about 11,000 years ago, as much as 30% of the continents had a coating of ice. New York City, Montréal, and many of the great cities of Europe now occupy land that once lay beneath hundreds of meters to a few kilometers of ice. The work of Louis Agassiz brought the subject of glaciers and ice ages into the realm of geologic study and led people to recognize that major climate changes happen in Earth history. In this chapter, after considering the nature of ice, we see how glaciers form and why they move. Next, we consider how glaciers modify landscapes by erosion and deposition. A substantial portion of the chapter concerns the Pleistocene ice ages, for their impact on the landscape can still be seen today, but we briefly introduce ice ages that happened earlier in Earth history too. We conclude by considering hypotheses to explain why ice ages happen.

22.2 ICE AND THE NATURE OF GLACIERS

What Is Ice?

Ice consists of solid water, formed when liquid water cools below its freezing point. We can consider a single ice crystal to be a mineral specimen: it is a naturally occurring, inorganic solid, with a definite chemical composition (H_2O) and a regular crystal structure. Ice crystals

(a)

(b)

FIGURE 22.2 **(a)** A piedmont glacier near the coast of Greenland. **(b)** A large valley glacier in Pakistan.

have a hexagonal form, so snowflakes grow into six-pointed stars (▶Fig. 22.3a). We can think of a layer of fresh snow as a layer of sediment, and a layer of snow that has been compacted so that the grains stick together as a layer of sedimentary rock (▶Fig. 22.3b). We can also think of the ice that appears on the surface of a pond as an igneous rock, for it forms when molten ice (liquid water) solidifies (▶Fig. 22.3c). Glacier ice, in effect, is a metamorphic rock. It develops when preexisting ice recrystallizes in the solid state, meaning that the molecules in solid water rearrange to form new crystals (▶Fig. 22.3d, e).

Pure ice has the transparency of glass, but if ice contains tiny air bubbles or cracks that disperse light, it becomes milky white. Like glass, ice has a high **albedo**, meaning that it reflects light well—so well, in fact, that if

you walk on ice without eye protection, you risk blindness from the glare. Ice differs from most other familiar materials in that its solid form is less dense than its liquid form—the architecture of an ice crystal holds water molecules apart. Ice, therefore, floats on water. This unusual characteristic has benefits; if ice didn't float, ice in oceans would sink, leaving room for new ice to form until the oceans froze solid.

Ice also has the unusual property of being slippery—that's why skaters can skate! Surprisingly, researchers still don't completely understand *why* ice is slippery. An older explanation—that skaters can glide on ice because a film of liquid water forms on the surface of the ice in response to frictional heating or to pressure-induced melting—can't explain how ice remains slippery even when it's so cold that water can't exist as liquid. Modern studies suggest that ice can remain slippery even at very low temperatures because the surface of ice consists of a layer of water molecules that are not completely fixed within a crystal lattice. Specifically, the molecules are chemically bonded to the solid ice below, but not to the air above. The existence of unattached bonds permits the surface molecules to behave somewhat like a liquid, even while attached to the solid.

Categories of Glaciers

Glaciers are streams or sheets of recrystallized ice that last all year long and flow under the influence of gravity. Today, they highlight coastal and mountain scenery in Alaska, western North America, the Alps of Europe, the Southern Alps of New Zealand, the Himalayas of Asia, and the Andes of South America, and they cover most of Greenland and Antarctica. Geologists distinguish between two main categories, mountain glaciers and continental glaciers.

Mountain glaciers (also called alpine glaciers) exist in or adjacent to mountainous regions (▶Fig. 22.4a). Topographical features of the mountains control their shape; overall, mountain glaciers flow from higher elevations to lower elevations. Mountain glaciers include cirque glaciers, which fill bowl-shaped depressions, or **cirques**, on the flank of a mountain; valley glaciers, rivers of ice that flow down valleys (▶Fig. 22.4b; Fig. 22.2b); mountain ice caps, mounds of ice that submerge peaks and ridges at the crest of a mountain range (▶Fig. 22.4c); and piedmont glaciers (Fig. 22.2a), fans or lobes of ice that form where a valley glacier emerges from a valley and spreads out into the adjacent plain (▶Fig. 22.4d). Mountain glaciers range in size from a few hundred meters to a few hundred kilometers long.

Continental glaciers are vast ice sheets that spread over thousands of square kilometers of continental crust. Continental glaciers now exist only on Antarctica and Greenland (▶Fig. 22.5a, b). Remember that Antarctica is a continent, so the ice beneath the South Pole rests mostly on solid ground. New research reveals that at least three lakes underlie the

FIGURE 22.3 **(a)** The hexagonal shape of snowflakes. No two are alike. **(b)** Snowdrifts carved by the wind in Antarctica. Note the person for scale. **(c)** Frost on a glass window, showing the large crystals of water. **(d)** Glacial ice, as seen here on the wall of a tunnel bored into a glacier, tends to be blue. **(e)** Thin section of glacial ice, showing the texture of ice crystals and air bubbles.

glacier. The largest, Lake Vostok, is 5,400 square km in area. In contrast, the ice beneath the North Pole forms part of a thin sheet of sea ice floating on the Arctic Ocean. Continental glaciers flow outward from their thickest point (up to 3.5 km thick) and thin toward their margins, where they may be only a few hundred meters thick. The front edge of the glacier may divide into several tongue-shaped lobes, because not all of the glacier flows at the same speed. Earth is not alone in hosting polar ice sheets—Mars has them too (▶Box 22.1).

Geologists also find it valuable to distinguish between types of glaciers on the basis of the thermal conditions in which the glaciers exist. **Temperate glaciers** occur in regions where atmospheric temperatures become high enough during a substantial portion of the year for the glacial ice to be at or near its melting temperature throughout much of the year. **Polar glaciers** occur in regions where atmospheric temperatures stay so low all year long that the glacial ice remains well below melting temperature throughout the entire year.

How a Glacier Forms

In order for a glacier to form, three conditions must be met. First, the local climate must be sufficiently cold that winter snow does not melt entirely away during the

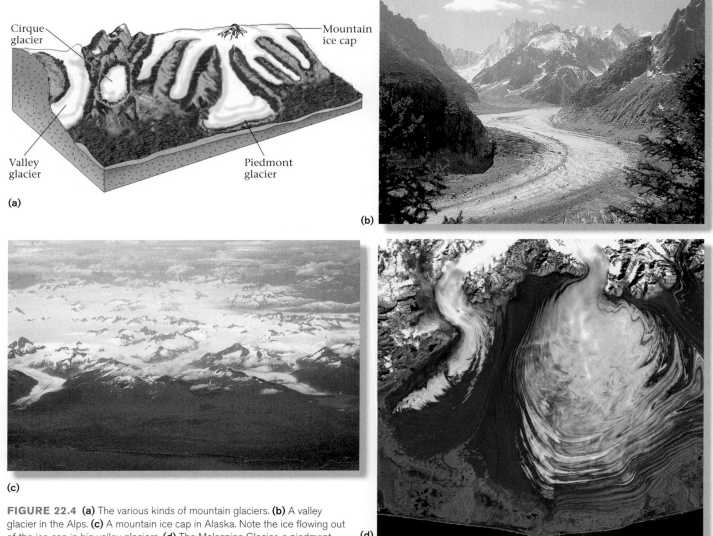

FIGURE 22.4 (a) The various kinds of mountain glaciers. (b) A valley glacier in the Alps. (c) A mountain ice cap in Alaska. Note the ice flowing out of the ice cap in big valley glaciers. (d) The Malaspina Glacier, a piedmont glacier along the coast of Alaska, as viewed by satellite. This is a "false color" image; red indicates rocky debris on the glacier surface, and gold indicates forest. Lines of debris trace out folds in the ice. The glacier is several kilometers across.

summer; second, there must be, or must have been, sufficient snowfall for a large amount to accumulate; and third, the slope of the surface on which the snow accumulates must be gentle enough that the snow does not slide away in avalanches, and must be protected enough that the snow doesn't blow away.

Glaciers develop in polar regions because even though relatively little snow falls today, temperatures remain so low that most ice and snow survive all year. Glaciers develop in mountains, even at low latitudes, because temperature decreases with elevation; at high elevations, the mean temperature stays low enough for ice and snow to survive all year. Since the temperature of a region depends on latitude, the specific elevation at which

mountain glaciers form also depends on latitude. In Earth's present-day climate, glaciers form only at elevations above 5 km between 0° and 30° latitude and down to sea level at 60° to 90° latitude (▶Fig. 22.7). Thus, you can see high-latitude glaciers from a cruise ship, but you have to climb the highest mountains of the Andes to find glaciers at the equator. Mountain glaciers tend to develop on the side of mountains that receives less wind, and on the side that receives less sunlight. Glaciers do not exist on slopes greater than about 30°, because avalanches clear such slopes.

The transformation of snow to glacier ice takes place slowly, as younger snow progressively buries older snow. Freshly fallen snow consists of delicate hexagonal crystals

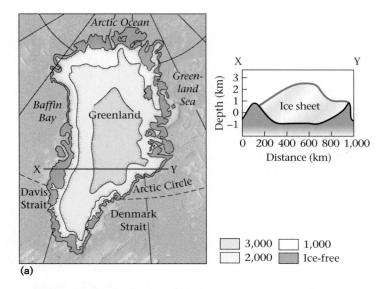

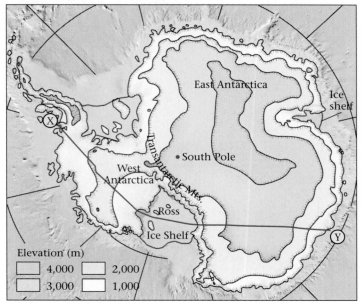

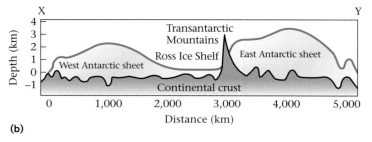

FIGURE 22.5 **(a)** A map and cross section of the Greenland ice sheet. **(b)** A map and cross section of the Antarctic ice sheet. Note that the Transantarctic Mountains divide the ice sheet in two. The Ross Ice Shelf lies between the East Antarctic and West Antarctic sheets. Valley glaciers carry ice from the ice sheets down to the shelf.

with sharp points. The crystals do not fit together tightly, so this snow contains about 90% air. With time, the points of the snowflakes become blunt because they either **sublimate** (evaporate directly into vapor) or melt, and the snow packs more tightly. As snow becomes buried, the weight of the overlying snow increases pressure, which causes remaining points of contact between snowflakes to melt. This process of melting at points of contact where the pressure is greatest is another example of pressure solution (see Chapter 11). Gradually, the snow transforms into a packed granular material called **firn**, which contains only about 25% air (▶Fig. 22.8a, b). Melting of firn grains at contact points produces water that crystallizes in the spaces between grains until eventually the firn transforms into a solid mass of glacial ice composed of interlocking ice crystals. Such glacial ice, which may still contain up to 20% air trapped in bubbles, tends to absorb red light and thus has a bluish color. The transformation of fresh snow to glacier ice can take as little as tens of years in regions with abundant snowfall, or as long as thousands of years in regions with little snowfall.

The Movement of Glacial Ice

When Louis Agassiz became fascinated by glaciers, he decided to find out how fast they flowed, so he hammered stakes into an Alpine glacier and watched the stakes move during the year. More recently, researchers have observed glacial movement with the aid of time-lapse photography, which shows the evolution of a glacier over several years in a movie that lasts a few minutes. In such movies, you can actually see a glacier flow across the screen, making its nickname "river of ice" seem perfectly appropriate.

How do glaciers move? Some move when meltwater accumulates at their bases, so that the mass of the glacier slides partially on a layer of water or on a slurry of water and sediment. During this kind of movement, known as basal sliding, the water or wet slurry holds the glacier above the underlying rock and reduces friction between the glacier and its substrate. Basal sliding is the dominant style of movement for **wet-bottom glaciers** (▶Fig. 22.9a). Recent research shows that basal sliding occurs episodically—the ice stays fixed for a while until stress builds up sufficiently to cause the ice to lurch forward suddenly. This "stick-slip" generates **ice quakes**, much as slip on a fault generates earthquakes. Measurements in Greenland show that larger ice quakes correspond to movement of a huge glacial mass by up to 12 m in less than a minute. The water forms at the base of wet-bottom glaciers because of pressure-induced melting, or because the glaciers exist in a climate that is warm enough for the ice to be at or near its melting temperature, or because the glacier ice acts like an insulating blanket and traps heat rising from the

BOX 22.1 THE REST OF THE STORY

Polar Ice Caps on Mars

The discovery that Mars has polar ice caps dates back to 1666, when the first telescopes allowed astronomers to resolve details of the red planet's surface. By 1719, astronomers had detected that Martian polar caps change in area with the season, suggesting that they partially melt and then refreeze. You can see these changes on modern images (▶Fig. 22.6a, b). The question of what the ice caps consist of remained a puzzle until fairly recently. Early studies revealed that the atmosphere of Mars consists mostly of carbon dioxide, so researchers first assumed that the ice caps consisted of frozen carbon dioxide. But data from modern spacecraft led to the conclusion that this initial assumption is false. It

now appears that the Martian ice caps consist mostly of water ice (mixed with dust) in layers from 1 to 3 km thick. During the winter, atmospheric carbon dioxide freezes and covers the north polar cap with a 1-m-thick layer of dry ice. During the summer this layer melts away. The south polar cap is different, for its dry ice blanket is 8 m thick and doesn't melt away entirely in the summer. The difference between the north and south poles may reflect elevation. The south pole is 6 km higher and therefore remains colder. The amount of melting appears to have increased as the years pass, hinting that Mars is now undergoing climate change.

High-resolution photographs reveal that distinctive canyons, up to 10 km wide and

1 km deep, spiral outward from the center of the north polar ice cap (▶Fig. 22.6c, d). Why did this pattern form? Recent calculations suggest that if the ice sublimates (transforms into gas) on the sunny side of a crack and refreezes on the shady side, the crack will migrate sideways over time. If the cracks migrate more slowly closer to the pole, where it's colder, than they do farther away, they will naturally evolve into spirals.

(c)

(a)

(b)

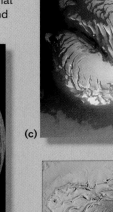

(d)

FIGURE 22.6 The ice caps of Mars. **(a)** During the winter, the ice caps expand to lower latitudes. **(b)** During the summer the ice cap shrinks. **(c)** A close up of the north polar cap in summer. **(d)** False-color imagery emphasizes the spiral canyons in the ice.

Earth below. Temperate glaciers are generally wet bottomed due primarily to melting by heat from the ground.

Glaciers also move by means of internal flow, during which the mass of ice slowly changes shape internally without breaking apart or completely melting. By studying ice deformation with a microscope, geologists have determined that internal flow involves two processes. First, ice crystals become plastically deformed by the rearrangement of water molecules within the crystal lattice. During this process, ice crystals change shape, old crystals disappear,

and new ones form. Second, if conditions allow very thin films of water to form on the surfaces of ice crystals, the crystals can slide past one another. Internal flow is the

FIGURE 22.7 The snow line depends on latitude.

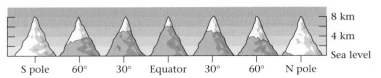

dominant style of movement for **dry-bottom glaciers**, those that are so cold that their base remains frozen to their substrate (▶Fig. 22.9b). Polar glaciers tend to be dry bottomed.

Note that the plastic deformation of ice in glaciers resembles the plastic deformation of metamorphic rock in mountain belts. However, metamorphic rock deforms plastically at depths greater than 10 to 15 km in the crust, whereas ice deforms plastically at depths below about 60 m (▶Fig. 22.9c). In other words, the "brittle-plastic transition" in ice lies at a depth of only about 60 m. Above this depth, large cracks can form in ice, whereas below this depth, cracks cannot form because ice is too plastic. A large crack that develops by brittle deformation in a glacier is called a **crevasse**. In large glaciers, crevasses can be

FIGURE 22.8 (a) Snow compacts and melts to form firn, which recrystallizes to make ice. **(b)** The size of ice crystals increases with depth in a glacier, where some crystals grow at the expense of others.

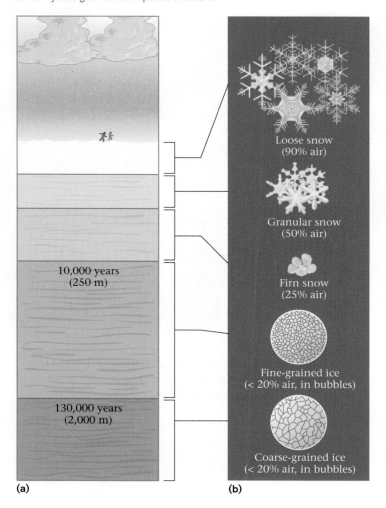

(a)   (b)

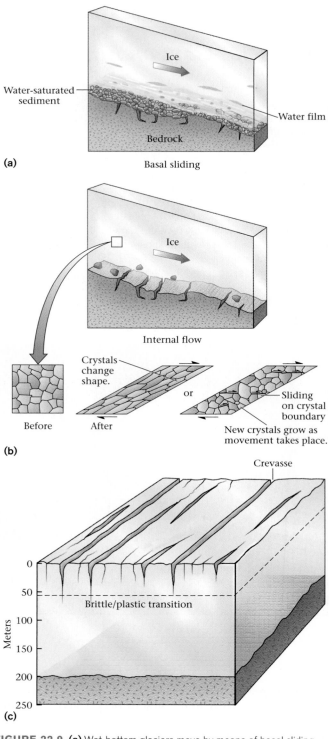

FIGURE 22.9 (a) Wet-bottom glaciers move by means of basal sliding. **(b)** Dry-bottom glaciers flow in the solid state. In some cases, the ice crystals stretch and rotate. In other cases, the ice recrystallizes and some crystals slide past each other, especially if there are thin films of water between grains. **(c)** Crevasses form in the brittle ice above the brittle-plastic transition in glaciers.

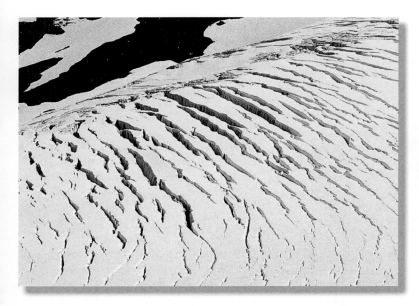

FIGURE 22.10 Crevasses in an Antarctic glacier. The crevasses are up to 10 m wide.

hundreds of meters long and tens of meters deep (▶Fig. 22.10). Tragically, several explorers have met their deaths by falling into crevasses.

Why do glaciers move? Ultimately, because the pull of gravity can cause weak ice to flow (▶Fig. 22.11a). Glaciers flow in the direction in which their surfaces slope. Thus, valley glaciers flow down their valleys, and continental ice sheets spread outward from their thickest point. To picture the movement of a continental ice sheet, imagine that a thick pile of ice builds up. Gravity causes the top of the pile to push down on the ice at the base. Eventually, the basal ice can no longer support the weight of the overlying ice and begins to deform plastically. When this happens, the basal ice starts squeezing out to the side, carrying the overlying ice with it. The greater the volume of ice that builds up, the wider the sheet of ice can become. You've seen a similar process of gravitational spreading if you've ever poured honey onto a plate. The honey can't build up into a narrow column because it's too weak; rather, it flows laterally away from the point where it lands to form a wide, thin layer (▶Fig. 22.11b, c).

Glaciers generally flow at rates of between 10 and 300 m per year—far slower than a river, but far faster than a silicate rock even under high-grade metamorphic conditions. The velocity of a particular glacier depends, in part, on the magnitude of the force driving its motion. For example, a glacier whose surface slopes steeply moves faster than one with a gently sloping surface. Flow velocity also depends on whether liquid water exists at places along the base of the glacier; wet-bottom glaciers tend to move faster than dry-bottom glaciers.

Not all parts of a glacier move at the same rate. For example, friction between rock and ice slows a glacier, so the center of a valley glacier moves faster than its margins, and the top of a glacier moves faster than its base (▶Fig. 22.12a, b). And because water at the base of a glacier allows it to travel more rapidly, portions of a continental glacier that flow over water or wet sediment become "ice streams" that travel 10 to 100 times faster than adjacent dry-bottom portions of the glacier. Similarly, if water builds up beneath a valley glacier to the point where it lifts the glacier off its substrate, the glacier undergoes a **surge** and flows much faster for a limited time (rarely more than a few months), until the water escapes. During surges, glaciers have been clocked at speeds of 20 to 110 m per day.

Glacial Advance and Retreat

Glaciers resemble bank accounts: snowfall adds to the account, while **ablation**—the removal of ice by sublimation (the evaporation of ice into water vapor), melting (the transformation of ice into liquid water, which flows away), and calving (the breaking off of chunks of ice at the edge of the glacier)—subtracts from the account (▶Fig. 22.13). Snowfall adds to the glacier in the **zone of accumulation**, whereas ablation subtracts in the **zone of ablation**; the boundary between these two zones is the **equilibrium line**. The zone of accumulation occurs where the temperature remains cold enough year round so that winter snow does not melt or sublimate away entirely during the summer. Therefore, elevation and latitude control the position of the equilibrium line.

The leading edge or margin of a glacier is called its **toe**, or terminus (▶Fig. 22.14a). If the rate at which ice builds up in the zone of accumulation exceeds the rate at which ablation occurs below the equilibrium line, then the toe moves forward into previously unglaciated regions. Such a change is called a **glacial advance** (▶Fig. 22.14b). In mountain glaciers, the position of a toe moves downslope during an advance, and in continental glaciers, the toe moves outward, away from the glacier's origin. If the rate of ablation below the equilibrium line equals the rate of accumulation, then the position of the toe remains fixed. But if the rate of ablation exceeds the rate of accumulation, then the position of the toe moves back toward the origin of the glacier; such a change is called a **glacial retreat** (▶Fig. 22.14c). During a mountain glacier's retreat, the position of the toe moves upslope. It's important to realize that when a glacier retreats, it's only the *position* of the toe that moves back toward the origin, for ice continues to flow toward the toe. Glacial ice cannot flow back toward the glacier's origin.

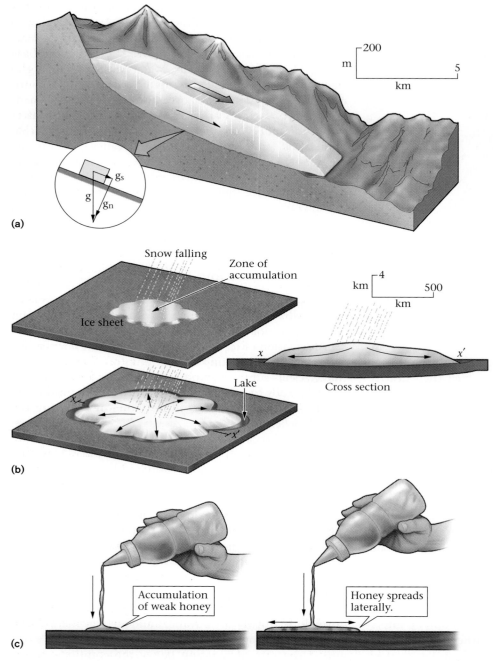

surface of the glacier, as overlying ice ablates. Thus, as a glacier flows, ice crystals follow curved trajectories (Fig. 22.14 a–c). For this reason, rocks picked up by ice at the base of the glacier may slowly move to the surface. The upward flow of ice where the Antarctic ice sheet collides with the Transantarctic Mountains brings up meteorites long buried in the ice (▶Fig. 22.15a, b).

Ice in the Sea

On the moonless night of April 14, 1912, the great ocean liner *Titanic* plowed through the calm but frigid waters of the North Atlantic on her maiden voyage from Southampton, England, to New York. Although radio broadcasts from other ships warned that icebergs, large blocks of ice floating in the water, had been sighted in the area and might pose a hazard, the ship sailed on, its crew convinced that they could see and avoid the biggest bergs, and that smaller ones would not be a problem for the steel hull of this "unsinkable" vessel. But in a story now retold countless times, their confidence was fatally wrong. At 11:40 P.M., while the first-class passengers danced, the *Titanic* struck a large iceberg. Lookouts had seen the ghostly mass of frozen water only minutes earlier and had alerted the ship's pilot, but the ship had been unable to turn fast enough to avoid disaster. The force of the blow split the steel hull that spanned five of the ship's sixteen water-tight compartments. The ship could stay afloat if four compartments flooded, but the

FIGURE 22.11 (a) Gravity drives the downslope motion of glaciers. Gravity can be portrayed by an arrow (vector) that points straight down. One component (g_s) is parallel to the slope and drives the flow, whereas the other (g_n) is perpendicular to the slope. **(b, c)** The gravitational spreading of an ice sheet resembles honey spreading across a table.

One final point before we leave the subject of glacial flow: beneath the zone of accumulation, a crystal of ice gradually moves down toward the base of the glacier as new ice accumulates above it, whereas beneath the zone of ablation, a crystal of ice gradually moves up toward the

flooding of five meant it would sink. At about 2:15 A.M., the bow disappeared below the water, and the stern rose until the ship protruded nearly vertically from the water. Without water to support its weight, the hull buckled and split in two. The stern section fell back down onto the

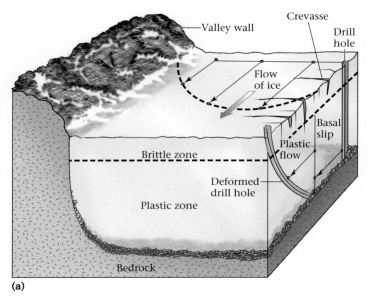

(a)

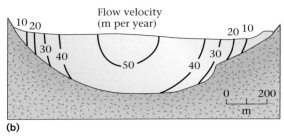

(b)

FIGURE 22.12 (a) Different parts of a glacier may flow at different velocities. The vector lengths indicate the velocity of flow. **(b)** This cross section of a glacier shows measured velocities of flow.

water and momentarily bobbed horizontally before following the bow, settling downward through over 3.5 km of water to the silent sea floor below. Because of an inadequate number of lifeboats and the inability of the crew to load passengers efficiently, only 705 passengers survived; 1,500 expired in the frigid waters of the Atlantic. The *Titanic* remained lost until 1985, when a team of oceanographers led by Robert Ballard located the sunken hull and photographed its eerie form where it lies upright in the mud.

Where do icebergs, such as the one responsible for the *Titanic*'s demise, originate? In high latitudes, mountain glaciers and continental ice sheets flow down to the sea, and they either stop at the shore or flow into the sea. Valley glaciers whose terminus lies in the water are tidewater glaciers. Some of these may protrude into the ocean to become ice tongues (▶Fig. 22.16a). Continental glaciers entering the sea become broad, flat sheets called **ice shelves** (▶Fig. 22.16b). Four-fifths of the ice lies below the sea surface, so in shallow water, it remains grounded (▶Fig. 22.17). But where the water becomes deep enough, the ice floats. At the boundary between glacier and ocean, blocks of ice calve off and tumble into the water with an impressive splash. If a free-floating chunk rises 6 m above the water and is at least 15 m long, it is formally called an **iceberg**. Smaller pieces, formed when ice blocks fragment before entering the water or after icebergs have had time to melt, include bergy bits, rising 1 to 5 m above the water and covering an area of 100 to 300 square m, and growlers, rising less than 1 m above the water and covering an area of about 20 square m—still big enough to damage a ship. Growlers get their name because of the sound they make as they bob in the sea and grind together.

Most large icebergs form along the western coast of Greenland or along the coast of Antarctica. Icebergs that calve off valley glaciers tend to be irregularly shaped with pointed peaks rising upward. Such glaciers are

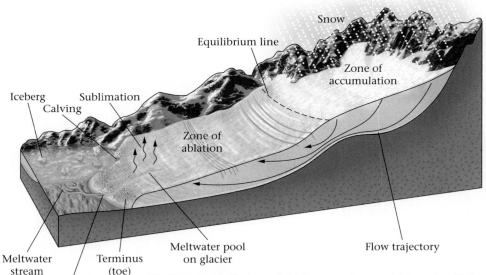

FIGURE 22.13 The zone of ablation, zone of accumulation, and equilibrium line. The arrows illustrate how a grain of ice flows down in the zone of accumulation and up in the zone of ablation, so that it follows a curved trajectory. Note that ice at the base of the glacier can flow up and over obstacles as long as the surface of the glacier has a downhill slope.

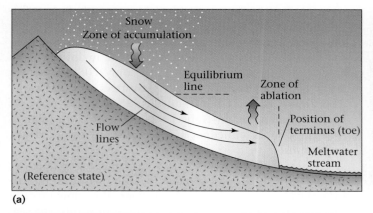

(a)

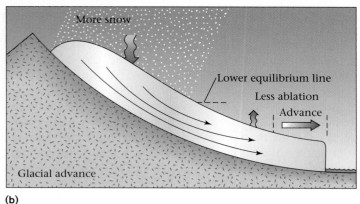

(b)

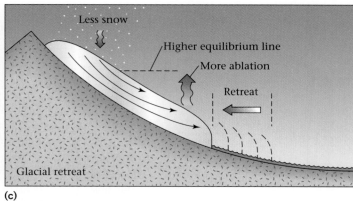

(c)

FIGURE 22.14 (a) The position of a glacier's toe represents the balance between the amount of ice that forms beneath the zone of accumulation and the amount of ice lost in the zone of ablation. (b) Glacial advance and (c) glacial retreat. Notice that ice always flows toward the toe of the glacier regardless of whether the toe advances or retreats, as indicated by the curving flow line.

called castle bergs or pinnacle bergs. One of the largest on record protruded about 180 m above the sea. Since four-fifths of the ice lies below the surface of the sea, the base of a large iceberg may actually be a few hundred meters below the surface (▶Fig. 22.18). Icebergs that originate in Greenland float into the "iceberg alley" region of the North Atlantic. These are the bergs that threaten ships, although the danger has diminished in modern times because of less ice and because of ice patrols that report the locations of floating ice. Blocks that calve off the vast ice shelves of Antarctica tend to have flat tops and nearly vertical sides—such glaciers are called tabular bergs. Some of the tabular

FIGURE 22.15 (a) Beneath the zone of ablation, ice crystals move up to the surface. Long-buried meteorites also collect at the surface. (b) A meteorite sitting on the surface of a glacier.

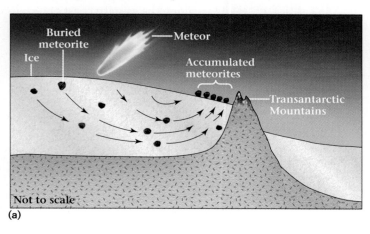

(a)

(b)

(a)

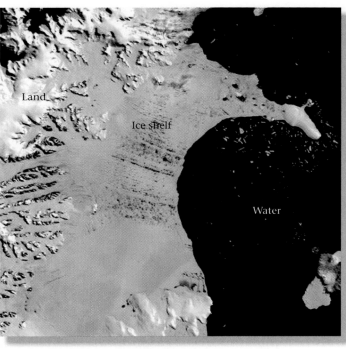

FIGURE 22.16 **(a)** An ice tongue protruding into the Ross Sea along the coast of Antarctica. **(b)** The Larsen Ice Shelf along the coast of Antarctica, as seen from a satellite in January 2002. Since then, much of the shelf has disintegrated.

(b)

FIGURE 22.18 An artist's rendition of an iceberg. This image is a composite of photographs spliced together. It conveys a sense of the relative proportions of ice above and below the surface of the sea.

FIGURE 22.17 Where a glacier reaches the sea, the ice stays grounded in shallow water and floats in deep water. Icebergs calve off the front of such tidewater glaciers. Sediment known as drop stones falls off the base of icebergs and collects on the sea floor.

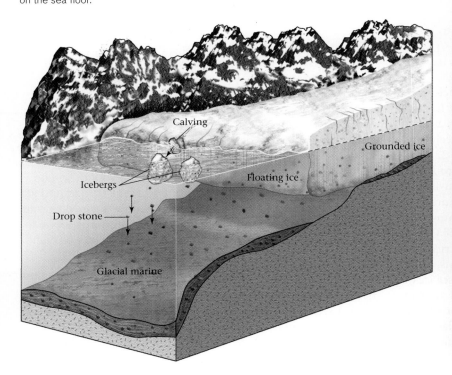

(a)

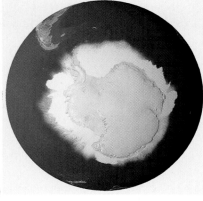

(b)

FIGURE 22.19 **(a)** Broken-up sea ice off the coast of Antarctica. **(b)** Sea ice covers most of the Arctic Ocean at the North Pole (top globe), and surounds Antarctica during the winter at the South Pole (bottom globe).

bergs in the Antarctic are truly immense; air photos have revealed bergs over 160 km across.

Not all ice floating in the sea originates as glaciers on land. In polar climates, the surface of the sea itself freezes, forming **sea ice** (▶Fig. 22.19a). Some sea ice, such as that covering the interior of the Arctic Ocean, floats freely, but some protrudes outward from the shore (▶Fig. 22.19b). Icebreakers can crunch through sea ice that is up to 2.5 m thick; the icebreaker rides up on the ice, then crushes it. Vast areas of ice shelves and of sea ice have been disintegrating in recent years, perhaps because of global warming. For example, open regions have been found in the Arctic Ocean during the summers, and the ice shelf in Antarctica has been decreasing rapidly in area. In some locations, large openings known as polynyas have developed in the sea ice of Antarctica. Some sea ice forms in winter and melts away in summer. But at high latitudes, sea ice may last for several years. For example, in the Arctic Ocean, sea ice may last long enough to make a 7- to 10-year voyage around the Arctic Ocean; this movement occurs in response to currents.

The existence of icebergs leaves a record in the stratigraphy of the sea floor, for icebergs carry ice-rafted sediment (▶Fig. 22.20). Larger rocks that drop from the ice to the sea floor are called **drop stones** (Fig. 22.17). In ancient glacial deposits, drop stones appear as isolated blocks surrounded by mud. Icebergs and smaller fragments also drop sand and gravel, derived by the erosion of continents, onto the sea floor. In cores extracted by drilling into sea-floor sediment, horizons of such land-derived sediment, sandwiched between layers of sediment formed from marine plankton shells, indicate times in Earth history when glaciers were breaking up and icebergs became particularly abundant.

> **Take-Home Message**
>
> Accumulation, burial, and recrystallization of snow forms glaciers. Continental ice sheets spread over broad areas of land and valley glaciers flow down valleys in mountain ranges in response to gravity. Sea ice covers portions of oceans.

(a)

FIGURE 22.20 A 7-m-long growler in Alaska containing now-tilted gravel-rich layers. The gravel fell on the ice and was buried during successive snowfalls when the ice was part of a glacier. The murky color of water is due to suspended glacial sediment.

(b)

22.3 CARVING AND CARRYING BY ICE

Glacial Erosion and Its Products

The Sierra Nevada mountains of California consist largely of granite that formed during the Mesozoic Era in the crust beneath a volcanic arc. During the past 10 to 20 million years, the land surface slowly rose, and erosion stripped away overlying rock to expose the granite. The style of erosion formed rounded domes. Then, during the last ice age, valley glaciers cut deep, steep-sided valleys into the range. In the process, some of the domes were cut in half, leaving a rounded surface on one side and a steep cliff on the other. Half Dome, in Yosemite National Park, formed in this way (▶Fig. 22.21a); its steep cliff has challenged many rock climbers. Such glacial erosion also produces the knife-edge ridges and pointed spires of high mountains (▶Fig. 22.21b), and the broad expanses where rock outcrops have been stripped of overlying sediment and polished smooth (▶Fig. 22.21c). Glacial erosion in the mountains can lower a valley floor by over 100 m, and continental glaciation during the last ice age stripped up to 30 m of material off the land in northern Canada.

Glaciers erode their substrate in several ways (▶Fig. 22.22a, b). During glacial incorporation, ice surrounds and incorporates debris. During glacial plucking (or glacial quarrying), a glacier breaks off and then carries away fragments of bedrock. Plucking occurs when ice freezes onto rock that has already started to crack and separate from its substrate; movement of the ice lifts off pieces of the rock. At the toe of a glacier, ice may actually bulldoze sediment and trees slightly before flowing over them.

As glaciers flow, clasts embedded in the ice act like the teeth of a giant rasp and grind away the substrate. This

(c)

FIGURE 22.21 Products of glacial erosion. **(a)** Half Dome, in Yosemite National Park, California. Before glaciation, the mountain was a complete dome; then glacial erosion truncated one side. **(b)** Glaciated mountains with sharp peaks. **(c)** A glacially polished surface on bedrock in New York City.

process, glacial abrasion, produces very fine sediment called rock flour, just as sanding wood produces sawdust. Rasping by embedded sand yields shiny **glacially polished surfaces**. Rasping by large clasts produces long

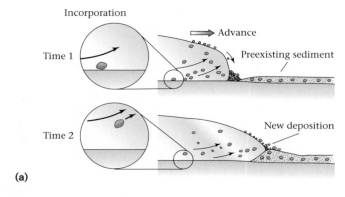

Incorporation

Time 1

Advance

Preexisting sediment

Time 2

New deposition

(a)

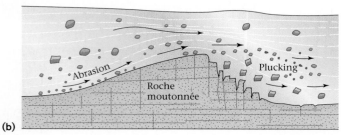

Abrasion

Plucking

Roche
moutonnée

(b)

(c)

(d)

FIGURE 22.22 The various kinds of glacial erosion. **(a)** Plowing and incorporation; **(b)** abrasion and plucking; **(c)** glacial striations in downtown Victoria, British Columbia (Canada); **(d)** chatter marks on a glacially eroded surface in Switzerland.

gouges, grooves, or scratches (1 cm to 1 m (0.4 in to 3 ft) across) called **glacial striations** (▶Fig. 22.22c). As you might expect, striations run parallel to the flow direction of the ice. When boulders entrained in the base of the ice strike bedrock below, as the ice moves, the impact may break off asymmetric wedges of bedrock, leaving behind indentations called chatter marks (▶Fig. 22.22d). In regions of wet-bottom glaciers, sediment-laden water rushes through tunnels at the base of the glaciers and can carve substantial channels.

Let's now look more closely at the erosional features associated with a mountain glacier (▶Fig. 22.23a–f). Freezing and thawing during the fall and spring help fracture the rock bordering the head of the glacier (the ice edge high in the mountains). This rock falls on the ice or gets picked up at the base of the ice, and moves downslope with the glacier. As a consequence, a bowl-shaped depression, or cirque, develops on the side of the mountain. If the ice later melts, a lake called a **tarn** may form at the base of the cirque. An **arête** (French for ridge), a residual knife-edge ridge of rock, separates two adjacent cirques. A pointed mountain peak surrounded by at least three cirques is called a **horn**. The Matterhorn, a peak in Switzerland, is a particularly beautiful example of a horn (Fig. 22.23e).

Glacial erosion severely modifies the shape of a valley. To see how, compare a river-eroded valley with a glacially eroded valley. If you look along the length of a river in unglaciated mountains, you'll see that it flows down a V-shaped valley, with the river channel forming the point of the V. The V develops because river erosion occurs only in the channel, and mass wasting causes the valley slopes to approach the angle of repose. But if you look down the length of a glacially eroded valley, you'll see that it resembles a U, with steep walls. A **U-shaped valley** forms because the combined processes of glacial abrasion and plucking not only lower the floor of the valley, but also bevel its sides.

Glacial erosion in mountains also modifies the intersections between tributaries and the trunk valley. In a river system, tributaries cut side valleys that merge with the trunk valley, such that the mouths of the tributary valleys lie at the same elevation as the trunk valley. The ridges (spurs) between valleys taper to a point when they join the trunk valley floor. During glaciation, tributary glaciers flow down side valleys into a trunk glacier. But the trunk glacier cuts the floor of its valley down to a depth that far exceeds the depth cut by the tributary glaciers. Thus, when the glaciers melt

FIGURE 22.23 **(a)** A mountain landscape before glaciation. The V-shaped valleys are the result of river erosion; the floors of the tributary valleys are at the same elevation as the trunk valley where they intersect. **(b)** During glaciation, the valleys fill with ice. **(c)** After glaciation, the region contains U-shaped valleys, hanging valleys, truncated spurs, and horns. **(d)** A U-shaped glacial valley. **(e)** The Matterhorn in Switzerland. **(f)** A waterfall spilling out of a hanging valley. Truncated spurs occur on both sides of the waterfall.

See for yourself...
Glacial Landscapes

On *Google Earth*™ imagery, you can easily spot glacial erosional and depositional features. And at high latitudes and/or elevations, you can still see glacial ice. The thumbnail images provided on this page are only to help identify tour sites. Go to wwnorton.com/studyspace to experience this flyover tour.

Continental Glacier, Antarctica
(Lat 89°59'52.49"S, Long 144°50'31.19"E)

At the southern end of the Earth, zoom 8,000 km (5,000 miles) and look down to see the continental glacier that covers Antarctica (Image G22.1). Rock crops out only along the Transantarctic Mountains and along the coast. The Antarctic Peninsula protrudes toward the southern tip of South America.

G22.1

Greenland and the Arctic Ocean
(Lat 85°4'55.25"N, Long 35°8'59.46"W)

At the northern end of the Earth, from 11,000 km (6,800 miles) out in space, you can see the Arctic Ocean. *Google Earth*™ does not generally show sea ice, so it's clear that, unlike the southern end, the northern end of the Earth is an ocean. A continental glacier covers the large island of Greenland (Image G22.2).

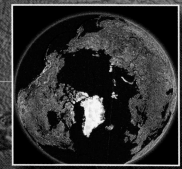

G22.2

G22.3

Southern Tip of Greenland (Lat 60°18'48.05"N, Long 44°28'1.45"W)

From an elevation of 250 km (155 miles), you can see the ice cap, valley glaciers, and fjords of southernmost Greenland (Image G22.3). Zoom to 15 km (9 miles), at Lat 60°9'20.86"N, Long 44°3'34.30"W, tilt your view to look north, and you can see the effects of glacial erosion in detail.

G22.4

G22.5

Baffin Island, Canada
(Lat 67°3'38.31"N, Long 65°33'2.88"W)

Zoom to 400 km (250 miles) and your view encompasses a portion of Baffin Island. You can see a small ice cap, and several valley glaciers ending in fjords (Image G22.4). Move to Lat 67°8'18.30"N, Long 65°0'56.99"W and zoom to 60 km (37 miles) to see valley glaciers "draining" the ice cap. Zoom down to 15 km (9 miles), tilt, and look upstream along a glacier (Image G22.5). Note how medial moraines form from the lateral moraines of valley glaciers that merge.

G22.6

Matterhorn, Switzerland (Lat 46°5'15.04"N, Long 7°38'2.72"E)

Ice-age glaciers sculpted the peaks that we see today. At these coordinates, zoom to 10 km (6 miles), tilt, and look south. You can see a glacially carved valley with the Matterhorn at its end. Multiple cirque formation produced the peak (Image G22.6).

Malaspina Glacier Area, Alaska (Lat 59°57'5.59"N, Long 140°32'54.35"W)

At the coordinates provided, zoom to 65 km (40 miles) and tilt your view to look NNE. You are gazing at the Malaspina Glacier, a piedmont glacier spreading out on a coastal plain (Image G22.7). A terminal moraine outlines the toe of the glacier. Note that the flow of the ice produced complex folds traced out by the pattern of debris on the glacier's surface. Zoom down to 8 km (5 miles) at Lat 59°50'35.76"N, Long 140°1'27.22"W. You can see knob-and-kettle topography and a braided outwash stream (Image G22.8).

G22.7

G22.8

Glaciated Peaks, Montana (Lat 48°54'53.37"N, Long 113°50'51.04"W)

At the coordinates provided, zoom to 7 km (4 miles), tilt your view, and look east. You can see the peak of Mt. Cleveland at the head of two cirques separated by a sharp arête (Image G22.9). Move a little south, to Lat 48°45'35.56"N, Long 113°37'47.02"W, and you are in Glacier National Park. Zoom to 8.5 km (5.3 miles) and tilt your view to look southwest, and you can see a U-shaped valley with a tarn at its head (Image G22.10).

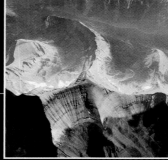

G22.9

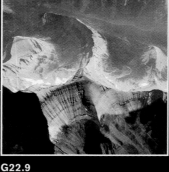

G22.10

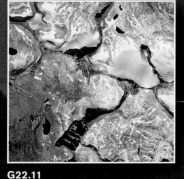

G22.11

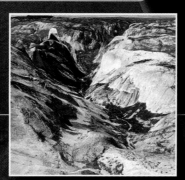

G22.12

Sierra Nevada, California (Lat 37°44'33.69"N, Long 119°16'21.06"W)

Fly to the coordinates provided and zoom to 7 km (4 miles). You can see several cirques and arêtes in the high peaks of the Sierra Nevada (Image G22.11). Fly to Lat 37°46'23.82"N, Long 119°30'10.68"W, zoom to 5 km (3 miles), and tilt to look SSW. You are looking down the Yosemite Valley, a U-shaped valley at the heart of Yosemite National Park. Waterfalls spill out of hanging valleys into the main valley (Image G22.12).

G22.13

Pluvial Lake Shore, Salt Lake City, Utah (Lat 40°31'18.45"N, Long 111°49'54.14"W)

Zoom to 3.5 km (2 miles), tilt, and look south to see the terrace of sediment that marks the beach of a pluvial lake, Lake Bonneville (Image G22.13). This lake covered the entire area of what is now Salt Lake City during the Pleistocene.

FIGURE 22.24 A roche moutonnée. The glacier flowed from right to left.

away, the mouths of the tributary valleys perch at a higher elevation than the floor of the trunk valley. Such side valleys are called **hanging valleys**. The water in post-glacial streams that flow down a hanging valley must cascade over a spectacular waterfall to reach the post-glacial trunk stream (Fig. 22.23f). As they erode, trunk glaciers also chop off the ends of spurs, creating truncated spurs.

Now let's look at the erosional features produced by continental ice sheets. To a large extent, these depend on the nature of the preglacial landscape. Where an ice sheet spreads over a region of low relief, such as the Canadian Shield, glacial erosion creates a vast region of polished, flat, striated surfaces. Where an ice sheet spreads over a hilly area, it deepens valleys and smooths hills. In central New York, for example, continental glaciers carved the deep valleys that now cradle the Finger Lakes, and in Maine, glaciers smoothed and streamlined the granite and metamorphic rock hills of Acadia National Park. Glacially eroded hills end up being elongate in the direction of flow and are asymmetric; glacial rasping smoothes and bevels the upstream part of the hill, creating a gentle slope, whereas glacial plucking eats away at the downstream part, making a steep slope. Ultimately, the hill's profile resembles that of a sheep lying in a meadow—such a hill is called a **roche moutonnée**, from the French for sheep rock (▶Fig. 22.24).

Fjords: Submerged Glacial Valleys

As noted earlier, where a valley glacier meets the sea, the glacier's base remains in contact with the ground until the water depth exceeds about four-fifths of the glacier's thickness, at which point the glacier floats. Thus, glaciers can continue carving U-shaped valleys even below sea level. In addition, during an ice age, water extracted from the sea becomes locked in the ice sheets on land, so sea level drops signifi-

cantly. Therefore, the floors of valleys cut by coastal glaciers during the last ice age were cut much deeper than present sea level. Today, the sea has flooded these deep valleys, producing **fjords** (see Chapter 18). In the spectacular fjord-land regions along the coasts of Norway, New Zealand, Chile, and Alaska, the walls of submerged U-shaped valleys rise straight from the sea as vertical cliffs up to 1,000 m high (▶Fig. 22.25). Fjords also develop where a glacial valley fills to become a lake.

The Glacial Conveyor: The Transport of Sediment by Ice

Glaciers can carry sediment of any size and, like a conveyor belt, transport it in the direction of flow (i.e., toward the toe). The sediment load either falls onto the surface of the glacier from bordering cliffs or gets plucked and lifted from the substrate and incorporated into the moving ice. Because of the curving flow lines of glacial ice, rocks plucked off the floor may eventually reach the surface.

Sediment dropped on the glacier's surface moves with the ice and becomes a stripe of debris. Stripes formed along the side edges of the glacier are **lateral moraines**. The word **moraine** was a local term used by Alpine farmers and shepherds for piles of rock and dirt. When the glacier finally melts, lateral moraines lie stranded along the side of the glacially carved valley, like bathtub rings. If flowing water runs along the edge of the glacier and sorts the sediment of a lateral moraine, a stratified sequence of sediment, called a **kame**, forms. Where two valley

> **Take-Home Message**
>
> A glacier scrapes up and plucks rock from its substrate, and carries debris that falls on its surface. Glacial erosion polishes and scratches rock and carves distinctive landforms, such as U-shaped valleys. Since ice is solid, moving ice does not sort sediment.

FIGURE 22.25 One of the many spectacular fjords of Norway.

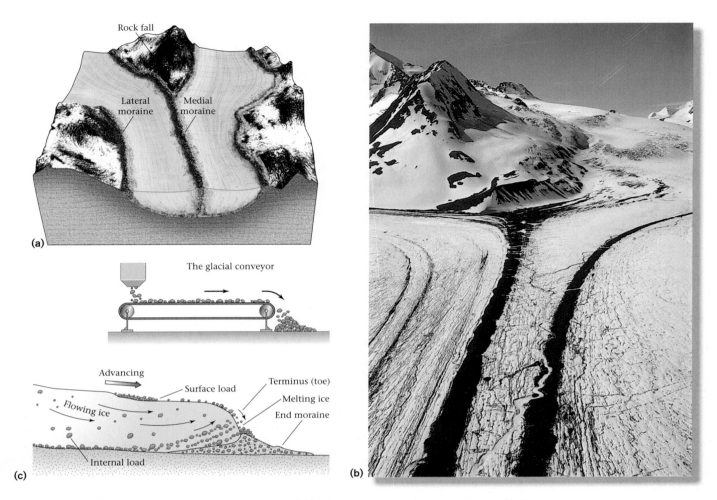

FIGURE 22.26 **(a)** Lateral and medial moraines on a glacier. **(b)** Medial moraines on a glacier near Alyeska, Alaska. The source of the moraine on the right is off the image. **(c)** The surface load plus the internal load accumulates at the toe of the glacier to constitute an end moraine; in effect, glaciers act like conveyor belts, constantly transporting more sediment toward the toe, regardless of whether the glacier advances or retreats.

glaciers merge, the debris constituting two lateral moraines merges to become a **medial moraine**, running as a stripe down the interior of the composite glacier (▶Fig. 22.26a, b). Trunk glaciers created by the merging of tributary glaciers contain several medial moraines. Glaciers passing through ranges with extremely high erosion rates may be completely buried by rocky debris. In some cases, a glacier incorporates so much rock that geologists refer to it as a rock glacier. Sediment transported to a glacier's toe by the glacial conveyor accumulates in a pile at the toe and builds up to form an **end moraine** (▶Fig. 22.26c).

So far, we've emphasized sediment moved by ice. But in temperate glacial environments, the flowing water at the base of the glacier, moving through channels, transports much of the sediment load, eventually depositing it beyond the terminus of the glacier.

22.4 DEPOSITION ASSOCIATED WITH GLACIATION

Types of Glacial Sedimentary Deposits

If you drill through the soil throughout much of the upper midwestern and northeastern United States and adjacent parts of Canada, the drill penetrates a layer of sediment deposited during the last ice age. A similar story holds true for much of northern Europe. Thus, many of the world's richest agricultural regions rely on soil derived from sediment deposited by glaciers during the ice age. This sediment buries a pre–ice age landscape, as frosting fills the irregularities on a cake. Preglacial valleys may be completely filled with sediment.

Glaciers and Glacial Landforms

Continental ice sheet

Crevasses

Ice shelf

Lower sea level

Drop stones

Iceberg

Mountain (alpine) glaciation

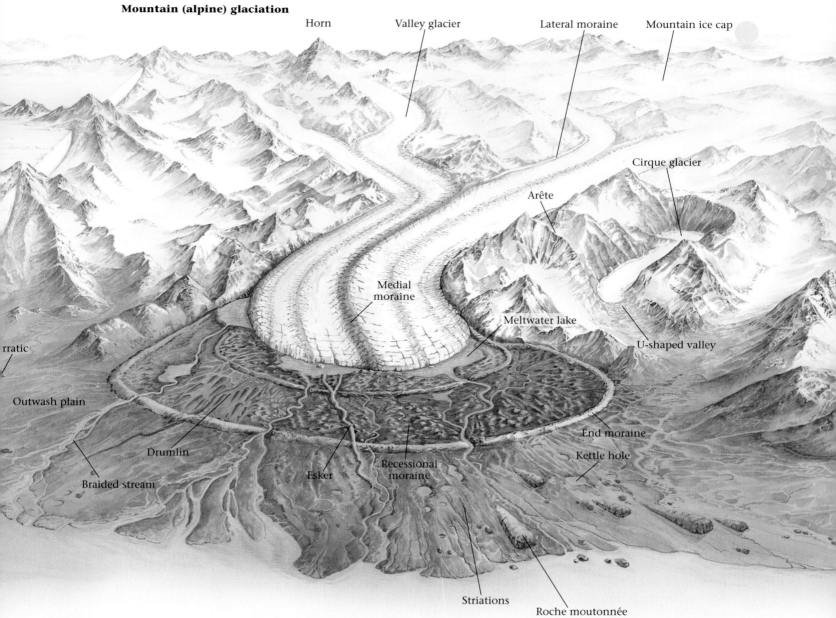

Horn   Valley glacier   Lateral moraine   Mountain ice cap

Cirque glacier

Arête

Medial moraine

Meltwater lake

U-shaped valley

rratic

Outwash plain

Drumlin

Braided stream

Esker

Recessional moraine

End moraine

Kettle hole

Striations

Roche moutonnée

Glaciers are rivers or sheets of ice that last all year and flow slowly. Continental glaciers, vast sheets of ice up to a few kilometers thick, covered extensive areas of land during times when Earth had a colder climate. Continental glaciers form when snow accumulates at high latitudes, then, when buried deeply enough, packs together and recrystallizes to make glacial ice. Ice, though solid, is weak, and thus ice sheets spread over the landscape, like syrup over a pancake. At the peak of the last ice age, ice sheets covered almost all of Canada, much of the United States, northern Europe, and parts of Russia.

The upper part of a sheet is brittle and may crack to form crevasses. Because ice sheets store so much of the Earth's water, sea level becomes lower during an ice age. When a glacier reaches the sea, it becomes an ice shelf. Rock that has been plucked up by the glacier along the way is carried out to sea with the ice; when the ice melts, the rocks fall to the sea floor as drop stones. At the edge of the shelf, icebergs calve off and float away.

A second class of glaciers, called mountain or alpine glaciers, form in mountainous areas because snow can last all year at high elevations. During the ice age, mountain glaciers grew and flowed out onto the land surface beyond the mountain front. The glacier at the

right has started to recede after formerly advancing and covering more of the land. Glacial recession may happen when the climate warms, so ice melts away faster at the toe (terminus) of the glacier than can be added at the source. In front of the glacier, you can find consequences of glacial erosion, such as striations on bedrock and roches moutonnées.

When the glacier pauses, till (unsorted glacial sediment) accumulates to form an end moraine. Meltwater lakes gather at the toe. Streams carry sediment and deposit it as glacial outwash. Sediment that accumulates in ice tunnels, exposed when the glacier melts, forms sinuous ridges called eskers. Even when the toe remains fixed in position for a while, the ice continues to flow, and thus molds underlying sediment into drumlins. Ice blocks buried in till melt to form kettle holes. (Though the examples shown here were left after the melting away of a piedmont glacier, most actually form during continental glaciation.)

In the mountains, the glacier is confined to a valley. Sediment falling on it from the mountains creates lateral and medial moraines. Glaciers carve distinct landforms in the mountains, such as cirques, arêtes, horns, and U-shaped valleys.

Several different types of sediment can be deposited in glacial environments; all of these types together constitute **glacial drift**. (The term dates from pre-Agassiz studies of glacial deposits, when geologists thought that the sediment had "drifted" into place during an immense flood.) Specifically, glacial drift includes the following:

- *Till*: Sediment transported by ice and deposited beneath or at the toe of a glacier is called **glacial till**. Glacial till is unsorted (it is a type of distinction), because the solid ice of glaciers can carry clasts of all sizes (▶Fig. 22.27a).

- *Erratics*: Glacial erratics are cobbles and boulders that have been dropped by a glacier. Some protrude from till piles, and others rest on glacially polished surfaces.

- *Glacial marine*: Where a sediment-laden glacier flows into the sea, icebergs calve off the toe and raft clasts out to sea. As the icebergs melt, they drop the clasts, which settle into the muddy sediment on the sea floor. Pebble- and larger-size clasts deposited in this way, as we have seen, are called drop stones. Sediment consisting of ice-rafted clasts mixed with marine sediment makes up glacial marine. Glacial marine can also consist of sediment carried into the sea by water flowing at the base of a glacier.

- *Glacial outwash*: Till deposited by a glacier at its toe may be picked up and transported by meltwater streams that sort the sediment. The clasts are deposited by a braided stream network in a broad area of gravel and sandbars called an outwash plain. This sediment is known as glacial outwash (▶Fig. 22.27b).

- *Glacial lake-bed sediment*: Streams transport fine clasts, including rock flour, away from the glacial front. This sediment eventually settles in meltwater lakes, forming a thick layer of glacial lake-bed sediment. This sediment commonly contains varves. A **varve** is a pair of thin layers deposited during a single year. One layer consists of silt brought in during spring floods, and the other of clay deposited in winter when the lake's surface freezes over and the water is still (▶Fig. 22.27c).

- *Loess*: When the warmer air above ice-free land beyond the toe of a glacier rises, the cold, denser air from above the glacier rushes in to take its place; a strong wind called catabatic wind therefore blows at the margin of a glacier. This wind picks up fine clay and silt and transports it away from the glacier's toe. Where the winds die down, the sediment settles and forms a thick layer. This sediment, called **loess**, sticks together because of the electrical charges on clay flakes; thus steep escarpments develop by erosion of loess deposits (▶Fig. 22.27d).

Till, which contains no layering, is sometimes called unstratified drift; glacial sediments that have been redistributed by flowing water are called stratified drift.

Depositional Landforms of Glacial Environments

Picture a hunter, dressed in deerskin, standing at the toe of a continental glacier in what is now southern Canada, waiting for an unwary woolly mammoth to wander by. It's a sunny summer day 12,000 years ago, and milky, sediment-laden streams gush from tunnels and channels at the base of the glacier and pour off the top as the ice melts. No mammoths venture by today, so the bored hunter climbs to the top of the glacier for a view. The climb isn't easy, partly because of the incessant catabatic wind, and partly because deep crevasses interrupt his path. Reaching the top of the ice sheet, the hunter looks northward, and the glare almost blinds him. Squinting, he sees the white of snow, and where the snow has blown away, he sees the rippled, glassy surface of bluish ice (Fig. 22.1). Here and there, a rock protrudes from the ice. Now looking southward, he surveys a stark landscape of low, sinuous ridges separated by hummocky (bumpy) plains. Braided streams, which carry meltwater out across this landscape, flow through the hummocky plains and supply a number of lakes. Dust fills the air because of the wind.

All of the landscape features that the hunter observes as he looks southward formed by deposition in glacial environments, both mountain and continental (▶Fig. 22.28a, b). The low, sinuous ridges, called end moraines, develop when the toe of a glacier stalls in one position for a while; the ice keeps flowing to the toe, and, like a giant conveyor belt, transports sediment with it; the sediment accumulates in a pile at the toe. The end moraine at the farthest limit of glaciation is called the **terminal moraine**. (The ridge of sediment that makes up Long Island, New York, and continues east-northeast into Cape Cod, Massachusetts, is part of the terminal moraine of the ice sheet that covered New England and eastern Canada during the last ice age; ▶Fig. 22.29.) The end moraines that form when a glacier stalls for a while as it recedes are **recessional moraines**.

Till that has been released at the base of a flowing glacier and remains after the glacier has melted away is called lodgment till. Clasts in lodgment till may be aligned and scratched during their movement or as ice

> ### Take-Home Message
>
> When ice melts, it drops unsorted sediment to form glacial till. Meltwater streams and wind can transport and sort the sediment to form outwash plains and loess deposits, respectively. Deposition by glaciers produces distinctive landforms, such as moraines.

(a)

(b)

(c)

(d)

FIGURE 22.27 (a) The unsorted sediment constituting glacial till in Ireland. (b) Braided streams choked with glacial outwash near the toe of a glacier in Alaska. (c) Note the alternating light and dark layers in this 20 cm-thick cross section of varved glacial lake-bed sediment. (d) A small escarpment cut into glacial loess deposits in Illinois.

flows over them. The flow of the glacier may mold till and other subglacial sediment into streamlined, elongate hills called **drumlins** (from the Gaelic word for hills). Drumlins tend to be asymmetric along their length, with a gentle downstream slope, tapered in the direction of flow, and a steeper upstream slope (▶Fig. 22.30a–c). The till left behind during rapid recession forms a thin, hummocky layer on the land surface; this till, together with lodgment till, forms a landscape feature known as **ground moraine**.

The hummocky surface of moraines reflects partly the variations in the amount of sediment supplied by the glacier, and partly the occurrence of **kettle holes**, circular depressions made when blocks of ice calve off the toe of the glacier, become buried by till, and then melt (▶Fig. 22.31a). A land surface with many kettle holes separated

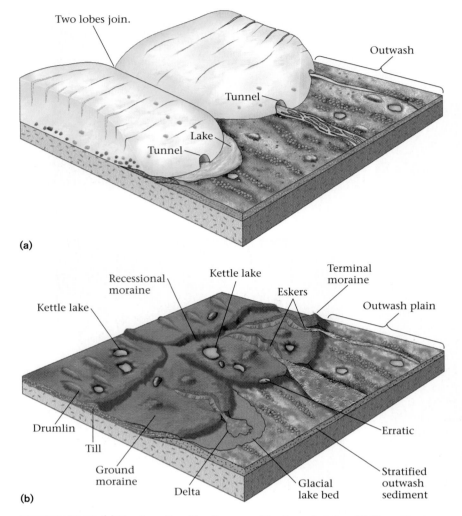

(a)

Two lobes join.

Outwash

Tunnel

Lake

Tunnel

(b)

Recessional
moraine

Kettle lake

Kettle lake

Terminal
moraine

Eskers

Outwash plain

Drumlin

Till

Ground
moraine

Delta

Glacial
lake bed

Erratic

Stratified
outwash
sediment

FIGURE 22.28 **(a)** The depositional landforms resulting from glaciation. **(b)** The setting in which various types of moraines form.

FIGURE 22.29 The moraines that constitute Long Island and Cape Cod.

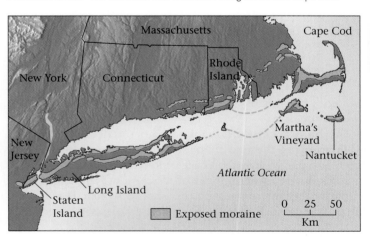

Massachusetts

Cape Cod

Rhode
Island

New York

Connecticut

New
Jersey

Martha's
Vineyard

Nantucket

Atlantic Ocean

Staten
Island

Long Island

☐ Exposed moraine

0   25   50
Km

by round hills of till displays knob-and-kettle topography (▶Fig. 22.31b, c).

Sediment-choked water pours out of tunnels in the ice. This water feeds large braided streams that sort till and redeposit it as glacial outwash, stratified layers, and lenses of gravel and sand. Thus, outwash plains form between recessional moraines and beyond the terminal moraine. Some of the meltwater pools in lowlands between recessional moraines, or in ice-margin lakes between the glacier's toe and the nearest recessional moraine. Even long after the glacier has melted away, the lowlands between recessional moraines persist as lakes or swamps. Glacial lake beds provide particularly fertile soil for agriculture. When a glacier eventually melts away, ridges of sorted sand and gravel, deposited in subglacial meltwater tunnels, snake across the ground moraine. These ridges are called **eskers** (▶Fig. 22.32). Sediments in glacial outwash plains and in eskers are important sources of sand and gravel for road building and construction.

22.5 OTHER CONSEQUENCES OF CONTINENTAL GLACIATION

Ice Loading and Glacial Rebound

When a large ice sheet (more than 50 km in diameter) grows on a continent, its weight causes the surface of the lithosphere to sink. In other words, ice loading causes **glacial subsidence**. Lithosphere, the relatively rigid outer shell of the Earth, can sink because the underlying asthenosphere is soft enough to flow slowly out of the way. You can see how this works by conducting a simple experiment. First, fill a bowl with honey and then place a thin rubber sheet over the honey (▶Fig. 22.33a). The rubber represents the lithosphere, and the honey represents the asthenosphere. If you place an ice cube on the rubber sheet, the sheet sinks because the weight of the ice pushes it down; the honey flows out of the way to make room. Because of ice loading, much of Antarctica and Greenland now lie below sea level (Fig. 22.5), so if their ice were instantly to melt away, these continents would be flooded by a shallow sea.

What happens when continental ice sheets do melt away? Gradually the surface of the underlying continent rises back up, a process called **glacial rebound**, and the asthenosphere flows back underneath to fill the space (▶Fig. 22.33b). In the honey and rubber example, when you remove the ice cube, the rubber sheet slowly returns to its original shape.

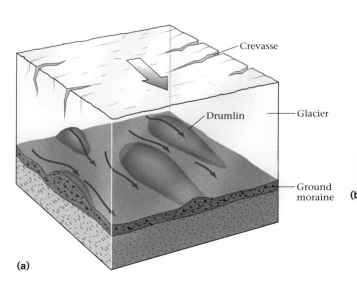

(a)

(b)

FIGURE 22.30 (a) The formation of a drumlin beneath a glacier. (b) Drumlins near Rochester, New York. (c) Topographic map emphasizing the shape of drumlins in New York State.

This process doesn't take place instantly, because the honey can only flow slowly. Similarly, because the asthenosphere flows *so* slowly (at rates of a few millimeters per year), it takes thousands of years for ice-depressed continents to rebound. Thus, glacial rebound is still taking place in some regions that were burdened by ice during the last ice age. Recently, researchers in North America have documented this movement by using GPS measurements (▶Fig. 22.34). Regions north of

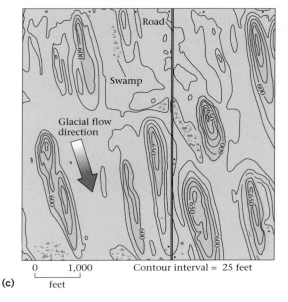

(c)

FIGURE 22.31 (a) When this ice block melts away, a kettle hole will form. (b, c) Knob-and-kettle topography, the hummocky surface of a moraine.

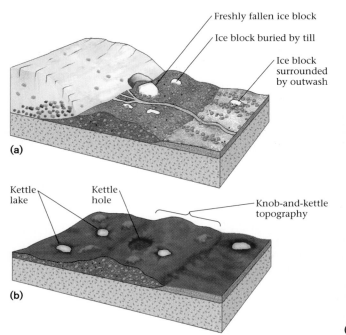

a line passing through the Great Lakes are rising, whereas regions south of this line are sinking. Where rebound affects coastal areas, beaches along the shoreline rise several meters above sea level and become terraces. Glacial rebound can also take place as mountain glaciers melt away.

(c)

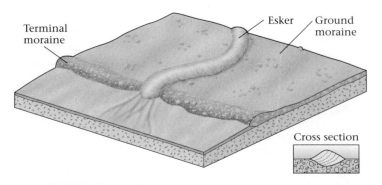

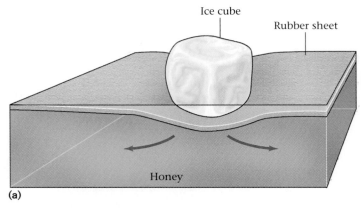

FIGURE 22.32 Eskers are snake-like ridges of sand and gravel that form when sediment fills meltwater tunnels at the base of a glacier.

Sea-Level Changes: The Glacial Reservoir in the Hydrologic Cycle

More of the Earth's surface and near-surface freshwater is stored in glacial ice than in any other reservoir. In fact, glacial ice accounts for 2.15% of Earth's total water supply, while lakes, rivers, soil, and the atmosphere together contain only 0.03%. During the last ice age, when glaciers covered almost 3 times as much land area as they do today, they held significantly more water (70 million cubic km, as opposed to 25 million cubic km today). In effect, water from the ocean reservoir transferred to the glacial reservoir and remained trapped on land. As a consequence, sea level dropped by as much as 100 m, and extensive areas of continental shelves became exposed as the coastline migrated seaward, in places by more than 100 km (▶Fig. 22.35a–c). People and animals migrated into the newly exposed coastal plains; in fact, fishermen dragging their nets along the Atlantic Ocean floor off New England today occasionally recover artifacts. The drop in sea level also created land bridges across the Bering Strait between North America and northeastern Asia and between Australia and Indonesia, providing convenient migration routes.

If today's ice sheets in Antarctica and Greenland were to melt, the crust of these continents would undergo glacial

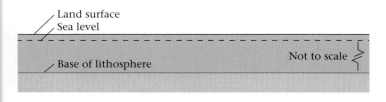

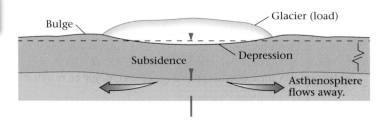

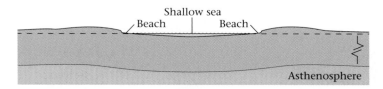

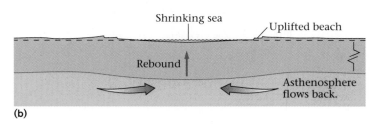

FIGURE 22.33 (a) An ice cube placed on the surface of a rubber sheet floating on honey illustrates the concept of glacial loading. **(b)** Cross sections illustrating the concept of glacial rebound. Top to bottom: Before glaciation, the surface of the lithosphere is flat. The weight of the glacier pushes the lithosphere down below sea level. The asthenosphere flows out of the way below, and lithosphere on either side bulges up. When the glacier melts, the depression fills with water. During glacial rebound, the floor of the shallow sea rises, and beaches along its shore rise.

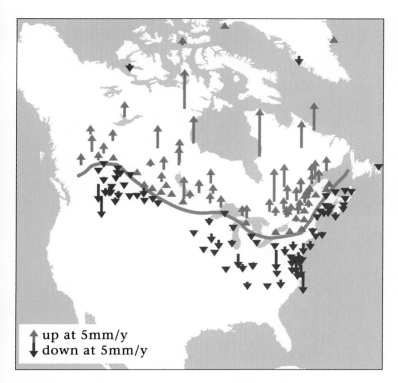

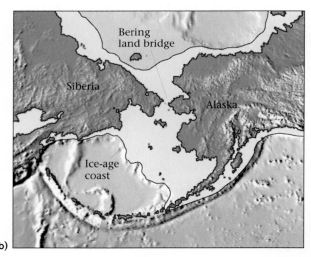

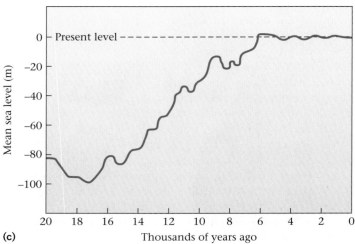

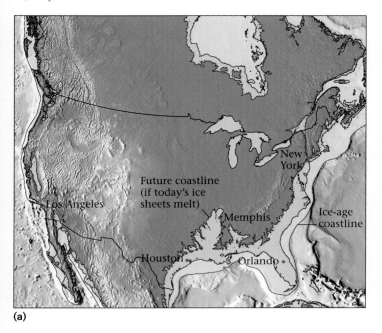

up at 5mm/y
down at 5mm/y

FIGURE 22.34 Rates of isostatic movement still taking place in response to the melting of the Pleistocene ice sheet, as indicated by satellite data. Red bars indicate upward movement; blue bars indicate downward movement.

FIGURE 22.35 (a) This map shows the coastline of North America during the last ice age, and the coastline should present-day ice sheets melt. Note that much of the continental shelf was exposed during the last ice age. (b) A land bridge across the Bering Strait connected Asia and North America during the last ice age. (c) The graph shows the change in sea level during the past 20,000 years.

rebound, global sea level would rise substantially, and low-lying areas of other continents would undergo flooding. In the United States, large areas of the coastal plain along the East Coast and Gulf Coast would flood, and cities such as Miami, Houston, New York, and Philadelphia would disappear beneath the waves. In Canada, substantial parts of the shield would flood.

Ice Dams, Drainage Reversals, and Lakes

When ice freezes over a sewer opening in a street, neither meltwater nor rain can enter the drain, and the street floods. Ice sheets play a similar role in glaciated environments. The ice may block the course of a river, leading to the formation of a lake. In addition, the weight of a glacier changes the tilt of the land surface and therefore the gradients of streams, and glacial sediment may fill preexisting valleys. In sum, continental glaciation destroys preexisting drainage networks. While the glacier exists, streams find different routes and carve out new valleys;

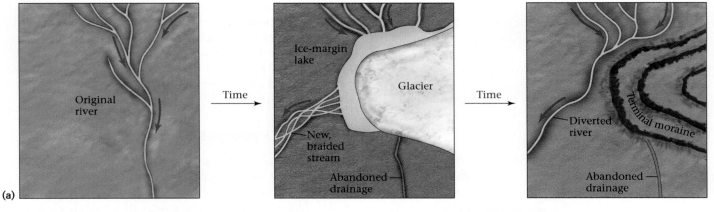

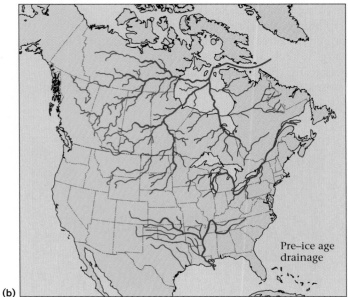

(b)

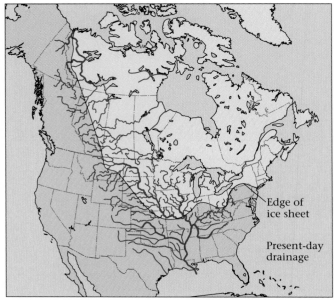

Edge of ice sheet

Present-day drainage

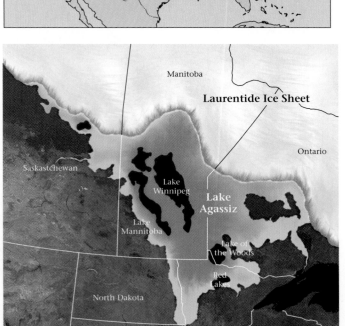

(c)

FIGURE 22.36 **(a)** When a glacier advances on the course of a river, the glacier blocks the drainage and causes a new stream to form. After the glacier melts away, the river remains diverted. **(b)** In North America, the major river systems that flowed northward before the last ice age have been destroyed. **(c)** Glacial Lake Agassiz was an ice-margin lake that formed near the end of the ice age. Lake Winnepeg in Manitoba is a small remnant.

by the time the glacier melts away, these new streams have become so well established that old river courses may remain abandoned.

Glaciation during the last ice age profoundly modified North America's drainage. Before the ice age, several major rivers drained much of the interior of the continent to the north, into the Arctic Ocean (▶Fig. 22.36a, b). The ice sheet buried this drainage network and diverted the flow into the Mississippi-Missouri network, which became larger. Then after the ice receded, regions covered by knob-and-kettle topography became spotted with thousands of small lakes, as now occur in central and southern Minnesota, and regions of the midwest covered by a smooth frosting of till became swampland, with little or no drainage at all. In the Canadian Shield, scouring left innumerable depressions that have now become lakes.

Inevitably, the ice dams that held back large ice-margin lakes melted and broke. In a matter of hours to days, the contents of the lakes drained, creating immense floodwaters that stripped the land of soil and left behind huge ripple marks. Glacial Lake Missoula, in Montana, filled when glaciers advanced and blocked the outlet of a large valley. When the glaciers retreated, the ice dam broke, releasing immense torrents that scoured eastern Washington, creating a barren, soil-free landscape called the channeled scablands (see Chapter 17). Recent evidence suggests that this process occurred many times.

The largest known ice-margin lake covered portions of Manitoba and Ontario, in south-central Canada, and North Dakota and Minnesota in the United States (▶Fig. 22.36c). This body of water, Glacial Lake Agassiz, existed between 11,700 and 9,000 years ago, a time during which the most recent phase of the last ice age came to a close and the continental glacier retreated north. At its largest, the lake covered over 250,000 square km (100,000 square miles), an area greater than that of all the present Great Lakes combined. Eventually, the ice sheet receded from the north shore of Glacial Lake Agassiz, so near the end of its life, the lake was surrounded by ice-free land. Field evidence suggests that the lake's demise came when it drained catastrophically. How did geologists reconstruct the history of Glacial Lake Agassiz? The present landscape holds the clues. Broad plains define the region that was once the lake floor, and beach terraces (now high and dry) define its former shoreline.

Pluvial Features

During ice ages, regions to the south of continental glaciers were wetter than they are today. Fed by enhanced rainfall, lakes accumulated in low-lying land even at a great distance from the ice front. The largest of these **pluvial lakes** (from the Latin *pluvia*, rain) in North America flooded interior basins of the Basin and Range Province in Utah and Nevada (▶Fig. 22.37a). These basins received drainage from the adjacent ranges but had no outlet to the sea, so they filled with water. The largest pluvial lake, Lake Bonneville, covered almost a third of western Utah (▶Fig. 22.37b). When this lake suddenly drained after a natural dam holding it back broke, it left a

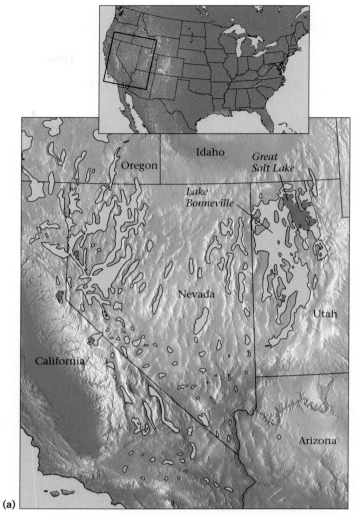

(a)

(b)

FIGURE 22.37 (a) The distribution of pluvial lakes in the Basin and Range Province during the last ice age. (b) The shoreline of Lake Bonneville along a mountain near Salt Lake City. The Great Salt Lake is a small remnant of this once-huge pluvial lake.

bathtub ring of shoreline rimming the Wasatch Mountains near Salt Lake City. Today's Great Salt Lake itself is but a small remnant of Lake Bonneville.

22.6 PERIGLACIAL ENVIRONMENTS

In polar latitudes today, and in regions adjacent to the fronts of continental glaciers during the last ice age, the mean annual temperature stays low enough (below −5°C) that soil moisture and groundwater freeze and, except in the upper few meters, stay solid all year. Such permanently frozen ground, or **permafrost**, may extend to depths of 1,500 m below the ground surface. Regions with widespread permafrost but without a blanket of snow or ice are called periglacial environments (the Greek *peri* means around, or encircling; periglacial environments appear around the edges of glacial environments) (▶Fig. 22.38a).

The upper few meters of permafrost may melt during the summer months, only to refreeze again when winter comes. As a consequence of the freeze-thaw process, the ground splits into pentagonal or hexagonal shapes, creating a landscape called **patterned ground** (▶Fig. 22.38b). Water fills the gaps between the cracks and freezes to create wedge-shaped walls of ice. In some places, freeze and thaw cycles in permafrost gradually push cobbles and pebbles up from the subsurface. Because the expansion of the ground is not even, the stones gradually collect between adjacent bulges to form stone rings (▶Fig. 22.38c). Some stone rings may also form when mud at depth pushes up from beneath a permafrost layer and forces stones aside.

Permafrost presents a unique challenge to people who live in polar regions or who work to extract resources from these regions. For example, heat from a building may warm and melt underlying permafrost, creating a mire into which the building settles. For this reason, buildings in permafrost regions must be placed on stilts, so that cold air can circulate beneath them to keep the ground frozen. When geologists discovered oil on the northern coast of Alaska, oil companies faced the challenge of shipping the oil to markets outside of Alaska. After much debate over the environmental impact, the trans-Alaska pipeline was built, and now it carries oil for 1,000 km to a seaport in southern Alaska (see Chapter 14). The oil must be warm during transport, or it would be too viscous to flow; thus, to prevent the warm pipeline from melting underlying permafrost, it had to be built on a frame that holds it above the ground for its entire length.

> **Take-Home Message**
>
> Extensive areas of permafrost—permanently frozen ground—form in high-latitude or high-elevation regions. Seasonal changes may cause the surface of permafrost to break into polygonal shapes, or to be covered by stone circles.

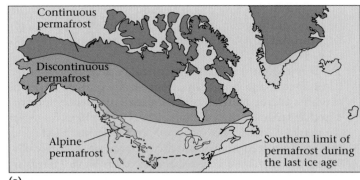

(a)

(b)

(c)

FIGURE 22.38 **(a)** The present-day distribution of periglacial environments in the Northern Hemisphere. **(b)** Patterned ground in the Northwest Territories, Canada, as viewed from a low-flying plane. The polygons are about 10 m across. The straight lines on the left side of the photo are caribou tracks. **(c)** Stone circles that are about 2 m across.

22.7 THE PLEISTOCENE ICE AGES

The Pleistocene Glaciers

Today, most of the land surface in New York City lies hidden beneath concrete and steel, but in Central Park it's still possible to see land in a seminatural state. If you stroll through the park and study the rock outcrops, you'll find that their top surfaces are smooth and polished, and in places have been grooved and scratched—you can also find erratics (▶Fig. 22.39). You are seeing evidence that an ice sheet once scraped along this ground. Geologists estimate that the ice sheet that overrode the New York City area may have been 250 m thick, enough to bury the Empire State Building up to the 80th floor.

Glacial features such as those on display in Central Park first led Louis Agassiz to propose the idea that vast continental glaciers advanced over substantial portions of North America, Europe, and Asia during a great ice age. Since Agassiz's day, thousands of geologists, by mapping out the distribution of glacial deposits and landforms, have gradually defined the extent of ice-age glaciers and a history of their movement (▶Box 22.2).

The fact that these glacial features decorate the surface of the Earth today means that the most recent ice age occurred fairly recently during Earth history. This ice age, responsible for the glacial landforms of North America and Eurasia, happened mostly during the Pleistocene Epoch, which began 1.8 million years ago (see Chapter 13), so it is commonly known as the Pleistocene ice age. This traditional title is a bit of a misnomer. Recent studies demonstrate that the glaciations of this ice age actually began between 3.0 and 2.5 million years ago, during the Pliocene Epoch, and continued through the Pleistocene. Further, there was not just a single ice advance, but, as we will see,

FIGURE 22.39 An erratic sitting on a glacially polished surface in Central Park, New York City.

there were many—probably over twenty. (Thus, all of these events together might better be called the Pleistocene ice ages.) Geologists use the name Holocene to refer to the time since the last glaciation (i.e., to the last 11,000 years).

From their mapping of glacial striations and deposits, geologists have determined where the great Pleistocene ice sheets originated and flowed. In North America, the Laurentide ice sheet started to grow over northeastern Canada, then merged with the Keewatin ice sheet, which originated in northwestern Canada. Together, these ice sheets eventually covered all of Canada east of the Rocky Mountains and extended southward across the border as far as southern Illinois (▶Fig. 22.40a, b). At their maximum, they attained a thickness of 2 to 3 km; each thinned toward its toe. In northeastern Canada, the ice sheet eroded the land surface. Farther south and west, it deposited sediment (▶Fig. 22.40c) These ice sheets also eventually merged with the Greenland ice sheet to the northeast and the Cordilleran ice sheet to the west; the Cordilleran covered the mountains of western Canada, as well as the southern third of Alaska.

During the Pleistocene ice age, mountain ice caps and valley glaciers also grew in the Rocky Mountains, the Sierra Nevada, and the Cascade Mountains. In Eurasia, a large ice sheet formed in northernmost Europe and adjacent Asia, and gradually covered all of Scandinavia and northern Russia. This ice sheet flowed southward across France until it reached the Alps and merged with Alpine mountain glaciers; it also covered all of Ireland and almost all of the United Kingdom. A smaller ice sheet grew in eastern Siberia and expanded in the mountains of central Asia. In the Southern Hemisphere, Antarctica remained ice-covered, and mountain ice caps expanded in the Andes, but there were no continental glaciers in South America, Africa, or Australia.

In addition to continental ice sheets, sea ice in the Northern Hemisphere expanded to cover all of the Arctic Ocean and parts of the North Atlantic. Sea ice surrounded Iceland and approached Scotland, and also fringed most of western Canada and southeastern Alaska.

Life and Climate in the Pleistocene World

During the Pleistocene ice age, all climatic belts shifted southward (▶Fig. 22.41a, b). Geologists can document this shift by examining fossil pollen, which can survive for thousands of years, preserved in the sediment of bogs. Presently, the southern boundary of North America's tundra, a treeless region supporting only low shrubs, moss, and lichen capable of living on permafrost, lies at a latitude of 68°N; during the Pleistocene ice age, it moved down to 48°N. Much of the interior of the United States, which now has temperate, deciduous forest, harbored cold-weather spruce and pine forest. Ice-age climates also

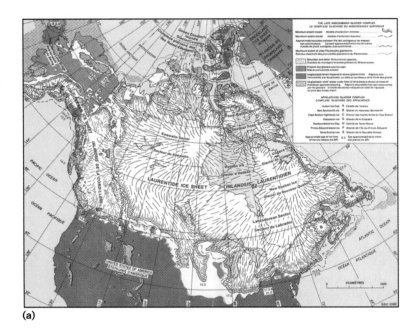

(a)

FIGURE 22.40 **(a)** A map of the major ice sheets in North America, as first compiled by V. K. Prest. The thin lines indicate the flow direction of the ice. Note that one major sheet originated to the west of Hudson Bay and another to the east. **(b)** A simplified map of the distribution of major ice sheets during the Pleistocene Epoch. The arrows indicate the flow trajectories of the ice sheets. **(c)** Top: A continental glacier scours and erodes the land surface beneath its center, while at the margins it deposits sediment. Map: The Laurentide ice sheet scoured and eroded the land in northern and eastern Canada, and deposited sediment in western Canada and the midwestern United States.

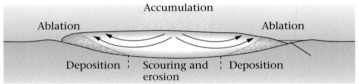

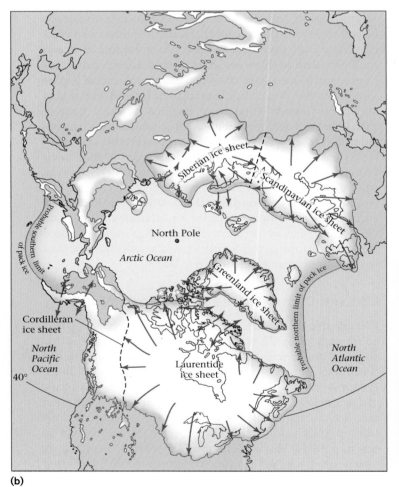

(b)

(c)

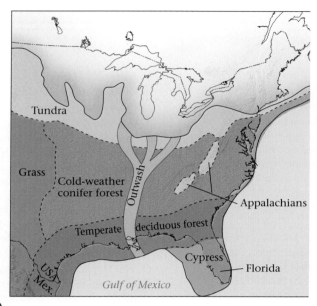

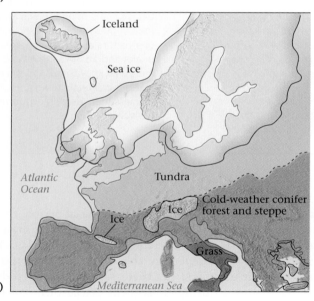

FIGURE 22.41 **(a)** Pleistocene climatic belts in North America; **(b)** Pleistocene climatic belts in Europe.

changed the distribution of rainfall on the planet: increased rainfall in North America led to the filling of pluvial lakes in Utah and Nevada, whereas decreased rainfall in equatorial regions led to shrinkage of the rain forest. Overall, the contrast between colder, glaciated regions and warmer, unglaciated regions created windier conditions worldwide. These winds sent glacial rock flour skyward, creating a dusty atmosphere (and, presumably, spectacular sunsets). The dust settled to create extensive deposits of loess. And because glaciers trapped so much water, as we have seen, sea level dropped.

Numerous species of now-extinct large mammals inhabited the Pleistocene world (▶Fig. 22.42). Giant mammoths and mastodons, relatives of the elephant, along with woolly rhinos, musk oxen, reindeer, giant ground sloths, bison, lions, saber-toothed cats, giant cave bears, and hyenas wandered forests and tundra in North America. Early human-like species were already foraging in the woods by the beginning of the Pleistocene Epoch, and by the end modern *Homo sapiens* lived on every continent except Antarctica, and had discovered fire and invented tools. Rapidly changing climates may have triggered a global migration of early humans, who gained access to the Americas, Indonesia, and Australia via land bridges that became exposed when sea level dropped.

Chronology of the Pleistocene Ice Ages

Louis Agassiz assumed that only one ice age had affected the planet. But close examination of the stratigraphy of glacial deposits on land revealed that paleosol (ancient soil preserved in the stratigraphic record), as well as beds containing fossils of warmer-weather animals and plants, separated distinct layers of glacial sediment. This observation suggested that between episodes of glacial deposition, glaciers receded and temperate climates prevailed. In the second half of the twentieth century, when modern methods for dating geological materials became available, the difference in ages between the different layers of glacial sediment could be confirmed. Clearly, ice-age glaciers had advanced and then retreated more than once. Times during which the glaciers grew and covered substantial areas of the continents are called glacial periods, or **glaciations**, and times between glacial periods are called interglacial periods, or **interglacials**.

Using the on-land sedimentary record, geologists recognized five Pleistocene glaciations in Europe (named, in order of increasing age: Würm, Riss, Mindel, Gunz, and

FIGURE 22.42 Examples of now-extinct large mammals that roamed the countryside during the Pleistocene Epoch.

BOX 22.2 THE HUMAN ANGLE

So You Want to See Glaciation?

Though the last of the Pleistocene continental ice sheet that once covered much of North America vanished about 6,000 years ago, you can find evidence of its power quite easily. The Great Lakes, along the U.S.-Canadian border, the Finger Lakes and drumlins of New York, the low-lying moraines and outwash plains of Illinois, and the polished outcrops of southern Canada all formed in response to the movement of this glacier. But if you want to see continental glaciers in action today, you must trek to Greenland or Antarctica.

Mountain glaciers are easier to reach. A trip to the mountains of western North America (including Alaska), the Alps of France or Switzerland, the Andes of South America, or the mountains of southern New Zealand will bring you in contact with live glaciers. You can even spot glaciers from the comfort of a cruise ship.

Some of the most spectacular glacial landscapes in North America formed during the Pleistocene Epoch, when mountain glaciers were more widespread. These are now on display in national parks.

- *Glacier National Park (Montana):* This park, which borders Waterton Lakes National Park in Canada, displays giant cirques, U-shaped valleys, hanging valleys, terminal moraines, and fifty small relicts of formerly larger glaciers, all in a mountainous terrain that reaches elevations of over 3 km. Unfortunately, these glaciers are melting away quickly!

- *Yosemite National Park (California):* A huge U-shaped valley, carved into the Sierra Nevada granite batholith, makes up the centerpiece of this park. Waterfalls spill out of hanging valleys bordering the valley.

- *Voyageurs National Park (Minnesota):* This park lacks the high peaks of mountainous parks, but shows the dramatic consequences of glacial scouring and deposition on the Canadian Shield. The low-lying landscape, dotted with lakes, contains abundant polished surfaces, glacial striations, and erratics, along with moraines, glacial lake beds, and outwash plains.

- *Acadia National Park (Maine):* During the last ice age, the continental ice sheet overrode low bedrock hills and flowed into the sea along the coast of Maine. This park provides some of the best examples of the consequences. Its hills were scoured and shaped into large roches moutonnées by glacial flow. Some of the deeper valleys have now become small fjords.

- *Glacier Bay National Park (Alaska):* In Glacier Bay, huge tidewater glaciers fringe the sea, creating immense ice cliffs from which icebergs calve off. Cruise ships bring tourists up to the toes of these glaciers. More adventurous visitors can climb the coastal peaks and observe lateral and medial moraines, crevasses, and the erosional and depositional consequences of glaciers that have already retreated up the valley.

Donau) and, traditionally four in the midwestern United States (Wisconsinan, Illinoian, Kansan, and Nebraskan, named after the southernmost states in which their till was deposited; ▶Fig. 22.43). Since the mid-1980s, geologists no longer recognize Nebraskan and Kansan; they are lumped together as "pre-Illinoian." With the advent of radiometric dating in the mid-twentieth century, the ages of the younger glaciations were determined by dating wood trapped in glacial deposits. Geologists estimate the ages of the older glaciations by identifying fossils in the deposits. Because of their greater age, these deposits have been thoroughly weathered and dissected.

The four-stage chronology of North American glaciation was turned on its head in the 1960s, when geologists began to study submarine sediment containing the fossilized shells of microscopic marine plankton. Because the assemblage of plankton species living in warm water is not the same as the assemblage living in cold water, geologists can track changes in the temperature of the ocean by studying plankton fossils. Researchers found that in sediment of the last 2 million years, assuming that cold water indicates a glacial period and warm water an interglacial period, there were twenty to thirty different glacial advances during the Pleistocene Epoch. The four traditionally recognized glaciations probably represent only the largest of these. Sediments deposited on land by other glaciations were eroded and redistributed during subsequent glaciations, or were eroded by streams and wind during interglacials.

Geologists refined their conclusions about the frequency of Pleistocene glaciations by examining the composition of fossil shells. Shells of many plankton species consist of calcite ($CaCO_3$). The oxygen in the shells includes two isotopes, a heavier one (^{18}O) and a lighter one (^{16}O). The ratio of these isotopes tells us about the water temperature in which the plankton grew; this is because as water gets colder, plankton incorporate a higher proportion of ^{18}O into their shells (see Chapter 23). Thus, intervals in the stratigraphic record during which plankton shells have a large ratio of ^{18}O to ^{16}O define times when Earth had a colder, glacial climate. The record also indicates that twenty to thirty of these events occurred during the last 3 million years (▶Fig. 22.44).

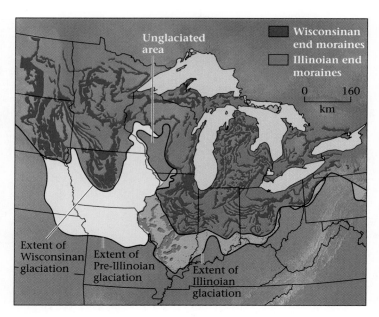

FIGURE 22.43 Pleistocene deposits in the United States.

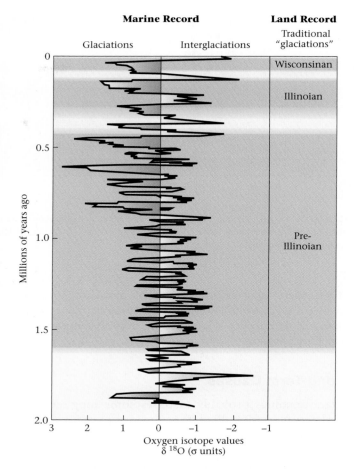

FIGURE 22.44 This time column shows the variations in oxygen-isotope ratios from marine sediment that define twenty to thirty glaciations and interglacials during the Pleistocene Epoch. The green bands represent the approximate boundaries of the principal glacial stages recognized on land in the midwestern United States. Note that the traditional names "Kansan" and "Nebraskan" are no longer used and have been replaced by "pre-Illinoian."

Older Ice Ages During Earth History

So far, we've focused on the Pleistocene ice age because of its importance in developing Earth's present landscape. Was this the only ice age during Earth history, or do ice ages happen frequently? To answer such questions, geologists study the stratigraphic record and search for ancient glacial deposits that have hardened into rock. These deposits, called **tillites**, consist of larger clasts distributed throughout a matrix of sandstone and mudstone. In many cases, tillites are deposited on glacially polished surfaces.

By using the stratigraphic principles described in Chapter 12, geologists have determined that tillites were deposited about 280 million years ago, in Permian time; these are the deposits Alfred Wegener studied when he argued in favor of continental drift (▶Fig. 22.45a, b). Tillites were also deposited about 600 to 700 million years ago (at the end of the Proterozoic Eon), about 2.2 billion years ago (near the beginning of the Proterozoic), and perhaps about 2.7 billion years ago (in the Archean Eon). Strata deposited at other times in Earth history do not contain tillites. Thus, it appears that glacial advances and retreats have not occurred steadily throughout Earth history, but rather are restricted to specific

> **Take-Home Message**
>
> During the last Ice Age, which began about 3 Ma, continental ice sheets advanced and retreated several times. This period includes the Pleistocene (which formally began 1.8 Ma, and ended at about 11 Ka). Other ice ages happened earlier in Earth history.

time intervals, or ice ages, of which there were four or five: Pleistocene, Permian, late Proterozoic, early Proterozoic, and perhaps Archean. Of particular note, some tillites of the late Proterozoic event were deposited at equatorial latitudes, suggesting that, for at least a short time, the continents worldwide were largely glaciated, and the sea may have been covered worldwide by ice. Geologists refer to the ice encrusted planet as **snowball Earth**.

22.8 THE CAUSES OF ICE AGES

Ice ages occur only during restricted intervals of Earth history, hundreds of millions of years apart. But within an ice age, glaciers advance and retreat with a frequency measured in tens of thousands to hundreds of thousands of years.

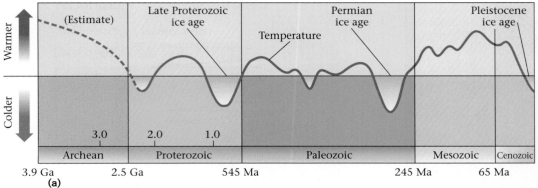

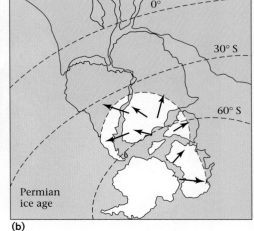

FIGURE 22.45 (a) This time column shows pre-Pleistocene glaciations during Earth history. **(b)** The distribution of Permian glacial features on a reconstruction of Gondwana, the southern part of the supercontinent that existed at the time.

Thus, there must be both long-term and short-term controls on glaciation. The nature of these controls emphasizes the complexity of interactions among components of the Earth System.

Long-Term Causes

Plate tectonics probably exercises some long-term control over glaciation. First, for an ice age to occur, substantial areas of continents must have drifted to high latitudes; if all continents sat along the equator, the land would be too warm for snow to accumulate. Second, glaciations can only take place when most continents lie well above sea level; sea-level changes may be controlled, in part, by changes in rates of sea-floor spreading. Finally, ice ages can't develop when oceanic currents carry heat to high latitudes; currents are controlled, in part, by positions of continents and volcanic arcs, as determined by plate motions.

The concentration of carbon dioxide in the atmosphere may also determine whether an ice age can occur. Carbon dioxide is a greenhouse gas—it traps infrared radiation rising from the Earth—so if the concentration of CO_2 increases, the atmosphere becomes warmer. Ice sheets cannot form during periods when the atmosphere has a relatively high concentration of CO_2, even if other factors favor glaciation. But what might cause long-term changes in CO_2 concentration? Possibilities include changes in the number of marine organisms that extract CO_2 to make shells; changes in the amount of chemical weathering on land (determined by the abundance of mountain ranges), for weathering absorbs CO_2; and changes in the amount of volcanic activity. Major stages in evolution may also affect CO_2 concentration. For example, the appearance of coal swamps at the end of the Paleozoic may have removed CO_2, for plants incorporate CO_2.

Short-Term Causes

Now we've seen how the stage could be set for an ice age to occur, but why do glaciers advance and retreat periodically *during* an ice age? In 1920, Milutin Milankovitch, a Serbian astronomer and geophysicist, came up with an explanation. Milankovitch studied how the Earth's orbit changes shape and how its axis changes orientation through time, and he calculated the frequency of these changes. In particular, he evaluated three aspects of Earth's movement around the Sun.

- *Orbital eccentricity*: Milankovitch showed that the Earth's orbit gradually changes from a more circular shape to a more elliptical shape. This eccentricity cycle takes around 100,000 years (▶Fig. 22.46a).

- *Tilt of Earth's axis*: We have seasons because the Earth's axis is not perpendicular to the plane of its orbit. Milankovitch calculated that over time, the tilt angle varies between 22.5° and 24.5°, with a frequency of 41,000 years (▶Fig. 22.46b).

- *Precession of Earth's axis*: If you've ever set a top spinning, you've probably noticed that its axis gradually traces a conical path. This motion, or wobble, is called precession (▶Fig. 22.46c). Milankovitch determined that the Earth's axis wobbles over the course of about 23,000 years. Right now, the Earth's axis points toward Polaris, making it the north star, but 12,000 years ago the axis pointed to Vega. Precession determines the relationship between the timing of the seasons and the position of Earth along its orbit around the Sun. If summer happens when Earth is closer to the Sun, then we have a warm summer, but if it happens when Earth is farther away from the Sun, then we have a cool summer.

Milankovitch showed that precession, along with variations in orbital eccentricity and tilt, combine to affect the

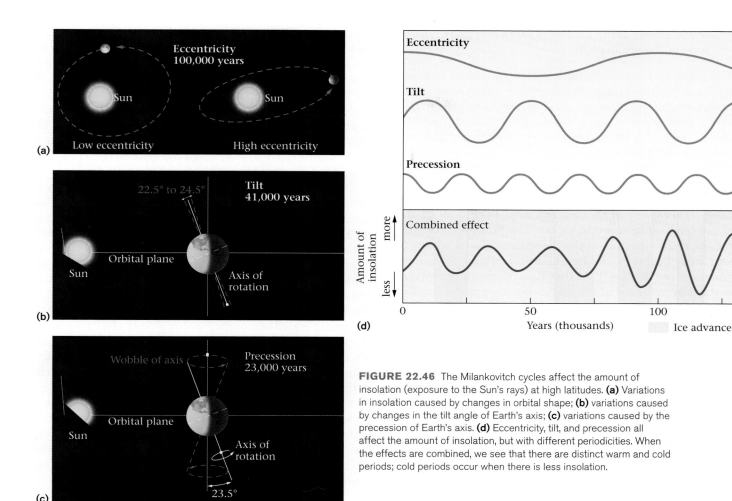

FIGURE 22.46 The Milankovitch cycles affect the amount of insolation (exposure to the Sun's rays) at high latitudes. **(a)** Variations in insolation caused by changes in orbital shape; **(b)** variations caused by changes in the tilt angle of Earth's axis; **(c)** variations caused by the precession of Earth's axis. **(d)** Eccentricity, tilt, and precession all affect the amount of insolation, but with different periodicities. When the effects are combined, we see that there are distinct warm and cold periods; cold periods occur when there is less insolation.

total annual amount of insolation (exposure to the Sun's rays) and the seasonal distribution of insolation that the Earth receives at the high latitudes (such as 65°N). For example, high-latitude regions receive more insolation when the Earth's axis is almost perpendicular to its orbital plane than when its axis is greatly tilted. According to Milankovitch, glaciers tend to advance during times of cool summers at 65°N, which occur roughly 100,000, 40,000, and 20,000 years apart. When geologists began to study the climate record, they found climate cycles with the frequency predicted by Milankovitch. These climate cycles are now called **Milankovitch cycles** (▶Fig. 22.46d).

The discovery of Milankovitch cycles in the geologic record strongly supports the contention that changes in the Earth's orbit and tilt help trigger short-term advances and retreats during an ice age. But orbit and tilt changes cannot be the whole story, because they could cause only about a 4°C temperature decrease (relative to today's temperature), and during glaciations the temperature decreased 5° to 7°C along coasts and 10° to 13°C inland. Geologists suggest

that several other factors may come into play in order to trigger a glacial advance.

- *A changing albedo*: When snow remains on land throughout the year, or clouds form in the sky, the albedo (reflectivity) of the Earth increases, so Earth's surface reflects incoming sunlight and thus becomes even cooler.

- *Interrupting the global heat conveyor*: As the climate cools, evaporation rates from the sea decrease, so seawater does not become as salty. And decreasing salinity might stop the system of thermohaline currents that brings warm water to high latitudes (see Chapter 18). Thus, the high latitudes become even colder than they would otherwise.

- *Biological processes that change CO_2 concentration*: Several kinds of biological processes may have amplified climate changes by altering the concentration of carbon dioxide in the atmosphere. For example, a greater amount of plankton growing in the oceans could absorb more carbon dioxide and thus remove it from the atmosphere.

The three processes described above are called positive-feedback mechanisms: they enhance the process that causes them. Because of positive feedback, the Earth could cool more than it would otherwise during the cooler stage of a Milankovitch cycle, and this could trigger a glacial advance.

A Model for Pleistocene-Ice-Age History

Long-term cooling in the Cenozoic Era. Taking all of the above causes into account, we can now propose a scenario for the events that led to the Pleistocene glacial advances. Our story begins in the Eocene Epoch, about 55 million years ago (▶Fig. 22.47). At that time, climates were warm and balmy, not only in the tropics but even above the Arctic Circle. At the end of the Middle Eocene (37 million years ago), the climate began to cool, and by Early Oligocene time (33 million years ago), Antarctica became glaciated. The Antarctic ice sheet came and went until the middle of the Miocene Epoch (15 million years ago), when an ice sheet formed that has lasted ever since. Ice sheets did not appear in the Arctic, however, until 2 to 3 million years ago, when the Pleistocene ice age began.

These long-term climate changes may have been caused, in part, by changes in the pattern of oceanic currents that happened, in turn, because of plate tectonics. For example, in the Eocene, the collision of India with Asia cut off warm equatorial currents that had been flowing in the Tethys Sea. And in the Miocene and Oligocene, Australia and South America drifted away from Antarctica, allowing the cold circum-Antarctic current to develop. This new current prevented warm, southward-flowing currents from reaching Antarctica, allowing ice to form and survive

in the region. With the loss of the warm currents, the climate of Antarctica overall underwent cooling. Changes to atmospheric circulation and temperature may also have happened at this time. Models suggest that the uplift of the Himalayas and Tibet diverted winds in a way that cooled the climate. Further, this uplift exposed more rock to chemical weathering, perhaps leading to extraction of CO_2 from the atmosphere (for chemical weathering reactions absorb CO_2); a decrease in the concentration of this greenhouse gas would contribute to atmospheric cooling.

So far, we've examined hypotheses that explain long-term cooling since 55 million years ago, but what caused the sudden appearance of the Laurentide ice sheet about 2–3 million years ago? This event coincides with another well-known plate-tectonic event, the closing of the gap between North and South America by the growth of the Isthmus of Panama. When this land bridge formed, it separated the waters of the Caribbean from those of the tropical Pacific for the first time, and when this happened, warm currents that previously flowed out of the Caribbean into the Pacific were blocked and diverted northward to merge with the Gulf Stream. This current transfers warm water from the Caribbean up the Atlantic Coast of North America and ultimately to the British Isles. As the warm water moves up the Atlantic Coast, it generates warm, moisture-laden air that provides a source for the snow that falls over New England, eastern Canada, and Greenland. In other words, the Arctic has long been cold enough for ice caps, but until the Gulf Stream was diverted northward by the growth of Panama, there was no source of moisture to make abundant snow and ice.

Short-term advances and retreats in the Pleistocene Epoch. Once the Earth's climate had cooled overall, short-term processes such as the Milankovitch cycles led to periodic advances and retreats of the glaciers. To understand how, let's look at a possible case history of a single advance and retreat of the Laurentide ice sheet. (Note that such models remain the subject of vigorous debate.)

- *Stage 1:* During the overall cooler climates of the late Cenozoic Era, the Earth reaches a point in the Milankovitch cycle when the average mean temperature in temperate latitudes drops. Because of glacial rebound, the ice-free surface of northern Canada has risen to an altitude of several hundred meters above sea level. With lower temperatures and higher elevations, not all of winter's snow melts away during the summer. Eventually, snow covers the entire region of northern Canada, even during the summer. Because of the snow's high albedo, it reflects sunlight, so the region grows still colder (a positive-feedback effect) and even more snow accumulates. Precipitation rates are high, because evaporation off the Gulf Stream provides moisture. Finally, the snow at the base of the pile turns

FIGURE 22.47 The graph shows the gradual cooling of Earth's atmosphere since the Cretaceous Period.

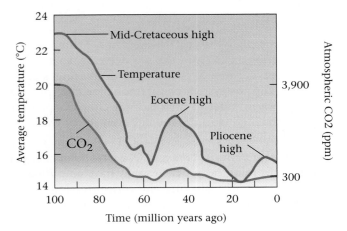

to ice, and the ice begins to spread outward under its own weight. A new continental glacier has been born.

- *Stage 2*: The ice sheet continues to grow as more snow piles up in the zone of accumulation. And as the ice sheet grows, the atmosphere continues to cool because of the albedo effect. But now, the weight of the ice loads the continent and makes it sink, so the elevation of the glacier decreases, and its surface approaches the equilibrium line. Also, the temperature becomes cold enough that in high latitudes the Atlantic Ocean begins to freeze. As the sea ice covers the ocean, the amount of evaporation decreases, so the source of snow is cut off and the amount of snowfall diminishes. The glacial advance pretty much chokes on its own success. The decrease in the glacier's elevation (leading to warmer summer temperatures) on the ice surface, as well as the decrease in snowfall, causes ablation to occur faster than accumulation, and the glacier begins to retreat.

> **Take-Home Message**
>
> Ice ages happen when the distribution of continents, ocean currents, and the concentration of atmospheric CO_2 are appropriate. Advances and retreats during an ice age are controlled by Milankovitch cycles (variations in Earth's orbit shape and in the orientation of Earth's rotation axis).

- *Stage 3*: As the glacier retreats, temperatures gradually increase, and the sea ice begins to melt. The supply of water to the atmosphere from evaporation increases once again, but with the warmer temperatures and lower elevations, this water precipitates as rain during the summer. The rain drastically accelerates the rate of ice melting, and the retreat progresses quite rapidly.

22.9 WILL THERE BE ANOTHER GLACIAL ADVANCE?

What does the future hold? Considering the periodicity of glacial advances and retreats during the Pleistocene Epoch, we may be living in an interglacial period. Pleistocene interglacials lasted about 10,000 years, and since the present interglacial began about 11,000 years ago, the time seems ripe for a new glaciation. If a glacier on the scale of the Laurentide ice sheet were to develop, major cities and agricultural belts would be overrun by ice, and their populations would have to migrate southward. Long before the ice front arrived, though, the climate would become so hostile that the cities would already be abandoned.

The Earth actually had a brush with ice-age conditions between the 1300s and the mid-1800s, when average annual temperatures in the Northern Hemisphere fell suffi-

FIGURE 22.48 Skaters (c. 1600) on the frozen canals of the Netherlands during the little ice age.

ciently for mountain glaciers to advance significantly. During this period, now known as the **little ice age**, sea ice surrounded Iceland, and canals froze in the Netherlands, leading to that country's tradition of skating (▶Fig. 22.48). Some researchers speculate that the depopulation of the western hemisphere, in the wake of European conquest, caused temporary reforestation, for without inhabitants, farmlands went untended. The new forests absorbed CO_2, and caused atmospheric concentrations of CO_2 to decrease, leading to the cooler conditions that triggered the little ice age. Others speculate that the change reflects increased cloud cover, not a change in CO_2 concentration. Researchers will likely propose additional ideas as work on this problem continues. During the past 150 years, temperatures have warmed, and most mountain glaciers have retreated significantly (▶Fig. 22.49). We no longer see the icebergs that once threatened Atlantic shipping lanes and sank the *Titanic*, and large slabs frequently calve off the Antarctic ice sheet. In fact, the Larsen B Ice Shelf of Antarctica, an area larger than Rhode Island, disintegrated in 2002. Some researchers suggest that this global-warming trend is due to the addition of CO_2 to the atmosphere from the burning of forests and the use of fossil fuels (see Chapter 23). Global warming could conceivably cause a "super-interglacial."

If the climate were to become significantly warmer than it is today, the ice sheet of West Antarctica might begin to float and then break up rapidly. If all of today's ice caps melted, global sea level would rise by 70 m (230 feet), extensive areas of coastal plains would be flooded, and major coastal cities such as New York, Miami, and

FIGURE 22.49 A couple of hundred years ago, glacial ice filled the cirque in the background and all of the valley up to the height of the lateral moraine in the foreground. In this 2003 photo, most of the ice of this Alaskan glacier has vanished.

London would be submerged (Fig. 22.37). Instead of protruding from ice, the tip of the Empire State Building would protrude from the sea. Icehouse or greenhouse? We may not know which scenario will play out in the future until it happens. However, researchers have voiced concern that, at least in the near term, glacial melting will be the order of the day, as global temperatures seem to be rising.

Chapter Summary

- Glaciers are streams or sheets of recrystallized ice that survive for the entire year and flow in response to gravity. Mountain glaciers exist in high regions and fill cirques and valleys. Continental glaciers (ice sheets) spread over substantial areas of the continents.

- Glaciers form when snow accumulates over a long period of time. With progressive burial, the snow first turns to firn, and then to ice.

- Wet-bottom glaciers move by basal sliding over water or wet sediment. Dry-bottom glaciers move by internal flow. In general, glaciers move tens of meters per year.

- Glaciers move because of gravitational pull, as long as the glaciers have a surface slope.

- Whether the toe of a glacier stays fixed in position, advances farther from the glacier's origin, or retreats back toward the origin depends on the balance between the rate at which snow builds up in the zone of accumulation and the rate at which glaciers melt or sublimate in the zone of ablation.

- Icebergs break off glaciers that flow into the sea. Continental glaciers that flow out into the sea along a coast make ice shelves. Sea ice forms where the ocean's surface freezes.

- Glacial ice can flow over sediment or incorporate sediment. The clasts embedded in glacial ice act like a rasp that abrades the substrate.

- Mountain glaciers carve numerous landforms, including cirques, arêtes, horns, U-shaped valleys, hanging valleys, and truncated spurs. Fjords are glacially carved valleys that fill with water when sea level rises after an ice age.

- Glaciers can transport sediment of all sizes. Glacial drift includes till, glacial marine, glacial outwash, lake-bed mud, and loess. Lateral moraines accumulate along the sides of valley glaciers, and medial moraines form down the middle of a glacier. End moraines accumulate at a glacier's toe.

- Glacial depositional landforms include moraines, knob-and-kettle topography, drumlins, kames, eskers, meltwater lakes, and outwash plains.

- Continental crust subsides as a result of ice loading. When the glacier melts away, the crust rebounds.

- When water is stored in continental glaciers, sea level drops. When glaciers melt, sea level rises.

- During past ice ages, the climate in regions south of the continental glaciers was wetter, and pluvial lakes formed. Permafrost (permanently frozen ground) exists in periglacial environments.

- During the Pleistocene ice age, large continental glaciers covered much of North America, Europe, and Asia.

- The stratigraphy of Pleistocene glacial deposits preserved on land records five European and four North American glaciations, times during which ice sheets advanced. The record preserved in marine sediments records twenty to thirty such events. The land record, therefore, is incomplete.

- Long-term causes of ice ages include plate tectonics and changes in the concentration of CO_2 in the atmosphere. Short-term causes include the Milankovitch cycles (caused by periodic changes in Earth's orbit and tilt).

Geopuzzle Revisited

Even if people had the ability to build large cities 12,000 years ago, they couldn't have built at the localities that are now New York or Edinburgh, for the land surface at these locations lay beneath hundreds of meters of ice. 12,000 years ago marked the end of the last glacial advance when huge ice sheets covered substantial areas of North America and Eurasia.

Key Terms

ablation (p. 688)
albedo (p. 683)
arête (p. 693)
cirque (p. 684)
continental glacier (ice sheet) (p. 684)
crevasse (p. 686)
drop stone (p. 693)
drumlin (p. 699)
dry-bottom glacier (p. 686)
equilibrium line (p. 688)
end moraine (p. 697)
erratic (p. 681)
esker (p. 700)
firn (p. 686)
fjord (p. 696)
glacial advance (p. 688)
glacial drift (p. 698)
glacial rebound (p. 705)
glacial retreat (p. 688)
glacial striation (p. 693)
glacial subsidence (p. 704)
glacial till (p. 698)
glacially polished surface (p. 693)
glaciation (p. 712)
glacier (p. 681)
ground moraine (p. 699)
hanging valley (p. 696)
horn (p. 693)
ice age (p. 682)
ice quake (p. 686)
ice sheet (p. 682)
interglacial (p. 712)

kame (p. 697)
kettle hole (p. 699)
lateral moraine (p. 697)
little ice age (p. 717)
loess (p. 698)
medial moraine (p. 697)
Milankovitch cycles (p. 715)
moraine (p. 697)
mountain (alpine) glacier (p. 684)
patterned ground (p. 707)
permafrost (p. 706)
pluvial lake (p. 706)
polar glacier (p. 685)
recessional moraine (p. 698)
roche moutonnée (p. 696)
sea ice (p. 691)
Snowball Earth (p. 714)
sublimation (p. 685)
surge (p. 688)
tarn (p. 693)
temperate glacier (p. 685)
terminal moraine (p. 698)
tillite (p. 713)
toe (p. 688)
U-shaped valley (p. 696)
varve (p. 698)
wet-bottom glacier (p. 686)
zone of ablation (p. 688)
zone of accumulation (p. 688)

Review Questions

1. What evidence did Louis Agassiz offer to support the idea of an ice age?
2. How do mountain glaciers and continental glaciers differ in terms of dimensions, thickness, and patterns of movement?
3. Describe the transformation from snow to ice.
4. Explain how arêtes, cirques, and horns form.
5. Describe the mechanisms that enable glaciers to move, and explain why they move.
6. How fast do glaciers normally move? How fast can they move during a surge?
7. Explain how the balance between ablation and accumulation determines whether a glacier advances or retreats.
8. How can a glacier continue to flow toward its toe even though its toe is retreating?
9. How does a glacier transform a V-shaped river valley into a U-shaped valley? Discuss how hanging valleys develop.
10. Describe the various kinds of glacial deposits. Be sure to note the materials from which the deposits are made and the landforms that result from deposition.
11. How do the crust and mantle respond to the weight of glacial ice?
12. How was the world different during the glacial advances of the Pleistocene ice ages? Be sure to mention the relation between glaciations and sea level.
13. How was the standard four-stage chronology of North American glaciations developed? Why was it so incomplete? How was it modified with the study of marine sediment?
14. Were there ice ages before the Pleistocene? If so, when?
15. What are some of the long-term causes that lead to ice ages? What are the short-term causes that trigger glaciations and interglacials?

On Further Thought

1. If you fly over the barren cornfields of central Illinois during the early spring, you will see slight differences in soil color due to variations in moisture content—wetter soil is darker. These variations outline the shapes of polygons that are tens of meters across. What do these patterns represent and how might they have formed? What do they tell us about the climate of central Illinois at the end of the last ice age?

2. Recent observations suggest that glaciers of southern Greenland have started to flow much faster in the past 10 years. Researchers suggest that this change might be a manifestation of the warming of the region's climate. What mechanism could account for the acceleration of the glaciers?

3. An unusual late Precambrian rock unit crops out in the Flinders and Mt. Lofty Range, a small mountain belt in South Australia, near Adelaide. Structures in the belt formed at the beginning of the Paleozoic. This unit consists of clasts of granite and gneiss, in a wide range of sizes, suspended through a matrix of slate. The clasts are now elliptical, with their long axes parallel to the plane of slaty cleavage. The rock unit lies unconformably above a basement of granite and gneiss, and if you dig out the unconformity surface, you will find that it is polished and striated. What is the "unusual rock," why does it have cleavage and elliptical clasts.

Suggested Reading

Alley, R. B. 2002. *The Two-Mile Time Machine: Ice Cores, Abrupt Climate Change, and Our Future.* Princeton, N.J.: Princeton University Press.

Anderson, B. G., and H. W. Borns, Jr. 1994. *The Ice Age World.* Oslo-Copenhagen-Stockholm: Scandinavian University Press.

Bennett, M. R., and N. F. Glasser. 1996. *Glacial Geology: Ice Sheets and Landforms.* New York: Wiley.

Dawson, A. G. 1992. *Ice Age Earth: Late Quaternary Geology and Climate.* New York: Routledge, Chapman, and Hall.

Erickson, J. 1996. *Glacial Geology: How Ice Shapes the Land.* New York: Facts on File.

Fagan, B. 2002. *The Little Ice Age: How Climate Made History, 1350–1850.* New York: Basic Books.

Hambrey, M. J., and J. Alean. 1992. *Glaciers.* Cambridge: Cambridge University Press.

MacDougall, J. D., 2004. *Frozen Earth: The Once and Future Story of Ice Ages.* Berkeley: University of California Press.

Menzies, J. 2002. *Modern and Past Glacial Environments.* Woburn, Mass.: Butterworth-Heinemann.

Paterson, W. S. B. 1999. *Physics of Glaciers.* Woburn, Mass.: Butterworth-Heinemann.

Post, A., and E. R. Lachapelle. 2000. *Glacier Ice.* Seattle: University of Washington Press.

THE VIEW FROM SPACE Glaciers and glacially carved features dominate the landscape of southeastern Alaska, as seen in this infrared image. Here we see Hubbard glacier as it enters Yakutat Bay. Where the glacier meets the sea, large blocks of ice calve off and float away. Many tributary glaciers can still be seen, but some glaciers have melted away, leaving behind long fjords.

Global Change in the Earth System

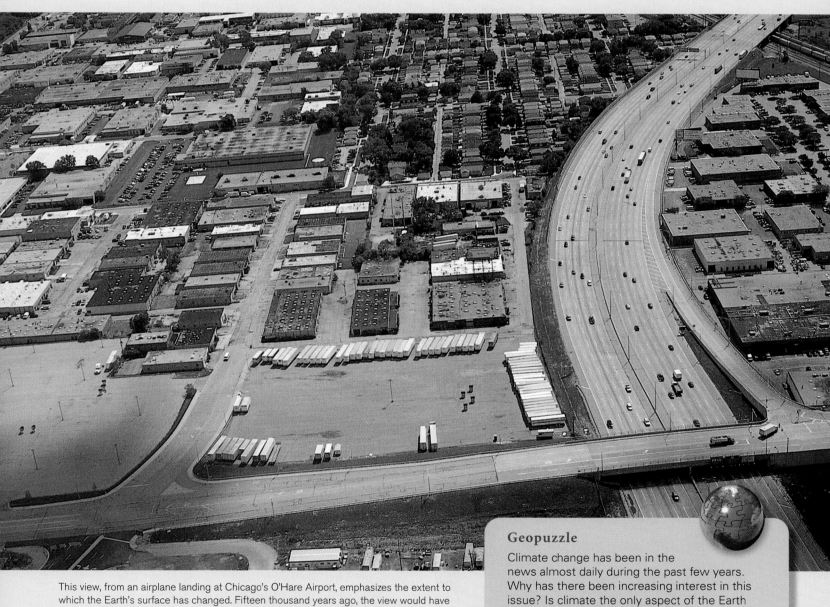

This view, from an airplane landing at Chicago's O'Hare Airport, emphasizes the extent to which the Earth's surface has changed. Fifteen thousand years ago, the view would have been the surface of a glacier. Five hundred years ago, it would have been a vast tall-grass prairie. One hundred fifty years ago, it would have been a checkerboard of farm fields. Today, most of the land has been covered by a layer of concrete or asphalt.

Geopuzzle

Climate change has been in the news almost daily during the past few years. Why has there been increasing interest in this issue? Is climate the only aspect of the Earth System that changes through time?

All we in one long caravan
are journeying since the world began,
we know not whither, we know . . . all must go.
—Bhartrihari (Indian poet, c. 500 C.E.)

23.1 INTRODUCTION

Would the Earth's surface have looked the same in the Jurassic Period as it does to a modern astronaut? Definitely not—the Earth of 200 million years ago differed from that of today in many ways. During the Jurassic, the North Atlantic Ocean was a narrow sea and the South Atlantic Ocean didn't exist at all, so most dry land connected to form a single vast continent (▶Fig. 23.1a). Today, both parts of the Atlantic are wide oceans, and the Earth has seven separate continents (▶Fig. 23.1b). Moreover, during the Jurassic, the call of the wild rumbled from the throats of dinosaurs, whereas today, the largest land animals are mammals. In essence, what we see of the Earth today is just

FIGURE 23.1 A comparison of **(a)** a map of the Earth 200 million years ago with **(b)** a map of today's Earth, emphasizing the change that has resulted from continental drift.

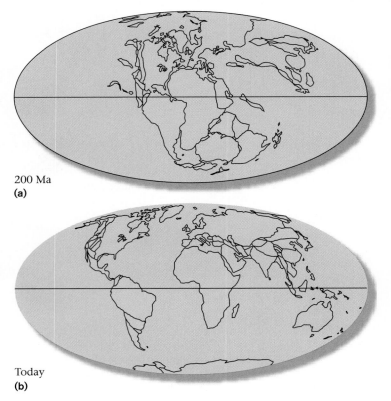

200 Ma
(a)

Today
(b)

a snapshot, an instant in the life story of a constantly changing planet. This idea arguably stands as geology's greatest philosophical contribution to humanity's understanding of its surroundings.

Why has the Earth changed so much over geologic time, and why can it continue to change? Ultimately, change happens because the Earth's internal heat makes the asthenosphere weak enough to flow, and because the Sun's heat keeps most of the Earth's surface at temperatures above the freezing point of water. Flow in the asthenosphere permits plate tectonics (sea-floor spreading, subduction, and transform motion), which, in turn, leads to continental drift, volcanism, and mountain building. These phenomena produce and modify rocks. In addition, volcanism provides raw materials from which the atmosphere and oceans form. The presence of liquid water and an atmosphere permits a variety of phenomena ranging from the evolution of life to weathering and erosion. Further, the interaction between internally driven processes (plate tectonics) and externally driven processes (atmospheric circulation) leads to the hydrologic cycle and the rock cycle, among many other phenomena. Biological and physical phenomena also interact. For example, photosynthetic organisms affect the composition of the atmosphere by producing oxygen, and atmospheric composition, in turn, determines the nature of chemical weathering in rocks.

No other object in our solar system has a mobile asthenosphere and a surface whose temperature straddles the freezing point of water (▶Box 23.1; ▶Fig. 23.2), so no other object of our solar system undergoes the kinds of changes that Earth does. The Moon, for example, changes so little through time that it looked essentially the same to a Jurassic dinosaur as it does to you. For purposes of discussion, we refer to the global interconnecting web of physical and biological phenomena on Earth as the **Earth System**, and we define **global change** as the transformations or modifications of both physical and biological components of the Earth System over time.

Geologists distinguish among different types of global change, on the basis of the rate or way in which change progresses with time. Gradual change takes place over long periods of geologic time (millions to billions of years) catastrophic change takes place relatively rapidly in the context of geologic time (seconds to millennia). Unidirectional change involves transformations that never repeat; cyclic change repeats the same steps over and over, though not necessarily with the same results. Some types of cyclic change are periodic in that the cycles happen with a definable frequency, others are not.

In this chapter, we begin by reviewing examples of global change, both unidirectional and cyclic, involving phenomena discussed earlier in the book. Then we look at an example of a **biogeochemical cycle**, the exchange of chemicals among living and nonliving reservoirs; some kinds of global change are due to changes in the

BOX 23.1 THE REST OF THE STORY

The Goldilocks Effect

Like Baby Bear's porridge in the tale of *Goldilocks and the Three Bears*, Earth is not too hot, and it's not too cold. . . . It's just right, as far as complex life is concerned (▶Fig. 23.2a–c). This condition is known as the **Goldilocks effect**.

Two factors play the key roles in determining Earth's surface temperature: the distance between the Earth and the Sun, and the concentration of carbon dioxide (CO_2) and other greenhouse gases in the atmosphere. If Earth were just 13% closer to the Sun, the heat would be so intense that liquid water could not exist. Without liquid water, CO_2 would not be stored as limestone and coal but rather would remain in the atmosphere. And without uplift and exposure of continental rocks, CO_2-absorbing chemical-weathering reactions could not take place. As a result, the atmosphere would contain so much CO_2 that it would become hot enough to melt lead, a condition that now exists on Venus. On the other hand, if Earth were significantly farther from the Sun so that sunlight was weaker, or if the atmosphere contained less CO_2 with which to trap heat, our planet's surface would be so cool that the oceans would freeze solid and complex life would die.

FIGURE 23.2 **(a)** Venus is too hot, **(b)** Mars is too cold, and **(c)** Earth is just right.

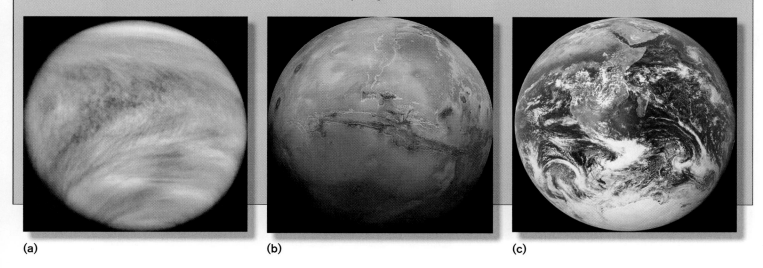

(a)   (b)   (c)

proportions of chemicals held in different reservoirs through time. Finally, we focus on **global climate change** (transformations or modifications in Earth's climate over time) and on anthropogenic (human-caused) contributions to global change. We conclude this chapter, and this book, by considering hypotheses that describe the ultimate global change—the end of the Earth.

23.2 UNIDIRECTIONAL CHANGES

The Evolution of the Solid Earth

Recall from Chapter 1 that Earth began as a fairly homogenous mass, formed by the coalescence of planetesimals. But the homogeneous proto-Earth did not last long—within about 100 million years of its birth the planet began to melt, yielding a liquid iron alloy that sank rapidly to the center to form the core (▶Fig. 23.3a, b). This process of differentiation represents major unidirectional change: it produced a layered, onion-like planet, with an iron alloy core surrounded by a rocky mantle.

Soon after differentiation, a Mars-sized proto-planet appears to have collided with the newborn Earth. This collision caused a catastrophic change—a significant portion of the Earth and the colliding object fragmented and vaporized, creating a ring of debris that coalesced to form the Moon (▶Fig. 23.3c–e). After the collision, the Earth's mantle was probably partially molten, and its surface became a sea of magma. The Earth endured intense bombardment by asteroids and comets between about 4.0 and 3.9 Ga, so any crust that had formed prior to 3.9 Ga was largely pulverized or melted. Eventually, however, bombardment

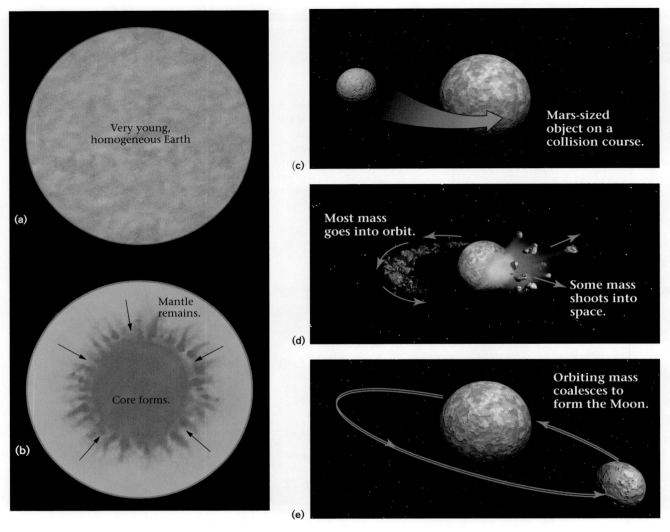

FIGURE 23.3 **(a)** When it first formed, the Earth was probably homogeneous. **(b)** Soon thereafter, the iron in the Earth melted and sank to the center. When this differention, a unidirectional change, was complete, the Earth had a distinct core and mantle. **(c–e)** Then, a Mars-sized body collided with Earth, sending off fragments that coalesced to form the Moon. All these phenomena radically changed the Earth.

ceased and our planet gradually cooled, permitting a crust to form at its surface and plate tectonics to begin operating. Subduction, and/or the rise of mantle plumes, produced relatively low-density rocks (e.g., granite). These rocks could not be subducted and thus remained buoyant blocks at the Earth's surface. Plate motion eventually caused these buoyant blocks to collide and suture together, forming the first continents. Overall, therefore, the transition from the Hadean Eon to the Archean Eon saw remarkable unidirectional change in the nature of the Earth. By early Archean time, our planet had distinct continents and ocean basins, and thus looked radically different from the other terrestrial planets (see Chapter 13).

The Evolution of the Atmosphere and Oceans

Like its surface, the Earth's atmosphere has also changed over time. Partial melting in the mantle produced magma and also released large quantities of gases that belched from volcanoes. More gases may have arrived when comets collided with our new planet. Eventually, Earth accumulated an early atmosphere composed dominantly of carbon dioxide (CO_2) and water (H_2O). Other gases, such as nitrogen (N_2), composed only a minor proportion of the early atmosphere. When the Earth's surface cooled, however, water condensed and fell as rain, collecting in low areas to form oceans. This may have happened before 4.0 Ga, but had certainly happened by 3.8 Ga. Gradually, CO_2 dissolved in the oceans and was absorbed by chemical-weathering reactions on land, so its concentration in the atmosphere decreased. Nitrogen, which doesn't react with other chemicals, was left behind. Thus, the atmosphere's composition changed to become dominated by nitrogen. Photosynthetic organisms appeared early in the Archean. But it probably wasn't until between 2.5 and 2.0 Ga (the early part of the Proterozoic) that oxygen (O_2) became a significant proportion of the atmosphere. Present concentrations of oxygen may have only existed for the past 400 million years.

The Evolution of Life

During most of the Hadean Eon, Earth's surface was probably lifeless, for carbon-based organisms could not survive the high temperatures of the time. The fossil record indicates that life had appeared at least by 3.8 billion years ago and has undergone unidirectional change (evolution) in fits and starts ever since (see Interlude E). Though simple organisms such as archaea and bacteria still exist, life evolution during the late Proterozoic and early Phanerozoic yielded multicellular plants and animals (▶Fig. 23.4). Life now inhabits regions from a few kilometers below the surface to a few kilometers above, yielding a diverse biosphere.

> **Take-Home Message**
>
> Since it first formed, the Earth has undergone major changes that are unidirectional, in that they will never repeat. Examples include the formation of the core, mantle, and moon; the evolution of the atmosphere and oceans; and the evolution of life.

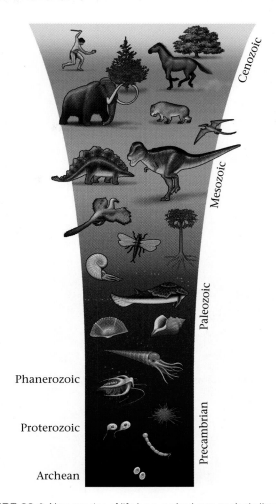

FIGURE 23.4 New species of life have evolved over geologic time. Though some of the simplest still exist, more complex organisms have appeared more recently.

23.3 PHYSICAL CYCLES

The Supercontinent Cycle

During Earth history, the map of the planet's surface has constantly changed. At times, almost all continental crust merged to form a single supercontinent, but usually the crust is distributed among several smaller continents. The process of change during which supercontinents form and later break apart is the **supercontinent cycle** (▶Fig. 23.5). Geologists have found evidence that at least three or four times during the past 3 billion years of Earth history, supercontinents existed. The most recent one, Pangaea, formed at the end of the Paleozoic Era. Others likely coalesced at 1.1 Ga, 2.1 Ga, and perhaps 2.7 Ga.

Plates move only 1 to 15 cm per year, so one passage through the supercontinent cycle takes at least a few hundred million years. Note that ocean basins do not simply open and close like accordions. In reality, plate motions are more complex, so the land never rearranges itself in exactly the same way through two supercontinent cycles.

The Sea-Level Cycle

Global sea level rose and fell by as much as 300 m during the Phanerozoic, and likely did the same in the Precambrian. When sea level rises, the shoreline migrates inland, and low-lying plains in the continents become submerged. During periods of particularly high sea level, more than half of Earth's continental area can be covered by shallow seas; at such times, sediment buries continental regions, thereby changing their surface (▶Fig. 23.6a). When sea level falls, the continents become dry again, and regional unconformities develop. For example, the sedimentary strata of the midwestern United States record at least six continent-wide advances and retreats of the sea, each of which left behind a blanket of sediment called a **sedimentary sequence**; unconformities define the boundaries between the sequences (▶Fig. 23.6b). The sequence deposited during the Pennsylvanian contains at least thirty shorter

The Earth System

External energy

Sun

Thunderhead

Lightning

Rain and snow

Mountain uplift

Continental glacier

City

Ocean

Rocky coastline

Desert

Valley

Arid mountains

Mining

Lakes

Deciduous forest

Beach

Forested mountains

Tropical rain forest

Coral reef

Shark

Internal energy

The Earth's surface is the interface among the solid Earth (the lithosphere); the ice and liquid water of oceans, lakes, streams, groundwater, and glaciers (the hydrosphere); and the planet's gaseous envelope (the atmosphere). Countless species of life, ranging from microscopic bacteria to giant whales and trees, make up the complex ecosystems of Earth's biosphere. All of these components—the lithosphere, hydrosphere, atmosphere, and biosphere—interact with each other. These components and the interactions among them constitute the Earth System.

Various materials cycle among living and nonliving components of the Earth System. In the hydrologic cycle, for example, water evaporates from the sea, rains on the land, and eventually flows back to the sea. During this process, water may be temporarily trapped in living organisms, clouds, subsurface pores, or ice sheets. Carbon dioxide can be stored in the air, dissolved in water, or trapped in plants, coal, or limestone. Some limestone forms when coral extracts ions from water. Meanwhile, over the vastness of geologic time, the atoms that make up minerals pass through the rock cycle. New elements from the mantle may enter the cycles of the Earth System at volcanoes or black smokers. Elements at the surface may be carried back into the mantle at subduction zones. Some atoms escape from the atmosphere into space.

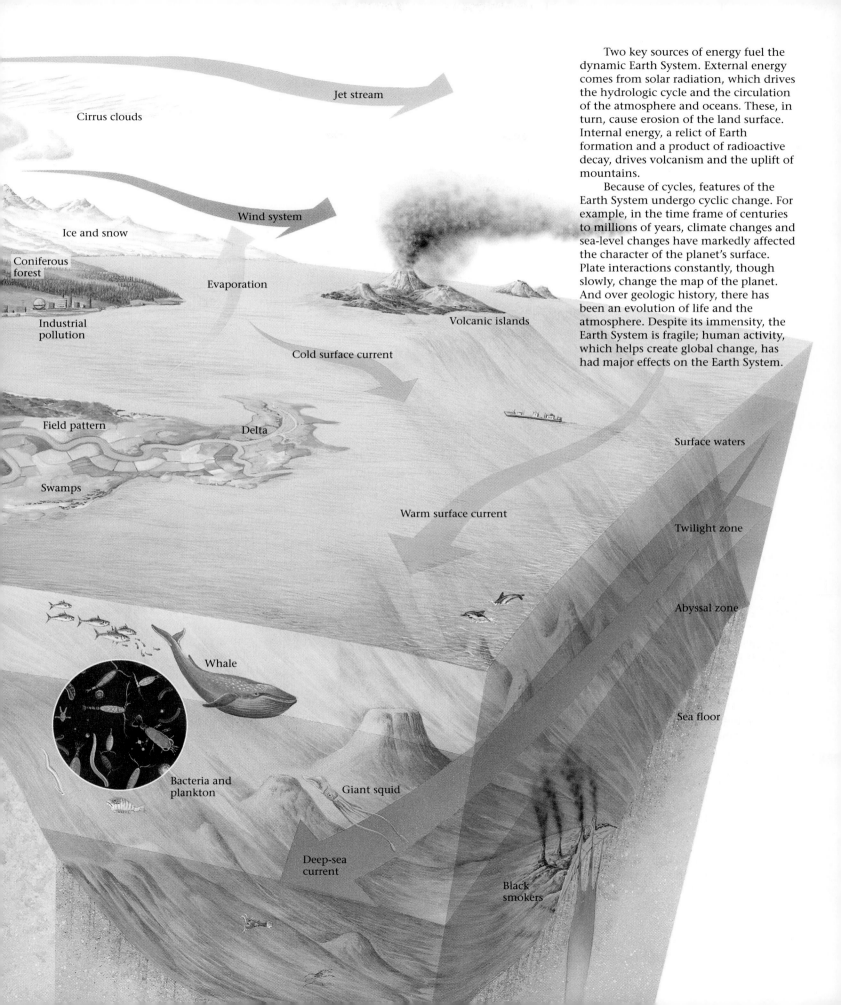

Jet stream

Cirrus clouds

Wind system

Ice and snow

Coniferous forest

Evaporation

Industrial pollution

Volcanic islands

Cold surface current

Field pattern

Delta

Surface waters

Swamps

Warm surface current

Twilight zone

Abyssal zone

Whale

Sea floor

Bacteria and plankton

Giant squid

Deep-sea current

Black smokers

Two key sources of energy fuel the dynamic Earth System. External energy comes from solar radiation, which drives the hydrologic cycle and the circulation of the atmosphere and oceans. These, in turn, cause erosion of the land surface. Internal energy, a relict of Earth formation and a product of radioactive decay, drives volcanism and the uplift of mountains.

Because of cycles, features of the Earth System undergo cyclic change. For example, in the time frame of centuries to millions of years, climate changes and sea-level changes have markedly affected the character of the planet's surface. Plate interactions constantly, though slowly, change the map of the planet. And over geologic history, there has been an evolution of life and the atmosphere. Despite its immensity, the Earth System is fragile; human activity, which helps create global change, has had major effects on the Earth System.

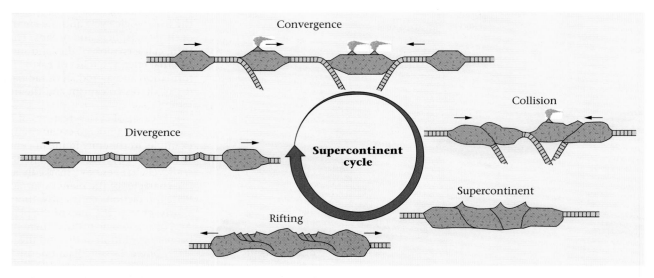

FIGURE 23.5 During the supercontinent cycle, smaller continents coalesce to form a supercontinent, which then later rifts and breaks apart, only to recombine later on. Since continents drift around the surface of the Earth, collisions do not necessarily bring previously adjacent continents back together again.

repeated intervals, called cyclothems, each of which contains a specific succession of sedimentary beds. At the base of each cyclothem you'll find sandstone, and in the middle you'll find coal (▶Fig. 23.6c). Cyclothems are short-term cycles of sea-level rise and fall.

After studying sedimentary sequences around the world, geologists at Exxon Corporation pieced together a chart defining the succession of global transgressions and regressions during the Phanerozoic Eon. The global sedimentary cycle chart may largely reflect the cycles of eustatic (worldwide) sea-level change (▶Fig. 23.6d, e). However, the chart probably does not give us an exact image of sea-level change, because the sedimentary record reflects other factors as well, such as changes in sediment supply. Eustatic sea-level changes may be due to a variety of factors, including advances and retreats of continental glaciers, changes in the volume of mid-ocean ridge systems, and changes in continental elevation and area.

> **Take-Home Message**
>
> Some changes are cyclic in that they have stages that may be repeated. Examples include the supercontinent cycle (accumulation of continents by collision, and then dispersal by rifting), the sea-level cycle (rise and fall of the sea), and the rock cycle.

The Rock Cycle

We learned early in this book that the crust of the Earth consists of three rock types: igneous, sedimentary, and metamorphic. Atoms making up the minerals of one rock type may later become part of another rock type. In effect, rocks are simply reservoirs of atoms, and the atoms move from reservoir to reservoir over time. As we learned in Interlude B, this process is the rock cycle. Each stage in the rock cycle changes the Earth by redistributing and modifying material.

23.4 BIOGEOCHEMICAL CYCLES

A **biogeochemical cycle** involves the passage of a chemical among nonliving and living reservoirs in the Earth System, mostly on or near the surface. Nonliving reservoirs include the atmosphere, the crust, and the ocean; living reservoirs include plants, animals, and microbes. Although a great variety of chemicals (water, carbon, oxygen, sulfur, ammonia, phosphorus, and nitrogen) participate in biogeochemical cycles, here we look at only two: water (H_2O) and carbon (C).

Some stages in a biogeochemical cycle may take only hours, some may take thousands of years, and others may take millions of years. Because chemicals can cycle rapidly, the transfer of a chemical from reservoir to reservoir during these cycles doesn't really seem like a change in the Earth in the way that the movement of continents or the metamorphism of rock seems like a change. In fact, for intervals of time, biogeochemical cycles attain a **steady-state condition**, meaning that the proportions of a chemical in different reservoirs remain fairly constant even though there is a constant flux (flow) of the chemical among reservoirs. When we speak of global change in a biogeochemical cycle, we mean a change in

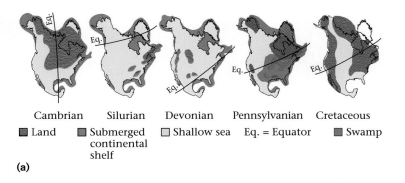

Cambrian Silurian Devonian Pennsylvanian Cretaceous

■ Land   ■ Submerged   □ Shallow sea   Eq. = Equator   ■ Swamp
              continental
              shelf

(a)

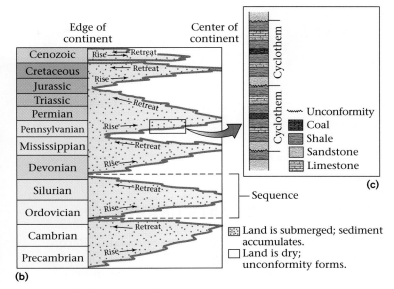

(b)

Land is submerged; sediment accumulates.

Land is dry; unconformity forms.

Unconformity
Coal
Shale
Sandstone
Limestone

Sequence

(c)

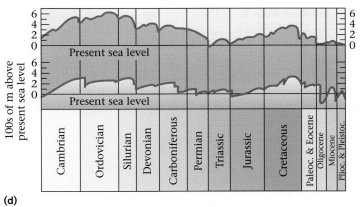

(d)

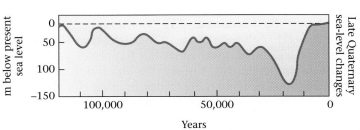

(e)

FIGURE 23.6 **(a)** Sea level has changed significantly over geologic time. For example, large parts of North America were once submerged by shallow seas. (The continental outline is shown for reference only—the continent did not have its present shape in the past.) **(b)** The stratigraphic record shows that sedimentary sequences are deposited during long-term transgressions (sea-level rise) and regressions (sea-level retreat). Transgression starts near the edge of a continent and then moves inland. **(c)** During the Pennsylvanian, short-term transgressions and regressions resulted in cyclothems. **(d)** To some extent, the transgressions and regressions indicated by this sedimentary sequence chart reflect global (eustatic) sea-level change. Here, two versions of the sequence chart are shown, each produced by a different author—the shape of the curve is still a subject for research. **(e)** A finer time scale shows many short-term ups and downs in sea level. This graph shows sea-level changes during the last 150,000 years.

the relative proportions of a chemical held in different reservoirs at a given time—in other words, a change in the steady-state condition.

The Hydrologic Cycle

As we learned in Interlude F, the hydrologic cycle involves the movement of water from reservoir to reservoir on or near the surface of the Earth. The hydrologic cycle is an example of a biogeochemical cycle in that a chemical (H_2O) passes through both nonliving and living entities—the oceans, the atmosphere, surface water, groundwater, glaciers, soil, and living organisms. Global change in the hydrologic cycle occurs when a change in global climate alters the ratio between the amount of water held in the ocean and the amount held in continental ice sheets. For example, during an ice age, water that had been stored in oceans moves into glacial reservoirs. Thus, the continents become covered with ice, and sea level drops. When the climate warms, water returns to the oceans, and sea level rises.

The Carbon Cycle: The Movement of a Greenhouse Gas

Most carbon in the near-surface realm of Earth originally bubbled out of the mantle in the form of CO_2 gas released by volcanoes (▶Fig. 23.7). Once it enters the atmosphere, it can be removed in various ways. Some dissolves in seawater to form bicarbonate (HCO_3^-) ions, whereas some is absorbed by photosynthetic organisms that convert it into sugar and other organic chemicals. This carbon enters the food chain and ultimately makes up the flesh, fat, and sinew of animals. In fact, about 63 billion tons of carbon move from the atmosphere into life forms every year. Some of the reactions that take place when rock undergoes chemical weathering incorporate atmospheric CO_2, and thus also remove carbon from the atmosphere.

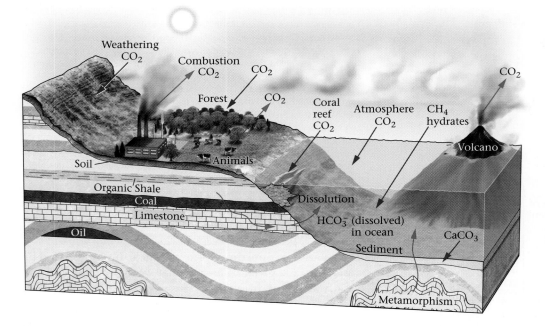

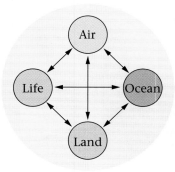

Carbon Exchange
in the Earth System

FIGURE 23.7 In the carbon cycle, carbon transfers among various reservoirs at or near the Earth's surface. Red arrows indicate release to the air, and green arrows indicate absorption from air.

Some carbon returns directly to the atmosphere through the respiration of animals (again as CO_2), by the flatulence of animals (as methane [CH_4]), or by the decay of dead organisms. But some can be stored for long periods of time in fossil fuels (oil and coal), in organic shale, in methane hydrates (see Chapter 14), or in limestone. Fossil fuel deposits, limestone, methane hydrates, and the organic portion of shale contain most of the carbon in the near-surface realm of Earth and can hold on to it for long periods of time. But this carbon either returns to the atmosphere in the form of CO_2, as a result of the burning of fossil fuels and the metamorphism of rocks containing carbonate, or returns to the sea after undergoing chemical weathering followed by dissolution as HCO_3^- in river water or groundwater. Melting of methane hydrates releases methane to the atmosphere.

The concentration of carbon dioxide and methane in the atmosphere play an essential role in controlling Earth's climate because, as we saw in Chapter 20, these gases, along with several other trace gases (such as water), are greenhouse gases. An increase in their concentration warms the atmosphere, whereas a decrease cools it down.

> **Take-Home Message**
>
> In the Earth System, chemicals—such as carbon and water—cycle through living and non-living reservoirs. For example, carbon can be stored in the atmosphere as CO_2, in sea water as bicarbonate ions, in rock as calcite, and in fossil fuels as oil, gas, and coal.

23.5 GLOBAL CLIMATE CHANGE

How often have you seen a newspaper proclaim "Record High Temperatures!" In August 2003, such a claim became reality for much of Europe, where thermometers registered weeks of temperatures as much as 8°C above "normal." The *New York Times* ran the headline "Europe Sizzles and Suffers in a Summer of Merciless Heat." Does this mean that the climate—the average range of weather conditions for a given region—is changing? As discussed in Chapter 20, the atmospheric conditions during a specific time interval in a given region define the region's *weather* for the time interval. The term *climate* indicates overall weather conditions as well as daily to seasonal variability of weather conditions over a period of many decades. So a newspaper's headline about a single hot spell or cold snap does not mean that the climate is changing. But if a new set of conditions—say, increased average temperature, rising snow line, or a longer growing season—becomes the new norm for a region, then climate change has occurred. The stratigraphic record clearly shows that global climate change (the transformation of Earth's climate over time) has happened repeatedly throughout Earth's history for natural reasons. Increasingly, measurements are indicating that global climate change is now occurring, and that human activities are playing a role in causing this change.

In this section, we describe how researchers study past climates, how the climate changes over geologic time, and why researchers have concluded that global climate change over the last two centuries may reflect human activities. We conclude by discussing the implications of current climate change. For purposes of discussion, we distinguish between long-term climate change, which takes place over millions to tens of millions of years, and short-term climate change, which takes place over tens to hundreds of thousands of years. If the average atmospheric and sea-surface temperature rises, we have **global warming**, and if it falls, we have **global cooling**. Some changes are great enough to cause oceanic islands and large regions of continents to be submerged by shallow seas or to be covered by ice, whereas others are subtle, creating only a slight latitudinal shift in vegetation belts and a sea-level change measured in meters or less.

Methods of Study

Geologists and climatologists are working hard to define the kinds of climate changes that can occur, the rates at which these changes take place, and the effects they may have on society. There are two basic approaches to studying global climate change: (1) researchers measure past climate change, as indicated by the stratigraphic record, to document the magnitude of change that is possible and the rate at which such change occurred; (2) researchers develop computer programs to calculate how factors such as atmospheric composition, topography, ocean currents, and Earth's orbit affect the climate. The resulting **climate-change models** provide insight into when and why changes took place in the past and whether they will happen in the future.

Let's look first at how geologists study the **paleoclimate** (past climate), so as to document climate changes throughout Earth history. Any feature whose character depends on the climate and whose age can be determined can be a clue to defining paleoclimate.

- *The stratigraphic record*: The nature of sedimentary strata deposited at a certain location reflects the climate at that location. For example, an outcrop exposing cross-bedded sandstone, overlain successively by coal and glacial till, indicates that the site of the outcrop has endured different climates (desert, then tropical, then glacial) over time.

- *Paleontological evidence*: Different assemblages of species survive in different climatic belts. Thus, the succession of species in a sedimentary sequence provides clues to the changes in climate at that site. For example, a record of short-term climate change can be obtained by studying the succession of plankton fossils in sea-floor sediments, for cold-water species of plankton are different from warm-water species. Fos-

sil pollen also yields clues to the paleoclimate. Pollen, tiny grains involved in plant reproduction, looks like dust to the unaided eye. But under a microscope, each grain has a distinctive structure, and grains of one species look different from grains of another species (▶Fig. 23.8a, b). Further, pollen grains have a tough coating and can survive burial. By studying pollen in sediment, palynologists (scientists who study pollen) can determine whether the sediment accumulated in a cold-climate coniferous forest or in a warm-climate deciduous forest. And by recording changes in the pollen assemblages found in successive layers of sediment, palynologists can track the movement of climate belts over the landscape (▶Fig. 23.8c). For example, studies of spruce pollen preserved in the mud of bogs show that spruce forests, indicative of cool climates, have slowly migrated north since the ice age (▶Fig. 23.8d, e).

- *Oxygen-isotope ratios*: Two isotopes of an element have the same atomic number but different atomic weights (see Appendix A). For instance, oxygen occurs as ^{16}O (8 protons and 8 neutrons) and ^{18}O (8 protons and 10 neutrons). Geologists have found that the ratio of ^{18}O to ^{16}O in glacial ice indicates the atmospheric temperature in which the snow that made up the ice formed: the ratio is larger in snow that forms in warmer air, but smaller in snow that forms in colder air. Because of this relationship, the isotope ratio measured in a succession of ice layers in a glacier indicates temperature change over time. Researchers have now obtained ice cores down to a depth of almost 3 km in Antarctica and in Greenland; this record spans up to 720,000 years (▶Fig. 23.9a). For a number of reasons, the $^{18}O/^{16}O$ ratio in the $CaCO_3$ making up plankton shells also gives geologists an indication of past temperatures. Thus, measurement of oxygen-isotope ratios in drill cores of marine sediment extends the record of temperature change back over millions of years (▶Fig. 23.9b, c).

- *Bubbles in ice*: Bubbles in ice trap the air present at the time the ice forms. By analyzing these bubbles, geologists can measure the concentration of CO_2 in the atmosphere back through time. This information can be used to correlate CO_2 concentration with past atmospheric temperature. The CO_2 record has been extended back through 240,000 years.

- *Growth rings:* If you've ever looked at a tree stump, you'll have noticed the concentric rings visible in the wood. Each ring represents one year of growth, and the thickness of the ring indicates the rate of growth in a given year. Trees grow faster during warmer, wetter years and more slowly during cold, dry years

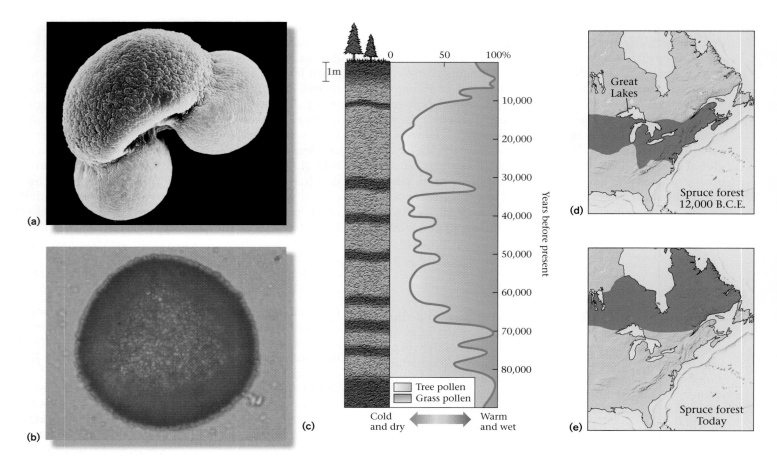

FIGURE 23.8 Changes in the assemblage of pollen in sediment indicate a shift in climate belts. **(a)** Spruce pollen from a cold-climate coniferous forest. **(b)** Hemlock pollen from a warm-climate deciduous forest. **(c)** This model shows how the proportion of tree pollen relative to grass pollen can change in a sedimentary sequence through time. Tree pollen indicates cooler and drier conditions, whereas grass pollen indicates warmer and wetter conditions. **(d)** Pollen data suggest that about 12,000 years ago, spruce forests (green areas) lay south of the Great Lakes. **(e)** Today, they are found north of the Great Lakes.

(▶Fig. 23.10a); thus, the succession of ring widths provides an easily calibrated record of climate during the lifetime of the tree. Bristlecone pines supply a record back through 4,000 years. To go further into the past, dendrochronologists (scientists who study tree rings) look at the record of rings in logs dated by the radiocarbon technique or in logs whose ages overlap with that of the oldest living tree. Growth rings in corals and shells can provide similar information.

- *Human history*: Researchers have been able to make careful, direct measurements of climate changes only for the past few decades. This record is not long enough to document long-term climate change. But history, both written and archaeological, contains important clues to climates at times in the past. Periods of unusual cold or drought leave an impression on people, who record them in paintings, stories, and records of crop success or failure (▶Fig. 23.10b; ▶Box 23.2).

Long-Term Climate Change

Using the variety of techniques described above, geologists have reconstructed an approximate record of global climate, represented by mean temperature and rainfall, for geologic time. The record shows that at some times in the past, the Earth's atmosphere was significantly warmer than it is today; whereas, at other times, it was significantly cooler. The warmer periods have come to be known as **greenhouse** (or hothouse) **periods** and the colder as **icehouse periods**. (The more familiar term, "ice age," refers to the times during an icehouse period when the Earth was cold enough for ice sheets to advance and cover substantial areas of the continents.) As the chart in ▶Figure 23.11a shows, there have been at least five major icehouse periods during geologic history.

Let's look a little more closely at the climate record of the last 100 million years, for this time interval includes the transition between a greenhouse and an icehouse

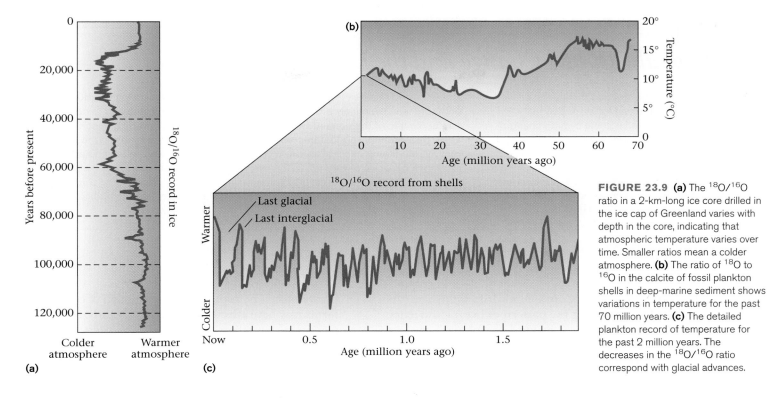

FIGURE 23.9 (a) The $^{18}O/^{16}O$ ratio in a 2-km-long ice core drilled in the ice cap of Greenland varies with depth in the core, indicating that atmospheric temperature varies over time. Smaller ratios mean a colder atmosphere. (b) The ratio of ^{18}O to ^{16}O in the calcite of fossil plankton shells in deep-marine sediment shows variations in temperature for the past 70 million years. (c) The detailed plankton record of temperature for the past 2 million years. The decreases in the $^{18}O/^{16}O$ ratio correspond with glacial advances.

period. Paleontological and other data suggest that the climate of the Mesozoic Era, the Age of Dinosaurs, was much warmer than the climate of today. At the equator, average annual temperatures may have been 2° to 6°C warmer, while at the poles, temperatures may have been 20° to 60°C warmer. In fact, during the Cretaceous Period, dinosaurs were able to live at high latitudes, and there were no polar ice caps on Earth. But starting about 80 million years ago, the Earth's atmosphere began to cool. We entered an icehouse period about 33 million years ago, and the climate reached its coldest condition about 2 million years ago, during the Pleistocene ice age.

What caused long-term global climate change? Discounting very long-term changes in the sun (▶Box 23.3; see Section 23.7) the answer probably lies in the complex relationships among the various geologic and biogeochemical cycles of the Earth system, as described earlier. For example:

FIGURE 23.10 (a) Tree rings provide a record of climate, for more growth happens in wet years than in dry years. (b) The stories of great floods recorded by artists and historians provide clues to the timing of climate changes. This woodcut depicts a flood in medieval England.

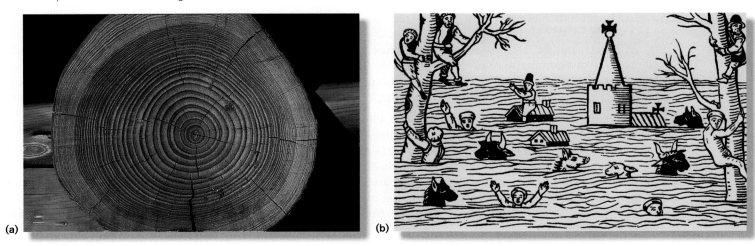

BOX 23.2 THE HUMAN ANGLE

Global Climate Change and the Birth of Legends

Some geologists argue that myths passed down from the early days of civilization may have their roots in global climate change. For example, recent evidence suggests that earlier than 7,600 years ago, the region that is now the Black Sea contained a much smaller freshwater lake surrounded by settlements. Subsequent to the most recent ice-age glacial advance, the ice sheets melted and sea level rose, and the Mediterranean eventually broke through a dam at the site of the present Bosporus Strait. Researchers suggest that seawater from the Mediterranean spilled into the Black Sea basin via a waterfall 200 times larger than Niagara Falls. This influx of water caused the lake level to rise by perhaps 10 cm per day, and within a year 155,000 square km (60,000 square miles) of populated land had become submerged beneath hundreds of meters of water. This flooding presumably led to a huge human migration, and its timing has suggested to some researchers that it may have inspired the Babylonian *Epic of Gilgamesh* (2000 B.C.E.) and, later, the biblical epic of Noah's Ark.

- *Positions of continents*: Continental drift influences the climate by controlling the pattern of oceanic currents, which redistribute heat around the planet's surface (►Fig. 23.11b). Drift also determines whether the land is at high or low latitudes (and thus how much solar radiation strikes it), and whether or not there are large continental interior regions where extremely cold winter temperatures can develop.

- *Volcanic activity*: A long-term global increase in volcanic activity may contribute to long-term global warming, because it increases the concentration of greenhouse gases in the atmosphere. For example, when Pangaea broke up during the Cretaceous Period, numerous rifts formed, and sea-floor-spreading rates were particularly high, so volcanoes were more abundant than they are today. Thus, volcanic activity may have triggered Cretaceous greenhouse conditions. The eruption of large igneous provinces (LIPs; see Chapter 6) at other times may have also caused cooling.

- *The uplift of land surfaces*: Tectonic events that lead to the long-term uplift of the land affect atmospheric CO_2 concentration, because such events expose land to weathering, and chemical-weathering reactions absorb CO_2. Thus, uplift decreases the greenhouse effect and causes global cooling. For example, uplift of the Himalayas and Tibet may have triggered Cenozoic icehouse conditions (►Fig. 23.11c). Such uplift will also affect atmospheric circulation and rainfall rates; see Chapter 20.

- *The formation of coal, oil, or organic shale*: At various times during Earth history, environments suitable for coal or oil formation have been particularly widespread. Such formation removes CO_2 from the atmosphere and stores it underground. For example, global cooling in the Carboniferous may correlate with the development of vast coal swamps on Pangaea.

- *Life evolution*: The appearance or extinction of certain life forms may have affected climate significantly. For example, some researchers speculate that the appearance of lichens in the Neoproterozoic may have decreased atmospheric CO_2 concentration, and thus could have triggered Neoproterozoic icehouse conditions. Similarly, the appearance of grass about 30 to 35 Ma may have triggered Cenozoic icehouse conditions.

Note that some of the effects described above add or subtract CO_2 from the atmosphere. Some researchers argue that change in the distribution of CO_2 among various reservoirs in the carbon cycle plays a particularly influential role in controlling climate because CO_2 is an important greenhouse gas. Factors that add CO_2 lead to warming, whereas those that decrease CO_2 lead to cooling. Others suggest that the greenhouse effect of CO_2 only amplifies the effects caused by other phenomena, such as changes in cloud cover. The relative importance of these different factors remains a focus of research.

It is important to note that feedback among components of the Earth System helps regulate the amount of CO_2 in the atmosphere. Negative feedback slows a process down or even reverses it. For example, as global temperature rises because of an increase in CO_2, rates of evaporation and therefore amounts of precipitation increase. As a consequence, weathering rates (which absorb CO_2) increase, so the concentration of CO_2 then goes down. Positive feedback, on the other hand, makes a process continue or even accelerate. For example, positive feedback on Venus may have led to a **runaway greenhouse effect**, through the following steps:

1. Because Venus orbits closer to the Sun than does Earth, it receives more solar radiation. In the past, the extreme heat caused any surface water to evaporate until the planet became enveloped in clouds.

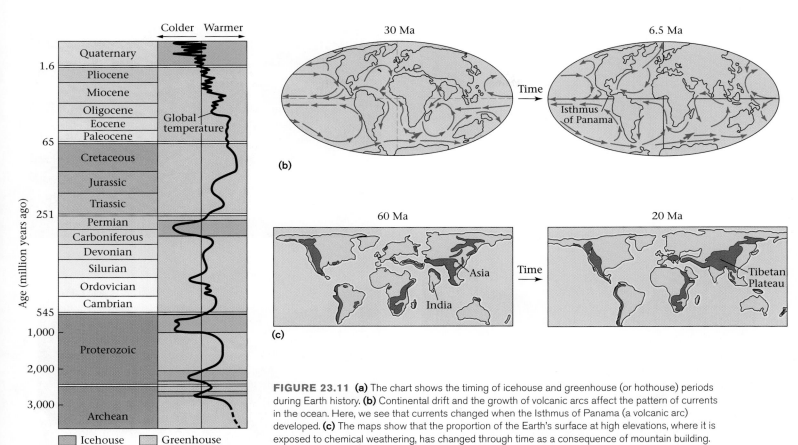

FIGURE 23.11 **(a)** The chart shows the timing of icehouse and greenhouse (or hothouse) periods during Earth history. **(b)** Continental drift and the growth of volcanic arcs affect the pattern of currents in the ocean. Here, we see that currents changed when the Isthmus of Panama (a volcanic arc) developed. **(c)** The maps show that the proportion of the Earth's surface at high elevations, where it is exposed to chemical weathering, has changed through time as a consequence of mountain building. For example, the collision of India with Asia about 40 Ma created a broad, high plateau.

2. The water vapor, a greenhouse gas, did not allow infrared radiation to escape, so the atmosphere became still hotter, approaching 1,500°C. At these extremely high temperatures, water molecules break apart, forming hydrogen and oxygen gas. The hydrogen escaped to space, and the oxygen reacted with surface rocks to oxidize (rust) them.

3. Without water, CO_2-absorbing chemical weathering ceased, so CO_2 built up in the atmosphere, making the temperature rise even more. This dense, hot atmosphere persists on Venus today.

Natural Short-Term Climate Change

The record of the past million years gives a sense of the magnitude and duration of short-term climate change. During this period, there have been about five major and twenty to thirty minor episodes of glaciation, separated by interglacial periods. If we focus on the last 15,000 years, we see that overall the temperature has increased, but there are still notable ups and downs (▶Fig. 23.12a). In fact, by studying evidence for icebergs in the ocean, researchers have identified cycles in this period that lasted for centuries to millennia. Some evidence suggests that significant shifts in climate can occur in only 10 years.

As a result of the warming that began 15,000 years ago, the glaciers retreated for about 4,500 years. Then colder conditions returned for a few thousand more years. This interval of cooler temperature is named the Younger Dryas, after an Arctic flower that became widespread during the time. The climate then warmed, reaching a peak at 5,000 to 6,000 years ago, a period called the Holocene maximum, when average temperatures were about 2°C above temperatures of today. This warming peak led to increased evaporation and therefore precipitation, making the Middle East unusually wet and fertile—conditions that may partially account for the rise of civilization in this part of the world.

The temperature dipped to a low about 3,000 years ago, before returning to a high during the Middle Ages, a time called the Medieval warm period. During this time, Vikings established self-supporting agricultural settlements along the coast of Greenland (▶Fig. 23.12b). The temperature dropped again from 1500 C.E. to about 1800 C.E., a period known as the "little ice age," when Alpine glaciers advanced and the canals of the Netherlands froze over

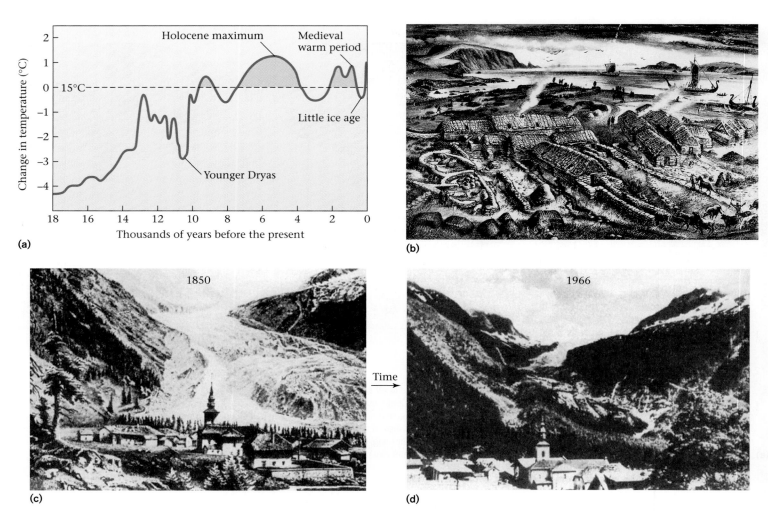

FIGURE 23.12 **(a)** The past 15,000 years (the Holocene Epoch) experienced several periods of warming and cooling. **(b)** Coastal Greenland was settled by the Vikings during one of the warmer periods, when the region could support agriculture. **(c)** Glaciers that formed during the little ice age persisted into the nineteenth century. Here we see a glacier in the French Alps as it appeared in 1850. **(d)** By the second half of the twentieth century (1966), the glacier had almost disappeared.

in winter (▶Fig. 23.12c, d; see Fig. 22.48). Overall, the climate has warmed since the end of the little ice age, and today it is as warm or warmer than it was during the Medieval warm period.

Geologists have focused on four factors to explain short-term climate change.

- *Fluctuations in solar radiation and cosmic rays*: The amount of energy produced by the Sun varies with the **sunspot cycle**. This cycle involves the appearance of large numbers of sunspots (black spots thought to be magnetic storms on the Sun's surface) about every 9 to 11.5 years (▶Fig. 23.13a, b). There may be longer-term cycles that have not yet been identified. This variation in energy may affect the climate. Recently, some researchers have speculated that changes in the rate of influx of cosmic

rays may affect climate, perhaps by generating clouds. Specifically, recent research suggests that cosmic rays striking the atmosphere produce clusters of ions that become condensation nuclei around which water molecules congregate, thus forming the mist droplets making up clouds. But how cloud formation changes climate remains uncertain. High-elevation clouds could reflect incoming solar radiaton and would cool the planet, whereas low-elevation clouds could absorb infrared rays rising from the Earth's surface and would warm the planet.

- *Changes in Earth's orbit and tilt*: As Milankovitch first recognized in 1920, the change in the tilt of Earth's axis over a period of 41,000 years, the Earth's 23,000-year precession cycle, and changes in the eccentricity of its orbit over a

BOX 23.3 THE REST OF THE STORY

The Faint Young Sun Paradox

The intensity of sunlight striking the Earth has likely changed substantially during the history of the solar system, because the composition of the Sun changes with time. The Sun's energy comes primarily from the fusion reaction that bonds four hydrogen atoms together to form one helium atom. Over time, the proportion of helium in the Sun increases. Calculations suggest that because one helium atom takes up less space than four separate hydrogen atoms, the production of helium should allow the inside of the Sun to contract. This contraction raises the Sun's temperature, which, in turn, increases the rate of fusion reactions and therefore the amount of energy the Sun produces. Overall, the Sun may be 30% brighter today than it was when the Earth first formed.

If this is the case, the Earth's average temperature *should have been* over 20°C less during the Archean Eon than it is today, and Archean oceans *should have been* frozen solid. But fossil evidence suggests that liquid water has existed on our planet's surface almost continuously since at least 3.8 billion years ago, so the Earth couldn't have been so cold. This apparent conflict between calculation and observation is called the "faint young Sun paradox."

Most researchers agree that the paradox can be explained by remembering that earlier in Earth history, before the widespread appearance of photosynthetic life, the atmosphere contained more carbon dioxide than it does today. The greenhouse effect caused by this excess CO_2 could have increased the temperature of the Earth sufficiently to counter the effect of the faint young Sun, so that surface temperatures remained above freezing.

FIGURE 23.13 **(a)** The appearance of sunspots correlates with increased energy production on the Sun. **(b)** There may be cycles in sunspot activity that could influence the climate.

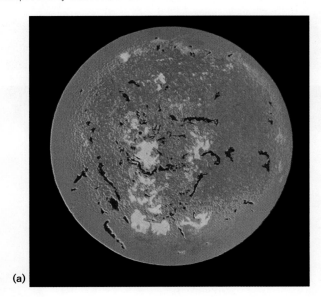

(a)

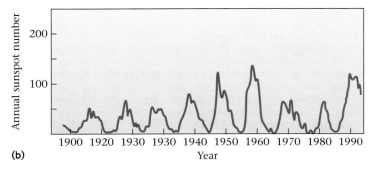

(b)

period of 100,000 years together cause the amount of summer heat in high latitudes to vary. They also cause the overall amount of heat reaching Earth to vary (see Chapter 22). These changes correlate with observed ups and downs in atmospheric and oceanic temperature.

- *Changes in volcanic emissions*: Not all of the sunlight that reaches the Earth penetrates its atmosphere and warms the ground. Some is reflected by the atmosphere. The degree of reflectivity, or **albedo**, of the atmosphere increases not only if cloud cover increases, as we have seen, but also if the concentration of volcanic aerosols in the atmosphere increases.

  The short-term effect of volcanism on global temperature is abundantly clear. For example, the year following the 1815 eruption of Mt. Tambora in the western Pacific became known as the "year without a summer," for aerosols that erupted encircled the Earth and blocked the Sun. Snow fell in Europe throughout the spring, and the entire summer was cold.

- *Changes in ocean currents*: Recent studies suggest that the configuration of currents can change quite quickly, and that this configuration affects the climate. The Younger Dryas may have resulted when a layer of freshwater from melting glaciers spread out over the North Atlantic and prevented thermohaline circulation in the ocean, thereby shutting down the Gulf Stream (see Chapter 18).

- *Changes in surface albedo*: Regional-scale changes in the nature of continental vegetation cover, and/or in the proportion of snow and ice on the Earth's surface, would affect our planet's albedo. Increasing albedo causes cooling, whereas decreasing albedo causes warming.

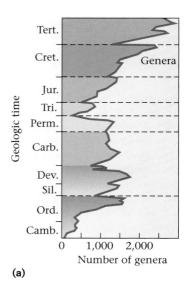

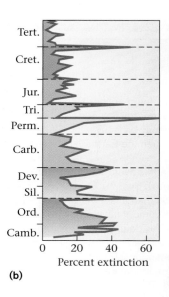

(a)
(b)

- *Abrupt changes in concentrations of greenhouse gases*: A sudden change in greenhouse gas concentration in the atmosphere could affect climate. One such change might happen if sea temperature warmed or sea level dropped, causing some of the methane hydrate that crystallized in sediment on the sea floor suddenly to melt. Such melting would release CH_4 to the atmosphere. Algal blooms and reforestation (or deforestation) conceivably could change CO_2 concentrations.

Catastrophic Climate Change and Mass-Extinction Events

Changes that happen on Earth almost instantaneously are called *catastrophic changes*. For example, a volcanic eruption, an earthquake, a tsunami, or a landslide can change a local landscape in seconds or minutes. But such events affect only relatively small areas. Can such catastrophes happen on a global scale? In the past decades, geoscientists have come to the conclusion that the answer is yes. The stratigraphic record shows that Earth history includes several **mass-extinction events**, when large numbers of species abruptly vanished (▶Fig. 23.14a). Some of these events define boundaries between geologic periods. A mass-extinction event decreases the biodiversity (the number of different species that exist at a given time) of life on Earth (▶Fig. 23.14b). It takes millions of years after a mass-extinction event for biodiversity to increase again, and the new species that appear differ from those that vanished, for evolution is unidirectional.

Geologists speculate that some mass-extinction events reflect a catastrophic change in the planet's climate, brought about by unusually voluminous volcanic eruptions or by the impact of a comet or an asteroid with the Earth (▶Fig. 23.14c). Either of these events could eject

FIGURE 23.14 (a) During a mass-extinction event, biodiversity on Earth, as indicated by the number of fossil species (grouped together in genera), suddenly decreases. **(b)** Paleontologists can calculate the percentage of species that became extinct during a given event. Notice that most existing species became extinct at the boundary between the Permian and Triassic periods. **(c)** The K-T (Cretaceous-Tertiary) boundary event may have been caused by the impact of a large comet or asteroid. Here, we see an artist's rendition of the impact (top) and its aftermath (middle and bottom).

enough debris into the atmosphere to block sunlight. Without the warmth of the Sun, winter-like or night-like conditions would last for weeks to years, long enough to disrupt the food chain. In addition, either event could eject aerosols that would turn into global acid rain, scatter hot debris that would ignite forest fires, or give off chemicals that, when dissolved in the ocean, would make the ocean either toxic or so nutritious that oxygen-consuming algae could thrive.

Let's examine possible causes for two of the more profound mass-extinction events in Earth history. The first event marks the boundary between the Permian and Triassic periods (i.e., between the Paleozoic and Mesozoic eras). During the Permian-Triassic extinction event, over two-thirds of the species on Earth became extinct. In fact, this boundary was first defined in the nineteenth century, precisely because the assemblage of fossils from rocks below the boundary differs so markedly from the assemblage in rocks above. Radiometric dating suggests that the extinction event roughly coincided with the eruption of vast quantities of basalt in Siberia, basalt that covered over 2.5 million square km of continental crust. So much basalt erupted that geologists attribute its source to a superplume, a mantle plume many times larger than the one currently beneath Hawaii. Because of the correlation between the time of the basalt eruptions and the time of the mass extinction, geologists suggest that the former caused the latter. But recently, researchers have found evidence suggesting that a large asteroid collided with the Earth (perhaps at a site that now lies off the northern coast of Australia) at the Permian-Triassic boundary. Thus, the mass extinction may have resulted, instead, from this collision.

The second event, called the K-T boundary event, caused the mass extinction that marks the boundary between the Cretaceous and Tertiary Periods (i.e., the boundary between the Mesozoic and Cenozoic Eras). As discussed in Chapter 13, the timing of this event correlates very well with the time at which an asteroid collided with the Earth at a site now called the Chicxulub crater in Yucatán, Mexico. Thus, most geologists suggest that the mass extinction is the aftermath of this collision. But it is interesting to note that the extinction is comparable in age to the eruption of extensive basalt flows in India, so the possibility remains that volcanic activity played a role. Some geologists are even speculating that the eruptions are related to the impact of the asteroid. This debate remains active.

> **Take-Home Message**
>
> Earth's past climate can be studied using fossils, isotopes, pollen assemblages, and growth rings in trees. The record shows that, over geologic time, climate alternates between warm (greenhouse) conditions and cold (icehouse) conditions.

23.6 HUMAN IMPACT ON THE EARTH SYSTEM

During the Stone Age, the human population worldwide had not yet topped 10 million. By the dawn of civilization, 4000 B.C.E., it was still, at most, a few tens of millions. But by the beginning of the nineteenth century, revolutions in industrial methods, agriculture, medicine, and hygiene had substantially lowered death rates and raised living standards, so that the population began to grow at accelerating rates. It took tens of thousands of years to grow from Stone Age populations to 1 billion people worldwide in 1850, but it took only 80 years for the population to double again, reaching 2 billion in 1930. The growth rate increased during the twentieth century, with the population reaching 4 billion in 1975. Now, the doubling time is only 44 years, and the population passed the 6 billion mark just before the year 2000 (▶Fig. 23.15).

As the population grows and the standard of living improves, per capita usage of resources increases. We use land for agriculture and grazing, forests for wood, rock and dirt for construction, oil and coal for energy or plastics, and ores for metals. Without a doubt, our usage of resources has affected the Earth System profoundly, and thus humanity has become a major agent of global change. Here, we examine some of these anthropogenic impacts.

The Modification of Landscapes

Every time we pick up a shovel and move a pile of dirt, we redistribute a portion of the Earth's crust, an activity that prior to humanity was accomplished only by rivers, the

FIGURE 23.15 The population has increased dramatically during the past two centuries. Currently, it doubles every 44 years. Occasional pandemics (widespread deadly disease) have caused abrupt drops, but the population quickly recovers.

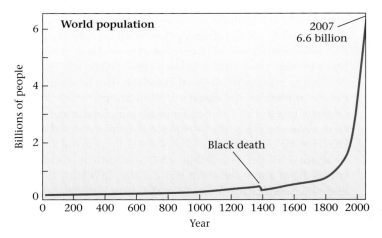

wind, rodents, and worms. In the last century, the pace of Earth movement has accelerated, for now we have shovels in coal mines that can move 300 cubic meters of coal in a single scoop, trucks that can carry 200 tons of ore in a single load, and tankers that can transport 50 million liters of oil during a single journey. In North America, human activity now moves more sediment each year than rivers do. The extraction of rock during mining, the building of levees and dams along rivers or of sea walls along the coast, and the construction of highways and cities all involve the redistribution of Earth materials (▶Fig. 23.16a). In addition, people clear and plow fields, drain and fill wetlands, and pave over the land surface (▶Fig. 23.16b; see also chapter-opening photo). All these activities change the landscape.

Landscape modification has side effects. For example, it may make the ground unstable and susceptible to landslides. And it may expose the land to erosion, thereby changing the volume of sediment transported by natural agents (such as running water and wind). Locally, flood-control projects may diminish the sediment supply downstream, also with unfortunate consequences; for example, the damming of the Nile by the Aswan High Dam has cut off the sediment supply to the Nile Delta, so ocean waves along the Mediterranean coast of the delta have begun to erode the coastline by more than 1 m per year.

The Modification of Ecosystems

In undisturbed areas, the **ecosystem** of a region (an interconnected network of organisms and the physical environment in which they live) is the product of evolution for an extended period of time. The ecosystem's flora (plant life) include species that have adapted to living together in that particular climate and on the substrate available, whereas its fauna (animal life) can survive local climate conditions and utilize local food supplies. Human-caused deforestation, overgrazing, agriculture, and urbanization disrupt ecosystems and lead to a decrease in biodiversity.

Archaeological studies have found that the earliest example of human modification of an ecosystem occurred in the Stone Age, when hunters played a major role in causing the mass extinction of many species of large mammals (mammoths, giant sloths, giant bears). Today, less than 5% of Europe retains its original habitats. The same number can be applied to the eastern United States, which lost its original forest and prairie. Tropical rain forests cover less than about half the area worldwide that they covered before the dawn of civilization, and they are disappearing at a rate of about 1.8% per year (▶Fig. 23.17a, b). Much of this loss comes from slash-and-burn agriculture, in which farmers and ranchers destroy forest to make open land for farming and grazing (▶Fig. 23.17c). Unfortunately, the heavy rainfall of tropical regions removes nutrients from the soil, making

(a)

(b)

FIGURE 23.16 (a) A giant, power-driven shovel can move more dirt and rock in a day than a stream can move in a decade. **(b)** Agriculture and development radically change the landscape of a region. The prairies of the midwestern United States have been replaced by giant farm fields.

the soil useless in just a few years. Overgrazing by domesticated animals can remove vegetation so completely that some grasslands have undergone desertification. And urbanization replaces the natural land surface with concrete or asphalt, a process that completely destroys an ecosystem and radically changes the amount of rain that infiltrates the land surface to become groundwater.

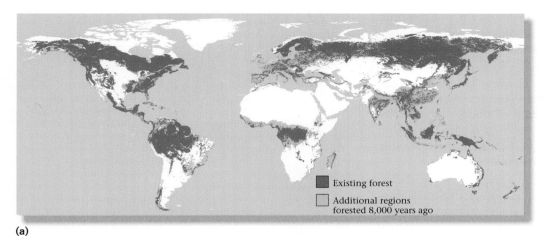

FIGURE 23.17 (a) The area of the Earth covered by rain forest shrank steadily during the past century. On this map, the dark-green areas indicate existing forest (including high-latitude scrub forest), whereas the yellow areas indicate regions that were forested 8,000 years ago. Note how tropical rain forests are shrinking. (b) A satellite image of Bolivia showing forest (dark-green) being replaced by fields (light-yellow). (c) Much of the loss is due to slash-and-burn agriculture.

■ Existing forest

□ Additional regions
forested 8,000 years ago

(a)

(b)

1990

2000

(c)

Human-caused changes to ecosystems affect the broader Earth System, because they modify biogeochemical cycles and Earth's albedo (surface reflectivity). For example, deforestation increases the CO_2 concentration in the atmosphere, for much of the carbon that was stored in trees is burned and rises. And the replacement of forest cover with concrete or fields increases Earth's albedo.

Pollution

The environment has always contained contaminants such as soot, dust, and the byproducts of organisms. But when human populations grew, urbanization; industrial and agricultural activity; the production of electricity; and modern modes of transportation greatly increased both the quantity and diversity of contaminants that entered the air, surface water, and groundwater. These contaminants, or **pollution**, include both natural and synthetic materials (in liquid, solid, or gaseous form). They have become a problem because they are produced at a higher rate than the Earth System can naturally absorb or modify. For example, small quantities of sewage can be absorbed by clay minerals in the soil or destroyed by bacterial metabolism, but large quantities overwhelm natural controls and can accumulate into destructive concentrations. Further, because many contaminants are not produced in nature, they are not easily removed by natural processes. Pollution of the Earth System is a type of global change, because it is a redistribution and reformulation of materials. Some key problems associated with this change include the following.

- *Smog*: The term was originally coined to refer to the dank, dark air that resulted when smoke from the burning of coal mixed with fog in London and other industrial cities. Another kind of smog, called photochemical smog, is the ozone-rich brown haze that blankets cities when exhaust from cars and trucks reacts with air in the presence of sunlight.

- *Water contamination*: We dump a great variety of chemicals into surface water and groundwater, including gasoline, other organic chemicals, radioactive waste, acids, fertilizers—the list could go on for pages. These chemicals affect biodiversity.

- *Acid runoff*: Dissolution of sulfide-containing minerals in ores or coal by groundwater or stream water makes the water acidic (it increases the concentration of hydrogen ions in the water) and toxic to life.

- *Acid rain*: When rain passes through air that contains sulfur-containing aerosols (emitted from power plants), the water dissolves the sulfur, creating sulfuric acid, or **acid rain**. Wind can carry aerosols far from a power plant, so acid rain can damage a broad region (▶Fig. 23.18).

- *Radioactive materials*: Nuclear weapons, nuclear energy, and medical waste transfer radioactive materials from rock to Earth's surface environment. Also, human-caused nuclear reactions produce new, nonnatural radioactive isotopes, some of which have relatively short half-lives. Thus, society has changed the distribution and composition of radioactive material worldwide.

- *Ozone depletion*: When emitted into the atmosphere, human-produced chemicals, most notably chlorofluorocarbons (CFCs), react with ozone in the stratosphere. This reaction, which happens most rapidly on the surfaces of tiny ice crystals in polar stratopheric clouds, destroys ozone molecules, thus creating an **ozone hole** over high-latitude regions, particularly during the spring (▶Fig. 23.19a, b). Note that the "hole" is not really an area where no ozone is present, but rather is a region where atmospheric ozone has been reduced substantially. The ozone hole is more prominent in the Antarctic than in the Arctic, because a current of air circulates around the land mass of Antarctica and traps the air, with its CFCs, above the continent, preventing it from mixing with air from elsewhere. Ozone holes have dangerous consequences, for they affect the ability of the atmosphere to shield the Earth's surface from harmful ultraviolet radiation. In 1987, a summit conference in Montréal proposed a global reduction of (ozone-destroying) CFC emissions. Reduction of such emissions may substantially reduce ozone depletion.

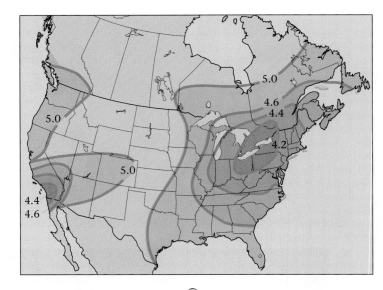

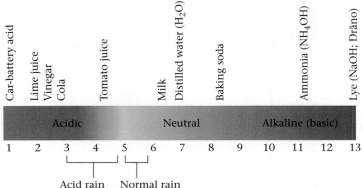

FIGURE 23.18 Acid rain has affected large portions of the United States. The map contours pH numbers, indicating the concentration of hydrogen ions in solution (according to the formula $pH = -\log[H^+]$; the square brackets mean "concentration"). Note that very acidic rain falls in the Northeast and Southwest. The scale gives a sense of what the numbers mean. A solution with a pH of 7 is neutral; acidic solutions have a pH less than 7, and alkaline solutions have a pH greater than 7.

Recent Global Warming

During the past two centuries, industry, energy production, and agriculture significantly altered the rate at which greenhouse gases, such as carbon dioxide (CO_2) and methane (CH_4), have been added to the atmosphere. In fact, the rate of addition has exceeded the rate at which these gases can be absorbed by dissolution in the ocean, by incorporation in plants, or by chemical-weathering reactions. Effectively, by burning fossil fuels at the rate that we do, we transfer CO_2 from underground reservoirs (oil and coal deposits) back into the atmosphere. In 1800, the mean concentration of CO_2 in the atmosphere was 280 parts per million (ppm); in 1900, it was 295 ppm; and by 2000, it had reached 370 ppm (▶Fig. 23.20a). At the same

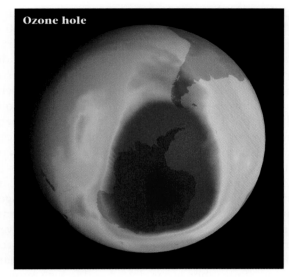

(a)

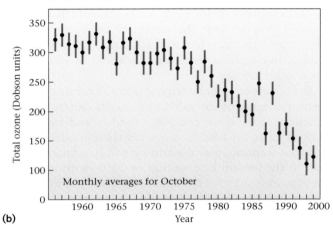

(b)

FIGURE 23.19 (a) The colors on this map show the quantity of ozone in the atmosphere when ozone levels are at a minimum. The darker blue area over Antarctica is the ozone hole. **(b)** The graph shows the decrease in ozone above Antarctica in October, specified in Dobson units (named for a scientist who helped identify the ozone hole). One Dobson unit is the amount of ozone needed to make a 0.01-mm-thick layer at the surface of the Earth. Note the increase in size between 1979 and 2000.

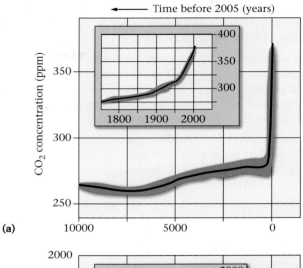

(a)

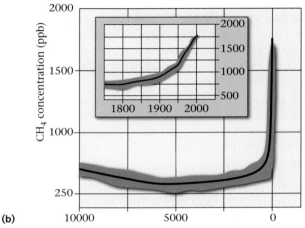

(b)

FIGURE 23.20 (a) By studying bubbles in ice cores from Antarctica, scientists can trace the record of carbon dioxide (CO_2) in the atmosphere back in time. Note that there has been a significant increase since the start of the Industrial Revolution. **(b)** The concentration of methane (CH_4) has also increased steadily during the past few centuries. In the graphs, the red band is the range of measurements.

time, the decay of organic material in rice paddies and the flatulence of cows have released enough methane to measurably change the concentration of this organic chemical in the atmosphere (▶Fig. 23.20b).

Have human-caused increases in greenhouse gases triggered global warming? To answer this question, researchers first need to determine if atmospheric and sea-surface temperature, weather patterns, the distribution of plants and animals, the dimensions of glaciers, and the extent of permafrost are different now than they were in the past. And if changes are observed, researchers must then determine if those changes are caused by a rise in greenhouse gas or by natural phenomena (e.g., sunspot cycles, changes in the

orbit or tilt of our planet, volcanic eruptions, or cosmic-ray intensity). In the past three decades, thousands of projects have been undertaken worldwide to collect relevant data, resulting in some startling discoveries:

- Large ice shelves, such as the Larsen B Ice Shelf, along the Antarctic Peninsula, and the Ayles Ice Shelf, along Ellesmere Island in northernmost Canada, are breaking up (▶Fig. 23.21a).

- The Greenland ice sheet is melting at an accelerating pace. Studies suggest that the rate of ice loss has increased from 90 to 220 square km per year in the last 10 years and that, in places, the sheet is thinning by about 1 m per year. In addition, the annual melt zone along the margins of the ice sheet has widened dramatically because the elevation of the equilibrium line (see Chapter 22) has risen. Valley glaciers draining the ice sheet moved 50% faster in 2003 than they did in 1992 (►Fig. 23.21b,c).

- The area covered by sea ice in the Arctic Ocean has decreased substantially. In fact, this ice covered 14% less area in 2005 than in 2004. This observation leads to the prediction that it may be possible to sail across the Arctic Ocean within decades (►Fig. 23.21d).

- Valley glaciers worldwide have been retreating rapidly, so that areas that were once ice covered are now bare. The change is truly dramatic in many locations (►Fig. 23.21e).

- Biological phenomena that are sensitive to climate are being disrupted. For example, the time at which sap in the maple trees of the northeastern United States starts to flow has changed, and the mosquito line (the elevation at which mosquitoes can survive) has risen substantially. Also, the average weight of polar bears has been decreasing, because the bears can no longer walk over pack ice to reach their hunting grounds in the sea.

- The area of permafrost in high latitutdes has substantially decreased, and melt ponds have replaced frozen land. Large regions that stayed frozen all year are now melting in the summer, so trees are tipping over.

To interpret these data from a broad perspective, a group of researchers founded the Intergovernmental Panel on Climate Change (IPCC) in 1988. This organization, sponsored by the World Meteorological Organization and the United Nations, reviews published research on climate change and, every 5 years, summarizes the conclusions in an assessment report. The language describing the likelihood that global warming is happening, and that humans have contributed significantly to causing it, has become progressively less equivocal in successive versions of the report. The Fourth Assessment Report, published in 2007, states:

> Warming of the climate system is unequivocal, as is now evident from observations of increases in global average air and ocean temperatures, widespread melting of snow and ice, and rising global average sea level. . . . The understanding of anthropogenic [human-caused] warming and cooling influences on climate has improved since the Third Assessment Report, leading to *very high confidence* that the globally averaged net effect of human activities since 1750 has been one of warming.

In other words, most climate researchers have concluded that global warming is real, and that the actions of people—burning fossil fuels, cutting down forests, paving over wetlands—have played a significant role in causing it. Specifically, global mean atmospheric temperature has risen by almost 1 °C during the last century,

and the rate of increase during the past 50 years appears to be greater than the rate for the previous 50 years (►Fig. 23.22a, b). This 1 °C temperature rise appears to be significantly more than would have taken place due to natural cycles (►Fig. 23.22c). Also, although such a change may seem small, the magnitude of temperature change between the ice age and now, by comparison, was only 3° to 5 °C. A small change in average temperature may have major consequences. Of note, sea-surface temperatures also appear to be rising in many parts of the world, though not everywhere.

Some researchers suggest that human impact on climate became noticeable as far back as 8,000 years ago, and that climate has been trending toward warmer conditions ever since. Deviations from the warming trend have been attributed, speculatively, to times when human population abruptly decreased (due to pandemics) so that production of greenhouse gas slowed and forests returned, to long-term sunspot cycles, or to volcanic events. It has been suggested that without the advent of agriculture, deforestation, and the burning of fossil fuels, Earth's climate would be in a cooling trend and we would be heading toward another ice age.

The effects of global warming over the coming decades to centuries remains the subject of intense debate, because predictions depend on computer models and not all researchers agree on how to construct these models. In the worst-case scenario, global warming will continue into the future at the present rate, so that by 2050—within the lifetime of many readers of this book—the average annual temperature will have increased in some parts of the world by 1.5° to 2.0 °C (Fig. 23.22c). At these rates, by the end of the century, temperatures could be almost 4 °C warmer depending on the model used (Fig. 23.22c), and by 2150, global temperatures may be 5° to 11 °C warmer than present—the warmest since the Eocene Epoch, 40 million years ago. The effects of such a change are controversial, but according to some climate models, the following events might happen.

- *A shift in climate belts*: Temperate climate belts would move to higher latitudes, and vegetation belts would follow this trend. As a result, desert regions would expand, and the soil would dry out in present agricultural regions (►Fig. 23.23a, b).

- *Rise in the snow line*: As temperatures increase, the snow line in mountainous areas rises, and globally, it moves to higher latitudes. This change impacts ecosystems and will even impact tourism by making it impossible for winter resorts to host snow-based activities.

- *Stronger storms*: An increase in average ocean temperatures would mean that more of the ocean could evaporate when a tropical depression passed over. This evaporation would nourish stronger hurricanes. In

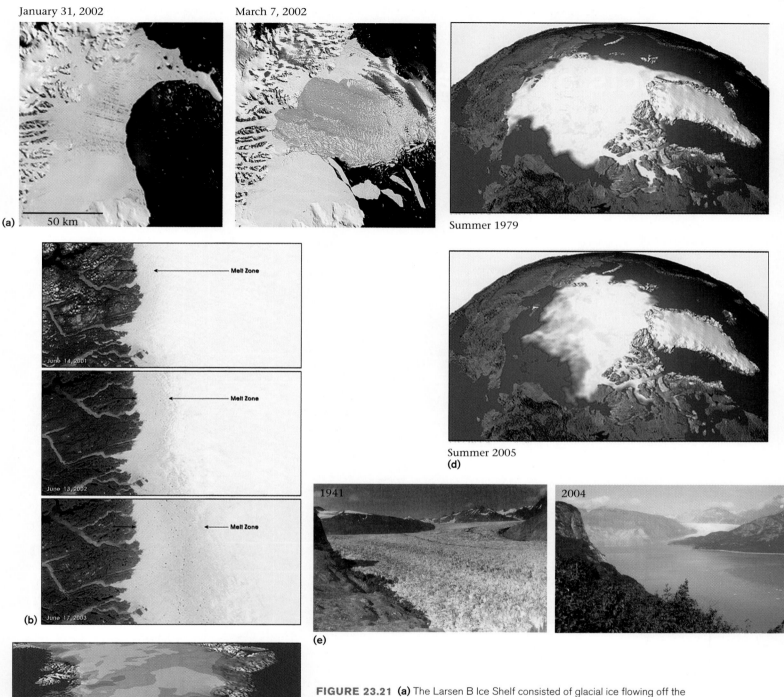

January 31, 2002

March 7, 2002

(a) 50 km

Melt Zone

June 14, 2001

Melt Zone

June 13, 2002

Melt Zone

June 17, 2003

(b)

(c)

Rate of Change in Icecap Height (cm/year)

-60   -20   -2 +2   +20   +60

Summer 1979

Summer 2005
(d)

1941

2004

(e)

FIGURE 23.21 (a) The Larsen B Ice Shelf consisted of glacial ice flowing off the Antarctic Penninsula. In a matter of weeks, during the Antarctic summer of 2002, the shelf cracked and disintegrated. Without the presence of the shelf to hold them back, the glaciers on land began to flow faster. **(b)** The summer melt line has risen notably along the edge of Greenland's ice sheet. The grayer area of ice, spotted with puddles, is the area undergoing melting during the summer. **(c)** The continental glacier that covers Greenland is melting at an accelerating rate. Along the southern edge, seen here, the glacier has been thinning at up to 60 cm per year. This change has led to modifications of weather, which have caused more snowfall and thickening of the glacier in the interior. **(d)** The north polar ice cap has been shrinking significantly, as can be seen by comparing an image of the cap from 1979 with an image from 2005. **(e)** A comparison of a photo from 1941 with a photo from 2004 shows the 12-km retreat of the Muir Glacier.

See for yourself . . .
Aspects of Global Change

Human activity has become a major agent of change on planet Earth. The consequences of this activity are clearly evident in the character of landscapes on the surface. The thumbnail images provided on this page are only to help identify tour sites. Go to wwnorton.com/studyspace to experience this flyover tour.

Clear Cutting the Amazon, Brazil
(Lat 4°3'51.60"S, Long 54°50'47.96"W)

At these coordinates, and an elevation of 100 km (62 miles), you can see swaths of clear cutting (Image G23.1). The pattern reflects access roads. Move east to Lat 3°51'7.92"S, Long 54°11'12.93"W, where the resolution is better, and from 5 km (3 miles), you can see clear-cut areas converted into rangeland (Image G23.2).

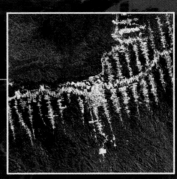

G23.1

G23.2

Long-Term Deforestation, Brazil
(Lat 16°41'18.46"S, Long 43°58'0.76"W)

At the coordinates provided, the view from 7 km (4 miles) shows hilly land (Image G23.3). Deforestation here took place decades ago. Forest remains only in valleys. Fly east to the Brazilian highlands at Lat 19°20'0.43"S, Long 41°14'25.05"W. From 25 km (15 miles), you can see that only the highest hills remain forested (Image G23.4). Yearly burning keeps the forest from regrowing.

G23.3

G23.4

Edge of the Everglades, Florida
(Lat 25°33'41.34"N, Long 80°35'2.75"W)

Looking down from 45 km (28 miles) at these coordinates, you see the transition between the natural Everglades on the west and intensively cultivated farmland on the east (Image G23.5). This radical change in landscape affects many aspects of the Earth System. Fly to Lat 25°53'27.04"N, Long 80°8'25.19"W, zoom to 6 km (4 miles) and look down at what was once a sandbar, but is now a forest of concrete (Image G23.6).

G23.5

G23.6

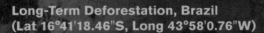

Urbanizing the Desert, Arizona
(Lat 33°33'48.47"N, Long 111°48'33.37"W)

At the coordinates provided, the view from 5 km (3 miles) shows the edge of a Phoenix suburb (Image G23.7). Note that construction has nearly eliminated natural drainage. The transformation of the Sonoran Desert of Arizona into cityscapes and farmland requires huge amounts of water. Some of the water comes from pumping underground aquifers, and some from canals. A canal traverses this image.

G23.7

G23.8

Receding Glacier, Switzerland
(Lat 46°34'21.13"N, Long 8°22'40.61"E)

Most glaciers have been retreating at accelerating paces in recent decades. Here, near the town of Gletsch, zoom to 7 km (4 miles), tilt, and look northeast (Image G23.8). You can see the empty U-shaped valley that was occupied by ice a century ago.

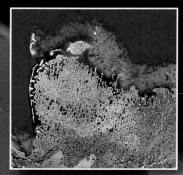

G23.9

Melting Permafrost, Siberia
(Lat 73°20'58.61"N, Long 124°43'59.76"E)

Here, on the Arctic coast of Siberia, a landscape that had been permafrost appears to be melting. Zoom to 200 km (125 miles), and you can see the pockmarks on land and beneath shallow coastal waters of the Arctic (Image G23.9). This image gives the impression that recent sea-level rise has submerged the landscape.

Intense Urbanization, Tokyo, Japan
(Lat 35°41'3.94"N, Long 139°48'35.69"E)

The view of Tokyo from an elevation of 1.5 km (1 mile) shows the results of complete transformation of a natural coastal area into a human-controlled environment (Image G23.10). Vehicles generate pollutants, the concrete absorbs and retains heat, and geologic materials have been transformed into buildings. Zoom out to 20 km (12 miles) to see how human activity has affected the coastline.

G23.10

G23.11

Village and Fields, China
(Lat 35°7'3.20"N, Long 114°27'16.86"E)

From an elevation of 5 km (3 miles), at these coordinates, you see a landscape in central China that has been transformed from forest to farmland (Image G23.11). Every square meter that does not lie beneath a house or road has been turned into a farm field.

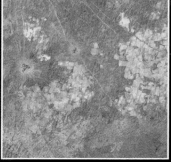

G23.12

Desertification of the Sahel, Africa
(Lat 13°1'46.67"N, Long 26°3'49.89"E)

Here, the view from 12 km (7 miles) reveals a dry landscape (Image G23.12). The radial pattern of paths on the left side of the image comes from cattle walking to a tiny watering hole. Cattle compact the land, preventing water infiltration, farming removes nutrients, and grazing causes devegetation. As a result, the landscape has become more desert-like.

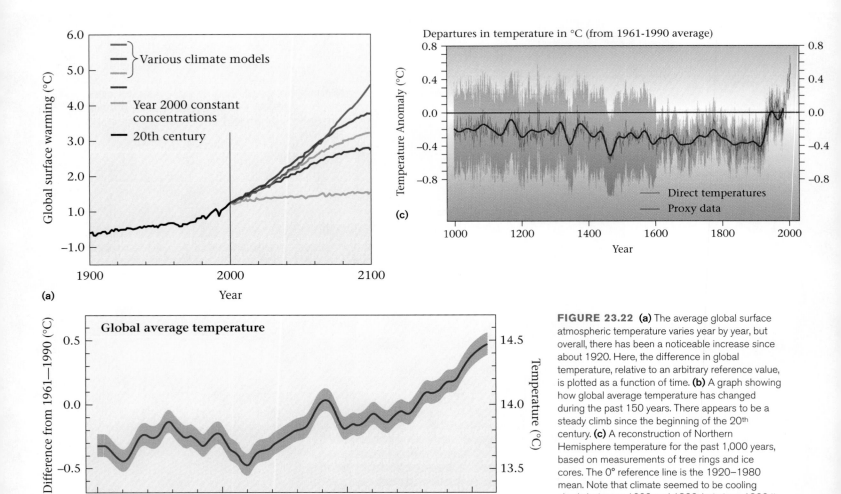

FIGURE 23.22 (a) The average global surface atmospheric temperature varies year by year, but overall, there has been a noticeable increase since about 1920. Here, the difference in global temperature, relative to an arbitrary reference value, is plotted as a function of time. **(b)** A graph showing how global average temperature has changed during the past 150 years. There appears to be a steady climb since the beginning of the 20th century. **(c)** A reconstruction of Northern Hemisphere temperature for the past 1,000 years, based on measurements of tree rings and ice cores. The 0° reference line is the 1920–1980 mean. Note that climate seemed to be cooling slowly between 1000 and 1900, but since 1900 it has warmed substantially.

FIGURE 23.23 The distribution of climate belts, indicated by vegetation type, will change if the global temperature rises. **(a)** Distribution of vegetation types today. **(b)** Distribution of vegetation types if the climate warms by just a couple of degrees.

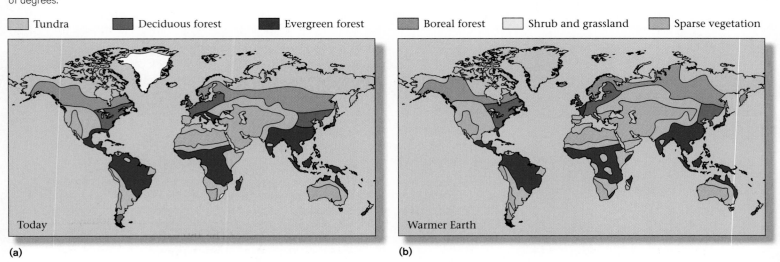

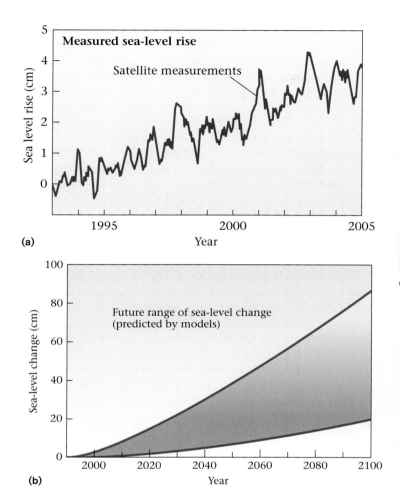

(a)

(b)

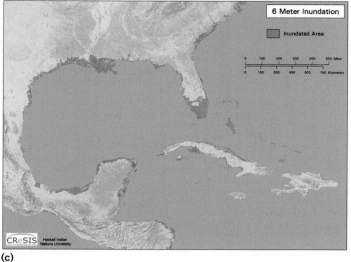

(c)

(d)

FIGURE 23.24 **(a)** Measurements of global sea-level change by satellite measurements show a gradual but noticeable rise during the past decade. The blue line indicates averages taken over 60 day periods. **(b)** All models of future sea-level change predict that there will be a rise. In the worst-case model, sea level could rise by almost 1 m in the next century. **(c)** This map shows the regions of the Gulf Coast that would be covered with seawater if sea level rose by 6 m. **(d)** Forest and grass fires rage in southern California with increasing frequency, partially due to drier overall conditions as shown in this satellite image.

nondesert areas, there might be more precipitation and, therefore, flooding.

- *A rise in sea level*: The melting of ice sheets in polar regions and the expansion of water in the sea as the water warms would make sea level rise enough to flood coastal wetlands and communities and damage deltas. Measurements suggest that there has been a rise of almost 12 cm in the past century (►Fig. 23.24a). Sea-level rise is already taking its toll in the South Pacific, where some islands have already become submerged. Even a change of only 1 m, which some models say may happen in little more than a century, could inundate regions of the world where 20% of the population currently lives (►Fig. 23.24b,c).

- *An increase in wildfires*: Warmer temperatures may lead to an increase in the frequency of wildfires because the moisture content of plants is lower (►Fig. 23.24d).

- *An interruption of the oceanic heat conveyor*: Oceanic currents play a major role in transferring heat across latitudes. If global warming melts polar ice, the resulting freshwater would dilute surface ocean water at high latitudes. This water could not sink, and thus thermohaline circulation would be shut off (see Chapter 18), preventing the water from conveying heat.

The potential changes described above, along with other studies that estimate the large economic cost of global warming, imply that the issue needs to be addressed seriously and soon. But what can be done? The 160 nations that signed the 1997 Kyoto Accord, at a summit meeting held in Japan, propose that the first step would be to slow the input of greenhouse gases into the atmosphere by decreasing the burning of fossil fuels and/or by forcing the CO_2 produced at power plants down wells into pore space underground. Motivating such actions, needless to say, involves controversial political decisions. Some researchers suggest that more aggressive solutions may be possible. Speculations include the intentional addition of sulfur dioxide to the atmosphere in order to increase the amount of sunlight reflected back to space, or the intentional addition of iron to the sea to encourage algal blooms that would absorb CO_2. But the feasibility of such approaches remains very far from certain.

> **Take-Home Message**
>
> Humans have become major agents of change in the Earth System. Human activities modify the landscape, disrupt ecosystems, and decrease forest cover. Evidence gathered in recent decades indicates that the release of greenhouse gases causes global warming.

23.7 THE FUTURE OF THE EARTH: A SCENARIO

Most of the discussion in this book has focused on the past, for the geologic record preserved in rocks tells us of earlier times. Let's now bring this book to a close by facing in the opposite direction and speculating what the world might look like in geologic time to come.

In the geologic near term, the future of the world depends largely on human activities. Whether the Earth System undergoes a major disruption and shifts to a new equilibrium, whether a catastrophic mass-extinction event takes place, or whether society achieves **sustainable growth** (an ability to prosper within the constraints of the Earth System) will depend on our own foresight and ingenuity. Projecting thousands of years into the future, we might well wonder if the Earth will return to ice-age conditions, with glaciers growing over major cities and the continental shelf becoming dry land, or if the ice age is over for good because of global warming. No one really knows for sure.

If we project millions of years into the future, it is clear that the map of the planet will change significantly because of the continuing activity of plate tectonics. For example, during the next 50 million years or so, the Atlantic Ocean will probably become bigger, the Pacific Ocean will shrink, and the western part of California will migrate northward. Eventually, Australia will crush against the southern margin of Asia, and the islands of Indonesia will be flattened in between.

Predicting the map of the Earth beyond that is hard, because we don't know for sure where new subduction zones will develop. Most likely, subduction of the Pacific Ocean will lead to the collision of the Americas with Asia, to produce a supercontinent ("Amasia"). A subduction zone eventually will form on one side of the Atlantic Ocean, and the ocean will be consumed. As a consequence, the eastern margin of the Americas will collide with the western margin of Europe and Africa. The sites of major cities—New York, Miami, Rio de Janeiro, Buenos Aires, and London—will be incorporated in a collisional mountain belt, and likely will be subjected to metamorphism and igneous intrusion before being uplifted and eroded. Shallow seas may once again cover the interiors of continents and then later retreat, and glaciers may once again cover the continents—it happened in the past, so it could happen again. And if the past is the key to the future, we *Homo sapiens* might not be around to watch our cities enter the rock cycle, for biological evolution may have introduced new species to the biosphere, and there is no way to predict what these species will be like.

And what of the end of the Earth? Geologic catastrophes resulting from asteroid and comet collisions will undoubtedly occur in the future as they have in the past. We can't predict when the next strike will come, but unless the object can be diverted, Earth is in for another radical readjustment of surface conditions. But it's not likely that such collisions will destroy our planet. Rather, astronomers predict that the end of the Earth will occur some 5 billion years from now, when the Sun begins to run out of nuclear fuel. When this happens, thermal pressure caused by fusion reactions will no longer be able to prevent the Sun from collapsing inward, because of the immense gravitational pull of its mass. Were the Sun a few times larger than it is, the collapse would trigger a supernova explosion that would blast matter of the Universe out into space to form a new nebula, perhaps surrounding a black hole. But since the Sun is not that large, the thermal energy

> **Take-Home Message**
>
> In the near term, Earth's surface will be affected by decisions of human society. Over longer time scales, the map of the Earth will change due to plate interactions and sea-level change. The end of the Earth will happen when the Sun becomes a red giant.

generated when its interior collapses inward will heat the gases of its outer layers sufficiently to cause them to expand. As a result, the Sun will become a red giant, a huge star whose radius would grow beyond the orbit of Earth (▶Fig. 23.25). Our planet will then vaporize, and its atoms will join an expanding ring of gas—the ultimate global change. If this happens, the atoms that once formed Earth and all its inhabitants through geologic time may eventually be incorporated in a future solar system, where the cycle of planetary formation and evolution will begin anew.

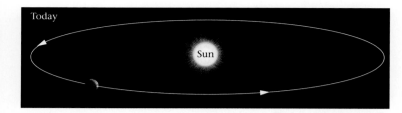

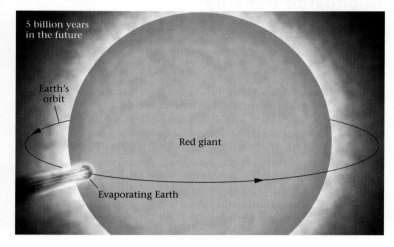

FIGURE 23.25 In about 5 billion years, the Sun will grow to become a red giant. When that happens, the Earth will first dry out, and then vaporize. Initially, it may look like a giant comet. Perhaps some of its atoms will become part of a nebula that someday condenses to form a new sun and planetary system.

Chapter Summary

- We refer to the global interconnecting web of physical and biological phenomena on Earth as the Earth System. Global change involves the transformations or modifications of physical and biological components of the Earth System through time. Unidirectional change results in transformations that never repeat, whereas cyclic change involves repetition of the same steps over and over.

- Examples of unidirectional change include the gradual evolution of the solid Earth from a homogeneous collection of planetesimals to a layered planet; the formation of the oceans and the gradual change in the composition of the atmosphere; and the evolution of life.

- Examples of physical cycles that take place on Earth include the supercontinent cycle, the sea-level cycle, and the rock cycle.

- A biogeochemical cycle involves the passage of a chemical among nonliving and living reservoirs. Examples include the hydrologic cycle and the carbon cycle. Global change occurs when factors change the relative proportions of the chemical in different reservoirs.

- Tools for documenting global climate change include the stratigraphic record, paleontology, oxygen-isotope ratios, bubbles in ice, growth rings in trees, and human history.

- Studies of long-term climate change show that, at times, in the past the Earth experienced greenhouse (warmer) periods; whereas, at other times there were icehouse (cooler) periods. Factors leading to long-term climate change include the positions of continents, volcanic activity, the uplift of land, and the formation of materials that remove CO_2, an important greenhouse gas.

- Short-term climate change can be seen in the record of the last million years. In fact, during only the past 15,000 years, we see that the climate has warmed and cooled a few times. Causes of short-term climate change include fluctuations in solar radiation and cosmic rays, changes in Earth's orbit and tilt, changes in reflectivity, and changes in ocean currents.

- Mass extinction, a catastrophic change in biodiversity, may be caused by the impact of a comet or asteroid or by intense volcanic activity associated with a superplume.

- During the last two centuries, humans have changed landscapes; modified ecosystems; and added pollutants to the land, air, and water at rates faster than the Earth System can process.

- The addition of CO_2 and CH_4 to the atmosphere may be causing global warming, which could shift climate belts and lead to a rise in sea level.

- In the future, in addition to climate change, the Earth will witness a continued rearrangement of continents resulting from plate tectonics, and will likely suffer the impact of asteroids and comets. The end of the Earth may come when the Sun runs out of fuel in about 5 billion years and becomes a red giant.

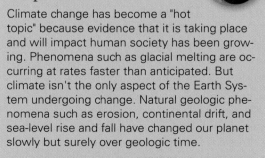

Geopuzzle Revisited

Climate change has become a "hot topic" because evidence that it is taking place and will impact human society has been growing. Phenomena such as glacial melting are occurring at rates faster than anticipated. But climate isn't the only aspect of the Earth System undergoing change. Natural geologic phenomena such as erosion, continental drift, and sea-level rise and fall have changed our planet slowly but surely over geologic time.

Key Terms

acid rain (p. 822)
albedo (p. 817)
biogeochemical cycle
    (pp. 802, 808)
climate-change model (p. 811)
Earth System (p. 802)
ecosystem (p. 820)
global change (p. 802)
global climate change (p. 803)
global cooling (p. 811)
global warming (p. 811)
Goldilocks effect (p. 803)
greenhouse (hothouse) period
    (p. 812)

icehouse period (p. 812)
mass-extinction event (p. 818)
ozone hole (p. 822)
paleoclimate (p. 811)
pollution (p. 821)
runaway greenhouse effect
    (p. 814)
sedimentary sequence
    (p. 805)
steady-state condition
    (p. 808)
sunspot cycle (p. 816)
supercontinent cycle (p. 805)
sustainable growth (p. 830)

Review Questions

1. Why do we use the term *Earth System* to describe the processes operating on this planet?

2. How have the Earth's crust and atmosphere changed since they first formed?

3. What processes control the rise and fall of sea level on Earth?

4. How does carbon cycle through the various Earth systems?

5. How do paleoclimatologists study ancient climate change?

6. Contrast icehouse and greenhouse conditions.

7. What are the possible causes of long-term climatic change?

8. What factors explain short-term climatic change?

9. Give some examples of events that cause catastrophic change.

10. Give some examples of how humans have changed the Earth.

11. What is the ozone hole, and how does it affect us?

12. Describe how carbon dioxide–induced global warming takes place, and how humans may be responsible. What effects might global warming have on the Earth System?

13. What are some of the likely scenarios for the long-term future of the Earth?

On Further Thought

1. If global warming continues, how will the distribution of grain crops change? Might this affect national economies? Why? How will the distribution of spruce forests change?

2. Currently, tropical rain forests are being cut down at a rate of 1.8% per year. At this rate, how many more years will the forests survive? In the eastern United States, the proportion of land with forest cover today has increased over the past century. In fact, most of the farmland that existed in New York State in 1850 is forestland today. Why? How might this change affect erosion rates in the region?

3. Using the library or the web, examine the change in the nature of world fisheries that has taken place in the last 50 years. Is the world's fish biomass sustainable if these patterns continue? What has happened to whale populations during the past 50 years?

Suggested Reading

Alvarez, W. 1997. *T. Rex and the Crater of Doom.* Princeton, N.J.: Princeton University Press.

Burroughs, W. J. 2001. *Climate Change: A Multidisciplinary Approach.* Cambridge: Cambridge University Press.

Collier, M., and R. H. Webb. 2002. *Floods, Droughts, and Climate Change.* Tucson: University of Arizona Press.

Ehler, J., W. C. Haneburg, and R. A. Larson, 2005. *Humans as Geologic Agents.* Boulder: Geological Society of America.

Flannery, T., 2002. *The Future Eaters: An Ecologic History of the Australian Lands and People.* New York: Grove Press.

Flannery, T., 2006. *The Weather Makers: How Man Is Changing the Climate and What it Means for Life on Earth.* Washington, D.C.: Atlantic Monthly Press.

Harvey, D. 1999. *Global Warming: The Hard Science.* Upper Saddle River, N.J.: Prentice-Hall.

Holland, H. D., and U. Petersen. 1995. *Living Dangerously: The Earth, Its Resources, and the Environment.* Princeton, N.J.: Princeton University Press.

Houghton, J. 2004. *Global Warming: The Complete Briefing, 3rd edition.* Cambridge: Cambridge University Press.

Intergovernmental Panel on Climate Change. 2007. *Climate Change 2007: The Physical Science Basis, Summary for Policymakers.* Geneva: IPCC Secretariat.

Kump, L. R., J. F. Kasting, and R. G. Crane. 2003. *The Earth System, 2nd ed.* Upper Saddle River, N.J.: Prentice-Hall.

Leggett, J. K. 2001. *The Carbon War: Global Warming and the End of the Oil Era.* New York: Routledge.

Lovejoy, T., and L. Hannah, eds. 2006. *Climate Change and Biodiversity.* New Haven: Yale University Press.

Mackay, A., et al., eds. 2003. *Global Change in the Holocene.* London: Edward Arnold.

MacKenzie, F. T. 2002. *Our Changing Planet: An Introduction to Earth System Science and Global Environmental Change, 3rd ed.* Upper Saddle River, N.J.: Prentice-Hall.

Officer, C., and J. Page. 1993. *Tales of the Earth: Paroxysms and Perturbations of the Blue Planet.* New York: Oxford University Press.

Ruddiman, W. F. 2005. *Plows, Plagues, and Petroleum: How Humans Took Control of Climate.* Princeton, N.J.: Princeton University Press.

Speth, J. G. 2004. *Red Sky at Morning: America and the Crisis of the Global Environment.* New Haven: Yale University Press.

Steffen, W., et al. 2005. *Global Change and the Earth System: A Planet Under Pressure.* New York: Springer.

Turco, R. P. 1997. *Earth under Siege: From Air Pollution to Global Change.* Oxford: Oxford University Press.

Van Andel, T. H. 1994. *New Views on an Old Planet: A History of Global Change.* Cambridge: Cambridge University Press.

Weart, S. R. 2003. *The Discovery of Global Warming.* Cambridge, Mass.: Harvard University Press.

Metric Conversion Chart

Length

1 kilometer (km) = 0.6214 mile (mi)
1 meter (m) = 1.094 yards = 3.281 feet
1 centimeter (cm) = 0.3937 inch
1 millimeter (mm) = 0.0394 inch
1 mile (mi) = 1.609 kilometers (km)
1 yard = 0.9144 meter (m)
1 foot = 0.3048 meter (m)
1 inch = 2.54 centimeters (cm)

Area

1 square kilometer (km^2) = 0.386 square mile (mi^2)
1 square meter (m^2) = 1.196 square yards (yd^2)
= 10.764 square feet (ft^2)
1 square centimeter (cm^2) = 0.155 square inch (in^2)
1 square mile (mi^2) = 2.59 square kilometers (km^2)
1 square yard (yd^2) = 0.836 square meter (m^2)
1 square foot (ft^2) = 0.0929 square meter (m^2)
1 square inch (in^2) = 6.4516 square centimeters
(cm^2)

Volume

1 cubic kilometer (km^3) = 0.24 cubic mile (mi^3)
1 cubic meter (m^3) = 264.2 gallons
= 35.314 cubic feet (ft^3)
1 liter (1) = 1.057 quarts
= 33.815 fluid ounces
1 cubic centimeter (cm^3) = 0.0610 cubic inch (in^3)
1 cubic mile (mi^3) = 4.168 cubic kilometers (km^3)
1 cubic yard (yd^3) = 0.7646 cubic meter (m^3)
1 cubic foot (ft^3) = 0.0283 cubic meter (m^3)
1 cubic inch (in^3) = 16.39 cubic centimeters (cm^3)

Mass

1 metric ton = 2,205 pounds
1 kilogram (kg) = 2.205 pounds
1 gram (g) = 0.03527 ounce
1 pound (lb) = 0.4536 kilogram (kg)
1 ounce (oz) = 28.35 grams (g)

Pressure

1 kilogram per square
centimeter (kg/cm^2)* = 0.96784 atmosphere (atm)
= 0.98066 bar
= 9.8067×10^4 pascals (Pa)
1 bar = 10 megapascals (Mpa)
= 1.0×10^5 pascals (Pa)
= 29.53 inches of mercury (in a barometer)
= 0.98692 atmosphere (atm)
= 1.02 kilograms per square centimeter (kg/cm^2)
1 pascal (Pa) = 1 $kg/m/s^2$
1 pound per square inch = 0.06895 bars
= 6.895×10^3 pascals (Pa)
= 0.0703 kilogram per square
centimeter

Temperature

To change from Fahrenheit (F) to Celsius (C):
$$°C = \frac{(°F - 32°)}{1.8}$$

To change from Celsius (C) to Fahrenheit (F):
$$°F = (°C \times 1.8) + 32°$$

To change from Celsius (C) to Kelvin (K):
$$K = °C + 273.15$$

To change from Fahrenheit (F) to Kelvin (K):
$$K = \frac{(°F - 32°)}{1.8} + 273.15$$

---

*Note: Because kilograms are a measure of mass whereas pounds are a unit of weight, pressure units incorporating kilograms assume a given gravitational constant (g) for Earth. In reality, the gravitational for Earth varies slightly with location.

APPENDIX A

Scientific Background: Matter and Energy

a.I INTRODUCTION

In order to understand the formation and evolution of the Universe, as well as descriptions of materials that constitute the Earth and processes that shape the Earth, we must first understand some basic facts about matter, energy, and heat (▶Fig. a.1). The following synopsis highlights key topics from physics and chemistry that serve as an essential background to the rest of the book.

a.1 DISCOVERING THE NATURE OF MATTER

Matter takes up space—you can feel it and see it. We use the word **matter** to refer to any material making up the universe. The amount of matter in an object is its **mass**. An object with a greater mass contains more matter—for example, a large tree contains more matter than a blade of grass. There's a subtle but important distinction between mass and weight. **Weight** depends on the amount of an object's mass but also on the strength of gravity. An astronaut has the same mass on both the Earth and the Moon, but weighs a different amount in each place. Since the amount of gravitational pull exerted by an object depends on its mass, and the Moon has about one-sixth the mass of the Earth, an astronaut weighing 68 kilograms (150 pounds) on Earth would weigh only about 11 kilograms (25 pounds) on the Moon. Thus, lunar explorers have no trouble jumping great distances even when burdened by a space suit and oxygen tanks.

What does matter consist of? Early philosophers deduced that the Earth and the plants and animals on it must be formed from simpler components, just as bread is made from a measured mixture of primary ingredients. They initially thought that the "primary ingredients" making up matter included only earth, air, fire, and water. Then a philosopher named Democritus (ca. 460–370 B.C.E.) argued that if you were able to keep dividing matter into progressively smaller pieces, you would eventually end up with nothing, and since it doesn't seem possible to make something out of nothing, there must be a smallest piece of matter that can't be subdivided further. He proposed the name **atom** for these smallest pieces, based on the Greek word *atomos,* which means indivisible.

Our modern understanding of matter didn't become established until the seventeenth century, when chemists such as Robert Boyle (1627–1691) recognized that certain substances, like hydrogen, oxygen, carbon, and sulfur, cannot break down into other substances, while others, like water and salt, *can* break down into other substances (▶Fig. a.2). For example, water breaks down into oxygen and hydrogen. Substances that can't be broken down came to be known as **elements**, while those that can be broken down came to be known as **compounds**. An English schoolteacher, John Dalton (1766–1844), adopted the word atom for the smallest piece of an

FIGURE a.1 A lightning storm over mountains. The solid rock, the air, and the clouds consist of matter. The lightning, a rapidly moving current of electrons, is a manifestation of energy.

Salt crystal

Sodium metal   Chlorine gas

FIGURE a.2 A compound, salt, can be subdivided to form two elements, sodium metal and chlorine gas. Neither sodium nor chlorine can be divided further.

element that maintains the property of the element, and suggested that compounds consisted of combinations of different atoms. Then in 1869, a Russian chemist named Dmitri Mendeléev (1834–1907) realized that groups of elements share similar characteristics. Mendeléev organized the elements into a chart we now call the **periodic table of the elements** (▶Fig. a.3). In the figure, elements within each column of the table behave similarly; for example, all elements in the right-hand column are inert gases, meaning that they can't combine with other elements to form compounds. But

FIGURE a.3 The modern periodic table of the elements. The columns group elements with related properties. For example, inert gases are listed in the column on the right. Metals are found in the central and left parts of the chart.

Alkali metals																	Inert gases
H 1 Hydrogen 1.007																	He 2 Helium 4.002
Li 3 Lithium 6.941	Be 4 Beryllium 9.0121											B 5 Boron 10.811	C 6 Carbon 12.011	N 7 Nitrogen 14.006	O 8 Oxygen 15.999	F 9 Fluorine 18.998	Ne 10 Neon 20.179
Na 11 Sodium 22.989	Mg 12 Magnesium 24.305											Al 13 Aluminum 26.981	Si 14 Silicon 28.085	P 15 Phosphorus 30.973	S 16 Sulfur 32.066	Cl 17 Chlorine 35.452	Ar 18 Argon 39.948
K 19 Potassium 39.098	Ca 20 Calcium 40.078	Sc 21 Scandium 44.955	Ti 22 Titanium 47.88	V 23 Vanadium 50.941	Cr 24 Chromium 51.996	Mn 25 Manganese 54.938	Fe 26 Iron 55.847	Co 27 Cobalt 58.933	Ni 28 Nickel 58.693	Cu 29 Copper 63.546	Zn 30 Zinc 65.39	Ga 31 Gallium 69.723	Ge 32 Germanium 72.61	As 33 Arsenic 74.921	Se 34 Selenium 78.96	Br 35 Bromine 79.904	Kr 36 Krypton 83.80
Rb 37 Rubidium 85.467	Sr 38 Strontium 87.62	Y 39 Yttrium 88.905	Zr 40 Zirconium 91.224	Nb 41 Niobium 92.906	Mo 42 Molybdenum 95.94	Tc 43 Technetium 98.907	Ru 44 Ruthenium 101.07	Rh 45 Rhodium 102.905	Pd 46 Palladium 106.42	Ag 47 Silver 107.868	Cd 48 Cadmium 112.411	In 49 Indium 114.82	Sn 50 Tin 118.710	Sb 51 Antimony 121.757	Te 52 Tellurium 127.60	I 53 Iodine 126.904	Xe 54 Xenon 131.29
Cs 55 Cesium 132.905	Ba 56 Barium 137.327	La 57 Lanthanum 138.905	Hf 72 Hafnium 178.49	Ta 73 Tantalum 180.947	W 74 Tungsten 183.85	Re 75 Rhenium 186.207	Os 76 Osmium 190.2	Ir 77 Iridium 192.22	Pt 78 Platinum 195.08	Au 79 Gold 196.966	Hg 80 Mercury 200.59	Tl 81 Thallium 204.383	Pb 82 Lead 207.2	Bi 83 Bismuth 208.980	Po 84 Polonium 208.982	At 85 Astatine 209.987	Rn 86 Radon 222.017
Fr 87 Francium 223.019	Ra 88 Radium 226.025	Ac 89 Actinium 227.027															

Symbol — He 2 Helium 4.002 — Atomic number, Name, Atomic weight

Transition elements (metals)

Nonmetals

Ce 58 Cerium 140.115	Pr 59 Praseodymium 140.907	Nd 60 Neodymium 144.24	Pm 61 Promethium 144.912	Sm 62 Samarium 150.36	Eu 63 Europium 151.965	Gd 64 Gadolinium 157.25	Tb 65 Terbium 158.925	Dy 66 Dysprosium 162.50	Ho 67 Holmium 164.930	Er 68 Erbium 167.26	Tm 69 Thulium 168.934	Yb 70 Ytterbium 173.04	Lu 71 Lutetium 174.967
Th 90 Thorium 232.038	Pa 91 Protactinium 231.035	U 92 Uranium 238.028	Np 93 Neptunium 237.048	Pu 94 Plutonium 244.064	Am 95 Americium 243.061	Cm 96 Curium 247.070	Bk 97 Berkelium 247.070	Cf 98 Californium 251.079	Es 99 Einsteinium 252.083	Fm 100 Fermium 257.095	Md 101 Mendelevium 258.10	No 102 Nobelium 259.100	Lr 103 Lawrencium 262.11

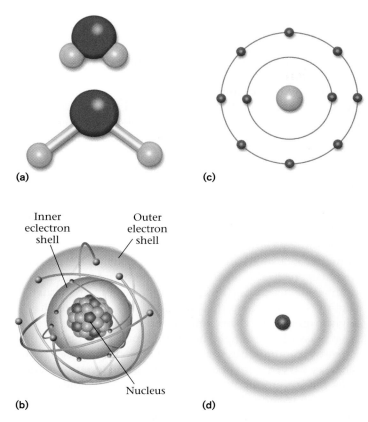

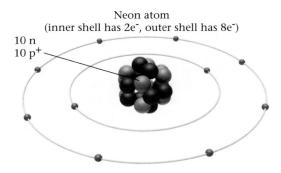

(a)

(c)

Inner eclectron shell   Outer electron shell

Nucleus

(b)

(d)

FIGURE a.4 Different ways of portraying atoms. **(a)** Two ways of portraying a water molecule. The large ball is oxygen, and the small ones are hydrogen. The "sticks" represent chemical bonds. **(b)** An image of an atom with a nucleus surrounded by electrons. **(c)** This diagram shows the number of electrons in the inner shells. **(d)** An alternative depiction of electron shells, implying that the electrons constitute a cloud. In reality, electrons do not follow simple circular orbits.

Mendeléev and his contemporaries didn't know what caused the similarities and differences. An understanding of the cause would have to wait until twentieth-century physicists discovered the internal structure of the atom.

a.2 A MODERN VIEW OF MATTER

Atoms

In modern terminology, an element is a substance composed only of the same kind of atoms. Ninety-two different elements occur naturally on Earth, but physicists have created more than a dozen new elements in the laboratory. Each element has a name (e.g., nitrogen, hydrogen, sulfur) and a **symbol**, an abbreviation of its English or Latin name (N = nitrogen, H = hydrogen, Fe = iron, Ag = silver).

Work in the late 1800s and early 1900s demonstrated that, contrary to the view of Democritus, atoms actually can be divided. Ernest Rutherford, a British physicist, made this key discovery in 1910 when he shot a beam of atoms at a gold foil and found, to his amazement, that only a tiny fraction of the atoms bounced back; most of the mass in the beam passed through the foil as if it were invisible. This result could mean only one thing. Most of the mass in an atom clusters in a dense ball at the atom's center, and this ball is surrounded by a cloud that contains very little mass, so an atom as a whole consists mostly of empty space.

Physicists now refer to the dense ball at the center of the atom as the **nucleus**, and the low-density cloud surrounding the nucleus as the **electron cloud**. Further study in the first half of the twentieth century led to the conclusion that the nucleus contains two types of subatomic particles: **neutrons**, which have a neutral electrical charge, and **protons**, which have a positive electrical charge (▶Fig. a.4a–d). The electron cloud consists of negatively charged particles, **electrons**, which are only about $\frac{1}{1,836}$ times as massive as protons. Remember that opposite charges attract; the positive charge on the nucleus attracts the negative charge of the electrons, so the nucleus holds on to the electron cloud. (For simplicity, think of a positive charge as the "+" end of a battery and a negative charge as the "−" end.) The mass of a neutron approximately equals the sum of the mass of a proton and the mass of an electron. In the past few decades, physicists have found that protons and neutrons are made up of a myriad of even smaller particles, the smallest of which is called a **quark**.

Electron clouds have a complex internal structure. Electron are grouped in intervals called **orbitals**, energy levels, or shells. Some shells have a spherical shape, whereas others resemble dumbbells, rings, or groups of balls—for simplicity, we portray shells as circles, in cross section (▶Fig. a.5). Electrons in inner shells are concentrated near the nucleus, whereas those of outer shells predominate farther

FIGURE a.5 This schematic drawing of a neon atom shows the two complete electron shells. The inner shell contains two electrons, the outer one eight. The "shells" merely represent the most likely location for an election to be.

Neon atom
(inner shell has 2e⁻, outer shell has 8e⁻)

10 n
10 p⁺

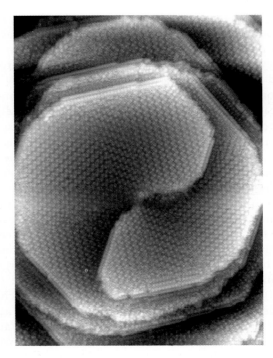

FIGURE a.6 An image of atoms taken with a special microscope that uses electrons to illuminate the surface.

from the nucleus. Successively higher shells lie at progressively greater distances from the nucleus. Each shell can only contain a specific number of electrons: the lowest shell can contain only two electrons, while the next several shells each can hold 8. Electrons fill the lowest shells first, so that atoms with a small number of electrons only have the innermost shells; the outer shells do not exist unless there are electrons to fill them.

The outermost shell of electrons, in effect, defines the outer edge of an atom, so an atom with many occupied electron shells is larger than one with few occupied shells (argon, for example, with 18 electrons, contains more occupied electron shells than helium, with 2 electrons, so an argon atom is bigger than a helium atom). If we picture the nucleus of a carbon atom as an orange, the electrons of the outermost shell would lie at a distance of over 1 km from the orange.

Atoms are so small that they can't be seen with even the strongest light microscopes. (However, by using some clever techniques, scientists have been able, in recent years, to create images of very large atoms; ▶Fig. a.6.) Atoms are so small, in fact, that 1 gram (0.036 ounces) of helium contains 6.02×10^{23} atoms; in other words, a quantity of helium weighing little more than a postage stamp contains 602,000,000,000,000,000,000,000 atoms! The number of atoms of helium in a small balloon is approximately the same as the number of balloons it would take to replace the entirety of Earth's atmosphere.

Atomic Number, Atomic Mass, and Isotopes

We distinguish atoms of different elements from one another by their **atomic number**, the number of protons in their nucleus. For example, hydrogen's atomic number is 1, oxygen's is 8, lead's is 82, and uranium's is 92. We write the atomic number as a subscript to the left of the element's symbol (e.g., $_1$H, $_8$O, $_{82}$Pb, $_{92}$U). With the exception of the most common hydrogen nuclei, all atomic nuclei also contain neutrons. In smaller atoms, the number of neutrons equals the number of protons, but in larger atoms, the number of neutrons exceeds the number of protons. Subatomic particles are held together in a nucleus by **nuclear bonds**.

Atomic weight (or **atomic mass**) defines the amount of matter in a single atom. For a given element, the atomic weight *approximately* equals the number of protons plus the number of neutrons. (Precise atomic weights are actually slightly greater than this sum, because neutrons have slightly more mass than protons, and because the electrons have mass.) For example, helium contains 2 protons and 2 neutrons, and thus has an atomic weight of about 4, whereas oxygen contains 8 protons and 8 neutrons and thus has an atomic weight of about 16. We indicate the weight of an atom by a superscript to the left of the element's symbol (^{4}He, ^{16}O).

Some elements occur in more than one form, and these differ in atomic weight. For example, uranium 235 ($^{235}_{92}$U) contains 92 protons and 143 neutrons, while uranium 238 ($^{238}_{92}$U) contains 92 protons and 146 neutrons. Note that both forms of uranium have the *same* atomic number—they must, if they are to be considered the same element. But they have different quantities of neutrons and thus differ in atomic weight. Multiple versions of the same element, which differ from one another in atomic weight, are called **isotopes** of the element (▶Fig. a.7).

FIGURE a.7 The three isotopes of hydrogen: regular hydrogen, deuterium, and tritium.

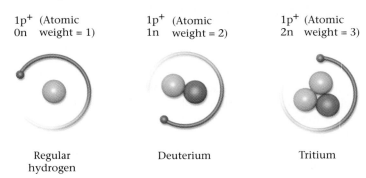

1p$^+$ (Atomic   1p$^+$ (Atomic   1p$^+$ (Atomic
0n weight = 1)   1n weight = 2)   2n weight = 3)

Regular   Deuterium   Tritium
hydrogen

Ions: Atoms with a Charge

If the number of electrons (negatively charged particles) exactly equals the number of protons (positively charged particles) in an atom, then the atom is electrically neutral. But if the numbers aren't equal, then the atom has a net electrical charge and is called an **ion**. For example, if an atom has two fewer electrons than protons, then we say it has a charge of +2, and if it has two more electrons than protons, it has a charge of −2. We write the charge as a superscript to the right of the symbol (e.g., Na^+). Ions with a negative charge are **anions** (pronounced ANN-eye-ons), and ions with a positive charge are **cations** (pronounced CAT-eye-ons). Ions form because atoms "prefer" to have a complete outer electron shell. Thus, an atom may give up or take on electrons in order for its outer shell to contain the proper number of electrons. As is the case with neutral atoms, the size of an ion depends on the number of shells containing electrons. Note that oxygen ions are larger than silicon ions, even though silicon has a larger atomic number, because oxygen has gained electrons and thus uses more electron shells than does silicon, for silicon has lost electrons (▶Fig. a.8).

Molecules and Chemical Bonds

Most of the materials we deal with in everyday life—oxygen, water, plastic—are not composed of isolated atoms. Rather, most atoms tend to stick, or bond, to other atoms; two or more atoms stuck together constitute a **molecule**. Hydrogen gas, for example, consists of H_2 molecules; note that a hydrogen molecule consists of two of the same kind of atom. But many materials contain different kinds of atoms bonded together. As we noted earlier, such materials are called compounds. For example, common table salt is a compound containing sodium and chlorine atoms bonded together. Compounds generally differ markedly from the elements that make them up—salt bears no resemblance at

all to pure sodium (a shiny metal) or pure chlorine (a noxious gas) (Fig. a.2). A molecule is the smallest identifiable piece of a compound, containing the correct proportion of the compound's elements. For example, a molecule of water consists of two atoms of hydrogen and one atom of oxygen. We represent water by the **chemical formula** H_2O, a concise recipe that indicates the relative proportions of different elements in the molecule (Fig. a.4a).

Chemical bonds act as the glue that holds atoms together to form molecules and holds molecules together to form larger pieces of a material. Chemical bonding results from the interaction among the electrons of nearby atoms, and can take place in four different ways. (Note that chemical bonds are not the same as the nuclear bonds that hold together protons and neutrons in a nucleus.)

- *Ionic bonds:* As an inviolate rule of nature, "like" electrical charges repel (two positive charges push each other away), while "unlike" electrical charges attract (a negative charge sticks to a positive charge). Bonds that form in this way are called **ionic bonds** (▶Fig. a.9). For example, in a molecule of salt, positively charged sodium ions (Na^+) attract negatively charged chloride (Cl^-) ions. (Chloride is the name given to ions of chlorine.)

FIGURE a.9 The sodium atom has an unfilled outer shell—the shell has room for eight electrons but only has one. The chlorine atom also has an unfilled outer shell—it's missing one electron. The sodium atom gives up its outer electron, and thus has one more proton than electron (i.e., a net positive charge), while the chlorine gains an electron and thus has one more electron than proton (a net negative charge). The two ions attract each other.

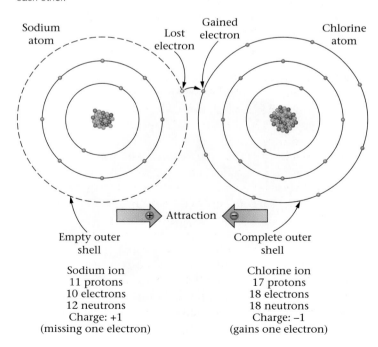

Sodium atom — Lost electron — Gained electron — Chlorine atom

⊕ Attraction ⊖

Empty outer shell — Complete outer shell

Sodium ion
11 protons
10 electrons
12 neutrons
Charge: +1
(missing one electron)

Chlorine ion
17 protons
18 electrons
18 neutrons
Charge: −1
(gains one electron)

FIGURE a.8 Relative sizes of common ions making up materials in rocks at the Earth's surface.

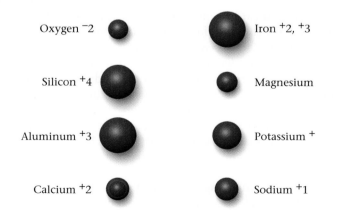

Oxygen $^-2$

Silicon $^{+4}$

Aluminum $^{+3}$

Calcium $^{+2}$

Iron $^{+2}$, $^{+3}$

Magnesium

Potassium $^+$

Sodium $^{+1}$

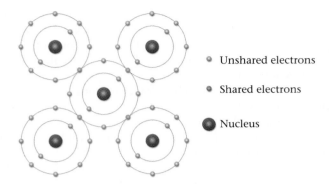

FIGURE a.10 The atoms in H_2 gas are covalently bonded. Each hydrogen atom alone has one electron, even though its outer shell could hold two electrons. When two atoms bond, they share electrons.

- *Covalent bonds*: The atoms of carbon making up a diamond do not transfer electrons to one another, but rather share electrons. Bonding that involves the sharing of electrons is called **covalent bonding** (▶Fig. a.10). Because of the sharing, the electron shells of all the carbon atoms in a diamond are complete, and all the carbon atoms have a neutral charge. Water molecules also exist because of covalent bonding: in a water molecule, two hydrogen atoms are covalently bonded to one oxygen atom.

- *Metallic bonds*: In metals, electrons of the outer shells move easily from atom to atom and bind the atoms to each other. We call this type of bonding **metallic bonding** (▶Fig. a.11). Because outer-shell electrons

FIGURE a.11 In a metallically bonded material, nuclei and their inner shells of electrons float in a "sea" of free electrons. Sometimes the electrons orbit the nuclei, but at other times they stream through the metal.

move so freely, metals conduct electricity easily—when you connect a metal wire to an electrical circuit, a current of electrons flows through the metal.

- *Bonds resulting from the polarity of atoms or molecules*: Chemists long recognized that some materials break or split so easily that they must be held together by particularly weak chemical bonds. Eventually, they realized that these bonds are due to the permanent or temporary polarity of molecules. **Polarity** means that the molecule has a positive charge on one side and a negative charge on the other. Polarity-related bonds form because the negative side of one molecule attracts the positive side of another.

Bonds resulting from the polarity of molecules are important in many geologic materials. For example, liquid water consists of molecules in which two hydrogen atoms covalently bond to one oxygen atom. The hydrogen atoms both lie on the same side of the oxygen atom. The oxygen nucleus, which contains 8 protons, exerts more attraction to the shared electrons than do the hydrogen nuclei, each of which contains only one proton. As a consequence, the shared electrons concentrate closer to the oxygen side of the molecule, making the oxygen side more negative than the hydrogen side. Thus, the water molecules are polar, and they attract each other. This attraction is called a **hydrogen bond** (▶Fig. a.12). The polarity of water molecules makes water a good solvent (meaning, it can dissolve substances), because the polar water molecules attract and surround ions of soluble materials and pull them apart.

Johannes van der Waals (1837–1923), a Dutch physicist, discovered another type of weak chemical bonding that depends on polarity. This type, now known as **van der Waals bonding**, links one covalently bonded molecule to another. The bonds exist because electrons temporarily cluster on one side of each molecule, giving it a polarity.

The Forms of Matter

Matter exists in one of four states: **solids**, which can maintain their shape for a long time without the restraint of a container; **liquids**, which flow fairly quickly and assume the shape of the container they fill (it's possible for a liquid to fill only part of its container); **gases**, which expand to fill the entire container they have been placed in and will disperse in all directions when not restrained; and **plasma**, an unfamiliar, gas-like mixture of positive ions and free electrons that exists only at very high temperatures (▶Fig. a.13a–d). Which state of matter exists at a location depends on the pressure and temperature (▶Fig. a.14).

To understand the differences at an atomic scale among solids, liquids, and gases—the common forms of matter on Earth—imagine a large group of students getting ready to take a 9:00 A.M. exam in a windowless auditorium.

In the figure a.10 legend:
- Unshared electrons
- Shared electrons
- Nucleus

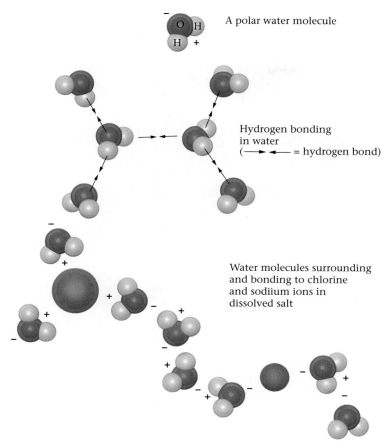
A polar water molecule

Hydrogen bonding in water
(⟶ ◄─ = hydrogen bond)

Water molecules surrounding and bonding to chlorine and sodiium ions in dissolved salt

FIGURE a.12 Water is a polar molecule, because the two hydrogen atoms lie on the same side of the oxygen atom. Thus, water molecules tend to attract one another (this attraction creates surface tension, and causes water to form drops). Similarly, the polar molecules surround chlorine and sodium ions when salt dissolves in water.

The students represent atoms. It's 8 o'clock, and some students are in their rooms, some are in the library, some are walking to the auditorium, and some are at the cafeteria. In other words, everyone is scattered all over the place; there is no order to the distribution of people, and they are too far apart to interact and bond to one another. Such a distribution is characteristic of a gas.

At ten minutes before nine, most students have arrived at the auditorium. Some are already in their seats, but many are walking in or out of the room. Such a distribution characterizes the state of a liquid, in which most of the particles are in the container, except for a slow exchange with the atmosphere (by evaporation or condensation). There is short-range order, represented by little clumps of students in their seats or chatting in the aisles, but overall there is still a lot of motion and no overall organization, or long-range order.

Suddenly the lights go off, when someone bumps against the switch, and everyone stops in place to avoid stumbling but people continue to fidget in place. Even though the migration of people has stopped, there is still no long-range order in the room. Such a situation characterizes a special kind of solid called a **glass**. Glass has a disordered arrangement of atoms, like a liquid, but cannot flow and thus behaves like solid.

The lights go on again, and the examiner finally enters. Everyone takes a seat according to a specified plan—alternate rows and alternate seats. Even though individual students are not motionless, because they are still fidgeting in place, they are no longer moving out of an orderly arrangement. The situation now corresponds to that in a solid.

a.3 THE FORCES OF NATURE

In our everyday experience, we constantly see or feel the effect of force. Forces can squash objects, speed them up and slow them down, tear, stretch, spin, and twist them, and make them float or sink. Isaac Newton, the great British scientist who effectively founded the field of physics, was the first to describe the way forces work. He defined a **force** as simply the push or pull that causes the velocity (speed) of a mass to change in magnitude and/or direction.

We can distinguish between two categories of force. The first, which includes force applied by the movement of a mass (a hand, a hammer, the wind, waves), is called **mechanical force**, or contact force. When you push a block across the floor, you are applying a mechanical force. The second category, which includes force resulting from the action of an invisible agent, is called a **field force**, or noncontact force. If you drop a book, the invisible force of gravity pulls the book toward the floor.

Physicists further recognize four types of field forces: gravity, electromagnetic, strong nuclear, and weak nuclear. **Gravity** is a force of attraction between any two masses, and can act over large distances. The magnitude of gravitational attraction depends on the size of the masses and the distance between them. We feel a much stronger gravitational pull to the planet we walk on than we do to a baseball. **Electromagnetic force**, the force associated with electricity and magnetism, is stronger than gravity, but only operates between materials that have electrical charges or are magnetic, and only operates over short distances. Like gravity, electromagnetic force depends on the distance between objects, but unlike gravity it can be either attractive or repulsive—as mentioned previously, like charges repel and unlike charges attract. Particles within atoms are subject to two kinds of nuclear force. We'll leave the discussion of nuclear forces to a physics text.

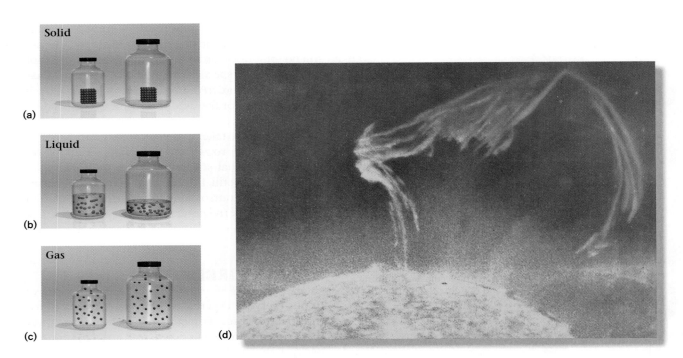

FIGURE a.13 The states of matter. **(a)** A solid block sitting in a bottle retains its shape regardless of the size of the container. **(b)** A liquid conforms to the shape of the container. If the container changes shape, the liquid also changes shape, so that it stays constant in density (mass per unit volume). **(c)** A gas will expand to fill whatever volume it occupies, and thus can change density if the volume changes. **(d)** The Sun contains plasma, a gas-like material that is so hot that electrons have been stripped from the nuclei.

a.4 ENERGY

To a physicist, **energy** is the ability to do work, or, in other words, the ability to apply a force that moves a mass by some distance (**work** = force × distance). According to this definition, gasoline serves as an energy source because we can burn gasoline to move heavy cars and trucks (a type of work).

FIGURE a.14 A material such as water can change state (from gas to liquid, liquid to solid, or solid to liquid), depending on the pressure and temperature. Water can exist simultaneously in all three states at a unique pressure and temperature called the triple point.

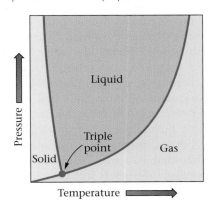

Kinetic and Potential Energy

Formally, we classify energy into two basic types: **kinetic energy**, the energy of motion, and **potential energy**, the energy stored in a material. A boulder sitting at the top of a hill has potential energy, because the force of gravity will do work on it as the boulder topples down. If the boulder falls down, part of its potential energy converts into kinetic energy (▶Fig. a.15). Some of this kinetic energy can be transferred to a second boulder if the original boulder strikes another boulder, causing a contact force that starts the second boulder moving. Similarly, the gas in a car's tank has potential energy, which converts to kinetic energy when the gas burns and molecules start moving quickly. This kinetic energy starts the cylinders in the engine moving up and down. Now we briefly look at where energy comes from.

Energy from Chemical Reactions

A **chemical reaction** is a process whereby one or more compounds (the reactants) come together or break apart to form new compounds and/or ions (the products). Many chemical reactions produce energy. In the case of burning gasoline, gasoline molecules react with oxygen molecules in the air to form carbon dioxide molecules

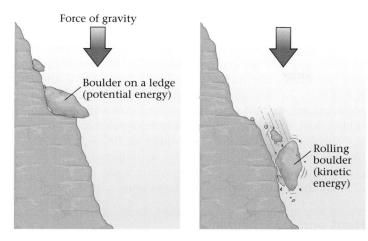

FIGURE a.15 A boulder sitting on a ledge in a gravitational field has potential energy. When the boulder starts to roll, this potential energy converts to kinetic energy.

and water molecules, plus energy. We can describe this reaction in a chemical formula, that uses symbols for the molecules:

$$2C_8H_{18} + 25O_2 \rightarrow 16CO_2 + 18H_2O + energy.$$

Note that *reactants* appear on the left side of the formula (C_8H_{18} is a typical compound in gasoline), and *products* on the right side. The arrow indicates the direction in which the reaction proceeds. The energy comes from the breaking of chemical bonds—more energy is stored in the chemical bonds of gasoline and oxygen than in the chemical bonds of carbon dioxide and water.

Energy from Field Forces

When an object you place in a field force, the force acts on the object, and the object, unless it is restrained, moves. In a gravity field, as described earlier, a boulder resting on top of a hill has potential energy (it is restrained by the ground), but when the boulder starts rolling, it has kinetic energy. Similarly, an iron bar taped to a table near a magnet has potential energy, but an unrestrained bar moving toward the magnet has kinetic energy.

Fission and Fusion: Energy from Breaking or Making Atoms

During **nuclear fission**, a large atomic nucleus splits into two or more smaller nuclei and other by-products, including free neutrons (▶Fig. a.16). Such fission is one form of nuclear decay. Other forms include: (1) the ejection of an electron from one of the neutrons in the nucleus, transforming the neutron into a proton; (2) the ejection of a

couple of protons or neutrons from the nucleus. Nuclear decay produces new atoms with different atomic numbers from that of the original atom. In other words, one element transforms into another. In effect, the ultimate goal of medieval alchemists, to turn lead into gold, can now be achieved in modern atom smashers, though at a significant cost.

When fission or other nuclear decay reactions occur, some of the matter constituting the original atom transforms into a huge amount of energy (heat and electromagnetic radiation), as defined by Einstein's famous equation: $E = mc^2$, where E is energy, m is mass, and c is the speed of light. The realization that fission releases vast quantities of energy led to the rush to build atomic weapons during and after World War II. Nuclear fission provides the energy in atomic bombs, nuclear power plants, and nuclear submarines.

Nuclear fusion results when two nuclei slam together at such high velocity that they get close enough for nuclear forces to bind them together, thereby creating a new and larger atom of a different element (▶Fig. a.17). In a manner of speaking, fusion is the opposite of fission. Fusion reactions also produce huge amounts of energy, because during fusion some of the matter converts into energy, according to Einstein's equation. Fusion reactions generate the heat in stars; in the Sun, for example, hydrogen atoms fuse together to form helium. Fusion reactions also generate the explosive energy of a hydrogen bomb—a thermonuclear device. Such reactions can only take place at extremely high temperatures (over 1,000,000°C). In fact, in order to get a hydrogen bomb to blow up, you need to use an atomic bomb as the trigger, because only an atomic explosion can generate high enough temperatures to start fusion. So far, no one has figured out how to economically sustain a controlled fusion reaction on Earth for the purpose of generating electricity.

FIGURE a.16 A uranium atom splits during nuclear fission.

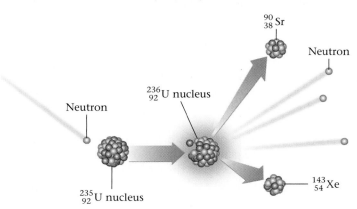

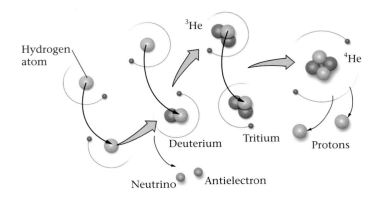

(a) Sun

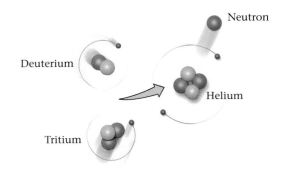

(b) Hydrogen bomb

FIGURE a.17 **(a)** In the Sun, four hydrogen atoms fuse in sequence to form a helium atom. **(b)** In a hydrogen bomb, a deuterium and a tritium atom fuse to form a helium atom plus a free neutron.

a.5 HOT AND COLD

Thermal Energy and Temperature

The atoms and molecules that make up an object do not stay rigidly fixed in place, but rather jiggle and jostle with respect to one another. This vibration creates **thermal energy**—the faster the atoms move, the greater the thermal energy and the hotter the object. In simple terms, the thermal energy in a substance represents the sum of the kinetic energy of all the substance's atoms (this includes the back-and-forth displacements that an atom makes as it vibrates as well as the movement of an atom from one place to another).

When we say that one object is hotter or colder than another, we are describing its temperature. It represents the average kinetic energy of atoms in the material. **Temperature** is a measure of warmth relative to some standard. In everyday life, we generally use the freezing or boiling point of water at sea level as the standard. When using the **Celsius (centigrade) scale**, we arbitrarily set the freezing point of water (at sea level) as 0 °C and the boiling point as 100 °C; whereas in the **Fahrenheit scale**, we set the freezing point as 32 °F and the boiling point as 212 °F.

The coldest a substance can be is the temperature at which its atoms or molecules stand still. We call this temperature **absolute zero**, or 0K (pronounced "zero kay"), where **K** stands for **Kelvin** (after Lord Kelvin [1824–1907], a British physicist), another unit of temperature. You simply can't get colder than absolute zero, meaning that you can't extract any thermal energy from a substance at 0K (−273.15 °C). Degrees in the Kelvin scale are the same increment as degrees in the Celsius scale. On the Kelvin scale, ice melts at 273.15K and water boils at 373.15K (▶Figure a.18).

Heat and Heat Transfer

Heat is the thermal energy transferred from one object to another. Heat can be measured in **calories**. One thousand calories can heat 1 kilogram of water by 1 °C. There are four ways in which heat transfer takes place in the Earth System: radiation, conduction, convection, and advection.

Radiation is the process by which electromagnetic waves transmit heat into a body or out of a body (▶Fig. a.19a). For example, when the Sun heats the ground during the day, radiative heating takes place. Similarly, when heat rises from the ground at night, radiative heating is occurring—in the opposite direction.

Conduction takes place when you stick the end of an iron bar in a fire (▶Fig. a.19b). The iron atoms at the fire-licked end of the bar start to vibrate more energetically;

FIGURE a.18 Physical phenomena that we observe occur at a wide range of temperatures (here, as specified by the Kelvin scale).

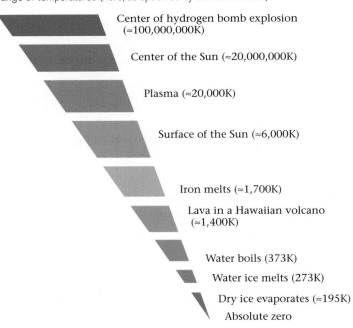

Center of hydrogen bomb explosion (≈100,000,000K)

Center of the Sun (≈20,000,000K)

Plasma (≈20,000K)

Surface of the Sun (≈6,000K)

Iron melts (≈1,700K)

Lava in a Hawaiian volcano (≈1,400K)

Water boils (373K)

Water ice melts (273K)

Dry ice evaporates (≈195K)

Absolute zero

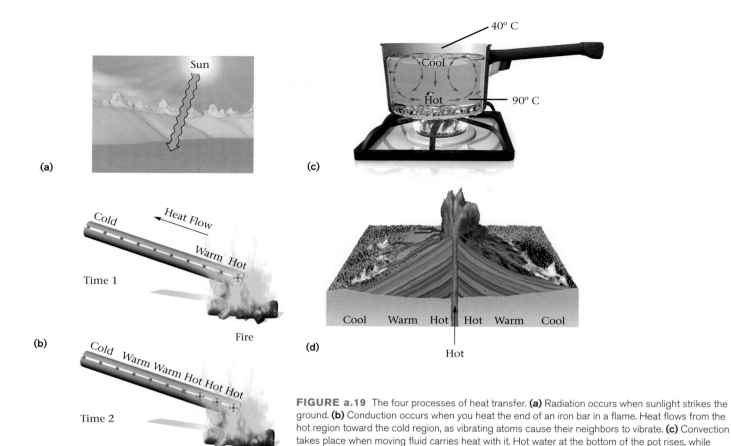

FIGURE a.19 The four processes of heat transfer. **(a)** Radiation occurs when sunlight strikes the ground. **(b)** Conduction occurs when you heat the end of an iron bar in a flame. Heat flows from the hot region toward the cold region, as vibrating atoms cause their neighbors to vibrate. **(c)** Convection takes place when moving fluid carries heat with it. Hot water at the bottom of the pot rises, while cool water sinks, setting up a convective cell. **(d)** During advection, a hot liquid (such as molten rock) rises into cooler material. Heat conducts from the hot liquid into the cooler material.

they gradually incite atoms farther from the flame to start jiggling, and these atoms in turn set atoms even farther along in motion. In this way, heat slowly flows down the bar until you feel it with your hand. Note that even though the iron atoms vibrate more as the bar heats up, the atoms remain locked in their position within the solid. Thus, conduction does not involve actual movement of atoms from one place to another. If you place two bars of different temperatures in contact with each other, heat conducts from the hot bar into the cold one until both end up at the same temperature.

Convection takes place when you set a pot of water on a stove (▶Fig. a.19c). The heat from the stove warms the water at the base of the pot (it makes the molecules of water vibrate faster and move around more). As a consequence, the density of the water at the base of the pot decreases, for as you heat a liquid, the atoms move away from each other and the liquid expands. For a time, cold water remains at the top of the pot, but eventually the warm, less dense water becomes buoyant relative to the cold, dense water. In a gravitational field, a buoyant material rises (like a styrofoam ball in a pool of water) if the material above is weak enough to flow out of the way. Since liquid water can easily flow, hot water rises. When this happens, cold water sinks to take its place. The new volume of cold water then heats up and then rises itself. Thus, during convection, by the actual flow of the material itself carries heat. The trajectory of flow defines **convective cells;** in a convective cell, water follows a loop, with warm water rising and cold water sinking. Convection occurs in the atmosphere, the ocean, and the interior of the Earth.

Advection, a less familiar process, happens when heat moves with a fluid, flowing through cracks and pores within a solid material (▶Fig. a.19d). The heat brought by the fluid conductively heats up the solid that the fluid passes through. Advection takes place, for example, if you pass hot water through a sponge and the sponge itself gets hot. In the Earth, advection occurs where molten rock rises through the crust beneath a volcano and heats up the crust in the process or where hot water circulates through cracks in rocks beneath geysers and hot springs.

Additional Maps and Charts

This Appendix contains several maps and charts for general reference. We list the purpose of each below.

Mineral Identification Flow Charts: Geologists use these charts to identify unknown mineral specimens (▶Fig. b.1a, b). A mineral flow chart is simply an organized series of questions concerning the mineral's physical properties. The questions are arranged in a sequence such that appropriate answers ultimately lead you on a path to a specific mineral. To understand this concept, let's imagine that we are trying to identify a shiny, bronze-colored, metallic-looking mineral specimen. We start by observing the specimen's luster. It is metallic, so we follow the path on the chart for metallic-luster minerals (Fig. b.1a). Next, we determine if the mineral is magnetic or nonmagnetic. If it is nonmagnetic, we follow the path for non-magnetic minerals. Then, we look at the mineral's color. Since it is bronze-colored, our path ends at pyrite.

Notice that one of the flow chart questions in Figure b.1b asks about the reaction of the specimen with hydrochloric acid (HCl). Only calcite and dolomite react, so the question allows definitive identification of these minerals. Another question pertains to striations, faint parallel lines on cleavage planes. Only plagioclase has striations.

World Magnetic Declination Map: This map shows the variation of magnetic declination with location on the surface of the Earth. Declination exists because the position of the Earth's magnetic pole does not coincide exactly with that of the geographic pole. In fact, the magnetic pole location constantly moves, currently at a rate of about 20 km per year. It now lies off the north coast of Canada, and in the not too distant future, it may lie along the coast of Siberia. In order for a compass to give an accurate indication of direction, it must be adjusted to accommodate for the declination at the location of measurement.

US Magnetic Declination Map: This map shows the magnetic declination for the United States.

Volcanoes of the Past Few Million Years: This map shows the location of volcanoes that have erupted in relatively recent time. The map also shows the location of earthquakes.

Earthquake Epicenter Map: This map shows the positions of epicenters for earthquakes that were large enough to be detected at seismic stations over a broad region. An epicenter is the point on the surface of the Earth above the earthquake's hypocenter (focus). The hypocenter is the place where the earthquake energy is generated. Different colors indicate the different depths of the earthquake hypocenters.

World Soils Map: This map displays areas of different types of soils, using one of the standard classification schemes for soils.

A Satellite Image of the Earth at Night: This image shows light sources on the surface of the planet. Light sources indicate both the density of human habitation, and the degree of industrialization. Note, for example, that most of the lights in Australia occur in coastal cities.

Mineral-Identification Flow Chart (metallic or dark-colored nonmetallic)

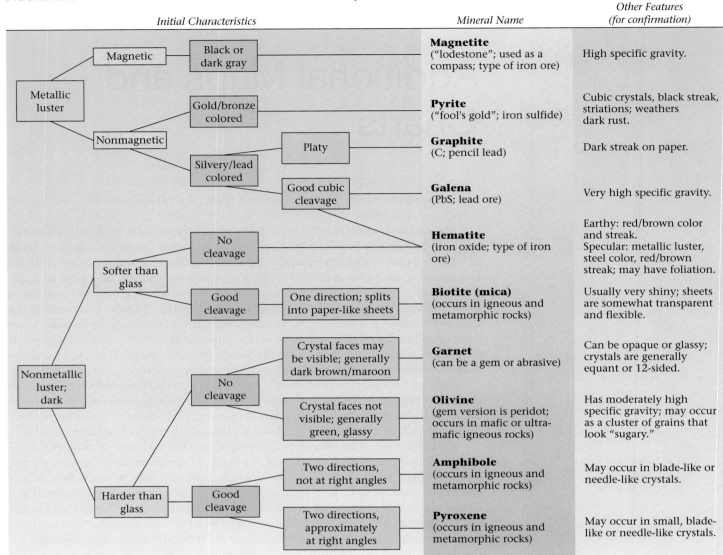

Initial Characteristics — *Mineral Name* — *Other Features (for confirmation)*

Metallic luster
- Magnetic → Black or dark gray → **Magnetite** ("lodestone"; used as a compass; type of iron ore) — High specific gravity.
- Nonmagnetic
  - Gold/bronze colored → **Pyrite** ("fool's gold"; iron sulfide) — Cubic crystals, black streak, striations; weathers dark rust.
  - Silvery/lead colored
    - Platy → **Graphite** (C; pencil lead) — Dark streak on paper.
    - Good cubic cleavage → **Galena** (PbS; lead ore) — Very high specific gravity.

Nonmetallic luster; dark
- Softer than glass
  - No cleavage → **Hematite** (iron oxide; type of iron ore) — Earthy: red/brown color and streak. Specular: metallic luster, steel color, red/brown streak; may have foliation.
  - Good cleavage → One direction; splits into paper-like sheets → **Biotite (mica)** (occurs in igneous and metamorphic rocks) — Usually very shiny; sheets are somewhat transparent and flexible.
- Harder than glass
  - No cleavage
    - Crystal faces may be visible; generally dark brown/maroon → **Garnet** (can be a gem or abrasive) — Can be opaque or glassy; crystals are generally equant or 12-sided.
    - Crystal faces not visible; generally green, glassy → **Olivine** (gem version is peridot; occurs in mafic or ultra-mafic igneous rocks) — Has moderately high specific gravity; may occur as a cluster of grains that look "sugary."
  - Good cleavage
    - Two directions, not at right angles → **Amphibole** (occurs in igneous and metamorphic rocks) — May occur in blade-like or needle-like crystals.
    - Two directions, approximately at right angles → **Pyroxene** (occurs in igneous and metamorphic rocks) — May occur in small, blade-like or needle-like crystals.

Mineral-Identification Flow Chart (light-colored nonmetallic)

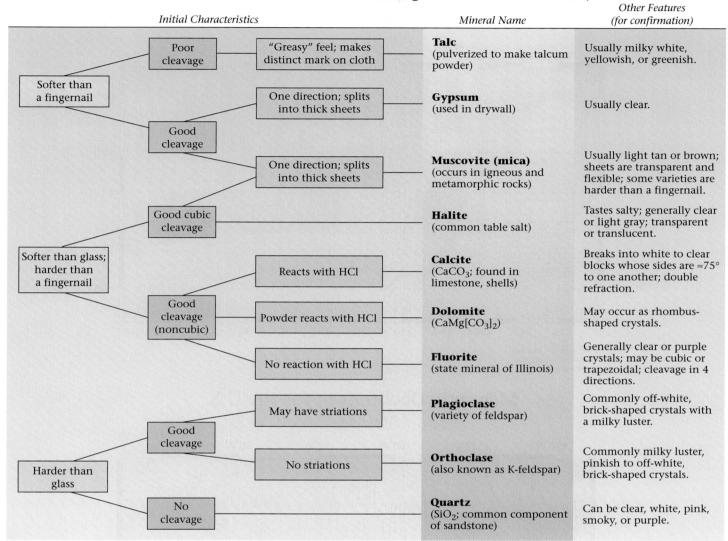

Initial Characteristics   *Mineral Name*   *Other Features (for confirmation)*

Softer than a fingernail

- Poor cleavage — "Greasy" feel; makes distinct mark on cloth — **Talc** (pulverized to make talcum powder) — Usually milky white, yellowish, or greenish.

- Good cleavage — One direction; splits into thick sheets — **Gypsum** (used in drywall) — Usually clear.

- One direction; splits into thick sheets — **Muscovite (mica)** (occurs in igneous and metamorphic rocks) — Usually light tan or brown; sheets are transparent and flexible; some varieties are harder than a fingernail.

Softer than glass; harder than a fingernail

- Good cubic cleavage — **Halite** (common table salt) — Tastes salty; generally clear or light gray; transparent or translucent.

- Good cleavage (noncubic)
  - Reacts with HCl — **Calcite** (CaCO₃; found in limestone, shells) — Breaks into white to clear blocks whose sides are ≈75° to one another; double refraction.
  - Powder reacts with HCl — **Dolomite** (CaMg[CO₃]₂) — May occur as rhombus-shaped crystals.
  - No reaction with HCl — **Fluorite** (state mineral of Illinois) — Generally clear or purple crystals; may be cubic or trapezoidal; cleavage in 4 directions.

Harder than glass

- Good cleavage
  - May have striations — **Plagioclase** (variety of feldspar) — Commonly off-white, brick-shaped crystals with a milky luster.
  - No striations — **Orthoclase** (also known as K-feldspar) — Commonly milky luster, pinkish to off-white, brick-shaped crystals.

- No cleavage — **Quartz** (SiO₂; common component of sandstone) — Can be clear, white, pink, smoky, or purple.

US/UK World Magnetic Chart—Epoch 2000
Declination - Main Field (D)

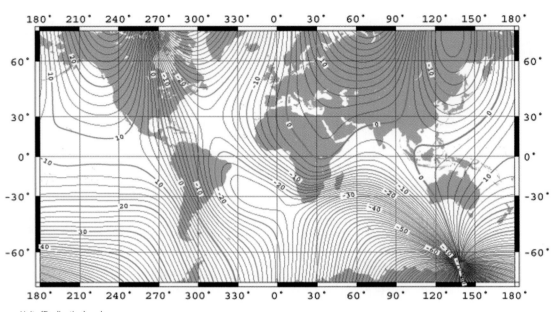

Units {Declination} : degrees
Contour Interval : 2 degrees
Map Projection : Mercator

(a)

Magnetic Declination for the U.S.
2004

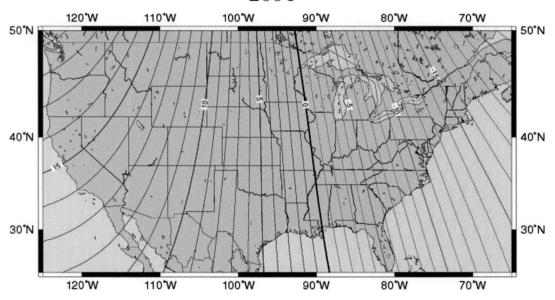

Mercator Projection

Contours of Declination of the Earth's magnetic
field. Contours are expressed in degrees.
Contour Interval: 1 Degree (Positive
declinations in blue, negative in red)

Produced by NOAA's National Geophysical
Data Center (NGDC), Boulder, Colorado

http://www.ngdc.noaa.gov

Based on International Geomagnetic
Reference Field (IGRF), Epoch 2000 updated to
December 31, 2004

The IGRF is developed by the International
Association of Geomagnetism and Aeronomy
(IAGA). Division V

(b)

Earthquake Locations 1990 - 1996 (Magnitudes 4 and greater)

Color indicates depth: Red 0-33 km, Orange 33-70 km, Green 70-300 km, Blue 300-700 km

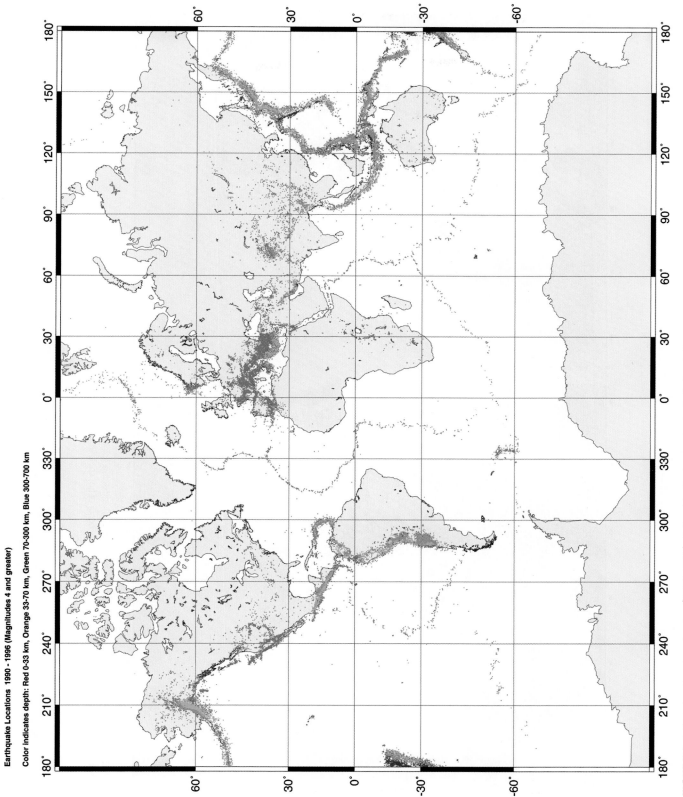

FIGURE b.3 Global earthquake epicenters (1990–1996), distinguished by hypocenter depth. Red = shallow; Green = intermediate; Blue = deep. Note that deep earthquakes only occur at convergent margins.

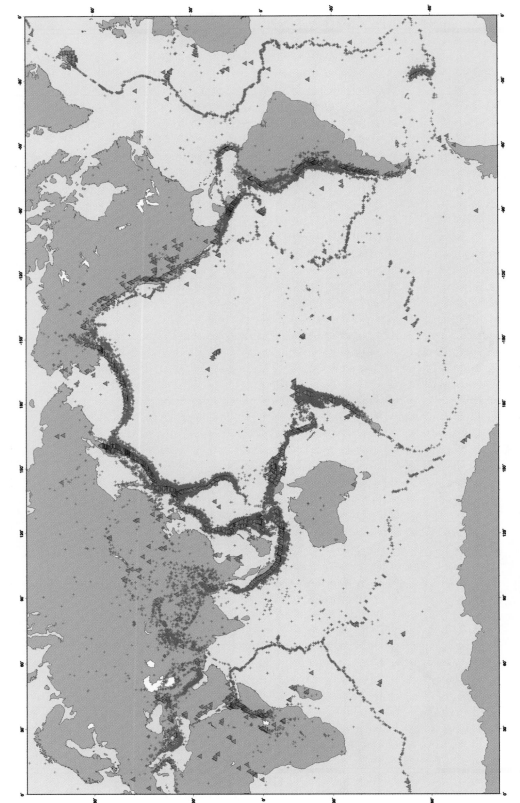

FIGURE b.4 Locations of earthquake epicenters (small purple dots) and active volcanoes (red triangles). Note that most volcanoes occur along plate boundaries, but some occur at hot spots in plate interiors.

Global Soil Regions

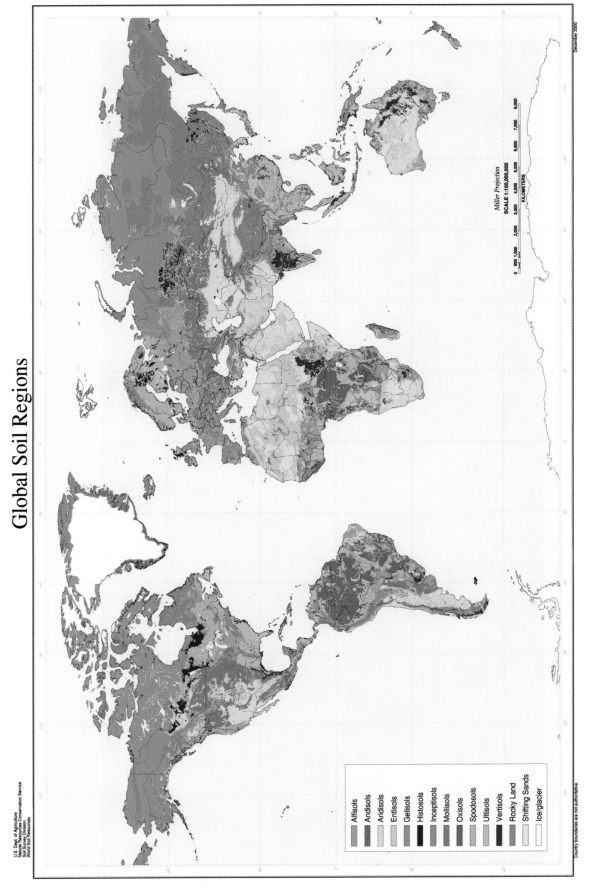

U.S. Dept. of Agriculture
Natural Resources Conservation Service
Soil Survey Division
World Soil Resources

Country boundaries are not authoritative

December 2000

Miller Projection
SCALE 1:100,000,000

KILOMETERS
0 500 1,000   2,000   3,000   4,000   5,000   6,000   7,000   8,000

Alfisols
Andisols
Aridisols
Entisols
Gelisols
Histosols
Inceptisols
Mollisols
Oxisols
Spodosols
Ultisols
Vertisols
Rocky Land
Shifting Sands
Ice/glacier

FIGURE b.5 Map of soil types around the world. The different colors indicate areas in which a particular soil type dominates.

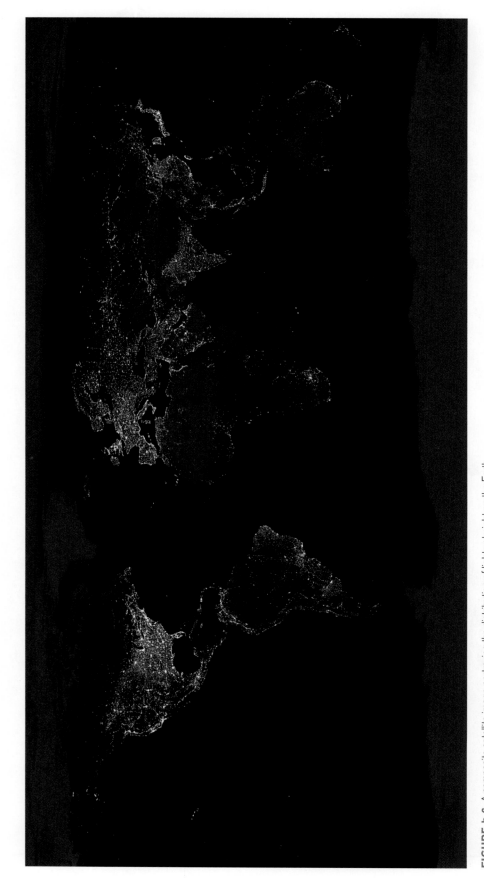

FIGURE b.6 A composite satellite image portraying the distribution of lights at night on the Earth. Light density represents a combination of population density and level of industrialization.

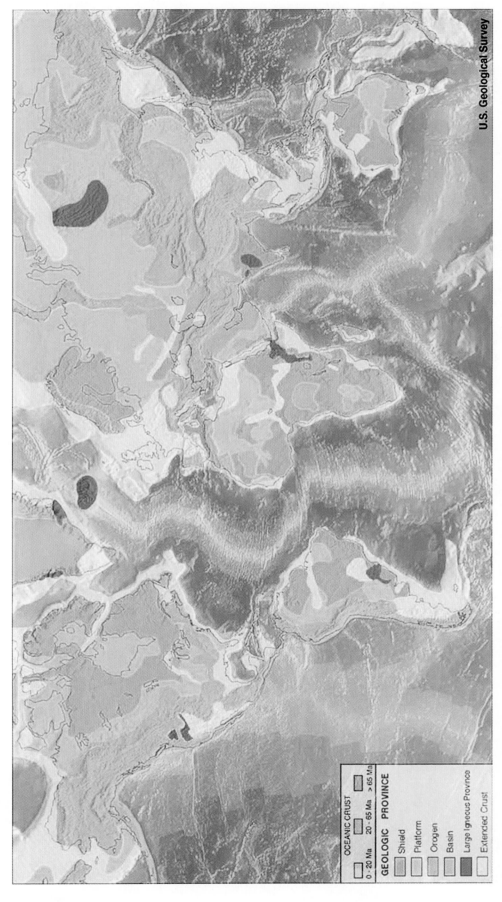

U.S. Geological Survey

OCEANIC CRUST

0 - 20 Ma 20 - 65 Ma > 65 Ma

GEOLOGIC PROVINCE

Shield
Platform
Orogen
Basin
Large Igneous Province
Extended Crust

FIGURE b.7 CAPTION TK

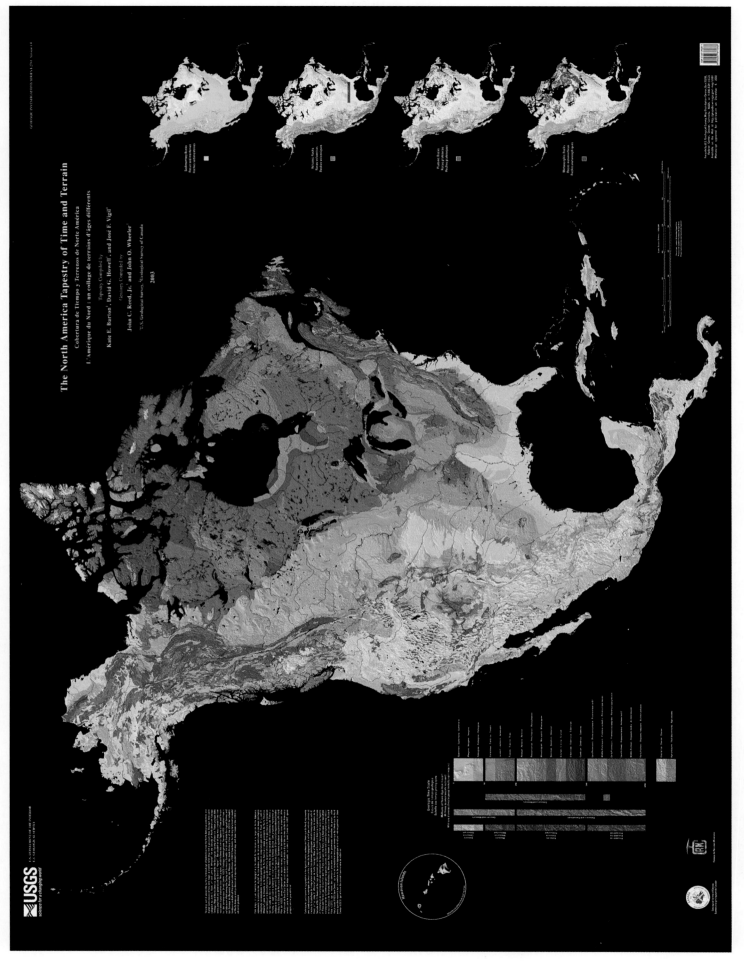

FIGURE b.8 CAPTION TK

Glossary

aa A lava flow with a rubbly surface.

abandoned meander A meander that dries out after it was cut off.

ablation The removal of ice at the toe of a glacier by melting, sublimation (the evaporation of ice into water vapor), and/or calving.

abrasion The process in which one material (such as sand-laden water) grinds away at another (such as a stream channel's floor and walls).

absolute age Numerical age (the age specified in years).

absolute plate velocity The movement of a plate relative to a fixed point in the mantle.

abyssal plain A broad, relatively flat region of the ocean that lies at least 4.5 km below sea level.

Acadian orogeny A convergent mountain-building event, that occurred around 400 million years ago, during which continental slivers accreted to the eastern edge of the North American continent.

accreted terrane A block of crust that collided with a continent at a convergent margin and stayed attached to the continent.

accretionary coast A coastline that receives more sediment than erodes away.

accretionary orogen An orogen formed by the attachment of numerous buoyant slivers of crust to an older, larger continental block.

accretionary prism A wedge-shaped mass of sediment and rock scraped off the top of a downgoing plate and accreted onto the overriding plate at a convergent plate margin.

acid mine runoff A dilute solution of sulfuric acid, produced when sulfur-bearing minerals in mines react with rainwater, that flows out of a mine.

acid rain Precipitation in which air pollutants react with water to make a weak acid that then falls from the sky.

active continental margin A continental margin that coincides with a plate boundary.

active fault A fault that has moved recently or is likely to move in the future.

active sand The top layer of beach sand, which moves daily because of wave action.

active volcano A volcano that has erupted within the past few centuries and will likely erupt again.

adiabatic cooling The cooling of a body of air or matter without the addition or subtraction of thermal energy (heat).

adiabatic heating The warming of a body of air or matter without the addition or subtraction of heat.

aerosols Tiny solid particles or liquid droplets that remain suspended in the atmosphere for a long time.

aftershocks The series of smaller earthquakes that follow a major earthquake.

air The mixture of gases that make up the Earth's atmosphere.

air-fall tuff Tuff formed when ash settles gently from the air.

air mass A body of air, about 1,500 km across, that has recognizable physical characteristics.

air pressure The push that air exerts on its surroundings.

albedo The reflectivity of a surface.

Alleghenian orogeny The orogenic event that occurred about 270 million years ago when Africa collided with North America.

alloy A metal containing more than one type of metal atom.

alluvial fan A gently sloping apron of sediment dropped by an ephemeral stream at the base of a mountain in arid or semi-arid regions.

alluvium Sorted sediment deposited by a stream.

alluvium-filled valley A valley whose floor fills with sediment.

amber Hardened (fossilized) ancient sap or resin.

amphibolite facies A set of metamorphic mineral assemblages formed under intermediate pressures and temperatures.

amplitude The height of a wave from crest to trough.

Ancestral Rockies The late Paleozoic uplifts of the Rocky Mountain region; they eroded away long before the present Rocky Mountains formed.

angiosperm A flowering plant.

angle of repose The angle of the steepest slope that a pile of uncemented material can attain without collapsing from the pull of gravity.

angularity The degree to which grains have sharp or rounded edges or corners.

angular unconformity An unconformity in which the strata below were tilted or folded before the unconformity developed; strata below the unconformity therefore have a different tilt than strata above.

anhedral grains Crystalline mineral grains without well-formed crystal faces.

Antarctic bottom water mass The mass of cold, dense water that sinks along the coast of Antarctica.

antecedent stream A stream that cuts across an uplifted mountain range; the stream must have existed before the range uplifted and must then have been able to downcut as fast as the land was rising.

anthracite coal Shiny black coal formed at temperatures between 200° and 300°C. A high-rank coal.

anticline A fold with an arch-like shape in which the limbs dip away from the hinge.

anticyclone The clockwise flow of air around a high-pressure mass.

Antler orogeny The Late Devonian mountain-building event in which slices of deep-marine strata were pushed eastward, up and over the shallow-water strata on the western coast of North America.

anvil cloud A large cumulonimbus cloud that spreads laterally at the tropopause to form a broad, flat top.

aphanitic A textural term for fine-grained igneous rock.

apparent polar-wander path A path on the globe along which a magnetic pole appears to have wandered over time; in fact, the continents drift, while the magnetic pole stays fairly fixed.

aquiclude Sediment or rock that transmits no water.

aquifer Sediment or rock that transmits water easily.

aquitard Sediment or rock that does not transmit water easily and therefore retards the motion of the water.

archaea A kingdom of "old bacteria," now commonly found in extreme environments like hot springs. (Also called "archaeobacteria.")

Archean The middle Precambrian Eon.

Archimedes' principle The mass of the water displaced by a block of material equals the mass of the whole block of material.

arête A residual knife-edge ridge of rock that separates two adjacent cirques.

argillaceous sedimentary rock Sedimentary rock that contains abundant clay.

arroyo The channel of an ephemeral stream; dry wash; wadi.

artesian well A well in which water rises on its own.

ash fall Ash that falls to the ground out of an ash cloud.

ash flow An avalanche of ash that tumbles down the side of an explosively erupting volcano.

assimilation The process of magma contamination in which blocks of wall rock fall into a magma chamber and dissolve.

asthenosphere The layer of the mantle that lies between 100–150 km and 350 km deep; the asthenosphere is relatively soft and can flow when acted on by force.

atm A unit of air pressure that approximates the pressure exerted by the atmosphere at sea level.

atmosphere A layer of gases that surrounds a planet.

atoll A coral reef that develops around a circular reef surrounding a lagoon.

atomic number The number of protons in the nucleus of a given element.

atomic weight The number of protons plus the number of neutrons in the nucleus of a given element. (Also known as atomic mass.)

aurora australis The same phenomenon as the aurora borealis, but in the Southern Hemisphere.

aurora borealis A ghostly curtain of varicolored light that appears across the night sky in the Northern Hemisphere when charged particles from the Sun interact with the ions in the ionosphere.

avalanche A turbulent cloud of debris mixed with air that rushes down a steep hill slope at high velocity; the debris can be rock and/or snow.

avalanche chute A downslope hillside pathway along which avalanches repeatedly fall, consequently clearing the pathway of mature trees.

avulsion The process in which a river overflows a natural levee and begins to flow in a new direction.

axial plane The imaginary surface that encompasses the hinges of successive layers of a fold.

axial trough A narrow depression that runs along a mid-ocean ridge axis.

backscattered light Atmospheric scattered sunlight that returns back to space.

backshore zone The zone of beach that extends from a small step cut by high-tide swash to the front of the dunes or cliffs that lie farther inshore.

backswamp The low marshy region between the bluffs and the natural levees of a floodplain.

backwash The gravity-driven flow of water back down the slope of a beach.

bajada An elongate wedge of sediment formed by the overlap of several alluvial fans emerging from adjacent valleys.

Baltica A Paleozoic continent that included crust that is now part of today's Europe.

banded-iron formation (BIF) Iron-rich sedimentary layers consisting of alternating gray beds of iron oxide and red beds of iron-rich chert.

bar (1) A sheet or elongate lens or mound of alluvium; (2) a unit of air-pressure measurement approximately equal to 1 atm.

barchan dune A crescent-shaped dune whose tips point downwind.

barrier island An offshore sand bar that rises above the mean high-water level, forming an island.

barrier reef A coral reef that develops offshore, separated from the coast by a lagoon.

basal sliding The phenomenon in which meltwater accumulates at the base of a glacier, so that the mass of the glacier slides on a layer of water or on a slurry of water and sediment.

basalt A fine-grained mafic igneous rock.

base level The lowest elevation a stream channel's floor can reach at a given locality.

basement Older igneous and metamorphic rocks making up the Earth's crust beneath sedimentary cover.

basement uplift Uplift of basement rock by faults that penetrate deep into the continental crust.

base metals Metals that are mined but not considered precious. Examples include copper, lead, zinc, and tin.

basin A fold or depression shaped like a right-side-up bowl.

Basin and Range Province A broad, Cenozoic continental rift that has affected a portion of the western United States in Nevada, Utah, and Arizona; in this province, tilted fault blocks form ranges, and alluvium-filled valleys are basins.

batholith A vast composite, intrusive, igneous rock body up to several hundred km long and 100 km wide, formed by the intrusion of numerous plutons in the same region.

bathymetric map A map illustrating the shape of the ocean floor.

bathymetric profile A cross section showing ocean depth plotted against location.

bathymetry Variation in depth.

bauxite A residual mineral deposit rich in aluminum.

baymouth bar A sandspit that grows across the opening of a bay.

beach drift The gradual migration of sand along a beach.

beach erosion The removal of beach sand caused by wave action and longshore currents.

beach face A steeply concave part of the foreshore zone formed where the swash of the waves actively scours the sand.

bedding Layering or stratification in sedimentary rocks.

bed load Large particles, such as sand, pebbles, or cobbles, that bounce or roll along a stream bed.

bedrock Rock still attached to the Earth's crust.

Bergeron process Precipitation involving the growth of ice crystals in a cloud at the expense of water droplets.

berm A horizontal or landward-sloping terrace in the backshore zone of a beach that receives sediment during a storm.

big bang A cataclysmic explosion that scientists suggest represents the formation of the Universe; before this event, all matter and all energy were packed into one volumeless point.

biochemical sedimentary rock Sedimentary rock formed from material (such as shells) produced by living organisms.

biodiversity The number of different species that exist at a given time.

biofuel Gas or liquid fuel made from plant material (biomass). Examples of biofuel include alcohol (from fermented sugar), bio diesel from vegetable oil, and wood.

biogeochemical cycle The exchange of chemicals between living and nonliving reservoirs in the Earth System.

bioremediation The injection of oxygen and nutrients into a contaminated aquifer to foster the growth of bacteria that will ingest or break down contaminants.

biosphere The region of the Earth and atmosphere inhabited by life; this region stretches from a few km below the Earth's surface to a few km above.

bioturbation The mixing of sediment by burrowing animals such as clams and worms.

bituminous coal Dull, black intermediate-rank coal formed at temperatures between 100° and 200°C.

black-lung disease Lung disease contracted by miners from the inhalation of too much coal dust.

black smoker The cloud of suspended minerals formed where hot water spews out of a vent along a mid-ocean ridge; the dissolved sulfide components of the hot water instantly precipitate when the water mixes with seawater and cools.

blind fault A fault that does not intersect the ground surface.

blocking temperature The temperature below which isotopes in a mineral are no longer free to move, so the radiometric clock starts.

blowout A deep, bowl-like depression scoured out of desert terrain by a turbulent vortex of wind.

blue shift The phenomenon in which a source of light moving toward you appears to have a higher frequency.

body waves Seismic waves that pass through the interior of the Earth.

bog A wetland dominated by moss and shrubs.

bolide A solid extraterrestrial object such as a meteorite, comet, or asteroid that explodes in the atmosphere.

bornhardt An inselberg with a loaf geometry, like that of Uluru (Ayers Rock) in central Australia.

Bowen's reaction series The sequence in which different silicate minerals crystallize during the progressive cooling of a melt.

braided stream A sediment-choked stream consisting of entwined subchannels.

breaker A water wave in which water at the top of the wave curves over the base of the wave.

breakwater An offshore wall, built parallel or at an angle to the beach, that prevents the full force of waves from reaching a harbor.

breccia Coarse sedimentary rock consisting of angular fragments; or rock broken into angular fragments by faulting.

breeder reactor A nuclear reactor that produces its own fuel.

brine Water that is not fresh but is less salty than seawater; brine may be found in estuaries.

brittle deformation The cracking and fracturing of a material subjected to stress.

brittle-ductile transition (brittle-plastic transition) The depth above which materials behave brittlely and below which materials behave ductilely (plastically); this transition typically lies between a depth of 10 and 15 km in continental crustal rock, and 60 m deep in glacial ice.

buoyancy The upward force acting on a less dense object immersed or floating in denser material.

butte A medium-size, flat-topped hill in an arid region.

caldera A large circular depression with steep walls and a fairly flat floor, formed after an eruption as the center of the volcano collapses into the drained magma chamber below.

caliche A solid mass created where calcite cements the soil together (also called calcrete).

calving The breaking off of chunks of ice at the edge of a glacier.

Cambrian explosion of life The remarkable diversification of life, indicated by the fossil record, that occurred at the beginning of the Cambrian Period.

Canadian Shield A broad, low-lying region of exposed Precambrian rock in the Canadian interior.

canyon A trough or valley with steeply sloping walls, cut into the land by a stream.

capillary fringe The thin subsurface layer in which water molecules seep up from the water table by capillary action to fill pores.

carbonate rocks Rocks containing calcite and/or dolomite.

carbon-14 dating A radiometric dating process that can tell us the age of organic material containing carbon originally extracted from the atmosphere.

cast Sediment that preserves the shape of a shell it once filled before the shell dissolved or mechanically weathered away.

catabatic winds Strong winds that form at the margin of a glacier where the warmer air above ice-free land rises and the cold, denser air from above the glaciers rushes in to take its place.

catastrophic change Change that takes place either instantaneously or rapidly in geologic time.

catchment *Drainage network.*

cement Mineral material that precipitates from water and fills the spaces between grains, holding the grains together.

cementation The phase of lithification in which cement, consisting of minerals that precipitate from groundwater, partially or completely fills the spaces between clasts and attaches each grain to its neighbor.

Cenozoic The most recent era of the Phanerozoic Eon, lasting from 65 Ma up until the present.

chalk Very fine-grained limestone consisting of weakly cemented plankton shells.

change of state The process in which a material changes from one phase (liquid, gas, or solid) to another.

channel A trough dug into the ground surface by flowing water.

channeled scablands A barren, soil-free landscape in eastern Washington, scoured clean by a flood unleashed when a large glacial lake drained.

chatter marks Wedge-shaped indentations left on rock surfaces by glacial plucking.

chemical sedimentary rocks Sedimentary rocks made up of minerals that precipitate directly from water solution.

chemical weathering The process in which chemical reactions alter or destroy minerals when rock comes in contact with water solutions and/or air.

chert A sedimentary rock composed of very fine-grained silica (cryptocrystalline quartz).

Chicxulub crater A circular excavation buried beneath younger sediment on the Yucután peninsula; geologists suggest that a meteorite landed there 65 Ma.

chimney (1) A conduit in a magma chamber in the shape of a long vertical pipe through which magma rises and erupts at the surface; (2) an isolated column of strata in an arid region.

cinder cone A subaerial volcano consisting of a cone-shaped pile of tephra whose slope approaches the angle of repose for tephra.

cinders Fragments of glassy rock ejected from a volcano.

cirque A bowl-shaped depression carved by a glacier on the side of a mountain.

cirrus cloud A wispy cloud that tapers into delicate, feather-like curls.

clastic (detrital) sedimentary rock   Sedimentary rock consisting of cemented-together detritus derived from the weathering of preexisting rock.

cleavage   (1) The tendency of a mineral to break along preferred planes; (2) a type of foliation in low-grade metamorphic rock.

cleavage planes   A series of surfaces on a crystal that form parallel to the weakest bonds holding the atoms of the crystal together.

cliff (or scarp) retreat   The change in the position of a cliff face caused by erosion.

climate   The average weather conditions, along with the range of conditions, of a region over a year.

cloud   A mist of tiny water droplets in the sky.

coal rank   A measurement of the carbon content of coal; higher-rank coal forms at higher temperatures.

coal reserve   The quantities of discovered, but not yet mined, coal in sedimentary rock of the continents.

coal swamp   A swamp whose oxygen-poor water allows thick piles of woody debris to accumulate; this debris transforms into coal upon deep burial.

coastal plain   Low-relief regions of land adjacent to the coast.

cold front   The boundary at which a cold air mass pushes underneath a warm air mass.

collision   The process of two buoyant pieces of lithosphere converging and squashing together.

columnar jointing   A type of fracturing that yields roughly hexagonal columns of basalt; columnar joints form when a dike, sill, or lava flow cools.

comet   A ball of ice and dust, probably remaining from the formation of the solar system, that orbits the Sun.

compaction   The phase of lithification in which the pressure of the overburden on the buried rock squeezes out water and air that was trapped between clasts, and the clasts press tightly together.

composite volcano   *Stratovolcano.*

compositional banding   A type of metamorphic foliation, found in gneiss, defined by alternating bands of light and dark minerals.

compressibility   The degree to which a material's volume changes in response to squashing.

compression   A push or squeezing felt by a body.

compressional waves   Waves in which particles of material move back and forth parallel to the direction in which the wave itself moves.

conchoidal fractures   Smoothly curving, clamshell-shaped surfaces along which materials with no cleavage planes tend to break.

condensation   The process of gas molecules linking together to form a liquid.

condensation nuclei   Preexisting solid or liquid particles, such as aerosols, onto which water condenses during cloud formation.

cone of depression   The downward-pointing, cone-shaped surface of the water table in a location where the water table is experiencing drawdown because of pumping at a well.

confined aquifer   An aquifer that is separated from the Earth's surface by an overlying aquitard.

conglomerate   Very coarse-grained sedimentary rock consisting of rounded clasts.

consuming boundary   *Convergent plate boundary.*

contact   The boundary surface between two rock bodies (as between two stratigraphic formations, between an igneous intrusion and adjacent rock, between two igneous rock bodies, or between rocks juxtaposed by a fault).

contact metamorphism   *Thermal metamorphism.*

contaminant plume   A cloud of contaminated groundwater that moves away from the source of the contamination.

continental crust   The crust beneath the continents.

continental divide   A highland separating drainage that flows into one ocean from drainage that flows into another.

continental-drift hypothesis   The idea that continents have moved and are still moving slowly across the Earth's surface.

continental glacier   A vast sheet of ice that spreads over thousands of square km of continental crust.

continental-interior desert   An inland desert that develops because by the time air masses reach the continental interior, they have lost all of their moisture.

continental lithosphere   Lithosphere topped by continental crust; this lithosphere reaches a thickness of 150 km.

continental margin   A continent's coastline.

continental rift   A linear belt along which continental lithosphere stretches and pulls apart.

continental rifting   The process by which a continent stretches and splits along a belt; if it is successful, rifting separates a larger continent into two smaller continents separated by a divergent boundary.

continental rise   The sloping sea floor that extends from the lower part of the continental slope to the abyssal plain.

continental shelf   A broad, shallowly submerged region of a continent along a passive margin.

continental slope   The slope at the edge of a continental shelf, leading down to the deep sea floor.

continental volcanic arc   A long curving chain of subaerial volcanoes on the margin of a continent adjacent to a convergent plate boundary.

contour lines   Lines on a map along which a parameter has a constant value; for example, all points along a contour line on a topographic map are at the same elevation.

control rod   Rods that absorb neutrons in a nuclear reactor and thus decrease the number of collisions between neutrons and radioactive atoms.

convection   Heat transfer that results when warmer, less dense material rises while cooler, denser material sinks.

convergence zone   A place where two surface air flows meet so that air has to rise.

convergent margin   *Convergent plate boundary.*

convergent plate boundary   A boundary at which two plates move toward each other so that one plate sinks (subducts) beneath the other; only oceanic lithosphere can subduct.

coral reef   A mound of coral and coral debris forming a region of shallow water.

core   The dense, iron-rich center of the Earth.

core-mantle boundary   An interface 2,900 km below the Earth's surface separating the mantle and core.

Coriolis effect   The deflection of objects, winds, and currents on the surface of the Earth owing to the planet's rotation.

cornice   A huge, overhanging drift of snow built up by strong winds at the crest of a mountain ridge.

correlation   The process of defining the age relations between the strata at one locality and the strata at another.

cosmic rays   Nuclei of hydrogen and other elements that bombard the Earth from deep space.

cosmology   The study of the overall structure of the Universe.

country rock (wall rock)   The preexisting rock into which magma intrudes.

crater   (1) A circular depression at the top of a volcanic mound; (2) a depression formed by the impact of a meteorite.

craton   A long-lived block of durable continental crust commonly found in the stable interior of a continent.

cratonic platform A province in the interior of a continent in which Phanerozoic strata bury most of the underlying Precambrian rock.

creep The gradual downslope movement of regolith.

crevasse A large crack that develops by brittle deformation in the top 60 m of a glacier.

critical mass A sufficiently dense and large mass of radioactive atoms in which a chain reaction happens so quickly that the mass explodes.

cross section A diagram depicting the geometry of materials underground as they would appear on an imaginary vertical slice through the Earth.

crude oil Oil extracted directly from the ground.

crust The rock that makes up the outermost layer of the Earth.

crustal root Low-density crustal rock that protrudes downward beneath a mountain range.

crystal A single, continuous piece of a mineral bounded by flat surfaces that formed naturally as the mineral grew.

crystal form The geometric shape of a crystal, defined by the arrangement of crystal faces.

crystal habit The general shape of a crystal or cluster of crystals that grew unimpeded.

crystal lattice The orderly framework within which the atoms or ions of a mineral are fixed.

crystalline Containing a crystal lattice.

cuesta An asymmetric ridge formed by tilted layers of rock, with a steep cliff on one side cutting across the layers and a gentle slope on the other side; the gentle slope is parallel to the layering.

cumulonimbus cloud A rain-producing puffy cloud.

cumulus cloud A puffy, cotton-ball-shaped cloud.

current (1) A well-defined stream of ocean water; (2) the moving flow of water in a stream.

cut bank The outside bank of the channel wall of a meander, which is continually undergoing erosion.

cutoff A straight reach in a stream that develops when erosion eats through a meander neck.

cyanobacteria Blue-green algae; a type of archaea.

cycle A series of interrelated events or steps that occur in succession and can be repeated, perhaps indefinitely.

cyclone (1) The counterclockwise flow of air around a low-pressure mass; (2) the equivalent of a hurricane in the Indian Ocean.

cyclothem A repeated interval within a sedimentary sequence that contains a specific succession of sedimentary beds.

Darcy's law A mathematical equation stating that a volume of water, passing through a specified area of material at a given time, depends on the material's permeability and hydraulic gradient.

daughter isotope The decay product of radioactive decay.

day The time it takes for the Earth to spin once on its axis.

debris avalanche An avalanche in which the falling debris consists of rock fragments and dust.

debris flow A downslope movement of mud mixed with larger rock fragments.

debris slide A sudden downslope movement of material consisting only of regolith.

decompression melting The kind of melting that occurs when hot mantle rock rises to shallower depths in the Earth so that pressure decreases while the temperature remains unchanged.

deep current An ocean current at a depth greater than 100 m.

deep-focus earthquake An earthquake that occurs at a depth between 300 and 670 km; below 670 km, earthquakes do not happen.

deflation The process of lowering the land surface by wind abrasion.

deformation A change in the shape, position, or orientation of a material, by bending, breaking, or flowing.

dehydration Loss of water.

delta A wedge of sediment formed at a river mouth when the running water of the stream enters standing water, the current slows, the stream loses competence, and sediment settles out.

delta plain The low, swampy land on the surface of a delta.

delta-plain flood A flood in which water submerges a delta plain.

dendritic network A drainage network whose interconnecting streams resemble the pattern of branches connecting to a deciduous tree.

dendrochronologist A scientist who analyzes tree rings to determine the geologic age of features.

density Mass per unit volume.

denudation The removal of rock and regolith from the Earth's surface.

deposition The process by which sediment settles out of a transporting medium.

depositional landform A landform resulting from the deposition of sediment where the medium carrying the sediment evaporates, slows down, or melts.

desert A region so arid that it contains no permanent streams except for those that bring water in from elsewhere, and has very sparse vegetation cover.

desertification The process of transforming nondesert areas into desert.

desert pavement A mosaic-like stone surface forming the ground in a desert.

desert varnish A dark, rusty-brown coating of iron oxide and magnesium oxide that accumulates on the surface of the rock.

detachment fault A nearly horizontal fault at the base of a fault system.

detritus The chunks and smaller grains of rock broken off outcrops by physical weathering.

dewpoint temperature The temperature at which air becomes saturated so that dew can form.

differential stress A condition causing a material to experience a push or pull in one direction of a greater magnitude than the push or pull in another direction; in some cases, differential stress can result in shearing.

differential weathering What happens when different rocks in an outcrop undergo weathering at different rates.

diffraction The splitting of light into many tiny beams that interfere with one another.

dike A tabular (wall-shaped) intrusion of rock that cuts across the layering of country rock.

dimension stone An intact block of granite or marble to be used for architectural purposes.

dipole A magnetic field with a north and south pole, like that of a bar magnet.

dipole field (for Earth) The part of the Earth's magnetic field, caused by the flow of liquid iron alloy in the outer core, that can be represented by an imaginary bar magnet with a north and south pole.

dip-slip fault A fault in which sliding occurs up or down the slope (dip) of the fault.

dip slope A hill slope underlain by bedding parallel to the slope.

disappearing stream A stream that intersects a crack or sinkhole leading to an underground cavern, so that the water disappears into the subsurface and becomes an underground stream.

discharge The volume of water in a conduit or channel passing a point in one second.

discharge area A location where groundwater flows back up to the surface, and may emerge at springs.

disconformity An unconformity parallel to the two sedimentary sequences it separates.

displacement (or **offset**) The amount of movement or slip across a fault plane.

disseminated deposit A hydrothermal ore deposit in which ore minerals are dispersed throughout a body of rock.

dissolution A process during which materials dissolve in water.

dissolved load Ions dissolved in a stream's water.

distillation column A vertical pipe in which crude oil is separated into several components.

distributaries The fan of small streams formed where a river spreads out over its delta.

divergence zone A place where sinking air separates into two flows that move in opposite directions.

divergent plate boundary A boundary at which two lithosphere plates move apart from each other; they are marked by mid-ocean ridges.

diversification The development of many different species.

DNA (deoxyribonucleic acid) The complex molecule, shaped like a double helix, containing the code that guides the growth and development of an organism.

doldrums A belt with very slow winds along the equator.

dome Folded or arched layers with the shape of an overturned bowl.

Doppler effect The phenomenon in which the frequency of wave energy appears to change when a moving source of wave energy passes an observer.

dormant volcano A volcano that has not erupted for hundreds to thousands of years but does have the potential to erupt again in the future.

downcutting The process in which water flowing through a channel cuts into the substrate and deepens the channel relative to its surroundings.

downdraft Downward-moving air.

downgoing plate (or **slab**) A lithosphere plate that has been subducted at a convergent margin.

downslope force The component of the force of gravity acting in the downslope direction.

downslope movement The tumbling or sliding of rock and sediment from higher elevations to lower ones.

downwelling zone A place where near-surface water sinks.

drag fold A fold that develops in layers of rock adjacent to a fault during or just before slip.

drainage divide A highland or ridge that separates one watershed from another.

drainage network (or **basin**) An array of interconnecting streams that together drain an area.

drawdown The phenomenon in which the water table around a well drops because the users are pumping water out of the well faster than it flows in from the surrounding aquifer.

drilling mud A slurry of water mixed with clay that oil drillers use to cool a drill bit and flush rock cuttings up and out of the hole.

dripstone Limestone (travertine in a cave) formed by the precipitation of calcium carbonate out of groundwater.

drop stone A rock that drops to the sea floor once the iceberg that was carrying the rock melts.

drumlin A streamlined, elongate hill formed when a glacier overrides glacial till.

dry-bottom (polar) glacier A glacier so cold that its base remains frozen to the substrate.

dry wash The channel of an ephemeral stream when empty of water.

dry well (1) A well that does not supply water because the well has been drilled into an aquitard or into rock that lies above the water table; (2) a well that does not yield oil, even though it has been drilled into an anticipated reservoir.

ductile (plastic) deformation The bending and flowing of a material (without cracking and breaking) subjected to stress.

dune A pile of sand generally formed by deposition from the wind.

dust storm An event in which strong winds hit unvegetated land, strip off the topsoil, and send it skyward to form rolling dark clouds that block out the Sun.

dynamic metamorphism Metamorphism that occurs as a consequence of shearing alone, with no change in temperature or pressure.

dynamo A power plant generator in which water or wind power spins an electrical conductor around a permanent magnet.

dynamothermal metamorphism Metamorphism that involves heat, pressure, and shearing.

earthquake A vibration caused by the sudden breaking or frictional sliding of rock in the Earth.

earthquake belt A relatively narrow, distinct belt of earthquakes that defines the position of a plate boundary.

earthquake engineering The design of buildings that can withstand shaking.

earthquake zoning The determination of where land is relatively stable and where it might collapse because of seismicity.

Earth System The global interconnecting web of physical and biological phenomena involving the solid Earth, the hydrosphere, and the atmosphere.

ebb tide The falling tide.

eccentricity cycle The cycle of the gradual change of the Earth's orbit from a more circular to a more elliptical shape; the cycle takes around 100,000 years.

ecliptic The plane defined by a planet's orbit.

ecosystem An environment and its inhabitants.

eddy An isolated, ring-shaped current of water.

effusive eruption An eruption that yields mostly lava, not ash.

Ekman spiral The change in flow direction of water with depth, caused by the Coriolis effect.

Ekman transport The overall movement of a mass of water, resulting from the Eckman spiral, in a direction 90° to the wind direction.

elastic strain A change in shape of a material that disappears instantly when stress is removed.

electromagnet An electrical device that produces a magnetic field.

electron microprobe A laboratory instrument that can focus a beam of electrons on a small part of a mineral grain in order to create a signal that defines its chemical composition.

El Niño The flow of warm water eastward from the Pacific Ocean that reverses the upwelling of cold water along the western coast of South America and causes significant global changes in weather patterns.

embayment   A low area of coastal land.

emergent coast   A coast where the land is rising relative to sea level or sea level is falling relative to the land.

end moraine (terminal moraine)   A low, sinuous ridge of till that develops when the terminus (toe) of a glacier stalls in one position for a while.

energy   The capacity to do work.

energy resource   Something that can be used to produce work; in a geologic context, a material (such as oil, coal, wind, flowing water) that can be used to produce energy.

eon   The largest subdivision of geologic time.

epeirogenic movement   The gradual uplift or subsidence of a broad region of the Earth's surface.

epeirogeny   An event of epeirogenic movement; the term is usually used in reference to the formation of broad mid-continent domes and basins.

ephemeral (intermittent) stream   A stream whose bed lies above the water table, so that the stream flows only when the rate at which water enters the stream from rainfall or meltwater exceeds the rate at which water infiltrates the ground below.

epicenter   The point on the surface of the Earth directly above the focus of an earthquake.

epicontinental sea   A shallow sea overlying a continent.

epoch   An interval of geologic time representing the largest subdivision of a period.

equant   A term for a grain that has the same dimensions in all directions.

equatorial low   The area of low pressure that develops over the equator because of the intertropical convergence zone.

equinox   One of two days out of the year (September 22 and March 21) in which the Sun is directly overhead at noon at the equator.

era   An interval of geologic time representing the largest subdivision of the Phanerozoic Eon.

erg   Sand seas formed by the accumulation of dunes in a desert.

erosion   The grinding away and removal of Earth's surface materials by moving water, air, or ice.

erosional coast   A coastline where sediment is not accumulating and wave action grinds away at the shore.

erosional landform   A landform that results from the breakdown and removal of rock or sediment.

erratic   A boulder or cobble that was picked up by a glacier and deposited hundreds of kilometers away from the outcrop from which it detached.

esker   A ridge of sorted sand and gravel that snakes across a ground moraine; the sediment of an esker was deposited in subglacial meltwater tunnels.

estuary   An inlet in which seawater and river water mix, created when a coastal valley is flooded because of either rising sea level or land subsidence.

Eubacteria   The kingdom of "true bacteria."

euhedral crystal   A crystal whose faces are well formed and whose shape reflects crystal form.

eukaryotic cell   A cell with a complex internal structure, capable of building multicellular organisms.

eustatic sea-level change   A global rising or falling of the ocean surface.

evaporate   To change from liquid to vapor.

evapotranspiration   The sum of evaporation from bodies of water and the ground surface and transpiration from plants and animals.

exfoliation   The process by which an outcrop of rock splits apart into onion-like sheets along joints that lie parallel to the ground surface.

exhumation   The process (involving uplift and erosion) that returns deeply buried rocks to the surface.

exotic terrane   A block of land that collided with a continent along a convergent margin and attached to the continent; the term "exotic" implies that the land was not originally part of the continent to which it is now attached.

expanding Universe theory   The theory that the whole Universe must be expanding because galaxies in every direction seem to be moving away from us.

explosive eruptions   Violent volcanic eruptions that produce clouds and avalanches of pyroclastic debris.

external process   A geomorphologic process—such as downslope movement, erosion, or deposition—that is the consequence of gravity or of the interaction between the solid Earth and its fluid envelope (air and water). Energy for these processes comes from gravity and sunlight.

extinction   The death of the last members of a species so that there are no parents to pass on their genetic traits to offspring.

extinct volcano   A volcano that was active in the past but has now shut off entirely and will not erupt in the future.

extraordinary fossil   A rare fossilized relict, or trace, of the soft part of an organism.

extrusive igneous rock   Rock that forms by the freezing of lava above ground, after it flows or explodes out (extrudes) onto the surface and comes into contact with the atmosphere or ocean.

eye   The relative calm in the center of a hurricane.

eye wall   A rotating vertical cylinder of clouds surrounding the eye of a hurricane.

facies   (1) Sedimentary: a group of rocks and primary structures indicative of a given depositional environment; (2) Metamorphic: a set of metamorphic mineral assemblages formed under a given range of pressures and temperatures.

fault   A fracture on which one body of rock slides past another.

fault-block mountains   An outdated term for a narrow, elongate range of mountains that develops in a continental-rift setting as normal faulting drops down blocks of crust, or tilts blocks.

fault breccia   Fragmented rock in which angular fragments were formed by brittle fault movement; fault breccia occurs along a fault.

fault creep   Gradual movement along a fault that occurs in the absence of an earthquake.

fault gouge   Pulverized rock consisting of fine powder that lies along fault surfaces; gouge forms by crushing and grinding.

faulting   Slip events along a fault.

fault scarp   A small step on the ground surface where one side of a fault has moved vertically with respect to the other.

fault system   A grouping of numerous related faults.

fault trace (or line)   The intersection between a fault and the ground surface.

felsic   An adjective used in reference to igneous rocks that are rich in elements forming feldspar and quartz.

Ferrel cells   The name given to the middle-latitude convection cells in the atmosphere.

fetch   The distance across a body of water along which a wind blows to build waves.

fine-grained   A textural term for rock consisting of many fine grains or clasts.

firn Compacted granular ice (derived from snow) that forms where snow is deeply buried; if buried more deeply, firn turns into glacial ice.

fission track A line of damage formed in the crystal lattice of a mineral by the impact of an atomic particle ejected during the decay of a radioactive isotope.

fissure A conduit in a magma chamber in the shape of a long crack through which magma rises and erupts at the surface.

fjord A deep, glacially carved, U-shaped valley flooded by rising sea level.

flank eruption An eruption that occurs when a secondary chimney, or fissure, breaks through the flank of a volcano.

flash flood A flood that occurs during unusually intense rainfall or as the result of a dam collapse, during which the floodwaters rise very fast.

flexing The process of folding in which a succession of rock layers bends and slip occurs between the layers.

flocculation The clumping together of clay suspended in river water into bunches that are large enough to settle out.

flood An event during which the volume of water in a stream becomes so great that it covers areas outside the stream's normal channel.

flood basalt Vast sheets of basalt that spread from a volcanic vent over an extensive surface of land; they may form where a rift develops above a continental hot spot, and where lava is particularly hot and has low viscosity.

floodplain The flat land on either side of a stream that becomes covered with water during a flood.

floodplain flood A flood during which a floodplain is submerged.

flood stage The stage when water reaches the top of a stream channel.

flood tide The rising tide.

floodway A mapped region likely to be flooded, in which people avoid constructing buildings.

flow fold A fold that forms when the rock is so soft that it behaves like weak plastic.

flowstone A sheet of limestone that forms along the wall of a cave when groundwater flows along the surface of the wall.

fluvial deposit Sediment deposited in a stream channel, along a stream bank, or on a floodplain.

flux Flow.

focus The location where a fault slips during an earthquake (hypocenter).

fog A cloud that forms at ground level.

fold A bend or wrinkle of rock layers or foliation; folds form as a consequence of ductile deformation.

fold axis An imaginary line that, when moved parallel to itself, can trace out the shape of a folded surface.

fold-thrust belt An assemblage of folds and related thrust faults that develop above a detachment fault.

foliation Layering formed as a consequence of the alignment of mineral grains, or of compositional banding in a metamorphic rock.

foraminifera Microscopic plankton with calcitic shells, components of some limestones.

foreland sedimentary basin A basin located under the plains adjacent to a mountain front, which develops as the weight of the mountains pushes the crust down, creating a depression that traps sediment.

foreshocks The series of smaller earthquakes that precede a major earthquake.

foreshore zone The zone of beach regularly covered and uncovered by rising and falling tides.

formation *Stratigraphic formation.*

fossil The remnant, or trace, of an ancient living organism that has been preserved in rock or sediment.

fossil assemblage A group of fossil species found in a specific sequence of sedimentary rock.

fossil correlation A determination of the stratigraphic relation between two sedimentary rock units, reached by studying fossils.

fossil fuel An energy resource such as oil or coal that comes from organisms that lived long ago, and thus stores solar energy that reached the Earth then.

fossiliferous limestone Limestone consisting of abundant fossil shells and shell fragments.

fossilization The process of forming a fossil.

fractional crystallization The process by which a magma becomes progressively more silicic as it cools, because early-formed crystals settle out.

fracture zone A narrow band of vertical fractures in the ocean floor; fracture zones lie roughly at right angles to a mid-ocean ridge, and the actively slipping part of a fracture zone is a transform fault.

fresh rock Rock whose mineral grains have their original composition and shape.

friction Resistance to sliding on a surface.

fringing reef A coral reef that forms directly along the coast.

front The boundary between two air masses.

frost wedging The process in which water trapped in a joint freezes, forces the joint open, and may cause the joint to grow.

fuel rod A metal tube that holds the nuclear fuel in a nuclear reactor.

Fujita scale A scale that distinguishes among tornadoes on the basis of wind speed, path dimensions, and possible damage.

Ga Billions of years ago (abbreviation).

gabbro A coarse-grained, intrusive mafic igneous rock.

Gaia The term used for the Earth System, with the implication that it resembles a complex living entity.

galaxy An immense system of hundreds of billions of stars.

gene An individual component of the DNA code that guides the growth and development of an organism.

genetics The study of genes and how they transmit information.

geocentric Universe concept An ancient Greek idea suggesting that the Earth sat motionless in the center of the Universe while stars and other planets and the Sun orbited around it.

geochronology The science of dating geologic events in years.

geode A cavity in which euhedral crystals precipitate out of water solutions passing through a rock.

geographical pole The locations (north and south) where the Earth's rotational axis intersects the planet's surface.

geologic column A composite stratigraphic chart that represents the entirety of the Earth's history.

geologic history The sequence of geologic events that has taken place in a region.

geologic map A map showing the distribution of rock units and structures across a region.

geologic time The span of time since the formation of the Earth.

geologic time scale A scale that describes the intervals of geologic time.

geology The study of the Earth, including our planet's composition, behavior, and history.

geotherm The change in temperature with depth in the Earth.

geothermal energy Heat and electricity produced by using the internal heat of the Earth.

geothermal gradient The rate of change in temperature with depth.

geothermal region A region of current or recent volcanism in which magma or very hot rock heats up groundwater, which may discharge at the surface in the form of hot springs and/or geysers.

geyser A fountain of steam and hot water that erupts periodically from a vent in the ground in a geothermal region.

glacial abrasion The process by which clasts embedded in the base of a glacier grind away at the substrate as the glacier flows.

glacial advance The forward movement of a glacier's toe when the supply of snow exceeds the rate of ablation.

glacial drift Sediment deposited in glacial environments.

glacial incorporation The process by which flowing ice surrounds and incorporates debris.

glacial marine Sediment consisting of ice-rafted clasts mixed with marine sediment.

glacial outwash Coarse sediment deposited on a glacial outwash plain by meltwater streams.

glacially polished surface A polished rock surface created by the glacial abrasion of the underlying substrate.

glacial plucking (or **quarrying**) The process by which a glacier breaks off and carries away fragments of bedrock.

glacial rebound The process by which the surface of a continent rises back up after an overlying continental ice sheet melts away and the weight of the ice is removed.

glacial retreat The movement of a glacier's toe back toward the glacier's origin; glacial retreat occurs if the rate of ablation exceeds the rate of supply.

glacial subsidence The sinking of the surface of a continent caused by the weight of an overlying glacial ice sheet.

glacial till Sediment transported by flowing ice and deposited beneath a glacier or at its toe.

glaciation A period of time during which glaciers grew and covered substantial areas of the continents.

glacier A river or sheet of ice that slowly flows across the land surface and lasts all year long.

glass A solid in which atoms are not arranged in an orderly pattern.

glassy igneous rock Igneous rock consisting entirely of glass, or of tiny crystals surrounded by a glass matrix.

glide horizon The surface along which a slump slips.

global change The transformations or modifications of both physical and biological components of the Earth System through time.

global circulation The movement of volumes of air in paths that ultimately take it around the planet.

global climate change Transformations or modifications in Earth's climate over time.

global cooling A fall in the average atmospheric temperature.

global positioning system (GPS) A satellite system people can use to measure rates of movement of the Earth's crust relative to one another, or simply to locate their position on the Earth's surface.

global warming A rise in the average atmospheric temperature.

gneiss A compositionally banded metamorphic rock typically composed of alternating dark- and light-colored layers.

Gondwana A supercontinent that consisted of today's South America, Africa, Antarctica, India, and Australia. Also called Gondwanaland.

graben A down-dropped crustal block bounded on either side by a normal fault dipping toward the basin.

gradualism The theory that evolution happens at a constant, slow rate.

grain A fragment of a mineral crystal or of a rock.

grain rotation The process by which rigid, inequant mineral grains distributed through a soft matrix may rotate into parallelism as the rock changes shape owing to differential stress.

granite A coarse-grained intrusive silicic igneous rock.

granulite facies A set of metamorphic mineral assemblages formed at very high pressures and temperatures.

gravitational spreading A process of lateral spreading that occurs in a material because of the weakness of the material; gravitational spreading causes continental glaciers to grow and mountain belts to undergo orogenic collapse.

graywacke An informal term used for sedimentary rock consisting of sand-size up to small-pebble-size grains of quartz and rock fragments all mixed together in a muddy matrix; typically, graywacke occurs at the base of a graded bed.

greenhouse conditions (greenhouse period) Relatively warm global climate leading to the rising of sea level for an interval of geologic time.

greenhouse effect The trapping of heat in the Earth's atmosphere by carbon dioxide and other greenhouse gases, which absorb infrared radiation; somewhat analogous to the effect of glass in a greenhouse.

greenhouse gases Atmospheric gases, such as carbon dioxide and methane, that regulate the Earth's atmospheric temperature by absorbing infrared radiation.

greenschist facies A set of metamorphic mineral assemblages formed under relatively low pressures and temperatures.

greenstone A low-grade metamorphic rock formed from basalt; if foliated, the rock is called greenschist.

Greenwich mean time (GMT) The time at the astronomical observatory in Greenwich, England; time in all other time zones is set in relation to GMT.

Grenville orogeny The orogeny that occurred about 1 billion years ago and yielded the belt of deformed and metamorphosed rocks that underlie the eastern fifth of the North American continent.

groin A concrete or stone wall built perpendicular to a shoreline in order to prevent beach drift from removing sand.

ground moraine A thin, hummocky layer of till left behind on the land surface during a rapid glacial recession.

groundwater Water that resides under the surface of the Earth, mostly in pores or cracks of rock or sediment.

group A succession of stratigraphic formations that have been lumped together, making a single, thicker stratigraphic entity.

growth ring A rhythmic layering that develops in trees, travertine deposits, and shelly organisms as a consequence of seasonal changes.

gusher A fountain of oil formed when underground pressure causes the oil to rise on its own out of a drilled hole.

guyot A seamount that had a coral reef growing on top of it, so that it is now flat-crested.

gymnosperm A plant whose seeds are "naked," not surrounded by a fruit.

gyre A large circular flow pattern of ocean surface currents.

Hadean The oldest of the Precambrian eons; the time between Earth's origin and the formation of the first rocks that have been preserved.

Hadley cells The name given to the low-latitude convection cells in the atmosphere.

hail Falling ice balls from the sky, formed when ice crystallizes in turbulent storm clouds.

hail streak An approximately 2-by-10-km stretch of ground, elongate in the direction of a storm, onto which hail has fallen.

half-graben A wedge-shaped basin in cross section that develops as the hanging-wall block above a normal fault slides down and rotates; the basin develops between the fault surface and the top surface of the rotated block.

half-life The time it takes for half of a group of a radioactive element's isotopes to decay.

halocline The boundary in the ocean between surface-water and deep-water salinities.

hamada Barren rocky highlands in a desert.

hanging valley A glacially carved tributary valley whose floor lies at a higher elevation than the floor of the trunk valley.

hanging wall The rock or sediment above an inclined fault plane.

hard water Groundwater that contains dissolved calcium and magnesium, usually after passing through limestone or dolomite.

head (1) The elevation of the water table above a reference horizon; (2) the edge of ice at the origin of a glacier.

headland A place where a hill or cliff protrudes into the sea.

head scarp The distinct step along the upslope edge of a slump where the regolith detached.

headward erosion The process by which a stream channel lengthens up its slope as the flow of water increases.

headwaters The beginning point of a stream.

heat Thermal energy resulting from the movement of molecules.

heat capacity A measure of the amount of heat that must be added to a material to change its temperature.

heat flow The rate at which heat rises from the Earth's interior up to the surface.

heat-transfer melting Melting that results from the transfer of heat from a hotter magma to a cooler rock.

heliocentric Universe concept An idea proposed by Greek philosophers around 250 B.C.E. suggesting that all heavenly objects including the Earth orbited the Sun.

Hercynian orogen The late Paleozoic orogen that affected parts of Europe; a continuation of the Alleghenian orogen.

heterosphere A term for the upper portion of the atmosphere in which gases separate into distinct layers on the basis of composition.

hiatus The interval of time between deposition of the youngest rock below an unconformity and deposition of the oldest rock above the unconformity.

high-altitude westerlies Westerly winds at the top of the troposphere.

high-grade metamorphic rocks Rocks that metamorphose under relatively high temperatures.

high-level waste Nuclear waste containing greater than 1 million times the safe level of radioactivity.

hinge The portion of a fold where curvature is greatest.

hogback A steep-sided ridge of steeply dipping strata.

Holocene The period of geologic time since the last glaciation.

Holocene climatic maximum The period from 5,000 to 6,000 years ago, when Holocene temperatures reached a peak.

homosphere The lower part of the atmosphere, in which the gases have stirred into a homogenous mixture.

hoodoo The local name for the brightly colored shale and sandstone chimneys found in Bryce Canyon National Park in Utah.

horn A pointed mountain peak surrounded by at least three cirques.

hornfels Rock that undergoes metamorphism simply because of a change in temperature, without being subjected to differential stress.

horse latitudes The region of the subtropical high in which winds are weak.

horst The high block between two grabens.

hot spot A location at the base of the lithosphere, at the top of a mantle plume, where temperatures can cause melting.

hot-spot track A chain of now-dead volcanoes transported off the hot spot by the movement of a lithosphere plate.

hot-spot volcano An isolated volcano not caused by movement at a plate boundary, but rather by the melting of a mantle plume.

hot spring A spring that emits water ranging in temperature from about 30° to 104°C.

hummocky surface An irregular and lumpy ground surface.

hurricane A huge rotating storm, resembling a giant spiral in map view, in which sustained winds blow over 119 km per hour.

hurricane track The path a hurricane follows.

hyaloclastite A rubbly extrusive rock consisting of glassy debris formed in a submarine or sub-ice eruption.

hydration The absorption of water into the crystal structure of minerals; a type of chemical weathering.

hydraulic conductivity The coefficient K in Darcy's law; hydraulic conductivity takes into account the permeability of the sediment or rock as well as the fluid's viscosity.

hydraulic gradient The slope of the water table.

hydrocarbon A chain-like or ring-like molecule made of hydrogen and carbon atoms; petroleum and natural gas are hydrocarbons.

hydrocarbon system The association of source rock, migration pathway, reservoir rock, seal, and trap geometry that leads to the occurrence of a hydrocarbon reserve.

hydrologic cycle The continual passage of water from reservoir to reservoir in the Earth System.

hydrolysis The process in which water chemically reacts with minerals and breaks them down.

hydrosphere The Earth's water, including surface water (lakes, rivers, and oceans), groundwater, and liquid water in the atmosphere.

hydrothermal deposit An accumulation of ore minerals precipitated from hot-water solutions circulating through a magma or through the rocks surrounding an igneous intrusion.

hypsometric curve A graph that plots surface elevation on the vertical axis and the percentage of the Earth's surface on the horizontal axis.

ice age An interval of time in which the climate was colder than it is today, glaciers occasionally advanced to cover large areas of the continents, and mountain glaciers grew; an ice age can include many glacials and interglacials.

iceberg A large block of ice that calves off the front of a glacier and drops into the sea.

icehouse period A period of time when the Earth's temperature was cooler than it is today and ice ages could occur.

ice-margin lake A meltwater lake formed along the edge of a glacier.

ice-rafted sediment Sediment carried out to sea by icebergs.

ice sheet A vast glacier that covers the landscape.

ice shelf A broad, flat region of ice along the edge of a continent formed where a continental glacier flowed into the sea.

ice stream A portion of a glacier that travels much more quickly than adjacent portions of the glacier.

ice tongue   The portion of a valley glacier that has flowed out into the sea.

igneous rock   Rock that forms when hot molten rock (magma or lava) cools and freezes solid.

ignimbrite   Rock formed when deposits of pyroclastic flows solidify.

inactive fault   A fault that last moved in the distant past and probably won't move again in the near future, yet is still recognizable because of displacement across the fault plane.

inactive sand   The sand along a coast that is buried beneath a layer of active sand and moves only during severe storms or not at all.

incised meander   A meander that lies at the bottom of a steep-walled canyon.

index minerals   Minerals that serve as good indicators of metamorphic grade.

induced seismicity   Seismic events caused by the actions of people (e.g. filling a reservoir, that lies over a fault, with water).

industrial minerals   Minerals that serve as the raw materials for manufacturing chemicals, concrete, and wallboard, among other products.

inequant   A term for a mineral grain whose length and width are not the same.

inertia   The tendency of an object at rest to remain at rest.

infiltrate   Seep down into.

injection well   A well in which a liquid is pumped down into the ground under pressure so that it passes from the well back into the pore space of the rock or regolith.

inner core   The inner section of the core 5,155 km deep to the Earth's center at 6,371 km, and consisting of solid iron alloy.

inselberg   An isolated mountain or hill in a desert landscape created by progressive cliff retreat, so that the hill is surrounded by a pediment or an alluvial fan.

insolation   Exposure to the Sun's rays.

interglacial   A period of time between two glaciations.

interior basin   A basin with no outlet to the sea.

interlocking texture   The texture of crystalline rocks in which mineral grains fit together like pieces of a jigsaw puzzle.

internal process   A process in the Earth System, such as plate motion, mountain building, or volcanism, ultimately caused by Earth's internal heat.

intertidal zone   The area of coastal land across which the tide rises and falls.

intertropical convergence zone   The equatorial convergence zone in the atmosphere.

intraplate earthquakes   Earthquakes that occur away from plate boundaries.

intrusive contact   The boundary between country rock and an intrusive igneous rock.

intrusive igneous rock   Rock formed by the freezing of magma underground.

ionosphere   The interval of Earth's atmosphere, at an elevation between 50 and 400 km, containing abundant positive ions.

iron catastrophe   The proposed event very early in Earth history when the Earth partly melted and molten iron sank to the center to form the core.

isobar   A line on a map along which the air has a specified pressure.

isograd   (1) A line on a pressure-temperature graph along which all points are taken to be at the same metamorphic grade; (2) A line on a map making the first appearance of a metamorphic index mineral.

isostasy (or isostatic equilibrium)   The condition that exists when the buoyancy force pushing lithosphere up equals the gravitational force pulling lithosphere down.

isostatic compensation   The process in which the surface of the crust slowly rises or falls to reestablish isostatic equilibrium after a geologic event changes the density or thickness of the lithosphere.

isotherm   Lines on a map or cross section along which the temperature is constant.

isotopes   Different versions of a given element that have the same atomic number but different atomic weights.

jet stream   A fast-moving current of air that flows at high elevations.

jetty   A manmade wall that protects the entrance to a harbor.

joints   Naturally formed cracks in rocks.

joint set   A group of systematic joints.

Jovian   A term used to describe the outer gassy, Jupiter-like planets (gas-giant planets).

kame   A stratified sequence of lateral-moraine sediment that's sorted by water flowing along the edge of a glacier.

karst landscape   A region underlain by caves in limestone bedrock; the collapse of the caves creates a landscape of sinkholes separated by higher topography, or of limestone spires separated by low areas.

kerogen   The waxy molecules into which the organic material in shale transforms on reaching about 100°C. At higher temperatures, kerogen transforms into oil.

kettle hole   A circular depression in the ground made when a block of ice calves off the toe of a glacier, becomes buried by till, and later melts.

knob-and-kettle topography   A land surface with many kettle holes separated by round hills of glacial till.

K-T boundary event   The mass extinction that happened at the end of the Cretaceous Period, 65 million years ago, possibly because of the collision of an asteroid with the Earth.

lag deposit   The coarse sediment left behind in a desert after wind erosion removes the finer sediment.

lagoon   A body of shallow seawater separated from the open ocean by a barrier island.

lahar   A thick slurry formed when volcanic ash and debris mix with water, either in rivers or from rain or melting snow and ice on the flank of a volcano.

landslide   A sudden movement of rock and debris down a nonvertical slope.

landslide-potential map   A map on which regions are ranked according to the likelihood that a mass movement will occur.

land subsidence   Sinking elevation of the ground surface; the process may occur over an aquifer that is slowly draining and decreasing in volume because of pore collapse.

La Niña   Years in which the El Niño event is not strong.

lapilli   Marble-to-plum-sized fragments of pyroclastic debris.

Laramide orogeny   The mountain-building event that lasted from about 80 Ma to 40 Ma, in western North America; in the United States, it formed the Rocky Mountains as a result of basement uplift and the warping of the younger overlying strata into large monoclines.

latent heat of condensation   The heat released during condensation, which comes only from a change in state.

lateral moraine   A strip of debris along the side margins of a glacier.

laterite soil Soil formed over iron-rich rock in a tropical environment, consisting primarily of a dark-red mass of insoluble iron and/or aluminum oxide.

Laurentia A continent in the early Paleozoic Era composed of today's North America and Greenland.

Laurentide ice sheet An ice sheet that spread over northeastern Canada during the Pleistocene ice age(s).

lava Molten rock that has flowed out onto the Earth's surface.

lava dome A dome-like mass of rhyolitic lava that accumulates above the eruption vent.

lava flows Sheets or mounds of lava that flow onto the ground surface or sea floor in molten form and then solidify.

lava lake A large pool of lava produced around a vent when lava fountains spew forth large amounts of lava in a short period of time.

lava tube The empty space left when a lava tunnel drains; this happens when the surface of a lava flow solidifies while the inner part of the flow continues to stream downslope.

leach To dissolve and carry away.

leader A conductive path stretching from a cloud toward the ground, along which electrons leak from the base of the cloud, and which provides the start for a lightning flash to the ground.

lightning flash A giant spark or pulse of current that jumps across a gap of charge separation.

light year The distance that light travels in one Earth year (about 6 trillion miles or 9.5 trillion km).

lignite Low-rank coal that consists of 50% carbon.

limb The side of a fold, showing less curvature than at the hinge.

limestone Sedimentary rock composed of calcite.

liquification The process by which wet sediment becomes a slurry; liquification may be triggered by earthquake vibrations.

lithification The transformation of loose sediment into solid rock through compaction and cementation.

lithologic correlation A correlation based on similarities in rock type.

lithosphere The relatively rigid, nonflowable, outer 100- to 150-km-thick layer of the Earth; constituting the crust and the top part of the mantle.

little ice age A period of cooler temperatures, between 1500 and 1800 C.E., during which many glaciers advanced.

local base level A base level upstream from a drainage network's mouth.

lodgment till A flat layer of till smeared out over the ground when a glacier overrides an end moraine as it advances.

loess Layers of fine-grained sediments deposited from the wind; large deposits of loess formed from fine-grained glacial sediment blown off outwash plains.

longitudinal (seif) dune A dune formed when there is abundant sand and a strong, steady wind, and whose axis lies parallel to the wind direction.

longitudinal profile A cross-sectional image showing the variation in elevation along the length of a river.

longshore current A current that flows parallel to a beach.

lower mantle The deepest section of the mantle, stretching from 670 km down to the core-mantle boundary.

low-grade metamorphic rocks Rocks that underwent metamorphism at relatively low temperatures.

low-velocity zone The asthenosphere underlying oceanic lithosphere in which seismic waves travel more slowly, probably because rock has partially melted.

luster The way a mineral surface scatters light.

L-waves (*love waves*) Surface seismic waves that cause the ground to ripple back and forth, creating a snake-like movement.

Ma Millions of years ago (abbreviation).

macrofossil A fossil large enough to be seen with the naked eye.

mafic A term used in reference to magmas or igneous rocks that are relatively poor in silica and rich in iron and magnesium.

magma Molten rock beneath the Earth's surface.

magma chamber A space below ground filled with magma.

magma contamination The process in which flowing magma incorporates components of the country rock through which it passes.

magmatic deposit An ore deposit formed when sulfide ore minerals accumulate at the bottom of a magma chamber.

magnetic anomaly The difference between the expected strength of the Earth's magnetic field at a certain location and the actual measured strength of the field at that location.

magnetic declination The angle between the direction a compass needle points at a given location and the direction of true north.

magnetic field The region affected by the force emanating from a magnet.

magnetic field lines The trajectories along which magnetic particles would align, or charged particles would flow, if placed in a magnetic field.

magnetic force The push or pull exerted by a magnet.

magnetic inclination The angle between a magnetic needle free to pivot on a horizontal axis and a horizontal plane parallel to the Earth's surface.

magnetic reversal The change of the Earth's magnetic polarity; when a reversal occurs, the field flips from normal to reversed polarity, or vice versa.

magnetic-reversal chronology The history of magnetic reversals through geologic time.

magnetization The degree to which a material can exert a magnetic force.

magnetometer An instrument that measures the strength of the Earth's magnetic field.

magnetosphere The region protected from the electrically charged particles of the solar winds by Earth's magnetic field.

magnetostratigraphy The comparison of the pattern of magnetic reversals in a sequence of strata, with a reference column showing the succession of reversals through time.

manganese nodules Lumpy accumulations of manganese-oxide minerals precipitated onto the sea floor.

mantle The thick layer of rock below the Earth's crust and above the core.

mantle plume A column of very hot rock rising up through the mantle.

marble A metamorphic rock composed of calcite and transformed from a protolith of limestone.

mare The broad darker areas on the Moon's surface, which consist of flood basalts that erupted over 3 billion years ago and spread out across the Moon's lowlands.

marginal sea A small ocean basin created when sea-floor spreading occurs behind an island arc.

maritime tropical air mass A mass of air that originates over tropical or subtropical oceanic regions.

marsh A wetland dominated by grasses.

mass-extinction event A time when vast numbers of species abruptly vanish.

mass movement (or **mass wasting**) The gravitationally caused downslope transport of rock, regolith, snow, or ice.

matrix Finer-grained material surrounding larger grains in a rock.

meander A snake-like curve along a stream's course.

meandering stream A reach of stream containing many meanders (snake-like curves).

meander neck A narrow isthmus of land separating two adjacent meanders.

mean sea level The average level between the high and low tide over a year at a given point.

mechanical weathering *Physical weathering.*

medial moraine A strip of sediment in the interior of a glacier, parallel to the flow direction of the glacier, formed by the lateral moraines of two merging glaciers.

Medieval warm period A period of high temperatures in the Middle Ages.

melt Molten (liquid) rock.

meltdown The melting of the fuel rods in a nuclear reactor that occurs if the rate of fission becomes too fast and the fuel rods become too hot.

melting curve The line defining the range of temperatures and pressures at which a rock melts.

melting temperature The temperature at which the thermal vibration of the atoms or ions in the lattice of a mineral is sufficient to break the chemical bonds holding them to the lattice, so a material transforms into a liquid.

meltwater lake A lake fed by glacial meltwater.

Mercalli intensity scale An earthquake characterization scale based on the amount of damage that the earthquake causes.

mesa A large, flat-topped hill (with a surface area of several square km) in an arid region.

mesopause The boundary that marks the top of the mesosphere of Earth's atmosphere.

mesosphere The cooler layer of atmosphere overlying the stratosphere.

Mesozoic The middle of the three Phanerozoic eras; it lasted from 245 Ma to 65 Ma.

metal A solid composed almost entirely of atoms of metallic elements; it is generally opaque, shiny, smooth, and malleable, and can conduct electricity.

metallic bond A chemical bond in which the outer atoms are attached to each other in such a way that electrons flow easily from atom to atom.

metamorphic aureole The region around a pluton, stretching tens to hundreds of meters out, in which heat transferred into the country rock and metamorphosed the country rock.

metamorphic facies A set of metamorphic mineral assemblages indicative of metamorphism under a specific range of pressures and temperatures.

metamorphic foliation A fabric defined by parallel surfaces or layers that develop in a rock as a result of metamorphism; schistocity and gneissic layering are examples.

metamorphic mineral assemblage A group of minerals that form in a rock as a result of metamorphism.

metamorphic rock Rock that forms when preexisting rock changes into new rock as a result of an increase in pressure and temperature and/or shearing under elevated temperatures; metamorphism occurs without the rock first becoming a melt or a sediment.

metamorphic zone The region between two metamorphic isograds, typically named after an index mineral found within the region.

metamorphism The process by which one kind of rock transforms into a different kind of rock.

metasomatism The process by which a rock's overall chemical composition changes during metamorphism because of reactions with hot water that bring in or remove elements.

meteoric water Water that falls to Earth from the atmosphere as either rain or snow.

meteorite A piece of rock or metal alloy that fell from space and landed on Earth.

micrite Limestone consisting of lime mud (i.e., very fine-grained limestone).

microfossil A fossil that can be seen only with a microscope or an electron microscope.

mid-latitude (wave) cyclone The circulation of air around large, low-pressure masses.

mid-ocean ridge A 2-km-high submarine mountain belt that forms along a divergent oceanic plate boundary.

migmatite A rock formed when gneiss is heated high enough so that it begins to partially melt, creating layers, or lenses, of new igneous rock that mix with layers of the relict gneiss.

Milankovitch cycles Climate cycles that occur over tens to hundreds of thousands of years, because of changes in Earth's orbit and tilt.

mine A site at which ore is extracted from the ground.

mineral A homogenous, naturally occurring, solid inorganic substance with a definable chemical composition and an internal structure characterized by an orderly arrangement of atoms, ions, or molecules in a lattice. Most minerals are inorganic.

mineral classes Groups of minerals distinguished from each other on the basis of chemical composition.

mineral resources The minerals extracted from the Earth's upper crust for practical purposes.

Mississippi Valley–type (MVT) ore An ore deposit, typically in dolostone, containing lead- and zinc-bearing minerals that precipitated from groundwater that had moved up from several km depth in the upper crust; such deposits occur in the upper Mississippi Valley.

Moho The seismic-velocity discontinuity that defines the boundary between the Earth's crust and mantle.

Mohs hardness scale A list of ten minerals in a sequence of relative hardness, with which other minerals can be compared.

mold A cavity in sedimentary rock left behind when a shell that once filled the space weathers out.

monocline A fold in the land surface whose shape resembles that of a carpet draped over a stair step.

monsoon A seasonal reversal in wind direction that causes a shift from a very dry season to a very rainy season in some regions of the world.

moraine A sediment pile composed of till deposited by a glacier.

mountain front The boundary between a mountain range and adjacent plains.

mountain (alpine) glacier A glacier that exists in or adjacent to a mountainous region.

mountain ice cap A mound of ice that submerges peaks and ridges at the crest of a mountain range.

mouth The outlet of a stream where it discharges into another stream, a lake, or a sea.

mudflow A downslope movement of mud at slow to moderate speed.

mud pot A viscous slurry that forms in a geothermal region when hot water or steam rises into soils rich in volcanic ash and clay.

mudstone Very fine-grained sedimentary rock that will not easily split into sheets.

mylonite Rock formed during dynamic metamorphism and characterized by foliation that lies roughly parallel to the fault (shear zone) involved in the shearing process; mylonites have very fine grains formed by the nonbrittle subdivision of larger grains.

native metal A naturally occurring pure mass of a single metal in an ore deposit.

natural arch An arch that forms when erosion along joints leaves narrow walls of rock; when the lower part of the wall erodes while the upper part remains, an arch results.

natural levees A pair of low ridges that appear on either side of a stream and develop as a result of the accumulation of sediment deposited naturally during flooding.

natural selection The process by which the fittest organisms survive to pass on their characteristics to the next generation.

neap tide An especially low tide that occurs when the angle between the direction of the Moon and the direction of the Sun is 90°.

nebula A cloud of gas or dust in space.

nebula theory of planet formation The concept that planets grow out of rings of gas, dust, and ice surrounding a new-born star.

negative anomaly An area where the magnetic field strength is less than expected.

negative feedback Feedback that slows a process down or reverses it.

Neocrystallization The growth of new crystals, not in the protolith, during metamorphism.

Nevadan orogeny A convergent-margin mountain-building event that took place in western North America during the Late Jurassic Period.

nonconformity A type of unconformity at which sedimentary rocks overlie basement (older intrusive igneous rocks and/or metamorphic rocks).

nonflowing artesian well An artesian well in which water rises on its own up to a level that lies below the ground surface.

nonfoliated metamorphic rock Rock containing minerals that recrystallized during metamorphism, but which has no foliation.

nonmetallic mineral resources Mineral resources that do not contain metals; examples include building stone, gravel, sand, gypsum, phosphate, and salt.

nonplunging fold A fold with a horizontal hinge.

nonrenewable resource A resource that nature will take a long time (hundreds to millions of years) to replenish or may never replenish.

nonsystematic joints Short cracks in rocks that occur in a range of orientations and are randomly placed and oriented.

nor'easter A large, mid-latitude North American cyclone; when it reaches the East Coast, it produces strong winds that come out of the northeast.

normal fault A fault in which the hanging-wall block moves down the slope of the fault.

normal force The component of the gravitational force acting perpendicular to a slope.

normal polarity Polarity in which the paleomagnetic dipole has the same orientation as it does today.

normal stress The push or pull that is perpendicular to a surface.

North Atlantic deep-water mass The mass of cold, dense water that sinks in the north polar regions.

northeast tradewinds Surface winds that come out of the northeast and occur in the region between the equator and 30°N.

nuclear fuel Pellets of concentrated uranium oxide or a comparable radioactive material that can provide energy in a nuclear reactor.

nuclear fusion The process by which the nuclei of atoms fuse together, thereby creating new, larger atoms.

nuclear reactor The part of a nuclear power plant where the fission reactions occur.

nuée ardente *Pyroclastic flow.*

oasis A verdant region surrounded by desert, occurring at a place where natural springs provide water at the surface.

oblique-slip fault A fault in which sliding occurs diagonally along the fault plane.

obsidian An igneous rock consisting of a solid mass of volcanic glass.

occluded front A front that no longer intersects the ground surface.

oceanic crust The crust beneath the oceans; composed of gabbro and basalt, overlain by sediment.

oceanic lithosphere Lithosphere topped by oceanic crust; it reaches a thickness of 100 km.

Oil Age The period of human history, including our own, so named because the economy depends on oil.

oil field A region containing a significant amount of accessible oil underground.

oil reserve The known supply of oil held underground.

oil shale Shale containing kerogen.

oil trap A geologic configuration that keeps oil underground in the reservoir rock and prevents it from rising to the surface.

oil window The narrow range of temperatures under which oil can form in a source rock.

olistotrome A large, submarine slump block, buried and preserved.

ophiolite A slice of oceanic crust that has been thrust onto continental crust.

ordinary well A well whose base penetrates below the water table, and can thus provide water.

ore Rock containing native metals or a concentrated accumulation of ore minerals.

ore deposit An economically significant accumulation of ore.

ore minerals Minerals that have metal in high concentrations and in a form that can be easily extracted.

organic carbon Carbon that has been incorporated in an organism.

organic chemical A carbon-containing compound that occurs in living organisms, or that resembles such compounds; it consists of carbon atoms bonded to hydrogen atoms along with varying amounts of oxygen, nitrogen, and other chemicals.

organic coast A coast along which living organisms control landforms along the shore.

organic sedimentary rock Sedimentary rock (such as coal) formed from carbon-rich relicts of organisms.

organic shale Lithified, muddy, organic-rich ooze that contains the raw materials from which hydrocarbons eventually form.

orogen (or orogenic belt) A linear range of mountains.

orogenic collapse The process in which mountains begin to collapse under their own weight and spread out laterally.

orogeny A mountain-building event.

orographic barrier A landform that diverts air flow upward or laterally.

outcrop An exposure of bedrock.

outer core The section of the core, between 2,900 and 5,150 km deep, that consists of liquid iron alloy.

outwash plain A broad area of gravel and sandbars deposited by a braided stream network, fed by the meltwater of a glacier.

overburden The weight of overlying rock on rock buried deeper in the Earth's crust.

overriding plate (or slab) The plate at a subduction zone that overrides the downgoing plate.

oversaturated solution A solution that contains so much solute (dissolved ions) that precipitation begins.

oversized stream valley A large valley with a small stream running through it; the valley formed earlier when the flow was greater.

oxbow lake A meander that has been cut off yet remains filled with water.

oxidation reaction A reaction in which an element loses electrons; an example is the reaction of iron with air to form rust.

ozone O_3, an atmospheric gas that absorbs harmful ultraviolet radiation from the Sun.

ozone hole An area of the atmosphere, over polar regions, from which ozone has been depleted.

pahoehoe A lava flow with a surface texture of smooth, glassy, rope-like ridges.

paleoclimate The past climate of the Earth.

paleomagnetism The record of ancient magnetism preserved in rock.

paleopole The supposed position of the Earth's magnetic pole in the past, with respect to a particular continent.

paleosol Ancient soil preserved in the stratigraphic record.

Paleozoic The oldest era of the Phanerozoic Eon.

Pangaea A supercontinent that assembled at the end of the Paleozoic Era.

Pannotia A supercontinent that may have existed sometime between 800 Ma and 600 Ma.

parabolic dunes Dunes formed when strong winds break through transverse dunes to make new dunes whose ends point upwind.

parallax The apparent movement of an object seen from two different points not on a straight line from the object (e.g., from your two different eyes).

parallax method A trigonometric method used to determine the distance from the Earth to a nearby star.

parent isotope A radioactive isotope that undergoes decay.

partial melting The melting in a rock of the minerals with the lowest melting temperatures, while other minerals remain solid.

passive margin A continental margin that is not a plate boundary.

passive-margin basin A thick accumulation of sediment along a tectonically inactive coast, formed over crust that stretched and thinned when the margin first began.

patterned ground A polar landscape in which the ground splits into pentagonal or hexagonal shapes.

pause An elevation in the atmosphere where temperature stops decreasing and starts increasing, or vice versa.

peat Compacted and partially decayed vegetation accumulating beneath a swamp.

pedalfer soil A temperate-climate soil characterized by well-defined soil horizons and an organic A-horizon.

pediment The broad, nearly horizontal bedrock surface at the base of a retreating desert cliff.

pedocal soil Thin soil, formed in arid climates. It contains very little organic matter, but significant precipitated calcite.

pegmatite A coarse-grained igneous rock containing crystals of up to tens of centimeters across and occurring in dike-shaped intrusions.

pelagic sediment Microscopic plankton shells and fine flakes of clay that settle out and accumulate on the deep-ocean floor.

Pelé's hair Droplets of basaltic lava that mold into long glassy strands as they fall.

Pelé's tears Droplets of basaltic lava that mold into tear-shaped glassy beads as they fall.

peneplain A nearly flat surface that lies at an elevation close to sea level; thought to be the product of long-term erosion.

perched water table A quantity of groundwater that lies above the regional water table because an underlying lens of impermeable rock or sediment prevents the water from sinking down to the regional water table.

percolation The process by which groundwater meanders through tiny, crooked channels in the surrounding material.

peridotite A coarse-grained ultramafic rock.

periglacial environment A region with widespread permafrost but without a blanket of snow or ice.

period An interval of geologic time representing a subdivision of a geologic era.

permafrost Permanently frozen ground.

permanent magnet A special material that behaves magnetically for a long time all by itself.

permanent stream A stream that flows year-round because its bed lies below the water table, or because more water is supplied from upstream than can infiltrate the ground.

permeability The degree to which a material allows fluids to pass through it via an interconnected network of pores and cracks.

permineralization The fossilization process in which plant material becomes transformed into rock by the precipitation of silica from groundwater.

petrified A term used by geologists to describe plant material that has transformed into rock by permineralization.

petroglyph Drawings formed by chipping into the desert varnish of rocks to reveal the lighter rock beneath.

petroleum *Oil.*

phaneritic A textural term used to describe coarse-grained igneous rock.

Phanerozoic Eon The most recent eon, an interval of time from 542 Ma to the present.

phenocryst A large crystal surrounded by a finer-grained matrix in an igneous rock.

photochemical smog Brown haze that blankets a city when exhaust from cars and trucks reacts in the presence of sunlight.

photosynthesis The process during which chlorophyll-containing plants remove carbon dioxide from the atmosphere, form tissues, and expel oxygen back to the atmosphere.

phreatomagmatic eruption An explosive eruption that occurs when water enters the magma chamber and turns into steam.

phyllite A fine-grained metamorphic rock with a foliation caused by the preferred orientation of very fine-grained mica.

phyllitic luster A silk-like sheen characteristic of phyllite, a result of the rock's fine-grained mica.

phylogenetic tree A chart representing the ideas of paleontologists showing which groups of organisms radiated from which ancestors.

physical weathering The process in which intact rock breaks into smaller grains or chunks.

piedmont glacier A fan or lobe of ice that forms where a valley glacier emerges from a valley and spreads out into the adjacent plain.

pillow basalt Glass-encrusted basalt blobs that form when magma extrudes on the sea floor and cools very quickly.

placer deposit Concentrations of metal grains in stream sediment that develop when rocks containing native metals erode and create a mixture of sand grains and metal fragments; the moving water of the stream carries away lighter mineral grains.

planetesimal Tiny, solid pieces of rock and metal that collect in a planetary nebula and eventually accumulate to form a planet.

plankton Tiny plants and animals that float in sea or lake water.

plastic deformation The deformational process in which mineral grains behave like plastic and, when compressed or sheared, become flattened or elongate without cracking or breaking.

plate One of about twenty distinct pieces of the relatively rigid lithosphere.

plate boundary The border between two adjacent lithosphere plates.

plate-boundary earthquakes The earthquakes that occur along and define plate boundaries.

plate-boundary volcano A volcanic arc or mid-ocean ridge volcano, formed as a consequence of movement along a plate boundary.

plate interior A region away from the plate boundaries that consequently experiences few earthquakes.

plate tectonics *Theory of plate tectonics.*

playa The flat, typically salty lake bed that remains when all the water evaporates in drier times; forms in desert regions.

Pleistocene ice age(s) The period of time from about 2 Ma to 14,000 years ago, during which the Earth experienced an ice age.

plunge pool A depression at the base of a waterfall scoured by the energy of the falling water.

plunging fold A fold with a tilted hinge.

pluton An irregular or blob-shaped intrusion; can range in size from tens of m across to tens of km across.

pluvial lake A lake formed to the south of a continental glacier as a result of enhanced rainfall during an ice age.

point bar A wedge-shaped deposit of sediment on the inside bank of a meander.

polar cell A high-latitude convection cell in the atmosphere.

polar easterlies Prevailing winds that come from the east and flow from the polar high to the subpolar low.

polar front The convergence zone in the atmosphere at latitude 60°.

polar glacier *Dry-bottom glacier.*

polar high The zone of high pressure in polar regions created by the sinking of air in the polar cells.

polarity The orientation of a magnetic dipole.

polarity chron The time interval between polarity reversals of Earth's magnetic field.

polarity subchron The time interval between magnetic reversals if the interval is of short duration (less than 200,000 years long).

polarized light A beam of filtered light waves that all vibrate in the same plane.

polar wander The phenomenon of the progressive changing through time of the position of the Earth's magnetic poles relative to a location on a continent; significant polar wander probably doesn't occur—in fact, poles seem to remain fairly fixed, while continents move.

polar-wander path The curving line representing the apparent progressive change in the position of the Earth's magnetic pole, relative to a locality X, assuming that the position of X on Earth has been fixed through time (in fact, poles stay fixed while continents move).

pollen Tiny grains involved in plant reproduction.

polymorphs Two minerals that have the same chemical composition but a different crystal lattice structure.

pore A small open space within sediment or rock.

pore collapse The closer packing of grains that occurs when groundwater is extracted from pores, thus eliminating the support holding the grains apart.

porosity The total volume of empty space (pore space) in a material, usually expressed as a percentage.

porphyritic A textural term for igneous rock that has phenocrysts distributed throughout a finer matrix.

positive anomaly An area where the magnetic field strength is stronger than expected.

positive-feedback mechanism A mechanism that enhances the process that causes the mechanism in the first place.

potentiometric surface The elevation to which water in an artesian system would rise if unimpeded; where there are flowing artesian wells, the potentiometric surface lies above ground.

pothole A bowl-shaped depression carved into the floor of a stream by a long-lived whirlpool carrying sand or gravel.

Precambrian The interval of geologic time between Earth's formation about 4.57 Ga and the beginning of the Phanerozoic Eon 542 Ma.

precession The gradual conical path traced out by Earth's spinning axis; simply put, it is the "wobble" of the axis.

precious metals Metals (like gold, silver, and platinum) that have high value.

precipitation (1) The process by which atoms dissolved in a solution come together and form a solid; (2) rainfall or snow.

preferred mineral orientation The metamorphic texture that exists where platy grains lie parallel to one another and/or elongate grains align in the same direction.

pressure Force per unit area, or the "push" acting on a material in cases where the push is the same in all directions.

pressure gradient The rate of pressure change over a given horizontal distance.

pressure solution The process of dissolution at points of contact, between grains, where compression is greatest, producing ions that then precipitate elsewhere, where compression is less.

prevailing winds Surface winds that generally flow in the same direction for long time periods.

primary porosity The space that remains between solid grains or crystals immediately after sediment accumulates or rock forms.

principal aquifer The geologic unit that serves as the primary source of groundwater in a region.

principle of baked contacts When an igneous intrusion "bakes" (metamorphoses) surrounding rock, the rock that has been baked must be older than the intrusion.

principle of cross-cutting relations If one geologic feature cuts across another, the feature that has been cut is older.

principle of fossil succession In a stratigraphic sequence, different species of fossil organisms appear in a definite order; once a fossil species disappears in a sequence of strata, it never reappears higher in the sequence.

principle of inclusions If a rock contains fragments of another rock, the fragments must be older than the rock containing them.

principle of original continuity Sedimentary layers, before erosion, formed fairly continuous sheets over a region.

principle of original horizontality Layers of sediment, when originally deposited, are fairly horizontal.

principle of superposition In a sequence of sedimentary rock layers, each layer must be younger than the one below, for a layer of sediment cannot accumulate unless there is already a substrate on which it can collect.

principle of uniformitariansim The physical processes we observe today also operated in the past in the same way, and at comparable rates.

prograde metamorphism Metamorphism that occurs as temperatures and pressures are increasing.

Proterozoic The most recent of the Precambrian eons.

protocontinent A block of crust composed of volcanic arcs and hot-spot volcanoes sutured together.

protolith The original rock from which a metamorphic rock formed.

protoplanet A body that grows by the accumulation of planetesimals but has not yet become big enough to be called a planet.

protoplanetary nebula A ring of gas and dust that surrounded the newborn Sun, from which the planets were formed.

protostar A dense body of gas that is collapsing inward because of gravitational forces and that may eventually become a star.

pumice A glassy igneous rock that forms from felsic frothy lava and contains abundant (over 50%) pore space.

punctuated equilibrium The hypothesis that evolution takes place in fits and starts; evolution occurs very slowly for quite a while and then, during a relatively short period, takes place very rapidly.

P-waves Compressional seismic waves that move through the body of the Earth.

P-wave shadow zone A band between 103° and 143° from an earthquake epicenter, as measured along the circumference of the Earth, inside which P-waves do not arrive at seismograph stations.

pycnocline The boundary between layers of water of different densities.

pyroclastic debris Fragmented material that sprayed out of a volcano and landed on the ground or sea floor in solid form.

pyroclastic flow A fast-moving avalanche that occurs when hot volcanic ash and debris mix with air and flow down the side of a volcano.

pyroclastic rock Rock made from fragments blown out of a volcano during an explosion that were then packed or welded together.

quarry A site at which stone is extracted from the ground.

quartzite A metamorphic rock composed of quartz and transformed from a protolith of quartz sandstone.

quenching A sudden cooling of molten material to form a solid.

quick clay Clay that behaves like a solid when still (because of surface tension holding the water-coated clay flakes together), but that flows like a liquid when shaken.

radial network A drainage network in which the streams flow outward from a cone-shaped mountain, and define a pattern resembling spokes on a wheel.

radioactive decay The process by which a radioactive atom undergoes fission or releases particles thereby transforming into a new element.

radioactive isotope An unstable isotope of a given element.

radiometric dating The science of dating geologic events in years by measuring the ratio of parent radioactive atoms to daughter product atoms.

rain band A spiraling arm of a hurricane radiating outward from the eye.

rain shadow The inland side of a mountain range, which is arid because the mountains block rain clouds from reaching the area.

range (for fossils) The interval of a sequence of strata in which a specific fossil species appears.

rapids A reach of a stream in which water becomes particularly turbulent; as a consequence, waves develop on the surface of the stream.

reach A specified segment of a stream's path.

recessional moraine The end moraine that forms when a glacier stalls for a while as it recedes.

recharge area A location where water enters the ground and infiltrates down to the water table.

recrystallization The process in which ions or atoms in minerals rearrange to form new minerals.

rectangular network A drainage network in which the streams join each other at right angles because of a rectangular grid of fractures that breaks up the ground and localizes channels.

recurrence interval The average time between successive geologic events.

red giant A huge red star that forms when Sun-sized stars start to die and expand.

red shift The phenomenon in which a source of light moving away from you very rapidly shifts to a lower frequency; that is, toward the red end of the spectrum.

reef bleaching The death and loss of color of a coral reef.

reflected ray A ray that bounces off a boundary between two different materials.

refracted ray A ray that bends as it passes through a boundary between two different materials.

refraction The bending of a ray as it passes through a boundary between two different materials.

reg A vast stony plain in a desert.

regional metamorphism *Dynamothermal metamorphism;* metamorphism of a broad region, usually the result of deep burial during an orogeny.

regolith Any kind of unconsolidated debris that covers bedrock.

regression The seaward migration of a shoreline caused by a lowering of sea level.

relative age The age of one geologic feature with respect to another.

relative humidity The ratio between the measured water content of air and the maximum possible amount of water the air can hold at a given condition.

relative plate velocity The movement of one lithosphere plate with respect to another.

relief The difference in elevation between adjacent high and low regions on the land surface.

renewable resource A resource that can be replaced by nature within a short time span relative to a human life span.

reservoir rock Rock with high porosity and permeability, so it can contain an abundant amount of easily accessible oil.

residence time The average length of time that a substance stays in a particular reservoir.

residual mineral deposit Soils in which the residuum left behind after leaching by rainwater is so concentrated in metals that the soil itself becomes an ore deposit.

resurgent dome The new mound, or cone, of igneous rock that grows within a caldera as an eruption begins anew.

retrograde metamorphism Metamorphism that occurs as pressures and temperatures are decreasing; for retrograde metamorphism to occur, water must be added.

return stroke An upward-flowing electric current from the ground that carries positive charges up to a cloud during a lightning flash.

reversed polarity Polarity in which the paleomagnetic dipole points north.

reverse fault A steeply dipping fault on which the hanging-wall block slides up.

Richter magnitude scale A scale that defines earthquakes on the basis of the amplitude of the largest ground motion recorded on a seismogram.

ridge axis The crest of a mid-ocean ridge; the ridge axis defines the position of a divergent plate boundary.

right-lateral strike-slip fault A strike-slip fault in which the block on the opposite fault plane from a fixed spot moves to the right of that spot.

rip current A strong, localized seaward flow of water perpendicular to a beach.

riprap Loose boulders or concrete piled together along a beach to absorb wave energy before it strikes a cliff face.

roche moutonnée A glacially eroded hill that becomes elongate in the direction of flow and asymmetric; glacial rasping smoothes the upstream part of the hill into a gentle slope, while glacial plucking erodes the downstream edge into a steep slope.

rock A coherent, naturally occurring solid, consisting of an aggregate of minerals or a mass of glass.

rock burst A sudden explosion of rock off the ceiling or wall of an underground mine.

rock cycle The succession of events that results in the transformation of Earth materials from one rock type to another, then another, and so on.

rock flour Fine-grained sediment produced by glacial abrasion of the substrate over which a glacier flows.

rock glacier A slow-moving mixture of rock fragments and ice.

rock slide A sudden downslope movement of rock.

rocky coast An area of coast where bedrock rises directly from the sea, so beaches are absent.

Rodinia A proposed Precambrian supercontinent that existed around 1 billion years ago.

rotational axis The imaginary line through the center of the Earth around which the Earth spins.

R-waves (*Rayleigh waves*) Surface seismic waves that cause the ground to ripple up and down, like water waves in a pond.

sabkah A region of formerly flooded coastal desert in which stranded seawater has left a salt crust over a mire of mud that is rich in organic material.

salinity The degree of concentration of salt in water.

saltation The movement of a sediment in which grains bounce along their substrate, knocking other grains into the water column (or air) in the process.

salt dome A rising bulbous dome of salt that bends up the adjacent layers of sedimentary rock.

salt wedging The process in arid climates by which dissolved salt in groundwater crystallizes and grows in open pore spaces in rocks and pushes apart the surrounding grains.

sand spit An area where the beach stretches out into open water across the mouth of a bay or estuary.

sandstone Coarse-grained sedimentary rock consisting almost entirely of quartz.

sand volcano (or sand blow) A small mound of sand produced when sand layers below the ground surface liquify as a result of seismic shaking, causing the sand to erupt onto the Earth's surface through cracks or holes in overlying clay layers.

saprolite A layer of rotten rock created by chemical weathering in warm, wet climates.

Sargasso Sea The center of North Atlantic Gyre, named for the tropical seaweed sargassum, which accumulates in its relatively noncirculating waters.

saturated solution Water that carries as many dissolved ions as possible under given environmental conditions.

saturated zone The region below the water table where pore space is filled with water.

scattering The dispersal of energy that occurs when light interacts with particles in the atmosphere.

schist A medium-to-coarse-grained metamorphic rock that possesses schistosity.

schistosity Foliation caused by the preferred orientation of large mica flakes.

scientific method A sequence of steps for systematically analyzing scientific problems in a way that leads to verifiable results.

scoria A glassy mafic igneous rock containing abundant air-filled holes.

scouring A process by which running water removes loose fragments of sediment from a stream bed.

sea arch An arch of land protruding into the sea and connected to the mainland by a narrow bridge.

sea-floor spreading The gradual widening of an ocean basin as new oceanic crust forms at a mid-ocean ridge axis and then moves away from the axis.

sea ice Ice formed by the freezing of the surface of the sea.

seal A relatively impermeable rock, such as shale, salt, or unfractured limestone, that lies above a reservoir rock and stops the oil from rising further.

seam A sedimentary bed of coal interlayered with other sedimentary rocks.

seamount An isolated submarine mountain.

seasonal floods Floods that appear almost every year during seasons when rainfall is heavy or when winter snows start to melt.

seasonal well A well that provides water only during the rainy season when the water table rises below the base of the well.

sea stack An isolated tower of land just offshore, disconnected from the mainland by the collapse of a sea arch.

seawall A wall of riprap built on the landward side of a backshore zone in order to protect shore cliffs from erosion.

second The basic unit of time measurement, now defined as the time it takes for the magnetic field of a cesium atom to flip polarity 9,192,631,770 times, as measured by an atomic clock.

secondary enrichment The process by which a new ore deposit forms from metals that were dissolved and carried away from preexisting ore minerals.

secondary porosity New pore space in rocks, created some time after a rock first forms.

secondary recovery technique A process used to extract the quantities of oil that will not come out of a reservoir rock with just simple pumping.

sediment An accumulation of loose mineral grains, such as boulders, pebbles, sand, silt, or mud, that are not cemented together.

sedimentary basin A depression, created as a consequence of subsidence, that fills with sediment.

sedimentary rock Rock that forms either by the cementing together of fragments broken off preexisting rock or by the precipitation of mineral crystals out of water solutions at or near the Earth's surface.

sedimentary sequence A grouping of sedimentary units bounded on top and bottom by regional unconformities.

sediment budget The proportion of sand supplied to sand removed from a depositional setting.

sediment load The total volume of sediment carried by a stream.

sediment maturity The degree to which a sediment has evolved from a crushed-up version of the original rock into a sediment that has lost its easily weathered minerals and become well sorted and rounded.

sediment sorting The segregation of sediment by size.

seep A place where oil-filled reservoir rock intersects the ground surface, or where fractures connect a reservoir to the ground surface, so that oil flows out onto the ground on its own.

seiche Rhythmic movement in a body of water caused by ground motion.

seismic belts (or **zones)** The relatively narrow strips of crust on Earth under which most earthquakes occur.

seismicity Earthquake activity.

seismic-moment magnitude scale A scale that defines earthquake size on the basis of calculations involving the amount of slip, length of rupture, depth of rupture, and rock strength.

seismic ray The changing position of an imaginary point on a wave front as the front moves through rock.

seismic-reflection profile A cross-sectional view of the crust made by measuring the reflection of artificial seismic waves off boundaries between different layers of rock in the crust.

seismic tomography Analysis by sophisticated computers of global seismic data in order to create a three-dimensional image of variations in seismic-wave velocities within the Earth.

seismic velocity The speed at which seismic waves travel.

seismic-velocity discontinuity A boundary in the Earth at which seismic velocity changes abruptly.

seismic (earthquake) waves Waves of energy emitted at the focus of an earthquake.

seismogram The record of an earthquake produced by a seismograph.

seismograph (seismometer) An instrument that can record the ground motion from an earthquake.

semipermanent pressure cell A somewhat elliptical zone of high or low atmospheric pressure that lasts much of the year; it forms because high-pressure zones tend to be narrower over land than over sea.

Sevier orogeny A mountain-building event that affected western North America between about 150 Ma and 80 Ma, a result of convergent margin tectonism; a fold-thrust belt formed during this event.

shale Very fine-grained sedimentary rock that breaks into thin sheets.

shatter cones Small, cone-shaped fractures formed by the shock of a meteorite impact.

shear strain A change in shape of an object that involves the movement of one part of a rock body sideways past another part so that angular relationships within the body change.

shear stress A stress that moves one part of a material sideways past another part.

shear waves Seismic waves in which particles of material move back and forth perpendicular to the direction in which the wave itself moves.

shear zone A fault in which movement has occurred ductilely.

sheetwash A film of water less than a few mm thick that covers the ground surface during heavy rains.

shield An older, interior region of a continent.

shield volcano A subaerial volcano with a broad, gentle dome, formed either from low-viscosity basaltic lava or from large pyroclastic sheets.

shocked quartz Grains of quartz that have been subjected to intense pressure such as occurs during a meteorite impact.

shoreline The boundary between the water and land.

shortening The process during which a body of rock or a region of crust becomes shorter.

short-term climate change Climate change that takes place over hundreds to thousands of years.

Sierran arc A large continental volcanic arc along western North America that initiated at the end of the Jurassic Period and lasted until about 80 million years ago.

silica SiO_2.

silicate minerals Minerals composed of silicon-oxygen tetrahedra linked in various arrangements; most contain other elements too.

silicate rock Rock composed of silicate minerals.

siliceous sedimentary rock Sedimentary rock that contains abundant quartz.

silicic Rich in silica with relatively little iron and magnesium.

sill A nearly horizontal table-top-shaped tabular intrusion that occurs between the layers of country rock.

siltstone Fine-grained sedimentary rock generally composed of very small quartz grains.

sinkhole A circular depression in the land that forms when an underground cavern collapses.

slab-pull force The force that downgoing plates (or slabs) apply to oceanic lithosphere at a convergent margin.

slate Fine-grained, low-grade metamorphic rock, formed by the metamorphism of shale.

slaty cleavage The foliation typical of slate, and reflective of the preferred orientation of slate's clay minerals, that allows slate to be split into thin sheets.

slickensides The polished surface of a fault caused by slip on the fault; lineated slickensides also have groves that indicate the direction of fault movement.

slip face The lee side of a dune, which sand slides down.

slip lineations Linear marks on a fault surface created during movement on the fault; some slip lineations are defined by grooves, some by aligned mineral fibers.

slope failure The downslope movement of material on an unstable slope.

slumping Downslope movement in which a mass of regolith detaches from its substrate along a spoon-shaped sliding surface and slips downward semicoherently.

smelting The heating of a metal-containing rock to high temperatures in a fire so that the rock will decompose to yield metal plus a nonmetallic residue (slag).

snotite A long gob of bacteria that slowly drips from the ceiling of a cave.

snow line The boundary above which snow remains all year.

soda straw A hollow stalactite in which calcite precipitates around the outside of a drip.

soil Sediment that has undergone changes at the surface of the Earth, including reaction with rainwater and the addition of organic material.

soil erosion The removal of soil by wind and runoff.

soil horizon Distinct zones within a soil, distinguished from each other by factors such as chemical composition and organic content.

soil moisture Underground water that wets the surface of the mineral grains and organic material making up soil, but lies above the water table.

soil profile   A vertical sequence of distinct zones of soil.

solar wind   A stream of particles with enough energy to escape from the Sun's gravity and flow outward into space.

solid-state diffusion   The slow movement of atoms or ions through a solid.

solifluction   The type of creep characteristic of tundra regions; during the summer, the uppermost layer of permafrost melts, and the soggy, weak layer of ground then flows slowly downslope in overlapping sheets.

solstice   A day on which the polar ends of the terminator (the boundary between the day hemisphere and the night hemisphere) lie 23.5° away from the associated geographic poles.

Sonoma orogeny   A convergent-margin mountain-building event that took place on the western coast of North America in the Late Permian and Early Triassic periods.

sorting   (1) The range of clast sizes in a collection of sediment; (2) the degree to which sediment has been separated by flowing currents into different-size fractions.

source rock   A rock (organic-rich shale) containing the raw materials from which hydrocarbons eventually form.

southeast tradewinds   Tradewinds in the Southern Hemisphere, which start flowing northward, deflect to the west, and end up flowing from southeast to northwest.

southern oscillation   The oscillating of atmospheric pressure cells back and forth across the Pacific Ocean, in association with El Niño.

specific gravity   A number representing the density of a mineral, as specified by the ratio between the weight of a volume of the mineral and the weight of an equal volume of water.

speleothem   A formation that grows in a limestone cave by the accumulation of travertine precipitated from water solutions dripping in a cave or flowing down the wall of a cave.

sphericity   The measure of the degree to which a clast approaches the shape of a sphere.

spreading boundary   *Divergent plate boundary.*

spreading rate   The rate at which sea floor moves away from a mid-ocean ridge axis, as measured with respect to the sea floor on the opposite side of the axis.

spring   A natural outlet from which groundwater flows up onto the ground surface.

spring tide   An especially high tide that occurs when the Sun is on the same side of the Earth as the Moon.

stable air   Air that does not have a tendency to rise rapidly.

stable slope   A slope on which downward sliding is unlikely.

stalactite   An icicle-like cone that grows from the ceiling of a cave as dripping water precipitates limestone.

stalagmite   An upward-pointing cone of limestone that grows when drips of water hit the floor of a cave.

standing wave   A wave whose crest and trough remain in place as water moves through the wave.

star dune   A constantly changing dune formed by frequent shifts in wind direction; it consists of overlapping crescent dunes pointing in many different directions.

stick-slip behavior   Stop-start movement along a fault plane caused by friction, which prevents movement until stress builds up sufficiently.

stone rings   Ridges of cobbles between adjacent bulges of permafrost ground.

stoping   A process by which magma intrudes; blocks of wall rock break off and then sink into the magma.

storm   An episode of severe weather in which winds, precipitation, and in some cases lightning become strong enough to be bothersome and even dangerous.

storm-center velocity   A storm's (hurricane's) velocity along its track.

storm surge   Excess seawater driven landward by wind during a storm; the low atmospheric pressure beneath the storm allows sea level to rise locally, increasing the surge.

strain   The change in shape of an object in response to deformation (i.e., as a result of the application of a stress).

stratified drift   Glacial sediment that has been redistributed and stratified by flowing water.

stratigraphic column   A cross-section diagram of a sequence of strata summarizing information about the sequence.

stratigraphic formation   A recognizable layer of a specific sedimentary rock type or set of rock types, deposited during a certain time interval, that can be traced over a broad region.

stratigraphic sequence   An interval of strata deposited during periods of relatively high sea level, and bounded above and below by regional unconformities.

stratopause   The temperature pause that marks the top of the stratosphere.

stratosphere   The stable, stratified layer of atmosphere directly above the troposphere.

stratovolcano   A large, cone-shaped subaerial volcano consisting of alternating layers of lava and tephra.

stratus cloud   A thin, sheet-like, stable cloud.

streak   The color of the powder produced by pulverizing a mineral on an unglazed ceramic plate.

stream   A ribbon of water that flows in a channel.

stream bed   The floor of a stream.

stream capacity   The total quantity of sediment a stream carries.

stream capture (or piracy)   The situation in which headward erosion causes one stream to intersect the course of another, previously independent stream, so that the intersected stream starts to flow down the channel of the first stream.

stream competence   The maximum particle size that a stream can carry.

stream gradient   The slope of a stream's channel in the downstream direction.

stream rejuvination   The renewed downcutting of a stream into a floodplain or peneplain, caused by a relative drop of the base level.

stress   The push, pull, or shear that a material feels when subjected to a force; formally, the force applied per unit area over which the force acts.

stretching   The process during which a layer of rock or a region of crust becomes longer.

striations   Linear scratches in rock.

strike-slip fault   A fault in which one block slides horizontally past another (and therefore parallel to the strike line), so there is no relative vertical motion.

strip mining   The scraping off of all soil and sedimentary rock above a coal seam in order to gain access to the seam.

stromatolite   Layered mounds of sediment formed by cyanobacteria; cyanobacteria secrete a mucuous-like substance to which sediment sticks, and as each layer of cyanobacteria gets buried by sediment, it colonizes the surface of the new sediment, building a mound upward.

structural control   The condition in which geologic structures, such as faults, affect the distribution and drainage of water or the shape of the land surface.

subaerial   Pertaining to land regions above sea level (i.e., under air).

subduction   The process by which one oceanic plate bends and sinks down into the asthenosphere beneath another plate.

subduction zone The region along a convergent boundary where one plate sinks beneath another.

sublimation The evaporation of ice directly into vapor without first forming a liquid.

submarine canyon A narrow, steep canyon that dissects a continental shelf and slope.

submarine fan A wedge-shaped accumulation of sediment at the base of a submarine slope; fans usually accumulate at the mouth of a submarine canyon.

submarine slump The underwater downslope movement of a semicoherent block of sediment along a weak mud detachment.

submergent coast A coast at which the land is sinking relative to sea level.

subpolar low The rise of air where the surface flow of a polar cell converges with the surface flow of a Ferrel cell, creating a low-pressure zone in the atmosphere.

subsidence The vertical sinking of the Earth's surface in a region, relative to a reference plane.

substrate A general term for material just below the ground surface.

subtropical high (subtropical divergence zone) A belt of high pressure in the atmosphere at 30° latitude formed where the Hadley cell converges with the Ferrel cell, causing cool, dense air to sink.

subtropics Desert climate regions that lie on either side of the equatorial tropics between the lines of 20° and 30° north or south of the equator.

summit eruption An eruption that occurs in the summit crater of a volcano.

sunspot cycle The cyclic appearance of large numbers of sunspots (black spots thought to be magnetic storms on the Sun's surface) every 9 to 11.5 years.

supercontinent cycle The process of change during which supercontinents develop and later break apart, forming pieces that may merge once again in geologic time to make yet another supercontinent.

supernova A short-lived, very bright object in space that results from the cataclysmic explosion marking the death of a very large star; the explosion ejects large quantities of matter into space to form new nebulae.

superplume A huge mantle plume.

superposed stream A stream whose geometry has been laid down on a rock structure and is not controlled by the structure.

surface current An ocean current in the top 100 m of water.

surface waves Seismic waves that travel along the Earth's surface.

surface westerlies The prevailing surface winds in North America and Europe, which come out of the west or southwest.

surf zone A region of the shore in which breakers crash onto the shore.

surge (glacial) A pulse of rapid flow in a glacier.

suspended load Tiny solid grains carried along by a stream without settling to the floor of the channel.

swamp A wetland dominated by trees.

swash The upward surge of water that flows up a beach slope when breakers crash onto the shore.

S-waves Seismic shear waves that pass through the body of the Earth.

S-wave shadow zone A band between 103° and 180° from the epicenter of an earthquake inside of which S-waves do not arrive at seismograph stations.

swelling clay Clay possessing a mineral structure that allows it to absorb water between its layers and thus swell to several times its original size.

symmetry The condition in which the shape of one part of an object is a mirror image of the other part.

syncline A trough-shaped fold whose limbs dip toward the hinge.

systematic joints Long planar cracks that occur fairly regularly throughout a rock body.

tabular intrusions Sheet intrusions that are planar and of roughly uniform thickness.

Taconic orogeny A convergent mountain-building event that took place around 400 million years ago, in which a volcanic island arc collided with eastern North America.

tailings pile A pile of waste rock from a mine.

talus A sloping apron of fallen rock along the base of a cliff.

tar Hydrocarbons that exist in solid form at room temperature.

tarn A lake that forms at the base of a cirque on a glacially eroded mountain.

tar sand Sandstone reservoir rock in which less viscous oil and gas molecules have either escaped or been eaten by microbes, so that only tar remains.

taxonomy The study and classification of the relationships among different forms of life.

tension A stress that pulls on a material and could lead to stretching.

tephra Unconsolidated accumulations of pyroclastic grains.

terminal moraine The end moraine at the farthest limit of glaciation.

terminator The boundary between the half of the Earth that has daylight and the half experiencing night.

terrace The elevated surface of an older floodplain into which a younger floodplain had cut down.

terrestrial A term used to describe the inner, Earth-like planets.

thalweg The deepest part of a stream's channel.

theory A scientific idea supported by an abundance of evidence that has passed many tests and failed none.

theory of plate tectonics The theory that the outer layer of the Earth (the lithosphere) consists of separate plates that move with respect to one another.

thermal metamorphism Metamorphism caused by heat conducted into country rock from an igneous intrusion.

thermocline A boundary between layers of water with differing temperatures.

thermohaline circulation The rising and sinking of water driven by contrasts in water density, which is due in turn to differences in temperature and salinity; this circulation involves both surface and deep-water currents in the ocean.

thermosphere The outermost layer of the atmosphere containing very little gas.

thin section A 3/100-mm-thick slice of rock that can be examined with a petrographic microscope.

thin-skinned deformation A distinctive style of deformation characterized by displacement on faults that terminate at depth along a subhorizontal detachment fault.

thrust fault A gently dipping reverse fault; the hanging-wall block moves up the slope of the fault.

tidal bore A visible wall of water that moves toward shore with the rising tide in quiet waters.

tidal flat A broad, nearly horizontal plain of mud and silt, exposed or nearly exposed at low tide but totally submerged at high tide.

tidal reach The difference in sea level between high tide and low tide at a given point.

tide The daily rising or falling of sea level at a given point on the Earth.

tide-generating force The force, caused in part by the gravitational attraction of the Sun and Moon, and in part by the centrifugal force created by the Earth's spin, that generates tides.

till A mixture of unsorted mud, sand, pebbles, and larger rocks deposited by glaciers.

tillite A rock formed from hardened ancient glacial deposits and consisting of larger clasts distributed through a matrix of sandstone and mudstone.

toe (terminus) The leading edge or margin of a glacier.

tombolo A narrow ridge of sand that links a sea stack to the mainland.

topographical map A map that uses contour lines to represent variations in elevation.

topography Variations in elevation.

topsoil The top soil horizons, which are typically dark and nutrient-rich.

tornado A near-vertical, funnel-shaped cloud in which air rotates extremely rapidly around the axis of the funnel.

tornado swarm Dozens of tornadoes produced by the same storm.

tower karst A karst landscape in which steep-sided residual bedrock towers remain between sinkholes.

transform fault A fault marking a transform plate boundary; along mid-ocean ridges, transform faults are the actively slipping segment of a fracture zone between two ridge segments.

transform plate boundary A boundary at which one lithosphere plate slips laterally past another.

transgression The inland migration of shoreline resulting from a rise in sea level.

transition zone The middle portion of the mantle, from 400 to 670 km deep, in which there are several jumps in seismic velocity.

transpiration The release of moisture as a metabolic byproduct.

transverse dune A simple, wave-like dune that appears when enough sand accumulates for the ground surface to be completely buried, but only moderate winds blow.

travel-time curve A graph that plots the time since an earthquake began on the vertical axis, and the distance to the epicenter on the horizontal axis.

trellis network A drainage system that develops across a landscape of parallel valleys and ridges so that major tributaries flow down the valleys and join a trunk stream that cuts through the ridge; the resulting map pattern resembles a garden trellis.

trench A deep elongate trough bordering a volcanic arc; a trench defines the trace of a convergent plate boundary.

triangulation The method for determining the map location of a point from knowing the distance between that point and three other points; this method is used to locate earthquake epicenters.

tributary A smaller stream that flows into a larger stream.

triple junction A point where three lithosphere plate boundaries intersect.

tropical depression A tropical storm with winds reaching up to 61 km per hour; such storms develop from tropical disturbances, and may grow to become hurricanes.

tropical disturbance Cyclonic winds that develop in the tropics.

tropopause The temperature pause marking the top of the troposphere.

troposphere The lowest layer of the atmosphere, where air undergoes convection and where most wind and clouds develop.

truncated spur A spur (elongate ridge between two valleys) whose end was eroded off by a glacier.

trunk stream The single larger stream into which an array of tributaries flow.

tsunami A large wave along the sea surface triggered by an earthquake or large submarine slump.

tuff A pyroclastic igneous rock composed of volcanic ash and fragmented pumice, formed when accumulations of the debris cement together.

tundra A cold, treeless region of land at high latitudes, supporting only species of shrubs, moss, and lichen capable of living on permafrost.

turbidite A graded bed of sediment built up at the base of a submarine slope and deposited by turbidity currents.

turbidity current A submarine avalanche of sediment and water that speeds down a submarine slope.

turbulence The chaotic twisting, swirling motion in flowing fluid.

typhoon The equivalent of a hurricane in the western Pacific Ocean.

ultimate base level Sea level; the level below which a trunk stream cannot cut.

ultramafic A term used to describe igneous rocks or magmas that are rich in iron and magnesium and very poor in silica.

unconfined aquifer An aquifer that intersects the surface of the Earth.

unconformity A boundary between two different rock sequences representing an interval of time during which new strata were not deposited and/or were eroded.

unconsolidated Consisting of unattached grains.

undercutting Excavation at the base of a slope that results in the formation of an overhang.

undersaturated A term used to describe a solution capable of holding more dissolved ions.

unsaturated zone The region of the subsurface above the water table.

unstable air Air that is significantly warmer than air above and has a tendency to rise quickly.

unstable ground Land capable of slumping or slipping down-slope in the near future.

unstable slope A slope on which sliding will likely happen.

updraft Upward-moving air.

upper mantle The uppermost section of the mantle, reaching down to a depth of 400 km.

upwelling zone A place where deep water rises in the ocean, or hot magma rises in the asthenosphere.

U-shaped valley A steep-walled valley shaped by glacial erosion into the form of a U.

vacuum Space that contains very little matter in a given volume (e.g., a region in which air has been removed).

valley A trough with sloping walls, cut into the land by a stream.

valley glacier A river of ice that flows down a mountain valley.

Van Allen radiation belts Belts of solar wind particles and cosmic rays that surround the Earth, trapped by Earth's magnetic field.

varve A pair of thin layers of glacial lake-bed sediment, one consisting of silt brought in during the spring floods, and the other of clay deposited during the winter when the lake's surface freezes over and the water is still.

vascular plant A plant with woody tissue and seeds and veins for transporting water and food.

vein A seam of minerals that forms when dissolved ions carried by water solutions precipitate in cracks.

vein deposit A hydrothermal deposit in which the ore minerals occur in veins that fill cracks in preexisting rocks.

velocity-versus-depth curve A graph that shows the variation in the velocity of seismic waves with increasing depth in the Earth.

ventifact (faceted rock) A desert rock whose surface has been faceted by the wind.

vesicles Open holes in igneous rock formed by the preservation of bubbles in magma as the magma cools into solid rock.

viscosity The resistance of material to flow.

volatiles Elements or compounds such as H_2O and CO_2 that evaporate easily and can exist in gaseous forms at the Earth's surface.

volatility A specification of the ease with which a material evaporates.

volcanic arc A curving chain of active volcanoes formed adjacent to a convergent plate boundary.

volcanic ash Tiny glass shards formed when a fine spray of exploded lava freezes instantly upon contact with the atmosphere.

volcanic bomb A large piece of pyroclastic debris thrown into the atmosphere during a volcanic eruption.

volcanic-danger-assessment map A map delineating areas that lie in the path of potential lava flows, lahars, debris flows, or pyroclastic flows of an active volcano.

volcanic gas Elements or compounds that bubble out of magma or lava in gaseous form.

volcanic island arc The volcanic island chain that forms on the edge of the overriding plate where one oceanic plate subducts beneath another oceanic plate.

volcano (1) A vent from which melt from inside the Earth spews out onto the planet's surface; (2) a mountain formed by the accumulation of extrusive volcanic rock.

V-shaped valley A valley whose cross-sectional shape resembles the shape of a V; the valley probably has a river running down the point of the V.

Wadati-Benioff zone A sloping band of seismicity defined by intermediate- and deep-focus earthquakes that occur in the downgoing slab of a convergent plate boundary.

wadi The name used in the Middle East and North Africa for a dry wash.

warm front A front in which warm air rises slowly over cooler air in the atmosphere.

waste rock Rock dislodged by mining activity yet containing no ore minerals.

waterfall A place where water drops over an escarpment.

water gap An opening in a resistant ridge where a trunk river has cut through the ridge.

watershed The region that collects water that feeds into a given drainage network.

water table The boundary, approximately parallel to the Earth's surface, that separates substrate in which groundwater fills the pores from substrate in which air fills the pores.

wave base The depth, approximately equal in distance to half a wavelength in a body of water, beneath which there is no wave movement.

wave-cut bench A platform of rock, cut by wave erosion, at the low-tide line that was left behind a retreating cliff.

wave-cut notch A notch in a coastal cliff cut out by wave erosion.

wave erosion The combined effects of the shattering, wedging, and abrading of a cliff face by waves and the sediment they carry.

wave front The boundary between the region through which a wave has passed and the region through which it has not yet passed.

wavelength The horizontal difference between two adjacent wave troughs or two adjacent crests.

wave refraction (ocean) The bending of waves as they approach a shore so that their crests make no more than a 5° angle with the shoreline.

weather Local-scale conditions as defined by temperature, air pressure, relative humidity, and wind speed.

weathered rock Rock that has reacted with air and/or water at or near the Earth's surface.

weathering The processes that break up and corrode solid rock, eventually transforming it into sediment.

weather system A specific set of weather conditions, reflecting the configuration of air movement in the atmosphere, that affects a region for a period of time.

welded tuff Tuff formed by the welding together of hot volcanic glass shards at the base of pyroclastic flows.

well A hole in the ground dug or drilled in order to obtain water.

Western Interior Seaway A north-south-trending seaway that ran down the middle of North America during the Late Cretaceous Period.

wet-bottom (temperate) glacier A glacier with a thin layer of water at its base, over which the glacier slides.

wetted perimeter The area in which water touches a stream channel's walls.

wind abrasion The grinding away at surfaces in a desert by windblown sand and dust.

wind gap An opening through a high ridge that developed earlier in geologic history by stream erosion, but that is now dry.

xenolith A relict of wall rock surrounded by intrusive rock when the intrusive rock freezes.

yardang A mushroom-like column with a resistant rock perched on an eroding column of softer rock; created by wind abrasion in deserts where a resistant rock overlies softer layers of rock.

yazoo stream A small tributary that runs parallel to the main river in a floodplain because the tributary is blocked from entering the main river by levees.

Younger Dryas An interval of cooler temperatures that took place 4,500 years ago during a general warming/glacier-retreat period.

zeolite facies The metamorphic facies just above diagenetic conditions, under which zeolite minerals form.

zone of ablation The area of a glacier in which ablation (melting, sublimation, calving) subtracts from the glacier.

zone of accumulation (1) The layer of regolith in which new minerals precipitate out of water passing through, thus leaving behind a load of fine clay; (2) the area of a glacier in which snowfall adds to the glacier.

zone of aeration *Unsaturated zone.*

zone of leaching The layer of regolith in which water dissolves ions and picks up very fine clay; these materials are then carried downward by infiltrating water.

Credits

UNNUMBERED PHOTOS AND ART

ii: © Mark Laricchia/CORBIS; **1**: Stephen Marshak; **12–13**: ImageState/Alamy; **14**: R. Williams (STScI), the Hubble Deep Field Team and NASA; **36**: Stephen Marshak; **48**: Fred Espenak, Photo Researchers, Inc.; **56**: Courtesy of Thomas N. Taylor; **72–73**: original artwork by Gary Hincks; **85**: U.S. Geological Survey; **117**: © 2006 Christoph Hormann; **118–19**: Robert Essel, NYC/Corbis; **120**: Javier Trueba/MSF/Photo Researchers, Inc.; **144**: Hal Lott/Corbis; **152**: Visuals Unlimited; **176–77**: original artwork by Gary Hincks; **182**: Tony Linck/SuperStock; **183**: William Manning/Corbis; **216–17**: original artwork by Gary Hincks; **228, 246**: Stephen Marshak; **250–51**: original artwork by Gary Hincks; **264–65**: National Geographic/Getty Images; **266**: U.S. Geological Survey; **278–79**: original artwork by Gary Hincks; **303**: Anjum Naveed/AP Images; **349**: Jacques Descloitres, MODIS Rapid Response Team, NASA/GSFC/Visible Earth; **362**: Stephen Marshak; **392–93**: original artwork by Gary Hincks; **398**: NationalAtlas.gov/USGS; **399**: U.S. Geological Survey; **400–01**: Stephen Marshak; **414, 415**: Stephen Marshak; **426–27**: original artwork by Gary Hincks; **447**: Images courtesy of USGS National Center for EROS and NASA Landsat Project Science office; **448**: Stephen Marshak; **476–77**: original artwork by Gary Hincks; **483**: Jacques Descloitres, MODIS Rapid Response Team, NASA/GSFC/Visible Earth; **484–85**: Bettmann/Corbis; **486**: Mario Tama/Getty Images; **514–15**: original artwork by Gary Hincks; **522**: Stephen Marshak; **528–29**: original artwork by Gary Hincks; **542**: Stephen Marshak; **544–45**: Stephen Marshak; **546**: Stephen Marshak; **554–55**: original artwork by Gary Hincks; **559**: Rob Varela/Ventura Country Star/Corbis; **574–75**: original artwork by Gary Hincks; **583**: NASA/Calvin J. Hamilton; **584**: Stephen Marshak; **610–11**: original artwork by Gary Hincks; **622**: Stephen Marshak; **654–55**: original artwork by Gary Hincks; **662**: Yann Arthus–Bertrand/Corbis; **684–85**: original artwork by Gary Hincks; **691**: Courtesy of Parris Lyew–Ayee of Mona Geo Informatics Institute, plyewayee@monainformatixltd.com; **694**: SVS/NASA; **731**: NASA/GSFC, USGS, Landsat7; **732**: Stephen Marshak; **750–51**: original artwork by Gary Hincks; **758**: U.S. Geological Survey; **759**: Stephen Marshak; **780–81**: original artwork by Gary Hincks; **802**: U. S. Geological Survey; **803**: Stephen Marshak; **808–09**: original artwork by Gary Hincks.

NUMBERED PHOTOS AND ART

Prelude: **P.1B**: Stephen Marshak; **P.1C**: Stephen Marshak; **P.2**: Stephen Marshak; **P.3**: AP Images; **P.7**: Courtesy Peter Fiske.

Chapter 1: **1.1**: Richard E. Hill/Visuals Unlimited; **1.3 A–B**: Rare Books Division, The New York Public Library, Astor, Lenox and Tilden Foundations; **1.5**: NASA; **1.9**: Moonrunner Design; **1.10**: J. Hester and P. Scowen/NASA; **1.11**: NASA;

1.12: J. Hester and P. Scowen/NASA; **1.14A**: NASA; **1.14A**: NASA; **1.14B**: Photograph by Pelisson, SaharaMet; **1.15**: NASA/JPL/Cal Tech; **1.16B**: NASA/NSSDC/GSFC.

Chapter 2: **2.1**: JPL/NASA; **2.4A**: NASA/Photo Researchers, Inc.; **2.5A**: Johnson Space Center/NASA; **2.10**: Tate Gallery, London/Art Resource, NY;

Chapter 3: **3.1**: Alfred Wegener Institute for Polar and Marine Research; **3.2**: Wegener A., *The Origin of the Continents and Oceans* (New York: Dover, 1966; trans. From 1929 German ed.); **3.4**: Modified from American Association of Petroleum Geologists; **3.6A**: Modified from Hurley; **3.11**: Rothé, J.R. 1954. La zone séismique median Indo–Atlantique: Proceedings of the Royal Society of London, Series A, V.222, p. 388; **3.13A–B**: Modified from Mason, 1955; **3.17**: Modified from Cox, Dalrymple, and Doell, in Hamblin and Christiansen, 1998; **3.19**: Modified from Cox, Dalrymple, and Doell.

Interlude A: **A.1**: Johnson Space Center/NASA

Chapter 4: **4.9**: University of Texas Institute of Geophysics; **4.10**: Photo by Dudley Foster, Woods Hole Oceanographic Institute; **4.18B**: R.E. Wallace (228), U.S. Geological Survey; **4.24B**: Johnson Space Center/NASA; **4.29**: Modified from Cox and Hardt, 1986; **4.31**: Michael Heflin/Cal Tech; **4.32**: Chris Scotese.

Chapter 5: **5.1**: © Crown copyright. Historic Royal Palaces; **5.2A**: Ken Lucas/Visuals Unlimited; **5.2B**: Stephen Marshak; **5.3A**: © Jay Schomer; **5.3B**: Wally McNamee/Corbis; **5.4A**: Modified from Wicander and Monroe; **5.4B**: Modified from Wicander and Monroe; **5.7B**: Charles O'Rear/Corbis; **5.7D**: Courtesy of John A. Jaszczak, Michigan Technological University; **5.10A**: Erich Schrempp/Photo Researchers, Inc.; **5.10D**: Courtesy of Prof. Huifang Xu, Department of Geology and Geophysics, University of Wisconsin, Madison; **5.14A**: © 1996 Jeff Scovil; **5.16**: Richard P. Jacobs/JLM Visuals; **5.17**: © Richard P. Jacobs/JLM Visuals; **5.18A**: © Breck P. Kent/JLM Visuals; **5.18B**: © Richard P. Jacobs/JLM Visuals; **5.19A**: © 1995–1998 by Amethyst Galleries, Inc., http://mineral.galleries.com; **5.19B**: © 1997 Jeff Scovil; **5.19C**: © 1993 Jeff Scovil; **5.20A**: © Richard P. Jacobs/JLM Visuals; **5.20B**: © 1992 Jeff Scovil; **5.20C**: © Marli Miller/Visuals Unlimited; **5.20D**: © Richard P. Jacobs/JLM Visuals; **5.20E**: © Richard P. Jacobs/JLM Visuals; **5.21**: © Richard P. Jacobs/JLM Visuals; **5.23**: © Richard P. Jacobs/JLM Visuals; **5.26**: © 1996 Smithsonian Institution; **5.27A**: © 1985 Darrel Plowes; **5.27B**: ©Ken Lucas/Visuals Unlimited; **5.28**: A.J. Copley/Visuals Unlimited; **5.30**: © Chip Clark.

Interlude B: **B.1A**: © Richard P. Jacobs/JLM Visuals; **B.1B**: Courtesy David W. Houseknecht, U.S. Geological Survey; **B.2A**: Stephen Marshak; **B.2B**: Courtesy of Dr. Kent Ratajeski,

Department of Geology and Geophysics, University of Wisconsin, Madison; **B.3A**: Stephen Marshak; **B.3B–D**: Stephen Marshak; **B.4A**: Stephen Marshak; **B.4B**: Philip Gostelow/Anzenberger Agency/Jupiterimages; **B.4.C**: David Muench/Corbis; **B.6A**: © Tom Bean; **B.6B**: H. Johnson; **B.8A**: Stephen Marshak; **B.8B**: Courtesy of Bradley Hacker, Department of Geology, University of California, Santa Barbara.

Chapter 6: **6.1A**: Jim Sugar Photography/Corbis; **6.1B**: Photo by J.D. Griggs/U.S. Geological Survey; **6.1C**: Roger Ressmeyer/Corbis; **6.8A**: Stephen Marshak; **6.8B**: Corbis; **6.9B**: Stephen Marshak; **6.11A**: Stephen Marshak; **6.11B**: © 1998 Tom Bean; **6.11D**: Paul Hoffman; **6.11E**: Stephen Marshak; **6.13A**: Stephen Marshak; **6.14B–C**: Stephen Marshak; **6.15E**: Stephen Marshak; **6.17A–B**: Courtesy of Dr. Kent Ratajeski, Department of Geology and Geophysics, University of Wisconsin, Madison; **6.17C–D**: © Richard P. Jacobs/JLM Visuals; **6.17E**: © John S. Shelton; **6.17F–G**: Stephen Marshak; **6.17H**: © Dane S. Johnson/Visuals Unlimited; **6.17I**: © Doug Sokell/Visuals Unlimited; **6.17J**: Stephen Marshak; **6.20**: Stephen Marshak; **6.21**: M.F Coffin; **6.22B**: Tony Stone Images/Getty Images; **6.23B**: Stephen Marshak; **6.23C**: © Peter Kresan.

Chapter 7: **7.2**: Stephen Marshak; **7.3**: Stephen Marshak; **7.5A**: © Marli Miller/Visuals Unlimited; **7.5B–C**: Stephen Marshak; **7.6B**: © 1998 Tom Bean; **7.6C**: Stephen Marshak; **7.7C**: Stephen Marshak; **7.7D**: Courtesy Carlo Giovanella; **7.7E**: PR/91–05C British Geological Survey. © NERC. All rights reserved; **7.10B–C**: Stephen Marshak; **7.11A–B**: Stephen Marshak; **7.13A**: © John D. Cunningham/Visuals Unlimited; **7.15**: Jim Richardson/Corbis; **7.19A–C**: Stephen Marshak; **7.19D**: Scottsdale Community College; **7.19E**: Duncan Heron; **7.19F–G**: Stephen Marshak; **7.19H**: E.R. Degginger/Color–Pic, Inc.; **7.19I–J**: Stephen Marshak; **7.19K**: R.L. Kugler and J.C. Pashin, Geological Survey of Alabama; **7.19L**: Earth Sciences Department, Oxford University; **7.21A–D**: Stephen Marshak; **7.22A**: Stephen Marshak; **7.22B**: © Steve McCutcheon/Visuals Unlimited; **7.23B**: © Gerald and Buff Corsi/Visuals Unlimited; **7.23D**: © John S. Shelton; **7.24A**: © Marli Miller/Visuals Unlimited; **7.24B**: Craig Lovell/Corbis; **7.24C**: © 1998 M.W. Schmidt; **7.25A–B**: Stephen Marshak; **7.27**: Stephen Marshak; **7.28C–D**; **7.29B**: Stephen Marshak; **7.30B**: © Marli Miller/Visuals Unlimited; **7.31A–B**: Stephen Marshak; **7.32A**: Emma Marshak; **7.32B**: Stephen Marshak; **7.32C**: © Marli Miller/Visuals Unlimited; **7.33A**: © John S. Shelton; **7.33C**; **7.34C**: Stephen Marshak; **7.35A**: Yann Arthus–Bertrand/Corbis; **7.36A**: Belgian Federal Science Policy Office; **7.36B**: G.R. Roberts ©NSIL.

Chapter 8: **8.1A**: Stephen Marshak; **8.1B**: L.S. Stephanowicz/Visuals Unlimited; **8.1C**: Stephen Marshak; **8.1D**: Courtesy Kurt Friehauf, Kutztown University of Pennsylvania; **8.1E–F**: Stephen Marshak; **8.7**: © 1996 Bruce Clendenning/Visuals Unlimited; **8.8B**: Ludovic Maisant/Corbis; **8.8C**: Emma Marshak; **8.10A–C**; **8–11**; **8.13**; **8.14**; **8.15A–B**: Stephen Marshak; **8.16A**: Hans Georg Roth/Corbis; **8.16B–C**: Stephen Marshak; **8.28B**: Stephen Marshak; **8.28C**: David Muench/Corbis.

Chapter 9: **9.1A, C**: Stephen Marshak; **9.3A**: U. S. Geological Survey; **9.3B–E**; **9.4A–B**: Stephen Marshak; **9.4C**: NOAA; **9.5A**: AP Images; **9.5C**: A.M. Sarna–Wojcicki/U.S. Geological Survey; **9.5D**: © Ken Wagner/Visuals Unlimited; **9.5E**:

Stephen Weaver; **9.5F**: Stephen Marshak; **9.6A**: Roger Ressmeyer/Corbis; **9.6B**: © 2001 Stephen and Donna O'Meara, Volcano Watch International; **9.6C**: Associated Press Antigua Sun; **9.7A**: Stephen Marshak; **9.7B**: U.S. Geological Survey; **9.8**: Stephen Marshak; **9.9C**: N. Banks/U.S. Geological Survey; **9.9D**: Corbis; **9.10D**: Yann Arthus-Bertrand/Corbis; **9.10E**: © Marli Miller/Visuals Unlimited; **9.11B**: © Marli Miller/Visuals Unlimited; **9.11C**: © 1985 Tom Bean; **9.11E**: Roger Ressmeyer/Corbis; **9.13A**: Stephen Marshak; **9.13B**: © S. Jonasson/FLPA; **9.16D**: U. S. Geological Survey; **9.15E**: Garu Braasch/Corbis; **9.18B**: Stephen Marshak; **9.18E**: Oxford University Press; **9.18F**; **9.20**: Stephen Marshak; **9.21B**: Courtesy of Dr. Robert M. Stesky/Pangaea Scientific; **9.22A**: Stephen Marshak; **9.22B**: J.D. Griggs/U.S. Geological Survey; **9.22C**: AP Images; **9.22D**: © Philippe Bourseiller; **9.22E**: AP Images; **9.22F**: Photo by J. Marso, U.S. Geological Survey; **9.23**: Peter Turnley/Corbis; **9.24**: Fact Sheet by Scott, K.M., Wolfe, E.W., and Driedger, C.L., Hawaiian Volcano Observatory/U.S. Geological Survey; **9.25**: photo by Sigurgeir Jonasson/Frank Lane Picture Agency/Corbis; **9.26D**: Stephen Marshak; **9.27**: Joseph Mallord William Turner. *Slave Ship*, 1840. Museum of Fine Arts, Boston. Reproduced with permission. © 2000 Museum of Fine Arts, Boston. All rights reserved; **9.28A**: Gail Mooney/Corbis; **9.28B**: © 1997 Birke Schreiber; **9.29**: Julian Baum/Photo Researchers, Inc.; **9.30A–D**: NASA/JPL.

Chapter 10: **10.2**: J. Dewey, U.S. Geological Survey; **10.3**: UC Berkeley; **10.6A**: National Geophysical Data Center/NOAA; **10.6C**: Photo courtesy of Paul "Kip" Otis–Diehl, USMC, 29 Palms, CA; **10.11D**: Peltzer, G., Crampe, F., and King, G., 1999, Evidence of nonlinear elasticity of the Crust. Science, v. 286, p. 272–276; **10.14E**: Center for Earthquake Research and Information; **10.18**: Adapted from Bolt, 1978; **10.22**: Bettmann/Corbis; **10.23A**: Courtesy of ABSG Consulting, Inc., Houston, Texas, modified by Stephen Marshak; **10.23B**: Patrick Robert/Corbis Sygma; **10.25B**: U.S. Geological Survey; **10.25C**: Corbis; **10.25D**: George Hall/Corbis; **10.25D**: George Hall/Corbis; **10.26B**: Courtesy State Historical Society of Missouri, Columbia; **10.29A**: AP Images; **10.29B**: J. Dewey, U.S Geological Survey; **10.29C**: Reuters; **10.29D**: M. Celebi, U.S. Geological Survey; **10.29E**: AP Images; **10.30A**: AP Images; **10.30B**: NGDC; **10.31C**: Karl V. Steinbrugge Collection, Earthquake Engineering Research Center; **10.32C**: National Information Service for Earthquake Engineering; **10.32D**: (c) James Mori, Research Center for Earthquake Prediction, Disaster Prevention Institute, Kyoto University; **10.33**: U.S. Geological Survey; **10.35A–C**: AFP/Getty Images; **10.35D**: David Rydevik; **10.35F–G**: Space Imaging; **10.35H**: Copyright© DigitalGlobe. All rights reserved; **10.37A**: NOAA; **10.37B**: National Geophysical Data Center; **10.38A**: NOAA/NOA Center for Tsunami Research; **10.38B**: Vasily V. Titov, Associate Director, Tsunami Inundation Mapping Efforts (TIME), NOAA/PMEL-UW/JISAO, USA; **10.40A**: Produced by the Global Seismic Hazard Assessment Program; **10.40A**: Produced by the Global Seismic Hazard Assessment Program; **10.40B**: Earthquake Hazards Program/U.S. Geological Survey; **10.40C**: Adapted from Wesson and Wallace, 1985; **10.40D**: Adapted from Nishenko, 1989/U.S. Geological Survey.

Interlude D: **D.3**: Jenifer Jackson and Jay Bass, University of Illinois; **D.16A**: ESA; **D.16B**: University of Texas Center for Space Research and NASA; **D17A–B**: © John Q Thompson, courtesy of Dawson Geophysical Company; **D17C**: Courtesy

Miller/Visuals Unlimited; **17.21B**: Stephen Marshak; **17.22A**: © 1998 Tom Bean; **17.22B**: Images provided by Google Earth™ mapping services/DigitalGlobe, Terra Metrics, NASA, Europa Technologies—copyright 2008; **17.24D**: NASA; **17.30D**: Dewey Moore; **17.31C**: © 1997 Tom Bean; **17.32A–B**: Courtesy Doug Hazelrigg, U.S. Geological Survey; **17.32C**: Stephen Marshak; **17.32D**: Peter Kresan; **17.32E**: Binsar Bakkara/AP Images; **17.33**: U.S. Geological Survey; **17.34B**; **17.35B**: Stephen Marshak; **17.40**: Photo courtesy of the Bureau of Reclamation.

Chapter 18: **18.1A**: © Rod Catanach, Woods Hole Oceanographic Institution; **18.1B**: © Topham/The Image Works; **18.2A**: NOAA Photo Library; **18.2B**: Courtesy Stephen Hurst, University of Illinois; **18.4**: Orbimage; **18.5A–B**: National Geophysical Data Center/NOAA; **18.5E**: U.S. Geological Survey; **18.6B**: University of New Brunswick Ocean Mapping Group; **18.8A**: Adapted from Time Atlas of the World; **18.8B**: Adapted from CLIMAP, 1981; **18.8C–D**: Adapted from Conte et al., 1994; **18.9**: NASA; **18.10**: Adapted from Skinner and Porter, 1995; **18.13**: Adapted from Davis, 1991; **18.14**: Adapted from Skinner and Porter, 1995; **18.15B**: Michael St. Maur Sheil/Corbis; **18.18B**: Karsten Petersen; **18.19B**: Stephen Marshak; **18.20B**: G.R. Roberts ©NSIL; **18.20D**; **18.22A–C**: Stephen Marshak; **18.22D**: Feary, D.A., Hine, A.C., Malone, M.J., et al., 2000. Proc. ODP, Init. Repts., 182 College Station, TX (Ocean Drilling Program); **18.24A–B**: Stephen Marshak; **18.26A**: NASA; **18.26B**: Stephen Marshak; **18.29C**: G.R. Roberts © NSIL; **18.29D**; **18.31A–B**; **18.32A–B**: Stephen Marshak; **18.34B**: G.R. Roberts © NSIL; **18.35A–B**: Stephen Marshak; **18.27**: Adapted from Skinner and Porter, 1995; **18.39**: Adapted from Kraft, 1973; **18.40A**: Annie Griffiths Belt/Corbis; **18.40B**; **18.41D**: Stephen Marshak.

Chapter 19: **19.1A–B**: © GeoPhoto Publishing Company; **19.12A**: Vince Streano/Corbis; **19.14A**: Photo courtesy of USDA Natural Resources Conservation Service; **19.15H**: Stephen Marshak; **19.16A**: JPL/NASA; **19.16C**: Courtesy of the Food and Agriculture Organization of the United Nations; **19.17**; **19.18A–D**; **19.19B**; **19.21F**: Stephen Marshak; **19.21G**: Photo by Richard O. Ireland, U.S. Geological Survey; **19.21H–I**: Stephen Marshak; **19.23D**: Courtesy of C Tech Development Corporation www.ctech.com; **19.27**: Stephen Marshak; **19.29A**: G.R. Roberts © NSIL; **19.29B**: Photo by David Parker, courtesy of the National Astronomy and Ionosphere Center–Arecibo Observatory, a facility of the NSF; **19.29C**: Lois Kent; **19.29D**: G.R. Roberts © NSIL; **19.31A–B**: Stephen Marshak; **19.32A**: © Kjell B. Sandyed/Visuals Unlimited; **19.32B**: Photo by Jim Pisarowicz.

Chapter 20: **20.1**: AFP/Getty Images; **20.3**: Reuters NewMedia, Inc./Corbis; **20.5A**: Stephen Marshak; **20.5B**: Kathryn Marshak; **20.8**: Bjorn Backe, Papilio/Corbis; **20.12B–C**: NASA; **20.19**: Stephen Marshak; **20.20D**: Stephen Marshak; **20.22B**: Stephen Marshak; **20.23**: Tim Thompson/Corbis; **20.26A**: Eric Nguyen/Corbis; **20.26B**: AFP/Getty Images; **20.27**: NASA/GSFC/METI/ERSDAC/JAROS, and U.S./Japan ASTER Science Team; **20.28**: Adapted from Coch, 1995; **20.29**: Adapted from NOAA; **20.30A**: Reuters NewMedia, Inc./Corbis; **20.30B**: Adapted from Coch, 1995; **20.30C**: Adapted from Lutgens and Tarbuck, 1998; **20.32**: NOAA Photo Library; **20.33**: Adapted from Getis et al., 1991; **20.34A**: NHC/NOAA; **20.34B**: NOAA; **20.34C**: AP Images;

20.34E: New York Times Graphics; photographs from DigitalGlove via Keyhole (satellite images) and the Associated Press; **20.34F**: Vincent Laforet/Pool/Reuters/Corbis; **20.34G**: Getty Images; **20.34H**: NOAA; **20.34I–K**: Stephen Marshak; **20.35**: Adapted from Lutgens and Tarbuck, 1998; **20.36A**: FAO–SDRN Agrometeorology Group; **20.36B**: NASA/GSFC; **20.37**: Modified from Lutgens and Tarbuck, 1998.

Chapter 21: **22.1**: Stephen Marshak; **22.2A–B**: © 1996 Galen Rowell; **22.3A**: Photos courtesy Kenneth G. Libbrecht, www.snowcrystals.com; **22.3B**: Stephen Marshak; **22.3C**: Lowell Georgia/Corbis; **22.3D**: Emma Marshak; **22.3E**: National Geographic Image Collection; **22.4B**: Emma Marshak; **22.4C**: Stephen Marshak; **22.4D**: USGS/EROS Data Center Satellite Systems Branch; **22.6A**: NASA Jet Propulsion Laboratory (NASA–JPL); **22.6B**: 2001–2002 by Wm. Robert Johnston; **22.6C**: NASA; **22.6D**: John Pelletier. University of Arizona; **22.10**: © Harry M. Walker; **22.12A–B**: Modified from Raymond, 1971; **22.15B**: Antarctic Search for Meteorites Program/Linda Martel; **22.16A**: Stephen Marshak; **22.16B**: Courtesy Ted Scambos, National Snow and Ice Data Center, University of Colorado, Boulder; **22.18**: Ralph A. Clevenger/Corbis; **22.19A**: Stephen Marshak; **22.19B**: ESA; **22.20**: Stephen Marshak; **22.21A**: © 1986 Jack Olson; **22.21B**: © Art Wolfe; **22.21C**: Stephen Marshak; **22.22C–D**: Stephen Marshak; **22.23D**: © Gerald and Buff Corsi/Visuals Unlimited; **22.23E**: Ric Ergenbright/Corbis; **22.23F**: © 1986 Keith S. Walklet/Quietworks; **22.24**: © Marti Miller/Visuals Unlimited; **22.25**: Stephen Marshak; **22.26B**: Stephen Marshak; **22.27A–B**: Stephen Marshak; **22.27C**: Courtesy Duncan Heron; **22.27D**: Stephen Marshak; **22.29**: Adapted from Tarbuck and Lutgens, 1996; **22.30B**: Stephen Marshak; **22.30C**: Modified from U.S. Geological Survey; **22.31C**: © Glenn Oliver/Visuals Unlimited; **22.32**: Tom Bean/Corbis; **22.34**: Giovanni Sella; **22.35C**: Stephen Marshak; **22.36B**: Adapted from Hamblin and Christiansen, 1998; **22.37A**: After Flint, 1971; **22.37B**: Scott T. Smith/Corbis; **22.38B**: Reproduced with permission of Dan Armstrong; **22.38C**: Mark A. Kessler, Complex Systems Laboratory, UCSD; **22.39**: Stephen Marshak; **22.40A**: Reproduced with permission of the Minister of Public Works and Government Services Canada, 2004 and Courtesy of Natural Resources Canada, Geological Survey of Canada; **22.40B**: After Flint, 1971; **22.41A–B**: Modified from Skinner and Porter, 1995; **22.42**: detail of mural by Charles R. Knight, American Museum of Natural History, #4950(5). Photo by Denis Finnin; **22.43**: Adapted from Glacial Map of the United States, Geological Society of America; **22.44**: Stephen Marshak; **22.47**: After American Geophysical Union; **22.48**: Hendrick Averkamp, *Winter Scene with Ice Skaters*, ca. 1600. courtesy Rijksmusuem, Amsterdam; **22.49**: Stephen Marshak.

Chapter 23: **23.2A–C**: NASA; **23.6A**: After Van Andel, 1994; **23.6B–C**: After Sloss, 1962; **23.6D**: After Boggs, 1995; **23.6E**: After Van Andel, 1994; **23.8A**: ISM/PhototakeUSA.com; **23.9A**: After Johnson, 1972; **23.9B–C**: After Van Andel, 1994; **23.10A**: Courtesy of the Climatic Research Unit, University of East Anglia; **23.11A**: After Kump et al., 1999; **23.11B**: After Van Andel, 1994; **23.12A**: After McKenzie, 1998; **23.12B**: © Crown Copyright. Reproduced courtesy of Historic Scotland, Edinburgh; **23.12C–D**: Reprinted with permission from Understanding Climate Change. Copyright 1975 by the National Academy of Sciences. Courtesy of the National Academy Press, Washington; **23.13A**: courtesy of SOHO/MDI consortium. SOHO is a project of international cooperation between ESA and NASA; **23.13B**: Adapted from Chaisson and McMillan, 1998; **23.14A–B**: After Kump et al., 1999; **23.14C**: © William K. Hartmann/Planetary Science Institute; **23.16A**: Courtesy P&H Mining Equipment; **23.16B**: Richard Hamilton Smith/Corbis; **23.17B**: M–SAT LTD/SPL/Photo Researchers, Inc.; **23.17C**: Kennan Ward/Corbis; **23.18**: After Turco, 1997; **23.19A**: ASA/Goddard Space Flight Center Scientific Visualization Studio/NASA; **23.20A–B**: Courtesy of GreenFacts asbl/vzw 2001–2007/IPCC 2007: WG1–AR4; **23.21A**: NASA/Goddard Space Flight Center Scientific Visualization Studio; **23.21B**: Courtesy Jacques Descloitres, MODIS Land Rapid Response Team at NASA GSFC; **23.21C–D**: Image courtesy Scientific Visualizations Studio, NASA GSFC; **23.21E**: 1941 USGS; 2004 USGS photograph by Bruce Molnia; **23.24D**: NASA/GSFC/METI/ERSDAC/JAROS, and U.S. Japan ASTER Science Team.

Appendix A: **a.1**: Robert Glusic/PhotoDisc; **a.6**: Image courtesy of Arthur Smith and Randall Feenstra, Carnegie Mellon University; **a.13D**: Visuals Unlimited.

Appendix B: **b.2A–B**: NGDC/NOAA; **b.3**: Dale Sawyer/Rice University; **b.4**: Lisa Gagagan/PLATES Project, University of Texas, Institute for Geophysics; **b.5**: National Resources Conservation Source/USDA; **b.6**: JPL/NASA; **b.7**: USGS; **b.8**: USGS.

Index

NOTE: Italicized page numbers refer to pictures, tables, and figures. Bold page numbers refer to key words.

aquifers, **664**, *664*, 665, *673*, 688
  confined vs. unconfined, 664, *664*
  contaminants in, 679
  hydraulic head in, 668
  for Sahara oasis, 673
  and wells, 670
aquitards, **664**, *664*, 688
  clay as, 679
  and wells, 670
Arabian Desert, 732
Arabian Peninsula, Empty Quarter of, 744
Arabian Plate, *81*, *114*
aragonite, 204, *205*, *207*
Aral Sea, 616, 750, *751*
Archaea, 409, *409*, 805
  as earliest life, 454–55
  fossils of, 411, 453
  in geothermal waters, 674, *674*
Archaeopteryx, 408, *408*, 470, 477
Archean crust, 61
  in North America, *457*
  remnants of, *456*
Archean Eon, **6**, *6*, 430, *432*, *433*, *443*, **453–56**, *476*, 480
  and BIF, 459
  changes during, 804
  Moon's distance during, 635
  photosynthesis in, 805
  temperature in, 817
  tillites from, 793
Archean rock:
  atmosphere of, 459
  bacteria in, 454–55
  in Canada, 254
  in North America, *457*
  in U.S. Midwest, 457
arches, natural, **743**
Arches National Park, Utah, 370, *371*, 388, 743
Archimedes, 88
Archimedes' principle of buoyancy, 88, 383, *383*
Arc rock, *454*
Arctic Circle, 703
Arctic Ocean, 474, *470*, 770, 774, 786, 824, 826
Arecibo Radio telescope, Puerto Rico, 685, 689
arête, **772**, **779**
Arfons, Art, 206
Argentina, Iguazu (Iguaçu) Falls on border of, *179*, *595*
argillaceous rocks, 198
aridisol, *196*, *197*
aridity, 731
Arizona, 173, 222, 826
  Colorado Plateau in, *376*
  desert erosion in, 753
  desert pavement in, 737, *738*
  desert varnish in, *736*
  flooded orchard in, *610*
  Meteor Crater in, *32*, 249
  mining claims in, 532
  Monument Valley in, *545*, *742*, 743
  national parks of, *434*
  Painted Desert of, *434*, 735, *736*
  sandstone canyon wall in, *590*
  Sonoran Desert in, *738*, *747*, *748*
  Sunset Crater National Monument in, *274*
  tilted beds of strata in, *366*
arkose, 200, **201**, **203**, 218
Armenia, earthquake in, *305*, *332*, 346
Armero, Colombia, lahar hits, 289, *290*, 561
Arno River, flooding of, 607
arrival time, **316**
arroyos, 737, *748*, *749*
arsenic, 516, 525
  in groundwater, 677
artesian springs, **672**, *672*
artesian well, **670**, *671*
Arthropoda, 410
artificial levees, 611–12, 613, *613*
asbestos, 535
ash, volcanic, *See* volcanic ash
ash fall, 154

ash flow, 154, 162
Asia:
  in collision with India, 106–7, *392–93*, *471*, *474*
    and chemical weathering, *815*
    warm ocean currents cut off by, 796
  ice sheet over, 789
  land bridge to Australia from, 475, 784, 791
  land bridge to North America from, 474–75, *480*, 784, *785*
  in late Cretaceous Period, *471*
  monsoons of, 722–23
  at Pangaea breakup, *469*
Asian Plate, *393*
as metallic mineral resource, in veins, 370
Aspen, Colo., 479
asperities, 308, *309*
asphalt, *801*
assay, 532
assimilation, **158**, *159*
asteroid belt, 30
asteroids, **30**, 31
  and mass extinction, 414, *414*, 818, *818*, 819
  material from to Earth, 48
asthenosphere, 53, **53**, *53*, **86**, *87*, *108*, *109*, *114*, 357–58, 802
  in continental rifting, *107*
  and forces driving plate motion, 112, *112*
  and geotherm, *155*
  in glacial loading and rebound, 782–83, *784*
  and igneous magma, 179
  and isostacy, 383
  in mid-ocean ridge formation, *92*, 93–94
  as mobile, 802
  and passive margin, *89*
  plasticity of, 802
  and plate tectonics, 87*n*
  and subduction, 94, *96*
asthenospheric mantle, *108–9*
astronauts, earth in perspective for, 13
astronomy, *See* cosmology; solar system; universe
Aswan High Dam, 673, 820
Atacama Desert, 733, *734*, 752
Atchafalaya River, 599, *603*
Atlantic Ocean:
  creation of, 394, 468, *471*, 473
  in future, 830
  growth of, 473
  and ice age, 797
  opening of, *605*
  water masses of, 628, 630, *631*, *632*
  See also North Atlantic Ocean; South Atlantic Ocean
Atlantis legend, 298
Atlas Mountains, *364*
atmosphere, **39**, 53, 119, 693–99, **693**, *806*
  and blueness of sky, 697, *697*
  convection in, 700–701, *702*, 802
  evolution of, 27, 804–5
    formation of, 693–94, *695*, 727
    during Hadean era, 452
    stages of, *695*
  layers of, 698–99, 727
  outer edge of, 696
  oxygen in, 458, *458*, 459, 694, *695*, 805
    in Archean Eon, 456, 480
    increase in, 456, 460, 693–94
    and photosynthesis, 6
    in thermosphere, 699
  residence time for, 550
  and Venus, 551
  and volcanic gases, 297
  as water reservoir, *550*, 553
  See also air pressure; storms; weather; winds
atmospheres of other planets, 39
atoll, 649, *650*
atomic clock, 416
atomic number, **123**, 433, **A-4**
atomic weight (atomic mass), **123**, 433, **A-4**
atoms, **123**, A-1–A-5, **A-1**
  of metals, 523
  passing through rock cycle, 258, 259, 262, *806*, 808
aureole, metamorphic (contact), *245*, **245–46**

aurora australis, **699**
aurora borealis, *39*, **699**, *700*
aurorae, 39
Australia, 173, 452
   ancient sandstone found in, 444
   Blue Mountains of, *545*
   in Cambrian Period, *461*
   chert beds of, 454
   and creation of Pannotia, 457, *458*, 461
   during Cretaceous Period, 471
   dirt road in, *202*
   in future, 830
   gold nuggets from, *524*
   in Gondwana, *474*
   Great Barrier Reef of, 204
   land bridge to Asia from, 475, 784, 791
   in late Cretaceous Period, *471*
   Nambung National Park, *118–119*
   Nullarbor Plain of, 638, *639*
   in Pangaea breakup, 473
   quartzite and schist in, *366*
   sand desert in, *202*
   sea stacks in, *645*
   separation of from Antarctica, 60, 796
   Uluru (Ayers Rock), in, 743, *745*
   underground coalbed fire in, 509
   zircon found in, 452
Australian Aborigines, 747
   and Uluru, *745*
Australian Desert, 732
Australian-Indian Plate, *89*
Australian Plate, *114*
Australian Shield, 252
Australopithecus, 403
Austria, avalanche in, 562–63
automobiles, gasoline produced for, 493
avalanches, **552–63**, 578, *579*
   from earthquakes, 330, 331
   glacier formation checked by, 761
   underwater, *261*
avalanche shed, *579*
Avalon microcontinent, *464*, *465*
avulsion, 599
axis (centerline) of ridge, 90
Ayles Ice Shelf, 823
azurite, *526*

backscattered light, 697
backwash, **637**, *637*, 641
   in hurricane, 654
bacteria:
   anaerobic, 500
   in caves, 687, *687*
   as earliest life, 454–55, 456
   and Earth history, *807*
   eubacteria, 409
   and evolution, 805
   fossils of, 411, 453, 454
      microfossils, 407
   in geothermal waters, 674, *674*
   in groundwater, 679
      for bioremediation, 679
   hydrocarbon-eating, 492–93, 497, 499
   weathering process supported by, 190
Badlands National Monument, S. Dak., *366*
Baffin Island, Canada, 774
Bahamas, 42, 223, *483*, 651, 722
Bahariya oasis, 673
bajada, **740**
baked contacts, principle of, **419**, 445
Ballard, Robert, 767
balloon travel, 693, *693*
Baltica, *458*, 461, *461*, *465*
Baltic Shield, *252*
Banda Aceh, Sumatra, 336–37, *337*
   tsunami damage of, 341
banded-iron formation (BIF), *458*, 459, *459*, **529**, 531, 536
banding, in gneiss, 236, *238*
Bangladesh, cyclone-flood deaths in, 607, 654, 718

bar, *216*, **591**, *591*
   baymouth, 641, *641*
   offshore, *640*
Baraboo, Wisc., *448*
barchan dunes, 744, *748*, *749*
barrier islands, **641**, 642, *642*
barrier reefs, 649
Barringer (Meteor) Crater, Ariz., 32, 249
Barton, Clara, 583
basal sliding, 762, *764*
basalt, **44**, *49*, 80–81, *149*, *156*, *171*, *176*, 172, 173, 178, 179
   beach sand from, 639
   in continental-rift eruptions, 288
   and cooling, 175
   flood, **179**, *179*, **286**, 299
   from lava flow, *153*, 173
   and marine anomalies, 70–71, *71*
   metamorphism of, 238, 241, 249
   from mid-ocean ridge eruptions, 287–88
   molten form of, *149*
   in oceanic crust, 47, 65, 179, 262
   and paleomagnetism, 81
   pillow, 91, **92**, **180**, 269, *284*, 286, 288, *623*
   sheared and unsheared, *374*
   soil formed on, 194
   weathering of, 258
   wet, *156*
basalt dike, *164*, 180, *426*
   in geologic history illustration, 419, *422*
basalt flows:
   in India, 819
   of sea floor, 180
basaltic eruptions, 269
basaltic lava, 175, 176, 267–73, 276, 280, 281, 284, 299, *427*
   freezing of, *295*
   threat from, 288
   at Yellowstone, 284
basaltic magma, *156*, 175, 176, 179, 280
basalt pillow, 269
basalt sill, *167*
base level of stream, **592**, *593*
basement, 185, **260**
   North American, *469*
   Precambrian, *254*, *468*
basement uplifts, **471**, *472*
base metals, **525**
basic (or mafic) metamorphic rocks, *241*
   See also mafic rocks
basin, **376**, **378**, 382, 390, 396
   alluvium-filled, *748*
   on geologic map of eastern U.S., *391*
   between mountains, *748*
   oceanic, 622
   regional, 390
   and tides, 632
   See also sedimentary basins
Basin and Range Province, 106, *111*, *166*, 179, 288, *326*, 327, 341,
      390, *390*, 437, **474**, *475*, 479, *480*, 665, **748**, 787, *787*
Bath, England, artesian hot spring at, *674*
batholiths, *163*, 165, **166**, 181
   granitic (Sierra Nevada), 471, 792
bathymetric map, 621, *625*
   of hot-spot tracks in Pacific Ocean, *106*
bathymetric profile, 63, *64*
bathymetric provinces, of sea floor, *623*
bathymetry, **63**, **622**, *652–53*
   of mid-Atlantic ridge, *93*
   of oceanic plate boundaries, 624
Bauer, George (Agricola, Georgius), 121
bauxite, 196, 531
baymouth bar, 641, *641*
Bay of Fundy, 632
Bay of Naples, 293
bayous, 602
bays, and tides, 632
beach drift, **641**, *641*
beach erosion, 654–56, *655*, *656*, 657
beaches, 218, 223, 638–43, **638–39**, *638*, *641*, 642–44, *642*, *643*,
   652–53

protection of, 654–56
sediment budget of, 642–43
beach face, **641**
beachfront homes, after hurricane, *655*
beach management, 654–56
beach nourishment, 656, **656**
beach profile, 641
bearing, **367**
Becquerel, Henri, 444
bed, **209**, 222
    coal seam as, 503–4
    and fold, 379
    thrust or reverse faults in, 371
bedding, **150**, **209**, **210**, 222, *235*, **381**
    disrupted, *343*
    and foliated marble, 240
    fossils in, *422*
bedding plane, 209, *235*, **279**
    as prone to become failure surfaces, 568, *568*
    vs. fault surfaces, 373
bedforms, 211
bed load, **589**, *590*, 617
bedrock, **146**, *147*, *197*
bed-surface markings, **213**
Bergeron, Tor, 709
Bergeron process, **709**, *709*
bergy bits, 767
Bering Strait, land bridge across, 474–75, *480*, 784, *785*
berms, **641**
beryl, 138, *141*
Berzelius, Jöns Jacob, 135
B-horizon, 194, **197**
BIF (banded-iron formation), *458*, 459, *459*, **529**, 531, 536
big bang, **21**, *21*, **22**, 34
    aftermath of, 21–23
    and geologic column, *433*
big bang nucleosynthesis, 22
Big Bend Ridge Caldera, *285*
Big Thompson River flood, 610–11, *611*
biochemical chert, 204
biochemical sedimentary rocks, **198**, 204, 226
biodegradation, **499**
biodiversity, 413, 818
    decrease in, 820
    and water contamination, 822
biofuels, **516**
biogenic minerals, 122, **122**
biogeochemical cycle, **802–3**, **808**
    carbon cycle, 809–10
    hydrologic cycle, *806*, 809
biography of Earth, *See* Earth history
biologic productivity, *624*
biomarkers, 453, 454
biomass, 489, 516
biomineralization, 129, 138
    as chemical weathering, 190
bioremediation, **679–80**
biosphere, 119, *806*
biotite, 123, 136, 146, *171*, *230*, 240
    and closure temperature, 439
    in gneiss, 236
    in granite, 525
    in granitic mountains, 185
    and igneous intrusion, *247*
    and metamorphism, 238, 240, 241, *243*
    in Onawa Pluton, *247*
    in schist, *251*
    stability of, *190*
bioturbation, 210, **642**
birds:
    appearance of, 430–31
    and dinosaurs, 470, 475
    in history of Earth, 445
    wind farms as danger to, 513
Bishop Tuff, 302
Bismarck Plate, *89*
Bissel, George, 493
bituminous coal, **503**, *504*
    in North America, *505*

bivalves, 410, *411*, 465
    fossil record of, 407
Black Canyon of the Gunnison River, *252*
Black Hills, 664
black hole, 830
black-lung disease, 508
Black Sea, 814
black smokers, **92**, **93**, 228, 297–98
    and Earth system, *806*
    first organisms at, 455
    and life on "snowball Earth," 460
    sulfide ore around, 528, *528*, 531, *538*
blasting of unstable slopes, 578
blind faults, 307, *308*
blocks (volcanic), **271**
blowouts, **738**
Blue Mountains, Australia, *545*
Blue Ridge, *468*
blueschist, *241*, **242**, 248, **250**, **251**, 255
blue shift, 20
body fossils, **405**
body-wave magnitude, 320
body waves, **313**, 351
bolides, 48, 472–73, *473*
Bolivia, *821*
bolting rock, 579
bombs (volcanic), 154, **271**, *272*, **278**, **279**, 280
bond, **A-5–A-6**
bonding in minerals, 125–26, *126*
    and hardness, *131*
Bonneville, Lake, 775, 787–88, *787*
Bonneville Salt Flats, 206
Boone, Daniel, 603
borax, 740
Borglum, Gutzon, *175*
bornhardts, 743, *745*
bornite, *526*
boron, 23, 516
bottled gas, carbon in, 489, *489*
bottomset beds, *220*
boulders, 186, 199
    in glacial abrasion, 772
    left by glaciers, 758
Bowen, Norman L., 160
Bowen's reaction series, **160**, **190**
Boyle, Robert, A-1
brachiopods, *366*, 410, *411*, *426*, 463, 535
Bragg, W. H., 121
Bragg, W. L., 121
braided stream, **597**, *597*, 780, *781*, *782*
Brazil, 193, 601, 826
    Iguazu (Iguaçu) Falls on border of, *179*, *595*
    Paraná Basin and Plateau of, 179, *179*, 286
    quarry wall in, *375*
    Rio de Janeiro mountains, 638, *639*
    Rio de Janeiro mudflows in, *561*
    rock and sand seascape along, *545*
    stream cut in, *147*
    tight fold in, *373*
    valleys in, *385*
Brazilde Ocean, *458*
Brazilian Shield, *252*
breakers, 637, *637*
breakwater, 655, *655*
breccia, **201**, 373, 395, 663
    collapsed, *682*
    crater, 33
    sedimentary, *202*
breeder reactors, 518
Breedlove, Craig, 206
Bretz, J. Harlan, 611
bricks, 535
Bright Angel Shale, *428*, *429*, *430*, *434*
Britain, *See* England; Scotland; United Kingdom; Wales
British Columbia, fjords of, 645
brittle deformation, 313, **366**, 368, 369–70, 395
brittle-ductile transition, 368
brittle-plastic transition, 764
Bronze Age, 524

China (*continued*)
    in Pangaea, 466
    squeezing of, *393*
    underground coalbed fire in, 509, *509*
    Yangtze River flood in, 607
Chinese Shield, *252*
Chinle Formation, 437
chloride, *127*
chlorine, and ozone breakdown, 696
chlorine atom, *A-5*
chlorite, 235, 242, *244*, 248
chloritic, *243*
chlorofluorocarbons (CFCs), 696, 822
chordates, 409
C-horizon, 194, *197*
chrome, *526*
chromite, 121, *526*
chromium, 537, *537*
chron, polarity, **69**, 71, **71**, 75
ciliate protozoans, 459
Cincinnati Arch, *391*
cinder cones, 173, 273, **275**, *277*, *278*, 280, 287, 288, 293, 300
cinders (volcanic), 169, *278*, 280
cinnabar, *526*
circum-Antarctic current, 628, 796
cirque glaciers, 759, *779*
cirques, **759**, 772, **779**, 792, *798*
cirrus clouds, **710**
city water tank system, 670, *671*
Clark, William, 591, 594
classification schemes, 135, 148
    of fossils, 410–11
    for life forms, 409, *409*
clastic deposits, shallow-marine, 218
clastic rocks, **146**
clastic (detrital) sedimentary rocks, 199–204, **200**, *216*, 226
clasts, **199**, 737
    lithic, 200
    and porosity, 662–63
    size of, 200
    See also detritus
clay, 136, *536*, *538*
    in A-horizon, 194
    and beaches, 639
    and bricks, 535, *538*
    and chemical weathering, **191**
    consumption of, *536*
    in desert, 737
        on playa, 740, *741*
    in desert varnish, *736*
    and ductile deformation, 380
    flakes of, *203*
    from glacier (in varve), 780
    in groundwater, 677, 679
        contamination prevented by, 679
    from hydrolysis, 189
    at K-T boundary, 473
    in lake environment, 215
    and liquefaction, 331
    in lithification, *200*
    in loess, 780
    and marble, 240
    and metamorphosis, 235, *243*
    and nuclear waste storage, 512
    on ocean floor, 65
        deep sea floor, 220
    in oil and gas creation, 490, *490*
    and organic-coast pollution, 656
    in pelagic sediment, 625
    potters', 246
    quick clay, 331, **569**, *569*
    in red shale, *230*
    in rock cycle, 258, 259
    in sandstone, 199
    and sedimentary rocks, 203
    and shale, 235, 243, 380, *662*
    and slate, *251*, 364
    and soil erosion, 198
    stability of, *190*

swelling clays, 569
    from weathering, 190, *191*
clay layers (wet), as prone to become failure surfaces, 568, *569*
clean energy, 519
cleavage, **133**, *134*, **381**
cleavage plane, **133**, 134, *134*, 141, *381*
    in slaty cleavage, 235, *236*
cliff:
    of jointed rock, 370
    stair-step, *743*
cliff retreat, **741**, *742*, 755
climate, **693**, 719, 727
    alternation of, 449
    and California mass movements, 576
    in Cenozoic Era, 474
    and coastal variability, 651
    in Cretaceous Period, 470, 471–72, 474
    cycles of (Milankovitch cycles), 795
    in deserts, 732–34
    and eruption of LIP, 178
    and evolution of genus *Homo*, 475, 480
    factors controlling, 719, *722*
    and glacier ice, 440–41
    global changes in, 3, **802**, 810–19
        and deserts, 755
        global cooling, *476*
        global warming, 519, **811**, 822–30; *See also* global warming
    of interior Pangaea, 466
    in Jurassic and Cretaceous, 468
    and landscape development, 549
    in Late Miocene Epoch, 474
    paleoclimate, **811**
    of Paleozoic Era, 466
    in Permian Period, 466
    pollen as indicator of, 407
    recognizing in past, 450–51
    shifts in (Proterozoic Eon), 460
    as soil-forming factor, 194, 196–97, *197*
    and transport of heat by currents, 628
    and tree rings, 440, *440*
    types of, 722, *722*, *723*
    variability of, 722–23, 726–27
    and volcanoes, 296–97, 301
    See also ice ages
climate belts, and global warming, 824, *828*
climate change, global, 810–19
    and extinction events, 413–14, 818–19, *818*
climate-change models, **811**
clinometer, 367
cloning, and extinct species, 408
Cloos, H., quoted, 267
closure temperature, **399**, **439**, 444
clouds, **698**, *698*, 707–8, *708*, *709*, 710–11, *710*, 727
coal, 139, **205**, *206*, 487, *487*, **500**, *501*, *502*, 520
    classification of, 503, *504*
    and climate change, 814
    consumption of, *505*, 536
    and continental drift, 59
    finding and mining, 503–8, *508*
        acid runoff from, 519
    formation of, 224, 466, 500, 503, *504*
    and Industrial Revolution, 487
    low-sulfur, 519
    sulfur dioxide from, 519
    supply of, *517*, 518
coalbed fires, underground, 508–9, *509*
coalbed methane, **508**
coal gasification, 487, **508**, 516
coal rank, 503, *504*
coal reserves, **503–4**
    distribution of, *505*
coal swamps, 466, *467*, *468*, **500–501**, *502*, *504*, 814
"coarse-grained" rocks, 168
coastal beach sands, 218
coastal deserts, 733, 755
coastal landforms, 638, *639*, *640*, *652*
    beaches and tidal flats, 638–43, *642*, *643*
    coastal wetlands, 643–44, *648*
    coral reefs, 648–49, *649*; *See also* coral reefs

estuaries, 644, *648*
fjords, 645, *648*
rocky coast, 643, *644, 645*
coastal plain (U.S.), *457, 468,* 471, **649–50, 651**
  east coast, *390*
  Gulf coast, *391,* 506
  as threatened by sea level rise, 785, *785*
coastal problems and solutions, 654–57
coastal variability, causes of, 649–52, *651*
coastal wetland, **643–44,** *648*
Coast Mountains, Canada, 185
Coast Ranges Batholith, *166*
coasts, **622,** *652–53*
  emergent, **650–51**
  pollution of, 654, 656–57
  submergent, **651,** *651*
cobalt, 537, *537*
cobbles, 186, 200, **201, 219**
Coconino Sandstone (Grand Canyon), *210, 428, 429, 434*
Cocos Plate, 88, *89, 114,* 330
coesite, 230, 232, 249
cold fronts, 706, *707*
Cold War, seismograph stations during, 357
collision, **85, 106,** 107, *111,* 396
  earthquakes from, *323*
  exhumation from, 249
  and fold-thrust belts, 396
  between India and Asia, 106–7, **392–93,** *471, 474*
  in late Paleozoic Era, 466
  and mountain building, 364, 386–87
  and orogens, 380
collisional mountain range (belt, orogen), 107, *108, 111, 245,* 382, 383, *383,* 386–87, *387, 457*
  of Appalachian region, 464
  Himalayas as, *392–93*
collision and coalescence, **708**
collision zones, 327
  and Canadian Shield, 390
  earthquakes at, *310, 322, 323, 326*
Colombia:
  earthquakes in, *305,* 339
  lahar in, *290,* 561
  Nevada del Ruiz in, 289
color, of mineral, **131**
Colorado, 479
  Black Canyon of Gunnison River in, *252*
  highway road cut in, *549*
  Mesa Verde dwellings in, 199
  Rocky Mountains of, *252, 372; See also* Rocky Mountains
  stone dam in, *549*
Colorado Plateau, 222, 341, *376,* 431, 436, 437, *457, 480,* 593, *593, 748*
Colorado River, 222, *590*
  as desert river, *748*
  Grand Canyon of, *185, 210,* 222, 436, 593, *593; See also* Grand Canyon
  water diverted from, 616
Columbia River:
  Grand Coulee Dam on, *513*
  saltwater wedges in, 644
Columbia River Plateau, 179, *179,* 268, *285,* 286
Columbia River Valley, and Great Missoula Flood, 611, *612*
columnar jointing, **269**
comets, 30
  and mass extinction, 414
  material from to earth, 48
  samples from, 30
compaction, **200**
compass needles, 77
competence (of stream), **589,** 617
  during flood, 607
competition, for minerals, 537
competitor, and extinction of species, 414
composite volcanoes, **275,** 288, 300
compositional banding, *251*
compositional layering, 236
compound, **123, A-1,** A-5
Comprehensive Soil Classification System, U.S., 196
compression, *232, 234, 343,* **369,** *370,* **381,** 386

buckle in response to, 380
  and mountain scenery, *362*
compressional waves, **313, 314**
concepts, geology as, 7
conchoidal fractures, **133,** 168
concrete, **534,** *539*
condensation, 550
  and temperature, 698
condensation nuclei, **708**
conduction, **A-10–A-11,** *A-11*
conduits, in permeable material, 663, 669, 675
cone of depression, **670,** *671,* 676–77, 688
conglomerate, **201,** 203, *217, 427*
  and alluvial-fan sediments, 218
  and bedding, *210*
  in Catskill Deltas, 465, *465*
  flattened-clast, **236,** *237*
  of stream gravel, *202*
Congo, 724
Congo River, 588
Congo volcano disaster, 288, *290*
conodonts, 463
consuming boundaries, *91,* **94**
  See also convergent plate boundaries
contact, geologic, **424,** 445
contact aureole, **245,** 382
contact metamorphism, *241,* **245, 251,** 255, *260, 261, 279,* 382, *382*
contaminant plume, **679,** *680*
contamination of water, 821
  of groundwater, 679, **679,** *680,* 821
continental arc, *97, 99,* **175**
continental collision, *See* collision
continental crust, 42, 47–48, *49,* 53, *108,* 350, *350*
  in Archean Eon, 453, *454,* 455
  in Asia-India collision, *392*
  brittle deformation in, 368
  earthquakes in, 313, 347
  formation of, 480
  igneous rocks in, 153
  and passive margin, *89*
  during Proterozoic Eon, 456, 460
  recognizing growth of, 450
  in Wegener's hypothesis, 62
continental divides, 586, *586*
continental drift, 57–58, **57,** 70–71, 74, 86, 113, 802
  in apparent polar-wander paths, 63
  change from, 802, *802*
  and climate change (long-term), 814, *815*
  criticism of, 62
  evidence for, 58–62, 70–71, 74, 86
    in apparent polar-wander paths, 63, *63,* 81–84
    in deep-sea drilling, 74, *75*
    in fossil distribution, *56,* 59, 61, *61*
  and paleomagnetism, 62
  and plate tectonics, 113
  and sea-floor spreading, 66
continental glaciers, *See* ice sheets
continental interior, *324*
continental-interior deserts, 733–34, 755
continental lithosphere, 53, *53,* 86, *87, 96, 107, 108,* 383, *392–93,* 622
continental margin, *64,* 87, *260*
  active, 50, **87, 623,** *625, 653*
    and earthquakes, 89
  and earthquakes, 89
  passive, **87,** 89, **623,** *652*
continental platform, **456**
continental rift, **106–7, 109**
  Basin and Range Province as, 474, *480*
  earthquakes at, 327
  and mountains, *387*
  volcanic eruptions in, 288
continental rifting, **106,** *106,* 115, 179
  and diamonds, 139
  igneous rock format at, 179
  mountains related to, 387, *387,* 390
  and passive margins, 623
  and salt precipitation, 206
continental rise, 623

rock deformation in, 364–69
and subsidence, 546
temperature of, 380
and uplift, *546*
See also continental crust; oceanic crust
crustal blocks, *390*
in Archean Eon, *454*
crustal fragments, in growth of North America, *469*
crustal rocks, in continental crust formation, 453
crustal root, 383, *383*
crust-mantle boundary, discovery of, 352–53, *353*
cryptocrystalline quartz, 204
crystal faces, **124**
crystal habit, **132–33**, *133*
crystal(line) lattice, **123**, 142
of metals, 523
quartz as, 124
crystalline (nonglassy) igneous rocks, 168, **168**, 169, *171*, 181
and cooling time, 169
crystalline rocks, **146**
crystalline solid, 123
crystal mush, *92*, 158, *159*
crystals, **44**, 124–27
arrangement of atoms in, 128
destruction of, 130
formation of, 127, 129–30
and stability of intact rock, 567
crystal structure, **125**
fractional, 158–59
of ice, 758–59
and weathering, 190
cuesta, 222, **385**, *385*, **743**, *743*
Cullinan Diamond, 139
Cumberland Gap, 603
cumulonimbus clouds, 710, *710*, 727
cumulus clouds, **710**
current meter, *587*
currents, oceanic, 627–30, *629*, *652*, 657
and climate change, 817
continental drift as cause, 814
and coastal deserts, 733
and El Niño, 723
and glaciation, 794, 795
global-warming effect on, 829
and Isthmus of Panama, 474
cut bank, 597
cutoff, 597
cyanobacteria, 454, 459, 476, 694
cycads, 466
cycle, 224
cycles in Earth history:
biogeochemical cycles, 808–10, 831
physical, 805–8, 830
cyclic change, 802, *807*, 831
cyclones, **707**, *708*
and nor'easters, 715
cyclones (Indian Ocean storms), *717*, 718
cyclothems, 808, *809*
Cynognathus, *61*

Dakota aquifer, 664, *664*, 665
Dakota Sandstone, 664, 665, 669
Dalton, John, A-1–A-2
dams, 614–15
and earthquake flooding danger, 347
environmental issues over, 614–15, 820
for hydroelectric power, 513, *513*
ice sheets as, 785–87, *786*
and Black Sea flooding, 814
and Great Missoula Flood, 611, *612*, 787
Johnstown Flood from collapse of, 583, *583*
on Nile River (Aswan High Dam), 820
and sediment carried downriver, 616
stone dam in Colorado, *549*
Vaiont Dam disaster, 562, *562*
Dante, 452
quoted, 451
Darcy, Henry, 669
Darcy's law, **669**, 688

"Darkness" (Byron), 297
Darwin, Charles, 444, 649
daughter isotope, *427*, **435**, *435*
Davenport, Iowa:
flood hazard map of region near, *614*
flooding of, 610
David Sandstone, *429*
Dayton Lakes, 601
Dead Sea, 40, *110*, 206
Death Valley, California, 222, *597*, 731, 753
debris:
and coastline, 652
in Colorado River, *595*
in desert environment, *260*
in desert pavements, *749*
on desert slopes, 385
exceeding angle of repose, *570*
Johnstown dam spillway blocked by, 583
as Midwest covering, 147
in Nile River canyon, 184, *184*
in orbit around Earth, 452
and rivers, *605*
scattered by tornado, *713*
submarine flows of, 288
unconsolidated, 193
volcanic (pyroclastic), 154, *154*, *156*, **269**, 271–73, *272*, *282*, *284*
from Yungay avalanches, 558
See also pyroclastic debris; regolith
debris avalanche, 563, *573*
debris falls, **563–64**, 578
debris flow, **561**, **573**, 578
submarine, **564**
debris slide, 562, 578
Deccan region, India, 179, 286
Deccan traps, *286*, 477
de Chelly Sandstone, 437
decompression melting, 155, *155*, *156*, 157, 179, 180
deeper-water settings, 223
deep-focus earthquakes, 322–23, *323*, 348
Deep Gorge, 600
Deep Impact spacecraft, 30
deep-marine deposits, 220
deep-ocean trenches, *50*, 64, *64*, 113
deep-sea drilling, 74
deep time, *416*
See also geologic time
deflation, **738**
deforestation, 824, 826
in tropical forests, 576
deformation, rock, *363*, 364–69, 380, 395
of cratonic platforms, 390
and deposition, 419
thin-skinned, 466
in western North America, *472*
See also brittle deformation; ductile deformation
Degas crater, 32
delta plain, 607
delta-plain floods, 607
of Ganges, 718
deltas, 218, **591**, 599–602, *602*, *603*, 607, 617
deposition on, 217
of Ganges, 718
global-warming damage to, 829
of Mississippi, 599, *602*, *603*
of Niger, *602*
of Nile, 218, 599, *602*, 820
and redbeds, *427*
soil deposition on, 203
swampy, *640*
See also Catskill Deltas
Democritus, A-1
Denali National Park region, Alaska, *306*, *591*, *597*
dendritic drainage network, 585, *586*, 601
dendrochronologists, 440
density (atmospheric), 39
Denver, Colo., 479
earthquake in, 327–39
deposition, *199*, *200*, *608*
alluvial-fan, 222

deposition (*continued*)
  deltas as, 599
  evaporite, 222
  from glaciation, 774–75, 777, 782, *782*
  in landscape evolution, 546
  and Scottish outcrops, 228
  of sediment, 546
depositional environment, **214**
  recognizing changes in, 450
depositional landform, **546**
  and passage of time, 549
depositional processes, 589–91
depositional sequence, 224
De Re Metallica (Agricola), 121
De revolutionibus (Copernicus), *17*
desertification, **750**, 754, 820
desert pavement, **737**, *738*, *749*, 754
desert plateau, *749*
deserts and desert regions, *722*, **731–32**, 732, 752–53, 755
  from atmospheric convection, *701*
  in Australia, *202*
  of Basin and Range Province, *748*
  and continental drift, 59
  depositional environments in, 739–41, *740*
  distribution of, *733*
  ecological balance of, 750
  extent of, 731–32
  groundwater irrigation of (Jordan), *660*
  landscapes of, 741–45, 752–53
  life in, *747*, 750, 751
  mountains in, 385
  Namib Desert, *756*
  and rock cycle, *260*
  Sahara Desert, *See* Sahara Desert
  Sonoran Desert, *730*, *747*, *748*
  types of, 732–34
  urbanization of, 753, 754
  weathering and erosional processes in, 734–39, *736*, *739*
desert varnish, **735**, *735*, *737*, 755
Des Moines, Iowa, flooding of, 610
detachment fault, **373**, *375*, 390
  and fold-thrust belt, 386
detrital (clastic) grains, age of, 439
detritus, 186, 225
  and weathering, 200
  See also Clasts
deuterium, *A-4*
developing countries, river pollution in, 614
development, and southern California mass movement, 576
Devil's Tower, Wyo., 294, 296, *296*
Devonian limestone, *230*
Devonian Period, *432*, *433*, *443*, 476
  animals in, 465
  Antler orogeny in, 465
  and Appalachian region, 464
  biodiversity in, *414*, *818*
  coal reserves in rocks of, *502*
  in correlation of strata, *434*
  day's length in (Middle), 635
  and eastern U.S. geologic features, *391*
  late (paleogeographical map of North America), *465*
  life forms in, 430
  sea level during, *463*, *809*
  and stratigraphic sequences, *463*
dewpoint temperature, **698**
diagenesis, **224**, 225, 228, *241*, 405, 407
  and metamorphism, 225
Diamantina, Brazil, 140
diamond anvil, 354–55
diamonds, 122, 126, *126*, 138, *138*, 139, *139*, 141, 354–55
  Cullinan Diamond, 139
  hardness of, 131, *132*
  Hope Diamond, 137, *137*
  mining of, 140
  placer deposits of, 531
  shape of, *126*
  in simulation of mantle, 354–55
    as ultra-high pressure metamorphic rock, 232

diatoms, 204
Dickinson, Emily, 621
Dickinsonia, *460*
Dietz, Robert, 58, 86
differential stress, **232**, **233**, **251**
differential weathering, 191
differentiation, **31**, 154, *238*, **452**, 803, *804*
diffraction, **128**, *128*
digital elevation maps, **548**, *548*
dike intrusion, *167*
dikes, 91, **92**, **162**, *163*, **164**, **165**, 180, *279*
  basalt, *164*, 180, 419, *422*, *426*
  composition of, 169
  and cross-cutting relations, *421*
  formation of, 167, *167*
  pegmatite, 169, *170*, 252
  on sea floor, *623*
  at Shiprock, *164*
  in volcano, *278*
  in Western Australia, 173
dimension stone, **533–34**
dinosaurs, 470, 472, 473, 475, *477*, 481
  bones of exposed, *406*
  and climate, 813
  extinction of, *414*, 430, 472, 477
  first appearance of, 430, 470, *477*
  footprints of in mudstone, *406*
  fossilization of, *404*
  during Jurassic Period, *469*, 470, 472
diorite, *171*
dip, **367**
dipole, **38**, **73**, **78**, **81**
  paleomagnetic, 68, 81
dip-slip faults, **371**, *373*
dip slope, 743, *743*
"dirty snowballs" (comets), 30
disappearing streams, *685*, **688**
disasters, *See* catastrophic change or event
discharge (groundwater), *668*
discharge area (groundwater), **668**, *668*
discharge of stream, **587–88**, *587*, 617
  and capacity, 589
disconformity, *424*, *425*, 445
discontinuous reaction series, **160**, *160*
disease, from earthquakes, 339
displacement of fault, **306**, 308, 331, 372, **372**
  on San Andreas Fault, *308*
disseminated deposit, 527, **527**
dissolution, 188, *189*, 191
  of minerals in stream, 589
dissolved load, **589**
distillation column (tower), **499**
distortion, 365
distributaries, **599**, *602*, *603*, *609*
divergence zone, **701**
divergent plate boundaries, 89, *93*, 98, *108–9*, 113
  and continental rifting, 106
  and sea-floor spreading, 90–92, *95*
  seismicity at, 323
diversification, 430
DNA (deoxyribonucleic acid), 408, 410, 411, 412
Dobson units, *823*
Dogger Epoch, *562*
doldrums, **704**
dolomite, 133, 136, **208**, 251, *526*, 528, 677
dolostone, **208**, **251**, 595, *596*
dome, **376**, *378*, 396
  on geologic map of eastern U.S., *391*
  regional, 390
Donau glaciation, 792
"doping", of silicon wafer, 516
Doppler, C. J., 19
Doppler effect, **19**, 31
Doppler radar, 715
dormant volcanoes, **291**, 294
Dover, England, chalk cliffs (White Cliffs) of, *221*, 449
downcutting, **584–85**
  and cave formation, *686*

in dry wash, *736*
downgoing plate (or slab), 95
down-slope movement, 546
downwelling zones, 628–30, *632*
dowsers, 670
drag folds, 372
drag lines, 504
drainage, and soil characteristics, 194
drainage basin, **586**, *592*
drainage divide, **586**, *586*, *604*, 617
drainage network, **585**, 601, 617
    and dam construction, 615
    and rock cycle, *260*
drainage reversal, **603**
drainage reversal from ice age, 785–87, *786*
Drake, Edwin, 493
Drake Passage, 628
drift, 780
drilling mud, 497
dripstone, 684–85, 687
drop stones, *769*, **770**, **779**
droughts, 198, 750, 754
drumlins, **779**, **781**, **783**, 792
dry-bottom glaciers, **764**, 765
dry wash, 587, *587*, **735**, *736*, *748*, *749*
ductile deformation, 313, *326*, **366**, 368, 380, 395
dunes, **211**, **741**
    coastal, *640*
    sand, *212*, **744–45**, *744*, 745, *748*, 755
Durant, Will, quoted, 2
dust, wind-blown, 754
"dust" (cosmic), *27*, 28–29, *29*, 34, 39, 48
"dust bowl" of Oklahoma, 198, 750, *751*, 754
dust devils, 735
dust storm, **750**
dwarf planets, 25
dynamic metamorphism, **246**
    along fault zone, *248*
dynamo, 79
dynamothermal (regional) metamorphism, **248**, *251*, 255

Earth, 13
    age of, 6, 442, 444, 451–52, 480
    atmosphere of, 39–40
    axis of
        precession of, 794, *795*
        tilt of, 35, *702*, 703, *703*, 794–95, *795*, 816–17
    basic components of, 119
    as blue, 622
    changes in from continental drift, 802, *802*
    circumference of, 17–18, 35
    climate belts of, 726–27
    density of, 45–46
    differentiation of, 31
    elemental composition of, 41, 54
    formation of, *26–27*
    future of, 829, 830, 831, *831*; *see also* global warming
    heat of (early life), 154–55
    history of in sedimentary rocks, *217*
    layers of, 45–46, 86, 350, *350*
    magnetic field of, 37–39, 63, 70, 78–79, *78*
        reversal of, 68, *68*, 75; *See also* magnetic reversals
    map of, *41*
    meteor bombardment of, 453
    as natural resource, 437
    new discoveries on, 357–59
    orbit of, 794, *795*, 816–17
    other-world explorers' view of, 37–38
    as planet, 5, 25, 31
    pressure in, 46, 54
    rotation of, 15, 42
    satellite view of, *624*
    shape of, 17, 31, 360
    stratigraphic record of, 447
    surface of, 40–41, 53; *see also* landforms; landscape
    temperature in, 46, 54
    tilt of axis of, *702*, 703, *703*
    topography of, *41*

Earth history, *448*, 449, 802–5, *804*
    biogeochemical cycles in, 808–10, 831
    causes of change in, 802
    end of, 830, 831, *831*
    in geologic column, 430–32, *433*; *See also* geologic time
    and Grand Canyon, 416
    and human history, 445, 449
    ice ages in, 58, 474, 758, 794, 812; *See also* ice ages
    methods for studying, 449–50
    physical cycles in, 831
        rock cycle, *806*, 808
        sea-level cycle, 805–8, *809*
        supercontinent cycle, 805, *808*
    in sedimentary rocks, 216, 226
    time periods of
        Archean Eon, **6**, *6*, 453–56
        Cenozoic Era, **6**, *6*, 473–75, 480
        Hadean Eon, **6**, *6*, 451–53
        Mesozoic Era, **6**, *6*, 468–73
        Paleozoic Era, **6**, *6*, 461–68
        Proterozoic Eon, **6**, *6*, 456–60
        See also specific periods
Earth-Moon system, **632**, **634**, *634*, 635
earthquake belts, 89, *90*
earthquake engineering, **346–47**
earthquakes, **46**, **304**, *304*, **311**, 347
    annual probability of, **342**
    causes of, 305–13
        and location, 322–29
    classes of, 323
    in continental crust, 313, 347
    damage from, 322, 329, 347
        in Turkey, *4*, *332*, 345
    deaths from, *305*
    engineering and zoning for, 346–47, 348
    household protections from, 346–47
    and induced seismicity, 328
    mass movement triggered by, 558, 568, 578
    measuring and locating, 315–22, *321*
    in Northridge, California, 304
    and plate tectonics, 113
    precursors of, 345, *345*
    predicting of, 339, 342–46, 348
    resonance in waves of, 330
    and sea-floor spreading, 66, *66*
    and seismic waves, 313
    and subducted plates, 95, *96*, 97
    in Turkey, *4*, *305*, *332*, 345, *345*
    as volcano threat, 289, 294
    See also seismic waves
earthquake waves, 46
    See also seismic waves
earthquake zoning, **346–47**
EarthScope, 359
Earth Summit, Rio de Janeiro (1992), 803
Earth System, 1, **5–6**, 449, **802**, **806–7**, 826, 830
    anthropogenic changes in, 819–22, *823*, 831
    climate of, and currents, 628
    deserts in, 731
    and global change, 802
    interplay of life and geology in, 459–60
    and life processes, 6
    physical-biological interaction in, 500
    rocks as insight into, 145
    sources of energy in, 488–89
    and transfer cycle, 258
East African Rift, 106, *110*, *174*, 179
    and continental rifting, 390
    on digital map, *364*
    and earthquakes, *326*, 327
    Olduvai Gorge in, 403, *403*
    and volcanoes, *174*, 288, 293
East Pacific Ridge, *64*
East Pacific Rise, 71
eccentricity cycle, 794, *795*
echinoderms, 463
echo sounding (sonar), 63
ecliptic, 31

gravel:
  and bedding, *210*
  conglomerate from, 203
  consumption of, *536*
  and deltas, 218
  in desert, *736*
  glacial sources of, *782*
  in icebergs, 770, *771*
  as nonmetallic mineral resource, 523, *536*
  in river, *220*
  from stream, *202*
gravestones:
  tilted by creep, *559*
  See also headstone
graveyard:
  salt wedging in, *188*
  weathering in, 192
  See also headstone
gravimeter, 384
gravitational energy, 546
gravitational spreading, 765, *766*
gravity (gravitational pull), 23, 154, 359, 360, 384, 632, 635, **A-7**
  energy from, 488
  in formation of stars, 23
  and glacier movement, 765, *766*
  and potential energy, 668
  and rock cycle, 262
  and running water, 588
  and tides, 631, *633*, 634–35
  and weight, A-1
gravity anomaly, 384
graywacke, *202*, 203, **453**, *454*
  See also wacke
Great Barrier Reef of Australia, 204
Great Britain, *See* England; Scotland; United Kingdom; Wales
Great Dividing Range, *364*
Great Exhuma, Bahamas, 223
Great Falls, Mont., flood damage in, *610*
Great Lakes, 792
Great Missoula Flood, 611, *612*, 787
Great Plains of North America:
  desertification of, 750, *751*, 754
  exposed rocks in, 364, *365*
Great Salt Lake, Utah, 206, 741, 787, *787*, 788
Greek philosophers:
  cosmological views of, 15
  and pre-Newtonian paradigms, 86
greenhouse effect, **464**, 468, 470, 474, 519
  from uplift, 814
  and young Sun, 817
greenhouse gases, 803, 810, 823–24
  carbon dioxide as, 460, 472, 694, 795, 810, 822
  and global warming, 822, 823, 824
  methane as, 694
  and solar energy, 694, 701
greenhouse (hot-house) periods, **812**, 814, *815*
Greenland, 42, 725, 774, *825*
  as below sea level, 782
  during Devonian Period, *465*
  glaciers and glacial ice in, 297, 550, 759, *759*, *761*, 762
    icebergs from, 768
    ice cores from, 811, *813*
    visit to, 792
  isotopic signatures of life in, 453–54
  mountain belts in, *62*
  in Pangaea, *467*
  and Pangaea breakup, 473
  Viking settlements on, 815, *816*
  Wegener's final expedition to, 62
Greenland ice cap, 42, 440–41
Greenland Shield, *252*
"Green Revolution," 615
greenschist, *241*
greenstone, 254, **453**, *454*
Greenwich Mean Time (GMT), 416
Grenoble, France, lake-bed shales in, *219*
Grenville orogeny, 391, *394*, 457, *457*
groins, 655
Gros Ventre slide, 569, *570*

Grotto Geyser, Yellowstone, *674*
ground moraine, **781**
ground shaking, as earthquake damage, 329–31, 348
groundwater, 40, 48, *579*, 584, **660**, 661–63, 687–88
  for agriculture, 754
  and archaea development, 411
  and caves, 681–84, *686*
  chemistry of, 677
  contamination of, 821
  and desert plants, 747
  extraction of, 670–72, *670*, *671*
  flow of, 667–69, 688
  and fossilization, 405
  and geothermal energy, 512, *512*
  global usage of, 675
  and hot springs or geysers, 672–75, *675*
  and hydrologic cycle, 551, 553
  and hydrothermal fluids, 233
  and induced seismicity, 328
  for irrigation (Jordan), *660*
  in joints, 370
  in lithification, *200*
  lowering level of (protecting against mass movement), 579
  as magma coolant, 168
  and nuclear waste storage, 512
  and oases, 673, *673*
  and permeability, 663, *663*
  and relative sea level, 650
  as reservoir, 550
  residence time for, 550
  and secondary-enrichment deposits, 528, *538*
  into stream, *584*
  in travertine formation, 207
  usage problems with, 675–80, 688
  and water in atmosphere, 694
  and water table, 665–67, *666*; *See also* water table
groundwater contamination, 679, **679**, *680*
group, 428
"growlers," 767, *771*
growth rings, **440**, *440*
  and climate change record, 440, 811–12, *813*
Guadeloupe, volcano on, 295
Guiana Shield, *252*
Guilin region of China, tower karst landscape in, *688*, 689
Gulf Coast, 654, *694*, *829*
Gulf of Aden, *110*
Gulf of California, *475*, 675
Gulf of Mexico, *470*, *694*, 722, *722*
  offshore drilling in, 506
  sediment along coast of, 471
Gulf of Suez, *110*
Gulf Stream, 627, *629*, *719*, 796, 817
Gulf War, Kuwaiti oil wells set afire after, *486*
Gunz glaciation, 791
gushers, 497
  at Spindletop, 496
Gutenberg, Beno, 355
guyot (flat-topped seamount), 65, 105, **626**, 649, *650*, *653*, 657
gymnosperms, 466, 470, 476
gypsum, *132*, 136, 533, *536*, *539*
  as evaporite, 206
  beneath Mediterranean Sea, 184
  as nonmetallic mineral resource, 523, *537*
  on playa, 740
gypsum beds, *426*
gypsum board, 535, *539*
gyres, **627–28**, *652*

Hadean Eon, **6**, *6*, *443*, 444, **451–53**, *476*, 480, 804, 805
Hades, 15
Hadley, George, 701
Hadley cells, **701**, *702*, 704
Hadrian's Wall, 162
hail, **711**
Hale-Bopp comet, *30*
Half Dome, 172, 771, *771*
half-graben, 374, *375*, *390*
half-life, **435**, *435*, 445
halide, **136**

at mid-ocean ridge, *92*
plates of, 58, 86–88; *See also* plate tectonics
and ridge-push force, 111
and sedimentary basins, 224
and slab-pull force, 111
and subsidence, *546*
and uplift, *546*
in Wegener's hypothesis, 62
See also continental lithosphere; oceanic lithosphere
lithosphere of Moon, 317
lithosphere plates, 88
See also plates
lithospheric mantle, *49, 89, 92, 108, 109, 324*
and Archimedes' principle of buoyancy, 383
in Asia-India collision, *392–93*
and crustal root, **383**
diamond in, 139
formation of at mid-ocean ridge, 93
and mid-ocean ridges, 93
lithospheric root, *381*
little ice age, **797**, *797*, 815–16, *816*
Lituya Bay, Alaska, landslide, 568
loading, *546*
loam, **196**
local magnitude, 320
Locke, John, 417
lodestone, 79
lodgment till, 780
loess, **741**, **780**, **781**
Loma Prieta, California, earthquake, *305*, 327, 346
London:
in future of Earth, 830
as threatened by sea-level rise, 798
Longfellow, Henry Wadsworth, 550
Long Island, N.Y., as terminal moraine, **780**, *782*
longitudinal dunes, 744–45, *746*
longitudinal profile, **591–92**, *592*, 617
longshore current, **637**, *638*, *642*
long-term climate change, 811, *812–15*, 831
long-term predictions of earthquakes, 339, 342, **342**
Long Valley Caldera, 302
Loop Current, 722
Los Angeles:
Colorado River water diverted to, 615
La Brea Tar Pits in, 405
Louisiana, 722
bayous in, 602
Louisiana Territory, 591
Louis IV (king of France), and Hope Diamond, 137
Love waves (L-waves), 313, *314*, 316, 329, *329*, 347, 351
Lowell, Percival, 554
lower mantle, **49**, 54, **354**
low-grade rocks, 241, 243, 255
low-sulfur coal, 519
low-velocity zone, 52, **353**
lubricating oil, *489*
Lucas, Anthony, 496
luster (mineral), **131**
L-waves (Love waves), **313**, *314*, 316, 329, *329*, 347, 351
Lyell, Charles, 412, 418
Lystrosaurus, 61, *61*

Maat Mons, *300*
McDermitt volcanic field, *285*
McMurdo Station, Antarctica, 2
macrofossils, **405**
Madison Canyon, Mont., landslide in, 568
mafic lava, 268
and viscosity, 161, *163*, 164
mafic magma, **158**, 160, 175
mafic melt, *156*
mafic minerals, stability of, *190*, 238
mafic rocks, 38, 54, 161, *161*
as crystalline rocks, 175, *176*
Magellan, Ferdinand, 17, 621
Magellan (spacecraft), *300*
magma, 44, 154, *154*, *156*, 179–80
andesitic, 179

in Archean Eon, *454*
basaltic, *156*, 175, 179, 280
composition of, 157, 179
cooling of, 168, *169*, 175, 181
Earth's surface as (Hadean Eon), 452, *452*
felsic (silicic), **160**, 172, 175, 179
formation of, 154–55, *156*
and hydrothermal fluids, 233
intermediate, **160**
mafic, **160**, 161, 175
major types of, 157–59, 162–63
materials in, 47
at mid-ocean ridge axis, 65, *66*
and migmatite, 237
movement of, 162–63, 181
of Mt. St. Helens, 282
rhyolitic, *156*, 286
stoping of, *163*
ultramafic, **158**
viscosity of, 161, 181
magma chamber, 91, **92**, 105, 156, 158, *163*, **274**, **278**, **279**, *390*
and hot springs, 674, 688
massive-sulfide deposit in, *527*, *538*
in shield volcano, *284*
magma contamination, *159*
magma mixing, 158
magmatic deposit, **527**, 531, 540
magnesite, *526*
magnesium, *52*, *127*, *526*
in basic metamorphic rocks, 241
in biotite, 123
in magma, 157
and weathering, 190
magnesium oxide, 158, 268
magnetic anomalies, **67**, *67*, 74–75
and plate movement, 112
magnetic-anomaly map, ore bodies shown on, *533*
magnetic declination, **79**, *80*
magnetic field, 38–39, **38**, *38*, 53, **77**
of Earth, 38–39, 63, 70, 78–79
as generated by rock flow, 52
reversal of, 68, *68*, 75; *See also* magnetic reversals
magnetic field lines, 38, 79, *80*, **81**
magnetic force, **77**
lines of, *51*
magnetic inclination, **79**, **81**
magnetic poles, *51*, **77**
magnetic-reversal chronology, 68, 69, 71, 74, 75
magnetic reversals, 67, 68–69, **68**
lava flows as recording, *73*
and sea-floor spreading, 68–69, *69*
magnetism, 77–78
magnetite, 70, 80, 135, 525, *526*, 529
magnetization, **77**
magnetometer, 67, *67*, 74
magnetosphere, 38
magnetostratigraphy, **441**, *441*
magnitude of earthquake, **320**
and Richter scale, 320
Mahomet aquifer, 664, *664*
Maili Glacier, avalanche from, 563
main bedding, 212
Maine, 646
Acadia National Park in, 776, 792
fjords of, 645
Onawa Pluton in, 245, *247*
Makran Range, Pakistan, 389
malachite, *122*, *526*
Malaspina Glacier, *761*
Malaspina Glacier Area, Alaska, 775
Mallard Lake resurgent dome, *285*
malleability, 523
Malm Epoch, *562*
mammals, 409, 475
in desert, 747
development of, 472
diversification of, 475
earliest ancestors of, 470

mid-ocean ridges (*continued*)
   and ridge-push force, 110, *112*
   and rifting, 107
   rise of, 93–94, *95*
   and rise of heat, 65, *65*
   and sea-floor spreading, *66*, *92*
   and sea-level changes, 462, 650, 808
   volcanic activity at, 174, 176, 284, 287
      hot spots, 103, 106
      plate-boundary volcanoes, 103
mid-stream bars, 601
Midway Island, 104
Midwestern United States:
   aquifers for, 664
   Archean rocks under, 457
   climate of, *722*
   Dust Bowl of, 750, *751*, 754
   floodplain flood in (1993), 607–10, *610*, 612
   higher-sulfur coals of, 519
   human impact on, 43
   ice-age results in, 786–87
   meteorite impact in, 7–8, *8*
   outcrops rare in, 147
   plutons in, 457
   and tornadoes, 714
   transported soil in, 193
migmatite, **237**, *239*, *241*, **251**, 255
migration (human), and Pleistocene ice age, 474–75, *480*, 784, 791
migration pathway (hydrocarbons), **492**
Milankovitch, Milutin, 794–95
Milankovitch cycles, **795**, *795*, 796, 816–17
Milford Sound, New Zealand, waterfall in, *595*
Milky Way, 19, *19*, 25, 31
Miller, Stanley, 454–55
Milton, John, 45
Mindel glaciation, 791
mine, 532–33
   deepest in world, 44, 350
mineral assemblage, in metamorphic rocks, 242
mineral classes, **135–36**
mineralogists, **121**
mineralogy, 5, **122**
mineral resources, **523**
   formation and processing of, 538–39
   metallic, 523–34, *536*, *531*, 540
   nonmetallic, 523, *536*, 533, *538*, 540, *542*
   as nonrenewable, 536–37
minerals, **44**, 121–24, **122**, 142, 540
   biogenic, **122**
   and Bowen's reaction series, 160
   classification of, 135–36, 142, 410
   criteria for, 122–24
   as crystals, 122, 124–31, **124**
   felsic, *190*
   as fossils, 405
   gems, 122, 137–41
   global needs for, 536–37
   in groundwater, 677
   ice as, 758
   industrial, 122
   mafic, *190*
   ore, 122
   and paleomagnetism, 63, 81
   physical properties of, 131–34, 142
   on playas, 740
   relative stability of, *190*
   rock as, 145
   silicate, 44, 136, **136**, 142, 146
   synthetic, **122**
miners, and faults terminology, 371
mining:
   for cement ingredients, 534
   of coal
      strip, 504, *505*, 508, *508*
      underground, 508, *508*
   and environment, 540, 549
   as landscape modification, *542*, 820
   new ways of, 540
   for ore minerals, 532–33, *538*

   open-pit, 530, *533*, 535, *538*, 540
   underground, 532–33, *538*, 540
   waste rock from, 540
   of sea-floor deposits, 531
   See also mine
Minnesota, 787
   ice-age lakes of, 786–87
   Voyageurs National Park in, 792
Minoan people, 298, *298*
Miocene Epoch, *432*, *433*, *443*, *477*
   and Antarctic ice sheet, 796
   Late (climate), 474
   and sea floor, *94*
   and sea level, *809*
mirage, **732**, *732*
Mississippi, 722, *722*
Mississippian Period, *432*, *443*
   in correlation of strata, *434*
   and eastern U.S. geologic features, *391*
   sea level during, *809*
Mississippi Delta, 599, *602*, *603*, 722
Mississippi drainage basin, 586, *586*
Mississippi Embayment, *391*
Mississippi-Missouri network, 786
Mississippi River, 600
   at Cape Girardeau, *613*
   discharge of, 587
   drainage basin of, *586*
   and Mark Twain, 597, 611
   as meander, 597
   Mississippi River Flood Control Act (1927), 611
   and New Orleans, *599*
   1993 flooding of, 607–610, *610*, 612
   peak annual discharge of, 615, *615*
   sediment load of, 536, 589
Mississippi Valley, Precambrian rift in, 328
Mississippi Valley-type (MVT) ores, **528**, 540
Missouri:
   cliff face in, *415*
   tornado in, 713
Missouri River, 1993 flooding of, 607–10, *610*, 612
mixture, **123**
Moenkopi Formation, *434*, 436
Moenkopi shale, 437
Moho, *47*, 52, **53**, **89**, *97*, 156, **353**, *353*, 357, *623*
Mohorovicic, Andrija, 47, 352–53
Mohs, Friedrich, 131
Mohs hardness scale, **131**, *132*
Mojave Desert, *167*, 545
mold of fossil shell, *406*
molecule, **123**, **A-5**
mollisol, *196*
Mollusca, 410
mollusks, 463
molybdenite, *526*
molybdenum, *526*
moment magnitude (M_w) scale, **321**
monoclines, **376**, *376*, *380*, 391
Mono Lake, California, 207, *208*, *302*
Mono Lakes volcanic area, 179
monsoons, 722, *722*, *726*, 727
   floods from, 607–10
Montana, 222
   Glacier National Park in, *373*, 792
   Madison Canyon landslide in, 568
Monte Cristo Limestone, 428–29, *430*
Monte Toc, rockslide from, 561–62, *562*
Montreal, Canada:
   earthquake in, 327
   during last ice age, 758
Montreal conference on CFC emissions, 822
Mont St. Michel, France, *420*, *633*
Montserrat (volcano), 273, *273*
Monument Valley, Ariz., 437, *545*, 743, *743*, 753
Moon, 25, 802
   changing distance from Earth of, *633*
   cratering on, 32, 35, 453
   Earth's distance to, 18
   formation of, *26*, 31, 34, 452, 803

in geocentric image, *17*
during Hadean Eon, *452*
knowledge of (vs. knowledge of ocean), 621
lack of change in, 802
landscape of, 551, *551*
layers of, *50*
material from to Earth, 48
phases of, 15
and tides, 488, 630, 631–32, *634*, 635, 657
volcanic activity on, 299, 301
moonquakes, 317
moon rock, *444*
radiometric dating of, 444, 446
moons, **25**
moraine, 647, **776**, **780**
setting of, *782*, *783*
Morgan, Jason, 104
Morley, Lawrence, 70
morphology, **410**
mortar, **534**
mountain belts or ranges (orogens), **363**, 395, 475
accretionary, **386**, **457**, *457*
collisional, 107, 108, *111*, *245*, 383, 386–87, *387*, 392–93, *457*
of Appalachian region, 463
Himalayas as, *392–93*
crustal roots of, 383, *383*
digital map of, *364*
identifying of, 450
life story of (Appalachians), 391, 394, *394*
metamorphic rocks in, 252–53, 255
and plate tectonics, 113
topography of, 383–84, 387, *387*, *450*
See also uplift; volcanoes
mountain building, *See* orogeny
mountain cliffs, *147*
mountain (alpine) glaciers, **759**, 761, *761*, 765, 771, *783*
visit to, 792
mountain ice cap, 759, *761*, *783*, 789
mountains, *41*
erosion of, 382, 384–85, 396
fascination of, 363
mountain stream environments, 215
Muav Limestone, *428*, *430*, *434*
mud:
in Peru landslide, 558
in river, *219*
mud cracks, **214**, *214*, 218
on floodplain, 218
and uniformitarianism, *420*
mudflow, *278*, **561**, **573**, 578
from hurricanes, 718
as volcano threat, 289, 291
mudstone, 200, **203**, 220
dinosaur footprints in, *406*
tillites in, 793
Muir, John, quoted, 363
Muir Glacier, *825*
muscovite, *134*, 136, 160, *251*
and igneous intrusion, *247*
and metamorphism, 233, 241, 243
in Onawa Pluton, *247*
stability of, *190*
mylonite, 246–47, *248*, *250*, *251*, 373, *374*

Namazu, 305, *306*
Namibia, 223
Namib Desert in, 223, *447*, 752, *756*
Nanaga Parbat, 397
Na-plagioclase, *190*
NASA World Wind, 9, 12, 42
Nashville Dome, *391*
Nassar, Lake, 673
National Weather Service, U.S., tornado warnings by, U.S., 714
Native Americans:
desert-varnished rock used by, 735, *736*
folklore of on earthquakes, 305
on Mesa Verde, Colo., 199, *202*
Onondaga, 208
native metals, **136**, **524**

natural arches, **743**
natural bridge, *682*, *685*, **688**
Natural Bridge, Va., *685*
natural gas, 487, *487*, **500**, 520
carbon in, *489*
consumption of, *536*
migration of, 492
world supply of, 517, 518
natural hazard, **558**
See also earthquakes; floods; mass movement; storms; volcanic
eruptions
natural levees, **598**, *599*, 607
natural selection, **412**, 444
nature, forces of, A-7
nautiloids, *464*
Navajo Sandstone, 432, *434*
Nazca Plate, 88, *89*, 97, *114*, 395
Neanderthal man, *477*, 480
Nebraska, sand hills region of, 745
Nebraskan glaciation, 792
nebulae, *23*, *23*, 34, 35
nebular hypothesis, *26–27*
nebular theory of solar system formation, **29**, 34, 154
negative anomalies, 67, *67*, 75
negative feedback, and CO_2 in atmosphere, 814
neocrystallization, 230, 231, *231*, 233, 236, *238*
of fossils, 407
neon atom, *A-3*
Neoproterozoic Era, *443*
Neptune, 25, 28
Neptunists, 146
Netherlands, little ice age in, 797, *797*, 816
neutrons, 123, **A-3**
Nevada, *436*, 479
mining claims in, 532
Yucca Mountain in, 512, 613
Nevadan orogeny, **468**
Nevado del Ruiz, 289
Nevado Huascarán Mountain, landslide on, 558
Newfoundland, submarine slide along coast of, 566
New Guinea, 473
New Madrid, Mo., earthquakes, 327–28, *328*, *334*
New Mexico:
Shiprock volcano in, *164*, 296
weathered sedimentary rock in, *192*
New Orleans, La., 598, *599*, 638, 722–23, *722*, *723*
Newport, Ind., 43
Newton, Sir Isaac, 16–17, 23, 86, 369, 417, A-7
New York City:
bedrock for skyscrapers of, 147
Central Park in, 789, *789*
concrete sidewalks of, 535
in future of Earth, 830
glacially polished surface in, *771*
during last ice age, 758
sea-level rise threat to, 654, *654*, 785, 797
New York State:
continental glaciers in, 776
drumlins in, *783*
New Zealand:
Alpine Fault in, 102, 325, 389, 576
fjords of, 645, 776
geothermal energy in, 512
glacier visit to, 792
Rotorua in, 675, *675*
Southern Alps of, 42
waterfall in, *595*
Niagara Escarpment, *596*
Niagara Falls, 595, *596*
Niagara Gorge, *596*
nickel, *526*, *537*
Niger Delta, 223, *602*
Nigeria, 223
Niger River, 223
Nile Delta, 218, 223, 599, *602*, 820
Nile River, 614, 673, *748*
damming of, 820
Nile River canyon, 184, *184*
"nimbus," 710

in rock cycle, 258, *259*
at Yucca Mountain, 512
rhyolitic eruptions, 269
rhyolitic lava, 269, 280, 299
rhyolitic lava flow, 269, *269*, 299
rhyolitic (fine-grained) magma, *156*, *171*, 280, 286
rhyolitic volcanoes, 288
rhyolitic welded tuff, *170*
rhythmic layering, **440**, *440*
Richter, Charles, 320
Richter scale, **320**, *321*, 347
ridge, 40
as divergent boundary, *92*
on map of relative velocities, *113*
and sea-floor spreading, *66*
on volcano map, *174*
See also Mid-Atlantic Ridge; mid-ocean ridges
ridge axis, 64, 98
ridge-push force, **110**, **112**, 116
rift basins, 224
in Pangaea, 468
rifts or rifting, 47
in Archean Eon, *454*
and earthquakes, *323*
in landscape evolution, 546
and mountains, 364, 377–78
salt layers in, *207*
rift volcanoes, 288
ring dikes, *177*
"Ring of Fire," 175, **288**
Rio de Janeiro:
in future of Earth, 830
mudflows in, 561, *561*
sugar-loaf mountains in, 639, *639*
Rio Grande Rift, 327, *480*
Rio Negro, 601
Rio Ucayali, 600
rip current, 637, *639*
ripples (ripple marks), **211**, 214
on floodplain, 218
riprap, **578**, **579**, *579*, 655, *655*
Riss glaciation, 791
river environments, 218, *220*
rivers, *582*
in Archean Eon, 453
deltas of, 218, 591, 599–602, *602*, *603*, *609*, 617;
See also deltas
in desert, 731
environmental issues over, 614–16, 617
erosion by, 362, 385, *385*, 395
in human history, 614
irrigation from, 750, 754
relocating of, 578, 579
residence time for, 550
and "stream," 583–84
undercutting by, 569
as water reservoir, 550, 553
and water table, 665
See also streams
river sediment, 218, *219*, 273
river systems, *608–9*
road cuts, *147*
roche moutonnée, **776**, *776*, **779**
Rochester, N.Y., drumlins near, *783*
rock assemblages, and continental drift, 61–62
rock avalanche, 563, *573*
rock bolts, *579*
rock bursts, 533
rock composition, *148*
rock cycle, **257–58**, *258*, *806*, 808
causes of, 263
and plate tectonics, 258, *259*, *261*, 262–263
rates of movement through, 262–263
and rock-forming environments, *260–61*
rock deformation, *See* deformation, rock
Rockefeller, John D., 494
rock exposure, Earth history told by, *401*
rock falls, **563–64**, *573*, 578
rock flour, 771

rock formation, during orogeny, 380–82
rock glaciers, **560**
in Alaska, *560*
"rock oil," 493
rocks, **44**, 119, **145–46**, 223
and carbon-14 dating, 439
classification of, 148–50
in desert, 732
determining age of, *415*
in Earth's history, 453
beginning of, 444, *444*
and Earth's magnetic field, 70
as flowing, 49, 53
as geological record, 145, 262
in glacier, 766, *779*
and groundwater, 613, 677
as insulators, 747
intact vs. fractured, 567, 578
intermediate, 44, 54, *160*, 161, *171*, 175
from lava flow, 267
mafic, 44, 54, 160, *161*, 175, *177*
magnetization of, 63
melting of, 154–56
modification of through plate tectonics, 53
from Moon, 444, *444*, 446
naming of, 150
pores of, 42, 491, **662**
and porosity, 662–63, *663*
silicate, 44, 54, 765
and soil, 194
study of
through high-tech equipment, 151
through outcrop observations, 150
through thin-section study, 150–51
surface occurrences of, 146–47
types of
igneous, 44, **148**, **150**, **153**
metamorphic, 44, **148**, **229**
sedimentary, 44, **148**, **184–85**
See also specific types
ultramafic, 44, 50, 54, 169, *171*
and water table, 665
See also boulders
"rock saws, 151, *151*
rock slide, 562, 578
rocky coasts, 643, *644*, *645*, 657
Rocky Erosion, 665
Rocky Mountains:
Ancestral Rockies, 466, *467*, 479
coal from, 503
formation of, 471
glaciers in, 789
metamorphic rocks of, *252*
road cut in, *372*
Rodinia, **457**, *458*, 461, *477*, 480
and Pangaea, 466
rogue waves, **636**
Romans, cement used by, 534
roots, 197
root wedging, **187**
Rosendale Formation, 535
Ross Ice Shelf, 2, 762
rotation, 365
rotational axism, 79
Rotorua, New Zealand, 675, *675*
Rub al Khali, 744
ruby, 138, *138*
Rufus Limestone, *429*
Rumi, Jalal-Uddin, 121
runaway greenhouse effect (Venus), **814**
running water, **583**
runoff, **584**
in hydrologic cycle, 584
Rushmore, Mt., *175*
Russia:
Caucasus Mountains avalanche in, 563
deepest drill hole in, 350
during ice age, *779*, 789
Rutherford, Ernest, 444, A-3

stony plains, 741, *749*
stoping, *163*, **167**
Storegga Slide, 566, *566*
storms, **711**
    hurricanes, 715–18, *717*, *718*, 727
    nor'easters, 715
    thunderstorms, **711–13**
    tornadoes, **713–15**, *715*
storm surge, 636, **718**
strain, 307–8, **365**, *368*
strata, 210
    tilted beds of, *366*
    and unconformities, 424
strategic metals, **537**
stratification, 209
stratified drift, 780
stratigrapher, 209
stratigraphic column, **424**, 445
stratigraphic formation, **210**, *210*, **424**, 445
    correlation of, 428–29, *430*, *434*
    and Grand Canyon, *424*, 428, *428*, *429*
stratigraphic sequence, **462**
stratigraphic trap, 495, *495*
stratigraphy, 5, 209
    of sea floor, 770
stratosphere, **698**
    volcanic materials in, 280, 297
stratovolcanoes, **275**, *277*, 293
stratus clouds, **710**
streak (mineral), **131**
stream cut, *147*
stream piracy, **603**, *604*
stream rejuvenation, **603**, 617
streams, **583–84**, 617
    antecedent, 604, **605**, *606*
    base level of, 592, *593*
    braided, **596**, *597*, 780, *781*, *782*
    deltas of, 591, 599–602, *602*, *603*, 607, 617
    depositional processes of, *217*, 589–91, *591*
    in deserts, *748*
    disappearing, *685*, **686**
    discharge of, **587–88**, *587*, *588*
    drainage network of, *605*
    ephemeral, **587**, *587*, 617, 735, 755
    and erosion, **588**, 589–9, *590*, 617
        headward erosion, **585**, *585*, 595, 596, 603, *609*, 617, *749*
        and meanders, 597, *598*
        from mountains, 378
        in stream creation, 584–85, *587*
    formation of, 584–86
    longitudinal profile of, **591–92**, *592*, 617
    meandering, 597, *598*, *608*, 617
    permanent, **587**, *587*, 617
    rapids in, 594, *595*
    sediment loads of, 589, *590*
    seepage into increased by paving, 549
    superposed, 604, **605**, *605*
    turbulence of, 588, *589*
    waterfalls in, 594–95, *595*
    as water reservoir, 553
    See also rivers
stress, 307–8, **369**, *370*, 391
    differential, **232**
    and earthquake building codes, 346–47
    normal, **232**
    shear, **232**, **369**, *370*
stress-triggering models, 345
stretching, 365, *366*
striations, 58
    glacial, **772**, **779**
strike, *367*, *367*
strike line, *367*, *367*
strike-slip displacement, 394
strike-slip faults, 306, 307, 309, **311**, **327**, *343*, 347, 365, 371, **371**, *373*, 389
strip mining, 504, *505*, 508
stromatolites, 454, *455*, 476
Strombolian eruptive style, *278*

subaerial volcanoes, **176**
    shield, *284*
subduction, **58**, **94**, **96**, 114, *469*
    along West Coast, *475*
    and carbon in mantle, 139
    and convergent plate boundaries, 94–100
    and dynamothermal metamorphism, *251*
    in early Earth history, 804
    and mountain building, 364
    and nuclear waste storage, 512
    and Redoubt Volcano eruption, *266*
    and volcanic activity, 176
subduction zone, *91*, 94, *625*
    and future, 830
    metamorphism in, 248
    and rock cycle, 262
    and subaerial zones, 287
sublimation, **762**, 765
submarine basaltic lava, 179, 269
submarine canyons, *217*, **623**, *625*, 657
submarine debris flows, 288, **564**
submarine fan, 105, *213*, *216*, *261*, 624, *625*
submarine landslides, 566
submarine plateaus, during Cretaceous Period, 471
submarine slumps, **564**, *564*
submarine volcanoes, 176, 269, 275, 281
submergent coasts, **651**, *651*
subsidence, **221**, **546**
    causes, *546*
    thermal, 224
substrate, 546
substrate composition, and landscape development, 549
substrate composition and soil characteristics, 194, *195*
subsurface water:
    reducing of, 578
    See also groundwater
subtropical deserts, 732–33, 755
subtropical high (subtropical divergence zone), 702
succulents, **747**
Sudbury, Ontario, smoke from smelters in, 540, *540*
sulfates, **136**
sulfides (sulfide minerals), 127, **136**, 524, 531
    around black smokers, 528, *528*, 531, *538*
    ore minerals as, 526
sulfur, 79, 536
    in groundwater, 613
    and iron, 525
sulfur dioxide, 452, 519
    in volcanic gas, 693
sulfuric-acid speleogenesis, 681, 684
Sumatra, earthquakes in, 305, 335, *336*
Sumeria, bronze discovered in, 524
summit eruptions, 275
Sun:
    burning of, 24
    and change in Earth, 802
    color of, 697
    Earth's distance to, 18
    and faint young sun paradox, 817
    formation of, *27*
    future of, 830, *831*
    in geocentric image, *17*
    mass of, 25
    nuclear fusion in, *A-10*
    plasma in, **A-6**
    and tides, 631, *633*
Sunbelt region of United States, 676
Sunda Trench, 335
sunlight, 697
Sunset Crater, 173, *270*, *274*, *277*
sunspot cycle, **816**, *817*
Supai Group, *210*, *428*, *430*, *434*
supercontinent cycle, **805**, *808*
supercontinents, 458
    Gondwana, 461, *461*, 466, *467*, 471, 473, *474*, 603, *794*
    Pangaea, **57**, 58, **466**, *467*, 480, 481; *See also* Pangaea
    Pannotia, **457–58**, *458*, 461, *477*, 480
    Rodinia, **457**, *458*, 461, *477*, 480